Function Notation

$f(x)$	Value of f at x
$(f \circ g)(x) = f[g(x)]$	Composite function
$f^{-1}(x)$	Value of inverse of f at x

Linear Equations and Functions

$y = mx + b$	Slope–intercept form
$(y - y_1) = m(x - x_1)$	Point–slope form
$f(x) = mx + b$	Linear function
$y = b$	Horizontal line
$x = a$	Vertical line

Polynomial and Rational Forms

$$f(x) = ax^2 + bx + c \qquad \text{Quadratic function}$$

$$f(x) = a_n x^n + a_{n-1} x^{n-1} + \cdots + a_1 x + a_0, \qquad \text{Polynomial function}$$
$$a_n \neq 0, n \text{ a nonnegative integer}$$

$$f(x) = \frac{p(x)}{q(x)}, \quad p \text{ and } q \text{ polynomial functions}, q(x) \neq 0 \qquad \text{Rational function}$$

Exponential and Logarithmic Functions

$f(x) = b^x, \quad b > 0, b \neq 1$	Exponential function
$f(x) = \log_b x, \quad b > 0, b \neq 1$	Logarithmic function
$y = \log_b x$ if and only if $x = b^y, \quad b > 0, b \neq 1$	

Matrices and Determinants

$$\begin{bmatrix} a & b & c \\ d & e & f \end{bmatrix} \quad \text{Matrix}$$

$$\begin{vmatrix} a & b & c \\ d & e & f \\ g & h & i \end{vmatrix} \quad \text{Determinant}$$

Arithmetic Sequence

$$a_1, a_2, \ldots, a_n, \ldots$$
$$a_n - a_{n-1} = d \qquad \text{Common difference}$$
$$a_n = a_1 + (n - 1)d \qquad n\text{th-term formula}$$

$$S_n = a_1 + \cdots + a_n = \frac{n}{2}[2a_1 + (n - 1)d] \qquad \text{Sum of } n \text{ terms}$$

$$S_n = \frac{n}{2}(a_1 + a_n)$$

Geometric Sequence

$$a_1, a_2, \ldots, a_n, \ldots$$

$$\frac{a_n}{a_{n-1}} = r \qquad \text{Common ratio}$$

$$a_n = a_1 r^{n-1} \qquad n\text{th-term formula}$$

$$S_n = a_1 + \cdots + a_n = \frac{a_1 - a_1 r^n}{1 - r}, \quad r \neq 1 \qquad \text{Sum of } n \text{ terms}$$

$$S_n = \frac{a_1 - r a_n}{1 - r}, \quad r \neq 1$$

$$S_\infty = a_1 + a_2 + \cdots = \frac{a_1}{1 - r}, \quad |r| < 1 \qquad \text{Sum of infinitely many terms}$$

Factorial and Binomial Formulas

$$n! = n(n - 1) \cdots 2 \cdot 1, \quad n \in N \qquad n \text{ factorial}$$
$$0! = 1$$

$$\binom{n}{r} = \frac{n!}{r!(n - r)!}, \quad 0 \leq r \leq n$$

$$(a + b)^n = \sum_{k=0}^{n} \binom{n}{k} a^{n-k} b^k, \quad n \geq 1 \qquad \text{Binomial formula}$$

Permutations and Combinations

For $0 \leq r \leq n$:

$$P_{n,r} = \frac{n!}{(n - r)!} \qquad \text{Permutation}$$

$$C_{n,r} = \binom{n}{r} = \frac{n!}{r!(n - r)!} \qquad \text{Combination}$$

(Continued on back endpaper)

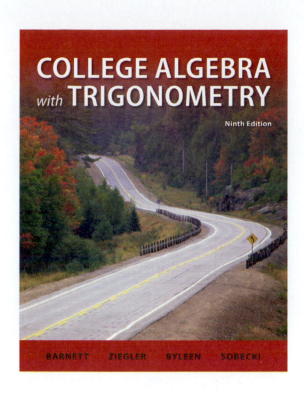

College Algebra
with Trigonometry

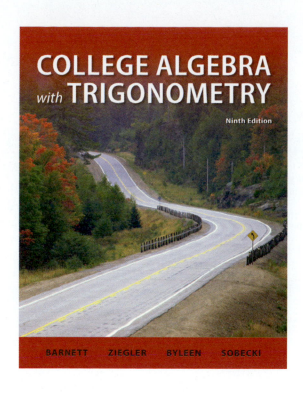

NINTH EDITION

College Algebra with Trigonometry

Raymond A. Barnett
Merritt College

Michael R. Ziegler
Marquette University

Karl E. Byleen
Marquette University

Dave Sobecki
Miami University Hamilton

Connect
Learn
Succeed™

The McGraw·Hill Companies

COLLEGE ALGEBRA WITH TRIGONOMETRY, NINTH EDITION

Published by McGraw-Hill, a business unit of The McGraw-Hill Companies, Inc., 1221 Avenue of the Americas, New York, NY 10020.

Some ancillaries, including electronic and print components, may not be available to customers outside the United States.

This book is printed on acid-free paper.

1 2 3 4 5 6 7 8 9 0 DOW/DOW 1 0 9 8 7 6 5 4 3 2 1 0

ISBN 978–0–07–351950–0
MHID 0–07–351950–2

ISBN 978–0–07–729720–6 (Annotated Instructor's Edition)
MHID 0–07–729720–2

Vice President, Editor-in-Chief: *Marty Lange*
Vice President, EDP: *Kimberly Meriwether David*
Editorial Director: *Stewart K. Mattson*
Sponsoring Editor: *John R. Osgood*
Director of Development: *Kristine Tibbetts*
Developmental Editor: *Christina A. Lane*
Marketing Manager: *Kevin M. Ernzen*
Lead Project Manager: *Sheila M. Frank*
Senior Production Supervisor: *Kara Kudronowicz*
Senior Media Project Manager: *Sandra M. Schnee*
Designer: *Tara McDermott*
Cover/Interior Designer: *Ellen Pettergell*
(USE) Cover Image: © *Comstock Images/Getty Images*
Senior Photo Research Coordinator: *Lori Hancock*
Supplement Producer: *Mary Jane Lampe*
Compositor: *Aptara®, Inc.*
Typeface: *10/12 Times Roman*
Printer: *R. R. Donnelley*

Chapter R Opener: © Corbis RF; **p. 31:** © The McGraw-Hill Companies, Inc./John Thoeming photographer. **Chapter 1 Opener:** © Corbis RF; **p. 56:** © Vol. 71/Getty RF; **p. 92:** © Getty RF. **Chapter 2 Opener:** © Vol. 88/Getty RF; **p. 142:** © Big Stock Photo; **p. 147:** © Corbis RF; **p. 151:** © Vol. 112/Getty RF. **Chapter 3 Opener, p. 170:** © Getty RF; **p. 187:** © Vol. 88/Getty RF; **p. 220:** © Corbis RF; **p. 250:** © The McGraw-Hill Companies, Inc./Andrew Resek photographer. **Chapter 4 Opener, p. 271:** © Corbis RF; **p. 272:** © Vol. 4/Getty RF. **Chapter 5 Opener:** © Getty RF; **p. 333:** © Vol. 68/Getty RF; **p. 345:** © Corbis RF. **Chapter 6 Opener:** © Corbis RF; **p. 395:** © Digital Vision Punchstock/RF; **p. 441:** © Corbis RF; **p. 460(left):** © Vol. DV289/Getty RF; **p. 460(right):** Vol. 44/Getty RF. **Chapter 7 Opener:** © Big Stock Photos; **p. 479:** © Corbis RF; **p. 492:** © Jacqui Hurst/Corbis. **Chapter 8 Opener:** © Digital Vision Punchstock/RF; **p. 553:** © Big Stock Photos. **Chapter 9 Opener:** © Brand X/Superstock RF; **p. 587:** © California Institute of Technology. **Chapter 10 Opener:** © Corbis RF; **p. 637:** Courtesy of Bill Tapenning, USDA; **p. 641:** © Vol. 5/Getty RF; **p. 658:** © Vol. 48/Getty RF; **p. 662:** © Getty RF. **Chapter 11 Opener:** © Vol. 6/Getty RF; **p. 733:** © ThinkStock/PictureQuest RF; **p. 745:** © Corbis RF.

Library of Congress Cataloging-in-Publication Data

Barnett, Raymond A.
 College algebra with trigonometry/Raymond A. Barnett ... [et al.]. — 9th ed.
 p. cm. — (Barnett, Ziegler & Byleen's precalculus series)
 Includes index.
 ISBN 978-0-07-351950-0 — ISBN 0-07-351950-2 (hard copy : alk. paper) 1. Algebra–Textbooks.
2. Trigonometry–Textbooks. I. Title
QA154.3.B368 2011
512.13–dc22
 2009019473

www.mhhe.com

The Barnett, Ziegler, Byleen, and Sobecki Precalculus Series

College Algebra, Ninth Edition
This book is the same as *Precalculus* without the three chapters on trigonometry.
ISBN 0-07-351949-9, ISBN 978-0-07-351-949-4

Precalculus, Seventh Edition
This book is the same as *College Algebra* with three chapters of trigonometry added. The trigonometric functions are introduced by a unit circle approach.
ISBN 0-07-351951-0, ISBN 978-0-07-351-951-7

College Algebra with Trigonometry, Ninth Edition
This book differs from *Precalculus* in that *College Algebra with Trigonometry* uses right triangle trigonometry to introduce the trigonometric functions.
ISBN 0-07-735010-3, ISBN 978-0-07-735010-9

College Algebra: Graphs and Models, Third Edition
This book is the same as *Precalculus: Graphs and Models* without the three chapters on trigonometry. This text assumes the use of a graphing calculator.
ISBN 0-07-305195-0, ISBN 978-0-07-305195-6

Precalculus: Graphs and Models, Third Edition
This book is the same as *College Algebra: Graphs and Models* with three additional chapters on trigonometry. The trigonometric functions are introduced by a unit circle approach. This text assumes the use of a graphing calculator.
ISBN 0-07-305196-9, ISBN 978-0-07-305-196-3

About the Authors

Raymond A. Barnett, a native of and educated in California, received his B.A. in mathematical statistics from the University of California at Berkeley and his M.A. in mathematics from the University of Southern California. He has been a member of the Merritt College Mathematics Department and was chairman of the department for four years. Associated with four different publishers, Raymond Barnett has authored or co-authored 18 textbooks in mathematics, most of which are still in use. In addition to international English editions, a number of the books have been translated into Spanish. Co-authors include Michael Ziegler, Marquette University; Thomas Kearns, Northern Kentucky University; Charles Burke, City College of San Francisco; John Fujii, Merritt College; Karl Byleen, Marquette University; and Dave Sobecki, Miami University Hamilton.

Michael R. Ziegler received his B.S. from Shippensburg State College and his M.S. and Ph.D. from the University of Delaware. After completing postdoctoral work at the University of Kentucky, he was appointed to the faculty of Marquette University where he held the rank of Professor in the Department of Mathematics, Statistics, and Computer Science. Dr. Ziegler published more than a dozen research articles in complex analysis and co-authored more than a dozen undergraduate mathematics textbooks with Raymond Barnett and Karl Byleen before passing away unexpectedly in 2008.

Karl E. Byleen received his B.S., M.A., and Ph.D. degrees in mathematics from the University of Nebraska. He is currently an Associate Professor in the Department of Mathematics, Statistics, and Computer Science of Marquette University. He has published a dozen research articles on the algebraic theory of semigroups and co-authored more than a dozen undergraduate mathematics textbooks with Raymond Barnett and Michael Ziegler.

Dave Sobecki earned a B.A. in math education from Bowling Green State University, then went on to earn an M.A. and a Ph.D. in mathematics from Bowling Green. He is an associate professor in the Department of Mathematics at Miami University in Hamilton, Ohio. He has written or co-authored five journal articles, eleven books and five interactive CD-ROMs. Dave lives in Fairfield, Ohio with his wife (Cat) and dogs (Macleod and Tessa). His passions include Ohio State football, Cleveland Indians baseball, heavy metal music, travel, and home improvement projects.

Dedicated to the memory of Michael R. Ziegler,
trusted author, colleague, and friend.

Brief Contents

Contents

Preface

Enhancing a Tradition of Success

The ninth edition of *College Algebra with Trigonometry* represents a substantial step forward in **student accessibility.** Every aspect of the revision of this classic text focuses on making the text more accessible to students, while retaining the precise presentation of the mathematics for which the Barnett name is renowned. Extensive work has been done to enhance the clarity of the exposition, improving to the overall presentation of the content. This in turn has decreased the length of the text.

Specifically, we concentrated on the areas of writing, exercises, worked examples, design, and technology. Based on numerous reviews, advice from expert consultants, and direct correspondence with the many users of previous editions, this edition is more relevant and accessible than ever before.

Writing Without sacrificing breadth or depth or coverage, we have rewritten explanations to make them clearer and more direct. As in previous editions, the text emphasizes computational skills, essential ideas, and problem solving rather than theory.

Exercises Over twenty percent of the exercises in the ninth edition are new. These exercises encompass both a variety of skill levels as well as increased content coverage, ensuring a gradual increase in difficulty level throughout. In addition, brand new writing exercises have been included at the beginning of each exercise set in order to encourage a more thorough understanding of key concepts for students.

Examples Color annotations accompany many examples, encouraging the learning process for students by explaining the solution steps in words. Each example is then followed by a similar matched problem for the student to solve. Answers to the matched problems are located at the end of each section for easy reference. This active involvement in learning while reading helps students develop a more thorough understanding of concepts and processes.

Technology Instructors who use technology to teach college algebra with trigonometry, whether it be exploring mathematics with a graphing calculator or assigning homework and quizzes online, will find the ninth edition to be much improved.

Refined "Technology Connections" boxes included at appropriate points in the text illustrate how problems previously introduced in an algebraic context may be solved using a graphing calculator. Exercise sets include calculator-based exercises marked with a calculator icon. Note, however, that the use of graphing technology is completely optional with this text. We understand that at many colleges a single text must serve the purposes of teachers with widely divergent views on the proper use of graphing and scientific calculators in college algebra with trigonometry, and this text remains flexible regarding the degree of calculator integration.

Additionally, McGraw-Hill's MathZone offers a complete online homework system for mathematics and statistics. Instructors can assign textbook-specific content as well as customize the level of feedback students receive, including the ability to have students show their work for any given exercise. Assignable content for the ninth edition of *College Algebra with Trigonometry* includes an array of videos and other multimedia along with algorithmic exercises, providing study tools for students with many different learning styles.

A Central Theme

In the Barnett series, the function concept serves as a unifying theme. A brief look at the table of contents reveals this emphasis. A major objective of this book is the development of a library of elementary functions, including their important properties and uses. Employing this library as a basic working tool, students will be able to proceed through this book with greater confidence and understanding.

Reflecting trends in the way college algebra with trigonometry is taught, the ninth edition emphasizes functions modeled in the real world more strongly than previous editions. In some cases, data are provided and the student is asked to produce an approximate corresponding function using regression on a graphing calculator. However, as with previous editions, the use of a graphing calculator remains completely optional and any such examples or exercises can be easily omitted without loss of continuity.

Key Features

The revised **full-color design** gives the book a more contemporary feel and will appeal to students who are accustomed to high production values in books, magazines, and nonprint media. The rich color palette, streamlined calculator explorations, and use of color to signify important steps in problem material work in conjunction to create a more visually appealing experience for students.

An **emphasis on mathematical modeling** is evident in section titles such as "Linear Equations and Models" and "Exponential Models." These titles reflect a focus on the relationship between functions and real-world phenomena, especially in examples and exercises. Modeling problems vary from those where only the function model is given (e.g., when the model is a physical law such as $F = ma$), through problems where a table of data and the function are provided, to cases where the student is asked to approximate a function from data using the regression function of a calculator or computer.

Matched problems following worked examples encourage students to practice problem solving immediately after reading through a solution. Answers to the matched problems are located at the end of each section for easy reference.

Interspersed throughout each section, **Explore-Discuss** boxes foster conceptual understanding by asking students to think about a relationship or process before a result is stated. Verbalization of mathematical concepts, results, and processes is strongly encouraged in these explanations and activities. Many Explore-Discuss boxes are appropriate for group work.

Refined **Technology Connections** boxes employ graphing calculators to show graphical and numerical alternatives to pencil-and-paper symbolic methods for problem solving—but the algebraic methods are not omitted. Screen shots are from the TI-84 Plus calculator, but the Technology Connections will interest users of any automated graphing utility.

Think boxes (color dashed boxes) are used to enclose steps that, with some experience, many students will be able to perform mentally.

Balanced exercise sets give instructors maximum flexibility in assigning homework. A wide variety of easy, moderate, and difficult level exercises presented in a range of problem types help to ensure a gradual increase in difficulty level throughout each exercise set. The division of exercise sets into A (routine, easy mechanics), B (more difficult mechanics), and C (difficult mechanics and some theory) is explicitly presented only in the Annotated Instructor's Edition. This is due to our attempt to avoid fueling students' anxiety about challenging exercises.

This book gives the student substantial experience in modeling and solving **applied problems.** Over 500 application exercises help convince even the most skeptical student that mathematics is relevant to life outside the classroom.

An **Applications Index** is included following the Guided Tour to help locate particular applications.

Most exercise sets include **calculator-based exercises** that are clearly marked with a calculator icon. These exercises may use real or realistic data, making them computationally heavy, or they may employ the calculator to explore mathematics in a way that would be impractical with paper and pencil.

As many students will use this book to **prepare for a calculus course,** examples and exercises that are especially pertinent to calculus are marked with an icon.

A **Group Activity** is located at the end of each chapter and involves many of the concepts discussed in that chapter. These activities require students to discuss and write about mathematical concepts in a complex, real-world context.

Changes to this Edition

A more modernized, casual, and student-friendly **writing style** has been infused throughout the chapters without radically changing the tone of the text overall. This directly works toward a goal of increasing motivation for students to actively engage with their textbooks, resulting in higher degrees of retention.

A significant revision to the **exercise sets** in the new edition has produced a variety of important changes for both students and instructors. As a result, over twenty percent of the exercises are new. These exercises encompass both a variety of skill levels as well as increased content coverage, ensuring a gradual increase in difficulty level throughout. In addition, brand new writing exercises have been included at the beginning of each exercise set in order to encourage a more thorough understanding of key concepts for students. Specific changes include:

- The addition of hundreds of new writing exercises to the beginning of each exercise set. These exercises encourage students to think about the key concepts of the sections before attempting the computational and application exercises, ensuring a more thorough understanding of the material.

- An update to the data in many application exercises to reflect more current statistics in topics that are both familiar and highly relevant to today's students.

- A significant increase the amount of moderate skill level problems throughout the text in response to the growing need expressed by instructors.

The number of **colored annotations** that guide students through worked examples has been increased throughout the text to add clarity and guidance for students who are learning critical concepts.

New instructional **videos on graphing calculator operations** posted on MathZone help students master the most essential calculator skills used in the college algebra course. The videos are closed-captioned for the hearing impaired, subtitled in Spanish, and meet the Americans with Disabilities Act Standards for Accessible Design. Though these are an entirely optional ancillary, instructors may use them as resources in a learning center, for online courses, and to provide extra help to students who require extra practice.

Chapter R, "Basic Algebraic Operations," has been extensively rewritten based upon feedback from reviewers to provide a streamlined review of basic algebra in four sections rather than six. Exponents and radicals are now covered in a single section (R-2), and the section covering operations on polynomials (R-3) now includes factoring.

Chapter 10, "Systems of Equations and Matrices," has been reorganized to focus on systems of linear equations, rather than on systems of inequalities or nonlinear systems. A section on determinants and Cramer's rule (10-5) has been added. Three additional sections on systems of nonlinear equations, systems of linear inequalities, and linear programming are also available online.

Design: A Refined Look with Your Students in Mind

The McGraw-Hill Mathematics Team has gathered a great deal of information about how to create a student-friendly textbook in recent years by going directly to the source—your students. As a result, two significant changes have been made to the design of the ninth edition based upon this feedback. First, example headings have been pulled directly out into the margins, making them easy for students to find. Additionally, we have modified the design of one of our existing features—the caution box—to create a more powerful tool for your students. Described by students as one of the most useful features in a math text, these boxes now demand attention with bold red headings pulled out into the margin, alerting students to avoid making a common mistake. These fundamental changes have been made entirely with the success of your students in mind and we are confident that they will improve your students' overall reaction to and enjoyment of the course.

Tegrity Campus, a service that makes class time available all the time by automatically capturing every lecture in a searchable format for students to review when they study and complete assignments, is an additional supplementary material available with the new edition. With a simple one-click start and stop process, you capture all computer screens and corresponding audio. Students can then replay any part of any class with easy-to-use browser-based viewing on a PC or Mac. With Tegrity Campus, students quickly recall key moments by using Tegrity Campus's unique search feature. This search helps students efficiently find what they need, when they need it across an entire semester of class recordings.

Features

Examples and Matched Problems

Integrated throughout the text, completely worked examples and practice problems are used to introduce concepts and demonstrate problem-solving techniques—algebraic, graphical, and numerical. Each example is followed by a similar Matched Problem for the student to work through while reading the material. Answers to the matched problems are located at the end of each section for easy reference. This active involvement in the learning process helps students develop a more thorough understanding of algebraic concepts and processes.

EXAMPLE 2 · Using the Distance Formula

Find the distance between the points $(-3, 5)$ and $(-2, -8)$.*

SOLUTION Let $(x_1, y_1) = (-3, 5)$ and $(x_2, y_2) = (-2, -8)$. Then,

$$d = \sqrt{[(-2) - (-3)]^2 + [(-8) - 5]^2}$$
$$= \sqrt{(-2 + 3)^2 + (-8 - 5)^2} = \sqrt{1^2 + (-13)^2} = \sqrt{1 + 169} = \sqrt{170}$$

Notice that if we choose $(x_1, y_1) = (-2, -8)$ and $(x_2, y_2) = (-3, 5)$, then

$$d = \sqrt{[(-3) - (-2)]^2 + [5 - (-8)]^2} = \sqrt{1 + 169} = \sqrt{170}$$

so it doesn't matter which point we designate as P_1 or P_2.

MATCHED PROBLEM 2 · Find the distance between the points $(6, -3)$ and $(-7, -5)$.

› **Midpoint of a Line Segment**

The **midpoint** of a line segment is the point that is equidistant from each of the endp
A formula for finding the midpoint is given in Theorem 2. The proof is discussed
exercises.

Exploration and Discussion

Would you like to incorporate more discovery learning in your course? Interspersed at appropriate places in every section, Explore-Discuss boxes encourage students to think critically about mathematics and to explore key concepts in more detail. Verbalization of mathematical concepts, results, and processes is encouraged in these Explore-Discuss boxes, as well as in some matched problems, and in problems marked with color numerals in almost every exercise set. Explore-Discuss material can be used in class or in an out-of-class activity.

⟫⟫ EXPLORE-DISCUSS 1

To graph the equation $y = -x^3 + 2x$, we use point-by-point plotting to obtain the graph in Figure 5.

(A) Do you think this is the correct graph of the equation? If so, why? If not, why?

(B) Add points on the graph for $x = -2$, $-0.5, 0.5$, and 2.

(C) Now, what do you think the graph looks like? Sketch your version of the graph, adding more points as necessary.

(D) Write a short statement explaining any conclusions you might draw from parts A, B, and C.

x	y
-1	-1
0	0
1	1

› Figure 5

Applications

One of the primary objectives of this book is to give the student substantial experience in modeling and solving real-world problems. Over 500 application exercises help convince even the most skeptical student that mathematics is relevant to everyday life. An Applications Index is included following the features to help locate particular applications.

15. $2^3 2^{-x} = 0.426$ **16.** $3^4 3^{-x} = 0.089$

In Problems 17–26, solve exactly.

17. $\log_5 x = 2$ **18.** $\log_3 y = 4$

19. $\log (t - 4) = -1$ **20.** $\ln (2x + 3) = 0$

21. $\log 5 + \log x = 2$ **22.** $\log x - \log 8 = 1$

23. $\log x + \log (x - 3) = 1$

24. $\log (x - 9) + \log 100x = 3$

25. $\log (x + 1) - \log (x - 1) = 1$

26. $\log (2x + 1) = 1 + \log (x - 2)$

In Problems 27–34, solve to three significant digits.

27. $2 = 1.05^x$ **28.** $3 = 1.06^x$

29. $e^{-1.4x} + 5 = 0$ **30.** $e^{0.32x} + 0.47 = 0$

31. $123 = 500e^{-0.12x}$ **32.** $438 = 200e^{0.25x}$

33. $e^{-x^2} = 0.23$ **34.** $e^{x^2} = 125$

In Problems 35–48, solve exactly.

35. $\log (5 - 2x) = \log (3x + 1)$

36. $\log (x + 3) = \log (6 + 4x)$

37. $\log x - \log 5 = \log 2 - \log (x - 3)$

38. $\log (6x + 5) - \log 3 = \log 2 - \log x$

39. $\ln x = \ln (2x - 1) - \ln (x - 2)$

40. $\ln (x + 1) = \ln (3x + 1) - \ln x$

41. $\log (2x + 1) = 1 - \log (x - 1)$

42. $1 - \log (x - 2) = \log (3x + 1)$

43. $\ln (x + 1) = \ln (3x + 3)$

44. $1 + \ln (x + 1) = \ln (x - 1)$

54. $L = 8.8 + 5.1 \log D$ for D (astronomy)

55. $I = \dfrac{E}{R}(1 - e^{-Rt/L})$ for t (circuitry)

56. $S = R\dfrac{(1 + i)^n - 1}{i}$ for n (annuity)

*The following combinations of exponential functions define four of six **hyperbolic functions**, a useful class of functions in calculus and higher mathematics. Solve Problems 57–60 for x in terms of y. The results are used to define **inverse hyperbolic functions**, another useful class of functions in calculus and higher mathematics.*

57. $y = \dfrac{e^x + e^{-x}}{2}$ **58.** $y = \dfrac{e^x - e^{-x}}{2}$

59. $y = \dfrac{e^x - e^{-x}}{e^x + e^{-x}}$ **60.** $y = \dfrac{e^x + e^{-x}}{e^x - e^{-x}}$

In Problems 61–68, use a graphing calculator to approximate to two decimal places any solutions of the equation in the interval $0 \le x \le 1$. None of these equations can be solved exactly using any step-by-step algebraic process.

61. $2^{-x} - 2x = 0$ **62.** $3^{-x} - 3x = 0$

63. $e^{-x} - x = 0$ **64.** $xe^{2x} - 1 = 0$

65. $\ln x + 2x = 0$ **66.** $\ln x + x^2 = 0$

67. $\ln x + e^x = 0$ **68.** $\ln x + x = 0$

APPLICATIONS

69. COMPOUND INTEREST How many years, to the nearest year, will it take a sum of money to double if it is invested at 7% compounded annually?

70. COMPOUND INTEREST How many years, to the nearest year, will it take money to quadruple if it is invested at 6% compounded

Technology Connections

Technology Connections boxes integrated at appropriate points in the text illustrate how concepts previously introduced in an algebraic context may be approached using a graphing calculator. Students always learn the algebraic methods first so that they develop a solid grasp of these methods and do not become calculator-dependent. The exercise sets contain calculator-based exercises that are clearly marked with a calculator icon. The use of technology is completely optional with this text. All technology features and exercises may be omitted without sacrificing content coverage.

Technology Connections

Figure 1 shows the details of constructing the logarithmic model of Example 5 on a graphing calculator.

(a) Entering the data (b) Finding the model (c) Graphing the data and the model

▶ Figure 1

62. $g(x) = 4e^{x+1} - 7$; $f(x) = e^x$

63. $g(x) = 3 - 4e^{2-x}$; $f(x) = e^x$

64. $g(x) = -2 - 5e^{4-x}$; $f(x) = e^x$

In Problems 65–68, simplify.

65. $\dfrac{-2x^3 e^{-2x} - 3x^2 e^{-2x}}{x^6}$ **66.** $\dfrac{5x^4 e^{5x} - 4x^3 e^{5x}}{x^8}$

67. $(e^x + e^{-x})^2 + (e^x - e^{-x})^2$

68. $e^x(e^{-x} + 1) - e^{-x}(e^x + 1)$

In Problems 69–76, use a graphing calculator to find local extrema, y intercepts, and x intercepts. Investigate the behavior as $x \to \infty$ and as $x \to -\infty$ and identify any horizontal asymptotes. Round any approximate values to two decimal places.

69. $f(x) = 2 + e^{x-2}$ **70.** $g(x) = -3 + e^{1+x}$

71. $s(x) = e^{-x^2}$ **72.** $r(x) = e^{x^2}$

Group Activities

A Group Activity is located at the end of each chapter and involves many of the concepts discussed in that chapter. These activities strongly encourage the verbalization of mathematical concepts, results, and processes. All of these special activities are highlighted to emphasize their importance.

Foundation for Calculus

As many students will use this book to prepare for a calculus course, examples and exercises that are especially pertinent to calculus are marked with an icon. ∫

Student Aids

Annotation of examples and explanations, in small colored type, is found throughout the text to help students through critical stages. **Think Boxes** are dashed boxes used to enclose steps that students may be encouraged to perform mentally.

> The domain of f is all x values except $-\frac{5}{2}$, or $(-\infty, -\frac{5}{2}) \cup (-\frac{5}{2}, \infty)$.
> The value of a fraction is 0 if and only if the numerator is zero:
>
> $$4 - 3x = 0 \qquad \text{Subtract 4 from both sides.}$$
> $$-3x = -4 \qquad \text{Divide both sides by } -3.$$
> $$x = \tfrac{4}{3}$$
>
> The x intercept of f is $\frac{4}{3}$.
>
> The y intercept is $f(0) = \dfrac{4 - 3(0)}{2(0) + 5} = \dfrac{4}{5}$.

Screen Boxes are used to highlight important definitions, theorems, results, and step-by-step processes.

> **COMPOUND INTEREST**
>
> If a **principal** P is invested at an annual **rate** r compounded m times a year, then the **amount** A in the account at the end of n compounding periods is given by
>
> $$A = P\left(1 + \frac{r}{m}\right)^n$$
>
> Note that the annual rate r must be expressed in decimal form, and that $n = mt$, where t is years.

> **DEFINITION 1** Increasing, Decreasing, and Constant Functions
>
> Let I be an interval in the domain of function f. Then,
>
> 1. f is **increasing** on I and the graph of f is **rising** on I if $f(x_1) < f(x_2)$ whenever $x_1 < x_2$ in I.
> 2. f is **decreasing** on I and the graph of f is **falling** on I if $f(x_1) > f(x_2)$ whenever $x_1 < x_2$ in I.
> 3. f is **constant** on I and the graph of f is **horizontal** on I if $f(x_1) = f(x_2)$ whenever $x_1 < x_2$ in I.

> **THEOREM 1** Tests for Symmetry
>
Symmetry with respect to the:	An equivalent equation results if:
> | y axis | x is replaced with $-x$ |
> | x axis | y is replaced with $-y$ |
> | Origin | x and y are replaced with $-x$ and $-y$ |

Caution Boxes appear throughout the text to indicate where student errors often occur.

> **»» CAUTION »»**
>
> A very common error occurs about now—students tend to confuse *algebraic expressions* involving fractions with *algebraic equations* involving fractions. Consider these two problems:
>
> **(A)** Solve: $\dfrac{x}{2} + \dfrac{x}{3} = 10$ **(B)** Add: $\dfrac{x}{2} + \dfrac{x}{3} + 10$
>
> The problems look very much alike but are actually very different. To solve the equation in (A) we multiply both sides by 6 (the LCD) to clear the fractions. This works so well for equations that students want to do the same thing for problems like (B). The only catch is that (B) is not an equation, and the multiplication property of equality does not apply. If we multiply (B) by 6, we simply obtain an expression 6 times as large as the original! Compare these correct solutions:
>
> **(A)** $\dfrac{x}{2} + \dfrac{x}{3} = 10$
>
> $6 \cdot \dfrac{x}{2} + 6 \cdot \dfrac{x}{3} = 6 \cdot 10$
>
> $3x + 2x = 60$
>
> $5x = 60$
>
> $x = 12$
>
> **(B)** $\dfrac{x}{2} + \dfrac{x}{3} + 10$
>
> $= \dfrac{3 \cdot x}{3 \cdot 2} + \dfrac{2 \cdot x}{2 \cdot 3} + \dfrac{6 \cdot 10}{6 \cdot 1}$
>
> $= \dfrac{3x}{6} + \dfrac{2x}{6} + \dfrac{60}{6}$
>
> $= \dfrac{5x + 60}{6}$

Chapter Review sections are provided at the end of each chapter and include a thorough review of all the important terms and symbols. This recap is followed by a comprehensive set of review exercises.

CHAPTER **5** **Review**

5-1 Exponential Functions

The equation $f(x) = b^x$, $b > 0$, $b \neq 1$, defines an **exponential function with base b**. The **domain** of f is $(-\infty, \infty)$ and the **range** is $(0, \infty)$. The **graph** of f is a continuous curve that has no sharp corners; passes through $(0, 1)$; lies above the x axis, which is a horizontal asymptote; increases as x increases if $b > 1$; decreases as x increases if $b < 1$; and intersects any horizontal line at most once. The function f is one-to-one and has an inverse. We often use the following **exponential function properties:**

1. $a^x a^y = a^{x+y}$ $(a^x)^y = a^{xy}$ $(ab)^x = a^x b^x$

 $\left(\dfrac{a}{b}\right)^x = \dfrac{a^x}{b^x}$ $\dfrac{a^x}{a^y} = a^{x-y}$

2. $a^x = a^y$ if and only if $x = y$.

3. For $x \neq 0$, $a^x = b^x$ if and only if $a = b$.

As x approaches ∞, the expression $[1 + (1/x)]^x$ approaches the irrational number $e \approx 2.718\ 281\ 828\ 459$. The function $f(x) = e^x$ is called the **exponential function with base e**. The growth of money in an account paying **compound interest** is described by $A = P(1 + r/m)^n$, where P is the **principal**, r is the annual **rate**, m is the number of compounding periods in 1 year, and A is the **amount** in the account after n compounding periods.

If the account pays **continuous compound interest**, the amount A in the account after t years is given by $A = Pe^{rt}$.

5-2 Exponential Models

Exponential functions are used to model various types of growth:

1. *Population growth* can be modeled by using the **doubling time growth model** $A = A_0 2^{t/d}$, where A is the population at time t, A_0 is the population at time $t = 0$, and d is the **doubling time**—

the time it takes for the population to double. Another model of population growth, $A = A_0 e^{kt}$, where A_0 is the population at time zero and k is a positive constant called the **relative growth rate**, uses the exponential function with base e. This model is used for many other types of quantities that exhibit exponential growth as well.

2. *Radioactive decay* can be modeled by using the **half-life decay model** $A = A_0(\frac{1}{2})^{t/h} = A_0 2^{-t/h}$, where A is the amount at time t, A_0 is the amount at time $t = 0$, and h is the **half-life**—the time it takes for half the material to decay. Another model of radioactive decay, $A = A_0 e^{-kt}$, where A_0 is the amount at time zero and k is a positive constant, uses the exponential function with base e. This model can be used for other types of quantities that exhibit negative exponential growth as well.

3. *Limited growth*—the growth of a company or proficiency at learning a skill, for example—can often be modeled by the equation $y = A(1 - e^{-kt})$, where A and k are positive constants.

 Logistic growth is another limited growth model that is useful for modeling phenomena like the spread of an epidemic, or sales of a new product. The logistic model is $A = M/(1 + ce^{-kt})$, where c, k, and M are positive constants. A good comparison of these different exponential models can be found in Table 3 at the end of Section 5-2.

 Exponential regression can be used to fit a function of the form $y = ab^x$ to a set of data points. Logistic regression can be used to find a function of the form $y = c/(1 + ae^{-bx})$.

5-3 Logarithmic Functions

The **logarithmic function with base b** is defined to be the inverse of the exponential function with base b and is denoted by $y = \log_b x$. So $y = \log_b x$ if and only if $x = b^y$, $b > 0$, $b \neq 1$. The **domain** of a logarithmic function is $(0, \infty)$ and the **range** is $(-\infty, \infty)$. The graph of a logarithmic function is a continuous curve that always passes

Cumulative Review Exercise Sets are provided in Appendix A for additional reinforcement of key concepts.

CHAPTERS **1–3** **Cumulative Review Exercises**

Work through all the problems in this cumulative review and check answers in the back of the book. Answers to all review problems are there, and following each answer is a number in italics indicating the section in which that type of problem is discussed. Where weaknesses show up, review appropriate sections in the text.

1. Solve for x: $\dfrac{7x}{5} - \dfrac{3 + 2x}{2} = \dfrac{x - 10}{3} + 2$

In Problems 2–4, solve and graph the inequality.

2. $2(3 - y) + 4 \leq 5 - y$ 3. $|x - 2| < 7$

4. $x^2 + 3x \geq 10$

5. Perform the indicated operations and write the answer in standard form:
 (A) $(2 - 3i) - (-5 + 7i)$
 (B) $(1 + 4i)(3 - 5i)$
 (C) $\dfrac{5 + i}{2 + 3i}$

In Problems 6–9, solve the equation.

6. $3x^2 = -12x$ 7. $4x^2 - 20 = 0$

8. $x^2 - 6x + 2 = 0$ 9. $x - \sqrt{12 - x} = 0$

10. Given the points $A = (3, 2)$ and $B = (5, 6)$, find:
 (A) Distance between A and B.
 (B) Slope of the line through A and B.
 (C) Slope of a line perpendicular to the line through A and B.

11. Find the equation of the circle with radius $\sqrt{2}$ and center:
 (A) $(0, 0)$ (B) $(-3, 1)$

12. Graph $2x - 3y = 6$ and indicate its slope and intercepts.

13. Indicate whether each set defines a function. Find the domain and range of each function.
 (A) $\{(1, 1), (2, 1), (3, 1)\}$

Problems 16–18 refer to the function f given by the graph:

16. Find the domain and range of f. Express answers in interval notation.

17. Is f an even function, an odd function, or neither? Explain.

18. Use the graph of f to sketch a graph of the following:
 (A) $y = -f(x + 1)$ (B) $y = 2f(x) - 2$

In Problems 19–21, solve the equation.

19. $\dfrac{x + 3}{2x + 2} + \dfrac{5x + 2}{3x + 3} = \dfrac{5}{6}$ 20. $\dfrac{3}{x} = \dfrac{6}{x + 1} - \dfrac{1}{x - 1}$

21. $2x + 1 = 3\sqrt{2x - 1}$

In Problems 22–24, solve and graph the inequality.

22. $|4x - 9| > 3$ 23. $\sqrt{(3m - 4)^2} \leq 2$

24. $\dfrac{x + 1}{2} > x - 2$

25. For what real values of x does the following expression represent a real number?

 $\dfrac{\sqrt{x - 2}}{x - 4}$

Experience Student Success!

ALEKS ALEKS is a unique online math tool that uses adaptive questioning and artificial intelligence to correctly place, prepare, and remediate students . . . all in one product! Institutional case studies have shown that **ALEKS has improved pass rates by over 20% versus traditional online homework and by over 30% compared to using a text alone.**

By offering each student an individualized learning path, ALEKS directs students to work on the math topics that they are ready to learn. Also, to help students keep pace in their course, instructors can correlate ALEKS to their textbook or syllabus in seconds.

To learn more about how ALEKS can be used to boost student performance, please visit www.aleks.com/highered/math or contact your McGraw-Hill representative.

ALEKS Pie
Each student is given their own individualized learning path.

Easy Graphing Utility!
Students can answer graphing problems with ease!

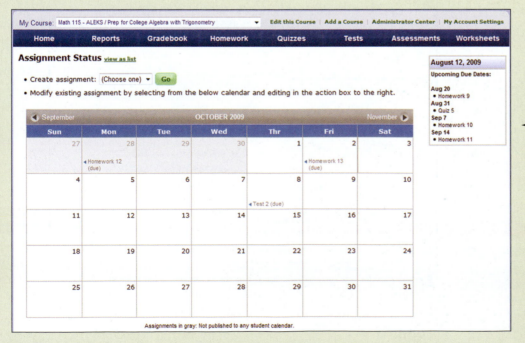

Course Calendar
Instructors can schedule assignments and reminders for students.

New ALEKS Instructor Module

Enhanced Functionality and Streamlined Interface Help to Save Instructor Time

ALEKS® The new ALEKS Instructor Module features enhanced functionality and streamlined interface based on research with ALEKS instructors and homework management instructors. Paired with powerful assignment driven features, textbook integration, and extensive content flexibility, the new ALEKS Instructor Module simplifies administrative tasks and makes ALEKS more powerful than ever.

Gradebook view for all students Gradebook view for an individual student

New Gradebook!

Instructors can seamlessly track student scores on automatically graded assignments. They can also easily adjust the weighting and grading scale of each assignment.

Track Student Progress Through Detailed Reporting

Instructors can track student progress through automated reports and robust reporting features.

Select topics for each assignment

Automatically Graded Assignments

Instructors can easily assign homework, quizzes, tests, and assessments to all or select students. Deadline extensions can also be created for select students.

Learn more about ALEKS by visiting www.aleks.com/highered/math or contact your McGraw-Hill representative.

Supplements

ALEKS

ALEKS (**A**ssessment and **L**earning in **K**nowledge **S**paces) is a dynamic online learning system for mathematics education, available over the Web 24/7. ALEKS assesses students, accurately determines their knowledge, and then guides them to the material that they are most ready to learn. With a variety of reports, Textbook Integration Plus, quizzes, and homework assignment capabilities, ALEKS offers flexibility and ease of use for instructors.

- ALEKS uses artificial intelligence to determine exactly what each student knows and is ready to learn. ALEKS remediates student gaps and provides highly efficient learning and improved learning outcomes

- ALEKS is a comprehensive curriculum that aligns with syllabi or specified textbooks. Used in conjunction with McGraw-Hill texts, students also receive links to text-specific videos, multimedia tutorials, and textbook pages.

- ALEKS offers a dynamic classroom management system that enables instructors to monitor and direct student progress towards mastery of course objectives.

ALEKS Prep/Remediation:

- Helps instructors meet the challenge of remediating under prepared or improperly placed students.

- Assesses students on their pre-requisite knowledge needed for the course they are entering (i.e. Calculus students are tested on Precalculus knowledge) and prescribes unique and efficient learning paths specific to their strengths and weaknesses.

- Students can address pre-requisite knowledge gaps outside of class freeing the instructor to use class time pursuing course outcomes.

McGraw-Hill's MathZone is a complete online homework system for mathematics and statistics. Instructors can assign textbook-specific content from over 40 McGraw-Hill titles as well as customize the level of feedback students receive, including the ability to have students show their work for any given exercise. Assignable content includes an array of videos and other multimedia along with algorithmic exercises, providing study tools for students with many different learning styles.

MathZone also helps ensure consistent assignment delivery across several sections through a course administration function and makes sharing courses with other instructors easy. In addition, instructors can also take advantage of a virtual whiteboard by setting up a Live Classroom for online office hours or a review session with students.

For more information, visit the book's website (www.mhhe.com/barnett) or contact your local McGraw-Hill sales representative (www.mhhe.com/rep).

McGraw Hill Tegrity campus

Tegrity Campus is a service that makes class time available all the time by automatically capturing every lecture in a searchable format for students to review when they study and complete assignments. With a simple one-click start and stop process, you capture all computer screens and corresponding audio. Students replay any part of any class with easy-to-use browser-based viewing on a PC or Mac.

Educators know that the more students can see, hear, and experience class resources, the better they learn. With Tegrity Campus, students quickly recall key moments by using

Tegrity Campus's unique search feature. This search helps students efficiently find what they need, when they need it across an entire semester of class recordings. Help turn all your students' study time into learning moments immediately supported by your lecture.

To learn more about Tegrity watch a 2 minute Flash demo at http://tegritycampus.mhhe.com.

Instructor Solutions Manual Prepared by Fred Safier of City College of San Francisco, this supplement provides detailed solutions to exercises in the text. The methods used to solve the problems in the manual are the same as those used to solve the examples in the textbook.

Student Solutions Manual Prepared by Fred Safier of City College of San Francisco, the *Student's Solutions Manual* provides complete worked-out solutions to odd-numbered exercises from the text. The procedures followed in the solutions in the manual match exactly those shown in worked examples in the text.

Lecture and Exercise Videos The video series is based on exercises from the textbook. J. D. Herdlick of St. Louis Community College-Meramec introduces essential definitions, theorems, formulas, and problem-solving procedures. Professor Herdlick then works through selected problems from the textbook, following the solution methodologies employed by the authors. The video series is available on DVD or online as part of MathZone. The DVDs are closed-captioned for the hearing impaired, subtitled in Spanish, and meet the Americans with Disabilities Act Standards for Accessible Design.

NetTutor Available through MathZone, NetTutor is a revolutionary system that enables students to interact with a live tutor over the web. NetTutor's web-based, graphical chat capabilities enable students and tutors to use mathematical notation and even to draw graphs as they work through a problem together. Students can also submit questions and receive answers, browse previously answered questions, and view previous sessions. Tutors are familiar with the textbook's objectives and problem-solving styles.

Computerized Test Bank (CTB) Online Available through the book's website, this computerized test bank, utilizing Brownstone Diploma® algorithm-based testing software, enables users to create customized exams quickly. This user-friendly program enables instructors to search for questions by topic, format, or difficulty level; to edit existing questions or to add new ones; and to scramble questions and answer keys for multiple versions of the same test. Hundreds of text-specific open-ended and multiple-choice questions are included in the question bank. Sample chapter tests and a sample final exam in Microsoft Word® and PDF formats are also provided.

Acknowledgments

In addition to the authors, many others are involved in the successful publication of a book. We wish to thank personally the following people who reviewed the text and offered invaluable advice for improvements:

Marwan Abu-Sawwa, Florida Community College at Jacksonville
Gerardo Aladro, Florida International University
Eugene Allevato, Woodbury University
Joy Becker, University of Wisconsin–Stout
Susan Bradley, Angelina College
Ellen Brook, Cuyahoga Community College, Eastern Campus
Kelly Brooks, Pierce College
Denise Brown, Collin County Community College
Cheryl Davids, Central Carolina Technical College
Timothy Delworth, Purdue University
Marcial Echenique, Broward Community College
Gay Ellis, Missouri State University
Jackie English, Northern Oklahoma College Enid
Mike Everett, Santa Ana College
Nicki Feldman, Pulaski Technical College
James Fightmaster, Virginia Western Community College
Perry Gillespie, University of North Carolina at Fayetteville
Vanetta Grier-Felix, Seminole Community College
David Gurney, Southeastern Louisiana University
Celeste Hernandez, Richland College
Fredrick Hoffman, Florida Atlantic University
Syed Hossain, University of Nebraska at Kearney
Glenn Jablonski, Triton College
Sarah Jackson, Pratt Community College
Charles Johnson, South Georgia College
Larry Johnson, Metropolitan State College of Denver
Cheryl Kane, University of Nebraska Lincoln
Raja Khoury, Collin County Community College
Betty Larson, South Dakota State University
Owen Mertens, Southwest Missouri State University
Dana Nimic, Southeast Community College, Lincoln Campus
Lyn Noble, Florida Community College at Jacksonville
Luke Papademas, DeVry University, DeVry Chicago Campus
David Phillips, Georgia State University
Margaret Rosen, Mercer County Community College
Patty Schovanec, Texas Tech University
Eleanor Storey, Front Range Community College, Westminster Campus
Linda Sundbye, Metropolitan State College of Denver
Cynthia Woodburn, Pittsburg State University
Martha Zimmerman, University of Louisville
Bob Martin, Tarrant County College
Susan Walker, Montana Tech of the University of Montana
Lynn Cleaveland, University of Arkansas
Michael Wodzak, Viterbo University
Ryan Kasha, Valencia Community College
Frank Juric, Brevard Community College
Jerry Mayfield, North Lake College
Andrew Shiers, Dakota State University
Richard Avery, Dakota State University

Mike Everett, Santa Ana College
Greg Boyd, Murray State College
Sarah Cook, Washburn University
Nga Wai Liu, Bowling Green State University
Donald Bennett, Murray State University
Sharon Suess, Asheville-Buncombe Technical Community College
Dale Rohm, University of Wisconsin at Stevens Point
George Anastassiou, The University of Memphis
Bill White, University of South Carolina Upstate
Linda Sundbye, Metropolitan State College of Denver
Khaled Hussein, University of Wisconsin
Diane Cook, Okaloosa Walton College
Celeste Hernandez, Richland College
Thomas Riedel, University of Louisville
Thomas English, College of the Mainland
Hayward Allan Edwards, West Virginia University at Parkersburg
Debra Lehman, State Fair Community College
Nancy Ressler, Oakton Community College
Marwan Zabdawi, Gordonn College
Ianna West, Nicholls State University
Tzu-Yi Alan Yang, Columbus State Community College
Patricia Jones, Methodist University
Kay Geving, Belmont University
Linda Horner, Columbia State Community College
Martha Zimmerman, University of Louisville
Faye Childress, Central Piedmont Community College
Bradley Thiessen, Saint Ambrose University
Pamela Lasher, Edinboro University of Pennsylvania

We also wish to thank

Carrie Green for providing a careful and thorough check of all the mathematical calculations in the book (a tedious but extremely important job).

Fred Safier for developing the supplemental manuals that are so important to the success of a text.

Mitchel Levy for scrutinizing our exercises in the manuscript and making recommendations that helped us to build balanced exercise sets.

Tony Palermino for providing excellent guidance in making the writing more direct and accessible to students.

Pat Steele for carefully editing and correcting the manuscript.

Christina Lane for editorial guidance throughout the revision process.

Sheila Frank for guiding the book smoothly through all publication details.

All the people at McGraw-Hill who contributed their efforts to the production of this book.

Producing this new edition with the help of all these extremely competent people has been a most satisfying experience.

APPLICATION INDEX

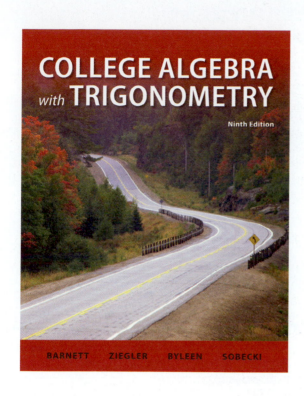

College Algebra
with Trigonometry

Basic Algebraic Operations

ALGEBRA is "generalized arithmetic." In arithmetic we add, subtract, multiply, and divide specific numbers. In algebra we use all that we know about arithmetic, but, in addition, we work with symbols that represent one or more numbers. In this chapter we review some important basic algebraic operations usually studied in earlier courses.

R-1 Algebra and Real Numbers

> › The Set of Real Numbers
> › The Real Number Line
> › Addition and Multiplication of Real Numbers
> › Further Operations and Properties

The numbers 14, -3, 0, $\frac{7}{3}$, $\sqrt{2}$, and $\sqrt[3]{6}$ are examples of *real numbers*. Because the symbols we use in algebra often stand for real numbers, we will discuss important properties of the real number system.

› The Set of Real Numbers

Informally, a **real number** is any number that has a decimal representation. So the real numbers are the numbers you have used for most of your life. The **set of real numbers,** denoted by R, is the collection of all real numbers. The notation $\sqrt{2} \in R$ (read "$\sqrt{2}$ is an **element** of R") expresses the fact that $\sqrt{2}$ is a real number. The set $Z = \{\ldots, -2, -1, 0, 1, 2, \ldots\}$ of the natural numbers, along with their negatives and zero, is called the **set of integers.** We write $Z \subset R$ (read "Z is a **subset** of R") to express the fact that every element of Z is an element of R; that is, that every integer is a real number. Table 1 describes the set of real numbers and some of its important subsets. Study Table 1 and note in particular that $N \subset Z \subset Q \subset R$.

No real number is both rational and irrational, so the intersection (overlap) of the sets Q and I is the **empty set** (or **null set**), denoted by $\varnothing$. The empty set contains no elements,

Table 1 The Set of Real Numbers

Symbol	Name	Description	Examples
N	Natural numbers	Counting numbers (also called positive integers)	1, 2, 3, . . .
Z	Integers	Natural numbers, their negatives, and 0 (also called whole numbers)	. . . , -2, -1, 0, 1, 2, . . .
Q	Rational numbers	Numbers that can be represented as a/b, where a and b are integers and $b \neq 0$; decimal representations are repeating or terminating	-4, 0, 1, 25, $-\frac{3}{5}$, $\frac{2}{3}$, 3.67, $-0.33\overline{3}$,* $5.27\overline{27}$
I	Irrational numbers	Numbers that can be represented as nonrepeating and nonterminating decimal numbers	$\sqrt{2}$, π, $\sqrt[3]{7}$, 1.414213 . . . ,† 2.71828182 . . .†
R	Real numbers	Rational numbers and irrational numbers	

*The overbar indicates that the number (or block of numbers) repeats indefinitely.
†Note that the ellipsis does *not* indicate that a number (or block of numbers) repeats indefinitely.

so it is true that every element of the empty set is an element of any given set. In other words, the empty set is a subset of every set.

Two sets are **equal** if they have exactly the same elements. The order in which the elements of a set are listed does not matter. For example,

$$\{1, 2, 3, 4\} = \{3, 1, 4, 2\}$$

› The Real Number Line

A one-to-one correspondence exists between the set of real numbers and the set of points on a line. That is, each real number corresponds to exactly one point, and each point to exactly one real number. A line with a real number associated with each point, and vice versa, as in Figure 1, is called a **real number line,** or simply a **real line.** Each number associated with a point is called the **coordinate** of the point. The point with coordinate 0 is called the **origin.** The arrow on the right end of the line indicates a positive direction. The coordinates of all points to the right of the origin are called **positive real numbers,** and those to the left of the origin are called **negative real numbers.** The real number 0 is neither positive nor negative.

> **Figure 1** A real number line.

› Addition and Multiplication of Real Numbers

How do you add or multiply two real numbers that have nonrepeating and nonterminating decimal expansions? The answer to this difficult question relies on a solid understanding of the arithmetic of rational numbers. The **rational numbers** are numbers that can be written in the form a/b, where a and b are integers and $b \neq 0$ (see Table 1 on page 2). The numbers 7/5 and $-2/3$ are rational, and any integer a is rational because it can be written in the form $a/1$. Two rational numbers a/b and c/d are **equal** if $ad = bc$; for example, $35/10 = 7/2$. Recall how the sum and product of rational numbers are defined:

> ### › DEFINITION 1 Addition and Multiplication of Rationals
>
> For rational numbers a/b and c/d, where a, b, c, and d are integers and $b \neq 0$, $d \neq 0$:
>
> **Addition:** $\dfrac{a}{b} + \dfrac{c}{d} = \dfrac{ad + bc}{bd}$
>
> **Multiplication:** $\dfrac{a}{b} \cdot \dfrac{c}{d} = \dfrac{ac}{bd}$

Addition and multiplication of rational numbers are **commutative;** changing the order in which two numbers are added or multiplied does not change the result.

$$\frac{3}{2} + \frac{5}{7} = \frac{5}{7} + \frac{3}{2} \qquad \text{Addition is commutative.}$$

$$\frac{3}{2} \cdot \frac{5}{7} = \frac{5}{7} \cdot \frac{3}{2} \qquad \text{Multiplication is commutative.}$$

Addition and multiplication of rational numbers is also **associative;** changing the grouping of three numbers that are added or multiplied does not change the result:

$$\frac{3}{2} + \left(\frac{5}{7} + \frac{9}{4}\right) = \left(\frac{3}{2} + \frac{5}{7}\right) + \frac{9}{4} \qquad \text{Addition is associative.}$$

$$\frac{3}{2} \cdot \left(\frac{5}{7} \cdot \frac{9}{4}\right) = \left(\frac{3}{2} \cdot \frac{5}{7}\right) \cdot \frac{9}{4} \qquad \text{Multiplication is associative.}$$

Furthermore, the operations of addition and multiplication are related in that multiplication **distributes** over addition:

$$\frac{3}{2} \cdot \left(\frac{5}{7} + \frac{9}{4}\right) = \frac{3}{2} \cdot \frac{5}{7} + \frac{3}{2} \cdot \frac{9}{4} \qquad \text{Left distributive law}$$

$$\left(\frac{5}{7} + \frac{9}{4}\right) \cdot \frac{3}{2} = \frac{5}{7} \cdot \frac{3}{2} + \frac{9}{4} \cdot \frac{3}{2} \qquad \text{Right distributive law}$$

The rational number 0 is an **additive identity;** adding 0 to a number does not change it. The rational number 1 is a **multiplicative identity;** multiplying a number by 1 does not change it. Every rational number r has an **additive inverse,** denoted $-r$; the additive inverse of $4/5$ is $-4/5$, and the additive inverse of $-3/2$ is $3/2$. The sum of a number and its additive inverse is 0. Every *nonzero* rational number r has a **multiplicative inverse,** denoted r^{-1}; the multiplicative inverse of $4/5$ is $5/4$, and the multiplicative inverse of $-3/2$ is $-2/3$. The product of a number and its multiplicative inverse is 1. The rational number 0 has no multiplicative inverse.

| EXAMPLE | 1 | **Arithmetic of Rational Numbers** |

Perform the indicated operations.

(A) $\dfrac{1}{3} + \dfrac{6}{5}$ (B) $\dfrac{8}{3} \cdot \dfrac{5}{4}$

(C) $(-17/9)^{-1}$ (D) $(-6 + 9/2)^{-1}$

SOLUTIONS

(A) $\dfrac{1}{3} + \dfrac{6}{5} = \dfrac{5 + 18}{15} = \dfrac{23}{15}$

(B) $\dfrac{8}{3} \cdot \dfrac{5}{4} = \dfrac{40}{12} = \dfrac{10}{3}$ $\dfrac{40}{12} = \dfrac{10}{3}$ because $40 \cdot 3 = 12 \cdot 10$

(C) $(-17/9)^{-1} = -9/17$

(D) $(-6 + 9/2)^{-1} = \left(\dfrac{-6}{1} + \dfrac{9}{2}\right)^{-1} = \left(\dfrac{-12 + 9}{2}\right)^{-1} = \left(\dfrac{-3}{2}\right)^{-1} = -\dfrac{2}{3}$

MATCHED PROBLEM 1[*] Perform the indicated operations.

(A) $-(5/2 + 7/3)$ (B) $-(8/17)^{-1}$

(C) $\dfrac{21}{20} \cdot \dfrac{15}{14}$ (D) $5 \cdot (1/2 + 1/3)$

Rational numbers have decimal expansions that are repeating or terminating. For example, using long division,

$$\frac{2}{3} = 0.66\overline{6} \qquad \text{The number 6 repeats indefinitely.}$$

$$\frac{22}{7} = 3.\overline{142857} \qquad \text{The block 142857 repeats indefinitely.}$$

$$\frac{13}{8} = 1.625 \qquad \text{Terminating expansion}$$

Conversely, any decimal expansion that is repeating or terminating represents a rational number (see Problems 49 and 50 in Exercise R-1).

The number $\sqrt{2}$ is *irrational* because it cannot be written in the form a/b, where a and b are integers, $b \neq 0$ (for an explanation, see Problem 89 in Section R-3). Similarly, $\sqrt{3}$ is irrational. But $\sqrt{4}$, which is equal to 2, is a rational number. In fact, if n is a positive integer, then $\sqrt{n}$ is irrational unless n belongs to the sequence of perfect squares 1, 4, 9, 16, 25, . . . (see Problem 90 in Section R-3).

We now return to our original question: how do you add or multiply two real numbers that have nonrepeating and nonterminating decimal expansions? Although we will not give a detailed answer to this question, the key idea is that every real number can be approximated to any desired precision by rational numbers. For example, the irrational number

$$\sqrt{2} \approx 1.414\,213\,562\ldots$$

is approximated by the rational numbers

$$\frac{14}{10} = 1.4$$

$$\frac{141}{100} = 1.41$$

$$\frac{1{,}414}{1{,}000} = 1.414$$

$$\frac{14{,}142}{10{,}000} = 1.4142$$

$$\frac{141{,}421}{100{,}000} = 1.41421$$

$$\vdots$$

Using the idea of approximation by rational numbers, we can extend the definitions of rational number operations to include real number operations. The following box summarizes the basic properties of real number operations.

[*]Answers to matched problems in a given section are found near the end of the section, before the exercise set.

> **BASIC PROPERTIES OF THE SET OF REAL NUMBERS**

Let R be the set of real numbers, and let x, y, and z be arbitrary elements of R.

Addition Properties

Closure: $x + y$ is a unique element in R.

Associative: $(x + y) + z = x + (y + z)$

Commutative: $x + y = y + x$

Identity: 0 is the additive identity; that is, $0 + x = x + 0 = x$ for all x in R, and 0 is the only element in R with this property.

Inverse: For each x in R, $-x$ is its unique additive inverse; that is, $x + (-x) = (-x) + x = 0$, and $-x$ is the only element in R relative to x with this property.

Multiplication Properties

Closure: xy is a unique element in R.

Associative: $(xy)z = x(yz)$

Commutative: $xy = yx$

Identity: 1 is the multiplicative identity; that is, for all x in R, $(1)x = x(1) = x$, and 1 is the only element in R with this property.

Inverse: For each x in R, $x \neq 0$, x^{-1} is its unique multiplicative inverse; that is, $xx^{-1} = x^{-1}x = 1$, and x^{-1} is the only element in R relative to x with this property.

Combined Property

Distributive: $x(y + z) = xy + xz$ $(x + y)z = xz + yz$

EXAMPLE 2 **Using Real Number Properties**

Which real number property justifies the indicated statement?

(A) $(7x)y = 7(xy)$
(B) $a(b + c) = (b + c)a$
(C) $(2x + 3y) + 5y = 2x + (3y + 5y)$
(D) $(x + y)(a + b) = (x + y)a + (x + y)b$
(E) If $a + b = 0$, then $b = -a$.

SOLUTIONS (A) Associative ($\cdot$)
(B) Commutative ($\cdot$)
(C) Associative ($+$)
(D) Distributive
(E) Inverse ($+$)

Which real number property justifies the indicated statement?

(A) $4 + (2 + x) = (4 + 2) + x$ (B) $(a + b) + c = c + (a + b)$

(C) $3x + 7x = (3 + 7)x$ (D) $(2x + 3y) + 0 = 2x + 3y$

(E) If $ab = 1$, then $b = 1/a$.

› Further Operations and Properties

Subtraction of real numbers can be defined in terms of addition and the additive inverse. If a and b are real numbers, then $a - b$ is defined to be $a + (-b)$. Similarly, division can be defined in terms of multiplication and the multiplicative inverse. If a and b are real numbers and $b \neq 0$, then $a \div b$ (also denoted a/b) is defined to be $a \cdot b^{-1}$.

> ### › DEFINITION 2 Subtraction and Division of Real Numbers
>
> For all real numbers a and b:
>
> **Subtraction:** $a - b = a + (-b)$ $5 - 3 = 5 + (-3) = 2$
>
> **Division:** $a \div b = a \cdot b^{-1}$ $b \neq 0$ $3 \div 2 = 3 \cdot 2^{-1} = 3 \cdot \dfrac{1}{2} = 1.5$

It is important to remember that

Division by 0 is never allowed.

(A) Give an example that shows that subtraction of real numbers is not commutative.

(B) Give an example that shows that division of real numbers is not commutative.

The basic properties of the set of real numbers, together with the definitions of subtraction and division, lead to the following properties of negatives and zero.

> ### › THEOREM 1 Properties of Negatives
>
> For all real numbers a and b:
>
> 1. $-(-a) = a$
> 2. $(-a)b = -(ab) = a(-b) = -ab$
> 3. $(-a)(-b) = ab$
> 4. $(-1)a = -a$
> 5. $\dfrac{-a}{b} = -\dfrac{a}{b} = \dfrac{a}{-b}$ $b \neq 0$
> 6. $\dfrac{-a}{-b} = \dfrac{-a}{b} = -\dfrac{a}{-b} = \dfrac{a}{b}$ $b \neq 0$

> › **THEOREM 2** Zero Properties
>
> For all real numbers a and b:
>
> 1. $a \cdot 0 = 0 \cdot a = 0$
> 2. $ab = 0$ if and only if* $a = 0$ or $b = 0$ or both

Note that if $b \neq 0$, then $\frac{0}{b} = 0 \cdot b^{-1} = 0$ by Theorem 2. In particular, $\frac{0}{3} = 0$; but the expressions $\frac{3}{0}$ and $\frac{0}{0}$ are undefined.

EXAMPLE	3

Using Negative and Zero Properties

Which real number property or definition justifies each statement?

(A) $3 - (-2) = 3 + [-(-2)] = 5$

(B) $-(-2) = 2$

(C) $-\dfrac{-3}{2} = \dfrac{3}{2}$

(D) $\dfrac{5}{-2} = -\dfrac{5}{2}$

(E) If $(x - 3)(x + 5) = 0$, then either $x - 3 = 0$ or $x + 5 = 0$.

SOLUTIONS

(A) Subtraction (Definition 1 and Theorem 1, part 1)

(B) Negatives (Theorem 1, part 1)

(C) Negatives (Theorem 1, part 6)

(D) Negatives (Theorem 1, part 5)

(E) Zero (Theorem 2, part 2)

MATCHED PROBLEM 3

Which real number property or definition justifies each statement?

(A) $\dfrac{3}{5} = 3\left(\dfrac{1}{5}\right)$ (B) $(-5)(2) = -(5 \cdot 2)$ (C) $(-1)3 = -3$

(D) $\dfrac{-7}{9} = -\dfrac{7}{9}$ (E) If $x + 5 = 0,$ then $(x - 3)(x + 5) = 0.$

››› EXPLORE-DISCUSS 2

A set of numbers is **closed** under an operation if performing the operation on numbers of the set always produces another number in the set. For example, the set of odd integers is closed under multiplication, but is not closed under addition.

(A) Give an example that shows that the set of irrational numbers is not closed under addition.

(B) Explain why the set of irrational numbers is closed under taking multiplicative inverses.

*Given statements P and Q, "P if and only if Q" stands for both "if P, then Q" and "if Q, then P."

If a and b are real numbers, $b \neq 0$, the quotient $a \div b$, when written in the form a/b, is called a **fraction.** The number a is the **numerator,** and b is the **denominator.** It can be shown that fractions satisfy the following properties. (Note that some of these properties, under the restriction that numerators and denominators are integers, were used earlier to define arithmetic operations on the rationals.)

> **THEOREM 3** Fraction Properties

For all real numbers a, b, c, d, and k (division by 0 excluded):

1. $\dfrac{a}{b} = \dfrac{c}{d}$ if and only if $ad = bc$

 $\dfrac{4}{6} = \dfrac{6}{9}$ since $4 \cdot 9 = 6 \cdot 6$

2. $\dfrac{ka}{kb} = \dfrac{a}{b}$ 3. $\dfrac{a}{b} \cdot \dfrac{c}{d} = \dfrac{ac}{bd}$ 4. $\dfrac{a}{b} \div \dfrac{c}{d} = \dfrac{a}{b} \cdot \dfrac{d}{c}$

 $\dfrac{7 \cdot 3}{7 \cdot 5} = \dfrac{3}{5}$ $\dfrac{3}{5} \cdot \dfrac{7}{8} = \dfrac{3 \cdot 7}{5 \cdot 8} = \dfrac{21}{40}$ $\dfrac{2}{3} \div \dfrac{5}{7} = \dfrac{2}{3} \cdot \dfrac{7}{5} = \dfrac{14}{15}$

5. $\dfrac{a}{b} + \dfrac{c}{b} = \dfrac{a + c}{b}$ 6. $\dfrac{a}{b} - \dfrac{c}{b} = \dfrac{a - c}{b}$ 7. $\dfrac{a}{b} + \dfrac{c}{d} = \dfrac{ad + bc}{bd}$

 $\dfrac{3}{6} + \dfrac{4}{6} = \dfrac{3 + 4}{6} = \dfrac{7}{6}$ $\dfrac{7}{8} - \dfrac{2}{8} = \dfrac{7 - 2}{8} = \dfrac{5}{8}$ $\dfrac{2}{3} + \dfrac{1}{5} = \dfrac{2 \cdot 5 + 3 \cdot 1}{3 \cdot 5} = \dfrac{13}{15}$

ANSWERS TO MATCHED PROBLEMS

1. (A) $-29/6$ (B) $-17/8$ (C) $9/8$ (D) $25/6$
2. (A) Associative $(+)$ (B) Commutative $(+)$ (C) Distributive
 (D) Identity $(+)$ (E) Inverse $(\cdot)$
3. (A) Division (Definition 1) (B) Negatives (Theorem 1, part 2)
 (C) Negatives (Theorem 1, part 4) (D) Negatives (Theorem 1, part 5)
 (E) Zero (Theorem 2, part 1)

R-1 Exercises

In Problems 1–16, perform the indicated operations, if defined. If the result is not an integer, express it in the form a/b, where a and b are integers.

1. $\dfrac{1}{3} + \dfrac{1}{5}$ **2.** $\dfrac{1}{2} + \dfrac{1}{7}$

3. $\dfrac{3}{4} - \dfrac{4}{3}$ **4.** $\dfrac{8}{9} - \dfrac{4}{5}$

5. $\dfrac{2}{3} \cdot \dfrac{4}{7}$ **6.** $\left(-\dfrac{1}{10}\right) \cdot \dfrac{3}{8}$

7. $\dfrac{11}{5} \div \dfrac{1}{3}$ **8.** $\dfrac{2}{9} \div \dfrac{7}{5}$

9. $100 \div 0$ **10.** $0 \div 0$

11. $\left(-\dfrac{3}{5}\right)\left(-\dfrac{5}{3}\right)$ **12.** $\dfrac{4}{7} \div \left(3 - \dfrac{6}{2}\right)$

13. $\dfrac{17}{8} \cdot \dfrac{2}{7}$ **14.** $\left(-\dfrac{2}{3}\right)\left(-\dfrac{5}{6}\right)$

15. $\left(\dfrac{3}{8}\right)^{-1} + 2^{-1}$ **16.** $-(4^{-1} + 3)$

In Problems 17–28, each statement illustrates the use of one of the following properties or definitions. Indicate which one.

Commutative (+)	Inverse (+)
Commutative (·)	Inverse (·)
Associative (+)	Subtraction
Associative (·)	Division
Distributive	Negatives (Theorem 1)
Identity (+)	Zero (Theorem 2)
Identity (·)	

17. $x + ym = x + my$ **18.** $7(3m) = (7 \cdot 3)m$

19. $7u + 9u = (7 + 9)u$ **20.** $-\dfrac{u}{-v} = \dfrac{u}{v}$

21. $(-2)(\frac{1}{-2}) = 1$ **22.** $8 - 12 = 8 + (-12)$

23. $w + (-w) = 0$ **24.** $5 \div (-6) = 5(\frac{1}{-6})$

25. $3(xy + z) + 0 = 3(xy + z)$

26. $ab(c + d) = abc + abd$

27. $\dfrac{-x}{-y} = \dfrac{x}{y}$ **28.** $(x + y) \cdot 0 = 0$

29. If $ab = 0$, does either a or b have to be 0?

30. If $ab = 1$, does either a or b have to be 1?

31. Indicate which of the following are true:
 (A) All natural numbers are integers.
 (B) All real numbers are irrational.
 (C) All rational numbers are real numbers.

32. Indicate which of the following are true:
 (A) All integers are natural numbers.
 (B) All rational numbers are real numbers.
 (C) All natural numbers are rational numbers.

33. Give an example of a rational number that is not an integer.

34. Give an example of a real number that is not a rational number.

In Problems 35 and 36, list the subset of S consisting of (A) natural numbers, (B) integers, (C) rational numbers, and (D) irrational numbers.

35. $S = \{-3, -\frac{2}{3}, 0, 1, \sqrt{3}, \frac{9}{5}, \sqrt{144}\}$

36. $S = \{-\sqrt{5}, -1, -\frac{1}{2}, 2, \sqrt{7}, 6, \sqrt{625/9}, \pi\}$

In Problems 37 and 38, use a calculator to express each number in decimal form. Classify each decimal number as terminating, repeating, or nonrepeating and nonterminating. Identify the pattern of repeated digits in any repeating decimal numbers.*

37. (A) $\frac{8}{9}$ (B) $\frac{3}{11}$ (C) $\sqrt{5}$ (D) $\frac{11}{8}$

38. (A) $\frac{13}{6}$ (B) $\sqrt{21}$ (C) $\frac{7}{16}$ (D) $\frac{29}{111}$

39. Indicate true (T) or false (F), and for each false statement find real number replacements for a and b that will provide a counterexample. For all real numbers a and b:
 (A) $a + b = b + a$
 (B) $a - b = b - a$
 (C) $ab = ba$
 (D) $a \div b = b \div a$

40. Indicate true (T) or false (F), and for each false statement find real number replacements for a, b, and c that will provide a counterexample. For all real numbers a, b, and c:
 (A) $(a + b) + c = a + (b + c)$
 (B) $(a - b) - c = a - (b - c)$
 (C) $a(bc) = (ab)c$
 (D) $(a \div b) \div c = a \div (b \div c)$

In Problems 41–48, indicate true (T) or false (F), and for each false statement give a specific counterexample.

41. The difference of any two natural numbers is a natural number.

42. The quotient of any two nonzero integers is an integer.

43. The sum of any two rational numbers is a rational number.

44. The sum of any two irrational numbers is an irrational number.

45. The product of any two irrational numbers is an irrational number.

46. The product of any two rational numbers is a rational number.

47. The multiplicative inverse of any irrational number is an irrational number.

48. The multiplicative inverse of any nonzero rational number is a rational number.

49. If $c = 0.151515\ldots$, then $100c = 15.1515\ldots$ and

$$100c - c = 15.1515\ldots - 0.151515\ldots$$
$$99c = 15$$
$$c = \tfrac{15}{99} = \tfrac{5}{33}$$

Proceeding similarly, convert the repeating decimal $0.090909\ldots$ into a fraction. (All repeating decimals are rational numbers, and all rational numbers have repeating decimal representations.)

50. Repeat Problem 49 for $0.181818\ldots$.

*Later in the book you will encounter optional exercises that require a graphing calculator. If you have such a calculator, you can certainly use it here. Otherwise, any scientific calculator will be sufficient for the problems in this chapter.

R-2 Exponents and Radicals

> Integer Exponents
> Scientific Notation
> Roots of Real Numbers
> Rational Exponents and Radicals
> Simplifying Radicals

The French philosopher/mathematician René Descartes (1596–1650) is generally credited with the introduction of the very useful exponent notation "x^n." This notation as well as other improvements in algebra may be found in his *Geometry*, published in 1637.

If n is a natural number, x^n denotes the product of n factors, each equal to x. In this section, the meaning of x^n will be expanded to allow the exponent n to be any rational number. Each of the following expressions will then represent a unique real number:

$$7^5 \qquad 5^{-4} \qquad 3.14^0 \qquad 6^{1/2} \qquad 14^{-5/3}$$

> Integer Exponents

If a is a real number, then

$$a^6 = a \cdot a \cdot a \cdot a \cdot a \cdot a \qquad \text{6 factors of } a$$

In the expression a^6, 6 is called an **exponent** and a is called the **base.**

Recall that a^{-1}, for $a \neq 0$, denotes the multiplicative inverse of a (that is, $1/a$). To generalize exponent notation to include negative integer exponents and 0, we define a^{-6} to be the multiplicative inverse of a^6, and we define a^0 to be 1.

> **DEFINITION 1** a^n, n an Integer and a a Real Number

For n a positive integer and a a real number:

$$a^n = a \cdot a \cdot \ldots \cdot a \qquad \text{n factors of } a$$

$$a^{-n} = \frac{1}{a^n} \qquad\qquad (a \neq 0)$$

$$a^0 = 1 \qquad\qquad (a \neq 0)$$

EXAMPLE 1 Using the Definition of Integer Exponents

Write parts (A) and (B) in decimal form and parts (C) and (D) using positive exponents. Assume all variables represent nonzero real numbers.

(A) $(u^3v^2)^0$ (B) 10^{-3}

(C) x^{-8} (D) $\dfrac{x^{-3}}{y^{-5}}$

SOLUTIONS

(A) $(u^3v^2)^0 = 1$ (B) $10^{-3} = \dfrac{1}{10^3} = \dfrac{1}{1,000} = 0.001$

(C) $x^{-8} = \dfrac{1}{x^8}$ (D) $\dfrac{x^{-3}}{y^{-5}} = \boxed{\dfrac{x^{-3}}{1} \cdot \dfrac{1}{y^{-5}} = \dfrac{1}{x^3} \cdot \dfrac{y^5}{1}}^{*} = \dfrac{y^5}{x^3}$

MATCHED PROBLEM 1

Write parts (A) and (B) in decimal form and parts (C) and (D) using positive exponents. Assume all variables represent nonzero real numbers.

(A) $(x^2)^0$ (B) 10^{-5} (C) $\dfrac{1}{x^{-4}}$ (D) $\dfrac{u^{-7}}{v^{-3}}$

To calculate with exponents, it is helpful to remember Definition 1. For example:

$$2^3 \cdot 2^4 = (2 \cdot 2 \cdot 2)(2 \cdot 2 \cdot 2 \cdot 2) = 2^{3+4} = 2^7$$
$$(2^3)^4 = (2 \cdot 2 \cdot 2)^4 = (2 \cdot 2 \cdot 2)(2 \cdot 2 \cdot 2)(2 \cdot 2 \cdot 2)(2 \cdot 2 \cdot 2) = 2^{3 \cdot 4} = 2^{12}$$

These are instances of Properties 1 and 2 of Theorem 1.

> **THEOREM 1** **Properties of Integer Exponents**

For n and m integers and a and b real numbers:

1. $a^m a^n = a^{m+n}$ $\quad a^5 a^{-7} \; \boxed{= a^{5+(-7)}} \; = a^{-2}$

2. $(a^n)^m = a^{mn}$ $\quad (a^3)^{-2} \; \boxed{= a^{(-2)3}} \; = a^{-6}$

3. $(ab)^m = a^m b^m$ $\quad (ab)^3 = a^3 b^3$

4. $\left(\dfrac{a}{b}\right)^m = \dfrac{a^m}{b^m}$ $\quad b \neq 0$ $\quad \left(\dfrac{a}{b}\right)^4 = \dfrac{a^4}{b^4}$

5. $\dfrac{a^m}{a^n} = \begin{cases} a^{m-n} \\ \dfrac{1}{a^{n-m}} \end{cases}$ $\quad a \neq 0$ $\quad \dfrac{a^3}{a^{-2}} = a^{3-(-2)} = a^5$

$\dfrac{a^3}{a^{-2}} = \dfrac{1}{a^{-2-3}} = \dfrac{1}{a^{-5}}$

EXAMPLE 2 **Using Exponent Properties**

Simplify using exponent properties, and express answers using positive exponents only.[†]

(A) $(3a^5)(2a^{-3})$ (B) $\dfrac{6x^{-2}}{8x^{-5}}$

(C) $-4y^3 - (-4y)^3$ (D) $(2a^{-3}b^2)^{-2}$

SOLUTIONS

(A) $(3a^5)(2a^{-3}) \; \boxed{= (3 \cdot 2)(a^5 a^{-3})} \; = 6a^2$

(B) $\dfrac{6x^{-2}}{8x^{-5}} \; \boxed{= \dfrac{3x^{-2-(-5)}}{4}} \; = \dfrac{3x^3}{4}$

*Throughout the book, dashed boxes—called **think boxes**—are used to represent steps that may be performed mentally.

[†]By "simplify" we mean eliminate common factors from numerators and denominators and reduce to a minimum the number of times a given constant or variable appears in an expression. We ask that answers be expressed using positive exponents only in order to have a definite form for an answer. Later (in this section and elsewhere) we will encounter situations where we will want negative exponents in a final answer.

(C) $-4y^3 - (-4y)^3 = -4y^3 - (-4)^3y^3 \quad \boxed{= -4y^3 - (-64)y^3}$

$$= -4y^3 + 64y^3 = 60y^3$$

(D) $(2a^{-3}b^2)^{-2} = 2^{-2}a^6b^{-4} = \dfrac{a^6}{4b^4}$

MATCHED PROBLEM 2

Simplify using exponent properties, and express answers using positive exponents only.

(A) $(5x^{-3})(3x^4)$ (B) $\dfrac{9y^{-7}}{6y^{-4}}$ (C) $2x^4 - (-2x)^4$ (D) $(3x^4y^{-3})^{-2}$

› Scientific Notation

Scientific work often involves the use of very large numbers or very small numbers. For example, the average cell contains about 200,000,000,000,000 molecules, and the diameter of an electron is about 0.000 000 000 0004 centimeter. It is generally troublesome to write and work with numbers of this type in standard decimal form. The two numbers written here cannot even be entered into most calculators as they are written. However, each can be expressed as the product of a number between 1 and 10 and an integer power of 10:

$$200{,}000{,}000{,}000{,}000 = 2 \times 10^{14}$$

$$0.000\ 000\ 000\ 0004 = 4 \times 10^{-13}$$

In fact, any positive number written in decimal form can be expressed in **scientific notation,** that is, in the form

$$a \times 10^n \qquad 1 \leq a < 10,\ n \text{ an integer, } a \text{ in decimal form}$$

EXAMPLE 3

Scientific Notation

(A) Write each number in scientific notation: 6,430; 5,350,000; 0.08; 0.000 32
(B) Write in standard decimal form: 2.7×10^2; 9.15×10^4; 5×10^{-3}; 8.4×10^{-5}

SOLUTIONS

(A) $6{,}430 = 6.43 \times 10^3$; $5{,}350{,}000 = 5.35 \times 10^6$; $0.08 = 8 \times 10^{-2}$;
$0.000\ 32 = 3.2 \times 10^{-4}$

(B) 270; 91,500; 0.005; 0.000 084

MATCHED PROBLEM 3

(A) Write each number in scientific notation: 23,000; 345,000,000; 0.0031; 0.000 000 683
(B) Write in standard decimal form: 4×10^3; 5.3×10^5; 2.53×10^{-2}; 7.42×10^{-6}

Most calculators express very large and very small numbers in scientific notation. Consult the manual for your calculator to see how numbers in scientific notation are entered in your calculator. Some common methods for displaying scientific notation on a calculator are shown here.

Number Represented	Typical Scientific Calculator Display	Typical Graphing Calculator Display
$5.427\ 493 \times 10^{-17}$	5.427493 – 17	5.427493E – 17
$2.359\ 779 \times 10^{12}$	2.359779 12	2.359779E12

EXAMPLE **4** **Using Scientific Notation on a Calculator**

Calculate $\dfrac{325,100,000,000}{0.000\ 000\ 000\ 000\ 0871}$ by writing each number in scientific notation and then using your calculator. (Refer to the user's manual accompanying your calculator for the procedure.) Express the answer to three significant digits* in scientific notation.

SOLUTION

$$\frac{325,100,000,000}{0.000\ 000\ 000\ 000\ 0871} = \frac{3.251 \times 10^{11}}{8.71 \times 10^{-14}}$$

$$= \boxed{3.732491389\text{E}24} \qquad \text{Calculator display}$$

$$= 3.73 \times 10^{24} \qquad \text{To three significant digits}$$

```
(3.251E11)/(8.71
E-14)
    3.732491389E24
325100000000/.00
0000000000871
    3.732491389E24
```

> **Figure 1**

Figure 1 shows two solutions to this problem on a graphing calculator. In the first solution we entered the numbers in scientific notation, and in the second we used standard decimal notation. Although the multiple-line screen display on a graphing calculator enables us to enter very long standard decimals, scientific notation is usually more efficient and less prone to errors in data entry. Furthermore, as Figure 1 shows, the calculator uses scientific notation to display the answer, regardless of the manner in which the numbers are entered.

MATCHED PROBLEM 4

Repeat Example 4 for:

$$\frac{0.000\ 000\ 006\ 932}{62,600,000,000}$$

› Roots of Real Numbers

The solutions of the equation $x^2 = 64$ are called **square roots** of 64 and the solutions of $x^3 = 64$ are the **cube roots** of 64. So there are two real square roots of 64 (-8 and 8) and one real cube root of 64 (4 is a cube root, but -4 is not). Note that -64 has no real square root ($x^2 = -64$ has no real solution because the square of a real number can't be negative), but -4 is a cube root of -64 because $(-4)^3 = -64$. In general:

> **DEFINITION 2** **Definition of an nth Root**
>
> For a natural number n and a and b real numbers:
>
> a is an **nth root** of b if $a^n = b$ 3 is a fourth root of 81, since $3^4 = 81$.

The number of real nth roots of a real number b is either 0, 1, or 2, depending on whether b is positive or negative, and whether n is even or odd. Theorem 2 gives the details, which are summarized in Table 1.

*For those not familiar with the meaning of *significant digits,* see Appendix A for a brief discussion of this concept.

Table 1 Number of Real
nth Roots of b

	n even	n odd
$b > 0$	2	1
$b = 0$	1	1
$b < 0$	0	1

> **THEOREM 2** Number of Real nth Roots of a Real Number b

Let n be a natural number and let b be a real number:

1. $b > 0$: If n is even, then b has two real nth roots, each the negative of the other; if n is odd, then b has one real nth root.

2. $b = 0$: 0 is the only nth root of $b = 0$.

3. $b < 0$: If n is even, then b has no real nth root; if n is odd, then b has one real nth root.

> ## Rational Exponents and Radicals

To denote nth roots, we can use rational exponents or we can use radicals. For example, the square root of a number b can be denoted by $b^{1/2}$ or $\sqrt{b}$. To avoid ambiguity, both expressions denote the *positive* square root when there are two real square roots. Furthermore, both expressions are undefined when there is no real square root. In general:

> **DEFINITION 3** Principal nth Root

For n a natural number and b a real number, the **principal nth root of b,** denoted by $b^{1/n}$ or $\sqrt[n]{b}$, is:

1. The real nth root of b if there is only one.

2. The positive nth root of b if there are two real nth roots.

3. Undefined if b has no real nth root.

In the notation $\sqrt[n]{b}$, the symbol $\sqrt{}$ is called a **radical,** n is called the **index,** and b is the **radicand.** If $n = 2$, we write $\sqrt{b}$ in place of $\sqrt[2]{b}$.

EXAMPLE 5 **Principal nth Roots**

Evaluate each expression:

(A) $9^{1/2}$ (B) $\sqrt{121}$ (C) $\sqrt[3]{-125}$ (D) $(-16)^{1/4}$ (E) $27^{1/3}$ (F) $\sqrt[5]{32}$

SOLUTIONS

(A) $9^{1/2} = 3$ (B) $\sqrt{121} = 11$

(C) $\sqrt[3]{-125} = -5$ (D) $(-16)^{1/4}$ is undefined (not a real number).

(E) $27^{1/3} = 3$ (F) $\sqrt[5]{32} = 2$

MATCHED PROBLEM 5

Evaluate each expression:

(A) $8^{1/3}$ (B) $\sqrt{-4}$ (C) $\sqrt[4]{10{,}000}$ (D) $(-1)^{1/5}$ (E) $\sqrt[3]{-27}$ (F) $0^{1/8}$

How should a symbol such as $7^{2/3}$ be defined? If the properties of exponents are to hold for rational exponents, then $7^{2/3} = (7^{1/3})^2$; that is, $7^{2/3}$ must represent the square of the cube root of 7. This leads to the following general definition:

> **DEFINITION 4** $b^{m/n}$ **and** $b^{-m/n}$, **Rational Number Exponent**

For m and n natural numbers and b any real number (except b cannot be negative when n is even):

$$b^{m/n} = (b^{1/n})^m \quad \text{and} \quad b^{-m/n} = \frac{1}{b^{m/n}}$$

$$4^{3/2} = (4^{1/2})^3 = 2^3 = 8 \qquad 4^{-3/2} = \frac{1}{4^{3/2}} = \frac{1}{8} \qquad (-4)^{3/2} \text{ is not real}$$

$$(-32)^{3/5} = [(-32)^{1/5}]^3 = (-2)^3 = -8$$

We have now discussed $b^{m/n}$ for all rational numbers m/n and real numbers b. It can be shown, though we will not do so, that all five properties of exponents listed in Theorem 1 continue to hold for rational exponents as long as we avoid even roots of negative numbers. With the latter restriction in effect, the following useful relationship is an immediate consequence of the exponent properties:

> **THEOREM 3** **Rational Exponent/Radical Property**

For m and n natural numbers and b any real number (except b cannot be negative when n is even):

$$(b^{1/n})^m = (b^m)^{1/n} \quad \text{and} \quad (\sqrt[n]{b})^m = \sqrt[n]{b^m}$$

⟩⟩⟩ EXPLORE-DISCUSS 1

Find the contradiction in the following chain of equations:

$$-1 = (-1)^{2/2} = [(-1)^2]^{1/2} = 1^{1/2} = 1 \tag{1}$$

Where did we try to use Theorem 3? Why was this not correct?

EXAMPLE 6 Using Rational Exponents and Radicals

Simplify and express answers using positive exponents only. All letters represent positive real numbers.

(A) $8^{2/3}$ (B) $\sqrt[4]{3^{12}}$ (C) $(3\sqrt[3]{x})(2\sqrt{x})$ (D) $\left(\dfrac{4x^{1/3}}{x^{1/2}}\right)^{1/2}$

SOLUTIONS

(A) $8^{2/3} = (8^{1/3})^2 = 2^2 = 4$ or $8^{2/3} = (8^2)^{1/3} = 64^{1/3} = 4$

(B) $\sqrt[4]{3^{12}} = (3^{12})^{1/4} = 3^3 = 27$

(C) $(3\sqrt[3]{x})(2\sqrt{x}) = (3x^{1/3})(2x^{1/2}) = 6x^{1/3+1/2} = 6x^{5/6}$

(D) $\left(\dfrac{4x^{1/3}}{x^{1/2}}\right)^{1/2} = \dfrac{4^{1/2}x^{1/6}}{x^{1/4}} = \dfrac{2}{x^{1/4-1/6}} = \dfrac{2}{x^{1/12}}$

MATCHED PROBLEM 6 Simplify and express answers using positive exponents only. All letters represent positive real numbers.

(A) $(-8)^{5/3}$ (B) $\sqrt[5]{32^4}$ (C) $(5\sqrt[4]{y^3})(2\sqrt[3]{y})$ (D) $\left(\dfrac{8x^{1/2}}{x^{2/3}}\right)^{1/3}$

› Simplifying Radicals

The exponent properties considered earlier lead to the following properties of radicals.

> **› THEOREM 4 Properties of Radicals**
>
> For n a natural number greater than 1, and x and y positive real numbers:
>
> 1. $\sqrt[n]{x^n} = x$ $\sqrt[3]{x^3} = x$
> 2. $\sqrt[n]{xy} = \sqrt[n]{x}\,\sqrt[n]{y}$ $\sqrt[5]{xy} = \sqrt[5]{x}\,\sqrt[5]{y}$
> 3. $\sqrt[n]{\dfrac{x}{y}} = \dfrac{\sqrt[n]{x}}{\sqrt[n]{y}}$ $\sqrt[4]{\dfrac{x}{y}} = \dfrac{\sqrt[4]{x}}{\sqrt[4]{y}}$

An algebraic expression that contains radicals is said to be in **simplified form** if all four of the conditions listed in the following definition are satisfied.

> **› DEFINITION 5 Simplified (Radical) Form**
>
> **1.** No radicand (the expression within the radical sign) contains a factor to a power greater than or equal to the index of the radical.
> For example, $\sqrt{x^5}$ violates this condition.
>
> **2.** No power of the radicand and the index of the radical have a common factor other than 1.
> For example, $\sqrt[6]{x^4}$ violates this condition.
>
> **3.** No radical appears in a denominator.
> For example, $y/\sqrt{x}$ violates this condition.
>
> **4.** No fraction appears within a radical.
> For example, $\sqrt{\frac{3}{5}}$ violates this condition.

EXAMPLE 7 **Finding Simplified Form**

Write in simplified radical form.

(A) $\sqrt{12x^5y^2}$ (B) $\sqrt[6]{16x^4y^2}$ (C) $\dfrac{6}{\sqrt{2x}}$ (D) $\sqrt[3]{\dfrac{8x^4}{y}}$

SOLUTIONS

(A) Condition 1 is violated. First we convert to rational exponent form.

$$\sqrt{12x^5y^2} = (12x^5y^2)^{1/2}$$ Use $(ab)^m = a^m b^m$ and $(a^n)^m = a^{mn}$.

$$= 12^{1/2}x^{5/2}y$$ $12 = 4 \cdot 3,\ x^{5/2} = x^2x^{1/2}$

$$= (4 \cdot 3)^{1/2}x^2x^{1/2}y$$ Write in radical form.

$$= 2\sqrt{3}\,x^2\sqrt{x}\,y$$ Use commutative property and radical property 2.

$$= 2x^2y\sqrt{3x}$$

(B) Condition 2 is violated. First we convert to rational exponent form.

$$\sqrt[6]{16x^4y^2} = (16x^4y^2)^{1/6}$$ Use $(ab)^m = a^m b^m$ and $(a^n)^m = a^{mn}$.

$$= 16^{1/6}x^{2/3}y^{1/3}$$ $16 = 2^4$

$$= 2^{2/3}x^{2/3}y^{1/3}$$ Write in radical form.

$$= \sqrt[3]{4x^2y}$$

(C) Condition 3 is violated. We multiply numerator and denominator by $\sqrt{2x}$; the effect is to multiply the expression by 1, so its value is unchanged, but the denominator is left free of radicals.

$$\frac{6}{\sqrt{2x}} = \frac{6}{\sqrt{2x}} \cdot \frac{\sqrt{2x}}{\sqrt{2x}} = \frac{6\sqrt{2x}}{2x} = \frac{3\sqrt{2x}}{x}$$

(D) Condition 4 is violated. First we convert to rational exponent form.

$$\sqrt[3]{\frac{8x^4}{y}} = \frac{8^{1/3}x^{4/3}}{y^{1/3}}$$ Multiply by $\frac{y^{2/3}}{y^{2/3}} = 1$.

$$= \frac{2x^{4/3}y^{2/3}}{y}$$ $x^{4/3} = xx^{1/3}$

$$= \frac{2xx^{1/3}y^{2/3}}{y}$$ Write in radical form.

$$= \frac{2x\sqrt[3]{xy^2}}{y}$$

MATCHED PROBLEM 7

Write in simplified radical form.

(A) $\sqrt{18x^4y^3}$ (B) $\sqrt[9]{8x^6y^3}$ (C) $\dfrac{30}{\sqrt[4]{16x}}$ (D) $\sqrt{\dfrac{5x^3}{y}}$

Eliminating a radical from a denominator [as in Example 7(C)] is called **rationalizing the denominator**. To rationalize the denominator, we multiply the numerator and denominator by a suitable factor that will leave the denominator free of radicals. This factor is called a **rationalizing factor**. If the denominator is of the form $\sqrt{a} + \sqrt{b}$, then $\sqrt{a} - \sqrt{b}$ is a rationalizing factor because

$$(\sqrt{a} + \sqrt{b})(\sqrt{a} - \sqrt{b}) = a - b$$

Similarly, if the denominator is of the form $\sqrt{a} - \sqrt{b}$, then $\sqrt{a} + \sqrt{b}$ is a rationalizing factor.

EXAMPLE 8 **Rationalizing Denominators**

Rationalize the denominator and write the answer in simplified radical form.

(A) $\dfrac{8}{\sqrt{6} + \sqrt{5}}$ (B) $\dfrac{\sqrt{x} + \sqrt{y}}{\sqrt{x} - \sqrt{y}}$

SOLUTIONS (A) Multiply numerator and denominator by the rationalizing factor $\sqrt{6} - \sqrt{5}$.

$$\frac{8}{\sqrt{6} + \sqrt{5}} = \frac{8}{\sqrt{6} + \sqrt{5}} \cdot \frac{\sqrt{6} - \sqrt{5}}{\sqrt{6} - \sqrt{5}} \qquad (\sqrt{a} + \sqrt{b})(\sqrt{a} - \sqrt{b}) = a - b$$

$$= \frac{8(\sqrt{6} - \sqrt{5})}{6 - 5} \qquad \text{Simplify.}$$

$$= 8(\sqrt{6} - \sqrt{5})$$

(B) Multiply numerator and denominator by the rationalizing factor $\sqrt{x} + \sqrt{y}$.

$$\frac{\sqrt{x} + \sqrt{y}}{\sqrt{x} - \sqrt{y}} = \frac{\sqrt{x} + \sqrt{y}}{\sqrt{x} - \sqrt{y}} \cdot \frac{\sqrt{x} + \sqrt{y}}{\sqrt{x} + \sqrt{y}} \qquad \text{Expand numerator and denominator.}$$

$$= \frac{x + \sqrt{x}\sqrt{y} + \sqrt{y}\sqrt{x} + y}{x - y} \qquad \text{Combine like terms.}$$

$$= \frac{x + 2\sqrt{xy} + y}{x - y}$$

MATCHED PROBLEM 8 Rationalize the denominator and write the answer in simplified radical form.

(A) $\dfrac{6}{1 - \sqrt{3}}$ (B) $\dfrac{2\sqrt{x} - 3\sqrt{y}}{\sqrt{x} + \sqrt{y}}$

ANSWERS TO MATCHED PROBLEMS

1. (A) 1 (B) 0.000 01 (C) x^4 (D) v^3/u^7
2. (A) $15x$ (B) $3/(2y^3)$ (C) $-14x^4$ (D) $y^6/(9x^8)$
3. (A) 2.3×10^4; 3.45×10^8; 3.1×10^{-3}; 6.83×10^{-7}
 (B) 4,000; 530,000; 0.0253; 0.000 007 42
4. 1.11×10^{-19}
5. (A) 2 (B) Not real (C) 10 (D) -1 (E) -3 (F) 0
6. (A) -32 (B) 16 (C) $10y^{13/12}$ (D) $2/x^{1/18}$
7. (A) $3x^2y\sqrt{2y}$ (B) $\sqrt[3]{2x^2y}$ (C) $\dfrac{15\sqrt[4]{x^3}}{x}$ (D) $\dfrac{x\sqrt{5xy}}{y}$
8. (A) $-3 - 3\sqrt{3}$ (B) $\dfrac{2x - 5\sqrt{xy} + 3y}{x - y}$

R-2 Exercises

All variables represent positive real numbers and are restricted to prevent division by 0.
In Problems 1–14, evaluate each expression. If the answer is not an integer, write it in fraction form.

1. 3^7 **2.** 5^6 **3.** $\left(\dfrac{1}{2}\right)^8$

4. $\left(\dfrac{3}{5}\right)^3$ **5.** 6^{-3} **6.** 2^{-6}

7. $(-5)^4$ **8.** $(-4)^5$ **9.** $(-3)^{-1}$

10. $(-7)^{-2}$ **11.** -7^{-2} **12.** -10^0

13. $\left(\dfrac{1}{3}\right)^0$ **14.** $\left(\dfrac{1}{10}\right)^{-1}$

In Problems 15–20, write the numbers in scientific notation.

15. 58,620,000

16. 4,390

17. 0.027

18. 0.11

19. 0.000 000 064

20. 0.000 0325

In Problems 21–26, write each number in standard decimal form.

21. 4×10^{-3}

22. 5×10^{-6}

23. 2.99×10^{5}

24. 7.75×10^{11}

25. 3.1×10^{-7}

26. 8.167×10^{-4}

In Problems 27–32, change to radical form. Do not simplify.

27. $32^{1/5}$

28. $625^{3/4}$

29. $4x^{-1/2}$

30. $32y^{-2/5}$

31. $x^{1/3} - y^{1/3}$

32. $(x - y)^{1/3}$

In Problems 33–38, change to rational exponent form. Do not simplify.

33. $\sqrt{361}$

34. $\sqrt[3]{17^2}$

35. $4x\sqrt[5]{y^3}$

36. $\sqrt[4]{7x^3y^2}$

37. $\sqrt[3]{x^2 + y^2}$

38. $\sqrt[3]{x^2} + \sqrt[3]{y^2}$

In Problems 39–50, evaluate each expression that represents a real number.

39. $100^{1/2}$

40. $169^{1/2}$

41. $\sqrt{121}$

42. $\sqrt{361}$

43. $125^{1/3}$

44. $27^{2/3}$

45. $\sqrt[3]{-27}$

46. $\sqrt[3]{64}$

47. $\sqrt[4]{-16}$

48. $\sqrt[6]{-1}$

49. $9^{-3/2}$

50. $64^{-4/3}$

In Problems 51–64, simplify and express answers using positive exponents only.

51. $x^5 x^{-2}$

52. $y^6 y^{-8}$

53. $(2y)(3y^2)(5y^4)$

54. $(6x^3)(4x^7)(x^{-5})$

55. $(a^2 b^3)^5$

56. $(2c^4 d^{-2})^{-3}$

57. $u^{1/3} u^{5/3}$

58. $v^{-1/5} v^{6/5}$

59. $(x^{-3})^{1/6}$

60. $(49a^4 b^{-2})^{1/2}$

61. $\left(\dfrac{m^{-2} n^3}{m^4 n^{-1}} \right)^2$

62. $\left(\dfrac{6mn^{-2}}{3m^{-1} n^2} \right)^{-3}$

63. $\left(\dfrac{w^4}{9x^{-2}} \right)^{-1/2}$

64. $\left(\dfrac{8a^{-4} b^3}{27a^2 b^{-3}} \right)^{1/3}$

In Problems 65–86, write in simplified radical form.

65. $-\sqrt{128}$

66. $-\sqrt{125}$

67. $\sqrt{27} - 5\sqrt{3}$

68. $2\sqrt{8} + \sqrt{18}$

69. $\sqrt[3]{5} - \sqrt[3]{25} + \sqrt[3]{625}$

70. $\sqrt{20} + \sqrt[3]{40} - \sqrt[3]{5}$

71. $\sqrt[3]{25}\sqrt[3]{10}$

72. $\sqrt{6}\sqrt{14}$

73. $\sqrt{16m^4 y^8}$

74. $\sqrt[4]{16m^4 n^8}$

75. $\dfrac{1}{2\sqrt{5}}$

76. $\dfrac{1}{\sqrt[3]{7}}$

77. $\dfrac{3}{\sqrt[3]{54}}$

78. $\dfrac{12y^2}{\sqrt{6y}}$

79. $\dfrac{4}{\sqrt{6} - 2}$

80. $\dfrac{\sqrt{2}}{\sqrt{6} + 2}$

81. $x\sqrt[5]{3^6 x^7 y^{11}}$

82. $2a\sqrt[3]{8a^8 b^{13}}$

83. $\dfrac{\sqrt{2m}\sqrt{5}}{\sqrt{20m}}$

84. $\dfrac{3\sqrt{y}}{2\sqrt{y} - 3}$

85. $\dfrac{2\sqrt{5} + 3\sqrt{2}}{5\sqrt{5} + 2\sqrt{2}}$

86. $\dfrac{3\sqrt{2} - 2\sqrt{3}}{3\sqrt{3} - 2\sqrt{2}}$

87. What is the result of entering 2^{3^2} on a calculator?

88. Refer to Problem 87. What is the difference between $2^{(3^2)}$ and $(2^3)^2$? Which agrees with the value of 2^{3^2} obtained with a calculator?

APPLICATIONS

89. ECONOMICS If in the United States in 2007 the national debt was about \$8,868,000,000,000 and the population was about 301,000,000, estimate to three significant digits each individual's share of the national debt. Write your answer in scientific notation and in standard decimal form.

90. ECONOMICS If in the United States in 2007 the gross domestic product (GDP) was about \$14,074,000,000,000 and the population was about 301,000,000, estimate to three significant digits the GDP per person. Write your answer in scientific notation and in standard decimal form.

91. ECONOMICS The number of units N of a finished product produced from the use of x units of labor and y units of capital for a particular Third World country is approximated by

$$N = 10x^{3/4} y^{1/4}$$

Estimate how many units of a finished product will be produced using 256 units of labor and 81 units of capital.

92. ECONOMICS The number of units N of a finished product produced by a particular automobile company where x units of labor and y units of capital are used is approximated by

$$N = 50x^{1/2} y^{1/2}$$

Estimate how many units will be produced using 256 units of labor and 144 units of capital.

93. BRAKING DISTANCE R. A. Moyer of Iowa State College found, in comprehensive tests carried out on 41 wet pavements, that the braking distance d (in feet) for a particular automobile traveling at v miles per hour was given approximately by

$$d = 0.0212v^{7/3}$$

Approximate the braking distance to the nearest foot for the car traveling on wet pavement at 70 miles per hour.

94. BRAKING DISTANCE Approximately how many feet would it take the car in Problem 93 to stop on wet pavement if it were traveling at 50 miles per hour? (Compute answer to the nearest foot.)

95. PHYSICS—RELATIVISTIC MASS The mass M of an object moving at a velocity v is given by

$$M = \frac{M_0}{\sqrt{1 - \dfrac{v^2}{c^2}}}$$

where M_0 = mass at rest and c = velocity of light. The mass of an object increases with velocity and tends to infinity as the velocity approaches the speed of light. Show that M can be written in the form

$$M = \frac{M_0 c \sqrt{c^2 - v^2}}{c^2 - v^2}$$

96. PHYSICS—PENDULUM A simple pendulum is formed by hanging a bob of mass M on a string of length L from a fixed support (see the figure). The time it takes the bob to swing from right to left

and back again is called the **period** T and is given by

$$T = 2\pi \sqrt{\frac{L}{g}}$$

where g is the gravitational constant. Show that T can be written in the form

$$T = \frac{2\pi \sqrt{gL}}{g}$$

R-3 — Polynomials: Basic Operations and Factoring

> Polynomials
> Addition and Subtraction
> Multiplication
> Factoring

In this section, we review the basic operations on *polynomials*. Polynomials are expressions such as $x^4 - 5x^2 + 1$ or $3xy - 2x + 5y + 6$ that are built from constants and variables using only addition, subtraction, and multiplication (the power x^4 is the product $x \cdot x \cdot x \cdot x$). Polynomials are used throughout mathematics to describe and approximate mathematical relationships.

> Polynomials

Algebraic expressions are formed by using constants and variables and the algebraic operations of addition, subtraction, multiplication, division, raising to powers, and taking roots. Some examples are

$$\sqrt[3]{x^3 + 5} \qquad 5x^4 + 2x^2 - 7$$

$$\frac{x - 5}{x^2 + 2x - 5} \qquad 1 + \frac{1}{1 + \dfrac{1}{x}}$$

An algebraic expression involving only the operations of addition, subtraction, multiplication, and raising to natural number powers is called a **polynomial.** (Note that raising to a natural number power is repeated multiplication.) Some examples are

$$2x - 3 \qquad 4x^2 - 3x + 7$$

$$x - 2y \qquad x^3 - 3x^2y + xy^2 + 2y^7$$

In a polynomial, a variable cannot appear in a denominator, as an exponent, or within a radical. Accordingly, a **polynomial in one variable** x is constructed by adding or subtracting constants and terms of the form ax^n, where a is a real number and n is a natural number. A **polynomial in two variables** x and y is constructed by adding and subtracting constants and terms of the form $ax^m y^n$, where a is a real number and m and n are natural numbers. Polynomials in three or more variables are defined in a similar manner.

Polynomials can be classified according to their *degree*. If a term in a polynomial has only one variable as a factor, then the **degree of that term** is the power of the variable. If two or more variables are present in a term as factors, then the degree of the term is the sum of the powers of the variables. The **degree of a polynomial** is the degree of the nonzero term with the highest degree in the polynomial. Any nonzero constant is defined to be a **polynomial of degree 0.** The number 0 is also a polynomial but is not assigned a degree.

EXAMPLE 1 **Polynomials and Nonpolynomials**

(A) Which of the following are polynomials?

$$2x + 5 - \frac{1}{x} \qquad x^2 - 3x + 2 \qquad \sqrt{x^3 - 4x + 1} \qquad x^4 + \sqrt{2}$$

(B) Given the polynomial $2x^3 - x^6 + 7$, what is the degree of the first term? The third term? The whole polynomial?

(C) Given the polynomial $x^3 y^2 + 2x^2 y + 1$, what is the degree of the first term? The second term? The whole polynomial?

SOLUTIONS (A) $x^2 - 3x + 2$ and $x^4 + \sqrt{2}$ are polynomials. (The others are not polynomials since a variable appears in a denominator or within a radical.)

(B) The first term has degree 3, the third term has degree 0, and the whole polynomial has degree 6.

(C) The first term has degree 5, the second term has degree 3, and the whole polynomial has degree 5. ◉

MATCHED PROBLEM 1 (A) Which of the following are polynomials?

$$3x^2 - 2x + 1 \qquad \sqrt{x - 3} \qquad x^2 - 2xy + y^2 \qquad \frac{x - 1}{x^2 + 2}$$

(B) Given the polynomial $3x^5 - 6x^3 + 5$, what is the degree of the first term? The second term? The whole polynomial?

(C) Given the polynomial $6x^4 y^2 - 3xy^3$, what is the degree of the first term? The second term? The whole polynomial? ◉

In addition to classifying polynomials by degree, we also call a single-term polynomial a **monomial,** a two-term polynomial a **binomial,** and a three-term polynomial a **trinomial.**

$$\tfrac{5}{2}x^2 y^3 \qquad \text{Monomial}$$

$$x^3 + 4.7 \qquad \text{Binomial}$$

$$x^4 - \sqrt{2}x^2 + 9 \qquad \text{Trinomial}$$

A constant in a term of a polynomial, including the sign that precedes it, is called the **numerical coefficient,** or simply, the **coefficient,** of the term. If a constant doesn't appear, or

only a $+$ sign appears, the coefficient is understood to be 1. If only a $-$ sign appears, the coefficient is understood to be -1. So given the polynomial

$$2x^4 - 4x^3 + x^2 - x + 5 \qquad 2x^4 + (-4)x^3 + 1x^2 + (-1)x + 5$$

the coefficient of the first term is 2, the coefficient of the second term is -4, the coefficient of the third term is 1, the coefficient of the fourth term is -1, and the coefficient of the last term is 5.

Two terms in a polynomial are called **like terms** if they have exactly the same variable factors to the same powers. The numerical coefficients may or may not be the same. Since constant terms involve no variables, all constant terms are like terms. If a polynomial contains two or more like terms, these terms can be combined into a single term by making use of distributive properties. Consider the following example:

$$
\begin{aligned}
5x^3y - 2xy - x^3y - 2x^3y \quad &= 5x^3y - x^3y - 2x^3y - 2xy & &\text{Group like terms.}\\
&= (5x^3y - x^3y - 2x^3y) - 2xy & &\text{Use the distributive property}\\
&= (5 - 1 - 2)x^3y - 2xy & &\text{Simplify.}\\
&= 2x^3y - 2xy
\end{aligned}
$$

It should be clear that free use has been made of the real number properties discussed earlier. The steps done in the dashed box are usually done mentally, and the process is quickly done as follows:

Like terms in a polynomial are combined by adding their numerical coefficients.

› Addition and Subtraction

Addition and subtraction of polynomials can be thought of in terms of removing parentheses and combining like terms. Horizontal and vertical arrangements are illustrated in the next two examples. You should be able to work either way, letting the situation dictate the choice.

EXAMPLE 2 **Adding Polynomials**

Add: $x^4 - 3x^3 + x^2$, $\quad -x^3 - 2x^2 + 3x$, $\quad$ and $\quad 3x^2 - 4x - 5$

SOLUTION Add horizontally:

$$
\begin{aligned}
(x^4 - 3x^3 + x^2) + (-x^3 - 2x^2 + 3x) + (3x^2 - 4x - 5) \qquad &\text{Remove parentheses.}\\
= x^4 - 3x^3 + x^2 - x^3 - 2x^2 + 3x + 3x^2 - 4x - 5 \qquad &\text{Combine like terms.}\\
= x^4 - 4x^3 + 2x^2 - x - 5
\end{aligned}
$$

Or vertically, by lining up like terms and adding their coefficients:

$$
\begin{array}{r}
x^4 - 3x^3 + x^2 \\
- x^3 - 2x^2 + 3x \\
3x^2 - 4x - 5 \\
\hline
x^4 - 4x^3 + 2x^2 - x - 5
\end{array}
$$

MATCHED PROBLEM 2 Add horizontally and vertically:

$$3x^4 - 2x^3 - 4x^2, \qquad x^3 - 2x^2 - 5x, \qquad \text{and} \qquad x^2 + 7x - 2$$

EXAMPLE **3** **Subtracting Polynomials**

Subtract: $4x^2 - 3x + 5$ from $x^2 - 8$

SOLUTION

$$(x^2 - 8) - (4x^2 - 3x + 5) \quad \text{or}$$
$$= x^2 - 8 - 4x^2 + 3x - 5$$
$$= -3x^2 + 3x - 13$$

$$\begin{array}{r} x^2 \qquad\quad - \;\; 8 \\ \underline{-4x^2 + 3x - \;\; 5} \\ -3x^2 + 3x - 13 \end{array}$$ ←Change signs and add.

MATCHED PROBLEM 3 Subtract: $2x^2 - 5x + 4$ from $5x^2 - 6$

⟩⟩⟩ CAUTION ⟩⟩

When you use a horizontal arrangement to subtract a polynomial with more than one term, you must enclose the polynomial in parentheses. For example, to subtract $2x + 5$ from $4x - 11$, you must write

$$4x - 11 - (2x + 5) \quad \text{and not} \quad 4x - 11 - 2x + 5$$

⟩ Multiplication

Multiplication of algebraic expressions involves extensive use of distributive properties for real numbers, as well as other real number properties.

EXAMPLE **4** **Multiplying Polynomials**

Multiply: $(2x - 3)(3x^2 - 2x + 3)$

SOLUTION

$$(2x - 3)(3x^2 - 2x + 3)$$
$$= 2x(3x^2 - 2x + 3) - 3(3x^2 - 2x + 3) \qquad \text{Distribute, multiply out parentheses.}$$
$$= 6x^3 - 4x^2 + 6x - 9x^2 + 6x - 9 \qquad \text{Combine like terms.}$$
$$= 6x^3 - 13x^2 + 12x - 9$$

Or, using a vertical arrangement,

$$\begin{array}{r} 3x^2 - 2x + 3 \\ \underline{2x \;\; - 3} \\ 6x^3 - \;\; 4x^2 + \;\; 6x \\ \underline{- \;\; 9x^2 + \;\; 6x - 9} \\ 6x^3 - 13x^2 + 12x - 9 \end{array}$$

MATCHED PROBLEM 4 Multiply:

$$(2x - 3)(2x^2 + 3x - 2)$$

To multiply two polynomials, multiply each term of one by each term of the other, and combine like terms.

› Factoring

A **factor of a number** is one of two or more numbers whose product is the given number. Similarly, a **factor of an algebraic expression** is one of two or more algebraic expressions whose product is the given algebraic expression. For example,

$$30 = 2 \cdot 3 \cdot 5$$ 2, 3, and 5 are each factors of 30.

$$x^2 - 4 = (x - 2)(x + 2)$$ (x − 2) and (x + 2) are each factors of $x^2 - 4$.

The process of writing a number or algebraic expression as the product of other numbers or algebraic expressions is called **factoring.** We start our discussion of factoring with the positive integers.

An integer such as 30 can be represented in a factored form in many ways. The products

$$6 \cdot 5 \qquad (\tfrac{1}{2})(10)(6) \qquad 15 \cdot 2 \qquad 2 \cdot 3 \cdot 5$$

all yield 30. A particularly useful way of factoring positive integers greater than 1 is in terms of *prime* numbers.

An integer greater than 1 is **prime** if its only positive integer factors are itself and 1. So 2, 3, 5, and 7 are prime, but 4, 6, 8, and 9 are not prime. An integer greater than 1 that is not prime is called a **composite number.** The integer 1 is neither prime nor composite.

A composite number is said to be **factored completely** if it is represented as a product of prime factors. The only factoring of 30 that meets this condition, except for the order of the factors, is $30 = 2 \cdot 3 \cdot 5$. This illustrates an important property of integers.

> **› THEOREM 1 The Fundamental Theorem of Arithmetic**
>
> Each integer greater than 1 is either prime or can be expressed uniquely, except for the order of factors, as a product of prime factors.

We can also write polynomials in completely factored form. A polynomial such as $2x^2 - x - 6$ can be written in factored form in many ways. The products

$$(2x + 3)(x - 2) \qquad 2(x^2 - \tfrac{1}{2}x - 3) \qquad 2(x + \tfrac{3}{2})(x - 2)$$

all yield $2x^2 - x - 6$. A particularly useful way of factoring polynomials is in terms of prime polynomials.

> **› DEFINITION 1 Prime Polynomials**
>
> A polynomial of degree greater than 0 is said to be **prime** relative to a given set of numbers if: (1) all of its coefficients are from that set of numbers; and (2) it cannot be written as a product of two polynomials (excluding constant polynomials that are factors of 1) having coefficients from that set of numbers.
>
> Relative to the set of integers:
>
> $x^2 - 2$ is prime
> $x^2 - 9$ is not prime, since $x^2 - 9 = (x - 3)(x + 3)$
>
> [*Note:* The set of numbers most frequently used in factoring polynomials is the set of integers.]

A nonprime polynomial is said to be **factored completely relative to a given set of numbers** if it is written as a product of prime polynomials relative to that set of numbers.

In Examples 5 and 6 we review some of the standard factoring techniques for polynomials with integer coefficients.

EXAMPLE 5 Factoring Out Common Factors

Factor out, relative to the integers, all factors common to all terms:

(A) $2x^3y - 8x^2y^2 - 6xy^3$ (B) $2x(3x - 2) - 7(3x - 2)$

SOLUTIONS (A) $2x^3y - 8x^2y^2 - 6xy^3 \quad = (2xy)x^2 - (2xy)4xy - (2xy)3y^2$ Factor out 2xy.

$$= 2xy(x^2 - 4xy - 3y^2)$$

(B) $2x(3x - 2) - 7(3x - 2) \quad = 2x(3x - 2) - 7(3x - 2)$ Factor out 3x − 2.

$$= (2x - 7)(3x - 2)$$

MATCHED PROBLEM 5 Factor out, relative to the integers, all factors common to all terms:

(A) $3x^3y - 6x^2y^2 - 3xy^3$ (B) $3y(2y + 5) + 2(2y + 5)$

The polynomials in Example 6 can be factored by first grouping terms to find a common factor.

EXAMPLE 6 Factoring by Grouping

Factor completely, relative to the integers, by grouping:

(A) $3x^2 - 6x + 4x - 8$ (B) $wy + wz - 2xy - 2xz$
(C) $3ac + bd - 3ad - bc$

SOLUTIONS (A) $3x^2 - 6x + 4x - 8$ Group the first two and last two terms.

$= (3x^2 - 6x) + (4x - 8)$ Remove common factors from each group.

$= 3x(x - 2) + 4(x - 2)$ Factor out the common factor (x − 2).

$= (3x + 4)(x - 2)$

(B) $wy + wz - 2xy - 2xz$ Group the first two and last two terms—be careful of signs.

$= (wy + wz) - (2xy + 2xz)$ Remove common factors from each group.

$= w(y + z) - 2x(y + z)$ Factor out the common factor (y + z).

$= (w - 2x)(y + z)$

(C) $3ac + bd - 3ad - bc$
In parts (A) and (B) the polynomials are arranged in such a way that grouping the first two terms and the last two terms leads to common factors. In this problem neither the first two terms nor the last two terms have a common factor. Sometimes rearranging terms will lead to a factoring by grouping. In this case, we interchange

the second and fourth terms to obtain a problem comparable to part (B), which can be factored as follows:

$$3ac - bc - 3ad + bd = (3ac - bc) - (3ad - bd) \qquad \text{Factor out } c, d.$$
$$= c(3a - b) - d(3a - b) \qquad \text{Factor out } 3a - b.$$
$$= (c - d)(3a - b)$$

MATCHED PROBLEM 6

Factor completely, relative to the integers, by grouping:

(A) $2x^2 + 6x + 5x + 15$ (B) $2pr + ps - 6qr - 3qs$

(C) $6wy - xz - 2xy + 3wz$

Example 7 illustrates an approach to factoring a second-degree polynomial of the form

$$2x^2 - 5x - 3 \qquad \text{or} \qquad 2x^2 + 3xy - 2y^2$$

into the product of two first-degree polynomials with integer coefficients.

EXAMPLE 7

Factoring Second-Degree Polynomials

Factor each polynomial, if possible, using integer coefficients:

(A) $2x^2 + 3xy - 2y^2$ (B) $x^2 - 3x + 4$ (C) $6x^2 + 5xy - 4y^2$

SOLUTIONS

(A) $2x^2 + 3xy - 2y^2 = (2x + y)(x - y)$

Put in what we know. Signs must be opposite. (We can reverse this choice if we get $-3xy$ instead of $+3xy$ for the middle term.)

$$\begin{array}{c}\uparrow \qquad \uparrow \\ ? \qquad ?\end{array}$$

Now, what are the factors of 2 (the coefficient of y^2)?

$$\dfrac{2}{1 \cdot 2}$$

$$1 \cdot 2 \qquad (2x + y)(x - 2y) = 2x^2 - 3xy - 2y^2$$

$$2 \cdot 1 \qquad (2x + 2y)(x - y) = 2x^2 - 2y^2$$

The first choice gives us $-3xy$ for the middle term—close, but not there—so we reverse our choice of signs to obtain

$$2x^2 + 3xy - 2y^2 = (2x - y)(x + 2y)$$

(B) $x^2 - 3x + 4 = (x - \;\;)(x - \;\;)$ Signs must be the same because the third term is positive and must be negative because the middle term is negative.

$$\dfrac{4}{2 \cdot 2}$$

$$2 \cdot 2 \qquad (x - 2)(x - 2) = x^2 - 4x + 4$$

$$1 \cdot 4 \qquad (x - 1)(x - 4) = x^2 - 5x + 4$$

$$4 \cdot 1 \qquad (x - 4)(x - 1) = x^2 - 5x + 4$$

No choice produces the middle term; so $x^2 - 3x + 4$ is not factorable using integer coefficients.

(C) $6x^2 + 5xy - 4y^2 = (\;\;x + \;\;y)(\;\;x - \;\;y)$

$$\begin{array}{cccc}\uparrow & \uparrow & \uparrow & \uparrow \\ ? & ? & ? & ?\end{array}$$

The signs must be opposite in the factors, because the third term is negative. We can reverse our choice of signs later if necessary. We now write all factors of 6 and of 4:

$$\begin{array}{cc} 6 & 4 \\ \hline 2 \cdot 3 & 2 \cdot 2 \\ 3 \cdot 2 & 1 \cdot 4 \\ 1 \cdot 6 & 4 \cdot 1 \\ 6 \cdot 1 & \end{array}$$

and try each choice on the left with each on the right—a total of 12 combinations that give us the first and last terms in the polynomial $6x^2 + 5xy - 4y^2$. The question is: Does any combination also give us the middle term, $5xy$? After trial and error and, perhaps, some educated guessing among the choices, we find that $3 \cdot 2$ matched with $4 \cdot 1$ gives us the correct middle term.

$$6x^2 + 5xy - 4y^2 = (3x + 4y)(2x - y)$$

If none of the 24 combinations (including reversing our sign choice) had produced the middle term, then we would conclude that the polynomial is not factorable using integer coefficients.

MATCHED PROBLEM 7 Factor each polynomial, if possible, using integer coefficients:

(A) $x^2 - 8x + 12$ (B) $x^2 + 2x + 5$

(C) $2x^2 + 7xy - 4y^2$ (D) $4x^2 - 15xy - 4y^2$

The special factoring formulas listed here will enable us to factor certain polynomial forms that occur frequently.

> **SPECIAL FACTORING FORMULAS**

1. $u^2 + 2uv + v^2 = (u + v)^2$ **Perfect Square**

2. $u^2 - 2uv + v^2 = (u - v)^2$ **Perfect Square**

3. $u^2 - v^2 = (u - v)(u + v)$ **Difference of Squares**

4. $u^3 - v^3 = (u - v)(u^2 + uv + v^2)$ **Difference of Cubes**

5. $u^3 + v^3 = (u + v)(u^2 - uv + v^2)$ **Sum of Cubes**

The formulas in the box can be established by multiplying the factors on the right.

⟩⟩⟩ EXPLORE-DISCUSS 1 Explain why there is no formula for factoring a sum of squares $u^2 + v^2$ into the product of two first-degree polynomials with real coefficients.

EXAMPLE	8	**Using Special Factoring Formulas**

Factor completely relative to the integers:

(A) $x^2 + 6xy + 9y^2$ (B) $9x^2 - 4y^2$ (C) $8m^3 - 1$ (D) $x^3 + y^3z^3$

SOLUTIONS (A) $x^2 + 6xy + 9y^2$ $= x^2 + 2(x)(3y) + (3y)^2$ $= (x + 3y)^2$ Perfect square

(B) $9x^2 - 4y^2$ $= (3x)^2 - (2y)^2$ $= (3x - 2y)(3x + 2y)$ Difference of squares

(C) $8m^3 - 1$ $= (2m)^3 - 1^3$ Difference of cubes
$= (2m - 1)[(2m)^2 + (2m)(1) + 1^2]$ Simplify.

$= (2m - 1)(4m^2 + 2m + 1)$

(D) $x^3 + y^3z^3$ $= x^3 + (yz)^3$ Sum of cubes

$= (x + yz)(x^2 - xyz + y^2z^2)$

MATCHED PROBLEM 8	Factor completely relative to the integers:

(A) $4m^2 - 12mn + 9n^2$ (B) $x^2 - 16y^2$ (C) $z^3 - 1$ (D) $m^3 + n^3$

ANSWERS TO MATCHED PROBLEMS

1. (A) $3x^2 - 2x + 1, x^2 - 2xy + y^2$ (B) 5, 3, 5 (C) 6, 4, 6
2. $3x^4 - x^3 - 5x^2 + 2x - 2$ 3. $3x^2 + 5x - 10$ 4. $4x^3 - 13x + 6$
5. (A) $3xy(x^2 - 2xy - y^2)$ (B) $(3y + 2)(2y + 5)$
6. (A) $(2x + 5)(x + 3)$ (B) $(p - 3q)(2r + s)$ (C) $(3w - x)(2y + z)$
7. (A) $(x - 2)(x - 6)$ (B) Not factorable using integers (C) $(2x - y)(x + 4y)$
 (D) $(4x + y)(x - 4y)$
8. (A) $(2m - 3n)^2$ (B) $(x - 4y)(x + 4y)$ (C) $(z - 1)(z^2 + z + 1)$
 (D) $(m + n)(m^2 - mn + n^2)$

R-3 Exercises

Problems 1–8 refer to the polynomials (a) $x^2 + 1$ and (b) $x^4 - 2x + 1$.

1. What is the degree of (a)?

2. What is the degree of (b)?

3. What is the degree of the sum of (a) and (b)?

4. What is the degree of the product of (a) and (b)?

5. Multiply (a) and (b).

6. Add (a) and (b).

7. Subtract (b) from (a).

8. Subtract (a) from (b).

In Problems 9–14, is the algebraic expression a polynomial? If so, give its degree.

9. $4 - x^2$ **10.** $x^3 - 5x^6 + 1$

11. $x^3 - 7x + 8\sqrt{x}$ **12.** $x^4 + 3x - \sqrt{5}$

13. $x^5 - 4x^2 + 6^{-2}$ **14.** $3x^4 - 2x^{-1} - 10$

In Problems 15–22, perform the indicated operations and simplify.

15. $2(x - 1) + 3(2x - 3) - (4x - 5)$

16. $2y - 3y[4 - 2(y - 1)]$ **17.** $(m - n)(m + n)$

18. $(5y - 1)(3 - 2y)$

19. $(3x + 2y)(x - 3y)$

20. $(4x - y)^2$

21. $(a + b)(a^2 - ab + b^2)$

22. $(a - b)(a^2 + ab + b^2)$

In Problems 23–28, factor out, relative to the integers, all factors common to all terms.

23. $6x^4 - 8x^3 - 2x^2$

24. $3x^5 + 6x^3 + 9x$

25. $x^2y + 2xy^2 + x^2y^2$

26. $8u^3v - 6u^2v^2 + 4uv^3$

27. $2w(y - 2z) - x(y - 2z)$

28. $2x(u - 3v) + 5y(u - 3v)$

In Problems 29–34, factor completely, relative to the integers.

29. $x^2 + 4x + x + 4$

30. $2y^2 - 6y + 5y - 15$

31. $x^2 - xy + 3xy - 3y^2$

32. $3a^2 - 12ab - 2ab + 8b^2$

33. $8ac + 3bd - 6bc - 4ad$

34. $3ux - 4vy + 3vx - 4uy$

In Problems 35–42, perform the indicated operations and simplify.

35. $2x - 3\{x + 2[x - (x + 5)] + 1\}$

36. $m - \{m - [m - (m - 1)]\}$

37. $(2x^2 - 3x + 1)(x^2 + x - 2)$

38. $(x^2 - 3xy + y^2)(x^2 + 3xy + y^2)$

39. $(3u - 2v)^2 - (2u - 3v)(2u + 3v)$

40. $(2a - b)^2 - (a + 2b)^2$

41. $(2m - n)^3$

42. $(3a + 2b)^3$

In Problems 43–62, factor completely, relative to the integers. If a polynomial is prime relative to the integers, say so.

43. $2x^2 + x - 3$

44. $3y^2 - 8y - 3$

45. $x^2 + 5xy - 14y^2$

46. $x^2 + 4y^2$

47. $4x^2 - 20x + 25$

48. $a^2b^2 - c^2$

49. $a^2b^2 + c^2$

50. $9x^2 - 4$

51. $4x^2 + 9$

52. $16x^2 - 25$

53. $6x^2 + 48x + 72$

54. $3z^2 - 28z + 48$

55. $2x^4 - 24x^3 + 40x^2$

56. $16x^2y - 8xy + y$

57. $6m^2 - mn - 12n^2$

58. $4u^3v - uv^3$

59. $3m^3 - 6m^2 + 15m$

60. $2x^3 - 2x^2 + 8x$

61. $m^3 + n^3$

62. $8x^3 - 125$

Problems 63–68 are calculus-related. Perform the indicated operations and simplify.

63. $3(x + h) - 7 - (3x - 7)$

64. $(x + h)^2 - x^2$

65. $2(x + h)^2 - 3(x + h) - (2x^2 - 3x)$

66. $-4(x + h)^2 + 6(x + h) - (-4x^2 + 6x)$

67. $(x + h)^3 - 2(x + h)^2 - (x^3 - 2x^2)$

68. $(x + h)^3 + 3(x + h) - (x^3 + 3x)$

Problems 69–74 are calculus-related. Factor completely, relative to the integers.

69. $2x(x + 1)^4 + 4x^2(x + 1)^3$

70. $(x - 1)^3 + 3x(x - 1)^2$

71. $6(3x - 5)(2x - 3)^2 + 4(3x - 5)^2(2x - 3)$

72. $2(x - 3)(4x + 7)^2 + 8(x - 3)^2(4x + 7)$

73. $5x^4(9 - x)^4 - 4x^5(9 - x)^3$

74. $3x^4(x - 7)^2 + 4x^3(x - 7)^3$

In Problems 75–86, factor completely, relative to the integers. In polynomials involving more than three terms, try grouping the terms in various combinations as a first step. If a polynomial is prime relative to the integers, say so.

75. $(a - b)^2 - 4(c - d)^2$

76. $(x + 2)^2 + 9$

77. $2am - 3an + 2bm - 3bn$

78. $15ac - 20ad + 3bc - 4bd$

79. $3x^2 - 2xy - 4y^2$

80. $5u^2 + 4uv - v^2$

81. $x^3 - 3x^2 - 9x + 27$ **82.** $t^3 - 2t^2 + t - 2$

83. $4(A + B)^2 - 5(A + B) - 5$

84. $x^4 + 6x^2 + 8$ **85.** $m^4 - n^4$

86. $y^4 - 3y^2 - 4$

87. Show by example that, in general, $(a + b)^2 \neq a^2 + b^2$. Discuss possible conditions on a and b that would make this a valid equation.

88. Show by example that, in general, $(a - b)^2 \neq a^2 - b^2$. Discuss possible conditions on a and b that would make this a valid equation.

89. To show that $\sqrt{2}$ is an irrational number, explain how the assumption that $\sqrt{2}$ is rational leads to a contradiction of Theorem 1, the fundamental theorem of arithmetic, by the following steps:
(A) Suppose that $\sqrt{2} = a/b$, where a and b are positive integers, $b \neq 0$. Explain why $a^2 = 2b^2$.
(B) Explain why the prime number 2 appears an even number of times (possibly 0 times) as a factor in the prime factorization of a^2.
(C) Explain why the prime number 2 appears an odd number of times as a factor in the prime factorization of $2b^2$.
(D) Explain why parts (B) and (C) contradict the fundamental theorem of arithmetic.

90. To show that $\sqrt{n}$ is an irrational number unless n is a perfect square, explain how the assumption that $\sqrt{n}$ is rational leads to a contradiction of the fundamental theorem of arithmetic by the following steps:

(A) Assume that n is not a perfect square, that is, does not belong to the sequence 1, 4, 9, 16, 25, Explain why some prime number p appears an odd number of times as a factor in the prime factorization of n.

(B) Suppose that $\sqrt{n} = a/b$, where a and b are positive integers, $b \neq 0$. Explain why $a^2 = nb^2$.

(C) Explain why the prime number p appears an even number of times (possibly 0 times) as a factor in the prime factorization of a^2.

(D) Explain why the prime number p appears an odd number of times as a factor in the prime factorization of nb^2.

(E) Explain why parts (C) and (D) contradict the fundamental theorem of arithmetic.

APPLICATIONS

91. GEOMETRY The width of a rectangle is 5 centimeters less than its length. If x represents the length, write an algebraic expression in terms of x that represents the perimeter of the rectangle. Simplify the expression.

92. GEOMETRY The length of a rectangle is 8 meters more than its width. If x represents the width of the rectangle, write an algebraic expression in terms of x that represents its area. Change the expression to a form without parentheses.

93. COIN PROBLEM A parking meter contains nickels, dimes, and quarters. There are 5 fewer dimes than nickels, and 2 more quarters than dimes. If x represents the number of nickels, write an algebraic expression in terms of x that represents the value of all the coins in the meter in cents. Simplify the expression.

94. COIN PROBLEM A vending machine contains dimes and quarters only. There are 4 more dimes than quarters. If x represents the number of quarters, write an algebraic expression in terms of x that represents the value of all the coins in the vending machine in cents. Simplify the expression.

95. PACKAGING A spherical plastic container for designer wristwatches has an inner radius of x centimeters (see the figure). If the plastic shell is

0.3 cm

x cm

Figure for 95

0.3 centimeters thick, write an algebraic expression in terms of x that represents the volume of the plastic used to construct the container. Simplify the expression. [*Recall:* The volume V of a sphere of radius r is given by $V = \frac{4}{3}\pi r^3$.]

96. PACKAGING A cubical container for shipping computer components is formed by coating a metal mold with polystyrene. If the metal mold is a cube with sides x centimeters long and the polystyrene coating is 2 centimeters thick, write an algebraic expression in terms of x that represents the volume of the polystyrene used to construct the container. Simplify the expression. [*Recall:* The volume V of a cube with sides of length t is given by $V = t^3$.]

97. CONSTRUCTION A rectangular open-topped box is to be constructed out of 20-inch-square sheets of thin cardboard by cutting x-inch squares out of each corner and bending the sides up as indicated in the figure. Express each of the following quantities as a polynomial in both factored and expanded form.

(A) The area of cardboard after the corners have been removed.

(B) The volume of the box.

98. CONSTRUCTION A rectangular open-topped box is to be constructed out of 9- by 16-inch sheets of thin cardboard by cutting x-inch squares out of each corner and bending the sides up. Express each of the following quantities as a polynomial in both factored and expanded form.

(A) The area of cardboard after the corners have been removed.

(B) The volume of the box.

R-4 Rational Expressions: Basic Operations

› Reducing to Lowest Terms
› Multiplication and Division
› Addition and Subtraction
› Compound Fractions

A quotient of two algebraic expressions, division by 0 excluded, is called a **fractional expression.** If both the numerator and denominator of a fractional expression are polynomials, the fractional expression is called a **rational expression.** Some examples of rational expressions are the following (recall that a nonzero constant is a polynomial of degree 0):

$$\frac{x-2}{2x^2-3x+5} \qquad \frac{1}{x^4-1} \qquad \frac{3}{x} \qquad \frac{x^2+3x-5}{1}$$

In this section, we discuss basic operations on rational expressions, including multiplication, division, addition, and subtraction.

Since variables represent real numbers in the rational expressions we are going to consider, the properties of real number fractions summarized in Section R-1 play a central role in much of the work that we will do.

Even though not always explicitly stated, we always assume that variables are restricted so that division by 0 is excluded.

› Reducing to Lowest Terms

We start this discussion by restating the fundamental property of fractions (from Theorem 3 in Section R-1):

> **› FUNDAMENTAL PROPERTY OF FRACTIONS**
>
> If a, b, and k are real numbers with $b, k \neq 0$, then
>
> $$\frac{ka}{kb} = \frac{a}{b} \qquad \frac{2 \cdot 3}{2 \cdot 4} = \frac{3}{4} \qquad \frac{(x-3)2}{(x-3)x} = \frac{2}{x}$$
> $$x \neq 0, x \neq 3$$

Using this property from left to right to eliminate all common factors from the numerator and the denominator of a given fraction is referred to as **reducing a fraction to lowest terms.** We are actually dividing the numerator and denominator by the same nonzero common factor.

Using the property from right to left—that is, multiplying the numerator and the denominator by the same nonzero factor—is referred to as **raising a fraction to higher terms.** We will use the property in both directions in the material that follows.

We say that a rational expression is **reduced to lowest terms** if the numerator and denominator do not have any factors in common. Unless stated to the contrary, factors will be relative to the integers.

EXAMPLE 1 **Reducing Rational Expressions**

Reduce each rational expression to lowest terms.

(A) $\dfrac{x^2 - 6x + 9}{x^2 - 9}$ (B) $\dfrac{x^3 - 1}{x^2 - 1}$

SOLUTIONS

(A) $\dfrac{x^2 - 6x + 9}{x^2 - 9} = \dfrac{(x-3)^2}{(x-3)(x+3)}$ Factor numerator and denominator completely. Divide numerator and denominator by $(x - 3)$; this is a valid operation as long as $x \ne 3$.

$\qquad\qquad = \dfrac{x - 3}{x + 3}$

(B) $\dfrac{x^3 - 1}{x^2 - 1} = \dfrac{\overset{1}{\cancel{(x - 1)}}(x^2 + x + 1)}{\underset{1}{\cancel{(x - 1)}}(x + 1)}$ Dividing numerator and denominator by $(x - 1)$ can be indicated by drawing lines through both $(x - 1)$'s and writing the resulting quotients, 1's.

$\qquad\qquad = \dfrac{x^2 + x + 1}{x + 1}$ $x \ne -1$ and $x \ne 1$

MATCHED PROBLEM 1 Reduce each rational expression to lowest terms.

(A) $\dfrac{6x^2 + x - 2}{2x^2 + x - 1}$ (B) $\dfrac{x^4 - 8x}{3x^3 - 2x^2 - 8x}$

⟩⟩⟩ CAUTION ⟩⟩⟩

Remember to always factor the numerator and denominator first, then divide out any *common factors*. Do not indiscriminately eliminate *terms* that appear in both the numerator and the denominator. For example,

$$\frac{2x^3 + y^2}{y^2} \ne \frac{2x^3 + \overset{1}{\cancel{y^2}}}{\underset{1}{\cancel{y^2}}}$$

$$\frac{2x^3 + y^2}{y^2} \ne 2x^3 + 1$$

Since the term y^2 is not a factor of the numerator, it cannot be eliminated. In fact, $(2x^3 + y^2)/y^2$ is already reduced to lowest terms.

⟩ Multiplication and Division

Since we are restricting variable replacements to real numbers, multiplication and division of rational expressions follow the rules for multiplying and dividing real number fractions (Theorem 3 in Section R-1).

⟩ MULTIPLICATION AND DIVISION

If a, b, c, and d are real numbers with $b, d \ne 0$, then:

1. $\dfrac{a}{b} \cdot \dfrac{c}{d} = \dfrac{ac}{bd}$ $\qquad\qquad$ $\dfrac{2}{3} \cdot \dfrac{x}{x-1} = \dfrac{2x}{3(x-1)}$

2. $\dfrac{a}{b} \div \dfrac{c}{d} = \dfrac{a}{b} \cdot \dfrac{d}{c}$ $\quad c \ne 0$ $\qquad$ $\dfrac{2}{3} \div \dfrac{x}{x-1} = \dfrac{2}{3} \cdot \dfrac{x-1}{x}$

EXAMPLE 2 ## Multiplying and Dividing Rational Expressions

Perform the indicated operations and reduce to lowest terms.

(A) $\dfrac{10x^3y}{3xy + 9y} \cdot \dfrac{x^2 - 9}{4x^2 - 12x}$

(B) $\dfrac{4 - 2x}{4} \div (x - 2)$

(C) $\dfrac{2x^3 - 2x^2y + 2xy^2}{x^3y - xy^3} \div \dfrac{x^3 + y^3}{x^2 + 2xy + y^2}$

SOLUTIONS

(A) $\dfrac{10x^3y}{3xy + 9y} \cdot \dfrac{x^2 - 9}{4x^2 - 12x} = \dfrac{\overset{5x^2}{\cancel{10x^3y}}}{\underset{3 \cdot 1}{3\cancel{y}(x + 3)}} \cdot \dfrac{\overset{1 \cdot 1}{(x - 3)(x + 3)}}{\underset{2 \cdot 1}{4\cancel{x}(x - 3)}}$

Factor numerators and denominators; then divide any numerator and any denominator with a like common factor.

$= \dfrac{5x^2}{6}$

(B) $\dfrac{4 - 2x}{4} \div (x - 2) = \dfrac{\overset{1}{2(2 - x)}}{\underset{2}{4}} \cdot \dfrac{1}{x - 2}$

$x - 2$ is the same as $\dfrac{x - 2}{1}$.

$= \dfrac{2 - x}{2(x - 2)} = \dfrac{\overset{-1}{-(x - 2)}}{2\underset{1}{(x - 2)}}$

$b - a = -(a - b)$, a useful change in some problems.

$= -\dfrac{1}{2}$

(C) $\dfrac{2x^3 - 2x^2y + 2xy^2}{x^3y - xy^3} \div \dfrac{x^3 + y^3}{x^2 + 2xy + y^2}$

$\dfrac{a}{b} \div \dfrac{c}{d} = \dfrac{a}{b} \cdot \dfrac{d}{c}$

$= \dfrac{\overset{2}{2x}(\overset{1}{x^2 - xy + y^2})}{\underset{y}{x\cancel{y}}(x + y)\underset{1}{(x - y)}} \cdot \dfrac{\overset{1}{(x + y)^2}}{\underset{1}{(x + y)}\underset{1}{(x^2 - xy + y^2)}}$

Divide out common factors.

$= \dfrac{2}{y(x - y)}$

MATCHED PROBLEM 2

Perform the indicated operations and reduce to lowest terms.

(A) $\dfrac{12x^2y^3}{2xy^2 + 6xy} \cdot \dfrac{y^2 + 6y + 9}{3y^3 + 9y^2}$

(B) $(4 - x) \div \dfrac{x^2 - 16}{5}$

(C) $\dfrac{m^3 + n^3}{2m^2 + mn - n^2} \div \dfrac{m^3n - m^2n^2 + mn^3}{2m^3n^2 - m^2n^3}$

› ## Addition and Subtraction

Again, because we are restricting variable replacements to real numbers, addition and subtraction of rational expressions follow the rules for adding and subtracting real number fractions (Theorem 3 in Section R-1).

> ADDITION AND SUBTRACTION

For a, b, and c real numbers with $b \neq 0$:

1. $\dfrac{a}{b} + \dfrac{c}{b} = \dfrac{a + c}{b}$ $\qquad$ $\dfrac{x}{x - 3} + \dfrac{2}{x - 3} = \dfrac{x + 2}{x - 3}$

2. $\dfrac{a}{b} - \dfrac{c}{b} = \dfrac{a - c}{b}$ $\qquad$ $\dfrac{x}{2xy^2} - \dfrac{x - 4}{2xy^2} = \dfrac{x - (x - 4)}{2xy^2}$

So we add rational expressions with the same denominators by adding or subtracting their numerators and placing the result over the common denominator. If the denominators are not the same, we raise the fractions to higher terms, using the fundamental property of fractions to obtain common denominators, and then proceed as described.

Even though any common denominator will do, our work will be simplified if the **least common denominator (LCD)** is used. Often, the LCD is obvious, but if it is not, the steps in the box describe how to find it.

> THE LEAST COMMON DENOMINATOR (LCD)

The LCD of two or more rational expressions is found as follows:

1. Factor each denominator completely.

2. Identify each different prime factor from all the denominators.

3. Form a product using each different factor to the highest power that occurs in any one denominator. This product is the LCD.

EXAMPLE | **3** | **Adding and Subtracting Rational Expressions**

Combine into a single fraction and reduce to lowest terms.

(A) $\dfrac{3}{10} + \dfrac{5}{6} - \dfrac{11}{45}$ $\qquad$ (B) $\dfrac{4}{9x} - \dfrac{5x}{6y^2} + 1$ $\qquad$ (C) $\dfrac{x + 3}{x^2 - 6x + 9} - \dfrac{x + 2}{x^2 - 9} - \dfrac{5}{3 - x}$

SOLUTIONS

(A) To find the LCD, factor each denominator completely:

$$\left. \begin{array}{l} 10 = 2 \cdot 5 \\ 6 = 2 \cdot 3 \\ 45 = 3^2 \cdot 5 \end{array} \right\} \text{LCD} = 2 \cdot 3^2 \cdot 5 = 90$$

Now use the fundamental property of fractions to make each denominator 90:

$$\frac{3}{10} + \frac{5}{6} - \frac{11}{45} = \frac{9 \cdot 3}{9 \cdot 10} + \frac{15 \cdot 5}{15 \cdot 6} - \frac{2 \cdot 11}{2 \cdot 45} \qquad \text{Multiply.}$$

$$= \boxed{\frac{27}{90} + \frac{75}{90} - \frac{22}{90}} \qquad \text{Combine into a single fraction.}$$

$$= \frac{27 + 75 - 22}{90} = \frac{80}{90} = \frac{8}{9}$$

(B) $\left.\begin{array}{l} 9x = 3^2x \\ 6y^2 = 2 \cdot 3y^2 \end{array}\right\}$ LCD $= 2 \cdot 3^2xy^2 = 18xy^2$

$$\frac{4}{9x} - \frac{5x}{6y^2} + 1 = \frac{2y^2 \cdot 4}{2y^2 \cdot 9x} - \frac{3x \cdot 5x}{3x \cdot 6y^2} + \frac{18xy^2}{18xy^2} \qquad \textit{Multiply, combine.}$$

$$= \frac{8y^2 - 15x^2 + 18xy^2}{18xy^2}$$

(C) $\dfrac{x+3}{x^2-6x+9} - \dfrac{x+2}{x^2-9} - \dfrac{5}{3-x} = \dfrac{x+3}{(x-3)^2} - \dfrac{x+2}{(x-3)(x+3)} + \dfrac{5}{x-3}$

$$\textit{Note: } -\frac{5}{3-x} = -\frac{5}{-(x-3)} = \frac{5}{x-3} \qquad \begin{array}{l}\textit{We have again used the fact} \\ \textit{that } a - b = -(b - a).\end{array}$$

The LCD $= (x-3)^2(x+3)$.

$$\frac{(x+3)^2}{(x-3)^2(x+3)} - \frac{(x-3)(x+2)}{(x-3)^2(x+3)} + \frac{5(x-3)(x+3)}{(x-3)^2(x+3)} \qquad \textit{Expand numerators.}$$

$$= \frac{(x^2+6x+9) - (x^2-x-6) + 5(x^2-9)}{(x-3)^2(x+3)} \qquad \begin{array}{l}\textit{Be careful of sign} \\ \textit{errors here.}\end{array}$$

$$= \frac{x^2+6x+9 - x^2+x+6 + 5x^2-45}{(x-3)^2(x+3)} \qquad \textit{Combine like terms.}$$

$$= \frac{5x^2+7x-30}{(x-3)^2(x+3)}$$

MATCHED PROBLEM 3 Combine into a single fraction and reduce to lowest terms.

(A) $\dfrac{5}{28} - \dfrac{1}{10} + \dfrac{6}{35}$ 　　　　　　　　(B) $\dfrac{1}{4x^2} - \dfrac{2x+1}{3x^3} + \dfrac{3}{12x}$

(C) $\dfrac{y-3}{y^2-4} - \dfrac{y+2}{y^2-4y+4} - \dfrac{2}{2-y}$

⟫⟫⟫ EXPLORE-DISCUSS 1

What is the result of entering $16 \div 4 \div 2$ on a calculator?

What is the difference between $16 \div (4 \div 2)$ and $(16 \div 4) \div 2$?

How could you use fraction bars to distinguish between these two cases when writing $\dfrac{\frac{16}{4}}{2}$?

› Compound Fractions

A fractional expression with fractions in its numerator, denominator, or both is called a **compound fraction**. It is often necessary to represent a compound fraction as a **simple fraction**—that is (in all cases we will consider), as the quotient of two polynomials. The process does not involve any new concepts. It is a matter of applying old concepts and processes in the right sequence. We will illustrate two approaches to the problem, each with its own merits, depending on the particular problem under consideration.

EXAMPLE	4	**Simplifying Compound Fractions**

Express as a simple fraction reduced to lowest terms:

$$\frac{\dfrac{2}{x} - 1}{\dfrac{4}{x^2} - 1}$$

SOLUTION | *Method 1.* Multiply the numerator and denominator by the LCD of all fractions in the numerator and denominator—in this case, x^2. (We are multiplying by $1 = x^2/x^2$.)

$$\frac{x^2\left(\dfrac{2}{x} - 1\right)}{x^2\left(\dfrac{4}{x^2} - 1\right)} = \frac{x^2\dfrac{2}{x} - x^2}{x^2\dfrac{4}{x^2} - x^2} = \frac{2x - x^2}{4 - x^2} = \frac{x(2 - x)}{(2 + x)(2 - x)}$$

$$= \frac{x}{2 + x}$$

Method 2. Write the numerator and denominator as single fractions. Then treat as a quotient.

$$\frac{\dfrac{2}{x} - 1}{\dfrac{4}{x^2} - 1} = \frac{\dfrac{2 - x}{x}}{\dfrac{4 - x^2}{x^2}} = \frac{2 - x}{x} \div \frac{4 - x^2}{x^2} = \frac{2 - x}{x} \cdot \frac{x^2}{(2 - x)(2 + x)}$$

$$= \frac{x}{2 + x}$$

MATCHED PROBLEM 4 | Express as a simple fraction reduced to lowest terms. Use the two methods described in Example 4.

$$\frac{1 + \dfrac{1}{x}}{x - \dfrac{1}{x}}$$

ANSWERS TO MATCHED PROBLEMS

1. (A) $\dfrac{3x + 2}{x + 1}$ (B) $\dfrac{x^2 + 2x + 4}{3x + 4}$ 2. (A) $2x$ (B) $\dfrac{-5}{x + 4}$ (C) mn

3. (A) $\dfrac{1}{4}$ (B) $\dfrac{3x^2 - 5x - 4}{12x^3}$ (C) $\dfrac{2y^2 - 9y - 6}{(y - 2)^2(y + 2)}$ 4. $\dfrac{1}{x - 1}$

R-4 Exercises

In Problems 1–10, reduce each rational expression to lowest terms.

1. $\dfrac{17}{85}$ 2. $\dfrac{91}{26}$ 3. $\dfrac{360}{288}$ 4. $\dfrac{63}{105}$ 5. $\dfrac{x + 1}{x^2 + 3x + 2}$ 6. $\dfrac{x^2 - 2x - 24}{x - 6}$ 7. $\dfrac{x^2 - 9}{x^2 + 3x - 18}$

8. $\dfrac{x^2 + 9x + 20}{x^2 - 16}$

9. $\dfrac{3x^2y^3}{x^4y}$

10. $\dfrac{2a^2b^4c^6}{6a^5b^3c}$

In Problems 11–36, perform the indicated operations and reduce answers to lowest terms. Represent any compound fractions as simple fractions reduced to lowest terms.

11. $\dfrac{5}{6} + \dfrac{11}{15}$

12. $\dfrac{7}{10} + \dfrac{19}{25}$

13. $\dfrac{1}{8} - \dfrac{1}{9}$

14. $\dfrac{9}{8} - \dfrac{8}{9}$

15. $\dfrac{1}{n} - \dfrac{1}{m}$

16. $\dfrac{m}{n} - \dfrac{n}{m}$

17. $\dfrac{5}{12} \div \dfrac{3}{4}$

18. $\dfrac{10}{3} \div \dfrac{5}{2}$

19. $\left(\dfrac{25}{8} \div \dfrac{5}{16}\right) \cdot \dfrac{4}{15}$

20. $\dfrac{25}{8} \div \left(\dfrac{5}{16} \cdot \dfrac{4}{15}\right)$

21. $\left(\dfrac{b^2}{2a} \div \dfrac{b}{a^2}\right) \cdot \dfrac{a}{3b}$

22. $\dfrac{b^2}{2a} \div \left(\dfrac{b}{a^2} \cdot \dfrac{a}{3b}\right)$

23. $\dfrac{x^2 - 1}{x + 2} \div \dfrac{x + 1}{x^2 - 4}$

24. $\dfrac{x^2 - 9}{x^2 - 1} \div \dfrac{x - 3}{x - 1}$

25. $\dfrac{1}{c} + \dfrac{1}{b} + \dfrac{1}{a}$

26. $\dfrac{1}{bc} + \dfrac{1}{ac} + \dfrac{1}{ab}$

27. $\dfrac{2a - b}{a^2 - b^2} - \dfrac{2a + 3b}{a^2 + 2ab + b^2}$

28. $\dfrac{x + 2}{x^2 - 1} - \dfrac{x - 2}{(x - 1)^2}$

29. $m + 2 - \dfrac{m - 2}{m - 1}$

30. $\dfrac{x + 1}{x - 1} + x$

31. $\dfrac{3}{x - 2} - \dfrac{2}{2 - x}$

32. $\dfrac{1}{a - 3} - \dfrac{2}{3 - a}$

33. $\dfrac{3}{y + 2} + \dfrac{2}{y - 2} - \dfrac{4y}{y^2 - 4}$

34. $\dfrac{4x}{x^2 - y^2} + \dfrac{3}{x + y} - \dfrac{2}{x - y}$

35. $\dfrac{\dfrac{x^2}{y^2} - 1}{\dfrac{x}{y} + 1}$

36. $\dfrac{\dfrac{4}{x} - x}{\dfrac{2}{x} - 1}$

Ⓤ *Problems 37–42 are calculus-related. Reduce each fraction to lowest terms.*

37. $\dfrac{6x^3(x^2 + 2)^2 - 2x(x^2 + 2)^3}{x^4}$

38. $\dfrac{4x^4(x^2 + 3) - 3x^2(x^2 + 3)^2}{x^6}$

39. $\dfrac{2x(1 - 3x)^3 + 9x^2(1 - 3x)^2}{(1 - 3x)^6}$

40. $\dfrac{2x(2x + 3)^4 - 8x^2(2x + 3)^3}{(2x + 3)^8}$

41. $\dfrac{-2x(x + 4)^3 - 3(3 - x^2)(x + 4)^2}{(x + 4)^6}$

42. $\dfrac{3x^2(x + 1)^3 - 3(x^3 + 4)(x + 1)^2}{(x + 1)^6}$

In Problems 43–54, perform the indicated operations and reduce answers to lowest terms. Represent any compound fractions as simple fractions reduced to lowest terms.

43. $\dfrac{y}{y^2 - 2y - 8} - \dfrac{2}{y^2 - 5y + 4} + \dfrac{1}{y^2 + y - 2}$

44. $\dfrac{x}{x^2 - 9x + 18} + \dfrac{x - 8}{x - 6} + \dfrac{x + 4}{x - 3}$

45. $\dfrac{16 - m^2}{m^2 + 3m - 4} \cdot \dfrac{m - 1}{m - 4}$

46. $\dfrac{x + 1}{x(1 - x)} \cdot \dfrac{x^2 - 2x + 1}{x^2 - 1}$

47. $\dfrac{x + 7}{ax - bx} + \dfrac{y + 9}{by - ay}$

48. $\dfrac{c + 2}{5c - 5} - \dfrac{c - 2}{3c - 3} + \dfrac{c}{1 - c}$

49. $\dfrac{x^2 - 16}{2x^2 + 10x + 8} \div \dfrac{x^2 - 13x + 36}{x^3 + 1}$

50. $\left(\dfrac{x^3 - y^3}{y^3} \cdot \dfrac{y}{x - y}\right) \div \dfrac{x^2 + xy + y^2}{y^2}$

51. $\left(\dfrac{x}{x^2 - 16} - \dfrac{1}{x + 4}\right) \div \dfrac{4}{x + 4}$

52. $\left(\dfrac{3}{x - 2} - \dfrac{1}{x + 1}\right) \div \dfrac{x + 4}{x - 2}$

53. $\dfrac{1 + \dfrac{2}{x} - \dfrac{15}{x^2}}{1 + \dfrac{4}{x} - \dfrac{5}{x^2}}$

54. $\dfrac{\dfrac{x}{y} - 2 + \dfrac{y}{x}}{\dfrac{x}{y} - \dfrac{y}{x}}$

Ⓤ *Problems 55–58 are calculus-related. Perform the indicated operations and reduce answers to lowest terms. Represent any compound fractions as simple fractions reduced to lowest terms.*

55. $\dfrac{\dfrac{1}{x + h} - \dfrac{1}{x}}{h}$

56. $\dfrac{\dfrac{1}{(x + h)^2} - \dfrac{1}{x^2}}{h}$

57. $\dfrac{\dfrac{(x + h)^2}{x + h + 2} - \dfrac{x^2}{x + 2}}{h}$

58. $\dfrac{\dfrac{2x + 2h + 3}{x + h} - \dfrac{2x + 3}{x}}{h}$

In Problems 59–62, perform the indicated operations and reduce answers to lowest terms. Represent any compound fractions as simple fractions reduced to lowest terms.

59. $\dfrac{y - \dfrac{y^2}{y - x}}{1 + \dfrac{x^2}{y^2 - x^2}}$

60. $\dfrac{\dfrac{s^2}{s - t} - s}{\dfrac{t^2}{s - t} + t}$

61. $2 - \dfrac{1}{1 - \dfrac{2}{a + 2}}$

62. $1 - \dfrac{1}{1 - \dfrac{1}{1 - \dfrac{1}{x}}}$

63. Show by example that, in general,

$$\frac{a + b}{b} \neq a + 1 \quad (\text{assume } b \neq 0)$$

Discuss possible conditions of a and b that would make this a valid equation.

64. Show by example that, in general,

$$\frac{a^2 + b^2}{a + b} \neq a + b \quad (\text{assume } a \neq -b)$$

Discuss possible conditions of a and b that would make this a valid equation.

CHAPTER **R** **Review**

R-1 Algebra and Real Numbers

A **real number** is any number that has a decimal representation. There is a one-to-one correspondence between the set of real numbers and the set of points on a line. Important subsets of the real numbers include the **natural numbers, integers,** and **rational numbers.** A rational number can be written in the form a/b, where a and b are integers and $b \neq 0$. A real number can be approximated to any desired precision by rational numbers. Consequently, arithmetic operations on rational numbers can be extended to operations on real numbers. These operations satisfy **basic real number properties,** including **associative properties:** $x + (y + z) = (x + y) + z$ and $x(yz) = (xy)z$; **commutative properties:** $x + y = y + x$ and $xy = yx$; **identities:** $0 + x = x + 0 = x$ and $(1)x = x(1) = x$; **inverses:** $-x$ is the **additive inverse** of x and, if $x \neq 0$, x^{-1} is the **multiplicative inverse** of x; and **distributive property:** $x(y + z) = xy + xz$. **Subtraction** is defined by $a - b = a + (-b)$ and **division** by $a/b = ab^{-1}$. Division by 0 is never allowed. Additional properties include **properties of negatives:**

1. $-(-a) = a$

2. $(-a)b = -(ab) = a(-b) = -ab$

3. $(-a)(-b) = ab$

4. $(-1)a = -a$

5. $\dfrac{-a}{b} = -\dfrac{a}{b} = \dfrac{a}{-b} \quad b \neq 0$

6. $\dfrac{-a}{-b} = -\dfrac{-a}{b} = -\dfrac{a}{-b} = \dfrac{a}{b} \quad b \neq 0$

zero properties:

1. $a \cdot 0 = 0$

2. $ab = 0$ if and only if $a = 0$ or $b = 0$ or both.

and **fraction properties** (division by 0 excluded):

1. $\dfrac{a}{b} = \dfrac{c}{d}$ if and only if $ad = bc$

2. $\dfrac{ka}{kb} = \dfrac{a}{b}$

3. $\dfrac{a}{b} \cdot \dfrac{c}{d} = \dfrac{ac}{bd}$

4. $\dfrac{a}{b} \div \dfrac{c}{d} = \dfrac{a}{b} \cdot \dfrac{d}{c}$

5. $\dfrac{a}{b} + \dfrac{c}{b} = \dfrac{a + c}{b}$

6. $\dfrac{a}{b} - \dfrac{c}{b} = \dfrac{a - c}{b}$

7. $\dfrac{a}{b} + \dfrac{c}{d} = \dfrac{ad + bc}{bd}$

R-2 Exponents and Radicals

The notation a^n, in which the **exponent** n is an integer, is defined as follows. For n a positive integer and a a real number:

$$a^n = a \cdot a \cdot \ldots \cdot a \quad (n \text{ factors of } a)$$

$$a^{-n} = \frac{1}{a^n} \quad (a \neq 0)$$

$$a^0 = 1 \quad (a \neq 0)$$

Properties of integer exponents (division by 0 excluded):

1. $a^m a^n = a^{m+n}$

2. $(a^n)^m = a^{mn}$

3. $(ab)^m = a^m b^m$

4. $\left(\dfrac{a}{b}\right)^m = \dfrac{a^m}{b^m}$

5. $\dfrac{a^m}{a^n} = a^{m-n} = \dfrac{1}{a^{n-m}}$

Any positive number written in decimal form can be expressed in **scientific notation,** that is, in the form $a \times 10^n$ $1 \leq a < 10$, n an integer, a in decimal form.

For n a natural number, a and b real numbers: a is an **nth root** of b if $a^n = b$. The number of real nth roots of a real number b is either 0, 1, or 2, depending on whether b is positive or negative, and whether n is even or odd. The **principal nth root of b,** denoted by $b^{1/n}$ or $\sqrt[n]{b}$, is the real nth root of b if there is only one, and the positive nth root of b if there are two real nth roots. In the notation $\sqrt[n]{b}$, the symbol $\sqrt{\ }$ is called a **radical,** n is called the **index,** and b is the **radicand.** If $n = 2$ we write $\sqrt{b}$ in place of $\sqrt[2]{b}$.

We extend exponent notation so that exponents can be rational numbers, not just integers, as follows. For m and n natural numbers and b any real number (except b can't be negative when n is even),

$$b^{m/n} = (b^{1/n})^m \quad \text{and} \quad b^{-m/n} = \frac{1}{b^{m/n}}$$

Rational exponent/radical property:

$$(b^{1/n})^m = (b^m)^{1/n} \quad \text{and} \quad (\sqrt[n]{b})^m = \sqrt[n]{b^m}$$

Properties of radicals $(x > 0, y > 0)$:

1. $\sqrt[n]{x^n} = x$ **2.** $\sqrt[n]{xy} = \sqrt[n]{x}\sqrt[n]{y}$ **3.** $\sqrt[n]{\dfrac{x}{y}} = \dfrac{\sqrt[n]{x}}{\sqrt[n]{y}}$

A radical is in **simplified form** if:

1. No radicand contains a factor to a power greater than or equal to the index of the radical.

2. No power of the radicand and the index of the radical have a common factor other than 1.

3. No radical appears in a denominator.

4. No fraction appears within a radical.

Eliminating a radical from a denominator is called **rationalizing the denominator.** To rationalize the denominator, we multiply the numerator and denominator by a suitable factor that will leave the denominator free of radicals. This factor is called a **rationalizing factor.** For example, if the denominator is of the form $\sqrt{a} + \sqrt{b}$, then $\sqrt{a} - \sqrt{b}$ is a rationalizing factor.

R-3 Polynomials: Basic Operations and Factoring

An **algebraic expression** is formed by using constants and variables and the operations of addition, subtraction, multiplication, division, raising to powers, and taking roots. A **polynomial** is an algebraic expression formed by adding and subtracting constants and terms of the form ax^n (one variable), ax^ny^m (two variables), and so on. The **degree of a term** is the sum of the powers of all variables in the term, and the **degree of a polynomial** is the degree of the nonzero term with highest degree in the polynomial. Polynomials with one, two, or three terms are called **monomials, binomials,** and **trinomials,** respectively. **Like terms** have exactly the same variable factors to the same powers and can be combined by adding their **coefficients.** Polynomials can be *added, subtracted,* and *multiplied* by repeatedly applying the distributive property and combining like terms.

A number or algebraic expression is **factored** if it is expressed as a product of other numbers or algebraic expressions, which are called **factors.** An integer greater than 1 is a **prime number** if its only positive integer factors are itself and 1, and a **composite**

number otherwise. Each composite number can be **factored uniquely into a product of prime numbers.** A polynomial is **prime** relative to a given set of numbers (usually the set of integers) if (1) all its coefficients are from that set of numbers, and (2) it cannot be written as a product of two polynomials of positive degree having coefficients from that set of numbers. A nonprime polynomial is **factored completely relative to a given set of numbers** if it is written as a product of prime polynomials relative to that set of numbers. *Common factors* can be factored out by applying the distributive properties. *Grouping* can be used to identify common factors. Second-degree polynomials can be factored by trial and error. The following special factoring formulas are useful:

1. $u^2 + 2uv + v^2 = (u + v)^2$ Perfect Square

2. $u^2 - 2uv + v^2 = (u - v)^2$ Perfect Square

3. $u^2 - v^2 = (u - v)(u + v)$ Difference of Squares

4. $u^3 - v^3 = (u - v)(u^2 + uv + v^2)$ Difference of Cubes

5. $u^3 + v^3 = (u + v)(u^2 - uv + v^2)$ Sum of Cubes

There is no factoring formula relative to the real numbers for $u^2 + v^2$.

R-4 Rational Expressions: Basic Operations

A **fractional expression** is the ratio of two algebraic expressions, and a **rational expression** is the ratio of two polynomials. The rules for adding, subtracting, multiplying, and dividing real number fractions (see Section R-1 in this review) all extend to fractional expressions with the understanding that **variables are always restricted to exclude division by zero.** Fractions can be **reduced to lowest terms** or **raised to higher terms** by using the **fundamental property of fractions:**

$$\frac{ka}{kb} = \frac{a}{b} \qquad \text{with } b, k \neq 0$$

A rational expression is **reduced to lowest terms** if the numerator and denominator do not have any factors in common relative to the integers. The **least common denominator** (LCD) is useful for adding and subtracting fractions with different denominators and for reducing **compound fractions** to **simple fractions.**

CHAPTER R Review Exercises

In Problems 1–6, perform the indicated operations, if defined. If the result is not an integer, express it in the form a/b, where a and b are integers.

1. $\dfrac{5}{6} + \dfrac{3}{4}$

2. $\dfrac{2}{3} - \dfrac{4}{9}$

3. $7^{-1}9^{-1}$

4. $\left(-\dfrac{10}{3}\right)\left(-\dfrac{6}{5}\right)$

5. $\dfrac{5}{7} \div \left(\dfrac{1}{3} - 3^{-1}\right)$

6. $\dfrac{11}{12} \div \left(-\dfrac{3}{4}\right)$

Problems 7–12 refer to the polynomials (a) $x^4 + 3x^2 + 1$ and (b) $4 - x^4$.

7. What is the degree of (a)?

8. What is the degree of (b)?

9. What is the degree of the sum of (a) and (b)?

10. What is the degree of the product of (a) and (b)?

11. Multiply (a) and (b).

12. Add (a) and (b).

In Problems 13–18, evaluate each expression that results in a rational number.

13. $289^{1/2}$

14. $216^{1/3}$

15. $8^{-2/3}$

16. $(-64)^{5/3}$

17. $\left(\dfrac{9}{16}\right)^{-1/2}$

18. $(121^{1/2} + 25^{1/2})^{-3/4}$

In Problems 19–22, perform the indicated operations and simplify.

19. $5x^2 - 3x[4 - 3(x - 2)]$

20. $(3m - 5n)(3m + 5n)$

21. $(2x + y)(3x - 4y)$

22. $(2a - 3b)^2$

In Problems 23–25, write each polynomial in a completely factored form relative to the integers. If the polynomial is prime relative to the integers, say so.

23. $9x^2 - 12x + 4$

24. $t^2 - 4t - 6$

25. $6n^3 - 9n^2 - 15n$

In Problems 26–29, perform the indicated operations and reduce to lowest terms. Represent all compound fractions as simple fractions reduced to lowest terms.

26. $\dfrac{2}{5b} - \dfrac{4}{3a^3} - \dfrac{1}{6a^2b^2}$

27. $\dfrac{3x}{3x^2 - 12x} + \dfrac{1}{6x}$

28. $\dfrac{y - 2}{y^2 - 4y + 4} \div \dfrac{y^2 + 2y}{y^2 + 4y + 4}$

29. $\dfrac{u - \dfrac{1}{u}}{1 - \dfrac{1}{u^2}}$

Simplify Problems 30–35, and write answers using positive exponents only. All variables represent positive real numbers.

30. $6(xy^3)^5$

31. $\dfrac{9u^8v^6}{3u^4v^8}$

32. $(2 \times 10^5)(3 \times 10^{-3})$

33. $(x^{-3}y^2)^{-2}$

34. $u^{5/3}u^{2/3}$

35. $(9a^4b^{-2})^{1/2}$

36. Change to radical form: $3x^{2/5}$

37. Change to rational exponent form: $-3\sqrt[3]{(xy)^2}$

Simplify Problems 38–42, and express answers in simplified form. All variables represent positive real numbers.

38. $3x\sqrt[3]{x^5y^4}$

39. $\sqrt{2x^2y^5}\sqrt{18x^3y^2}$

40. $\dfrac{6ab}{\sqrt{3a}}$

41. $\dfrac{\sqrt{5}}{3 - \sqrt{5}}$

42. $\sqrt[8]{y^6}$

In Problems 43–48, each statement illustrates the use of one of the following real number properties or definitions. Indicate which one.

Commutative (+) Identity (+)
Commutative (·) Identity (·)
Division Associative (+)
Inverse (+) Associative (·)
Inverse (·) Zero
Distributive Subtraction
Negatives

43. $(-3) - (-2) = (-3) + [-(-2)]$

44. $3y + (2x + 5) = (2x + 5) + 3y$

45. $(2x + 3)(3x + 5) = (2x + 3)3x + (2x + 3)5$

46. $3 \cdot (5x) = (3 \cdot 5)x$

47. $\dfrac{a}{-(b - c)} = -\dfrac{a}{b - c}$

48. $3xy + 0 = 3xy$

49. Indicate true (T) or false (F):
 (A) An integer is a rational number and a real number.
 (B) An irrational number has a repeating decimal representation.

50. Give an example of an integer that is not a natural number.

51. Given the algebraic expressions:
 (a) $2x^2 - 3x + 5$ (b) $x^2 - \sqrt{x} - 3$
 (c) $x^{-3} + x^{-2} - 3x^{-1}$ (d) $x^2 - 3xy - y^2$
 (A) Identify all second-degree polynomials.
 (B) Identify all third-degree polynomials.

In Problems 52–55, perform the indicated operations and simplify.

52. $(2x - y)(2x + y) - (2x - y)^2$

53. $(m^2 + 2mn - n^2)(m^2 - 2mn - n^2)$

54. $5(x + h)^2 - 7(x + h) - (5x^2 - 7x)$

55. $-2x\{(x^2 + 2)(x - 3) - x[x - x(3 - x)]\}$

In Problems 56–61, write in a completely factored form relative to the integers.

56. $(4x - y)^2 - 9x^2$

57. $2x^2 + 4xy - 5y^2$

58. $6x^3y + 12x^2y^2 - 15xy^3$

59. $(y - b)^2 - y + b$

60. $y^3 + 2y^2 - 4y - 8$

61. $2x(x - 4)^3 + 3x^2(x - 4)^2$

In Problems 62–65, perform the indicated operations and reduce to lowest terms. Represent all compound fractions as simple fractions reduced to lowest terms.

62. $\dfrac{3x^2(x + 2)^2 - 2x(x + 2)^3}{x^4}$

63. $\dfrac{m - 1}{m^2 - 4m + 4} + \dfrac{m + 3}{m^2 - 4} + \dfrac{2}{2 - m}$

64. $\dfrac{y}{x^2} \div \left(\dfrac{x^2 + 3x}{2x^2 + 5x - 3} \div \dfrac{x^3y - x^2y}{2x^2 - 3x + 1}\right)$

65. $\dfrac{1 - \dfrac{1}{1 + \dfrac{x}{y}}}{1 - \dfrac{1}{1 - \dfrac{x}{y}}}$

66. Convert to scientific notation and simplify:
$$\dfrac{0.000\ 000\ 000\ 52}{(1,300)(0.000\ 002)}$$

In Problems 67–75, perform the indicated operations and express answers in simplified form. All radicands represent positive real numbers.

67. $-2x\sqrt[5]{3^6x^7y^{11}}$

68. $\dfrac{2x^2}{\sqrt[3]{4x}}$

69. $\sqrt[5]{\dfrac{3y^2}{8x^2}}$

70. $\sqrt[9]{8x^6y^{12}}$

71. $\sqrt{\sqrt{\sqrt[3]{4x^4}}}$

72. $(2\sqrt{x} - 5\sqrt{y})(\sqrt{x} + \sqrt{y})$

73. $\dfrac{3\sqrt{x}}{2\sqrt{x} - \sqrt{y}}$

74. $\dfrac{2\sqrt{u} - 3\sqrt{v}}{2\sqrt{u} + 3\sqrt{v}}$

75. $\dfrac{y^2}{\sqrt{y^2 + 4} - 2}$

APPLICATIONS

76. CONSTRUCTION A circular fountain in a park includes a concrete wall that is 3 ft high and 2 ft thick (see the figure). If the inner radius of the wall is x feet, write an algebraic expression in terms of x that represents the volume of the concrete used to construct the wall. Simplify the expression.

2 feet

x feet

3 feet

77. ECONOMICS If in the United States in 2007 the total personal income was about $11,580,000,000,000 and the population was about 301,000,000, estimate to three significant digits the average personal income. Write your answer in scientific notation and in standard decimal form.

78. ECONOMICS The number of units N produced by a petroleum company from the use of x units of capital and y units of labor is approximated by

$$N = 20x^{1/2}y^{1/2}$$

(A) Estimate the number of units produced by using 1,600 units of capital and 900 units of labor.
(B) What is the effect on production if the number of units of capital and labor are doubled to 3,200 units and 1,800 units, respectively?

(C) What is the effect on production of doubling the units of labor and capital at any production level?

79. ELECTRIC CIRCUIT If three electric resistors with resistances R_1, R_2, and R_3 are connected in parallel, then the total resistance R for the circuit shown in the figure is given by

$$R = \dfrac{1}{\dfrac{1}{R_1} + \dfrac{1}{R_2} + \dfrac{1}{R_3}}$$

Represent this compound fraction as a simple fraction.

80. CONSTRUCTION A box with a hinged lid is to be made out of a piece of cardboard that measures 16 by 30 inches. Six squares, x inches on a side, will be cut from each corner and the middle, and then the ends and sides will be folded up to form the box and its lid (see the figure). Express each of the following quantities as a polynomial in both factored and expanded form.
(A) The area of cardboard after the corners have been removed.
(B) The volume of the box.

30 in.

16 in.

x

x

Equations and Inequalities

SOLVING equations and inequalities is one of the most important skills in algebra because it can be applied to solving a boundless supply of real-world problems. In this chapter, we will begin with a look at techniques for solving linear equations and inequalities. After a study of complex numbers, we'll return to equations, learning how to solve a variety of nonlinear equations. For each type of equation and inequality we solve, we will look at some real-world problems that can be solved using those solution techniques. This doesn't close the book on solving equations, though—we will learn how to solve new types of equations in many of the remaining chapters.

1-1 Linear Equations and Applications

› Understanding Basic Terms
› Solving Linear Equations
› Solving Number and Geometric Problems
› Solving Rate–Time Problems
› Solving Mixture Problems

We begin this section with a quick look at what an equation is and what it means to solve one. After solving some linear equations, we move on to the main topic: using linear equations to solve word problems.

› Understanding Basic Terms

An **algebraic equation** is a mathematical statement that two algebraic expressions are equal. Some examples of equations with variable x are

$$3x - 2 = 7 \qquad \frac{1}{1 + x} = \frac{x}{x - 2}$$

$$2x^2 - 3x + 5 = 0 \qquad \sqrt{x + 4} = x - 1$$

The **replacement set,** or **domain,** for a variable is defined to be the set of numbers that are permitted to replace the variable.

› **ASSUMPTION** On Domains of Variables

Unless stated to the contrary, we assume that the domain for a variable in an algebraic expression or equation is the set of those real numbers for which the algebraic expressions involving the variable are real numbers.

For example, the domain for the variable x in the expression

$$2x - 4$$

is R, the set of all real numbers, since $2x - 4$ represents a real number for all replacements of x by real numbers. The domain of x in the equation

$$\frac{1}{x} = \frac{2}{x - 3}$$

is the set of all real numbers except 0 and 3. These values are excluded because the expression on the left is not defined for $x = 0$ and the expression on the right is not defined for $x = 3$. Both expressions represent real numbers for all other replacements of x by real numbers.

The **solution set** for an equation is defined to be the set of all elements in the domain of the variable that make the equation true. Each element of the solution set is called a **solution,** or **root,** of the equation. To **solve an equation** is to find the solution set for the equation.

An equation is called an **identity** if the equation is true for all elements from the domain of the variable. An equation is called a **conditional equation** if it is true for certain domain values and false for others. For example,

$$2x - 4 = 2(x - 2) \quad \text{and} \quad \frac{5}{x^2 - 3x} = \frac{5}{x(x - 3)}$$

are identities, since both equations are true for all elements from the respective domains of their variables. On the other hand, the equations

$$3x - 2 = 5 \quad \text{and} \quad \frac{2}{x - 1} = \frac{1}{x}$$

are conditional equations, since, for example, neither equation is true for the domain value 2.

Knowing what we mean by the solution set of an equation is one thing; finding it is another. We introduce the idea of equivalent equations to help us find solutions. We will call two equations **equivalent** if they both have the same solution set. To solve an equation, we perform operations on the equation to produce simpler equivalent equations. We stop when we find an equation whose solution is obvious. Then we check this obvious solution in the original equation. Any of the properties of equality given in Theorem 1 can be used to produce equivalent equations.

> **THEOREM 1** Properties of Equality

For a, b, and c any real numbers:

1. If $a = b$, then $a + c = b + c$. **Addition Property**

2. If $a = b$, then $a - c = b - c$. **Subtraction Property**

3. If $a = b$ and $c \neq 0$, then $ca = cb$. **Multiplication Property**

4. If $a = b$ and $c \neq 0$, then $\dfrac{a}{c} = \dfrac{b}{c}$. **Division Property**

5. If $a = b$, then either may replace the other in any statement without changing the truth or falsity of the statement. **Substitution Property**

> ## Solving Linear Equations

We now turn our attention to methods of solving *first-degree*, or *linear, equations* in one variable.

> **DEFINITION 1** Linear Equation in One Variable

Any equation that can be written in the form

$$ax + b = 0 \qquad a \neq 0 \qquad \text{Standard Form}$$

where a and b are real constants and x is a variable, is called a **linear**, or **first-degree, equation** in one variable.

$5x - 1 = 2(x + 3)$ is a linear equation because after simplifying, it can be written in the standard form $3x - 7 = 0$.

EXAMPLE 1 **Solving a Linear Equation**

Solve $5x - 9 = 3x + 7$ and check.

SOLUTION We will use the properties of equality to transform the given equation into an equivalent equation whose solution is obvious.

$$5x - 9 = 3x + 7 \qquad \text{Add 9 to both sides.}$$
$$5x - 9 + 9 = 3x + 7 + 9 \qquad \text{Combine like terms.}$$
$$5x = 3x + 16 \qquad \text{Subtract } 3x \text{ from both sides.}$$
$$5x - 3x = 3x + 16 - 3x \qquad \text{Combine like terms.}$$
$$2x = 16 \qquad \text{Divide both sides by 2.}$$
$$\frac{2x}{2} = \frac{16}{2} \qquad \text{Simplify.}$$
$$x = 8$$

The solution set for this last equation is obvious:

$$\text{Solution set: } \{8\}$$

And since the equation $x = 8$ is equivalent to all the preceding equations in our solution, $\{8\}$ is also the solution set for all these equations, including the original equation. [*Note:* If an equation has only one element in its solution set, we generally use the last equation (in this case, $x = 8$) rather than set notation to represent the solution.]

CHECK

$$5x - 9 = 3x + 7 \qquad \text{Substitute } x = 8.$$
$$5(8) - 9 \overset{?}{=} 3(8) + 7 \qquad \text{Simplify each side.}$$
$$40 - 9 \overset{?}{=} 24 + 7$$
$$31 \overset{\checkmark}{=} 31 \qquad \text{A true statement}$$

MATCHED PROBLEM 1 Solve and check: $7x - 10 = 4x + 5$

We often encounter equations involving more than one variable. For example, if l and w are the length and width of a rectangle, respectively, the area of the rectangle is given by $A = lw$ (see Fig. 1).

Depending on the situation, we may want to solve this equation for l or w. To solve for w, we simply consider A and l to be constants and w to be a variable. Then the equation $A = lw$ becomes a linear equation in w that can be solved easily by dividing both sides by l:

$$w = \frac{A}{l} \qquad l \neq 0$$

$A = lw$ w

l

> **Figure 1** Area of a rectangle.

EXAMPLE 2 **Solving an Equation with More Than One Variable**

Solve for P in terms of the other variables: $A = P + Prt$

SOLUTION

$$A = P + Prt \qquad \text{Factor to isolate } P.$$
$$A = P(1 + rt) \qquad \text{Divide both sides by } 1 + rt.$$
$$\frac{A}{1 + rt} = P$$
$$P = \frac{A}{1 + rt} \qquad \text{Restriction: } 1 + rt \neq 0$$

MATCHED PROBLEM 2 Solve for F in terms of C: $C = \frac{5}{9}(F - 32)$

A great many practical problems can be solved using algebraic techniques—so many, in fact, that there is no one method of attack that will work for all. However, we can put together a strategy that will help you organize your approach.

> **STRATEGY FOR SOLVING WORD PROBLEMS**

1. Read the problem slowly and carefully, more than once if necessary. Write down information as you read the problem the first time to help you get started. Identify what it is that you are asked to find.

2. Use a variable to represent an unknown quantity in the problem, usually what you are asked to find. Then try to represent any other unknown quantities in terms of that variable. It's pretty much impossible to solve a word problem without this step.

3. If it helps to visualize a situation, draw a diagram and label known and unknown parts.

4. Write an equation relating the quantities in the problem. Often, you can accomplish this by finding a formula that connects those quantities. Try to write the equation in words first, then translate to symbols.

5. Solve the equation, then answer the question in a sentence by rephrasing the question. Make sure that you're answering all of the questions asked.

6. Check to see if your answers make sense in the original problem, not just the equation you wrote.

>>> **EXPLORE-DISCUSS 1**

Translate each of the following sentences involving two numbers into an equation.

(A) The first number is 10 more than the second number.

(B) The first number is 15 less than the second number.

(C) The first number is half the second number.

(D) The first number is three times the second number.

(E) Ten times the first number is 15 more than the second number.

The remaining examples in this section contain solutions to a variety of word problems illustrating both the process of setting up word problems and the techniques used to solve the resulting equations. As you read an example, try covering up the solution and working the problem yourself. If you need a hint, uncover just part of the solution and try to work out the rest. After you successfully solve an example problem, try the matched problem. If you work through the remainder of the section in this way, you will already have experience with a wide variety of word problems.

> ## Solving Number and Geometric Problems

Example 3 introduces the process of setting up and solving word problems in a simple mathematical context. Examples 4–8 are more realistic.

| EXAMPLE | 3 | ### Setting Up and Solving a Word Problem |

Find four consecutive even integers so that the sum of the first three is 8 more than the fourth.

SOLUTION Let $x =$ the first even integer; then

$$x \qquad x + 2 \qquad x + 4 \qquad \text{and} \qquad x + 6$$

represent four consecutive even integers starting with the even integer x. (Remember, even integers are separated by 2.) The phrase "the sum of the first three is 8 more than the fourth" translates into an equation:

$$\text{Sum of the first three} = \text{Fourth} + 8$$

$$x + (x + 2) + (x + 4) = (x + 6) + 8 \qquad \text{Combine like terms.}$$

$$3x + 6 = x + 14 \qquad \text{Subtract 6 and } x \text{ from both sides.}$$

$$2x = 8 \qquad \text{Divide both sides by 2.}$$

$$x = 4$$

The first even integer is 4, so the four consecutive integers are 4, 6, 8, and 10.

CHECK $4 + 6 + 8 = 18 = 10 + 8$ Sum of first three is 8 more than the fourth.

MATCHED PROBLEM 3 Find three consecutive odd integers so that 3 times their sum is 5 more than 8 times the middle one.

>>> EXPLORE-DISCUSS 2

According to Part 3 of Theorem 1, multiplying both sides of an equation by a nonzero number always produces an equivalent equation. By what number would you choose to multiply both sides of the following equation to eliminate all the fractions?

$$\frac{x + 1}{3} - \frac{x}{4} = \frac{1}{2}$$

If you did not choose 12, the LCD of all the fractions in this equation, you could still solve the resulting equation, but with more effort. (For a discussion of LCDs and how to find them, see Section R-4.)

| EXAMPLE | 4 | ### Using a Diagram in the Solution of a Word Problem |

A landscape designer plans a series of small triangular gardens outside a new office building. Her plans call for one side to be one-third of the perimeter, and another side to be one-fifth of the perimeter. The space allotted for each will allow the third side to be 7 meters. Find the perimeter of the triangle.

SOLUTION Draw a triangle, and label one side 7 meters. Let $p =$ the perimeter: then the remaining sides are one-third p, or $p/3$, and one-fifth p, or $p/5$ (see Fig. 2).

$$\text{Perimeter} = \text{Sum of the side lengths}$$

$$p = \frac{p}{3} + \frac{p}{5} + 7 \qquad \text{Multiply both sides by 15, the LCD. Make sure to multiply every term by 15!}$$

> Figure 2

$$15 \cdot p = 15 \cdot \left(\frac{p}{3} + \frac{p}{5} + 7 \right) \quad *$$

$$15p = 15 \cdot \frac{p}{3} + 15 \cdot \frac{p}{5} + 15 \cdot 7$$

$15p = 5p + 3p + 105$ Combine like terms.

$15p = 8p + 105$ Subtract $8p$ from both sides.

$7p = 105$ Divide both sides by 7.

$p = 15$

The perimeter is 15 meters.

CHECK

$$\frac{p}{3} = \frac{15}{3} = 5 \qquad \text{Side 1}$$

$$+\frac{p}{5} = \frac{15}{5} = 3 \qquad \text{Side 2}$$

$$+ \qquad\qquad 7 \qquad \text{Side 3}$$

$$\overline{\qquad\qquad 15 \text{ meters}} \qquad \text{Perimeter}$$

MATCHED PROBLEM 4

If one side of a triangle is one-fourth the perimeter, the second side is 7 centimeters, and the third side is two-fifths the perimeter, what is the perimeter?

››› CAUTION ›››

A very common error occurs about now—students tend to confuse *algebraic expressions* involving fractions with *algebraic equations* involving fractions.

Consider these two problems:

(A) Solve: $\dfrac{x}{2} + \dfrac{x}{3} = 10$ **(B)** Add: $\dfrac{x}{2} + \dfrac{x}{3} + 10$

The problems look very much alike but are actually very different. To solve the equation in (A) we multiply both sides by 6 (the LCD) to clear the fractions. This works so well for equations that students want to do the same thing for problems like (B). The only catch is that (B) is not an equation, and the multiplication property of equality does not apply. If we multiply (B) by 6, we simply obtain an expression 6 times as large as the original! Compare these correct solutions:

(A) $\qquad\qquad \dfrac{x}{2} + \dfrac{x}{3} = 10$ **(B)** $\dfrac{x}{2} + \dfrac{x}{3} + 10$

$$6 \cdot \frac{x}{2} + 6 \cdot \frac{x}{3} = 6 \cdot 10 \qquad = \frac{3 \cdot x}{3 \cdot 2} + \frac{2 \cdot x}{2 \cdot 3} + \frac{6 \cdot 10}{6 \cdot 1}$$

$$3x + 2x = 60 \qquad\qquad\qquad = \frac{3x}{6} + \frac{2x}{6} + \frac{60}{6}$$

$$5x = 60 \qquad\qquad\qquad\qquad = \frac{5x + 60}{6}$$

$$x = 12$$

*Throughout the book, dashed boxes—called **think boxes**—are used to represent steps that may be performed mentally.

There are many problems in which a rate plays a key role. For example, if you're losing weight at the rate of 2 lb per week, you can use that rate to find a total weight loss for some period of time. Rate problems can often be solved using the following basic formula:

> **› QUANTITY-RATE-TIME FORMULA**
>
> The change in a quantity is the rate at which it changes times the time passed: Quantity = Rate × Time, or $Q = RT$. If the quantity is distance, then $D = RT$. The formulas can be solved for R or T to get a related formula to find the rate or the time. [*Note:* R is an average or uniform rate.]

››› EXPLORE-DISCUSS 3

(A) If you drive at an average rate of 65 miles per hour, how far do you go in 3 hours?

(B) If you make $750 for 2 weeks of part-time work, what is your weekly rate of pay?

(C) If you eat at the rate of 1,900 calories per day, how long will it take you to eat 7,600 calories?

EXAMPLE 5 **A Distance–Rate–Time Problem**

The distance along a shipping route between San Francisco and Honolulu is 2,100 nautical miles. If one ship leaves San Francisco at the same time another leaves Honolulu, and if the former travels at 15 knots* and the latter at 20 knots, how long will it take the two ships to rendezvous? How far will they be from Honolulu and San Francisco at that time?

SOLUTION Let T = number of hours until both ships meet. Draw a diagram and label known and unknown parts. Both ships will have traveled the same amount of time when they meet.

$D_1 = 20T$ $D_2 = 15T$

20 knots 15 knots

H Meeting SF

$D = RT$
$D_1 = 20$ knots · Time
$D_2 = 15$ knots · Time

$$\begin{pmatrix} \text{Distance ship 1} \\ \text{from Honolulu} \\ \text{travels to} \\ \text{meeting point} \end{pmatrix} + \begin{pmatrix} \text{Distance ship 2} \\ \text{from San Francisco} \\ \text{travels to} \\ \text{meeting point} \end{pmatrix} = \begin{pmatrix} \text{Total distance} \\ \text{from Honolulu} \\ \text{to San Francisco} \end{pmatrix}$$

D_1	+	D_2	=	2,100
$20T$	+	$15T$	=	2,100
		$35T$	=	2,100
		T	=	60

Therefore, it takes 60 hours, or 2.5 days, for the ships to meet.

Distance from Honolulu = $20 \cdot 60 = 1{,}200$ nautical miles

Distance from San Francisco = $15 \cdot 60 = 900$ nautical miles

CHECK $1{,}200 + 900 = 2{,}100$ nautical miles

*15 knots means 15 nautical miles per hour. There are 6,076.1 feet in 1 nautical mile, and 5,280 feet in 1 statute mile.

MATCHED PROBLEM 5 An old piece of equipment can print, stuff, and label 38 mailing pieces per minute. A newer model can handle 82 per minute. How long will it take for both pieces of equipment to prepare a mailing of 6,000 pieces? [*Hint:* Use Quantity = Rate × Time for each machine.]

Some equations involving variables in a denominator can be transformed into linear equations. We can proceed in essentially the same way as in Example 5; however, we need to exclude any value of the variable that will make a denominator 0. With these values excluded, we can multiply through by the LCD even though it contains a variable, and, according to Theorem 1, the new equation will be equivalent to the old.

EXAMPLE 6 **A Distance–Rate–Time Problem**

An excursion boat takes 1.5 times as long to go 360 miles up a river as to return. If the boat cruises at 15 miles per hour in still water, what is the rate of the current?

SOLUTION Let

360 miles

$$x = \text{Rate of current (in miles per hour)} \qquad \text{What we were asked to find.}$$

$$15 - x = \text{Rate of boat upstream}$$

$$15 + x = \text{Rate of boat downstream} \qquad \text{Faster downstream.}$$

$$\text{Time upstream} = (1.5)(\text{Time downstream})$$

$$\frac{\text{Distance upstream}}{\text{Rate upstream}} = (1.5)\frac{\text{Distance downstream}}{\text{Rate downstream}} \qquad \text{Because } D = RT,\ T = \frac{D}{R}$$

$$\frac{360}{15 - x} = (1.5)\frac{360}{15 + x} \qquad x \text{ cannot be 15 or } -15$$

$$\frac{360}{15 - x} = \frac{540}{15 + x} \qquad \begin{array}{l}\text{Multiply both sides by the LCD,}\\ (15 - x)(15 + x).\end{array}$$

$$360(15 + x) = 540(15 - x) \qquad \text{Multiply out parentheses.}$$

$$5,400 + 360x = 8,100 - 540x \qquad \text{Add 540x to both sides.}$$

$$5,400 + 900x = 8,100 \qquad \text{Subtract 5,400 from both sides.}$$

$$900x = 2,700 \qquad \text{Divide both sides by 900.}$$

$$x = 3$$

The rate of the current is 3 miles per hour. The check is left to the reader.

MATCHED PROBLEM 6 A jetliner takes 1.2 times as long to fly from Paris to New York (3,600 miles) as to return. If the jet cruises at 550 miles per hour in still air, what is the average rate of the wind blowing in the direction of Paris from New York?

EXAMPLE 7 **A Quantity–Rate–Time Problem**

An advertising firm has an old computer that can prepare a whole mailing in 6 hours. With the help of a newer model the job is complete in 2 hours. How long would it take the newer model to do the job alone?

SOLUTION Let x = time (in hours) for the newer model to do the whole job alone.

$$\left(\begin{array}{c}\text{Part of job completed}\\ \text{in a given length of time}\end{array}\right) = (\text{Rate})(\text{Time})$$

$$\text{Rate of old model} = \frac{1}{6} \text{ job per hour}$$

$$\text{Rate of new model} = \frac{1}{x} \text{ job per hour}$$

$$\left(\begin{array}{c}\text{Part of job completed}\\ \text{by old model}\\ \text{in 2 hours}\end{array}\right) + \left(\begin{array}{c}\text{Part of job completed}\\ \text{by new model}\\ \text{in 2 hours}\end{array}\right) = 1 \text{ whole job}$$

$$\left(\begin{array}{c}\text{Rate of}\\ \text{old model}\end{array}\right)\left(\begin{array}{c}\text{Time of}\\ \text{old model}\end{array}\right) + \left(\begin{array}{c}\text{Rate of}\\ \text{new model}\end{array}\right)\left(\begin{array}{c}\text{Time of}\\ \text{new model}\end{array}\right) = 1 \qquad \color{red}{\text{Recall: } Q = RT}$$

$$\frac{1}{6}(2) \qquad + \qquad \frac{1}{x}(2) \qquad = 1 \qquad \color{red}{x \text{ cannot be zero.}}$$

$$\frac{1}{3} \qquad + \qquad \frac{2}{x} \qquad = 1 \qquad \color{red}{\substack{\text{Multiply both sides}\\ \text{by } 3x, \text{ the LCD.}}}$$

$$3x\left(\frac{1}{3}\right) \qquad + \qquad 3x\left(\frac{2}{x}\right) \qquad = 3x$$

$$x \qquad + \qquad 6 \qquad = 3x \qquad \color{red}{\substack{\text{Subtract } x \text{ from}\\ \text{both sides.}}}$$

$$6 \qquad = 2x \qquad \color{red}{\substack{\text{Divide both}\\ \text{sides by 2.}}}$$

$$3 \qquad = x$$

Therefore, the new computer could do the job alone in 3 hours.

CHECK Part of job completed by old model in 2 hours $= 2(\frac{1}{6}) = \frac{1}{3}$

$+$ Part of job completed by new model in 2 hours $= 2(\frac{1}{3}) = \frac{2}{3}$

Part of job completed by both models in 2 hours $= 1$

MATCHED PROBLEM 7 Two pumps are used to fill a water storage tank at a resort. One pump can fill the tank by itself in 9 hours, and the other can fill it in 6 hours. How long will it take both pumps operating together to fill the tank?

> ## Solving Mixture Problems

A variety of applications can be classified as mixture problems. Even though the problems come from different areas, their mathematical treatment is essentially the same.

EXAMPLE 8 ## A Mixture Problem

How many liters of a mixture containing 80% alcohol should be added to 5 liters of a 20% solution to yield a 30% solution?

SOLUTION Let x = amount of 80% solution used.

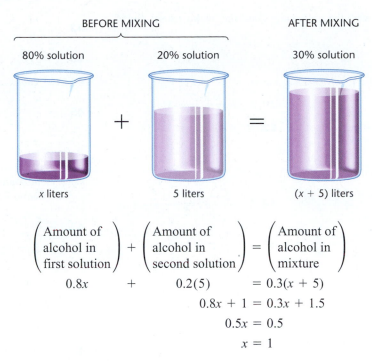

BEFORE MIXING AFTER MIXING

80% solution 20% solution 30% solution

+ =

x liters 5 liters $(x + 5)$ liters

$$\begin{pmatrix} \text{Amount of} \\ \text{alcohol in} \\ \text{first solution} \end{pmatrix} + \begin{pmatrix} \text{Amount of} \\ \text{alcohol in} \\ \text{second solution} \end{pmatrix} = \begin{pmatrix} \text{Amount of} \\ \text{alcohol in} \\ \text{mixture} \end{pmatrix}$$

$$0.8x + 0.2(5) = 0.3(x + 5)$$
$$0.8x + 1 = 0.3x + 1.5$$
$$0.5x = 0.5$$
$$x = 1$$

Add 1 liter of the 80% solution.

CHECK	Liters of solution	Liters of alcohol	Percent alcohol
First solution	1	$0.8(1) = 0.8$	80 or 0.8/1
Second solution	+ 5	+ 0.2(5) = 1	20 or 1/5
Mixture	6	1.8	1.8/6 = 0.3, or 30%

MATCHED PROBLEM 8

A chemical storeroom has a 90% acid solution and a 40% acid solution. How many centiliters of the 90% solution should be added to 50 centiliters of the 40% solution to yield a 50% solution?

ANSWERS TO MATCHED PROBLEMS

1. $x = 5$ **2.** $F = \frac{9}{5}C + 32$ **3.** 3, 5, 7 **4.** 20 centimeters
5. 50 minutes **6.** 50 miles per hour **7.** 3.6 hours **8.** 12.5 centiliters

1-1 Exercises

1. What does it mean to solve an equation?

2. Describe the difference between an equation and an expression.

3. How can you tell if an equation is linear?

4. In one or two sentences, describe what parts 1–4 in Theorem 1 say about working with equations.

5. How can you check your solution to an equation?

6. How do you check your solution to a word problem?

7. Explain why the following does not make sense: Solve the equation $P = 2l + 2w$.

8. Explain why the following does not make sense: Solve $\dfrac{y}{4} - \dfrac{y}{5} + 1$.

In Problems 9–34, solve each equation.

9. $10x - 7 = 4x - 25$

10. $11 + 3y = 5y - 5$

11. $3(x + 2) = 5(x - 6)$

12. $3(y - 4) + 2y = 18$

13. $5 + 4(t - 2) = 2(t + 7) + 1$

14. $4 - 3(t + 2) + t = 5(t - 1) - 7t$

15. $5 - \dfrac{3a - 4}{5} = \dfrac{7 - 2a}{2}$

16. $5 - \dfrac{2x - 1}{4} = \dfrac{x + 2}{3}$

17. $\dfrac{x + 3}{4} - \dfrac{x - 4}{2} = \dfrac{3}{8}$

18. $\dfrac{x}{5} + \dfrac{3x - 1}{2} = \dfrac{6x + 5}{4}$

19. $0.1(t + 0.5) + 0.2t = 0.3(t - 0.4)$

20. $0.1(w + 0.5) + 0.2w = 0.2(w - 0.4)$

21. $0.35(s + 0.34) + 0.15s = 0.2s - 1.66$

22. $0.35(u + 0.34) - 0.15u = 0.2u - 1.66$

23. $\dfrac{2}{y} + \dfrac{5}{2} = 4 - \dfrac{2}{3y}$

24. $\dfrac{3 + w}{6w} = \dfrac{1}{2w} + \dfrac{4}{3}$

25. $\dfrac{z}{z - 1} = \dfrac{1}{z - 1} + 2$

26. $\dfrac{t}{t - 1} = \dfrac{2}{t - 1} + 2$

27. $\dfrac{y}{3} + \dfrac{y - 10}{5} = \dfrac{2y - 2}{4} - 3$

28. $\dfrac{z + 4}{7} + \dfrac{z}{6} = \dfrac{z + 8}{3} + 5$

29. $1 - \dfrac{x - 3}{x - 2} = \dfrac{2x - 3}{x - 2}$

30. $\dfrac{2x - 3}{x + 1} = 2 - \dfrac{3x - 1}{x + 1}$

31. $\dfrac{6}{y + 4} + 1 = \dfrac{5}{2y + 8}$

32. $\dfrac{4y}{y - 3} + 5 = \dfrac{12}{y - 3}$

33. $\dfrac{3a - 1}{a^2 + 4a + 4} - \dfrac{3}{a^2 + 2a} = \dfrac{3}{a}$

34. $\dfrac{1}{b - 5} - \dfrac{10}{b^2 - 5b + 25} = \dfrac{1}{b + 5}$

*In Problems 35–38, use a calculator to solve each equation to three significant digits.**

35. $3.142x - 0.4835(x - 4) = 6.795$

36. $1.73y + 0.279(y - 3) = 2.66y$

37. $\dfrac{2.32x}{x - 2} - \dfrac{3.76}{x} = 2.32$

38. $\dfrac{2.34}{x} + 5.67 = \dfrac{5.67x}{x + 4}$

In Problems 39–46, solve for the indicated variable in terms of the other variables.

39. $a_n = a_1 + (n - 1)d$ for d (arithmetic progressions)

40. $F = \frac{9}{5}C + 32$ for C (temperature scale)

**Appendix A contains a brief discussion of significant digits.*

41. $\dfrac{1}{f} = \dfrac{1}{d_1} + \dfrac{1}{d_2}$ for f (simple lens formula)

42. $\dfrac{1}{R} = \dfrac{1}{R_1} + \dfrac{1}{R_2}$ for R_1 (electric circuit)

43. $A = 2ab + 2ac + 2bc$ for a (surface area of a rectangular solid)

44. $A = 2ab + 2ac + 2bc$ for c

45. $y = \dfrac{2x - 3}{3x + 5}$ for x

46. $x = \dfrac{3y + 2}{y - 3}$ for y

In Problems 47 and 48, imagine that the indicated "solutions" were given to you by a student whom you were tutoring in this class. Is the solution right or wrong? If the solution is wrong, explain what is wrong and show a correct solution.

47.
$$\dfrac{x}{x - 3} + 4 = \dfrac{2x - 3}{x - 3}$$
$$x + 4x - 12 = 2x - 3$$
$$x = 3$$

48.
$$\dfrac{x^2 + 1}{x - 1} = \dfrac{x^2 + 4x - 3}{x - 1}$$
$$x^2 + 1 = x^2 + 4x - 3$$
$$x = 1$$

In Problems 49–51, solve the equation.

49. $\dfrac{x - \dfrac{1}{x}}{1 + \dfrac{1}{x}} = 3$

50. $\dfrac{x - \dfrac{1}{x}}{x + 1 - \dfrac{2}{x}} = 1$

51. $\dfrac{x + 1 - \dfrac{2}{x}}{1 - \dfrac{1}{x}} = x + 2$

52. Solve for y in terms of x: $\dfrac{y}{1 - y} = \left(\dfrac{x}{1 - x}\right)^3$

53. Solve for x in terms of y: $y = \dfrac{a}{1 + \dfrac{b}{x + c}}$

54. Let m and n be real numbers with m larger than n. Then there exists a positive real number p such that $m = n + p$. Find the fallacy in the following argument:

$$m = n + p$$
$$(m - n)m = (m - n)(n + p)$$
$$m^2 - mn = mn + mp - n^2 - np$$
$$m^2 - mn - mp = mn - n^2 - np$$
$$m(m - n - p) = n(m - n - p)$$
$$m = n$$

APPLICATIONS

These problems are grouped according to subject area.

Numbers

55. Find a number so that 10 less than two-thirds the number is one-fourth the number.

56. Find a number so that 6 more than one-half the number is two-thirds the number.

57. Find four consecutive even integers so that the sum of the first three is 2 more than twice the fourth.

58. Find three consecutive even integers so that the first plus twice the second is twice the third.

Geometry

59. Find the perimeter of a triangle if one side is 16 feet, another side is two-sevenths the perimeter, and the third side is one-third the perimeter.

60. Find the perimeter of a triangle if one side is 11 centimeters, another side is two-fifths the perimeter, and the third side is one-half the perimeter.

61. A new game show requires a playing field with a perimeter of 54 yards and length 3 yards less than twice the width. What are the dimensions?

62. A celebrity couple wants to have a rectangular pool put in the backyard of their vacation home. They want it to be 24 meters long, and they insist that it have at least as much area as the neighbor's pool, which is a square 12 meters on a side. Find the dimensions of the smallest pool that meets these criteria.

Business and Economics

63. The sale price of an MP3 player after a 30% discount was $140. What was the original price?

64. A sporting goods store marks up each item it sells 60% above wholesale price. What is the wholesale price on a snowboard that sells for $144?

65. One employee of a computer store is paid a base salary of $2,150 a month plus an 8% commission on all sales over $7,000 during the month. How much must the employee sell in 1 month to earn a total of $3,170 for the month?

66. A second employee of the computer store in Problem 65 is paid a base salary of $1,175 a month plus a 5% commission on all sales during the month.
(A) How much must this employee sell in 1 month to earn a total of $3,170 for the month?
(B) Determine the sales level where both employees receive the same monthly income. If employees can select either of these payment methods, how would you advise an employee to make this selection?

Earth Science

67. In 1970, Russian geologists began drilling a very deep borehole in the Kola Peninsula. Their goal was to reach a depth of 15 kilometers, but high temperatures in the borehole forced them to stop in 1994 after reaching a depth of 12 kilometers. They found that below 3 kilometers the temperature T increased 2.5°C for each additional 100 meters of depth.
(A) If the temperature at 3 kilometers is 30°C and x is the depth of the hole in kilometers, write an equation using x that will give the temperature T in the hole at any depth beyond 3 kilometers.
(B) What would the temperature be at 12 kilometers?
(C) At what depth (in kilometers) would they reach a temperature of 200°C?

68. An earthquake emits a primary wave and a secondary wave. Near the surface of the Earth the primary wave travels at about 5 miles per second, and the secondary wave travels at about 3 miles per second. From the time lag between the two waves arriving at a given seismic station, it is possible to estimate the distance to the quake. Suppose a station measures a time difference of 12 seconds between the arrival of the two waves. How far is the earthquake from the station? (The *epicenter* can be located by obtaining distance bearings at three or more stations.)

Life Science

69. The kangaroo rat is an endangered species native to California. In order to keep track of their population size in a state nature preserve, a conservation biologist trapped, tagged, and released 80 individuals from the population. After waiting 2 weeks for the animals to mix back in with the general population, she again caught 80 individuals and found that 22 of them were tagged. Assuming that the ratio of tagged animals to total animals in the second sample is the same as the ratio of all tagged animals to the total population in the preserve, estimate the total number of kangaroo rats in the preserve.

70. Repeat Problem 69 with a first (marked) sample of 70 and a second sample of 30 with only 11 marked animals.

Chemistry

71. How many gallons of distilled water must be mixed with 50 gallons of 30% alcohol solution to obtain a 25% solution?

72. How many gallons of hydrochloric acid must be added to 12 gallons of a 30% solution to obtain a 40% solution?

73. A chemist mixes distilled water with a 90% solution of sulfuric acid to produce a 50% solution. If 5 liters of distilled water are used, how much 50% solution is produced?

74. A fuel oil distributor has 120,000 gallons of fuel with 0.9% sulfur content, which exceeds pollution control standards of 0.8% sulfur content. How many gallons of fuel oil with a 0.3% sulfur content must be added to the 120,000 gallons to obtain fuel oil that will comply with the pollution control standards?

Rate–Time

75. An old computer can do the weekly payroll in 5 hours. A newer computer can do the same payroll in 3 hours. The old computer starts on the payroll, and after 1 hour the newer computer is brought on-line to work with the older computer until the job is finished. How long will it take both computers working together to finish the job? (Assume the computers operate independently.)

76. One pump can fill a gasoline storage tank in 8 hours. With a second pump working simultaneously, the tank can be filled in 3 hours. How long would it take the second pump to fill the tank operating alone?

77. The cruising speed of an airplane is 150 miles per hour (relative to the ground). You plan to hire the plane for a 3-hour sightseeing trip. You instruct the pilot to fly north as far as she can and still return to the airport at the end of the allotted time.
(A) How far north should the pilot fly if the wind is blowing from the north at 30 miles per hour?
(B) How far north should the pilot fly if there is no wind?

78. Suppose you are at a river resort and rent a motor boat for 5 hours starting at 7 A.M. You are told that the boat will travel at 8 miles per hour upstream and 12 miles per hour returning. You decide that you would like to go as far up the river as you can and still be back at noon. At what time should you turn back, and how far from the resort will you be at that time?

79. A two-woman rowing team can row 1,200 meters with the current in a river in the same amount of time it takes them to row 1,000 meters against that same current. In each case, their average rowing speed without the effect of the current is 3 meters per second. Find the speed of the current.

80. The winners of the men's 1,000-meter double sculls event in the 2008 Olympics rowed at an average of 11.3 miles per hour. If this team were to row this speed for a half mile with a current in 80% of the time they were able to row that same distance against the current, what would be the speed of the current?

Music

81. A major chord in music is composed of notes whose frequencies are in the ratio 4:5:6. If the first note of a chord has a frequency of 264 hertz (middle C on the piano), find the frequencies of the other two notes. [*Hint:* Set up two proportions using 4:5 and 4:6.]

82. A minor chord is composed of notes whose frequencies are in the ratio 10:12:15. If the first note of a minor chord is A, with a frequency of 220 hertz, what are the frequencies of the other two notes?

Psychology

83. In an experiment on motivation, Professor Brown trained a group of rats to run down a narrow passage in a cage to receive food in a goal box. He then put a harness on each rat and connected it to an overhead wire attached to a scale. In this way, he could place the rat different distances from the food and measure the pull (in grams) of the rat toward the food. He found that the relationship between motivation (pull) and position was given approximately by the equation

$$p = -\tfrac{1}{5}d + 70 \qquad 30 \le d \le 170$$

where pull p is measured in grams and distance d in centimeters. When the pull registered was 40 grams, how far was the rat from the goal box?

84. Professor Brown performed the same kind of experiment as described in Problem 83, except that he replaced the food in the goal box with a mild electric shock. With the same kind of apparatus, he was able to measure the avoidance strength relative to the distance from the object to be avoided. He found that the avoidance strength a (measured in grams) was related to the distance d that the rat was from the shock (measured in centimeters) approximately by the equation

$$a = -\tfrac{4}{3}d + 230 \qquad 30 \le d \le 170$$

If the same rat were trained as described in this problem and in Problem 83, at what distance (to one decimal place) from the goal box would the approach and avoidance strengths be the same? (What do you think the rat would do at this point?)

1-2 # Linear Inequalities

> **Understanding Inequality and Interval Notation**
> **Solving Linear Inequalities**
> **Applying Linear Inequalities**

An equation is a statement that two expressions are equal. Sometimes it is useful to find when one expression is more or less than another, so in this section we turn our attention to linear inequalities in one variable, like

$$3x + 5 > x - 10 \qquad \text{and} \qquad -4 < 3 - 2x < 7$$

› ## Understanding Inequality and Interval Notation

The preceding mathematical statements use the **inequality,** or **order, relations,** more commonly known as "greater than" and "less than." Just as we use the symbol "=" to replace the words "is equal to," we use the **inequality symbols** < and > to replace "is less than" and "is greater than," respectively.

You probably have a natural understanding of how to compare numbers using these symbols, but to be precise about using inequality symbols, we should have a clear definition of what they mean.

> ### › DEFINITION 1 $a < b$ and $b > a$

> For two real numbers a and b, we say that **a is less than b,** and write $a < b$, if there is a positive real number p so that $a + p = b$. The statement $b > a$, read **b is greater than a,** means exactly the same as $a < b$.

This definition basically says that if you add a positive number to any number, the sum is larger than the original number.

When we write $a \leq b$ we mean $a < b$ or $a = b$ and say **a is less than or equal to b.** When we write $a \geq b$ we mean $a > b$ or $a = b$ and say **a is greater than or equal to b.**

The inequality symbols < and > have a very clear geometric interpretation on the real number line. If $a < b$, then a is to the left of b; if $c > d$, then c is to the right of d (Fig. 1). This is called a **line graph.**

› **Figure 1** $a < b, c > d$.

If we want to state that some number x is between a and b, we could use two inequalities: $x > a$ and $x < b$. Instead, we will write one double inequality, $a < x < b$. For example, the inequality $-2 < x \leq 5$ indicates that x is between -2 and 5, and could be equal to 5, but not -2. The set of all real numbers that satisfy this inequality is called an **interval,** and is commonly represented by $(-2, 5]$. In general,

$$(a, b] = \{x \mid a < x \leq b\}*$$

The number a is called the **left endpoint** of the interval, and the symbol "(" indicates that a is not included in the interval. The number b is called the **right endpoint** of the interval, and the symbol "]" indicates that b is included in the interval. An interval is **closed** if it contains its endpoint(s) and **open** if it does not contain any endpoint. Other types of intervals of real numbers are shown in Table 1.

Note that the symbol "∞," read "infinity," used in Table 1 is not a numeral. When we write $[b, \infty)$, we are simply referring to the interval starting at b and continuing indefinitely to the right. We would never write $[b, \infty]$ or $b \leq x \leq \infty$, because ∞ cannot be used as an endpoint of an interval. The interval $(-\infty, \infty)$ represents the set of real numbers R, since its graph is the entire real number line.

*In general, $\{x \mid P(x)\}$ represents the set of all x such that statement $P(x)$ is true. To express this set verbally, just read the vertical bar as "such that."

Table 1 Interval Notation

Interval notation	Inequality notation	Line graph	Type
$[a, b]$	$a \leq x \leq b$		Closed
$[a, b)$	$a \leq x < b$		Half-open
$(a, b]$	$a < x \leq b$		Half-open
(a, b)	$a < x < b$		Open
$[b, \infty)$	$x \geq b$		Closed*
(b, ∞)	$x > b$		Open
$(-\infty, a]$	$x \leq a$		Closed*
$(-\infty, a)$	$x < a$		Open

*These intervals are closed because they contain all of their endpoints; they have only one endpoint.

>>> **CAUTION** >>>

It is important to note that

$$5 > x \geq -3 \qquad \text{is equivalent to } [-3, 5) \text{ and not to } (5, -3]$$

In interval notation, the smaller number is always written to the left. It may be useful to rewrite the inequality as $-3 \leq x < 5$ before rewriting it in interval notation. The symbol $(5, -3]$ is meaningless.

EXAMPLE 1

Graphing Intervals and Inequalities

Write each of the following in inequality notation and graph on a real number line:

(A) $[-2, 3)$ (B) $(-4, 2)$ (C) $[-2, \infty)$ (D) $(-\infty, 3)$

SOLUTIONS

(A) $-2 \leq x < 3$

(B) $-4 < x < 2$

(C) $x \geq -2$

(D) $x < 3$

MATCHED PROBLEM 1

Write each of the following in interval notation and graph on a real number line:

(A) $-3 < x \leq 3$ (B) $2 \geq x \geq -1$ (C) $x > 1$ (D) $x \leq 2$

>>> **EXPLORE-DISCUSS 1**

Example 1C shows the graph of the inequality $x \geq -2$. What is the graph of $x < -2$? What is the corresponding interval? Describe the relationship between these sets.

Since intervals are sets of real numbers, the set operations of *union* and *intersection* are often useful when working with intervals. The **union** of sets A and B, denoted by $A \cup B$, is the set formed by combining all the elements of A and all the elements of B. The **intersection** of sets A and B, denoted by $A \cap B$, is the set of elements of A that are also in B. Symbolically:

> **DEFINITION 2 Union and Intersection**
>
> Union: $A \cup B = \{x \mid x \text{ is in } A \text{ **or** } x \text{ is in } B\}$
> $\quad\quad\quad\quad$ {1, 2, 3} ∪ {2, 3, 4, 5} = {1, 2, 3, 4, 5}
>
> Intersection: $A \cap B = \{x \mid x \text{ is in } A \text{ **and** } x \text{ is in } B\}$
> $\quad\quad\quad\quad$ {1, 2, 3} ∩ {2, 3, 4, 5} = {2, 3}

EXAMPLE 2 Graphing the Union and Intersection of Intervals

If $A = (-2, 5]$ and $B = (1, \infty)$, graph the sets $A \cup B$ and $A \cap B$ and write them in interval notation.

SOLUTION

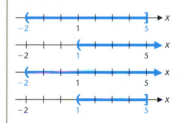

$A = (-2, 5]$

$B = (1, \infty)$

$A \cup B = (-2, \infty)$

$A \cap B = (1, 5]$

MATCHED PROBLEM 2

If $C = [-4, 3)$ and $D = (-\infty, -1]$, graph the sets $C \cup D$ and $C \cap D$ and write them in interval notation.

⫸ EXPLORE-DISCUSS 2

Replace ? with $<$ or $>$ in each of the following.

(A) $-1 \; ? \; 3$ $\quad\quad$ and $\quad\quad$ $2(-1) \; ? \; 2(3)$

(B) $-1 \; ? \; 3$ $\quad\quad$ and $\quad\quad$ $-2(-1) \; ? \; -2(3)$

(C) $12 \; ? \; -8$ $\quad\quad$ and $\quad\quad$ $\dfrac{12}{4} \; ? \; \dfrac{-8}{4}$

(D) $12 \; ? \; -8$ $\quad\quad$ and $\quad\quad$ $\dfrac{12}{-4} \; ? \; \dfrac{-8}{-4}$

Based on your results, describe verbally the effect of multiplying or dividing both sides of an inequality by a number.

> ## Solving Linear Inequalities

We now turn to the problem of solving linear inequalities in one variable, such as

$$2(2x + 3) < 6(x - 2) + 10 \quad\quad \text{and} \quad\quad -3 < 2x + 3 \leq 9$$

The **solution set** for an inequality is the set of all values of the variable that make the inequality a true statement. Each element of the solution set is called a **solution** of the inequality. To **solve an inequality** is to find its solution set. Two inequalities are **equivalent**

if they have the same solution set. Just as with equations, we perform operations on inequalities that produce simpler equivalent inequalities, and continue the process until an inequality is reached whose solution is obvious. The properties of inequalities given in Theorem 1 can be used to produce equivalent inequalities.

> › **THEOREM 1** Inequality Properties
>
> An equivalent inequality will result and the **sense (or direction) will remain the same** if each side of the original inequality
>
> • Has the same real number added to or subtracted from it
> • Is multiplied or divided by the same positive number
>
> An equivalent inequality will result and the **sense (or direction) will reverse** if each side of the original inequality
>
> • Is multiplied or divided by the same negative number
>
> *Note:* Multiplication by 0 and division by 0 are not permitted.

Theorem 1 tells us that we can perform essentially the same operations on inequalities that we perform on equations, with the exception that *the sense (or direction) of the inequality reverses if we multiply or divide both sides by a negative number:* Otherwise the sense of the inequality does not change.

Now let's see how the inequality properties are used to solve linear inequalities. Examples 3, 4, and 5 will illustrate the process.

EXAMPLE 3 **Solving a Linear Inequality**

Solve and graph: $2(2x + 3) - 10 < 6(x - 2)$

SOLUTION

$$2(2x + 3) - 10 < 6(x - 2)$$ Multiply out parentheses.
$$4x + 6 - 10 < 6x - 12$$ Combine like terms.
$$4x - 4 < 6x - 12$$ Add 4 to both sides.
$$4x - 4 + 4 < 6x - 12 + 4$$
$$4x < 6x - 8$$ Subtract 6x from both sides.
$$4x - 6x < 6x - 8 - 6x$$
$$-2x < -8$$ Divide both sides by −2. Note that direction reverses because −2 is negative.
$$\frac{-2x}{-2} > \frac{-8}{-2}$$
$$x > 4 \qquad \text{or} \qquad (4, \infty) \qquad \text{Solution set}$$

Graph of solution set

MATCHED PROBLEM 3 Solve and graph: $3(x - 1) \geq 5(x + 2) - 5$

EXAMPLE 4

Solving a Linear Inequality Involving Fractions

Solve and graph: $\dfrac{2x-3}{4} + 6 \geq 2 + \dfrac{4x}{3}$

SOLUTION

$$\dfrac{2x-3}{4} + 6 \geq 2 + \dfrac{4x}{3}$$ Multiply both sides by 12, the LCD.

$$12 \cdot \dfrac{2x-3}{4} + 12 \cdot 6 \geq 12 \cdot 2 + 12 \cdot \dfrac{4x}{3}$$ Direction doesn't change: we multiplied by a positive number.

$$3(2x-3) + 72 \geq 24 + 4(4x)$$ Multiply out parentheses.

$$6x - 9 + 72 \geq 24 + 16x$$ Combine like terms.

$$6x + 63 \geq 24 + 16x$$ Subtract 63 from both sides.

$$6x \geq -39 + 16x$$ Subtract 16x from both sides.

$$-10x \geq -39$$ Order reverses when both sides are divided by −10, a negative number.

$$x \leq 3.9 \quad \text{or} \quad (-\infty, 3.9]$$

3.9 Graph of solution set

MATCHED PROBLEM 4

Solve and graph: $\dfrac{4x-3}{3} + 8 < 6 + \dfrac{3x}{2}$

EXAMPLE 5

Solving a Double Inequality

Solve and graph: $-3 \leq 4 - 7x < 18$

SOLUTION We proceed as before, except we try to isolate x in the middle with a coefficient of 1, being sure to perform operations on all three parts of the inequality.

$$-3 \leq 4 - 7x < 18$$ Subtract 4 from each member.

$$-3 - 4 \leq 4 - 7x - 4 < 18 - 4$$

$$-7 \leq -7x < 14$$ Divide each member by −7 and reverse each inequality.

$$\dfrac{-7}{-7} \geq \dfrac{-7x}{-7} > \dfrac{14}{-7}$$

$$1 \geq x > -2 \quad \text{or} \quad -2 < x \leq 1 \quad \text{or} \quad (-2, 1]$$

−2 1 Graph of solution set

MATCHED PROBLEM 5

Solve and graph: $-3 < 7 - 2x \leq 7$

› **Applying Linear Inequalities to Chemistry**

EXAMPLE 6

Chemistry

In a chemistry experiment, a solution of hydrochloric acid is to be kept between 30°C and 35°C—that is, $30 \leq C \leq 35$. What is the range in temperature in degrees Fahrenheit if the Celsius/Fahrenheit conversion formula is $C = \frac{5}{9}(F - 32)$?

SOLUTION

$$30 \leq C \leq 35$$

Replace C with $\frac{5}{9}(F - 32)$.

$$30 \leq \frac{5}{9}(F - 32) \leq 35$$

Multiply each member by $\frac{9}{5}$ to clear fractions.

$$\frac{9}{5} \cdot 30 \leq \frac{9}{5} \cdot \frac{5}{9}(F - 32) \leq \frac{9}{5} \cdot 35$$

$$54 \leq F - 32 \leq 63$$

Add 32 to each member.

$$54 + 32 \leq F - 32 + 32 \leq 63 + 32$$

$$86 \leq F \leq 95$$

The range of the temperature is from 86°F to 95°F, inclusive.

MATCHED PROBLEM 6

A film developer is to be kept between 68°F and 77°F—that is, $68 \leq F \leq 77$. What is the range in temperature in degrees Celsius if the Celsius/Fahrenheit conversion formula is $F = \frac{9}{5}C + 32$?

ANSWERS TO MATCHED PROBLEMS

1. (A) $(-3, 3]$

(B) $[-1, 2]$

(C) $(1, \infty)$

(D) $(-\infty, 2]$

2.

$$C \cup D = (-\infty, 3)$$

$$C \cap D = [-4, -1]$$

3. $x \leq -4$ or $(-\infty, -4]$

4. $x > 6$ or $(6, \infty)$

5. $5 > x \geq 0$ or $0 \leq x < 5$ or $[0, 5)$

6. $20 \leq C \leq 25$: the range in temperature is from 20°C to 25°C

1-2 Exercises

1. Explain in your own words what it means to solve an inequality.

2. Explain why the "interval" $[5, -3)$ is meaningless.

3. What is the main difference between the procedures for solving linear equations and linear inequalities?

4. Describe how to graph the solution set of an inequality.

In Problems 5–10, rewrite in inequality notation and graph on a real number line.

5. $[-8, 7]$ **6.** $(-4, 8)$ **7.** $[-6, 6)$

8. $(-3, 3]$ **9.** $[-6, \infty)$ **10.** $(-\infty, 7)$

In Problems 11–16, rewrite in interval notation and graph on a real number line.

11. $-2 < x \le 6$ **12.** $-5 \le x \le 5$ **13.** $-7 < x < 8$

14. $-4 \le x < 5$ **15.** $x \le -2$ **16.** $x > 3$

In Problems 17–20, write in interval and inequality notation.

17.

18.

19.

20.

In Problems 21–28, replace each ? with $>$ or $<$ to make the resulting statement true.

21. $12 \; ? \; 6$ and $12 + 5 \; ? \; 6 + 5$

22. $-4 \; ? \; -2$ and $-4 - 7 \; ? \; -2 - 7$

23. $-6 \; ? \; -8$ and $-6 - 3 \; ? \; -8 - 3$

24. $4 \; ? \; 9$ and $4 + 2 \; ? \; 9 + 2$

25. $2 \; ? \; -1$ and $-2(2) \; ? \; -2(-1)$

26. $-3 \; ? \; 2$ and $4(-3) \; ? \; 4(2)$

27. $2 \; ? \; 6$ and $\dfrac{2}{2} \; ? \; \dfrac{6}{2}$

28. $-10 \; ? \; -15$ and $\dfrac{-10}{5} \; ? \; \dfrac{-15}{5}$

In Problems 29–42, solve and graph.

29. $7x - 8 < 4x + 7$ **30.** $5x - 21 \ge 3x + 5$

31. $12 - y \ge 2(9 - 2y)$ **32.** $4(y + 1) - 7 < -9 - 2y$

33. $\dfrac{N}{-2} > 4$ **34.** $\dfrac{Z}{-10} \le 3$

35. $-5t < -10$ **36.** $-20m \ge 100$

37. $3 - m < 4(m - 3)$ **38.** $6(5 - 2k) \ge 6 - 8k$

39. $-2 - \dfrac{B}{4} \le \dfrac{1 + B}{3}$ **40.** $\dfrac{t - 2}{5} + 2 > \dfrac{t}{3}$

41. $-4 < 5t + 6 \le 21$ **42.** $-2 \le 4r - 14 < 2$

In Problems 43–54, graph the indicated set and write as a single interval, if possible.

43. $(-5, 5) \cup [4, 7]$ **44.** $(-5, 5) \cap [4, 7]$

45. $[-1, 4) \cap (2, 6]$ **46.** $[-1, 4) \cup (2, 6]$

47. $(-\infty, 1) \cup (-2, \infty)$ **48.** $(-\infty, 1) \cap (2, \infty)$

49. $(-\infty, -1) \cup [3, 7)$ **50.** $(1, 6] \cup [9, \infty)$

51. $[2, 3] \cup (1, 5)$ **52.** $[2, 3] \cap (1, 5)$

53. $(-\infty, 4) \cup (-1, 6]$ **54.** $(-3, 2) \cup [0, \infty)$

In Problems 55–70, solve and graph.

55. $\dfrac{q}{7} - 3 > \dfrac{q - 4}{3} + 1$ **56.** $\dfrac{p}{3} - \dfrac{p - 2}{2} \le \dfrac{p}{4} - 4$

57. $\dfrac{2x}{5} - \dfrac{1}{2}(x - 3) \le \dfrac{2x}{3} - \dfrac{3}{10}(x + 2)$

58. $\dfrac{2}{3}(x + 7) - \dfrac{x}{4} > \dfrac{1}{2}(3 - x) + \dfrac{x}{6}$

59. $-4 \le \dfrac{9}{5}x + 32 \le 68$ **60.** $2 \le \dfrac{4}{5}z + 6 < 18$

61. $-20 < \dfrac{5}{2}(4 - x) < -5$

62. $24 \le \dfrac{2}{3}(x - 5) < 36$

63. $16 < 7 - 3x \le 31$

64. $19 \le 7 - 6x < 49$

65. $-8 \le -\dfrac{1}{4}(2 - x) + 3 < 10$

66. $0 < \dfrac{1}{3}(4 - x) - 10 \le 16$

67. $0.1(x - 7) < 0.8 - 0.05x$

68. $0.4(x + 5) > 0.3x + 17$

69. $0.3x - 2.04 \ge 0.04(x + 1)$

70. $0.02x - 5.32 \le 0.5(x - 2)$

Problems 71–76 are calculus-related. For what real number(s) x does each expression represent a real number?

71. $\sqrt{1 - x}$ **72.** $\sqrt{x + 5}$

73. $\sqrt{3x + 5}$ **74.** $\sqrt{7 - 2x}$

75. $\dfrac{1}{\sqrt[4]{2x + 3}}$ **76.** $\dfrac{1}{\sqrt[4]{5 - 6x}}$

77. What can be said about the signs of the numbers a and b in each case?

(A) $ab > 0$ (B) $ab < 0$

(C) $\dfrac{a}{b} > 0$ (D) $\dfrac{a}{b} < 0$

78. What can be said about the signs of the numbers a, b, and c in each case?

(A) $abc > 0$ (B) $\dfrac{ab}{c} < 0$

(C) $\dfrac{a}{bc} > 0$ (D) $\dfrac{a^2}{bc} < 0$

79. Replace each question mark with $<$ or $>$, as appropriate:
 (A) If $a - b = 1$, then a ? b.
 (B) If $u - v = -2$, then u ? v.

80. For what p and q is $p + q < p - q$?

81. If both a and b are negative numbers and b/a is greater than 1, then is $a - b$ positive or negative?

82. If both a and b are positive numbers and b/a is greater than 1, then is $a - b$ positive or negative?

83. Indicate true (T) or false (F):
 (A) If $p > q$ and $m > 0$, then $mp < mq$.
 (B) If $p < q$ and $m < 0$, then $mp > mq$.
 (C) If $p > 0$ and $q < 0$, then $p + q > q$.

84. Assume that $m > n > 0$; then

$$mn > n^2$$
$$mn - m^2 > n^2 - m^2$$
$$m(n - m) > (n + m)(n - m)$$
$$m > n + m$$
$$0 > n$$

But it was assumed that $n > 0$. Find the error.

Prove each inequality property in Problems 85–88, given a, b, and c are arbitrary real numbers.

85. If $a < b$, then $a + c < b + c$.

86. If $a < b$, then $a - c < b - c$.

87. (A) If $a < b$ and c is positive, then $ca < cb$.
 (B) If $a < b$ and c is negative, then $ca > cb$.

88. (A) If $a < b$ and c is positive, then $\dfrac{a}{c} < \dfrac{b}{c}$.

 (B) If $a < b$ and c is negative, then $\dfrac{a}{c} > \dfrac{b}{c}$.

APPLICATIONS

Write all your answers using inequality notation.

89. EARTH SCIENCE In 1970, Russian geologists began drilling a very deep borehole in the Kola Peninsula. Their goal was to reach a depth of 15 kilometers, but high temperatures in the borehole forced them to stop in 1994 after reaching a depth of 12 kilometers. They found that the approximate temperature x kilometers below the surface of the Earth is given by

$$T = 30 + 25(x - 3) \qquad 3 \le x \le 12$$

where T is temperature in degrees Celsius. At what depth is the temperature between 150°C and 250°C, inclusive?

90. EARTH SCIENCE As dry air moves upward it expands, and in so doing it cools at a rate of about 5.5°F for each 1,000-foot rise up to about 40,000 feet. If the ground temperature is 70°F, then the temperature T at height h is given approximately by $T = 70 - 0.0055h$.

For what range in altitude will the temperature be between 26°F and −40°F, inclusive?

91. BUSINESS AND ECONOMICS An electronics firm is planning to market a new graphing calculator. The fixed costs are $650,000 and the variable costs are $47 per calculator. The wholesale price of the calculator will be $63. For the company to make a profit, it is clear that revenues must be greater than costs.
(A) How many calculators must be sold for the company to make a profit?
(B) How many calculators must be sold for the company to break even?
(C) Discuss the relationship between the results in parts A and B.

92. BUSINESS AND ECONOMICS A video game manufacturer is planning to market a handheld version of its game machine. The fixed costs are $550,000 and the variable costs are $120 per machine. The wholesale price of the machine will be $140.
(A) How many game machines must be sold for the company to make a profit?
(B) How many game machines must be sold for the company to break even?
(C) Discuss the relationship between the results in parts A and B.

93. BUSINESS AND ECONOMICS The electronics firm in Problem 91 finds that rising prices for parts increases the variable costs to $50.50 per calculator.
(A) Discuss possible strategies the company might use to deal with this increase in costs.
(B) If the company continues to sell the calculators for $63, how many must they sell now to make a profit?
(C) If the company wants to start making a profit at the same production level as before the cost increase, how much should they increase the wholesale price?

94. BUSINESS AND ECONOMICS The video game manufacturer in Problem 92 finds that unexpected programming problems increases the fixed costs to $660,000.
(A) Discuss possible strategies the company might use to deal with this increase in costs.
(B) If the company continues to sell the game machines for $140, how many must they sell now to make a profit?
(C) If the company wants to start making a profit at the same production level as before the cost increase, how much should they increase the wholesale price?

95. ENERGY If the power demands in a 110-volt electric circuit in a home vary between 220 and 2,750 watts, what is the range of current flowing through the circuit? ($W = EI$, where $W = $ Power in watts, $E = $ Pressure in volts, and $I = $ Current in amperes.)

96. PSYCHOLOGY A person's IQ is given by the formula

$$IQ = \frac{MA}{CA} 100$$

where MA is mental age and CA is chronological age. If

$$80 \le IQ \le 140$$

for a group of 12-year-old children, find the range of their mental ages.

1-3 Absolute Value in Equations and Inequalities

> Relating Absolute Value and Distance
> Solving Absolute Value Equations and Inequalities
> Using Absolute Value to Solve Radical Inequalities

We can express the distance between two points on a number line using the concept of *absolute value.* As a result, absolute values often appear in equations and inequalities that are associated with distance. In this section, we define absolute value and we show how to solve equations and inequalities that involve absolute value.

› Relating Absolute Value and Distance

We start with a geometric definition of absolute value. If a is the coordinate of a point on a real number line, then the distance from the origin to a is represented by $|a|$ and is referred to as the **absolute value** of a. So $|5| = 5$, since the point with coordinate 5 is five units from the origin, and $|-6| = 6$, since the point with coordinate -6 is six units from the origin (Fig. 1).

> **Figure 1** Absolute value.

We can use symbols to write a formal definition of absolute value:

> **DEFINITION 1** Absolute Value
>
> $$|x| = \begin{cases} -x & \text{if } x < 0 \\ x & \text{if } x \geq 0 \end{cases}$$
>
> For example, $|-3| = -(-3) = 3$
> For example, $|4| = 4$
>
> [*Note:* $-x$ is positive if x is negative.]

Both the geometric and algebraic definitions of absolute value are useful, as will be seen in the material that follows. Remember:

The absolute value of a number is never negative.

| EXAMPLE | 1 | **Finding Absolute Value** |

Write without the absolute value sign:

(A) $|\pi - 3|$ (B) $|3 - \pi|$

SOLUTIONS

(A) $|\pi - 3| = \pi - 3$ Because $\pi \approx 3.14$, $\pi - 3$ is positive.

(B) $|3 - \pi| = -(3 - \pi) = \pi - 3$ Because $3 - \pi$ is negative.

MATCHED PROBLEM 1

Write without the absolute value sign:

(A) $|8|$ (B) $|\sqrt[3]{9} - 2|$ (C) $|-\sqrt{2}|$ (D) $|2 - \sqrt[3]{9}|$

Notice that the solution in both parts of Example 1 was the same. This suggests Theorem 1, which will be proved in Problem 81.

> **THEOREM 1** For All Real Numbers a and b,
>
> $$|b - a| = |a - b|$$

To find the distance between two numbers, we subtract, larger minus smaller. But if we don't know which is larger, we can use absolute value; Theorem 1 tells us that the order is immaterial.

> **DEFINITION 2** Distance Between Points A and B
>
> Let A and B be two points on a real number line with coordinates a and b, respectively. The **distance between A and B** is given by
>
> $$d(A, B) = |b - a|$$
>
> This distance is also called the **length of the line segment** joining A and B.

It will come in very handy to observe that an expression like $|b - a|$ can always be interpreted as the distance between two numbers a and b, and that the order of the subtraction doesn't matter.

› Solving Absolute Value Equations and Inequalities

The connection between algebra and geometry is an important tool when working with equations and inequalities involving absolute value. For example, the algebraic statement

$$|x - 1| = 2$$

can be interpreted geometrically as stating that the distance from x to 1 is 2.

››› EXPLORE-DISCUSS 1

Write geometric interpretations of the following algebraic statements:

(A) $|x - 1| < 2$ (B) $0 < |x - 1| < 2$ (C) $|x - 1| > 2$

EXAMPLE	2	**Solving Absolute Value Problems Geometrically**

Interpret geometrically, solve, and graph. For inequalities, write solutions in both inequality and interval notation.

(A) $|x - 3| = 5$ (B) $|x - 3| < 5$

(C) $0 < |x - 3| < 5$ (D) $|x - 3| > 5$

SOLUTIONS

(A) The expression $|x - 3|$ represents the distance between x and 3, so the solutions to $|x - 3| = 5$ are all numbers that are exactly 5 units away from 3 on a number line.

$$x = 3 \pm 5 = -2 \text{ or } 8$$

The solution set is $\{-2, 8\}$. This *is not* interval notation.

(B) Solutions to $|x - 3| < 5$ are all numbers whose distance from 3 is less than 5. These are the numbers between -2 and 8:

$$-2 < x < 8$$

The solution set is $(-2, 8)$. This *is* interval notation.

(C) Expressions like $0 < |x - 3| < 5$ are important in calculus. The solutions are all numbers whose distance from 3 is less than 5, and is not zero. This excludes 3 itself from the solution set:

$$-2 < x < 8 \qquad x \neq 3 \qquad \text{or} \qquad (-2, 3) \cup (3, 8)$$

(D) The solutions to $|x - 3| > 5$ are all numbers whose distance from 3 is greater than 5; that is,

$$x < -2 \qquad \text{or} \qquad x > 8 \qquad (-\infty, -2) \cup (8, \infty)$$

››› CAUTION ›››	The pair of inequalities $-2 < x$ and $x < 8$ can be written as a double inequality:

$$-2 < x < 8 \text{ or in interval notation } (-2, 8)$$

But the pair $x < -2$ or $x > 8$ from Example 2(D) cannot be written as a double inequality, or as a single interval: no number is both less than -2 and greater than 8.

| MATCHED PROBLEM 2 | Interpret geometrically, solve, and graph. For inequalities, write solutions in both inequality and interval notation. [*Hint:* $|x + 2| = |x - (-2)|$.] |

(A) $|x + 2| = 6$ (B) $|x + 2| < 6$

(C) $0 < |x + 2| < 6$ (D) $|x + 2| > 6$

The preceding results are summarized in Table 1.

Table 1 Geometric Interpretation of Absolute Value Equations and Inequalities

Form ($d > 0$)	Geometric interpretation	Solution	Graph
$\|x - c\| = d$	Distance between x and c is equal to d.	$\{c - d, c + d\}$	
$\|x - c\| < d$	Distance between x and c is less than d.	$(c - d, c + d)$	
$0 < \|x - c\| < d$	Distance between x and c is less than d, but $x \neq c$.	$(c - d, c) \cup (c, c + d)$	
$\|x - c\| > d$	Distance between x and c is greater than d.	$(-\infty, c - d) \cup (c + d, \infty)$	

| EXAMPLE 3 | **Interpreting Verbal Statements Algebraically** |

Express each verbal statement as an absolute value equation or inequality.

(A) x is 4 units from 2.

(B) y is less than 3 units from -5.

(C) t is no more than 5 units from 7.

(D) w is no less than 2 units from -1.

SOLUTIONS

(A) $d(x, 2) \;=\; |x - 2| = 4$ The distance from x to 2 is 4.

(B) $d(y, -5) =\; |y + 5| < 3$ The distance from y to -5 is < 3.

(C) $d(t, 7) \;=\; |t - 7| \leq 5$ The distance from t to 7 is ≥ 5.

(D) $d(w, -1) =\; |w + 1| \geq 2$ The distance from w to -1 is ≥ 2.

| MATCHED PROBLEM 3 | Express each verbal statement as an absolute value equation or inequality. |

(A) x is 6 units from 5.

(B) y is less than 7 units from -6.

(C) w is no less than 3 units from -2.

(D) t is no more than 4 units from 3.

>>> EXPLORE-DISCUSS 2

Describe the set of numbers that satisfies each of the following:

(A) $2 > x > 1$ **(B)** $2 > x < 1$

(C) $2 < x > 1$ **(D)** $2 < x < 1$

Explain why we never write double inequalities with inequality symbols pointing in different directions.

The results of Example 2 can be generalized as Theorem 2. [*Note:* $|x| = |x - 0|$.]

> **THEOREM 2 Properties of Equations and Inequalities Involving $|x|$**

For $p > 0$: *p has to be positive!*

1. $|x| = p$ is equivalent to $x = p$ or $x = -p$.
 The distance from x to zero is p.

2. $|x| < p$ is equivalent to $-p < x < p$.
 The distance from x to zero is less than p.

3. $|x| > p$ is equivalent to $x < -p$ or $x > p$.
 The distance from x to zero is greater than p.

If we replace x in Theorem 2 with $ax + b$, we obtain the more general Theorem 3.

> **THEOREM 3 Properties of Equations and Inequalities Involving $|ax + b|$**

For $p > 0$: *p has to be positive!*

1. $|ax + b| = p$ is equivalent to $ax + b = p$ or $ax + b = -p$.*

2. $|ax + b| < p$ is equivalent to $-p < ax + b < p$.

3. $|ax + b| > p$ is equivalent to $ax + b < -p$ or $ax + b > p$.

| EXAMPLE | 4 | **Solving Absolute Value Problems** |

Solve each equation or inequality. For inequalities, write solutions in both inequality and interval notation.

(A) $|3x + 5| = 4$ (B) $|x| < 5$

(C) $|2x - 1| < 3$ (D) $|7 - 3x| \leq 2$

*This can be more concisely written as $ax + b = \pm p$.

SOLUTIONS

(A) $|3x + 5| = 4$ Use Theorem 3, part 1

$$3x + 5 = \pm 4$$
$$3x = -5 \pm 4$$
$$x = \frac{-5 \pm 4}{3}$$
$$x = -3, -\tfrac{1}{3}$$

or $\{-3, -\tfrac{1}{3}\}$

(B) $|x| < 5$ Use Theorem 2, part 2

$$-5 < x < 5$$

or $(-5, 5)$

(C) $|2x - 1| < 3$ Use Theorem 3, part 2

$$-3 < 2x - 1 < 3$$
$$-2 < 2x < 4$$
$$-1 < x < 2$$

or $(-1, 2)$

(D) $|7 - 3x| \leq 2$ Use Theorem 3, part 2

$$-2 \leq 7 - 3x \leq 2$$
$$-9 \leq -3x \leq -5$$
$$3 \geq x \geq \tfrac{5}{3}$$
$$\tfrac{5}{3} \leq x \leq 3$$

or $[\tfrac{5}{3}, 3]$

MATCHED PROBLEM 4

Solve each equation or inequality. For inequalities, write solutions in both inequality and interval notation.

(A) $|2x - 1| = 8$ (B) $|x| \leq 7$ (C) $|3x + 3| \leq 9$ (D) $|5 - 2x| < 9$

EXAMPLE **5** **Solving Absolute Value Inequalities**

Solve, and write solutions in both inequality and interval notation.

(A) $|x| > 3$ (B) $|2x - 1| \geq 3$ (C) $|7 - 3x| > 2$

SOLUTIONS

(A) $|x| > 3$ Use Theorem 2, part 3.

$\quad x < -3 \quad$ or $\quad x > 3$ Solution in inequality notation

$\quad (-\infty, -3) \cup (3, \infty)$ Solution in interval notation

(B) $|2x - 1| \geq 3$ Use Theorem 3, part 3.

$\quad 2x - 1 \leq -3 \quad$ or $\quad 2x - 1 \geq 3$ Add 1 to both sides.

$\quad\quad 2x \leq -2 \quad$ or $\quad\quad 2x \geq 4$ Divide both sides by 2.

$\quad\quad\quad x \leq -1 \quad$ or $\quad\quad\quad x \geq 2$ Solution in inequality notation

$\quad (-\infty, -1] \cup [2, \infty)$ Solution in interval notation

(C) $|7 - 3x| > 2$ Use Theorem 3, part 3.

$\quad 7 - 3x < -2 \quad$ or $\quad 7 - 3x > 2$ Subtract 7 from both sides.

$\quad\quad -3x < -9 \quad$ or $\quad\quad -3x > -5$ Divide both sides by -3 and reverse the direction of the inequality.

$\quad\quad\quad x > 3 \quad$ or $\quad\quad\quad x < \tfrac{5}{3}$ Solution in inequality notation

$\quad (-\infty, \tfrac{5}{3}) \cup (3, \infty)$ Solution in interval notation

MATCHED PROBLEM 5

Solve, and write solutions in both inequality and interval notation.

(A) $|x| \geq 5$ (B) $|4x - 3| > 5$ (C) $|6 - 5x| > 16$

| EXAMPLE | 6 | **An Absolute Value Problem with Two Cases** |

Solve: $|x + 4| = 3x - 8$

SOLUTION We can't use Theorem 3 directly, because we don't know that $3x - 8$ is positive. However, we can use the definition of absolute value and two cases: $x + 4 \geq 0$ and $x + 4 < 0$.

Case 1. $x + 4 \geq 0$ (in which case, $x \geq -4$)
For this case, the only acceptable values of x are greater than or equal to -4.

$$|x + 4| = 3x - 8 \quad \text{If } x + 4 \text{ is positive, } |x + 4| = x + 4.$$
$$x + 4 = 3x - 8 \quad \text{Subtract 3x and 4 from both sides.}$$
$$-2x = -12 \quad \text{Divide both sides by } -2.$$
$$x = 6 \quad \text{A solution, because 6 is among the acceptable values of x } (6 \geq -4).$$

Case 2. $x + 4 < 0$ (in which case, $x < -4$)
In this case, the only acceptable values of x are less than -4.

$$|x + 4| = 3x - 8 \quad \text{If } x + 4 \text{ is negative, } |x + 4| = -(x + 4).$$
$$-(x + 4) = 3x - 8 \quad \text{Distribute } -1.$$
$$-x - 4 = 3x - 8 \quad \text{Subtract 3x and add 4 to both sides.}$$
$$-4x = -4 \quad \text{Divide both sides by } -4.$$
$$x = 1 \quad \text{Not a solution, since 1 is not among the acceptable values of x } (1 > -4).$$

Combining both cases, we see that the only solution is $x = 6$.

As a final check, we substitute $x = 6$ and $x = 1$ in the original equation.

$$|x + 4| = 3x - 8 \qquad |x + 4| = 3x - 8$$
$$|6 + 4| \stackrel{?}{=} 3(6) - 8 \qquad |1 + 4| \stackrel{?}{=} 3(1) - 8$$
$$10 \stackrel{\checkmark}{=} 10 \qquad\qquad 5 \neq -5$$

MATCHED PROBLEM 6 Solve: $|3x - 4| = x + 5$

› Using Absolute Value to Solve Radical Inequalities

In Section R-2, we found that if x is positive or zero, $\sqrt{x^2} = x$. But what if x is negative? Let's look at an example:

$$\sqrt{(-2)^2} = \sqrt{4} = 2$$

We see that for negative x, $\sqrt{x^2} = -x$. So for any real number,

$$\sqrt{x^2} = \begin{cases} -x & \text{if } x < 0 \\ x & \text{if } x \geq 0 \end{cases}$$

But this is exactly how we defined $|x|$ at the beginning of this section (see Definition 1). So for any real number x,

$$\sqrt{x^2} = |x| \tag{1}$$

EXAMPLE	7	**Solving a Radical Inequality**

Solve the inequality. Write your answer in both inequality and interval notation.

$$\sqrt{(x-2)^2} \le 5$$

SOLUTION

$$\sqrt{(x-2)^2} \le 5 \qquad \text{Use equation (1): } \sqrt{(x-2)^2} = |x-2|$$
$$|x-2| \le 5 \qquad \text{Use Theorem 3, part 2.}$$
$$-5 \le x - 2 \le 5 \qquad \text{Add 2 to each member.}$$
$$-3 \le x \le 7 \qquad \text{Solution in inequality notation}$$
$$\text{or } [-3, 7] \qquad \text{Solution in interval notation}$$

MATCHED PROBLEM 7

Solve the inequality. Write your answers in both inequality and interval notation.

$$\sqrt{(x+2)^2} < 3$$

ANSWERS TO MATCHED PROBLEMS

1. (A) 8 (B) $\sqrt[3]{9} - 2$ (C) $\sqrt{2}$ (D) $\sqrt[3]{9} - 2$

2. (A) x is a number whose distance from -2 is 6.
 $x = -8, 4$ or $\{-8, 4\}$

 (B) x is a number whose distance from -2 is less than 6.
 $-8 < x < 4$ or $(-8, 4)$

 (C) x is a number whose distance from -2 is less than 6, but x cannot equal -2.
 $-8 < x < 4, x \ne -2$, or $(-8, -2) \cup (-2, 4)$

 (D) x is a number whose distance from -2 is greater than 6.
 $x < -8$ or $x > 4$, or $(-\infty, -8) \cup (4, \infty)$

3. (A) $|x - 5| = 6$ (B) $|y + 6| < 7$ (C) $|w + 2| \ge 3$ (D) $|t - 3| \le 4$

4. (A) $x = -\frac{7}{2}, \frac{9}{2}$ or $\{-\frac{7}{2}, \frac{9}{2}\}$ (B) $-7 \le x \le 7$ or $[-7, 7]$ (C) $-4 \le x \le 2$ or $[-4, 2]$
 (D) $-2 < x < 7$ or $(-2, 7)$

5. (A) $x \le -5$ or $x \ge 5$, or $(-\infty, -5] \cup [5, \infty)$ (B) $x < -\frac{1}{2}$ or $x > 2$, or $(-\infty, -\frac{1}{2}) \cup (2, \infty)$
 (C) $x < -2$ or $x > \frac{22}{5}$, or $(-\infty, -2) \cup (\frac{22}{5}, \infty)$

6. $x = -\frac{1}{4}, \frac{9}{2}$ or $\{-\frac{1}{4}, \frac{9}{2}\}$ **7.** $-5 < x < 1$ or $(-5, 1)$

1-3 Exercises

1. Describe how to find the absolute value of a number, then explain how your description matches Definition 1.

2. Explain what the expression $|x - 5|$ represents geometrically, and why.

3. Describe the equation $|x - 5| = 10$ in terms of your answer to Problem 2, then explain how that helps you to solve it.

4. Repeat Problem 3 for the inequalities $|x - 5| < 10$ and $|x - 5| > 10$.

5. Explain why it is incorrect to say that $\sqrt{x^2} = x$.

6. Why can't the following be a legitimate solution to an inequality? $x < 1$ and $x > 5$.

In Problems 7–14, simplify, and write without absolute value signs. Do not replace radicals with decimal approximations.

7. $|\sqrt{5}|$

8. $|-\frac{3}{4}|$

9. $|(-6) - (-2)|$

10. $|(-2) - (-6)|$

11. $|5 - \sqrt{5}|$

12. $|\sqrt{7} - 2|$

13. $|\sqrt{5} - 5|$

14. $|2 - \sqrt{7}|$

In Problems 15–20, use the number line shown to find the indicated distances.

15. $d(B, O)$

16. $d(A, B)$

17. $d(O, B)$

18. $d(B, A)$

19. $d(B, C)$

20. $d(D, C)$

Write each of the statements in Problems 21–30 as an absolute value equation or inequality.

21. x is 4 units from 3.

22. y is 3 units from 1.

23. m is 5 units from -2.

24. n is 7 units from -5.

25. x is less than 5 units from 3.

26. z is less than 8 units from -2.

27. p is more than 6 units from -2.

28. c is no greater than 7 units from -3.

29. q is no less than 2 units from 1.

30. d is no more than 4 units from 5.

In Problems 31–42, solve, interpret geometrically, and graph. When applicable, write answers using both inequality notation and interval notation.

31. $|y - 5| = 3$

32. $|t - 3| = 4$

33. $|y - 5| < 3$

34. $|t - 3| < 4$

35. $|y - 5| > 3$

36. $|t - 3| > 4$

37. $|u + 8| = 3$

38. $|x + 1| = 5$

39. $|u + 8| \leq 3$

40. $|x + 1| \leq 5$

41. $|u + 8| \geq 3$

42. $|x + 1| \geq 5$

In Problems 43–60, solve the equation or inequality. Write solutions to inequalities using both inequality and interval notation.

43. $|2x - 11| \leq 13$

44. $|5x + 20| \geq 5$

45. $|100 - 40t| > 60$

46. $|150 - 20y| < 10$

47. $|4x - 7| = 13$

48. $|-8x + 3| \leq 91$

49. $|\frac{1}{2}w - \frac{3}{4}| < 2$

50. $|\frac{1}{3}z + \frac{5}{6}| = 1$

51. $|0.2u + 1.7| \geq 0.5$

52. $|0.5v - 2.5| > 1.6$

53. $|\frac{9}{5}C + 32| < 31$

54. $|\frac{5}{9}(F - 32)| < 40$

55. $\sqrt{x^2} < 2$

56. $\sqrt{m^2} > 3$

57. $\sqrt{(1 - 3t)^2} \leq 2$

58. $\sqrt{(3 - 2x)^2} < 5$

59. $\sqrt{(2t - 3)^2} > 3$

60. $\sqrt{(3m + 5)^2} \geq 4$

In Problems 61–64, solve and write answers in inequality notation. Round decimals to three significant digits.

61. $|2.25 - 1.02x| \leq 1.64$

62. $|0.962 - 0.292x| \leq 2.52$

63. $|21.7 - 11.3x| = 15.2$

64. $|195 - 55.5x| = 315$

Problems 65–68 involve expressions that are important in the study of limits in calculus. First, provide a verbal translation of the inequality. Then solve and graph, writing your solution in interval notation.

65. $0 < |x - 3| < 0.1$

66. $0 < |x + 5| < 0.5$

67. $0 < |x - a| < \dfrac{1}{10}$

68. $0 < |x - 8| < d$

In Problems 69–76, for what values of x does each hold?

69. $|x - 2| = 2x - 7$

70. $|x + 4| = 3x + 8$

71. $|3x + 5| = 2x + 6$

72. $|7 - 2x| = 5 - x$

73. $|x| + |x + 3| = 3$

74. $|x| - |x - 5| = 5$

75. $|3 - x| = 2(4 + x)$

76. $|5 - 2x| = 4(x - 5)$

77. What are the possible values of $\dfrac{x}{|x|}$?

78. What are the possible values of $\dfrac{|x - 1|}{x - 1}$?

79. Explain why $|ax + b| < -3$ has no solution for any values of a and b.

80. Explain why $|ax + b| > -3$ has solution all real numbers for any values of a and b.

81. Prove that $|b - a| = |a - b|$ for all real numbers a and b. [*Hint:* Apply Definition 1 and use cases.]

82. Prove that $|x|^2 = x^2$ for all real numbers x.

83. Prove that the average of two numbers is between the two numbers; that is, if $m < n$, then

$$m < \frac{m + n}{2} < n$$

84. Prove that for $m < n$,

$$d\left(m, \frac{m + n}{2}\right) = d\left(\frac{m + n}{2}, n\right)$$

85. Prove that $|-m| = |m|$.

86. Prove that $|m| = |n|$ if and only if $m = n$ or $m = -n$.

87. Prove that for $n \neq 0$,

$$\left| \frac{m}{n} \right| = \frac{|m|}{|n|}$$

88. Prove that $|mn| = |m||n|$.

89. Prove that $-|m| \leq m \leq |m|$.

90. Prove the **triangle inequality:**

$$|m + n| \leq |m| + |n|$$

Hint: Use Problem 89 to show that

$$-|m| - |n| \leq m + n \leq |m| + |n|$$

APPLICATIONS

91. STATISTICS Inequalities of the form

$$\left| \frac{x - m}{s} \right| < n$$

occur frequently in statistics. If $m = 45.4$, $s = 3.2$, and $n = 1$, solve for x.

92. STATISTICS Repeat Problem 91 for $m = 28.6$, $s = 6.5$, and $n = 2$.

93. BUSINESS The daily production P in an automobile assembly plant is always within 20 units of 500 units. Write the daily production as an absolute value inequality, then solve to find the range of daily productions possible.

94. CHEMISTRY In order to manufacture a polymer for soft drink containers, a chemical reaction must take place within 10°C of 200°C. Write this temperature restriction as an absolute value inequality, then solve to find the acceptable temperatures.

95. APPROXIMATION The area A of a region is approximately equal to 12.436. The error in this approximation is less than 0.001. Describe the possible values of this area both with an absolute value inequality and with interval notation.

96. APPROXIMATION The volume V of a solid is approximately equal to 6.94. The error in this approximation is less than 0.02. Describe the possible values of this volume both with an absolute value inequality and with interval notation.

97. SIGNIFICANT DIGITS If $N = 2.37$ represents a measurement, then we assume an accuracy of 2.37 ± 0.005. Express the accuracy assumption using an absolute value inequality.

98. SIGNIFICANT DIGITS If $N = 3.65 \times 10^{-3}$ is a number from a measurement, then we assume an accuracy of $3.65 \times 10^{-3} \pm 5 \times 10^{-6}$. Express the accuracy assumption using an absolute value inequality.

1-4　　Complex Numbers

> Understanding Complex Number Terminology
> Performing Operations with Complex Numbers
> Relating Complex Numbers and Radicals
> Solving Equations Involving Complex Numbers

The idea of inventing new numbers might seem odd to you, but think about this example: A group of mathematicians known as the Pythagoreans proved over 2,000 years ago that the equation $x^2 = 2$ has no solutions that are rational numbers. You may be thinking that the solutions are $\sqrt{2}$ and $-\sqrt{2}$, but at the time, those numbers had not been defined, so the Pythagoreans invented a new kind of number—irrational numbers, like $\sqrt{2}$ and $-\sqrt{2}$.

　　Now consider the similar equation $x^2 = -1$. To be a solution, a number has to result in -1 when squared. But we know that the square of any real number cannot be negative, so again a new type of number is invented—a number whose square is -1. The concept of square roots of negative numbers had been kicked around for a few centuries, but in 1748, the Swiss mathematician Leonhard Euler (pronounced "Oiler") used the letter i to represent a square root of -1. From this simple beginning, it is possible to build a new system of numbers called the complex number system.

> Understanding Complex Number Terminology

The number i, whose square is -1, is called the **imaginary unit.** Complex numbers are defined in terms of the imaginary unit.

> › **DEFINITION 1** Complex Number
>
> A **complex number** is a number of the form $a + bi$, where a and b are real numbers, and i is the imaginary unit (a square root of -1). A complex number written this way is said to be in **standard form.** The real number a is called the **real part,** and bi is called the **imaginary part.**

Some examples of complex numbers are

$$3 - 2i \qquad \tfrac{1}{2} + 5i \qquad 2 - \tfrac{1}{3}i$$
$$0 + 3i \qquad 5 + 0i \qquad 0 + 0i$$

The notation $3 - 2i$ is shorthand for $3 + (-2)i$.

Particular kinds of complex numbers are given special names as follows:

> › **DEFINITION 2** Special Terms
>
> | $a + bi \qquad b \neq 0$ | **Imaginary Number** |
> | $0 + bi = bi \qquad b \neq 0$ | **Pure Imaginary Number** |
> | $a + 0i = a$ | **Real Number** |
> | $0 = 0 + 0i$ | **Zero** |
> | $a - bi$ | **Conjugate of $a + bi$** |

EXAMPLE 1 **Complex Numbers**

Identify the real part, the imaginary part, and the conjugate of each of the following numbers:

(A) $3 - 2i$ (B) $2 + 5i$ (C) $7i$ (D) 6

SOLUTIONS (A) Real part: 3; imaginary part: $-2i$; conjugate: $3 + 2i$

(B) Real part: 2; imaginary part: $5i$; conjugate: $2 - 5i$

(C) Real part: 0; imaginary part: $7i$; conjugate: $-7i$

(D) Real part: 6; imaginary part: 0; conjugate: 6

MATCHED PROBLEM 1 Identify the real part, the imaginary part, and the conjugate of each of the following numbers:

(A) $6 + 7i$ (B) $-3 - 8i$ (C) $-4i$ (D) -9

We will identify a complex number of the form $a + 0i$ with the real number a, a complex number of the form $0 + bi$, $b \neq 0$, with the **pure imaginary number** bi, and the complex number $0 + 0i$ with the real number 0. So a real number is also a complex number, just as a rational number is also a real number. Any complex number that is not a real number is called an **imaginary number.** If we combine the set of all real numbers with the set of all imaginary numbers, we obtain **C, the set of complex numbers.** The relationship of the complex number system to the other number systems we have studied is shown in Figure 1.

> Figure 1

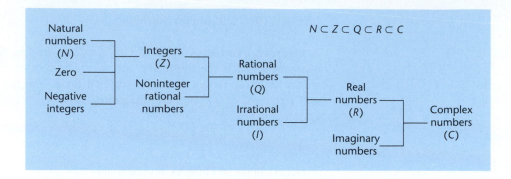

$$N \subset Z \subset Q \subset R \subset C$$

› Performing Operations with Complex Numbers

To work with complex numbers, we will need to know how to add, subtract, multiply, and divide them. We start by defining equality, addition, and multiplication.

> ### › DEFINITION 3 Equality and Basic Operations
>
> **1. Equality:** $a + bi = c + di$ if and only if $a = c$ and $b = d$
>
> **2. Addition:** $(a + bi) + (c + di) = (a + c) + (b + d)i$
>
> **3. Multiplication:** $(a + bi)(c + di) = (ac - bd) + (ad + bc)i$

In Section R-1 we listed the basic properties of the real number system. Using Definition 3, it can be shown that the complex number system possesses the same properties. That is,

1. Addition and multiplication of complex numbers are commutative and associative operations.

2. There is an additive identity and a multiplicative identity for complex numbers.

3. Every complex number has an additive inverse or negative.

4. Every nonzero complex number has a multiplicative inverse or reciprocal.

5. Multiplication distributes over addition.

This is actually really good news: it tells us that we don't have to memorize the formulas for adding and multiplying complex numbers in Definition 3. Instead:

We can treat complex numbers of the form $a + bi$ exactly as we treat algebraic expressions of the form $a + bx$. We just need to remember that in this case, i stands for the imaginary unit; it is not a variable that represents a real number.

The first two arithmetic operations we consider are *addition* and *subtraction*.

| EXAMPLE | 2 | ### Addition and Subtraction of Complex Numbers |

Carry out each operation and express the answer in standard form:

(A) $(2 - 3i) + (6 + 2i)$ (B) $(-5 + 4i) + (0 + 0i)$

(C) $(7 - 3i) - (6 + 2i)$ (D) $(-2 + 7i) + (2 - 7i)$

SOLUTIONS (A) We could apply the definition of addition directly, but it is easier to use complex number properties.

$$(2 - 3i) + (6 + 2i) = 2 - 3i + 6 + 2i \qquad \text{Use the commutative property.}$$
$$= (2 + 6) + (-3 + 2)i \qquad \text{Combine like terms.}$$
$$= 8 - i$$

(B) $(-5 + 4i) + (0 + 0i) = -5 + 4i + 0 + 0i$
$$= -5 + 4i$$

(C) $(7 - 3i) - (6 + 2i) = 7 - 3i - 6 - 2i \qquad \text{Make sure you distribute the negative sign!}$
$$= 1 - 5i$$

(D) $(-2 + 7i) + (2 - 7i) = -2 + 7i + 2 - 7i = 0$

MATCHED PROBLEM 2 Carry out each operation and express the answer in standard form:

(A) $(3 + 2i) + (6 - 4i)$ (B) $(0 + 0i) + (7 - 5i)$

(C) $(3 - 5i) - (1 - 3i)$ (D) $(-4 + 9i) + (4 - 9i)$

Example 2, part B, illustrates the following general property: For any complex number $a + bi$,

$$(a + bi) + (0 + 0i) = a + bi \quad \text{and} \quad (0 + 0i) + (a + bi) = a + bi$$

That is, $0 + 0i$ is the **additive identity** or **zero** for the complex numbers. This is why we identify $0 + 0i$ with the real number zero in Definition 2.

Example 2, part D, illustrates a different result: In general, the **additive inverse** or **negative** of $a + bi$ is $-a - bi$ because

$$(a + bi) + (-a - bi) = 0 \quad \text{and} \quad (-a - bi) + (a + bi) = 0$$

Now we turn our attention to multiplication. Just like addition and subtraction, **multiplication of complex numbers** can be carried out by treating $a + bi$ in the same way we treat the algebraic expression $a + bx$. The key difference is that we replace i^2 with -1 each time it occurs.

EXAMPLE 3 Multiplying Complex Numbers

Carry out each operation and express the answer in standard form:

(A) $(2 - 3i)(6 + 2i)$ (B) $1(3 - 5i)$

(C) $i(1 + i)$ (D) $(3 + 4i)(3 - 4i)$

SOLUTIONS (A) $(2 - 3i)(6 + 2i) = 12 + 4i - 18i - 6i^2 \qquad \text{Replace } i^2 \text{ with } -1.$
$$= 12 - 14i - 6(-1) \qquad -6(-1) = 6; \text{ combine like terms.}$$
$$= 18 - 14i$$

(B) $1(3 - 5i) = 1 \cdot 3 - 1 \cdot 5i = 3 - 5i$

(C) $i(1 + i) = i + i^2 = i - 1 = -1 + i \qquad \text{Answer in standard form.}$

(D) $(3 + 4i)(3 - 4i) = 9 - 12i + 12i - 16i^2 \qquad -16i^2 = -16(-1) = 16$
$$= 9 + 16 = 25$$

MATCHED PROBLEM 3

Carry out each operation and express the answer in standard form:

(A) $(5 + 2i)(4 - 3i)$ (B) $3(-2 + 6i)$

(C) $i(2 - 3i)$ (D) $(2 + 3i)(2 - 3i)$

For any complex number $a + bi$,

$$1(a + bi) = a + bi \quad \text{and} \quad (a + bi)1 = a + bi$$

(see Example 3, part B). This indicates that 1 is the **multiplicative identity** for complex numbers, just as it is for real numbers.

Earlier we stated that every nonzero complex number has a multiplicative inverse or reciprocal. We will denote this as a fraction, just as we do with real numbers:

$$\frac{1}{a + bi} \quad \text{is the \textbf{reciprocal} of} \quad a + bi \qquad a + bi \neq 0$$

The following important property of the conjugate of a complex number is used to express reciprocals and quotients in standard form. (See Example 3, part D)

> **› THEOREM 1 Product of a Complex Number and Its Conjugate**
>
> $$(a + bi)(a - bi) = a^2 + b^2 \qquad \text{A real number}$$

To divide complex numbers, multiply the numerator and denominator by the conjugate of the denominator.

EXAMPLE 4 Reciprocals and Quotients

Write each expression in standard form:

(A) $\dfrac{1}{2 + 3i}$ (B) $\dfrac{7 - 3i}{1 + i}$

SOLUTIONS

(A) Multiply numerator and denominator by the conjugate of the denominator:

$$\frac{1}{2 + 3i} = \frac{1}{2 + 3i} \cdot \frac{2 - 3i}{2 - 3i} = \frac{2 - 3i}{4 - 9i^2} = \frac{2 - 3i}{4 + 9}$$

$$= \frac{2 - 3i}{13} = \frac{2}{13} - \frac{3}{13}i \qquad \text{Answer in standard form.}$$

This answer can be checked by multiplication:

CHECK

$$(2 + 3i)\left(\frac{2}{13} - \frac{3}{13}i\right) = \frac{4}{13} - \frac{6}{13}i + \frac{6}{13}i - \frac{9}{13}i^2$$

$$= \frac{4}{13} + \frac{9}{13} = 1 \quad \checkmark$$

(B) $\dfrac{7 - 3i}{1 + i} = \dfrac{7 - 3i}{1 + i} \cdot \dfrac{1 - i}{1 - i} = \dfrac{7 - 7i - 3i + 3i^2}{1 - i^2}$ $3i^2 = -3$

$= \dfrac{4 - 10i}{2} = 2 - 5i$ Answer in standard form.

CHECK $(1 + i)(2 - 5i) = 2 - 5i + 2i - 5i^2 = 7 - 3i$

MATCHED PROBLEM 4 Carry out each operation and express the answer in standard form:

(A) $\dfrac{1}{4 + 2i}$ (B) $\dfrac{6 + 7i}{2 - i}$

EXAMPLE 5 **Combined Operations**

Carry out the indicated operations and write each answer in standard form:

(A) $(3 - 2i)^2 - 6(3 - 2i) + 13$ (B) $\dfrac{2 - 3i}{2i}$

SOLUTIONS (A) $(3 - 2i)^2 - 6(3 - 2i) + 13 = 9 - 12i + 4i^2 - 18 + 12i + 13$

$= 9 - 12i - 4 - 18 + 12i + 13$

$= 0$

(B) If a complex number is divided by a pure imaginary number, we can make the denominator real by multiplying numerator and denominator by i. (We could also multiply by the conjugate of $2i$, which is $-2i$.)

$$\dfrac{2 - 3i}{2i} \cdot \dfrac{i}{i} = \dfrac{2i - 3i^2}{2i^2} = \dfrac{2i + 3}{-2} = -\dfrac{3}{2} - i$$

MATCHED PROBLEM 5 Carry out the indicated operations and write each answer in standard form:

(A) $(3 + 2i)^2 - 6(3 + 2i) + 13$ (B) $\dfrac{4 - i}{3i}$

>>> EXPLORE-DISCUSS 1

Natural number powers of i take on particularly simple forms:

i $i^5 = i^4 \cdot i = (1)i = i$

$i^2 = -1$ $i^6 = i^4 \cdot i^2 = 1(-1) = -1$

$i^3 = i^2 \cdot i = (-1)i = -i$ $i^7 = i^4 \cdot i^3 = 1(-i) = -i$

$i^4 = i^2 \cdot i^2 = (-1)(-1) = 1$ $i^8 = i^4 \cdot i^4 = 1 \cdot 1 = 1$

In general, what are the possible values for i^n, n a natural number? Explain how you could easily evaluate i^n for any natural number n. Then evaluate each of the following:

(A) i^{17} **(B)** i^{24} **(C)** i^{38} **(D)** i^{47}

› Relating Complex Numbers and Radicals

Recall that we say that a is a square root of b if $a^2 = b$. If x is a positive real number, then x has two square roots, the principal square root, denoted by $\sqrt{x}$, and its negative, $-\sqrt{x}$ (Section R-2). If x is a negative real number, then x still has two square roots, but now these square roots are imaginary numbers.

> **DEFINITION 4 Principal Square Root of a Negative Real Number**

The **principal square root of a negative real number,** denoted by $\sqrt{-a}$, where a is positive, is defined by

$$\sqrt{-a} = i\sqrt{a} \qquad \text{For example } \sqrt{-3} = i\sqrt{3}; \ \sqrt{-9} = i\sqrt{9} = 3i$$

The other square root of $-a$, $a > 0$, is $-\sqrt{-a} = -i\sqrt{a}$.

Note in Definition 4 that we wrote $i\sqrt{a}$ and $i\sqrt{3}$ in place of the standard forms $\sqrt{a}i$ and $\sqrt{3}i$. We follow this convention to avoid confusion over whether the i should or should not be under the radical. (Notice that $\sqrt{3}i$ and $\sqrt{3i}$ look a lot alike, but are not the same number.)

EXAMPLE 6 Complex Numbers and Radicals

Write in standard form:

(A) $\sqrt{-4}$ (B) $4 + \sqrt{-5}$

(C) $\dfrac{-3 - \sqrt{-5}}{2}$ (D) $\dfrac{1}{1 - \sqrt{-9}}$

SOLUTIONS

(A) $\sqrt{-4} = i\sqrt{4} = 2i$

(B) $4 + \sqrt{-5} = 4 + i\sqrt{5}$

(C) $\dfrac{-3 - \sqrt{-5}}{2} = \dfrac{-3 - i\sqrt{5}}{2} = -\dfrac{3}{2} - \dfrac{\sqrt{5}}{2}i$ Answer in standard form.

(D) $\dfrac{1}{1 - \sqrt{-9}} = \dfrac{1}{1 - 3i} = \dfrac{1 \cdot (1 + 3i)}{(1 - 3i) \cdot (1 + 3i)}$

$= \dfrac{1 + 3i}{1 - 9i^2} = \dfrac{1 + 3i}{10} = \dfrac{1}{10} + \dfrac{3}{10}i$ Standard form

MATCHED PROBLEM 6

Write in standard form:

(A) $\sqrt{-16}$ (B) $5 + \sqrt{-7}$

(C) $\dfrac{-5 - \sqrt{-2}}{2}$ (D) $\dfrac{1}{3 - \sqrt{-4}}$

>>> **EXPLORE-DISCUSS 2**

From Theorem 4 in Section R-2, we know that if a and b are positive real numbers, then

$$\sqrt{a}\,\sqrt{b} = \sqrt{ab} \tag{1}$$

So we can evaluate expressions like $\sqrt{9}\,\sqrt{4}$ two ways:

$$\sqrt{9}\,\sqrt{4} = \sqrt{(9)(4)} = \sqrt{36} = 6 \quad \text{and} \quad \sqrt{9}\,\sqrt{4} = (3)(2) = 6$$

Evaluate each of the following two ways. Is equation (1) a valid property to use in all cases?

(A) $\sqrt{9}\,\sqrt{-4}$ **(B)** $\sqrt{-9}\,\sqrt{4}$ **(C)** $\sqrt{-9}\,\sqrt{-4}$

>>> **CAUTION** >>>

Note that in Example 6, part D, we wrote $1 - \sqrt{-9} = 1 - 3i$ before proceeding with the simplification. This is a necessary step because some of the properties of radicals that are true for real numbers turn out not to be true for complex numbers. In particular, for positive real numbers a and b,

$$\sqrt{a}\,\sqrt{b} = \sqrt{ab} \quad \text{but} \quad \sqrt{-a}\,\sqrt{-b} \neq \sqrt{(-a)(-b)}$$

(See Explore-Discuss 2.)

To avoid having to worry about this, always convert expressions of the form $\sqrt{-a}$ to the equivalent form in terms of i before performing any operations.

> ## Solving Equations Involving Complex Numbers

EXAMPLE **7** **Equations Involving Complex Numbers**

(A) Solve for real numbers x and y:

$$(3x + 2) + (2y - 4)i = -4 + 6i$$

(B) Solve for complex number z:

$$(3 + 2i)z - 3 + 6i = 8 - 4i$$

SOLUTIONS (A) This equation is really a statement that two complex numbers are equal: $(3x + 2) + (2y - 4)i$, and $-4 + 6i$. In order for these numbers to be equal, the real parts must be the same, and the imaginary parts must be the same as well.

$$
\begin{array}{ll}
3x + 2 = -4 \qquad \text{and} & 2y - 4 = 6 \\
3x = -6 & 2y = 10 \\
x = -2 & y = 5
\end{array}
$$

(B) Solve for z, then write the answer in standard form.

$$(3 + 2i)z - 3 + 6i = 8 - 4i \qquad \text{Add 3 and subtract } 6i \text{ from both sides.}$$

$$(3 + 2i)z = 11 - 10i \qquad \text{Divide both sides by } 3 + 2i.$$

$$z = \frac{11 - 10i}{3 + 2i} \qquad \text{Multiply numerator and denominator by the conjugate of the denominator.}$$

$$= \frac{(11 - 10i)(3 - 2i)}{(3 + 2i)(3 - 2i)} \qquad \text{Simplify.}$$

$$= \frac{13 - 52i}{13} \qquad \text{Write in standard form.}$$

$$= 1 - 4i$$

A check is left to the reader.

MATCHED PROBLEM 7

(A) Solve for real numbers x and y:

$$(2y - 7) + (3x + 4)i = 1 + i$$

(B) Solve for complex number z:

$$(1 + 3i)z + 4 - 5i = 3 + 2i$$

The truth is that the numbers we studied in this section weren't received very well when they were invented, as you can guess from the names they were given: *complex* and *imaginary*. These names are not exactly ringing endorsements.

Still, complex numbers eventually came into widespread use in areas like electrical engineering, physics, chemistry, statistics, and aeronautical engineering. Our first application of complex numbers will be in solving second-degree equations in Section 1-5.

ANSWERS TO MATCHED PROBLEMS

1. (A) Real part: 6; imaginary part: $7i$; conjugate: $6 - 7i$ (B) Real part: -3; imaginary part: $-8i$; conjugate: $-3 + 8i$ (C) Real part: 0; imaginary part: $-4i$; conjugate: $4i$ (D) Real part: -9; imaginary part: 0; conjugate: -9
2. (A) $9 - 2i$ (B) $7 - 5i$ (C) $2 - 2i$ (D) 0
3. (A) $26 - 7i$ (B) $-6 + 18i$ (C) $3 + 2i$ (D) 13
4. (A) $\frac{1}{5} - \frac{1}{10}i$ (B) $1 + 4i$ 5. (A) 0 (B) $-\frac{1}{3} - \frac{4}{3}i$
6. (A) $4i$ (B) $5 + i\sqrt{7}$ (C) $-\frac{5}{2} - (\sqrt{2}/2)i$ (D) $\frac{3}{13} + \frac{2}{13}i$
7. (A) $x = -1, y = 4$ (B) $z = 2 + i$

1-4 Exercises

1. Do negative real numbers have square roots? Explain.

2. Arrange the following sets of numbers so that each one contains the one that comes before it in the list: rational numbers, complex numbers, integers, real numbers, natural numbers.

3. Is it possible to square an imaginary number and get a real number? Explain.

4. What is the conjugate of a complex number? How do we use conjugates?

5. Which statement is false, and which is true? Justify your response.
 (A) Every real number is a complex number.
 (B) Every complex number is a real number.

6. Is it possible to add a real number and an imaginary number? If so, what kind of number is the result?

For each number in Problems 7–18, find the (A) real part, (B) imaginary part, and (C) conjugate.

7. $2 - 9i$

8. $-6i + 4$

9. $-\frac{3}{2} + \frac{5}{6}i$

10. $4.2 - 9.7i$

11. $6.5 + 2.1i$

12. $\frac{3}{5} + \frac{4}{5}i$

13. $i\pi$

14. 6π

15. 4π

16. $-2\pi i$

17. $-5 + i\sqrt{2}$

18. $4 - i\sqrt{7}$

In Problems 19–44, perform the indicated operations and write each answer in standard form.

19. $(3 + 5i) + (2 + 4i)$

20. $(4 + i) + (5 + 3i)$

21. $(8 - 3i) + (-5 + 6i)$

22. $(-1 + 2i) + (4 - 7i)$

23. $(9 + 5i) - (6 + 2i)$

24. $(3 + 7i) - (2 + 5i)$

25. $(3 - 4i) - (-5 + 6i)$

26. $(-4 - 2i) - (1 + i)$

27. $2 + (3i + 5)$

28. $(2i + 7) - 4i$

29. $(2i)(4i)$ 30. $(3i)(5i)$

31. $-2i(4 - 6i)$

32. $(-4i)(2 - 3i)$

33. $(1 + 2i)(3 - 4i)$

34. $(2 - i)(-5 + 6i)$

35. $(3 - i)(4 + i)$

36. $(5 + 2i)(4 - 3i)$

37. $(2 + 9i)(2 - 9i)$

38. $(3 + 8i)(3 - 8i)$

39. $\dfrac{1}{2 + 4i}$

40. $\dfrac{i}{3 + i}$ **41.** $\dfrac{4 + 3i}{1 + 2i}$ **42.** $\dfrac{3 - 5i}{2 - i}$

43. $\dfrac{7 + i}{2 + i}$ **44.** $\dfrac{-5 + 10i}{3 + 4i}$

In Problems 45–52, evaluate and express results in standard form.

45. $\sqrt{2}\,\sqrt{8}$ **46.** $\sqrt{3}\,\sqrt{12}$

47. $\sqrt{2}\,\sqrt{-8}$ **48.** $\sqrt{-3}\,\sqrt{12}$

49. $\sqrt{-2}\,\sqrt{8}$ **50.** $\sqrt{3}\,\sqrt{-12}$

51. $\sqrt{-2}\,\sqrt{-8}$ **52.** $\sqrt{-3}\,\sqrt{-12}$

In Problems 53–62, convert imaginary numbers to standard form, perform the indicated operations, and express answers in standard form.

53. $(2 - \sqrt{-4}) + (5 - \sqrt{-9})$

54. $(3 - \sqrt{-4}) + (-8 + \sqrt{-25})$

55. $(9 - \sqrt{-9}) - (12 - \sqrt{-25})$

56. $(-2 - \sqrt{-36}) - (4 + \sqrt{-49})$

57. $(3 - \sqrt{-4})(-2 + \sqrt{-49})$

58. $(2 - \sqrt{-1})(5 + \sqrt{-9})$

59. $\dfrac{5 - \sqrt{-4}}{7}$ **60.** $\dfrac{6 - \sqrt{-64}}{2}$

61. $\dfrac{1}{2 - \sqrt{-9}}$ **62.** $\dfrac{1}{3 - \sqrt{-16}}$

In Problems 63–68, write the complex number in standard form.

63. $-\dfrac{5}{i}$ **64.** $\dfrac{1}{10i}$

65. $(2i)^2 - 5(2i) + 6$ **66.** $(i\sqrt{3})^4 + 2(i\sqrt{3})^2 + 15$

67. $(5 + 2i)^2 - 4(5 + 2i) - 1$

68. $(7 - 3i)^2 + 8(7 - 3i) - 30$

69. Evaluate $x^2 - 2x + 2$ for $x = 1 - i$.

70. Evaluate $x^2 - 2x + 2$ for $x = 1 + i$.

In Problems 71–74, for what real values of x does each expression represent an imaginary number?

71. $\sqrt{3 - x}$ **72.** $\sqrt{5 + x}$

73. $\sqrt{2 - 3x}$ **74.** $\sqrt{3 + 2x}$

In Problems 75–78, solve for x and y.

75. $(2x - 1) + (3y + 2)i = 5 - 4i$

76. $3x + (y - 2)i = (5 - 2x) + (3y - 8)i$

77. $\dfrac{(1 + x) + (y - 2)i}{1 + i} = 2 - i$

78. $\dfrac{(2 + x) + (y + 3)i}{1 - i} = -3 + i$

In Problems 79–82, solve for z and write your answer in standard form.

79. $(10 - 2i)z + (5 + i) = 2i$

80. $(3 - 2i)z + (4i + 6) = 8i$

81. $(4 + 2i)z + (7 - 2i) = (4 - i)z + (3 + 5i)$

82. $(-2 + 3i) + (4 + 5i)z = (1 + i) - (-4 + 2i)z$

83. Show that $2 - i$ and $-2 + i$ are square roots of $3 - 4i$.

84. Show that $-3 + 2i$ and $3 - 2i$ are square roots of $5 - 12i$.

85. Explain what is wrong with the following "proof" that $-1 = 1$:
$$-1 = i^2 = \sqrt{-1}\,\sqrt{-1} = \sqrt{(-1)(-1)} = \sqrt{1} = 1$$

86. Explain what is wrong with the following "proof" that $1/i = i$. What is the correct value of $1/i$?
$$\frac{1}{i} = \frac{1}{\sqrt{-1}} = \frac{\sqrt{1}}{\sqrt{-1}} = \sqrt{\frac{1}{-1}} = \sqrt{-1} = i$$

87. Show that $i^{4k} = 1$, k a natural number

88. Show that $i^{4k+1} = i$, k a natural number

Supply the reasons in the proofs for the theorems stated in Problems 89 and 90.

89. *Theorem:* The complex numbers are commutative under addition.

Proof: Let $a + bi$ and $c + di$ be two arbitrary complex numbers; then:

Statement
1. $(a + bi) + (c + di) = (a + c) + (b + d)i$
2. $\qquad\qquad\qquad = (c + a) + (d + b)i$
3. $\qquad\qquad\qquad = (c + di) + (a + bi)$

Reason
1.
2.
3.

90. *Theorem:* The complex numbers are commutative under multiplication.

Proof: Let $a + bi$ and $c + di$ be two arbitrary complex numbers; then:

Statement
1. $(a + bi) \cdot (c + di) = (ac - bd) + (ad + bc)i$
2. $\qquad\qquad\qquad = (ca - db) + (da + cb)i$
3. $\qquad\qquad\qquad = (c + di)(a + bi)$

Reason
1.
2.
3.

Letters z and w are often used as complex variables, where z = x + yi, w = u + vi, and x, y, u, v are real numbers. The conjugates of z and w, denoted by $\bar{z}$ and $\bar{w}$, respectively, are given by $\bar{z} = x - yi$ and $\bar{w} = u - vi$. In Problems 91–98, express each property of conjugates verbally and then prove the property.

91. $z\bar{z}$ is a real number.

92. $z + \bar{z}$ is a real number.

93. $\bar{z} = z$ if and only if z is real.

94. $\bar{\bar{z}} = z$

95. $\overline{z + w} = \bar{z} + \bar{w}$

96. $\overline{z - w} = \bar{z} - \bar{w}$

97. $\overline{zw} = \bar{z} \cdot \bar{w}$

98. $\overline{z/w} = \bar{z}/\bar{w}$

1-5 Quadratic Equations and Applications

> Using Factoring to Solve Quadratic Equations
> Using the Square Root Property to Solve Quadratic Equations
> Using Completing the Square to Solve Quadratic Equations
> Using the Quadratic Formula to Solve Quadratic Equations
> Solving Applications Involving Quadratic Equations

The next class of equations we consider are the second-degree polynomial equations in one variable, called *quadratic equations*.

> ### DEFINITION 1 Quadratic Equation

A **quadratic equation** in one variable is any equation that can be written in the form

$$ax^2 + bx + c = 0 \qquad a \neq 0 \qquad \textbf{Standard Form}$$

where x is a variable and a, b, and c are constants.

Now that we have discussed the complex number system, we can use complex numbers when solving equations. Recall that a solution of an equation is also called a *root* of the equation. A real number solution of an equation is called a **real root,** and an imaginary number solution is called an **imaginary root.** In this section, we develop methods for finding all real and imaginary roots of a quadratic equation.

> ### Using Factoring to Solve Quadratic Equations

There is one single reason why factoring is so important in solving equations. It's called the **zero product property.**

>>> **EXPLORE-DISCUSS 1**

(A) Write down a pair of numbers whose product is zero. Is one of them zero? Can you think of two nonzero numbers whose product is zero?

(B) Choose any number other than zero and call it a. Write down a pair of numbers whose product is a. Is one of them a? Can you think of a pair, neither of which is a, whose product is a?

> **ZERO PRODUCT PROPERTY**

If m and n are complex numbers, then

$$m \cdot n = 0 \qquad \text{if and only if} \qquad m = 0 \text{ or } n = 0 \text{ (or both)}$$

It is very helpful to think about what this says in words: If the product of two factors is zero, then *at least one of those factors has to be zero*. It's also helpful to observe that *zero is the only number for which this is true.*

EXAMPLE 1 **Solving Quadratic Equations by Factoring**

Solve by factoring:

(A) $(x - 5)(x + 3) = 0$ (B) $6x^2 - 19x - 7 = 0$
(C) $x^2 - 6x + 5 = -4$ (D) $2x^2 = 3x$

SOLUTIONS (A) The product of two factors is zero, so by the zero product property, one of the two must be zero. This enables us to write two easier equations to solve.

$$(x - 5)(x + 3) = 0$$
$$x - 5 = 0 \qquad \text{or} \qquad x + 3 = 0$$
$$x = 5 \qquad\qquad\qquad x = -3 \qquad \text{Solution set: } \{-3, 5\}.$$

(B) $\quad 6x^2 - 19x - 7 = 0 \qquad$ Factor the left side.
$\quad (2x - 7)(3x + 1) = 0 \qquad$ Use the zero product property.
$\qquad\qquad 2x - 7 = 0 \qquad \text{or} \qquad 3x + 1 = 0$
$\qquad\qquad\qquad x = \frac{7}{2} \qquad\qquad\qquad x = -\frac{1}{3} \qquad \text{Solution set: } \{-\frac{1}{3}, \frac{7}{2}\}.$

(C) $\quad x^2 - 6x + 5 = -4 \qquad$ Add 4 to both sides.
$\quad x^2 - 6x + 9 = 0 \qquad$ Factor left side.
$\quad (x - 3)(x - 3) = 0 \qquad$ Use the zero product property.
$\qquad\qquad x - 3 = 0$
$\qquad\qquad\qquad x = 3 \qquad$ Solution set: $\{3\}$.

The equation has one root, 3. But because it came from two factors, we call 3 a *double root* or a *root of multiplicity 2.*

(D) $\qquad\quad 2x^2 = 3x \qquad$ Subtract 3x from both sides.
$\quad 2x^2 - 3x = 0 \qquad$ Factor the left side.
$\quad x(2x - 3) = 0 \qquad$ Use the zero product property.
$\qquad\quad x = 0 \qquad \text{or} \qquad 2x - 3 = 0$
$\qquad\qquad\qquad\qquad x = \frac{3}{2} \qquad$ Solution set: $\{0, \frac{3}{2}\}$

MATCHED PROBLEM 1 Solve by factoring:

(A) $(2x + 4)(x - 7) = 0$ (B) $3x^2 + 7x - 20 = 0$
(C) $4x^2 + 12x + 9 = 0$ (D) $4x^2 = 5x$

>>> **CAUTION** >>>

1. One side of an equation must be 0 before the zero product property can be applied. So

$$x^2 - 6x + 5 = -4$$
$$(x - 1)(x - 5) = -4$$

does not mean that $x - 1 = -4$ or $x - 5 = -4$. See Example 1, part C, for the correct solution of this equation.

2. The equations

$$2x^2 = 3x \qquad \text{and} \qquad 2x = 3$$

are not equivalent. The first has solution set $\{0, \frac{3}{2}\}$, but the second has solution set $\{\frac{3}{2}\}$. The root $x = 0$ is lost when each member of the first equation is divided by the variable x. See Example 1, part D, for the correct solution of this equation.

Never divide both sides of an equation by an expression containing the variable for which you are solving. You may be dividing by 0, which of course is not allowed.

> ## Using the Square Root Property to Solve Quadratic Equations

We now turn our attention to quadratic equations that do not have the first-degree term—that is, equations of the special form

$$ax^2 + c = 0 \qquad a \neq 0$$

The method of solution of this special form makes direct use of the square root property:

> **SQUARE ROOT PROPERTY**
>
> If $A^2 = C$, then $A = \pm\sqrt{C}$.

The use of the square root property is illustrated in Example 2.

EXAMPLE **2** **Using the Square Root Property**

Solve using the square root property:

(A) $9x^2 - 7 = 0$ (B) $3x^2 + 27 = 0$ (C) $(x + \frac{1}{2})^2 = \frac{5}{4}$

SOLUTIONS

(A) $9x^2 - 7 = 0$ Add 7 to both sides.

$\quad 9x^2 = 7$ Divide both sides by 9.

$\quad x^2 = \dfrac{7}{9}$ Apply the square root property; don't forget the ±!

$\quad x = \pm\sqrt{\dfrac{7}{9}} = \pm\dfrac{\sqrt{7}}{3}$ Solution set: $\left\{\dfrac{\sqrt{7}}{3}, \dfrac{-\sqrt{7}}{3}\right\}$

(B) $3x^2 + 27 = 0$ *Solve for x^2.*

$$x^2 = -9$$ *Apply the square root property.*

$$x = \pm\sqrt{-9} = \pm 3i$$ *Solution set: $\{-3i, 3i\}$*

(C) $(x + \frac{1}{2})^2 = \frac{5}{4}$ *Apply the square root property.*

$$x + \frac{1}{2} = \pm\sqrt{\frac{5}{4}}$$ *Subtract $\frac{1}{2}$ from both sides, and simplify $\sqrt{\frac{5}{4}}$.*

$$x = -\frac{1}{2} \pm \frac{\sqrt{5}}{2}$$ *Combine fractions with common denominators.*

$$= \frac{-1 \pm \sqrt{5}}{2}$$

MATCHED PROBLEM 2

Solve using the square root property:

(A) $9x^2 - 5 = 0$ (B) $2x^2 + 8 = 0$ (C) $(x + \frac{1}{3})^2 = \frac{2}{9}$

Note: It is common practice to represent solutions of quadratic equations informally by the last equation (Example 2, part C) rather than by writing a solution set using set notation (Example 2, parts A and B). From now on, we will follow this practice unless we need to make a special point.

› Using Completing the Square to Solve Quadratic Equations

The methods of square root property and factoring are generally fast when they apply; however, there are equations, such as $x^2 + 6x - 2 = 0$, that cannot be solved directly by these methods. A more general procedure must be developed to take care of this type of equation. One is called the method of completing the square.* This method is based on the process of transforming the standard quadratic equation $ax^2 + bx + c = 0$ into the form

$$(x + A)^2 = B$$

where A and B are constants. Equations of this form can easily be solved by using the square root property. But how do we transform the first equation into the second? We will need to find a way to make the left side factor as a perfect square.

›››EXPLORE-DISCUSS 2

Replace ? in each of the following with a number that makes the equation valid.

(A) $(x + 1)^2 = x^2 + 2x + ?$ **(B)** $(x + 2)^2 = x^2 + 4x + ?$

(C) $(x + 3)^2 = x^2 + 6x + ?$ **(D)** $(x + 4)^2 = x^2 + 8x + ?$

Replace ? in each of the following with a number that makes the expression a perfect square of the form $(x + h)^2$.

(E) $x^2 + 10x + ?$ **(F)** $x^2 + 12x + ?$

(G) $x^2 + bx + ?$

Given the quadratic expression

$$x^2 + bx$$

*We will find many other uses for this important method.

what number should be added to this expression to make it a perfect square? To find out, consider the square of the following expression:

$$(x + m)^2 = x^2 + \underbrace{2mx} + \underbrace{m^2} \quad\quad m^2 \text{ is the square of one-half the coefficient of } x.$$

We see that the third term on the right side of the equation is the square of one-half the coefficient of x in the second term on the right; that is, m^2 is the square of $\frac{1}{2}(2m)$. This observation leads to the following rule:

> ### › COMPLETING THE SQUARE
>
> To complete the square of a quadratic expression of the form $x^2 + bx$, add the square of one-half the coefficient of x; that is, add $(b/2)^2$, or $b^2/4$. The resulting expression factors as a perfect square,
>
> $$x^2 + bx \quad\quad\quad\quad\quad \text{For example, } x^2 + 5x$$
>
> $$x^2 + bx + \left(\frac{b}{2}\right)^2 = \left(x + \frac{b}{2}\right)^2 \quad x^2 + 5x + \left(\frac{5}{2}\right)^2 = \left(x + \frac{5}{2}\right)^2$$

EXAMPLE 3 Completing the Square

Complete the square for each of the following:

(A) $x^2 - 3x$ (B) $x^2 - bx$

SOLUTIONS

(A) $x^2 - 3x$ Add $\left(\dfrac{-3}{2}\right)^2$; that is, $\dfrac{9}{4}$ and factor.

$x^2 - 3x + \frac{9}{4} = (x - \frac{3}{2})^2$

(B) $x^2 - bx$ Add $\left(\dfrac{-b}{2}\right)^2$; that is, $\dfrac{b^2}{4}$ and factor.

$$x^2 - bx + \frac{b^2}{4} = \left(x - \frac{b}{2}\right)^2$$

MATCHED PROBLEM 3

Complete the square for each of the following:

(A) $x^2 - 5x$ (B) $x^2 + mx$

You should note that the rule for completing the square applies only if the coefficient of the second-degree term is 1. This causes little trouble, however, as you will see. To solve equations by completing the square, we will add $b^2/4$ to both sides after moving the constant term to the right side.

EXAMPLE 4 Solution by Completing the Square

Solve by completing the square:

(A) $x^2 + 6x - 2 = 0$ (B) $2x^2 - 4x + 3 = 0$

SOLUTIONS

(A) $x^2 + 6x - 2 = 0$ Add 2 to both sides to obtain the form $x^2 + bx$ on the left side.

$x^2 + 6x = 2$ Complete the square on the left side and add $\left(\frac{b}{2}\right)^2 = \left(\frac{6}{2}\right)^2 = 9$ to both sides.

$x^2 + 6x + 9 = 2 + 9$ Factor the left side; add on the right.

$(x + 3)^2 = 11$ Use the square root property. Don't forget the ±!

$x + 3 = \pm\sqrt{11}$ Add −3 to both sides.

$x = -3 \pm \sqrt{11}$

(B) $2x^2 - 4x + 3 = 0$ Make the leading coefficient 1 by dividing both sides by 2.

$x^2 - 2x + \frac{3}{2} = 0$ Subtract $-\frac{3}{2}$ from both sides.

$x^2 - 2x = -\frac{3}{2}$ Complete the square on the left side and add $\left(\frac{b}{2}\right)^2 = \left(\frac{-2}{2}\right)^2 = 1$ to both sides.

$x^2 - 2x + 1 = -\frac{3}{2} + 1$ Factor the left side; add on the right.

$(x - 1)^2 = -\frac{1}{2}$ Use the square root property.

$x - 1 = \pm\sqrt{-\frac{1}{2}}$ Add 1 to both sides and simplify $\sqrt{-\frac{1}{2}}$.

$x = 1 \pm i\sqrt{\frac{1}{2}}$

$= 1 \pm \frac{\sqrt{2}}{2}i$ Answer in $a + bi$ form.

MATCHED PROBLEM 4

Solve by completing the square:

(A) $x^2 + 8x - 3 = 0$ (B) $3x^2 - 12x + 13 = 0$

› Using the Quadratic Formula to Solve Quadratic Equations

If we solve a generic quadratic equation using the method of completing the square, the result will be a formula for solving *any* quadratic equation.

$ax^2 + bx + c = 0 \qquad a \neq 0$ Make the leading coefficient 1 by dividing by a.

$x^2 + \frac{b}{a}x + \frac{c}{a} = 0$ Subtract $\frac{c}{a}$ from both sides.

$x^2 + \frac{b}{a}x = -\frac{c}{a}$ Complete the square on the left side and add $\left(\frac{b}{2a}\right)^2 = \frac{b^2}{4a^2}$ to both sides.

$x^2 + \frac{b}{a}x + \frac{b^2}{4a^2} = \frac{b^2}{4a^2} - \frac{c}{a}$ Factor the left side and combine terms on the right side, getting a common denominator.

$\left(x + \frac{b}{2a}\right)^2 = \frac{b^2 - 4ac}{4a^2}$ Use the square root property.

$x + \frac{b}{2a} = \pm\sqrt{\frac{b^2 - 4ac}{4a^2}}$ Add $-\frac{b}{2a}$ to both sides and simplify $\sqrt{\frac{b^2 - 4ac}{4a^2}}$ (see Problem 75 in Exercises 1-5).

$x = -\frac{b}{2a} \pm \frac{\sqrt{b^2 - 4ac}}{2a}$ Combine terms on the right side.

$x = \frac{-b \pm \sqrt{b^2 - 4ac}}{2a}$

The result is known as the **quadratic formula:**

> **› THEOREM 1** Quadratic Formula
>
> If $ax^2 + bx + c = 0$, $a \neq 0$, then
>
> $$x = \frac{-b \pm \sqrt{b^2 - 4ac}}{2a}$$

The quadratic formula should be memorized and used to solve quadratic equations when other methods fail, or are more difficult to apply.

EXAMPLE 5 Using the Quadratic Formula

Solve $2x + \frac{3}{2} = x^2$ using the quadratic formula. Leave the answer in simplest radical form.

SOLUTION

$$2x + \tfrac{3}{2} = x^2 \qquad \text{Multiply both sides by 2.}$$

$$4x + 3 = 2x^2 \qquad \text{Write in standard form.}$$

$$2x^2 - 4x - 3 = 0 \qquad \text{Identify } a, b, \text{ and } c \text{ and use the quadratic}$$
$$\text{formula: } a = 2, b = -4, c = -3$$

$$x = \frac{-b \pm \sqrt{b^2 - 4ac}}{2a}$$

$$= \frac{-(-4) \pm \sqrt{(-4)^2 - 4(2)(-3)}}{2(2)}$$

$$= \frac{4 \pm \sqrt{40}}{4} = \frac{4 \pm 2\sqrt{10}}{4} = \frac{2 \pm \sqrt{10}}{2}$$

››› CAUTION ›››

1. $-4^2 \neq (-4)^2$ $-4^2 = -16$ and $(-4)^2 = 16$

2. $2 + \dfrac{\sqrt{10}}{2} \neq \dfrac{2 + \sqrt{10}}{2}$ $2 + \dfrac{\sqrt{10}}{2} = \dfrac{4 + \sqrt{10}}{2}$

3. $\dfrac{4 \pm 2\sqrt{10}}{4} \neq \pm 2\sqrt{10}$ $\dfrac{4 \pm 2\sqrt{10}}{4} = \dfrac{2(2 \pm \sqrt{10})}{4} = \dfrac{2 \pm \sqrt{10}}{2}$

MATCHED PROBLEM 5

Solve $x^2 - \frac{5}{2} = -3x$ by use of the quadratic formula. Leave the answer in simplest radical form.

The expression under the square root in the quadratic formula, $b^2 - 4ac$, is called the **discriminant.** It gives us useful information about the corresponding roots, as shown in Table 1.

Table 1 Discriminant and Roots

Discriminant $b^2 - 4ac$	Roots of $ax^2 + bx + c = 0$ a, b, and c real numbers, $a \neq 0$
Positive	Two distinct real roots
0	One real root (a double root)
Negative	Two imaginary roots, one the conjugate of the other

| EXAMPLE | 6 | **Using the Discriminant** |

Find the number of real roots of each quadratic equation.

(A) $2x^2 - 4x + 1 = 0$ (B) $2x^2 - 4x + 2 = 0$ (C) $2x^2 - 4x + 3 = 0$

SOLUTIONS

(A) $b^2 - 4ac = (-4)^2 - 4(2)(1) = 8 > 0$; two real roots

(B) $b^2 - 4ac = (-4)^2 - 4(2)(2) = 0$; one real (double) root

(C) $b^2 - 4ac = (-4)^2 - 4(2)(3) = -8 < 0$; no real roots (two imaginary roots)

MATCHED PROBLEM 6

Find the number of real roots of each quadratic equation.

(A) $3x^2 - 6x + 5 = 0$ (B) $3x^2 - 6x + 1 = 0$ (C) $3x^2 - 6x + 3 = 0$

› Solving Applications Involving Quadratic Equations

Now that we're good at solving quadratic equations, we can use them to solve many applied problems. It would be a good idea to review the problem-solving strategy on page 47 before beginning.

| EXAMPLE | 7 | **Setting Up and Solving a Word Problem** |

The sum of a number and its reciprocal is $\frac{13}{6}$. Find all such numbers.

SOLUTION Let $x =$ the number we're asked to find; then its reciprocal is $\frac{1}{x}$.

$$x + \frac{1}{x} = \frac{13}{6}$$ Multiply both sides by the LCD, 6x. [*Note: x cannot be zero.*]

$$(6x)x + (6x)\frac{1}{x} = (6x)\frac{13}{6}$$ Make sure to multiply every term by 6x.

$$6x^2 + 6 = 13x$$ Subtract 13x from both sides.

$$6x^2 - 13x + 6 = 0$$ Factor the left side.

$$(2x - 3)(3x - 2) = 0$$ Use the zero product property.

$$2x - 3 = 0 \quad \text{or} \quad 3x - 2 = 0$$ Solve each equation for x.

$$x = \tfrac{3}{2} \qquad\qquad x = \tfrac{2}{3}$$

These are two such numbers: $\frac{3}{2}$ and $\frac{2}{3}$.

CHECK $\frac{3}{2} + \frac{2}{3} = \frac{13}{6}$ $\frac{2}{3} + \frac{3}{2} = \frac{13}{6}$

MATCHED PROBLEM 7

The sum of two numbers is 23 and their product is 132. Find the two numbers. [*Hint:* If one number is x, then the other number is $23 - x$.]

EXAMPLE | **8** **A Distance–Rate–Time Problem**

A casino boat takes 1.6 hours longer to go 36 miles up a river than to return. If the rate of the current is 4 miles per hour, what is the speed of the boat in still water?

SOLUTION Let

$$x = \text{Speed of boat in still water}$$
$$x + 4 = \text{Speed downstream}$$
$$x - 4 = \text{Speed upstream}$$

$$\left(\begin{array}{c}\text{Time}\\\text{upstream}\end{array}\right) - \left(\begin{array}{c}\text{Time}\\\text{downstream}\end{array}\right) = 1.6 \qquad \text{Use Time} = \frac{\text{Distance}}{\text{Rate}}.$$

$$\frac{36}{x-4} - \frac{36}{x+4} = 1.6 \qquad \begin{array}{l}\text{Multiply both sides by}\\(x-4)(x+4), \text{ the LCD.}\end{array}$$

$$36(x+4) - 36(x-4) = 1.6(x-4)(x+4) \qquad \text{Multiply out parentheses.}$$

$$36x + 144 - 36x + 144 = 1.6x^2 - 25.6 \qquad \begin{array}{l}\text{Combine like terms and isolate}\\1.6x^2 \text{ on one side of the equation.}\end{array}$$

$$1.6x^2 = 313.6 \qquad \text{Divide both sides by 1.6.}$$

$$x^2 = 196 \qquad \text{Use the square root property.}$$

$$x = \pm\sqrt{196} = \pm 14$$

The speed in still water is 14 miles per hour. (The negative answer is thrown out, because it doesn't make sense in the problem to have a negative speed.)

CHECK
$$\text{Time upstream} = \frac{D}{R} = \frac{36}{14-4} = 3.6$$

$$\underline{- \text{ Time downstream} = \frac{D}{R} = \frac{36}{14+4} = 2}$$

$$1.6 \qquad \text{Difference of times}$$

MATCHED PROBLEM 8 Two boats travel at right angles to each other after leaving a dock at the same time. One hour later they are 25 miles apart. If one boat travels 5 miles per hour faster than the other, what is the rate of each? [*Hint:* Use the Pythagorean theorem,* remembering that distance equals rate times time.]

In Example 9, we introduce some concepts from economics that will be used throughout this book. The quantity of a product that people are willing to buy during some period

*Pythagorean theorem: In a right triangle, the square of the length of the longest side is equal to the sum of the squares of the lengths of the two shorter sides.

of time is called the **demand** for that product. The price p of a product and the demand q for that product are often related by a **price–demand equation** of the following form:

$$q = a - bp \qquad \text{\small{q is the number of items that can be sold at \$$p$ per item.}}$$

The constants a and b in a price–demand equation are usually determined by using historical data and statistical analysis.

The amount of money received from the sale of q items at $\$p$ per item is called the **revenue** and is given by

$$R = (\text{Number of items sold}) \times (\text{Price per item}) \qquad \text{\small{Using the price-demand equation}}$$
$$= qp = (a - bp)p$$

EXAMPLE 9 Price and Demand

The daily price–demand equation for whole milk in a chain of supermarkets is

$$q = 5,600 - 800p$$

where p is the price per gallon and q is the number of gallons sold per day. Find the price(s) that will produce a revenue of $9,500. Round answer(s) to two decimal places.

SOLUTION The revenue equation is

$$R = qp = (5,600 - 800p)p$$
$$= 5,600p - 800p^2$$

To get a revenue of $9,500, we substitute 9,500 for R:

$$5,600p - 800p^2 = 9,500 \qquad \text{\small{Subtract 9,500 from both sides.}}$$
$$-9,500 + 5,600p - 800p^2 = 0 \qquad \text{\small{Divide both sides by -800.}}$$
$$p^2 - 7p + 11.875 = 0 \qquad \text{\small{Use the quadratic formula with $a = 1$,}}$$
$$ \qquad \text{\small{$b = -7$, and $c = 11.875$.}}$$
$$p = \frac{7 \pm \sqrt{1.5}}{2}$$
$$= 2.89, 4.11$$

Selling whole milk for either $2.89 per gallon or $4.11 per gallon will produce a revenue of $9,500.

MATCHED PROBLEM 9 If the price–demand equation for milk is $q = 4,800 - 600p$, find the price that will produce revenues of

(A) $9,300 (B) $10,500

ANSWERS TO MATCHED PROBLEMS

1. (A) $x = -2, 7$ (B) $x = -4, \frac{5}{3}$ (C) $x = -\frac{3}{2}$ (a double root) (D) $x = 0, \frac{5}{4}$
2. (A) $x = \pm\sqrt{5}/2$ (B) $x = \pm 2i$ (C) $x = (-1 \pm \sqrt{2})/3$
3. (A) $x^2 - 5x + \frac{25}{4} = (x - \frac{5}{2})^2$ (B) $x^2 + mx + (m^2/4) = [x + (m/2)]^2$
4. (A) $x = -4 \pm \sqrt{19}$ (B) $x = (6 \pm i\sqrt{3})/3$ or $2 \pm (\sqrt{3}/3)i$
5. $x = (-3 \pm \sqrt{19})/2$ 6. (A) No real roots (two imaginary roots) (B) Two real roots
(C) One real (double) root 7. 11 and 12 8. 15 and 20 miles per hour
9. (A) $3.29 or $4.71 (B) Not possible

1-5 Exercises

Leave all answers involving radicals in simplified radical form unless otherwise stated.

1. How can you tell when an equation is quadratic?

2. What do a, b, and c in the quadratic formula stand for?

3. Explain what the zero product property says in your own words.

4. Explain what the square root property says in your own words.

5. If you could only use one of factoring, completing the square, and quadratic formula on an important test featuring a variety of quadratic equations, which would you choose, and why?

6. Does every quadratic equation have two solutions? Explain.

In Problems 7–12, solve by factoring.

7. $2x^2 = 8x$

8. $3y^2 = y + 10$

9. $-8 = 22t - 6t^2$

10. $25z^2 = -10z$

11. $3w^2 + 13w = 10$

12. $36x^2 = -12x - 1$

In Problems 13–24, solve by using the square root property.

13. $m^2 - 25 = 0$

14. $n^2 + 16 = 0$

15. $c^2 + 9 = 0$

16. $d^2 - 36 = 0$

17. $4y^2 + 9 = 0$

18. $9x^2 - 25 = 0$

19. $25z^2 - 32 = 0$

20. $16w^2 + 27 = 0$

21. $(2k - 5)^2 = 16$

22. $(t - 2)^2 = -3$

23. $(n - 3)^2 = -4$

24. $(5m - 6)^2 = 7$

In Problems 25–32, use the discriminant to determine the number of real roots of each equation and then solve each equation using the quadratic formula.

25. $x^2 - 2x - 1 = 0$

26. $y^2 - 4y + 7 = 0$

27. $x^2 - 2x + 3 = 0$

28. $y^2 - 4y + 1 = 0$

29. $2t^2 + 8 = 6t$

30. $9s^2 + 2 = 12s$

31. $2t^2 + 1 = 6t$

32. $9s^2 + 7 = 12s$

In Problems 33–40, solve by completing the square.

33. $x^2 - 4x - 1 = 0$

34. $y^2 + 4y - 3 = 0$

35. $2r^2 + 10r + 11 = 0$

36. $2s^2 - 6s + 7 = 0$

37. $4u^2 + 8u + 15 = 0$

38. $4v^2 + 16v + 23 = 0$

39. $3w^2 + 4w + 3 = 0$

40. $3z^2 - 8z + 1 = 0$

In Problems 41–56, solve by any method.

41. $12x^2 + 7x = 10$

42. $9x^2 + 9x = 4$

43. $(2y - 3)^2 = 5$

44. $(3m + 2)^2 = -4$

45. $x^2 = 3x + 1$

46. $x^2 + 2x = 2$

47. $7n^2 = -4n$

48. $8u^2 + 3u = 0$

49. $1 + \dfrac{8}{x^2} = \dfrac{4}{x}$

50. $\dfrac{2}{u} = \dfrac{3}{u^2} + 1$

51. $\dfrac{24}{10 + m} + 1 = \dfrac{24}{10 - m}$

52. $\dfrac{1.2}{y - 1} + \dfrac{1.2}{y} = 1$

53. $\dfrac{2}{x - 2} = \dfrac{4}{x - 3} - \dfrac{1}{x + 1}$

54. $\dfrac{3}{x - 1} - \dfrac{2}{x + 3} = \dfrac{4}{x - 2}$

55. $\dfrac{x + 2}{x + 3} - \dfrac{x^2}{x^2 - 9} = 1 - \dfrac{x - 1}{3 - x}$

56. $\dfrac{11}{x^2 - 4} + \dfrac{x + 3}{2 - x} = \dfrac{2x - 3}{x + 2}$

In Problems 57–60, solve for the indicated variable in terms of the other variables. Use positive square roots only.

57. $s = \frac{1}{2}gt^2$ for t

58. $a^2 + b^2 = c^2$ for a

59. $P = EI - RI^2$ for I

60. $A = P(1 + r)^2$ for r

61. Consider the quadratic equation

$$x^2 + 4x + c = 0$$

where c is a real number. Discuss the relationship between the values of c and the three types of roots listed in Table 1.

62. Consider the quadratic equation

$$x^2 - 2x + c = 0$$

where c is a real number. Discuss the relationship between the values of c and the three types of roots listed in Table 1.

Solve the equation in Problems 63–66 and leave answers in simplified radical form (i is the imaginary unit).

63. $x^2 + 3ix - 2 = 0$

64. $x^2 - 7ix - 10 = 0$

65. $x^2 + 2ix = 3$

66. $x^2 = 2ix - 3$

In Problems 67 and 68, find all solutions.

67. $x^3 - 1 = 0$

68. $x^4 - 1 = 0$

69. Prove that when the discriminant of a quadratic equation with real coefficients is negative, the equation has two imaginary solutions.

70. Prove that when the discriminant of a quadratic equation with real coefficients is zero, the equation has one real solution.

71. Can a quadratic equation with rational coefficients have one rational root and one irrational root? Explain.

72. Can a quadratic equation with real coefficients have one real root and one imaginary root? Explain.

73. Show that if r_1 and r_2 are the two roots of $ax^2 + bx + c = 0$, then $r_1 r_2 = c/a$.

74. For r_1 and r_2 in Problem 73, show that $r_1 + r_2 = -b/a$.

75. In one stage of the derivation of the quadratic formula, we replaced the expression

$$\pm\sqrt{(b^2 - 4ac)/4a^2}$$

with

$$\pm\sqrt{b^2 - 4ac}/2a$$

What justifies using $2a$ in place of $|2a|$?

76. Find the error in the following "proof" that two arbitrary numbers are equal to each other: Let a and b be arbitrary numbers such that $a \neq b$. Then

$$(a - b)^2 = a^2 - 2ab + b^2 = b^2 - 2ab + a^2$$
$$(a - b)^2 = (b - a)^2$$
$$a - b = b - a$$
$$2a = 2b$$
$$a = b$$

77. Find two numbers such that their sum is 21 and their product is 104.

78. Find all numbers with the property that when the number is added to itself the sum is the same as when the number is multiplied by itself.

79. Find two consecutive positive even integers whose product is 168.

80. The sum of a number and its reciprocal is $\frac{10}{3}$. Find the number.

APPLICATIONS

81. ALCOHOL CONSUMPTION The beer consumption by Americans for the years 1960–2005 can be modeled by the equation $y = -0.0665x^2 + 3.58x + 122$, where x is the number of years after 1960, and y is the number of ounces of beer consumed per person in that year. Find the per person consumption in 1960, then find in what year the model predicts that consumption will return to that level.

82. ALCOHOL CONSUMPTION The wine consumption by Americans for the years 1985–2005 can be modeled by the equation $y = 0.0951x^2 - 2.06x + 49.0$, where x is the number of years after 1985, and y is the number of ounces of wine consumed per person in that year. In what year does the model predict that consumption will reach the 1960 level of beer consumption (see Problem 81)?

83. CONSTRUCTION A gardener has a 30 foot by 20 foot rectangular plot of ground. She wants to build a brick walkway of uniform width on the border of the plot (see the figure). If the gardener wants to have 400 square feet of ground left for planting, how wide (to two decimal places) should she build the walkway?

84. CONSTRUCTION Refer to Problem 83. The gardener buys enough bricks to build 160 square feet of walkway. Is this sufficient to build the walkway determined in Problem 83? If not, how wide (to two decimal places) can she build the walkway with these bricks?

85. CONSTRUCTION A 1,200 square foot rectangular garden is enclosed with 150 feet of fencing. Find the dimensions of the garden to the nearest tenth of a foot.

86. CONSTRUCTION The intramural fields at a small college will cover a total area of 140,000 square feet, and the administration has budgeted for 1,600 feet of fence to enclose the rectangular field. Find the dimensions of the field.

87. PRICE AND DEMAND The daily price–demand equation for hamburgers at a fast-food restaurant is

$$q = 1,600 - 200p$$

where q is the number of hamburgers sold daily and p is the price of one hamburger (in dollars). Find the demand and the revenue when the price of a hamburger is $3.

88. PRICE AND DEMAND The weekly price–demand equation for medium pepperoni pizzas at a fast-food restaurant is

$$q = 8,000 - 400p$$

where q is the number of pizzas sold weekly and p is the price of one medium pepperoni pizza (in dollars). Find the demand and the revenue when the price is $8.

89. PRICE AND DEMAND Refer to Problem 87. Find the price p that will produce each of the following revenues. Round answers to two decimal places.
(A) $2,800 (B) $3,200 (C) $3,400

90. PRICE AND DEMAND Refer to Problem 88. Find the price p that will produce each of the following revenues. Round answers to two decimal places.
(A) $38,000 (B) $40,000 (C) $42,000

91. NAVIGATION Two planes travel at right angles to each other after leaving the same airport at the same time. One hour later they are 260 miles apart. If one travels 140 miles per hour faster than the other, what is the rate of each?

92. NAVIGATION A speedboat takes 1 hour longer to go 24 miles up a river than to return. If the boat cruises at 10 miles per hour in still water, what is the rate of the current?

93. AIR SEARCH A search plane takes off from an airport at 6:00 A.M. and travels due north at 200 miles per hour. A second plane leaves that airport at the same time and travels due east at 170 miles

per hour. The planes carry radios with a maximum range of 500 miles. When (to the nearest minute) will these planes no longer be able to communicate with each other?

94. AIR SEARCH If the second plane in Problem 93 leaves at 6:30 A.M. instead of 6 A.M., when (to the nearest minute) will the planes lose communication with each other?

95. ENGINEERING One pipe can fill a tank in 5 hours less than another. Together they can fill the tank in 5 hours. How long would it take each alone to fill the tank? Compute the answer to two decimal places.

96. ENGINEERING Two gears rotate so that one completes 1 more revolution per minute than the other. If it takes the smaller gear 1 second less than the larger gear to complete $\frac{1}{5}$ revolution, how many revolutions does each gear make in 1 minute?

97. PHYSICS—ENGINEERING For a car traveling at a speed of v miles per hour, under the best possible conditions the shortest distance d necessary to stop it (including reaction time) is given by the formula $d = 0.044v^2 + 1.1v$, where d is measured in feet. Estimate the speed of a car that requires 165 feet to stop in an emergency.

98. PHYSICS—ENGINEERING If a projectile is shot vertically into the air (from the ground) with an initial velocity of 176 feet per second, its distance y (in feet) above the ground t seconds after it is shot is given by $y = 176t - 16t^2$ (neglecting air resistance).
(A) Find the times when y is 0, and interpret the results physically.
(B) Find the times when the projectile is 16 feet off the ground. Compute answers to two decimal places.

99. ARCHITECTURE A developer wants to erect a rectangular building on a triangular-shaped piece of property that is 200 feet wide and 400 feet long (see the figure).

(A) Building codes require that industrial buildings on lots that size have a floor area of at least 15,000 square feet. Find the dimensions that will yield the smallest building that meets code. [*Hint:* Use Euclid's theorem* to find a relationship between the length and width of the building.]

Euclid's theorem: If two triangles are similar, their corresponding sides are proportional:

$$\frac{a}{a'} = \frac{b}{b'} = \frac{c}{c'}$$

(B) A potential buyer for the building needs to have a floor area of 25,000 square feet. Can the builder accommodate them?

100. ARCHITECTURE An architect is designing a small A-frame cottage for a resort area. A cross section of the cottage is an isosceles triangle with an area of 98 square feet. The front wall of the cottage must accommodate a sliding door that is 6 feet wide and 8 feet high (see the figure). Find the width and height of the cross section of the cottage. [*Recall:* The area of a triangle with base b and altitude h is $bh/2$.]

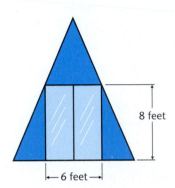

8 feet

6 feet

101. TRANSPORTATION A delivery truck leaves a warehouse and travels north to factory A. From factory A the truck travels east to factory B and then returns directly to the warehouse (see the figure). The driver recorded the truck's odometer reading at the warehouse at both the beginning and the end of the trip and also at factory B, but forgot to record it at factory A (see the table). The driver does recall that it was farther from the warehouse to factory A than it was from factory A to factory B. Since delivery charges are based on distance from the warehouse, the driver needs to know how far factory A is from the warehouse. Find this distance.

Factory A

Factory B

Warehouse

Odometer readings	
Warehouse	5 2 8 4 6
Factory *A*	5 2 ? ? ?
Factory *B*	5 2 9 3 7
Warehouse	5 3 0 0 2

102. CONSTRUCTION A $\frac{1}{4}$-mile track for racing stock cars consists of two semicircles connected by parallel straightaways (see the figure). In order to provide sufficient room for pit crews, emergency vehicles, and spectator parking, the track must enclose an area of 100,000 square feet. Find the length of the straightaways and the diameter of the semicircles to the nearest foot. [*Recall:* The area A and circumference C of a circle of diameter d are given by $A = \pi d^2/4$ and $c = \pi d$.]

100,000 square feet

1-6 Additional Equation-Solving Techniques

> Solving Equations Involving Radicals
> Revisiting Equations Involving Absolute Value
> Solving Equations of Quadratic Type

In this section, we'll study equations that are not quadratic but can be transformed into quadratic equations. We can then solve the quadratic equation, and with a little bit of interpretation, use the solutions to solve the original equation.

› Solving Equations Involving Radicals

In solving an equation involving a radical, like

$$x = \sqrt{x + 2}$$

it seems reasonable that we can remove the radical by squaring each side and then proceed to solve the resulting quadratic equation. Let's give it a try:

$$x = \sqrt{x + 2} \qquad \text{Square both sides.}$$
$$x^2 = (\sqrt{x + 2})^2 \qquad \text{Recall that } (\sqrt{a})^2 = a \text{ if } a \geq 0.$$
$$x^2 = x + 2 \qquad \text{Subtract } x + 2 \text{ from both sides.}$$
$$x^2 - x - 2 = 0 \qquad \text{Factor the left side.}$$
$$(x - 2)(x + 1) = 0 \qquad \text{Use the zero product property.}$$
$$x - 2 = 0 \quad \text{or} \quad x + 1 = 0$$
$$x = 2 \quad \text{or} \quad x = -1$$

Now we check these results in the original equation.

$$\text{Check: } x = 2 \qquad \text{Check: } x = -1$$
$$x = \sqrt{x + 2} \qquad x = \sqrt{x + 2}$$
$$2 \overset{?}{=} \sqrt{2 + 2} \qquad -1 \overset{?}{=} \sqrt{-1 + 2}$$
$$2 \overset{?}{=} \sqrt{4} \qquad -1 \overset{?}{=} \sqrt{1}$$
$$2 \overset{\checkmark}{=} 2 \qquad -1 \neq 1$$

That's interesting: 2 is a solution, but −1 is not. These results are a special case of Theorem 1.

> **THEOREM 1** Squaring Operation on Equations

If both sides of an equation are squared, then the solution set of the original equation is a subset of the solution set of the new equation.

Equation	Solution Set
$x = 3$	$\{3\}$
$x^2 = 9$	$\{-3, 3\}$

This theorem provides us with a method of solving some equations involving radicals. It is important to remember that any new equation obtained by raising both sides of an equation to the same power may have solutions, called **extraneous solutions,** that are not solutions of the original equation. Fortunately though, any solution of the original equation must be among those of the new equation.

When raising both sides of an equation to a power, checking solutions is not just a good idea—it is essential to identify any extraneous solutions.

>>> **EXPLORE-DISCUSS 1**

Squaring both sides of the equations $x = \sqrt{x}$ and $x = -\sqrt{x}$ produces the new equation $x^2 = x$. Find the solutions to the new equation and then check for extraneous solutions in each of the original equations.

EXAMPLE 1 Solving Equations Involving Radicals

Solve:

(A) $x + \sqrt{x - 4} = 4$ (B) $\sqrt{2x + 3} - \sqrt{x - 2} = 2$

SOLUTIONS (A) $x + \sqrt{x - 4} = 4$ Isolate radical on one side.

$\sqrt{x - 4} = 4 - x$ Square both sides.

$(\sqrt{x - 4})^2 = (4 - x)^2$ See the upcoming caution on squaring the right side.

$x - 4 = 16 - 8x + x^2$ Write in standard form.

$x^2 - 9x + 20 = 0$ Factor left side.

$(x - 5)(x - 4) = 0$ Use the zero product property.

$x - 5 = 0$ or $x - 4 = 0$

$x = 5$ or $x = 4$

CHECK $x = 5$ $x = 4$

$x + \sqrt{x - 4} = 4$ $x + \sqrt{x - 4} = 4$

$5 + \sqrt{5 - 4} \overset{?}{=} 4$ $4 + \sqrt{4 - 4} \overset{?}{=} 4$

$6 \neq 4$ $4 \overset{\checkmark}{=} 4$

This shows that 4 is a solution to the original equation and 5 is extraneous. The only solution is $x = 4$.

(B) To solve an equation that contains more than one radical, isolate one radical at a time and square both sides to eliminate the isolated radical. Repeat this process until all the radicals are eliminated.

$$\sqrt{2x + 3} - \sqrt{x - 2} = 2$$ Isolate one of the radicals.

$$\sqrt{2x + 3} = \sqrt{x - 2} + 2$$ Square both sides.

$$(\sqrt{2x + 3})^2 = (\sqrt{x - 2} + 2)^2$$ See the upcoming caution on squaring the right side.

$$2x + 3 = x - 2 + 4\sqrt{x - 2} + 4$$ Isolate the remaining radical.

$$x + 1 = 4\sqrt{x - 2}$$ Square both sides.

$$(x + 1)^2 = (4\sqrt{x - 2})^2$$

$$x^2 + 2x + 1 = 16(x - 2)$$ Write in standard form.

$$x^2 - 14x + 33 = 0$$ Factor left side.

$$(x - 3)(x - 11) = 0$$ Use the zero property.

$$x - 3 = 0 \quad \text{or} \quad x - 11 = 0$$

$$x = 3 \quad \text{or} \quad x = 11$$

CHECK

$$x = 3 \qquad\qquad x = 11$$

$$\sqrt{2x + 3} - \sqrt{x - 2} = 2 \qquad \sqrt{2x + 3} - \sqrt{x - 2} = 2$$

$$\sqrt{2(3) + 3} - \sqrt{3 - 2} \overset{?}{=} 2 \qquad \sqrt{2(11) + 3} - \sqrt{11 - 2} \overset{?}{=} 2$$

$$2 \overset{\checkmark}{=} 2 \qquad\qquad\qquad 2 \overset{\checkmark}{=} 2$$

Both solutions check, so there are two solutions: $x = 3, 11$.

MATCHED PROBLEM 1

Solve:

(A) $x - 5 = \sqrt{x - 3}$ (B) $\sqrt{2x + 5} + \sqrt{x + 2} = 5$

≫ CAUTION ≫

1. When squaring both sides, it is very important to isolate the radical first.
2. Be sure to square binomials like $(4 - x)$ by first writing as $(4 - x)(4 - x)$ and then multiplying. Remember: $(4 - x)^2 \neq 4^2 - x^2$.

› Revisiting Equations Involving Absolute Value

Squaring both sides of an equation can be a useful operation even if the equation does not involve any radicals. Because $|x|^2 = x^2$ for any x, squaring can be helpful in some absolute value equations.

EXAMPLE 2 **Absolute Value Equations Revisited**

Solve the following equation by squaring both sides:

$$|x + 4| = 3x - 8$$

SOLUTION

$$|x + 4| = 3x - 8$$ Square both sides.

$$|x + 4|^2 = (3x - 8)^2$$ Use $|x + 4|^2 = (x + 4)^2$ and expand each side.

$$x^2 + 8x + 16 = 9x^2 - 48x + 64$$ Write in standard form.

$$8x^2 - 56x + 48 = 0$$ Divide both sides by 8.

$$x^2 - 7x + 6 = 0$$ Factor the left side.

$$(x - 1)(x - 6) = 0$$ Use the zero product property.

$$x - 1 = 0 \quad \text{or} \quad x - 6 = 0$$

$$x = 1 \quad \text{or} \quad x = 6$$

CHECK

$$x = 1 \qquad\qquad x = 6$$

$$|x + 4| = 3x - 8 \qquad |x + 4| = 3x - 8$$

$$|1 + 4| \overset{?}{=} 3(1) - 8 \qquad |6 + 4| \overset{?}{=} 3(6) - 8$$

$$|5| \overset{?}{=} -5 \qquad\qquad |10| \overset{?}{=} 10$$

$$5 \neq -5 \qquad\qquad 10 \overset{\checkmark}{=} 10$$

The only solution is $x = 6$.

Compare this solution with the solution of Example 6, Section 1-3. Squaring both sides eliminates the need to consider two separate cases.

MATCHED PROBLEM 2

Solve the following equation by squaring both sides:

$$|3x - 4| = x + 4$$

› Solving Equations of Quadratic Type

Quadratic equations in standard form have two terms with the variable; one has power 2, the other power 1. When equations have two variable terms where the larger power is twice the smaller, we can use quadratic solving techniques.

EXAMPLE 3 Solving an Equation of Quadratic Type

Solve $x^{2/3} - x^{1/3} - 6 = 0$.

SOLUTIONS

Method I. Direct solution:

Note that the larger power (2/3) is twice the smaller. Using the properties of exponents from basic algebra, we can write $x^{2/3}$ as $(x^{1/3})^2$ and solve by factoring.

$$(x^{1/3})^2 - x^{1/3} - 6 = 0$$ Factor left side.

$$(x^{1/3} - 3)(x^{1/3} + 2) = 0$$ Use the zero product property.

$$x^{1/3} = 3 \quad \text{or} \quad x^{1/3} = -2$$ Cube both sides.

$$(x^{1/3})^3 = 3^3 \qquad (x^{1/3})^3 = (-2)^3$$

$$x = 27 \qquad\qquad x = -8$$

The solution is $x = 27, -8$.

Method II. Using substitution:

Replace $x^{1/3}$ (the smaller power) with a new variable u. Then the larger power $x^{2/3}$ is u^2. This gives us a quadratic equation with variable u.

$$u^2 - u - 6 = 0 \qquad \text{Factor.}$$
$$(u - 3)(u + 2) = 0 \qquad \text{Use the zero product property.}$$
$$u = 3, -2$$

This is not the solution! We still need to find the values of x that correspond to $u = 3$ and $u = -2$.

Replacing u with $x^{1/3}$, we obtain

$$x^{1/3} = 3 \qquad \text{or} \qquad x^{1/3} = -2 \qquad \text{Cube both sides.}$$
$$x = 27 \qquad\qquad\quad x = -8$$

The solution is $x = 27, -8$.

MATCHED PROBLEM 3

Solve algebraically using both Method I and Method II: $x^{1/2} - 5x^{1/4} + 6 = 0$.

In general, if an equation that is not quadratic can be transformed to the form

$$au^2 + bu + c = 0$$

where u is an expression in some other variable, then the equation is called an **equation of quadratic type.** Equations of quadratic type often can be solved using quadratic methods.

⟩⟩⟩ EXPLORE-DISCUSS 2

Which of the following can be transformed into a quadratic equation by making a substitution of the form $u = x^n$? What is the resulting quadratic equation?

(A) $3x^{-4} + 2x^{-2} + 7 = 0$ **(B)** $7x^5 - 3x^2 + 3 = 0$

(C) $2x^5 + 4x^2\sqrt{x} - 6 = 0$ **(D)** $8x^{-2}\sqrt{x} - 5x^{-1}\sqrt{x} - 2 = 0$

In general, if a, b, c, m, and n are nonzero real numbers, when can an equation of the form $ax^m + bx^n + c = 0$ be transformed into an equation of quadratic type?

EXAMPLE **4** **Solving an Equation of Quadratic Type**

Solve: $3x^4 - 5x^2 + 1 = 0$

SOLUTION

If we let $u = x^2$, then $u^2 = x^4$, and the equation becomes

$$3u^2 - 5u + 1 = 0 \qquad \text{Use the quadratic formula with } a = 3, b = -5, c = 1.$$

$$u = \frac{5 \pm \sqrt{13}}{6} \qquad \text{Substitute } x^2 \text{ back in for } u.$$

$$x^2 = \frac{5 \pm \sqrt{13}}{6} \qquad \text{Use the square root property to solve for } x.$$

$$x = \pm\sqrt{\frac{5 \pm \sqrt{13}}{6}} \qquad \text{There are four solutions.}$$

MATCHED PROBLEM 4

Solve: $2x^4 + 3x^2 - 4 = 0$

Many applied problems result in equations that can be solved using the techniques in this section.

EXAMPLE	5

An Application: Court Design

A hardcourt version of the game broomball becomes popular on college campuses because it enables people to hit each other with a stick. The court is a rectangle with diagonal 30 feet and area 400 square feet. Find the dimensions to one decimal place.

SOLUTION Draw a rectangle and label the dimensions as shown in Figure 1. The area is given by $A = xy$. Also, $x^2 + y^2 = 30^2$ (Pythagorean theorem), and we can solve for y to get $y = \sqrt{900 - x^2}$. Now substitute in for y in our area equation, then set area equal to 400 and solve.

> **Figure 1**

$$x\sqrt{900 - x^2} = 400 \qquad \text{Square both sides.}$$
$$x^2(900 - x^2) = 160{,}000 \qquad \text{Multiply out parentheses.}$$
$$900x^2 - x^4 = 160{,}000 \qquad \text{Write in standard quadratic form.}$$
$$(x^2)^2 - 900x^2 + 160{,}000 = 0 \qquad \text{Use quadratic formula with } a = 1,\ b = -900,\text{ and } c = 160{,}000.$$
$$x^2 = \frac{900 \pm \sqrt{(-900)^2 - 4(1)(160{,}000)}}{2} \qquad \text{Simplify inside the square root.}$$
$$= \frac{900 \pm \sqrt{170{,}000}}{2} \qquad \text{Use a calculator.}$$
$$x^2 \approx 656.2,\ 243.8 \qquad \text{Use square root property;}$$
$$x = \sqrt{656.2} \approx 25.6 \text{ or } \sqrt{243} \approx 15.6 \qquad \text{discard negative solutions.}$$

If $x = 25.6$, then $y = \sqrt{900 - 25.6^2} \approx 15.6$.
If $x = 15.6$, then $y = \sqrt{900 - 15.6^2} \approx 25.6$.

In either case, the dimensions are 25.6 feet by 15.6 feet.

CHECK Area: $25.6 \times 15.6 = 399.36 \approx 400$
Diagonal: $\sqrt{25.6^2 + 15.6^2} \approx 30$

MATCHED PROBLEM 5

If the area of a right triangle is 24 square inches and the hypotenuse is 12 inches, find the lengths of the legs of the triangle correct to one decimal place.

ANSWERS TO MATCHED PROBLEMS

1. (A) $x = 7$ (B) $x = 2$ **2.** $x = 0, 4$ **3.** $x = 16, 81$
4. $x = \dfrac{\pm\sqrt{-3 \pm \sqrt{41}}}{2}$ **5.** 11.2 inches by 4.3 inches

1-6 Exercises

1. What is meant by the term "extraneous solution"?

2. When is it necessary to check for extraneous solutions?

3. How can squaring both sides help in solving absolute value equations?

4. How can you recognize when an equation is of quadratic type?

In Problems 5–12, determine the validity of each statement. If a statement is false, explain why.

5. If $x^2 = 5$, then $x = \pm\sqrt{5}$. **6.** $\sqrt{25} = \pm 5$

7. $(x + 5)^2 = x^2 + 25$

8. $(2x - 1)^2 = 4x^2 - 1$

9. $(\sqrt{x - 1} + 1)^2 = x$

10. $(\sqrt{x - 1})^2 + 1 = x$

11. If $x^3 = 2$, then $x = 8$.

12. If $x^{1/3} = 8$, then $x = 2$.

In Problems 13–26, solve the equation.

13. $\sqrt{x + 2} = 4$

14. $\sqrt{x - 4} = 2$

15. $\sqrt{3y - 5} + 10 = 0$

16. $\sqrt{4 - x} + 5 = 0$

17. $\sqrt{3y - 2} = y - 2$

18. $\sqrt{4y + 1} = 5 - y$

19. $\sqrt{5w + 6} - w = 2$

20. $\sqrt{2w - 3} + w = 1$

21. $|2x + 1| = x + 2$

22. $|2x + 2| = 5 - x$

23. $|x - 5| = 7 - 2x$

24. $|x + 7| = 1 - 2x$

25. $|3x - 4| = 2x - 5$

26. $|3x - 1| = x - 1$

In Problems 27–32, transform each equation of quadratic type into a quadratic equation in u and state the substitution used in the transformation. If the equation is not an equation of quadratic type, say so.

27. $2x^{-6} - 4x^{-3} = 0$

28. $\dfrac{4}{7} - \dfrac{3}{x} + \dfrac{6}{x^2} = 0$

29. $3x^3 - 4x + 9 = 0$

30. $7x^{-1} + 3x^{-1/2} + 2 = 0$

31. $\dfrac{10}{9} + \dfrac{4}{x^2} - \dfrac{7}{x^4} = 0$

32. $3x^{3/2} - 5x^{1/2} + 12 = 0$

In Problems 33–56, solve the equation.

33. $\sqrt{3t - 2} = 1 - 2\sqrt{t}$

34. $\sqrt{5t + 4} - 2\sqrt{t} = 1$

35. $m^4 + 2m^2 - 15 = 0$

36. $m^4 + 4m^2 - 12 = 0$

37. $3x = \sqrt{x^2 - 2}$

38. $x = \sqrt{5x^2 + 9}$

39. $2y^{2/3} + 5y^{1/3} - 12 = 0$

40. $3y^{2/3} + 2y^{1/3} - 8 = 0$

41. $(m^2 - 2m)^2 + 2(m^2 - 2m) = 15$

42. $(m^2 + 2m)^2 - 6(m^2 + 2m) = 16$

43. $\sqrt{2t + 3} + 2 = \sqrt{t - 2}$

44. $\sqrt{2x - 1} - \sqrt{x - 5} = 3$

45. $\sqrt{w + 3} + \sqrt{2 - w} = 3$

46. $\sqrt{w + 7} = 2 + \sqrt{3 - w}$

47. $\sqrt{8 - z} = 1 + \sqrt{z + 5}$

48. $\sqrt{3z + 1} + 2 = \sqrt{z - 1}$

49. $\sqrt{4x^2 + 12x + 1} - 6x = 9$

50. $6x - \sqrt{4x^2 - 20x + 17} = 15$

51. $y^{-2} - 2y^{-1} + 3 = 0$

52. $y^{-2} - 3y^{-1} + 4 = 0$

53. $2t^{-4} - 5t^{-2} + 2 = 0$

54. $15t^{-4} - 23t^{-2} + 4 = 0$

55. $3z^{-1} - 3z^{-1/2} + 1 = 0$

56. $2z^{-1} - 3z^{-1/2} + 2 = 0$

Solve Problems 57–60 two ways: by squaring and by substitution.

57. $m - 7\sqrt{m} + 12 = 0$

58. $y - 6 + \sqrt{y} = 0$

59. $t - 11\sqrt{t} + 18 = 0$

60. $x = 15 - 2\sqrt{x}$

In Problems 61–68, solve the equation.

61. $\sqrt{7 - 2x} - \sqrt{x + 2} = \sqrt{x + 5}$

62. $\sqrt{1 + 3x} - \sqrt{2x - 1} = \sqrt{x + 2}$

63. $3 + x^{-4} = 5x^{-2}$

64. $2 + 4x^{-4} = 7x^{-2}$

65. $2\sqrt{x + 5} = 0.01x + 2.04$

66. $3\sqrt{x - 1} = 0.05x + 2.9$

67. $2x^{-2/5} - 5x^{-1/5} + 1 = 0$

68. $x^{-2/5} - 3x^{-1/5} + 1 = 0$

69. Explain why the following "solution" is incorrect:

$$\sqrt{x + 3} + 5 = 12$$
$$x + 3 + 25 = 144$$
$$x = 116$$

70. Explain why the following "solution" is incorrect.

$$\sqrt{x^2 - 16} = 2x + 3$$
$$x - 4 = 2x + 3$$
$$-7 = x$$

APPLICATIONS

71. PHYSICS—WELL DEPTH When a stone is dropped into a deep well, the number of seconds until the sound of a splash is heard is given by the formula $t = \dfrac{\sqrt{x}}{4} + \dfrac{x}{1,100}$, where x is the depth of the well in feet. For one particular well, the splash is heard 14 seconds after the stone is released. How deep (to the nearest foot) is the well?

72. PHYSICS—WELL DEPTH Refer to Problem 71. For a different well, the sound of the splash is heard 2 seconds after the stone is released. How deep (to the nearest foot) is the well?

73. GEOMETRY The diagonal of a rectangle is 10 inches and the area is 45 square inches. Find the dimensions of the rectangle, correct to one decimal place.

74. GEOMETRY The hypotenuse of a right triangle is 12 inches and the area is 24 square inches. Find the dimensions of the triangle, correct to one decimal place.

75. MANUFACTURING A lumber mill cuts rectangular beams from circular logs (see the figure). If the diameter of the log is 16 inches and the cross-sectional area of the beam is 120 square inches, find the dimensions of the cross section of the beam correct to one decimal place.

76. DESIGN A food-processing company packages an assortment of their products in circular metal tins 12 inches in diameter. Four identically sized rectangular boxes are used to divide the tin into eight compartments (see the figure). If the cross-sectional area of each box is 15 square inches, find the dimensions of the boxes correct to one decimal place.

77. CONSTRUCTION A water trough is constructed by bending a 4- by 6-foot rectangular sheet of metal down the middle and attaching triangular ends (see the figure). If the volume of the trough is 9 cubic feet, find the width correct to two decimal places.

6 feet

2 feet

78. DESIGN A paper drinking cup in the shape of a right circular cone is constructed from 125 square centimeters of paper (see the figure). If the height of the cone is 10 centimeters, find the radius correct to two decimal places.

Lateral surface area:

$$S = \pi r \sqrt{r^2 + h^2}$$

CHAPTER 1 Review

1-1 Linear Equations and Applications

Solving an equation is the process of finding all values of the variable that make the equation a true statement. An equation that is true for some values of the variable is called a **conditional equation.** An equation that is true for all permissible values of the variable is called an **identity.** An equation that is false for all permissible values of the variable is called a **contradiction,** and has no solution.

An equation that can be written in the **standard form** $ax + b = 0$, $a \neq 0$, is a **linear** or **first-degree equation.** Linear equations are solved by performing algebraic steps that result in equivalent equations until the result is an equation whose solution is obvious. When an equation has fractions, begin by multiplying both sides by the least common denominator of all the fractions. The formula

$$\text{Quantity} = \text{Rate} \times \text{Time}$$

is useful in modeling problems that involve a rate of change, like speed.

> **STRATEGY FOR SOLVING WORD PROBLEMS**

1. Read the problem slowly and carefully, more than once if necessary. Write down information as you read the problem the first time to help you get started. Identify what it is that you are asked to find.

2. Use a variable to represent an unknown quantity in the problem, usually what you are asked to find. Then try to represent any other unknown quantities in terms of that variable. It's pretty much impossible to solve a word problem without this step.

3. If it helps to visualize a situation, draw a diagram and label known and unknown parts.

4. Write an equation relating the quantities in the problem. Often, you can accomplish this by finding a formula that connects those quantities. Try to write the equation in words first, then translate to symbols.

5. Solve the equation, then answer the question in a sentence by rephrasing the question. Make sure that you're answering all of the questions asked.

6. Check to see if your answers make sense in the original problem, not just the equation you wrote.

1-2 Linear Inequalities

The **inequality symbols** $<$, $>$, $\leq$, $\geq$ are used to express **inequality relations**. **Line graphs, interval notation,** and the set operations of **union** and **intersection** are used to describe inequality relations. A **solution** of a linear inequality in one variable is a value of the variable that makes the inequality a true statement. Two inequalities are **equivalent** if they have the same **solution set.**

Linear inequalities can be solved using the same basic procedure as linear equations, with one important difference: *the direction of an inequality reverses if we multiply or divide both sides by a negative number.*

1-3 Absolute Value in Equations and Inequalities

The **absolute value** of a number x is the distance on a real number line from the origin to the point with coordinate x and is given by

$$|x| = \begin{cases} -x & \text{if } x < 0 \\ x & \text{if } x \geq 0 \end{cases}$$

The **distance between points A and B** with coordinates a and b, respectively, is $d(A, B) = |b - a|$, which has the following *geometric interpretations:*

$	x - c	= d$	Distance between x and c is equal to d.
$	x - c	< d$	Distance between x and c is less than d.
$0 <	x - c	< d$	Distance between x and c is less than d, but $x \neq c$.
$	x - c	> d$	Distance between x and c is greater than d.

Equations and inequalities involving absolute values are solved using the following relationships for $p > 0$:

1. $|x| = p$ is equivalent to $x = p$ or $x = -p$.

2. $|x| < p$ is equivalent to $-p < x < p$.

3. $|x| > p$ is equivalent to $x < -p$ or $x > p$.

These relationships also hold if x is replaced with $ax + b$. For x any real number, $\sqrt{x^2} = |x|$.

1-4 Complex Numbers

A **complex number** in **standard form** is a number in the form $a + bi$ where a and b are real numbers and i denotes a square root of -1. The number i is known as the **imaginary unit.** For a complex number $a + bi$, a is the **real part** and bi is the **imaginary part.**

If $b \neq 0$ then $a + bi$ is also called an **imaginary number.** If $a = 0$ then $0 + bi = bi$ is also called a **pure imaginary number.** If $b = 0$ then $a + 0i = a$ is a real number. The complex **zero** is $0 + 0i = 0$. The **conjugate** of $a + bi$ is $a - bi$. **Equality, addition,** and **multiplication** are defined as follows:

1. $a + bi = c + di$ if and only if $a = c$ and $b = d$

2. $(a + bi) + (c + di) = (a + c) + (b + d)i$

3. $(a + bi)(c + di) = (ac - bd) + (ad + bc)i$

Because complex numbers obey the same commutative, associative, and distributive properties as real numbers, most operations with complex numbers are performed by using these properties in the same way that algebraic operations are performed on the expression $a + bx$. Keep in mind that $i^2 = -1$.

The **property of conjugates,**

$$(a + bi)(a - bi) = a^2 + b^2$$

can be used to find **reciprocals** and **quotients.** To divide by a complex number, we multiply the numerator and denominator by the conjugate of the denominator. This enables us to write the result in $a + bi$ form. If $a > 0$, then the **principal square root of the negative real number** $-a$ is $\sqrt{-a} = i\sqrt{a}$.

To solve equations involving complex numbers, set the real and imaginary parts equal to each other and solve.

1-5 Quadratic Equations and Applications

A **quadratic equation** is an equation that can be written in the **standard form**

$$ax^2 + bx + c = 0 \qquad a \neq 0$$

where x is a variable and a, b, and c are constants. Methods of solution include:

1. Factoring and using the **zero product property:**

$$m \cdot n = 0 \qquad \text{if and only if} \qquad m = 0 \text{ or } n = 0 \text{ (or both)}$$

2. Using the **square root property:**

$$\text{If } A^2 = C, \text{ then } A = \pm\sqrt{C}$$

3. Completing the square:

$$x^2 + bx + \left(\frac{b}{2}\right)^2 = \left(x + \frac{b}{2}\right)^2$$

4. Using the **quadratic formula:**

$$x = \frac{-b \pm \sqrt{b^2 - 4ac}}{2a}$$

If the **discriminant $b^2 - 4ac$** is positive, the equation has two distinct **real roots;** if the discriminant is 0, the equation has one real double root; and if the discriminant is negative, the equation has two **imaginary roots,** each the conjugate of the other.

1-6 Additional Equation-Solving Techniques

A *square root radical* can be eliminated from an equation by isolating the radical on one side of the equation and squaring both sides of the equation. The new equation formed by squaring both sides may have **extraneous solutions.** Consequently, *every solution of the new equation must be checked in the original equation to eliminate extraneous solutions.* If an equation contains more than one radical, then the process of isolating a radical and squaring both sides can be repeated until all radicals are eliminated. If a substitution transforms an equation into the form $au^2 + bu + c = 0$, where u is an expression in some other variable, then the equation is an **equation of quadratic type** that can be solved by quadratic methods.

CHAPTER 1 Review Exercises

Work through all the problems in this chapter review and check answers in the back of the book. Answers to all review problems are there, and following each answer is a number in italics indicating the section in which that type of problem is discussed. Where weaknesses show up, review appropriate sections in the text.

In Problems 1–3, solve the equation.

1. $8x + 10 = 4x - 30$

2. $4 - 3(x + 2) = 5x - 7(4 - x)$

3. $\dfrac{y + 10}{15} - \dfrac{1}{5} = \dfrac{y + 1}{6} - \dfrac{1}{10}$

Solve and graph the inequality in Problems 4–6.

4. $3(2 - x) - 2 \le 2x - 1$ **5.** $|y + 9| < 5$

6. $|3 - 2x| \le 5$

7. Find the real part, the imaginary part, and the conjugate:
(A) $9 - 4i$ (B) $5i$ (C) -10

8. Perform the indicated operations and write the answer in standard form.
(A) $(4 + 7i) + (-2 - 3i)$
(B) $(-3 + 5i) - (4 - 8i)$
(C) $(1 - 2i)(3 + 4i)$
(D) $\dfrac{21 + 9i}{5 - 2i}$

Solve the equation in Problems 9–15.

9. $2x^2 - 7 = 0$ **10.** $5x^2 + 20 = 0$

11. $2x^2 = 4x$ **12.** $2x^2 = 7x - 3$

13. $m^2 + m + 1 = 0$ **14.** $y^2 = \frac{3}{2}(y + 1)$

15. $\sqrt{5x - 6} - x = 0$

16. For what values of x does the expression $\sqrt{15 + 6x}$ represent a real number?

Solve the equation in Problems 17 and 18.

17. $\dfrac{7}{2 - x} = \dfrac{10 - 4x}{x^2 + 3x - 10}$ **18.** $\dfrac{u - 3}{2u - 2} = \dfrac{1}{6} - \dfrac{1 - u}{3u - 3}$

Solve and graph the inequality in Problems 19–21.

19. $\dfrac{x + 3}{8} \le 5 - \dfrac{2 - x}{3}$ **20.** $|3x - 8| > 2$

21. $\sqrt{(1 - 2m)^2} \le 3$

22. If points A, B, and C have coordinates on a number line of 5, 20, and -8 respectively, find
(A) $d(A, B)$ (B) $d(A, C)$ (C) $d(B, C)$

23. Perform the indicated operations and write the final answers in standard form:
(A) $(3 + i)^2 - 2(3 + i) + 3$ (B) i^{27}

24. Convert to $a + bi$ forms, perform the indicated operations, and write the final answers in standard form:
(A) $(2 - \sqrt{-4}) - (3 - \sqrt{-9})$
(B) $\dfrac{2 - \sqrt{-1}}{3 + \sqrt{-4}}$ (C) $\dfrac{4 + \sqrt{-25}}{\sqrt{-4}}$

Solve the equation in Problems 25–30.

25. $\left(y + \dfrac{11}{3}\right)^2 = 20$

26. $1 + \dfrac{3}{u^2} = \dfrac{2}{u}$

27. $\dfrac{x}{x^2 - x - 6} - \dfrac{2}{x - 3} = 3$ **28.** $2x^{2/3} - 5x^{1/3} - 12 = 0$

29. $m^4 + 5m^2 - 36 = 0$ **30.** $\sqrt{y - 2} - \sqrt{5y + 1} = -3$

Solve the equation or inequality in Problems 31–35, and round answers to three significant digits if necessary.

31. $2.15x - 3.73(x - 0.930) = 6.11x$

32. $-1.52 \le 0.770 - 2.04x \le 5.33$

33. $|9.71 - 3.62x| > 5.48$

34. $\left|\dfrac{8}{3} - \dfrac{4}{5}t\right| \le \dfrac{1}{2}$

35. $6.09x^2 + 4.57x - 8.86 = 0$

Solve the equation in Problems 36–38 for the indicated variable in terms of the other variables.

36. $P = M - Mdt$ for M (mathematics of finance)

37. $P = EI - RI^2$ for I (electrical engineering)

38. $x = \dfrac{4y + 5}{2y + 1}$ for y

39. Find the error in the following "solution" and then find the correct solution.

$$\dfrac{4}{x^2 - 4x + 3} = \dfrac{3}{x^2 - 3x + 2}$$
$$4x^2 - 12x + 8 = 3x^2 - 12x + 9$$
$$x^2 = 1$$
$$x = -1 \quad \text{or} \quad x = 1$$

40. Consider the quadratic equation $x^2 - 8x + c = 0$, where c is a real number. Describe the number and type of solutions for $c = -16$, 16, and 32. Use your result to make a general statement about the number and type of solutions for certain values of c, then use an inequality to prove your statement.

41. For what values of a and b is the inequality $a + b < b - a$ true?

42. If a and b are negative numbers and $a > b$, then is a/b greater than 1 or less than 1?

43. Solve for x in terms of y: $y = \dfrac{1}{1 - \dfrac{1}{1 - x}}$

44. Solve and graph: $0 < |x - 6| < d$

Solve the equation in Problems 45–47.

45. $2x^2 = \sqrt{3}x - \frac{1}{2}$

46. $4 = 8x^{-2} - x^{-4}$

47. $2ix^2 + 3ix - 5i = 0$

48. Evaluate: $(a + bi)\left(\dfrac{a}{a^2 + b^2} - \dfrac{b}{a^2 + b^2}i\right)$, $a, b \neq 0$

APPLICATIONS

49. NUMBERS Find a number such that subtracting its reciprocal from the number gives $\frac{16}{15}$.

50. SPORTS MEDICINE The following quotation was found in a sports medicine handout: "The idea is to raise and sustain your heart rate to 70% of its maximum safe rate for your age. One way to determine this is to subtract your age from 220 and multiply by 0.7."
(A) If H is the maximum safe sustained heart rate (in beats per minute) for a person of age A (in years), write a formula relating H and A.
(B) What is the maximum safe sustained heart rate for a 20-year-old?
(C) If the maximum safe sustained heart rate for a person is 126 beats per minute, how old is the person?

51. CHEMISTRY A chemical storeroom has an 80% alcohol solution and a 30% alcohol solution. How many milliliters of each should be used to obtain 50 milliliters of a 60% solution?

52. RATE–TIME A student group flies to Cancun for spring break, a distance of 1,200 miles. The plane used for both trips has an average cruising speed of 300 miles per hour in still air. The trip down is with the prevailing winds and takes $1\frac{1}{2}$ hours less than the trip back, against the same strength wind. What is the wind speed?

53. RATE–TIME A crew of four practices by rowing up a river for a fixed distance and then returning to their starting point. The river has a current of 3 km/h.
(A) Currently the crew can row 15 km/h in still water. If it takes them 25 minutes to make the round-trip, how far upstream did they row?
(B) After some additional practice the crew cuts the round-trip time to 23 minutes. What is their still-water speed now? Round answers to one decimal place.

(C) If the crew wants to increase their still-water speed to 18 km/h, how fast must they make the round-trip? Express answer in minutes rounded to one decimal place.

54. COST ANALYSIS Cost equations for manufacturing companies are often quadratic in nature. If the cost equation for manufacturing inexpensive calculators is $C = x^2 - 10x + 31$, where C is the cost of manufacturing x units per week (both in thousands), find:
(A) The output for a $15 thousand weekly cost
(B) The output for a $6 thousand weekly cost

55. BREAK-EVEN ANALYSIS The manufacturing company in Problem 54 sells its calculators to wholesalers for $3 each. So its revenue equation is $R = 3x$, where R is revenue and x is the number of units sold per week (both in thousands). Find the break-even point(s) for the company—that is, the output at which revenue equals cost.

56. POLITICS Before the 2008 presidential election, one news outlet estimated that the percentage of voters casting a vote for Barack Obama would be within 1.2% of 54%. Express this range as an absolute value inequality, then solve the inequality.

57. DESIGN The pages of a textbook have uniform margins of 2 centimeters on all four sides (see the figure). If the area of the entire page is 480 square centimeters and the area of the printed portion is 320 square centimeters, find the dimensions of the page.

Figure for 57.

58. DESIGN A landscape designer uses 8-foot timbers to form a pattern of isosceles triangles along the wall of a building (see the figure). If the area of each triangle is 24 square feet, find the base correct to two decimal places.

8 feet

CHAPTER 1

››› GROUP ACTIVITY Solving a Cubic Equation

If a, b, and c are real numbers with $a \neq 0$, then the quadratic equation $ax^2 + bx + c = 0$ can be solved by a variety of methods, including the quadratic formula. How can we solve the cubic equation

$$ax^3 + bx^2 + cx + d = 0, \qquad a \neq 0 \qquad (1)$$

and is there a formula for the roots of this equation?

The first published solution of equation (1) is generally attributed to the Italian mathematician Girolamo Cardano (1501–1576) in 1545. His work led to a complicated formula for the roots of equation (1) that involves topics that are discussed later in this text. For now, we will use Cardano's method to find a real solution in special cases of equation (1). Note that because a is nonzero, we can always multiply both sides of (1) by $1/a$ to make the coefficient of x^3 equal to 1.

CARDANO'S METHOD FOR SOLVING A CUBIC EQUATION

Let $x^3 + bx^2 + cx + d = 0$

Example problem: $x^3 - 6x^2 + 6x - 5 = 0$. Steps will be in red.

Step 1. Substitute $x = y - b/3$ to obtain the *reduced cubic* $y^3 + my = n$.

$x = y - \dfrac{-6}{3}$ or $x = y + 2$. The equation becomes

$$(y + 2)^3 - 6(y + 2) + 6(y + 2) - 5 = 0,$$

which simplifies to $y^3 - 6y = 9$: $m = -6$, $n = 9$.

Step 2. Define u and v by $m = 3uv$ and $n = u^3 - v^3$. Use $v = \frac{m}{3u}$ to write

$$n = u^3 - \left(\frac{m}{3u}\right)^3$$

Multiply both sides by u^3 to obtain an equation quadratic in u^3. Solve for u^3 by factoring or by using the quadratic formula. Then solve for u, and find the associated value of v.

$v = \dfrac{-6}{3u}$ or $v = -\dfrac{2}{u}$; $9 = u^3 - \left(-\dfrac{2}{u}\right)^3 = u^3 + \dfrac{8}{u^3}$ Multiply both sides by u^3 to obtain $u^6 - 9u^3 + 8 = 0$; solve by factoring to get $u = 2$ (in which case $v = -1$) or $u = 1$ (in which case $v = -2$).

Step 3. Using either of the solutions found in step 2,

$$x = y - \frac{b}{3} = u - v - \frac{b}{3}$$

is a solution to $x^3 + bx^2 + cx + d = 0$

For $u = 2$, $v = -1$, $x = 2 - (-1) - \dfrac{-6}{3} = 5$ (Solution)

(A) The key to Cardano's method is to recognize that if u and v are defined as in step 2, then $y = u - v$ is a solution of the reduced cubic. Verify this by substituting $y = u - v$, $m = 3uv$, and $n = u^3 - v^3$ in $y^3 + my = n$ and show that the result is an identity.

(B) Use Cardano's method to solve

$$x^3 - 6x^2 - 3x - 8 = 0 \qquad (2)$$

Use a calculator to find a decimal approximation of your solution and check your answer by substituting this approximate value in equation (2).

(C) Use Cardano's method to solve

$$x^3 - 6x^2 + 9x - 6 = 0 \qquad (3)$$

Use a calculator to find a decimal approximation of your solution and check your answer by substituting this approximate value in equation (3).

(D) In step 2 of Cardano's method, show that u^3 is real if $\left(\dfrac{n}{2}\right)^2 \geq \left(\dfrac{-m}{3}\right)^3$.

Graphs

EQUATIONS and inequalities are algebraic objects. A graph, on the other hand, is a geometric object such as a line, circle, or parabola. The idea of visualizing an equation or inequality by means of a graph was crucial to the development of analytic geometry, a subject that combines algebra and geometry. In this chapter, we study the fundamentals of analytic geometry: The Cartesian coordinate system, named after the French mathematician and philosopher René Descartes (1596–1650); the calculation of distances in the plane; and equations of lines and circles. We conclude the chapter by applying linear models to solve real-world problems.

2-1 Cartesian Coordinate Systems

› Reviewing Cartesian Coordinate Systems
› Graphing: Point by Point
› Using Symmetry as an Aid in Graphing

In Chapter 1, we discussed algebraic methods for solving equations. In this section we show how to find a geometric representation (*graph*) of an equation. Examining the graph of an equation often results in additional insight into the nature of the equation's solutions.

› Reviewing Cartesian Coordinate Systems

› **Figure 1** Cartesian coordinate system.

Just as a real number line is formed by establishing a one-to-one correspondence between the points on a line and the elements in the set of real numbers, we can form a *real plane* by establishing a one-to-one correspondence between the points in a plane and elements in the set of all ordered pairs of real numbers. This can be done by means of a Cartesian coordinate system.

To form a **Cartesian** or **rectangular coordinate system,** we select two real number lines, one horizontal and one vertical, and let them cross through their origins, as indicated in Figure 1. Up and to the right are the usual choices for the positive directions. These two number lines are called the **horizontal axis** and the **vertical axis,** or, together, the **coordinate axes.** The horizontal axis is usually referred to as the *x* **axis** and the vertical axis as the *y* **axis,** and each is labeled accordingly. Other labels may be used in certain situations. The coordinate axes divide the plane into four parts called **quadrants,** which are numbered counterclockwise from I to IV (see Fig. 1).

› **Figure 2** Coordinates in a plane.

Given an arbitrary point *P* in the plane, pass horizontal and vertical lines through the point (Fig. 2). The vertical line will intersect the horizontal axis at a point with coordinate *a*, and the horizontal line will intersect the vertical axis at a point with coordinate *b*. These two numbers written as the ordered pair* (*a*, *b*) form the **coordinates** of the point *P*. The first coordinate *a* is called the **abscissa** of *P*; the second coordinate *b* is called the **ordinate** of *P*. The abscissa of *Q* in Figure 2 is −10, and the ordinate of *Q* is 5. The coordinates of a point can also be referenced in terms of the axis labels. The *x* **coordinate** of *R* in Figure 2 is 5, and the *y* **coordinate** of *R* is 10. The point with coordinates (0, 0) is called the **origin.**

The procedure we have just described assigns to each point *P* in the plane a unique pair of real numbers (*a*, *b*). Conversely, if we are given an ordered pair of real numbers (*a*, *b*), then, reversing this procedure, we can determine a unique point *P* in the plane.

There is a one-to-one correspondence between the points in a plane and the elements in the set of all ordered pairs of real numbers.

This correspondence is often referred to as the *fundamental theorem of analytic geometry*. Because of this correspondence, we regularly speak of the point (*a*, *b*) when we are referring to the point with coordinates (*a*, *b*). We also write *P* = (*a*, *b*) to identify the coordinates of the point *P*. In Figure 2, referring to *Q* as the point (−10, 5) and writing *R* = (5, 10) are both acceptable statements.

*An **ordered pair** of real numbers is a pair of numbers in which the order is specified. We now use (*a*, *b*) as the coordinates of a point in a plane. In Chapter 1, we used (*a*, *b*) to represent an interval on a real number line. These concepts are not the same. You must always interpret the symbol (*a*, *b*) in terms of the context in which it is used.

> ## Graphing: Point by Point

Given any set of ordered pairs of real numbers S, the **graph** of S is the set of points in the plane corresponding to the ordered pairs in S. The fundamental theorem of analytic geometry enables us to look at an algebraic object (a set of ordered pairs) geometrically and to look at a geometric object (a graph) algebraically. We begin by considering an equation in two variables:

$$y = x^2 - 4 \qquad\qquad (1)$$

A **solution** to equation (1) is an ordered pair of real numbers (a, b) such that $b = a^2 - 4$. The **solution set** of equation (1) is the set of all its solutions.

To find a solution to equation (1) we simply replace one of the variables with a number and solve for the other variable. For example, if $x = 2$, then $y = 2^2 - 4 = 0$, and the ordered pair $(2, 0)$ is a solution. Similarly, if $y = 5$, then $5 = x^2 - 4$, $x^2 = 9$, $x = \pm 3$, and the ordered pairs $(3, 5)$ and $(-3, 5)$ are solutions.

Sometimes replacing one variable with a number and solving for the other variable will introduce imaginary numbers. For example, if $y = -5$ in equation (1), then

$$-5 = x^2 - 4$$
$$x^2 = -1$$
$$x = \pm\sqrt{-1} = \pm i$$

So $(-i, -5)$ and $(i, -5)$ are solutions to $y = x^2 - 4$. However, the coordinates of a point in a rectangular coordinate system must be real numbers.

For that reason, when graphing an equation, we consider only those values of the variables that produce real solutions to the equation.

The **graph of an equation in two variables** is the graph of its solution set. In equation (1), we find that its solution set will have infinitely many elements and its graph will extend off any paper we might choose, no matter how large. *To sketch the graph of an equation,* we include enough points from its solution set so that the total graph is apparent. This process is called **point-by-point plotting.**

EXAMPLE	1

Graphing an Equation Using Point-by-Point Plotting

Sketch a graph of $y = x^2 - 4$.

SOLUTION We make a table of solutions—ordered pairs of real numbers that satisfy the given equation.

x	-4	-3	-2	-1	0	1	2	3	4
y	12	5	0	-3	-4	-3	0	5	12

After plotting these solutions, if there are any portions of the graph that are unclear, we plot additional points until the shape of the graph is apparent. Then we join all these plotted points with a smooth curve, as shown in Figure 3. Arrowheads are used to indicate that the graph continues beyond the portion shown here with no significant changes in shape.

The resulting figure is called a *parabola*. Notice that if we fold the paper along the y axis, the right side will match the left side. We say that the graph is *symmetric with respect to the y axis* and call the y axis the *axis of the parabola*. More will be said about parabolas later in the text.

> Figure 3

MATCHED PROBLEM 1 Sketch a graph of $y^2 = x$.

This book contains a number of activities that use a graphing calculator or computer with appropriate software. All of these activities are clearly marked and easily omitted if no such device is available.

Technology Connections

To graph the equation in Example 1 on a graphing calculator, we first enter the equation in the calculator's **equation editor*** [Fig. 4(a)]. Using Figure 3 as a guide, we next enter values for the **window variables** [Fig. 4(b)], and then we **graph** the equation [Fig. 4(c)]. The values of the window variables, shown in red in Figure 4(c), are not displayed on the calculator screen. We add them to give you additional insight into the graph.

Compare the graphs in Figure 3 and Figure 4(c). They are similar in shape, but they are not identical. The discrepancy is due to the difference in the axes scales. In Figure 3, one unit on the x axis is equal to one unit on the y axes. In Figure 4(c), one unit on the x axis is equal to about three units on the y axis. We will have more to say about axes scales later in this section.

Enter the equation.
(a)

Enter the window variables.
(b)

Graph the equation.
(c)

> **Figure 4**

*See the Technology Index for a list of graphing calculator terms used in this book.

⟩⟩⟩ EXPLORE-DISCUSS 1

To graph the equation $y = -x^3 + 2x$, we use point-by-point plotting to obtain the graph in Figure 5.

(A) Do you think this is the correct graph of the equation? If so, why? If not, why?

(B) Add points on the graph for $x = -2$, -0.5, 0.5, and 2.

(C) Now, what do you think the graph looks like? Sketch your version of the graph, adding more points as necessary.

(D) Write a short statement explaining any conclusions you might draw from parts A, B, and C.

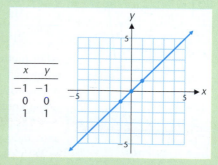

x	y
-1	-1
0	0
1	1

> **Figure 5**

Graphs illustrate the relationship between two quantities, one represented by x coordinates and the other by y coordinates. If no equation for the graph is available, we can find specific examples of this relationship by estimating coordinates of points on the graph. Example 2 illustrates this process.

EXAMPLE	2	### Ozone Levels

The ozone level during a 12-hour period in a suburb of Milwaukee, Wisconsin, on a particular summer day is given in Figure 6, where L is ozone in parts per billion and t is time in hours. Use this graph to estimate the following ozone levels to the nearest integer and times to the nearest quarter hour.

(A) The ozone level at 6 P.M.

(B) The highest ozone level and the time when it occurs.

(C) The time(s) when the ozone level is 90 ppb.

› **Figure 6** Ozone level.

SOLUTIONS

(A) The L coordinate of the point on the graph with t coordinate 6 is approximately 97 ppb.

(B) The highest ozone level is approximately 109 ppb at 3 P.M.

(C) The ozone level is 90 ppb at about 12:30 P.M. and again at 10 P.M.

MATCHED PROBLEM 2	Use Figure 6 to estimate the following ozone levels to the nearest integer and times to the nearest quarter hour.

(A) The ozone level at 7 P.M.

(B) The time(s) when the ozone level is 100 ppb.

› Using Symmetry as an Aid in Graphing

We noticed that the graph of $y = x^2 - 4$ in Example 1 is *symmetric with respect to the y axis;* that is, the two parts of the graph coincide if the paper is folded along the y axis. Similarly, we say that a graph is symmetric with respect to the x axis if the parts above and

below the x axis coincide when the paper is folded along the x axis. To make the intuitive idea of folding a graph along a line more concrete, we introduce two related concepts—reflection and symmetry.

> › **DEFINITION 1** Reflection
>
> 1. The **reflection through the y axis** of the point (a, b) is the point $(-a, b)$.
> 2. The **reflection through the x axis** of the point (a, b) is the point $(a, -b)$.
> 3. The **reflection through the origin** of the point (a, b) is the point $(-a, -b)$.
> 4. To **reflect a graph** just reflect each point on the graph.

EXAMPLE 3 **Reflections**

In a Cartesian coordinate system, plot the point $P = (4, -2)$ along with its reflection through (A) the y axis, (B) the x axis, (C) and the origin.

SOLUTION

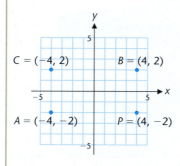

MATCHED PROBLEM 3

In a Cartesian coordinate system, plot the point $P = (-3, 5)$ along with its reflection through (A) the x axis, (B) the y axis, and (C) the origin.

> › **DEFINITION 2** Symmetry
>
> A graph is **symmetric with respect to**
>
> 1. **The x axis** if $(a, -b)$ is on the graph whenever (a, b) is on the graph—reflecting the graph through the x axis does not change the graph.
> 2. **The y axis** if $(-a, b)$ is on the graph whenever (a, b) is on the graph—reflecting the graph through the y axis does not change the graph.
> 3. **The origin** if $(-a, -b)$ is on the graph whenever (a, b) is on the graph—reflecting the graph through the origin does not change the graph.

Figure 7 illustrates these three types of symmetry.

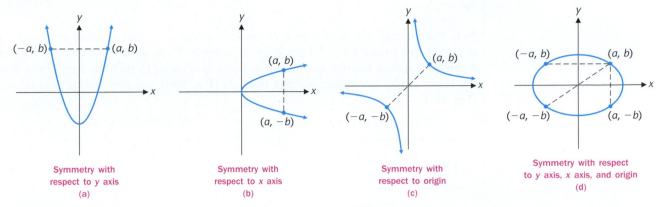

Symmetry with
respect to *y* axis
(a)

Symmetry with
respect to *x* axis
(b)

Symmetry with
respect to origin
(c)

Symmetry with respect
to *y* axis, *x* axis, and origin
(d)

› **Figure 7** Symmetry.

>>> **EXPLORE-DISCUSS 2** If a graph possesses two of the three types of symmetry in Definition 1, must it also possess the third? Explain.

Given an equation, if we can determine the symmetry properties of its graph ahead of time, we can save a lot of time and energy in sketching the graph. For example, we know that the graph of $y = x^2 - 4$ in Example 1 is symmetric with respect to the *y* axis, so we can carefully sketch only the right side of the graph; then reflect the result through the *y* axis to obtain the whole sketch—the point-by-point plotting is cut in half!

The tests for symmetry are given in Theorem 1. These tests are easily applied and are very helpful aids to graphing. Recall, two equations are equivalent if they have the same solution set.

› **THEOREM 1 Tests for Symmetry**

Symmetry with respect to the:	An equivalent equation results if:
y axis	*x* is replaced with $-x$
x axis	*y* is replaced with $-y$
Origin	*x* and *y* are replaced with $-x$ and $-y$

EXAMPLE 4 Using Symmetry as an Aid to Graphing

Test the equation $y = x^3$ for symmetry and sketch its graph.

SOLUTION

Test *y* Axis
Replace *x* with $-x$:

$$y = (-x)^3$$
$$y = -x^3$$

Test *x* Axis
Replace *y* with $-y$:

$$-y = x^3$$
$$y = -x^3$$

Test Origin
Replace *x* with $-x$
and *y* with $-y$:

$$-y = (-x)^3$$
$$-y = -x^3$$
$$y = x^3$$

The only test that produces an equivalent equation is replacing x with $-x$ and y with $-y$. So the only symmetry property for the graph of $y = x^3$ is symmetry with respect to the origin.

Note that positive values of x produce positive values for y and negative values of x produce negative values for y. So the graph is in the first and third quadrants. First we make a careful sketch in the first quadrant [Fig. 8(a)]. It is easier to perform a reflection through the origin if you first reflect through one axis [Fig. 8(b)] and then through the other axis [Fig. 8(c)].

x	0	1	2
y	0	1	8

(a)

(b)

(c)

› **Figure 8**

MATCHED PROBLEM 4

Test the equation $y = x$ for symmetry and sketch its graph.

EXAMPLE 5

Using Symmetry as an Aid to Graphing

Test the equation $y = |x|$ for symmetry and sketch its graph.

SOLUTION

Test y Axis	**Test x Axis**	**Test Origin**						
Replace x with $-x$:	Replace y with $-y$:	Replace x with $-x$ and y with $-y$:						
$y =	-x	$	$-y =	x	$	$-y =	-x	$
$y =	x	$	$y = -	x	$	$-y =	x	$
		$y = -	x	$				

The only symmetry property for the graph of $y = |x|$ is symmetry with respect to the y axis.

Since $|x|$ is never negative, this graph is in the first and second quadrants. We make a careful sketch in the first quadrant; then reflect this graph through the y axis to obtain the complete sketch shown in Figure 9.

x	0	2	4
y	0	2	4

> Figure 9

MATCHED PROBLEM 5

Test the equation $y = -|x|$ for symmetry and sketch its graph.

EXAMPLE **6** **Using Symmetry as an Aid to Graphing**

Test the equation $y^2 - x^2 = 4$ for symmetry and sketch its graph.

SOLUTION Since $(-x)^2 = x^2$ and $(-y)^2 = y^2$, the equation $y^2 - x^2 = 4$ will be unchanged if x is replaced with $-x$ or if y is replaced with $-y$. So the graph is symmetric with respect to the y axis, the x axis, and the origin. We need to make a careful sketch in only the first quadrant, reflect this graph through the y axis, and then reflect everything through the x axis. To find quadrant I solutions, we solve the equation for either y in terms of x or x in terms of y. We choose to solve for y.

$$y^2 - x^2 = 4$$
$$y^2 = x^2 + 4$$
$$y = \pm\sqrt{x^2 + 4}$$

To obtain the quadrant I portion of the graph, we sketch $y = \sqrt{x^2 + 4}$ for $x = 0, 1, 2, \ldots$. The final graph is shown in Figure 10.

x	0	1	2	3	4
y	2	$\sqrt{5} \approx 2.2$	$\sqrt{8} \approx 2.8$	$\sqrt{13} \approx 3.6$	$\sqrt{20} \approx 4.5$

> Figure 10

MATCHED PROBLEM 6

Test the equation $2y^2 - x^2 = 2$ for symmetry and sketch its graph.

Technology Connections

To graph $y^2 - x^2 = 4$ on a graphing calculator, we enter both $\sqrt{x^2 + 4}$ and $-\sqrt{x^2 + 4}$ in the equation editor [Fig. 11(a)] and graph.

(a)

(b)

> **Figure 11**

ANSWERS TO MATCHED PROBLEMS

1.

2. (A) 96 ppb (B) 1:45 P.M. and 5 P.M.

3.

4. Symmetric with respect to the origin

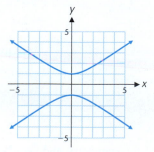

5. Symmetric with respect to the y axis

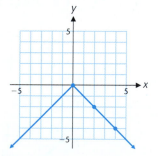

6. Symmetric with respect to the x axis, the y axis, and the origin

2-1 Exercises

1. Describe the one-to-one correspondence between points in the plane and ordered pairs of real numbers.

2. Explain how to graph an equation in two variables using point-by-point plotting.

3. Explain how to sketch the reflection of a graph through the y axis.

4. How can you tell whether the graph of an equation is symmetric with respect to the origin?

In Problems 5–14, give a verbal description of the indicated subset of the plane in terms of quadrants and axes.

5. $\{(x, y) \mid x = 0\}$

6. $\{(x, y) \mid x > 0, y > 0\}$

7. $\{(x, y) \mid x < 0, y < 0\}$

8. $\{(x, y) \mid y = 0\}$

9. $\{(x, y) \mid x > 0, y < 0\}$

10. $\{(x, y) \mid y < 0, x \neq 0\}$

11. $\{(x, y) \mid x > 0, y \neq 0\}$

12. $\{(x, y) \mid x < 0, y > 0\}$

13. $\{(x, y) \mid xy < 0\}$

14. $\{(x, y) \mid xy > 0\}$

[*Hint:* In Problems 13 and 14, consider two cases.]

In Problems 15–18, plot the given points in a rectangular coordinate system.

15. $(5, 0), (3, -2), (-4, 2), (4, 4)$

16. $(0, 4), (-3, 2), (5, -1), (-2, -4)$

17. $(0, -2), (-1, -3), (4, -5), (-2, 1)$

18. $(-2, 0), (3, 2), (1, -4), (-3, 5)$

In Problems 19–22, find the coordinates of points A, B, C, and D and the coordinates of the indicated reflections.

19. Reflect A, B, C, and D through the y axis.

20. Reflect A, B, C, and D through the x axis.

21. Reflect A, B, C, and D through the origin.

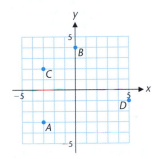

22. Reflect A, B, C, and D through the x axis and then through the y axis.

Test each equation in Problems 23–30 for symmetry with respect to the x axis, y axis, and the origin. Sketch the graph of the equation.

23. $y = 2x - 4$

24. $y = \frac{1}{2}x + 1$

25. $y = \frac{1}{2}x$

26. $y = 2x$

27. $|y| = x$

28. $|y| = -x$

29. $|x| = |y|$

30. $y = -x$

In Problems 31–34, use the graph to estimate to the nearest integer the missing coordinates of the indicated points. (Be sure you find all possible answers.)

31. (A) (8, ?) (B) (−5, ?) (C) (0, ?)
 (D) (?, 6) (E) (?, −5) (F) (?, 0)

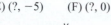

32. (A) (3, ?) (B) (−5, ?) (C) (0, ?)
 (D) (?, 3) (E) (?, −4) (F) (?, 0)

33. (A) (1, ?) (B) (−8, ?) (C) (0, ?)
 (D) (?, −6) (E) (?, 4) (F) (?, 0)

34. (A) (6, ?) (B) (−6, ?) (C) (0, ?)
 (D) (?, −2) (E) (?, 1) (F) (?, 0)

The figures in Problems 35 and 36 show a portion of a graph. Extend the given graph to one that exhibits the indicated type of symmetry.

35. (A) *x* axis only
 (B) *y* axis only
 (C) origin only
 (D) *x* axis, *y* axis, and origin

36. (A) *x* axis only
 (B) *y* axis only
 (C) origin only
 (D) *x* axis, *y* axis, and origin

Test each equation in Problems 37–46 for symmetry with respect to the x axis, the y axis, and the origin. Do not sketch the graph.

37. $2x + 7y = 0$

38. $x^2 + 6y + y^2 = 25$

39. $x^2 - 4xy^2 = 3$

40. $3x - 5y = 2$

41. $x^4 - 5x^2y + y^4 = 1$

42. $x^4 - y^4 = 16$

43. $x^3 - y^3 = 8$

44. $x^2 + 2xy + 3y^2 = 12$

45. $x^4 - 4x^2y^2 + y^4 = 81$

46. $x^3 - 4y^2 = 1$

Test each equation in Problems 47–58 for symmetry with respect to the x axis, the y axis, and the origin. Sketch the graph of the equation.

47. $y^2 = x + 2$ **48.** $y^2 = x - 2$

49. $y = x^2 + 1$ **50.** $y + 2 = x^2$

51. $4y^2 - x^2 = 1$ **52.** $4x^2 - y^2 = 1$

53. $y^3 = x$ **54.** $y = x^4$

55. $y = 0.6x^2 - 4.5$ **56.** $x = 0.8y^2 - 3.5$

57. $y = x^{2/3}$ **58.** $y^{2/3} = x$

59. (A) Graph the triangle with vertices $A = (1, 1)$, $B = (7, 2)$, and $C = (4, 6)$.
 (B) Now graph the triangle with vertices $A' = (1, -1)$, $B' = (7, -2)$, and $C' = (4, -6)$ in the same coordinate system.
 (C) How are these two triangles related? How would you describe the effect of changing the sign of the *y* coordinate of all the points on a graph?

60. (A) Graph the triangle with vertices $A = (1, 1)$, $B = (7, 2)$, and $C = (4, 6)$.

(B) Now graph the triangle with vertices $A' = (-1, 1)$, $B' = (-7, 2)$, and $C' = (-4, 6)$ in the same coordinate system.

(C) How are these two triangles related? How would you describe the effect of changing the sign of the x coordinate of all the points on a graph?

61. (A) Graph the triangle with vertices $A = (1, 1)$, $B = (7, 2)$, and $C = (4, 6)$.

(B) Now graph the triangle with vertices $A' = (-1, -1)$, $B' = (-7, -2)$, and $C' = (-4, -6)$ in the same coordinate system.

(C) How are these two triangles related? How would you describe the effect of changing the signs of the x and y coordinates of all the points on a graph?

62. (A) Graph the triangle with vertices $A = (1, 2)$, $B = (1, 4)$, and $C = (3, 4)$.

(B) Now graph $y = x$ and the triangle obtained by reversing the coordinates for each vertex of the original triangle: $A' = (2, 1)$, $B' = (4, 1)$, $B' = (4, 3)$.

(C) How are these two triangles related? How would you describe the effect of reversing the coordinates of each point on a graph?

 In Problems 63–66, solve for y, producing two equations, and then graph both of these equations in the same viewing window.

63. $2x + y^2 = 3$ **64.** $x^3 + y^2 = 8$

65. $x^2 - (y + 1)^2 = 4$ **66.** $(y - 2)^2 - x^2 = 9$

Test each equation in Problems 67–76 for symmetry with respect to the x axis, the y axis, and the origin. Sketch the graph of the equation.

67. $y^3 = |x|$ **68.** $|y| = x^3$ **69.** $xy = 1$

70. $xy = -1$ **71.** $y = 6x - x^2$ **72.** $y = x^2 - 6x$

73. $y^2 = |x| + 1$ **74.** $y^2 = 4|x| + 1$

75. $|xy| + 2|y| = 6$ **76.** $|xy| + |y| = 4$

77. If a graph is symmetric with respect to the x axis and to the origin, must it be symmetric with respect to the y axis? Explain.

78. If a graph is symmetric with respect to the y axis and to the origin, must it be symmetric with respect to the x axis? Explain.

79. If a graph is symmetric with respect to the origin, must it be symmetric with respect to the x axis? Explain.

80. If a graph is symmetric with respect to the origin, must it be symmetric with respect to the y axis? Explain.

APPLICATIONS

81. BUSINESS After extensive surveys, the marketing research department of a producer of popular compact discs developed the demand equation

$$n = 10 - p \qquad 5 \le p \le 10$$

where n is the number of units (in thousands) retailers are willing to buy per day at \$$p$ per disc. The company's daily revenue R (in thousands of dollars) is given by

$$R = np = (10 - p)p \qquad 5 \le p \le 10$$

Graph the revenue equation for the indicated values of p.

82. BUSINESS Repeat Problem 81 for the demand equation

$$n = 8 - p \qquad 4 \le p \le 8$$

83. PRICE AND DEMAND The quantity of a product that consumers are willing to buy during some period of time depends on its price. The price p and corresponding weekly demand q for a particular brand of diet soda in a city are shown in the figure. Use this graph to estimate the following demands to the nearest 100 cases.

(A) What is the demand when the price is \$6.00 per case?

(B) Does the demand increase or decrease if the price is increased from \$6.00 to \$6.30 per case? By how much?

(C) Does the demand increase or decrease if the price is decreased from \$6.00 to \$5.70? By how much?

(D) Write a brief description of the relationship between price and demand illustrated by this graph.

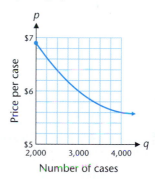

Number of cases

84. PRICE AND SUPPLY The quantity of a product that suppliers are willing to sell during some period of time depends on its price. The price p and corresponding weekly supply q for a particular brand of diet soda in a city are shown in the figure. Use this graph to estimate the following supplies to the nearest 100 cases.

(A) What is the supply when the price is \$5.60 per case?

(B) Does the supply increase or decrease if the price is increased from \$5.60 to \$5.80 per case? By how much?

(C) Does the supply increase or decrease if the price is decreased from \$5.60 to \$5.40 per case? By how much?

(D) Write a brief description of the relationship between price and supply illustrated by this graph.

Number of cases

85. TEMPERATURE The temperature during a spring day in the Midwest is given in the figure. Use this graph to estimate the following temperatures to the nearest degree and times to the nearest hour.
(A) The temperature at 9:00 A.M.
(B) The highest temperature and the time when it occurs.
(C) The time(s) when the temperature is 49°F.

86. TEMPERATURE Use the graph in Problem 85 to estimate the following temperatures to the nearest degree and times to the nearest half hour.
(A) The temperature at 7:00 P.M.
(B) The lowest temperature and the time when it occurs.
(C) The time(s) when the temperature is 52°F.

87. PHYSICS The speed (in meters per second) of a ball swinging at the end of a pendulum is given by

$$v = 0.5\sqrt{2 - x}$$

where x is the vertical displacement (in centimeters) of the ball from its position at rest (see the figure).

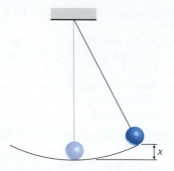

(A) Graph v for $0 \le x \le 2$.
(B) Describe the relationship between this graph and the physical behavior of the ball as it swings back and forth.

88. PHYSICS The speed (in meters per second) of a ball oscillating at the end of a spring is given by

$$v = 4\sqrt{25 - x^2}$$

where x is the vertical displacement (in centimeters) of the ball from its position at rest (positive displacement measured downward—see the figure).

(A) Graph v for $-5 \le x \le 5$.
(B) Describe the relationship between this graph and the physical behavior of the ball as it oscillates up and down.

2-2 Distance in the Plane

> **Distance Between Two Points**
> **Midpoint of a Line Segment**
> **Circles**

Two basic problems studied in analytic geometry are

1. Given an equation, find its graph.

2. Given a figure (line, circle, parabola, ellipse, etc.) in a coordinate system, find its equation.

The first problem was discussed in Section 2-1. In this section, we introduce some tools that are useful when studying the second problem.

› **Distance Between Two Points**

Given two points P_1 and P_2 in a rectangular coordinate system, we denote the **distance** between P_1 and P_2 by $d(P_1, P_2)$. We begin with an example.

EXAMPLE	1

Distance Between Two Points

Find the distance between the points $P_1 = (1, 2)$ and $P_2 = (4, 6)$.

SOLUTION

Connecting the points P_1, P_2, and $P_3 = (4, 2)$ with straight line segments forms a right triangle (Fig. 1).

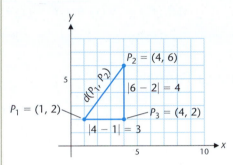

› **Figure 1**

From the figure, we see that the lengths of the legs of the triangle are

$$d(P_1, P_3) = |4 - 1| = 3$$

and

$$d(P_3, P_2) = |6 - 2| = 4$$

The length of the hypotenuse is $d(P_1, P_2)$, the distance we are seeking. Applying the Pythagorean theorem (see Appendix B), we get

$$[d(P_1, P_2)]^2 = [d(P_1, P_3)]^2 + [d(P_3, P_2)]^2$$
$$= 3^2 + 4^2$$
$$= 9 + 16$$
$$= 25$$

Therefore, $d(P_1, P_2) = \sqrt{25} = 5$

MATCHED PROBLEM 1	Find the distance between the points $P_1 = (1, 2)$ and $P_2 = (13, 7)$.

The ideas used in Example 1 can be applied to any two distinct points in the plane. If $P_1 = (x_1, y_1)$ and $P_2 = (x_2, y_2)$ are two points in a rectangular coordinate system (Fig. 2), then

$$[d(P_1, P_2)]^2 = |x_2 - x_1|^2 + |y_2 - y_1|^2$$
$$= (x_2 - x_1)^2 + (y_2 - y_1)^2 \qquad \text{Because } |N|^2 = N^2$$

Taking square roots gives the distance formula.

> **Figure 2** Distance between two points.

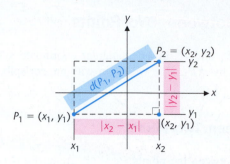

> **THEOREM 1** Distance Formula

The distance between $P_1 = (x_1, y_1)$ and $P_2 = (x_2, y_2)$ is

$$d(P_1, P_2) = \sqrt{(x_2 - x_1)^2 + (y_2 - y_1)^2}$$

EXAMPLE 2 Using the Distance Formula

Find the distance between the points $(-3, 5)$ and $(-2, -8)$.*

SOLUTION Let $(x_1, y_1) = (-3, 5)$ and $(x_2, y_2) = (-2, -8)$. Then,

$$d = \sqrt{[(-2) - (-3)]^2 + [(-8) - 5]^2}$$
$$= \sqrt{(-2 + 3)^2 + (-8 - 5)^2} = \sqrt{1^2 + (-13)^2} = \sqrt{1 + 169} = \sqrt{170}$$

Notice that if we choose $(x_1, y_1) = (-2, -8)$ and $(x_2, y_2) = (-3, 5)$, then

$$d = \sqrt{[(-3) - (-2)]^2 + [5 - (-8)]^2} = \sqrt{1 + 169} = \sqrt{170}$$

so it doesn't matter which point we designate as P_1 or P_2.

MATCHED PROBLEM 2 Find the distance between the points $(6, -3)$ and $(-7, -5)$.

> ## Midpoint of a Line Segment

The **midpoint** of a line segment is the point that is equidistant from each of the endpoints. A formula for finding the midpoint is given in Theorem 2. The proof is discussed in the exercises.

> **THEOREM 2** Midpoint Formula

The midpoint of the line segment joining $P_1 = (x_1, y_1)$ and $P_2 = (x_2, y_2)$ is

$$M = \left(\frac{x_1 + x_2}{2}, \frac{y_1 + y_2}{2} \right)$$

The point M is the unique point satisfying

$$d(P_1, M) = d(M, P_2) = \frac{1}{2}d(P_1, P_2)$$

*We often speak of the point (a, b) when we are referring to the point with coordinates (a, b). This shorthand, though not technically accurate, causes little trouble, and we will continue the practice.

Note that the coordinates of the midpoint are simply the averages of the respective coordinates of the two given points.

| EXAMPLE | 3 | **Using the Midpoint Formula** |

Find the midpoint M of the line segment joining $A = (-3, 2)$ and $B = (4, -5)$. Plot A, B, and M and verify that $d(A, M) = d(M, B) = \frac{1}{2}d(A, B)$.

SOLUTION

We use the midpoint formula with $(x_1, y_1) = (-3, 2)$ and $(x_2, y_2) = (4, -5)$ to obtain the coordinates of the midpoint M.

$$M = \left(\frac{x_1 + x_2}{2}, \frac{y_1 + y_2}{2} \right) \qquad \text{Substitute } x_1 = -3, y_1 = 2, x_2 = 4, \text{ and } y_2 = -5.$$

$$= \left(\frac{-3 + 4}{2}, \frac{2 + (-5)}{2} \right) \qquad \text{Simplify.}$$

$$= \left(\frac{1}{2}, \frac{-3}{2} \right)$$

$$= (0.5, -1.5)$$

We plot the three points (Fig. 3) and compute the distances $d(A, M)$, $d(M, B)$, and $d(A, B)$:

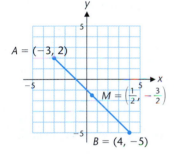

$A = (-3, 2)$

$M = \left(\frac{1}{2}, -\frac{3}{2} \right)$

$B = (4, -5)$

> **Figure 3**

$$d(A, M) = \sqrt{(-3 - 0.5)^2 + [2 - (-1.5)]^2} = \sqrt{12.25 + 12.25} = \sqrt{24.5}$$

$$d(M, B) = \sqrt{(0.5 - 4)^2 + [-1.5 - (-5)]^2} = \sqrt{12.25 + 12.25} = \sqrt{24.5}$$

$$d(A, B) = \sqrt{(-3 - 4)^2 + [2 - (-5)]^2} = \sqrt{49 + 49} = \sqrt{98}$$

$$\frac{1}{2}d(A, B) = \frac{1}{2}\sqrt{98} = \sqrt{\frac{98}{4}} = \sqrt{24.5} = d(A, M) = d(M, B)$$

This verifies that M is the midpoint of the line segment joining A and B.

| MATCHED PROBLEM 3 |

Find the midpoint M of the line segment joining $A = (4, 1)$ and $B = (-3, -5)$. Plot A, B, and M and verify that $d(A, M) = d(M, B) = \frac{1}{2}d(A, B)$.

| EXAMPLE | 4 | **Using the Midpoint Formula** |

If $M = (1, 1)$ is the midpoint of the line segment joining $A = (-3, -1)$ and $B = (x, y)$, find the coordinates of B.

SOLUTION

From the midpoint formula, we have

$$M = (1, 1) = \left(\frac{-3 + x}{2}, \frac{-1 + y}{2} \right)$$

We equate the corresponding coordinates and solve the resulting equations for x and y:

$$1 = \frac{-3 + x}{2} \qquad\qquad 1 = \frac{-1 + y}{2}$$

$$2 = -3 + x \qquad\qquad 2 = -1 + y$$

$$\boxed{2 + 3 = -3 + x + 3}^{*} \qquad \boxed{2 + 1 = -1 + y + 1}$$

$$5 = x \qquad\qquad\qquad 3 = y$$

Therefore, $B = (5, 3)$.

MATCHED PROBLEM 4

If $M = (1, -1)$ is the midpoint of the line segment joining $A = (-1, -5)$ and $B = (x, y)$, find the coordinates of B.

› Circles

The distance formula would be helpful if its only use were to find actual distances between points, such as in Example 2. However, its more important use is in finding equations of figures in a rectangular coordinate system. We start with an example.

EXAMPLE 5 ## Equations and Graphs of Circles

Write an equation for the set of all points that are 5 units from the origin. Graph your equation.

SOLUTION

The distance between a point (x, y) and the origin is

$$d = \sqrt{(x - 0)^2 + (y - 0)^2} = \sqrt{x^2 + y^2}$$

So, an equation for the set of points that are 5 units from the origin is

$$\sqrt{x^2 + y^2} = 5$$

We square both sides of this equation to obtain an equation that does not contain any radicals.

$$x^2 + y^2 = 25$$

Because $(-x)^2 = x^2$ and $(-y)^2 = y^2$, the graph will be symmetric with respect to the x axis, y axis, and origin. We make up a table of solutions, sketch the curve in the first quadrant, and use symmetry properties to produce a familiar geometric object—a circle (Fig. 4).

x	y
0	5
3	4
4	3
5	0

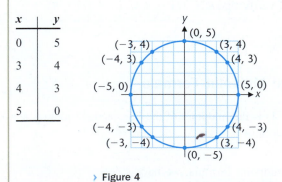

> **Figure 4**

MATCHED PROBLEM 5

Write an equation for the set of all points that are three units from the origin. Graph your equation.

*Throughout the book, dashed boxes—called **think boxes**—are used to represent steps that may be performed mentally.

Technology Connections

Refer to Example 5. To graph this circle on a graphing calculator, first we solve $x^2 + y^2 = 25$ for y:

$$x^2 + y^2 = 25$$
$$y^2 = 25 - x^2$$
$$y = \pm\sqrt{25 - x^2}$$

Next we enter $y = \sqrt{25 - x^2}$ and $y = -\sqrt{25 - x^2}$ in the equation editor of a graphing calculator [Fig. 5(a)], enter appropriate window variables [Fig. 5(b)], and graph [Fig. 5(c)].

The graph in Figure 5(c) doesn't look like a circle. (A circle is as wide as it is tall.) This distortion is caused by the difference between axes scales. One unit on the x axis appears to be longer than one unit on the y axis. Most graphing calculators have an option called **ZSquare** under the **zoom menu** [Fig. 6(a)] that automatically adjusts the x axis scale [Fig. 6(b)] to produce a **squared viewing window**. The graph of a circle in a squared viewing window is not distorted [Fig. 6(c)].

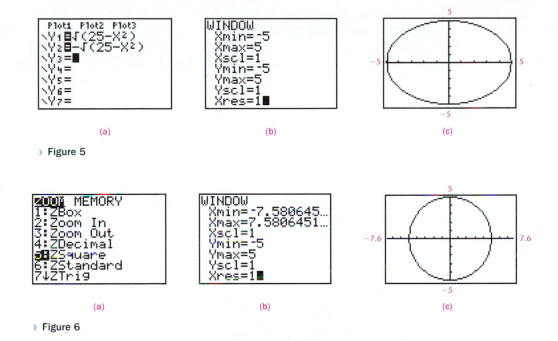

(a) (b) (c)

> **Figure 5**

(a) (b) (c)

> **Figure 6**

In Example 5, we began with a verbal description of a set of points, produced an algebraic equation that these points must satisfy, constructed a numerical table listing some of these points, and then drew a graphical representation of this set of points. The interplay between verbal, algebraic, numerical, and graphical concepts is one of the central themes of this book.

Now we generalize the ideas introduced in Example 5.

> **DEFINITION 1** Circle

A **circle** is the set of all points in a plane equidistant from a fixed point. The fixed distance is called the **radius,** and the fixed point is called the **center.**

> **Figure 7** Circle.

Let's find the equation of a circle with radius r $(r > 0)$ and center C at (h, k) in a rectangular coordinate system (Fig. 7). The circle consists of all points $P = (x, y)$ satisfying $d(P, C) = r$; that is, all points satisfying

$$\sqrt{(x - h)^2 + (y - k)^2} = r \quad r > 0$$

or, equivalently,

$$(x - h)^2 + (y - k)^2 = r^2 \quad r > 0$$

> **THEOREM 3 Standard Form of the Equation of a Circle**

The **standard form** of a circle with radius r and center at (h, k) is:

$$(x - h)^2 + (y - k)^2 = r^2 \quad r > 0$$

EXAMPLE 6 **Equations and Graphs of Circles**

Find the equation of a circle with radius 4 and center at $C = (-3, 6)$. Graph the equation.

SOLUTION

$$C = (h, k) = (-3, 6) \text{ and } r = 4$$
$$(x - h)^2 + (y - k)^2 = r^2 \quad \text{Substitute } h = -3, k = 6$$
$$[x - (-3)]^2 + (y - 6)^2 = 4^2 \quad \text{Simplify}$$
$$(x + 3)^2 + (y - 6)^2 = 16$$

To graph the equation, plot the center and a few points on the circle (the easiest points to plot are those located 4 units from the center in either the horizontal or vertical direction), then draw a circle of radius 4 (Fig. 8).

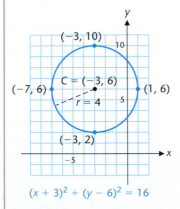

$(x + 3)^2 + (y - 6)^2 = 16$

> **Figure 8**

MATCHED PROBLEM 6 Find the equation of a circle with radius 3 and center at $C = (3, -2)$. Graph the equation.

>>> EXPLORE-DISCUSS 1 Explain how to find the equation of the circle with diameter AB, if $A = (3, 8)$ and $B = (11, 12)$.

EXAMPLE **7** **Finding the Center and Radius of a Circle**

Find the center and radius of the circle with equation $x^2 + y^2 + 6x - 4y = 23$.

SOLUTION We transform the equation into the form $(x - h)^2 + (y - k)^2 = r^2$ by completing the square relative to x and relative to y (see Section 1-5). From this standard form we can determine the center and radius.

$$x^2 + y^2 + 6x - 4y = 23$$

Group together the terms involving x and those involving y.

$$(x^2 + 6x \quad) + (y^2 - 4y \quad) = 23$$

Complete the squares.

$$(x^2 + 6x + 9) + (y^2 - 4y + 4) = 23 + 9 + 4$$

Factor each trinomial.

$$(x + 3)^2 + (y - 2)^2 = 36$$

Write $+3$ as $-(-3)$ to identify h.

$$[x - (-3)]^2 + (y - 2)^2 = 6^2$$

Center: $(h, k) = (-3, 2)$

Radius: $r = \sqrt{36} = 6$

MATCHED PROBLEM 7 Find the center and radius of the circle with equation $x^2 + y^2 - 8x + 10y = -25$.

ANSWERS TO MATCHED PROBLEMS

1. 13

2. $\sqrt{173}$

3. $M = (\frac{1}{2}, -2) = (0.5, -2)$; $d(A, B) = \sqrt{85}$; $d(A, M) = \sqrt{21.25} = d(M, B) = \frac{1}{2} d(A, B)$

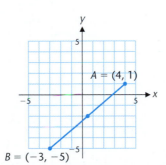

4. $B = (3, 3)$

5. $x^2 + y^2 = 9$ **6.** $(x - 3)^2 + (y + 2)^2 = 9$

7. $(x - 4)^2 + (y + 5)^2 = 16$; radius: 4, center: $(4, -5)$

2-2 Exercises

1. State the Pythagorean theorem.

2. Explain how to calculate the distance between two points in the plane if you know their coordinates.

3. Explain how to calculate the midpoint of a line segment if you know the coordinates of the endpoints.

4. Explain how to find the standard form of the equation of the circle with center $(1, 5)$ and radius $\sqrt{2}$.

In Problems 5–12, find the distance between each pair of points and the midpoint of the line segment joining the points. Leave distance in radical form, if applicable.

5. $(1, 0), (4, 4)$ **6.** $(0, 1), (3, 5)$

7. $(0, -2), (5, 10)$ **8.** $(3, 0), (-2, -3)$

9. $(-6, -4), (3, 4)$ **10.** $(-5, 4), (6, -1)$

11. $(-6, -3), (-2, -1)$ **12.** $(-5, -2), (-1, 2)$

In Problems 13–20, write the equation of a circle with the indicated center and radius.

13. $C = (0, 0), r = 7$ **14.** $C = (0, 0), r = 5$

15. $C = (2, 3), r = 6$ **16.** $C = (5, 6), r = 2$

17. $C = (-4, 1), r = \sqrt{7}$ **18.** $C = (-5, 6), r = \sqrt{11}$

19. $C = (-3, -4), r = \sqrt{2}$ **20.** $C = (4, -1), r = \sqrt{5}$

In Problems 21–26, write an equation for the given set of points. Graph your equation.

21. The set of all points that are two units from the origin.

22. The set of all points that are four units from the origin.

23. The set of all points that are one unit from $(1, 0)$.

24. The set of all points that are one unit from $(0, -1)$.

25. The set of all points that are three units from $(-2, 1)$.

26. The set of all points that are two units from $(3, -2)$.

27. Let M be the midpoint of A and B, where

$$A = (a_1, a_2), B = (1, 3), \text{ and } M = (-2, 6).$$

(A) Use the fact that -2 is the average of a_1 and 1 to find a_1.
(B) Use the fact that 6 is the average of a_2 and 3 to find a_2.
(C) Find $d(A, M)$ and $d(M, B)$.

28. Let M be the midpoint of A and B, where

$$A = (-3, 5), B = (b_1, b_2), \text{ and } M = (4, -2).$$

(A) Use the fact that 4 is the average of -3 and b_1 to find b_1.
(B) Use the fact that -2 is the average of 5 and b_2 to find b_2.
(C) Find $d(A, M)$ and $d(M, B)$.

29. Find x such that $(x, 7)$ is 10 units from $(-4, 1)$.

30. Find x such that $(x, 2)$ is 4 units from $(3, -3)$.

31. Find y such that $(2, y)$ is 3 units from $(-1, 4)$.

32. Find y such that $(3, y)$ is 13 units from $(-9, 2)$.

In Problems 33–36, write a verbal description of the graph and then write an equation that would produce the graph.

33.

34.

35.

36.

In Problems 37–42, M is the midpoint of A and B. Find the indicated point. Verify that d(A, M) = d(M, B) = ½d(A, B).

37. $A = (-4.3, 5.2)$, $B = (9.6, -1.7)$, $M = ?$

38. $A = (2.8, -3.5)$, $B = (-4.1, 7.6)$, $M = ?$

39. $A = (25, 10)$, $M = (-5, -2)$, $B = ?$

40. $M = (2.5, 3.5)$, $B = (12, 10)$, $A = ?$

41. $M = (-8, -6)$, $B = (2, 4)$, $A = ?$

42. $A = (-4, -2)$, $M = (-1.5, -4.5)$, $B = ?$

In Problems 43–52, find the center and radius of the circle with the given equation. Graph the equation.

43. $x^2 + (y + 2)^2 = 9$

44. $(x - 5)^2 + y^2 = 16$

45. $(x + 4)^2 + (y - 2)^2 = 7$

46. $(x - 5)^2 + (y + 7)^2 = 15$

47. $x^2 + 6x + y^2 = 16$

48. $x^2 + y^2 - 8y = 9$

49. $x^2 + y^2 - 6x - 4y = 36$

50. $x^2 + y^2 - 2x - 10y = 55$

51. $3x^2 + 3y^2 + 24x - 18y + 24 = 0$

52. $2x^2 + 2y^2 + 8x + 20y + 30 = 0$

 In Problems 53–56, solve for y, producing two equations, and then graph both of these equations in the same viewing window.

53. $x^2 + y^2 = 3$

54. $x^2 + y^2 = 5$

55. $(x + 3)^2 + (y + 1)^2 = 2$

56. $(x - 2)^2 + (y - 1)^2 = 3$

In Problems 57 and 58, show that the given points are the vertices of a right triangle (see the Pythagorean theorem in Appendix B). Find the length of the line segment from the midpoint of the hypotenuse to the opposite vertex.

57. $(-3, 2), (1, -2), (8, 5)$

58. $(-1, 3), (3, 5), (5, 1)$

Find the perimeter (to two decimal places) of the triangle with the vertices indicated in Problems 59 and 60.

59. $(-3, 1), (1, -2), (4, 3)$

60. $(-2, 4), (3, 1), (-3, -2)$

61. If $P_1 = (x_1, y_1)$, $P_2 = (x_2, y_2)$ and $M = \left(\dfrac{x_1 + x_2}{2}, \dfrac{y_1 + y_2}{2} \right)$, show that $d(P_1, M) = d(M, P_2) = \frac{1}{2}d(P_1, P_2)$. (This is one step in the proof of Theorem 2.)

62. A parallelogram *ABCD* is shown in the figure.
(A) Find the midpoint of the line segment joining *A* and *C*.
(B) Find the midpoint of the line segment joining *B* and *D*.
(C) What can you conclude about the diagonals of the parallelogram?

In Problems 63–68, find the standard form of the equation of the circle that has a diameter with the given endpoints.

63. $(-4, 3), (6, 3)$

64. $(5, -1), (5, 7)$

65. $(4, 0), (0, 10)$

66. $(-6, 0), (0, -8)$

67. $(11, -2), (3, -4)$

68. $(-8, 9), (12, 15)$

In Problems 69–72, find the standard form of the equation of the circle with the given center that passes through the given point.

69. Center: $(0, 5)$; point on circle: $(2, -4)$

70. Center: $(-3, 0)$; point on circle: $(6, 1)$

71. Center: $(-2, 9)$; point on circle: $(8, -7)$

72. Center: $(7, -12)$; point on circle: $(13, 8)$

APPLICATIONS

73. SPORTS A singles court for lawn tennis is a rectangle 27 feet wide and 78 feet long (see the figure). Points *B* and *F* are the midpoints of the end lines of the court.

(A) Sketch a graph of the court with *A* at the origin of your coordinate system, *C* on the positive *y* axis, and *G* on the positive *x* axis. Find the coordinates of points *A* through *G*.
(B) Find $d(B, D)$ and $d(F, C)$ to the nearest foot.

74. SPORTS Refer to Problem 73. Find $d(A, D)$ and $d(C, G)$ to the nearest foot.

75. ARCHITECTURE An arched doorway is formed by placing a circular arc on top of a rectangle (see the figure). If the doorway is 4 feet wide and the height of the arc above its ends is 1 foot, what is the radius of the circle containing the arc? [*Hint:* Note that $(2, r - 1)$ must satisfy $x^2 + y^2 = r^2$.]

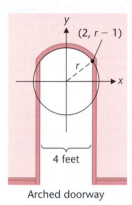

Arched doorway

76. ENGINEERING The cross section of a rivet has a top that is an arc of a circle (see the figure). If the ends of the arc are 12 millimeters apart and the top is 4 millimeters above the ends, what is the radius of the circle containing the arc?

Rivet

77. CONSTRUCTION Town B is located 36 miles east and 15 miles north of town A (see the figure). A local telephone company wants to position a relay tower so that the distance from the tower to town B is twice the distance from the tower to town A.
(A) Show that the tower must lie on a circle, find the center and radius of this circle, and graph.
(B) If the company decides to position the tower on this circle at a point directly east of town A, how far from town A should they place the tower? Compute answer to one decimal place.

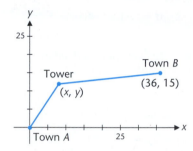

78. CONSTRUCTION Repeat Problem 77 if the distance from the tower to town A is twice the distance from the tower to town B.

2-3 Equations of a Line

> Graphing Lines
> Finding the Slope of a Line
> Determining Special Forms of the Equation of a Line
> Finding Slopes of Parallel or Perpendicular Lines

In this section, we consider one of the most basic geometric figures—a line. When we use the term *line* in this book, we mean *straight line*. We will learn how to recognize and graph a line and how to use information concerning a line to find its equation.

> Graphing Lines

With your past experience in graphing equations in two variables, you probably remember that first-degree equations in two variables, such as

$$y = -3x + 5 \qquad 3x - 4y = 9 \qquad y = -\tfrac{2}{3}x$$

have graphs that are lines. This fact is stated in Theorem 1.

> **THEOREM 1** The Equation of a Line

If A, B, and C are constants, with A and B not both 0, and x and y are variables, then the graph of the equation

$$Ax + By = C \qquad \textbf{Standard Form} \qquad (1)$$

is a line. Any line in a rectangular coordinate system has an equation of this form.

Also, the graph of any equation of the form

$$y = mx + b \qquad\qquad (2)$$

where m and b are constants, is a line. Equation (2), which we will discuss in detail later, is simply a special case of equation (1) for $B \neq 0$. This can be seen by solving equation (1) for y in terms of x:

$$y = -\frac{A}{B}x + \frac{C}{B} \qquad B \neq 0$$

To graph either equation (1) or (2), we plot any two points from the solution set and use a straightedge to draw a line through these two points. The points where the line crosses the axes are convenient to use and easy to find. The **y intercept*** is the y coordinate of the point where the graph crosses the y axis, and the **x intercept** is the x coordinate of the point where the graph crosses the x axis. To find the y intercept, let $x = 0$ and solve for y; to find the x intercept, let $y = 0$ and solve for x. It is often advisable to find a third point as a checkpoint. All three points must lie on the same line or a mistake has been made.

EXAMPLE 1 **Using Intercepts to Graph a Line**

Graph the equation $3x - 4y = 12$.

SOLUTION Find intercepts, a third checkpoint (optional), and draw a line through the two (three) points (Fig. 1).

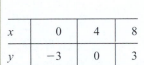

x	0	4	8
y	-3	0	3

> **Figure 1**

MATCHED PROBLEM 1 Graph the equation $4x + 3y = 12$.

*If the x intercept is a and the y intercept is b, then the graph of the line passes through the points $(a, 0)$ and $(0, b)$. It is common practice to refer to both the numbers a and b and the points $(a, 0)$ and $(0, b)$ as the x and y intercepts of the line.

Technology Connections

To solve Example 1 on a graphing calculator, we first solve the equation for y:

$$3x - 4y = 12$$

$$-4y = -3x + 12$$

$$y = 0.75x - 3$$

To find the y intercept of this line, we graph the preceding equation, press **TRACE**, and then enter 0 for x [Fig. 2(a)]. The displayed y value is the y intercept.

The x intercept can be found by using the **zero option** on the CALC menu. After selecting the zero option, you will be asked to provide three x values: a **left bound** (a number less than the zero), a **right bound** (a number greater than the zero), and a **guess** (a number between the left and right bounds). You can enter the three values from the keypad, but most find it easier to use the cursor. The zero or x intercept is displayed at the bottom of the screen [Fig. 2(b)].

(a) y intercept

(b) x intercept

› **Figure 2**

› Finding the Slope of a Line

If we take two different points $P_1 = (x_1, y_1)$ and $P_2 = (x_2, y_2)$ on a line, then the ratio of the change in y to the change in x as we move from point P_1 to point P_2 is called the **slope** of the line. Roughly speaking, slope is a measure of the "steepness" of a line. Sometimes the change in x is called the **run** and the change in y is called the **rise.**

> **DEFINITION 1** Slope of a Line

If a line passes through two distinct points $P_1 = (x_1, y_1)$ and $P_2 = (x_2, y_2)$, then its slope m is given by the formula

$$m = \frac{y_2 - y_1}{x_2 - x_1} \qquad x_1 \neq x_2$$

$$= \frac{\text{Vertical change (rise)}}{\text{Horizontal change (run)}}$$

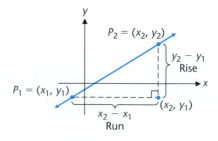

For a horizontal line, y doesn't change as x changes, so its slope is 0. On the other hand, for a vertical line, x doesn't change as y changes, so its slope is not defined:

$$\frac{y_2 - y_1}{x_2 - x_1} = \frac{y_2 - y_1}{0} \qquad \text{For a vertical line, slope is not defined.}$$

In general, the slope of a line may be positive, negative, 0, or not defined. Each of these cases is interpreted geometrically as shown in Table 1.

Table 1 Geometric Interpretation of Slope

Line	Slope	Example
Rising as x moves from left to right y values are increasing	Positive	
Falling as x moves from left to right y values are decreasing	Negative	
Horizontal y values are constant	0	
Vertical x values are constant	Not defined	

In using the formula to find the slope of the line through two points, it doesn't matter which point is labeled P_1 or P_2, because changing the labeling will change the sign in both the numerator and denominator of the slope formula:

$$\frac{y_2 - y_1}{x_2 - x_1} = \frac{y_1 - y_2}{x_1 - x_2}$$

For example, the slope of the line through the points $(3, 2)$ and $(7, 5)$ is

$$\frac{5 - 2}{7 - 3} = \frac{3}{4} = \frac{-3}{-4} = \frac{2 - 5}{3 - 7}$$

In addition, it is important to note that the definition of slope doesn't depend on the two points chosen on the line as long as they are distinct. This follows from the fact that the ratios of corresponding sides of similar triangles are equal (Fig. 3).

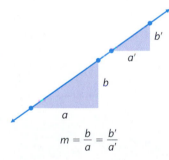

$$m = \frac{b}{a} = \frac{b'}{a'}$$

› **Figure 3**

EXAMPLE | **2** | **Finding Slopes**

For each line in Figure 4, find the run, the rise, and the slope. (All the horizontal and vertical line segments have integer lengths.)

(a) (b)

› **Figure 4**

SOLUTION In Figure 4(a), the run is 3, the rise is 6 and the slope is $\frac{6}{3} = 2$. In Figure 4(b), the run is 6, the rise is -4 and the slope is $\frac{-4}{6} = -\frac{2}{3}$.

MATCHED PROBLEM 2 For each line in Figure 5, find the run, the rise, and the slope. (All the horizontal and vertical line segments have integer lengths.)

(a) (b)

› Figure 5

EXAMPLE 3 Finding Slopes

Sketch a line through each pair of points and find the slope of each line.

(A) $(-3, -4)$, $(3, 2)$ (B) $(-2, 3)$, $(1, -3)$

(C) $(-4, 2)$, $(3, 2)$ (D) $(2, 4)$, $(2, -3)$

SOLUTIONS (A)

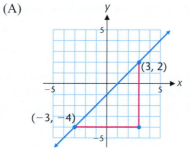

$$m = \frac{2 - (-4)}{3 - (-3)} = \frac{6}{6} = 1$$

(B)

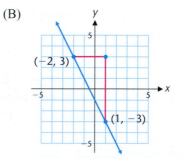

$$m = \frac{-3 - 3}{1 - (-2)} = \frac{-6}{3} = -2$$

(C)

$$m = \frac{2 - 2}{3 - (-4)} = \frac{0}{7} = 0$$

(D)

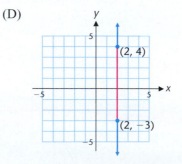

$$m = \frac{-3 - 4}{2 - 2} = \frac{-7}{0};$$
slope is not defined

| MATCHED PROBLEM 3 | Find the slope of the line through each pair of points. Do not graph. |

(A) $(-3, -3), (2, -3)$ (B) $(-2, -1), (1, 2)$

(C) $(0, 4), (2, -4)$ (D) $(-3, 2), (-3, -1)$

› Determining Special Forms of the Equation of a Line

We start by investigating why $y = mx + b$ is called the *slope–intercept form* for a line.

| ››› EXPLORE-DISCUSS 1 | **(A)** Graph $y = x + b$ for $b = -5, -3, 0, 3,$ and 5 simultaneously in the same coordinate system. Verbally describe the geometric significance of b.

(B) Graph $y = mx - 1$ for $m = -2, -1, 0, 1,$ and 2 simultaneously in the same coordinate system. Verbally describe the geometric significance of m. |

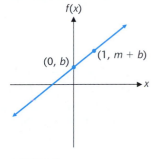

› Figure 6

As you see from the preceding exploration, constants m and b in $y = mx + b$ have special geometric significance.

If we let $x = 0$, then $y = b$ and the graph of $y = mx + b$ crosses the y axis at $(0, b)$. So the constant b is the y intercept. For example, the y intercept of the graph of $y = 2x - 7$ is -7.

We have already seen that the point $(0, b)$ is on the graph of $y = mx + b$. If we let $x = 1$, then it follows that the point $(1, m + b)$ is also on the graph (Fig. 6). Because the graph of $y = mx + b$ is a line, we can use these two points to compute the slope:

$$\text{Slope} = \frac{y_2 - y_1}{x_2 - x_1} = \frac{(m + b) - b}{1 - 0} = m \qquad \begin{array}{l}(x_1, y_1) = (0, b)\\(x_2, y_2) = (1, m + b)\end{array}$$

So m is the slope of the line with equation $y = mx + b$.

> **THEOREM 2** Slope–Intercept Form
>
> An equation of the line with slope m and y intercept b is
>
> $$y = mx + b$$
>
> which is called the **slope–intercept form**.
>
>

| EXAMPLE | 4 | Using the Slope–Intercept Form |

(A) Write the slope–intercept form of a line with slope $\frac{2}{3}$ and y intercept -5.

(B) Find the slope and y intercept, and graph $y = \frac{3}{4}x - 1$.

SOLUTIONS (A) Substitute $m = \frac{2}{3}$ and $b = -5$ in $y = mx + b$ to obtain $y = \frac{2}{3}x - 5$.

(B) The y intercept of $y = \frac{3}{4}x - 1$ is -1 and the slope is $\frac{3}{4}$. If we start at the point $(0, -1)$ and move four units to the right (run), then the y coordinate of a point on the line must move up three units (rise) to the point $(4, 2)$. Drawing a line through these two points produces the graph shown in Figure 7.

> **Figure 7**

MATCHED PROBLEM 4 Write the slope–intercept form of the line with slope $\frac{5}{4}$ and y intercept -2. Graph the equation.

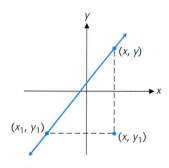

> **Figure 8**

In Example 4 we found the equation of a line with a given slope and y intercept. It is also possible to find the equation of a line passing through a given point with a given slope or to find the equation of a line containing two given points.

Suppose a line has slope m and passes through the point (x_1, y_1). If (x, y) is any other point on the line (Fig. 8), then

$$\frac{y - y_1}{x - x_1} = m$$

that is,

$$y - y_1 = m(x - x_1) \tag{3}$$

Because the point (x_1, y_1) also satisfies equation (3), we can conclude that equation (3) is an equation of a line with slope m that passes through (x_1, y_1).

> **THEOREM 3** Point–Slope Form

An equation of the line with slope m that passes through (x_1, y_1) is

$$y - y_1 = m(x - x_1)$$

which is called the **point–slope form.**

If we are given the coordinates of two points on a line, we can use the given coordinates to find the slope and then use the point–slope form with either of the given points to find the equation of the line.

| **EXAMPLE** | **5** | **Point–Slope Form** |

(A) Find an equation for the line that has slope $\frac{2}{3}$ and passes through the point $(-2, 1)$. Write the final answer in the form $Ax + By = C$.

(B) Find an equation for the line that passes through the two points $(4, -1)$ and $(-8, 5)$. Write the final answer in the form $y = mx + b$.

SOLUTIONS (A) If $m = \frac{2}{3}$ and $(x_1, y_1) = (-2, 1)$, then

$$y - y_1 = m(x - x_1) \qquad \text{Substitute } y_1 = 1, x_1 = -2, \text{ and } m = \frac{2}{3}.$$

$$y - 1 = \frac{2}{3}[x - (-2)] \qquad \text{Multiply both sides by 3.}$$

$$3(y - 1) = 2(x + 2) \qquad \text{Distribute.}$$

$$3y - 3 = 2x + 4 \qquad \text{Write in standard form.}$$

$$-2x + 3y = 7 \qquad \text{or} \qquad 2x - 3y = -7$$

(B) First use the slope formula to find the slope of the line:

$$m = \frac{y_2 - y_1}{x_2 - x_1} = \frac{5 - (-1)}{-8 - 4} = \frac{6}{-12} = -\frac{1}{2} \qquad \begin{array}{l}\text{Substitute } x_1 = 4, y_1 = -1, x_2 = -8, \\ \text{and } y_2 = 5 \text{ in the slope formula.}\end{array}$$

Now we choose $(x_1, y_1) = (4, -1)$ and proceed as in part A:

$$y - y_1 = m(x - x_1) \qquad \text{Substitute } x_1 = 4, y_1 = -1, \text{ and } m = -\frac{1}{2}.$$

$$y - (-1) = -\frac{1}{2}(x - 4) \qquad y - (-1) = y + 1; \text{ Distribute on right side.}$$

$$y + 1 = -\frac{1}{2}x + 2 \qquad \text{Subtract 1 from both sides.}$$

$$y = -\frac{1}{2}x + 1$$

You may want to verify that choosing $(x_1, y_1) = (-8, 5)$, the other given point, produces the same equation.

| **MATCHED PROBLEM 5** |

(A) Find an equation for the line that has slope $-\frac{2}{5}$ and passes through the point $(3, -2)$. Write the final answer in the form $Ax + By = C$.

(B) Find an equation for the line that passes through the two points $(-3, 1)$ and $(7, -3)$. Write the final answer in the form $y = mx + b$.

 The simplest equations of lines are those for horizontal and vertical lines. Consider the following two equations:

$$x + 0y = a \qquad \text{or} \qquad x = a \qquad (4)$$
$$0x + y = b \qquad \text{or} \qquad y = b \qquad (5)$$

In equation (4), y can be any number as long as $x = a$. So the graph of $x = a$ is a vertical line crossing the x axis at $(a, 0)$. In equation (5), x can be any number as long as $y = b$.

So the graph of $y = b$ is a horizontal line crossing the y axis at $(0, b)$. We summarize these results as follows:

> **THEOREM 4 Vertical and Horizontal Lines**

Equation		**Graph**
$x = a$	(short for $x + 0y = a$)	Vertical line through $(a, 0)$ (Slope is undefined.)
$y = b$	(short for $0x + y = b$)	Horizontal line through $(0, b)$ (Slope is 0.)

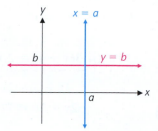

EXAMPLE 6 Graphing Horizontal and Vertical Lines

Graph the line $x = -2$ and the line $y = 3$.

SOLUTION

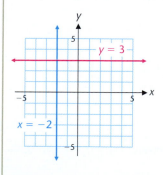

MATCHED PROBLEM 6 Graph the line $x = 4$ and the line $y = -2$.

The various forms of the equation of a line that we have discussed are summarized in Table 2 for convenient reference.

Table 2 Equations of a Line

Standard form	$Ax + By = C$	A and B not both 0
Slope–intercept form	$y = mx + b$	Slope: m; y intercept: b
Point–slope form	$y - y_1 = m(x - x_1)$	Slope: m; Point: (x_1, y_1)
Horizontal line	$y = b$	Slope: 0
Vertical line	$x = a$	Slope: Undefined

› **Finding Slopes of Parallel or Perpendicular Lines**

From geometry, we know that two vertical lines are parallel to each other and that a horizontal line and a vertical line are perpendicular to each other. How can we tell when two nonvertical lines are parallel or perpendicular to each other? Theorem 5, which we state without proof, provides a convenient test.

› **THEOREM 5** Parallel and Perpendicular Lines

Given two nonvertical lines L_1 and L_2 with slopes m_1 and m_2, respectively, then

$$L_1 \parallel L_2 \qquad \text{if and only if} \qquad m_1 = m_2$$
$$L_1 \perp L_2 \qquad \text{if and only if} \qquad m_1 m_2 = -1$$

The symbols $\parallel$ and $\perp$ mean, respectively, "is parallel to" and "is perpendicular to." In the case of perpendicularity, the condition $m_1 m_2 = -1$ also can be written as

$$m_2 = -\frac{1}{m_1} \qquad \text{or} \qquad m_1 = -\frac{1}{m_2}$$

Therefore,

Two nonvertical lines are perpendicular if and only if their slopes are the negative reciprocals of each other.

EXAMPLE 7 **Parallel and Perpendicular Lines**

Given the line $L: 3x - 2y = 5$ and the point $P = (-3, 5)$, find an equation of a line through P that is

(A) Parallel to L (B) Perpendicular to L

Write the final answers in the slope–intercept form $y = mx + b$.

SOLUTIONS First, find the slope of L by writing $3x - 2y = 5$ in the equivalent slope–intercept form $y = mx + b$:

$$3x - 2y = 5$$
$$-2y = -3x + 5$$
$$y = \tfrac{3}{2}x - \tfrac{5}{2}$$

So the slope of L is $\tfrac{3}{2}$. The slope of a line parallel to L is the same, $\tfrac{3}{2}$, and the slope of a line perpendicular to L is $-\tfrac{2}{3}$. We now can find the equations of the two lines in parts A and B using the point–slope form.

(A) Parallel ($m = \tfrac{3}{2}$): (B) Perpendicular ($m = -\tfrac{2}{3}$):

$y - y_1 = m(x - x_1)$	$y - y_1 = m(x - x_1)$ Substitute for x_1, y_1, and m.
$y - 5 = \tfrac{3}{2}(x + 3)$	$y - 5 = -\tfrac{2}{3}(x + 3)$ Distribute.
$y - 5 = \tfrac{3}{2}x + \tfrac{9}{2}$	$y - 5 = -\tfrac{2}{3}x - 2$ Add 5 to both sides.
$y = \tfrac{3}{2}x + \tfrac{19}{2}$	$y = -\tfrac{2}{3}x + 3$

MATCHED PROBLEM 7

Given the line L: $4x + 2y = 3$ and the point $P = (2, -3)$, find an equation of a line through P that is

(A) Parallel to L (B) Perpendicular to L

Write the final answers in the slope–intercept form $y = mx + b$.

EXAMPLE 8

Cost Analysis

A hot dog vendor pays $25 per day to rent a pushcart and $1.25 for the ingredients in one hot dog.

(A) Find the cost of selling x hot dogs in 1 day.

(B) What is the cost of selling 200 hot dogs in 1 day?

(C) If the daily cost is $355, how many hot dogs were sold that day?

SOLUTIONS

(A) The rental charge of $25 is the vendor's **fixed cost**—a cost that is accrued every day and does not depend on the number of hot dogs sold. The cost of the ingredients for x hot dogs is $1.25x$. This is the vendor's **variable cost**—a cost that depends on the number of hot dogs sold. The total cost for selling x hot dogs is

$$C(x) = 1.25x + 25 \qquad \text{Total Cost = Variable Cost + Fixed Cost}$$

(B) The cost of selling 200 hot dogs in 1 day is

$$C(200) = 1.25(200) + 25 = \$275$$

(C) The number of hot dogs that can be sold for $355 is the solution of the equation

$$\begin{aligned}
1.25x + 25 &= 355 &&\text{Subtract 25 from each side.}\\
1.25x &= 330 &&\text{Divide both sides by 1.25.}\\
x &= \frac{330}{1.25} &&\text{Simplify.}\\
&= 264 \text{ hot dogs}
\end{aligned}$$

MATCHED PROBLEM 8

It costs a pretzel vendor $20 per day to rent a cart and $0.75 for each pretzel.

(A) Find the cost of selling x pretzels in 1 day.

(B) What is the cost of selling 150 pretzels in 1 day?

(C) If the daily cost is $275, how many pretzels were sold that day?

Technology Connections

A graphing calculator can be used to solve equations like $1.25x + 25 = 355$ (see Example 8). First enter both sides of the equation in the equation editor [Fig. 9(a)] and choose window variables [Fig. 9(b)] so that the graphs of both equations appear on the screen. There is no "right" choice for the window variables. Any choice that displays the intersection point will do. (Here is how we chose our window variables: We chose $Y_{max} = 600$ to place the graph of the horizontal line below the top of the window. We chose $Y_{min} = -200$ to place the graph of the x axis above the text displayed at the bottom of the screen. Since x cannot be negative, we chose $X_{min} = 0$. We used trial and error to determine a reasonable choice for X_{max}.) Now choose **intersect** on the CALC menu, and respond to the prompts from the calculator. The coordinates of the intersection point of the two graphs are shown at the bottom of the screen [Fig. 9(c)].

(a)

(b)

(c)

> **Figure 9**

ANSWERS TO MATCHED PROBLEMS

1.

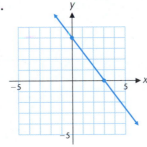

2. (A) Run = 5, rise = 4, slope = $\frac{4}{5}$

 (B) Run = 3, rise = -6, slope = $\frac{-6}{3} = -2$

3. (A) $m = 0$ (B) $m = 1$

 (C) $m = -4$ (D) m is not defined

4. $y = \frac{5}{4}x - 2$

5. (A) $2x + 5y = -4$ (B) $y = -\frac{2}{5}x - \frac{1}{5}$

6.

7. (A) $y = -2x + 1$ (B) $y = \frac{1}{2}x - 4$

8. (A) $C(x) = 0.75x + 20$ (B) \$132.50 (C) 340 pretzels

2-3 Exercises

1. Explain how to find the x and y intercepts of a line if its equation is written in standard form.

2. Given the graph of a line, explain how to determine whether the slope is negative.

3. Explain why $y = mx + b$ is called the slope–intercept form.

4. Explain why $y - y_1 = m(x - x_1)$ is called the point–slope form.

5. Given the equations of two lines in standard form, explain how to determine whether the lines are parallel.

6. Given the equations of two lines in standard form, explain how to determine whether the lines are perpendicular.

In Problems 7–12, use the graph of each line to find the rise, run, and slope. Write the equation of each line in the standard form $Ax + By = C$, $A \geq 0$. (All the horizontal and vertical line segments have integer lengths.)

7.

8.

9.

10.

11.

12.

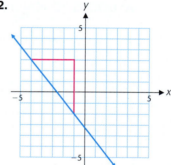

In Problems 13–18, use the graph of each line to find the x intercept, y intercept, and slope, if they exist. Write the equation of each line, using the slope–intercept form whenever possible.

13.

14.

15.

16.

17.

18.

Graph each equation in Problems 19–32, and indicate the slope, if it exists.

19. $y = -\frac{3}{5}x + 4$ **20.** $y = -\frac{3}{2}x + 6$

21. $y = -\frac{3}{4}x$ **22.** $y = \frac{2}{3}x - 3$

23. $4x + 2y = 0$ **24.** $6x - 2y = 0$

25. $4x - 5y = -24$ **26.** $6x - 7y = -49$

27. $\frac{y}{8} - \frac{x}{4} = 1$ **28.** $\frac{y}{6} - \frac{x}{5} = 1$

29. $x = -3$ **30.** $y = -2$

31. $y = 3.5$ **32.** $x = 2.5$

In Problems 33–38, find an equation of the line with the indicated slope and y intercept, and write it in the form $Ax + By = C$, $A \geq 0$, where A, B, and C are integers.

33. Slope $= -3$; y intercept $= 7$

34. Slope $= 4$; y intercept $= -10$

35. Slope $= \frac{7}{2}$; y intercept $= -\frac{1}{3}$

36. Slope $= -\frac{5}{4}$; y intercept $= \frac{11}{5}$

37. Slope $= 0$; y intercept $= \frac{2}{3}$

38. Slope $= 0$; y intercept $= 0$

In Problems 39–44, find the equation of the line passing through the given point with the given slope. Write the final answer in the slope–intercept form $y = mx + b$.

39. $(0, 3)$; $m = -2$ **40.** $(4, 0)$; $m = 3$

41. $(-5, 4)$; $m = \frac{3}{2}$ **42.** $(2, -3)$; $m = -\frac{4}{5}$

43. $(-2, -3)$; $m = -\frac{1}{2}$ **44.** $(2, 1)$; $m = \frac{4}{3}$

In Problem 45–58, write the equation of the line that contains the indicated point(s), and/or has the given slope or intercepts; use either the slope–intercept form $y = mx + b$, or the form $x = c$.

45. $(0, 4)$; $m = -3$ **46.** $(2, 0)$; $m = 2$

47. $(-5, 4)$; $m = -\frac{2}{5}$ **48.** $(-4, -2)$; $m = \frac{1}{2}$

49. $(1, 6)$; $(5, -2)$ **50.** $(-3, 4)$; $(6, 1)$

51. $(-4, 8)$; $(2, 0)$ **52.** $(2, -1)$; $(10, 5)$

53. $(-3, 4)$; $(5, 4)$ **54.** $(0, -2)$; $(4, -2)$

55. $(4, 6)$; $(4, -3)$ **56.** $(-3, 1)$; $(-3, -4)$

57. x intercept -4; **58.** x intercept -4;
 y intercept 3 y intercept -5

In Problems 59–66, write an equation of the line that contains the indicated point and meets the indicated condition(s). Write the final answer in the standard form $Ax + By = C$, $A \geq 0$.

59. $(-3, 4)$; parallel to $y = 3x - 5$

60. $(-4, 0)$; parallel to $y = -2x + 1$

61. $(2, -3)$; perpendicular to $y = -\frac{1}{3}x$

62. $(-2, -4)$; perpendicular to $y = \frac{2}{3}x - 5$

63. $(5, 0)$; parallel to $3x - 2y = 4$

64. $(3, 5)$; parallel to $3x + 4y = 8$

65. $(0, -4)$; perpendicular to $x + 3y = 9$

66. $(-2, 4)$; perpendicular to $4x + 5y = 0$

Problems 67–72 refer to the quadrilateral with vertices $A = (0, 2)$, $B = (4, -1)$, $C = (1, -5)$, and $D = (-3, -2)$.

67. Show that $AB \parallel DC$.

68. Show that $DA \parallel CB$.

69. Show that $AB \perp BC$.

70. Show that $AD \perp DC$.

71. Find an equation of the perpendicular bisector* of AD.

72. Find an equation of the perpendicular bisector of AB.

73. Prove that if a line L has x intercept $(a, 0)$ and y intercept $(0, b)$, then the equation of L can be written in the **intercept form**

$$\frac{x}{a} + \frac{y}{b} = 1 \qquad a, b \neq 0$$

74. Prove that if a line L passes through $P_1 = (x_1, y_1)$ and $P_2 = (x_2, y_2)$, then the equation of L can be written in the **two-point form**

$$(y - y_1)(x_2 - x_1) = (y_2 - y_1)(x - x_1)$$

Problems 75–80 are calculus related. Recall that a line tangent to a circle at a point is perpendicular to the radius drawn to that point (see the figure). Find the equation of the line tangent to the circle at the indicated point. Write the final answer in the standard form $Ax + By = C, A \geq 0$. Graph the circle and the tangent line on the same coordinate system.

75. $x^2 + y^2 = 25, (3, 4)$

76. $x^2 + y^2 = 100, (-8, 6)$

77. $x^2 + y^2 = 50, (5, -5)$

78. $x^2 + y^2 = 80, (-4, -8)$

79. $(x - 3)^2 + (y + 4)^2 = 169, (8, -16)$

80. $(x + 5)^2 + (y - 9)^2 = 289, (-13, -6)$

APPLICATIONS

81. BOILING POINT OF WATER At sea level, water boils when it reaches a temperature of 212°F. At higher altitudes, the atmospheric pressure is lower and so is the temperature at which water boils. The boiling point B in degrees Fahrenheit at an altitude of x feet is given approximately by

$$B = 212 - 0.0018x$$

(A) Complete Table 3.

Table 3

x	0	5,000	10,000	15,000	20,000	25,000	30,000
B							

(B) Based on the information in the table, write a brief verbal description of the relationship between altitude and the boiling point of water.

82. AIR TEMPERATURE As dry air moves upward, it expands and cools. The air temperature A in degrees Celsius at an altitude of x kilometers is given approximately by

$$A = 25 - 9x$$

*The perpendicular bisector of a line segment is a line perpendicular to the segment and passing through its midpoint.

(A) Complete Table 4.

Table 4

x	0	1	2	3	4	5
A						

(B) Based on the information in the table, write a brief verbal description of the relationship between altitude and air temperature.

83. COST ANALYSIS A doughnut shop has a fixed cost of $124 per day and a variable cost of $0.12 per doughnut. Find the total daily cost of producing x doughnuts. How many doughnuts can be produced for a total daily cost of $250?

84. COST ANALYSIS A small company manufactures picnic tables. The weekly fixed cost is $1,200 and the variable cost is $45 per table. Find the total weekly cost of producing x picnic tables. How many picnic tables can be produced for a total weekly cost of $4,800?

85. PHYSICS Hooke's law states that the relationship between the stretch s of a spring and the weight w causing the stretch is linear (a principle upon which all spring scales are constructed). For a particular spring, a 5-pound weight causes a stretch of 2 inches, while with no weight the stretch of the spring is 0.
(A) Find a linear equation that expresses s in terms of w.
(B) What is the stretch for a weight of 20 pounds?
(C) What weight will cause a stretch of 3.6 inches?

86. PHYSICS The distance d between a fixed spring and the floor is a linear function of the weight w attached to the bottom of the spring. The bottom of the spring is 18 inches from the floor when the weight is 3 pounds and 10 inches from the floor when the weight is 5 pounds.
(A) Find a linear equation that expresses d in terms of w.
(B) Find the distance from the bottom of the spring to the floor if no weight is attached.
(C) Find the smallest weight that will make the bottom of the spring touch the floor. (Ignore the height of the weight.)

87. PHYSICS The two most widespread temperature scales are Fahrenheit* (F) and Celsius† (C). It is known that water freezes at 32°F or 0°C and boils at 212°F or 100°C.
(A) Find a linear equation that expresses F in terms of C.
(B) If a European family sets its house thermostat at 20°C, what is the setting in degrees Fahrenheit? If the outside temperature in Milwaukee is 86°F, what is the temperature in degrees Celsius?

88. PHYSICS Two other temperature scales, used primarily by scientists, are Kelvin‡ (K) and Rankine** (R). Water freezes at 273 K or 492°R and boils at 373 K or 672°R. Find a linear equation that expresses R in terms of K.

89. OCEANOGRAPHY After about 9 hours of a steady wind, the height of waves in the ocean is approximately linearly related to

*Invented in 1724 by Daniel Gabriel Fahrenheit (1686–1736), a German physicist.
†Invented in 1742 by Anders Celsius (1701–1744), a Swedish astronomer.
‡Invented in 1848 by Lord William Thompson Kelvin (1824–1907), a Scottish mathematician and physicist. Note that the degree symbol "°" is not used with degrees Kelvin.
**Invented in 1859 by John Maquorn Rankine (1820–1872), a Scottish engineer and physicist.

the duration of time the wind has been blowing. During a storm with 50-knot winds, the wave height after 9 hours was found to be 23 feet, and after 24 hours it was 40 feet.

(A) If t is time after the 50-knot wind started to blow and h is the wave height in feet, write a linear equation that expresses height h in terms of time t.

(B) How long will the wind have been blowing for the waves to be 50 feet high?

Express all calculated quantities to three significant digits.

90. OCEANOGRAPHY Refer to Problem 89. A steady 25-knot wind produces a wave 7 feet high after 9 hours and 11 feet high after 25 hours.

(A) Write a linear equation that expresses height h in terms of time t.

(B) How long will the wind have been blowing for the waves to be 20 feet high?

91. DEMOGRAPHICS Life expectancy in the United States has increased from about 49.2 years in 1900 to about 77.3 years in 2000. The growth in life expectancy is approximately linear with respect to time.

(A) If L represents life expectancy and t represents the number of years since 1900, write a linear equation that expresses L in terms of t.

(B) What is the predicted life expectancy in the year 2020?

Express all calculated quantities to three significant digits.

92. DEMOGRAPHICS The average number of persons per household in the United States has been shrinking steadily for as long as statistics have been kept and is approximately linear with respect to time. In 1900, there were about 4.76 persons per household and in 2000, about 2.59.

(A) If N represents the average number of persons per household and t represents the number of years since 1900, write a linear equation that expresses N in terms of t.

(B) What is the predicted household size in the year 2025?

Express all calculated quantities to three significant digits.

93. CITY PLANNING The design of a new subdivision calls for three parallel streets connecting First Street with Main Street (see the figure). Find the distance d_1 (to the nearest foot) from Avenue A to Avenue B.

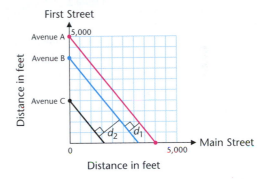

94. CITY PLANNING Refer to Problem 93. Find the distance d_2 (to the nearest foot) from Avenue B to Avenue C.

2-4 Linear Equations and Models

> Slope as a Rate of Change
> Linear Models
> Linear Regression

Mathematical modeling is the process of using mathematics to solve real-world problems. This process can be broken down into three steps (Fig. 1):

Step 1. *Construct* the **mathematical model,** a mathematics problem that, when solved, will provide information about the real-world problem.

Step 2. *Solve* the mathematical model.

Step 3. *Interpret* the solution to the mathematical model in terms of the original real-world problem.

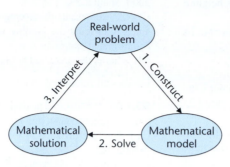

> **Figure 1**

In more complex problems, this cycle may have to be repeated several times to obtain the required information about the real-world problem. In this section, we discuss one of the simplest mathematical models, a linear equation. With the aid of a graphing calculator, we also learn how to analyze a linear model based on real-world data.

› Slope as a Rate of Change

If x and y are related by the equation $y = mx + b$, where m and b are constants with $m \neq 0$, then x and y are **linearly related.** If (x_1, y_1) and (x_2, y_2) are two distinct points on this line, then the slope of the line is

$$m = \frac{y_2 - y_1}{x_2 - x_1} = \frac{\text{Change in } y}{\text{Change in } x} \tag{1}$$

In applications, ratio (1) is called the **rate of change** of y with respect to x. Since the slope of a line is unique, *the rate of change of two linearly related variables is constant.* Here are some examples of familiar rates of change: miles per hour, revolutions per minute, price per pound, passengers per plane, etc. If y is distance and x is time, then the rate of change is also referred to as **speed** or **velocity.** If the relationship between x and y is not linear, ratio (1) is called the **average rate of change** of y with respect to x.

| EXAMPLE | 1 | **Estimating Body Surface Area** |

Appropriate doses of medicine for both animals and humans are often based on body surface area (BSA). Since weight is much easier to determine than BSA, veterinarians use the weight of an animal to estimate BSA. The following linear equation expresses BSA for canines in terms of weight*:

$$a = 16.21w + 375.6$$

where a is BSA in square inches and w is weight in pounds.

(A) Interpret the slope of the BSA equation.

(B) What is the effect of a 1-pound increase in weight?

SOLUTIONS (A) The rate of change BSA with respect to weight is 16.21 square inches per pound.

(B) Since slope is the ratio of rise to run, increasing w by 1 pound (run) increases a by 16.21 square inches (rise).

*Based on data from Veterinary Oncology Consultants, PTY LTD.

MATCHED PROBLEM 1

The following linear equation expresses BSA for felines in terms of weight:

$$a = 28.55w + 118.7$$

where a is BSA in square inches and w is weight in pounds.

(A) Interpret the slope of the BSA equation.

(B) What is the effect of a 1-pound increase in weight?

❯ Linear Models

We can use our experience with lines in Section 2-3 to construct linear models for applications involving linearly related quantities. This process is best illustrated through examples.

EXAMPLE 2

Business Markup Policy

A sporting goods store sells a fishing rod that cost $60 for $82 and a pair of cross-country ski boots that cost $80 for $106.

(A) If the markup policy of the store for items that cost more than $30 is assumed to be linear, find a linear model that express the retail price P in terms of the wholesale cost C.

(B) What is the effect on the price of a $1 increase in cost for any item costing over $30?

(C) Use the model to find the retail price for a pair of running shoes that cost $40.

SOLUTIONS

(A) If price P is linearly related to cost C, then we are looking for the equation of a line whose graph passes through $(C_1, P_1) = (60, 82)$ and $(C_2, P_2) = (80, 106)$. We find the slope, and then use the point–slope form to find the equation.

$$m = \frac{P_2 - P_1}{C_2 - C_1} = \frac{106 - 82}{80 - 60} = \frac{24}{20} = 1.2 \qquad \text{Substitute } C_1 = 60, P_1 = 82, C_2 = 80, \text{ and } P_2 = 106 \text{ into the slope formula.}$$

$$P - P_1 = m(C - C_1) \qquad \text{Substitute } P_1 = 82, C_1 = 60, \text{ and } m = 1.2 \text{ into the point–slope formula.}$$

$$P - 82 = 1.2(C - 60) \qquad \text{Distribute}$$

$$P - 82 = 1.2C - 72 \qquad \text{Add 82 to both sides.}$$

$$P = 1.2C + 10 \qquad C > 30 \qquad \text{Linear model}$$

(B) If the cost is increased by $1, then the price will increase by $1.2(1) = \$1.20$.

(C) $P = 1.2(40) + 10 = \$58$.

MATCHED PROBLEM 2

The sporting goods store in Example 2 is celebrating its twentieth anniversary with a 20% off sale. The sale price of a mountain bike is $380. What was the presale price of the bike? How much did the bike cost the store?

❯❯❯ EXPLORE-DISCUSS 1

The wholesale supplier for the sporting goods store in Example 2 offers the store a 15% discount on all items. The store decides to pass on the savings from this discount to the consumer. Which of the following markup policies is better for the consumer?

1. Apply the store's markup policy to the discounted cost.

2. Apply the store's markup policy to the original cost and then reduce this price by 15%.

Support your choice with examples.

EXAMPLE 3 Mixing Antifreeze

Ethylene glycol and propylene glycol are liquids used in antifreeze and deicing solutions. Ethylene glycol is listed as a hazardous chemical by the Environmental Protection Agency, while propylene glycol is generally regarded as safe. Table 1 lists solution concentration percentages and the corresponding freezing points for each chemical.

Table 1

Concentration	Ethylene Glycol	Propylene Glycol
20%	15°F	17°F
50%	−36°F	−28°F

(A) Assume that the concentration and the freezing point for ethylene glycol are linearly related. Construct a linear model for the freezing point.

(B) Interpret the slope in part (A).

(C) What percentage (to one decimal place) of ethylene glycol will result in a freezing point of −10°F?

SOLUTIONS

(A) We begin by defining appropriate variables:

Let

p = percentage of ethylene glycol in the antifreeze solution

f = freezing point of the antifreeze solution

From Table 1, we see that $(20, 15)$ and $(50, -36)$ are two points on the line relating p and f. The slope of this line is

$$m = \frac{f_2 - f_1}{p_2 - p_1} = \frac{15 - (-36)}{20 - 50} = \frac{51}{-30} = -1.7$$

and its equation is

$$f - 15 = -1.7(p - 20)$$
$$f = -1.7p + 49 \qquad \text{Linear model}$$

(B) The rate of change of the freezing point with respect to the percentage of ethylene glycol in the antifreeze solution is −1.7 degrees per percentage of ethylene glycol. Increasing the amount of ethylene glycol by 1% will lower the freezing point by 1.7°F.

(C) We must find p when f is −10°.

$$f = -1.7p + 49$$
$$-10 = -1.7p + 49 \qquad \text{Add } 10 + 1.7p \text{ to both sides.}$$
$$1.7p = 59 \qquad \text{Divide both sides by 1.7.}$$
$$p = \frac{59}{1.7} = 34.7\%$$

MATCHED PROBLEM 3

Refer to Table 1.

(A) Assume that the concentration and the freezing point for propylene glycol are linearly related. Construct a linear model for the freezing point.

(B) Interpret the slope in part (A).

(C) What percentage (to one decimal place) of propylene glycol will result in a freezing point of −15°F?

EXAMPLE	4	**Underwater Pressure**

The pressure at sea level is 14.7 pounds per square inch. As you descend into the ocean, the pressure increases linearly at a rate of about 0.445 pounds per square foot.

(A) Find the pressure p at a depth of d feet.

(B) If a diver's equipment is rated to be safe up to a pressure of 40 pounds per square foot, how deep (to the nearest foot) is it safe to use this equipment?

SOLUTIONS

(A) Let $p = md + b$. At the surface, $d = 0$ and $p = 14.7$, so $b = 14.7$. The slope m is the given rate of change, $m = 0.445$. So the pressure at a depth of d feet is

$$p = 0.445d + 14.7$$

(B) The safe depth is the solution of the equation

$0.445d + 14.7 = 40$	Subtract 14.7 from each side.
$0.445d = 25.3$	Divide both sides by 0.445.
$d = \dfrac{25.3}{0.445}$	Simplify.
≈ 57 feet	

MATCHED PROBLEM 4

The rate of change of pressure in fresh water is 0.432 pounds per square foot. Repeat Example 4 for a body of fresh water.

Technology Connections

Figure 2 shows the solution of Example 4(B) on a graphing calculator.

```
Intersection
X=56.853933   Y=40
```

> **Figure 2**
> $y_1 = 0.445x + 14.7$, $y_2 = 40$

› Linear Regression

In real-world applications we often encounter numerical data in the form of a table. The very powerful mathematical tool, *regression analysis,* can be used to analyze numerical data. In general, **regression analysis** is a process for finding an equation that provides a useful model for a set of data points. Graphs of equations are often called **curves** and regression analysis is also referred to as **curve fitting.** In Example 5, we use a linear model obtained by using *linear regression* on a graphing calculator.

EXAMPLE	5

Diamond Prices

Table 2 Round-Shaped Diamond Prices

Weight (Carats)	Price
0.5	$1,340
0.6	$1,760
0.7	$2,540
0.8	$3,350
0.9	$4,130
1.0	$4,920

Source: www.tradeshop.com

Prices for round-shaped diamonds taken from an online trader are given in Table 2.

(A) A linear model for the data in Table 2 is given by

$$p = 7,380c - 2,530 \qquad (2)$$

where p is the price of a diamond weighing c carats. (We will discuss the source of models like this later in this section.) Plot the points in Table 2 on a Cartesian coordinate system, producing a **scatter plot,** and graph the model on the same axes.

(B) Interpret the slope of the model in equation (2).

(C) Use the model to estimate the cost of a 0.85-carat diamond and the cost of a 1.2-carat diamond. Round answers to the nearest dollar.

(D) Use the model to estimate the weight of a diamond that sells for $3,000. Round the answer to two significant digits.

SOLUTIONS

(A) A scatter plot is simply a plot of the points in Table 2 [Fig. 3(a)]. To add the graph of the model to the scatter plot, we find any two points that satisfy equation (2) [we choose (0.4, 422) and (1.4, 7,802)]. Plotting these points and drawing a line through them gives us Figure 3(b).

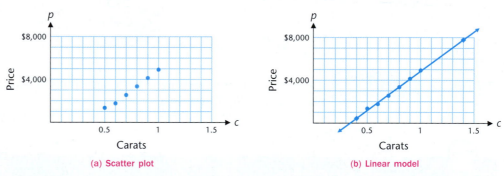

(a) Scatter plot (b) Linear model

› **Figure 3**

(B) The rate of change of the price of a diamond with respect to its weight is 7,380. Increasing the weight by 1 carat will increase the price by about $7,380.

(C) The graph of the model [Fig. 3(b)] does not pass through any of the points in the scatter plot, but it comes close to all of them. [Verify this by evaluating equation (2) at $c = 0.5$, 0.6, . . . , 1.] So we can use equation (2) to approximate points not in Table 2.

$$c = 0.85 \qquad\qquad c = 1.2$$
$$p = 7,380(0.85) - 2,530 \qquad p = 7,380(1.2) - 2,530$$
$$= \$3,743 \qquad\qquad = \$6,326$$

A 0.85-carat diamond will cost about $3,743 and a 1.2-carat diamond will cost about $6,326.

(D) To find the weight of a $3,000 diamond, we solve the following equation for c:

$$7,380c - 2,530 = 3,000$$
$$7,380c = 3,000 + 2,530$$
$$= 5,530$$
$$c = \frac{5,530}{7,380} = 0.75 \qquad \text{To two significant digits}$$

A $3,000 diamond will weigh about 0.75 carats.

MATCHED **PROBLEM 5**

Table 3 Emerald-Shaped Diamond Prices

Weight (Carats)	Price
0.5	$1,350
0.6	$1,740
0.7	$2,610
0.8	$3,320
0.9	$4,150
1.0	$4,850

Source: www.tradeshop.com

Prices for emerald-shaped diamonds taken from an online trader are given in Table 3. Repeat Example 5 for this data with the linear model

$$p = 7{,}270c - 2{,}450$$

where p is the price of an emerald-shaped diamond weighing c carats.

The model we used in Example 5 was obtained by using a technique called **linear regression** and the model is called the **regression line.** This technique produces a line that is the best fit for a given data set. We will not discuss the theory behind this technique, nor the meaning of "best fit." Although you can find a linear regression line by hand, we prefer to leave the calculations to a graphing calculator or a computer. Don't be concerned if you don't have either of these electronic devices. We will supply the regression model in the applications we discuss, as we did in Example 5.

Technology Connections

If you want to use a graphing calculator to construct regression lines, you should consult your user's manual.* The process varies from one calculator to another. Figure 4 shows three of the screens related to the construction of the model in Example 5 on a Texas Instruments TI-84 Plus.

(a) Entering the data.

(b) Finding the model.

(c) Graphing the data and the model.

› **Figure 4**

*User's manuals for the most popular graphing calculators are readily available on the Internet.

In Example 5, we used the regression line to approximate points that were not given in Table 2, but would fit between points in the table. This process is called **interpolation.** In the next example we use a regression model to approximate points outside the given data set. This process is called **extrapolation** and the approximations are often referred to as **predictions.**

EXAMPLE	6

Telephone Expenditures

Table 4 gives information about expenditures for residential and cellular phone service. The linear regression model for residential service is

$$r = 722 - 33.1t$$

where r is the average annual expenditure (in dollars per consumer unit) on residential service and t is time in years with $t = 0$ corresponding to 2000.

(A) Interpret the slope of the regression line as a rate of change.

(B) Use the regression line to predict expenditures for residential service in 2018.

Table 4 Average Annual Telephone Expenditures (dollars per consumer unit)

	2001	2003	2005	2007
Residential	686	620	570	482
Cellular	210	316	455	608

Source: Bureau of Labor Statistics

SOLUTIONS

(A) The slope $m = -33.1$ is the rate of change of expenditures with respect to time. Because the slope is negative, the expenditures for residential service are decreasing at a rate of $33.10 per year.

(B) If $t = 18$, then

$$r = 722 - 33.1(18) = \$126$$

So the model predicts that expenditures for residential phone service will be approximately $126 in 2018.

MATCHED PROBLEM 6

Repeat Example 6 using the following linear regression model for cellular service:

$$c = 66.7t + 131$$

where c is the average annual expenditure (in dollars per consumer unit) on cellular service and t is time in years with $t = 0$ corresponding to 2000.

ANSWERS TO MATCHED PROBLEMS

1. (A) The rate of change of BSA with respect to weight is 28.55 square inches per pound.
 (B) Increasing w by 1 pound increases a by 28.55 square inches.
2. Presale price is $475. Cost is $387.50
3. (A) $f = -1.5p + 47$
 (B) The rate of change of the freezing point with respect to the percentage of propylene glycol in the antifreeze solution is -1.5. Increasing the percentage of propylene glycol by 1% will lower the freezing point by 1.5°F.
 (C) 41.3%
4. (A) $p = 0.432d + 14.7$
 (B) 59 ft
5. (A)

Carats

(B) The rate of change of the price of a diamond with respect to the size is 7,270. Increasing the size by 1 carat will increase the price by about $7,270.
 (C) $3,730; $6,274
 (D) 0.75 carats
6. (A) The expenditures for cellular service are increasing at a rate of $66.70 per year.
 (B) $1,332.

2-4 Exercises

1. Explain the steps that are involved in the process of mathematical modeling.

2. If two variables x and y are linearly related, explain how to calculate the rate of change.

3. If two variables x and y are not linearly related, explain how to calculate the average rate of change from $x = x_1$ to $x = x_2$.

4. Explain the difference between interpolation and extrapolation in the context of regression analysis.

APPLICATIONS

5. COST ANALYSIS A plant can manufacture 80 golf clubs per day for a total daily cost of $8,147 and 100 golf clubs per day for a total daily cost of $9,647.
(A) Assuming that daily cost and production are linearly related, find the total daily cost of producing x golf clubs.
(B) Interpret the slope of this cost equation.
(C) What is the effect of a 1 unit increase in production?

6. COST ANALYSIS A plant can manufacture 50 tennis rackets per day for a total daily cost of $4,174 and 60 tennis rackets per day for a total daily cost of $4,634.
(A) Assuming that daily cost and production are linearly related, find the total daily cost of producing x tennis rackets.
(B) Interpret the slope of this cost equation.
(C) What is the effect of a 1 unit increase in production?

7. FORESTRY Forest rangers estimate the height of a tree by measuring the tree's diameter at breast height (DBH) and then using a model constructed for a particular species.* A model for white spruce trees is

$$h = 4.06d + 24.1$$

where d is the DBH in inches and h is the tree height in feet.
(A) Interpret the slope of this model.
(B) What is the effect of a 1-inch increase in DBH?
(C) How tall is a white spruce with a DBH of 12 inches? Round answer to the nearest foot.
(D) What is the DBH of a white spruce that is 100 feet tall? Round answer to the nearest inch.

8. FORESTRY A model for black spruce trees is

$$h = 2.27d + 33.1$$

where d is the DBH in inches and h is the tree height in feet.
(A) Interpret the slope of this model.
(B) What is the effect of a 1-inch increase in DBH?
(C) How tall is a black spruce with a DBH of 12 inches? Round answer to the nearest foot.
(D) What is the DBH of a black spruce that is 100 feet tall? Round answer to the nearest inch.

*Models in Problems 7 and 8 are based on data found at
http://flash.lakeheadu.ca/~fluckai/htdbh04.xls

9. Dr. J. D. Robinson and Dr. D. R. Miller published the following models for estimating the weight of a woman:

$$\text{Robinson: } w = 108 + 3.7h$$
$$\text{Miller: } w = 117 + 3.0h$$

where w is weight (in pounds) and h is height over 5 feet (in inches).
(A) Interpret the slope of each model.
(B) If a woman is 5′6″ tall, what does each model predict her weight to be?
(C) If a woman weighs 140 pounds, what does each model predict her height to be?

10. Dr. J. D. Robinson and Dr. D. R. Miller also published the following models for estimating the weight of a man:

$$\text{Robinson: } w = 115 + 4.2h$$
$$\text{Miller: } w = 124 + 3.1h$$

where w is weight (in pounds) and h is height over 5 feet (in inches).
(A) Interpret the slope of each model.
(B) If a man is 5′10″ tall, what does each model predict his weight to be?
(C) If a man weighs 160 pounds, what does each model predict his height to be?

11. SPEED OF SOUND The speed of sound through the air near sea level is linearly related to the temperature of the air. If sound travels at 741 mph at 32°F and at 771 mph at 72°F, construct a linear model relating the speed of sound (s) and the air temperature (t). Interpret the slope of this model.

12. SPEED OF SOUND The speed of sound through the air near sea level is linearly related to the temperature of the air. If sound travels at 337 mps (meters per second) at 10°C and at 343 mps at 20°C, construct a linear model relating the speed of sound (s) and the air temperature (t). Interpret the slope of this model.

13. SMOKING STATISTICS The percentage of male cigarette smokers in the United States declined from 25.7% in 2000 to 23.9% in 2006. Find a linear model relating the percentage m of male smokers to years t since 2000. Use the model to predict the first year for which the percentage of male smokers will be less than or equal to 18%.

14. SMOKING STATISTICS The percentage of female cigarette smokers in the United States declined from 21.0% in 2000 to 18.0% in 2006. Find a linear model relating the percentage f of female smokers to years t since 2000. Use the model to predict the first year for which the percentage of female smokers will be less than or equal to 10%.

15. BUSINESS—DEPRECIATION A farmer buys a new tractor for $142,000 and assumes that it will have a trade-in value of $67,000 after 10 years. The farmer uses a constant rate of depreciation (commonly called **straight-line depreciation**—one of several methods permitted by the IRS) to determine the annual value of the tractor.
(A) Find a linear model for the depreciated value V of the tractor t years after it was purchased.

(B) Interpret the slope of this model.

(C) What is the depreciated value of the tractor after 6 years?

16. BUSINESS—DEPRECIATION A charter fishing company buys a new boat for $154,900 and assumes that it will have a trade-in value of $46,100 after 16 years.

(A) Use straight-line depreciation (See Problem 15) to find a linear model for the depreciated value V of the boat t years after it was purchased.

(B) Interpret the slope of this model.

(C) In which year will the depreciated value of the boat fall below $100,000?

17. BUSINESS—MARKUP POLICY A drugstore sells a drug costing $85 for $112 and a drug costing $175 for $238.

(A) If the markup policy of the drugstore is assumed to be linear, write an equation that expresses retail price R in terms of cost C (wholesale price).

(B) What is the slope of the graph of the equation found in part A? Interpret verbally.

(C) What does a store pay (to the nearest dollar) for a drug that retails for $185?

18. BUSINESS—MARKUP POLICY A clothing store sells a shirt costing $20 for $33 and a jacket costing $60 for $93.

(A) If the markup policy of the store for items costing over $10 is assumed to be linear, write an equation that expresses retail price R in terms of cost C (wholesale price).

(B) What is the slope of the equation found in part A? Interpret verbally.

(C) What does a store pay for a suit that retails for $240?

19. FLIGHT CONDITIONS In stable air, the air temperature drops about 5°F for each 1,000-foot rise in altitude.

(A) If the temperature at sea level is 70°F and a commercial pilot reports a temperature of −20°F at 18,000 feet, write a linear equation that expresses temperature T in terms of altitude A (in thousands of feet).

(B) How high is the aircraft if the temperature is 0°F?

20. FLIGHT NAVIGATION An airspeed indicator on some aircraft is affected by the changes in atmospheric pressure at different altitudes. A pilot can estimate the true airspeed by observing the indicated airspeed and adding to it about 2% for every 1,000 feet of altitude.

(A) If a pilot maintains a constant reading of 200 miles per hour on the airspeed indicator as the aircraft climbs from sea level to an altitude of 10,000 feet, write a linear equation that expresses true airspeed T (miles per hour) in terms of altitude A (thousands of feet).

(B) What would be the true airspeed of the aircraft at 6,500 feet?

21. RATE OF DESCENT—PARACHUTES At low altitudes, the altitude of a parachutist and time in the air are linearly related. A jump at 2,880 ft using the U.S. Army's T-10 parachute system lasts 120 seconds.

(A) Find a linear model relating altitude a (in feet) and time in the air t (in seconds).

(B) The **rate of descent** is the speed at which the jumper falls. What is the rate of descent for a T-10 system?

22. RATE OF DESCENT—PARACHUTES The U.S. Army is considering a new parachute, the ATPS system. A jump at 2,880 ft using the ATPS system lasts 180 seconds.

(A) Find a linear model relating altitude a (in feet) and time in the air t (in seconds).

(B) What is the rate of descent for an ATPS system parachute?

23. LICENSED DRIVERS Table 5 contains the state population and the number of licensed drivers in the state (both in millions) for the states with population under 1 million. The regression model for this data is

$$y = 0.72x + 0.03$$

where x is the state population and y is the number of licensed drivers in the state.

Table 5 Licensed Drivers in 2006

State	Population	Licensed Drivers
Alaska	0.67	0.49
Delaware	0.85	0.62
Montana	0.94	0.72
North Dakota	0.64	0.47
South Dakota	0.78	0.58
Vermont	0.62	0.53
Wyoming	0.52	0.39

Source: Bureau of Transportation Statistics

(A) Plot the data in Table 5 and the model on the same axes.

(B) If the population of New Hampshire in 2006 was about 1.3 million, use the model to estimate the number of licensed drivers in New Hampshire.

(C) If the population of Nebraska in 2006 was about 1.8 million, use the model to estimate the number of licensed drivers in Nebraska.

24. LICENSED DRIVERS Table 6 contains the state population and the number of licensed drivers in the state (both in millions) for several states with population over 10 million. The regression model for this data is

$$y = 0.60x + 1.15$$

where x is the state population and y is the number of licensed drivers in the state.

Table 6 Licensed Drivers in 2006

State	Population	Licensed Drivers
California	36	23
Florida	18	14
Illinois	13	8
Michigan	10	7
New York	19	11
Ohio	11	8
Pennsylvania	12	9
Texas	24	15

Source: Bureau of Transportation Statistics

(A) Plot the data in Table 6 and the model on the same axes.

(B) If the population of Georgia in 2006 was about 9.4 million, use the model to estimate the number of licensed drivers in Georgia.

(C) If the population of New Jersey in 2006 was about 8.7 million, use the model to estimate the number of licensed drivers in New Jersey.

Problems 25–28 require a graphing calculator or a computer that can calculate the linear regression line for a given data set.

25. OLYMPIC GAMES Find a linear regression model for the men's 100-meter freestyle data given in Table 7, where x is years since 1968 and y is winning time (in seconds). Do the same for the women's 100-meter freestyle data. (Round regression coefficients to four significant digits.) Do these models indicate that the women will eventually catch up with the men?

Table 7 Winning Times in Olympic Swimming Events

	100-Meter Freestyle		200-Meter Backstroke	
	Men	**Women**	**Men**	**Women**
1968	52.20	60.0	2:09.60	2:24.80
1976	49.99	55.65	1:59.19	2:13.43
1984	49.80	55.92	2:00.23	2:12.38
1992	49.02	54.65	1:58.47	2:07.06
2000	48.30	53.83	1:56.76	2:08.16
2008	47.21	53.12	1:53.94	2:05.24

Source: www.infoplease.com

26. OLYMPIC GAMES Find a linear regression model for the men's 200-meter backstroke data given in Table 7 where x is years since 1968 and y is winning time (in seconds). Do the same for the women's 200-meter backstroke data. (Round regression coefficients to five significant digits.) Do these models indicate that the women will eventually catch up with the men?

27. SUPPLY AND DEMAND Table 8 contains price–supply data and price–demand data for corn. Find a linear regression model for the price–supply data where x is supply (in billions of bushels) and y is price (in dollars). Do the same for the price–demand data. (Round regression coefficients to three significant digits.) Find the price at which supply and demand are equal. (In economics, this price is referred to as the **equilibrium price.**)

Table 8 Supply and Demand for U.S. Corn

Price ($/bu.)	Supply (Billion bu.)	Price ($/bu.)	Demand (Billion bu.)
2.15	6.29	2.07	9.78
2.29	7.27	2.15	9.35
2.36	7.53	2.22	8.47
2.48	7.93	2.34	8.12
2.47	8.12	2.39	7.76
2.55	8.24	2.47	6.98

Source: www.usda.gov/nass/pubs/histdata.htm

28. SUPPLY AND DEMAND Table 9 contains price–supply data and price–demand data for soybeans. Find a linear regression model for the price–supply data where x is supply (in billions of bushels) and y is price (in dollars). Do the same for the price–demand data. (Round regression coefficients to three significant digits.) Find the equilibrium price for soybeans.

Table 9 Supply and Demand for U.S. Soybeans

Price ($/bu.)	Supply (Billion bu.)	Price ($/bu.)	Demand (Billion bu.)
5.15	1.55	4.93	2.60
5.79	1.86	5.48	2.40
5.88	1.94	5.71	2.18
6.07	2.08	6.07	2.05
6.15	2.15	6.40	1.95
6.25	2.27	6.66	1.85

Source: www.usda.gov/nass/pubs/histdata.htm

CHAPTER 2 Review

2-1 Cartesian Coordinate System

A **Cartesian** or **rectangular coordinate system** is formed by the intersection of a horizontal real number line and a vertical real number line at their origins. These lines are called the **coordinate axes.** The **horizontal axis** is often referred to as the x **axis** and the **vertical axis** as the y **axis.** These axes divide the plane into four **quadrants.** Each point in the plane corresponds to its **coordinates**—an ordered pair (a, b) determined by passing horizontal and vertical lines through the point. The **abscissa** or x **coordinate** a is the coordinate of the intersection of the vertical line with the horizontal axis, and the **ordinate** or y **coordinate** b is the coordinate of the intersection of the horizontal line with the vertical axis. The point $(0, 0)$ is

called the **origin.** A **solution** of an equation in two variables is an ordered pair of real numbers that makes the equation a true statement. The **solution set** of an equation is the set of all its solutions. The **graph of an equation in two variables** is the graph of its solution set formed using **point-by-point plotting** or with the aid of a graphing calculator. The **reflection** of the point (a, b) through the y axis is the point $(-a, b)$, through the x axis is the point $(a, -b)$, and through the **origin** is the point $(-a, -b)$. The **reflection** of a graph is the reflection of each point on the graph. If reflecting a graph through the y axis, x axis, or origin does not change its shape, the graph is said to be symmetric with respect to the y **axis, x axis,** or **origin,** respectively. To **test** an equation for symmetry, determine if the equation is unchanged when y is replaced with $-y$ (x axis symmetry), x is replaced

with $-x$ (y axis symmetry), or both x and y are replaced with $-x$ and $-y$ (origin symmetry).

2-2 Distance in the Plane

The **distance** between the two points $P_1 = (x_1, y_1)$ and $P_2 = (x_2, y_2)$ is

$$d(P_1, P_2) = \sqrt{(x_2 - x_1)^2 + (y_2 - y_1)^2}$$

and the **midpoint** of the line segment joining $P_1 = (x_1, y_1)$ and $P_2 = (x_2, y_2)$ is

$$M = \left(\frac{x_1 + x_2}{2}, \frac{y_1 + y_2}{2} \right)$$

The **standard form** for the equation of a **circle** with **radius** r and **center** at (h, k) is

$$(x - h)^2 + (y - k)^2 = r^2, \qquad r > 0$$

2-3 Equations of a Line

The standard form for the equation of a line is $Ax + By = C$, where A, B, and C are constants, A and B not both 0. The **y intercept** is the y coordinate of the point where the graph crosses the y axis, and the **x intercept** is the x coordinate of the point where the graph crosses the x axis. The slope of the line through the points (x_1, y_1) and (x_2, y_2) is

$$m = \frac{y_2 - y_1}{x_2 - x_1} \qquad \text{if } x_1 \neq x_2$$

The slope is not defined for a vertical line where $x_1 = x_2$. Two lines with slopes m_1 and m_2 are parallel if and only if $m_1 = m_2$ and perpendicular if and only if $m_1 m_2 = -1$.

Equations of a Line

Standard form	$Ax + By = C$	A and B not both 0
Slope–intercept form	$y = mx + b$	Slope: m; y intercept: b
Point–slope form	$y - y_1 = m(x - x_1)$	Slope: m; Point: (x_1, y_1)
Horizontal line	$y = b$	Slope: 0
Vertical line	$x = a$	Slope: Undefined

2-4 Linear Equations and Models

A **mathematical model** is a mathematics problem that, when solved, will provide information about a real-world problem. If $y = mx + b$, then the variables x and y are **linearly related** and the **rate of change** of y with respect to x is the constant m. If x and y are not linearly related, the ratio $(y_2 - y_1)/(x_2 - x_1)$ is called the **average rate of change** of y with respect to x. **Regression analysis** produces an equation whose graph is a **curve** that **fits** (approximates) a set of data points. A **scatter plot** is the graph of the points in a data set. **Linear regression** produces a **regression line** that is the best fit for a given data set. Graphing calculators or other electronic devices are frequently used to find regression lines.

CHAPTER 2 Review Exercises

Work through all the problems in this chapter review and check answers in the back of the book. Answers to all review problems are there, and following each answer is a number in italics indicating the section in which that type of problem is discussed. Where weaknesses show up, review appropriate sections in the text.

1. Plot $A = (-4, 1)$, $B = (2, -3)$, and $C = (-1, -2)$ in a rectangular coordinate system.

2. Refer to Problem 1. Plot the reflection of A through the x axis, the reflection of B through the y axis, and the reflection of C through the origin.

3. Test each equation for symmetry with respect to the x axis, y axis, and origin and sketch its graph.
(A) $y = 2x$ (B) $y = 2x - 1$
(C) $y = 2|x|$ (D) $|y| = 2x$

4. Use the following graph to estimate to the nearest integer the missing coordinates of the indicated points. (Be sure you find all possible answers.)
(A) $(0, ?)$ (B) $(?, 0)$ (C) $(?, 4)$

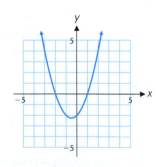

5. Given the points $A = (-2, 3)$ and $B = (4, 0)$, find:
(A) Distance between A and B
(B) Slope of the line through A and B
(C) Slope of a line perpendicular to the line through A and B

6. Write the equation of a circle with radius $\sqrt{7}$ and center:
(A) $(0, 0)$ (B) $(3, -2)$

7. Find the center and radius of the circle given by

$$(x + 3)^2 + (y - 2)^2 = 5$$

8. Let M be the midpoint of A and B, where $A = (a_1, a_2)$, $B = (2, -5)$, and $M = (-4, 3)$.
(A) Use the fact that -4 is the average of a_1 and 2 to find a_1.
(B) Use the fact that 3 is the average of a_2 and -5 to find a_2.
(C) Find $d(A, M)$ and $d(M, B)$.

9. (A) Graph the triangle with vertices $A = (-1, -2)$, $B = (4, 3)$, and $C = (1, 4)$.
(B) Find the perimeter to two decimal places.
(C) Use the Pythagorean theorem to determine if the triangle is a right triangle.
(D) Find the midpoint of each side of the triangle.

10. Use the graph of the linear function in the figure to find the rise, run, and slope. Write the equation of the line in the form $Ax + By = C$, where A, B, and C are integers with $A > 0$. (The horizontal and vertical line segments have integer lengths.)

11. Graph $3x + 2y = 9$ and indicate its slope.

12. Write an equation of a line with x intercept 6 and y intercept 4. Write the final answer in the standard form $Ax + By = C$, where A, B, and C are integers.

13. Write the slope–intercept form of the equation of the line with slope $-\frac{2}{3}$ and y intercept 2.

14. Write the equations of the vertical and horizontal lines passing through the point $(-3, 4)$. What is the slope of each?

Test each equation in Problems 15–18 for symmetry with respect to the x axis, y axis, and the origin. Sketch the graph of the equation.

15. $y = x^2 - 2$

16. $y^2 = x - 2$

17. $9y^2 + 4x^2 = 36$

18. $9y^2 - 4x^2 = 36$

19. Write a verbal description of the graph shown in the figure and then write an equation that would produce the graph.

20. (A) Find an equation of the line through $P = (-4, 3)$ and $Q = (0, -3)$. Write the final answer in the standard form $Ax + By = C$, where A, B, and C are integers with $A > 0$.
(B) Find $d(P, Q)$.

21. Write the slope–intercept form of the equation of the line that passes through the point $(-2, 1)$ and is
(A) parallel to the line $6x + 3y = 5$
(B) perpendicular to the line $6x + 3y = 5$

22. Find the equation of a circle that passes through the point $(-1, 4)$ with center at $(3, 0)$.

23. Find the center and radius of the circle given by
$$x^2 + y^2 + 4x - 6y = 3$$

24. Find the equation of the set of points equidistant from $(3, 3)$ and $(6, 0)$. What is the name of the geometric figure formed by this set?

25. Are the graphs of $mx - y = b$ and $x + my = b$ parallel, perpendicular, or neither? Justify your answer.

26. Use completing the square to find the center and radius of the circle with equation:
$$x^2 - 4x + y^2 - 2y - 3 = 0$$

27. Refer to Problem 26. Find the equation of the line tangent to the circle at the point $(4, 3)$. Graph the circle and the line on the same coordinate system.

28. Find the equation of a circle with center $(4, -3)$ whose graph passes through the point $(1, 2)$.

29. Extend the following graph to one that exhibits the indicated symmetry:
(A) x axis only (B) y axis only
(C) origin only (D) x axis, y axis, and origin

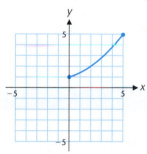

Problems 30 and 31 refer to a triangle with base b and height h (see the figure). Write a mathematical expression in terms of b and h for each of the verbal statements in Problems 30 and 31.

30. The base is five times the height.

31. The height is one-fourth of the base.

APPLICATIONS

32. LINEAR DEPRECIATION A computer system was purchased by a small company for $12,000 and is assumed to have a depreciated value of $2,000 after 8 years. If the value is depreciated linearly from $12,000 to $2,000:
(A) Find the linear equation that relates value V (in dollars) to time t (in years).
(B) What would be the depreciated value of the system after 5 years?

33. COST ANALYSIS A video production company is planning to produce an instructional CD. The producer estimates that it will cost $24,900 to produce the CD and $5 per unit to copy and distribute the CD. The budget for this project is $62,000. How many CDs can be produced without exceeding the budget?

34. FORESTRY Forest rangers estimate the height of a tree by measuring the tree's diameter at breast height (DBH) and then using a model constructed for a particular species. A model for sugar maples is

$$h = 2.9d + 30.2$$

where d is the DBH in inches and h is the tree height in feet.
(A) Interpret the slope of this model.
(B) What is the effect of a 1-inch increase in DBH?
(C) How tall is a sugar maple with a DBH of 3 inches? Round answer to the nearest foot.
(D) What is the DBH of a sugar maple that is 45 feet tall? Round answer to the nearest inch.

35. ESTIMATING BODY SURFACE AREA An important criterion for determining drug dosage for children is the patient's body surface area (BSA). John D. Current published the following useful model for estimating BSA*:

$$BSA = 1,321 + 0.3433 \times Wt$$

where BSA is given in square centimeters and Wt in grams.
(A) Interpret the slope of this model.
(B) What is the effect of a 100-gram increase in weight?
(C) What is the BSA for a child that weighs 15 kilograms?

*"Body Surface Area in Infants and Children," *The Internet Journal of Anesthesiology,* 1998, Volume 2, Number 2.

36. ARCHITECTURE A circular arc forms the top of an entryway with 6-foot vertical sides 8 feet apart. If the top of the arc is 2 feet above the ends, what is the radius of the arc?

37. SPORTS MEDICINE The following quotation was found in a sports medicine handout: "The idea is to raise and sustain your heart rate to 70% of its maximum safe rate for your age. One way to determine this is to subtract your age from 220 and multiply by 0.7."
(A) If H is the maximum safe sustained heart rate (in beats per minute) for a person of age A (in years), write a formula relating H and A.
(B) What is the maximum safe sustained heart rate for a 20-year-old?
(C) If the maximum safe sustained heart rate for a person is 126 beats per minute, how old is the person?

38. DATA ANALYSIS Winning times in the men's Olympic 400-meter freestyle event in minutes for selected years are given in Table 1. A mathematical model for these data is

$$y = -0.021x + 5.57$$

where x is years since 1900.
(A) Compare the model and the data graphically and numerically.
(B) Estimate (to three decimal places) the winning time in 2024.

Table 1

Year	Time
1912	5.41
1932	4.81
1952	4.51
1972	4.00
1992	3.75

CHAPTER 2

››› GROUP ACTIVITY Average Speed

If you score 40 on the first exam and 80 on the second, then your average score for the two exams is $(40 + 80)/2 = 60$. The number 60 is the *arithmetic* **average** of 40 and 80.

On the other hand, if you drive 100 miles at a speed of 40 mph, and then drive an additional 100 miles at 80 mph, your average speed for the entire trip is *not* 60 mph. **Average speed** is defined to be the constant speed at which you could drive the same distance in the same length of time. So to calculate average speed, total distance (200 miles) must be divided by total time: The time t_1 it takes to drive 100 miles at 40 mph is $t_1 = (100 \text{ miles})/(40 \text{ mph}) = 2.5$ hours. Similarly, the time t_2 it takes to drive 100 miles at 80 mph is $t_2 = (100 \text{ miles})/(80 \text{ mph}) = 1.25$ hours. Therefore, your average speed is

$$\frac{200 \text{ miles}}{t_1 + t_2} = \frac{200}{2.5 + 1.25} = \frac{200}{3.75} = 53.\overline{3} \text{ mph}$$

(A) You bicycle 15 miles at 21 mph, then 20 miles at 18 mph, and finally 30 miles at 12 mph. Find the average speed.
(B) You bicycle for 2 hours at 18 mph, then 2 more hours at 12 mph. Find the average speed.
(C) You run a 10-mile race by running at a pace of 8 minutes per mile for 1 hour, and after that at a pace of 9 minutes per mile. Define **average pace,** find it (to the nearest second) for the 10-mile race, and discuss the connection between average pace (in minutes per mile) and average speed (in miles per hour).

Functions

THE function concept is one of the most important ideas in mathematics. To study math beyond the elementary level, you absolutely need to have a solid understanding of functions and their graphs. In this chapter, you'll learn the fundamentals of what functions are all about, and how to apply them. As you work through this and subsequent chapters, this will pay off as you study specific types of functions in depth. Everything you learn in this chapter will increase your chance of success in this course, and in almost any other course you may take that involves mathematics.

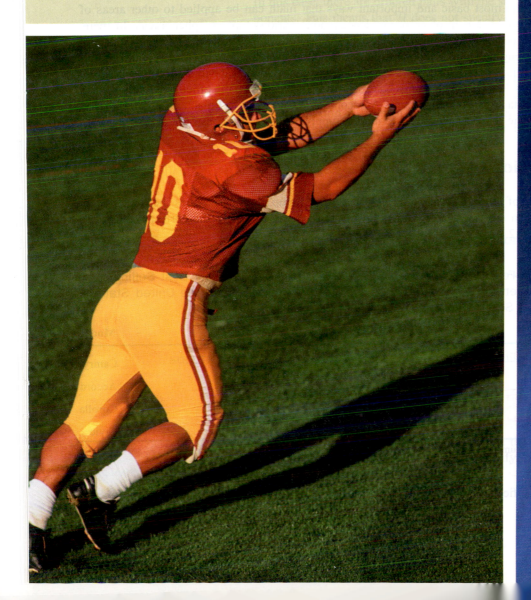

EXAMPLE 1 Vertical and Horizontal Shifts

(A) How are the graphs of $y = x^2 + 2$ and $y = x^2 - 3$ related to the graph of $y = x^2$? Confirm your answer by graphing all three functions in the same coordinate system.

(B) How are the graphs of $y = (x + 2)^2$ and $y = (x - 3)^2$ related to the graph of $y = x^2$? Confirm your answer by graphing all three functions in the same coordinate system.

SOLUTIONS

(A) Note that the output of $y = x^2 + 2$ is always exactly two more than the output of $y = x^2$. Consequently, the graph of $y = x^2 + 2$ is the same as the graph of $y = x^2$ shifted upward two units, and the graph of $y = x^2 - 3$ is the same as the graph of $y = x^2$ shifted downward three units. Figure 3 confirms these conclusions. (It appears that the graph of $y = f(x) + k$ is the graph of $y = f(x)$ shifted up if k is positive and down if k is negative.)

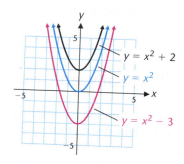

> **Figure 3** Vertical shifts.

(B) Note that the output of $y = (x + 2)^2$ is zero for $x = -2$, while the output of $y = x^2$ is zero for $x = 0$. This suggests that the graph of $y = (x + 2)^2$ is the same as the graph of $y = x^2$ shifted to the left two units, and the graph of $y = (x - 3)^2$ is the same as the graph of $y = x^2$ shifted to the right three units. Figure 4 confirms these conclusions. It appears that the graph of $y = f(x + h)$ is the graph of $y = f(x)$ shifted right if h is negative and left if h is positive.

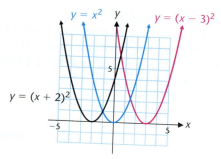

> **Figure 4** Horizontal shifts.

MATCHED PROBLEM 1

(A) How are the graphs of $y = \sqrt{x} + 3$ and $y = \sqrt{x} - 1$ related to the graph of $y = \sqrt{x}$? Confirm your answer by graphing all three functions in the same coordinate system.

(B) How are the graphs of $y = \sqrt{x + 3}$ and $y = \sqrt{x - 1}$ related to the graph of $y = \sqrt{x}$? Confirm your answer by graphing all three functions in the same coordinate system.

To summarize our experiences in Explore-Discuss 1 and Example 1: We can graph $y = f(x) + k$ by **vertically** shifting the graph of $y = f(x)$ upward k units if k is positive

and downward $|k|$ units if k is negative. We can graph $y = f(x + h)$ by **horizontally shifting** the graph of $y = f(x)$ left h units if h is positive and right $|h|$ units if h is negative.

EXAMPLE	2	### Vertical and Horizontal Shifts

The graphs in Figure 5 are either horizontal or vertical shifts of the graph of $f(x) = |x|$. Write appropriate equations for functions H, G, M, and N in terms of f.

> **Figure 5** Vertical and horizontal shifts.

SOLUTION

The graphs of functions H and G are 3 units lower and 1 unit higher, respectively, than the graph of f, so H and G are vertical shifts given by

$$H(x) = |x| - 3 \qquad G(x) = |x| + 1$$

The graphs of functions M and N are 2 units to the left and 3 units to the right, respectively, of the graph of f, so M and N are horizontal shifts given by

$$M(x) = |x + 2| \qquad N(x) = |x - 3|$$

MATCHED PROBLEM 2

The graphs in Figure 6 are either horizontal or vertical shifts of the graph of $f(x) = x^3$. Write appropriate equations for functions H, G, M, and N in terms of f.

> **Figure 6** Vertical and horizontal shifts.

> ## Reflecting Graphs

In Section 2-1, we discussed reflections of graphs and developed symmetry properties that we used as an aid in graphing equations. Now we will consider reflection as an operation that transforms the graph of a function.

>>> EXPLORE-DISCUSS 2

The following activities refer to the graph of f shown in Figure 7 and the corresponding points on the graph shown in Table 2.

(A) Construct a similar table for $y = -f(x)$ and then sketch the graph of $y = -f(x)$. Describe the relationship between the graph of $y = f(x)$ and the graph of $y = -f(x)$ in terms of reflections.

(B) Construct a similar table for $y = f(-x)$ and then sketch the graph of $y = f(-x)$. [*Hint:* Choose x values so that $-x$ is in Table 2.] Describe the relationship between the graph of $y = f(x)$ and the graph of $y = f(-x)$ in terms of reflections.

(C) Construct a similar table for $y = -f(-x)$ and then sketch the graph of $y = -f(-x)$. [*Hint:* Choose x values so that $-x$ is in Table 2.] Describe the relationship between the graph of $y = f(x)$ and the graph of $y = -f(-x)$ in terms of reflections.

> **Figure 7**

Table 2

	x	$f(x)$
A	-2	5
B	-1	0
C	1	-4
D	3	0
E	4	5

EXAMPLE	3	**Reflecting the Graph of a Function**

Let $f(x) = (x - 1)^2$.

(A) How are the graphs of $y = f(x)$ and $y = -f(x)$ related? Confirm your answer by graphing both functions in the same coordinate system.

(B) How are the graphs of $y = f(x)$ and $y = f(-x)$ related? Confirm your answer by graphing both functions in the same coordinate system.

(C) How are the graphs of $y = f(x)$ and $y = -f(-x)$ related? Confirm your answer by graphing both functions in the same coordinate system.

SOLUTIONS Refer to Definition 1 in Section 2-1.

(A) The graph of $y = -f(x)$ can be obtained from the graph of $y = f(x)$ by changing the sign of each y coordinate. This has the effect of moving every point to the opposite side of the x axis. So the graph of $y = -f(x)$ is the **reflection through the x axis** of the graph of $y = f(x)$ [Fig. 8(a)].

(B) The graph of $y = f(-x)$ can be obtained from the graph of $y = f(x)$ by changing the sign of each x coordinate. This has the effect of moving every point to the opposite side of the y axis. So the graph of $y = f(-x)$ is the **reflection through the y axis** of the graph of $y = f(x)$ [Fig. 8(b)].

(C) The graph of $y = -f(-x)$ can be obtained from the graph of $y = f(x)$ by changing the sign of each x and y coordinate. So the graph of $y = -f(-x)$ is the **reflection through the origin** of the graph of $y = f(x)$ [Fig. 8(c)].

> **Figure 8**

(a)
$y = f(x)$ and $y = -f(x)$;
reflection through the x axis

(b)
$y = f(x)$ and $y = f(-x)$; reflection
through the y axis

(c)
$y = f(x)$ and $y = -f(-x)$; reflection
through the origin

| **MATCHED PROBLEM 3** | Repeat Example 3 for $f(x) = |x + 2|$. |
|---|---|

› Stretching and Shrinking Graphs

Horizontal shifts, vertical shifts, and reflections are called **rigid transformations** because they do not change the shape of a graph, only its location. Now we consider some **non-rigid transformations** that change the shape by stretching or shrinking a graph.

››› EXPLORE-DISCUSS 3

› **Figure 9**

Table 3

	x	$f(x)$
A	-2	5
B	0	3
C	2	1
D	4	3
E	6	5

The following activities refer to the graph of f shown in Figure 9 and the corresponding points on the graph shown in Table 3.

(A) Use the points in Table 3 to construct a similar table and sketch a graph for each of the following functions: $y = 2f(x)$ and $y = \frac{1}{2}f(x)$. If $A > 1$, does multiplying f by A stretch or shrink the graph of $y = f(x)$ in the vertical direction? What happens if $0 < A < 1$?

(B) Use the points in Table 3 to complete the following tables and then sketch a graph of $y = f(2x)$ and of $y = f(\frac{1}{2}x)$:

x	$2x$	$f(2x)$	x	$\frac{1}{2}x$	$f(\frac{1}{2}x)$
-1			-4		
0			0		
1			4		
2			8		
3			12		

If $A > 1$, is the graph of $y = f(Ax)$ a horizontal stretch or a horizontal shrink of the graph of $y = f(x)$? What if $0 < A < 1$?

In general, the graph of $y = Af(x)$ can be obtained from the graph of $y = f(x)$ by multiplying the y coordinate of each point on the graph f by A. This **vertically stretches** the graph of $y = f(x)$ if $A > 1$ and **vertically shrinks** the graph if $0 < A < 1$.

The graph of $y = f(Ax)$ can be obtained from the graph of $y = f(x)$ by multiplying the x coordinate of each point on the graph f by $1/A$. This **horizontally stretches** the graph of $y = f(x)$ if $0 < A < 1$ and **horizontally shrinks** the graph if $A > 1$.

Another common name for a stretch is an **expansion** and another common name for a shrink is a **contraction.**

EXAMPLE 4 Stretching or Shrinking a Graph

Let $f(x) = 1 + x^2$.

(A) How are the graphs of $y = 2f(x)$ and $y = \frac{1}{2}f(x)$ related to the graph of $y = f(x)$? Confirm your answer by graphing all three functions in the same coordinate system.

(B) How are the graphs of $y = f(2x)$ and $y = f(\frac{1}{2}x)$ related to the graph of $y = f(x)$? Confirm your answer by graphing all three functions in the same coordinate system.

SOLUTIONS

(A) The graph of $y = 2f(x) = 2 + 2x^2$ can be obtained from the graph of f by multiplying each y value by 2. This stretches the graph of f vertically (away from the x axis) by a factor of 2. The graph of $y = \frac{1}{2}f(x) = \frac{1}{2} + \frac{1}{2}x^2$ can be obtained from the graph of f by multiplying each y value by $\frac{1}{2}$. This shrinks the graph of f vertically (toward the x axis) by a factor of $\frac{1}{2}$ [Fig. 10(a)].

(B) The graph of $y = f(2x) = 1 + 4x^2$ can be obtained from the graph of f by multiplying each x value by $\frac{1}{2}$. This shrinks the graph of f horizontally (toward the y axis) by a factor of $\frac{1}{2}$. The graph of $y = f(\frac{1}{2}x) = 1 + \frac{1}{4}x^2$ can be obtained from the graph of f by multiplying each x value by 2. This stretches the graph of f horizontally (away from the y axis) by a factor of 2 [Fig. 10(b)].

(a) Vertical stretching and shrinking

(b) Horizontal stretching and shrinking

› Figure 10

MATCHED PROBLEM 4

Let $f(x) = 4 - x^2$.

(A) How are the graphs of $y = 2f(x)$ and $y = \frac{1}{2}f(x)$ related to the graph of $y = f(x)$? Confirm your answer by graphing all three functions in the same coordinate system.

(B) How are the graphs of $y = f(2x)$ and $y = f(\frac{1}{2}x)$ related to the graph of $y = f(x)$? Confirm your answer by graphing all three functions in the same coordinate system.

Plotting points with the same x coordinate will help you recognize vertical stretches and shrinks [Fig. 10(a)]. And plotting points with the same y coordinate will help you recognize horizontal stretches and shrinks [Fig. 10(b)].

Note that for some functions, a horizontal stretch or shrink can also be interpreted as a vertical stretch or shrink. For example, if $y = f(x) = x^2$, then

$$y = 4f(x) = 4x^2 = (2x)^2 = f(2x)$$

So the graph of $y = 4x^2$ is both a vertical stretch and a horizontal shrink of the graph of $y = x^2$.

The transformations we've studied are summarized next for easy reference.

> **Figure 11** Graph transformations.

(a)
Vertical translation
$g(x) = f(x) + 2$
$h(x) = f(x) - 3$

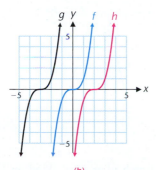

(b)
Horizontal translation
$g(x) = f(x + 3)$
$h(x) = f(x - 2)$

> **GRAPH TRANSFORMATIONS (SUMMARY)**

Vertical Shift [Fig. 11(a)]:

$$y = f(x) + k \qquad \begin{cases} k > 0 & \text{Shift graph of } y = f(x) \text{ up } k \text{ units} \\ k < 0 & \text{Shift graph of } y = f(x) \text{ down } |k| \text{ units} \end{cases}$$

Horizontal Shift [Fig. 11(b)]:

$$y = f(x + h) \qquad \begin{cases} h > 0 & \text{Shift graph of } y = f(x) \text{ left } h \text{ units} \\ h < 0 & \text{Shift graph of } y = f(x) \text{ right } |h| \text{ units} \end{cases}$$

Vertical Stretch and Shrink [Fig. 11(c)]:

$$y = Af(x) \qquad \begin{cases} A > 1 & \text{Vertically stretch the graph of } y = f(x) \\ & \text{by multiplying each } y \text{ value by } A \\ 0 < A < 1 & \text{Vertically shrink the graph of } y = f(x) \\ & \text{by multiplying each } y \text{ value by } A \end{cases}$$

Horizontal Stretch and Shrink [Fig. 11(d)]:

$$y = f(Ax) \qquad \begin{cases} A > 1 & \text{Horizontally shrink the graph of } y = f(x) \\ & \text{by multiplying each } x \text{ value by } \frac{1}{A} \\ 0 < A < 1 & \text{Horizontally stretch the graph of } y = f(x) \\ & \text{by multiplying each } x \text{ value by } \frac{1}{A} \end{cases}$$

Reflection [Fig. 11(e)]:

$$y = -f(x) \qquad \text{Reflect the graph of } y = f(x) \text{ through the } x \text{ axis}$$
$$y = f(-x) \qquad \text{Reflect the graph of } y = f(x) \text{ through the } y \text{ axis}$$
$$y = -f(-x) \qquad \text{Reflect the graph of } y = f(x) \text{ through the origin}$$

(c)
Vertical expansion and contraction
$g(x) = 2f(x)$
$h(x) = \frac{1}{2}f(x)$

(d)
Horizontal expansion and contraction
$g(x) = f(2x)$
$h(x) = f(\frac{1}{2}x)$

(e)
Reflection
$g(x) = f(-x)$
$h(x) = -f(x)$
$k(x) = -f(-x)$

EXAMPLE 5 Combining Graph Transformations

The graph of $y = g(x)$ in Figure 12 is a transformation of the graph of $y = x^2$. Find an equation for the function g.

> **Figure 12**

SOLUTION To transform the graph of $y = x^2$ [Fig. 13(a)] into the graph of $y = g(x)$, we first reflect the graph of $y = x^2$ through the x axis [Fig. 13(b)], then shift it to the right two units [Fig. 13(c)]. An equation for the function g is

$$g(x) = -(x - 2)^2$$

(a) $y = x^2$

(b) $y = -x^2$

(c) $y = -(x - 2)^2$

> **Figure 13**

MATCHED PROBLEM 5 The graph of $y = h(x)$ in Figure 14 is a transformation of the graph of $y = x^3$. Find an equation for the function h.

> **Figure 14**

> ## Even and Odd Functions

Certain transformations leave the graphs of some functions unchanged. For example, reflecting the graph of $y = x^2$ through the y axis does not change the graph. Functions with this property are called *even functions*. Similarly, reflecting the graph of $y = x^3$ through the origin does not change the graph. Functions with this property are called *odd functions*. More formally, we have the following definitions.

> **EVEN AND ODD FUNCTIONS**

If $f(x) = f(-x)$ for all x in the domain of f, then f is an **even function.**

If $f(-x) = -f(x)$ for all x in the domain of f, then f is an **odd function.**

The graph of an even function is **symmetric with respect to the y axis** and the graph of an odd function is **symmetric with respect to the origin** (Fig. 15).

> **Figure 15** Even and odd functions.

EXAMPLE 6 Testing for Even and Odd Functions

Determine whether the functions f, g, and h are even, odd, or neither.

(A) $f(x) = x^4 + 1$ (B) $g(x) = x^3 + 1$ (C) $h(x) = x^5 + x$

SOLUTIONS It will be useful to note the following: if n is an even integer, then $(-x)^n = (-1)^n x^n = x^n$ because $(-1)^n = 1$ if n is even. But if n is an odd integer, $(-x)^n = (-1)^n x^n = -x^n$ because $(-1)^n = -1$ when n is odd.

(A) $f(x) = x^4 + 1$

 $f(-x) = (-x)^4 + 1$ $(-x)^4 = x^4$ because 4 is even.

 $= x^4 + 1$

 $= f(x)$

This shows that f is even.

(B) $g(x) = x^3 + 1$

 $g(-x) = (-x)^3 + 1$ $(-x)^3 = -x^3$ because 3 is odd.

 $= -x^3 + 1$

 $-g(x) = -(x^3 + 1)$ Distribute the negative.

 $= -x^3 - 1$

The function $g(-x)$ is neither $g(x)$ nor $-g(x)$, so g is neither even nor odd.

(C) $h(x) = x^5 + x$

 $h(-x) = (-x)^5 + (-x)$ $(-x)^5 = -x^5$ because 5 is odd.

 $= -x^5 - x$

 $-h(x) = -(x^5 + x)$ Distribute the negative.

 $= -x^5 - x$

Since $h(-x) = -h(x)$, h is odd.

MATCHED PROBLEM 6 Determine whether the functions F, G, and H are even, odd, or neither:

(A) $F(x) = x^3 - 2x$ (B) $G(x) = x^2 + 1$ (C) $H(x) = 2x + 4$

ANSWERS TO MATCHED PROBLEMS

1. (A) The graph of $y = \sqrt{x} + 3$ is the same as the graph of $y = \sqrt{x}$ shifted upward 3 units, and the graph of $y = \sqrt{x} - 1$ is the same as the graph of $y = \sqrt{x}$ shifted downward 1 unit. The figure confirms these conclusions.

(B) The graph of $y = \sqrt{x + 3}$ is the same as the graph of $y = \sqrt{x}$ shifted to the left 3 units, and the graph of $y = \sqrt{x - 1}$ is the same as the graph of $y = \sqrt{x}$ shifted to the right 1 unit. The figure confirms these conclusions.

2. $G(x) = (x + 3)^3$, $H(x) = (x - 1)^3$, $M(x) = x^3 + 3$, $N(x) = x^3 - 4$

3. (A) The graph of $y = -f(x)$ is the reflection through the x axis of the graph of $y = f(x)$.

(B) The graph of $y = f(-x)$ is the reflection through the y axis of the graph of $y = f(x)$.

(C) The graph of $y = -f(-x)$ is the reflection through the origin of the graph of $y = f(x)$.

4. (A) The graph of $y = 2f(x)$ is a vertical stretch of the graph of $y = f(x)$ by a factor of 2. The graph of $y = \frac{1}{2}f(x)$ is a vertical shrink of the graph of $y = f(x)$ by a factor of $\frac{1}{2}$.

(B) The graph of $y = f(2x)$ is a horizontal shrink of the graph of $y = f(x)$ by a factor of $\frac{1}{2}$. The graph of $y = f(\frac{1}{2}x)$ is a horizontal stretch of the graph of $y = f(x)$ by a factor of 2.

5. The graph of function h is a reflection through the x axis and a horizontal translation of three units to the left of the graph of $y = x^3$. An equation for h is $h(x) = -(x + 3)^3$.

6. (A) Odd (B) Even (C) Neither

3-3 Exercises

1. Explain why the graph of $y = f(x) + k$ is the same as the graph of $y = f(x)$ moved upward k units when k is positive.

2. Explain why the graph of $y = Af(x)$ is a vertical stretch of the graph of $y = f(x)$ when $A > 1$, and a vertical shrink when $A < 1$.

3. Explain why the graph of $y = -f(x)$ is a reflection of the graph of $y = f(x)$ about the x axis, and why the graph of $y = f(-x)$ is a reflection about the y axis.

4. Is every function either even or odd? Explain your answer.

In Problems 5–10, without looking back in the text, indicate the domain and range of each of the following functions. (Making rough sketches on scratch paper may help.)

5. $h(x) = -\sqrt{x}$ **6.** $m(x) = -\sqrt[3]{x}$

7. $g(x) = -2x^2$ **8.** $f(x) = -0.5|x|$

9. $F(x) = -0.5x^3$ **10.** $G(x) = 4x^3$

Problems 11–26 refer to the functions f and g given by the graphs below. The domain of each function is $[-2, 2]$, the range of f is $[-2, 2]$, and the range of g is $[-1, 1]$. Use the graph of f or g, as required, to graph the function h and state the domain and range of h.

11. $h(x) = f(x) + 2$

12. $h(x) = g(x) - 1$

13. $h(x) = g(x) + 2$

14. $h(x) = f(x) - 1$

15. $h(x) = f(x - 2)$

16. $h(x) = g(x - 1)$

17. $h(x) = g(x + 2)$

18. $h(x) = f(x - 1)$

19. $h(x) = -f(x)$

20. $h(x) = -g(x)$

21. $h(x) = 2g(x)$

22. $h(x) = \frac{1}{2}f(x)$

23. $h(x) = g(2x)$

24. $h(x) = f\left(\frac{1}{2}x\right)$

25. $h(x) = f(-x)$

26. $h(x) = -g(-x)$

Indicate whether each function in Problems 27–36 is even, odd, or neither.

27. $g(x) = x^3 + x$ **28.** $f(x) = x^5 - x$

29. $m(x) = x^4 + 3x^2$ **30.** $h(x) = x^4 - x^2$

31. $F(x) = x^5 + 1$ **32.** $f(x) = x^5 - 3$

33. $G(x) = x^4 + 2$ **34.** $P(x) = x^4 - 4$

35. $q(x) = x^2 + x - 3$ **36.** $n(x) = 2x - 3$

In Problems 37–44, the graph of the function g is formed by applying the indicated sequence of transformations to the given function f. Find an equation for the function g. Check your work by graphing f and g in a standard viewing window.

37. The graph of $f(x) = \sqrt[3]{x}$ is shifted four units to the left and five units down.

38. The graph of $f(x) = x^3$ is shifted five units to the right and four units up.

39. The graph of $f(x) = \sqrt{x}$ is shifted six units up, reflected in the x axis, and vertically shrunk by a factor of 0.5.

40. The graph of $f(x) = \sqrt{x}$ is shifted two units down, reflected in the x axis, and vertically stretched by a factor of 4.

41. The graph of $f(x) = x^2$ is reflected in the x axis, vertically stretched by a factor of 2, shifted four units to the left, and shifted two units down.

42. The graph of $f(x) = |x|$ is reflected in the x axis, vertically shrunk by a factor of 0.5, shifted three units to the right, and shifted four units up.

43. The graph of $f(x) = \sqrt{x}$ is horizontally stretched by a factor of 0.5, reflected in the y axis, and shifted two units to the left.

44. The graph of $f(x) = \sqrt[3]{x}$ is horizontally shrunk by a factor of 2, shifted three units up, and reflected in the y axis.

Use graph transformations to sketch the graph of each function in Problems 45–62.

45. $f(x) = 4x^2$

46. $g(x) = \dfrac{1}{3}\sqrt{x}$

47. $h(x) = |x + 2|$

48. $k(x) = |x - 4|$

49. $m(x) = -|4x - 8|$

50. $n(x) = -|9 + 3x|$

51. $p(x) = 3 - \sqrt{x}$

52. $q(x) = -2 + \sqrt{x + 3}$

53. $r(x) = 3\sqrt{x - 1} + 2$

54. $s(x) = \sqrt{x - 1} + 2$

55. $h(x) = x^2 + 3$

56. $h(x) = 4 - x^2$

57. $k(x) = 2x^3 + 1$

58. $h(x) = 3x^3 - 1$

59. $n(x) = (x + 2)^2$

60. $m(x) = (x - 4)^2$

61. $q(x) = 4 - \dfrac{1}{2}(x - 2)^2$

62. $p(x) = 5 - \dfrac{2}{3}(x + 3)^2$

Each graph in Problems 63–78 is a transformation of one of the six basic functions in Figure 1. Find an equation for the given graph.

63.

64.

65.

66.

67.

68.

69.

70.

71.

72.

73.

74.

75.

76.

77.

78.

79. Consider the graphs of $f(x) = \sqrt[3]{8x}$ and $g(x) = 2\sqrt[3]{x}$.
(A) Describe each as a stretch or shrink of $y = \sqrt[3]{x}$.
(B) Graph both functions in the same viewing window on a graphing calculator. What do you notice?
(C) Rewrite the formula for f algebraically to show that f and g are in fact the same function. (This shows that for some functions, a horizontal stretch or shrink can also be interpreted as a vertical stretch or shrink.)

80. Consider the graphs of $f(x) = (3x)^3$ and $g(x) = 27x^3$.
(A) Describe each as a stretch or shrink of $y = x^3$.
(B) Graph both functions in the same viewing window on a graphing calculator. What do you notice?
(C) Rewrite the formula for f algebraically to show that f and g are in fact the same function. (This shows that for some functions, a horizontal stretch or shrink can also be interpreted as a vertical stretch or shrink.)

81. (A) Starting with the graph of $y = x^2$, apply the following transformations.
(i) Shift downward 5 units, then reflect in the x axis.
(ii) Reflect in the x axis, then shift downward 5 units.
What do your results indicate about the significance of order when combining transformations?
(B) Write a formula for the function corresponding to each of the above transformations. Discuss the results of part A in terms of order of operations.

82. (A) Starting with the graph of $y = |x|$, apply the following transformations.

 (i) Stretch vertically by a factor of 2, then shift upward 4 units.

 (ii) Shift upward 4 units, then stretch vertically by a factor of 2.

 What do your results indicate about the significance of order when combining transformations?

 (B) Write a formula for the function corresponding to each of the above transformations. Discuss the results of part A in terms of order of operations.

83. Based on the graphs of the six elementary functions in Figure 1, which are odd, which are even, and which are neither? Use the definitions of odd and even functions to prove your answers.

84. Based on the results of Example 6, why do you think the terms "even" and "odd" are used to describe functions with particular symmetry properties?

Changing the order in a sequence of transformations may change the final result. Investigate each pair of transformations in Problems 85–90 to determine if reversing their order can produce a different result. Support your conclusions with specific examples and/or mathematical arguments.

85. Vertical shift, horizontal shift

86. Vertical shift, reflection in y axis

87. Vertical shift, reflection in x axis

88. Vertical shift, expansion

89. Horizontal shift, reflection in x axis

90. Horizontal shift, contraction

Problems 91–94 refer to two functions f and g with domain $[-5, 5]$ and partial graphs as shown here.

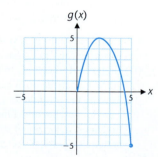

91. Complete the graph of f over the interval $[-5, 0]$, given that f is an even function.

92. Complete the graph of f over the interval $[-5, 0]$, given that f is an odd function.

93. Complete the graph of g over the interval $[-5, 0]$, given that g is an odd function.

94. Complete the graph of g over the interval $[-5, 0]$, given that g is an even function.

95. Let f be any function with the property that $-x$ is in the domain of f whenever x is in the domain of f, and let E and O be the functions defined by

$$E(x) = \tfrac{1}{2}[f(x) + f(-x)]$$

and

$$O(x) = \tfrac{1}{2}[f(x) - f(-x)]$$

 (A) Show that E is always even.

 (B) Show that O is always odd.

 (C) Show that $f(x) = E(x) + O(x)$. What is your conclusion?

96. Let f be any function with the property that $-x$ is in the domain of f whenever x is in the domain of f, and let $g(x) = xf(x)$.

 (A) If f is even, is g even, odd, or neither?

 (B) If f is odd, is g even, odd, or neither?

APPLICATIONS

97. PRODUCTION COSTS Total production costs for a product can be broken down into fixed costs, which do not depend on the number of units produced, and variable costs, which do depend on the number of units produced. So, the total cost of producing x units of the product can be expressed in the form

$$C(x) = K + f(x)$$

where K is a constant that represents the fixed costs and $f(x)$ is a function that represents the variable costs. Use the graph of the variable-cost function $f(x)$ shown in the figure to graph the total cost function if the fixed costs are $30,000.

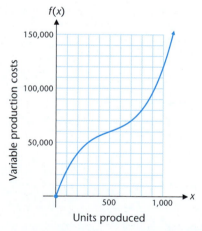

98. COST FUNCTIONS Refer to the variable-cost function $f(x)$ in Problem 97. Suppose construction of a new production facility results in a 25% decrease in the variable cost at all levels of output. If F is the new variable-cost function, use the graph of f to graph $y = F(x)$, then graph the total cost function for fixed costs of $30,000.

99. TIMBER HARVESTING To determine when a forest should be harvested, forest managers often use formulas to estimate the number of board feet a tree will produce. A board foot equals 1 square foot of wood, 1 inch thick. Suppose that the number of board feet y yielded by a tree can be estimated by

$$y = f(x) = C + 0.004(x - 10)^3$$

where x is the diameter of the tree in inches measured at a height of 4 feet above the ground and C is a constant that depends on the species being harvested. Graph $y = f(x)$ for $C = 10$, 15, and 20 simultaneously in the viewing window with Xmin $= 10$, Xmax $= 25$, Ymin $= 10$, and Ymax $= 35$. Write a brief verbal description of this collection of functions.

 100. SAFETY RESEARCH If a person driving a vehicle slams on the brakes and skids to a stop, the speed v in miles per hour at the time the brakes are applied is given approximately by

$$v = f(x) = C\sqrt{x}$$

where x is the length of the skid marks and C is a constant that depends on the road conditions and the weight of the vehicle. The table lists values of C for a midsize automobile and various road conditions. Graph $v = f(x)$ for the values of C in the table simultaneously in the viewing window with Xmin $= 0$, Xmax $= 100$, Ymin $= 0$, and Ymax $= 60$. Write a brief verbal description of this collection of functions.

Road Condition	C
Wet (concrete)	3.5
Wet (asphalt)	4
Dry (concrete)	5
Dry (asphalt)	5.5

 101. FLUID FLOW A cubic tank is 4 feet on a side and is initially full of water. Water flows out an opening in the bottom of the tank at a rate proportional to the square root of the depth (see the figure). Using advanced concepts from mathematics and physics, it can be shown that the volume of the water in the tank t minutes after the water begins to flow is given by

$$V(t) = \frac{64}{C^2}(C - t)^2 \quad 0 \le t \le C$$

where C is a constant that depends on the size of the opening. Sketch by hand the graphs of $y = V(t)$ for $C = 1, 2, 4$, and 8. Write

a brief verbal description of this collection of functions. Based on the graphs, do larger values of C correspond to a larger or smaller opening?

Figure for 101

 102. EVAPORATION A water trough with triangular ends is 9 feet long, 4 feet wide, and 2 feet deep (see the figure). Initially, the trough is full of water, but due to evaporation, the volume of the water in the trough decreases at a rate proportional to the square root of the volume. Using advanced concepts from mathematics and physics, it can be shown that the volume after t hours is given by

$$V(t) = \frac{1}{C^2}(t + 6C)^2 \quad 0 \le t \le 6|C|$$

where C is a constant. Sketch by hand the graphs of $y = V(t)$ for $C = -4, -5$, and -6. Write a brief verbal description of this collection of functions. Based on the graphs, do values of C with a larger absolute value correspond to faster or slower evaporation?

3-4

Quadratic Functions

> **Graphing Quadratic Functions**
> **Modeling with Quadratic Functions**
> **Solving Quadratic Inequalities**
> **Modeling with Quadratic Regression**

The graph of the squaring function $h(x) = x^2$ is shown in Figure 1 on page 204. Notice that h is an even function; that is, the graph of h is symmetric with respect to the y axis. Also, the lowest point on the graph is $(0, 0)$. Let's explore the effect of applying a sequence of basic transformations to the graph of h.

Indicate how the graph of each function is related to the graph of $h(x) = x^2$. Discuss the symmetry of the graphs and find the highest or lowest point, whichever exists, on each graph.

(A) $f(x) = (x - 3)^2 - 7 = x^2 - 6x + 2$

(B) $g(x) = 0.5(x + 2)^2 + 3 = 0.5x^2 + 2x + 5$

(C) $m(x) = -(x - 4)^2 + 8 = -x^2 + 8x - 8$

(D) $n(x) = -3(x + 1)^2 - 1 = -3x^2 - 6x - 4$

> **Graphing Quadratic Functions**

Graphing the functions in Explore-Discuss 1 produces figures similar in shape to the graph of the squaring function in Figure 1. These figures are called *parabolas*. The functions that produced these parabolas are examples of the important class of *quadratic functions*, which we will now define.

> **DEFINITION 1** Quadratic Functions

If a, b, and c are real numbers with $a \neq 0$, then the function

$$f(x) = ax^2 + bx + c$$

is called a **quadratic function** and its graph is called a **parabola.** This is known as the **general form** of a quadratic function.

> **Figure 1** Squaring function $h(x) = x^2$.

Because the expression $ax^2 + bx + c$ represents a real number no matter what number we substitute for x,

the domain of a quadratic function is the set of all real numbers.

We will discuss methods for determining the range of a quadratic function later in this section. Typical graphs of quadratic functions are illustrated in Figure 2.

> **Figure 2** Graphs of quadratic functions.

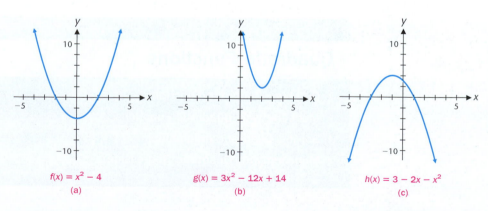

$f(x) = x^2 - 4$ $g(x) = 3x^2 - 12x + 14$ $h(x) = 3 - 2x - x^2$
(a) (b) (c)

We will begin our detailed study of quadratic functions by examining some in a special form, which we will call the **vertex form:***

$$f(x) = a(x - h)^2 + k$$

*In Problem 75 of Exercises 3-4, you will be asked to show that any function of this form fits the definition of quadratic function in Definition 1.

We'll see where the name comes from in a bit. For now, refer to Explore-Discuss 1. Any function of this form is a transformation of the basic squaring function $g(x) = x^2$, so we can use transformations to analyze the graph.

EXAMPLE	1

The Graph of a Quadratic Function

Use transformations of $g(x) = x^2$ to graph the function $f(x) = 2(x - 3)^2 + 4$. Use your graph to determine the graphical significance of the constants 2, 3, and 4 in this function.

SOLUTION Multiplying by 2 vertically stretches the graph by a factor of 2. Subtracting 3 inside the square moves the graph 3 units to the right. Adding 4 outside the square moves the graph 4 units up. The graph of f is shown in Figure 3, along with the graph of $g(x) = x^2$.

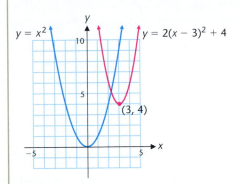

> **Figure 3**

The lowest point on the graph of f is $(3, 4)$, so $h = 3$ and $k = 4$ determine the key point where the graph changes direction. The constant $a = 2$ affects the width of the parabola. ◉

MATCHED PROBLEM 1 Use transformations of $g(x) = x^2$ to graph the function $f(x) = -\frac{1}{2}(x - 2)^2 + 5$. Use your graph to determine the significance of the constants $-\frac{1}{2}$, 2, and 5 in this function.

Every parabola has a point where the graph reaches a maximum or minimum and changes direction. We will call that point the **vertex** of the parabola. Finding the vertex is key to many of the things we'll do with parabolas. Example 1 and Explore-Discuss 1 demonstrate that

if a quadratic function is in the form $f(x) = a(x - h)^2 + k$, then the vertex is the point (h, k).

Next, notice that the graph of $h(x) = x^2$ is symmetric about the y axis. As a result, the transformation $f(x) = 2(x - 3)^2 + 4$ is symmetric about the vertical line $x = 3$ (which runs through the vertex). We will call this vertical line of symmetry the **axis**, or **axis of symmetry** of a parabola. If the page containing the graph of f is folded along the line $x = 3$, the two halves of the graph will match exactly.

Finally, Explore-Discuss 1 illustrates the significance of the constant a in $f(x) = a(x - h)^2 + k$. If a is positive, the graph has a minimum and opens upward. But if a is negative, the graph will be a vertical reflection of $h(x) = x^2$ and will have a maximum and open downward. The size of a determines the width of the parabola: if $|a| > 1$, the graph is narrower than $h(x) = x^2$, and if $|a| < 1$, it is wider.

These properties of a quadratic function in vertex form are summarized next.

> PROPERTIES OF A QUADRATIC FUNCTION AND ITS GRAPH

Given a quadratic function in vertex form

$$f(x) = a(x - h)^2 + k \qquad a \neq 0$$

we summarize general properties as follows:

1. The graph of f is a parabola:

2. Vertex: (h, k) (parabola rises on one side of the vertex and falls on the other).

3. Axis (of symmetry): $x = h$ (parallel to y axis).

4. $f(h) = k$ is the minimum if $a > 0$ and the maximum if $a < 0$.

5. Domain: all real numbers; range: $(-\infty, k]$ if $a < 0$ or $[k, \infty)$ if $a > 0$.

6. The graph of f is the graph of $g(x) = ax^2$ translated horizontally h units and vertically k units.

Now that we can recognize the key properties of quadratic functions in vertex form, the obvious question is "What if a quadratic function is *not* in vertex form?" More often than not, the quadratic functions we will encounter will be in the form $f(x) = ax^2 + bx + c$. The method of completing the square, which we studied in Section 1-5, can be used to find the vertex form in this case.

| EXAMPLE | 2 | **Finding the Vertex Form of a Parabola** |

Find the vertex form of $f(x) = 2x^2 - 8x + 4$ by completing the square, then write the vertex and the axis.

SOLUTION

We will begin by separating the first two terms with parentheses; then we will complete the square to factor part of f as a perfect square.

$$f(x) = 2x^2 - 8x + 4 \qquad \text{Group first two terms.}$$
$$= (2x^2 - 8x) + 4 \qquad \text{Factor out 2.}$$
$$= 2(x^2 - 4x) + 4 \qquad (b/a)^2 = (-2)^2 = 4$$
$$= 2(x^2 - 4x + ?) + 4 \qquad \text{Add 4 inside parentheses; because of the 2 in front, we really added 8, so subtract 8 as well.}$$
$$= 2(x^2 - 4x + 4) + 4 - 8 \qquad \text{Factor inside parentheses; simplify } 4 - 8.$$
$$= 2(x - 2)^2 - 4$$

The vertex form is $f(x) = 2(x - 2)^2 - 4$; the vertex is $(2, -4)$ and the axis is $x = 2$.

MATCHED PROBLEM 2

Find the vertex form of $g(x) = 3x^2 - 18x + 2$ by completing the square, then write the vertex and axis.

EXAMPLE 3

Graphing a Quadratic Function

Let $f(x) = -0.5x^2 - x + 2$.

(A) Use completing the square to find the vertex form of f. State the vertex and the axis of symmetry.

(B) Graph f and find the maximum or minimum of $f(x)$, the domain, the range, and the intervals where f is increasing or decreasing.

SOLUTIONS

(A) Complete the square:

$$f(x) = -0.5x^2 - x + 2 \qquad \text{Group first two terms}$$
$$= (-0.5x^2 - x) + 2 \qquad \text{Factor out } -0.5$$
$$= -0.5(x^2 + 2x + ?) + 2 \qquad \text{Add 1 inside the parentheses to complete the square and 0.5 outside the parentheses.}$$
$$= -0.5(x^2 + 2x + \mathbf{1}) + 2 + 0.5 \qquad \text{Factor the trinomial and combine like terms.}$$
$$= -0.5(x + 1)^2 + 2.5$$

From this last form we see that $h = -1$ and $k = 2.5$, so the vertex is $(-1, 2.5)$ and the axis of symmetry is $x = -1$.

(B) To graph f, locate the axis and vertex; then plot several points on either side of the axis

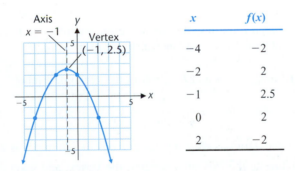

x	$f(x)$
-4	-2
-2	2
-1	2.5
0	2
2	-2

The domain of f is $(-\infty, \infty)$. From the graph we see that the maximum value is $f(-1) = 2.5$ and that f is increasing on $(-\infty, -1]$ and decreasing on $[-1, \infty)$. Also, $y = f(x)$ can be any number less than or equal to 2.5; the range of f is $y \leq 2.5$ or $(-\infty, 2.5]$.

MATCHED PROBLEM 3

Let $f(x) = -x^2 + 4x + 2$.

(A) Use completing the square to find the vertex form of f. State the vertex and the axis of symmetry.

(B) Graph f and find the maximum or minimum of $f(x)$, the domain, the range, and the intervals where f is increasing or decreasing.

We can develop a simple formula for finding the vertex of a parabola if we apply completing the square to $f(x) = ax^2 + bx + c$.

$$f(x) = ax^2 + bx + c$$ Factor *a* out of the first two terms.

$$= a\left(x^2 + \frac{b}{a} + ?\right) + c$$ Add $\left(\dfrac{b}{2a}\right)^2$ inside the parentheses and

subtract $a\left(\dfrac{b}{2a}\right)^2 = \dfrac{b^2}{4a}$ outside the parentheses.

$$= a\left(x^2 + \frac{b}{a} + \frac{b^2}{4a^2}\right) + c - \frac{b^2}{4a}$$ Factor the trinomial.

$$= a\left(x + \frac{b}{2a}\right)^2 + c - \frac{b^2}{4a}$$

This is in vertex form, and the x coordinate of the vertex is $-b/2a$.

> **FINDING THE VERTEX OF A PARABOLA**
>
> When a quadratic function is written in the form $f(x) = ax^2 + bx + c$, the first coordinate of the vertex can be found using the formula
>
> $$x = -\frac{b}{2a}$$
>
> The second coordinate can then be found by evaluating f at the first coordinate.

EXAMPLE 4 Graphing a Quadratic Function

Let $f(x) = x^2 - 6x + 4$.

(A) Use the vertex formula to find the vertex and the axis of symmetry of f.

(B) Graph f and find the maximum or minimum of $f(x)$, the domain, the range, and the intervals where f is increasing or decreasing.

SOLUTIONS (A) Using $a = 1$ and $b = -6$ in the vertex formula,

$$x = -\frac{b}{2a} = -\frac{-6}{2} = 3; \; f(3) = 3^2 - 6(3) + 4 = -5$$

The vertex is $(3, -5)$ and the axis of symmetry is $x = 3$.

(B) Locate the axis of symmetry, the vertex, and several points on either side of the axis of symmetry, and graph f.

x	$f(x)$
0	4
2	-4
3	-5
4	-4
6	4

Vertex
$(3, -5)$

The minimum of $f(x)$ is -5, the domain is $(-\infty, \infty)$, the range is $[-5, \infty)$, f is decreasing on $(-\infty, 3]$ and increasing on $[3, \infty)$.

MATCHED PROBLEM 4

Let $f(x) = \frac{1}{4}x^2 + \frac{1}{2}x - 5$.

(A) Use the vertex formula to find the vertex and the axis of symmetry of f.

(B) Graph f and find the maximum or minimum of $f(x)$, the domain, the range, and the intervals where f is increasing or decreasing.

EXAMPLE 5 Finding the Equation of a Parabola

Find the equation of the parabola with vertex $(3, -2)$ and x intercept 4.

SOLUTION Since the vertex is $(3, -2)$, the vertex form for the equation is

$$f(x) = a(x - 3)^2 - 2 \qquad h = 3,\ k = -2 \text{ in } a(x - h)^2 + k$$

Since 4 is an x intercept, $f(4) = 0$. Substituting $x = 4$ and $f(x) = 0$ into the vertex formula, we have

$$f(4) = a(4 - 3)^2 - 2 = 0 \qquad \text{Add 2 to both sides.}$$
$$a = 2$$

The equation of this parabola is

$$f(x) = 2(x - 3)^2 - 2 = 2x^2 - 12x + 16$$

MATCHED PROBLEM 5 Find the equation of the parabola with vertex $(4, -2)$ and y intercept 2.

We have presented two methods for locating the vertex of a parabola: completing the square and evaluating the vertex formula. You may prefer to use the completing the square process or to remember the formula. Unless directed otherwise, we will leave this choice to you. If you have a graphing calculator, there is a third approach.

Technology Connections

The **maximum** and **minimum** options on the CALC menu of a graphing calculator can be used to find the vertex of a parabola. After selecting the appropriate option (maximum or minimum), you will be asked to provide three x values: a left bound, a right bound, and a guess. The maximum or minimum is displayed at the bottom of the screen. Figure 4(a) locates the vertex of the parabola in Example 1 and Figure 4(b) locates the vertex of the parabola in Example 4.

(a) $f(x) = -0.5x^2 - x + 2$

(b) $f(x) = x^2 - 6x + 4$

▸ Figure 4

> ## Modeling with Quadratic Functions

We will now look at some applications that can be modeled using quadratic functions.

| EXAMPLE | 6 | **Maximum Area** |

A dairy farm has a barn that is 150 feet long and 75 feet wide. The owner has 240 feet of fencing and plans to use all of it in the construction of two identical adjacent outdoor pens, with part of the long side of the barn as one side of the pens, and a common fence between the two (Fig. 5). The owner wants the pens to be as large as possible.

150 feet

75 feet

> **Figure 5**

(A) Construct a mathematical model for the combined area of both pens in the form of a function $A(x)$ (see Fig. 5) and state the domain of A.

(B) Find the value of x that produces the maximum combined area.

(C) Find the dimensions and the area of each pen.

SOLUTIONS (A) The combined area of the two pens is

$$A = xy$$

Adding up the lengths of all four segments of fence, we find that building the pens will require $3x + y$ feet of fencing. We have 240 feet of fence to use, so

$$3x + y = 240$$
$$y = 240 - 3x$$

Because the distances x and y must be nonnegative, x and y must satisfy $x \geq 0$ and $y = 240 - 3x \geq 0$. It follows that $0 \leq x \leq 80$. Substituting for y in the combined area equation, we have the following model for this problem:

$$A(x) = x(240 - 3x) = 240x - 3x^2 \qquad 0 \leq x \leq 80$$

(B) The function $A(x) = 240x - 3x^2$ is a parabola that opens downward, so the maximum value of area will occur at the vertex.

$$x = -\frac{b}{2a} = -\frac{240}{2(-3)} = 40;$$
$$A(40) = 240(40) - 3(40)^2 = 4,800$$

A value of $x = 40$ gives a maximum area of 4,800 square feet.

(C) When $x = 40$, $y = 240 - 3(40) = 120$. Each pen is x by $y/2$, or 40 feet by 60 feet. The area of each pen is 40 feet × 60 feet = 2,400 square feet.

MATCHED PROBLEM 6

Repeat Example 6 with the owner constructing three identical adjacent pens instead of two.

The great sixteenth-century astronomer and physicist Galileo was the first to discover that the distance an object falls is proportional to the square of the time it has been falling. This makes quadratic functions a natural fit for modeling falling objects. Neglecting air resistance, the quadratic function

$$h(t) = h_0 - 16t^2$$

represents the *height of an object* t seconds after it is dropped from an initial height of h_0 feet. The constant -16 is related to the force of gravity and is dependent on the units used. That is, -16 only works for distances measured in feet and time measured in seconds. If the object is thrown either upward or downward, the quadratic model will also have a term involving t. (See Problems 93 and 94 in Exercises 3-4.)

EXAMPLE **7** **Projectile Motion**

As a publicity stunt, a late-night talk show host drops a pumpkin from a rooftop that is 200 feet high. When will the pumpkin hit the ground? Round your answer to two decimal places.

SOLUTION

Because the initial height is 200 feet, the quadratic model for the height of the pumpkin is

$$h(t) = 200 - 16t^2$$

Because $h(t) = 0$ when the pumpkin hits the ground, we must solve this equation for t.

$$h(t) = 200 - 16t^2 = 0 \qquad \text{Add } 16t^2 \text{ to both sides.}$$
$$16t^2 = 200 \qquad \text{Divide both sides by 16.}$$
$$t^2 = \frac{200}{16} = 12.5 \qquad \text{Take the square root of both sides.}$$
$$t = \sqrt{12.5} \qquad \text{Only the positive solution is relevant.}$$
$$\approx 3.54 \text{ seconds}$$

MATCHED PROBLEM 7

A watermelon is dropped from a rooftop that is 300 feet high. When will the melon hit the ground? Round your answer to two decimal places.

› **Solving Quadratic Inequalities**

Given a quadratic function $f(x) = ax^2 + bx + c$, $a \neq 0$, the zeros of f are the solutions of the quadratic equation

$$ax^2 + bx + c = 0 \qquad (1)$$

(see Section 1-5). If the equal sign in equation (1) is replaced with $<$, $>$, $\leq$, or $\geq$, the result is a **quadratic inequality** in **standard form.** Just as was the case with linear inequalities (see Section 1-2), the **solution set** for a quadratic inequality is the subset of the real number line that makes the inequality a true statement. We can identify this subset by examining the graph of a quadratic function. We begin with a specific example and then generalize the results.

The graph of

$$f(x) = x^2 - 2x - 3 = (x - 3)(x + 1)$$

is shown in Figure 6. Information obtained from the graph is listed in Table 1.

> **Figure 6**
$y = f(x) = x^2 - 2x - 3 = (x - 3)(x + 1)$

Table 1

x	$f(x)$
$-\infty < x < -1$	Positive
$x = -1$	Zero
$-1 < x < 3$	Negative
$x = 3$	Zero
$3 < x < \infty$	Positive

Because we now know where the output of f is positive, negative, and zero, we can use the graph or the table to solve a number of related inequalities involving f. For example,

$$f(x) > 0 \text{ on } (-\infty, -1) \cup (3, \infty) \quad \text{and} \quad f(x) \le 0 \text{ on } [-1, 3]$$

The key steps in the preceding process are summarized in the box.

> **SOLVING A QUADRATIC INEQUALITY**

1. Write the inequality in standard form (a form where one side of the inequality defines a quadratic function f and the other side is 0).

2. Find the zeros of f.

3. Graph f and plot its zeros.

4. Use the graph to identify the intervals on the x axis that satisfy the original inequality.

EXAMPLE 8 Solving a Quadratic Inequality

Solve: $x^2 - 4x \ge 14$

SOLUTION **Step 1.** Write in standard form.

$$x^2 - 4x \ge 14 \qquad \text{Subtract 14 from both sides.}$$
$$x^2 - 4x - 14 \ge 0 \qquad \text{Write using function notation.}$$
$$f(x) = x^2 - 4x - 14 \ge 0 \qquad \text{Standard form}$$

Step 2. Solve: $f(x) = x^2 - 4x - 14 = 0$ Use the quadratic formula with $a = 1$, $b = -4$, and $c = -14$.

$$x = \frac{-b \pm \sqrt{b^2 - 4ac}}{2a}$$
$$= \frac{-(-4) \pm \sqrt{(-4)^2 - 4(1)(-14)}}{2(1)}$$
$$= \frac{4 \pm \sqrt{72}}{2} = \frac{4 \pm 6\sqrt{2}}{2} \qquad \text{Divide } both \text{ terms in numerator by 2.}$$
$$= 2 \pm 3\sqrt{2}$$

The zeros of f are $2 - 3\sqrt{2} \approx -2.24$ and $2 + 3\sqrt{2} \approx 6.24$.

Step 3. Plot these zeros, along with a few other points, and graph f (Figure 7).

> **Figure 7**

Step 4. We need to identify intervals where $f(x) \geq 0$. From the graph we see that $f(x) \geq 0$ for $x \leq 2 - 3\sqrt{2}$ and for $x \geq 2 + 3\sqrt{2}$. Returning to the original inequality, the solution to

$$x^2 - 4x \geq 14 \quad \text{is} \quad (-\infty, 2 - 3\sqrt{2}] \cup [2 + 3\sqrt{2}, \infty)$$

MATCHED PROBLEM 8 Solve: $x^2 + 6x \leq 6$

| **EXAMPLE** | **9** |

Break-Even, Profit, and Loss

Table 2 Price–Demand Data

Weekly Sales (in gallons)	Price per Gallon
1,400	$43.00
2,550	$37.25
3,475	$32.60
4,856	$25.72
5,625	$21.88
6,900	$15.50

Table 2 contains price–demand data for a paint manufacturer. A linear regression model for this data is

$$p = 50 - 0.005x \qquad \text{Price–demand equation}$$

where x is the weekly sales (in gallons) and p is the price per gallon. The manufacturer has weekly fixed costs of \$58,500 and variable costs of \$3.50 per gallon produced.

(A) Find the weekly revenue function R and weekly cost function C as functions of the sales x. What is the domain of each function?

(B) Graph R and C on the same coordinate axes and find the level of sales for which the company will break even.

(C) Describe verbally and graphically the sales levels that result in a profit and those that result in a loss.

SOLUTIONS

(A) If x gallons of paint are sold weekly at a price of p per gallon, then the weekly revenue is

$$R = xp = x(50 - 0.005x) = 50x - 0.005x^2$$

Since the sales x and the price p cannot be negative, x must satisfy

$$x \geq 0 \quad \text{and} \quad p = 50 - 0.005x \geq 0 \qquad \text{Subtract 50 from both sides.}$$
$$-0.005x \geq -50 \qquad \text{Divide both sides by } -0.005 \text{ and reverse the inequality.}$$
$$x \leq \frac{-50}{-0.005} = 10,000 \qquad \text{Simplify.}$$

The revenue function and its domain are

$$R(x) = 50x - 0.005x^2 \qquad 0 \leq x \leq 10,000$$

The cost of producing x gallons of paint weekly is

$$C(x) = 58,500 + 3.5x \qquad x \geq 0 \qquad \text{Fixed costs} + \$3.50 \text{ times number of gallons}$$

(B) The graph of C is a line and the graph of R is a parabola opening downward. Using the vertex formula,

$$x = -\frac{b}{2a} = -\frac{50}{2(-0.005)} = 5,000$$
$$R(5,000) = 50(5,000) - 0.005(5,000)^2 = 125,000$$

The vertex is (5,000, 125,000).

After plotting a few points (Table 3), we sketch the graphs of R and C (Fig. 8).

Table 3

x	$R(x)$	$C(x)$
0	0	58,500
5,000	125,000	76,000
10,000	0	93,500

> **Figure 8** Profit when $R > C$; loss when $R < C$

The company breaks even if cost equals revenue:

$$C(x) = R(x)$$
$$58,500 + 3.5x = 50x - 0.005x^2$$
$$0.005x^2 - 46.5x + 58,000 = 0$$

Use the quadratic formula with $a = 0.005$, $b = -46.5$, and $c = 58,000$.

$$x = \frac{46.5 \pm \sqrt{46.5^2 - 4(0.005)(58,500)}}{2(0.005)} \qquad = \frac{46.5 \pm \sqrt{992.25}}{0.01} = \frac{46.5 \pm 31.5}{0.01}$$

$$= 1,500 \text{ or } 7,800$$

Now we find the corresponding points on the graph:

$$C(1,500) = R(1,500) = \$63,750$$
$$C(7,800) = R(7,800) = \$85,800$$

The graphs of C and R intersect at the points $(1,500, 63,750)$ and $(7,800, 85,800)$ (see Figure 8). These intersection points are called the **break-even points.**

(C) If the company produces and sells between 1,500 and 7,800 gallons of paint weekly, then $R > C$ and the company will make a profit. These sales levels are shown in blue in Figure 8. If it produces and sells between 0 and 1,500 gallons or between 7,800 and 10,000 gallons of paint, then $R < C$ and the company will lose money. These sales levels are shown in red in Figure 8.

MATCHED PROBLEM 9

Refer to Example 9.

(A) Find the profit function P and state its domain.

(B) Find the sales levels for which $P(x) > 0$ and those for which $P(x) < 0$.

(C) Find the maximum profit and the sales level at which it occurs.

> ## Modeling with Quadratic Regression

We obtained the linear model for the price–demand data in Example 9 by applying linear regression to the data in Table 2. Regression is not limited to just linear functions. In Example 10 we will use a quadratic model obtained by applying quadratic regression to a data set.

EXAMPLE **10** **Stopping Distance**

Automobile accident investigators often use the length of skid marks to approximate the speed of vehicles involved in an accident. The skid mark length depends on a number of factors, including the make and weight of the vehicle, the road surface, and the road

Table 4

Speed (mph)	Length of Skid Marks (in feet)	
	Wet Asphalt	Dry Concrete
20	22	16
30	49	33
40	84	61
50	137	94
60	197	133

conditions at the time of the accident. Investigators conduct tests to determine skid mark length for various vehicles under varying conditions. Some of the test results for a particular vehicle are listed in Table 4.

Using the quadratic regression feature on a graphing calculator, (see the Technology Connections following this example) we find a model for the skid mark length on wet asphalt:

$$L(x) = 0.06x^2 - 0.42x + 6.6$$

where x is speed in miles per hour.

(A) Graph $y = L(x)$ and the data for skid mark length on wet asphalt on the same axes.

(B) How fast (to the nearest mile) was the vehicle traveling if it left skid marks 100 feet long?

SOLUTIONS

(A)

(B) To approximate the speed from the skid mark length, we solve

$$L(x) = 100$$

$$0.06x^2 - 0.42x + 6.6 = 100 \qquad \text{Subtract 100 from both sides.}$$

$$0.06x^2 - 0.42x - 93.4 = 0 \qquad \text{Use the quadratic formula.}$$

$$x = \frac{-(-0.42) \pm \sqrt{(-0.42)^2 - 4(0.06)(-93.4)}}{2(0.06)}$$

$$= \frac{0.42 \pm \sqrt{22.5924}}{0.12}$$

$$x \approx 43 \text{ mph} \qquad \text{The negative root was discarded.}$$

MATCHED PROBLEM 10

A model for the skid mark length on dry concrete in Table 4 is

$$M(x) = 0.035x^2 + 0.15x - 1.6$$

where x is speed in miles per hour.

(A) Graph $y = L(x)$ and the data for skid mark length on dry concrete on the same axes.

(B) How fast (to the nearest mile) was the vehicle traveling if it left skid marks 100 feet long?

Technology Connections

Figure 9 shows three of the screens related to the construction of the quadratic model in Example 10 on a Texas Instruments TI-84 Plus.

The use of regression to construct mathematical models is not limited to just linear and quadratic models. As we expand our library of functions, we will see that regression can be used to construct models involving these new functions.

(a) Enter the data.

(b) Use the QuadReg option on a calculator.

(c) Graph the data and the model.

> Figure 9

ANSWERS TO MATCHED PROBLEMS

1.

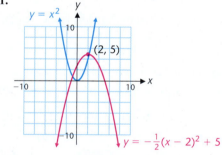

The $-\frac{1}{2}$ makes the graph open downward and vertically shrinks it by a factor of $\frac{1}{2}$, the 2 moves it 2 units right, and the 5 moves it 5 units up.

2. $g(x) = 3(x - 3)^2 - 25$; vertex: $(3, -25)$; axis: $x = 3$

3. (A) Vertex form: $f(x) = -(x - 2)^2 + 6$; vertex: $(2, 6)$; axis of symmetry: $x = 2$.

 (B)

 Max $f(x) = f(2) = 6$; domain: $(-\infty, \infty)$; range: $(-\infty, 6]$; increasing on $(-\infty, 2]$; decreasing on $[2, \infty)$

4. (A) Vertex: $(-1, -\frac{21}{4})$; axis of symmetry: $x = -1$

(B)

Min $f(x) = f(-1) = -\frac{21}{4}$; domain: $(-\infty, \infty)$; range: $[-\frac{21}{4}, \infty)$; decreasing on $(-\infty, -1]$; increasing on $[-1, \infty)$

5. $y = \frac{1}{4}(x - 4)^2 - 2 = 0.25x^2 - 2x + 2$

6. (A) $A(x) = (240 - 4x)x$, $0 \le x \le 60$

(B) The maximum combined area of 3,600 ft.2 occurs at $x = 30$ feet.

(C) Each pen is 30 feet by 40 feet with area 1,200 ft.2

7. 4.33 seconds

8. $[-3 - \sqrt{15}, -3 + \sqrt{15}]$

9. (A) $P(x) = 46.5x - 0.005x^2 - 58,500$, $0 \le x \le 10,000$

(B) Profit is positive for sales between 1,500 and 7,800 gallons per week and negative for sales less than 1,500 or for sales between 7,800 and 10,000.

(C) The maximum profit is $49,612.50 at a sales level of 4,650 gallons.

10. (A)

(B) 52 mph

3-4 Exercises

1. Describe the graph of any quadratic function.

2. How can you tell from a quadratic function whether its graph opens up or down?

3. True or False: Every quadratic function has a maximum. Explain.

4. Using transformations, explain why the vertex of $f(x) = a(x - h)^2 + k$ is (h, k).

5. What information does the constant a provide about the graph of a function of the form $f(x) = ax^2 + bx + c$?

6. Explain how to find the maximum or minimum value of a quadratic function.

In Problems 7–12, find the vertex and axis of the parabola, then draw the graph.

7. $f(x) = (x + 3)^2 - 4$

8. $f(x) = (x + 2)^2 - 2$

9. $f(x) = -\left(x - \frac{3}{2}\right)^2 - 5$

10. $f(x) = -\left(x - \frac{11}{2}\right)^2 + 3$

11. $f(x) = 2(x + 10)^2 + 20$

12. $f(x) = -\frac{1}{2}(x + 8)^2 + 12$

In Problems 13–18, write a brief verbal description of the relationship between the graph of the indicated function and the graph of $y = x^2$.

13. $f(x) = (x - 2)^2 + 1$

14. $g(x) = -(x + 1)^2 - 2$

15. $h(x) = -(x + 1)^2$ **16.** $k(x) = (x - 2)^2$

17. $m(x) = (x - 2)^2 - 3$ **18.** $n(x) = -(x + 1)^2 + 4$

In Problems 19–24, match each graph with one of the functions in Problems 13–18.

19.

20.

21.

22.

23.

24.

In Problems 25–34, complete the square and find the vertex form of each quadratic function, then write the vertex and the axis and draw the graph.

25. $f(x) = x^2 - 4x + 5$ **26.** $g(x) = x^2 - 6x + 1$

27. $h(x) = -x^2 - 2x - 3$ **28.** $k(x) = -x^2 - 10x + 3$

29. $m(x) = 2x^2 - 12x + 22$ **30.** $n(x) = 3x^2 + 6x - 2$

31. $f(x) = \dfrac{1}{2}x^2 + 3x - \dfrac{7}{2}$ **32.** $g(x) = -\dfrac{3}{2}x^2 + 9x + \dfrac{11}{2}$

33. $f(x) = 2x^2 - 24x + 90$ **34.** $g(x) = 3x^2 + 24x + 30$

In Problems 35–46, use the formula $x = -b/2a$ to find the vertex. Then write a description of the graph using all of the following words: axis, increases, decreases, range, and maximum or minimum. Finally, draw the graph.

35. $f(x) = x^2 + 8x + 8$

36. $f(x) = x^2 + 10x + 10$

37. $f(x) = -x^2 - 7x + 4$

38. $f(x) = -x^2 - 11x + 1$

39. $f(x) = 4x^2 - 18x + 25$

40. $f(x) = 5x^2 + 30x - 17$

41. $f(x) = -10x^2 + 50x + 12$

42. $f(x) = -8x^2 - 24x + 16$

43. $f(x) = x^2 + 3x$

44. $f(x) = 4x - x^2$

45. $f(x) = 0.5x^2 - 2x - 7$

46. $f(x) = 0.4x^2 + 4x + 4$

In Problems 47–60, solve and write the answer using interval notation.

47. $x^2 < 10 - 3x$ **48.** $x^2 + x < 12$

49. $x^2 + 21 > 10x$ **50.** $x^2 + 7x + 10 > 0$

51. $x^2 \le 8x$ **52.** $x^2 + 6x \ge 0$

53. $x^2 + 5x \le 0$ **54.** $x^2 \le 4$

55. $x^2 + 1 < 2x$ **56.** $x^2 + 25 < 10x$

57. $x^2 < 3x - 3$ **58.** $x^2 + 3 > 2x$

59. $x^2 - 1 \ge 4x$ **60.** $2x + 2 > x^2$

In Problems 61–68, find the standard form of the equation for the quadratic function whose graph is shown.

61.

62.

63.

64.

65.

66.

67.

68.

In Problems 69–74, find the equation of a quadratic function whose graph satisfies the given conditions.

69. Vertex: $(4, 8)$; x intercept: 6

70. Vertex: $(-2, -12)$; x intercept: -4

71. Vertex: $(-4, 12)$; y intercept: 4

72. Vertex: $(5, 8)$; y intercept: -2

73. Vertex: $(-5, -25)$; additional point on graph: $(-2, 20)$

74. Vertex: $(6, -40)$; additional point on graph: $(3, 50)$

75. For $f(x) = a(x - h)^2 + k$, expand the parentheses and simplify to write in the form $f(x) = ax^2 + bx + c$. This proves that any function in vertex form is a quadratic function as defined in Definition 1.

76. Find a general formula for the constant term c when expanding $f(x) = a(x - h)^2 + k$ into the form $f(x) = ax^2 + bx + c$.

77. Let $g(x) = x^2 + kx + 1$. Graph g for several different values of k and discuss the relationship between these graphs.

78. Confirm your conclusions in Problem 77 by finding the vertex form for g.

79. Let $f(x) = (x - 1)^2 + k$. Discuss the relationship between the values of k and the number of x intercepts for the graph of f. Generalize your comments to any function of the form

$$f(x) = a(x - h)^2 + k, \; a > 0$$

80. Let $f(x) = -(x - 2)^2 + k$. Discuss the relationship between the values of k and the number of x intercepts for the graph of f. Generalize your comments to any function of the form

$$f(x) = a(x - h)^2 + k, \; a < 0$$

81. Find the minimum product of two numbers whose difference is 30. Is there a maximum product? Explain.

82. Find the maximum product of two numbers whose sum is 60. Is there a minimum product? Explain.

APPLICATIONS

83. PROFIT ANALYSIS A consultant hired by a small manufacturing company informs the company owner that their annual profit can be modeled by the function $P(x) = -1.2x^2 + 62.5x - 491$, where x represents the number of employees and P is profit in thousands of dollars. How many employees should the company have to maximize annual profit? What is the maximum annual profit they can expect in that case?

84. PROFIT ANALYSIS The annual profits (in thousands of dollars) from 2000 to 2009 for the company in Problem 83 can be modeled by the function $P(t) = 6.8t^2 - 80.5t + 427.3, \; 0 \le t \le 9$, where t is years after 2000. How much profit did the company make in their worst year?

85. MOVIE INDUSTRY REVENUE The annual U.S. box office revenue in billions of dollars for a span of years beginning in 2002 can be modeled by the function $B(x) = -0.19x^2 + 1.2x + 7.6$, $0 \le x \le 7$, where x is years after 2002.
(A) In what year was box office revenue at its highest in that time span?
(B) Explain why you should not use the exact vertex in answering part A in this problem.

86. GAS MILEAGE The speed at which a car is driven can have a big effect on gas mileage. Based on EPA statistics for compact cars, the function $m(x) = -0.025x^2 + 2.45x - 30, \; 30 \le x \le 65$, models the average miles per gallon for compact cars in terms of the speed driven x (in miles per hour).
(A) At what speed should the owner of a compact car drive to maximize miles per gallon?
(B) If one compact car has a 14-gallon gas tank, how much farther could you drive it on one tank of gas driving at the speed you found in part A than if you drove at 65 miles per hour?

87. CONSTRUCTION A horse breeder plans to construct a corral next to a horse barn that is 50 feet long, using all of the barn as one side of the corral (see the figure). He has 250 feet of fencing available and wants to use all of it.
(A) Express the area $A(x)$ of the corral as a function of x and indicate its domain.
(B) Find the value of x that produces the maximum area.
(C) What are the dimensions of the corral with the maximum area?

88. CONSTRUCTION Repeat Problem 87 if the horse breeder has only 140 feet of fencing available for the corral. Does the maximum value of the area function still occur at the vertex? Explain.

Problems 89–92 use the falling object function described on page 211.

89. FALLING OBJECT A sandbag is dropped off a high-altitude balloon at an altitude of 10,000 ft. When will the sandbag hit the ground?

90. FALLING OBJECT A prankster drops a water balloon off the top of a 144-ft.-high building. When will the balloon hit the ground?

91. FALLING OBJECT A cliff diver hits the water 2.5 seconds after diving off the cliff. How high is the cliff?

92. FALLING OBJECT A forest ranger drops a coffee cup off a fire watchtower. If the cup hits the ground 1.5 seconds later, how high is the tower?

93. PROJECTILE FLIGHT An arrow shot vertically into the air reaches a maximum height of 484 feet after 5.5 seconds of flight. Let the quadratic function $d(t)$ represent the distance above ground (in feet) t seconds after the arrow is released. (If air resistance is neglected, a quadratic model provides a good approximation for the flight of a projectile.)
(A) Find $d(t)$ and state its domain.
(B) At what times (to two decimal places) will the arrow be 250 feet above the ground?

94. PROJECTILE FLIGHT Repeat Problem 93 if the arrow reaches a maximum height of 324 feet after 4.5 seconds of flight.

95. ENGINEERING The arch of a bridge is in the shape of a parabola 14 feet high at the center and 20 feet wide at the base (see the figure).

(A) Express the height of the arch $h(x)$ in terms of x and state its domain.

(B) Can a truck that is 8 feet wide and 12 feet high pass through the arch?

(C) What is the tallest 8-ft.-wide truck that can pass through the arch?

(D) What (to two decimal places) is the widest 12-ft.-high truck that can pass through the arch?

96. ENGINEERING The roadbed of one section of a suspension bridge is hanging from a large cable suspended between two towers that are 200 feet apart (see the figure). The cable forms a parabola that is 60 feet above the roadbed at the towers and 10 feet above the roadbed at the lowest point.

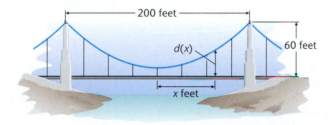

(A) Express the vertical distance $d(x)$ (in feet) from the roadbed to the suspension cable in terms of x and state the domain of d.

(B) The roadbed is supported by seven equally spaced vertical cables (see the figure). Find the combined total length of these supporting cables.

 97. STOPPING DISTANCE Table 5 contains data related to the length of the skid marks left by two different cars when making emergency stops.

Table 5

Speed (mph)	Length of Skid Marks (in feet)	
	Car *A*	Car *B*
20	26	38
30	45	62
40	73	102
50	118	158
60	171	230

(A) Use the quadratic regression feature on a graphing calculator to find a quadratic model $L(x)$ for the skid mark length for Car *A*, where x is speed in miles per hour. (Round to two significant digits.)

(B) Graph $y = L(x)$ and the data for skid mark length on the same axes.

(C) How fast (to the nearest mile per hour) was the car traveling if it left skid marks 150 feet long?

98. STOPPING DISTANCE (A) Use the quadratic regression feature on a graphing calculator to find a quadratic model $M(x)$ for the skid mark length for Car *B*, where x is speed in miles per hour. (Round to two significant digits.)

(B) Graph $y = M(x)$ and the data for skid mark length on the same axes.

(C) How fast (to the nearest mile) was the car traveling if it left skid marks 100 feet long?

99. ALCOHOL CONSUMPTION Table 6 contains data related to the per capita ethanol consumption in the United States from 1960 to 2000 (Source: NIAAA). A quadratic regression model for the per capita beer consumption is

$$B(x) = -0.0006x^2 + 0.03x + 1$$

(A) If beer consumption continues to follow the trend exhibited in Table 6, when (to the nearest year) would the consumption return to the 1960 level?

(B) What does this model predict for beer consumption in the year 2005? Use the Internet or a library to compare the predicted results with the actual results.

Table 6 Per Capita Alcohol Consumption (in gallons)

Year	Beer	Wine
1960	0.99	0.22
1970	1.14	0.27
1980	1.38	0.34
1990	1.34	0.33
2000	1.22	0.31

100. ALCOHOL CONSUMPTION Refer to Table 6. A quadratic regression model for the per capita wine consumption is

$$W(x) = -0.00016x^2 + 0.009x + 0.2$$

(A) If wine consumption continues to follow the trend exhibited in Table 6, when (to the nearest year) would the consumption return to the 1960 level?

(B) What does this model predict for wine consumption in the year 2005? Use the Internet or a library to compare the predicted results with the actual results.

101. PROFIT ANALYSIS A screen printer produces custom silk-screen apparel. The cost $C(x)$ of printing x custom T-shirts and the revenue $R(x)$ from the sale of x T-shirts (both in dollars) are given by

$$C(x) = 245 + 1.6x$$
$$R(x) = 10x - 0.04x^2$$

Find the break-even points and determine the sales levels x (to the nearest integer) that will result in the printer showing a profit.

102. PROFIT ANALYSIS Refer to Problem 101. Determine the sales levels x (to the nearest integer) that will result in the printer showing a profit of at least $60.

103. MAXIMIZING REVENUE A company that manufactures beer mugs has collected the price–demand data in Table 7. A linear regression model for this data is

$$p = d(x) = 9.3 - 0.15x$$

where x is the number of mugs (in thousands) that the company can sell at a price of p. Find the price that maximizes the company's revenue from the sale of beer mugs.

Table 7

Demand	Price
45,800	$2.43
40,500	$3.23
37,900	$3.67
34,700	$4.10
30,400	$4.74
28,900	$4.97
25,400	$5.49

104. MAXIMIZING REVENUE A company that manufactures inexpensive flash drives has collected the price–demand data in Table 8. A linear regression model for this data is

$$p = d(x) = 12.3 - 0.15x$$

where x is the number of drives (in thousands) that the company can sell at a price of p. Find the price that maximizes the company's revenue from the sale of flash drives.

Table 8

Demand	Price
47,800	$5.13
45,600	$5.46
42,700	$5.90
39,600	$6.36
34,700	$7.10
31,600	$7.56
27,800	$8.13

105. BREAK-EVEN ANALYSIS Table 9 contains weekly price–demand data for orange juice for a fruit-juice producer. The producer has weekly fixed cost of $24,500 and variable cost of $0.35 per gallon of orange juice produced. A linear regression model for the data in Table 9 is

$$p = d(x) = 3.5 - 0.00007x$$

where x is the number of gallons of orange juice that can be sold at a price of p.
(A) Find the revenue and cost functions as functions of the sales x. What is the domain of each function?
(B) Graph R and C on the same coordinate axes and find the sales levels for which the company will break even.
(C) Describe verbally and graphically the sales levels that result in a profit and those that result in a loss.
(D) Find the sales and the price that will produce the maximum profit. Find the maximum profit.

Table 9

Orange Juice	
Demand	Price
21,800	$1.97
24,300	$1.80
26,700	$1.63
28,900	$1.48
29,700	$1.42
33,700	$1.14
34,800	$1.06

106. BREAK-EVEN ANALYSIS Table 10 contains weekly price–demand data for grapefruit juice for a fruit-juice producer. The producer has weekly fixed cost of $4,500 and variable cost of $0.15 per gallon of grapefruit juice produced. A linear regression model for the data in Table 10 is

$$p = d(x) = 3 - 0.0003x$$

where x is the number of gallons of grapefruit juice that can be sold at a price of p.
(A) Find the revenue and cost functions as functions of the sales x. What is the domain of each function?
(B) Graph R and C on the same coordinate axes and find the sales levels for which the company will break even.
(C) Describe verbally and graphically the sales levels that result in a profit and those that result in a loss.
(D) Find the sales and the price that will produce the maximum profit. Find the maximum profit.

Table 10

Grapefruit Juice	
Demand	Price
2,130	$2.36
2,480	$2.26
2,610	$2.22
2,890	$2.13
3,170	$2.05
3,640	$1.91
4,350	$1.70

3-5

Operations on Functions; Composition

> **Performing Operations on Functions**
> **Composition**
> **Mathematical Modeling**

Perhaps the most basic thing you've done in math classes is operations on numbers: things like addition, subtraction, multiplication, and division. In this section, we will explore the concept of operations on functions. In many cases, combining functions will enable us to model more complex and useful situations.

If two functions f and g are both defined at some real number x, then $f(x)$ and $g(x)$ are both real numbers, so it makes sense to perform the four basic arithmetic operations with $f(x)$ and $g(x)$. Furthermore, if $g(x)$ is a number in the domain of f, then it is also possible to evaluate f at $g(x)$. We will see that operations on the outputs of the functions can be used to define operations on the functions themselves.

> Performing Operations on Functions

The functions f and g given by

$$f(x) = 2x + 3 \text{ and } g(x) = x^2 - 4$$

are both defined for all real numbers. Note that $f(3) = 9$ and $g(3) = 5$, so it would seem reasonable to assign the value $9 + 5$, or 14, to a new function $(f + g)(x)$. Based on this idea, for any real x we can perform the operation

$$f(x) + g(x) = (2x + 3) + (x^2 - 4) = x^2 + 2x - 1$$

Similarly, we can define other operations on functions:

$$f(x) - g(x) = (2x + 3) - (x^2 - 4) = -x^2 + 2x + 7$$
$$f(x)g(x) = (2x + 3)(x^2 - 4) = 2x^3 + 3x^2 - 8x - 12$$

For $x \neq \pm 2$ (to avoid zero in the denominator) we can also form the quotient

$$\frac{f(x)}{g(x)} = \frac{2x + 3}{x^2 - 4} \qquad x \neq \pm 2$$

Notice that the result of each operation is a new function. So, we have

$$(f + g)(x) = f(x) + g(x) = x^2 + 2x - 1 \qquad \text{Sum}$$
$$(f - g)(x) = f(x) - g(x) = -x^2 + 2x + 7 \qquad \text{Difference}$$
$$(fg)(x) = f(x)g(x) = 2x^3 + 3x^2 - 8x - 12 \qquad \text{Product}$$
$$\left(\frac{f}{g}\right)(x) = \frac{f(x)}{g(x)} = \frac{2x + 3}{x^2 - 4} \qquad x \neq \pm 2 \qquad \text{Quotient}$$

The sum, difference, and product functions are defined for all values of x, as were the original functions f and g, but the domain of the quotient function must be restricted to exclude those values where $g(x) = 0$.

> **DEFINITION 1** Operations on Functions

The **sum, difference, product,** and **quotient** of the functions f and g are the functions defined by

$$(f + g)(x) = f(x) + g(x) \qquad \text{Sum function}$$
$$(f - g)(x) = f(x) - g(x) \qquad \text{Difference function}$$
$$(fg)(x) = f(x)g(x) \qquad \text{Product function}$$
$$\left(\frac{f}{g}\right)(x) = \frac{f(x)}{g(x)} \qquad g(x) \neq 0 \qquad \text{Quotient function}$$

The **domain** of each function consists of all elements in the domains of *both f* and *g*, with the exception that the values of x where $g(x) = 0$ must be excluded from the domain of the quotient function.

>>> **EXPLORE-DISCUSS 1**

The following activities refer to the graphs of f and g shown in Figure 1 and the corresponding points on the graph shown in Table 1.

> **Figure 1**

Table 1

x	$f(x)$	$g(x)$
0	8	0
2	7	2
4	6	3
6	5	3
8	4	2
10	3	0

For each of the following functions, construct a table of values, sketch a graph, and state the domain and range.

(A) $(f + g)(x)$ **(B)** $(f - g)(x)$ **(C)** $(fg)(x)$ **(D)** $\left(\dfrac{f}{g}\right)(x)$

EXAMPLE 1 **Finding the Sum, Difference, Product, and Quotient Functions**

Let $f(x) = \sqrt{4 - x}$ and $g(x) = \sqrt{3 + x}$. Find the functions $f + g$, $f - g$, fg, and f/g, and find their domains.

SOLUTION

$$(f + g)(x) = f(x) + g(x) = \sqrt{4 - x} + \sqrt{3 + x}$$
$$(f - g)(x) = f(x) - g(x) = \sqrt{4 - x} - \sqrt{3 + x}$$
$$(fg)(x) = f(x)g(x) = \sqrt{4 - x}\,\sqrt{3 + x}$$
$$= \sqrt{(4 - x)(3 + x)}$$
$$= \sqrt{12 + x - x^2}$$

$$\left(\frac{f}{g}\right)(x) = \frac{f(x)}{g(x)} = \frac{\sqrt{4 - x}}{\sqrt{3 + x}} = \sqrt{\frac{4 - x}{3 + x}}$$

Domain of *f*

(a)

Domain of *g*

(b)

Domain of *f* + *g*, *f* − *g*, and *fg*

(c)

› **Figure 2**

The domains of *f* and *g* are

$$\text{Domain of } f: x \leq 4 \text{ or } (-\infty, 4] \text{ [Fig. 2(a)]}$$
$$\text{Domain of } g: x \geq -3 \text{ or } [-3, \infty) \text{ [Fig. 2(b)]}$$

The intersection of these domains is shown in Figure 2(c):

$$(-\infty, 4] \cap [-3, \infty) = [-3, 4]$$

This is the domain of the functions *f* + *g*, *f* − *g*, and *fg*. Since $g(-3) = 0$, $x = -3$ must be excluded from the domain of the quotient function, and

$$\text{Domain of } \frac{f}{g}: (-3, 4]$$

MATCHED PROBLEM 1

Let $f(x) = \sqrt{x}$ and $g(x) = \sqrt{10 - x}$. Find the functions *f* + *g*, *f* − *g*, *fg*, and *f/g*, and find their domains.

Technology Connections

A graphing calculator can be used to check the domains in the solution of Example 1. To check the domain of *f* + *g*, we enter $y_1 = \sqrt{4 - x}$, $y_2 = \sqrt{3 + x}$, and $y_3 = y_1 + y_2$ in the equation editor of a graphing calculator and graph y_3 (Fig. 3).

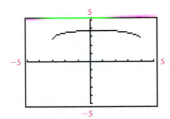

› **Figure 3**

Next we press TRACE and enter −3 (Fig. 4). Pressing the left cursor indicates that y_3 is not defined for $x < -3$ (Fig. 5).

› **Figure 4**

› **Figure 5**

Figures 6 and 7 indicate that y_3 is not defined for $x > 4$. This confirms that the domain of $y_3 = f + g$ is $[-3, 4]$.

› **Figure 6**

› **Figure 7**

EXAMPLE **2**

Finding the Quotient of Two Functions

Let $f(x) = \dfrac{x}{x - 1}$ and $g(x) = \dfrac{x - 4}{x + 3}$. Find the function $\dfrac{f}{g}$ and find its domain.

SOLUTION

Because division by 0 must be excluded, the domain of f is all x except $x = 1$ and the domain of g is all x except $x = -3$. Now we find f/g.

$$\left(\frac{f}{g}\right)(x) = \frac{f(x)}{g(x)} = \frac{\dfrac{x}{x - 1}}{\dfrac{x - 4}{x + 3}}$$

$$= \frac{x}{x - 1} \cdot \frac{x + 3}{x - 4}$$

$$= \frac{x(x + 3)}{(x - 1)(x - 4)} \tag{1}$$

The fraction in equation (1) indicates that 1 and 4 must be excluded from the domain of f/g to avoid division by 0. But equation (1) does not indicate that -3 must be excluded also. Although the fraction in equation (1) is defined at $x = -3$, -3 was excluded from the domain of g, so it must be excluded from the domain of f/g also. The domain of f/g is all real numbers x except -3, 1, and 4.

MATCHED PROBLEM 2

Let $f(x) = \dfrac{1}{x + 2}$ and $g(x) = \dfrac{x - 5}{x}$. Find the function $\dfrac{f}{g}$ and find its domain.

› Composition

Consider the functions f and g given by

$$f(x) = \sqrt{x} \quad \text{and} \quad g(x) = 4 - 2x$$

Note that $g(0) = 4 - 2(0) = 4$ and $f(4) = \sqrt{4} = 2$. So if we apply these two functions *consecutively*, we get

$$f(g(0)) = f(4) = 2$$

In a diagram, this would look like

When two functions are applied consecutively, we call the result the **composition** of functions. We will use the symbol $f \circ g$ to represent the composition of f and g, which we formally define now.

> **DEFINITION 2 Composition of Functions**

The composition of a function f with another function g is denoted by $f \circ g$ (read "f composed with g") and is defined by

$$(f \circ g)(x) = f(g(x))$$

EXAMPLE **3** **Computing Composition From a Table**

Functions f and g are defined by Table 2. Find $(f \circ g)(2)$, $(f \circ g)(5)$, and $(f \circ g)(-3)$.

Table 2

x	$f(x)$	$g(x)$
-5	-8	11
-3	-6	2
0	-1	-6
2	5	-3
5	12	0

SOLUTION We will use the formula provided by Definition 2.

$$(f \circ g)(2) = f(g(2)) = f(-3) = -6$$
$$(f \circ g)(5) = f(g(5)) = f(0) = -1$$
$$(f \circ g)(-3) = f(g(-3)) = f(2) = 5$$

MATCHED PROBLEM 3 Functions h and k are defined by Table 3. Find $(h \circ k)(10)$, $(h \circ k)(-8)$, and $(h \circ k)(0)$.

Table 3

x	$h(x)$	$k(x)$
-8	12	0
-4	18	22
0	40	-4
10	52	-8
20	70	-30

⟫ CAUTION ⟫ When computing $f \circ g$, it's important to keep in mind that the first function that appears in the notation (f, in this case) is actually the second function that is applied. For this reason, some people read $f \circ g$ as "f following g."

Refer to the functions f and g on page 226, and let $h(x) = (f \circ g)(x)$. Complete Table 4 and graph h.

Table 4

x	$g(x)$	$h(x) = f(g(x))$
0	$g(0) = 4$	$h(0) = f(g(0)) = f(4) = 2$
1		
2		
3		
4		

The domain of f is $\{x \mid x \geq 0\}$ and the domain of g is the set of all real numbers. What is the domain of h?

So far, we have looked at composition on a point-by-point basis. Using algebra, we can find a formula for the composition of two functions.

EXAMPLE 4 **Finding the Composition of Two Functions**

Find $(f \circ g)(x)$ for $f(x) = x^2 - x$ and $g(x) = 3 + 2x$.

SOLUTION We again use the formula in Definition 2.

$$(f \circ g)(x) = f(g(x))$$
$$= f(3 + 2x)$$
$$= (3 + 2x)^2 - (3 + 2x)$$
$$= 9 + 12x + 4x^2 - 3 - 2x$$
$$= 4x^2 + 10x + 6$$

MATCHED PROBLEM 4 Find $(h \circ k)(x)$ for $h(x) = 11 + x^2$ and $k(x) = 4x - 1$.

(A) For $f(x) = x - 10$ and $g(x) = 3 + 7x$, find $(f \circ g)(x)$ and $(g \circ f)(x)$. Based on this result, what do you think is the relationship between $f \circ g$ and $g \circ f$ in general?

(B) Repeat for $f(x) = 2x + 1$ and $g(x) = \dfrac{x - 1}{2}$. Does this change your thoughts on the relationship between $f \circ g$ and $g \circ f$?

Explore-Discuss 3 tells us that order is important in composition. Sometimes $f \circ g$ and $g \circ f$ are equal, but more often they are not.

Finding the domain of a composition of functions can sometimes be a bit tricky. Based on the definition $(f \circ g)(x) = f(g(x))$, we can see that for an x value to be in the domain of $f \circ g$, two things must occur. First, x must be in the domain of g so that $g(x)$ is defined. Second, $g(x)$ must be in the domain of f, so that $f(g(x))$ is defined.

| EXAMPLE | 5 | **Finding the Composition of Two Functions** |

Find $(f \circ g)(x)$ and $(g \circ f)(x)$ and their domains for $f(x) = x^{10}$ and $g(x) = 3x^4 - 1$.

SOLUTION

$$(f \circ g)(x) = f(g(x)) = f(3x^4 - 1) = (3x^4 - 1)^{10}$$
$$(g \circ f)(x) = g(f(x)) = g(x^{10}) = 3(x^{10})^4 - 1 = 3x^{40} - 1$$

Note that the functions f and g are both defined for all real numbers. If x is any real number, then x is in the domain of g, so $g(x)$ is a real number. This then tells us that $g(x)$ is in the domain of f, which means that $f(g(x))$ is a real number. In other words, every real number is in the domain of $f \circ g$. Using similar reasoning, we can conclude that the domain of $g \circ f$ is also the set of all real numbers.

| MATCHED PROBLEM 5 | Find $(f \circ g)(x)$ and $(g \circ f)(x)$ and their domains for $f(x) = \sqrt[3]{x}$ and $g(x) = 7x + 5$.

The line of reasoning used in Example 5 can be used to deduce the following fact:

If two functions are both defined for all real numbers, then so is their composition.

If either function in a composition is not defined for some real numbers, then, as Example 6 illustrates, the domain of the composition may not be what you first think it should be.

| EXAMPLE | 6 | **Finding the Composition of Two Functions** |

Find $(f \circ g)(x)$ for $f(x) = \sqrt{4 - x^2}$ and $g(x) = \sqrt{3 - x}$, then find the domain of $f \circ g$.

SOLUTION We begin by stating the domains of f and g, which is a good idea in any composition problem:

$$\text{Domain } f: -2 \le x \le 2 \quad \text{or} \quad [-2, 2]$$
$$\text{Domain } g: x \le 3 \quad \text{or} \quad (-\infty, 3]$$

Next we find the composition:

$$(f \circ g)(x) = f(g(x))$$
$$= f(\sqrt{3 - x}) \qquad \text{Substitute } \sqrt{3-x} \text{ for } g(x).$$
$$= \sqrt{4 - (\sqrt{3 - x})^2} \qquad \text{Square: } (\sqrt{t})^2 = t \text{ as long as } t \ge 0.$$
$$= \sqrt{4 - (3 - x)} \qquad \text{Subtract.}$$
$$= \sqrt{1 + x}$$

Although $\sqrt{1 + x}$ is defined for all $x \ge -1$, we must restrict the domain of $f \circ g$ to those values that also are in the domain of g.

$$\text{Domain } f \circ g: x \ge -1 \text{ and } x \le 3 \quad \text{or} \quad [-1, 3]$$

| MATCHED PROBLEM 6 | Find $f \circ g$ for $f(x) = \sqrt{9 - x^2}$ and $g(x) = \sqrt{x - 1}$, then find the domain of $f \circ g$.

>>> CAUTION >>> The domain of $f \circ g$ cannot always be determined simply by examining the final form of $(f \circ g)(x)$. Any numbers that are excluded from the domain of g must also be excluded from the domain of $f \circ g$.

In calculus, it is not only important to be able to find the composition of two functions, but also to recognize when a given function is the composition of simpler functions.

EXAMPLE 7 **Recognizing Composition Forms**

 Express h as a composition of two simpler functions for
$$h(x) = \sqrt{1 + 3x^4}$$

SOLUTION If we were to evaluate this function for some x value, say, $x = 1$, we would do so in two stages. First, we would find the value of $1 + 3(1)^4$, which is 4. Then we would apply the square root to get 2. This shows that h can be thought of as two consecutive functions: First, $g(x) = 1 + 3x^4$, then $f(x) = \sqrt{x}$. So $h(x) = f(g(x))$, and we have written h as $f \circ g$. ●

MATCHED PROBLEM 7 Express h as the composition of two simpler functions for $h(x) = (4x^3 - 7)^4$. ●

The answers to Example 7 and Matched Problem 7 are not unique. For example, if $f(x) = \sqrt{1 + 3x}$ and $g(x) = x^4$, then
$$f(g(x)) = \sqrt{1 + 3g(x)} = \sqrt{1 + 3x^4} = h(x)$$

› Mathematical Modeling

The operations discussed in this section can be applied in many different situations. Example 8 shows how they are used to construct a model in economics.

EXAMPLE 8 **Modeling Profit**

The research department for an electronics firm estimates that the weekly demand for a certain brand of headphones is given by
$$x = f(p) = 20{,}000 - 1{,}000p \qquad 0 \le p \le 20 \qquad \text{Demand function}$$

This function describes the number x of pairs of headphones retailers are likely to buy per week at p dollars per pair. The research department also has determined that the total cost (in dollars) of producing x pairs per week is given by
$$C(x) = 25{,}000 + 3x \qquad \text{Cost function}$$

and the total weekly revenue (in dollars) obtained from the sale of these headphones is given by
$$R(x) = 20x - 0.001x^2 \qquad \text{Revenue function}$$

Express the firm's weekly profit as a function of the price p and find the price that produces the largest profit. What is the largest possible profit?

SOLUTION The basic economic principle we are using is that profit is revenue minus cost. So the profit function P is the difference of the revenue function R and the cost function C.

$$P(x) = (R - C)(x)$$
$$= R(x) - C(x)$$
$$= (20x - 0.001x^2) - (25{,}000 + 3x)$$
$$= 17x - 0.001x^2 - 25{,}000$$

This is a function of the demand x. We were asked to find the profit P as a function of the price p; we can accomplish this using composition, because $x = f(p)$.

$$(P \circ f)(p) = P(f(p))$$
$$= P(20{,}000 - 1{,}000p)$$
$$= 17(20{,}000 - 1{,}000p) - 0.001(20{,}000 - 1{,}000p)^2 - 25{,}000$$
$$= 340{,}000 - 17{,}000p - 400{,}000 + 40{,}000p - 1{,}000p^2 - 25{,}000$$
$$= -85{,}000 + 23{,}000p - 1{,}000p^2$$

Technically, $P \circ f$ and P are different functions, because the first has independent variable p and the second has independent variable x. However, because both functions represent the same quantity (the profit), it is customary to use the same symbol to name each function. So

$$P(p) = -85{,}000 + 23{,}000p - 1{,}000p^2$$

expresses the weekly profit P as a function of price p. Now we can use the vertex formula to find the maximum.

$$p = -\frac{b}{2a} = -\frac{23{,}000}{-2{,}000} = 11.5$$

$$P(11.5) = -85{,}000 + 23{,}000(11.5) - 1{,}000(11.5)^2 = 47{,}250$$

Since $a < 0$, the parabola opens downward, and the maximum value of P occurs at the vertex. So the largest profit is $47,250 and it will occur when the price of the headphones is $11.50.

MATCHED PROBLEM 8 Repeat Example 8 for the functions

$$x = f(p) = 10{,}000 - 1{,}000p \qquad 0 \le p \le 10$$
$$C(x) = 10{,}000 + 2x \qquad R(x) = 10x - 0.001x^2$$

ANSWERS TO MATCHED PROBLEMS

1. $(f + g)(x) = \sqrt{x} + \sqrt{10 - x}$, $(f - g)(x) = \sqrt{x} - \sqrt{10 - x}$, $(fg)(x) = \sqrt{10x - x^2}$, $(f/g)(x) = \sqrt{x/(10 - x)}$; the functions $f + g$, $f - g$, and fg have domain: $[0, 10]$, the domain of f/g is $[0, 10)$

2. $\left(\dfrac{f}{g}\right)(x) = \dfrac{x}{(x + 2)(x - 5)}$; domain: all real numbers x except -2, 0, and 5

3. $(h \circ k)(10) = 12$; $(h \circ k)(-8) = 40$; $(h \circ k)(0) = 18$

4. $(h \circ k)(x) = 16x^2 - 8x + 12$

5. $(f \circ g)(x) = \sqrt[3]{7x + 5}$, domain: $(-\infty, \infty)$
 $(g \circ f)(x) = 7\sqrt[3]{x} + 5$, domain: $(-\infty, \infty)$

6. $(f \circ g)(x) = \sqrt{10 - x}$; domain: $x \ge 1$ and $x \le 10$ or $[1, 10]$

7. $h(x) = (f \circ g)(x)$ where $f(x) = x^4$ and $g(x) = 4x^3 - 7$

8. $P(p) = -30{,}000 + 12{,}000p - 1{,}000p^2$. The largest profit is $6,000 and occurs when the price is $6.

3-5 Exercises

1. Explain how to find the sum of two functions.

2. Explain how to find the product of two functions.

3. Describe in your own words what the composition of two functions means. Don't focus on how to find composition, but rather on what it really means.

4. Is the domain of f/g always the same as the intersection of the domains of f and g? Explain.

5. When composing two functions, why can't you always find the domain by simply looking at the simplified form of the composition?

6. Describe a real-world situation where the composition of two functions would have significance.

Problems 7–18 refer to functions f and g whose graphs are shown below.

 In Problems 7–10 use the graphs of f and g to construct a table of values and sketch the graph of the indicated function.

7. $(f + g)(x)$ **8.** $(g - f)(x)$

9. $(fg)(x)$ **10.** $(f - g)(x)$

 In Problems 11–18, use the graphs of f and g to find each of the following:

11. $(f \circ g)(-1)$ **12.** $(f \circ g)(2)$

13. $(g \circ f)(-2)$ **14.** $(g \circ f)(3)$

15. $f(g(1))$ **16.** $f(g(0))$

17. $g(f(2))$ **18.** $g(f(-3))$

In Problems 19–26, find the indicated function value, if it exists, given $f(x) = 2 - x$ and $g(x) = \sqrt{3 - x}$.

19. $(f + g)(-3)$ **20.** $(g - f)(-5)$

21. $(fg)(-1)$ **22.** $\left(\dfrac{f}{g}\right)(3)$

23. $(f \circ g)(-2)$ **24.** $(f \circ g)(1)$

25. $(g \circ f)(1)$ **26.** $(g \circ g)(-7)$

27. Functions f and g are defined by Table 5. Find $(f \circ g)(-7)$, $(f \circ g)(0)$, and $(f \circ g)(4)$.

28. Functions h and k are defined by Table 6. Find $(h \circ k)(-15)$, $(h \circ k)(-10)$, and $(h \circ k)(15)$.

Table 5

x	$f(x)$	$g(x)$
-7	5	4
-2	9	10
0	0	-2
4	3	6
6	-10	-3

Table 6

x	$h(x)$	$k(x)$
-20	-100	30
-15	-200	5
-10	-300	15
5	-150	8
15	-90	-10

In Problems 29–42, for the indicated functions f and g, find the functions $f + g$, $f - g$, fg, and f/g, and find their domains.

29. $f(x) = 4x;$ $g(x) = x + 1$

30. $f(x) = 3x;$ $g(x) = x - 2$

31. $f(x) = 2x^2;$ $g(x) = x^2 + 1$

32. $f(x) = 3x;$ $g(x) = x^2 + 4$

33. $f(x) = 3x + 5;$ $g(x) = x^2 - 1$

34. $f(x) = 2x - 7;$ $g(x) = 9 - x^2$

35. $f(x) = \sqrt{2 - x};$ $g(x) = \sqrt{x + 3}$

36. $f(x) = \sqrt{x + 4};$ $g(x) = \sqrt{3 - x}$

37. $f(x) = \sqrt{x} + 2;$ $g(x) = \sqrt{x} - 4$

38. $f(x) = 1 - \sqrt{x};$ $g(x) = 2 - \sqrt{x}$

39. $f(x) = \sqrt{x^2 + x - 6};$ $g(x) = \sqrt{7 + 6x - x^2}$

40. $f(x) = \sqrt{8 + 2x - x^2};$ $g(x) = \sqrt{x^2 - 7x + 10}$

41. $f(x) = x + \dfrac{1}{x};$ $g(x) = x - \dfrac{1}{x}$

42. $f(x) = x - 1;$ $g(x) = x - \dfrac{6}{x - 1}$

In Problems 43–60, for the indicated functions f and g, find the functions $f \circ g$, and $g \circ f$, and find their domains.

43. $f(x) = x^3;$ $g(x) = x^2 - x + 1$

44. $f(x) = x^2;$ $g(x) = x^3 + 2x + 4$

45. $f(x) = |x + 1|;$ $g(x) = 2x + 3$

46. $f(x) = |x - 4|$; $g(x) = 3x + 2$

47. $f(x) = x^{1/3}$; $g(x) = 2x^3 + 4$

48. $f(x) = x^{2/3}$; $g(x) = 8 - x^3$

49. $f(x) = \sqrt{x}$; $g(x) = x - 4$

50. $f(x) = \sqrt{x}$; $g(x) = 2x + 5$

51. $f(x) = x + 2$; $g(x) = \dfrac{1}{x}$

52. $f(x) = x - 3$; $g(x) = \dfrac{1}{x^2}$

53. $f(x) = \sqrt{4 - x}$; $g(x) = x^2$

54. $f(x) = \sqrt{x - 1}$; $g(x) = x^2$

55. $f(x) = \dfrac{x + 5}{x}$; $g(x) = \dfrac{x}{x - 2}$

56. $f(x) = \dfrac{x}{x - 1}$; $g(x) = \dfrac{2x - 4}{x}$

57. $f(x) = \dfrac{2x + 1}{x}$; $g(x) = \dfrac{1}{x - 2}$

58. $f(x) = \dfrac{2}{x + 3}$; $g(x) = \dfrac{2 - 3x}{x}$

59. $f(x) = \sqrt{25 - x^2}$; $g(x) = \sqrt{9 + x^2}$

60. $f(x) = \sqrt{x^2 - 9}$; $g(x) = \sqrt{x^2 + 25}$

Use the graphs of functions f and g shown below to match each function in Problems 61–64 with one of graphs (a)–(d).

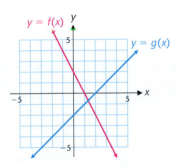

61. $(f + g)(x)$ **62.** $(f - g)(x)$

63. $(g - f)(x)$ **64.** $(fg)(x)$

(a)

(b)

(c)

(d)

In Problems 65–72, find $f \circ g$ and $g \circ f$. Graph $f, g, f \circ g,$ and $g \circ f$ in the same coordinate system and describe any apparent symmetry between these graphs.

65. $f(x) = \tfrac{1}{2}x + 1$; $g(x) = 2x - 2$

66. $f(x) = 3x + 2$; $g(x) = \tfrac{1}{3}x - \tfrac{2}{3}$

67. $f(x) = -\tfrac{2}{3}x - \tfrac{5}{3}$; $g(x) = -\tfrac{3}{2}x - \tfrac{5}{2}$

68. $f(x) = -2x + 3$; $g(x) = -\tfrac{1}{2}x + \tfrac{3}{2}$

69. $f(x) = \dfrac{x^3}{8}$; $g(x) = 2\sqrt[3]{x}$

70. $f(x) = 3\sqrt[3]{x}$; $g(x) = \dfrac{x^3}{27}$

71. $f(x) = \sqrt[3]{x - 2}$; $g(x) = x^3 + 2$

72. $f(x) = x^3 - 3$; $g(x) = \sqrt[3]{x + 3}$

In Problems 73–80, express h as a composition of two simpler functions f and g.

73. $h(x) = (2x - 7)^4$

74. $h(x) = (3 - 5x)^7$

75. $h(x) = \sqrt{4 + 2x}$

76. $h(x) = \sqrt{3x - 11}$

77. $h(x) = 3x^7 - 5$

78. $h(x) = 5x^6 + 3$

79. $h(x) = \dfrac{4}{\sqrt{x}} + 3$

80. $h(x) = -\dfrac{2}{\sqrt{x}} + 1$

81. Are the functions fg and gf identical? Justify your answer.

82. Are the functions $f \circ g$ and $g \circ f$ identical? Justify your answer.

83. Is there a function g that satisfies $f \circ g = g \circ f = f$ for all functions f? If so, what is it?

84. Is there a function g that satisfies $fg = gf = f$ for all functions f? If so, what is it?

In Problems 85–88, for the indicated functions f and g, find the functions f + g, f − g, fg, and f/g, and find their domains.

85. $f(x) = x + \dfrac{1}{x}$; $g(x) = x - \dfrac{1}{x}$

86. $f(x) = x - 1$; $g(x) = x - \dfrac{6}{x - 1}$

87. $f(x) = 1 - \dfrac{x}{|x|}$; $g(x) = 1 + \dfrac{x}{|x|}$

88. $f(x) = x + |x|$; $g(x) = x - |x|$

APPLICATIONS

89. MARKET RESEARCH The demand x and the price p (in dollars) for new release CDs for a large online retailer are related by

$$x = f(p) = 4{,}000 - 200p \qquad 0 \le p \le 20$$

The revenue (in dollars) from the sale of x units is given by

$$R(x) = 20x - \frac{1}{200}x^2$$

and the cost (in dollars) of producing x units is given by

$$C(x) = 2x + 8{,}000$$

Express the profit as a function of the price p and find the price that produces the largest profit.

90. MARKET RESEARCH The demand x and the price p (in dollars) for portable iPod speakers at a national electronics store are related by

$$x = f(p) = 5{,}000 - 100p \qquad 0 \le p \le 50$$

The revenue (in dollars) from the sale of x units and the cost (in dollars) of producing x units are given, respectively, by

$$R(x) = 50x - \frac{1}{100}x^2 \quad \text{and} \quad C(x) = 20x + 40{,}000$$

Express the profit as a function of the price p and find the price that produces the largest profit.

91. POLLUTION An oil tanker aground on a reef is leaking oil that forms a circular oil slick about 0.1 foot thick (see the figure). The radius of the slick (in feet) t minutes after the leak first occurred is given by

$$r(t) = 0.4t^{1/3}$$

Express the volume of the oil slick as a function of t.

$A = \pi r^2$
$V = 0.1A$

92. WEATHER BALLOON A weather balloon is rising vertically. An observer is standing on the ground 100 meters from the point where the weather balloon was released.
(A) Express the distance d between the balloon and the observer as a function of the balloon's distance h above the ground.
(B) If the balloon's distance above ground after t seconds is given by $h = 5t$, express the distance d between the balloon and the observer as a function of t.

93. FLUID FLOW A conical paper cup with diameter 4 inches and height 4 inches is initially full of water. A small hole is made in the bottom of the cup and the water begins to flow out of the cup. Let h and r be the height and radius, respectively, of the water in the cup t minutes after the water begins to flow.

$V = \frac{1}{3}\pi r^2 h$

(A) Express r as a function of h.
(B) Express the volume V as a function of h.
(C) If the height of the water after t minutes is given by

$$h(t) = 4 - 0.5\sqrt{t}$$

express V as a function of t.

94. EVAPORATION A water trough with triangular ends is 6 feet long, 4 feet wide, and 2 feet deep. Initially, the trough is full of water, but due to evaporation, the volume of the water is decreasing. Let h and w be the height and width, respectively, of the water in the tank t hours after it began to evaporate.

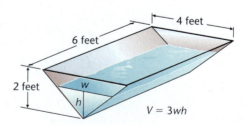

$V = 3wh$

(A) Express w as a function of h.
(B) Express V as a function of h.
(C) If the height of the water after t hours is given by

$$h(t) = 2 - 0.2\sqrt{t}$$

express V as a function of t.

3-6 Inverse Functions

> One-to-One Functions
> Finding the Inverse of a Function
> Mathematical Modeling
> Graphing Inverse Functions

We have seen that many important mathematical relationships can be expressed in terms of functions. For example,

$$C = \pi d$$ The circumference of a circle is a function of the diameter d.

$$V = s^3$$ The volume of a cube is a function of length s of the edges.

$$d = 1{,}000 - 100p$$ The demand for a product is a function of the price p.

$$F = \frac{9}{5}C + 32$$ Temperature measured in °F is a function of temperature in °C.

In many cases, we are interested in *reversing* the correspondence determined by a function. For our examples,

$$d = \frac{C}{\pi}$$ The diameter of a circle is a function of the circumference C.

$$s = \sqrt[3]{V}$$ The length of the edge of a cube is a function of the volume V.

$$p = 10 - \frac{1}{100}d$$ The price of a product is a function of the demand d.

$$C = \frac{5}{9}(F - 32)$$ Temperature measured in °C is a function of temperature in °F.

As these examples illustrate, reversing the correspondence between two quantities often produces a new function. This new function is called the *inverse* of the original function. Later in this text we will see that many important functions are actually defined as the inverses of other functions.

In this section, we develop techniques for determining whether the inverse of a function exists, some general properties of inverse functions, and methods for finding the rule of correspondence that defines the inverse function. A review of function basics in Section 3-1 would be very helpful at this point.

> One-to-One Functions

Recall the set form of the definition of function:

A function is a set of ordered pairs with the property that no two ordered pairs have the same first component and different second components.

However, it is possible that two ordered pairs in a function could have the same second component and different first components. If this does not happen, then we call the function a *one-to-one function.*

In other words, a function is one-to-one if there are no duplicates among the second components.

> **DEFINITION 1** One-to-One Function

A function is **one-to-one** if no two ordered pairs in the function have the same second component and different first components.

›› EXPLORE-DISCUSS 1

Given the following sets of ordered pairs:

$$f = \{(0, 1), (0, 2), (1, 1), (1, 2)\}$$
$$g = \{(0, 1), (1, 1), (2, 2), (3, 2)\}$$
$$h = \{(0, 1), (1, 2), (2, 3), (3, 0)\}$$

(A) Which of these sets represent functions?

(B) Which of the functions are one-to-one functions?

(C) For each set that is a function, form a new set by reversing each ordered pair in the set. Which of these new sets represent functions?

(D) What do these results tell you about the result of reversing the ordered pairs for functions that are one-to-one, and for functions that are not one-to-one?

Explore-Discuss 1 illustrates an important idea that we will examine later: Only one-to-one functions have inverses.

EXAMPLE **1** **Determining Whether a Function Is One-to-One**

Determine whether f is a one-to-one function for

(A) $f(x) = x^2$ (B) $f(x) = 2x - 1$

SOLUTIONS

(A) To show that a function is *not* one-to-one, all we have to do is find two different ordered pairs in the function with the same second component and different first components. Because

$$f(2) = 2^2 = 4 \quad \text{and} \quad f(-2) = (-2)^2 = 4$$

the ordered pairs $(2, 4)$ and $(-2, 4)$ both belong to f, and f is not one-to-one. (Note that there's nothing special about 2 and -2 here: Any real number and its negative can be used in the same way.)

(B) To show that a function *is* one-to-one, we have to show that no two ordered pairs have the same second component and different first components. To do this, we'll show that if any two ordered pairs $(a, f(a))$ and $(b, f(b))$ in f have the same second components, then the first components must also be the same. That is, we show that $f(a) = f(b)$ implies $a = b$. We proceed as follows:

$$f(a) = f(b)$$ Assume second components are equal. Evaluate $f(a)$ and $f(b)$.
$$2a - 1 = 2b - 1$$
$$2a = 2b$$ Simplify.
$$a = b$$ Conclusion: f is one-to-one.

By Definition 1, f is a one-to-one function.

MATCHED PROBLEM 1 Determine whether f is a one-to-one function for

(A) $f(x) = 4 - x^2$ (B) $f(x) = 4 - 2x$

The methods used in the solution of Example 1 can be stated as a theorem.

> **THEOREM 1 One-to-One Functions**

1. If $f(a) = f(b)$ for at least one pair of domain values a and b, $a \neq b$, then f is not one-to-one.
2. If the assumption $f(a) = f(b)$ always implies that the domain values a and b are equal, then f is one-to-one.

Applying Theorem 1 is not always easy—try testing $f(x) = x^3 + 2x + 3$, for example. (Good luck!) However, the graph of a function can help us develop a simple procedure for determining if a function is one-to-one. If any horizontal line intersects the graph in more than one point [as shown in Fig. 1(a)], then there is a second component (height) that corresponds to two different first components (x values). This shows that the function is not one-to-one.

On the other hand, if every horizontal line intersects the graph in just one point or not at all [as shown in Fig. 1(b)], the function is one-to-one. These observations form the basis of the *horizontal line test*.

> **Figure 1** Intersections of graphs and horizontal lines.

> **THEOREM 2 Horizontal Line Test**

A function is one-to-one if and only if every horizontal line intersects the graph of the function in at most one point.

The graphs of the functions considered in Example 1 are shown in Figure 2 on page 238. Applying the horizontal line test to each graph confirms the results we obtained in Example 1.

A function that is increasing throughout its domain or decreasing throughout its domain will always pass the horizontal line test [Figs. 3(a) and 3(b)]. This gives us the following theorem.

$f(x) = x^2$ does not pass
the horizontal line test;
f is not one-to-one
(a)

$f(x) = 2x - 1$ passes
the horizontal line test;
f is one-to-one
(b)

› **Figure 2** Applying the horizontal line test.

> › **THEOREM 3** Increasing and Decreasing Functions
>
> If a function f is increasing throughout its domain or decreasing throughout its domain, then f is a one-to-one function.

An increasing function
is always one-to-one
(a)

A decreasing function
is always one-to-one
(b)

A one-to-one function is not
always increasing or decreasing
(c)

› **Figure 3** Increasing, decreasing, and one-to-one functions.

Figure 3(c) shows that a function can still be one-to-one even if it is neither increasing nor decreasing. The function illustrated is increasing on $[-\infty, 0]$ and decreasing on $(0, \infty)$.

› Finding the Inverse of a Function

Now we will demonstrate how we can form a new function by reversing the correspondence determined by a given function. Let g be the function defined as follows:

$$g = \{(-3, 9), (0, 0), (3, 9)\} \qquad g \text{ is not one-to-one.}$$

Notice that g is not one-to-one because the domain elements -3 and 3 both correspond to the range element 9. We can reverse the correspondence determined by function g simply by reversing the components in each ordered pair in g, producing the following set:

$$G = \{(9, -3), (0, 0), (9, 3)\} \qquad G \text{ is not a function.}$$

But the result is not a function because the domain element 9 corresponds to two different range elements, -3 and 3. On the other hand, if we reverse the ordered pairs in the function

$$f = \{(1, 2), (2, 4), (3, 9)\} \qquad \text{\textcolor{red}{f is one-to-one; all second components are distinct.}}$$

we obtain

$$F = \{(2, 1), (4, 2), (9, 3)\} \qquad \text{\textcolor{red}{F is a function.}}$$

This time f is a one-to-one function, and the set F turns out to be a function also. This new function F, formed by reversing all the ordered pairs in f, is called the *inverse* of f and is usually denoted by f^{-1} (this is read as "inverse f" or "the inverse of f"):

$$f^{-1} = \{(2, 1), (4, 2), (9, 3)\} \qquad \text{\textcolor{red}{The inverse of f}}$$

Notice that f^{-1} is also a one-to-one function and that the following relationships hold:

$$\text{Domain of } f^{-1} = \{2, 4, 9\} = \text{Range of } f$$
$$\text{Range of } f^{-1} = \{1, 2, 3\} = \text{Domain of } f$$

We conclude that reversing all the ordered pairs in a one-to-one function forms a new one-to-one function and reverses the domain and range in the process. We are now ready to present a formal definition of the inverse of a function.

> **DEFINITION 2 Inverse of a Function**

If f is a one-to-one function, then the **inverse** of f, denoted f^{-1}, is the function formed by reversing all the ordered pairs in f. That is,

$$f^{-1} = \{(y, x) \mid (x, y) \text{ is in } f\}$$

If f is not one-to-one, then f *does not have an inverse* and f^{-1} *does not exist.*

»» CAUTION »»

Be careful not to confuse inverse notation and reciprocal notation. For numbers, a superscript of -1 means reciprocal: $2^{-1} = \dfrac{1}{2}$. For functions, a superscript of -1 means inverse: $f^{-1}(x)$ is the inverse of $f(x)$, which is not the same as $\dfrac{1}{f(x)}$.

The following properties of inverse functions follow directly from the definition.

> **THEOREM 4 Properties of Inverse Functions**

For a given function f, if f^{-1} exists, then

1. f^{-1} is a one-to-one function.
2. The domain of f^{-1} is the range of f.
3. The range of f^{-1} is the domain of f.

>>> EXPLORE-DISCUSS 2

(A) For the function $f = \{(3, 5), (7, 11), (11, 17)\}$, find f^{-1}.

(B) What do you think would be the result of composing f with f^{-1}? Justify your answer using Definition 2.

(C) Check your conjecture from part B by finding both $f \circ f^{-1}$ and $f^{-1} \circ f$. Were you correct?

Explore-Discuss 2 brings up an important point: If you apply a function to any number in its domain, then apply the inverse of that function to the result, you'll get right back where you started. This leads to the following theorem.

> **› THEOREM 5** Inverse Functions and Composition

> If f^{-1} exists, then

> 1. $f(f^{-1}(x)) = x$ for all x in the domain of f^{-1}.
> 2. $f^{-1}(f(x)) = x$ for all x in the domain of f.

> If f and g are one-to-one functions satisfying

> $$f(g(x)) = x \text{ for all } x \text{ in the domain of } g \text{ and}$$
> $$g(f(x)) = x \text{ for all } x \text{ in the domain of } f$$

> then f and g are inverses of one another.

We can use Theorem 5 to see if two functions defined by equations are inverses.

| EXAMPLE | 2 | **Deciding If Two Functions Are Inverses** |

Use Theorem 5 to decide if these two functions are inverses.

$$f(x) = 3x - 7 \qquad g(x) = \frac{x + 7}{3}$$

SOLUTION The domain of both functions is all real numbers. For any x,

$$f(g(x)) = f\left(\frac{x + 7}{3}\right) \qquad \text{Substitute into } f(x).$$

$$= 3\left(\frac{x + 7}{3}\right) - 7 \qquad \text{Multiply.}$$

$$= x + 7 - 7 \qquad \text{Add.}$$

$$= x$$

$$g(f(x)) = g(3x - 7) \qquad \text{Substitute into } g(x).$$

$$= \frac{3x - 7 + 7}{3} \qquad \text{Add.}$$

$$= \frac{3x}{3} \qquad \text{Simplify.}$$

$$= x$$

By Theorem 5, f and g are inverses.

MATCHED PROBLEM 2 Use Theorem 5 to decide if these two functions are inverses.

$$f(x) = \frac{2}{5}(11 - x) \qquad g(x) = -\frac{5}{2}x + 11$$

There is one obvious question that remains: when a function is defined by an equation, how can we find the inverse? Given a function $y = f(x)$, the first coordinates of points on the graph are represented by x, and the second coordinates are represented by y. Finding the inverse by reversing the order of the coordinates would then correspond to switching the variables x and y. This leads us to the following procedure, which can be applied whenever it is possible to solve $y = f(x)$ for x in terms of y.

> ### ❯ FINDING THE INVERSE OF A FUNCTION f
>
> **Step 1.** Find the domain of f and verify that f is one-to-one. If f is not one-to-one, then stop, because f^{-1} does not exist.
>
> **Step 2.** If the function is written with function notation, like $f(x)$, replace the function symbol with the letter y. Then interchange x and y.
>
> **Step 3.** Solve the resulting equation for y. The result is $f^{-1}(x)$.
>
> **Step 4.** Find the domain of f^{-1}. Remember, the domain of f^{-1} must be the same as the range of f.
>
> You can check your work using Theorem 5.

EXAMPLE 3 ### Finding the Inverse of a Function

Find f^{-1} for $f(x) = \sqrt{x - 1}$.

SOLUTION **Step 1.** Find the domain of f and verify that f is one-to-one. Since $\sqrt{x - 1}$ is defined only for $x - 1 \geq 0$, the domain of f is $[1, \infty)$. The graph of f in Figure 4 shows that f is one-to-one, so f^{-1} exists.

Step 2. Replace $f(x)$ with y, then interchange x and y.

$$y = \sqrt{x - 1} \qquad \text{Interchange } x \text{ and } y.$$
$$x = \sqrt{y - 1}$$

Step 3. Solve the equation for y.

$$x = \sqrt{y - 1} \qquad \text{Square both sides.}$$
$$x^2 = y - 1 \qquad \text{Add 1 to each side.}$$
$$x^2 + 1 = y$$

The inverse is $f^{-1}(x) = x^2 + 1$.

Step 4. Find the domain of f^{-1}.

The equation we found for f^{-1} is defined for all x, but the domain should be the range of f. From Figure 4, we see that the range of f is $[0, \infty)$ so that is the domain of f^{-1}. Therefore,

$$f^{-1}(x) = x^2 + 1 \qquad x \geq 0$$

$f(x) = \sqrt{x - 1}, x \geq 1$

> Figure 4

CHECK Find the composition of f with the alleged inverse (in both orders!).
For x in $[1, \infty)$, the domain of f, we have

$$f^{-1}(f(x)) = f^{-1}(\sqrt{x-1}) \qquad \text{Substitute } \sqrt{x-1} \text{ into } f^{-1}.$$
$$= (\sqrt{x-1})^2 + 1 \qquad \text{Square } \sqrt{x-1}.$$
$$= x - 1 + 1 \qquad \text{Add.}$$
$$\overset{\checkmark}{=} x$$

For x in $[0, \infty)$, the domain of f^{-1}, we have

$$f(f^{-1}(x)) = f(x^2 + 1) \qquad \text{Substitute } x^2 + 1 \text{ into } f.$$
$$= \sqrt{(x^2 + 1) - 1} \qquad \text{Add.}$$
$$= \sqrt{x^2} \qquad \sqrt{x^2} = |x| \text{ for any real number } x.$$
$$= |x| \qquad |x| = x \text{ for } x \geq 0.$$
$$\overset{\checkmark}{=} x$$

MATCHED PROBLEM 3 Find f^{-1} for $f(x) = \sqrt{x + 2}$.

The technique of finding an inverse by interchanging x and y leads to the following property of inverses that comes in very handy later in the course.

> **THEOREM 6** A Property of Inverses

If f^{-1} exists, then $x = f^{-1}(y)$ if and only if $y = f(x)$.

> ## Mathematical Modeling

Example 4 shows how an inverse function is used in constructing a revenue model. It is based on Example 8 in Section 3-5.

EXAMPLE 4 **Modeling Revenue**

The research department for an electronics firm estimates that the weekly demand for a certain brand of headphones is given by

$$x = f(p) = 20{,}000 - 1{,}000p \qquad \text{Demand function}$$

where x is the number of pairs retailers are likely to buy per week at p dollars per pair. Express the revenue as a function of the demand x and state its domain.

SOLUTION If x pairs of headphones are sold at p dollars each, the total revenue is

$$\text{Revenue} = (\text{Number of pairs})(\text{price of each pair})$$
$$R = xp$$

To express the revenue as a function of the demand x, we need to express the price in terms of x. That is, we must find the inverse of the demand function.

Step 1. Find the domain of f and verify that f is one-to-one. Price and demand are never negative, so $p \geq 0$ and

$$x = 20{,}000 - 1{,}000p \qquad \text{\color{magenta}Factor.}$$
$$= 1{,}000(20 - p) \geq 0 \qquad \text{\color{magenta}Divide both sides by 1,000.}$$
$$20 - p \geq 0 \qquad \text{\color{magenta}Add } p \text{ to both sides.}$$
$$20 \geq p \qquad \text{or} \qquad p \leq 20$$

Since p must satisfy both $p \geq 0$ and $p \leq 20$, the domain of f is $[0, 20]$. The graph of f (Fig. 5) shows that f is one-to-one.

> **Figure 5**

Step 2. Since x and p have specific meaning in the context of this problem, interchanging them does not apply here.

Step 3. Solve the equation $x = 20{,}000 - 1{,}000p$ for p.

$$x = 20{,}000 - 1{,}000p \qquad \text{\color{magenta}Subtract 20,000 from both sides.}$$
$$x - 20{,}000 = -1{,}000p \qquad \text{\color{magenta}Divide both sides by -1,000.}$$
$$-0.001x + 20 = p$$

The inverse of the demand function is

$$p = f^{-1}(x) = 20 - 0.001x$$

Step 4. From Figure 5, we see that the range of f is $[0, 20{,}000]$, so this must also be the domain of f^{-1}.

$$p = f^{-1}(x) = 20 - 0.001x \qquad 0 \leq x \leq 20{,}000$$

We should check that $f(f^{-1}(x)) = x$ and $f^{-1}(f(p)) = p$, but we will leave that to the reader.

The revenue R is given by

$$R = xp$$
$$R(x) = x(20 - 0.001x)$$
$$= 20x - 0.001x^2$$

and the domain of R is $[0, 20{,}000]$.

MATCHED PROBLEM 4

Repeat Example 3 for the demand function

$$x = f(p) = 10{,}000 - 1{,}000p \qquad 0 \leq p \leq 10$$

The demand function in Example 4 was defined with independent variable p and dependent variable x. When we found the inverse function, we did not rewrite it with independent variable p. Because p represents price and x represents number of players, to interchange these variables would be confusing. In most applications, the variables have specific meaning and should not be interchanged as part of the inverse process.

> ## Graphing Inverse Functions

The following activities refer to the graph of f in Figure 6 and Tables 1 and 2.

> **Figure 6**

Table 1	
x	$f(x)$
-4	
-2	
0	
2	

Table 2	
x	$f^{-1}(x)$

(A) Complete the second column in Table 1.

(B) Reverse the ordered pairs in Table 1 and list the results in Table 2.

(C) Add the points in Table 2 to Figure 6 (or a copy of the figure) and sketch the graph of f^{-1}.

(D) Discuss any symmetry you observe between the graphs of f and f^{-1}.

Explore-Discuss 3 is based on an important relationship between the graph of any function and its inverse. In a rectangular coordinate system, the points (a, b) and (b, a) are symmetric with respect to the line $y = x$ [Fig. 7(a)]. Theorem 6 is an immediate consequence of this observation.

> **Figure 7** Symmetry with respect to the line $y = x$.

(a, b) and (b, a)
are symmetric with
respect to the line $y = x$
(a)

$f(x) = 2x - 1$
$f^{-1}(x) = \frac{1}{2}x + \frac{1}{2}$
(b)

$f(x) = \sqrt{x - 1}$
$f^{-1}(x) = x^2 + 1, \ x \geq 0$
(c)

> **THEOREM 7 Symmetry Property for the Graphs of f and f^{-1}**

The graphs of $y = f(x)$ and $y = f^{-1}(x)$ are symmetric with respect to the line $y = x$.

Knowledge of this symmetry property allows us to graph f^{-1} if the graph of f is known, and vice versa. Figures 7(b) and 7(c) illustrate this property for the two inverse functions we found earlier.

If a function is not one-to-one, we can usually restrict the domain of the function to produce a new function that is one-to-one. Then we can find an inverse for the restricted function. Suppose we start with $f(x) = x^2 - 4$. Because f is not one-to-one, f^{-1} does not exist [Fig. 8(a)]. But there are many ways the domain of f can be restricted to obtain a one-to-one function. Figures 8(b) and 8(c) illustrate two such restrictions. In essence, we are "forcing" the function to be one-to-one by throwing out a portion of the graph that would make it fail the horizontal line test.

> **Figure 8** Restricting the domain of a function.

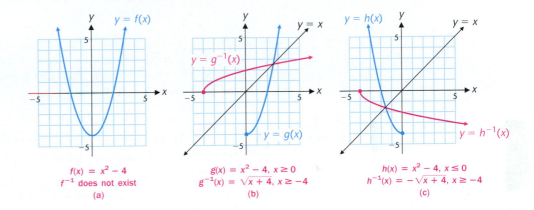

$f(x) = x^2 - 4$
f^{-1} does not exist
(a)

$g(x) = x^2 - 4, x \geq 0$
$g^{-1}(x) = \sqrt{x + 4}, x \geq -4$
(b)

$h(x) = x^2 - 4, x \leq 0$
$h^{-1}(x) = -\sqrt{x + 4}, x \geq -4$
(c)

Recall from Theorem 3 that increasing and decreasing functions are always one-to-one. This provides the basis for a convenient method of restricting the domain of a function:

If the domain of a function f is restricted to an interval on the x axis over which f is increasing (or decreasing), then the new function determined by this restriction is one-to-one and has an inverse.

We used this method to form the functions g and h in Figure 8.

| EXAMPLE | 5 | **Finding the Inverse of a Function** |

Find the inverse of $f(x) = 4x - x^2$, $x \leq 2$. Graph f, f^{-1}, and the line $y = x$ in the same coordinate system.

SOLUTION **Step 1.** Find the domain of f and verify that f is one-to-one. We are given that the domain of f is $(-\infty, 2]$. The graph of $y = 4x - x^2$ is a parabola opening downward with vertex $(2, 4)$ (Fig. 9). The graph of f is the left side of this parabola (Fig. 10). From the graph of f, we see that f is increasing and one-to-one on $(-\infty, 2]$.

> **Figure 9**

> **Figure 10**

Step 2. Replace $f(x)$ with y, then interchange x and y.

$$y = 4x - x^2$$
$$x = 4y - y^2$$

Step 3. Solve the equation for y.

$$x = 4y - y^2 \qquad \text{Rewrite so that the coefficient of } y^2 \text{ is } +1.$$
$$y^2 - 4y = -x \qquad \text{Add 4 to both sides to complete the square.}$$
$$y^2 - 4x + 4 = -x + 4 \qquad \text{Factor the left side.}$$
$$(y - 2)^2 = 4 - x \qquad \text{Take the square root of both sides.}$$
$$y - 2 = \pm\sqrt{4 - x} \qquad \text{Add 2 to both sides.}$$
$$y = 2 \pm\sqrt{4 - x}$$

Now we have two possible solutions. The domain of f was $(-\infty, 2]$, and this should be the range of f^{-1}. In other words, the output of the inverse is never greater than 2. But $y = 2 + \sqrt{4 - x}$ would always be greater than or equal to 2, so we must instead choose $y = 2 - \sqrt{4 - x}$.

$$f^{-1}(x) = 2 - \sqrt{4 - x}$$

Step 4. The domain of f^{-1} is the range of f. We can see from Figure 10 that this is $(-\infty, 4]$. Notice that the equation we found for $f^{-1}(x)$ is defined for these values. Our final answer is

$$f^{-1}(x) = 2 - \sqrt{4 - x} \qquad x \le 4$$

The check is again left for the reader.

The graphs of f, f^{-1}, and $y = x$ are shown in Figure 11. To aid in graphing f^{-1}, we plotted several points on the graph of f and then reflected these points in the line $y = x$.

> **Figure 11**

MATCHED PROBLEM 5 Find the inverse of $f(x) = 4x - x^2$, $x \ge 2$. Graph f, f^{-1}, and $y = x$ in the same coordinate system.

Technology Connections

To reproduce Figure 11 on a graphing calculator, first enter

$$y_1 = (4x - x^2)/(x \le 2)$$

in the equation editor (Fig. 12) and graph (Fig. 13). (For graphs involving both f and f^{-1} it is best to use a squared viewing window.) The Boolean expression $(x \le 2)$ is assigned the value 1 if the inequality is true and 0 if it is false. The calculator recognizes that division by 0 is an undefined operation and no graph is drawn for $x > 2$. Now enter

$$y_2 = 2 - \sqrt{4 - x} \qquad \text{and} \qquad y_3 = x$$

in the equation editor and graph (Fig. 14).

> **Figure 12**

> **Figure 13**

> **Figure 14**

ANSWERS TO MATCHED PROBLEMS

1. (A) Not one-to-one
 (B) One-to-one
2. They are inverses.
3. $f^{-1}(x) = x^2 - 2, x \geq 0$
4. $R(x) = 10x - 0.001x^2$

5. $f^{-1}(x) = 2 + \sqrt{4 - x}, x \leq 4$

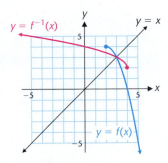

3-6 Exercises

1. When a function is defined by ordered pairs, how can you tell if it is one-to-one?

2. When you have the graph of a function, how can you tell if it is one-to-one?

3. Why does a function fail to have an inverse if it is not one-to-one? Give an example using ordered pairs to illustrate your answer.

4. True or False: Any function whose graph changes direction is not one-to-one. Explain.

5. What is the result of composing a function with its inverse? Why does this make sense?

6. What is the relationship between the graphs of two functions that are inverses?

For each set of ordered pairs in Problems 7–12, determine if the set is a function, a one-to-one function, or neither. Reverse all the ordered pairs in each set and determine if this new set is a function, a one-to-one function, or neither.

7. $\{(1, 2), (2, 1), (3, 4), (4, 3)\}$

8. $\{(-1, 0), (0, 1), (1, -1), (2, 1)\}$

9. $\{(5, 4), (4, 3), (3, 3), (2, 4)\}$

10. $\{(5, 4), (4, 3), (3, 2), (2, 1)\}$

11. $\{(1, 2), (1, 4), (-3, 2), (-3, 4)\}$

12. $\{(0, 5), (-4, 5), (-4, 2), (0, 2)\}$

In Problems 13–30, determine if the function is one-to-one.

13.
Domain	Range
-2	-4
-1	-2
0	0
1	1
2	5

14.
Domain	Range
-2	
-1	-3
0	7
1	
2	9

15.
Domain	Range
1	
2	
3	7
4	
5	

16.
Domain	Range
1	5
2	3
3	1
4	2
5	4

17.

18.

19.

20.

21.

22.

23.

24.

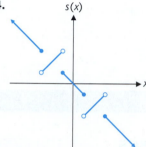

25. $F(x) = \frac{1}{2}x + 2$ **26.** $G(x) = -\frac{1}{3}x + 1$

27. $H(x) = 4 - x^2$ **28.** $K(x) = \sqrt{4 - x}$

29. $M(x) = \sqrt{x + 1}$ **30.** $N(x) = x^2 - 1$

In Problems 31–40, determine if g is the inverse of f.

31. $f(x) = 3x + 5$; $g(x) = \frac{1}{3}x - \frac{5}{3}$

32. $f(x) = 2x - 4$; $g(x) = \frac{1}{2}x - 2$

33. $f(x) = 2 - (x + 1)^3$; $g(x) = \sqrt[3]{3 - x} - 1$

34. $f(x) = (x - 3)^3 + 4$; $g(x) = \sqrt[3]{x - 4} + 3$

35. $f(x) = \dfrac{2x - 3}{x + 4}$; $g(x) = \dfrac{3 + 4x}{2 - x}$

36. $f(x) = \dfrac{x + 1}{2x - 3}$; $g(x) = \dfrac{3x + 1}{2x + 1}$

37. $f(x) = 4 + x^2, x \geq 0$; $g(x) = \sqrt{x - 4}$

38. $f(x) = \sqrt{x + 2}$; $g(x) = x^2 - 2, x \geq 0$

39. $f(x) = 1 - x^2, x \geq 0$; $g(x) = -\sqrt{1 - x}$

40. $f(x) = -\sqrt{x - 2}$; $g(x) = x^2 + 2, x \leq 0$

In Problems 41–44, find the domain and range of f, sketch the graph of f^{-1}, and find the domain and range of f^{-1}.

41.

42.

43.

44.

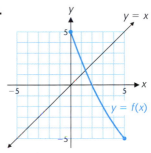

In Problems 45–74, graph f and verify that f is a one-to-one function. Find f^{-1} and add the graph of f^{-1} and the line $y = x$ to the graph of f. State the domain and range of f and the domain and range of f^{-1}.

45. $f(x) = 3x$

46. $f(x) = \dfrac{1}{2}x$

47. $f(x) = 4x - 3$

48. $f(x) = -\dfrac{1}{3}x + \dfrac{5}{3}$

49. $f(x) = 0.2x + 0.4$

50. $f(x) = 0.25x + 2.25$

51. $f(x) = \sqrt{x} + 3$

52. $f(x) = 2 - \sqrt{x}$

53. $f(x) = \dfrac{1}{2}\sqrt{16 - x}$

54. $f(x) = \dfrac{1}{3}\sqrt{36 - x}$

55. $f(x) = 3 - \sqrt{x - 1}$

56. $f(x) = 2 + \sqrt{5 - x}$

57. $f(x) = x^2 + 5, x \geq 0$

58. $f(x) = x^2 + 5, x \leq 0$

59. $f(x) = 4 - x^2, x \leq 0$

60. $f(x) = 4 - x^2, x \geq 0$

61. $f(x) = x^2 + 8x, x \geq -4$

62. $f(x) = x^2 + 8x, x \leq -4$

63. $f(x) = (2 - x)^2, x \leq 2$

64. $f(x) = (2 - x)^2, x \geq 2$

65. $f(x) = (x - 1)^2 + 2, x \geq 1$

66. $f(x) = 3 - (x - 2)^2, x \leq 2$

67. $f(x) = x^2 + 2x - 2, x \leq -1$

68. $f(x) = x^2 + 8x + 7, x \geq -4$

69. $f(x) = -\sqrt{9 - x^2}, 0 \leq x \leq 3$

70. $f(x) = \sqrt{9 - x^2}, 0 \leq x \leq 3$

71. $f(x) = \sqrt{9 - x^2}, -3 \leq x \leq 0$

72. $f(x) = -\sqrt{9 - x^2}, -3 \leq x \leq 0$

73. $f(x) = 1 - \sqrt{1 - x^2}, -1 \leq x \leq 0$

74. $f(x) = 1 + \sqrt{1 - x^2}, -1 \leq x \leq 0$

The functions in Problems 75–84 are one-to-one. Find f^{-1}.

75. $f(x) = 3 - \dfrac{2}{x}$

76. $f(x) = 5 + \dfrac{4}{x}$

77. $f(x) = \dfrac{2}{x - 1}$

78. $f(x) = \dfrac{3}{x + 4}$

79. $f(x) = \dfrac{2x}{x + 1}$

80. $f(x) = \dfrac{4x}{2 - x}$

81. $f(x) = \dfrac{2x + 5}{3x - 4}$

82. $f(x) = \dfrac{5 - 3x}{7 - 4x}$

83. $f(x) = 4 - \sqrt[5]{x + 2}$

84. $f(x) = \sqrt[3]{x + 3} - 2$

85. How are the x and y intercepts of a function and its inverse related?

86. Does a constant function have an inverse? Explain.

87. Are the functions $f(x) = x^2$ and $g(x) = \sqrt{x}$ inverses? Why or why not?

88. Are the functions $f(x) = x^3$ and $g(x) = \sqrt[3]{x}$ inverses? Why or why not?

In Problems 89–92, the given function is not one-to-one. Find a way to restrict the domain so that the function is one-to-one, then find the inverse of the function with that domain.

89. $f(x) = (2 - x)^2$

90. $f(x) = (1 + x)^2$

91. $f(x) = \sqrt{4x - x^2}$

92. $f(x) = \sqrt{6x - x^2}$

APPLICATIONS

93. BODY WEIGHT Two formulas for estimating body weight as a function of height that are commonly used are

$$\text{Women: } p = W(h) = 100 + 5h$$
$$\text{Men: } p = M(h) = 110 + 5h$$

where p is weight in pounds and h is height over 5 feet (in inches). Find $h = W^{-1}(p)$ and state its domain.

94. BODY WEIGHT Refer to Problem 93. Find $h = M^{-1}(p)$ and state its domain.

95. PRICE AND DEMAND The number q of CD players consumers are willing to buy per week from a retail chain at a price of \$p is given approximately by (see the figure)

$$q = d(p) = \frac{3,000}{0.2p + 1} \qquad 10 \leq p \leq 70$$

form of a parabola can be used to find the equation when the vertex and one other point on the graph are known.

Replacing the equal sign in a quadratic equation with $<$, $>$, $\leq$, or $\geq$ produces a **quadratic inequality.** The set of all values of the variable that make the inequality a true statement is the **solution set.**

3-5 Combining Functions; Composition

The **sum, difference, product,** and **quotient** of the functions f and g are defined by

$$(f + g)(x) = f(x) + g(x) \qquad (f - g)(x) = f(x) - g(x)$$

$$(fg)(x) = f(x)g(x) \qquad \left(\frac{f}{g}\right)(x) = \frac{f(x)}{g(x)} \qquad g(x) \neq 0$$

The **domain** of each function is the intersection of the domains of f and g, with the exception that values of x where $g(x) = 0$ must be excluded from the domain of f/g.

The **composition** of functions f and g is defined by $(f \circ g)(x) = f(g(x))$. The *domain* of $f \circ g$ is the set of all real numbers x in the domain of g such that $g(x)$ is in the domain of f. The domain of $f \circ g$ is always a subset of the domain of g.

3-6 Inverse Functions

A function is **one-to-one** if no two ordered pairs in the function have the same second component and different first components. According to the **horizontal line test,** a horizontal line will intersect the graph of a one-to-one function in at most one point. A function that is increasing (or decreasing) throughout its domain is one-to-one. The **inverse** of the one-to-one function f is the function f^{-1} formed by reversing all the ordered pairs in f.

If f is a one-to-one function, then:

1. f^{-1} is one-to-one.

2. Domain of f^{-1} = Range of f.

3. Range of f^{-1} = Domain of f.

4. $x = f^{-1}(y)$ if and only if $y = f(x)$.

5. $f^{-1}(f(x)) = x$ for all x in the domain of f.

6. $f(f^{-1}(x)) = x$ for all x in the domain of f^{-1}.

7. To find f^{-1}, solve the equation $y = f(x)$ for x. Interchanging x and y at this point is an option.

8. The graphs of $y = f(x)$ and $y = f^{-1}(x)$ are symmetric with respect to the line $y = x$.

CHAPTER **3** Review Exercises

Work through all the problems in this review and check answers in the back of the book. Answers to most review problems are there, and following each answer is a number in italics indicating the section in which that type of problem is discussed. Where weaknesses show up, review appropriate sections in the text.

1. Indicate whether each table defines a function.

(A)
Domain	Range
1	4
3	6
5	8

(B)
Domain	Range
7	0
8	
9	

(C)
Domain	Range
5	1
10	2
20	

2. Indicate whether each set defines a function. Indicate whether any of the functions are one-to-one. Find the domain and range of each function. Find the inverse of any one-to-one functions. Find the domain and range of any inverse functions.
 (A) $\{(1, 1), (2, 4), (3, 9)\}$
 (B) $\{(1, 1), (1, -1), (2, 2), (2, -2)\}$
 (C) $\{(\text{Albany, New York}), (\text{Utica, New York}), (\text{Akron, Ohio}), (\text{Dayton, Ohio})\}$
 (D) $\{(\text{Albany, New York}), (\text{Akron, Ohio}), (\text{Tucson, Arizona}), (\text{Atlanta, Georgia}), (\text{Muncie, Indiana})\}$

3. Let T be the set of teams in the National Football League that have won at least one Super Bowl, and let Y be the set of years during which a Super Bowl was played. If each team corresponds to the year or years in which they won the Super Bowl, does this correspondence define a function? Explain your answer.

4. Indicate whether each graph specifies a function:

(A)

(B)

(C)

(D)

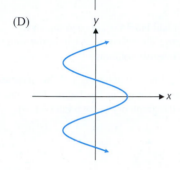

5. Which of the following equations define functions?
(A) $y = x$ (B) $y^2 = x$
(C) $y^3 = x$ (D) $|y| = x$

Problems 6–15 refer to the functions f, g, k, and m given by:

$$f(x) = 3x + 5 \quad g(x) = 4 - x^2$$
$$k(x) = 5$$
$$m(x) = 2|x| - 1$$

Find the indicated quantities or expressions.

6. $f(2) + g(-2) + k(0)$ **7.** $\dfrac{m(-2) + 1}{g(2) + 4}$

8. $\dfrac{f(2 + h) - f(2)}{h}$ **9.** $\dfrac{g(a + h) - g(a)}{h}$

10. $(f + g)(x)$ **11.** $(f - g)(x)$

12. $(fg)(x)$ **13.** $\left(\dfrac{f}{g}\right)(x)$

14. $(f \circ g)(x)$ **15.** $(g \circ f)(x)$

16. For $f(x) = x^2 - 2x$, find
(A) $f(1)$ (B) $f(-4)$ (C) $f(2) \cdot f(-1)$ (D) $\dfrac{f(0)}{f(3)}$

Problems 17–21 refer to the function f given by the following graph.

17. Find $f(-4)$, $f(0)$, $f(3)$, and $f(5)$.

18. Find all values of x for which $f(x) = -2$.

19. Find the domain and range of f.

20. Find the intervals over which f is increasing and decreasing.

21. Find any points of discontinuity.

Problems 22–29 refer to the graphs of f and g shown here.

22. Construct a table of values of $(f - g)(x)$ for $x = -3, -2, -1, 0, 1, 2,$ and 3, and sketch the graph of $f - g$.

23. Construct a table of values of $(fg)(x)$ for $x = -3, -2, -1, 0, 1, 2,$ and 3, and sketch the graph of fg.

In Problems 24–27, use the graphs of f and g to find:

24. $(f \circ g)(-1)$ **25.** $(g \circ f)(-2)$

26. $f[g(1)]$ **27.** $g[f(-3)]$

28. Is f a one-to-one function?

29. Is g a one-to-one function?

30. Indicate whether each function is even, odd, or neither:
(A) $f(x) = x^5 + 6x$
(B) $g(t) = t^4 + 3t^2$
(C) $h(z) = z^5 + 4z^2$

Problems 31–36 refer to the graph of the function f used in Problems 17–21.
Sketch the graph of each of the following.

31. $f(x) + 1$ **32.** $f(x + 1)$ **33.** $-f(x)$

34. $0.5f(x)$ **35.** $f(2x)$ **36.** $-f(-x)$

37. Match each equation with a graph of one of the functions $f, g, m,$ or n in the figure. Each graph is a graph of one of the equations.
(A) $y = (x - 2)^2 - 4$ (B) $y = -(x + 2)^2 + 4$
(C) $y = -(x - 2)^2 + 4$ (D) $y = (x + 2)^2 - 4$

38. Referring to the graph of function f in the figure for Problem 37 and using known properties of quadratic functions, find each of the following to the nearest integer:
(A) Intercepts (B) Vertex
(C) Maximum or minimum (D) Range
(E) Interval of increase (F) Interval of decrease

39. Let $f(x) = x^2 - 4$ and $g(x) = x + 3$. Find each of the following functions and find their domains.
(A) f/g (B) g/f (C) $f \circ g$ (D) $g \circ f$

40. For each function, find the maximum or minimum value without graphing. Then write the coordinates of the vertex.
(A) $f(x) = -2(x + 4)^2 - 10$
(B) $f(x) = x^2 - 6x + 11$

41. Complete the square to write the quadratic function in vertex form: $q(x) = 2x^2 - 14x + 3$

42. How are the graphs of the following related to the graph of $y = x^2$?
(A) $y = -x^2$
(B) $y = x^2 - 3$
(C) $y = (x + 3)^2$

Problems 43–49 refer to the function q given by the following graph.

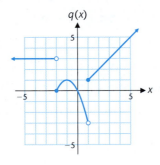

$q(x)$

43. Find y to the nearest integer:
(A) $y = q(0)$ (B) $y = q(1)$
(C) $y = q(2)$ (D) $y = q(-2)$

44. Find x to the nearest integer:
(A) $q(x) = 0$ (B) $q(x) = 1$
(C) $q(x) = -3$ (D) $q(x) = 3$

45. Find the domain and range of q.

46. Find the intervals over which q is increasing, decreasing, and constant.

47. Identify any points of discontinuity.

48. The function f multiplies the cube of the domain element by 4 and then subtracts the square root of the domain element. Write an algebraic definition of f.

49. Write a verbal description of the function $f(x) = 3x^2 + 4x - 6$.

In Problems 50 and 51, determine if the indicated equation defines a function. Justify your answer.

50. $x + 2y = 10$ **51.** $x + 2y^2 = 10$

In Problems 52–57, find the domain, y intercept (if it exists), and any x intercepts.

52. $m(x) = x^2 - 4x + 5$ **53.** $r(x) = 2 + 3\sqrt{x}$

54. $p(x) = \dfrac{1 - x^2}{x^3}$ **55.** $f(x) = \dfrac{x}{\sqrt{3 - x}}$

56. $g(x) = \dfrac{2x + 3}{x^2 - 4}$ **57.** $h(x) = \dfrac{1}{4 - \sqrt{x}}$

58. Let $f(x) = 0.5x^2 - 4x + 5$.
(A) Sketch the graph of f and label the axis and the vertex.
(B) Where is f increasing? Decreasing? What is the range? (Express answers in interval notation.)

59. Find the equations of the linear function g and the quadratic function f whose graphs are shown in the figure. This line is called the tangent line to the graph of f at the point $(-1, 0)$.

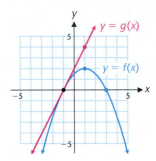

60. Let
$$f(x) = \begin{cases} -x - 5 & \text{for } -4 \le x < 0 \\ 0.2x^2 & \text{for } 0 \le x \le 5 \end{cases}$$
(A) Find $f(-4), f(-2), f(0), f(2),$ and $f(5)$.
(B) Sketch the graph of $y = f(x)$.
(C) Find the domain and range.
(D) Find any points of discontinuity.
(E) Find the intervals over which f is increasing, decreasing, and constant.

61. Given $f(x) = \sqrt{x} - 8$ and $g(x) = |x|$:
(A) Find $f \circ g$ and $g \circ f$.
(B) Find the domains of $f \circ g$ and $g \circ f$.

62. Which of the following functions are one-to-one?
(A) $f(x) = x^3$
(B) $g(x) = (x - 2)^2$
(C) $h(x) = 2x - 3$
(D) $F(x) = (x + 3)^2, x \ge -3$

63. Is $u(x) = 4x - 8$ the inverse of $v(x) = 0.25x + 2$?

64. The function $f(x) = 2(x - 3)^2$ is not one-to-one.
(A) Graph f using transformations of $y = x^2$.
(B) Restrict the domain of f to make it a one-to-one function.
(C) Find the inverse of the one-to-one function.

65. Given $f(x) = 3x - 7$:
(A) Find $f^{-1}(x)$.
(B) Find $f^{-1}(5)$.
(C) Find $f^{-1}[f(x)]$.
(D) Is f increasing, decreasing, or constant on $(-\infty, \infty)$?

66. The following graph is the result of applying a sequence of transformations to the graph of $y = x^2$. Describe the transformations verbally and write an equation for the given graph.

 Check by graphing your equation on a graphing calculator.

67. The graph of $f(x) = |x|$ is vertically stretched by a factor of 3, reflected through the x axis, and shifted 2 units to the right and 5 units up to form the graph of the function g. Find an equation for the function g and graph g.

68. Write an equation for the following graph in the form $y = a(x - h)^2 + k$, where a is either -1 or $+1$ and h and k are integers.

 Check by graphing your equation on a graphing calculator.

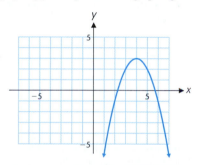

69. The following graph is the result of applying a sequence of transformations to the graph of $y = \sqrt[3]{x}$. Describe the transformations verbally, and write an equation for the given graph.

 Check by graphing your equation on a graphing calculator.

70. How is the graph of $f(x) = -(x - 2)^2 - 1$ related to the graph of $g(x) = x^2$?

71. Each of the following graphs is the result of applying one or more transformations to the graph of one of the six basic functions in Figure 1, Section 3-3. Find an equation for the graph. Check by graphing the equation on a graphing calculator.

(A)

(B)

72. The graph of $f(x) = |x|$ is stretched vertically by a factor of 3, reflected through the x axis, shifted four units to the right and eight units up to form the graph of the function g. Find an equation for the function g and graph g.

73. The graph of $m(x) = x^2$ is stretched horizontally by a factor of 2, shifted two units to the left and four units down to form the graph of the function t. Find an equation for the function t and graph t.

Use graph transformations to sketch the graph of each equation in Problems 74–81:

74. $y = |x + 1|$

75. $y = 1 + \sqrt[3]{1 - x}$

76. $y = |x| - 2$

77. $y = 9 - 3\sqrt{x}$

78. $y = \frac{1}{2}|x|$

79. $y = \sqrt[3]{4 - 0.5x}$

80. $y = 2 - 3(x - 1)^3$

81. $y = -|x + 1| - 1$

Solve Problems 82 and 83. Express answers in interval notation.

82. $x^2 + x < 20$

83. $x^2 > 4x + 12$

84. Find the domain of $f(x) = \sqrt{25 - x^2}$.

85. Given $f(x) = x^2$ and $g(x) = \sqrt{1 - x}$, find each function and its domain.

(A) fg (B) f/g (C) $f \circ g$ (D) $g \circ f$

86. For the one-to-one function f given by

$$f(x) = \frac{x + 2}{x - 3}$$

(A) Find $f^{-1}(x)$.
(B) Find $f^{-1}(3)$.
(C) Find $f^{-1}[f(x)]$.

87. Given $f(x) = \sqrt{x - 1}$:
(A) Find $f^{-1}(x)$.
(B) Find the domain and range of f and f^{-1}.
(C) Graph f, f^{-1}, and $y = x$ on the same coordinate system.

Check by graphing f, f^{-1}, and $y = x$ in a squared window on a graphing calculator.

88. Given $f(x) = x^2 - 1, x \geq 0$:
(A) Find the domain and range of f and f^{-1}.
(B) Find $f^{-1}(x)$.
(C) Find $f^{-1}(3)$.
(D) Find $f^{-1}[f(4)]$.
(E) Find $f^{-1}[f(x)]$.

Check by graphing f, f^{-1}, and $y = x$ in a squared window on a graphing calculator.

89. A partial graph of the function f is shown in the figure. Complete the graph of f over the interval [0, 5] given that:
(A) f is symmetric with respect to the y axis.
(B) f is symmetric with respect to the origin.

90. The function f is decreasing on $[-5, 5]$ with $f(-5) = 4$ and $f(5) = -3$.
(A) If f is continuous on $[-5, 5]$, how many times can the graph of f cross the x axis? Support your conclusion with examples and/or verbal arguments.
(B) Repeat part A if the function does not have to be continuous.

APPLICATIONS

91. INCOME Megan works 20 hours per week at an electronics store to help pay for tuition and rent. She gets a base salary of $6 per hour, a commission of 10% on all sales over $2,000 for the week, and a bonus of $250 if her weekly sales are over $5,000.
(A) Write a function that describes Megan's weekly earnings, where x represents her weekly sales.
(B) Find Megan's weekly earnings if her sales are $2,000, $4,000, and $6,000.
(C) If Megan needs to average at least $400 per week to cover her tuition and rent, how much does she need to sell on average each week?

92. On the set of a movie, a stuntman will be jumping from a helicopter that is hovering at a height of 120 feet, and landing in a moving truck full of chicken feathers. How many seconds after he jumps does the truck need to be in position?

93. BUSINESS—MARKUP POLICY A sporting goods store sells tennis shorts that cost $30 for $48 and sunglasses that cost $20 for $32.
(A) If the markup policy of the store for items that cost over $10 is assumed to be linear and is reflected in the pricing of these two items, find a function $r = f(c)$ that expresses retail price r as a function of cost c.
(B) What should be the retail price of a pair of skis that cost $105?
(C) Find $c = f^{-1}(r)$ and find its domain and range.
(D) What is the cost of a box of golf balls that retail for $39.99?

94. STOPPING DISTANCE Table 1 contains data related to the length of the skid marks left by an automobile when making an emergency stop. A model for the skid mark length L (in feet) of the auto is

$$L = f(s) = 0.06s^2 - 2.4s + 50, s \geq 20$$

where s is speed in miles per hour.

Table 1

Speed (mph)	Length of Skid Marks (feet)
20	26
30	32
40	49
50	80
60	122
70	176
80	242

(A) Graph $L = f(s)$ and the data for skid mark length on the same axes.
(B) Find $s = f^{-1}(L)$ and find its domain and range.
(C) How fast (to the nearest mile) was the auto traveling if it left skid marks 200 feet long?

95. PRICE AND DEMAND The price $p per hot dog at which q hot dogs can be sold during a baseball game is given approximately by

$$p = g(q) = \frac{9}{1 + 0.002q} \qquad 1,000 \leq q \leq 4,000$$

(A) Find the range of g.
(B) Find $q = g^{-1}(p)$ and find its domain and range.
(C) Express the revenue as a function of p.
(D) Express the revenue as a function of q.

96. MARKET RESEARCH A market research firm is hired to study demand for a new blanket that looks an awful lot like a bathrobe worn backwards. They determine that if x units are produced each week and sold at a price of $p per unit, then the weekly demand, revenue, and cost equations are, respectively

$$x = 500 - 10p$$
$$R(x) = 50x - 0.1x^2$$
$$C(x) = 10x + 1,500$$

Express the weekly profit as a function of the price p and find the price that produces the largest profit.

97. CONSTRUCTION A farmer has 120 feet of fencing to be used in the construction of two identical rectangular pens sharing a common side (see the figure).

(A) Express the total area $A(x)$ enclosed by both pens as a function of the width x.

(B) From physical considerations, what is the domain of the function A?

(C) Find the dimensions of the pens that will make the total enclosed area maximum.

98. COMPUTER SCIENCE In computer programming, it is often necessary to check numbers for certain properties (even, odd, perfect square, etc.). The greatest integer function provides a convenient method for determining some of these properties. Consider the function

$$f(x) = x - (\llbracket \sqrt{x} \rrbracket)^2$$

(A) Evaluate f for $x = 1, 2, \ldots, 16$.

(B) Find $f(n^2)$, where n is a positive integer.

(C) What property of x does this function determine?

99. Use the schedule in Table 2 to construct a piecewise-defined model for the taxes due for a single taxpayer in Virginia with a taxable income of x dollars. Find the tax on the following incomes: $2,000, $4,000, $10,000, $30,000.

Table 2 Virginia Tax Rate Schedule

Status	Taxable Income Over	But Not Over	Tax Is	Of the Amount Over
Single	$ 0	$ 3,000	2%	$ 0
	$ 3,000	$ 5,000	$ 60 + 3%	$ 3,000
	$ 5,000	$17,000	$120 + 5%	$ 5,000
	$17,000	—	$720 + 5.75%	$17,000

CHAPTER **3**

»» GROUP ACTIVITY Mathematical Modeling: Choosing a Cell Phone Plan

The number of companies offering cellular telephone service has grown rapidly in recent years. The plans they offer vary greatly and it can be difficult to select the plan that is best for you. Here are five typical plans:

Plan 1: A flat fee of $50 per month for unlimited calls.

Plan 2: A $30 per month fee for a total of 30 hours of calls and an additional charge of $0.01 per minute for all minutes over 30 hours.

Plan 3: A $5 per month fee and a charge of $0.04 per minute for all calls.

Plan 4: A $2 per month fee and a charge of $0.045 per minute for all calls; the fee is waived if the charge for calls is $20 or more.

Plan 5: A charge of $0.05 per minute for all calls; there are no additional fees.

(A) Construct a mathematical model for each plan that gives the total monthly cost in terms of the total number of minutes of calls placed in a month.

(B) Compare plans 1 and 2. Determine how many minutes per month would make plan 1 cheaper and how many would make plan 2 cheaper.

(C) Repeat part (B) for plans 1 and 3; plans 1 and 4; plans 1 and 5.

(D) Repeat part (B) for plans 2 and 3; plans 2 and 4; plans 2 and 5.

(E) Repeat part (B) for plans 3 and 4; plans 3 and 5.

(F) Repeat part (B) for plans 4 and 5.

(G) Is there one plan that is always better than all the others? Based on your personal calling history, which plan would you choose and why?

SOLUTIONS

(A) The real zeros are the x intercepts: $-4, -2, 0,$ and 3.

(B) Note first that $P(x)$ is a polynomial because it can be written in the form of Definition 1 (it is not necessary to actually multiply out $P(x)$ to find that form). The zeros of $P(x)$ are the solutions to the equation $P(x) = 0$. Because a product equals 0 if and only if one of the factors equals 0, we can find the zeros by solving each of the following equations (the last was solved using the quadratic formula):

$$x - 4 = 0 \qquad (x + 7)^3 = 0 \qquad x^2 + 9 = 0 \qquad x^2 - 2x + 2 = 0$$
$$x = 4 \qquad\qquad x = -7 \qquad\quad x = \pm 3i \qquad\qquad x = 1 \pm i$$

Therefore, the zeros of $P(x)$, are $4, -7, 3i, -3i, 1 + i,$ and $1 - i$. Only two of the six zeros are real numbers and therefore x intercepts: 4 and -7.

MATCHED PROBLEM 1

(A) Figure 4 shows the graph of a polynomial function of degree 4. List its real zeros.

› Figure 4

(B) List all zeros of the polynomial function

$$P(x) = (x + 5)(x^2 - 4)(x^2 + 4)(x^2 + 2x + 5)$$

Which zeros of $P(x)$ are x intercepts?

A point on a continuous graph that separates an increasing portion from a decreasing portion, or vice versa, is called a **turning point.** The vertex of a parabola, for example, is a turning point. Linear functions with real coefficients have exactly one real zero and no turning points; quadratic functions with real coefficients have at most two real zeros and exactly one turning point.

≫ EXPLORE-DISCUSS 1

Examine Figures 2(a), 2(b), 3, and 4, which show the graphs of polynomial functions of degree 2, 2, 5, and 4, respectively. In each figure, all real zeros and all turning points of the function appear in the given viewing window.

(A) Is the number of real zeros ever less than the degree? Equal to the degree? Greater than the degree? How is the number of real zeros of a polynomial related to its degree?

(B) Is the number of turning points ever less than the degree? Equal to the degree? Greater than the degree? How is the number of turning points of a polynomial related to its degree?

Explore-Discuss 1 suggests that graphs of polynomial functions with real coefficients have the properties listed in Theorem 1, which we accept now without proof. Property 3 is proved later in this section. The other properties are established in calculus.

> **THEOREM 1** Properties of Graphs of Polynomial Functions

Let $P(x)$ be a polynomial of degree $n > 0$ with real coefficients. Then the graph of $P(x)$:

1. Is continuous for all real numbers
2. Has no sharp corners
3. Has at most n real zeros
4. Has at most $n - 1$ turning points
5. Increases or decreases without bound as $x \to \infty$ and as $x \to -\infty$*

Figure 5 shows graphs of representative polynomial functions of degrees 1 through 6, illustrating the five properties of Theorem 1.

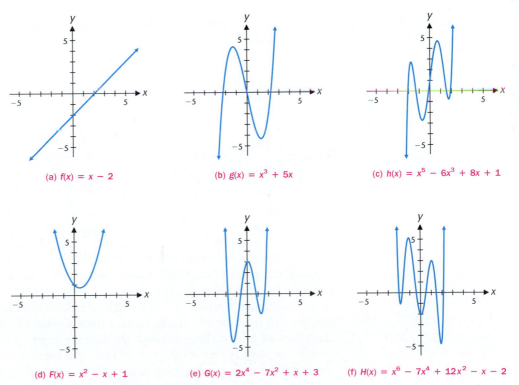

(a) $f(x) = x - 2$ (b) $g(x) = x^3 + 5x$ (c) $h(x) = x^5 - 6x^3 + 8x + 1$

(d) $F(x) = x^2 - x + 1$ (e) $G(x) = 2x^4 - 7x^2 + x + 3$ (f) $H(x) = x^6 - 7x^4 + 12x^2 - x - 2$

> **Figure 5** Graphs of polynomial functions.

*Remember that ∞ and $-\infty$ are not real numbers. The statement *the graph of $P(x)$ increases without bound as $x \to -\infty$* means that for any horizontal line $y = b$ there is some interval $(-\infty, a] = \{x \mid x \le a\}$ on which the graph of $P(x)$ is above the horizontal line.

| EXAMPLE | 2 | **Properties of Graphs of Polynomials** |

Explain why each graph is *not* the graph of a polynomial function by listing the properties of Theorem 1 that it *fails* to satisfy.

(A) (B) (C)

 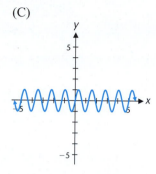

SOLUTIONS

(A) The graph has a sharp corner when $x = 0$. Property 2 fails.

(B) There are no points on the graph with x coordinate less then or equal to 0, so properties 1 and 5 fail.

(C) There are an infinite number of zeros and an infinite number of turning points, so properties 3 and 4 fail. Furthermore, the graph is bounded by the horizontal lines $y = \pm 1$, so property 5 fails.

| MATCHED PROBLEM 2 |

Explain why each graph is *not* the graph of a polynomial function by listing the properties of Theorem 1 that it *fails* to satisfy.

(A) (B) (C)

 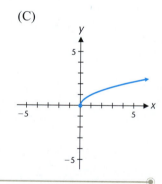

The shape of the graph of a polynomial function with real coefficients is similar to the shape of the graph of the **leading term,** that is, the term of highest degree. Figure 6 compares the graph of the polynomial $h(x) = x^5 - 6x^3 + 8x + 1$ from Figure 5 with the graph of its leading term $p(x) = x^5$. The graphs are dissimilar near the origin, but as we zoom out, the shapes of the two graphs become quite similar. The leading term in the polynomial dominates all other terms combined. Because the graph of $p(x)$ increases without bound as $x \to \infty$, the same is true of the graph of $h(x)$. And because the graph of $p(x)$ decreases without bound as $x \to -\infty$, the same is true of the graph of $h(x)$.

› **Figure 6** $p(x) = x^5$,
$h(x) = x^5 - 6x^3 + 8x + 1$.

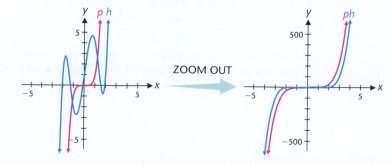

The left and right behavior of a polynomial function with real coefficients is determined by the left and right behavior of its leading term (see Fig. 6). Property 5 of Theorem 1 can therefore be refined. The various possibilities are summarized in Theorem 2.

> › **THEOREM 2** Left and Right Behavior of Polynomial Functions
>
> Let $P(x) = a_n x^n + a_{n-1} x^{n-1} + \ldots + a_1 x + a_0$ be a polynomial function with real coefficients, $a_n \neq 0$, $n > 0$.
>
> 1. $a_n > 0$, n even: The graph of $P(x)$ increases without bound as $x \to \infty$ and increases without bound as $x \to -\infty$ (like the graphs of x^2, x^4, x^6, etc.).
>
> 2. $a_n > 0$, n odd: The graph of $P(x)$ increases without bound as $x \to \infty$ and decreases without bound as $x \to -\infty$ (like the graphs of x, x^3, x^5, etc.).
>
> 3. $a_n < 0$, n even: The graph of $P(x)$ decreases without bound as $x \to \infty$ and decreases without bound as $x \to -\infty$ (like the graphs of $-x^2$, $-x^4$, $-x^6$, etc.).
>
> 4. $a_n < 0$, n odd: The graph of $P(x)$ decreases without bound as $x \to \infty$ and increases without bound as $x \to -\infty$ (like the graphs of $-x$, $-x^3$, $-x^5$, etc.).

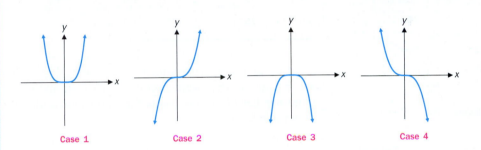

Case 1 Case 2 Case 3 Case 4

It is convenient to write $P(x) \to \infty$ as an abbreviation for the phrase *the graph of $P(x)$ increases without bound*. Using this notation, the left and right behavior in Case 4 of Theorem 2, for example, is $P(x) \to -\infty$ as $x \to \infty$ and $P(x) \to \infty$ as $x \to -\infty$.

EXAMPLE	3

Left and Right Behavior of Polynomials

Determine the left and right behavior of each polynomial.

(A) The degree of $P(x) = 3 - x^2 + 4x^3 - x^4 - 2x^6$

(B) The degree of $Q(x) = 4x^5 + 8x^3 + 5x - 1$

SOLUTIONS

(A) The degree $P(x)$ is 6 (even) and the coefficient a_6 is -2 (negative), so the left and right behavior is the same as that of $-x^6$ (Case 3 of Theorem 2): $P(x) \to -\infty$ as $x \to \infty$ and $P(x) \to -\infty$ as $x \to -\infty$.

(B) The degree $Q(x)$ is 5 (odd) and the coefficient a_5 is 4 (positive), so the left and right behavior is the same as that of x^5 (Case 2 of Theorem 2): $P(x) \to \infty$ as $x \to \infty$ and $P(x) \to -\infty$ as $x \to -\infty$.

MATCHED PROBLEM 3

Determine the left and right behavior of each polynomial.

(A) $P(x) = 4x^9 - 3x^{11} + 5$

(B) $Q(x) = 1 - 2x^{50} + x^{100}$

EXAMPLE 4 Graphing a Polynomial

Graph the polynomial $P(x) = x^3 - 12x - 16$, $-5 \leq x \leq 5$. List the real zeros and turning points.

SOLUTION First we construct a table of values by calculating $P(x)$ for each integer x, $-5 \leq x \leq 5$. For example,

$$P(-5) = (-5)^3 - 12(-5) - 16 = -81$$

x	$P(x)$	x	$P(x)$
-5	-81	1	-27
-4	-32	2	-32
-3	-7	3	-25
-2	0	4	0
-1	-5	5	49
0	-16		

> **Figure 7** $P(x) = x^3 - 12x - 16$.

Then we plot the points in the table and join them with a smooth curve (Fig. 7). The zeros are -2 and 4. The turning points are $(-2, 0)$ and $(2, -32)$. Note that $P(x)$ has the maximum number of turning points for a polynomial of degree 3, but one fewer than the maximum number of real zeros.

MATCHED PROBLEM 4 Graph $P(x) = x^4 - 6x^2 - 8x - 3$, $-4 \leq x \leq 4$. List the real zeros and turning points.

>>> CAUTION >>>

Finding the real zeros and turning points of a polynomial is usually more difficult than suggested by Example 4. In Example 4, how did we know that the real zeros were between -5 and 5 rather than between, say, 95 and 105? Could there be another real zero just to the left or right of -2? How do we know that $(-2, 0)$ and $(2, -32)$, rather than nearby points having noninteger coordinates, are the turning points? To answer such questions we must view polynomials from an algebraic perspective. Polynomials can be factored. So next we will study the division and factorization of polynomials.

> Polynomial Division

We can find quotients of polynomials by a long-division process similar to the one used in arithmetic. Example 5 will illustrate the process.

EXAMPLE 5 Polynomial Long Division

Divide $P(x) = 3x^3 - 5 + 2x^4 - x$ by $2 + x$.

SOLUTION First, rewrite the dividend $P(x)$ in descending powers of x, inserting 0 as the coefficient for any missing terms of degree less than 4:

$$P(x) = 2x^4 + 3x^3 + 0x^2 - x - 5$$

Similarly, rewrite the divisor $2 + x$ in the form $x + 2$. Then divide the first term x of the divisor into the first term $2x^4$ of the dividend. Multiply the result, $2x^3$, by the divisor, obtaining $2x^4 + 4x^3$. Line up like terms, subtract as in arithmetic, and bring down $0x^2$. Repeat the process until the degree of the remainder is less than the degree of the divisor.

$$
\begin{array}{r}
\underline{2x^3 - x^2 + 2x - 5} \qquad \text{Quotient} \\
x + 2\overline{)2x^4 + 3x^3 + 0x^2 - x - 5} \qquad \text{Dividend} \\
\underline{2x^4 + 4x^3} \qquad\qquad\qquad\qquad \text{Subtract} \\
-x^3 + 0x^2 \\
\underline{-x^3 - 2x^2} \qquad\qquad\qquad \text{Subtract} \\
2x^2 - x \\
\underline{2x^2 + 4x} \qquad\qquad \text{Subtract} \\
-5x - 5 \\
\underline{-5x - 10} \qquad \text{Subtract} \\
5 \qquad \text{Remainder}
\end{array}
$$

Therefore,

$$\frac{2x^4 + 3x^3 - x - 5}{x + 2} = 2x^3 - x^2 + 2x - 5 + \frac{5}{x + 2}$$

CHECK You can always check division using multiplication:

$$(x + 2)\left[2x^3 - x^2 + 2x - 5 + \frac{5}{x + 2}\right]$$

$$= (x + 2)(2x^3 - x^2 + 2x - 5) + 5 \qquad \text{Multiply and collect like terms}$$

$$= 2x^4 + 3x^3 - x - 5$$

MATCHED PROBLEM 5 Divide $6x^2 - 30 + 9x^3$ by $x - 2$.

The procedure illustrated in Example 5 is called the **division algorithm.** The concluding equation of Example 5 (before the check) may be multiplied by the divisor $x + 2$ to give the following form:

Dividend $\quad=\quad$ Divisor $\cdot$ Quotient $\quad+$ Remainder

$$2x^4 + 3x^3 - x - 5 = (x + 2)(2x^3 - x^2 + 2x - 5) + 5$$

This last equation is an *identity:* it is true for all replacements of x by real or complex numbers including $x = -2$. Theorem 3, which we state without proof, gives the general result of applying the division algorithm when the divisor has the form $x - r$.

> **THEOREM 3 Division Algorithm**

For each polynomial $P(x)$ of degree greater than 0 and each number r, there exists a unique polynomial $Q(x)$ of degree 1 less than $P(x)$ and a unique number R such that

$$P(x) = (x - r)Q(x) + R$$

The polynomial $Q(x)$ is called the **quotient,** $x - r$ is the **divisor,** and R is the **remainder.** Note that R may be 0.

There is a shortcut called **synthetic division** for the long division of Example 5. First write the coefficients of the dividend and the *negative* of the constant term of the divisor in the format shown below at the left. Bring down the 2 as indicated next on the right, multiply by -2, and record the product -4. Add 3 and -4, bringing down their sum -1. Repeat the process until the coefficients of the quotient and the remainder are obtained.

Compare the preceding synthetic division to the long division shown below, in which the essential numerals appear in color, to convince yourself that synthetic division produces the correct quotient and remainder. (In synthetic division we use the negative of the constant term of the divisor so we can add rather than subtract.)

$$
\begin{array}{r}
\,2x^3 - 1x^2 + 2x - 5 \qquad \text{Quotient}\\
x+2 \overline{)\,2x^4 + 3x^3 + 0x^2 - 1x - 5} \qquad \text{Dividend}\\
\underline{2x^4 + 4x^3}\\
-1x^3 + 0x^2\\
\underline{-1x^3 - 2x^2}\\
2x^2 - 1x\\
\underline{2x^2 + 4x}\\
-5x - 5\\
\underline{-5x - 10}\\
5 \qquad \text{Remainder}
\end{array}
$$

Divisor

› **KEY STEPS IN THE SYNTHETIC DIVISION PROCESS**

To divide the polynomial $P(x)$ by $x - r$:

Step 1. Arrange the coefficients of $P(x)$ in order of descending powers of x. Write 0 as the coefficient for each missing power.

Step 2. After writing the divisor in the form $x - r$, use r to generate the second and third rows of numbers as follows. Bring down the first coefficient of the dividend and multiply it by r; then add the product to the second coefficient of the dividend. Multiply this sum by r, and add the product to the third coefficient of the dividend. Repeat the process until a product is added to the constant term of $P(x)$.

Step 3. The last number to the right in the third row of numbers is the remainder. The other numbers in the third row are the coefficients of the quotient, which is of degree 1 less than $P(x)$.

| EXAMPLE | 6 | **Synthetic Division** |

Use synthetic division to divide $P(x) = 4x^5 - 30x^3 - 50x - 2$ by $x + 3$. Find the quotient and remainder. Write the conclusion in the form $P(x) = (x - r)Q(x) + R$ of Theorem 3.

SOLUTION

Because $x + 3 = x - (-3)$, we have $r = -3$, and

$$
\begin{array}{r|rrrrrr}
 & 4 & 0 & -30 & 0 & -50 & -2 \\
 & & -12 & 36 & -18 & 54 & -12 \\
\hline
-3 & 4 & -12 & 6 & -18 & 4 & -14 \\
\end{array}
$$

The quotient is $4x^4 - 12x^3 + 6x^2 - 18x + 4$ with a remainder of -14. So

$$4x^5 - 30x^3 - 50x - 2 = (x + 3)(4x^4 - 12x^3 + 6x^2 - 18x + 4) - 14$$

| MATCHED PROBLEM 6 | Repeat Example 6 with $P(x) = 3x^4 - 11x^3 - 18x + 8$ and divisor $x - 4$. |

› Remainder and Factor Theorems

››› EXPLORE-DISCUSS 2

Let $P(x) = x^3 - 3x^2 - 2x + 8$.

(A) Evaluate $P(x)$ for
 (i) $x = -2$ (ii) $x = 1$ (iii) $x = 3$

(B) Use synthetic division to find the remainder when $P(x)$ is divided by
 (i) $x + 2$ (ii) $x - 1$ (iii) $x - 3$

What conclusion does a comparison of the results in parts A and B suggest?

Explore-Discuss 2 suggests that when a polynomial $P(x)$ is divided by $x - r$, the remainder is equal to $P(r)$, the value of the polynomial $P(x)$ at $x = r$. In Problem 87 of Exercises 4-1, you are asked to complete a proof of this fact, which is called the *remainder theorem*.

> **THEOREM 4 Remainder Theorem**

If R is the remainder after dividing the polynomial $P(x)$ by $x - r$, then

$$P(r) = R$$

| EXAMPLE | 7 | **Two Methods for Evaluating Polynomials** |

If $P(x) = 4x^4 + 10x^3 + 19x + 5$, find $P(-3)$ by

(A) Using the remainder theorem and synthetic division
(B) Evaluating $P(-3)$ directly

SOLUTIONS

(A) Use synthetic division to divide $P(x)$ by $x - (-3)$.

$$
\begin{array}{r|rrrrr}
 & 4 & 10 & 0 & 19 & 5 \\
 & & -12 & 6 & -18 & -3 \\
\hline
-3 & 4 & -2 & 6 & 1 & 2 = R = P(-3) \\
\end{array}
$$

(B) $P(-3) = 4(-3)^4 + 10(-3)^3 + 19(-3) + 5$

$= 2$

MATCHED PROBLEM 7

Repeat Example 7 for $P(x) = 3x^4 - 16x^2 - 3x - 7$ and $x = -2$.

You might think the remainder theorem is not a very effective tool for evaluating polynomials. But let's consider the number of operations performed in parts A and B of Example 7. Synthetic division requires only four multiplications and four additions to find $P(-3)$, whereas the direct evaluation requires ten multiplications and four additions. [Note that evaluating $4(-3)^4$ actually requires five multiplications.] The difference becomes even larger as the degree of the polynomial increases. Computer programs that involve numerous polynomial evaluations often use synthetic division because of its efficiency. We will find synthetic division and the remainder theorem to be useful tools later in this chapter.

The remainder theorem shows that the division algorithm equation,

$$P(x) = (x - r)Q(x) + R$$

can be written in the form where R is replaced by $P(r)$:

$$P(x) = (x - r)Q(x) + P(r)$$

Therefore, $x - r$ is a factor of $P(x)$ if and only if $P(r) = 0$, that is, if and only if r is a zero of the polynomial $P(x)$. This result is called the *factor theorem*.

> **THEOREM 5 Factor Theorem**

If r is a zero of the polynomial $P(x)$, then $x - r$ is a factor of $P(x)$. Conversely, if $x - r$ is a factor of $P(x)$, then r is a zero of $P(x)$.

EXAMPLE 8 Factors of Polynomials

Use the factor theorem to show that $x + 1$ is a factor of $P(x) = x^{25} + 1$ but is not a factor of $Q(x) = x^{25} - 1$.

SOLUTION Because

$$P(-1) = (-1)^{25} + 1 = -1 + 1 = 0$$

$x - (-1) = x + 1$ is a factor of $x^{25} + 1$. On the other hand,

$$Q(-1) = (-1)^{25} - 1 = -1 - 1 = -2$$

and $x + 1$ is not a factor of $x^{25} - 1$.

MATCHED PROBLEM 8

Use the factor theorem to show that $x - i$ is a factor of $P(x) = x^8 - 1$ but is not a factor of $Q(x) = x^8 + 1$.

One consequence of the factor theorem is Theorem 6 (a proof is outlined in Problem 88 in Exercises 4-1).

> **THEOREM 6** Zeros of Polynomials

A polynomial of degree n has at most n zeros.

Theorem 6 says that the graph of a polynomial of degree n with real coefficients has at most n real zeros (Property 3 of Theorem 1). The polynomial

$$H(x) = x^6 - 7x^4 + 12x^2 - x - 2$$

for example, has degree 6 and the maximum number of zeros [see Fig. 5(f), p. 263]. Of course, polynomials of degree 6 may have fewer than six real zeros. In fact, $p(x) = x^6 + 1$ has no real zeros. However, it can be shown that the polynomial $p(x) = x^6 + 1$ has exactly six complex zeros.

› Mathematical Modeling and Data Analysis

In Chapters 2 and 3 we saw that linear and quadratic functions can be useful models for certain sets of data. For some data, however, no linear function and no quadratic function can provide a reasonable model. In that case, we investigate the suitability of polynomial models of degree greater than 2. In Examples 9 and 10 we discuss cubic and quartic models, respectively, for the given data.

EXAMPLE	9

Table 1 Sturgeon

Length (in.) x	Weight (oz.) y
18	13
22	26
26	46
30	75
34	115
38	166
44	282
52	492
60	796

Source: www.thefishernet.com

Estimating the Weight of Fish

Scientists and fishermen often estimate the weight of a fish from its length. The data in Table 1 give the average weight of North American sturgeon for certain lengths.

Because weight is associated with volume, which involves three dimensions, we might expect that weight would be associated with the cube of the length. A cubic model for the data is given by

$$y = 0.00526x^3 - 0.117x^2 + 1.43x - 5.00$$

where y is the weight (in ounces) of a sturgeon that has length x (in inches).

(A) Use the model to estimate the weight of a sturgeon of length 56 inches.

(B) Compare the weight of a sturgeon of length 44 inches as given by Table 1 with the weight given by the model.

SOLUTIONS

(A) If $x = 56$, then

$$y = 0.00526(56)^3 - 0.117(56)^2 + 1.43(56) - 5.00 \approx 632 \text{ ounces}$$

(B) If $x = 44$, then

$$y = 0.00526(44)^3 - 0.117(44)^2 + 1.43(44) - 5.00 \approx 279 \text{ ounces}$$

The weight given by the table, 282 ounces, is 3 ounces greater than the weight given by the model.

Technology Connections

Figure 8 shows the details of constructing the cubic model of Example 9 on a graphing calculator.

(a) Entering the data

(b) Finding the model

(c) Graphing the data and the model

> Figure 8

MATCHED PROBLEM 9

Use the cubic model of Example 9.

(A) Estimate the weight of a sturgeon of length 65 inches.

(B) Compare the weight of a sturgeon of length 30 inches as given by Table 1 with the weight given by the model.

EXAMPLE 10 Hydroelectric Power

Table 2

Year	U.S. Consumption of Hydroelectric Power (Quadrillion BTU)
1983	3.90
1985	3.40
1987	3.12
1989	2.99
1991	3.14
1993	3.13
1995	3.48
1997	3.88
1999	3.47
2001	2.38
2003	2.53
2005	2.61

Source: U.S. Department of Energy

The data in Table 2 gives the annual consumption of hydroelectric power (in quadrillion BTU) in the United States for selected years since 1983. From Table 2 it appears that a polynomial model of the data would have three turning points—near 1989, 1997, and 2001. Because a polynomial with three turning points must have degree at least four, we can model the data with a quartic (fourth-degree) polynomial:

$$y = 0.00013x^4 - 0.0067x^3 + 0.107x^2 - 0.59x + 4.03$$

where y is the consumption (in quadrillion BTU) and x is time in years with $x = 0$ representing 1983.

(A) Use the model to predict the consumption of hydroelectric power in 2018.

(B) Compare the consumption of hydroelectric power in 2003 (as given by Table 2) to the consumption given by the model.

SOLUTIONS

(A) If $x = 35$ (which represents the year 2018), then

$$y = 0.00013(35)^4 - 0.0067(35)^3 + 0.107(35)^2 - 0.59(35) + 4.03 \approx 22.3$$

The model predicts a consumption of 22.3 quadrillion BTU in 2018. However, because the predicted consumption for 2018 is so dramatically greater than earlier consumption levels, it is unlikely to be accurate. This brings up an important point: A model that fits a set of data points well is not automatically a good model for predicting future trends.

(B) If $x = 20$ (which represents 2003), then

$$y = 0.00013(20)^4 - 0.0067(20)^3 + 0.107(20)^2 - 0.59(20) + 4.03 = 2.23$$

The consumption reported in the table, 2.53 quadrillion BTU, is 0.30 quadrillion BTU greater than the consumption given by the model.

Technology Connections

Figure 9 shows the details of constructing the quartic model of Example 10 on a graphing calculator.

(a) (b) (c)

> **Figure 9**

MATCHED PROBLEM 10

Use the quartic model of Example 10.

(A) Estimate the consumption of hydroelectric power in 2000.

(B) Compare the consumption of hydroelectric power in 1991 (as given by Table 2) to the consumption given by the model.

ANSWERS TO MATCHED PROBLEMS

1. (A) $-1, 1, 2$
 (B) The zeros are $-5, -2, 2, 2i, -2i, -1 + 2i$, and $-1 - 2i$; the x intercepts are $-5, -2$, and 2.
2. (A) Properties 1 and 5
 (B) Property 5
 (C) Properties 1 and 5
3. (A) $P(x) \to -\infty$ as $x \to \infty$ and $P(x) \to \infty$ as $x \to -\infty$.
 (B) $Q(x) \to \infty$ as $x \to \infty$ and $Q(x) \to \infty$ as $x \to -\infty$.
4.

zeros: $-1, 3$; turning point; $(2, -27)$

5. $9x^2 + 24x + 48 + \dfrac{66}{x - 2}$

6. $3x^4 - 11x^3 - 18x + 8 = (x - 4)(3x^3 + x^2 + 4x - 2)$
7. $P(-2) = -3$ for both parts, as it should
8. $P(i) = 0$, so $x - i$ is a factor of $x^8 - 1$;
 $Q(i) = 2$, so $x - i$ is not a factor of $x^8 + 1$
9. (A) 1,038 in.
 (B) The weight given in the table is 0.38 oz greater than the weight given by the model.
10. (A) 2.86 quadrillion BTU
 (B) The consumption given in the table is 0.12 quadrillion BTU less than the consumption given by the model.

4-1 Exercises

1. What is a polynomial function?

2. Explain the connection between the zeros of a polynomial and its linear factors.

3. Explain what is wrong with the following setup for dividing $x^4 + 5x^2 - 2x + 6$ by $x - 2$ using synthetic division.

$$\begin{array}{r} 1 \quad 5 \quad -2 \quad 6 \\ 2\overline{} \end{array}$$

4. Explain what is wrong with the following setup for dividing $3x^3 - x^2 + 8x + 9$ by $x + 4$ using synthetic division.

$$\begin{array}{r} 3 \quad -1 \quad 8 \quad 9 \\ 4\overline{} \end{array}$$

In Problems 5–8, decide whether the statement is true or false, and explain your answer.

5. Every quadratic function is a polynomial function.

6. Every polynomial of degree 3 has three x intercepts.

7. If a polynomial has no x intercepts, then it has no zeros.

8. Every polynomial function is continuous.

In Problems 9–12, a is a positive real number. Match each function with one of graphs (a)–(d).

9. $f(x) = ax^3$ **10.** $g(x) = -ax^4$

11. $h(x) = ax^6$ **12.** $k(x) = -ax^5$

(a)

(b)

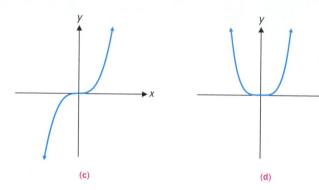

(c)

(d)

In Problems 13–16, list the real zeros and turning points, and state the left and right behavior, of the polynomial function P(x) that has the indicated graph.

13.

14.

15.

16.

In Problems 17–20, explain why each graph is not the graph of a polynomial function.

17.

18.

19.

20.

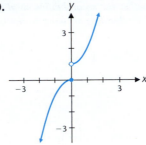

In Problems 21–24, list all zeros of each polynomial function, and specify those zeros that are x intercepts.

21. $P(x) = x(x^2 - 9)(x^2 + 4)$

22. $P(x) = (x^2 - 4)(x^4 - 1)$

23. $P(x) = (x + 5)(x^2 + 9)(x^2 + 16)$

24. $P(x) = (x^2 - 5x + 6)(x^2 - 5x + 7)$

In Problems 25–34, use algebraic long division to find the quotient and the remainder.

25. $(3x^2 + 5x + 6) \div (x + 1)$

26. $(2x^2 - 7x + 4) \div (x - 2)$

27. $(4m^2 - 1) \div (m - 1)$

28. $(y^2 - 9) \div (y + 3)$

29. $(6 - 6x + 8x^2) \div (x + 1)$

30. $(11x - 2 + 12x^2) \div (3x + 2)$

31. $\dfrac{x^3 - 1}{x - 1}$ **32.** $\dfrac{a^3 + 27}{a + 3}$

33. $(3y - y^2 + 2y^3 - 1) \div (y + 2)$

34. $(3 + x^3 - x) \div (x - 3)$

In Problems 35–40, divide using synthetic division.

35. $(x^2 + 3x - 7) \div (x - 2)$

36. $(x^2 + 3x - 3) \div (x - 3)$

37. $(4x^2 + 10x - 9) \div (x + 3)$

38. $(2x^2 + 7x - 5) \div (x + 4)$

39. $\dfrac{2x^3 - 3x + 1}{x - 2}$ **40.** $\dfrac{x^3 + 2x^2 - 3x - 4}{x + 2}$

In Problems 41–44, is the given number a zero of the polynomial? Use synthetic division.

41. $x^2 + 4x - 221; -17$

42. $x^2 - 7x - 551; 29$

43. $2x^3 + 38x^2 - x + 19; -19$

44. $2x^3 - 397x + 70; 14$

In Problems 45–48, determine whether the second polynomial is a factor of the first polynomial without dividing or using synthetic division.

45. $x^{18} - 1; x - 1$ **46.** $x^{18} - 1; x + 1$

47. $3x^3 - 7x^2 - 8x + 2; x + 1$

48. $3x^4 - 2x^3 + 5x - 6; x - 1$

Use synthetic division and the remainder theorem in Problems 49–54.

49. Find $P(-2)$, given $P(x) = 3x^2 - x - 10$.

50. Find $P(-3)$, given $P(x) = 4x^2 + 10x - 8$.

51. Find $P(2)$, given $P(x) = 2x^3 - 5x^2 + 7x - 7$.

52. Find $P(5)$, given $P(x) = 2x^3 - 12x^2 - x + 30$.

53. Find $P(-4)$, given $P(x) = x^4 - 10x^2 + 25x - 2$.

54. Find $P(-7)$, given $P(x) = x^4 + 5x^3 - 13x^2 - 30$.

In Problems 55–62, use synthetic division to find the quotient and the remainder. As coefficients get more involved, a calculator should prove helpful. Do not round off.

55. $(3x^4 - x - 4) \div (x + 1)$

56. $(5x^4 - 2x^2 - 3) \div (x - 1)$

57. $(x^5 + 1) \div (x + 1)$

58. $(x^4 - 16) \div (x - 2)$

59. $(3x^4 + 2x^3 - 4x - 1) \div (x + 3)$

60. $(x^4 - 3x^3 - 5x^2 + 6x - 3) \div (x - 4)$

61. $(2x^6 - 13x^5 + 75x^3 + 2x^2 - 50) \div (x - 5)$

62. $(4x^6 + 20x^5 - 24x^4 - 3x^2 - 13x + 30) \div (x + 6)$

In Problems 63–68, without graphing, state the left and right behavior, the maximum number of x intercepts, and the maximum number of local extrema.

63. $P(x) = x^3 - 5x^2 + 2x + 6$

64. $P(x) = x^3 + 2x^2 - 5x - 3$

65. $P(x) = -x^3 + 4x^2 + x + 5$

66. $P(x) = -x^3 - 3x^2 + 4x - 4$

67. $P(x) = x^4 + x^3 - 5x^2 - 3x + 12$

68. $P(x) = -x^4 + 6x^2 - 3x - 16$

In Problems 69–72, either give an example of a polynomial with real coefficients that satisfies the given conditions or explain why such a polynomial cannot exist.

69. $P(x)$ is a third-degree polynomial with one x intercept.

70. $P(x)$ is a fourth-degree polynomial with no x intercepts.

71. $P(x)$ is a third-degree polynomial with no x intercepts.

72. $P(x)$ is a fourth-degree polynomial with no turning points.

In Problems 73 and 74, divide, using synthetic division.

73. $(x^3 - 3x^2 + x - 3) \div (x - i)$

74. $(x^3 - 2x^2 + x - 2) \div (x + i)$

75. Let $P(x) = x^2 + 2ix - 10$. Use synthetic division to find:
(A) $P(2 - i)$
(B) $P(5 - 5i)$
(C) $P(3 - i)$
(D) $P(-3 - i)$

76. Let $P(x) = x^2 - 4ix - 13$. Use synthetic division to find:
(A) $P(5 + 6i)$
(B) $P(1 + 2i)$
(C) $P(3 + 2i)$
(D) $P(-3 + 2i)$

In Problems 77–82, approximate (to two decimal places) the x intercepts and the local extrema.

77. $P(x) = 40 + 50x - 9x^2 - x^3$

78. $P(x) = 40 + 70x + 18x^2 + x^3$

79. $P(x) = 0.04x^3 - 10x + 5$

80. $P(x) = -0.01x^3 + 2.8x - 3$

81. $P(x) = 0.1x^4 + 0.3x^3 - 23x^2 - 23x + 90$

82. $P(x) = 0.1x^4 + 0.2x^3 - 19x^2 + 17x + 100$

83. (A) What is the least number of turning points that a polynomial function of degree 4, with real coefficients, can have? The greatest number? Explain and give examples.
(B) What is the least number of x intercepts that a polynomial function of degree 4, with real coefficients, can have? The greatest number? Explain and give examples.

84. (A) What is the least number of turning points that a polynomial function of degree 3, with real coefficients, can have? The greatest number? Explain and give examples.
(B) What is the least number of x intercepts that a polynomial function of degree 3, with real coefficients, can have? The greatest number? Explain and give examples.

85. Is every polynomial of even degree an even function? Explain.

86. Is every polynomial of odd degree an odd function? Explain.

87. Prove the remainder theorem (Theorem 4):
(A) Write the result of the division algorithm if a polynomial $P(x)$ is divided by $x - r$.
(B) Evaluate both sides of the equation from part (A) when $x = r$. What can you conclude?

88. In this problem, we will prove that a polynomial of degree n has at most n zeros (Theorem 6). Give a reason for each step. Let $P(x)$ be a polynomial of degree n, and suppose that P has n distinct zeros $r_1, r_2, \ldots, r_n$. We will show that it is impossible for P to have any other zeros.

Step 1: We can write $P(x)$ in the form $P(x) = (x - r_1)Q_1(x)$, where the degree of $Q_1(x)$ is $n - 1$.

Step 2: r_2 is a zero of $Q_1(x)$.

Step 3: We can write $Q_1(x)$ in the form $Q_1(x) = (x - r_2)Q_2(x)$, where the degree of $Q_2(x)$ is $n - 2$.

Step 4: $P(x) = (x - r_1)(x - r_2)Q_2(x)$

Step 5: $P(x) = (x - r_1)(x - r_2) \cdots (x - r_n)Q_n(x)$, where the degree of $Q_n(x)$ is 0.

Step 6: The only zeros of P are $r_1, r_2, \ldots, r_n$.

APPLICATIONS

 89. REVENUE The price–demand equation for 8,000-BTU window air conditioners is given by

$$p = 0.0004x^2 - x + 569 \qquad 0 \le x \le 800$$

where x is the number of air conditioners that can be sold at a price of p dollars each.
(A) Find the revenue function.
(B) Find the number of air conditioners that must be sold to maximize the revenue, the corresponding price to the nearest dollar, and the maximum revenue to the nearest dollar.

 90. PROFIT Refer to Problem 89. The cost of manufacturing 8,000-BTU window air conditioners is given by

$$C(x) = 10,000 + 90x$$

where $C(x)$ is the total cost in dollars of producing x air conditioners.
(A) Find the profit function.
(B) Find the number of air conditioners that must be sold to maximize the profit, the corresponding price to the nearest dollar, and the maximum profit to the nearest dollar.

91. CONSTRUCTION A rectangular container measuring 1 foot by 2 feet by 4 feet is covered with a layer of lead shielding of uniform thickness (see the figure).
(A) Find the volume of lead shielding V as a function of the thickness x (in feet) of the shielding.
(B) Find the volume of the lead shielding if the thickness of the shielding is 0.05 feet.

Lead shielding

92. MANUFACTURING A rectangular storage container measuring 2 feet by 2 feet by 3 feet is coated with a protective coating of plastic of uniform thickness.
(A) Find the volume of plastic V as a function of the thickness x (in feet) of the coating.

(B) Find the volume of the plastic coating to four decimal places if the thickness of the shielding is 0.005 feet.

 Problems 93–96 require a graphing calculator or a computer that can calculate cubic regression polynomials for a given data set.

93. HEALTH CARE Table 3 shows the total national health care expenditures (in billion dollars) and the per capita expenditures (in dollars) for selected years since 1960.

Table 3 National Health Care Expenditures

Year	Total Expenditures (Billion $)	Per Capita Expenditures ($)
1960	28	148
1970	75	356
1980	253	1,100
1990	714	2,814
2000	1,353	4,789
2007	2,241	7,421

Source: U.S. Census Bureau.

(A) Let x represent the number of years since 1960 and find a cubic regression polynomial for the total national expenditures.
(B) Use the polynomial model from part A to estimate the total national expenditures (to the nearest billion) for 2018.

94. HEALTH CARE Refer to Table 3.
(A) Let x represent the number of years since 1960 and find a cubic regression polynomial for the per capita expenditures.
(B) Use the polynomial model from part A to estimate the per capita expenditures (to the nearest dollar) for 2018.

95. MARRIAGE Table 4 shows the marriage and divorce rates per 1,000 population for selected years since 1950.

Table 4 Marriages and Divorces (per 1,000 Population)

Year	Marriages	Divorces
1950	11.1	2.6
1960	8.5	2.2
1970	10.6	3.5
1980	10.6	5.2
1990	9.8	4.7
2000	8.2	4.1

Source: U.S. Census Bureau.

(A) Let x represent the number of years since 1950 and find a cubic regression polynomial for the marriage rate.
(B) Use the polynomial model from part A to estimate the marriage rate (to one decimal place) for 2016.

96. DIVORCE Refer to Table 4.
(A) Let x represent the number of years since 1950 and find a cubic regression polynomial for the divorce rate.
(B) Use the polynomial model from part A to estimate the divorce rate (to one decimal place) for 2016.

4-2 Real Zeros and Polynomial Inequalities

> Upper and Lower Bounds for Real Zeros
> Location Theorem and Bisection Method
> Approximating Real Zeros at Turning Points
> Polynomial Inequalities
> Mathematical Modeling

The real zeros of a polynomial $P(x)$ with real coefficients are just the x intercepts of the graph of $P(x)$. So an obvious strategy for finding the real zeros consists of two steps:

1. Graph $P(x)$.

2. Approximate each x intercept.

In this section, we develop important tools for carrying out this strategy: the *upper and lower bound theorem,* which determines an interval $[a, b]$ that is guaranteed to contain all x intercepts of $P(x)$, and the *bisection method,* which permits approximation of x intercepts to any desired accuracy. We emphasize the approximation of real zeros in this section; the problem of finding zeros exactly, when possible, is considered in Section 4-3.

> Upper and Lower Bounds for Real Zeros

On which interval should you graph a polynomial $P(x)$ in order to see all of its x intercepts? The answer is provided by the upper and lower bound theorem. This theorem explains how to find two numbers: a *lower bound,* which is less than or equal to all real zeros of the polynomial, and an *upper bound,* which is greater than or equal to all real zeros of the polynomial. A proof of Theorem 1 is outlined in Problems 67 and 68 of Exercises 4-2.

> ### THEOREM 1 Upper and Lower Bound Theorem

Let $P(x)$ be a polynomial of degree $n > 0$ with real coefficients, $a_n > 0$:

1. Upper bound: A number $r > 0$ is an upper bound for the real zeros of $P(x)$ if, when $P(x)$ is divided by $x - r$ by synthetic division, all numbers in the quotient row, including the remainder, are nonnegative.

2. Lower bound: A number $r < 0$ is a lower bound for the real zeros of $P(x)$ if, when $P(x)$ is divided by $x - r$ by synthetic division, all numbers in the quotient row, including the remainder, alternate in sign.

[*Note:* In the lower bound test, if 0 appears in one or more places in the quotient row, including the remainder, the sign in front of it can be considered either positive or negative, but not both. For example, the numbers 1, 0, 1 can be considered to alternate in sign, whereas 1, 0, −1 cannot.]

EXAMPLE 1 Bounding Real Zeros

Let $P(x) = x^4 - 2x^3 - 10x^2 + 40x - 90$. Find the smallest positive integer and the largest negative integer that, by Theorem 1, are upper and lower bounds, respectively, for the real zeros of $P(x)$.

SOLUTION We perform synthetic division for $r = 1, 2, 3, \ldots$ until the quotient row turns nonnegative; then repeat this process for $r = -1, -2, -3, \ldots$ until the quotient row alternates in sign. We organize these results in the *synthetic division table* shown below. In a synthetic division table we dispense with writing the product of r with each coefficient in the quotient and simply list the results in the table.

		1	-2	-10	40	-90
	1	1	-1	-11	29	-61
	2	1	0	-10	20	-50
	3	1	1	-7	19	-33
	4	1	2	-2	32	38
UB	5	1	3	5	65	235
	-1	1	-3	-7	47	-137
	-2	1	-4	-2	44	-178
	-3	1	-5	5	25	-165
	-4	1	-6	14	-16	-26
LB	-5	1	-7	25	-85	335

← {This quotient row is nonnegative; 5 is an upper bound (UB).

← {This quotient row alternates in sign; -5 is a lower bound (LB).

The graph of $P(x) = x^4 - 2x^3 - 10x^2 + 40x - 90$ for $-5 \le x \le 5$ is shown in Figure 1. Theorem 1 guarantees that all the real zeros of $P(x)$ are between -5 and 5. We can be certain that the graph does not change direction and cross the x axis somewhere outside the viewing window in Figure 1.

> **Figure 1** $P(x) = x^4 - 2x^3 - 10x^2 + 40x - 90$.

MATCHED PROBLEM 1 Let $P(x) = x^4 - 5x^3 - x^2 + 40x - 70$. Find the smallest positive integer and the largest negative integer that, by Theorem 1, are upper and lower bounds, respectively, for the real zeros of $P(x)$.

EXAMPLE 2 Bounding Real Zeros

Let $P(x) = x^3 - 30x^2 + 275x - 720$. Find the smallest positive integer multiple of 10 and the largest negative integer multiple of 10 that, by Theorem 1, are upper and lower bounds, respectively, for the real zeros of $P(x)$.

SOLUTION We construct a synthetic division table to search for bounds for the zeros of $P(x)$. The size of the coefficients in $P(x)$ indicates that we can speed up this search by choosing larger increments between test values.

		1	-30	275	-720
	10	1	-20	75	30
	20	1	-10	75	780
UB	30	1	0	275	7,530
LB	-10	1	-40	675	$-7,470$

> **Figure 2** $P(x) = x^3 - 30x^2 + 275x - 720$.

Therefore, all real zeros of $P(x) = x^3 - 30x^2 + 275x - 720$ must lie between -10 and 30, as confirmed by Figure 2.

MATCHED PROBLEM 2 Let $P(x) = x^3 - 25x^2 + 170x - 170$. Find the smallest positive integer multiple of 10 and the largest negative integer multiple of 10 that, by Theorem 1, are upper and lower bounds, respectively, for the real zeros of $P(x)$.

Technology Connections

How do you determine the correct viewing window for graphing a function? This is one of the most frequently asked questions about graphing calculators. For polynomial functions, the upper and lower bound theorem gives an answer: let Xmin and Xmax be the lower and upper bounds, respectively, of Theorem 1 (appropriate values for Ymin and Ymax can then be found using TRACE). We can approximate the zeros, all of which appear in the chosen viewing window, using the ZERO command. The upper and lower bound theorem and the ZERO command on a graphing calculator are two important mathematical tools that work very well together.

› **Figure 3** $P(x) = x^5 + 3x - 1$.

› # Location Theorem and Bisection Method

The graph of every polynomial function is continuous. Because the polynomial function $P(x) = x^5 + 3x - 1$ is negative when $x = 0$ [$P(0) = -1$] and positive when $x = 1$ [$P(1) = 3$], the graph of $P(x)$ must cross the x axis at least once between $x = 0$ and $x = 1$ (Fig. 3). This observation is the basis for Theorem 2 and leads to a simple method for approximating zeros.

> ### › THEOREM 2 Location Theorem*

> Suppose that a function f is continuous on an interval I that contains numbers a and b. If $f(a)$ and $f(b)$ have opposite signs, then the graph of f has at least one x intercept between a and b.

The conclusion of Theorem 2 says that at least one zero of the function is "located" between a and b. There may be more than one zero between a and b: if $g(x) = x^3 + x^2 - 2x - 1$, then $g(-2)$ and $g(2)$ have opposite signs and there are three zeros between $x = -2$ and $x = 2$ [Fig. 4(a)]. The converse of Theorem 2 is false: $h(x) = x^2$ has an x intercept at $x = 0$ but does not change sign [Fig. 4(b)].

(a)

(b)

› **Figure 4** Polynomials may or may not change sign at a zero.

>>> **EXPLORE-DISCUSS 1**

When synthetic division is used to divide a polynomial $P(x)$ by $x - 3$ the remainder is -33. When the same polynomial is divided by $x - 4$ the remainder is 38. Must $P(x)$ have a zero between 3 and 4? Explain.

*The location theorem is a formulation of the important intermediate value theorem of calculus.

Explore-Discuss 2 will provide an introduction to the repeated systematic application of the location theorem (Theorem 2) called the *bisection method*. This method forms the basis for the zero approximation routines in many graphing calculators.

>>> **EXPLORE-DISCUSS 2**

Let $P(x) = x^5 + 3x - 1$. Because $P(0)$ is negative and $P(1)$ is positive, the location theorem guarantees that $P(x)$ must have at least one zero in the interval $(0, 1)$.

(A) Is $P(0.5)$ positive or negative? Does the location theorem guarantee a zero of $P(x)$ in the interval $(0, 0.5)$ or in $(0.5, 1)$?

(B) Let m be the midpoint of the interval from part A that contains a zero of $P(x)$. Is $P(m)$ positive or negative? What does this tell you about the location of the zero?

(C) Explain how this process could be used repeatedly to approximate a zero to any desired accuracy.

The **bisection method** is a systematic application of the procedure suggested in Explore-Discuss 2: Let $P(x)$ be a polynomial with real coefficients. If $P(x)$ has opposite signs at the endpoints of an interval (a, b), then by the location theorem $P(x)$ has a zero in (a, b). Bisect this interval (that is, find the midpoint $m = \frac{a+b}{2}$), check the sign of $P(m)$, and select the interval (a, m) or (m, b) that has opposite signs at the endpoints. We repeat this bisection procedure (producing a set of intervals, each contained in and half the length of the previous interval, and each containing a zero) until the desired accuracy is obtained. If at any point in the process $P(m) = 0$, we stop, because a real zero m has been found. Example 3 illustrates the procedure, and clarifies when the procedure is finished.

EXAMPLE	3

The Bisection Method

The polynomial $P(x) = x^4 - 2x^3 - 10x^2 + 40x - 90$ of Example 2 has a zero between 3 and 4. Use the bisection method to approximate it to one-decimal-place accuracy.

SOLUTION

We organize the results of our calculations in Table 1. Because the sign of $P(x)$ changes at the endpoints of the interval $(3.5625, 3.625)$, we conclude that a real zero lies in this interval and is given by $r = 3.6$ to one-decimal place accuracy (each endpoint rounds to 3.6).

Table 1 Bisection Approximation

Sign Change Interval (a, b)	Midpoint m	Sign of P		
		$P(a)$	$P(m)$	$P(b)$
(3, 4)	3.5	–	–	+
(3.5, 4)	3.75	–	+	+
(3.5, 3.75)	3.625	–	+	+
(3.5, 3.625)	3.5625	–	–	+
(3.5625, 3.625)	We stop here	–		+

Figure 5 illustrates the nested intervals produced by the bisection method in Table 1. Match each step in Table 1 with an interval in Figure 5. Note how each interval that contains a zero gets smaller and smaller and is contained in the preceding interval that contained the zero.

> **Figure 5** Nested intervals produced by the bisection method in Table 1.

If we had wanted two-decimal-place accuracy, we would have continued the process in Table 1 until the endpoints of a sign change interval rounded to the same two-decimal-place number.

MATCHED PROBLEM 3

The polynomial $P(x) = x^4 - 2x^3 - 10x^2 + 40x - 90$ of Example 1 has a zero between -5 and -4. Use the bisection method to approximate it to one-decimal-place accuracy.

› Approximating Real Zeros at Turning Points

The bisection method for approximating zeros fails if a polynomial has a turning point at a zero, because the polynomial does not change sign at such a zero. Most graphing calculators use methods that are more sophisticated than the bisection method. Nevertheless, it is not unusual to get an error message when using the zero command to approximate a zero that is also a turning point. In this case, we can use the maximum or minimum command, as appropriate, to approximate the turning point, and the zero.

EXAMPLE 4 **Approximating Zeros at Turning Points**

Let $P(x) = x^5 + 6x^4 + 4x^3 - 24x^2 - 16x + 32$. Find the smallest positive integer and the largest negative integer that, by Theorem 1, are upper and lower bounds, respectively, for the real zeros of $P(x)$. Approximate the zeros to two decimal places, using maximum or minimum commands to approximate any zeros at turning points.

SOLUTION

The pertinent rows of a synthetic division table show that 2 is the upper bound and -6 is the lower bound:

	1	6	4	-24	-16	32
1	1	7	11	-13	-29	3
2	1	8	20	16	16	64
-5	1	1	-1	-19	79	-363
-6	1	0	4	-48	272	-1600

Examining the graph of $P(x)$ we find three zeros: the zero -3.24, found using the MAXIMUM command [Fig. 6(a)]; the zero -2, found using the ZERO command [Fig. 6(b)]; and the zero 1.24, found using the MINIMUM command [Fig. 6(c)].

> **Figure 6** Zeros of $P(x) = x^5 + 6x^4 + 4x^3 - 24x^2 - 16x + 32$.

MATCHED PROBLEM 4

Let $P(x) = x^5 - 6x^4 + 40x^2 - 12x - 72$. Find the smallest positive integer and the largest negative integer that, by Theorem 1, are upper and lower bounds, respectively, for the real zeros of $P(x)$. Approximate the zeros to two decimal places, using maximum or minimum commands to approximate any zeros at turning points.

› Polynomial Inequalities

We can apply the techniques we have introduced for finding real zeros to solve polynomial inequalities. Consider, for example, the inequality

$$x^3 - 2x^2 - 5x + 6 > 0$$

The real zeros of $P(x) = x^3 - 2x^2 - 5x + 6$ are easily found to be -2, 1, and 3. They partition the x axis into four intervals

$$(-\infty, -2), (-2, 1), (1, 3), \quad \text{and} \quad (3, \infty)$$

On any one of these intervals, the graph of P is either above the x axis or below the x axis, because, by the location theorem, *a continuous function can change sign only at a zero.*

One way to decide whether the graph of P is above or below the x axis on a given interval, say $(-2, 1)$, is to choose a "test number" that belongs to the interval, 0, for example, and evaluate P at the test number. Because $P(0) = 6 > 0$, the graph of P is above the x axis throughout the interval $(-2, 1)$. A second way to decide whether the graph of P is above or below the x axis on $(-2, 1)$ is to simply inspect the graph of P. Each technique has its advantages, and both are illustrated in the solutions to Examples 5 and 6.

EXAMPLE 5

Solving Polynomial Inequalities

Solve the inequality $x^3 - 2x^2 - 5x + 6 > 0$.

SOLUTION

Let $P(x) = x^3 - 2x^2 - 5x + 6$. Then

$$P(1) = 1^3 - 2(1^2) - 5 + 6 = 0$$

so 1 is a zero of P and $x - 1$ is a factor. Dividing $P(x)$ by $x - 1$ (details omitted) gives the quotient $x^2 - x - 6$. Therefore,

$$P(x) = (x - 1)(x^2 - x - 6) = (x - 1)(x + 2)(x - 3)$$

The zeros of P are -2, 1, and 3. They partition the x axis into the four intervals shown in the table on page 284. A test number is chosen from each interval as indicated to determine whether $P(x)$ is positive (above the x axis) or negative (below the x axis) on that interval.

Interval	$(-\infty, -2)$	$(-2, 1)$	$(1, 3)$	$(3, \infty)$
Test number x	-3	0	2	4
$P(x)$	-24	6	-4	18
Sign of P	$-$	$+$	$-$	$+$

We conclude that the solution set of the inequality is the intervals where $P(x)$ is positive:

$$(-2, 1) \cup (3, \infty)$$

MATCHED PROBLEM 5 Solve the inequality $x^3 + x^2 - x - 1 < 0$.

EXAMPLE 6

Solving Polynomial Inequalities with a Graphing Calculator

Solve $3x^2 + 12x - 4 \geq 2x^3 - 5x^2 + 7$ to three decimal places.

SOLUTION Subtracting the right-hand side gives the equivalent inequality

$$P(x) = -2x^3 + 8x^2 + 12x - 11 \geq 0$$

The zeros of $P(x)$, to three decimal places, are -1.651, 0.669, and 4.983 [Fig. 7(a)].
 By inspecting the graph of P we see that P is above the x axis on the intervals $(-\infty, -1.651)$ and $(0.669, 4.983)$. So the solution set of the inequality is

$$(-\infty, -1.651] \cup [0.669, 4.983]$$

The square brackets indicate that the endpoints of the intervals—the zeros of the polynomial—also satisfy the inequality.
 An alternative to inspecting the graph of P is to inspect the graph of

$$f(x) = \frac{P(x)}{|P(x)|}$$

The function $f(x)$ has the value 1 if $P(x)$ is positive, because then the absolute value of $P(x)$ is equal to $P(x)$. Similarly, $f(x)$ has the value -1 if $P(x)$ is negative. This technique makes it easy to identify the solution set of the original inequality [Fig. 7(b)] and often eliminates difficulties in choosing appropriate window variables.

(a) $P(x) = -2x^3 + 8x^2 + 12x - 11$

(b) $f(x) = \dfrac{P(x)}{|P(x)|}$

> **Figure 7**

MATCHED PROBLEM 6 Solve to three decimal places $5x^3 - 13x < 4x^2 + 10x - 5$.

› Mathematical Modeling

EXAMPLE	7	**Construction**

An oil tank is in the shape of a right circular cylinder with a hemisphere at each end (Fig. 8). The cylinder is 55 inches long, and the volume of the tank is $11{,}000\pi$ cubic inches (approximately 20 cubic feet). Let x denote the common radius of the hemispheres and the cylinder.

—— 55 inches ——

› **Figure 8**

(A) Find a polynomial equation that x must satisfy.

(B) Approximate x to one decimal place.

SOLUTIONS (A) If x is the common radius of the hemispheres and the cylinder in inches, then

$$\begin{pmatrix}\text{Volume}\\\text{of}\\\text{tank}\end{pmatrix} = \begin{pmatrix}\text{Volume}\\\text{of two}\\\text{hemispheres}\end{pmatrix} + \begin{pmatrix}\text{Volume}\\\text{of}\\\text{cylinder}\end{pmatrix}$$

$$11{,}000\pi = \tfrac{4}{3}\pi x^3 + 55\pi x^2 \qquad \text{Multiply by } 3/\pi.$$

$$33{,}000 = 4x^3 + 165x^2 \qquad \text{Subtract 33,000 from both sides.}$$

$$0 = 4x^3 + 165x^2 - 33{,}000$$

The radius we are looking for (x) must be a positive zero of

$$P(x) = 4x^3 + 165x^2 - 33{,}000$$

(B) Because the coefficients of $P(x)$ are large, we use larger increments in the synthetic division table:

		4	165	0	$-33{,}000$
	10	4	205	2,050	$-12{,}500$
UB	20	4	245	4,900	65,000

Applying the bisection method to the interval [10, 20] (nine midpoints are calculated; details omitted) or graphing $y = P(x)$ for $0 \le x \le 20$ (Fig. 9), we see that $x = 12.4$ inches (to one decimal place).

$P(x) = 4x^3 + 165x^2 - 33{,}000.$
› **Figure 9**

MATCHED PROBLEM 7

Repeat Example 7 if the volume of the tank is $44{,}000\pi$ cubic inches.

ANSWERS TO MATCHED PROBLEMS

1. Lower bound: -3; upper bound: 6 **2.** Lower bound: -10; upper bound: 30
3. $x = -4.1$ **4.** Lower bound: -2; upper bound: 6; $-1.65, 2, 3.65$
5. $(-\infty, -1) \cup (-1, 1)$ **6.** $(-\infty, -1.899) \cup (0.212, 2.488)$
7. (A) $P(x) = 4x^3 + 165x^2 - 132{,}000 = 0$ (B) 22.7 inches

4-2 Exercises

1. Given a polynomial of degree $n > 0$, explain why there must exist an upper bound and a lower bound for its real zeros.

2. State the location theorem in your own words.

3. A polynomial P has degree 6 and leading coefficient 1. If synthetic division by $x - 5$ results in all positive numbers in the quotient row, is 10 an upper bound for the real zeros of P? Explain.

4. A polynomial has degree 12 and leading coefficient 1. If synthetic division by $x + 5$ results in numbers that alternate in sign in the quotient row, is -10 a lower bound for the real zeros of P? Explain.

5. Explain the basic steps in the bisection method.

6. If you use the bisection method to approximate a real root to three decimal place accuracy, explain how you can tell when the method is finished.

In Problems 7–10, approximate the real zeros of each polynomial to three decimal places.

7. $P(x) = x^2 + 5x - 2$

8. $P(x) = 3x^2 - 7x + 1$

9. $P(x) = 2x^3 - 5x + 2$

10. $P(x) = x^3 - 4x^2 - 8x + 3$

In Problems 11–14, use the graph of P(x) to write the solution set for each inequality.

11. $P(x) \geq 0$ **12.** $P(x) < 0$

13. $P(x) > 0$ **14.** $P(x) \leq 0$

In Problems 15–18, solve each polynomial inequality to three decimal places (note the connection with Problems 7–10).

15. $x^2 + 5x - 2 > 0$ **16.** $3x^2 - 7x + 1 \geq 0$

17. $2x^3 - 5x + 2 \leq 0$ **18.** $x^3 - 4x^2 - 8x + 3 < 0$

Find the smallest positive integer and largest negative integer that, by Theorem 1, are upper and lower bounds, respectively, for the real zeros of each of the polynomials given in Problems 19–24.

19. $P(x) = x^3 - 3x + 1$ **20.** $P(x) = x^3 - 4x^2 + 4$

21. $P(x) = x^4 - 3x^3 + 4x^2 + 2x - 9$

22. $P(x) = x^4 - 4x^3 + 6x^2 - 4x - 7$

23. $P(x) = x^5 - 3x^3 + 3x^2 + 2x - 2$

24. $P(x) = x^5 - 3x^4 + 3x^2 + 2x - 1$

In Problems 25–30, (A) use the location theorem to explain why the polynomial function has a zero in the indicated interval; and (B) determine the number of additional intervals required by the bisection method to obtain a one-decimal-place approximation to the zero and state the approximate value of the zero.

25. $P(x) = x^3 - 2x^2 - 5x + 4;\ (3, 4)$

26. $P(x) = x^3 + x^2 - 4x - 1;\ (1, 2)$

27. $P(x) = x^3 - 2x^2 - x + 5;\ (-2, -1)$

28. $P(x) = x^3 - 3x^2 - x - 2;\ (3, 4)$

29. $P(x) = x^4 - 2x^3 - 7x^2 + 9x + 7;\ (3, 4)$

30. $P(x) = x^4 - x^3 - 9x^2 + 9x + 4;\ (2, 3)$

In Problems 31–36, (A) find the smallest positive integer and largest negative integer that, by Theorem 1, are upper and lower bounds, respectively, for the real zeros of P(x); and (B) use the bisection method to approximate a real zero of each polynomial to one decimal place.

31. $P(x) = x^3 - 2x^2 + 3x - 8$

32. $P(x) = x^3 + 3x^2 + 4x + 5$

33. $P(x) = 2x^3 + x^2 + 2x + 1$

34. $P(x) = 2x^3 - x^2 + 4x - 2$

35. $P(x) = x^4 + x^2 - 6$ **36.** $P(x) = x^4 - 2x^2 - 3$

Problems 37–40, refer to the polynomial

$$P(x) = (x - 1)^2(x - 2)(x - 3)^4$$

37. Can the zero at $x = 1$ be approximated by the bisection method? Explain.

38. Can the zero at $x = 2$ be approximated by the bisection method? Explain.

39. Can the zero at $x = 3$ be approximated by the bisection method? Explain.

40. Which of the zeros can be approximated by a maximum approximation routine? By a minimum approximation routine? By the zero approximation routine on your graphing calculator?

 In Problems 41–46, approximate the zeros of each polynomial function to two decimal places, using maximum or minimum commands to approximate any zeros at turning points.

41. $P(x) = x^4 - 4x^3 - 10x^2 + 28x + 49$

42. $P(x) = x^4 + 4x^3 - 4x^2 - 16x + 16$

43. $P(x) = x^5 - 6x^4 + 4x^3 + 24x^2 - 16x - 32$

44. $P(x) = x^5 - 6x^4 + 2x^3 + 28x^2 - 15x + 2$

45. $P(x) = x^5 - 6x^4 + 11x^3 - 4x^2 - 3.75x - 0.5$

46. $P(x) = x^5 + 12x^4 + 47x^3 + 56x^2 - 15.75x + 1$

In Problems 47–52, solve each polynomial inequality.

47. $x^2 > 9$

48. $1 - x^2 \leq 0$

49. $x^3 \leq 16x$

50. $2x > x^2 + x^3$

51. $x^4 + 4 \geq 5x^2$

52. $2 + x + x^2 + x^3 < x^4$

 In Problems 53–58, solve each polynomial inequality to three decimal places.

53. $x^2 + 7x - 3 \leq x^3 + x + 4$

54. $x^4 + 1 > 3x^2$

55. $x^4 < 8x^3 - 17x^2 + 9x - 2$

56. $x^3 + 5x \geq 2x^3 - 4x^2 + 6$

57. $(x^2 + 2x - 2)^2 \geq 2$

58. $5 + 2x < (x^2 - 4)^2$

 In Problems 59–64, (A) find the smallest positive integer multiple of 10 and largest negative integer multiple of 10 that, by Theorem 1, are upper and lower bounds, respectively, for the real zeros of each polynomial; and (B) approximate the real zeros of each polynomial to two decimal places.

59. $P(x) = x^3 - 24x^2 - 25x + 10$

60. $P(x) = x^3 - 37x^2 + 70x - 20$

61. $P(x) = x^4 + 12x^3 - 900x^2 + 5,000$

62. $P(x) = x^4 - 12x^3 - 425x^2 + 7,000$

63. $P(x) = x^4 - 100x^2 - 1,000x - 5,000$

64. $P(x) = x^4 - 5x^3 - 50x^2 - 500x + 7,000$

65. When synthetic division is used to divide a polynomial $P(x)$ by $x + 4$ the remainder is 10. When the same polynomial is divided by $x + 5$ the remainder is -8. Must $P(x)$ have a zero between -5 and -4? Explain.

66. When synthetic division is used to divide a polynomial $Q(x)$ by $x + 4$ the remainder is 10. When the same polynomial is divided by $x + 5$ the remainder is 8. Could $Q(x)$ have a zero between -5 and -4? Explain.

67. Give a reason for each step in the proof of the upper bound case of Theorem 1 on page 278.
Step 1: $P(x)$ can be written in the form $P(x) = (x - r)Q(x) + R$, where the coefficients of $Q(x)$ and R are positive.
Step 2: Suppose $s > r > 0$. Then $P(s) > 0$.
Step 3: r is an upper bound for the real zeros of $P(x)$.

68. Give a reason for each step in the proof of the lower bound case of Theorem 1 on page 278.
Step 1: $P(x)$ can be written in the form $P(x) = (x - r)Q(x) + R$, where the coefficients of $Q(x)$ and R alternate in sign.
Step 2: Suppose $s < r < 0$. If P has even degree, then $P(s) > 0$; if P has odd degree, then $P(s) < 0$.
Step 3: r is a lower bound for the real zeros of $P(x)$.

Problems 69 and 70 explore the cases in which 0 is an upper bound or lower bound for the real zeros of a polynomial. These cases are not covered by Theorem 1, the upper and lower bound theorem, as formulated on page 278.

69. Let $P(x)$ be a polynomial of degree $n > 0$ such that all of the coefficients of $P(x)$ are nonnegative. Explain why 0 is an upper bound for the real zeros of $P(x)$.

70. Let $P(x)$ be a polynomial of degree $n > 0$ such that $a_n > 0$ and the coefficients of $P(x)$ alternate in sign (as in Theorem 1, a coefficient 0 can be considered either positive or negative, but not both). Explain why 0 is a lower bound for the real zeros of $P(x)$.

APPLICATIONS

Express the solutions to Problems 71–76 as the roots of a polynomial equation of the form P(x) = 0 and approximate these solutions to one decimal place.

71. GEOMETRY Find all points on the graph of $y = x^2$ that are one unit away from the point $(1, 2)$. [*Hint:* Use the distance formula from Section 2-2.]

72. GEOMETRY Find all points on the graph of $y = x^2$ that are one unit away from the point $(2, 1)$.

73. MANUFACTURING A box is to be made out of a piece of cardboard that measures 18 by 24 inches. Squares, x inches on a side, will be cut from each corner, and then the ends and sides will be folded up (see the figure). Find the value of x that would result in a box with a volume of 600 cubic inches.

74. MANUFACTURING A box with a hinged lid is to be made out of a piece of cardboard that measures 20 by 40 inches. Six squares, x inches on a side, will be cut from each corner and the middle, and then the ends and sides will be folded up to form the box and its lid

(see the figure). Find the value of x that would result in a box with a volume of 500 cubic inches.

75. CONSTRUCTION A propane gas tank is in the shape of a right circular cylinder with a hemisphere at each end (see the figure). If the overall length of the tank is 10 feet and the volume is 20π cubic feet, find the common radius of the hemispheres and the cylinder.

76. SHIPPING A shipping box is reinforced with steel bands in all three directions (see the figure). A total of 20.5 feet of steel tape is to be used, with 6 inches of waste because of a 2-inch overlap in each direction. If the box has a square base and a volume of 2 cubic feet, find the side length of the base.

4-3

Complex Zeros and Rational Zeros of Polynomials

> **The Fundamental Theorem of Algebra**
> **Factors of Polynomials with Real Coefficients**
> **Graphs of Polynomials with Real Coefficients**
> **Rational Zeros**

The graph of the polynomial function $P(x) = x^2 + 4$ does not cross the x axis, so $P(x)$ has no real zeros. It does, however, have complex zeros, $2i$ and $-2i$; by the factor theorem, $x^2 + 4 = (x - 2i)(x + 2i)$. The *fundamental theorem of algebra* guarantees that *every* non-constant polynomial with real or complex coefficients has a complex zero; as a result, such a polynomial can be factored as a product of linear factors. In Section 4-3, we study the fundamental theorem and its implications, including results on the graphs of polynomials with real coefficients. Finally, we consider a problem that has led to important advances in mathematics and its applications: When can zeros of a polynomial be found *exactly*?

> The Fundamental Theorem of Algebra

The fundamental theorem of algebra was proved by Karl Friedrich Gauss (1777–1855), one of the greatest mathematicians of all time, in his doctoral thesis. A proof of the theorem is beyond the scope of this book, so we will state and use it without proof.

> **THEOREM 1 Fundamental Theorem of Algebra**

Every polynomial of degree $n > 0$ with complex coefficients has a complex zero.

If $P(x)$ is a polynomial of degree $n > 0$ with complex coefficients, then by Theorem 1 it has a zero r_1. So $x - r_1$ is a factor of $P(x)$ by Theorem 5 of Section 4-1, and

$$P(x) = (x - r_1)Q(x), \deg Q(x) = n - 1$$

Now, if $\deg Q(x) > 0$, then, applying the fundamental theorem to $Q(x)$, $Q(x)$ has a root r_2 and therefore a factor $x - r_2$. (It is possible that r_2 is equal to r_1.) By continuing this reasoning we obtain a proof of Theorem 2.

> **THEOREM 2 n Linear Factors Theorem**

Every polynomial of degree $n > 0$ with complex coefficients can be factored as a product of n linear factors.

Suppose that a polynomial $P(x)$ is factored as a product of n linear factors. Any zero r of $P(x)$ must be a zero of one or more of the factors. The number of linear factors that have zero r is said to be the **multiplicity** of r. For example, the polynomial

$$P(x) = (x - 5)^3(x + 1)^2(x - 6i)(x + 2 + 3i) \tag{1}$$

has degree 7 and is written as a product of seven linear factors. $P(x)$ has just four zeros, namely 5, -1, $6i$, and $-2 - 3i$. Because the factor $x - 5$ appears to the power 3, we say that the zero 5 has *multiplicity* 3. Similarly, -1 has *multiplicity* 2, $6i$ has *multiplicity* 1, and $-2 - 3i$ has *multiplicity* 1. A zero of multiplicity 2 is called a **double zero,** and a zero of multiplicity 3 is called **triple zero.** Note that the sum of the multiplicities is always equal to the degree of the polynomial: for $P(x)$ in equation (1), $3 + 2 + 1 + 1 = 7$.

EXAMPLE 1 **Multiplicities of Zeros**

Find the zeros and their multiplicities:

(A) $P(x) = (x + 2)^7(x - 4)^8(x^2 + 1)$

(B) $Q(x) = (x + 1)^3(x^2 - 1)(x + 1 - i)$

SOLUTIONS (A) Note that $x^2 + 1 = 0$ has the solutions i and $-i$. The zeros of $P(x)$ are -2 (multiplicity 7), 4 (multiplicity 8), i and $-i$ (each multiplicity 1).

(B) Note that $x^2 - 1 = (x - 1)(x + 1)$, so $x + 1$ appears four times as a factor of $Q(x)$. The zeros of $Q(x)$ are -1 (multiplicity 4), 1 (multiplicity 1), and $-1 + i$ (multiplicity 1).

MATCHED PROBLEM 1 Find the zeros and their multiplicities:

(A) $P(x) = (x - 5)^3(x + 3)^2(x^2 + 16)$

(B) $Q(x) = (x^2 - 25)^3(x + 5)(x - i)$

› Factors of Polynomials with Real Coefficients

If $p + qi$ is a zero of $P(x) = ax^2 + bx + c$, where a, b, c, p, and q are real numbers, then

$$P(p + qi) = 0$$

$$a(p + qi)^2 + b(p + qi) + c = 0 \qquad \text{Take the conjugate of both sides.}$$

$$\overline{a(p + qi)^2 + b(p + qi) + c} = \overline{0} \qquad \overline{z + w} = \overline{z} + \overline{w}, \; \overline{zw} = \overline{z}\,\overline{w}$$

$$\overline{a}\,(\overline{p + qi})^2 + \overline{b}\,(\overline{p + qi}) + \overline{c} = \overline{0} \qquad \overline{z} = z \text{ if } z \text{ is real}, \; \overline{p + qi} = p - qi$$

$$a(p - qi)^2 + b(p - qi) + c = 0$$

$$P(p - qi) = 0$$

Therefore, $p - qi$ is also a zero of $P(x)$. This method of proof can be applied to any polynomial $P(x)$ of degree $n > 0$ with real coefficients, justifying Theorem 3.

> **THEOREM 3** Imaginary Zeros of Polynomials with Real Coefficients

Imaginary zeros of polynomials with real coefficients, if they exist, occur in conjugate pairs.

If a polynomial $P(x)$ of degree $n > 0$ has real coefficients and a linear factor of the form $x - (p + qi)$ where $q \neq 0$, then, by Theorem 3, $P(x)$ also has the linear factor $x - (p - qi)$. But

$$[x - (p + qi)][x - (p - qi)] = x^2 - 2px + p^2 - q^2$$

which is a quadratic factor of $P(x)$ with real coefficients and imaginary zeros. By this reasoning we can prove Theorem 4.

> **THEOREM 4** Linear and Quadratic Factors Theorem*

If $P(x)$ is a polynomial of degree $n > 0$ with real coefficients, then $P(x)$ can be factored as a product of linear factors (with real coefficients) and quadratic factors (with real coefficients and imaginary zeros).

| EXAMPLE | 2 | **Factors of Polynomials** |

Factor $P(x) = x^3 + x^2 + 4x + 4$ in two ways:

(A) As a product of linear factors (with real coefficients) and quadratic factors (with real coefficients and imaginary zeros)

(B) As a product of linear factors with complex coefficients

SOLUTIONS (A) Note that $P(-1) = 0$, so -1 is a zero of $P(x)$ (or graph $P(x)$ and note that -1 is an x intercept). Therefore, $x + 1$ is a factor of $P(x)$. Using synthetic division, the quotient is $x^2 + 4$, which has imaginary roots. Therefore,

$$P(x) = (x + 1)(x^2 + 4)$$

*Theorem 4 underlies the technique of decomposing a rational function into partial fractions, which is useful in calculus. See Appendix B, Section B-2.

An alternative solution is to factor by grouping:

$$x^3 + x^2 + 4x + 4 = x^2(x + 1) + 4(x + 1)$$
$$= (x^2 + 4)(x + 1)$$

(B) Because $x^2 + 4$ has roots $2i$ and $-2i$,

$$P(x) = (x + 1)(x - 2i)(x + 2i)$$

MATCHED PROBLEM 2

Factor $P(x) = x^5 - x^4 - x + 1$ in two ways:

(A) As a product of linear factors (with real coefficients) and quadratic factors (with real coefficients and imaginary zeros)

(B) As a product of linear factors with complex coefficients

> ## Graphs of Polynomials with Real Coefficients

The factorization described in Theorem 4 gives additional information about the graphs of polynomial functions with real coefficients. For certain polynomials the factorization of Theorem 4 will involve only linear factors; for others, only quadratic factors. Of course if only quadratic factors are present, then the degree of the polynomial $P(x)$ must be even. In other words, a polynomial $P(x)$ of odd degree with real coefficients must have a linear factor with real coefficients. This proves Theorem 5.

> **THEOREM 5 Real Zeros and Polynomials of Odd Degree**
>
> Every polynomial of odd degree with real coefficients has at least one real zero, and consequently at least one x intercept.

>>> **EXPLORE-DISCUSS 1**

The graph of the polynomial $P(x) = x(x - 1)^2(x + 1)^4(x - 2)^3$ is shown in Figure 1. Find the real zeros of $P(x)$ and their multiplicities. How can a real zero of even multiplicity be distinguished from a real zero of odd multiplicity using only the graph?

> **Figure 1** Graph of
$P(x) = x(x - 1)^2(x + 1)^4(x - 2)^3$.

For polynomials with real coefficients, as suggested by Explore-Discuss 1, you can easily distinguish real zeros of even multiplicity from those of odd multiplicity using only the graph. Theorem 6, which we state without proof, tells how to do that.

> **THEOREM 6 Zeros of Even or Odd Multiplicity**
>
> Let $P(x)$ be a polynomial with real coefficients:
>
> 1. If r is a real zero of $P(x)$ of even multiplicity, then $P(x)$ has a turning point at r and does not change sign at r. (The graph just touches the x axis, then changes direction.)
>
> 2. If r is a real zero of $P(x)$ of odd multiplicity, then $P(x)$ does not have a turning point at r and changes sign at r. (The graph continues through to the opposite side of the x axis.)

EXAMPLE 3 ## Multiplicities from Graphs

Figure 2 shows the graph of a polynomial function of degree 6. Find the real zeros and their multiplicities.

> **Figure 2**

SOLUTION The numbers -2, -1, 1, and 2 are real zeros (x intercepts). The graph has turning points at $x = \pm 1$ but not at $x = \pm 2$. Therefore, by Theorem 6, the zeros -1 and 1 have even multiplicity, and -2 and 2 have odd multiplicity. Because the sum of the multiplicities must equal 6 (the degree), the zeros -1 and 1 each have multiplicity 2, and the zeros -2 and 2 each have multiplicity 1.

MATCHED PROBLEM 3 Figure 3 shows the graph of a polynomial function of degree 7. Find the real zeros and their multiplicities.

> **Figure 3**

> **Figure 4** $P(x) = x^2 - (4 + 10^{-9})$.

› Rational Zeros

From a graphical perspective, finding a zero of a polynomial means finding a good approximation to an actual zero. A graphing calculator, for example, might give 2 as a zero of $P(x) = x^2 - (4 + 10^{-9})$ even though $P(2)$ is equal to -10^{-9}, not 0 (Fig. 4).

It is natural, however, to want to find zeros *exactly*. Although this is impossible in general, we will adopt an algebraic strategy to find exact zeros in a special case, that of rational zeros of polynomials with rational coefficients. We will find a graphing calculator to be helpful in carrying out the algebraic strategy.

First note that a polynomial with rational coefficients can always be written as a constant times a polynomial with integer coefficients. For example,

$$P(x) = \frac{1}{2}x^3 - \frac{2}{3}x^2 + \frac{7}{4}x + 5$$

$$= \frac{1}{12}(6x^3 - 8x^2 + 21x + 60)$$

Because the zeros of $P(x)$ are the zeros of $6x^3 - 8x^2 + 21x + 60$, it is sufficient, for the purpose of finding rational zeros of polynomials with rational coefficients, to study just the polynomials with integer coefficients.

We introduce the rational zero theorem by examining the following quadratic polynomial whose zeros can be found easily by factoring:

$$P(x) = 6x^2 - 13x - 5 = (2x - 5)(3x + 1)$$

Zeros of $P(x)$: $\dfrac{5}{2}$ and $-\dfrac{1}{3} = \dfrac{-1}{3}$

Notice that the numerators 5 and -1 of the zeros are both integer factors of -5, the constant term in $P(x)$. The denominators 2 and 3 of the zeros are both integer factors of 6, the coefficient of the highest-degree term in $P(x)$. These observations are generalized in Theorem 7 (a proof is outlined in Problem 89 of Exercises 4-3).

> **THEOREM 7** Rational Zero Theorem
>
> If the rational number b/c, in lowest terms, is a zero of the polynomial
>
> $$P(x) = a_n x^n + a_{n-1} x^{n-1} + \cdots + a_1 x + a_0 \qquad a_n \neq 0$$
>
> with integer coefficients, then b must be an integer factor of a_0 and c must be an integer factor of a_n.
>
>

Theorem 7 enables us to construct a finite list of possible rational zeros of $P(x)$. Each number in the list must then be tested to determine whether or not it is actually a zero. As Example 4 illustrates, a graphing calculator can often reduce the effort required to locate rational zeros.

EXAMPLE 4 Finding Rational Zeros

Find all the rational zeros for $P(x) = 2x^3 + 9x^2 + 7x - 6$.

SOLUTION If b/c in lowest terms is a rational zero of $P(x)$, then b must be a factor of -6 and c must be a factor of 2.

$$\text{Possible values of } b \text{ are the integer factors of } -6\text{: } \pm 1, \pm 2, \pm 3, \pm 6 \qquad (2)$$

$$\text{Possible values of } c \text{ are the integer factors of } 2\text{: } \pm 1, \pm 2 \qquad (3)$$

Writing all possible fractions b/c where b is from (2) and c is from (3), we have

$$\text{Possible rational zeros for } P(x)\text{: } \pm 1, \pm 2, \pm 3, \pm 6, \pm \tfrac{1}{2}, \pm \tfrac{3}{2} \qquad (4)$$

[Note that all fractions are in lowest terms and duplicates like $\pm 6/\pm 2 = \pm 3$ are not repeated.] If $P(x)$ has any rational zeros, they must be in list (4). We can test each number r in this list simply by evaluating $P(r)$. However, exploring the graph of $y = P(x)$ first will usually indicate which numbers in the list are the most likely candidates for zeros. Examining a graph of $P(x)$, we see that there are zeros near -3, near -2, and between 0 and 1, so we begin by evaluating $P(x)$ at -3, -2, and $\tfrac{1}{2}$ (Fig. 5).

(a)

(b)

(c)

› Figure 5

Therefore, -3, -2, and $\frac{1}{2}$ are rational zeros of $P(x)$. Because a third-degree polynomial can have at most three zeros, we have found all the rational zeros. There is no need to test the remaining candidates in list (4).

MATCHED PROBLEM 4 Find all rational zeros for $P(x) = 2x^3 + x^2 - 11x - 10$.

As we saw in the solution of Example 4, rational zeros can be located by simply evaluating the polynomial. However, if we want to find multiple zeros, imaginary zeros, or exact values of irrational zeros, we need to consider *reduced polynomials*. If r is a zero of a polynomial $P(x)$, then we can write

$$P(x) = (x - r)Q(x)$$

where $Q(x)$ is a polynomial of degree one less than the degree of $P(x)$. The quotient polynomial $Q(x)$ is called a **reduced polynomial** for $P(x)$. In Example 4, after determining that -3 is a zero of $P(x)$, we can write

$$
\begin{array}{r}
\;\;2\quad\;\;9\quad\;\;7\quad -6 \\
\;\,-6\quad -9\quad\;\;6 \\
\hline
-3\,|\;\;2\quad\;\;3\quad -2\quad\;\;0
\end{array}
$$

$$
\begin{aligned}
P(x) &= 2x^3 + 9x^2 + 7x - 6 \\
&= (x + 3)(2x^2 + 3x - 2) \\
&= (x + 3)Q(x)
\end{aligned}
$$

Because the reduced polynomial $Q(x) = 2x^2 + 3x - 2$ is a quadratic, we can find its zeros by factoring or the quadratic formula. We get

$$P(x) = (x + 3)(2x^2 + 3x - 2) = (x + 3)(x + 2)(2x - 1)$$

and we see that the zeros of $P(x)$ are -3, -2, and $\frac{1}{2}$, as before.

EXAMPLE 5 Finding Rational and Irrational Zeros

Find all zeros exactly for $P(x) = 2x^3 - 7x^2 + 4x + 3$.

SOLUTION First, list the possible rational zeros:

$$\pm 1, \pm 3, \pm\tfrac{1}{2}, \pm\tfrac{3}{2}$$

Examining the graph of $y = P(x)$ (Fig. 6), we see that there is a zero between -1 and 0, another between 1 and 2, and a third between 2 and 3. We test the only likely candidates, $-\frac{1}{2}$ and $\frac{3}{2}$:

$$P(-\tfrac{1}{2}) = -1 \qquad \text{and} \qquad P(\tfrac{3}{2}) = 0$$

So $\frac{3}{2}$ is a zero, but $-\frac{1}{2}$ is not. Using synthetic division (details omitted), we can write

$$P(x) = (x - \tfrac{3}{2})(2x^2 - 4x - 2)$$

Because the reduced polynomial is quadratic, we can use the quadratic formula to find the exact values of the remaining zeros:

$$2x^2 - 4x - 2 = 0 \qquad\qquad \text{\color{red}Divide both sides by 2.}$$

$$x^2 - 2x - 1 = 0 \qquad\qquad \text{\color{red}Use the quadratic formula.}$$

$$x = \frac{2 \pm \sqrt{4 - 4(1)(-1)}}{2}$$

$$= \frac{2 \pm 2\sqrt{2}}{2} = 1 \pm \sqrt{2}$$

> Figure 6

So the exact zeros of $P(x)$ are $\frac{3}{2}$ and $1 \pm \sqrt{2}$.*

MATCHED PROBLEM 5 Find all zeros exactly for $P(x) = 3x^3 - 10x^2 + 5x + 4$.

EXAMPLE 6 Finding Rational and Imaginary Zeros

Find all zeros exactly for $P(x) = x^4 - 6x^3 + 14x^2 - 14x + 5$.

SOLUTION The possible rational zeros are ±1 and ±5. Examining the graph of $P(x)$ (Fig. 7), we see that 1 is a zero. Because the graph of $P(x)$ does not appear to change sign at 1, this may be a multiple root. Using synthetic division (details omitted), we find that

$$P(x) = (x - 1)(x^3 - 5x^2 + 9x - 5)$$

> Figure 7

The possible rational zeros of the reduced polynomial

$$Q(x) = x^3 - 5x^2 + 9x - 5$$

are ±1 and ±5. Examining the graph of $Q(x)$ (Fig. 8), we see that 1 is a rational zero. After a division, we have a quadratic reduced polynomial:

$$Q(x) = (x - 1)Q_1(x) = (x - 1)(x^2 - 4x + 5)$$

We use the quadratic formula to find the zeros of $Q_1(x)$:

> Figure 8

$$x^2 - 4x + 5 = 0$$
$$x = \frac{4 \pm \sqrt{16 - 4(1)(5)}}{2}$$
$$= \frac{4 \pm \sqrt{-4}}{2} = 2 \pm i$$

So the exact zeros of $P(x)$ are 1 (multiplicity 2), $2 - i$, and $2 + i$.

MATCHED PROBLEM 6 Find all zeros exactly for $P(x) = x^4 + 4x^3 + 10x^2 + 12x + 5$.

REMARK

We were successful in finding all the zeros of the polynomials in Examples 5 and 6 because we could find sufficient rational zeros to reduce the original polynomial to a quadratic. This is not always possible. For example, the polynomial

$$P(x) = x^3 + 6x - 2$$

has no rational zeros, but does have an irrational zero at $x \approx 0.32748$ (Fig. 9). The other two zeros are imaginary. The techniques we have developed will not find the exact value of these roots.

> Figure 9 $P(x) = x^3 + 6x - 2$.

*By analogy with Theorem 3 (imaginary zeros of polynomials with real coefficients occur in conjugate pairs), it can be shown that if $r + s\sqrt{t}$ is a zero of a polynomial with rational coefficients, where r, s, and t are rational but t is not the square of a rational, then $r - s\sqrt{t}$ is also a zero.

1. (A) 5 (multiplicity 3), -3 (multiplicity 2), $4i$ and $-4i$ (each multiplicity 1)
(B) -5 (multiplicity 4), 5 (multiplicity 3), i (multiplicity 1)
2. (A) $(x + 1)(x - 1)^2(x^2 + 1)$ (B) $(x + 1)(x - 1)^2(x + i)(x - i)$
3. -3 (multiplicity 2), -2 (multiplicity 1), -1 (multiplicity 1), 0 (multiplicity 2), 1 (multiplicity 1)
4. $-2, -1, \frac{5}{2}$ **5.** $\frac{4}{3}, 1 - \sqrt{2}, 1 + \sqrt{2}$ **6.** -1 (multiplicity 2), $-1 - 2i, -1 + 2i$

4-3 Exercises

1. Explain in your own words what the fundamental theorem of algebra says.

2. Does every polynomial of degree > 0 with real coefficients have a real zero? Explain.

3. What is meant by the *multiplicity* of a zero of a polynomial?

4. If $P(x)$ is a polynomial with integer coefficients and leading coefficient 1, explain why every rational zero of $P(x)$ is actually an integer.

Write the zeros of each polynomial in Problems 5–12, and indicate the multiplicity of each. What is the degree of each polynomial?

5. $P(x) = (x + 8)^3(x - 6)^2$

6. $P(x) = (x - 5)(x + 7)^2$

7. $P(x) = 3(x + 4)^3(x - 3)^2(x + 1)$

8. $P(x) = 5(x - 2)^3(x + 3)^2(x - 1)$

9. $P(x) = x^3(2x + 1)^2$

10. $P(x) = 6x^2(5x - 4)(3x + 2)$

11. $P(x) = (x^2 + 4)^3(x^2 - 4)^5(x + 2i)$

12. $P(x) = (x^2 + 7x + 10)^2(x^2 + 6x + 10)^3$

In Problems 13–18, find a polynomial P(x) of lowest degree, with leading coefficient 1, that has the indicated set of zeros. Write P(x) as a product of linear factors. Indicate the degree of P(x).

13. 3 (multiplicity 2) and -4

14. -2 (multiplicity 3) and 1 (multiplicity 2)

15. -7 (multiplicity 3), $-3 + \sqrt{2}, -3 - \sqrt{2}$

16. $\frac{1}{3}$ (multiplicity 2), $5 + \sqrt{7}, 5 - \sqrt{7}$

17. $(2 - 3i), (2 + 3i), -4$ (multiplicity 2)

18. $i\sqrt{3}$ (multiplicity 2), $-i\sqrt{3}$ (multiplicity 2), and 4 (multiplicity 3)

In Problems 19–24, find a polynomial of lowest degree, with leading coefficient 1, that has the indicated graph. Assume all

zeros are integers. Write the polynomial as a product of linear factors. Indicate the degree of the polynomial.

19.

20.

21.

22.

23.

24.

In Problems 25–28, factor each polynomial in two ways:
(A) as a product of linear factors (with real coefficients) and

quadratic factors (with real coefficients and imaginary zeros); and (B) as a product of linear factors with complex coefficients.

25. $P(x) = x^4 + 5x^2 + 4$

26. $P(x) = x^4 + 18x^2 + 81$

27. $P(x) = x^3 - x^2 + 25x - 25$

28. $P(x) = x^5 + x^4 - x - 1$

In Problems 29–34, list all possible rational zeros (Theorem 7) of a polynomial with integer coefficients that has the given leading coefficient a_n and constant term a_0.

29. $a_n = 1, a_0 = -4$ **30.** $a_n = 1, a_0 = 9$

31. $a_n = 10, a_0 = 1$ **32.** $a_n = 6, a_0 = -1$

33. $a_n = 7, a_0 = -2$ **34.** $a_n = 3, a_0 = 8$

 When searching for zeros of a polynomial, a graphing calculator often can be helpful in eliminating from consideration certain candidates for rational zeros.

In Problems 35–40, write $P(x)$ as a product of linear factors.

35. $P(x) = x^3 + 9x^2 + 24x + 16$; -1 is a zero

36. $P(x) = x^3 - 4x^2 - 3x + 18$; 3 is a double zero

37. $P(x) = x^4 + 2x^2 + 1$; i is a double zero

38. $P(x) = x^4 - 1$; 1 and -1 are zeros

39. $P(x) = 2x^3 - 17x^2 + 90x - 41$; $\frac{1}{2}$ is a zero

40. $P(x) = 3x^3 - 10x^2 + 31x + 26$; $-\frac{2}{3}$ is a zero

In Problems 41–48, find all roots exactly (rational, irrational, and imaginary) for each polynomial equation.

41. $2x^3 - 5x^2 + 1 = 0$ **42.** $2x^3 - 10x^2 + 12x - 4 = 0$

43. $x^4 + 4x^3 - x^2 - 20x - 20 = 0$

44. $x^4 - 4x^2 - 4x - 1 = 0$

45. $x^4 - 2x^3 - 5x^2 + 8x + 4 = 0$

46. $x^4 - 2x^2 - 16x - 15 = 0$

47. $x^4 + 10x^2 + 9 = 0$ **48.** $x^4 + 29x^2 + 100 = 0$

In Problems 49–54, find all zeros exactly (rational, irrational, and imaginary) for each polynomial.

49. $P(x) = x^3 - 19x + 30$ **50.** $P(x) = x^3 - 7x^2 + 36$

51. $P(x) = x^4 - \frac{21}{10}x^3 + \frac{2}{5}x$ **52.** $P(x) = x^4 + \frac{7}{6}x^3 - \frac{7}{3}x^2 - \frac{5}{2}x$

53. $P(x) = x^4 - 5x^3 + \frac{15}{2}x^2 - 2x - 2$

54. $P(x) = x^4 - \frac{13}{4}x^2 - \frac{5}{2}x - \frac{1}{4}$

In Problems 55–60, write each polynomial as a product of linear factors.

55. $P(x) = 6x^3 + 13x^2 - 4$ **56.** $P(x) = 6x^3 - 17x^2 - 4x + 3$

57. $P(x) = x^3 + 2x^2 - 9x - 4$

58. $P(x) = x^3 - 8x^2 + 17x - 4$

59. $P(x) = 4x^4 - 4x^3 - 9x^2 + x + 2$

60. $P(x) = 2x^4 + 3x^3 - 4x^2 - 3x + 2$

In Problems 61–68, find a polynomial $P(x)$ that satisfies all of the given conditions. Write the polynomial using only real coefficients.

61. $2 - 5i$ is a zero; leading coefficient 1; degree 2

62. $4 + 3i$ is a zero; leading coefficient 1; degree 2

63. $6 + i$ is a zero; $P(0) = 74$; degree 2

64. $1 - 4i$ is a zero; $P(0) = 51$; degree 2

65. -5 and $8i$ are zeros; leading coefficient 1; degree 3

66. 7 and $-2i$ are zeros; leading coefficient 1; degree 3

67. i and $1 - i$ are zeros; $P(1) = 10$; degree 4

68. $-i$ and $3 + i$ are zeros; $P(1) = 20$; degree 4

In Problems 69–74, multiply.

69. $[x - (4 - 5i)][x - (4 + 5i)]$

70. $[x - (2 - 3i)][x - (2 + 3i)]$

71. $[x - (3 + 4i)][x - (3 - 4i)]$

72. $[x - (5 + 2i)][x - (5 - 2i)]$

73. $[x - (a + bi)][x - (a - bi)]$

74. $(x - bi)(x + bi)$

In Problems 75–80, find all other zeros of $P(x)$, given the indicated zero.

75. $P(x) = x^3 - 5x^2 + 4x + 10$; $3 - i$ is one zero

76. $P(x) = x^3 + x^2 - 4x + 6$; $1 + i$ is one zero

77. $P(x) = x^3 - 3x^2 + 25x - 75$; $-5i$ is one zero

78. $P(x) = x^3 + 2x^2 + 16x + 32$; $4i$ is one zero

79. $P(x) = x^4 - 4x^3 + 3x^2 + 8x - 10$; $2 + i$ is one zero

80. $P(x) = x^4 - 2x^3 + 7x^2 - 18x - 18$; $-3i$ is one zero

In Problems 81–86, final all zeros (rational, irrational, and imaginary) exactly.

81. $P(x) = 3x^3 - 37x^2 + 84x - 24$

82. $P(x) = 2x^3 - 9x^2 - 2x + 30$

83. $P(x) = 4x^4 + 4x^3 + 49x^2 + 64x - 240$

84. $P(x) = 6x^4 + 35x^3 + 2x^2 - 233x - 360$

85. $P(x) = 4x^4 - 44x^3 + 145x^2 - 192x + 90$

86. $P(x) = x^5 - 6x^4 + 6x^3 + 28x^2 - 72x + 48$

87. The solutions to the equation $x^3 - 1 = 0$ are all the cube roots of 1.
 (A) 1 is obviously a cube root of 1; find all others.
 (B) How many distinct cube roots of 1 are there?

88. The solutions to the equation $x^3 - 8 = 0$ are all the cube roots of 8.
 (A) 2 is obviously a cube root of 8; find all others.
 (B) How many distinct cube roots of 8 are there?

89. Give a reason for each step in the proof of the rational zero theorem, assuming that $P(x)$ has degree two.
 Step 1: $a_2 \left(\frac{b}{c}\right)^2 + a_1\left(\frac{b}{c}\right) + a_0 = 0$
 Step 2: $a_2 b^2 + a_1 bc + a_0 c^2 = 0$
 Step 3: $a_2 b^2 + a_1 bc = -a_0 c^2$
 Step 4: b is a factor of $-a_0 c^2$, so b is a factor of a_0.
 Step 5: Modify steps 3 and 4 to conclude that c is a factor of a_2.

90. Explain how the ideas in Problem 89 can be adapted to give a proof of the rational zero theorem for $P(x)$ of degree n.

91. Given $P(x) = x^2 + 2ix - 5$ with $2 - i$ a zero, show that $2 + i$ is not a zero of $P(x)$. Does this contradict Theorem 3? Explain.

92. If $P(x)$ and $Q(x)$ are two polynomials of degree n, and if $P(x) = Q(x)$ for more than n values of x, then how are $P(x)$ and $Q(x)$ related? [*Hint:* Consider the polynomial $D(x) = P(x) - Q(x)$.]

APPLICATIONS

Find all rational solutions exactly, and find irrational solutions to one decimal place.

93. **STORAGE** A rectangular storage unit has dimensions 1 by 2 by 3 feet. If each dimension is increased by the same amount, how much should this amount be to create a new storage unit with volume 10 times the old?

94. **CONSTRUCTION** A rectangular box has dimensions 1 by 1 by 2 feet. If each dimension is increased by the same amount, how much should this amount be to create a new box with volume six times the old?

95. **PACKAGING** An open box is to be made from a rectangular piece of cardboard that measures 8 by 5 inches, by cutting out squares of the same size from each corner and bending up the sides (see the figure). If the volume of the box is to be 14 cubic inches, how large a square should be cut from each corner? [*Hint:* Determine the domain of x from physical considerations before starting.]

96. **FABRICATION** An open metal chemical tank is to be made from a rectangular piece of stainless steel that measures 10 by 8 feet, by cutting out squares of the same size from each corner and bending up the sides (see the figure for Problem 95). If the volume of the tank is to be 48 cubic feet, how large a square should be cut from each corner?

4-4 Rational Functions and Inequalities

> Rational Functions and Properties of Their Graphs
> Vertical and Horizontal Asymptotes
> Analyzing the Graph of a Rational Function
> Rational Inequalities

In Section 4-4, we will apply our knowledge of graphs and zeros of polynomial functions to study the graphs of *rational functions*, that is, functions that are quotients of polynomials. Our goal will be to produce hand sketches that clearly show all of the important features of the graph.

> Rational Functions and Properties of Their Graphs

The number $\frac{7}{13}$ is called a *rational number* because it is a quotient (or ratio) of integers. The function

$$f(x) = \frac{x + 1}{x^2 - x - 6}$$

is called a *rational function* because it is a quotient of polynomials.

(a) $f(x) = \dfrac{(x-1)(x^2-3)}{x-1}$

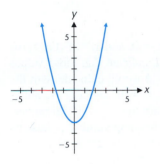

(b) $f(x) = x^2 - 3$

> Figure 1

> **DEFINITION 1 Rational Function**

A function f is a **rational function** if it can be written in the form

$$f(x) = \frac{p(x)}{q(x)}$$

where $p(x)$ and $q(x)$ are polynomials.

When working with rational functions, we will assume that the coefficients of $p(x)$ and $q(x)$ are real numbers, and that the domain of f is the set of all real numbers x such that $q(x) \neq 0$.

If a real number c is a zero of both $p(x)$ and $q(x)$, then, by the factor theorem, $x - c$ is a factor of both $p(x)$ and $q(x)$. The graphs of

$$f(x) = \frac{p(x)}{q(x)} = \frac{(x-c)p_r(x)}{(x-c)q_r(x)} \qquad \text{and} \qquad f_r(x) = \frac{p_r(x)}{q_r(x)}$$

are then identical, except possibly for a "hole" at $x = c$ (Fig. 1).

Later in this section we will explain how to handle the minor complication caused by common real zeros of $p(x)$ and $q(x)$. But to avoid that complication now,

unless stated to the contrary, we will assume that for any rational function f we consider, $p(x)$ and $q(x)$ have no real zero in common.

Because a polynomial $q(x)$ of degree n has at most n real zeros, there are at most n real numbers that are not in the domain of $f(x) = P(x)/q(x)$. Because a fraction equals 0 only if its numerator is 0, the x intercepts of the graph of f are the real zeros of a polynomial $p(x)$ of degree m. So the number of x intercepts of f is at most m.

EXAMPLE 1 Domain and x Intercepts

Find the domain and x intercepts for $f(x) = \dfrac{2x^2 - 2x - 4}{x^2 - 9}$.

SOLUTION

$$f(x) = \frac{p(x)}{q(x)} = \frac{2x^2 - 2x - 4}{x^2 - 9} = \frac{2(x-2)(x+1)}{(x-3)(x+3)}$$

Because $q(x) = 0$ for $x = 3$ and $x = -3$, the domain of f is

$$x \neq \pm 3 \qquad \text{or} \qquad (-\infty, -3) \cup (-3, 3) \cup (3, \infty)$$

Because $p(x) = 0$ for $x = 2$ and $x = -1$, the zeros of f, and the x intercepts of f, are -1 and 2.

MATCHED PROBLEM 1

Find the domain and x intercepts for $f(x) = \dfrac{3x^2 - 12}{x^2 + 2x - 3}$.

The graph of the rational function

$$f(x) = \frac{x^2 - 1.44}{x^3 - x}$$

is shown in Figure 2 on the next page.

› **Figure 2** $f(x) = \dfrac{x^2 - 1.44}{x^3 - x}$.

The domain of f consists of all real numbers except $x = -1$, $x = 0$, and $x = 1$ (the zeros of the denominator $x^3 - x$). The dotted vertical lines at $x = \pm 1$ indicate that those values of x are excluded from the domain (a dotted vertical line at $x = 0$ would coincide with the y axis and is omitted). The graph is discontinuous at $x = -1$, $x = 0$, and $x = 1$, but is continuous elsewhere and has no sharp corners. The zeros of f are the zeros of the numerator $x^2 - 1.44$, namely $x = -1.2$ and $x = 1.2$. The graph of f has four turning points. Its left and right behavior is the same as that of the function $g(x) = \frac{1}{x}$ (the graph is close to the x axis for very large and very small values of x). The graph of f illustrates the general properties of rational functions that are listed in Theorem 1. We have already justified Property 3; the other properties are established in calculus.

› **THEOREM 1** **Properties of Rational Functions**

Let $f(x) = p(x)/q(x)$ be a rational function where $p(x)$ and $q(x)$ are polynomials of degrees m and n, respectively. Then the graph of $f(x)$:

1. Is continuous with the exception of at most n real numbers
2. Has no sharp corners
3. Has at most m real zeros
4. Has at most $m + n - 1$ turning points
5. Has the same left and right behavior as the quotient of the leading terms of $p(x)$ and $q(x)$

Figure 3 shows graphs of several rational functions, illustrating the properties of Theorem 1.

› **Figure 3** Graphs of rational functions.

(a) $f(x) = \dfrac{1}{x}$ (b) $g(x) = \dfrac{1}{x^2 - 1}$ (c) $h(x) = \dfrac{1}{x^2 + 1}$

> **Figure 3** (continued)

(d) $F(x) = \dfrac{x^2 + 3x}{x - 1}$

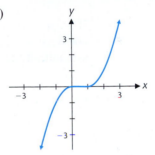

(e) $G(x) = \dfrac{-x - 1}{x^3 - 4x}$

(f) $H(x) = \dfrac{x^2 + x + 1}{x^2 + 1}$

EXAMPLE 2 Properties of Graphs of Rational Functions

Use Theorem 1 to explain why each graph is not the graph of a rational function.

(A)

(B)

(C)

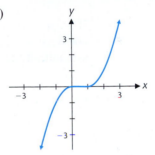

SOLUTIONS (A) The graph has a sharp corner when $x = 0$, so Property 2 is not satisfied.

(B) The graph has an infinite number of turning points, so Property 4 is not satisfied.

(C) The graph has an infinite number of zeros (all values of x between 0 and 1, inclusive, are zeros), so Property 3 is not satisfied.

MATCHED PROBLEM 2 Use Theorem 1 to explain why each graph is not the graph of a rational function.

(A)

(B)

(C)

(a) $f(x) = \dfrac{1}{x}$

> **Figure 3(a)** Graphs of rational functions.

› Vertical and Horizontal Asymptotes

The graphs of Figure 3 on pages 300–301 exhibit similar behaviors near points of discontinuity that can be described using the concept of *vertical asymptote*. Consider, for example, the rational function $f(x) = \frac{1}{x}$ of Figure 3(a). As x approaches 0 from the right, the points $(x, \frac{1}{x})$ on the graph have larger and larger y coordinates—that is, $\frac{1}{x}$ increases without bound—as confirmed by Table 1. We write this symbolically as

$$\frac{1}{x} \to \infty \qquad \text{as} \qquad x \to 0^+$$

and say that the line $x = 0$ (the y axis) is a *vertical asymptote* for the graph of f.

Table 1 Behavior of $1/x$ as $x \to 0^+$

x	1	0.1	0.01	0.001	0.0001	0.000 01	0.000 001	. . .	x approaches 0 from the right $(x \to 0^+)$
$1/x$	1	10	100	1,000	10,000	100,000	1,000,000	. . .	$1/x$ increases without bound $(1/x \to \infty)$

If x approaches 0 from the left, the points $(x, \frac{1}{x})$ on the graph have smaller and smaller y coordinates—that is, $\frac{1}{x}$ decreases without bound—as confirmed by Table 2. We write this symbolically as

$$\frac{1}{x} \to -\infty \qquad \text{as} \qquad x \to 0^-$$

Table 2 Behavior of $1/x$ as $x \to 0^-$

x	-1	-0.1	-0.01	-0.001	-0.0001	$-0.000\ 01$	$-0.000\ 001$	. . .	x approaches 0 from the left $(x \to 0^-)$
$1/x$	-1	-10	-100	$-1,000$	$-10,000$	$-100,000$	$-1,000,000$	. . .	$1/x$ decreases without bound $(1/x \to -\infty)$

>>> **EXPLORE-DISCUSS 1**

Construct tables similar to Tables 1 and 2 for $g(x) = \frac{1}{x^2}$ and discuss the behavior of the graph of $g(x)$ near $x = 0$.

› DEFINITION 2 Vertical Asymptote

The vertical line $x = a$ is a **vertical asymptote** for the graph of $y = f(x)$ if

$$f(x) \to \infty \quad \text{or} \quad f(x) \to -\infty \quad \text{as} \quad x \to a^+ \quad \text{or as} \quad x \to a^-$$

(that is, if $f(x)$ either increases or decreases without bound as x approaches a from the right or from the left).

› THEOREM 2 Vertical Asymptotes of Rational Functions

Let $f(x) = p(x)/q(x)$ be a rational function. If a is a zero of $q(x)$, then the line $x = a$ is a vertical asymptote of the graph of f.*

*Recall that we are assuming that $p(x)$ and $q(x)$ have no real zero in common. Theorem 2 is not valid without this assumption.

For example,

$$f(x) = \frac{x^2 - 1.44}{x^3 - x} = \frac{x^2 - 1.44}{x(x - 1)(x + 1)}$$

has three vertical asymptotes, $x = -1$, $x = 0$, and $x = 1$ (see Fig. 2 on p. 300).

The left and right behavior of some, but not all, rational functions can be described using the concept of *horizontal asymptote*. Consider $f(x) = \frac{1}{x}$. As values of x get larger and larger—that is, as x increases without bound—the points $(x, \frac{1}{x})$ have y coordinates that are positive and approach 0, as confirmed by Table 3. Similarly, as values of x get smaller and smaller—that is, as x decreases without bound—the points $(x, \frac{1}{x})$ have y coordinates that are negative and approach 0, as confirmed by Table 4. We write these facts symbolically as

$$\frac{1}{x} \to 0 \quad \text{as} \quad x \to \infty \quad \text{and as} \quad x \to -\infty$$

and say that the line $y = 0$ (the x axis) is a *horizontal asymptote* for the graph of f.

Table 3 Behavior of $1/x$ as $x \to \infty$

x	1	10	100	1,000	10,000	100,000	1,000,000	. . .	x increases without bound ($x \to \infty$)
$1/x$	1	0.1	0.01	0.001	0.0001	0.000 01	0.000 001	. . .	$1/x$ approaches 0 ($1/x \to 0$)

Table 4 Behavior of $1/x$ as $x \to -\infty$

x	-1	-10	-100	$-1,000$	$-10,000$	$-100,000$	$-1,000,000$	. . .	x decreases without bound ($x \to -\infty$)
$1/x$	-1	-0.1	-0.01	-0.001	-0.0001	$-0.000\ 01$	$-0.000\ 001$	. . .	$1/x$ approaches 0 ($1/x \to 0$)

⟩⟩⟩ EXPLORE-DISCUSS 2

Construct tables similar to Tables 3 and 4 for each of the following functions, and discuss the behavior of each as $x \to \infty$ and as $x \to -\infty$:

(A) $f(x) = \dfrac{3x}{x^2 + 1}$ **(B)** $g(x) = \dfrac{3x^2}{x^2 + 1}$ **(C)** $h(x) = \dfrac{3x^3}{x^2 + 1}$

⟩ DEFINITION 3 Horizontal Asymptote

The horizontal line $y = b$ is a **horizontal asymptote** for the graph of $y = f(x)$ if

$$f(x) \to b \quad \text{as} \quad x \to -\infty \quad \text{or as} \quad x \to \infty$$

(that is, if $f(x)$ approaches b as x increases without bound or as x decreases without bound).

A rational function $f(x) = p(x)/q(x)$ has the same left and right behavior as the quotient of the leading terms of $p(x)$ and $q(x)$ (Property 5 of Theorem 1). Consequently, *a rational function has at most one horizontal asymptote.* Moreover, we can determine easily whether a rational function has a horizontal asymptote, and if it does, find its equation. Theorem 3 gives the details.

> **THEOREM 3** Horizontal Asymptotes of Rational Functions

Consider the rational function

$$f(x) = \frac{a_m x^m + \ldots + a_1 x + a_0}{b_n x^n + \ldots + b_1 x + b_0}$$

where $a_m \neq 0$, $b_n \neq 0$.

1. If $m < n$, the line $y = 0$ (the x axis) is a horizontal asymptote.
2. If $m = n$, the line $y = a_m / b_n$ is a horizontal asymptote.
3. If $m > n$, there is no horizontal asymptote.

In 1 and 2, the graph of f approaches the horizontal asymptote both as $x \rightarrow \infty$ and as $x \rightarrow -\infty$.

EXAMPLE 3

Finding Vertical and Horizontal Asymptotes for a Rational Function

Find all vertical and horizontal asymptotes for

$$f(x) = \frac{p(x)}{q(x)} = \frac{2x^2 - 2x - 4}{x^2 - 9}$$

SOLUTION Because $q(x) = x^2 - 9 = (x - 3)(x + 3)$, the graph of $f(x)$ has vertical asymptotes at $x = 3$ and $x = -3$ (Theorem 1). Because $p(x)$ and $q(x)$ have the same degree, the line

$$y \boxed{= \frac{a_2}{b_2}}^* = \frac{2}{1} = 2 \qquad a_2 = 2, b_2 = 1$$

is a horizontal asymptote (Theorem 3, part 2).

MATCHED PROBLEM 3 Find all vertical and horizontal asymptotes for

$$f(x) = \frac{3x^2 - 12}{x^2 + 2x - 3}$$

> ## Analyzing the Graph of a Rational Function

We now use the techniques for locating asymptotes, along with other graphing aids discussed in the text, to graph several rational functions. First, we outline a systematic approach to the problem of graphing rational functions.

*Throughout the book, dashed boxes—called think boxes—are used to represent steps that may be performed mentally.

> **ANALYZING AND SKETCHING THE GRAPH OF A RATIONAL FUNCTION:**
> $f(x) = p(x)/q(x)$

Step 1. *Intercepts.* Find the real solutions of the equation $p(x) = 0$ and use these solutions to plot any x intercepts of the graph of f. Evaluate $f(0)$, if it exists, and plot the y intercept.

Step 2. *Vertical Asymptotes.* Find the real solutions of the equation $q(x) = 0$ and use these solutions to determine the domain of f, the points of discontinuity, and the vertical asymptotes. Sketch any vertical asymptotes as dashed lines.

Step 3. *Horizontal Asymptotes.* Determine whether there is a horizontal asymptote and, if so, sketch it as a dashed line.

Step 4. *Complete the Sketch.* For each interval in the domain of f, plot additional points and join them with a smooth continuous curve.

| EXAMPLE | 4 | **Graphing a Rational Function** |

Graph $f(x) = \dfrac{2x}{x - 3}$.

SOLUTION

$$f(x) = \frac{2x}{x - 3} = \frac{p(x)}{q(x)}$$

Step 1. *Intercepts.* Find real zeros of $p(x) = 2x$ and find $f(0)$:

$$2x = 0$$

$$x = 0 \qquad \text{x intercept}$$

$$f(0) = 0 \qquad \text{y intercept}$$

The graph crosses the coordinate axes only at the origin. Plot this intercept, as shown in Figure 4.

> **Figure 4**

Step 2. *Vertical Asymptotes.* Find real zeros of $q(x) = x - 3$:

$$x - 3 = 0$$

$$x = 3$$

The domain of f is $(-\infty, 3) \cup (3, \infty)$, f is discontinuous at $x = 3$, and the graph has a vertical asymptote at $x = 3$. Sketch this asymptote, as shown in Figure 4.

Step 3. *Horizontal Asymptote.* Because $p(x)$ and $q(x)$ have the same degree, the line $y = 2$ is a horizontal asymptote, as shown in Figure 4.

Step 4. *Complete the Sketch.* By plotting a few additional points, we obtain the graph in Figure 5. Notice that the graph is a smooth continuous curve over the interval

$$f(x) = \frac{2x}{x - 3}$$

> **Figure 5**

$(-\infty, 3)$ and over the interval $(3, \infty)$. As expected, there is a break in the graph at $x = 3$.

MATCHED PROBLEM 4 Proceed as in Example 4 and sketch the graph of $f(x) = \dfrac{3x}{x + 2}$.

Technology Connections

Refer to Example 4. When $f(x) = 2x/(x - 3)$ is graphed on a graphing calculator [Fig. 6(a)], it appears that the graphing calculator has also drawn the vertical asymptote at $x = 3$, but this is not the case. Many graphing calculators, when set in *connected mode*, calculate points on a graph and connect these points with line segments. The last point plotted to the left of the asymptote and the first plotted to the right of the asymptote will usually have very large y coordinates. If these y coordinates have opposite signs, then the graphing calculator may connect the two points with a nearly vertical line segment, which gives the appearance of an asymptote. If you wish, you can set the calculator in *dot mode* to plot the points without the connecting line segments [Fig. 6(b)].

Depending on the scale, a graph may even appear to be continuous at a vertical asymptote [Fig. 6(c)]. It is important to always locate the vertical asymptotes as we did in step 2 before turning to the graphing calculator graph to complete the sketch.

(a) Connected mode

(b) Dot mode

(c) Connected mode

> **Figure 6** Graphing calculator graphs of $f(x) = \dfrac{2x}{x - 3}$.

In Examples 5 and 6 we will just list the results of each step in the graphing strategy and omit the computational details.

EXAMPLE 5 **Graphing a Rational Function**

Graph $f(x) = \dfrac{x^2 - 6x + 9}{x^2 + x - 2}$.

SOLUTION
$$f(x) = \frac{x^2 - 6x + 9}{x^2 + x - 2} = \frac{(x - 3)^2}{(x + 2)(x - 1)}$$

x intercept: $x = 3$

y intercept: $f(0) = -\frac{9}{2} = -4.5$

Domain: $(-\infty, -2) \cup (-2, 1) \cup (1, \infty)$

Points of discontinuity: $x = -2$ and $x = 1$

Vertical asymptotes: $x = -2$ and $x = 1$

Horizontal asymptote: $y = 1$

Locate the intercepts, draw the asymptotes, and plot additional points in each interval of the domain of f to complete the graph (Fig. 7).

> **Figure 7**

MATCHED PROBLEM 5 Graph $f(x) = \dfrac{x^2}{x^2 - 7x + 10}$.

>>> **CAUTION** >>>

The graph of a function cannot cross a vertical asymptote, but the same statement is not true for horizontal asymptotes. The rational function

$$f(x) = \frac{2x^6 + x^5 - 5x^3 + 4x + 2}{x^6 + 1}$$

has the line $y = 2$ as a horizontal asymptote. The graph of f in Figure 8 clearly shows that *the graph of a function can cross a horizontal asymptote.* The definition of a horizontal asymptote requires $f(x)$ to approach b as x increases or decreases without bound, but it does not preclude the possibility that $f(x) = b$ for one or more values of x.

> **Figure 8** Multiple intersections of a graph and a horizontal asymptote.

EXAMPLE **6** **Graphing a Rational Function**

Graph $f(x) = \dfrac{x^2 - 3x - 4}{x - 2}$.

SOLUTION
$$f(x) = \frac{x^2 - 3x - 4}{x - 2} = \frac{(x + 1)(x - 4)}{x - 2}$$

x intercepts: $x = -1$ and $x = 4$

y intercept: $f(0) = 2$

Domain: $(-\infty, 2) \cup (2, \infty)$

Points of discontinuity: $x = 2$

Vertical asymptote: $x = 2$

No horizontal asymptote

Although the graph of f does not have a horizontal asymptote, we can still gain some useful information about the behavior of the graph as $x \to -\infty$ and as $x \to \infty$ if we first perform a long division:

$$
\begin{array}{r}
x - 1 \qquad\quad \text{Quotient}\\
x - 2\overline{)x^2 - 3x - 4}\\
\underline{x^2 - 2x}\qquad\\
-x - 4\quad\\
\underline{-x + 2}\quad\\
-6 \qquad \text{Remainder}
\end{array}
$$

This shows that

$$
f(x) = \frac{x^2 - 3x - 4}{x - 2} = x - 1 - \frac{6}{x - 2}
$$

As $x \to -\infty$ or $x \to \infty$, $6/(x - 2) \to 0$ and the graph of f approaches the line $y = x - 1$. This line is called an **oblique asymptote** for the graph of f. The asymptotes and intercepts are shown in Figure 9, and the graph of f is sketched in Figure 10.

Oblique asymptote $y = x - 1$

Intercepts and asymptotes

> Figure 9

$y = x - 1$

$f(x) = \dfrac{x^2 - 3x - 4}{x - 2}$

> Figure 10

Generalizing the results of Example 6, we have Theorem 4.

> **THEOREM 4 Oblique Asymptotes and Rational Functions**

If $f(x) = p(x)/q(x)$, where $p(x)$ and $q(x)$ are polynomials and the degree of $p(x)$ is 1 more than the degree of $q(x)$, then $f(x)$ can be expressed in the form

$$
f(x) = mx + b + \frac{r(x)}{q(x)}
$$

where the degree of $r(x)$ is less than the degree of $q(x)$. The line

$$
y = mx + b
$$

is an oblique asymptote for the graph of f. That is,

$$
[f(x) - (mx + b)] \to 0 \qquad \text{as} \qquad x \to -\infty \qquad \text{or} \qquad x \to \infty
$$

MATCHED PROBLEM 6 Graph, including any oblique asymptotes, $f(x) = \dfrac{x^2 + 5}{x + 1}$.

At the beginning of this section we made the assumption that for a rational function $f(x) = p(x)/q(x)$, the polynomials $p(x)$ and $q(x)$ have no common real zero. *Now we abandon that assumption.* Suppose that $p(x)$ and $q(x)$ have one or more real zeros in common. Then, by the factor theorem, $p(x)$ and $q(x)$ have one or more linear factors in common, so $f(x)$ can be "reduced." We proceed to divide out common linear factors in

$$f(x) = \frac{p(x)}{q(x)}$$

until we obtain a rational function

$$f_r(x) = \frac{p_r(x)}{q_r(x)}$$

in which $p_r(x)$ and $q_r(x)$ have no common real zero. We analyze and graph $f_r(x)$, then insert "holes" as required in the graph of f_r to obtain the graph of f. Example 7 illustrates the details.

EXAMPLE	7

Graphing Arbitrary Rational Functions

Graph $f(x) = \dfrac{2x^5 - 4x^4 - 6x^3}{x^5 - 3x^4 - 3x^3 + 7x^2 + 6x}$.

SOLUTION The real zeros of

$$p(x) = 2x^5 - 4x^4 - 6x^3$$

(obtained by graphing or factoring) are -1, 0, and 3.
 The real zeros of

$$q(x) = x^5 - 3x^4 - 3x^3 + 7x^2 + 6x$$

are -1, 0, 2, and 3. The common zeros are -1, 0, and 3. Factoring and dividing out common linear factors gives

$$f(x) = \frac{2x^3(x + 1)(x - 3)}{x(x + 1)^2(x - 2)(x - 3)} \qquad \text{and} \qquad f_r(x) = \frac{2x^2}{(x + 1)(x - 2)}$$

We analyze $f_r(x)$ as usual:
 x intercept: $x = 0$
 y intercept: $f_r(0) = 0$
 Domain: $(-\infty, -1) \cup (-1, 2) \cup (2, \infty)$
 Points of discontinuity: $x = -1$, $x = 2$
 Vertical asymptotes: $x = -1$, $x = 2$
 Horizontal asymptote: $y = 2$

The graph of f is identical to the graph of f_r except possibly at the common real zeros -1, 0, and 3. We consider each common zero separately.

 $x = -1$: Both f and f_r are undefined (no difference in their graphs).

 $x = 0$: f is undefined but $f_r(0) = 0$, so the graph of f has a hole at $(0, 0)$.

 $x = 3$: f is undefined but $f_r(3) = 4.5$, so the graph of f has a hole at $(3, 4.5)$.

Therefore, $f(x)$ has the following analysis:
 x intercepts: none
 y intercepts: none
 Domain: $(-\infty, -1) \cup (-1, 0) \cup (0, 2) \cup (2, 3) \cup (3, \infty)$
 Points of discontinuity: $x = -1$, $x = 0$, $x = 2$, $x = 3$

Vertical asymptotes: $x = -1$, $x = 2$
Horizontal asymptote: $y = 2$
Holes: $(0, 0)$, $(3, 4.5)$

Figure 11 shows the graphs of f and f_r.

(a) $f(x) = \dfrac{2x^5 - 4x^4 - 6x^3}{x^5 - 3x^4 - 3x^3 + 7x^2 + 6x}$

(b) $f_r(x) = \dfrac{2x^2}{(x + 1)(x - 2)}$

› **Figure 11**

MATCHED PROBLEM 7 Graph $f(x) = \dfrac{x^3 - x}{x^4 - x^2}$.

› Rational Inequalities

A rational function $f(x) = p(x)/q(x)$ can change sign at a real zero of $p(x)$ (where f has an x intercept) or at a real zero of $q(x)$ (where f is discontinuous), but nowhere else (because f is continuous except where it is not defined). Rational inequalities can therefore be solved in the same way as polynomial inequalities, except that the partition of the x axis is determined by the zeros of $p(x)$ and the zeros of $q(x)$.

EXAMPLE 8 **Solving Rational Inequalities**

Solve $\dfrac{x^3 + 4x^2}{x^2 - 4} < 0$.

SOLUTION Let

$$f(x) = \frac{p(x)}{q(x)} = \frac{x^3 + 4x^2}{x^2 - 4}$$

The zeros of

$$p(x) = x^3 + 4x^2 = x^2(x + 4)$$

are 0 and -4. The zeros of

$$q(x) = x^2 - 4 = (x + 2)(x - 2)$$

are -2 and 2. These four zeros partition the x axis into the five intervals shown in the table. A test number is chosen from each interval as indicated to determine whether $f(x)$ is positive or negative.

Interval	Test number x	$f(x)$	Sign of f
$(-\infty, -4)$	-5	$-25/21$	$-$
$(-4, -2)$	-3	$9/5$	$+$
$(-2, 0)$	-1	-1	$-$
$(0, 2)$	1	$-5/3$	$-$
$(2, \infty)$	3	$63/5$	$+$

We conclude that the solution set of the inequality is

$$(-\infty, -4) \cup (-2, 0) \cup (0, 2)$$

MATCHED PROBLEM 8 Solve $\dfrac{x^2 - 1}{x^2 - 9} \geq 0$.

EXAMPLE 9 Solving Rational Inequalities with a Graphing Calculator

Solve $1 \geq \dfrac{9x - 9}{x^2 + x - 3}$ to three decimal places.

SOLUTION First we convert the inequality to an equivalent inequality in which one side is 0:

$$1 \geq \frac{9x - 9}{x^2 + x - 3} \qquad \text{Subtract } \frac{9x-9}{x^2+x-3} \text{ from both sides.}$$

$$1 - \frac{9x - 9}{x^2 + x - 3} \geq 0 \qquad \text{Find a common denominator.}$$

$$\frac{x^2 + x - 3}{x^2 + x - 3} - \frac{9x - 9}{x^2 + x - 3} \geq 0 \qquad \text{Simplify.}$$

$$\frac{x^2 - 8x + 6}{x^2 + x - 3} \geq 0$$

(a) $f(x) = \dfrac{x^2 - 8x + 6}{x^2 + x - 3}$

(b) $g(x) = \dfrac{f(x)}{|f(x)|}$

▸ Figure 12

The zeros of $x^2 - 8x + 6$, to three decimal places, are 0.838 and 7.162. The zeros of $x^2 + x - 3$ are -2.303 and 1.303. These four zeros partition the x axis into five intervals:

$$(-\infty, -2.303), \ (-2.303, 0.838), \ (0.838, 1.303), \ (1.303, 7.162), \text{ and } (7.162, \infty)$$

We graph

$$f(x) = \frac{x^2 - 8x + 6}{x^2 + x - 3} \qquad \text{and} \qquad g(x) = \frac{f(x)}{|f(x)|}$$

(Fig. 12) and observe that the graph of f is above the x axis on the intervals $(-\infty, -2.303)$, $(0.838, 1.303)$, and $(7.162, \infty)$. So the solution set of the inequality is

$$(-\infty, -2.303) \cup [0.838, 1.303) \cup [7.162, \infty)$$

Note that the endpoints that are zeros of f are included in the solution set of the inequality, but not the endpoints at which f is undefined.

MATCHED PROBLEM 9 Solve $\dfrac{x^3 + 4x^2 - 7}{x^2 - 5x - 1} \geq 0$ to three decimal places.

ANSWERS TO MATCHED PROBLEMS

1. Domain: $(-\infty, -3) \cup (-3, 1) \cup (1, \infty)$; x intercepts: $x = -2$, $x = 2$

2. (A) Properties 3 and 4 are not satisfied.
 (B) Property 1 is not satisfied.
 (C) Properties 1 and 3 are not satisfied.

3. Vertical asymptotes: $x = -3$, $x = 1$; horizontal asymptote: $y = 3$

4.

$$f(x) = \frac{3x}{x + 2}$$

5.

$$f(x) = \frac{x^2}{x^2 - 7x + 10}$$

6.

$$f(x) = \frac{x^2 + 5}{x + 1}$$

7.

$$f(x) = \frac{x^3 - x}{x^4 - x^2}$$

8. $(-\infty, -3) \cup [-1, 1] \cup (3, \infty)$

9. $[-3.391, -1.773] \cup (-0.193, 1.164] \cup (5.193, \infty)$

4-4 Exercises

1. Is every polynomial function a rational function? Explain.

2. Is every rational function a polynomial function? Explain.

3. Explain in your own words what a vertical asymptote is.

4. Explain in your own words what a horizontal asymptote is.

5. Explain in your own words what an oblique asymptote is.

6. Explain why a rational function can't have both a horizontal asymptote and an oblique asymptote.

In Problems 7–10, match each graph with one of the following functions:

$$f(x) = \frac{2x - 4}{x + 2} \qquad g(x) = \frac{2x + 4}{2 - x}$$

$$h(x) = \frac{2x + 4}{x - 2} \qquad k(x) = \frac{4 - 2x}{x + 2}$$

7.

8.

9.

10.

11. Let $f(x) = \dfrac{2x - 4}{x + 2}$. Complete each statement:

(A) As $x \to -2^-$, $f(x) \to$?
(B) As $x \to -2^+$, $f(x) \to$?
(C) As $x \to -\infty$, $f(x) \to$?
(D) As $x \to \infty$, $f(x) \to$?

12. Let $g(x) = \dfrac{2x + 4}{2 - x}$. Complete each statement:

(A) As $x \to 2^-$, $g(x) \to$?
(B) As $x \to 2^+$, $g(x) \to$?
(C) As $x \to -\infty$, $g(x) \to$?
(D) As $x \to \infty$, $g(x) \to$?

13. Let $h(x) = \dfrac{2x + 4}{x - 2}$. Complete each statement:

(A) As $x \to 2^-$, $h(x) \to$?
(B) As $x \to 2^+$, $h(x) \to$?
(C) As $x \to -\infty$, $h(x) \to$?
(D) As $x \to \infty$, $h(x) \to$?

14. Let $k(x) = \dfrac{4 - 2x}{x + 2}$. Complete each statement:

(A) As $x \to 2^-$, $k(x) \to$?
(B) As $x \to 2^+$, $k(x) \to$?
(C) As $x \to -\infty$, $k(x) \to$?
(D) As $x \to \infty$, $k(x) \to$?

In Problems 15–22, find the domain and x intercepts.

15. $f(x) = \dfrac{3x - 9}{x}$ **16.** $g(x) = \dfrac{2x + 10}{x + 1}$

17. $h(x) = \dfrac{x + 6}{x^2 - 4}$ **18.** $k(x) = \dfrac{x^2 - 9}{x}$

19. $r(x) = \dfrac{x^2 - 3x - 4}{x^2 + 1}$ **20.** $s(x) = \dfrac{x^2 + 4x - 5}{x^2 + 4}$

21. $F(x) = \dfrac{x^4 + 16}{x^2 - 36}$ **22.** $G(x) = \dfrac{x^4 + x^2 + 1}{x^2 - 25}$

In Problems 23–30, find all vertical and horizontal asymptotes.

23. $f(x) = \dfrac{5x + 1}{x + 2}$ **24.** $g(x) = \dfrac{7x - 2}{x - 3}$

25. $s(x) = \dfrac{2x - 3}{x^2 - 16}$ **26.** $t(x) = \dfrac{3x + 4}{x^2 - 49}$

27. $p(x) = \dfrac{x^2 + 2x + 1}{x}$ **28.** $q(x) = \dfrac{x^3 - 1}{x + 1}$

29. $h(x) = \dfrac{3x^2 + 8}{2x^2 + 6x}$ **30.** $k(x) = \dfrac{6x^2 - 5x + 1}{7x^2 - 28x}$

In Problems 31–34, explain why each graph is not the graph of a rational function.

31.

32.

33.

34.

In Problems 35–38, explain how the graph of f differs from the graph of g.

35. $f(x) = \dfrac{x^2 + 2x}{x}$; $g(x) = x + 2$

36. $f(x) = \dfrac{x + 5}{x^2 - 25}$; $g(x) = \dfrac{1}{x - 5}$

37. $f(x) = \dfrac{x + 2}{x^2 + 10x + 16}$; $g(x) = \dfrac{1}{x + 8}$

38. $f(x) = \dfrac{x^2 - x - 12}{x - 4}$; $g(x) = x + 3$

In Problems 39–52, use the graphing strategy outlined in the text to sketch the graph of each function.

39. $f(x) = \dfrac{1}{x - 4}$

40. $g(x) = \dfrac{1}{x + 3}$

41. $f(x) = \dfrac{x}{x + 1}$

42. $f(x) = \dfrac{3x}{x - 3}$

43. $g(x) = \dfrac{1 - x^2}{x^2}$

44. $f(x) = \dfrac{x^2 + 1}{x^2}$

45. $f(x) = \dfrac{9}{x^2 - 9}$

46. $g(x) = \dfrac{6}{x^2 - x - 6}$

47. $f(x) = \dfrac{x}{x^2 - 1}$

48. $p(x) = \dfrac{x}{1 - x^2}$

49. $g(x) = \dfrac{2}{x^2 + 1}$

50. $f(x) = \dfrac{x}{x^2 + 1}$

51. $f(x) = \dfrac{12x^2}{(3x + 5)^2}$

52. $f(x) = \dfrac{7x^2}{(2x - 3)^2}$

In Problems 53–56, give an example of a rational function that satisfies the given conditions.

53. Real zeros: $-2, -1, 1, 2$; vertical asymptotes: none; horizontal asymptote: $y = 3$

54. Real zeros: none; vertical asymptotes: $x = 4$; horizontal asymptote: $y = -2$

55. Real zeros: none; vertical asymptotes: $x = 10$; oblique asymptote: $y = 2x + 5$

56. Real zeros: $1, 2, 3$; vertical asymptotes: none; oblique asymptote: $y = 2 - x$

In Problems 57–64, solve each rational inequality.

57. $\dfrac{x}{x - 2} \le 0$

58. $\dfrac{2x - 1}{x + 3} > 0$

59. $\dfrac{x^2 - 16}{5x - 2} > 0$

60. $\dfrac{x - 4}{x^2 - 9} \le 0$

61. $\dfrac{x^2 + 4x - 20}{3x} \ge 4$

62. $\dfrac{3x - 7}{x^2 + 6x} < 2$

63. $\dfrac{5x}{x^2 - 1} < \dfrac{9}{x}$

64. $\dfrac{1}{x^2 + 8x + 12} \ge \dfrac{1}{x}$

 In Problems 65–72, solve each rational inequality to three decimal places.

65. $\dfrac{x^2 + 7x + 3}{x + 2} > 0$

66. $\dfrac{x^3 + 4}{x^2 + x - 3} \le 0$

67. $\dfrac{9}{x} - \dfrac{5}{x^2} \le 1$

68. $\dfrac{x + 4}{x^2 + 1} > 2$

69. $\dfrac{3x + 2}{x - 5} > 10$

70. $\dfrac{x}{x^2 + 5x - 6} \le 0.5$

71. $\dfrac{4}{x + 1} \ge \dfrac{7}{x}$

72. $\dfrac{1}{x^2 - 1} < \dfrac{x^2}{x^4 + 1}$

In Problems 73–78, find all vertical, horizontal, and oblique asymptotes.

73. $f(x) = \dfrac{2x^2}{x - 1}$

74. $g(x) = \dfrac{3x^2}{x + 2}$

75. $p(x) = \dfrac{x^3}{x^2 + 1}$

76. $q(x) = \dfrac{x^5}{x^3 - 8}$

77. $r(x) = \dfrac{2x^2 - 3x + 5}{x}$

78. $s(x) = \dfrac{-3x^2 + 5x + 9}{x}$

In Problems 79–84, use the graphing strategy outlined in the text to sketch the graph of each function. Write the equations of all vertical, horizontal, and oblique asymptotes.

79. $f(x) = \dfrac{x^2 + 1}{x}$

80. $g(x) = \dfrac{x^2 - 1}{x}$

81. $k(x) = \dfrac{x^2 - 4x + 3}{2x - 4}$

82. $h(x) = \dfrac{x^2 + x - 2}{2x - 4}$

83. $F(x) = \dfrac{8 - x^3}{4x^2}$

84. $G(x) = \dfrac{x^4 + 1}{x^3}$

In calculus, it is often necessary to consider rational functions that are not in lowest terms, such as the functions given in Problems 85–88. For each function, state the domain. Write the equations of all vertical and horizontal asymptotes, and sketch the graph.

85. $f(x) = \dfrac{x^2 - 4}{x - 2}$

86. $g(x) = \dfrac{x^2 - 1}{x + 1}$

87. $r(x) = \dfrac{x + 2}{x^2 - 4}$

88. $s(x) = \dfrac{x - 1}{x^2 - 1}$

APPLICATIONS

89. EMPLOYEE TRAINING A company producing electronic components used in television sets has established that on the average, a new employee can assemble $N(t)$ components per day after t days of on-the-job training, as given by

$$N(t) = \dfrac{50t}{t + 4} \qquad t \ge 0$$

Sketch the graph of N, including any vertical or horizontal asymptotes. What does N approach as $t \to \infty$?

90. PHYSIOLOGY In a study on the speed of muscle contraction in frogs under various loads, researchers W. O. Fems and J. Marsh found that the speed of contraction decreases with increasing loads. More precisely, they found that the relationship between

speed of contraction S (in centimeters per second) and load w (in grams) is given approximately by

$$S(w) = \frac{26 + 0.06w}{w} \qquad w \geq 5$$

Sketch the graph of S, including any vertical or horizontal asymptotes. What does S approach as $w \to \infty$?

91. RETENTION An experiment on retention is conducted in a psychology class. Each student in the class is given 1 day to memorize the same list of 40 special characters. The lists are turned in at the end of the day, and for each succeeding day for 20 days each student is asked to turn in a list of as many of the symbols as can be recalled. Averages are taken, and it is found that a good approximation of the average number of symbols, $N(t)$, retained after t days is given by

$$N(t) = \frac{5t + 30}{t} \qquad t \geq 1$$

Sketch the graph of N, including any vertical or horizontal asymptotes. What does N approach as $t \to \infty$?

92. LEARNING THEORY In 1917, L. L. Thurstone, a pioneer in quantitative learning theory, proposed the function

$$f(x) = \frac{a(x + c)}{(x + c) + b}$$

to describe the number of successful acts per unit time that a person could accomplish after x practice sessions. Suppose that for a particular person enrolling in a typing class,

$$f(x) = \frac{50(x + 1)}{x + 5} \qquad x \geq 0$$

where $f(x)$ is the number of words per minute the person is able to type after x weeks of lessons. Sketch the graph of f, including any vertical or horizontal asymptotes. What does f approach as $x \to \infty$?

In Problems 93–96, use the fact from calculus that a function of the form

$$q(x) = ax + b + \frac{c}{x}, a > 0, c > 0, x > 0$$

has its minimum value when $x = \sqrt{c/a}$.

93. REPLACEMENT TIME A desktop office copier has an initial price of \$2,500. A maintenance/service contract costs \$200 for the first year and increases \$50 per year thereafter. It can be shown that the total cost of the copier after n years is given by

$$C(n) = 2{,}500 + 175n + 25n^2$$

The average cost per year for n years is $\overline{C}(n) = C(n)/n$.
(A) Find the rational function $\overline{C}$.
(B) When is the average cost per year a minimum? (This is frequently referred to as the *replacement time* for this piece of equipment.)
(C) Sketch the graph of $\overline{C}$, including any asymptotes.

94. AVERAGE COST The total cost of producing x units of a certain product is given by

$$C(x) = \tfrac{1}{5}x^2 + 2x + 2{,}000$$

The average cost per unit for producing x units is $\overline{C}(x) = C(x)/x$.
(A) Find the rational function $\overline{C}$.
(B) At what production level will the average cost per unit be minimal?
(C) Sketch the graph of $\overline{C}$, including any asymptotes.

95. CONSTRUCTION A rectangular dog pen is to be made to enclose an area of 225 square feet.
(A) If x represents the width of the pen, express the total length L of the fencing material required for the pen in terms of x.
(B) Considering the physical limitations, what is the domain of the function L?
(C) Find the dimensions of the pen that will require the least amount of fencing material.
(D) Graph the function L, including any asymptotes.

96. CONSTRUCTION Rework Problem 95 with the added assumption that the pen is to be divided into two sections, as shown in the figure. (Approximate dimensions to three decimal places.)

4-5 Variation and Modeling

> Direct Variation

> Inverse Variation

> Joint and Combined Variation

If you work more hours at a part-time job, then you will get more pay. If you increase your average speed in a bicycle race, then you will decrease the time required to finish. The relationship between hours and pay in the first instance, and between average speed and finishing time in the second, are expressed by saying "Pay is directly proportional to

hours worked, but average speed is inversely proportional to finishing time." Such statements, which describe how one quantity varies with respect to another, have a precise mathematical meaning. The purpose of this section is to explain the terminology of variation and how it is used in engineering and the sciences.

› Direct Variation

The perimeter of a square is a constant multiple of the side length, and the circumference of a circle is a constant multiple of the radius. These are examples of *direct variation*.

> ### › DEFINITION 1 Direct Variation
>
> Let x and y be variables. The statement **y is directly proportional to x** (or **y varies directly as x**) means
>
> $$y = kx$$
>
> for some nonzero constant k, called the **constant of proportionality** (or **constant of variation**).

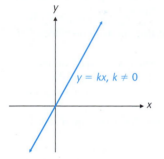

$y = kx, k \neq 0$

› **Figure 1** Direct variation.

The perimeter P of a square is directly proportional to the side length x; the constant of proportionality is 4 and the equation of variation is $P = 4x$. Similarly, the circumference C of a circle is directly proportional to the radius r; the constant of proportionality is 2π and the equation of variation is $C = 2\pi r$.

Note that the equation of direct variation $y = kx$, $k \neq 0$, gives a linear model with nonzero slope that passes through the origin (Fig. 1).

EXAMPLE 1 Direct Variation

The force F exerted by a spring is directly proportional to the distance x that it is stretched (Hooke's law). Find the constant of proportionality and the equation of variation if $F = 12$ pounds when $x = \frac{1}{3}$ foot.

SOLUTION The equation of variation has the form $F = kx$. To find the constant of proportionality, substitute $F = 12$ and $x = \frac{1}{3}$ and solve for k.

$$F = kx \qquad \text{Let } F = 12 \text{ and } x = \tfrac{1}{3}.$$
$$12 = k\left(\tfrac{1}{3}\right) \qquad \text{Multiply both sides by 3.}$$
$$k = 36$$

Therefore, the constant of proportionality is $k = 36$ and the equation of variation is

$$F = 36x$$

MATCHED PROBLEM 1 Find the constant of proportionality and the equation of variation if p is directly proportional to v, and $p = 200$ when $v = 8$.

› Inverse Variation

If variables x and y are *inversely proportional,* the functional relationship between them is a constant multiple of the rational function $y = 1/x$.

> **DEFINITION 2** Inverse Variation

Let x and y be variables. The statement **y is inversely proportional to x** (or **y varies inversely as x**) means

$$y = \frac{k}{x}$$

for some nonzero constant k, called the **constant of proportionality** (or **constant of variation**).

The rate r and time t it takes to travel a distance of 100 miles are inversely proportional (recall that distance equals rate times time, $d = rt$). The equation of variation is

$$t = \frac{100}{r}$$

and the constant of proportionality is 100.

The equation of inverse variation, $y = k/x$, determines a rational function having the y axis as a vertical asymptote and the x axis as a horizontal asymptote (Fig. 2). In most applications, the constant k of proportionality will be positive, and only the portion of the graph in Quadrant I will be relevant. If x is very large, then y is close to 0; if x is close to 0, then y is very large.

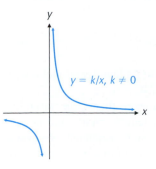

$y = k/x,\ k \neq 0$

> **Figure 2** Inverse variation.

EXAMPLE	2	**Inverse Variation**

The note played by each pipe in a pipe organ is determined by the frequency of vibration of the air in the pipe. The fundamental frequency f of vibration of air in an organ pipe is inversely proportional to the length L of the pipe. (This is why the low frequency notes come from the long pipes.)

(A) Find the constant of proportionality and the equation of variation if the fundamental frequency of an 8-foot pipe is 64 vibrations per second.

(B) Find the fundamental frequency of a 1.6-foot pipe.

SOLUTIONS (A) The equation has the form $f = k/L$. To find the constant of proportionality, substitute $L = 8$ and $f = 64$ and solve for k.

$$f = \frac{k}{L} \qquad \text{Let } f = 64 \text{ and } L = 8.$$

$$64 = \frac{k}{8} \qquad \text{Multiply both sides by 8.}$$

$$k = 512$$

The constant of proportionality is $k = 512$ and the equation of variation is

$$f = \frac{512}{L}$$

(B) If $L = 1.6$, then $f = \frac{512}{1.6} = 320$ vibrations per second.

MATCHED PROBLEM 2 Find the constant of proportionality and the equation of variation if P is inversely proportional to V, and $P = 56$ when $V = 3.5$.

> ## Joint and Combined Variation

The area of a rectangle is the product of its length and width. This is an example of *joint variation*.

> ### › DEFINITION 3 Joint Variation
>
> Let x, y, and w be variables. The statement w **is jointly proportional to x and y** (or w **varies jointly as x and y**) means
>
> $$w = kxy$$
>
> for some nonzero constant k, called the **constant of proportionality** (or **constant of variation**).

The area of a rectangle, for example, is jointly proportional to its length and width with constant of proportionality 1; the equation of variation is $A = LW$.

The concept of joint variation can be extended to apply to more than three variables. For example, the volume of a box is jointly proportional to its length, width, and height: $V = LWH$. Similarly, the concepts of direct and inverse variation can be extended. For example, the area of a circle is directly proportional to the square of its radius; the constant of proportionality is π and the equation of variation is $A = \pi r^2$.

The three basic types of variation also can be combined. For example, Newton's law of gravitation, "The force of attraction F between two objects is jointly proportional to their masses m_1 and m_2 and inversely proportional to the square of the distance d between them," has the equation

$$F = k\frac{m_1 m_2}{d^2}$$

EXAMPLE 3 | **Joint Variation**

The volume V of a right circular cone is jointly proportional to the square of its radius r and its height h. Find the constant of proportionality and the equation of variation if a cone of height 8 inches and radius 3 inches has a volume of 24π cubic inches.

SOLUTION | The equation of variation has the form $V = kr^2h$. To find the constant of proportionality k, substitute $V = 24\pi$, $r = 3$, and $h = 8$.

$$V = kr^2h \qquad \text{Let } V = 24\pi, \; r = 3, \text{ and } h = 8.$$
$$24\pi = k(3)^2 8 \qquad \text{Simplify.}$$
$$24\pi = 72k \qquad \text{Divide both sides by 72.}$$
$$k = \frac{\pi}{3}$$

The constant of proportionality is $k = \dfrac{\pi}{3}$ and the equation of variation is

$$V = \frac{\pi}{3}r^2h$$

MATCHED PROBLEM 3 | The volume V of a box with a square base is jointly proportional to the square of a side x of the base and the height h. Find the constant of proportionality and the equation of variation.

EXAMPLE	4	**Combined Variation**

The frequency f of a vibrating guitar string is directly proportional to the square root of the tension T and inversely proportional to the length L. What is the effect on the frequency if the length is doubled and the tension is quadrupled?

SOLUTION The equation of variation has the form

$$f = k\frac{\sqrt{T}}{L}$$

Let f_1, T_1, and L_1 denote the initial frequency, tension, and length, respectively. Then $L_2 = 2L_1$ and $T_2 = 4T_1$. Therefore,

$$f_2 = k\frac{\sqrt{T_2}}{L_2} \qquad \text{Let } L_2 = 2L_1, \text{ and } T_2 = 4T_1.$$

$$= k\frac{\sqrt{4T_1}}{2L_1} \qquad \text{Simplify the radical.}$$

$$= k\frac{2\sqrt{T_1}}{2L_1} \qquad \text{Cancel and use the equation of variation.}$$

$$= f_1$$

We conclude that there is no effect on the frequency—the pitch remains the same.

MATCHED PROBLEM 4 Refer to Example 4. What is the effect on the frequency if the tension is increased by a factor of 4 and the length is cut in half?

>>> **EXPLORE-DISCUSS 1** Refer to the equation of variation in Example 4. Explain why the frequency f, for fixed T, is a rational function of L, but f is *not*, for fixed L, a rational function of T.

ANSWERS TO MATCHED PROBLEMS

1. $k = 25$; $p = 25v$ **2.** $k = 196$; $P = \dfrac{196}{V}$ **3.** $k = 1$; $V = x^2h$

4. The frequency is increased by a factor of 4.

4-5 Exercises

1. Suppose that y is directly proportional to x and that the constant of proportionality is positive. If x increases, what happens to y? Explain.

2. Suppose that y is directly proportional to x and that the constant of proportionality is negative. If x increases, what happens to y? Explain.

3. Suppose that y is inversely proportional to x and that the constant of proportionality is positive. If x increases, what happens to y? Explain.

4. Explain what it means for w to be jointly proportional to x and y.

5. Suppose that y varies directly with x. What is the value of y when $x = 0$? Explain.

6. Suppose that y varies inversely with x. What is the value of y when $x = 1$? Explain.

In Problems 7–22, translate each statement into an equation using k as the constant of proportionality.

7. F is inversely proportional to x.

8. y is directly proportional to the square of x.

9. R is jointly proportional to S and T.

10. u is inversely proportional to v.

11. L is directly proportional to the cube of m.

12. W is jointly proportional to X, Y, and Z.

13. A varies jointly as the square of c and d.

14. q varies inversely as t.

15. P varies directly as x.

16. f varies directly as the square of b.

17. h varies inversely as the square root of s.

18. C varies jointly as the square of x and cube of y.

19. R varies directly as m and inversely as the square of d.

20. T varies jointly as p and q and inversely as w.

21. D is jointly proportional to x and the square of y and inversely proportional to z.

22. S is directly proportional to the square root of u and inversely proportional to v.

23. u varies directly as the square root of v. If $u = 3$ when $v = 4$, find u when $v = 10$.

24. y varies directly as the cube of x. If $y = 48$ when $x = 4$, find y when $x = 8$.

25. L is inversely proportional to the square of M. If $L = 9$ when $M = 9$, find L when $M = 6$.

26. I is directly proportional to the cube root of y. If $I = 5$ when $y = 64$, find I when $y = 8$.

27. Q varies jointly as m and the square of n, and inversely as P. If $Q = 2$ when $m = 3$, $n = 6$, and $P = 12$, find Q when $m = 4$, $n = 18$, and $P = 2$.

28. w varies jointly as x, y, and z. If $w = 36$ when $x = 2$, $y = 8$, and $z = 12$, find w when $x = 1$, $y = 2$, and $z = 4$.

In Problems 29–34, translate each statement into an equation using k as the constant of variation.

29. The biologist René Réaumur suggested in 1735 that the length of time t that it takes fruit to ripen is inversely proportional to the sum T of the average daily temperatures during the growing season.

30. The erosive force P of a swiftly flowing stream is directly proportional to the sixth power of the velocity v of the water.

31. The maximum safe load L for a horizontal beam varies jointly as its width w and the square of its height h, and inversely as its length x.

32. The number N of long-distance phone calls between two cities varies jointly as the populations P_1 and P_2 of the two cities, and inversely as the distance d between the two cities.

33. The f-stop numbers N on a camera, known as focal ratios, are directly proportional to the focal length F of the lens and inversely proportional to the diameter d of the effective lens opening.

34. The time t required for an elevator to lift a weight is jointly proportional to the weight w and the distance d through which it is lifted, and inversely proportional to the power P of the motor.

35. Suppose that f varies directly as x. Show that the ratio x_1/x_2 of two values of x is equal to f_1/f_2, the ratio of the corresponding values of f.

36. Suppose that f varies inversely as x. Show that the ratio x_1/x_2 of two values of x is equal to f_2/f_1, the reciprocal of the ratio of corresponding values of f.

APPLICATIONS

37. PHYSICS The weight w of an object on or above the surface of the Earth varies inversely as the square of the distance d between the object and the center of Earth. If a girl weighs 100 pounds on the surface of Earth, how much would she weigh (to the nearest pound) 400 miles above Earth's surface? (Assume the radius of Earth is 4,000 miles.)

38. PHYSICS A child was struck by a car in a crosswalk. The driver of the car had slammed on his brakes and left skid marks 160 feet long. He told the police he had been driving at 30 miles/hour. The police know that the length of skid marks L (when brakes are applied) varies directly as the square of the speed of the car v, and that at 30 miles/hour (under ideal conditions) skid marks would be 40 feet long. How fast was the driver actually going before he applied his brakes?

39. ELECTRICITY Ohm's law states that the current I in a wire varies directly as the electromotive forces E and inversely as the resistance R. If $I = 22$ amperes when $E = 110$ volts and $R = 5$ ohms, find I if $E = 220$ volts and $R = 11$ ohms.

40. ANTHROPOLOGY Anthropologists, in their study of race and human genetic groupings, often use an index called the *cephalic index*. The cephalic index C varies directly as the width w of the head and inversely as the length l of the head (both when viewed from the top). If an Indian in Baja California (Mexico) has measurements of $C = 75$, $w = 6$ inches, and $l = 8$ inches, what is C for an Indian in northern California with $w = 8.1$ inches and $l = 9$ inches?

41. PHYSICS If the horsepower P required to drive a speedboat through water is directly proportional to the cube of the speed v of the boat, what change in horsepower is required to double the speed of the boat?

42. ILLUMINATION The intensity of illumination E on a surface is inversely proportional to the square of its distance d from a light source. What is the effect on the total illumination on a book if the distance between the light source and the book is doubled?

43. MUSIC The frequency of vibration f of a musical string is directly proportional to the square root of the tension T and inversely proportional to the length L of the string. If the tension of the string is increased by a factor of 4 and the length of the string is doubled, what is the effect on the frequency?

44. PHYSICS In an automobile accident the destructive force F of a car is (approximately) jointly proportional to the weight w of the car and the square of the speed v of the car. (This is why accidents at high speed are generally so serious.) What would be the effect on the destructive forces of a car if its weight were doubled and its speed were doubled?

45. SPACE SCIENCE The length of time t a satellite takes to complete a circular orbit of Earth varies directly as the radius r of the orbit and inversely as the orbital velocity v of the satellite. If $t = 1.42$ hours when $r = 4,050$ miles and $v = 18,000$ miles/hour (Sputnik I), find t to two decimal places for $r = 4,300$ miles and $v = 18,500$ miles/hour.

46. GENETICS The number N of gene mutations resulting from x-ray exposure varies directly as the size of the x-ray dose r. What is the effect on N if r is quadrupled?

47. BIOLOGY In biology there is an approximate rule, called the *bioclimatic rule* for temperate climates, which states that the difference d in time for fruit to ripen (or insects to appear) varies directly as the change in altitude h. If $d = 4$ days when $h = 500$ feet, find d when $h = 2,500$ feet.

48. PHYSICS Over a fixed distance d, speed r varies inversely as time t. Police use this relationship to set up speed traps. If in a given speed trap $r = 30$ miles/hour when $t = 6$ seconds, what would be the speed of a car if $t = 4$ seconds?

49. PHYSICS The length L of skid marks of a car's tires (when the brakes are applied) is directly proportional to the square of the speed v of the car. How is the length of skid marks affected by doubling the speed?

50. PHOTOGRAPHY In taking pictures using flashbulbs, the lens opening (f-stop number) N is inversely proportional to the distance d from the object being photographed. What adjustment should you make on the f-stop number if the distance between the camera and the object is doubled?

51. ENGINEERING The total pressure P of the wind on a wall is jointly proportional to the area of the wall A and the square of the velocity of the wind v. If $P = 120$ pounds when $A = 100$ square feet

and $v = 20$ miles/hour, find P if $A = 200$ square feet and $v = 30$ miles/hour.

52. ENGINEERING The thrust T of a given type of propeller is jointly proportional to the fourth power of its diameter d and the square of the number of revolutions per minute n it is turning. What happens to the thrust if the diameter is doubled and the number of revolutions per minute is cut in half?

53. PSYCHOLOGY In early psychological studies on sensory perception (hearing, seeing, feeling, and so on), the question was asked: "Given a certain level of stimulation S, what is the minimum amount of added stimulation ΔS that can be detected?" A German physiologist, E. H. Weber (1795–1878) formulated, after many experiments, the famous law that now bears his name: "The amount of change ΔS that will be just noticed varies directly as the magnitude S of the stimulus."
(A) Write the law as an equation of variation.
(B) If a person lifting weights can just notice a difference of 1 ounce at the 50-ounce level, what will be the least difference she will be able to notice at the 500-ounce level?
(C) Determine the just noticeable difference in illumination a person is able to perceive at 480 candlepower if he is just able to perceive a difference of 1 candlepower at the 60-candle-power level.

54. PSYCHOLOGY Psychologists in their study of intelligence often use an index called IQ. IQ varies directly as mental age MA and inversely as chronological age CA (up to the age of 15). If a 12-year-old boy with a mental age of 14.4 has an IQ of 120, what will be the IQ of an 11-year-old girl with a mental age of 15.4?

55. GEOMETRY The volume of a sphere varies directly as the cube of its radius r. What happens to the volume if the radius is doubled?

56. GEOMETRY The surface area S of a sphere varies directly as the square of its radius r. What happens to the area if the radius is cut in half?

57. MUSIC The frequency of vibration of air in an open organ pipe is inversely proportional to the length of the pipe. If the air column in an open 32-foot pipe vibrates 16 times per second (low C), then how fast would the air vibrate in a 16-foot pipe?

58. MUSIC The frequency of pitch f of a musical string is directly proportional to the square root of the tension T and inversely proportional to the length l and the diameter d. Write the equation of variation using k as the constant of variation. (It is interesting to note that if pitch depended on only length, then pianos would have to have strings varying from 3 inches to 38 feet.)

CHAPTER 4 Review

4-1 Polynomial Functions and Models

A function that can be written in the form

$$P(x) = a_n x^n + a_{n-1} x^{n-1} + \ldots + a_1 x + a_0, \ a_n \neq 0,$$

is a **polynomial function of degree n**. In this chapter, when not specified otherwise, the **coefficients** $a_n, a_{n-1}, \ldots, a_1, a_0$ are complex numbers and the domain of P is the set of complex numbers. A number r is said to be a **zero** (or **root**) of a function $P(x)$ if $P(r) = 0$.

The real zeros of $P(x)$ are just the x intercepts of the graph of $P(x)$. A point on a continuous graph that separates an increasing portion from a decreasing portion, or vice versa, is called a **turning point**. If $P(x)$ is a polynomial of degree $n > 0$ with real coefficients, then the graph of $P(x)$:

1. Is continuous for all real numbers

2. Has no sharp corners

3. Has at most n real zeros

4. Has at most $n - 1$ turning points

5. Increases or decreases without bound as $x \to \infty$ and as $x \to -\infty$

The left and right behavior of such a polynomial $P(x)$ is determined by its highest degree or **leading term:** As $x \to \pm\infty$, both $a_n x^n$ and $P(x)$ approach $\pm\infty$, with the sign depending on n and the sign of a_n.

For any polynomial $P(x)$ of degree n, we have the following important results:

Division Algorithm
$P(x) = (x - r)Q(x) + R$ where the **quotient** $Q(x)$ and **remainder** R are unique; $x - r$ is the **divisor.**

Remainder Theorem
$P(r) = R$

Factor Theorem
$x - r$ is a factor of $P(x)$ if and only if $R = 0$.

Zeros of Polynomials
$P(x)$ has at most n zeros.

Synthetic division is an efficient method for dividing polynomials by linear terms of the form $x - r$.

4-2 Real Zeros and Polynomial Inequalities

The following theorems are useful in locating and approximating all real zeros of a polynomial $P(x)$ of degree $n > 0$ with real coefficients, $a_n > 0$:

Upper and Lower Bound Theorem
1. Upper bound: A number $r > 0$ is an upper bound for the real zeros of $P(x)$ if, when $P(x)$ is divided by $x - r$ using synthetic division, all numbers in the quotient row, including the remainder, are nonnegative.

2. Lower bound: A number $r < 0$ is a lower bound for the real zeros of $P(x)$ if, when $P(x)$ is divided by $x - r$ using synthetic division, all numbers in the quotient row, including the remainder, alternate in sign.

Location Theorem
Suppose that a function f is continuous on an interval I that contains numbers a and b. If $f(a)$ and $f(b)$ have opposite signs, then the graph of f has at least one x intercept between a and b.

The **bisection method** uses the location theorem repeatedly to approximate real zeros to any desired accuracy.

Polynomial inequalities can be solved by finding the zeros and inspecting the graph of an appropriate polynomial with real coefficients.

4-3 Complex Zeros and Rational Zeros of Polynomials

If $P(x)$ is a polynomial of degree $n > 0$ we have the following important theorems:

Fundamental Theorem of Algebra
$P(x)$ has at least one zero.

n Linear Factors Theorem
$P(x)$ can be factored as a product of n linear factors.

If $P(x)$ is factored as a product of linear factors, the number of linear factors that have zero r is said to be the **multiplicity** of r.

Imaginary Zeros Theorem
Imaginary zeros of polynomials with real coefficients, if they exist, occur in conjugate pairs.

Linear and Quadratic Factors Theorem
If $P(x)$ has real coefficients, then $P(x)$ can be factored as a product of linear factors (with real coefficients) and quadratic factors (with real coefficients and imaginary zeros).

Real Zeros and Polynomials of Odd Degree
If $P(x)$ has odd degree and real coefficients, then the graph of P has at least one x intercept.

Zeros of Even or Odd Multiplicity
Let $P(x)$ have real coefficients:

1. If r is a real zero of $P(x)$ of even multiplicity, then $P(x)$ has a turning point at r and does not change sign at r.

2. If r is a real zero of $P(x)$ of odd multiplicity, then $P(x)$ does not have a turning point at r and changes sign at r.

Rational Zero Theorem
If the rational number b/c, in lowest terms, is a zero of the polynomial

$$P(x) = a_n x^n + a_{n-1} x^{n-1} + \cdots + a_1 x + a_0 \qquad a_n \neq 0$$

with integer coefficients, then b must be an integer factor of a_0 and c must be an integer factor of a_n.

If $P(x) = (x - r)Q(x)$, then $Q(x)$ is called a **reduced polynomial** for $P(x)$.

4-4 Rational Functions and Inequalities

A function f is a **rational function** if it can be written in the form

$$f(x) = \frac{p(x)}{q(x)}$$

where $p(x)$ and $q(x)$ are polynomials of degrees m and n, respectively. The graph of a rational function $f(x)$:

1. Is continuous with the exception of at most n real numbers

2. Has no sharp corners

3. Has at most m real zeros

4. Has at most $m + n - 1$ turning points

5. Has the same left and right behavior as the quotient of the leading terms of $p(x)$ and $q(x)$

The vertical line $x = a$ is a **vertical asymptote** for the graph of $y = f(x)$ if $f(x) \to \infty$ or $f(x) \to -\infty$ as $x \to a^+$ or as $x \to a^-$. The horizontal line $y = b$ is a **horizontal asymptote** for the graph of $y = f(x)$ if $f(x) \to b$ as $x \to -\infty$ or as $x \to \infty$. The line $y = mx + b$ is an **oblique asymptote** if $[f(x) - (mx + b)] \to 0$ as $x \to -\infty$ or as $x \to \infty$.

Let $f(x) = \dfrac{a_m x^m + \cdots + a_1 x + a_0}{b_n x^n + \cdots + b_1 x + b_0}$, $a_m \neq 0, b_n \neq 0$.

1. If $m < n$, the line $y = 0$ (the x axis) is a horizontal asymptote.

2. If $m = n$, the line $y = a_m/b_n$ is a horizontal asymptote.

3. If $m > n$, there is no horizontal asymptote.

Analyzing and Sketching the Graph of a Rational Function: $f(x) = p(x)/q(x)$

Step 1. *Intercepts.* Find the real solutions of the equation $p(x) = 0$ and use these solutions to plot any x intercepts of the graph of f. Evaluate $f(0)$, if it exists, and plot the y intercept.

Step 2. *Vertical Asymptotes.* Find the real solutions of the equation $q(x) = 0$ and use these solutions to determine the domain of f, the points of discontinuity, and the vertical asymptotes. Sketch any vertical asymptotes as dashed lines.

Step 3. *Horizontal Asymptotes.* Determine whether there is a horizontal asymptote and, if so, sketch it as a dashed line.

Step 4. *Complete the Sketch.* For each interval in the domain of f, plot additional points and join them with a smooth continuous curve.

Rational inequalities can be solved by finding the zeros of $p(x)$ and $q(x)$, for an appropriate rational function $f(x) = p(x)/q(x)$, and inspecting the graph of f.

4-5 Variation and Modeling

Let x and y be variables. The statement:

1. **y is directly proportional to x (or y varies directly as x) means**

$$y = kx$$

for some nonzero constant k;

2. **y is inversely proportional to x (or y varies inversely as x) means**

$$y = \frac{k}{x}$$

for some nonzero constant k;

3. **w is jointly proportional to x and y (or w varies jointly as x and y) means**

$$w = kxy$$

for some nonzero constant k.

In each case the nonzero constant k is called the **constant of proportionality** (or **constant of variation**).

The three basic types of variation also can be combined. For example, Newton's law of gravitation, "The force of attraction F between two objects is jointly proportional to their masses m_1 and m_2 and inversely proportional to the square of the distance d between them" has the equation

$$F = k\frac{m_1 m_2}{d^2}$$

CHAPTER **4** **Review Exercises**

Work through all the problems in this chapter review, and check answers in the back of the book. Answers to all review problems are there, and following each answer is a number in italics indicating the section in which that type of problem is discussed. Where weaknesses show up, review appropriate sections in the text.

1. List the real zeros and turning points, and state the left and right behavior, of the polynomial function that has the indicated graph.

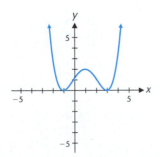

2. Use synthetic division to divide $P(x) = 2x^3 + 3x^2 - 1$ by $D(x) = x + 2$, and write the answer in the form $P(x) = D(x)Q(x) + R$.

3. If $P(x) = x^5 - 4x^4 + 9x^2 - 8$, find $P(3)$ using the remainder theorem and synthetic division.

4. What are the zeros of $P(x) = 3(x - 2)(x + 4)(x + 1)$?

5. If $P(x) = x^2 - 2x + 2$ and $P(1 + i) = 0$, find another zero of $P(x)$.

6. Let $P(x)$ be the polynomial whose graph is shown in the following figure.
(A) Assuming that $P(x)$ has integer zeros and leading coefficient 1, find the lowest-degree equation that could produce this graph.
(B) Describe the left and right behavior of $P(x)$.

7. According to the upper and lower bound theorem, which of the following are upper or lower bounds of the zeros of $P(x) = x^3 - 4x^2 + 2$?

$$-2, -1, 3, 4$$

8. How do you know that $P(x) = 2x^3 - 3x^2 + x - 5$ has at least one real zero between 1 and 2?

9. List all possible rational zeros of a polynomial with integer coefficients that has leading coefficient 5 and constant term -15.

10. Find all rational zeros for $P(x) = 5x^2 + 74x - 15$.

11. Find the domain and x intercepts for:

(A) $f(x) = \dfrac{6x}{x - 5}$

(B) $g(x) = \dfrac{7x + 3}{x^2 + 2x - 8}$

12. Find the horizontal and vertical asymptotes for the functions in Problem 11.

13. Explain why the graph is not the graph of a polynomial function.

In Problems 14–19, translate each statement into an equation using k as the constant of proportionality.

14. F is directly proportional to the square root of x.

15. G is jointly proportional to x and the square of y.

16. H is inversely proportional to the cube of z.

17. R varies jointly as the square of x and the square of y.

18. S varies inversely as the square of u.

19. T varies directly as v and inversely as w.

20. Let $P(x) = x^3 - 3x^2 - 3x + 4$.
(A) Graph $P(x)$ and describe the graph verbally, including the number of x intercepts, the number of turning points, and the left and right behavior.
(B) Approximate the largest x intercept to two decimal places.

21. If $P(x) = 8x^4 - 14x^3 - 13x^2 - 4x + 7$, find $Q(x)$ and R such that $P(x) = (x - \frac{1}{4})Q(x) + R$. What is $P(\frac{1}{4})$?

22. If $P(x) = 4x^3 - 8x^2 - 3x - 3$, find $P(-\frac{1}{2})$ using the remainder theorem and synthetic division.

23. Use the quadratic formula and the factor theorem to factor $P(x) = x^2 - 2x - 1$.

24. Is $x + 1$ a factor of $P(x) = 9x^{26} - 11x^{17} + 8x^{11} - 5x^4 - 7$? Explain, without dividing or using synthetic division.

25. Determine all rational zeros of $P(x) = 2x^3 - 3x^2 - 18x - 8$.

26. Factor the polynomial in Problem 25 into linear factors.

27. Find all rational zeros of $P(x) = x^3 - 3x^2 + 5$.

28. Find all zeros (rational, irrational, and imaginary) exactly for $P(x) = 2x^4 - x^3 + 2x - 1$.

29. Factor the polynomial in Problem 28 into linear factors.

30. If $P(x) = (x - 1)^2(x + 1)^3(x^2 - 1)(x^2 + 1)$, what is its degree? Write the zeros of $P(x)$, indicating the multiplicity of each if greater than 1.

31. Factor $P(x) = x^4 + 5x^2 - 36$ in two ways:
(A) As a product of linear factors (with real coefficients) and quadratic factors (with real coefficients and imaginary zeros)
(B) As a product of linear factors with complex coefficients

32. Let $P(x) = x^5 - 10x^4 + 30x^3 - 20x^2 - 15x - 2$.
(A) Approximate the zeros of $P(x)$ to two decimal places and state the multiplicity of each zero.
(B) Can any of these zeros be approximated with the bisection method? A maximum command? A minimum command? Explain.

33. Let $P(x) = x^4 - 2x^3 - 30x^2 - 25$.
(A) Find the smallest positive and largest negative integers that, by Theorem 1 in Section 4-2, are upper and lower bounds, respectively, for the real zeros of $P(x)$.
(B) If $(k, k + 1)$, k an integer, is the interval containing the largest real zero of $P(x)$, determine how many additional intervals are required in the bisection method to approximate this zero to one decimal place.

(C) Approximate the real zeros of $P(x)$ of two decimal places.

34. Let $f(x) = \dfrac{x - 1}{2x + 2}$.
(A) Find the domain and the intercepts for f.
(B) Find the vertical and horizontal asymptotes for f.
(C) Sketch a graph of f. Draw vertical and horizontal asymptotes with dashed lines.

35. Solve each polynomial inequality to three decimal places:
(A) $x^3 - 5x + 4 < 0$
(B) $x^3 - 5x + 4 < 2$

36. Explain why the graph is not the graph of a rational function.

37. B varies inversely as the square root of c. If $B = 5$ when $c = 4$, find B when $c = 25$.

38. D is jointly proportional to x and y. If $D = 10$ when $x = 3$ and $y = 2$, find D when $x = 9$ and $y = 8$.

39. Use synthetic division to divide $P(x) = x^3 + 3x + 2$ by $[x - (1 + i)]$. Write the answer in the form $P(x) = D(x)Q(x) + R$.

40. Find a polynomial of lowest degree with leading coefficient 1 that has zeros $-\frac{1}{2}$ (multiplicity 2), -3, and 1 (multiplicity 3). (Leave the answer in factored form.) What is the degree of the polynomial?

41. Repeat Problem 40 for a polynomial $P(x)$ with zeros $-5, 2 - 3i$, and $2 + 3i$.

42. Find all zeros (rational, irrational, and imaginary) exactly for $P(x) = 2x^5 - 5x^4 - 8x^3 + 21x^2 - 4$.

43. Factor the polynomial in Problem 42 into linear factors.

 44. Let $P(x) = x^4 + 16x^3 + 47x^2 - 137x + 73$. Approximate (to two decimal places) the x intercepts and the local extrema.

45. What is the minimal degree of a polynomial $P(x)$, given that $P(-1) = -4, P(0) = 2, P(1) = -5$, and $P(2) = 3$? Justify your conclusion.

46. If $P(x)$ is a cubic polynomial with integer coefficients and if $1 + 2i$ is a zero of $P(x)$, can $P(x)$ have an irrational zero? Explain.

47. The solutions to the equation $x^3 - 27 = 0$ are the cube roots of 27.
(A) How many cube roots of 27 are there?
(B) 3 is obviously a cube root of 27; find all others.

48. Let $P(x) = x^4 + 2x^3 - 500x^2 - 4{,}000$.
(A) Find the smallest positive integer multiple of 10 and the largest negative integer multiple of 10 that, by Theorem 1 in Section 4-2, are upper and lower bounds, respectively, for the real zeros of $P(x)$.
 (B) Approximate the real zero of $P(x)$ to two decimal places.

49. Graph

$$f(x) = \frac{x^2 + 2x + 3}{x + 1}$$

Indicate any vertical, horizontal, or oblique asymptotes with dashed lines.

 50. Use a graphing calculator to find any horizontal asymptotes for

$$f(x) = \frac{2x}{\sqrt{x^2 + 3x + 4}}$$

51. Solve each rational inequality:
(A) $\dfrac{x - 2}{5 - x} \le 0$ (B) $\dfrac{17}{x + 3} > \dfrac{5}{x}$

52. Solve each rational inequality to three decimal places:
(A) $\dfrac{x^2 - 3}{x^3 - 3x + 1} \le 0$
(B) $\dfrac{x^2 - 3}{x^3 - 3x + 1} > \dfrac{5}{x^2}$

53. If $P(x) = x^3 - x^2 - 5x + 4$, determine the number of real zeros of $P(x)$ and explain why $P(x)$ has no rational zeros.

54. Give an example of a rational function $f(x)$ that satisfies the following conditions: the real zeros of f are $-3, 0$, and 2; the vertical asymptotes of f are the line $x = -1$ and $x = 4$; and the line $y = 5$ is a horizontal asymptote.

APPLICATIONS

In Problems 55–58, express the solutions as the roots of a polynomial equation of the form $P(x) = 0$. Find rational solutions exactly and irrational solutions to one decimal place.

55. ARCHITECTURE An entryway is formed by placing a rectangular door inside an arch in the shape of the parabola with graph $y = 16 - x^2$, x and y in feet (see the figure). If the area of the door is 48 square feet, find the dimensions of the door.

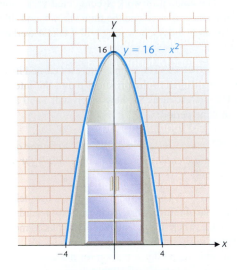

56. CONSTRUCTION A grain silo is formed by attaching a hemisphere to the top of a right circular cylinder (see the figure). If the cylinder is 18 feet high and the volume of the silo is 486π cubic feet, find the common radius of the cylinder and the hemisphere.

57. MANUFACTURING A box is to be made out of a piece of cardboard that measures 15 by 20 inches. Squares, x inches on a side, will be cut from each corner, and then the ends and sides will be folded up (see the figure). Find the value of x that would result in a box with a volume of 300 cubic inches.

5-1 Exponential Functions

> Defining Exponential Functions
> Graphs of Exponential Functions
> Additional Exponential Properties
> The Exponential Function with Base e
> Compound Interest
> Interest Compounded Continuously

Many of the functions we've studied so far have included exponents. But in every case, the exponent was a constant, and the base was often a variable. In this section, we will reverse those roles. In an *exponential function,* the variable appears in an exponent. As we'll see, this has a significant effect on the properties and graphs of these functions. A review of the basic properties of exponents in Section R-2, would be very helpful before moving on.

> Defining Exponential Functions

Let's start by noting that the functions f and g given by

$$f(x) = 2^x \qquad \text{and} \qquad g(x) = x^2$$

are not the same function. Whether a variable appears as an exponent with a constant base or as a base with a constant exponent makes a big difference. The function g is a quadratic function, which we have already discussed. The function f is an *exponential function.*

The graphs of f and g are shown in Figure 1. As expected, they are very different.

We know how to define the values of 2^x for many types of inputs. For positive integers, it's simply repeated multiplication:

$$2^2 = 2 \cdot 2 = 4; \quad 2^3 = 2 \cdot 2 \cdot 2 = 8; \quad 2^4 = 2 \cdot 2 \cdot 2 \cdot 2 = 16$$

For negative integers, we use properties of negative exponents:

$$2^{-1} = \frac{1}{2}; \quad 2^{-2} = \frac{1}{2^2} = \frac{1}{4}; \quad 2^{-3} = \frac{1}{2^3} = \frac{1}{8}$$

For rational numbers, a calculator comes in handy:

$$2^{\frac{1}{2}} = \sqrt{2} \approx 1.4; \quad 2^{\frac{3}{2}} = \sqrt{2^3} \approx 2.8; \quad 2^{\frac{9}{4}} = \sqrt[4]{2^9} \approx 4.8$$

(a)

(b)

> Figure 1

The only catch is that we don't know how to define 2^x for *all* real numbers. For example, what does

$$2^{\sqrt{2}}$$

mean? Your calculator can give you a decimal approximation, but where does it come from? That question is not easy to answer at this point. In fact, a precise definition of $2^{\sqrt{2}}$ must wait for more advanced courses. For now, we will simply state that for any positive real number b, the expression b^x is defined for all real values of x, and the output is a real number as well. This enables us to draw the continuous graph for $f(x) = 2^x$ in Figure 1. In Problems 79 and 80 in Exercises 5-1, we will explore a method for defining b^x for irrational x values like $\sqrt{2}$.

> **DEFINITION 1** Exponential Function

The equation

$$f(x) = b^x \qquad b > 0, b \neq 1$$

defines an **exponential function** for each different constant b, called the **base**. The independent variable x can assume any real value.

The domain of f is the set of all real numbers, and it can be shown that the range of f is the set of all positive real numbers. We require the base b to be positive to avoid imaginary numbers such as $(-2)^{1/2}$. Problems 53 and 54 in Exercises 5-1 explore why $b = 0$ and $b = 1$ are excluded.

> ## Graphs of Exponential Functions

>>> **EXPLORE-DISCUSS 1**

Compare the graphs of $f(x) = 3^x$ and $g(x) = 2^x$ by plotting both functions on the same coordinate system. Find all points of intersection of the graphs. For which values of x is the graph of f above the graph of g? Below the graph of g? Are the graphs of f and g close together as $x \to \infty$? As $x \to -\infty$? Discuss.

> **Figure 2** $y = b^x$ for $b = 2, 3, 5$.

The graphs of $y = b^x$ for $b = 2, 3,$ and 5 are shown in Figure 2. Note that all three have the same basic shape, and pass through the point $(0, 1)$. Also, the x axis is a horizontal asymptote for each graph, but only as $x \to -\infty$. The main difference between the graphs is their steepness.

Next, let's look at the graphs of $y = b^x$ for $b = \frac{1}{2}, \frac{1}{3},$ and $\frac{1}{5}$ (Fig. 3). Again, all three have the same basic shape, pass through $(0, 1)$, and have horizontal asymptote $y = 0$, but we can see that for $b < 1$, the asymptote is only as $x \to \infty$. In general, for bases less than 1, the graph is a reflection through the y axis of the graphs for bases greater than 1.

The graphs in Figures 2 and 3 suggest that the graphs of exponential functions have the properties listed in Theorem 1, which we state without proof.

> **Figure 3** $y = b^x$ for $b = \frac{1}{2}, \frac{1}{3}, \frac{1}{5}$.

> **THEOREM 1** Properties of Graphs of Exponential Functions

Let $f(x) = b^x$ be an exponential function, $b > 0, b \neq 1$. Then the graph of $f(x)$:

1. Is continuous for all real numbers

2. Has no sharp corners

3. Passes through the point $(0, 1)$

4. Lies above the x axis, which is a horizontal asymptote either as $x \to \infty$ or $x \to -\infty$, but not both

5. Increases as x increases if $b > 1$; decreases as x increases if $0 < b < 1$

6. Intersects any horizontal line at most once (that is, f is one-to-one)

These properties indicate that the graphs of exponential functions are distinct from the graphs we have already studied. (Actually, property 4 is enough to ensure that graphs of exponential functions are different from graphs of polynomials and rational functions.) Property 6 is important because it guarantees that exponential functions have inverses. Those inverses, called *logarithmic functions*, are the subject of Section 5-3.

Transformations of exponential functions are very useful in modeling real-world phenomena, like population growth and radioactive decay. These are among the applications we'll study in Section 5-2. It is important to understand how the graphs of those functions are related to the graphs of the exponential functions in this section. In Example 1, we will use the transformations we studied in Section 3-3 to examine this relationship.

EXAMPLE 1 Transformations of Exponential Functions

For the function $g(x) = \frac{1}{4}(2^x)$, use transformations to explain how the graph of g is related to the graph of $f(x) = 2^x$ in Figure 1(b). Find the intercepts and asymptotes, and draw the graph of g.

SOLUTION The graph of g is a vertical shrink of the graph of f by a factor of $\frac{1}{4}$. So like f, $g(x) > 0$ for all real numbers x, and $g(x) \to 0$ as $x \to -\infty$. In other words, there are no x intercepts, and the x axis is a horizontal asymptote. Since $g(0) = \frac{1}{4}(2^0) = \frac{1}{4}$, $\frac{1}{4}$ is the y intercept. Plotting the intercept and a few more points, we obtain the graph of g shown in the figure, with a portion magnified to illustrate the behavior better.

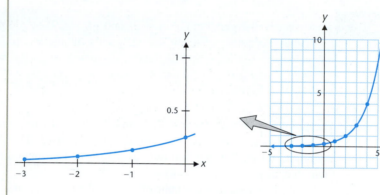

MATCHED PROBLEM 1 Let $g(x) = \frac{1}{2}(4^{-x})$. Use transformations to explain how the graph of g is related to the graph of the exponential function $f(x) = 4^x$. Find the intercepts and asymptotes, and sketch the graph of g.

› Additional Exponential Properties

Exponential functions whose domains include irrational numbers obey the familiar laws of exponents for rational exponents. We summarize these exponent laws here and add two other useful properties.

> **EXPONENTIAL FUNCTION PROPERTIES**

For a and b positive, $a \neq 1$, $b \neq 1$, and x and y real:

1. Exponent laws:

$$a^x a^y = a^{x+y} \qquad (a^x)^y = a^{xy} \qquad (ab)^x = a^x b^x$$

$$\left(\frac{a}{b}\right)^x = \frac{a^x}{b^x} \qquad \frac{a^x}{a^y} = a^{x-y} \qquad \frac{2^{5x}}{2^{7x}} \; \boxed{= 2^{5x-7x}}^* = 2^{-2x}$$

2. $a^x = a^y$ if and only if $x = y$. If $6^{4x} = 6^{2x+4}$, then $4x = 2x + 4$, and $x = 2$.

3. For $x \neq 0$, $a^x = b^x$ if and only if $a = b$. If $a^4 = 3^4$, then $a = 3$.

*Throughout the book, dashed boxes—called think boxes—are used to represent steps that may be performed mentally.

Property 2 is another way to express the fact that the exponential function $f(x) = a^x$ is one-to-one (see property 6 of Theorem 1). Because all exponential functions of the form $f(x) = a^x$ pass through the point $(0, 1)$ (see property 3 of Theorem 1), property 3 indicates that the graphs of exponential functions with different bases do not intersect at any other points.

EXAMPLE | **2** | ### Using Exponential Function Properties

Solve $4^{x-3} = 8$ for x.

SOLUTION | Express both sides in terms of the same base, and use property 2 to equate exponents.

$$4^{x-3} = 8 \qquad \text{Express 4 and 8 as powers of 2.}$$
$$(2^2)^{x-3} = 2^3 \qquad \text{Use the property } (a^x)^y = a^{xy}.$$
$$2^{2x-6} = 2^3 \qquad \text{Use property 2 to set exponents equal.}$$
$$2x - 6 = 3 \qquad \text{Add 6 to both sides.}$$
$$2x = 9 \qquad \text{Divide both sides by 2.}$$
$$x = \tfrac{9}{2}$$

CHECK | $4^{(9/2)-3} = 4^{3/2} = (\sqrt{4})^3 = 2^3 \overset{\checkmark}{=} 8$

Technology Connections

As an alternative to the algebraic method of Example 2, you can use a graphing calculator to solve the equation $4^{x-3} = 8$. Graph $y_1 = 4^{x-3}$ and $y_2 = 8$, then use the intersect command to obtain $x = 4.5$ (Fig. 4).

> Figure 4

MATCHED PROBLEM 2 | Solve $27^{x+1} = 9$ for x.

> ## The Exponential Function with Base e

Surprisingly, among the exponential functions it is not the function $g(x) = 2^x$ with base 2 or the function $h(x) = 10^x$ with base 10 that is used most frequently in mathematics. Instead, the most commonly used base is a number that you may not be familiar with.

>>> **EXPLORE-DISCUSS 2**

(A) Calculate the values of $[1 + (1/x)]^x$ for $x = 1, 2, 3, 4,$ and 5. Are the values increasing or decreasing as x gets larger?

(B) Graph $y = [1 + (1/x)]^x$ and discuss the behavior of the graph as x increases without bound.

Table 1

x	$\left(1 + \dfrac{1}{x}\right)^x$
1	2
10	2.593 74 ...
100	2.704 81 ...
1,000	2.716 92 ...
10,000	2.718 14 ...
100,000	2.718 27 ...
1,000,000	2.718 28 ...

By calculating the value of $[1 + (1/x)]^x$ for larger and larger values of x (Table 1), it looks like $[1 + (1/x)]^x$ approaches a number close to 2.7183. In a calculus course, we can show that as x increases without bound, the value of $[1 + (1/x)]^x$ approaches an irrational number that we call e. Just as irrational numbers such as π and $\sqrt{2}$ have unending, nonrepeating decimal representations, e also has an unending, nonrepeating decimal representation. To 12 decimal places,

$$e = 2.718\ 281\ 828\ 459$$

Don't let the symbol "e" intimidate you! It's just a number.

Exactly who discovered e is still being debated. It is named after the great Swiss mathematician Leonhard Euler (1707–1783), who computed e to 23 decimal places using $[1 + (1/x)]^x$.

The constant e turns out to be an ideal base for an exponential function because in calculus and higher mathematics many operations take on their simplest form using this base. This is why you will see e used extensively in expressions and formulas that model real-world phenomena.

> **DEFINITION 2** Exponential Function with Base e
>
> For x a real number, the equation
>
> $$f(x) = e^x$$
>
> defines the **exponential function with base e.**

> **Figure 5** Exponential functions.

The exponential function with base e is used so frequently that it is often referred to as *the* exponential function. The graphs of $y = e^x$ and $y = e^{-x}$ are shown in Figure 5.

EXAMPLE 3 Analyzing a Graph

Let $g(x) = 4 - e^{x/2}$. Use transformations to explain how the graph of g is related to the graph of $f_1(x) = e^x$. Determine whether g is increasing or decreasing, find any asymptotes, and sketch the graph of g.

SOLUTION The graph of g can be obtained from the graph of f_1 by a sequence of three transformations:

$$f_1(x) = e^x \quad \rightarrow \quad f_2(x) = e^{x/2} \quad \rightarrow \quad f_3(x) = -e^{x/2} \quad \rightarrow \quad g(x) = 4 - e^{x/2}$$

<div style="text-align:center">Horizontal stretch Reflection in x axis Vertical translation</div>

[See Fig. 6(a) for the graphs of f_1, f_2, and f_3, and Fig. 6(b) for the graph of g.] The function g is decreasing for all x. Because $e^{x/2} \rightarrow 0$ as $x \rightarrow -\infty$, it follows that $g(x) = 4 - e^{x/2} \rightarrow 4$ as $x \rightarrow -\infty$. Therefore, the line $y = 4$ is a horizontal asymptote [indicated by the dashed line in Fig. 6(b)]; there are no vertical asymptotes. [To check that the graph of g (as obtained by graph transformations) is correct, plot a few points.]

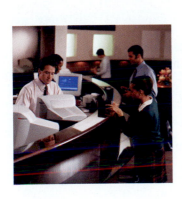

> **Figure 6**

MATCHED PROBLEM 3

Let $g(x) = 2e^{x/2} - 5$. Use transformations to explain how the graph of g is related to the graph of $f_1(x) = e^x$. Describe the increasing/decreasing behavior, find any asymptotes, and sketch the graph of g.

› Compound Interest

The fee paid to use someone else's money is called **interest**. It is usually computed as a percentage, called the **interest rate**, of the original amount (or **principal**) over a given period of time. At the end of the payment period, the interest paid is usually added to the principal amount, so the interest in the next period is earned on both the original amount, as well as the interest previously earned. Interest paid on interest previously earned and reinvested in this manner is called **compound interest.**

Suppose you deposit $1,000 in a bank that pays 8% interest compounded semiannually. How much will be in your account at the end of 2 years? "Compounded semiannually" means that the interest is paid to your account at the end of each 6-month period, and the interest will in turn earn more interest. To calculate the **interest rate per period**, we take the annual rate r, 8% (or 0.08), and divide by the number m of compounding periods per year, in this case 2. If A_1 represents the amount of money in the account after one compounding period (6 months), then

<center>Principal + 4% of principal</center>

$$A_1 = \$1,000 + \$1,000\left(\frac{0.08}{2}\right) \qquad \text{Factor out \$1,000.}$$

$$= \$1,000(1 + 0.04)$$

We will next use $A_2, A_3,$ and A_4 to represent the amounts at the end of the second, third, and fourth periods. (Note that the amount we're looking for is A_4.) A_2 is calculated by multiplying the amount at the beginning of the second compounding period (A_1) by 1.04.

$$A_2 = A_1(1 + 0.04) \qquad \text{Substitute our expression for } A_1.$$
$$= [\$1,000(1 + 0.04)](1 + 0.04) \qquad \text{Multiply.}$$
$$= \$1,000(1 + 0.04)^2 \qquad P\left(1 + \frac{r}{m}\right)^2$$

$$A_3 = A_2(1 + 0.04) \qquad \text{Substitute our expression for } A_2.$$
$$= [\$1,000(1 + 0.04)^2](1 + 0.04) \qquad \text{Multiply.}$$
$$= \$1,000(1 + 0.04)^3 \qquad P\left(1 + \frac{r}{m}\right)^3$$

$$A_4 = A_3(1 + 0.04) \qquad \text{Substitute our expression for } A_3.$$
$$= [\$1,000(1 + 0.04)^3](1 + 0.04) \qquad \text{Multiply.}$$
$$= \$1,000(1 + 0.04)^4 \qquad P\left(1 + \frac{r}{m}\right)^4$$
$$= \$1,169.86$$

What do you think the savings and loan will owe you at the end of 6 years (12 compounding periods)? If you guessed

$$A = \$1{,}000(1 + 0.04)^{12}$$

you have observed a pattern that is generalized in the following compound interest formula:

> › COMPOUND INTEREST
>
> If a **principal** P is invested at an annual **rate** r compounded m times a year, then the **amount** A in the account at the end of n compounding periods is given by
>
> $$A = P\left(1 + \frac{r}{m}\right)^{n}$$
>
> Note that the annual rate r must be expressed in decimal form, and that $n = mt$, where t is years.

EXAMPLE 4 Compound Interest

If you deposit $5,000 in an account paying 9% compounded daily,* how much will you have in the account in 5 years? Compute the answer to the nearest cent.

SOLUTION We will use the compound interest formula with $P = 5{,}000$, $r = 0.09$, (which is 9% written as a decimal), $m = 365$, and $n = 5(365) = 1{,}825$:

$$A = P\left(1 + \frac{r}{m}\right)^{n}$$ Let $P = 5{,}000$, $r = 0.09$, $m = 365$, $n = 5(365)$, or 1,825

$$= 5{,}000\left(1 + \frac{0.09}{365}\right)^{1{,}825}$$ Calculate to nearest cent.

$$= \$7{,}841.13$$

MATCHED PROBLEM 4 If $1,000 is invested in an account paying 10% compounded monthly, how much will be in the account at the end of 10 years? Compute the answer to the nearest cent.

EXAMPLE 5 Comparing Investments

If $1,000 is deposited into an account earning 10% compounded monthly and, at the same time, $2,000 is deposited into an account earning 4% compounded monthly, will the first account ever be worth more than the second? If so, when?

SOLUTION Let y_1 and y_2 represent the amounts in the first and second accounts, respectively, then

$$y_1 = 1{,}000(1 + 0.10/12)^{x}$$ $P = 1{,}000$, $r = 0.10$, $m = 12$
$$y_2 = 2{,}000(1 + 0.04/12)^{x}$$ $P = 2{,}000$, $r = 0.04$, $m = 12$

where x is the number of compounding periods (months). Examining the graphs of y_1 and y_2 [Fig. 7(a)], we see that the graphs intersect at $x \approx 139.438$ months. Because compound

*In all problems involving interest that is compounded daily, we assume a 365-day year.

interest is paid at the end of each compounding period, we compare the amount in the accounts after 139 months and after 140 months [Fig. 7(b)]. The first account is worth more than the second for $x \geq 140$ months, or after 11 years and 8 months.

> Figure 7

MATCHED PROBLEM 5 If $4,000 is deposited into an account earning 10% compounded quarterly and, at the same time, $5,000 is deposited into an account earning 6% compounded quarterly, when will the first account be worth more than the second?

› Interest Compounded Continuously

If $1,000 is deposited in an account that earns compound interest at an annual rate of 8% for 2 years, how will the amount A change if the number of compounding periods is increased? If m is the number of compounding periods per year, then

$$A = 1{,}000\left(1 + \frac{0.08}{m}\right)^{2m}$$

The amount A is computed for several values of m in Table 2. Notice that the largest gain appears in going from annually to semiannually. Then, the gains slow down as m increases. In fact, it appears that A might be approaching something close to $1,173.50 as m gets larger and larger.

Table 2 Effect of Compounding Frequency

Compounding Frequency	m	$A = 100\left(1 + \dfrac{0.08}{m}\right)^{2m}$
Annually	1	$1,166.400
Semiannually	2	1,169.859
Quarterly	4	1,171.659
Weekly	52	1,173.367
Daily	365	1,173.490
Hourly	8,760	1,173.501

We now return to the general problem to see if we can determine what happens to $A = P[1 + (r/m)]^{mt}$ as m increases without bound. A little algebraic manipulation of the

compound interest formula will lead to an answer and a significant result in the mathematics of finance:

$$A = P\left(1 + \frac{r}{m}\right)^{mt}$$ Replace $\frac{r}{m}$ with $\frac{1}{m/r}$, and mt with $\frac{m}{r} \cdot rt$.

$$= P\left(1 + \frac{1}{m/r}\right)^{(m/r)rt}$$ Replace $\frac{m}{r}$ with variable x.

$$= P\left[\left(1 + \frac{1}{x}\right)^{x}\right]^{rt}$$

Does the expression within the square brackets look familiar? Recall from the first part of this section that

$$\left(1 + \frac{1}{x}\right)^{x} \to e \qquad \text{as} \qquad x \to \infty$$

Because the interest rate r is fixed, $x = m/r \to \infty$ as $m \to \infty$. So $(1 + \frac{1}{x})^{x} \to e$, and

$$P\left(1 + \frac{r}{m}\right)^{mt} = P\left[\left(1 + \frac{1}{x}\right)^{x}\right]^{rt} \to Pe^{rt} \qquad \text{as} \qquad m \to \infty$$

This is known as the **continuous compound interest formula**, a very important and widely used formula in business, banking, and economics.

> **› CONTINUOUS COMPOUND INTEREST FORMULA**
>
> If a principal P is invested at an annual rate r compounded continuously, then the amount A in the account at the end of t years is given by
>
> $$A = Pe^{rt}$$
>
> The annual rate r must be expressed as a decimal.

EXAMPLE | **6** | **Continuous Compound Interest**

If \$1,000 is invested at an annual rate of 8% compounded continuously, what amount, to the nearest cent, will be in the account after 2 years?

SOLUTION Use the continuous compound interest formula to find A when $P = \$1,000$, $r = 0.08$, and $t = 2$:

$$A = Pe^{rt}$$ 8% is equivalent to $r = 0.08$.

$$= \$1,000e^{(0.08)(2)}$$ Calculate to nearest cent.

$$= \$1,173.51$$

Notice that the values calculated in Table 2 get closer to this answer as m gets larger. ●

MATCHED PROBLEM 6 What amount will an account have after 5 years if \$1,000 is invested at an annual rate of 12% compounded annually? Quarterly? Continuously? Compute answers to the nearest cent.

ANSWERS TO MATCHED PROBLEMS

1. The graph of g is the same as the graph of f reflected in the y axis and vertically shrunk by a factor of $\frac{1}{2}$.
 x intercepts: none
 y intercept: $\frac{1}{2}$
 horizontal asymptote: $y = 0$ (x axis)
 vertical asymptotes: none

2. $x = -\frac{1}{3}$

3. The graph of g is the same as the graph of f_1 stretched horizontally by a factor of 2, stretched vertically by a factor of 2, and shifted 5 units down; g is increasing.
 horizontal asymptote: $y = -5$
 vertical asymptote: none

4. $2,707.04 5. After 23 quarters

6. Annually: $1,762.34; quarterly: $1,806.11; continuously: $1,822.12

5-1 Exercises

1. What is an exponential function?

2. What is the significance of the symbol e in the study of exponential functions?

3. For a function $f(x) = b^x$, explain how you can tell if the graph increases or decreases without looking at the graph.

4. Explain why $f(x) = (1/4)^x$ and $g(x) = 4^{-x}$ are really the same function. Can you use this fact to add to your answer for Problem 3?

5. How do we know that the equation $e^x = 0$ has no solution?

6. Define the following terms related to compound interest: principal, interest rate, compounding period.

7. Match each equation with the graph of f, g, m, or n in the figure.
(A) $y = (0.2)^x$ (B) $y = 2^x$
(C) $y = (\frac{1}{3})^x$ (D) $y = 4^x$

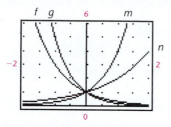

8. Match each equation with the graph of f, g, m, or n in the figure.
(A) $y = e^{-1.2x}$ (B) $y = e^{0.7x}$
(C) $y = e^{-0.4x}$ (D) $y = e^{1.3x}$

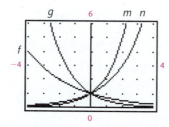

In Problems 9–16, use a calculator to compute answers to four significant digits.

9. $5^{\sqrt{3}}$ **10.** $3^{-\sqrt{2}}$

11. $e^2 + e^{-2}$ **12.** $e - e^{-1}$

13. $\sqrt{e}$ **14.** $e^{\sqrt{2}}$

15. $\dfrac{2^\pi + 2^{-\pi}}{2}$ **16.** $\dfrac{3^\pi - 3^{-\pi}}{2}$

In Problems 17–24, simplify.

17. $10^{3x-1}10^{4-x}$ **18.** $(4^{3x})^{2y}$

19. $\dfrac{3^x}{3^{1-x}}$ **20.** $\dfrac{5^{x-3}}{5^{x-4}}$

21. $\left(\dfrac{4^x}{5^y}\right)^{3z}$ **22.** $(2^x 3^y)^z$

23. $\dfrac{e^{5x}}{e^{2x+1}}$ **24.** $\dfrac{e^{4-3x}}{e^{2-5x}}$

In Problems 25–32, use transformations to explain how the graph of g is related to the graph of $f(x) = e^x$. Determine whether g is increasing or decreasing, find the asymptotes, and sketch the graph of g.

25. $g(x) = 3e^x$ **26.** $g(x) = 2e^{-x}$

27. $g(x) = \frac{1}{3}e^{-x}$ **28.** $g(x) = \frac{1}{5}e^x$

29. $g(x) = 2 + e^x$ **30.** $g(x) = -4 + e^x$

31. $g(x) = e^{x+2}$ **32.** $g(x) = e^{x-1}$

In Problems 33–50, solve for x.

33. $5^{3x} = 5^{4x-2}$ **34.** $10^{2-3x} = 10^{5x-6}$

35. $7^{x^2} = 7^{2x+3}$ **36.** $4^{5x-x^2} = 4^{-6}$

37. $(\frac{4}{5})^{6x+1} = \frac{5}{4}$ **38.** $(\frac{7}{3})^{2-x} = \frac{3}{7}$

39. $(1-x)^5 = (2x-1)^5$ **40.** $5^3 = (x+2)^3$

41. $2xe^{-x} = 0$ **42.** $(x-3)e^x = 0$

43. $x^2 e^x - 5xe^x = 0$ **44.** $3xe^{-x} + x^2 e^{-x} = 0$

45. $9^{x^2} = 3^{3x-1}$ **46.** $4^{x^2} = 2^{x+3}$

47. $25^{x+3} = 125^x$ **48.** $4^{5x+1} = 16^{2x-1}$

49. $4^{2x+7} = 8^{x+2}$ **50.** $100^{2x+3} = 1,000^{x+5}$

51. Find all real numbers a such that $a^2 = a^{-2}$. Explain why this does not violate the second exponential function property in the box on page 330.

52. Find real numbers a and b such that $a \neq b$ but $a^4 = b^4$. Explain why this does not violate the third exponential function property in the box on page 330.

53. Evaluate $y = 1^x$ for $x = -3, -2, -1, 0, 1, 2,$ and 3. Why is $b = 1$ excluded when defining the exponential function $y = b^x$?

54. Evaluate $y = 0^x$ for $x = -3, -2, -1, 0, 1, 2,$ and 3. Why is $b = 0$ excluded when defining the exponential function $y = b^x$?

In Problems 55–64, use transformations to explain how the graph of g is related to the graph of the given exponential function f. Determine whether g is increasing or decreasing, find any asymptotes, and sketch the graph of g.

55. $g(x) = -(\frac{1}{2})^x$; $f(x) = (\frac{1}{2})^x$

56. $g(x) = -(\frac{1}{3})^{-x}$; $f(x) = (\frac{1}{3})^x$

57. $g(x) = (\frac{1}{4})^{x/2} + 3$; $f(x) = (\frac{1}{4})^x$

58. $g(x) = 5 - (\frac{2}{3})^{3x}$; $f(x) = (\frac{2}{3})^x$

59. $g(x) = 500(1.04)^x$; $f(x) = 1.04^x$

60. $g(x) = 1,000(1.03)^x$; $f(x) = 1.03^x$

61. $g(x) = 1 + 2e^{x-3}$; $f(x) = e^x$

62. $g(x) = 4e^{x+1} - 7$; $f(x) = e^x$

63. $g(x) = 3 - 4e^{2-x}$; $f(x) = e^x$

64. $g(x) = -2 - 5e^{4-x}$; $f(x) = e^x$

In Problems 65–68, simplify.

65. $\dfrac{-2x^3 e^{-2x} - 3x^2 e^{-2x}}{x^6}$ **66.** $\dfrac{5x^4 e^{5x} - 4x^3 e^{5x}}{x^8}$

67. $(e^x + e^{-x})^2 + (e^x - e^{-x})^2$

68. $e^x(e^{-x} + 1) - e^{-x}(e^x + 1)$

In Problems 69–76, use a graphing calculator to find local extrema, y intercepts, and x intercepts. Investigate the behavior as $x \to \infty$ and as $x \to -\infty$ and identify any horizontal asymptotes. Round any approximate values to two decimal places.

69. $f(x) = 2 + e^{x-2}$ **70.** $g(x) = -3 + e^{1+x}$

71. $s(x) = e^{-x^2}$ **72.** $r(x) = e^{x^2}$

73. $F(x) = \dfrac{200}{1 + 3e^{-x}}$

74. $G(x) = \dfrac{100}{1 + e^{-x}}$

75. $f(x) = \dfrac{2^x + 2^{-x}}{2}$

76. $g(x) = \dfrac{3^x + 3^{-x}}{2}$

 77. Use a graphing calculator to investigate the behavior of $f(x) = (1 + x)^{1/x}$ as x approaches 0.

 78. Use a graphing calculator to investigate the behavior of $f(x) = (1 + x)^{1/x}$ as x approaches ∞.

79. The irrational number $\sqrt{2}$ is approximated by 1.414214 to six decimal places. Each of $x = 1.4, 1.41, 1.414, 1.4142, 1.41421,$ and 1.414214 is a rational number, so we know how to define 2^x for each. Compute the value of 2^x for each of these x values, and use your results to estimate the value of $2^{\sqrt{2}}$. Then compute $2^{\sqrt{2}}$ using your calculator to check your estimate.

80. The irrational number $\sqrt{3}$ is approximated by 1.732051 to six decimal places. Each of $x = 1.7, 1.73, 1.732, 1.7321, 1.73205,$ and 1.732051 is a rational number, so we know how to define 3^x for each. Compute the value of 3^x for each of these x values, and use your results to estimate the value of $3^{\sqrt{3}}$. Then compute $3^{\sqrt{3}}$ using your calculator to check your estimate.

 It is common practice in many applications of mathematics to approximate nonpolynomial functions with appropriately selected polynomials. For example, the polynomials in Problems 81–84, called Taylor polynomials, can be used to approximate the exponential function $f(x) = e^x$. To illustrate this approximation graphically, in each problem graph $f(x) = e^x$ and the indicated polynomial in the same viewing window, $-4 \le x \le 4$ and $-5 \le y \le 50$.

81. $P_1(x) = 1 + x + \frac{1}{2}x^2$

82. $P_2(x) = 1 + x + \frac{1}{2}x^2 + \frac{1}{6}x^3$

83. $P_3(x) = 1 + x + \frac{1}{2}x^2 + \frac{1}{6}x^3 + \frac{1}{24}x^4$

84. $P_4(x) = 1 + x + \frac{1}{2}x^2 + \frac{1}{6}x^3 + \frac{1}{24}x^4 + \frac{1}{120}x^5$

 85. Investigate the behavior of the functions $f_1(x) = x/e^x$, $f_2(x) = x^2/e^x$, and $f_3(x) = x^3/e^x$ as $x \to \infty$ and as $x \to -\infty$, and find any horizontal asymptotes. Generalize to functions of the form $f_n(x) = x^n/e^x$, where n is any positive integer.

 86. Investigate the behavior of the functions $g_1(x) = xe^x$, $g_2(x) = x^2e^x$, and $g_3(x) = x^3e^x$ as $x \to \infty$ and as $x \to -\infty$, and find any horizontal asymptotes. Generalize to functions of the form $g_n(x) = x^ne^x$, where n is any positive integer.

APPLICATIONS*

87. FINANCE A couple just had a new child. How much should they invest now at 6.25% compounded daily to have $100,000 for the child's education 17 years from now? Compute the answer to the nearest dollar.

88. FINANCE A person wants to have $25,000 cash for a new car 5 years from now. How much should be placed in an account now if the account pays 4.75% compounded weekly? Compute the answer to the nearest dollar.

*Round monetary amounts to the nearest cent unless specified otherwise. In all problems involving interest that is compounded daily, assume a 365-day year.

89. MONEY GROWTH If you invest $5,250 in an account paying 6.38% compounded continuously, how much money will be in the account at the end of
(A) 6.25 years? (B) 17 years?

90. MONEY GROWTH If you invest $7,500 in an account paying 5.35% compounded continuously, how much money will be in the account at the end of
(A) 5.5 years? (B) 12 years?

 91. FINANCE If $3,000 is deposited into an account earning 8% compounded daily and, at the same time, $5,000 is deposited into an account earning 5% compounded daily, will the first account ever be worth more than the second? If so, when?

 92. FINANCE If $4,000 is deposited into an account earning 9% compounded weekly and, at the same time, $6,000 is deposited into an account earning 7% compounded weekly, will the first account ever be worth more than the second? If so, when?

93. FINANCE Will an investment of $10,000 at 4.9% compounded daily ever be worth more at the end of any quarter than an investment of $10,000 at 5% compounded quarterly? Explain.

94. FINANCE A sum of $5,000 is invested at 7% compounded semiannually. Suppose that a second investment of $5,000 is made at interest rate r compounded daily. Both investments are held for 1 year. For which values of r, to the nearest tenth of a percent, is the second investment better than the first? Discuss.

95. PRESENT VALUE A promissory note will pay $30,000 at maturity 10 years from now. How much should you pay for the note now if the note gains value at a rate of 6% compounded continuously?

96. PRESENT VALUE A promissory note will pay $50,000 at maturity $5\frac{1}{2}$ years from now. How much should you pay for the note now if the note gains value at a rate of 5% compounded continuously?

97. MONEY GROWTH The website Bankrate.com publishes a weekly list of the top savings deposit yields. In the category of 3-year certificates of deposit, the following were listed:

Flagstar Bank, FSB	3.12% (CQ)
UmbrellaBank.com	3.00% (CD)
Allied First Bank	2.96% (CM)

where CQ represents compounded quarterly, CD compounded daily, and CM compounded monthly. Find the value of $5,000 invested in each account at the end of 3 years.

98. Refer to Problem 97. In the 1-year certificate of deposit category, the following accounts were listed:

GMAC Bank	2.91% (CD)
UFBDirect.com	2.86% (CM)

Find the value of $10,000 invested in each account at the end of 1 year.

99. FINANCE Suppose $4,000 is invested at 6% compounded weekly. How much money will be in the account in
(A) $\frac{1}{2}$ year? (B) 10 years?

100. FINANCE Suppose $2,500 is invested at 4% compounded quarterly. How much money will be in the account in
(A) $\frac{3}{4}$ year? (B) 15 years?

5-2 Exponential Models

> Mathematical Modeling
> Data Analysis and Regression
> A Comparison of Exponential Growth Phenomena

One of the best reasons for studying exponential functions is the fact that many things that occur naturally in our world can be modeled accurately by these functions. In this section, we will study a wide variety of applications, including growth of populations of people, animals, and bacteria; radioactive decay; spread of epidemics; propagation of rumors; light intensity; atmospheric pressure; and electric circuits. The regression techniques we used in Chapter 1 to construct linear and quadratic models will be extended to construct exponential models.

> Mathematical Modeling

Populations tend to grow exponentially and at different rates. A convenient and easily understood measure of growth rate is the **doubling time**—that is, the time it takes for a population to double. Over short periods the **doubling time growth model** is often used to model population growth:

$$A = A_0 2^{t/d}$$

where A = Population at time t

A_0 = Population at time $t = 0$

d = Doubling time

Note that when $t = d$,

$$A = A_0 2^{d/d} = A_0 2$$

and the population is double the original, as it should be. We will use this model to solve a population growth problem in Example 1.

| EXAMPLE | 1 | **Population Growth** |

According to a 2008 estimate, the population of Nicaragua was about 5.7 million, and that population is growing due to a high birth rate and relatively low mortality rate. If the population continues to grow at the current rate, it will double in 37 years. If the growth remains steady, what will the population be in

(A) 15 years? (B) 40 years?

Calculate answers to three significant digits.

SOLUTIONS We can use the doubling time growth model, $A = A_0(2)^{t/d}$ with $A_0 = 5.7$ and $d = 37$:

$$A = 5.7(2)^{t/37} \text{See Figure 1.}$$

> **Figure 1** $A = 5.7(2)^{t/37}$

(A) Find A when $t = 15$ years:

$$A = 5.7(2)^{15/37} = 7.55 \text{ million}$$ To 3 significant digits

(B) Find A when $t = 40$ years:

$$A = 5.7(2)^{40/37} = 12.1 \text{ million}$$ To 3 significant digits

MATCHED PROBLEM 1 Before the great housing bust, Palm Coast, Florida, was the fastest-growing city in America. Its population was about 34,000 in 2000, and it doubled in 6.6 years. If the population had continued growing at that rate, what would it be in

(A) 2010? (B) 2020?

Calculate answers to three significant digits.

>>> EXPLORE-DISCUSS 1 The doubling time growth model would *not* be expected to give accurate results over long periods. According to the doubling time growth model of Example 1, what was the population of Nicaragua 500 years ago when it was settled as a Spanish colony? What will the population of Nicaragua be 200 years from now? Explain why these results are unrealistic. Discuss factors that affect human populations that are not taken into account by the doubling time growth model.

The doubling time model is not the only one used to model populations. An alternative model based on the continuous compound interest formula will be used in Example 2. In this case, the formula is written as

$$A = A_0 e^{kt}$$

where A = Population at time t

A_0 = Population at time $t = 0$

k = Relative growth rate

The **relative growth rate** is written as a percentage in decimal form. For example, if a population is growing so that at any time the population is increasing at 3% of the current population per year, the relative growth rate k would be 0.03.

EXAMPLE 2 Medicine—Bacteria Growth

Cholera, an intestinal disease, is caused by a cholera bacterium that multiplies exponentially by cell division as modeled by

$$A = A_0 e^{1.386t}$$

where A is the number of bacteria present after t hours and A_0 is the number of bacteria present at $t = 0$. If we start with 1 bacterium, how many bacteria will be present in

(A) 5 hours? (B) 12 hours?

Calculate the answers to three significant digits.

SOLUTIONS (A) Use $A_0 = 1$ and $t = 5$:

$$
\begin{aligned}
A &= A_0 e^{1.386t} & \text{Let } A_0 = 1 \text{ and } t = 5. \\
&= e^{1.386(5)} & \text{Calculate to three significant digits.} \\
&\approx 1{,}020
\end{aligned}
$$

(B) Use $A_0 = 1$ and $t = 12$:

$$
\begin{aligned}
A &= A_0 e^{1.386t} & \text{Let } A_0 = 1 \text{ and } t = 12. \\
&= e^{1.386(12)} & \text{Calculate to three significant digits.} \\
&= 16{,}700{,}000
\end{aligned}
$$

MATCHED PROBLEM 2 Repeat Example 2 if $A = A_0 e^{0.783t}$ and all other information remains the same.

Exponential functions can also be used to model radioactive decay, which is sometimes referred to as **negative growth.** Radioactive materials are used extensively in medical diagnosis and therapy, as power sources in satellites, and as power sources in many countries. If we start with an amount A_0 of a particular radioactive substance, the amount declines exponentially over time. The rate of decay varies depending on the particular radioactive substance. A convenient and easily understood measure of the rate of decay is the **half-life** of the material—that is, the time it takes for half of a particular material to decay. We can use the following **half-life decay model:**

$$
\begin{aligned}
A &= A_0 \left(\tfrac{1}{2}\right)^{t/h} \\
&= A_0 2^{-t/h}
\end{aligned}
$$

where $A =$ Amount at time t

$A_0 =$ Amount at time $t = 0$

$h =$ Half-life

Note that when the amount of time passed is equal to the half-life ($t = h$),

$$
A = A_0 2^{-h/h} = A_0 2^{-1} = A_0 \cdot \tfrac{1}{2}
$$

and the amount of radioactive material is half the original amount, as it should be.

EXAMPLE 3 **Radioactive Decay**

The radioactive isotope gallium 67 (^{67}Ga), used in the diagnosis of malignant tumors, has a biological half-life of 46.5 hours. If we start with 100 milligrams of the isotope, how many milligrams will be left after

(A) 24 hours? (B) 1 week?

Calculate answers to three significant digits.

SOLUTIONS We can use the half-life decay model:

$$A = A_0\left(\tfrac{1}{2}\right)^{t/h} = A_0 2^{-t/h}$$

Using $A_0 = 100$ and $h = 46.5$, we obtain

$$A = 100(2^{-t/46.5})$$ See Figure 2.

A (milligrams)

100

50

100 200 *t*

Hours

> **Figure 2** $A = 100(2^{-t/46.5})$.

(A) Find A when $t = 24$ hours:

$$A = 100(2^{-24/46.5})$$ Calculate to three significant digits.

$$= 69.9 \text{ milligrams}$$

(B) Find A when $t = 168$ hours Be careful about units! Half-life was given in hours.
 (1 week $= 168$ hours):

$$A = 100(2^{-168/46.5})$$ Calculate to three significant digits.

$$= 8.17 \text{ milligrams}$$

MATCHED PROBLEM 3 Radioactive gold 198 (^{198}Au), used in imaging the structure of the liver, has a half-life of 2.67 days. If we start with 50 milligrams of the isotope, how many milligrams will be left after:

(A) $\tfrac{1}{2}$ day? (B) 1 week?

Calculate answers to three significant digits.

In Example 2, we saw that a base e exponential function can be used as an alternative to the doubling time model. Not surprisingly, the same can be said for the half-life model. In this case, the formula will be

$$A = A_0 e^{-kt}$$

where $A =$ the amount of radioactive material at time t

$A_0 =$ the amount at time $t = 0$

$k =$ a positive constant specific to the type of material

Our atmosphere is constantly being bombarded with cosmic rays. These rays produce neutrons, which in turn react with nitrogen to produce radioactive carbon-14. Radioactive carbon-14 enters all living tissues through carbon dioxide, which is first absorbed by plants. As long as a plant or animal is alive, carbon-14 is maintained in the living organism at a constant level. Once the organism dies, however, carbon-14 decays according to the equation

$$A = A_0 e^{-0.000124t}$$ **Carbon-14 decay equation**

where A is the amount of carbon-14 present after t years and A_0 is the amount present at time $t = 0$. This can be used to calculate the approximate age of fossils.

EXAMPLE 4 **Carbon-14 Dating**

If 1,000 milligrams of carbon-14 are present in the tissue of a recently deceased animal, how many milligrams will be present in

(A) 10,000 years? (B) 50,000 years?

Calculate answers to three significant digits.

SOLUTIONS

Substituting $A_0 = 1,000$ in the decay equation, we have

$$A = 1,000e^{-0.000124t}$$ See Figure 3.

> Figure 3

(A) Solve for A when $t = 10,000$:

$$A = 1,000e^{-0.000124(10,000)}$$ Calculate to three significant digits.

$$= 289 \text{ milligrams}$$

(B) Solve for A when $t = 50,000$:

$$A = 1,000e^{-0.000124(50,000)}$$ Calculate to three significant digits.

$$= 2.03 \text{ milligrams}$$

More will be said about carbon-14 dating in Exercises 5-5, where we will be interested in solving for t after being given information about A and A_0.

MATCHED PROBLEM 4

Referring to Example 4, how many milligrams of carbon-14 would have to be present at the beginning to have 10 milligrams present after 20,000 years? Compute the answer to four significant digits.

One of the problems with using exponential functions to model things like population is that the growth is completely unlimited in the long term. But in real life, there is often some reasonable maximum value, like the largest population that space and resources allow. We can use modified versions of exponential functions to model such phenomena more realistically.

One such type of function is called a *learning curve* since it can be used to model the performance improvement of a person learning a new task. **Learning curves** are functions of the form $A = c(1 - e^{-kt})$, where c and k are positive constants.

EXAMPLE 5

> Figure 4 $A = 40(1 - e^{-0.12t})$.

Learning Curve

People assigned to assemble circuit boards for a computer manufacturing company undergo on-the-job training. From past experience, it was found that the learning curve for the average employee is given by

$$A = 40(1 - e^{-0.12t})$$

where A is the number of boards assembled per day after t days of training (Fig. 4).

(A) How many boards can an average employee produce after 3 days of training? After 5 days of training? Round answers to the nearest integer.

(B) Does A approach a limiting value as t increases without bound? Explain.

SOLUTION

(A) When $t = 3$,

$$A = 40(1 - e^{-0.12(3)}) = 12$$ Rounded to nearest integer

so the average employee can produce 12 boards after 3 days of training. Similarly, when $t = 5$,

$$A = 40(1 - e^{-0.12(5)}) = 18$$ Rounded to nearest integer

(B) Because $e^{-0.12t} = \dfrac{1}{e^{0.12t}}$ approaches 0 as t increases without bound,

$$A = 40(1 - e^{-0.12t}) \to 40(1 - 0) = 40$$

So the limiting value of A is 40 boards per day. (Note the horizontal asymptote with equation $A = 40$ that is indicated by the dashed line in Fig. 4.)

MATCHED PROBLEM 5

A company is trying to expose as many people as possible to a new product through television advertising in a large metropolitan area with 2 million potential viewers. A model for the number of people A, in millions, who are aware of the product after t days of advertising was found to be

$$A = 2(1 - e^{-0.037t})$$

(A) How many viewers are aware of the product after 2 days? After 10 days? Express answers as integers, rounded to three significant digits.

(B) Does A approach a limiting value as t increases without bound? Explain.

Another limited-growth model is useful for phenomena such as the spread of an epidemic or the propagation of a rumor. It is called the *logistic equation,* and is given by

$$A = \frac{M}{1 + ce^{-kt}}$$

where M, c, and k are positive constants. Logistic growth, illustrated in Example 6, also approaches a limiting value as t increases without bound.

EXAMPLE 6

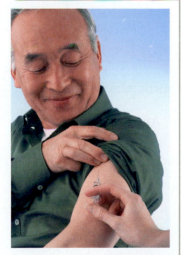

Logistic Growth in an Epidemic

A certain community consists of 1,000 people. One individual who has just returned from another community has a particularly contagious strain of influenza. Assume the community has not had influenza shots and all are susceptible. The spread of the disease in the community is predicted to be given by the logistic curve

$$A(t) = \frac{1,000}{1 + 999e^{-0.3t}}$$

where A is the number of people who have contracted the flu after t days.

(A) How many people have contracted the flu after 10 days? After 20 days?

(B) Does A approach a limiting value as t increases without bound? Explain.

SOLUTIONS

(A) When $t = 10$,

$$A = \frac{1,000}{1 + 999e^{-0.3(10)}} = 20 \qquad \text{Rounded to nearest integer}$$

so 20 people have contracted the flu after 10 days. Similarly, when $t = 20$,

$$A = \frac{1,000}{1 + 999e^{-0.3(20)}} = 288 \qquad \text{Rounded to nearest integer}$$

so 288 people have contracted the flu after 20 days.

(B) Because $e^{-0.3t}$ approaches 0 as t increases without bound,

$$A = \frac{1,000}{1 + 999e^{-0.3t}} \rightarrow \frac{1,000}{1 + 999(0)} = 1,000$$

So the limiting value is 1,000 individuals (everyone in the community will eventually get the flu). (Note the horizontal asymptote with equation $A = 1,000$ that is indicated by the dashed line in Fig. 5.)

▸ **Figure 5** $A = \dfrac{1,000}{1 + 999e^{-0.3t}}.$

MATCHED PROBLEM 6 A group of 400 parents, relatives, and friends are waiting anxiously at Kennedy Airport for a charter flight returning students after a year in Europe. It is stormy and the plane is late. A particular parent thought he heard that the plane's radio had gone out and related this news to some friends, who in turn passed it on to others. The propagation of this rumor is predicted to be given by

$$A(t) = \frac{400}{1 + 399e^{-0.4t}}$$

where A is the number of people who have heard the rumor after t minutes.

(A) How many people have heard the rumor after 10 minutes? After 20 minutes? Round answers to the nearest integer.

(B) Does A approach a limiting value as t increases without bound? Explain.

▸ Data Analysis and Regression

Many graphing calculators have options for exponential and logistic regression. We can use exponential regression to fit a function of the form $y = ab^x$ to a set of data points, and logistic regression to fit a function of the form

$$y = \frac{c}{1 + ae^{-bx}}$$

to a set of data points. The techniques are similar to those introduced in Chapters 2 and 3 for linear and quadratic functions.

| EXAMPLE | 7 | **Infectious Diseases** |

The U.S. Department of Health and Human Services published the data in Table 1.

Table 1 Reported Cases of Infectious Diseases

Year	Mumps	Rubella
1970	104,953	56,552
1980	8,576	3,904
1990	5,292	1,125
1995	906	128
2000	323	152
2005	314	11

An exponential model for the data on mumps is given by

$$A = 81,082(0.844)^t$$

where A is the number of reported cases of mumps and t is time in years with $t = 0$ representing 1970.

(A) Use the model to predict the number of reported cases of mumps in 2010.

(B) Compare the actual number of cases of mumps reported in 1980 to the number given by the model.

SOLUTIONS

(A) The year 2010 is represented by $t = 40$. Evaluating $A = 81,082(0.844)^t$ at $t = 40$ gives a prediction of 92 cases of mumps in 2010.

(B) The year 1980 is represented by $t = 10$. Evaluating $A = 81,082(0.844)^t$ at $t = 10$ gives 14,871 cases in 1980. The actual number of cases reported in 1980 was 8,576, nearly 6,300 less than the number given by the model.

Technology Connections

Figure 6 shows the details of constructing the exponential model of Example 7 on a graphing calculator.

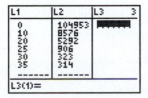

(a) Entering the data

ExpReg
y=a*b^x
a=81082.26436
b=.8444294513
r²=.9525693763
r=-.9759966067

(b) Finding the model

(c) Graphing the data and the model

› Figure 6

MATCHED PROBLEM 7

An exponential model for the data on rubella in Table 1 is given by

$$A = 54{,}988(0.799)^t$$

where A is the number of reported cases of rubella and t is time in years with $t = 0$ representing 1970.

(A) Use the model to predict the number of reported cases of rubella in 2010.

(B) Compare the actual number of cases of rubella reported in 1980 to the number given by the model.

EXAMPLE 8

AIDS Cases and Deaths

The U.S. Department of Health and Human Services published the data in Table 2.

Table 2 Acquired Immunodeficiency Syndrome (AIDS) Cases and Deaths in the United States

Year	Cases Diagnosed to Date	Known Deaths to Date
1985	23,185	12,648
1988	107,755	62,468
1991	261,259	159,294
1994	493,713	296,507
1997	672,970	406,179
2000	774,467	447,648
2005	944,306	529,113

A logistic model for the data on AIDS cases is given by

$$A = \frac{947{,}000}{1 + 17.3e^{-0.313t}}$$

where A is the number of AIDS cases diagnosed by year t with $t = 0$ representing 1985.

(A) Use the model to predict the number of AIDS cases diagnosed by 2010.

(B) Compare the actual number of AIDS cases diagnosed by 2005 to the number given by the model.

SOLUTIONS

(A) The year 2010 is represented by $t = 25$. Evaluating

$$A = \frac{947{,}000}{1 + 17.3e^{-0.313t}}$$

at $t = 25$ gives a prediction of approximately 940,000 cases of AIDS diagnosed by 2010.

(B) The year 2005 is represented by $t = 20$. Evaluating

$$A = \frac{947,000}{1 + 17.3e^{-0.313t}}$$

at $t = 20$ gives 916,690 cases in 2005. The actual number of cases diagnosed by 2005 was 944,306, nearly 28,000 greater than the number given by the model.

Technology Connections

Figure 7 shows the details of constructing the logistic model of Example 8 on a graphing calculator.

(a) Entering the data (b) Finding the model (c) Graphing the data and the model

› Figure 7

MATCHED PROBLEM 8

A logistic model for the data on deaths from AIDS in Table 2 is given by

$$A = \frac{521,000}{1 + 18.8e^{-0.349t}}$$

where A is the number of known deaths from AIDS by year t with $t = 0$ representing 1985.

(A) Use the model to predict the number of known deaths from AIDS by 2010.

(B) Compare the actual number of known deaths from AIDS by 2005 to the number given by the model.

› A Comparison of Exponential Growth Phenomena

The equations and graphs given in Table 3 compare several widely used growth models. These are divided basically into two groups: unlimited growth and limited growth. Following each equation and graph is a short, incomplete list of areas in which the models are used. We have only touched on a subject that has been extensively developed and that you are likely to study in greater depth in the future.

Table 3 Exponential Growth and Decay

Description	Equation	Graph	Short List of Uses
Unlimited growth	$A = A_0 e^{kt}$ $k > 0$		Short-term population growth (people, bacteria, etc.); growth of money at continuous compound interest
Exponential decay	$A = A_0 e^{-kt}$ $k > 0$		Radioactive decay; light absorption in water, glass, and the like; atmospheric pressure; electric circuits
Limited growth	$A = c(1 - e^{-kt})$ $c, k > 0$		Learning skills; sales fads; company growth; electric circuits
Logistic growth	$A = \dfrac{M}{1 + ce^{-kt}}$ $c, k, M > 0$		Long-term population growth; epidemics; sales of new products; spread of rumors; company growth

ANSWERS TO MATCHED PROBLEMS

1. (A) 97,200
 (B) 278,000
2. (A) 50 bacteria
 (B) 12,000 bacteria
3. (A) 43.9 milligrams
 (B) 8.12 milligrams
4. 119.4 milligrams
5. (A) 143,000 viewers; 619,000 viewers
 (B) A approaches an upper limit of 2 million, the number of potential viewers
6. (A) 48 individuals; 353 individuals
 (B) A approaches an upper limit of 400, the number of people in the entire group.
7. (A) 7 cases
 (B) The actual number of cases was 1,927 less than the number given by the model.
8. (A) 519,000 deaths
 (B) The actual number of known deaths was approximately 17,000 greater than the number given by the model.

5-2 Exercises

1. Define the terms "doubling time" and "half-life" in your own words.

2. One of the models below represents positive growth, and the other represents negative growth. Classify each, and explain how you decided on your answer. (Assume that $k > 0$.)

$$A = A_0 e^{-kt} \qquad A = A_0 e^{kt}$$

3. Explain the difference between exponential growth and limited growth.

4. Explain why a limited growth model would be more accurate than regular exponential growth in modeling the long-term population of birds on an island in Lake Erie.

In Problems 5–8, write an exponential equation describing the given population at any time t.

5. Initial population 200; doubling time 5 months

6. Initial population 5,000; doubling time 3 years

7. Initial population 2,000; continuous growth at 2% per year

8. Initial population 500; continuous growth at 3% per week

In Problems 9–12, write an exponential equation describing the amount of radioactive material present at any time t.

9. Initial amount 100 grams; half-life 6 hours

10. Initial amount 5 pounds; half-life 1,300 years

11. Initial amount 4 kilograms; continuous decay at 12.4% per year

12. Initial amount 50 milligrams; continuous decay at 0.03% per year

APPLICATIONS

13. **GAMING** A person bets on red and black on a roulette wheel using a *Martingale strategy.* That is, a $2 bet is placed on red, and the bet is doubled each time until a win occurs. The process is then repeated. If black occurs n times in a row, then $L = 2^n$ dollars is lost on the nth bet. Graph this function for $1 \le n \le 10$. Although the function is defined only for positive integers, points on this type of graph are usually joined with a smooth curve as a visual aid.

14. **BACTERIAL GROWTH** If bacteria in a certain culture double every $\frac{1}{2}$ hour, write an equation that gives the number of bacteria A in the culture after t hours, assuming the culture has 100 bacteria at the start. Graph the equation for $0 \le t \le 5$.

15. **POPULATION GROWTH** Because of its short life span and frequent breeding, the fruit fly *Drosophila* is used in some genetic studies. Raymond Pearl of Johns Hopkins University, for example, studied 300 successive generations of descendants of a single pair

of *Drosophila* flies. In a laboratory situation with ample food supply and space, the doubling time for a particular population is 2.4 days. If we start with 5 male and 5 female flies, how many flies should we expect to have in
(A) 1 week?
(B) 2 weeks?

16. **POPULATION GROWTH** It was estimated in 2008 that Kenya had a population of about 38,000,000 people, and a doubling time of 25 years. If growth continues at the same rate, find the population in
(A) 2012
(B) 2040
Calculate answers to two significant digits.

17. **COMPUTER DESIGN** In 1965, Gordon Moore, founder of Intel, predicted that the number of transistors that could be placed on a computer chip would double every 2 years. This has come to be known as *Moore's law.* In 1970, 2,200 transistors could be placed on a chip. Use Moore's law to predict the number of transistors in
(A) 1990
(B) 2005

18. **HISTORY OF TECHNOLOGY** The earliest mechanical clocks appeared around 1350 in Europe, and would gain or lose an average of 30 minutes per day. After that, accuracy roughly doubled every 30 years. Find the predicted accuracy of clocks in
(A) 1700
(B) 2000

19. **INSECTICIDES** The use of the insecticide DDT is no longer allowed in many countries because of its long-term adverse effects. If a farmer uses 25 pounds of active DDT, assuming its half-life is 12 years, how much will still be active after
(A) 5 years?
(B) 20 years?
Compute answers to two significant digits.

20. **RADIOACTIVE TRACERS** The radioactive isotope technetium-99m (^{99m}Tc) is used in imaging the brain. The isotope has a half-life of 6 hours. If 12 milligrams are used, how much will be present after
(A) 3 hours?
(B) 24 hours?
Compute answers to three significant digits.

21. **POPULATION GROWTH** According to the CIA World Factbook, the population of the world was estimated to be about 6.8 billion people in 2008, and the population was growing continuously at a relative growth rate of 1.188%. If this growth rate continues, what would the population be in 2020 to two significant digits?

22. **POPULATION GROWTH** According to the CIA World Factbook, the population of Mexico was about 100 million in 2008, and was growing continuously at a relative growth rate of 1.142%. If that growth continues, what will the population be in 2015 to three significant digits?

23. POPULATION GROWTH In 2005 the population of Russia was 143 million and the population of Nigeria was 129 million. If the populations of Russia and Nigeria grow continuously at relative growth rates of −0.37% and 2.56%, respectively, in what year did Nigeria have a greater population than Russia?
Use the Internet to find if the prediction was accurate.

24. POPULATION GROWTH In 2005 the population of Germany was 82 million and the population of Egypt was 78 million. If the populations of Germany and Egypt grow continuously at relative growth rates of 0% and 1.78%, respectively, in what year did Egypt have a greater population than Germany?
Use the Internet to find if the prediction was accurate.

25. SPACE SCIENCE Radioactive isotopes, as well as solar cells, are used to supply power to space vehicles. The isotopes gradually lose power because of radioactive decay. On a particular space vehicle the nuclear energy source has a power output of P watts after t days of use as given by

$$P = 75e^{-0.0035t}$$

Graph this function for $0 \le t \le 100$.

26. EARTH SCIENCE The atmospheric pressure P, in pounds per square inch, decreases exponentially with altitude h, in miles above sea level, as given by

$$P = 14.7e^{-0.21h}$$

Graph this function for $0 \le h \le 10$.

27. MARINE BIOLOGY Marine life is dependent upon the microscopic plant life that exists in the *photic zone,* a zone that goes to a depth where about 1% of the surface light still remains. Light intensity I relative to depth d, in feet, for one of the clearest bodies of water in the world, the Sargasso Sea in the West Indies, can be approximated by

$$I = I_0 e^{-0.00942d}$$

where I_0 is the intensity of light at the surface. To the nearest percent, what percentage of the surface light will reach a depth of
(A) 50 feet?
(B) 100 feet?

28. MARINE BIOLOGY Refer to Problem 27. In some waters with a great deal of sediment, the photic zone may go down only 15 to 20 feet. In some murky harbors, the intensity of light d feet below the surface is given approximately by

$$I = I_0 e^{-0.23d}$$

What percentage of the surface light will reach a depth of
(A) 10 feet?
(B) 20 feet?

29. AIDS EPIDEMIC The World Health Organization estimated that there were 33.2 million people worldwide living with the HIV infection in 2007, and that the number had been growing continuously at a relative growth rate of 2.37%. If the growth

continues at the rate, find the number of people that will be living with HIV in
(A) 2014
(B) 2020

30. AIDS EPIDEMIC The World Health Organization estimated that there were 3.25 million deaths from AIDS in 2007, and that the number had been growing continuously at a relative growth rate of 3.0%. If the growth continues at this rate, find the number of expected deaths from AIDS in
(A) 2012
(B) 2030

31. NEWTON'S LAW OF COOLING This law states that the rate at which an object cools is proportional to the difference in temperature between the object and its surrounding medium. The temperature T of the object t hours later is given by

$$T = T_m + (T_0 - T_m)e^{-kt}$$

where T_m is the temperature of the surrounding medium and T_0 is the temperature of the object at $t = 0$. Suppose a bottle of wine at a room temperature of 72°F is placed in the refrigerator to cool before a dinner party. If the temperature in the refrigerator is kept at 40°F and $k = 0.4$, find the temperature of the wine, to the nearest degree, after 3 hours. (In Exercises 5-5 we will find out how to determine k.)

32. NEWTON'S LAW OF COOLING Refer to Problem 31. What is the temperature, to the nearest degree, of the wine after 5 hours in the refrigerator?

33. PHOTOGRAPHY An electronic flash unit for a camera is activated when a capacitor is discharged through a filament of wire. After the flash is triggered, and the capacitor is discharged, the circuit (see the figure) is connected and the battery pack generates a current to recharge the capacitor. The time it takes for the capacitor to recharge is called the *recycle time*. For a particular flash unit using a 12-volt battery pack, the charge q, in coulombs, on the capacitor t seconds after recharging has started is given by

$$q = 0.0009(1 - e^{-0.2t})$$

Find the value that q approaches as t increases without bound and interpret.

34. MEDICINE An electronic heart pacemaker uses the same type of circuit as the flash unit in Problem 33, but it is designed so that the capacitor discharges 72 times a minute. For a particular pacemaker, the charge on the capacitor t seconds after it starts recharging is given by

$$q = 0.000\,008(1 - e^{-2t})$$

Find the value that q approaches as t increases without bound and interpret.

35. WILDLIFE MANAGEMENT A herd of 20 white-tailed deer is introduced to a coastal island where there had been no deer before. Their population is predicted to increase according to the logistic curve

$$A = \frac{100}{1 + 4e^{-0.14t}}$$

where A is the number of deer expected in the herd after t years.
(A) How many deer will be present after 2 years? After 6 years? Round answers to the nearest integer.
(B) How many years will it take for the herd to grow to 50 deer? Round answer to the nearest integer.
(C) Does A approach a limiting value as t increases without bound? Explain.

36. TRAINING A trainee is hired by a computer manufacturing company to learn to test a particular model of a personal computer after it comes off the assembly line. The learning curve for an average trainee is given by

$$A = \frac{200}{4 + 21e^{-0.1t}}$$

where A is the number of computers an average trainee can test per day after t days of training.
(A) How many computers can an average trainee be expected to test after 3 days of training? After 6 days? Round answers to the nearest integer.
(B) How many days will it take until an average trainee can test 30 computers per day? Round answer to the nearest integer.
(C) Does A approach a limiting value as t increases without bound? Explain.

Problems 37–40 require a graphing calculator or a computer that can calculate exponential and logistic regression models for a given data set.

37. DEPRECIATION Table 4 gives the market value of a minivan (in dollars) x years after its purchase. Find an exponential regression model of the form $y = ab^x$ for this data set. Round to four significant digits. Estimate the purchase price of the van. Estimate the value of the van 10 years after its purchase. Round answers to the nearest dollar.

Table 4

x	Value ($)
1	12,575
2	9,455
3	8,115
4	6,845
5	5,225
6	4,485

Source: Kelley Blue Book

38. DEPRECIATION Table 5 gives the market value of an SUV (in dollars) x years after its purchase. Find an exponential regression model of the form $y = ab^x$ for this data set. Estimate the purchase price of the SUV. Estimate the value of the SUV 10 years after its purchase. Round answers to the nearest dollar.

Table 5

x	Value ($)
1	23,125
2	19,050
3	15,625
4	11,875
5	9,450
6	7,125

Source: Kelley Blue Book

39. NUCLEAR POWER Table 6 gives data on nuclear power generation by region for the years 1980–2005.

Table 6 Nuclear Power Generation

	(Billion Kilowatt-Hours)	
Year	North America	Central and South America
1980	287.0	2.2
1985	440.8	8.4
1990	649.0	9.0
1995	774.4	9.5
2000	830.9	10.9
2005	879.7	16.3

Source: U.S. Energy Information Administration

(A) Let x represent time in years with $x = 0$ representing 1980. Find a logistic regression model ($y = \frac{c}{1 + ae^{-bx}}$) for the generation of nuclear power in North America. (Round the constants a, b, and c to three significant digits.)
(B) Use the logistic regression model to predict the generation of nuclear power in North America in 2010 and 2020.

40. NUCLEAR POWER Refer to Table 6.
(A) Let x represent time in years with $x = 0$ representing 1980. Find a logistic regression model ($y = \frac{c}{1 + ae^{-bx}}$) for the generation of nuclear power in Central and South America. (Round the constants a, b, and c to three significant digits.)
(B) Use the logistic regression model to predict the generation of nuclear power in Central and South America in 2010 and 2020.

5-3 Logarithmic Functions

> Defining Logarithmic Functions
> Converting Between Logarithmic Form and Exponential Form
> Properties of Logarithmic Functions
> Common and Natural Logarithms
> The Change-of-Base Formula

Solving an equation like $3^x = 9$ is easy: We know that $3^2 = 9$, so $x = 2$ is the solution. But what about an equation like $3^x = 20$? There probably is an exponent x between 2 and 3 for which 3^x is 20, but its exact value is not at all clear.

Compare this situation to an equation like $x^2 = 9$. This is easy to solve because we know that 3^2 and $(-3)^2$ are both 9. But what about $x^2 = 20$? To solve this equation, we needed to introduce a new function to be the opposite of the squaring function. This, of course, is the function $f(x) = \sqrt{x}$.

In this section, we will do something very similar with exponential functions. In the first section of this chapter, we learned that exponential functions are one-to-one, so we can define their inverses. These are known as the *logarithmic functions*.

> Defining Logarithmic Functions

The exponential function $f(x) = b^x$ for $b > 0$, $b \neq 1$, is a one-to-one function, and therefore has an inverse. Its inverse, denoted $f^{-1}(x) = \log_b x$ (read "log to the base b of x") is called the *logarithmic function with base b*. Just like exponentials, there are different logarithmic functions for each positive base other than 1. A point (x, y) is on the graph of $f^{-1} = \log_b x$ if and only if the point (y, x) is on the graph of $f = b^x$. In other words,

$$y = \log_b x \text{ if and only if } x = b^y$$

In a specific example,

$$y = \log_2 x \text{ if and only if } x = 2^y, \text{ and}$$

$\log_2 x$ is the power to which 2 must be raised to obtain x: $2^{\log_2 x} = 2^y = x$.

We can use this fact to learn some things about the logarithmic functions from our knowledge of exponential functions. For example, the graph of $f^{-1}(x) = \log_b x$ is the graph of $f(x) = b^x$ reflected through the line $y = x$. Also, the domain of $f^{-1}(x) = \log_b x$ is the range of $f(x) = b^x$, and vice versa.

In Example 1, we will use information about $f(x) = 2^x$ to graph its inverse, $f^{-1}(x) = \log_2 x$.

| EXAMPLE | 1 | Graphing a Logarithmic Function |

Make a table of values for $f(x) = 2^x$ and reverse the ordered pairs to obtain a table of values for $f^{-1}(x) = \log_2 x$. Then use both tables to graph $f(x)$ and $f^{-1}(x)$ on the same set of axes.

SOLUTION We chose to evaluate f for integer values from -3 to 3. The tables are shown here, along with the graph (Fig. 1). Note the important comments about domain and range below the graph.

x	$y = 2^x$	x	$y = \log_2 x$
-3	$\frac{1}{8}$	$\frac{1}{8}$	-3
-2	$\frac{1}{4}$	$\frac{1}{4}$	-2
-1	$\frac{1}{2}$	$\frac{1}{2}$	-1
0	1	1	0
1	2	2	1
2	4	4	2
3	8	8	3

Ordered pairs reversed

DOMAIN of $f = (-\infty, \infty)$ = RANGE of f^{-1}
RANGE of $f = (0, \infty)$ = DOMAIN of f^{-1}

› **Figure 1** Logarithmic function with base 2.

MATCHED PROBLEM 1

Repeat Example 1 for $f(x) = \left(\frac{1}{2}\right)^x$ and $f^{-1}(x) = \log_{1/2} x$.

› **DEFINITION 1** Logarithmic Function

For $b > 0, b \neq 1$, the inverse of $f(x) = b^x$, denoted $f^{-1}(x) = \log_b x$, is the **logarithmic function** with base b.

<div style="text-align:center">

Logarithmic form Exponential form

$y = \log_b x$ is equivalent to $x = b^y$

</div>

The log to the base b of x is the exponent to which b must be raised to obtain x. For example,

$$y = \log_{10} x \quad \text{is equivalent to} \quad x = 10^y$$
$$y = \log_e x \quad \text{is equivalent to} \quad x = e^y$$

Remember: A logarithm is an exponent.

$y = \log_b x$
$0 < b < 1$

DOMAIN $= (0, \infty)$
RANGE $= (-\infty, \infty)$

(a)

$y = \log_b x$
$b > 1$

DOMAIN $= (0, \infty)$
RANGE $= (-\infty, \infty)$

(b)

› **Figure 2** Typical logarithmic graphs.

It is very important to remember that the equations $y = \log_b x$ and $x = b^y$ define the same function, and as such can be used interchangeably.

Because the domain of an exponential function includes all real numbers and its range is the set of positive real numbers, the domain of a logarithmic function is the set of all positive real numbers and its range is the set of all real numbers. For example, $\log_{10} 3$ is defined, but $\log_{10} 0$ and $\log_{10} (-5)$ are not defined.

In short, the function $y = \log_b x$ for any b is only defined for positive x values. Typical logarithmic curves are shown in Figure 2. Notice that in each case, the y axis is a vertical asymptote for the graph.

The graphs in Example 1 and Figure 2 suggest that logarithmic graphs share some common properties. Several of these properties are listed in Theorem 1. It might be helpful in understanding them to review Theorem 1 in Section 5-1. Each of these properties is a consequence of a corresponding property of exponential graphs.

> **THEOREM 1** Properties of Graphs of Logarithmic Functions

Let $f(x) = \log_b x$ be a logarithmic function, $b > 0$, $b \neq 1$. Then the graph of $f(x)$:

1. Is continuous on its domain $(0, \infty)$
2. Has no sharp corners
3. Passes through the point $(1, 0)$
4. Lies to the right of the y axis, which is a vertical asymptote
5. Is increasing as x increases if $b > 1$; is decreasing as x increases if $0 < b < 1$
6. Intersects any horizontal line exactly once, so is one-to-one

>>> **EXPLORE-DISCUSS 1**

For the exponential function $f(x) = (\frac{2}{3})^x$, graph f and $y = x$ on the same coordinate system. Then sketch the graph of f^{-1}. Discuss the domains and ranges of f and its inverse. By what other name is f^{-1} known?

> ## Converting Between Logarithmic Form and Exponential Form

We now look into the matter of converting logarithmic forms to equivalent exponential forms, and vice versa. Throughout the remainder of the chapter, it will be useful to sometimes convert a logarithmic expression into the equivalent exponential form. At other times, it will be useful to do the reverse.

EXAMPLE 2 **Logarithmic–Exponential Conversions**

Change each logarithmic form to an equivalent exponential form.

(A) $\log_2 8 = 3$ (B) $\log_{25} 5 = \frac{1}{2}$ (C) $\log_2 (\frac{1}{4}) = -2$

SOLUTIONS

(A) $\log_2 8 = 3$ is equivalent to $8 = 2^3$.

(B) $\log_{25} 5 = \frac{1}{2}$ is equivalent to $5 = 25^{1/2}$.

(C) $\log_2 (\frac{1}{4}) = -2$ is equivalent to $\frac{1}{4} = 2^{-2}$.

Note that in each case, the base of the logarithm matches the base of the corresponding exponent.

MATCHED PROBLEM 2 Change each logarithmic form to an equivalent exponential form.

(A) $\log_3 27 = 3$ (B) $\log_{36} 6 = \frac{1}{2}$ (C) $\log_3 (\frac{1}{9}) = -2$

EXAMPLE 3 **Logarithmic–Exponential Conversions**

Change each exponential form to an equivalent logarithmic form.

(A) $49 = 7^2$ (B) $3 = \sqrt{9}$ (C) $\frac{1}{5} = 5^{-1}$

SOLUTIONS

(A) $49 = 7^2$ is equivalent to $\log_7 49 = 2$.

(B) $3 = \sqrt{9}$ is equivalent to $\log_9 3 = \frac{1}{2}$. Recall that $\sqrt{9} = 9^{1/2}$.

(C) $\frac{1}{5} = 5^{-1}$ is equivalent to $\log_5 \left(\frac{1}{5}\right) = -1$.

Again, the bases match.

MATCHED PROBLEM 3

Change each exponential form to an equivalent logarithmic form.

(A) $64 = 4^3$ (B) $2 = \sqrt[3]{8}$ (C) $\frac{1}{16} = 4^{-2}$

To gain a little deeper understanding of logarithmic functions and their relationship to the exponential functions, we will consider a few problems where we want to find x, b, or y in $y = \log_b x$, given the other two values. All values were chosen so that the problems can be solved without a calculator. In each case, converting to the equivalent exponential form is useful.

EXAMPLE 4 **Solutions of the Equation $y = \log_b x$**

Find x, b, or y as indicated.

(A) Find y: $y = \log_4 8$. (B) Find x: $\log_3 x = -2$. (C) Find b: $\log_b 81 = 4$.

SOLUTIONS

(A) Write $y = \log_4 8$ in equivalent exponential form.

$$8 = 4^y \qquad \text{\color{magenta}{Write each number to the same base 2.}}$$
$$2^3 = 2^{2y} \qquad \text{\color{magenta}{Recall that $b^m = b^n$ if and only if $m = n$.}}$$
$$2y = 3$$
$$y = \tfrac{3}{2}$$

We conclude that $\frac{3}{2} = \log_4 8$.

(B) Write $\log_3 x = -2$ in equivalent exponential form.

$$x = 3^{-2}$$
$$= \frac{1}{3^2} = \frac{1}{9}$$

We conclude that $\log_3 \left(\frac{1}{9}\right) = -2$.

(C) Write $\log_b 81 = 4$ in equivalent exponential form:

$$81 = b^4 \qquad \text{\color{magenta}{Write 81 as a fourth power.}}$$
$$3^4 = b^4 \qquad \text{\color{magenta}{b could be 3 or -3, but the base of a logarithm must be positive.}}$$
$$b = 3$$

We conclude that $\log_3 81 = 4$.

MATCHED PROBLEM 4

Find x, b, or y as indicated.

(A) Find y: $y = \log_9 27$. (B) Find x: $\log_2 x = -3$. (C) Find b: $\log_b 100 = 2$.

> ## Properties of Logarithmic Functions

Some of the properties of exponential functions that we studied in Section 5-1 can be used to develop corresponding properties of logarithmic functions. Several of these important properties of logarithmic functions are listed in Theorem 2. We will justify them individually.

> ### THEOREM 2 Properties of Logarithmic Functions
>
> If b, M, and N are positive real numbers, $b \neq 1$, and p and x are real numbers, then
>
> 1. $\log_b 1 = 0$ 5. $\log_b M = \log_b N$ if and only if $M = N$
> 2. $\log_b b = 1$ 6. $\log_b MN = \log_b M + \log_b N$
> 3. $\log_b b^x = x$ 7. $\log_b \dfrac{M}{N} = \log_b M - \log_b N$
> 4. $b^{\log_b x} = x, x > 0$ 8. $\log_b M^p = p \log_b M$

>>> CAUTION >>>

1. In properties 3 and 4, it's essential that the base of the exponential and the base of the logarithm are the same.

2. Properties 6 and 7 are often misinterpreted, so you should examine them carefully.

$$\frac{\log_b M}{\log_b N} \neq \log_b M - \log_b N$$

$\log_b M - \log_b N = \log_b \dfrac{M}{N}$;

$\dfrac{\log_b M}{\log_b N}$ cannot be simplified.

$$\log_b (M + N) \neq \log_b M + \log_b N$$

$\log_b M + \log_b N = \log_b MN$;

$\log_b (M + N)$ cannot be simplified.

Now we will justify properties in Theorem 2.

1. $\log_b 1 = 0$ because $b^0 = 1$.
2. $\log_b b = 1$ because $b^1 = b$.

3 and **4.** These are simply another way to state that $f(x) = b^x$ and $f^{-1}(x) = \log_b x$ are inverse functions. Property 3 can be written as $f^{-1}(f(x)) = x$ for all x in the domain of f. Property 4 can be written as $f(f^{-1}(x)) = x$ for all x in the domain of f^{-1}. This matches our characterization of inverse functions in Theorem 5, Section 3-6. Together, these properties say that if you apply an exponential function and a logarithmic function with the same base consecutively (in either order) you end up with the same value you started with.

5. This follows from the fact that logarithmic functions are one-to-one.

Properties 6, 7, and 8 are used often in manipulating logarithmic expressions. We will justify them in Problems 111 and 112 in Exercises 5-3, and Problem 69 in the Chapter 5 Review Exercises.

EXAMPLE 5 Using Logarithmic Properties

Simplify, using the properties in Theorem 2.

(A) $\log_e 1$ (B) $\log_{10} 10$ (C) $\log_e e^{2x+1}$

(D) $\log_{10} 0.01$ (E) $10^{\log_{10} 7}$ (F) $e^{\log_e x^2}$

SOLUTIONS (A) $\log_e 1 = 0$ Property 1 (B) $\log_{10} 10 = 1$ Property 2

(C) $\log_e e^{2x+1} = 2x + 1$ Property 3 (D) $\log_{10} 0.01 = \log_{10} 10^{-2} = -2$ Property 3

(E) $10^{\log_{10} 7} = 7$ Property 4 (F) $e^{\log_e x^2} = x^2$ Property 4

MATCHED PROBLEM 5 Simplify, using the properties in Theorem 2.

(A) $\log_{10} 10^{-5}$ (B) $\log_5 25$ (C) $\log_{10} 1$

(D) $\log_e e^{m+n}$ (E) $10^{\log_{10} 4}$ (F) $e^{\log_e (x^4 + 1)}$

› Common and Natural Logarithms

To work with logarithms effectively, we will need to be able to calculate (or at least approximate) the logarithms of any positive number to a variety of bases. Historically, tables were used for this purpose, but now calculators are used because they are faster and can find far more values than any table can possibly include.

 Of all possible bases, there are two that are used most often. **Common logarithms** are logarithms with base 10. **Natural logarithms** are logarithms with base e. Most calculators have a function key labeled "log" and a function key labeled "ln." The former represents the common logarithmic function and the latter the natural logarithmic function. In fact, "log" and "ln" are both used in most math books, and whenever you see either used in this book without a base indicated, they should be interpreted as follows:

› LOGARITHMIC FUNCTIONS

$$y = \log x = \log_{10} x \qquad \text{Common logarithmic function}$$
$$y = \ln x = \log_e x \qquad \text{Natural logarithmic function}$$

››› EXPLORE-DISCUSS 2 **(A)** Sketch the graph of $y = 10^x$, $y = \log x$, and $y = x$ in the same coordinate system and state the domain and range of the common logarithmic function.

(B) Sketch the graph of $y = e^x$, $y = \ln x$, and $y = x$ in the same coordinate system and state the domain and range of the natural logarithmic function.

EXAMPLE 6 **Calculator Evaluation of Logarithms**

Use a calculator to evaluate each to six decimal places.

(A) $\log 3{,}184$ (B) $\ln 0.000\,349$ (C) $\log(-3.24)$

SOLUTIONS (A) $\log 3{,}184 = 3.502\,973$ (B) $\ln 0.000\,349 = -7.960\,439$

(C) $\log(-3.24) = \text{Error}$

Why is an error indicated in part C? Because -3.24 is not in the domain of the log function. [*Note:* Calculators display error messages in various ways. Some calculators use a more advanced definition of logarithmic functions that involves complex numbers. They will

display an ordered pair, representing a complex number, as the value of log (−3.24), rather than an error message. You should interpret such a display as indicating that the number entered is not in the domain of the logarithmic function as we have defined it.]

MATCHED PROBLEM 6 Use a calculator to evaluate each to six decimal places.

(A) log 0.013 529 (B) ln 28.693 28 (C) ln (−0.438)

When working with common and natural logarithms, we will follow the common practice of using the equal sign "=" where it might be technically correct to use the approximately equal sign "≈." No harm is done as long as we keep in mind that in a statement such as log 3.184 = 0.503, the number on the right is only assumed accurate to three decimal places and is not exact.

>>> EXPLORE-DISCUSS 3

Graphs of the functions $f(x) = \log x$ and $g(x) = \ln x$ are shown in the graphing calculator display of Figure 3. Which graph belongs to which function? It appears from the display that one of the functions might be a constant multiple of the other. Is that true? Find and discuss the evidence for your answer.

› Figure 3

EXAMPLE 7 Calculator Evaluation of Logarithms

Use a calculator to evaluate each expression to three decimal places.

(A) $\dfrac{\log 2}{\log 1.1}$ (B) $\log \dfrac{2}{1.1}$ (C) $\log 2 - \log 1.1$

SOLUTIONS (A) $\dfrac{\log 2}{\log 1.1} = 7.273$ Enter as (log 2) ÷ (log 1.1).

(B) $\log \dfrac{2}{1.1} = 0.260$ Enter as log (2 ÷ 1.1).

(C) $\log 2 - \log 1.1 = 0.260$. Note that $\dfrac{\log 2}{\log 1.1} \neq \log 2 - \log 1.1$, but

$\log \dfrac{2}{1.1} = \log 2 - \log 1.1$ (see Theorem 2).

MATCHED PROBLEM 7 Use a calculator to evaluate each to three decimal places.

(A) $\dfrac{\ln 3}{\ln 1.08}$ (B) $\ln \dfrac{3}{1.08}$ (C) $\ln 3 - \ln 1.08$

We now turn to the opposite problem: Given the logarithm of a number, find the number. To solve this problem, we make direct use of the logarithmic–exponential relationships, and change logarithmic expressions into exponential form.

> LOGARITHMIC–EXPONENTIAL RELATIONSHIPS

$$\log x = y \quad \text{is equivalent to} \quad x = 10^y.$$
$$\ln x = y \quad \text{is equivalent to} \quad x = e^y.$$

EXAMPLE 8

Solving $\log_b x = y$ for x

Find x to three significant digits, given the indicated logarithms.

(A) $\log x = -9.315$ (B) $\ln x = 2.386$

SOLUTIONS

(A) $\log x = -9.315$ Change to exponential form (Definition 1).
$\quad\quad\; x = 10^{-9.315}$
$\quad\quad\quad = 4.84 \times 10^{-10}$

Notice that the answer is displayed in scientific notation in the calculator.

(B) $\ln x = 2.386$ Change to exponential form (Definition 1).
$\quad\quad\; x = e^{2.386}$
$\quad\quad\quad = 10.9$

MATCHED PROBLEM 8

Find x to four significant digits, given the indicated logarithms.

(A) $\ln x = -5.062$ (B) $\log x = 12.0821$

>>> EXPLORE-DISCUSS 4

Example 8 was solved algebraically using logarithmic–exponential relationships. Use the INTERSECT command on a graphing calculator to solve this problem graphically. Discuss the relative merits of the two approaches.

› The Change-of-Base Formula

How would you find the logarithm of a positive number to a base other than 10 or e? For example, how would you find $\log_3 5.2$? In Example 9 we evaluate this logarithm using several properties of logarithms. Then we develop a change-of-base formula to find such logarithms more easily.

EXAMPLE 9

Evaluating a Base 3 Logarithm

Evaluate $\log_3 5.2$ to four decimal places.

SOLUTION

Let $y = \log_3 5.2$ and proceed as follows:

$\log_3 5.2 = y$ Change to exponential form.

$\quad\quad 5.2 = 3^y$ Apply the natural log (or common log) to each side.

$\ln 5.2 = \ln 3^y$ Use $\log_b M^p = p \log_b M$, which brings the exponent y in front of $\ln 3$ as a factor.

$\ln 5.2 = y \ln 3$ Solve for y.

$\quad\quad\; y = \dfrac{\ln 5.2}{\ln 3}$

Replace y with $\log_3 5.2$ from the first step, and use a calculator to evaluate the right side:

$$\log_3 5.2 = \frac{\ln 5.2}{\ln 3} = 1.5007$$

MATCHED PROBLEM 9

Evaluate $\log_{0.5} 0.0372$ to four decimal places.

If we repeat the process we used in Example 9 on a generic logarithm, something interesting happens. The goal is to evaluate $\log_b N$, where b is any acceptable base, and N is any positive real number. As in Example 9, let $y = \log_b N$.

$$\log_b N = y \qquad \text{Write in exponential form.}$$
$$N = b^y \qquad \text{Apply natural log to each side.}$$
$$\ln N = \ln b^y \qquad \text{Use } \ln b^y = y \ln b \text{ (property 8, Theorem 2).}$$
$$\ln N = y \ln b \qquad \text{Solve for } y.$$
$$y = \frac{\ln N}{\ln b}$$

This provides a formula for evaluating a logarithm to any base by using natural log:

$$\log_b N = \frac{\ln N}{\ln b}$$

We could also have used log base 10 rather than natural log, and developed an alternative formula:

$$\log_b N = \frac{\log N}{\log b}$$

In fact, the same approach would enable us to rewrite $\log_b N$ in terms of a logarithm with any base we choose!

> **THE CHANGE-OF-BASE FORMULA**

For any $b > 0$, $b \neq 1$, and any positive real number N,

$$\log_b N = \frac{\log_a N}{\log_a b}$$

where a is any positive number other than 1.

>>> **EXPLORE-DISCUSS 5**

If b is any positive real number different from 1, the change-of-base formula shows that the function $y = \log_b x$ is a constant multiple of the natural logarithmic function; that is, $\log_b x = k \ln x$ for some k.

(A) Graph the functions $y = \ln x$, $y = 2 \ln x$, $y = 0.5 \ln x$, and $y = -3 \ln x$.

(B) Write each function of part A in the form $y = \log_b x$ by finding the base b to two decimal places.

(C) Is every exponential function $y = b^x$ a constant multiple of $y = e^x$? Explain.

ANSWERS TO MATCHED PROBLEMS

1.

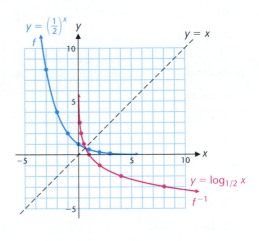

f		f^{-1}	
x	$y = \left(\frac{1}{2}\right)^x$	x	$y = \log_{1/2} x$
-3	8	8	-3
-2	4	4	-2
-1	2	2	-1
0	1	1	0
1	$\frac{1}{2}$	$\frac{1}{2}$	1
2	$\frac{1}{4}$	$\frac{1}{4}$	2
3	$\frac{1}{8}$	$\frac{1}{8}$	3

2. (A) $27 = 3^3$ (B) $6 = 36^{1/2}$ (C) $\frac{1}{9} = 3^{-2}$
3. (A) $\log_4 64 = 3$ (B) $\log_8 2 = \frac{1}{3}$ (C) $\log_4 \left(\frac{1}{16}\right) = -2$
4. (A) $y = \frac{3}{2}$ (B) $x = \frac{1}{8}$ (C) $b = 10$
5. (A) -5 (B) 2 (C) 0 (D) $m + n$ (E) 4 (F) $x^4 + 1$
6. (A) $-1.868\ 734$ (B) $3.356\ 663$ (C) Not possible
7. (A) 14.275 (B) 1.022 (C) 1.022
8. (A) $x = 0.006\ 333$ (B) $x = 1.208 \times 10^{12}$ **9.** 4.7486

5-3 Exercises

1. Describe the relationship between logarithmic functions and exponential functions in your own words.

2. Explain why there are infinitely many different logarithmic functions.

3. Why are logarithmic functions undefined for zero and negative inputs?

4. Why is $\log_b 1 = 0$ for any base?

5. Explain how to calculate $\log_5 3$ on a calculator that only has log buttons for base 10 and base e.

6. Using the word "inverse," explain why $\log_b b^x = x$ for any x and any acceptable base b.

Rewrite Problems 7–12 in equivalent exponential form.

7. $\log_3 81 = 4$

8. $\log_5 125 = 3$

9. $\log_{10} 0.001 = -3$

10. $\log_{10} 1,000 = 3$

11. $\log_6 \frac{1}{36} = -2$

12. $\log_2 \frac{1}{64} = -6$

Rewrite Problems 13–18 in equivalent logarithmic form.

13. $8 = 4^{3/2}$

14. $9 = 27^{2/3}$

15. $\frac{1}{2} = 32^{-1/5}$

16. $\frac{1}{8} = 2^{-3}$

17. $\left(\frac{2}{3}\right)^3 = \frac{8}{27}$

18. $\left(\frac{5}{2}\right)^{-2} = 0.16$

In Problems 19–22, make a table of values similar to the one in Example 1, then use it to graph both functions by hand.

19. $f(x) = 3^x$ $f^{-1}(x) = \log_3 x$

20. $f(x) = \left(\frac{1}{3}\right)^x$ $f^{-1}(x) = \log_{1/3} x$

21. $f(x) = \left(\frac{2}{3}\right)^x$ $f^{-1}(x) = \log_{2/3} x$

22. $f(x) = 10^x$ $f^{-1}(x) = \log x$

In Problems 23–38, simplify each expression using Theorem 2.

23. $\log_{16} 1$

24. $\log_{25} 1$

25. $\log_{0.5} 0.5$

26. $\log_7 7$

27. $\log_e e^4$

28. $\log_{10} 10^5$

29. $\log_{10} 0.01$ **30.** $\log_{10} 100$ **31.** $\log_3 27$

32. $\log_4 256$ **33.** $\log_{1/2} 2$ **34.** $\log_{1/5}\left(\frac{1}{25}\right)$

35. $e^{\log_e 5}$ **36.** $e^{\log_e 10}$ **37.** $\log_5 \sqrt[3]{5}$

38. $\log_2 \sqrt{8}$

In Problems 39–46, evaluate to four decimal places.

39. $\log 49{,}236$ **40.** $\log 691{,}450$

41. $\ln 54.081$ **42.** $\ln 19.722$

43. $\log_7 13$ **44.** $\log_9 78$

45. $\log_5 120.24$ **46.** $\log_{17} 304.66$

In Problems 47–54, evaluate x to four significant digits.

47. $\log x = 5.3027$ **48.** $\log x = 1.9168$

49. $\log x = -3.1773$ **50.** $\log x = -2.0411$

51. $\ln x = 3.8655$ **52.** $\ln x = 5.0884$

53. $\ln x = -0.3916$ **54.** $\ln x = -4.1083$

Find x, y, or b, as indicated in Problems 55–72.

55. $\log_2 x = 2$ **56.** $\log_3 x = 3$

57. $\log_4 16 = y$ **58.** $\log_8 64 = y$

59. $\log_b 16 = 2$ **60.** $\log_b 10^{-3} = -3$

61. $\log_b 1 = 0$ **62.** $\log_b b = 1$

63. $\log_4 x = \frac{1}{2}$ **64.** $\log_8 x = \frac{1}{3}$

65. $\log_{1/3} 9 = y$ **66.** $\log_{49}\left(\frac{1}{7}\right) = y$

67. $\log_b 1{,}000 = \frac{3}{2}$ **68.** $\log_b 4 = \frac{2}{3}$

69. $\log_8 x = -\frac{4}{3}$ **70.** $\log_{25} x = -\frac{3}{2}$

71. $\log_{16} 8 = y$ **72.** $\log_9 27 = y$

In Problems 73–78, evaluate to three decimal places.

73. $\dfrac{\log 2}{\log 1.15}$ **74.** $\dfrac{\log 2}{\log 1.12}$

75. $\dfrac{\ln 3}{\ln 1.15}$ **76.** $\dfrac{\ln 4}{\ln 1.2}$

77. $\dfrac{\ln 150}{2 \ln 3}$ **78.** $\dfrac{\log 200}{3 \log 2}$

In Problems 79–82, rewrite the expression in terms of log x and log y.

79. $\log\left(\dfrac{x}{y}\right)$ **80.** $\log(xy)$

81. $\log(x^4 y^3)$ **82.** $\log\left(\dfrac{x^2}{\sqrt{y}}\right)$

In Problems 83–86, rewrite the expression as a single log.

83. $\ln x - \ln y$ **84.** $\log_3 x + \log_3 y$

85. $2 \ln x + 5 \ln y - \ln z$ **86.** $\log a - 2 \log b + 3 \log c$

In Problems 87–90, given that log x = −2 and log y = 3, find:

87. $\log(xy)$ **88.** $\log\left(\dfrac{x}{y}\right)$

89. $\log\left(\dfrac{\sqrt{x}}{y^3}\right)$ **90.** $\log(x^5 y^3)$

In Problems 91–98, use transformations to explain how the graph of g is related to the graph of the given logarithmic function f. Determine whether g is increasing or decreasing, find its domain and asymptote, and sketch the graph of g.

91. $g(x) = 3 + \log_2 x; f(x) = \log_2 x$

92. $g(x) = -4 + \log_3 x; f(x) = \log_3 x$

93. $g(x) = \log_{1/3}(x - 2); f(x) = \log_{1/3} x$

94. $g(x) = \log_{1/2}(x + 3); f(x) = \log_{1/2} x$

95. $g(x) = -1 - \log x; f(x) = \log x$

96. $g(x) = 2 - \log x; f(x) = \log x$

97. $g(x) = 5 - 3 \ln x; f(x) = \ln x$

98. $g(x) = -3 - 2 \ln x; f(x) = \ln x$

In Problems 99–102, find f^{-1}.

99. $f(x) = \log_5 x$ **100.** $f(x) = \log_{1/3} x$

101. $f(x) = 4 \log_3(x + 3)$ **102.** $f(x) = 2 \log_2(x - 5)$

103. Let $f(x) = \log_3(2 - x)$.
(A) Find f^{-1}. (B) Graph f^{-1}.
(C) Reflect the graph of f^{-1} in the line $y = x$ to obtain the graph of f.

104. Let $f(x) = \log_2(-3 - x)$.
(A) Find f^{-1}. (B) Graph f^{-1}.
(C) Reflect the graph of f^{-1} in the line $y = x$ to obtain the graph of f.

105. What is wrong with the following "proof" that 3 is less than 2?

$$1 < 3 \qquad \text{Divide both sides by 27.}$$
$$\tfrac{1}{27} < \tfrac{3}{27}$$
$$\tfrac{1}{27} < \tfrac{1}{9}$$
$$\left(\tfrac{1}{3}\right)^3 < \left(\tfrac{1}{3}\right)^2$$
$$\log\left(\tfrac{1}{3}\right)^3 < \log\left(\tfrac{1}{3}\right)^2$$
$$3 \log \tfrac{1}{3} < 2 \log \tfrac{1}{3} \qquad \text{Divide both sides by } \log \tfrac{1}{3}.$$
$$3 < 2$$

106. What is wrong with the following "proof" that 1 is greater than 2?

$$3 > 2 \qquad \text{Multiply both sides by } \log \tfrac{1}{2}.$$

$$3 \log \tfrac{1}{2} > 2 \log \tfrac{1}{2}$$

$$\log \left(\tfrac{1}{2}\right)^3 > \log \left(\tfrac{1}{2}\right)^2$$

$$\left(\tfrac{1}{2}\right)^3 > \left(\tfrac{1}{2}\right)^2$$

$$\tfrac{1}{8} > \tfrac{1}{4} \qquad \text{Multiply both sides by } 8.$$

$$1 > 2$$

*The polynomials in Problems 107–110, called **Taylor polynomials**, can be used to approximate the function g(x) = ln (1 + x). To illustrate this approximation graphically, in each problem, graph*

$g(x) = ln(1 + x)$ *and the indicated polynomial in the same viewing window, $-1 \le x \le 3$ and $-2 \le y \le 2$.*

107. $P_1(x) = x - \tfrac{1}{2}x^2$ **108.** $P_2(x) = x - \tfrac{1}{2}x^2 + \tfrac{1}{3}x^3$

109. $P_3(x) = x - \tfrac{1}{2}x^2 + \tfrac{1}{3}x^3 - \tfrac{1}{4}x^4$

110. $P_4(x) = x - \tfrac{1}{2}x^2 + \tfrac{1}{3}x^3 - \tfrac{1}{4}x^4 + \tfrac{1}{5}x^5$

111. Prove that for any positive M, N, and b ($b \ne 1$), $\log_b \left(\tfrac{M}{N}\right) = \log_b M - \log_b N$. (*Hint:* Start by writing $u = \log_b M$ and $v = \log_b N$ and changing each to exponential form.)

112. Prove that for any positive integer p and any positive b and M ($b \ne 1$), $\log_b M^p = p \log_b M$. [*Hint:* Write M^p as $M \cdot M \cdot \cdots M$ (p factors).]

5-4 Logarithmic Models

› **Logarithmic Scales**
› **Data Analysis and Regression**

Logarithmic functions occur naturally as the inverses of exponential functions. But that's not to say that they are not useful in their own right. Some of these uses are probably familiar to you, but you might not have realized that they involved logarithmic functions.

In this section, we will study logarithmic scales that are used to compare the intensity of sounds, the severity of earthquakes, and the brightness of distant stars. We will also look at using regression to model data with a logarithmic function, and discuss what sort of data is likely to fit such a model.

› Logarithmic Scales

SOUND INTENSITY: The human ear is able to hear sound over a very wide range of intensities. The loudest sound a healthy person can hear without damage to the eardrum has an intensity 1 trillion (1,000,000,000,000) times that of the softest sound a person can hear. If we were to use these intensities as a scale for measuring volume, we would be stuck using numbers from zero all the way to the trillions, which seems cumbersome, if not downright silly. In the last section, we saw that logarithmic functions increase very slowly. We can take advantage of this to create a scale for sound intensity that is much more condensed, and therefore more manageable.

The decibel scale for sound intensity is an example of such a scale. The **decibel,** named after the inventor of the telephone, Alexander Graham Bell (1847–1922), is defined as follows:

$$D = 10 \log \frac{I}{I_0} \qquad \text{Decibel scale} \qquad (1)$$

where D is the **decibel level** of the sound, I is the **intensity** of the sound measured in watts per square meter (W/m^2), and I_0 is the intensity of the least audible sound that an average healthy young person can hear. The latter is standardized to be $I_0 = 10^{-12}$ watts per square meter. Table 1 lists some typical sound intensities from familiar sources. In Example 1 and Problems 5 and 6 in Exercises 5-4, we will calculate the decibel levels for these sounds.

Table 1 Typical Sound Intensities

Sound Intensity (W/m^2)	Sound
1.0×10^{-12}	Threshold of hearing
5.2×10^{-10}	Whisper
3.2×10^{-6}	Normal conversation
8.5×10^{-4}	Heavy traffic
3.2×10^{-3}	Jackhammer
1.0×10^{0}	Threshold of pain
8.3×10^{2}	Jet plane

| EXAMPLE | 1 | **Sound Intensity** |

(A) Find the number of decibels from a whisper with sound intensity 5.2×10^{-10} watts per square meter, then from heavy traffic at 8.5×10^{-4} watts per square meter. Round your answers to two decimal places.

(B) How many times larger is the sound intensity of heavy traffic compared to a whisper?

SOLUTIONS (A) We can use the decibel formula (1) with $I_0 = 10^{-12}$. First, we use $I = 5.2 \times 10^{-10}$:

$$D = 10 \log \frac{I}{I_0} \qquad \text{Substitute } I = 5.2 \times 10^{-10}, I_0 = 10^{-12}.$$

$$= 10 \log \frac{5.2 \times 10^{-10}}{10^{-12}} \qquad \text{Simplify the fraction.}$$

$$= 10 \log 520$$

$$= 27.16 \text{ decibels}$$

Next, for heavy traffic:

$$D = 10 \log \frac{I}{I_0} \qquad \text{Substitute } I = 8.5 \times 10^{-4}, I_0 = 10^{-12}.$$

$$= 10 \log \frac{8.5 \times 10^{-4}}{10^{-12}} \qquad \text{Simplify the fraction.}$$

$$= 10 \log 850,000,000$$

$$= 89.29 \text{ decibels}$$

(B) Dividing the larger intensity by the smaller,

$$\frac{8.5 \times 10^{-4}}{5.2 \times 10^{-10}} = 1,634,615.4$$

we see that the sound intensity of heavy traffic is more than 1.6 million times as great as the intensity of a whisper!

MATCHED PROBLEM 1 Find the number of decibels from a jackhammer with sound intensity 3.2×10^{-3} watts per square meter. Compute the answer to two decimal places.

⟩⟩⟩ EXPLORE-DISCUSS 1

Suppose that you are asked to draw a graph of the data in Table 1, with sound intensities on the x axis, and the corresponding decibel levels on the y axis.

(A) What would be the coordinates of the point corresponding to a jackhammer (see Matched Problem 1)?

(B) Suppose the axes of this graph are labeled as follows: Each tick mark on the x axis corresponds to the intensity of the least audible sound (10^{-12} watts per square meter), and each tick mark on the y axis corresponds to 1 decibel. If there is $\frac{1}{8}$ inch between all tick marks, how far away from the x axis is the point you found in part A? From the y axis? (Give the first answer in inches and the second in miles!) Discuss your result.

EARTHQUAKE INTENSITY: The energy released by the largest earthquake recorded, measured in joules, is about 100 billion (100,000,000,000) times the energy released by a small earthquake that is barely felt. In 1935 the California seismologist Charles Richter devised a logarithmic

Table 2 The Richter Scale

Magnitude on Richter Scale	Destructive Power
$M < 4.5$	Small
$4.5 < M < 5.5$	Moderate
$5.5 < M < 6.5$	Large
$6.5 < M < 7.5$	Major
$7.5 < M$	Great

scale that bears his name and is still widely used in the United States. The **magnitude** of an earthquake M on the **Richter scale*** is given as follows:

$$M = \frac{2}{3} \log \frac{E}{E_0} \qquad \text{Richter scale} \qquad (2)$$

where E is the energy released by the earthquake, measured in joules, and E_0 is the energy released by a very small reference earthquake, which has been standardized to be

$$E_0 = 10^{4.40} \text{ joules}$$

The destructive power of earthquakes relative to magnitudes on the Richter scale is indicated in Table 2.

EXAMPLE 2 Earthquake Intensity

The 1906 San Francisco earthquake released approximately 5.96×10^{16} joules of energy. Another quake struck the Bay Area just before game 3 of the 1989 World Series, releasing 1.12×10^{15} joules of energy.

(A) Find the magnitude of each earthquake on the Richter scale. Round your answers to two decimal places.

(B) How many times more energy did the 1906 earthquake release than the one in 1989?

SOLUTIONS (A) We can use the magnitude formula (2) with $E_0 = 10^{4.40}$. First, for the 1906 earthquake, $E = 5.96 \times 10^{16}$:

$$M = \frac{2}{3} \log \frac{E}{E_0} \qquad \text{Substitute } E = 5.96 \times 10^{16}, E_0 = 10^{4.40}.$$

$$= \frac{2}{3} \log \frac{5.96 \times 10^{16}}{10^{4.40}}$$

$$= 8.25$$

Next, for the 1989 earthquake, $E = 1.12 \times 10^{15}$

$$M = \frac{2}{3} \log \frac{E}{E_0} \qquad \text{Substitute } E = 1.12 \times 10^{15}, E_0 = 10^{4.40}.$$

$$= \frac{2}{3} \log \frac{1.12 \times 10^{15}}{10^{4.40}}$$

$$= 7.1$$

(B) Dividing the larger energy release by the smaller,

$$\frac{5.96 \times 10^{16}}{1.12 \times 10^{15}} = 53.2$$

we see that the 1906 earthquake released 53.2 times as much energy as the 1989 quake.

MATCHED PROBLEM 2 A 1985 earthquake in central Chile released approximately 1.26×10^{16} joules of energy. What was its magnitude on the Richter scale? Compute the answer to two decimal places.

*Originally, Richter defined the magnitude of an earthquake in terms of logarithms of the maximum seismic wave amplitude, in thousandths of a millimeter, measured on a standard seismograph. Equation (2) gives essentially the same magnitude that Richter obtained for a given earthquake but in terms of logarithms of the energy released by the earthquake.

| EXAMPLE | 3 | **Earthquake Intensity** |

If the energy release of one earthquake is 1,000 times that of another, how much larger is the Richter scale reading of the larger than the smaller?

SOLUTION Let

$$M_1 = \frac{2}{3} \log \frac{E_1}{E_0} \quad \text{and} \quad M_2 = \frac{2}{3} \log \frac{E_2}{E_0}$$

be the Richter equations for the smaller and larger earthquakes, respectively. Since the larger earthquake released 1,000 times as much energy, we can write $E_2 = 1,000E_1$.

$$M_2 = \frac{2}{3} \log \frac{E_2}{E_0} \qquad \text{Substitute } 1{,}000E_1 \text{ for } E_2.$$

$$= \frac{2}{3} \log \frac{1{,}000E_1}{E_0} \qquad \text{Use } \log(MN) = \log M + \log N;\ \frac{1{,}000E_1}{E_0} = 1{,}000 \cdot \frac{E_1}{E_0}$$

$$= \frac{2}{3} \left(\log 1{,}000 + \log \frac{E_1}{E_0} \right) \qquad \log 1{,}000 = \log 10^3 = 3$$

$$= \frac{2}{3} \left(3 + \log \frac{E_1}{E_0} \right) \qquad \text{Distribute.}$$

$$= \frac{2}{3}(3) + \frac{2}{3} \log \frac{E_1}{E_0} \qquad \frac{2}{3} \log \frac{E_1}{E_0} \text{ is } M_1!$$

$$= 2 + M_1$$

An earthquake with 1,000 times the energy of another has a Richter scale reading of 2 more than the other.

| MATCHED PROBLEM 3 |

If the energy release of one earthquake is 10,000 times that of another, how much larger is the Richter scale reading of the larger than the smaller?

ROCKET FLIGHT: The theory of rocket flight uses advanced mathematics and physics to show that the **velocity** v of a rocket at burnout (depletion of fuel supply) is given by

$$v = c \ln \frac{W_t}{W_b} \qquad \textbf{Rocket equation} \tag{3}$$

where c is the exhaust velocity of the rocket engine, W_t is the takeoff weight (fuel, structure, and payload), and W_b is the burnout weight (structure and payload).

Because of the Earth's atmospheric resistance, a launch vehicle velocity of at least 9.0 kilometers per second is required to achieve the minimum altitude needed for a stable orbit. Formula (3) indicates that to increase velocity v, either the weight ratio W_t/W_b must be increased or the exhaust velocity c must be increased. The weight ratio can be increased by the use of solid fuels, and the exhaust velocity can be increased by improving the fuels, solid or liquid.

| EXAMPLE | 4 | **Rocket Flight Theory** |

A typical single-stage, solid-fuel rocket may have a weight ratio $W_t/W_b = 18.7$ and an exhaust velocity $c = 2.38$ kilometers per second. Would this rocket reach a launch velocity of 9.0 kilometers per second?

SOLUTION We can use the rocket equation (3) with $c = 2.38$ and $W_t/W_b = 18.7$:

$$v = c \ln \frac{W_t}{W_b}$$
$$= 2.38 \ln 18.7$$
$$= 6.97 \text{ kilometers per second}$$

The velocity of the launch vehicle is far short of the 9.0 kilometers per second required to achieve orbit. This is why multiple-stage launchers are used—the deadweight from a preceding stage can be jettisoned into the ocean when the next stage takes over.

MATCHED PROBLEM 4 A launch vehicle using liquid fuel, such as a mixture of liquid hydrogen and liquid oxygen, can produce an exhaust velocity of $c = 4.7$ kilometers per second. However, the weight ratio W_t/W_b must be low—around 5.5 for some vehicles—because of the increased structural weight to accommodate the liquid fuel. How much more or less than the 9.0 kilometers per second required to reach orbit will be achieved by this vehicle?

> ## Data Analysis and Regression

Based on the logarithmic graphs we studied in the last section, when a quantity increases relatively rapidly at first, but then levels off and increases very slowly, it might be a good candidate to be modeled by a logarithmic function. Most graphing calculators with regression commands can fit functions of the form $y = a + b \ln x$ to a set of data points using the same techniques we used earlier for other types of regression.

EXAMPLE 5 Home Ownership Rates

Table 3 Home Ownership Rates

Year	Home Ownership Rate (%)
1940	43.6
1950	55.0
1960	61.9
1970	62.9
1980	64.4
1990	64.2
2000	67.4

The U.S. Census Bureau published the data in Table 3 on home ownership rates. A logarithmic model for the data is given by

$$R = -36.7 + 23.0 \ln t$$

where R is the home ownership rate and t is time in years with $t = 0$ representing 1900.

(A) Use the model to predict the home ownership rate in 2015.

(B) Compare the actual home ownership rate in 1950 to the rate given by the model.

SOLUTIONS (A) The year 2015 is represented by $t = 115$. Evaluating

$$R = -36.7 + 23.0 \ln t$$

at $t = 115$ predicts a home ownership rate of 72.4% in 2015.

(B) The year 1950 is represented by $t = 50$. Evaluating

$$R = -36.7 + 23.0 \ln t$$

at $t = 50$ gives a home ownership rate of 53.3% in 1950. The actual home ownership rate in 1950 was 55%, approximately 1.7% greater than the number given by the model.

Technology Connections

Figure 1 shows the details of constructing the logarithmic model of Example 5 on a graphing calculator.

 (a) Entering the data (b) Finding the model (c) Graphing the data and the model

› **Figure 1**

MATCHED PROBLEM 5

Refer to Example 5. The home ownership rate in 2008 was 67.8%. If this data is added to Table 3, a logarithmic model for the expanded data is given by

$$R = -30.6 + 21.5 \ln t$$

where R is the home ownership rate and t is time in years with $t = 0$ representing 1900.

(A) Use the model to predict the home ownership rate in 2015.

(B) Compare the actual home ownership rate in 1950 to the rate given by the model.

ANSWERS TO MATCHED PROBLEMS

1. 95.05 decibels **2.** 7.80 **3.** 2.67 **4.** 1 kilometer per second less
5. (A) 70.5% (B) The actual rate was 1.5% greater than the rate given by the model.

5-4 Exercises

1. Describe the decibel scale in your own words.

2. Describe the Richter scale in your own words.

3. Explain why logarithms are a good choice for describing sound intensity and earthquake magnitude.

4. Think of a real-life quantity that is likely to be modeled well by a logarithmic function, and explain your reasoning.

APPLICATIONS

5. SOUND What is the decibel level of
(A) The threshold of hearing, 1.0×10^{-12} watts per square meter?

(B) The threshold of pain, 1.0 watt per square meter?
Compute answers to two significant digits.

6. SOUND What is the decibel level of
(A) A normal conversation, 3.2×10^{-6} watts per square meter?
(B) A jet plane with an afterburner, 8.3×10^2 watts per square meter?
Compute answers to two significant digits.

7. SOUND If the intensity of a sound from one source is 1,000 times that of another, how much more is the decibel level of the louder sound than the quieter one?

8. SOUND If the intensity of a sound from one source is 10,000 times that of another, how much more is the decibel level of the louder sound than the quieter one?

9. EARTHQUAKES One of the strongest recorded earthquakes to date was in Colombia in 1906, with an energy release of 1.99×10^{17} joules. What was its magnitude on the Richter scale? Compute the answer to one decimal place.

10. EARTHQUAKES Anchorage, Alaska, had a major earthquake in 1964 that released 7.08×10^{16} joules of energy. What was its magnitude on the Richter scale? Compute the answer to one decimal place.

11. EARTHQUAKES The 1933 Long Beach, California, earthquake had a Richter scale reading of 6.3, and the 1964 Anchorage, Alaska, earthquake had a Richter scale reading of 8.3. How many times more powerful was the Anchorage earthquake than the Long Beach earthquake?

12. EARTHQUAKES Generally, an earthquake requires a magnitude of over 5.6 on the Richter scale to inflict serious damage. How many times more powerful than this was the great 1906 Colombia earthquake, which registered a magnitude of 8.6 on the Richter scale?

13. EXPLOSIVE ENERGY The atomic bomb dropped on Nagasaki, Japan, on August 9, 1945, released about 1.34×10^{14} joules of energy. What would be the magnitude of an earthquake that released that much energy?

14. EXPLOSIVE ENERGY The largest and most powerful nuclear weapon ever detonated was tested by the Soviet Union on October 30, 1961, on an island in the Arctic Sea. The blast was so powerful there were reports of windows breaking in Finland, over 700 miles away. The detonation released about 2.1×10^{17} joules of energy. What would be the magnitude of an earthquake that released that much energy?

15. ASTRONOMY A moderate-size solar flare observed on the sun on July 9, 1996, released enough energy to power the United States for almost 23,000 years at 2001 consumption levels, 2.38×10^{21} joules. What would be the magnitude of an earthquake that released that much energy?

16. CONSTRUCTION The energy released by a typical construction site explosion is about 7.94×10^5 joules. What would be the magnitude of an earthquake that released that much energy?

17. SPACE VEHICLES A new solid-fuel rocket has a weight ratio $W_t/W_b = 19.8$ and an exhaust velocity $c = 2.57$ kilometers per second. What is its velocity at burnout? Compute the answer to two decimal places.

18. SPACE VEHICLES A liquid-fuel rocket has a weight ratio $W_t/W_b = 6.2$ and an exhaust velocity $c = 5.2$ kilometers per second. What is its velocity at burnout? Compute the answer to two decimal places.

19. CHEMISTRY The hydrogen ion concentration of a substance is related to its acidity and basicity. Because hydrogen ion concentrations vary over a very wide range, logarithms are used to create a compressed **pH scale,** which is defined as follows:

$$\text{pH} = -\log\, [\text{H}^+]$$

where $[\text{H}^+]$ is the hydrogen ion concentration, in moles per liter. Pure water has a pH of 7, which means it is neutral. Substances with a pH less than 7 are acidic, and those with a pH greater than 7 are basic. Compute the pH of each substance listed, given the indicated hydrogen ion concentration. Also, indicate whether each substance is acidic or basic. Compute answers to one decimal place.
(A) Seawater, 4.63×10^{-9}
(B) Vinegar, 9.32×10^{-4}

20. CHEMISTRY Refer to Problem 19. Compute the pH of each substance below, given the indicated hydrogen ion concentration. Also, indicate whether it is acidic or basic. Compute answers to one decimal place.
(A) Milk, 2.83×10^{-7}
(B) Garden mulch, 3.78×10^{-6}

21. ECOLOGY Refer to Problem 19. Many lakes in Canada and the United States will no longer sustain some forms of wildlife because of the increase in acidity of the water from acid rain and snow caused by sulfur dioxide emissions from industry. If the pH of a sample of rainwater is 5.2, what is its hydrogen ion concentration in moles per liter? Compute the answer to two significant digits.

22. ECOLOGY Refer to Problem 19. If normal rainwater has a pH of 5.7, what is its hydrogen ion concentration in moles per liter? Compute the answer to two significant digits.

23. ASTRONOMY The brightness of stars is expressed in terms of magnitudes on a numerical scale that increases as the brightness decreases. The magnitude m is given by the formula

$$m = 6 - 2.5 \log \frac{L}{L_0}$$

where L is the light flux of the star and L_0 is the light flux of the dimmest stars visible to the naked eye.
(A) What is the magnitude of the dimmest stars visible to the naked eye?
(B) How many times brighter is a star of magnitude 1 than a star of magnitude 6?

24. ASTRONOMY An optical instrument is required to observe stars beyond the sixth magnitude, the limit of ordinary vision. However, even optical instruments have their limitations. The limiting magnitude L of any optical telescope with lens diameter D, in inches, is given by

$$L = 8.8 + 5.1 \log D$$

(A) Find the limiting magnitude for a homemade 6-inch reflecting telescope.
(B) Find the diameter of a lens that would have a limiting magnitude of 20.6.
Compute answers to three significant digits.

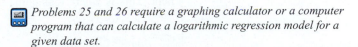

Problems 25 and 26 require a graphing calculator or a computer program that can calculate a logarithmic regression model for a given data set.

25. INTERNET ACCESS Table 4 on page 372 shows the percentage of Americans that had access to the Internet either at home or at work between 2000 and 2006. Let x represent years since 1995.

Table 4 Internet Access in the United States

Year	Percentage with Home Access	Percentage with Work Access
2000	46.9	35.2
2001	58.4	37.5
2002	59.3	40.2
2003	65.1	49.6
2005	66.2	55.1
2006	68.1	55.8

(A) Find a logarithmic regression model ($y = a + b \ln x$) for the percentage with home access. Round a and b to three significant digits. Use your model to estimate the percentage in 2008 and 2015.

(B) Examine the model for larger and larger values of x. Does it remain reasonable in the long term?

26. INTERNET ACCESS Refer to Table 4.

(A) Find a logarithmic regression model ($y = a + b \ln x$) for the percentage with work access. (Keep in mind that x represents years since 1995.) Round a and b to three significant digits. Use your model to estimate the percentage in 2008 and 2015.

(B) Examine the model for larger and larger values of x. Does it remain reasonable in the long term?

5-5 Exponential and Logarithmic Equations

› Solving Exponential Equations
› Solving Logarithmic Equations

We have seen that many quantities can be modeled by exponential or logarithmic functions. So it's not surprising that equations involving exponential or logarithmic expressions, like those shown next, are useful in studying those quantities.

$$2^{3x-2} = 5 \quad \text{and} \quad \log (x + 3) + \log x = 1$$

Equations like these are called **exponential** and **logarithmic** equations, respectively. The properties of logarithms that we studied in Section 5-3 will play a key role in solving both types of equations.

› Solving Exponential Equations

The distinguishing feature of exponential equations is that the variable appears in an exponent. Before defining logarithms, we didn't have a reliable method for removing variables from an exponent: Now we do. We'll illustrate how these properties are helpful in Examples 1-4.

EXAMPLE 1 Solving an Exponential Equation

Find all solutions to $2^{3x-2} = 5$ to four decimal places.

SOLUTION In order to have any chance of solving for x, we will first need to get x out of the exponent. This is where logs come in very handy.

$$2^{3x-2} = 5 \qquad \text{Take the common or natural log of both sides.}$$

$$\log 2^{3x-2} = \log 5 \qquad \text{Use } \log_b N^p = p \log_b N \text{ to get } 3x - 2 \text{ out of the exponent position.}$$

$$(3x - 2) \log 2 = \log 5 \qquad \text{Divide both sides by log 2.}$$

$$3x - 2 = \frac{\log 5}{\log 2} \qquad \text{Add 2 to both sides.}$$

$$3x = 2 + \frac{\log 5}{\log 2} \qquad \text{\color{magenta}Divide both sides by 3, or multiply both sides by } \frac{1}{3}.$$

$$x = \frac{1}{3}\left(2 + \frac{\log 5}{\log 2}\right) \qquad \text{\color{magenta}Use a calculator.}$$

$$= 1.4406 \qquad \text{\color{magenta}Solution to four decimal places}$$

MATCHED PROBLEM 1

Solve $35^{1-2x} = 7$ for x to four decimal places.

EXAMPLE 2 Compound Interest

Recall that when an amount of money P (principal) is invested at an annual rate r compounded annually, the amount of money A in the account after n years, assuming no withdrawals, is given by

$$A = P\left(1 + \frac{r}{m}\right)^n = P(1 + r)^n \qquad \text{\color{magenta}} m = 1 \text{ for annual compounding.}$$

How many years to the nearest year will it take the money to double if it is invested at 6% compounded annually?

SOLUTION The interest rate is $r = 0.06$, and we want the amount A to be twice the principal, or $2P$. So we substitute $r = 0.06$ and $A = 2P$, and solve for n.

$$2P = P(1.06)^n \qquad \text{\color{magenta}Divide both sides by } P \text{ to isolate } (1.06)^n.$$

$$2 = 1.06^n \qquad \text{\color{magenta}Take the common or natural log of both sides.}$$

$$\log 2 = \log 1.06^n \qquad \text{\color{magenta}Note how log properties are used to get } n \text{ out of the exponent position.}$$

$$\log 2 = n \log 1.06 \qquad \text{\color{magenta}Divide both sides by } \log 1.06 \text{ (which is just a number!).}$$

$$n = \frac{\log 2}{\log 1.06} \qquad \text{\color{magenta}Calculate to the nearest year.}$$

$$= 12 \text{ years}$$

MATCHED PROBLEM 2

Repeat Example 2, changing the interest rate to 9% compounded annually.

EXAMPLE 3 Atmospheric Pressure

The atmospheric pressure P, in pounds per square inch, at x miles above sea level is given approximately by

$$P = 14.7e^{-0.21x}$$

At what height will the atmospheric pressure be half the sea-level pressure? Compute the answer to two significant digits.

SOLUTION Since x is miles above sea level, sea-level pressure is the pressure at $x = 0$, which is $14.7e^0$, or 14.7.

One-half of sea level pressure is $14.7/2 = 7.35$. Now our problem is to find x so that $P = 7.35$; that is, we solve $7.35 = 14.7e^{-0.21x}$ for x:

$$7.35 = 14.7e^{-0.21x} \qquad \text{\color{magenta}Divide both sides by 14.7 to isolate the exponential.}$$

$$0.5 = e^{-0.21x} \qquad \text{\color{magenta}Because the base is } e, \text{ take the natural log of both sides.}$$

$$\ln 0.5 = \ln e^{-0.21x} \qquad \text{\color{magenta}} \ln e^a = a, \text{ so } \ln e^{-0.21x} = -0.21x$$

$$\ln 0.5 = -0.21x \qquad \text{Divide both sides by } -0.21.$$

$$x = \frac{\ln 0.5}{-0.21} \qquad \text{Calculate to two significant digits.}$$

$$= 3.3 \text{ miles}$$

MATCHED PROBLEM 3

Using the formula in Example 3, find the altitude in miles so that the atmospheric pressure will be one-eighth that at sea level. Compute the answer to two significant digits.

$y = \dfrac{e^x + e^{-x}}{2}$

> **Figure 1** Catenary.

The graph of

$$y = \frac{e^x + e^{-x}}{2} \tag{1}$$

is a curve called a **catenary** (Fig. 1). A uniform cable suspended between two fixed points is a physical example of such a curve, which resembles a parabola, but isn't.

EXAMPLE 4 **Solving an Exponential Equation**

In equation (1), find x when $y = 2.5$. Compute the answer to four decimal places.

SOLUTION

$$y = \frac{e^x + e^{-x}}{2} \qquad \text{Let } y = 2.5.$$

$$2.5 = \frac{e^x + e^{-x}}{2} \qquad \text{Multiply both sides by 2.}$$

$$5 = e^x + e^{-x} \qquad \text{Multiply both sides by } e^x.$$

$$5e^x = e^{2x} + 1 \qquad \text{Subtract } 5e^x \text{ from both sides.}$$

$$e^{2x} - 5e^x + 1 = 0 \qquad \text{This is a quadratic in } e^x.$$

Let $u = e^x$; then

$$u^2 - 5u + 1 = 0 \qquad \text{Use the quadratic formula.}$$

$$u = \frac{5 \pm \sqrt{25 - 4(1)(1)}}{2} \qquad \text{Simplify.}$$

$$= \frac{5 \pm \sqrt{21}}{2} \qquad \text{Replace } u \text{ with } e^x \text{ and solve for } x.$$

$$e^x = \frac{5 \pm \sqrt{21}}{2} \qquad \text{Take the natural log of both sides}$$
$$\text{(both values on the right are positive).}$$

$$\ln e^x = \ln \frac{5 \pm \sqrt{21}}{2} \qquad \log_b b^x = x, \text{ so } \ln e^x = x.$$

$$x = \ln \frac{5 \pm \sqrt{21}}{2} \qquad \text{Exact solutions}$$

$$= -1.5668, \ 1.5668 \qquad \text{Rounded to four decimal places.}$$

Note that the method produces exact solutions, an important consideration in certain calculus applications (see Problems 57–60 of Exercises 5-5).

MATCHED PROBLEM 4

Given $y = (e^x - e^{-x})/2$, find x for $y = 1.5$. Compute the answer to three decimal places.

› **Solving Logarithmic Equations**

We will now illustrate the solution of several types of logarithmic equations.

| EXAMPLE | 5 | **Solving a Logarithmic Equation** |

Solve $\log (x + 3) + \log x = 1$, and check.

SOLUTION First use properties of logarithms to express the left side as a single logarithm, then convert to exponential form and solve for x.

$$\log (x + 3) + \log x = 1 \qquad \text{Combine left side using log } M + \log N = \log MN.$$
$$\log [x(x + 3)] = 1 \qquad \text{Change to equivalent exponential form (the base is 10).}$$
$$x(x + 3) = 10^1 \qquad \text{Write in } ax^2 + bx + c = 0 \text{ form and solve.}$$
$$x^2 + 3x - 10 = 0 \qquad \text{Factor.}$$
$$(x + 5)(x - 2) = 0 \qquad \text{If } ab = 0, \text{ then } a = 0 \text{ or } b = 0.$$
$$x = -5, 2$$

CHECK $x = -5$: $\log (-5 + 3) + \log (-5)$ is not defined
because the domain of the log function is $(0, \infty)$.
$x = 2$: $\log (2 + 3) + \log 2 = \log 5 + \log 2$
$= \log (5 \cdot 2) = \log 10 \overset{\checkmark}{=} 1$

The only solution to the original equation is $x = 2$. Extraneous solutions are common in log equations, so answers should always be checked in the original equation to see whether any should be discarded.

| MATCHED PROBLEM 5 | Solve $\log (x - 15) = 2 - \log x$, and check. |

| EXAMPLE | 6 | **Solving a Logarithmic Equation** |

Solve $(\ln x)^2 = \ln x^2$.

SOLUTION There are no logarithmic properties for simplifying $(\ln x)^2$. However, we can simplify $\ln x^2$, obtaining an equation involving $\ln x$ and $(\ln x)^2$.

$$(\ln x)^2 = \ln x^2 \qquad \text{In } M^p = p \ln M, \text{ so } \ln x^2 = 2 \ln x.$$
$$(\ln x)^2 = 2 \ln x \qquad \text{This is a quadratic equation in ln } x. \text{ Move all nonzero terms to the left.}$$
$$(\ln x)^2 - 2 \ln x = 0 \qquad \text{Factor out ln } x.$$
$$(\ln x)(\ln x - 2) = 0 \qquad \text{If } ab = 0, \text{ then } a = 0 \text{ or } b = 0.$$
$$\ln x = 0 \quad \text{or} \quad \ln x - 2 = 0 \qquad \text{If ln } x = a, x = e^a.$$
$$x = e^0 \qquad\qquad \ln x = 2$$
$$= 1 \qquad\qquad\quad x = e^2$$

Checking that both $x = 1$ and $x = e^2$ are solutions to the original equation is left to you.

| MATCHED PROBLEM 6 | Solve $\log x^2 = (\log x)^2$. |

>>> CAUTION >>>

Note that

$$(\log_b x)^2 \neq \log_b x^2$$

$(\log_b x)^2 = (\log_b x)(\log_b x)$
$\log_b x^2 = 2 \log_b x$

You might find it helpful to keep these straight by writing $\log_b x^2$ as $\log_b (x^2)$.

EXAMPLE 7 Earthquake Intensity

Recall from Section 5-4 that the magnitude of an earthquake on the Richter scale is given by

$$M = \frac{2}{3} \log \frac{E}{E_0}$$

Solve for E in terms of the other symbols.

SOLUTION

$$M = \frac{2}{3} \log \frac{E}{E_0}$$ Multiply both sides by $\frac{3}{2}$ and switch sides.

$$\log \frac{E}{E_0} = \frac{3M}{2}$$ Change to exponential form with base 10.

$$\frac{E}{E_0} = 10^{3M/2}$$ Multiply both sides by E_0.

$$E = E_0 10^{3M/2}$$

MATCHED PROBLEM 7 Solve the rocket equation from Section 5-4 for W_b in terms of the other symbols:

$$v = c \ln \frac{W_t}{W_b}$$

ANSWERS TO MATCHED PROBLEMS

1. $x = 0.2263$ 2. More than double in 9 years, but not quite double in 8 years
3. 9.9 miles 4. $x = 1.195$ 5. $x = 20$ 6. $x = 1{,}100$ 7. $W_b = W_t e^{-v/c}$

5-5 Exercises

1. Which property of logarithms do you think is most useful in solving exponential equations? Explain.

2. Which properties of logarithms do you think are most useful in solving equations with more than one logarithm? Explain.

3. If u and v represent expressions with variable x, how can you solve equations of the form $\log_b u = \log_b v$ for x? Explain why this works.

4. Why is it especially important to check answers when solving logarithmic equations?

5. Explain the difference between $(\ln x)^2$ and $\ln x^2$.

6. Can you use a logarithm with the same base to solve both equations below? Explain.

$$e^x = 10 \qquad \text{and} \qquad 5^x = 8$$

In Problems 7–16, solve to three significant digits.

7. $10^{-x} = 0.0347$ 8. $10^x = 14.3$

9. $10^{3x+1} = 92$ 10. $10^{5x-2} = 348$

11. $e^x = 3.65$ 12. $e^{-x} = 0.0142$

13. $e^{2x-1} + 68 = 207$ 14. $13 + e^{3x+5} = 23$

15. $2^3 2^{-x} = 0.426$ **16.** $3^4 3^{-x} = 0.089$

In Problems 17–26, solve exactly.

17. $\log_5 x = 2$ **18.** $\log_3 y = 4$

19. $\log (t - 4) = -1$ **20.** $\ln (2x + 3) = 0$

21. $\log 5 + \log x = 2$ **22.** $\log x - \log 8 = 1$

23. $\log x + \log (x - 3) = 1$

24. $\log (x - 9) + \log 100x = 3$

25. $\log (x + 1) - \log (x - 1) = 1$

26. $\log (2x + 1) = 1 + \log (x - 2)$

In Problems 27–34, solve to three significant digits.

27. $2 = 1.05^x$ **28.** $3 = 1.06^x$

29. $e^{-1.4x} + 5 = 0$ **30.** $e^{0.32x} + 0.47 = 0$

31. $123 = 500e^{-0.12x}$ **32.** $438 = 200e^{0.25x}$

33. $e^{-x^2} = 0.23$ **34.** $e^{x^2} = 125$

In Problems 35–48, solve exactly.

35. $\log (5 - 2x) = \log (3x + 1)$

36. $\log (x + 3) = \log (6 + 4x)$

37. $\log x - \log 5 = \log 2 - \log (x - 3)$

38. $\log (6x + 5) - \log 3 = \log 2 - \log x$

39. $\ln x = \ln (2x - 1) - \ln (x - 2)$

40. $\ln (x + 1) = \ln (3x + 1) - \ln x$

41. $\log (2x + 1) = 1 - \log (x - 1)$

42. $1 - \log (x - 2) = \log (3x + 1)$

43. $\ln (x + 1) = \ln (3x + 3)$

44. $1 + \ln (x + 1) = \ln (x - 1)$

45. $(\ln x)^3 = \ln x^4$ **46.** $(\log x)^3 = \log x^4$

47. $\ln (\ln x) = 1$ **48.** $\log (\log x) = 1$

Solve Problems 49–56 for the indicated variable in terms of the remaining symbols. Use the natural log for solving exponential equations.

49. $A = Pe^{rt}$ for r (finance)

50. $A = P\left(1 + \dfrac{r}{n}\right)^{nt}$ for t (finance)

51. $D = 10 \log \dfrac{I}{I_0}$ for I (sound)

52. $t = \dfrac{-1}{k} (\ln A - \ln A_0)$ for A (decay)

53. $M = 6 - 2.5 \log \dfrac{I}{I_0}$ for I (astronomy)

54. $L = 8.8 + 5.1 \log D$ for D (astronomy)

55. $I = \dfrac{E}{R} (1 - e^{-Rt/L})$ for t (circuitry)

56. $S = R\dfrac{(1 + i)^n - 1}{i}$ for n (annuity)

*The following combinations of exponential functions define four of six **hyperbolic functions,** a useful class of functions in calculus and higher mathematics. Solve Problems 57–60 for x in terms of y. The results are used to define **inverse hyperbolic functions,** another useful class of functions in calculus and higher mathematics.*

57. $y = \dfrac{e^x + e^{-x}}{2}$ **58.** $y = \dfrac{e^x - e^{-x}}{2}$

59. $y = \dfrac{e^x - e^{-x}}{e^x + e^{-x}}$ **60.** $y = \dfrac{e^x + e^{-x}}{e^x - e^{-x}}$

In Problems 61–68, use a graphing calculator to approximate to two decimal places any solutions of the equation in the interval $0 \le x \le 1$. None of these equations can be solved exactly using any step-by-step algebraic process.

61. $2^{-x} - 2x = 0$ **62.** $3^{-x} - 3x = 0$

63. $e^{-x} - x = 0$ **64.** $xe^{2x} - 1 = 0$

65. $\ln x + 2x = 0$ **66.** $\ln x + x^2 = 0$

67. $\ln x + e^x = 0$ **68.** $\ln x + x = 0$

APPLICATIONS

69. COMPOUND INTEREST How many years, to the nearest year, will it take a sum of money to double if it is invested at 7% compounded annually?

70. COMPOUND INTEREST How many years, to the nearest year, will it take money to quadruple if it is invested at 6% compounded annually?

71. COMPOUND INTEREST At what annual rate compounded continuously will $1,000 have to be invested to amount to $2,500 in 10 years? Compute the answer to three significant digits.

72. COMPOUND INTEREST How many years will it take $5,000 to amount to $8,000 if it is invested at an annual rate of 9% compounded continuously? Compute the answer to three significant digits.

73. IMMIGRATION According to the U.S. Office of Immigration Statistics, there were 10.5 million illegal immigrants in the United States in May 2005, and that number had grown to 11.3 million by May 2007.
(A) Find the relative growth rate if we use the $P = P_0 e^{rt}$ model for population growth. Round to three significant digits.
(B) Use your answer from part A to write a function describing the illegal immigrant population in millions in terms of years after May 2005, and use it to predict when the illegal immigrant population should reach 20 million.

74. POPULATION GROWTH According to U.S. Census Bureau estimates, the population of the United States was 227.2 million on July 1, 1980, and 249.5 million on July 1, 1990.
(A) Find the relative growth rate if we use the $P = P_0 e^{rt}$ model for population growth. Round to three significant digits.
(B) Use your answer from part A to write a function describing the population of the United States in millions in terms of years after July 1980, and use it to predict when the population should reach 400 million.
(C) Use your function from part B to estimate the population of the United States today, then compare your estimate to the one found at www.census.gov/population/www/popclockus.html.

75. WORLD POPULATION A mathematical model for world population growth over short periods is given by

$$P = P_0 e^{rt}$$

where P is the population after t years, P_0 is the population at $t = 0$, and the population is assumed to grow continuously at the annual rate r. How many years, to the nearest year, will it take the world population to double if it grows continuously at an annual rate of 1.14%?

76. WORLD POPULATION Refer to Problem 75. Starting with a world population of 6.8 billion people (the estimated population in March 2009) and assuming that the population grows continuously at an annual rate of 1.14%, how many years, to the nearest year, will it be before there is only 1 square yard of land per person? Earth contains approximately 1.7×10^{14} square yards of land.

77. MEDICAL RESEARCH A medical researcher is testing a radioactive isotope for use in a new imaging process. She finds that an original sample of 5 grams decays to 1 gram in 6 hours. Find the half-life of the sample to three significant digits. [Recall that the half-life model is $A = A_0(\frac{1}{2})^{t/h}$, where A_0 is the original amount and h is the half-life.]

78. CARBON-14 DATING If 90% of a sample of carbon-14 remains after 866 years, what is the half-life of carbon-14? (See Problem 77 for the half-life model.)

As long as a plant or animal remains alive, carbon-14 is maintained in a constant amount in its tissues. Once dead, however, the plant or animal ceases taking in carbon, and carbon-14 diminishes by radioactive decay. The amount remaining can be modeled by the equation $A = A_0 e^{-0.000124t}$, where A is the amount after t years, and A_0 is the amount at time $t = 0$. Use this model to solve Problems 79–82.

79. CARBON-14 DATING In 2003, Japanese scientists announced the beginning of an effort to bring the long-extinct woolly mammoth back to life using modern cloning techniques. Their efforts were focused on an especially well-preserved specimen discovered frozen in the Siberian ice. Nearby samples of plant material were found to have 28.9% of the amount of carbon-14 in a living sample. What was the approximate age of these samples?

80. CARBON-14 DATING In 2004, archaeologist Al Goodyear discovered a site in South Carolina that contains evidence of the earliest human settlement in North America. Carbon dating of burned plant material indicated 0.2% of the amount of carbon-14 in a live sample. How old was that sample?

81. CARBON-14 DATING Many scholars believe that the earliest nonnative settlers of North America were Vikings who sailed from Iceland.

If a fragment of a wooden tool found and dated in 2004 had 88.3% of the amount of carbon-14 in a living sample, when was this tool made?

82. CARBON-14 DATING In 1998, the Shroud of Turin was examined by researchers, who found that plant fibers in the fabric had 92.1% of the amount of carbon-14 in a living sample. If this is accurate, when was the fabric made?

83. PHOTOGRAPHY An electronic flash unit for a camera is activated when a capacitor is discharged through a filament of wire. After the flash is triggered and the capacitor is discharged, the circuit (see the figure) is connected and the battery pack generates a current to recharge the capacitor. The time it takes for the capacitor to recharge is called the *recycle time*. For a particular flash unit using a 12-volt battery pack, the charge q, in coulombs, on the capacitor t seconds after recharging has started is given by

$$q = 0.0009(1 - e^{-0.2t})$$

How many seconds will it take the capacitor to reach a charge of 0.0007 coulomb? Compute the answer to three significant digits.

84. ADVERTISING A company is trying to expose as many people as possible to a new product through television advertising in a large metropolitan area with 2 million possible viewers. A model for the number of people N, in millions, who are aware of the product after t days of advertising was found to be

$$N = 2(1 - e^{-0.037t})$$

How many days, to the nearest day, will the advertising campaign have to last so that 80% of the possible viewers will be aware of the product?

85. NEWTON'S LAW OF COOLING This law states that the rate at which an object cools is proportional to the difference in temperature between the object and its surrounding medium. The temperature T of the object t hours later is given by

$$T = T_m + (T_0 - T_m)e^{-kt}$$

where T_m is the temperature of the surrounding medium and T_0 is the temperature of the object at $t = 0$. Suppose a bottle of wine at a room temperature of 72°F is placed in a refrigerator at 40°F to cool before a dinner party. After an hour the temperature of the wine is found to be 61.5°F. Find the constant k, to two decimal places, and the time, to one decimal place, it will take the wine to cool from 72 to 50°F.

86. MARINE BIOLOGY Marine life is dependent upon the microscopic plant life that exists in the *photic zone*, a zone that goes to a depth where about 1% of the surface light still remains. Light intensity is reduced according to the exponential function

$$I = I_0 e^{-kd}$$

where I is the intensity d feet below the surface and I_0 is the intensity at the surface. The constant k is called the *coefficient of extinction*. At Crystal Lake in Wisconsin it was found that half the surface light remained at a depth of 14.3 feet. Find k, and find the depth of the photic zone. Compute answers to three significant digits.

Problems 87–90 are based on the Richter scale equation from Section 5-4, $M = \frac{2}{3} \log \frac{E}{10^{4.40}}$, where M is the magnitude and E is the amount of energy in joules released by the earthquake. Round all calculations to three significant digits.

87. EARTHQUAKES There were 12 earthquakes recorded worldwide in 2008 with magnitude at least 7.0.
(A) How much energy is released by a magnitude 7.0 earthquake?
(B) The total average daily consumption of energy for the entire United States in 2006 was 2.88×10^{14} joules. How many days could the energy released by a magnitude 7.0 earthquake power the United States?

88. EARTHQUAKES On December 26, 2004, a magnitude 9.0 earthquake struck in the Indian Ocean, causing a massive tsunami that resulted in over 230,000 deaths.
(A) How much energy was released by this earthquake?

(B) The total average daily consumption of energy for the entire United States in 2006 was 2.88×10^{14} joules. How many days could the energy released by a magnitude 9.0 earthquake power the United States?

89. EARTHQUAKES There were 12 earthquakes worldwide in 2008 with magnitudes between 7.0 and 7.9. Assume that these earthquakes had an average magnitude of 7.5. How long could the total energy released by these 12 earthquakes power the United States, which had a total energy consumption of 1.05×10^{17} joules in 2006?

90. EARTHQUAKES There were 166 earthquakes worldwide in 2008 with magnitudes between 6.0 and 6.9. Assume that these earthquakes had an average magnitude of 6.5. How long could the total energy released by these 166 earthquakes power the United States, which had a total energy consumption of 1.05×10^{17} joules in 2006?

CHAPTER 5 Review

5-1 Exponential Functions

The equation $f(x) = b^x$, $b > 0$, $b \neq 1$, defines an **exponential function** with **base b**. The **domain** of f is $(-\infty, \infty)$ and the **range** is $(0, \infty)$. The **graph** of f is a continuous curve that has no sharp corners; passes through $(0, 1)$; lies above the x axis, which is a horizontal asymptote; increases as x increases if $b > 1$; decreases as x increases if $b < 1$; and intersects any horizontal line at most once. The function f is one-to-one and has an inverse. We often use the following **exponential function properties:**

1. $a^x a^y = a^{x+y}$ $(a^x)^y = a^{xy}$ $(ab)^x = a^x b^x$

$$\left(\frac{a}{b}\right)^x = \frac{a^x}{b^x} \qquad \frac{a^x}{a^y} = a^{x-y}$$

2. $a^x = a^y$ if and only if $x = y$.

3. For $x \neq 0$, $a^x = b^x$ if and only if $a = b$.

As x approaches ∞, the expression $[1 + (1/x)]^x$ approaches the irrational number $e \approx 2.718\,281\,828\,459$. The function $f(x) = e^x$ is called the **exponential function with base e.** The growth of money in an account paying **compound interest** is described by $A = P(1 + r/m)^n$, where P is the **principal**, r is the annual **rate**, m is the number of compounding periods in 1 year, and A is the **amount** in the account after n compounding periods.

If the account pays **continuous compound interest,** the amount A in the account after t years is given by $A = Pe^{rt}$.

5-2 Exponential Models

Exponential functions are used to model various types of growth:

1. *Population growth* can be modeled by using the **doubling time growth model** $A = A_0 2^{t/d}$, where A is the population at time t, A_0 is the population at time $t = 0$, and d is the **doubling time**—

the time it takes for the population to double. Another model of population growth, $A = A_0 e^{kt}$, where A_0 is the population at time zero and k is a positive constant called the **relative growth rate,** uses the exponential function with base e. This model is used for many other types of quantities that exhibit exponential growth as well.

2. *Radioactive decay* can be modeled by using the **half-life decay model** $A = A_0 (\frac{1}{2})^{t/h} = A_0 2^{-t/h}$, where A is the amount at time t, A_0 is the amount at time $t = 0$, and h is the **half-life**—the time it takes for half the material to decay. Another model of radioactive decay, $A = A_0 e^{-kt}$, where A_0 is the amount at time zero and k is a positive constant, uses the exponential function with base e. This model can be used for other types of quantities that exhibit negative exponential growth as well.

3. *Limited growth*—the growth of a company or proficiency at learning a skill, for example—can often be modeled by the equation $y = A(1 - e^{-kt})$, where A and k are positive constants.

Logistic growth is another limited growth model that is useful for modeling phenomena like the spread of an epidemic, or sales of a new product. The logistic model is $A = M/(1 + ce^{-kt})$, where c, k, and M are positive constants. A good comparison of these different exponential models can be found in Table 3 at the end of Section 5-2.

Exponential regression can be used to fit a function of the form $y = ab^x$ to a set of data points. Logistic regression can be used to find a function of the form $y = c/(1 + ae^{-bx})$.

5-3 Logarithmic Functions

The **logarithmic function with base b** is defined to be the inverse of the exponential function with base b and is denoted by $y = \log_b x$. So $y = \log_b x$ if and only if $x = b^y$, $b > 0$, $b \neq 1$. The **domain** of a logarithmic function is $(0, \infty)$ and the **range** is $(-\infty, \infty)$. The graph of a logarithmic function is a continuous curve that always passes

through the point $(1, 0)$ and has the y axis as a vertical asymptote. The following **properties of logarithmic functions** are useful:

1. $\log_b 1 = 0$

2. $\log_b b = 1$

3. $\log_b b^x = x$

4. $b^{\log_b x} = x, x > 0$

5. $\log_b MN = \log_b M + \log_b N$

6. $\log_b \dfrac{M}{N} = \log_b M - \log_b N$

7. $\log_b M^p = p \log_b M$

8. $\log_b M = \log_b N$ if and only if $M = N$

Logarithms to the base 10 are called **common logarithms** and are denoted by $\log x$. Logarithms to the base e are called **natural logarithms** and are denoted by $\ln x$. So $\log x = y$ is equivalent to $x = 10^y$, and $\ln x = y$ is equivalent to $x = e^y$.

The **change-of-base formula**, $\log_b N = (\log_a N)/(\log_a b)$, relates logarithms to two different bases and can be used, along with a calculator, to evaluate logarithms to bases other than e or 10.

5-4 Logarithmic Models

The following applications involve logarithmic functions:

1. The **decibel** is defined by $D = 10 \log (I/I_0)$, where D is the **decibel level** of the sound, I is the **intensity** of the sound, and $I_0 = 10^{-12}$ watts per square meter is a standardized sound level.

2. The **magnitude** M of an earthquake on the **Richter scale** is given by $M = \frac{2}{3} \log (E/E_0)$, where E is the energy released by the earthquake and $E_0 = 10^{4.40}$ joules is a standardized energy level.

3. The **velocity** v of a rocket at burnout is given by the **rocket equation** $v = c \ln (W_t/W_b)$, where c is the exhaust velocity, W_t is the takeoff weight, and W_b is the burnout weight.

Logarithmic regression can be used to fit a function of the form $y = a + b \ln x$ to a set of data points.

5-5 Exponential and Logarithmic Equations

Exponential equations are equations in which the variable appears in an exponent. If the exponential expression is isolated, applying a logarithmic function to both sides and using the property $\log_b N^p = p \log_b N$ will enable you to remove the variable from the exponent. If the exponential expression is not isolated, we can use previously developed techniques to first solve for the exponential, then solve as above.

Logarithmic equations are equations in which the variable appears inside a logarithmic function. In most cases, the key to solving them is to change the equation to the equivalent exponential expression. For equations with multiple log expressions, properties of logarithms can be used to combine the expressions before solving.

CHAPTER 5 Review Exercises

Work through all the problems in this chapter review and check answers in the back of the book. Answers to all review problems are there, and following each answer is a number in italics indicating the section in which that type of problem is discussed. Where weaknesses show up, review appropriate sections in the text.

1. Match each equation with the graph of f, g, m, or n in the figure.
 (A) $y = \log_2 x$ (B) $y = 0.5^x$
 (C) $y = \log_{0.5} x$ (D) $y = 2^x$

2. Write in logarithmic form using base 10: $m = 10^n$.

3. Write in logarithmic form using base e: $x = e^y$.

Write the expression in Problems 4 and 5 in exponential form.

4. $\log x = y$ **5.** $\ln y = x$

6. (A) Plot at least five points, then draw a hand sketch of the graph of $y = (\frac{4}{3})^x$.
 (B) Use your result from part A to sketch the graph of $y = \log_{4/3} x$.

In Problems 7 and 8, simplify.

7. $\dfrac{7^{x+2}}{7^{2-x}}$ **8.** $\left(\dfrac{e^x}{e^{-x}}\right)^x$

In Problems 9–11, solve for x exactly.

9. $\log_2 x = 3$ **10.** $\log_x 25 = 2$ **11.** $\log_3 27 = x$

In Problems 12–15, solve for x to three significant digits.

12. $10^x = 17.5$ **13.** $e^x = 143{,}000$

14. $\ln x = -0.015\,73$ **15.** $\log x = 2.013$

Evaluate the expression in Problems 16–19 to four significant digits using a calculator.

16. $\ln \pi$ **17.** $\log(-e)$

18. $\pi^{\ln 2}$ **19.** $\dfrac{e^{\pi} + e^{-\pi}}{2}$

20. Write as a single log: $2 \log a - \dfrac{1}{3} \log b + \log c$

21. Write in terms of $\ln a$ and $\ln b$: $\ln \dfrac{a^5}{\sqrt{b}}$

In Problems 22–35, solve for x exactly.

22. $3^x = 120$ **23.** $10^{2x} = 500$

24. $\log_2(4x - 5) = 5$ **25.** $\ln(x - 5) = 0$

26. $\ln(2x - 1) = \ln(x + 3)$

27. $\log(x^2 - 3) = 2 \log(x - 1)$

28. $e^{x^2 - 3} = e^{2x}$ **29.** $4^{x-1} = 2^{1-x}$

30. $2x^2 e^{-x} = 18 e^{-x}$ **31.** $\log_{1/4} 16 = x$

32. $\log_x 9 = -2$ **33.** $\log_{16} x = \frac{3}{2}$

34. $\log_x e^5 = 5$ **35.** $10^{\log_{10} x} = 33$

In Problems 36–45, solve for x to three significant digits.

36. $x = 2(10^{1.32})$ **37.** $x = \log_5 23$

38. $\ln x = -3.218$ **39.** $x = \log(2.156 \times 10^{-7})$

40. $x = \dfrac{\ln 4}{\ln 2.31}$ **41.** $25 = 5(2^x)$

42. $4{,}000 = 2{,}500(e^{0.12x})$ **43.** $0.01 = e^{-0.05x}$

44. $5^{2x-3} = 7.08$ **45.** $\dfrac{e^x - e^{-x}}{2} = 1$

In Problems 46–51, solve for x exactly.

46. $\log 3x^2 - \log 9x = 2$

47. $\log x - \log 3 = \log 4 - \log(x + 4)$

48. $\ln(x + 3) - \ln x = 2 \ln 2$

49. $\ln(2x + 1) - \ln(x - 1) = \ln x$

50. $(\log x)^3 = \log x^9$ **51.** $\ln(\log x) = 1$

In Problems 52 and 53, simplify.

52. $(e^x + 1)(e^{-x} - 1) - e^x(e^{-x} - 1)$

53. $(e^x + e^{-x})(e^x - e^{-x}) - (e^x - e^{-x})^2$

In Problems 54–57, use transformations to explain how the graph of g is related to the graph of the given logarithmic function f. Determine whether g is increasing or decreasing, find its domain and any asymptotes, and sketch the graph of g.

54. $g(x) = 3 - \frac{1}{3} 2^x; f(x) = 2^x$

55. $g(x) = 2e^x - 4; f(x) = e^x$

56. $g(x) = -2 + \log_4 x; f(x) = \log_4 x$

57. $g(x) = 1 + 2 \log_{1/3} x; f(x) = \log_{1/3} x$

58. If the graph of $y = e^x$ is reflected in the line $y = x$, the graph of the function $y = \ln x$ is obtained. Discuss the functions that are obtained by reflecting the graph of $y = e^x$ in the x axis and the y axis.

59. (A) Explain why the equation $e^{-x/3} = 4 \ln(x + 1)$ has exactly one solution.

 (B) Find the solution of the equation to three decimal places.

60. Approximate all real zeros of $f(x) = 4 - x^2 + \ln x$ to three decimal places.

61. Find the coordinates of the points of intersection of $f(x) = 10^{x-3}$ and $g(x) = 8 \log x$ to three decimal places.

In Problems 62–65, solve for the indicated variable in terms of the remaining symbols.

62. $D = 10 \log \dfrac{I}{I_0}$ for I (sound intensity)

63. $y = \dfrac{1}{\sqrt{2\pi}} e^{-x^2/2}$ for x (probability)

64. $x = -\dfrac{1}{k} \ln \dfrac{I}{I_0}$ for I (X-ray intensity)

65. $r = P \dfrac{i}{1 - (1 + i)^{-n}}$ for n (finance)

66. Write $\ln y = -5t + \ln c$ in an exponential form free of logarithms; then solve for y in terms of the remaining symbols.

67. For $f = \{(x, y) \mid y = \log_2 x\}$, graph f and f^{-1} on the same coordinate system. What are the domains and ranges for f and f^{-1}?

68. Explain why 1 cannot be used as a logarithmic base.

69. Prove that $\log_b(MN) = \log_b M + \log_b N$.

APPLICATIONS

70. **POPULATION GROWTH** Many countries have a population growth rate of 3% (or more) per year. At this rate, how many years will it take a population to double? Use the annual compounding growth model $P = P_0(1 + r)^t$. Compute the answer to three significant digits.

71. **POPULATION GROWTH** Repeat Problem 70 using the continuous compounding growth model $P = P_0 e^{rt}$.

72. **CARBON 14-DATING** How many years will it take for carbon-14 to diminish to 1% of the original amount after the death of a plant or animal? Use the formula $A = A_0 e^{-0.000124t}$. Compute the answer to three significant digits.

73. MEDICINE One leukemic cell injected into a healthy mouse will divide into two cells in about $\frac{1}{2}$ day. At the end of the day these two cells will divide into four. This doubling continues until 1 billion cells are formed; then the animal dies with leukemic cells in every part of the body.
(A) Write an equation that will give the number N of leukemic cells at the end of t days.
(B) When, to the nearest day, will the mouse die?

74. MONEY GROWTH Assume $1 had been invested at an annual rate of 3% compounded continuously in the year A.D. 1. What would be the value of the account in the year 2011? Compute the answer to two significant digits.

75. PRESENT VALUE Solving $A = Pe^{rt}$ for P, we obtain $P = Ae^{-rt}$, which is the **present value** of the amount A due in t years if money is invested at a rate r compounded continuously.
(A) Graph $P = 1,000(e^{-0.08t})$, $0 \leq t \leq 30$.
(B) What does it appear that P tends to as t tends to infinity? [*Conclusion:* The longer the time until the amount A is due, the smaller its present value, as we would expect.]

76. EARTHQUAKES The 1971 San Fernando, California, earthquake released 1.99×10^{14} joules of energy. Compute its magnitude on the Richter scale using the formula $M = \frac{2}{3} \log (E/E_0)$, where $E_0 = 10^{4.40}$ joules. Compute the answer to one decimal place.

77. EARTHQUAKES Refer to Problem 76. If the 1906 San Francisco earthquake had a magnitude of 8.3 on the Richter scale, how much energy was released? Compute the answer to three significant digits.

78. SOUND If the intensity of a sound from one source is 100,000 times that of another, how much more is the decibel level of the louder sound than the softer one? Use the formula $D = 10 \log (I/I_0)$.

79. MARINE BIOLOGY The intensity of light entering water is reduced according to the exponential function

$$I = I_0 e^{-kd}$$

where I is the intensity d feet below the surface, I_0 is the intensity at the surface, and k is the coefficient of extinction. Measurements in the Sargasso Sea in the West Indies have indicated that half the surface light reaches a depth of 73.6 feet. Find k, and find the depth at which 1% of the surface light remains. Compute answers to three significant digits.

80. WILDLIFE MANAGEMENT A lake formed by a newly constructed dam is stocked with 1,000 fish. Their population is expected to increase according to the logistic curve

$$N = \frac{30}{1 + 29e^{-1.35t}}$$

where N is the number of fish, in thousands, expected after t years. The lake will be open to fishing when the number of fish reaches 20,000. How many years, to the nearest year, will this take?

Problems 81 and 82 require a graphing calculator or a computer that can calculate exponential, logarithmic, and logistic regression models for a given data set.

81. MEDICARE The annual expenditures for Medicare (in billions of dollars) by the U.S. government for selected years since 1980 are shown in Table 1. Let x represent years since 1980.
(A) Find an exponential regression model of the form $y = ab^x$ for these data. Round to three significant digits. Estimate (to the nearest billion) the total expenditures in 2010 and in 2020.
(B) When (to the nearest year) will the total expenditures reach $900 billion?

Table 1 Medicare Expenditures

Year	Billion $
1980	37
1985	72
1990	111
1995	181
2000	225
2005	342

Source: U.S. Bureau of the Census

82. Table 2 lists the number of cell phone subscribers in the United States for selected years from 1994 to 2006. Let $x = 0$ correspond to 1990 and round all coefficients to four significant digits.
(A) Find a logarithmic regression model of the form $y = a + b \ln x$ for the data, then use the model to predict the number of subscribers in 2015.
(B) Repeat part A, this time finding a logistic regression model of the form $y = c/(1 + ae^{-bx})$.
(C) Which of the models do you think models the data better? Explain. Consider how well it fits the points from the table, as well as how well you think it predicts long-term trends.

Table 2 Cell Phone Subscribers in the U.S.

Year	Subscribers in millions
1994	24.13
1997	55.31
2000	109.5
2003	158.8
2006	233.0

Source: CTIA—The Wireless Association

CHAPTER 5

»»» GROUP ACTIVITY Comparing Regression Models

We have used polynomial, exponential, and logarithmic regression models to fit curves to data sets. How can we determine which equation provides the best fit for a given set of data? There are two principal ways to select models. The first is to use information about the type of data to help make a choice. For example, we expect the weight of a fish to be related to the cube of its length. And we expect most populations to grow exponentially, at least over the short term. The second method for choosing among equations involves developing a measure of how closely an equation fits a given data set. This is best introduced through an example. Consider the data set in Figure 1, where L1 represents the x coordinates and L2 represents the y coordinates. The graph of this data set is shown in Figure 2. Suppose we arbitrarily choose the equation $y_1 = 0.6x + 1$ to model these data (Fig. 3).

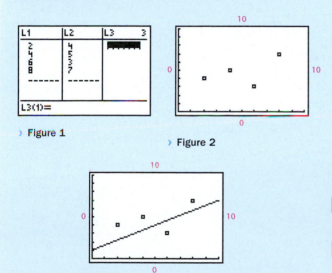

> Figure 1

> Figure 2

> Figure 3 $y_1 = 0.6x + 1.$

To measure how well the graph of y_1 fits these data, we examine the difference between the y coordinates in the data set and the corresponding y coordinates on the graph of y_1 (L3 in Figs. 4 and 5).

> Figure 4

> Figure 5 Here + is L2 and □ is L3.

Each of these differences is called a **residual.** Note that three of the residuals are positive and one is negative (three of the points lie above the line, one lies below). The most commonly accepted measure of the fit provided by a given model is the **sum of the squares of the residuals (SSR).** When squared, each residual (whether positive or negative or zero) makes a nonnegative contribution to the SSR.

$$SSR = (4 - 2.2)^2 + (5 - 3.4)^2 + (3 - 4.6)^2 \\ + (7 - 5.8)^2 = 9.8$$

(A) A linear regression model for the data in Figure 1 is given by

$$y_2 = 0.35x + 3$$

Compute the SSR for the data and y_2, and compare it to the one we computed for y_1.

It turns out that among all possible linear polynomials, *the linear regression model minimizes the sum of the squares of the residuals.* For this reason, the linear regression model is often called the **least-squares line.** A similar statement can be made for polynomials of any fixed degree. That is, the quadratic regression model minimizes the SSR over all quadratic polynomials, the cubic regression model minimizes the SSR over all cubic polynomials, and so on. The same statement cannot be made for exponential or logarithmic regression models. Nevertheless, the SSR can still be used to compare exponential, logarithmic, and polynomial models.

(B) Find the exponential and logarithmic regression models for the data in Figure 1, compute their SSRs, and compare with the linear model.

(C) National annual advertising expenditures for selected years since 1950 are shown in Table 1 where x is years since 1950 and y is total expenditures in billions of dollars. Which regression model would fit this data best: a quadratic model, a cubic model, or an exponential model? Use the SSRs to support your choice.

Table 1 Annual Advertising Expenditures, 1950–2000

x (years)	0	10	20	30	40	50
y (billion $)	5.7	12.0	19.6	53.6	128.6	247.5

Source: U.S. Bureau of the Census.

Trigonometric Functions

TRIGONOMETRIC functions seem to have had their origins with the Greeks' investigation of the indirect measurement of distances and angles in the "celestial sphere." (The ancient Egyptians had used some elementary geometry to build the pyramids and remeasure lands flooded by the Nile, but neither they nor the ancient Babylonians had developed the concept of angle measure.) The word *trigonometry,* based on the Greek words for "triangle measurement," was first used as the title for a text by the German mathematician Pitiscus in A.D. 1600.

Originally the trigonometric functions were restricted to angles and their applications to the indirect measurement of angles and distances. These functions gradually broke free of these restrictions, and we now have trigonometric functions of real numbers. Modern applications range over many types of problems that have little or nothing to do with angles or triangles—applications involving periodic phenomena such as sound, light, and electrical waves; business cycles; and planetary motion.

In our approach to the subject, we define the trigonometric functions both in terms of angles, and coordinates of points on the unit circle.

6-1 Angles and Their Measure

> Angles
> Degree and Radian Measure
> Converting Degrees to Radians and Vice Versa
> Linear and Angular Speed

> **Figure 1** Angle θ or angle *PVQ* or ∠ *V*.

In Section 6-1, we introduce the concept of angle and two measures of angles, *degree* and *radian*.

> Angles

The study of trigonometry depends on the concept of angle. An **angle** is formed by rotating (in a plane) a ray *m,* called the **initial side** of the angle, around its endpoint until it coincides with a ray *n,* called the **terminal side** of the angle. The common endpoint *V* of *m* and *n* is called the **vertex** (Fig. 1).

A counterclockwise rotation produces a **positive angle,** and a clockwise rotation produces a **negative angle,** as shown in Figures 2(a) and 2(b). The amount of rotation in either direction is not restricted. Two different angles may have the same initial and terminal sides, as shown in Figure 2(c). Such angles are said to be **coterminal.**

θ positive θ negative α and β coterminal

(a) (b) (c)

> **Figure 2** Angles and rotation.

An angle in a rectangular coordinate system is said to be in **standard position** if its vertex is at the origin and the initial side is along the positive *x* axis. If the terminal side of an angle in standard position lies along a coordinate axis, the angle is said to be a **quadrantal angle.** If the terminal side does not lie along a coordinate axis, then the angle is often referred to in terms of the quadrant in which the terminal side lies (Fig. 3).

θ is a quadrantal angle θ is a third-quadrant angle θ is a second-quadrant angle

(a) (b) (c)

> **Figure 3** Angles in standard positions.

› Degree and Radian Measure

Just as line segments are measured in centimeters, meters, inches, or miles, angles are measured in different units. The two most commonly used units for angle measure are *degree* and *radian*.

> ### › DEFINITION 1 Degree Measure
>
> A positive angle formed by one complete rotation is said to have a measure of *360 degrees* (360°). A positive angle formed by $\frac{1}{360}$ of a complete rotation is said to have a measure of **1 degree** (1°). The symbol ° denotes degrees.

Definition 1 is extended to all angles, not just the positive (counterclockwise) ones, in the obvious way. So, for example, a negative angle formed by $\frac{1}{4}$ of a complete clockwise rotation has a measure of $-90°$, and an angle for which the initial and terminal sides coincide, without rotation, has a measure of 0°.

Certain angles have special names that indicate their degree measure. Figure 4 shows a **straight angle**, a **right angle**, an **acute angle**, and an **obtuse angle**.

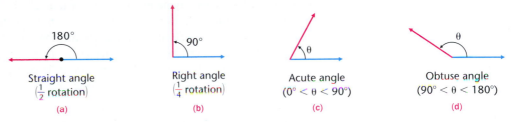

180°	90°	θ	θ
Straight angle $(\frac{1}{2}$ rotation$)$	Right angle $(\frac{1}{4}$ rotation$)$	Acute angle $(0° < θ < 90°)$	Obtuse angle $(90° < θ < 180°)$
(a)	(b)	(c)	(d)

> **Figure 4** Types of angles.

Two positive angles are **complementary** if their sum is 90°; they are **supplementary** if their sum is 180°.

A degree can be divided further using decimal notation. For example, 42.75° represents an angle of degree measure 42 plus three-quarters of 1 degree. A degree can also be divided further using minutes and seconds just as an hour is divided into minutes and seconds. Each degree is divided into 60 equal parts called **minutes,** and each minute is divided into 60 equal parts called **seconds.** Symbolically, minutes are represented by ′ and seconds by ″. So

$$12°23'14''$$

is a concise way of writing 12 degrees, 23 minutes, and 14 seconds.

Decimal degrees (DD) are useful in some instances and degrees–minutes–seconds (DMS) are useful in others. You should be able to go from one form to the other as demonstrated in Example 1.

> ### › CONVERSION ACCURACY
>
> If an angle is measured to the nearest second, the converted decimal form should not go beyond three decimal places, and vice versa.

EXAMPLE 1 From DMS to DD and Back

(A) Convert 21°47′12″ to decimal degrees.

(B) Convert 105.183° to degree–minute–second form.

SOLUTIONS

(A) $21°47′12″ = \left(21 + \dfrac{47}{60} + \dfrac{12}{3{,}600}\right)^{\circ} = 21.787°$

(B) $105.183°$ ⌐ $= 105° \ (0.183 \cdot 60)′$ ⌐ *

$= 105° \ 10.98′$

⌐ $= 105° \ 10′ \ (0.98 \cdot 60)″$ ⌐

$= 105° \ 10′ \ 59″$

MATCHED PROBLEM 1

(A) Convert 193°17′34″ to DD form.

(B) Convert 237.615° to DMS form.

Some scientific and some graphing calculators can convert the DD and DMS forms automatically, but the process differs significantly among the various types of calculators. Check your owner's manual for your particular calculator. The conversion methods outlined in Example 1 show you the reasoning behind the process, and are sometimes easier to use than the "automatic" methods for some calculators.

Degree measure of angles is used extensively in engineering, surveying, and navigation. Another unit of angle measure, called the *radian,* is better suited for certain mathematical developments, scientific work, and engineering applications.

> › **DEFINITION 2** Radian Measure
>
> A positive angle θ formed by a central angle of a circle has measure **1 radian** if the length s of the arc opposite θ is equal to the radius r of the circle.
>
> More generally, if θ is any positive angle formed by the central angle of a circle, then the **radian measure** of θ is given by
>
> $$\theta = \frac{s}{r} \text{ radians}$$
>
> where s is the length of the arc opposite θ and r is the radius of the circle. [*Note:* s and r must be measured in the same units.]

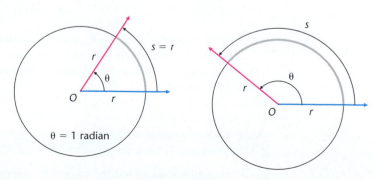

*Throughout the book, dashed boxes—called think boxes—are used to represent steps that may be performed mentally.

The circumference of a circle of radius r is $2\pi r$, so the radian measure of a positive angle formed by one complete rotation is

$$\theta = \frac{s}{r} = \frac{2\pi r}{r} = 2\pi \approx 6.283 \text{ radians}$$

Just as for degree measure, the definition is extended to apply to all angles; if θ is a negative angle, its radian measure is given by $\theta = -\frac{s}{r}$. Note that in the preceding sentence, as well as in Definition 2, the symbol θ is used in two ways: as the name of the angle and as the measure of the angle. The context indicates the meaning.

| EXAMPLE | 2 | **Computing Radian Measure** |

What is the radian measure of a central angle θ opposite an arc of 24 meters in a circle of radius 6 meters?

SOLUTION

$$\theta = \frac{s}{r} = \frac{24 \text{ meters}}{6 \text{ meters}} = 4 \text{ radians}$$

MATCHED PROBLEM 2

What is the radian measure of a central angle θ opposite an arc of 60 feet in a circle of radius 12 feet?

⟩⟩⟩ EXPLORE-DISCUSS 1

Discuss why the radian measure of an angle is independent of the size of the circle having the angle as a central angle.

⟩ Converting Degrees to Radians and Vice Versa

What is the radian measure of an angle of 180°? Let θ be a central angle of 180° in a circle of radius r. Then the length s of the arc opposite θ is $\frac{1}{2}$ the circumference C of the circle. Therefore,

$$s = \frac{C}{2} = \frac{2\pi r}{2} = \pi r \qquad \text{and} \qquad \theta = \frac{s}{r} = \frac{\pi r}{r} = \pi \text{ radians}$$

So, 180° corresponds to π radians. This is important to remember, because the radian measures of many special angles can be obtained from this correspondence. For example, 90° is 180°/2; therefore, 90° corresponds to $\pi/2$ radians.

⟩⟩⟩ EXPLORE-DISCUSS 2

Write the radian measure of each of the following angles in the form $\frac{a}{b}\pi$, where a and b are positive integers and fraction $\frac{a}{b}$ is reduced to lowest terms: 15°, 30°, 45°, 60°, 75°, 90°, 105°, 120°, 135°, 150°, 165°, 180°.

Some key results from Explore-Discuss 2 are summarized in Figure 5 for easy reference. These correspondences and multiples of them will be used extensively in work that follows.

⟩ **Figure 5** Radian–degree correspondences.

In general, the following proportion can be used to convert degree measure to radian measure and vice versa.

> **RADIAN–DEGREE CONVERSION FORMULAS**

$$\frac{\theta_{deg}}{180°} = \frac{\theta_{rad}}{\pi \text{ radians}} \qquad \text{Basic proportion}$$

$$\theta_{deg} = \frac{180°}{\pi \text{ radians}} \theta_{rad} \qquad \text{Radians to degrees}$$

$$\theta_{rad} = \frac{\pi \text{ radians}}{180°} \theta_{deg} \qquad \text{Degrees to radians}$$

[*Note:* The basic proportion is usually easier to remember. Also we will omit units in calculations until the final answer. If your calculator does not have a key labeled π, use $\pi \approx 3.14159$.]

Some scientific and graphing calculators can automatically convert radian measure to degree measure, and vice versa. Check the owner's manual for your particular calculator.

| EXAMPLE | 3 | **Radian–Degree Conversions** |

(A) Find the radian measure, exact and to three significant digits, of an angle of 75°.

(B) Find the degree measure, exact and to four significant digits, of an angle of 5 radians.

(C) Find the radian measure to two decimal places of an angle of 41°12′.

SOLUTIONS

$$\text{(A)} \quad \theta_{rad} = \frac{\pi \text{ radians}}{180°} \theta_{deg} = \frac{\pi}{180}(75) = \overset{\text{Exact}}{\frac{5\pi}{12}} = \overset{\text{Three significant digits}}{1.31}$$

$$\text{(B)} \quad \theta_{deg} = \frac{180°}{\pi \text{ radians}} \theta_{rad} = \frac{180}{\pi}(5) = \overset{\text{Exact}}{\frac{900}{\pi}} = \overset{\text{Four significant digits}}{286.5°}$$

$$\text{(C)} \quad 41°12′ = \left(41 + \frac{12}{60}\right)° = 41.2° \qquad \text{Change 41°12′ to DD first.}$$

$$\theta_{rad} = \frac{\pi \text{ radians}}{180°} \theta_{deg} = \frac{\pi}{180}(41.2) = 0.72 \qquad \text{To two decimal places}$$

```
75°
                  1.31
5ʳ
                  286.5
(41°12')°
                   .72
```

> **Figure 6** Automatic conversion.

Figure 6 shows the three preceding conversions done automatically on a graphing calculator by selecting the appropriate angle mode.

| MATCHED PROBLEM 3 |

(A) Find the radian measure, exact and to three significant digits, of an angle of 240°.

(B) Find the degree measure, exact and to three significant digits, of an angle of 1 radian.

(C) Find the radian measure to three decimal places of an angle of 125°23′.

EXAMPLE 4 **Engineering**

A belt connects a pulley of 2-inch radius with a pulley of 5-inch radius. If the larger pulley turns through 10 radians, through how many radians will the smaller pulley turn?

SOLUTION First we draw a sketch (Fig. 7).

> Figure 7

When the larger pulley turns through 10 radians, the point P on its circumference will travel the same distance s (arc length) that point Q on the smaller circle travels. For the larger pulley,

$$\theta = \frac{s}{r}$$

$$s = r\theta = (5)(10) = 50 \text{ inches}$$

For the smaller pulley,

$$\theta = \frac{s}{r} = \frac{50}{2} = 25 \text{ radians}$$

MATCHED PROBLEM 4 In Example 4, through how many radians will the larger pulley turn if the smaller pulley turns through 4 radians?

› Linear and Angular Speed

The *average speed v* of an object that travels a distance $d = 30$ meters in time $t = 3$ seconds is given by

$$v = \frac{d}{t} = \frac{30 \text{ meters}}{3 \text{ seconds}} = 10 \text{ meters per second}$$

Suppose that a point P moves an arc length of $s = 30$ meters in $t = 3$ seconds on the circumference of a circle of radius $r = 20$ meters (Fig. 8). Then, in those 3 seconds, the point P has moved through an angle of

$$\theta = \frac{s}{r} = \frac{30}{20} = 1.5 \text{ radians}$$

We call the average speed of point P, given by

$$v = \frac{s}{t} = 10 \text{ meters per second}$$

> Figure 8

the (average) *linear speed* to distinguish it from the (average) *angular speed* that is given by

$$\omega = \frac{\theta}{t} = \frac{1.5}{3} = 0.5 \text{ radians per second}$$

The simple formula, $v = r\omega$, which relates linear and angular speed, is obtained from the definition of radian measure as follows:

$$\frac{s}{r} = \theta \qquad \text{Multiply both sides by } r.$$

$$s = r\theta \qquad \text{Divide both sides by } t.$$

$$\frac{s}{t} = r\frac{\theta}{t} \qquad \text{Substitute } v = \frac{s}{t} \text{ and } \omega = \frac{\theta}{t}$$

$$v = r\omega$$

These concepts are summarized in the box.

> ### › LINEAR SPEED AND ANGULAR SPEED
>
> Suppose a point P moves through an angle θ and arc length s, in time t, on the circumference of a circle of radius r. The (average) **linear speed** of P is
>
> $$v = \frac{s}{t}$$
>
> and the (average) **angular speed** is
>
> $$\omega = \frac{\theta}{t}$$
>
> Furthermore, $v = r\omega$.

EXAMPLE 5 **Wind Power**

A wind turbine of rotor diameter 15 meters makes 62 revolutions per minute. Find the angular speed (in radians per second) and the linear speed (in meters per second) of the rotor tip.

SOLUTION The radius of the rotor is $15/2 = 7.5$ meters. In 1 minute the rotor moves through an angle of $62(2\pi) = 124\pi$ radians. Therefore, the angular speed is

$$\omega = \frac{\theta}{t} = \frac{124\pi \text{ radians}}{60 \text{ seconds}} \approx 6.49 \text{ radians per second}$$

and the linear speed of the rotor tip is

$$v = r\omega = 7.5\frac{124\pi}{60} \approx 48.69 \text{ meters per second}$$

MATCHED PROBLEM 5 A wind turbine of rotor diameter 12 meters has a rotor tip speed of 34.2 meters per second. Find the angular speed of the rotor (in radians per second) and the number of revolutions per minute.

6-1 Exercises

In all problems, if angle measure is expressed by a number that is not in degrees, it is assumed to be in radians.

1. Explain the difference between a positive angle and a negative angle.

2. Explain the difference between complementary angles and supplementary angles.

3. Would it be better to measure angles by dividing the circumference of a circle into 100 equal parts rather than 360 equal parts as in degree measure? Explain.

4. Explain the connection between an angle of 1 *radian* and the *radius* of a circle.

5. You are watching your nieces ride a Ferris wheel. Explain how you could do a mental calculation to estimate their angular speed.

6. Refer to Problem 5. Explain how you could do a mental calculation to estimate their linear speed.

Find the degree measure of each of the angles in Problems 7–12, keeping in mind that an angle of one complete rotation corresponds to 360°.

7. $\frac{1}{9}$ rotation **8.** $\frac{1}{5}$ rotation **9.** $\frac{3}{4}$ rotation

10. $\frac{3}{8}$ rotation **11.** $\frac{9}{8}$ rotations **12.** $\frac{7}{6}$ rotations

Find the radian measure of a central angle θ opposite an arc s in a circle of radius r, where r and s are as given in Problems 13–16.

13. $r = 4$ centimeters, $s = 24$ centimeters

14. $r = 8$ inches, $s = 16$ inches

15. $r = 12$ feet, $s = 30$ feet

16. $r = 18$ meters, $s = 27$ meters

Find the radian measure of each angle in Problems 17–22, keeping in mind that an angle of one complete rotation corresponds to 2π radians.

17. $\frac{1}{8}$ rotation **18.** $\frac{1}{6}$ rotation **19.** $\frac{3}{4}$ rotation

20. $\frac{5}{12}$ rotation **21.** $\frac{13}{12}$ rotations **22.** $\frac{11}{8}$ rotations

Find the exact radian measure, in terms of π, of each angle in Problems 23–26.

23. 30°, 60°, 90°, 120°, 150°, 180°

24. 60°, 120°, 180°, 240°, 300°, 360°

25. −45°, −90°, −135°, −180°

26. −90°, −180°, −270°, −360°

Find the exact degree measure of each angle in Problems 27–30.

27. $\dfrac{\pi}{3}, \dfrac{2\pi}{3}, \pi, \dfrac{4\pi}{3}, \dfrac{5\pi}{3}, 2\pi$ **28.** $\dfrac{\pi}{6}, \dfrac{\pi}{3}, \dfrac{\pi}{2}, \dfrac{2\pi}{3}, \dfrac{5\pi}{6}, \pi$

29. $-\dfrac{\pi}{2}, -\pi, -\dfrac{3\pi}{2}, -2\pi$ **30.** $-\dfrac{\pi}{4}, -\dfrac{\pi}{2}, -\dfrac{3\pi}{4}, -\pi$

In Problems 31–36, determine whether the statement is true or false. If true, explain why. If false, give a counterexample.

31. If two angles in standard position have the same measure, then they are coterminal.

32. If two angles in standard position are coterminal, then they have the same measure.

33. If two positive angles are complementary, then both are acute.

34. If two positive angles are supplementary, then one is obtuse and the other is acute.

35. If the terminal side of an angle in standard position lies in quadrant I, then the angle is positive.

36. If the initial and terminal sides of an angle coincide, then the measure of the angle is zero.

Convert each angle in Problems 37–40 to decimal degrees to three decimal places.

37. 5°51′33″ **38.** 14°18′37″

39. 354°8′29″ **40.** 184°31′7″

Convert each angle in Problems 41–44 to degree–minute–second form.

41. 3.042° **42.** 49.715°

43. 403.223° **44.** 156.808°

Find the radian measure to three decimal places for each angle in Problems 45–50.

45. 64° **46.** 25° **47.** 108.413°

48. 203.097° **49.** 13°25′14″ **50.** 56°11′52″

Find the degree measure to two decimal places for each angle in Problems 51–56.

51. 0.93 **52.** 0.08 **53.** 1.13

54. 3.07 **55.** −2.35 **56.** −1.72

Indicate whether each angle in Problems 57–68 is a first-, second-, third-, or fourth-quadrant angle or a quadrantal angle. All angles are in standard position in a rectangular coordinate system. (A sketch may be of help in some problems.)

57. 250° **58.** 150° **59.** 275°

60. 195° **61.** $\frac{3\pi}{4}$ **62.** $\frac{\pi}{3}$

63. $\frac{3\pi}{2}$ **64.** $\frac{7\pi}{4}$ **65.** −330°

66. −450° **67.** −1.5 **68.** −4

69. Verbally describe the meaning of a central angle in a circle with radian measure 1.

70. Verbally describe the meaning of an angle with degree measure 1.

In Problems 71–74, find all angles θ in degree measure that satisfy the given conditions.

71. 360° ≤ θ ≤ 720° and θ is coterminal with 210°

72. 360° ≤ θ ≤ 720° and θ is coterminal with 120°

73. −360° ≤ θ ≤ 0° and θ is coterminal with 45°

74. −360° ≤ θ ≤ 0° and θ is coterminal with 280°

In Problems 75–78, find all angles θ in radian measure that satisfy the given conditions.

75. 2π ≤ θ ≤ 6π and θ is coterminal with π/3

76. 2π ≤ θ ≤ 6π and θ is coterminal with 5π/4

77. −4π ≤ θ ≤ 0 and θ is coterminal with π

78. −4π ≤ θ ≤ 0 and θ is coterminal with π/6

APPLICATIONS

79. CIRCUMFERENCE OF THE EARTH The early Greeks used the proportion $s/C = \theta°/360°$, where s is an arc length on a circle, $\theta°$ is degree measure of the corresponding central angle, and C is the circumference of the circle ($C = 2\pi r$). Eratosthenes (240 B.C.), in his famous calculation of the circumference of the Earth, reasoned as follows: He knew at Syene (now Aswan) during the summer solstice the noon sun was directly overhead and shined on the water straight down a deep well. On the same day at the same time, 5,000 stadia (approx. 500 miles) due north in Alexandria, sun rays crossed a vertical pole at an angle of 7.5° as indicated in the figure. Carry out Eratosthenes' calculation for the circumference of the Earth to the nearest thousand miles. (The current calculation for the equatorial circumference is 24,902 miles.)

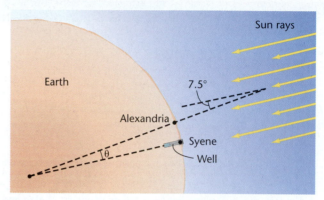

80. CIRCUMFERENCE OF THE EARTH Repeat Problem 79 with the sun crossing the vertical pole in Alexandria at 7°12′.

81. CIRCUMFERENCE OF THE EARTH In Problem 79, verbally explain how θ in the figure was determined.

82. CIRCUMFERENCE OF THE EARTH Verbally explain how the radius, surface area, and volume of the Earth can be determined from the result of Problem 79.

83. ANGULAR SPEED A wheel with diameter 6 feet makes 200 revolutions per minute. Find the angular speed (in radians per second) and the linear speed (in feet per second) of a point on the rim.

84. ANGULAR SPEED A point on the rim of a wheel with diameter 6 feet has a linear speed of 100 feet per second. Find the angular speed (in radians per second) and the number of revolutions per minute.

85. RADIAN MEASURE What is the radian measure of the larger angle made by the hands of a clock at 4:30? Express the answer exactly in terms of π.

86. RADIAN MEASURE What is the radian measure of the smaller angle made by the hands of a clock at 1:30? Express the answer exactly in terms of π.

87. ENGINEERING Through how many radians does a pulley of 10-centimeter diameter turn when 10 meters of rope are pulled through it without slippage?

88. ENGINEERING Through how many radians does a pulley of 6-inch diameter turn when 4 feet of rope are pulled through it without slippage?

89. ASTRONOMY A line from the sun to the Earth sweeps out an angle of how many radians in 1 week? Assume the Earth's orbit is circular and there are 52 weeks in a year. Express the answer in terms of π and as a decimal to two decimal places.

90. ASTRONOMY A line from the center of the Earth to the equator sweeps out an angle of how many radians in 9 hours? Express the answer in terms of π and as a decimal to two decimal places.

91. ENGINEERING A trail bike has a front wheel with a diameter of 40 centimeters and a back wheel of diameter 60 centimeters. Through what angle in radians does the front wheel turn if the back wheel turns through 8 radians?

92. ENGINEERING In Problem 91, through what angle in radians will the back wheel turn if the front wheel turns through 15 radians?

93. ANGULAR SPEED If the trail bike of Problem 91 travels at a speed of 10 kilometers per hour, find the angular speed (in radians per second) of each wheel.

94. ANGULAR SPEED If a car travels at a speed of 60 miles per hour, find the angular speed (in radians per second) of a tire that has a diameter of 2 feet.

The arc length on a circle is easy to compute if the corresponding central angle is given in radians and the radius of the circle is known ($s = r\theta$). If the radius of a circle is large and a central angle is small, then an arc length is often used to approximate the length of the corresponding chord as shown in the figure. If an angle is given in degree measure, converting to radian measure first may be helpful in certain problems. This information will be useful in Problems 95–98.

$$c \approx s = r\theta$$

95. ASTRONOMY The sun is about 9.3×10^7 mi from the Earth. If the angle subtended by the diameter of the sun on the surface of the Earth is 9.3×10^{-3} rad, approximately what is the diameter of the sun to the nearest thousand miles in standard decimal notation?

96. ASTRONOMY The moon is about 381,000 kilometers from the Earth. If the angle subtended by the diameter of the moon on the surface of the Earth is 0.0092 radians, approximately what is the diameter of the moon to the nearest hundred kilometers?

97. PHOTOGRAPHY The angle of view of a 1,000-millimeter telephoto lens is 2.5°. At 750 feet, what is the width of the field of view to the nearest foot?

98. PHOTOGRAPHY The angle of view of a 300-millimeter lens is 8°. At 500 feet, what is the width of the field of view to the nearest foot?

6-2

Right Triangle Trigonometry

› **Trigonometric Ratios**
› **Evaluation of Trigonometric Ratios**
› **Solving Right Triangles**

In Section 6-2 we introduce the six trigonometric ratios associated with an acute angle of a right triangle. We also use the trigonometric ratios to solve a variety of problems that require the indirect measurement of distances and angles.

› Trigonometric Ratios

A **right triangle** is a triangle with one 90° angle (Fig. 1).

If only the angles of a right triangle are known, it is impossible to solve for the sides. (Why?) But if we are given two sides, or one acute angle and a side, then it is possible to solve for the remaining three quantities. This process is called *solving the right triangle*. We use *trigonometric ratios* to solve right triangles.

If two right triangles have the same acute angle θ, then the triangles are similar and ratios of corresponding sides are equal (see Figure 2 on the next page).

› **Figure 1**

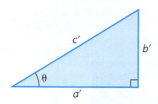

> Figure 2

Therefore,

$$\frac{b}{c} = \frac{b'}{c'} \qquad \frac{c}{b} = \frac{c'}{b'}$$

$$\frac{a}{c} = \frac{a'}{c'} \qquad \frac{c}{a} = \frac{c'}{a'}$$

$$\frac{b}{a} = \frac{b'}{a'} \qquad \frac{a}{b} = \frac{a'}{b'}$$

So the six possible ratios of the sides of a right triangle depend only on the acute angle θ, not on the size of the triangle. These six ratios, the **trigonometric ratios**, are called the **sine, cosine, tangent, cotangent, secant,** and **cosecant** of the angle θ. The values of these ratios at an angle θ are denoted by **sin θ, cos θ, tan θ, cot θ, sec θ,** and **csc θ,** respectively, as displayed in the box.

> **TRIGONOMETRIC RATIOS**

Side b is often referred to as the **side opposite** angle θ, a as the **side adjacent** to angle θ, and c as the **hypotenuse.** Using these designations for an arbitrary right triangle removed from a coordinate system, we have the following:

> **RIGHT TRIANGLE RATIOS**

>>> **EXPLORE-DISCUSS 1**

(A) Explain why sin θ < 1 and csc θ > 1 for $0° < \theta < 90°$.

(B) Explain why cos θ < 1 and sec θ > 1 for $0° < \theta < 90°$.

(C) Does there exist an angle θ, $0° < \theta < 90°$, such that tan θ > 100? Such that tan θ < 0.01? Explain.

Each of the six trigonometric ratios is the reciprocal of another. For example, csc θ is the reciprocal of sin θ. These **reciprocal relationships** make it easy to compute csc θ, sec θ, and cot θ from sin θ, cos θ, and tan θ, respectively.

> **RECIPROCAL RELATIONSHIPS**

For $0° < θ < 90°$: $\csc θ = \dfrac{1}{\sin θ}$ $\sec θ = \dfrac{1}{\cos θ}$ $\cot θ = \dfrac{1}{\tan θ}$

The names of the six trigonometric ratios suggest a second set of relationships among them. The prefix *co-* (in cosine, cotangent, cosecant) refers to "complement." If θ is an acute angle of a right triangle, then the other acute angle is its complement, $90° − θ$. (Why?) Because the side opposite angle θ is the side adjacent to $90° − θ$, the sine of θ equals the cosine of its complement. This establishes the first of the following **complementary relationships.** The second and third are justified similarly.

> **COMPLEMENTARY RELATIONSHIPS**

For $0° < θ < 90°$:
$$\sin θ = \cos (90° − θ)$$
$$\tan θ = \cot (90° − θ)$$
$$\sec θ = \csc (90° − θ)$$

So the sine of θ is equal to the cosine of the complement of θ, the tangent of θ is equal to the cotangent of the complement of θ, and the secant of θ is equal to the cosecant of the complement of θ. The trigonometric ratios cosine, cotangent, and cosecant are sometimes referred to as the **cofunctions** of sine, tangent, and secant, respectively.

> ## Evaluation of Trigonometric Ratios

Trigonometric ratios of certain angles can be evaluated using simple geometric facts. For example, by the Pythagorean theorem, the diagonal of a square with sides of length 1 has length $\sqrt{2}$ (Fig. 3). Therefore,

$$\sin 45° = \frac{1}{\sqrt{2}} \qquad \csc 45° = \sqrt{2}$$

$$\cos 45° = \frac{1}{\sqrt{2}} \qquad \sec 45° = \sqrt{2}$$

$$\tan 45° = 1 \qquad \cot 45° = 1$$

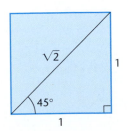

> Figure 3

Similarly, by the Pythagorean theorem, an equilateral triangle with sides of length 2 has height $\sqrt{3}$ (Fig. 4). Therefore,

$$\sin 60° = \frac{\sqrt{3}}{2} \qquad \csc 60° = \frac{2}{\sqrt{3}}$$

$$\cos 60° = \frac{1}{2} \qquad \sec 60° = 2$$

$$\tan 60° = \sqrt{3} \qquad \cot 60° = \frac{1}{\sqrt{3}}$$

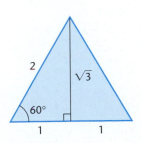

> Figure 4

For most angles, however, simple geometric arguments fail to give exact values of the trigonometric ratios. But a calculator, set in **degree mode,** can be used to give approximations. (Calculations using trigonometric ratios are valid if the angle is measured in either degrees or radians, provided the calculator is set in the correct mode. In Section 6-2, we restrict our interest to degree mode; in Section 6-3, radian mode is emphasized.) For example, $\sin 60° \approx 0.866$, corresponding to the exact value $\sqrt{3}/2$ obtained earlier [Fig. 5(a)]. Similarly, $\sin 50° \approx 0.766$ and $\cos 50° \approx 0.643$ [Fig. 5(b)].

> Figure 5

(a)

(b)

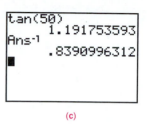

(c)

Most calculators have function keys for the sine, cosine, and tangent, but not for the cosecant, secant, and tangent. We use the reciprocal relationships to evaluate the last three. For example, $\cot 50° = 1/\tan 50° \approx 0.839$ [Fig. 5(c)]. Do not use the calculator function keys marked $\sin^{-1}$, $\cos^{-1}$, or $\tan^{-1}$ for this purpose—these keys are used to evaluate the inverse trigonometric functions of Section 6-6, not reciprocals.

EXAMPLE 1

Calculator Evaluation

Evaluate to four significant digits.

(A) $\tan 85°$ (B) $\cos 26°42'$ (C) $\csc 34°$

SOLUTIONS First, make certain that the calculator is set in degree mode.

(A) $\tan 85° = 11.43$

(B) $\cos 26°42' = 0.8934$ [Note that minutes can be converted to degrees, or the degree–second notation can be used, as shown in Figure 6(a)].

(C) $\csc 34° = 1/\sin 34° = 1.788$ [Figure 6(b)].

(a)

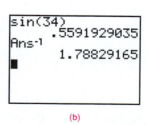

(b)

> Figure 6

MATCHED PROBLEM 1

Evaluate to four significant digits.

(A) $\sin 9°$ (B) $\tan 14°26'$ (C) $\sec 78°$

Table 1

Angle to Nearest	Significant Digits for Side Measure
1°	2
10′ or 0.1°	3
1′ or 0.01°	4
10″ or 0.001°	5

Solving Right Triangles

If we are given two sides of a right triangle, or an acute angle and a side, then it is possible to solve for the remaining three quantities. This process is called **solving the right triangle.** It is the key to solving many practical problems in engineering, surveying, navigation, and astronomy. Several examples will make the process clear.

Regarding computational accuracy, we use Table 1 as a guide. We will use = rather than ≈ in many places, realizing that the accuracy indicated in Table 1 is all that is assumed. Another word of caution: Be sure your calculator is set in degree mode.

EXAMPLE 2 Right Triangle Solution

Solve the right triangle with $c = 6.25$ feet and $\beta = 32.2°$.

SOLUTION First draw a figure and label the parts (Fig. 7):

SOLVE FOR α

$$\alpha = 90° - 32.2° = 57.8° \qquad \text{α and β are complementary.}$$

SOLVE FOR b

$$\sin \beta = \frac{b}{c} \qquad \text{Or use csc } \beta = \frac{c}{b}.$$

$$\sin 32.2° = \frac{b}{6.25} \qquad \text{Multiply both sides by 6.25.}$$

$$b = 6.25 \sin 32.2° \qquad \text{Calculate.}$$

$$= 3.33 \text{ feet}$$

SOLVE FOR a

$$\cos \beta = \frac{a}{c} \qquad \text{Or use sec } \beta = \frac{c}{a}.$$

$$\cos 32.2° = \frac{a}{6.25} \qquad \text{Multiply both sides by 6.25.}$$

$$a = 6.25 \cos 32.2° \qquad \text{Calculate.}$$

$$= 5.29 \text{ feet}$$

> **Figure 7**

6.25 ft α b

32.2°

a

MATCHED PROBLEM 2 Solve the right triangle with $c = 27.3$ meters and $\alpha = 47.8°$.

In Example 3 we are confronted with a problem of the type: Find θ given

$$\sin \theta = 0.4196$$

We know how to find (or approximate) $\sin \theta$ given θ, but how do we reverse the process? How do we find θ given $\sin \theta$? First, we note that the solution to the problem can be written symbolically as either

$$\theta = \arcsin 0.4196 \qquad \text{or} \qquad \theta = \sin^{-1} 0.4196 \qquad \text{"arcsin" and "sin}^{-1}\text{" both represent the same thing.}$$

Both expressions are read "θ is the angle whose sine is 0.4196."

>>> CAUTION >>>

It is important to note that $\sin^{-1} 0.4196$ does not mean $1/(\sin 0.4196)$. The superscript $^{-1}$ is part of a function symbol, and *$\sin^{-1}$ represents the inverse sine function.* Inverse trigonometric functions are developed in detail in Section 6-6.

Fortunately, we can find θ directly using a calculator. Most calculators of the type used in this book have the function keys $\boxed{\sin^{-1}}$, $\boxed{\cos^{-1}}$, and $\boxed{\tan^{-1}}$ or their equivalents (check your manual). These function keys take us from a trigonometric ratio back to the corresponding acute angle in degree measure when the calculator is in degree mode. So if $\sin \theta = 0.4196$, we can write either $\theta = \arcsin 0.4196$ or $\theta = \sin^{-1} 0.4196$. We choose the latter and proceed as follows:

$$\theta = \sin^{-1} 0.4196$$
$$= 24.81° \qquad \text{To the nearest hundredth degree.}$$
$$\text{or } 24°49' \qquad \text{To the nearest minute.}$$

CHECK $\sin 24.81° = 0.4196$

>>> EXPLORE-DISCUSS 2

Solve each of the following for θ to the nearest hundredth of a degree using a calculator. Explain why an error message occurs in one of the problems.

(A) $\cos \theta = 0.2044$ **(B)** $\tan \theta = 1.4138$ **(C)** $\sin \theta = 1.4138$

EXAMPLE 3 **Right Triangle Solution**

Solve the right triangle with $a = 4.32$ centimeters and $b = 2.62$ centimeters. Compute the angle measures to the nearest 10′.

SOLUTION Draw a figure and label the known parts (Fig. 8):

> **Figure 8**

SOLVE FOR β

$$\tan \beta = \frac{2.62}{4.32} \qquad \text{Use } \tan^{-1} \text{ to solve for } \beta.$$

$$\beta = \tan^{-1} \frac{2.62}{4.32} \qquad \text{Calculate.}$$

$$= 31.2° \text{ or } 31°10' \qquad 0.2° = [(0.2)(60)]' = 12' \approx 10' \text{ to nearest } 10'.$$

SOLVE FOR α

$$\alpha = 90° - 31°10' \quad \boxed{= 89°60' - 31°10'} \quad = 58°50'$$

SOLVE FOR c

$$\sin \beta = \frac{2.62}{c} \qquad \text{Or use } \csc \beta = \frac{c}{2.62}.$$

$$c = \frac{2.62}{\sin 31.2°} = 5.06 \text{ centimeters}$$

or, using the Pythagorean theorem,

$$c = \sqrt{4.32^2 + 2.62^2} = 5.05 \text{ centimeters}$$

Note the slight difference in the values obtained for c (5.05 versus 5.06). This was caused by rounding β to the nearest 10′ in the first calculation for c.

MATCHED PROBLEM 3 Solve the right triangle with $a = 1.38$ kilometers and $b = 6.73$ kilometers.

EXAMPLE 4 **Geometry**

If a regular pentagon (a five-sided regular polygon) is inscribed in a circle of radius 5.35 centimeters, find the length of one side of the pentagon.

SOLUTION Sketch a figure and insert triangle ACB with C at the center (Fig. 9). Add the auxiliary line CD as indicated. We will find AD and double it to find the length of the side wanted.

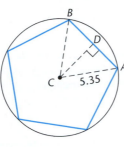

> **Figure 9**

$$\text{Angle } ACB = \frac{360°}{5} = 72° \qquad \text{Exact}$$

$$\text{Angle } ACD = \frac{72°}{2} = 36° \qquad \text{Exact}$$

$$\sin (\text{angle } ACD) = \frac{AD}{AC}$$

$$AD = AC \sin (\text{angle } ACD)$$

$$= 5.35 \sin 36°$$

$$= 3.14 \text{ centimeters} \qquad \text{To three significant digits}$$

$$AB = 2AD = 6.28 \text{ centimeters}$$

MATCHED PROBLEM 4 If a square of side 43.6 meters is inscribed in a circle, what is the radius of the circle?

EXAMPLE 5 **Architecture**

In designing a house an architect wishes to determine the amount of overhang of a roof so that it shades the entire south wall at noon during the summer solstice when the angle of elevation of the sun is 81° (Fig. 10). Minimally, how much overhang should be provided for this purpose?

SOLUTION Using Figure 10, we consider the right triangle with angle θ and sides x (the overhang) and 11 feet, and solve for x:

> **Figure 10**

$$\theta = 90° - 81° = 9°$$

$$\tan \theta = \frac{x}{11}$$

$$x = 11 \tan 9° = 1.7 \text{ feet}$$

MATCHED PROBLEM 5 With the overhang found in Example 5, how far will the shadow of the overhang come down the wall at noon during the winter solstice when the angle of elevation of the sun is 32°?

ANSWERS TO MATCHED PROBLEMS

1. (A) 0.1564 (B) 0.2574 (C) 4.810 2. $\beta = 42.2°$, $a = 20.2$ meters, $b = 18.3$ meters
3. $\alpha = 11°40'$, $\beta = 78°20'$, $c = 6.87$ kilometers 4. 30.8 meters 5. 1.1 feet

6-2 Exercises

1. Explain what it means to solve a right triangle.

2. Does every triangle have a hypotenuse? Explain.

3. Can every rectangle be partitioned into two right triangles? Explain.

4. Can every triangle be partitioned into two right triangles? Explain.

5. Explain why it is not possible to solve for the sides of a triangle if only its angles are known.

6. Explain why the cosine of an acute angle of a right triangle is equal to the sine of the complementary angle.

In Problems 7–12, use the figure to write the ratio of sides that corresponds to each trigonometric function.

7. $\sin \theta$ 8. $\cot \theta$ 9. $\csc \theta$

10. $\cos \theta$ 11. $\tan \theta$ 12. $\sec \theta$

Figure for Problems 7-18

In Problems 13–18, use the figure to write the trigonometric function that corresponds to each ratio.

13. $\dfrac{24}{25}$ 14. $\dfrac{7}{24}$ 15. $\dfrac{25}{24}$

16. $\dfrac{7}{25}$ 17. $\dfrac{24}{7}$ 18. $\dfrac{25}{7}$

In Problems 19–24, find each acute angle θ in degree measure to two decimal places using a calculator.

19. $\cos \theta = 0.4917$ 20. $\sin \theta = 0.0859$

21. $\theta = \tan^{-1} 8.031$ 22. $\theta = \cos^{-1} 0.5097$

23. $\sin \theta = 0.6031$ 24. $\tan \theta = 1.993$

In Problems 25–36, use the figure and the given information to solve each triangle.

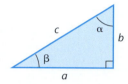

25. $\beta = 17.8°$, $c = 3.45$ 26. $\beta = 33.7°$, $b = 22.4$

27. $\beta = 43°20'$, $a = 123$ 28. $\beta = 62°30'$, $c = 42.5$

29. $\alpha = 23°0'$, $a = 54.0$ 30. $\alpha = 54°$, $c = 4.3$

31. $\alpha = 53.21°$, $b = 23.82$ 32. $\alpha = 35.73°$, $b = 6.482$

33. $a = 6.00$, $b = 8.46$ 34. $a = 22.0$, $b = 46.2$

35. $b = 10.0$, $c = 12.6$ 36. $b = 50.0$, $c = 165$

In Problems 37–42, determine whether the statement is true or false. If true, explain why. If false, give a counterexample.

37. If any two angles of a right triangle are known, then it is possible to solve for the remaining angle and the three sides.

38. If any two sides of a right triangle are known, then it is possible to solve for the remaining side and the three angles.

39. If α and β are the acute angles of a right triangle, then $\sin \alpha = \sin \beta$.

40. If α and β are the acute angles of a right triangle, then $\tan \alpha = \cot \beta$.

41. If α and β are the acute angles of a right triangle, then $\sec \alpha = \cos \beta$.

42. If α and β are the acute angles of a right triangle, then $\csc \alpha = \sec \beta$.

In Problems 43–48, find the degree measure to one decimal place of the acute angle between the given line and the x axis.

43. $y = \dfrac{1}{2}x + 3$ 44. $y = \dfrac{1}{3}x - \dfrac{1}{4}$

45. $y = 5x + 21$ 46. $y = 4x - 16$

47. $y = -2x + 7$ 48. $y = -3x - 1$

In Problems 49–54, find the slope to two decimal places of each line for which there is an angle of measure θ between the line and the x axis. [Hint: Note that there is an angle of measure 45° between the line y = x and the x axis, and also between the line y = −x and the x axis.]

49. θ = 20° **50.** θ = 40° **51.** θ = 80°

52. θ = 70° **53.** θ = π/30 **54.** θ = π/20

*Problems 55–60 give a **geometric interpretation of the trigonometric ratios.** Refer to the figure, where O is the center of a circle of radius 1, θ is the acute angle AOD, D is the intersection point of the terminal side of angle θ with the circle, and EC is tangent to the circle at D.*

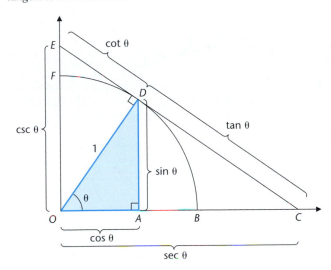

55. Explain why
(A) cos θ = OA (B) cot θ = DE (C) sec θ = OC

56. Explain why
(A) sin θ = AD (B) tan θ = DC (C) csc θ = OE

57. Explain what happens to each of the following as the acute angle θ approaches 90°.
(A) cos θ (B) cot θ (C) sec θ

58. Explain what happens to each of the following as the acute angle θ approaches 90°.
(A) sin θ (B) tan θ (C) csc θ

59. Explain what happens to each of the following as the acute angle θ approaches 0°.
(A) sin θ (B) tan θ (C) csc θ

60. Explain what happens to each of the following as the acute angle θ approaches 0°.
(A) cos θ (B) cot θ (C) sec θ

61. Show that

$$h = \frac{d}{\cot \alpha - \cot \beta}$$

62. Show that

$$h = \frac{d}{\cot \alpha + \cot \beta}$$

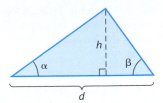

APPLICATIONS

63. SURVEYING Find the height of a tree (growing on level ground) if at a point 105 feet from the base of the tree the angle to its top relative to the horizontal is found to be 65.3°.

64. AIR SAFETY To measure the height of a cloud ceiling over an airport, a searchlight is directed straight upward to produce a lighted spot on the clouds. Five hundred meters away an observer reports the angle of the spot relative to the horizontal to be 32.2°. How high (to the nearest meter) are the clouds above the airport?

65. ENGINEERING If a train climbs at a constant angle of 1°23′, how many vertical feet has it climbed after going 1 mile? (1 mile = 5,280 feet)

66. AIR SAFETY If a jet airliner climbs at an angle of 15°30′ with a constant speed of 315 miles per hour, how long will it take (to the nearest minute) to reach an altitude of 8.00 miles? Assume there is no wind.

67. ASTRONOMY Find the diameter of the moon (to the nearest mile) if at 239,000 miles from Earth it produces an angle of 32′ relative to an observer on Earth.

68. ASTRONOMY If the sun is 93,000,000 miles from Earth and its diameter is opposite an angle of 32′ relative to an observer on Earth, what is the diameter of the sun (to two significant digits)?

69. GEOMETRY If a circle of radius 4 centimeters has a chord of length 3 centimeters, find the central angle that is opposite this chord (to the nearest degree).

70. GEOMETRY Find the length of one side of a nine-sided regular polygon inscribed in a circle of radius 4.06 inches.

71. PHYSICS In a course in physics it is shown that the velocity v of a ball rolling down an inclined plane (neglecting air resistance and friction) is given by

$$v = gt \sin \theta$$

where g is a gravitational constant (acceleration due to gravity), t is time, and θ is the angle of inclination of the plane (see the following figure). Galileo (1564–1642) used this equation in the form

$$g = \frac{v}{t \sin \theta}$$

to estimate g after measuring v experimentally. (At that time, no timing devices existed to measure the velocity of a free-falling body, so Galileo used the inclined plane to slow the motion down.) A steel ball is rolled down a glass plane inclined at 8.0°. Approximate g to one decimal place if at the end of 3.0 seconds the ball has a measured velocity of 4.2 meters per second.

72. PHYSICS Refer to Problem 71. A steel ball is rolled down a glass plane inclined at 4.0°. Approximate g to one decimal place if at the end of 4.0 seconds the ball has a measured velocity of 9.0 feet per second.

73. ENGINEERING—COST ANALYSIS A cable television company wishes to run a cable from a city to a resort island 3 miles offshore. The cable is to go along the shore, then to the island underwater, as indicated in the accompanying figure. The cost of running the cable along the shore is $15,000 per mile and underwater, $25,000 per mile.

(A) Referring to the figure, show that the cost in terms of θ is given by

$$C(\theta) = 75{,}000 \sec \theta - 45{,}000 \tan \theta + 300{,}000$$

(B) Calculate a table of costs, each cost to the nearest dollar, for the following values of θ: 10°, 20°, 30°, 40°, and 50°. (Notice how the costs vary with θ. In a course in calculus, students are asked to find θ so that the cost is minimized.)

74. ENGINEERING—COST ANALYSIS Refer to Problem 73. Suppose the island is 4 miles offshore and the cost of running the cable along the shore is $20,000 per mile and underwater, $30,000 per mile.

(A) Referring to the figure for Problem 73 with appropriate changes, show that the cost in terms of θ is given by

$$C(\theta) = 120{,}000 \sec \theta - 80{,}000 \tan \theta + 400{,}000$$

(B) Calculate a table of costs, each cost to the nearest dollar, for the following values of θ: 10°, 20°, 30°, 40°, and 50°.

75. GEOMETRY Find r in the accompanying figure (to two significant digits) so that the circle is tangent to all three sides of the isosceles triangle. [*Hint:* The radius of a circle is perpendicular to a tangent line at the point of tangency.]

76. GEOMETRY Find r in the accompanying figure (to two significant digits) so that the smaller circle is tangent to the larger circle and the two sides of the angle. [See the hint in Problem 75.]

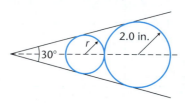

6-3 Trigonometric Functions: A Unit Circle Approach

> **The Wrapping Function**
> **Defining the Trigonometric Functions**
> **Graphing the Trigonometric Functions**

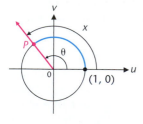

> Figure 1

In Section 6-3 we introduce the six trigonometric functions in terms of the coordinates of points on the unit circle.

> The Wrapping Function

Consider a positive angle θ in standard position, and let P denote the point of intersection of the terminal side of θ with the unit circle $u^2 + v^2 = 1$ (Fig. 1).* Let x denote the length

*We use the variables u and v instead of x and y so that x can be used without ambiguity as an independent variable in defining the wrapping function and the trigonometric functions.

of the arc opposite θ on the unit circle. Because the unit circle has radius $r = 1$, the radian measure of θ is given by

$$\theta = \frac{x}{r} = \frac{x}{1} = x \text{ radians}$$

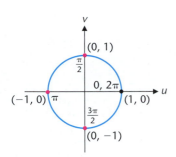

> **Figure 2** Circular points on the coordinate axes.

In other words, on the unit circle, the radian measure of a positive angle is equal to the length of the intercepted arc; similarly, on the unit circle, the radian measure of a negative angle is equal to the negative of the length of the intercepted arc. Because $\theta = x$, we may consider the real number x to be the name of the angle θ, when convenient. The function W that associates with each real number x the point $W(x) = P$ is called the **wrapping function.** The point P is called a **circular point.**

Consider, for example, the angle in standard position that has radian measure $\pi/2$. Its terminal side intersects the unit circle at the point $(0, 1)$. Therefore, $W(\pi/2) = (0, 1)$. Similarly, we can find the circular point associated with any angle that is an integer multiple of $\pi/2$ (Fig. 2).

$$W(0) = (1, 0)$$

$$W\left(\frac{\pi}{2}\right) = (0, 1)$$

$$W(\pi) = (-1, 0)$$

$$W\left(\frac{3\pi}{2}\right) = (0, -1)$$

$$W(2\pi) = (1, 0)$$

>>> **EXPLORE-DISCUSS 1**

The name *wrapping function* stems from visualizing the correspondence as a wrapping of the real number line, with origin at $(1, 0)$, around the unit circle—the positive real axis is wrapped counterclockwise, and the negative real axis is wrapped clockwise—so that each real number is paired with a unique circular point (Fig. 3).

> **Figure 3** The wrapping function.

(A) Explain why the wrapping function is not one-to-one.

(B) In which quadrant is the circular point $W(1)$? $W(-10)$? $W(100)$?

> **Figure 4**

Given a real number x, it is difficult, in general, to find the coordinates (a, b) of the circular point $W(x)$ that is associated with x. (It is trigonometry that overcomes this difficulty.) For certain real numbers x, however, we can find the coordinates (a, b) of $W(x)$ by using simple geometric facts. For example, consider $x = \pi/6$ and let P denote the circular point $W(x) = (a, b)$ that is associated with x. Let P' be the reflection of P through the u axis (Fig. 4).

Then triangle $0PP'$ is equiangular (each angle has measure $\pi/3$ radians or 60°) and therefore equilateral. So $b = 1/2$. Because (a, b) lies on the unit circle, we solve for a:

$$a^2 + b^2 = 1 \qquad \text{Substitute } b = \tfrac{1}{2}.$$

$$a^2 + \left(\frac{1}{2}\right)^2 = 1 \qquad \text{Subtract } \tfrac{1}{4} \text{ from both sides.}$$

$$a^2 = \frac{3}{4} \qquad \text{Take square roots of both sides.}$$

$$a = \pm\frac{\sqrt{3}}{2} \qquad a = -\frac{\sqrt{3}}{2} \text{ must be discarded. (Why?)}$$

Therefore,

$$W\left(\frac{\pi}{6}\right) = \left(\frac{\sqrt{3}}{2}, \frac{1}{2}\right)$$

EXAMPLE 1 **Coordinates of Circular Points**

Find the coordinates of the following circular points:

(A) $W(-\pi/2)$　　(B) $W(5\pi/2)$　　(C) $W(\pi/3)$　　(D) $W(7\pi/6)$　　(E) $W(\pi/4)$

SOLUTIONS

(A) Because the circumference of the unit circle is 2π, $-\pi/2$ is the radian measure of a negative angle that is $\tfrac{1}{4}$ of a complete clockwise rotation. So $W(-\pi/2) = (0, -1)$ (Fig. 5).

(B) Starting at $(1, 0)$ and proceeding counterclockwise, we count quarter-circle steps, $\pi/2, 2\pi/2, 3\pi/2, 4\pi/2$, and end at $5\pi/2$. So the circular point is on the positive vertical axis, and $W(5\pi/2) = (0, 1)$ (see Fig. 5).

(C) The circular point $W(\pi/3)$ is the reflection of the point $W(\pi/6) = (\sqrt{3}/2, 1/2)$ through the line $u = v$. So $W(\pi/3) = (1/2, \sqrt{3}/2)$ (Fig. 6).

(D) The circular point $W(7\pi/6)$ is the reflection of the point $W(\pi/6) = (\sqrt{3}/2, 1/2)$ through the origin. So $W(7\pi/6) = (-\sqrt{3}/2, -1/2)$ (see Fig. 6).

(E) The circular point $W(\pi/4)$ lies on the line $u = v$, so $a = b$.

$$a^2 + b^2 = 1 \qquad \text{Substitute } b = a.$$

$$2a^2 = 1 \qquad \text{Divide both sides by 2.}$$

$$a^2 = \frac{1}{2} \qquad \text{Take square roots of both sides.}$$

$$a = \frac{1}{\sqrt{2}} \qquad a = -\frac{1}{\sqrt{2}} \text{ is impossible. (Why?)}$$

So $W(\pi/4) = (1/\sqrt{2}, 1/\sqrt{2})$ (Fig. 7).

> **Figure 5**

> **Figure 6**

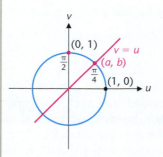

> **Figure 7**

MATCHED PROBLEM 1 Find the coordinates of the following circular points:

(A) $W(3\pi)$ (B) $W(-7\pi/2)$ (C) $W(5\pi/6)$

(D) $W(-\pi/3)$ (E) $W(5\pi/4)$

Some key results from Example 1 are summarized in Figure 8. If x is any integer multiple of $\pi/6$ or $\pi/4$, then the coordinates of $W(x)$ can be determined easily from Figure 8 by using symmetry properties. For example, change the sign of the first coordinate of the three points in Quadrant I to obtain the coordinates of their reflections through the v axis in Quadrant II.

> **COORDINATES OF KEY CIRCULAR POINTS**

> **Figure 8**

> ## Defining the Trigonometric Functions

We define the trigonometric functions in terms of the coordinates of points on the unit circle. This suggests that the trigonometric functions are useful in analyzing circular motion, for example, of satellites, DVD players, generators, wheels, and propellers. While true, we will also discover that these functions have many applications that are apparently unrelated to rotary motion. There are six **trigonometric functions: sine, cosine, tangent, cotangent, secant,** and **cosecant.** The values of these functions at a real number x are denoted by **sin x, cos x, tan x, cot x, sec x,** and **csc x,** respectively.

> **DEFINITION 1 Trigonometric Functions**

Let x be a real number and let (a, b) be the coordinates of the circular point $W(x)$ that lies on the terminal side of the angle with radian measure x. Then:

$$\sin x = b \qquad\qquad \csc x = \frac{1}{b} \quad b \neq 0$$

$$\cos x = a \qquad\qquad \sec x = \frac{1}{a} \quad a \neq 0$$

$$\tan x = \frac{b}{a} \quad a \neq 0 \qquad \cot x = \frac{a}{b} \quad b \neq 0$$

REMARKS:

1. Note that $\sin x$ and $\cos x$ are the second and first coordinates, respectively, of the point (a, b) on the unit circle.

2. We assume in Definition 1 that (a, b) is the point on the *unit circle* that lies on the terminal side of the angle with radian measure x. More generally, however, if (a, b) is the point on that terminal side that lies on the *circle of radius* $r > 0$, then:

$$a^2 + b^2 = r^2 \quad \text{Divide both sides by } r^2.$$

$$\left(\frac{a}{r}\right)^2 + \left(\frac{b}{r}\right)^2 = 1 \quad \text{So } \left(\frac{a}{r}, \frac{b}{r}\right) \text{ lies on the unit circle.}$$

Therefore, $(a/r, b/r)$ is the point on the terminal side of the angle with radian measure x (see Problems 89 and 90 in Exercises 6-3) that lies on the unit circle (Fig. 9).

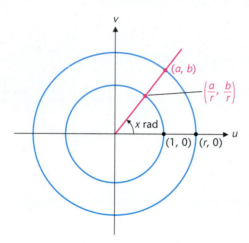

› **Figure 9**

By Definition 1,

$$\sin x = \frac{b}{r} \qquad\qquad \csc x = \frac{r}{b} \quad b \neq 0$$

$$\cos x = \frac{a}{r} \qquad\qquad \sec x = \frac{r}{a} \quad a \neq 0$$

$$\tan x = \frac{b}{a} \quad a \neq 0 \qquad \cot x = \frac{a}{b} \quad b \neq 0$$

Note that these formulas coincide with those of Definition 1 when $r = 1$.

EXAMPLE **2** **Evaluating Trigonometric Functions**

Find the values of all six trigonometric functions of the angle x if

(A) $W(x) = \left(\dfrac{3}{5}, -\dfrac{4}{5}\right)$.

(B) The terminal side of x contains the point $(-60, -11)$.

SOLUTIONS (A) Note that $W(x)$ is indeed on the unit circle because

$$\sqrt{\left(\frac{3}{5}\right)^2 + \left(-\frac{4}{5}\right)^2} = \sqrt{\frac{9}{25} + \frac{16}{25}} = \sqrt{\frac{25}{25}} = 1$$

Using Definition 1, with $a = \frac{3}{5}$ and $b = -\frac{4}{5}$,

$$\sin x = b = -\frac{4}{5} \qquad\qquad \csc x = \frac{1}{b} = -\frac{5}{4}$$

$$\cos x = a = \frac{3}{5} \qquad\qquad \sec x = \frac{1}{a} = \frac{5}{3}$$

$$\tan x = \frac{b}{a} = \frac{-4/5}{3/5} = -\frac{4}{3} \qquad \cot x = \frac{a}{b} = \frac{3/5}{-4/5} = -\frac{3}{4}$$

(B) The distance r from $(-60, -11)$ to $(0, 0)$ is

$$\sqrt{(-60)^2 + (-11)^2} = \sqrt{3{,}600 + 121} = \sqrt{3{,}721} = 61$$

Using Remark 2 following Definition 1, with $a = -60$, $b = -11$, and $r = 61$:

$$\sin x = \frac{b}{r} = -\frac{11}{61} \qquad\qquad \csc x = \frac{r}{b} = -\frac{61}{11}$$

$$\cos x = \frac{a}{r} = -\frac{60}{61} \qquad\qquad \sec x = \frac{r}{a} = -\frac{61}{60}$$

$$\tan x = \frac{b}{a} = \frac{11}{60} \qquad\qquad \cot x = \frac{a}{b} = \frac{60}{11}$$

»» CAUTION »»

Always check that values of $\sin x$ and $\cos x$ are numbers that are between (or equal to) -1 and 1, as implied by Definition 1. Note in particular that this is the case in Example 2.

MATCHED PROBLEM 2

Find the values of all six trigonometric functions of the angle x if

(A) $W(x) = \left(-\dfrac{12}{13}, \dfrac{5}{13} \right)$.

(B) The terminal side of x contains the point $(13, 84)$.

The domain of both the sine and cosine functions is the set of real numbers R. The range of both the sine and cosine functions is $[-1, 1]$. This is the set of numbers assumed by b, for sine, and a, for cosine, as the circular point (a, b) moves around the unit circle. The domain of cosecant is the set of real numbers x such that b in $W(x) = (a, b)$ is not 0. Similar restrictions are made on the domains of the other three trigonometric functions. We will have more to say about the domains and ranges of all six trigonometric functions in subsequent sections.

Note from Definition 1 that $\csc x$ is the reciprocal of $\sin x$, provided that $\sin x \neq 0$. Therefore, $\sin x$ is the reciprocal of $\csc x$. Similarly, $\cos x$ and $\sec x$ are reciprocals of each other, as are $\tan x$ and $\cot x$. We call these useful facts the *reciprocal identities*.

› RECIPROCAL IDENTITIES

For x any real number:

$$\csc x = \frac{1}{\sin x} \qquad \sin x \neq 0$$

$$\sec x = \frac{1}{\cos x} \qquad \cos x \neq 0$$

$$\cot x = \frac{1}{\tan x} \qquad \tan x \neq 0$$

In Example 1 we were able to give a simple geometric argument to find, for example, that the coordinates of $W(7\pi/6)$ are $(-\sqrt{3}/2, -1/2)$. Therefore, $\sin(7\pi/6) = -1/2$ and $\cos(7\pi/6) = -\sqrt{3}/2$. These exact values correspond to the approximations given by a calculator [Fig. 10(a)]. For most values of x, however, simple geometric arguments fail to give the exact coordinates of $W(x)$. But a calculator, set in radian mode, can be used to give approximations. For example, if $x = \pi/7$, then $W(\pi/7) \approx (0.901, 0.434)$ [Fig. 10(b)].

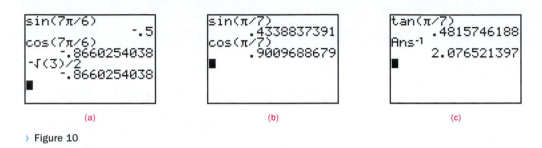

(a) (b) (c)

› **Figure 10**

Most calculators have function keys for the sine, cosine, and tangent functions, but not for the cotangent, secant, and cosecant. Because the cotangent, secant, and cosecant are the reciprocals of the tangent, cosine, and sine, respectively, they can be evaluated easily. For example, $\cot(\pi/7) = 1/\tan(\pi/7) \approx 2.077$ [Fig. 10(c)]. Do not use the calculator function keys marked $\sin^{-1}$, $\cos^{-1}$, or $\tan^{-1}$ for this purpose—these keys are used to evaluate the inverse trigonometric functions of Section 6-6, not reciprocals.

EXAMPLE 3 **Calculator Evaluation**

Evaluate to four significant digits.

(A) $\tan 1.5$ (B) $\csc(-6.27)$ (C) $\sec(11\pi/12)$

(D) The coordinates (a, b) of $W(1)$

SOLUTIONS (A) $\tan 1.5 = 14.10$

(B) $\csc(-6.27) = 1/\sin(-6.27) = 75.84$ [Fig. 11(a)]

(C) $\sec(11\pi/12) = 1/\cos(11\pi/12) = -1.035$ [Fig. 11(b)]

(D) $W(1) = (\cos 1, \sin 1) = (0.5403, 0.8415)$

(a) (b)

› **Figure 11**

MATCHED PROBLEM 3 Evaluate to four significant digits.

(A) $\cot(-8.25)$ (B) $\sec(7\pi/8)$ (C) $\csc(4.67)$

(D) The coordinates (a, b) of $W(100)$

› **Graphing the Trigonometric Functions**

The graph of $y = \sin x$ is the set of all ordered pairs (x, y) of real numbers that satisfy the equation. Because $\sin x$, by Definition 1, is the second coordinate of the circular point $W(x)$, our knowledge of the coordinates of certain circular points (Table 1) gives the following solutions to $y = \sin x$: $(0, 0)$, $(\pi/2, 1)$, $(\pi, 0)$, and $(3\pi/2, -1)$.

Table 1

x	0	$\pi/2$	π	$3\pi/2$
$W(x)$	$(0, 0)$	$(0, 1)$	$(-1, 0)$	$(0, -1)$
$\sin x$	0	1	0	-1

As x increases from 0 to $\pi/2$, the circular point $W(x)$ moves on the circumference of the unit circle from $(0, 0)$ to $(0, 1)$, and so $\sin x$ [the second coordinate of $W(x)$] increases from 0 to 1. Similarly, as x increases from $\pi/2$ to π, the circular point $W(x)$ moves on the circumference of the unit circle from $(0, 1)$ to $(-1, 0)$, and so $\sin x$ decreases from 1 to 0. These observations are in agreement with the graph of $y = \sin x$, obtained from a graphing calculator in radian mode [Fig. 12(a)].

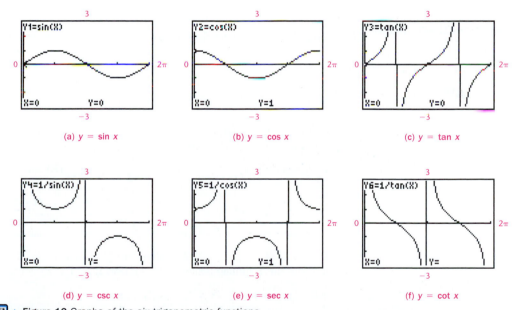

(a) $y = \sin x$ (b) $y = \cos x$ (c) $y = \tan x$

(d) $y = \csc x$ (e) $y = \sec x$ (f) $y = \cot x$

› **Figure 12** Graphs of the six trigonometric functions.

Figure 12 shows the graphs of all six trigonometric functions from $x = 0$ to $x = 2\pi$. Because the circular point $W(2\pi)$ coincides with the circular point $W(0)$, the graphs of the six trigonometric functions from $x = 2\pi$ to $x = 4\pi$ would be identical to the graphs shown in Figure 12. The functions $y = \sin x$ and $y = \cos x$ are bounded; their maximum values are 1 and their minimum values are -1. The functions $y = \tan x$, $y = \cot x$, $y = \sec x$, and $y = \csc x$ are unbounded; they have vertical asymptotes at the values of x for which they are undefined. It is instructive to study and compare the graphs of reciprocal pairs, for example, $y = \cos x$ and $y = \sec x$. Note that $\sec x$ is undefined when $\cos x$ equals 0, and that because the maximum positive value of $\cos x$ is 1, the minimum positive value of $\sec x$ is 1. We will study the properties of trigonometric functions and their graphs in Section 6-4.

| EXAMPLE | 4 | **Zeros and Turning Points** |

Find the zeros and turning points of $y = \cos x$ on the interval $[\pi/2, 5\pi/2]$.

SOLUTION

Recall that a *turning point* is a point on a graph that separates an increasing portion from a decreasing portion, or vice versa. As x increases from $\pi/2$ to $3\pi/2$, the first coordinate of the circular point $W(x)$ (that is, $\cos x$) decreases from 0 to a minimum value of -1 (when $x = \pi$), then increases to a value of 0 (when $x = 3\pi/2$) (Fig. 13). Similarly, as x increases from $3\pi/2$ to $5\pi/2$, $\cos x$ increases from 0 to a maximum value of 1 (when $x = 2\pi$), then decreases to a value of 0. Therefore, the graph of $y = \cos x$ has turning points $(\pi, -1)$ and $(2\pi, 1)$, and zeros $\pi/2$, $3\pi/2$, and $5\pi/2$. These conclusions are confirmed by the graph of $y = \cos x$ in Figure 12(b) on page 411.

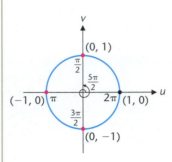

› Figure 13

MATCHED PROBLEM 4

Find all zeros and turning points of $y = \csc x$ on the interval $(0, 4\pi)$.

ANSWERS TO MATCHED PROBLEMS

1. (A) $(-1, 0)$ (B) $(0, 1)$ (C) $(-\sqrt{3}/2, 1/2)$
 (D) $(1/2, -\sqrt{3}/2)$ (E) $(-1/\sqrt{2}, -1/\sqrt{2})$

2. (A) $\sin x = \dfrac{5}{13}$ $\csc x = \dfrac{13}{5}$ (B) $\sin x = \dfrac{84}{85}$ $\csc x = \dfrac{85}{84}$

 $\cos x = -\dfrac{12}{13}$ $\sec x = -\dfrac{13}{12}$ $\cos x = \dfrac{13}{85}$ $\sec x = \dfrac{85}{13}$

 $\tan x = -\dfrac{5}{12}$ $\cot x = -\dfrac{12}{5}$ $\tan x = \dfrac{84}{13}$ $\cot x = \dfrac{13}{84}$

3. (A) 0.4181 (B) -1.082 (C) -1.001 (D) $(0.8623, -0.5064)$

4. Zeros: none; turning points: $(\pi/2, 1)$, $(3\pi/2, -1)$, $(5\pi/2, 1)$, $(7\pi/2, -1)$

6-3 Exercises

1. What is the unit circle?

2. Describe the wrapping function, including its domain and range.

3. Explain the connection between points on the unit circle and the six trigonometric functions.

4. Explain why the function $y = \sec x$ is undefined for certain values of x.

5. Explain why the graph of $y = \tan x$ has vertical asymptotes at $x = \pm\frac{\pi}{2}, \pm\frac{3\pi}{2}, \pm\frac{5\pi}{2}, \ldots$.

6. Explain why every point on the graph of $y = \cos x$ lies on or between the lines $y = -1$ and $y = 1$.

In Problems 7–18, find the coordinates of each circular point.

7. $W(\pi)$ **8.** $W(2\pi)$

9. $W(-\pi/2)$ **10.** $W(5\pi/2)$

11. $W(7\pi/4)$ **12.** $W(-3\pi/4)$

13. $W(-\pi/3)$ **14.** $W(2\pi/3)$

15. $W(11\pi/6)$ **16.** $W(5\pi/6)$

17. $W(-13\pi)$ **18.** $W(-9\pi/2)$

In Problems 19–30, use your answers to Problems 7–18 to give the exact value of the expression if it is defined.

19. $\sin \pi$ **20.** $\csc 2\pi$

21. $\sec (-\pi/2)$ **22.** $\cos (5\pi/2)$

23. $\tan (7\pi/4)$ **24.** $\cot (-3\pi/4)$

25. $\cos (-\pi/3)$ **26.** $\tan (2\pi/3)$

27. $\csc (11\pi/6)$ **28.** $\sec (5\pi/6)$

29. $\cot (-13\pi)$ **30.** $\sin (-9\pi/2)$

In Problems 31–36, in which quadrants must $W(x)$ lie so that:

31. $\cos x < 0$ **32.** $\tan x > 0$

33. $\sin x > 0$ **34.** $\sec x > 0$

35. $\cot x < 0$ **36.** $\csc x < 0$

Evaluate Problems 37–42 to four significant digits using a calculator set in radian mode.

37. $\cos 2.288$ **38.** $\sin 3.104$

39. $\tan (-4.644)$ **40.** $\sec (-1.555)$

41. $\csc 1.571$ **42.** $\cot 0.7854$

Evaluate Problems 43–48 to four significant digits using a calculator. Make sure your calculator is in the correct mode (degree or radian) for each problem.

43. $\sin 25°$ **44.** $\tan 89°$

45. $\cot 12$ **46.** $\csc 13$

47. $\sin 113°27'13''$ **48.** $\cos 235°12'47''$

In Problems 49–52, determine whether the statement about the wrapping function W is true or false. Explain.

49. The domain of the wrapping function is the set of all points on the unit circle.

50. The domain of the wrapping function is the set of all real numbers.

51. If $W(x) = W(y)$, then $x = y$.

52. If $x = y$, then $W(x) = W(y)$.

In Problems 53–58, determine whether the statement about the trigonometric functions is true or false. Explain.

53. If x is a real number, then $\cos x$ is the reciprocal of $\sin x$.

54. If x is a real number, then $(\cot x)(\tan x) = 1$.

55. If $\sec x = \sec y$, then $x = y$.

56. If $x = y$, then $\cos x = \cos y$.

57. The functions $\sin x$ and $\csc x$ have the same domain.

58. The functions $\sin x$ and $\cos x$ have the same domain.

In Problems 59–62, find all zeros and turning points of each function on $[0, 4\pi]$.

59. $y = \sec x$ **60.** $y = \sin x$

61. $y = \tan x$ **62.** $y = \cot x$

Determine the signs of a and b for the coordinates (a, b) of each circular point indicated in Problems 63–72. First determine the quadrant in which each circular point lies. [Note:$\pi/2 \approx 1.57$, $\pi \approx 3.14$, $3\pi/2 \approx 4.71$, and $2\pi \approx 6.28$.]

63. $W(2)$ **64.** $W(1)$

65. $W(3)$ **66.** $W(4)$

67. $W(5)$ **68.** $W(7)$

69. $W(-2.5)$ **70.** $W(-4.5)$

71. $W(-6.1)$ **72.** $W(-1.8)$

In Problems 73–76, for each equation find all solutions for $0 \le x < 2\pi$, then write an expression that represents all solutions for the equation without any restrictions on x.

73. $W(x) = (1, 0)$

74. $W(x) = (-1, 0)$

75. $W(x) = (-1/\sqrt{2}, 1/\sqrt{2})$

76. $W(x) = (1/\sqrt{2}, -1/\sqrt{2})$

77. Describe in words why $W(x) = W(x + 4\pi)$ for every real number x.

78. Describe in words why $W(x) = W(x - 6\pi)$ for every real number x.

In Problems 79–82, in which quadrants are the statements true and why?

79. $\sin x < 0$ and $\cot x < 0$

80. $\cos x > 0$ and $\tan x < 0$

81. $\cos x < 0$ and $\sec x > 0$

82. $\sin x > 0$ and $\csc x < 0$

For which values of x, $0 \leq x \leq 2\pi$, is each of Problems 83–88 not defined?

83. $\cos x$ **84.** $\sin x$

85. $\tan x$ **86.** $\cot x$

87. $\sec x$ **88.** $\csc x$

In Problems 89 and 90, consider the point $P = (a, b)$, where a and b are not both zero, and let $O = (0, 0)$. Ray $\overrightarrow{OP}$ is defined by

$$\overrightarrow{OP} = \{(ka, kb) \mid k \geq 0\}$$

89. Show that $bx - ay = 0$ is the equation of the line through O and P.

90. Refer to Problem 89. Show that every point on $\overrightarrow{OP}$ satisfies the equation of the line through O and P.

APPLICATIONS

If an n-sided regular polygon is inscribed in a circle of radius r, then it can be shown that the area of the polygon is given by

$$A = \frac{1}{2} nr^2 \sin \frac{2\pi}{n}$$

Compute each area exactly and then to four significant digits using a calculator if the area is not an integer.

91. $n = 12$, $r = 5$ meters

92. $n = 4$, $r = 3$ inches

93. $n = 3$, $r = 4$ inches

94. $n = 8$, $r = 10$ centimeters

APPROXIMATING π *Problems 95 and 96 refer to a sequence of numbers generated as follows:*

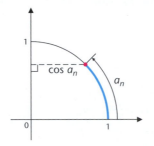

$$a_1$$
$$a_2 = a_1 + \cos a_1$$
$$a_3 = a_2 + \cos a_2$$
$$\vdots$$
$$a_{n+1} = a_n + \cos a_n$$

95. Let $a_1 = 0.5$, and compute the first five terms of the sequence to six decimal places and compare the fifth term with $\pi/2$ computed to six decimal places.

96. Repeat Problem 95, starting with $a_1 = 1$.

6-4 Properties of Trigonometric Functions

> Basic Identities
> Sign Properties
> Reference Triangles
> Periodic Functions

In Section 6-4, we study properties of the trigonometric functions that distinguish them from the polynomial, rational, exponential, and logarithmic functions. The trigonometric functions are *periodic*, and as a consequence, have infinitely many zeros, or infinitely many turning points, or both.

> ## Basic Identities

The definition of trigonometric functions provides several useful relationships among these functions. For convenience, we restate that definition.

> **DEFINITION 1** Trigonometric Functions

Let x be a real number and let (a, b) be the coordinates of the circular point $W(x)$ that lies on the terminal side of the angle with radian measure x. Then:

$$\sin x = b \qquad\qquad \csc x = \frac{1}{b} \quad b \neq 0$$

$$\cos x = a \qquad\qquad \sec x = \frac{1}{a} \quad a \neq 0$$

$$\tan x = \frac{b}{a} \quad a \neq 0 \qquad \cot x = \frac{a}{b} \quad b \neq 0$$

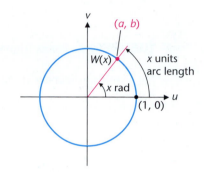

Because $\sin x = b$ and $\cos x = a$, we obtain the following equations:

$$\csc x = \frac{1}{b} = \frac{1}{\sin x} \tag{1}$$

$$\sec x = \frac{1}{a} = \frac{1}{\cos x} \tag{2}$$

$$\cot x = \frac{a}{b} = \frac{1}{b/a} = \frac{1}{\tan x} \tag{3}$$

$$\tan x = \frac{b}{a} = \frac{\sin x}{\cos x} \tag{4}$$

$$\cot x = \frac{a}{b} = \frac{\cos x}{\sin x} \tag{5}$$

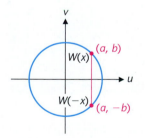

> **Figure 1** Symmetry property.

Because the circular points $W(x)$ and $W(-x)$ are symmetrical with respect to the horizontal axis (Fig. 1), we have the following sign properties:

$$\sin(-x) = -b = -\sin x \tag{6}$$

$$\cos(-x) = a = \cos x \tag{7}$$

$$\tan(-x) = \frac{-b}{a} = -\frac{b}{a} = -\tan x \tag{8}$$

Finally, because $(a, b) = (\cos x, \sin x)$ is on the unit circle $u^2 + v^2 = 1$, it follows that

$$(\cos x)^2 + (\sin x)^2 = 1$$

which is usually written as

$$\sin^2 x + \cos^2 x = 1 \tag{9}$$

where $\sin^2 x$ and $\cos^2 x$ are concise ways of writing $(\sin x)^2$ and $(\cos x)^2$, respectively.

››› **CAUTION** ›››

$$(\sin x)^2 \neq \sin x^2$$
$$(\cos x)^2 \neq \cos x^2$$

Equations (1)–(9) are called **basic identities.** They hold true for all replacements of x by real numbers for which both sides of an equation are defined. These basic identities must be memorized along with the definitions of the six trigonometric functions, because the material is used extensively in developments that follow. Note that most of Chapter 7 is devoted to trigonometric identities.

We summarize the basic identities for convenient reference in Theorem 1.

> › **THEOREM 1** Basic Trigonometric Identities
>
> For x any real number (in all cases restricted so that both sides of an equation are defined),
>
> **Reciprocal identities**
>
> (1) $\qquad$ (2) $\qquad$ (3)
>
> $$\csc x = \frac{1}{\sin x} \qquad \sec x = \frac{1}{\cos x} \qquad \cot x = \frac{1}{\tan x}$$
>
> **Quotient identities**
>
> (4) $\qquad\qquad$ (5)
>
> $$\tan x = \frac{\sin x}{\cos x} \qquad \cot x = \frac{\cos x}{\sin x}$$
>
> **Identities for negatives**
>
> (6) $\qquad\qquad$ (7) $\qquad\qquad$ (8)
>
> $$\sin(-x) = -\sin x \qquad \cos(-x) = \cos x \qquad \tan(-x) = -\tan x$$
>
> **Pythagorean identity**
>
> (9)
>
> $$\sin^2 x + \cos^2 x = 1$$

EXAMPLE 1 Using Basic Identities

Use the basic identities to find the values of the other five trigonometric functions given $\sin x = -\frac{1}{2}$ and $\tan x > 0$.

SOLUTION We first note that the circular point $W(x)$ is in Quadrant III, because that is the only quadrant in which $\sin x < 0$ and $\tan x > 0$. We next find $\cos x$ using identity (9):

$$\sin^2 x + \cos^2 x = 1 \qquad \text{Substitute } \sin x = -\tfrac{1}{2}.$$
$$(-\tfrac{1}{2})^2 + \cos^2 x = 1 \qquad \text{Subtract } \tfrac{1}{4} \text{ from both sides.}$$
$$\cos^2 x = \tfrac{3}{4} \qquad \text{Take square roots of both sides.}$$
$$\cos x = -\frac{\sqrt{3}}{2} \qquad \text{Because } W(x) \text{ is in Quadrant III.}$$

Now, because we have values for $\sin x$ and $\cos x$, we can find values for the other four trigonometric functions using identities (1), (2), (4), and (5):

$$\csc x = \frac{1}{\sin x} = \frac{1}{-\frac{1}{2}} = -2 \qquad \text{Reciprocal identity (1)}$$

$$\sec x = \frac{1}{\cos x} = \frac{1}{-\sqrt{3}/2} = -\frac{2}{\sqrt{3}} \qquad \text{Reciprocal identity (2)}$$

$$\tan x = \frac{\sin x}{\cos x} = \frac{-\frac{1}{2}}{-\sqrt{3}/2} = \frac{1}{\sqrt{3}} \qquad \text{Quotient identity (4)}$$

$$\cot x = \frac{\cos x}{\sin x} = \frac{-\sqrt{3}/2}{-\frac{1}{2}} = \sqrt{3} \qquad \text{Quotient identity (5)}$$
[Note: We could also use identity (3).]

It is important to note that we were able to find the values of the other five trigonometric functions without finding x.

MATCHED PROBLEM 1

Use the basic identities to find the values of the other five trigonometric functions given $\cos x = 1/\sqrt{2}$ and $\cot x < 0$.

⟩⟩⟩ EXPLORE-DISCUSS 1

Suppose that $\sin x = -\frac{1}{2}$ and $\tan x > 0$, as in Example 1. Using basic identities and the results in Example 1, find each of the following:

(A) $\sin(-x)$ **(B)** $\sec(-x)$ **(C)** $\tan(-x)$

Verbally justify each step in your solution process.

⟩ Sign Properties

As a circular point $W(x)$ moves from quadrant to quadrant, its coordinates (a, b) undergo sign changes. So the trigonometric functions also undergo sign changes. It is important to know the sign of each trigonometric function in each quadrant. Table 1 shows the sign behavior for each function. It is not necessary to memorize Table 1, because the sign of each function for each quadrant is easily determined from its definition (which *should* be memorized).

Table 1 Sign Properties

Trigonometric Function	Sign in Quadrant I	II	III	IV
$\sin x = b$	+	+	−	−
$\csc x = 1/b$	+	+	−	−
$\cos x = a$	+	−	−	+
$\sec x = 1/a$	+	−	−	+
$\tan x = b/a$	+	−	+	−
$\cot x = a/b$	+	−	+	−

› Reference Triangles

Consider an angle θ in standard position. Let $P = (a, b)$ be the point of intersection of the terminal side of θ with a circle of radius $r > 0$ (Fig. 2). Then, by Definition 1 and the remarks on pp. 407 and 408,

$$
\begin{array}{ll}
\sin \theta = \dfrac{b}{r} & \csc \theta = \dfrac{r}{b} \quad b \neq 0 \\[2mm]
\cos \theta = \dfrac{a}{r} & \sec \theta = \dfrac{r}{a} \quad a \neq 0 \qquad (10) \\[2mm]
\tan \theta = \dfrac{b}{a} \quad a \neq 0 & \cot \theta = \dfrac{a}{b} \quad b \neq 0
\end{array}
$$

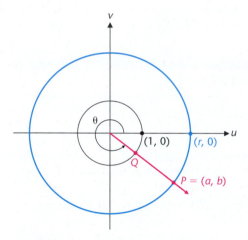

› Figure 2

To simplify the use of formulas (10), it is often convenient to associate a reference triangle and reference angle with θ, and to label the horizontal side, vertical side, and hypotenuse of the reference triangle with a, b, and r, respectively, to easily obtain the values of the trigonometric functions of θ.

› REFERENCE TRIANGLE AND REFERENCE ANGLE

1. To form a **reference triangle** for θ, draw a perpendicular from a point $P = (a, b)$ on the terminal side of θ to the horizontal axis.

2. The **reference angle** α is the acute angle (always taken positive) between the terminal side of θ and the horizontal axis.

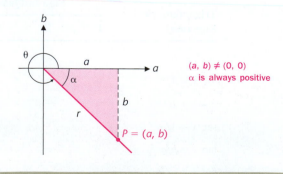

$(a, b) \neq (0, 0)$
α is always positive

If Adj and Opp denote the labels a and b (possibly negative) on the horizontal and vertical sides of the reference triangle, and Hyp denotes the length r of the hypotenuse, then

$$\sin \theta = \frac{\text{Opp}}{\text{Hyp}} \qquad \csc \theta = \frac{\text{Hyp}}{\text{Opp}}$$

$$\cos \theta = \frac{\text{Adj}}{\text{Hyp}} \qquad \sec \theta = \frac{\text{Hyp}}{\text{Adj}} \qquad \qquad (11)$$

$$\tan \theta = \frac{\text{Opp}}{\text{Adj}} \qquad \cot \theta = \frac{\text{Adj}}{\text{Opp}}$$

Formulas (11) are easy to remember, because, if the signs of Adj and Opp are ignored, the formulas coincide with the right triangle ratios (see Section 6.2) for the angle α of the reference triangle.

In Example 2, a reference triangle is used to find values of the trigonometric functions. This method provides a simple alternative to the approach of Example 1, which emphasized basic identities.

| EXAMPLE | 2 | **Values of the Trigonometric Functions** |

If $\sin \theta = 4/7$ and $\cos \theta < 0$, find the values of each of the other five trigonometric functions of θ.

SOLUTION Because the sine of θ is positive and the cosine is negative, the angle θ is in quadrant II. We sketch a reference triangle (Fig. 3) and use the Pythagorean theorem to calculate the length of the horizontal side:

$$\sqrt{7^2 - 4^2} = \sqrt{33}$$

Therefore, $\text{Adj} = -\sqrt{33}, \text{Opp} = 4, \text{Hyp} = 7.$ The values of the other five trigonometric functions are:

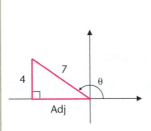

> **Figure 3**

$$\cos \theta = -\frac{\sqrt{33}}{7}$$

$$\tan \theta = -\frac{4}{\sqrt{33}}$$

$$\csc \theta = \frac{7}{4}$$

$$\sec \theta = -\frac{7}{\sqrt{33}}$$

$$\cot \theta = -\frac{\sqrt{33}}{4}$$

| MATCHED PROBLEM 2 | If $\tan \theta = 10$ and $\sin \theta < 0$, find the values of each of the other five trigonometric functions of θ.

》》 CAUTION 》》

When using a reference triangle, the label Hyp on the hypotenuse is always positive. The label Adj on the horizontal leg of the reference triangle is positive or negative depending on whether that leg lies on the positive or negative horizontal axis, respectively. The label Opp on the vertical leg is positive or negative depending on whether that leg is above or below the horizontal axis, respectively.

› **Periodic Functions**

Because the unit circle has a circumference of 2π, we find that for a given value of x (Fig. 4) we will return to the circular point $W(x) = (a, b)$ if we add any integer multiple of 2π to x. Think of a point P moving around the unit circle in either direction. Every time P covers a distance of 2π, the circumference of the circle, it is back at the point where it started. So for any real number x,

$$\sin (x + 2k\pi) = \sin x \qquad k \text{ any integer}$$
$$\cos (x + 2k\pi) = \cos x \qquad k \text{ any integer}$$

Functions with this kind of repetitive behavior are called *periodic functions*. In general, we have Definition 2.

› **Figure 4**

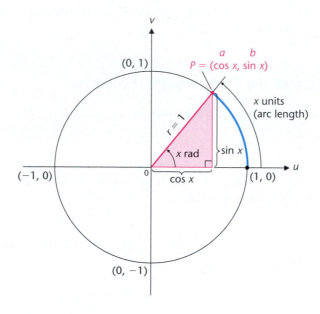

> **DEFINITION 2** Periodic Functions

A function f is **periodic** if there exists a positive real number p such that

$$f(x + p) = f(x)$$

for all x in the domain of f. The smallest such positive p, if it exists, is called the **fundamental period of f** (or often just the **period of f**).

Both the sine and cosine functions are periodic with period 2π. Once the graph for one period is known, the entire graph is obtained by repetition. The domain of both functions is the set of all real numbers, and the range of both is $[-1, 1]$. Because $b = 0$ at the circular points $(1, 0)$ and $(-1, 0)$, the zeros of the sine function are $k\pi$, k any integer. Because $a = 0$ at the circular points $(0, 1)$ and $(0, -1)$, the zeros of the cosine function are $\pi/2 + k\pi$, k any integer. Both the sine and cosine functions possess symmetry properties (see Section 3-3). By the basic identity $\sin (-x) = -\sin x$, the sine function is symmetric with respect to the origin, so it is an odd function. Because $\cos (-x) = \cos x$, the cosine function is symmetric with respect to the y axis, so it is an even function. Figures 5 and 6 summarize these properties and show the graphs of the sine and cosine functions, respectively.

› GRAPH OF y = sin x

› **Figure 5**

Period: 2π
Domain: All real numbers
Range: $[-1, 1]$

Symmetric with respect to the origin

› GRAPH OF y = cos x

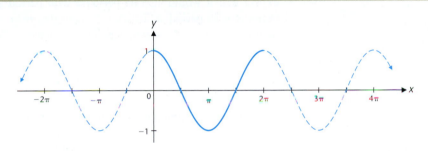

› **Figure 6**

Period: 2π
Domain: All real numbers
Range: $[-1, 1]$

Symmetric with respect to the y axis

EXAMPLE 3 **Symmetry**

Determine whether the function $f(x) = \dfrac{\sin x}{x}$ is even, odd, or neither.

SOLUTION

$$f(-x) = \frac{\sin(-x)}{-x} \qquad \text{Sine function is odd.}$$

$$= \frac{-\sin x}{-x}$$

$$= \frac{\sin x}{x}$$

$$= f(x)$$

> **Figure 7** $y_1 = \dfrac{\sin x}{x}$

Therefore, $f(x)$ is symmetric with respect to the y axis and is an even function. This fact is confirmed by the graph of $f(x)$ (Fig. 7). Note that although $f(x)$ is undefined at $x = 0$, it appears that $f(x)$ approaches 1 as x approaches 0 from either side.

MATCHED PROBLEM 3 Determine whether the function $g(x) = \dfrac{\cos x}{x}$ is even, odd, or neither.

Because the tangent function is the quotient of the sine and cosine functions, you might expect that it would also be periodic with period 2π. Surprisingly, the tangent function is periodic with period π. To see this, note that if (a, b) is the circular point associated with x, then $(-a, -b)$ is the circular point associated with $x + \pi$. Therefore,

$$\tan (x + \pi) = \frac{-b}{-a} = \frac{b}{a} = \tan x$$

The tangent function is symmetric with respect to the origin because

$$\tan (-x) = \frac{\sin (-x)}{\cos (-x)} = \frac{-\sin x}{\cos x} = -\tan x$$

Because $\tan x = \sin x/\cos x$, the zeros of the tangent function are the zeros of the sine function, namely, $k\pi$, k any integer, and the tangent function is undefined at the zeros of the cosine function, namely, $\pi/2 + k\pi$, k any integer. What does the graph of the tangent function look like near one of the values of x, say $\pi/2$, at which it is undefined? If $x < \pi/2$ but x is close to $\pi/2$, then b is close to 1 and a is positive and close to 0, so the ratio b/a is large and positive. So,

$$\tan x \to \infty \qquad \text{as} \qquad x \to (\pi/2)^-$$

Similarly, if $x > \pi/2$ but x is close to $\pi/2$, then b is close to 1 and a is negative and close to 0, so the ratio b/a is large in absolute value and negative. So,

$$\tan x \to -\infty \qquad \text{as} \qquad x \to (\pi/2)^+$$

Therefore, the line $x = \pi/2$ is a vertical asymptote for the tangent function and, by periodicity, so are the vertical lines $x = \pi/2 + k\pi$, k any integer. Figure 8 summarizes these properties of the tangent function and shows its graph. The analogous properties of the cotangent function and its graph are shown in Figure 9.

› **GRAPH OF** $y = \tan x$

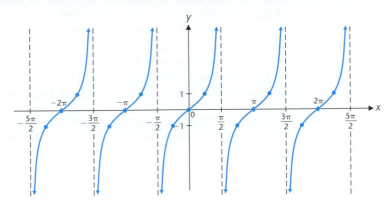

› **Figure 8**

Period: π
Domain: All real numbers except $\pi/2 + k\pi$, k an integer
Range: All real numbers

Symmetric with respect to the origin
Increasing function between consecutive asymptotes
Discontinuous at $x = \pi/2 + k\pi$, k an integer

› **GRAPH OF** $y = \cot x$

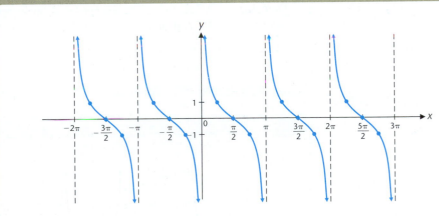

› **Figure 9**

Period: π
Domain: All real numbers except $k\pi$, k an integer
Range: All real numbers

Symmetric with respect to the origin
Decreasing function between consecutive asymptotes
Discontinuous at $x = k\pi$, k an integer

›› **EXPLORE-DISCUSS 2**

(A) Discuss how the graphs of the tangent and cotangent functions are related.

(B) How would you shift and/or reflect the tangent graph to obtain the cotangent graph?

(C) Is either the graph of $y = \tan(x - \pi/2)$ or $y = -\tan(x - \pi/2)$ the same as the graph of $y = \cot x$? Explain in terms of shifts and/or reflections.

Note that for a particular value of x, the y value on the graph of $y = \cot x$ is the reciprocal of the y value on the graph of $y = \tan x$. The vertical asymptotes of $y = \cot x$ occur at the zeros of $y = \tan x$, and vice versa.

The graphs of $y = \csc x$ and $y = \sec x$ can be obtained by taking the reciprocals of the y values of the graphs of $y = \sin x$ and $y = \cos x$, respectively. Vertical asymptotes occur at the zeros of $y = \sin x$ or $y = \cos x$. Figures 10 and 11 summarize the properties and show the graphs of $y = \csc x$ and $y = \sec x$. To emphasize the reciprocal relationships, the graphs of $y = \sin x$ and $y = \cos x$ are indicated in broken lines.

› **GRAPH OF** $y = \csc x$

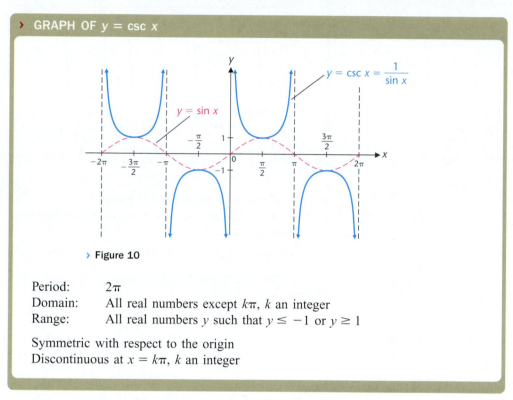

› Figure 10

Period: 2π
Domain: All real numbers except $k\pi$, k an integer
Range: All real numbers y such that $y \leq -1$ or $y \geq 1$

Symmetric with respect to the origin
Discontinuous at $x = k\pi$, k an integer

› **GRAPH OF** $y = \sec x$

› Figure 11

Period: 2π
Domain: All real numbers except $\pi/2 + k\pi$, k an integer
Range: All real numbers y such that $y \leq -1$ or $y \geq 1$

Symmetric with respect to the y axis
Discontinuous at $x = \pi/2 + k\pi$, k an integer

6-4 Exercises

The figure will be useful in many of the problems in this exercise.

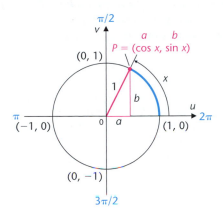

Figure for Problems 7–16.

1. When is an equation an identity?

2. Explain the meaning of the expressions $\sin^2 x$, $\sin x^2$, and $(\sin x)^2$.

3. How would you use your calculator to evaluate the expressions in Problem 2 for $x = \frac{\pi}{6}$?

4. Explain how to form a reference triangle.

5. How can you tell from the graph of a function whether it is periodic?

6. Explain why a periodic function either has no zeros, or infinitely many zeros, and give an example of each case.

Try to answer Problems 7–16 without looking back in the text or using a calculator. You can refer to the figure.

7. What are the periods of the sine, cotangent, and cosecant functions?

8. What are the periods of the cosine, tangent, and secant functions?

9. How far does the graph of each function deviate from the x axis?
(A) $y = \cos x$ (B) $y = \tan x$ (C) $y = \csc x$

10. How far does the graph of each function deviate from the x axis?
(A) $y = \sin x$ (B) $y = \cot x$ (C) $y = \sec x$

11. What are the x intercepts for the graph of each function over the interval $-2\pi \le x \le 2\pi$?
(A) $y = \sin x$ (B) $y = \cot x$ (C) $y = \csc x$

12. What are the x intercepts for the graph of each function over the interval $-2\pi \le x \le 2\pi$?
(A) $y = \cos x$ (B) $y = \tan x$ (C) $y = \sec x$

13. For what values of x, $-2\pi \le x \le 2\pi$, are the following functions not defined?
(A) $y = \cos x$ (B) $y = \tan x$ (C) $y = \csc x$

14. For what values of x, $-2\pi \le x \le 2\pi$, are the following functions not defined?
(A) $y = \sin x$ (B) $y = \cot x$ (C) $y = \sec x$

15. At what points, $-2\pi \le x \le 2\pi$, do the vertical asymptotes for the following functions cross the x axis?
(A) $y = \cos x$ (B) $y = \tan x$ (C) $y = \csc x$

16. At what points, $-2\pi \le x \le 2\pi$, do the vertical asymptotes for the following functions cross the x axis?
(A) $y = \sin x$ (B) $y = \cot x$ (C) $y = \sec x$

17. (A) Describe a shift and/or reflection that will transform the graph of $y = \csc x$ into the graph of $y = \sec x$.
(B) Is either the graph of $y = -\csc (x + \pi/2)$ or $y = -\csc (x - \pi/2)$ the same as the graph of $y = \sec x$? Explain in terms of shifts and/or reflections.

18. (A) Describe a shift and/or reflection that will transform the graph of $y = \sec x$ into the graph of $y = \csc x$.
(B) Is either the graph of $y = -\sec (x - \pi/2)$ or $y = -\sec (x + \pi/2)$ the same as the graph of $y = \csc x$? Explain in terms of shifts and/or reflections.

In Problems 19–30, determine whether each function is even, odd, or neither.

19. $y = \dfrac{\tan x}{x}$

20. $y = \dfrac{\sec x}{x}$

21. $y = \dfrac{\csc x}{x}$

22. $y = \dfrac{\cot x}{x}$

23. $y = \sin x \cos x$

24. $y = x \sin x \cos x$

25. $y = \sin x - \cos x$ **26.** $y = x - \sin x$

27. $y = x^2 + \sin^2 x$ **28.** $y = x + \sin^2 x$

29. $y = x^2 \sin x$ **30.** $y = x^3 \sin x$

Find the value of each of the six trigonometric functions for an angle θ that has a terminal side containing the point indicated in Problems 31–34.

31. $(6, 8)$ **32.** $(-3, 4)$

33. $(-1, \sqrt{3})$ **34.** $(\sqrt{3}, 1)$

Find the reference angle α for each angle θ in Problems 35–40.

35. $\theta = 300°$ **36.** $\theta = 135°$

37. $\theta = \dfrac{7\pi}{6}$ **38.** $\theta = \dfrac{\pi}{4}$

39. $\theta = -\dfrac{5\pi}{3}$ **40.** $\theta = -\dfrac{5\pi}{4}$

In Problems 41–46, find the smallest positive θ in degree and radian measure for which

41. $\cos = -\dfrac{1}{2}$ **42.** $\sin = -\dfrac{\sqrt{3}}{2}$

43. $\sin = -\dfrac{1}{2}$ **44.** $\tan \theta = -\sqrt{3}$

45. $\csc = -\dfrac{2}{\sqrt{3}}$ **46.** $\sec \theta = -\sqrt{2}$

Find the value of each of the other five trigonometric functions for an angle θ, without finding θ, given the information indicated in Problems 47–50. Sketching a reference triangle should be helpful.

47. $\sin \theta = \frac{3}{5}$ and $\cos \theta < 0$

48. $\tan \theta = -\frac{4}{3}$ and $\sin \theta < 0$

49. $\cos \theta = -\sqrt{5}/3$ and $\cot \theta > 0$

50. $\cos \theta = -\sqrt{5}/3$ and $\tan \theta > 0$

51. Which trigonometric functions are not defined when the terminal side of an angle lies along the vertical axis. Why?

52. Which trigonometric functions are not defined when the terminal side of an angle lies along the horizontal axis? Why?

53. Find exactly, all θ, $0° \le \theta < 360°$, for which $\cos \theta = -\sqrt{3}/2$.

54. Find exactly, all θ, $0° \le \theta < 360°$, for which $\cot \theta = -1/\sqrt{3}$.

55. Find exactly, all θ, $0 \le \theta < 2\pi$, for which $\tan \theta = 1$.

56. Find exactly, all θ, $0 \le \theta < 2\pi$, for which $\sec \theta = -\sqrt{2}$.

In Problems 57–62, determine whether the statement is true or false. Explain.

57. If the function f is not even, then it is odd.

58. The constant function with value 0 is both even and odd.

59. If f and g are each periodic with period p, then the function f/g is periodic with period p.

60. If f and g are each periodic with period p, then the function f/g is periodic.

61. If f and g are both odd, then the function fg is even.

62. If f and g are both even, then the function fg is odd.

63. Find all functions of the form $f(x) = ax + b$ that are periodic.

64. Find all functions of the form $f(x) = ax^2 + bx + c$ that are periodic.

65. Find all functions of the form $f(x) = ax + b$ that are even.

66. Find all functions of the form $f(x) = ax + b$ that are odd.

Problems 67–72 offer a preliminary investigation into the relationships of the graphs of $y = \sin x$ and $y = \cos x$ with the graphs of $y = A \sin x$, $y = A \cos x$, $y = \sin Bx$, $y = \cos Bx$, $y = \sin (x + C)$, and $y = \cos (x + C)$. This important topic is discussed in detail in Section 6-5.

67. (A) Graph $y = A \cos x$, $(-2\pi \le x \le 2\pi, -3 \le y \le 3)$, for $A = 1, 2$, and -3, all in the same viewing window.

(B) Do the x intercepts change? If so, where?

(C) How far does each graph deviate from the x axis? (Experiment with additional values of A.)

(D) Describe how the graph of $y = \cos x$ is changed by changing the values of A in $y = A \cos x$.

68. (A) Graph $y = A \sin x$, $(-2\pi \le x \le 2\pi, -3 \le y \le 3)$, for $A = 1, 3$, and -2, all in the same viewing window.

(B) Do the x intercepts change? If so, where?

(C) How far does each graph deviate from the x axis? (Experiment with additional values of A.)

(D) Describe how the graph of $y = \sin x$ is changed by changing the values of A in $y = A \sin x$.

69. (A) Graph $y = \sin Bx$ $(-\pi \le x \le \pi, -2 \le y \le 2)$, for $B = 1, 2$, and 3, all in the same viewing window.

(B) How many periods of each graph appear in this viewing rectangle? (Experiment with additional positive integer values of B.)

(C) Based on the observations in part B, how many periods of the graph of $y = \sin nx$, n a positive integer, would appear in this viewing window?

70. (A) Graph $y = \cos Bx$ $(-\pi \le x \le \pi, -2 \le y \le 2)$, for $B = 1, 2$, and 3, all in the same viewing window.

(B) How many periods of each graph appear in this viewing rectangle? (Experiment with additional positive integer values of B.)

(C) Based on the observations in part B, how many periods of the graph of $y = \cos nx$, n a positive integer, would appear in this viewing window?

71. (A) Graph $y = \cos (x + C)$, $-2\pi \le x \le 2\pi$, $-1.5 \le y \le 1.5$, for $C = 0$, $-\pi/2$, and $\pi/2$, all in the same viewing window. (Experiment with additional values of C.)

(B) Describe how the graph of $y = \cos x$ is changed by changing the values of C in $y = \cos (x + C)$.

72. (A) Graph $y = \sin (x + C)$, $-2\pi \le x \le 2\pi$, $-1.5 \le y \le 1.5$, for $C = 0$, $-\pi/2$, and $\pi/2$, all in the same viewing window. (Experiment with additional values of C.)

(B) Describe how the graph of $y = \sin x$ is changed by changing the values of C in $y = \sin (x + C)$.

73. Try to calculate each of the following on your calculator. Explain the results.
 (A) $\sec (\pi/2)$ (B) $\tan (-\pi/2)$ (C) $\cot (-\pi)$

74. Try to calculate each of the following on your calculator. Explain the results.
 (A) $\csc \pi$ (B) $\tan (\pi/2)$ (C) $\cot 0$

75. Graph $f(x) = \sin x$ and $g(x) = x$ in the same viewing window $(-1 \le x \le 1, -1 \le y \le 1)$.
 (A) What do you observe about the two graphs when x is close to 0, say $-0.5 \le x \le 0.5$?
 (B) Complete the table to three decimal places (use the table feature on your graphing calculator if it has one):

x	-0.3	-0.2	-0.1	0.0	0.1	0.2	0.3
$\sin x$							

(In applied mathematics certain derivations, formulas, and calculations are simplified by replacing $\sin x$ with x for small values of $|x|$.)

76. Graph $h(x) = \tan x$ and $g(x) = x$ in the same viewing window $(-1 \le x \le 1, -1 \le y \le 1)$.
 (A) What do you observe about the two graphs when x is close to 0, say $-0.5 \le x \le 0.5$?
 (B) Complete the table to three decimal places (use the table feature on your graphing calculator if it has one):

x	-0.3	-0.2	-0.1	0.0	0.1	0.2	0.3
$\tan x$							

(In applied mathematics certain derivations, formulas, and calculations are simplified by replacing $\tan x$ with x for small values of $|x|$.)

Use the following figure for Problems 77 and 78:

77. If the coordinates of A are $(4, 0)$ and arc length s is 7 units, find
 (A) The exact radian measure of θ
 (B) The coordinates of P to three decimal places

78. If the coordinates of A are $(2, 0)$ and arc length s is 8 units, find
 (A) The exact radian measure of θ
 (B) The coordinates of P to three decimal places

79. In a rectangular coordinate system, a circle with center at the origin passes through the point $(6\sqrt{3}, 6)$. What is the length of the arc on the circle in quadrant I between the positive horizontal axis and the point $(6\sqrt{3}, 6)$?

80. In a rectangular coordinate system, a circle with center at the origin passes through the point $(2, 2\sqrt{3})$. What is the length of the arc on the circle in quadrant I between the positive horizontal axis and the point $(2, 2\sqrt{3})$?

APPLICATIONS

81. SOLAR ENERGY The intensity of light I on a solar cell changes with the angle of the sun and is given by the formula $I = k \cos \theta$, where k is a constant (see the figure). Find light intensity I in terms of k for $\theta = 0°$, $\theta = 30°$, and $\theta = 60°$.

Solar cell

82. SOLAR ENERGY Refer to Problem 81. Find light intensity I in terms of k for $\theta = 20°$, $\theta = 50°$, and $\theta = 90°$.

83. PHYSICS—ENGINEERING The figure below illustrates a piston connected to a wheel that turns 3 revolutions per second; so the angle θ is being generated at $3(2\pi) = 6\pi$ radians per second, or $\theta = 6\pi t$, where t is time in seconds. If P is at $(1, 0)$ when $t = 0$, show that

$$y = b + \sqrt{4^2 - a^2}$$
$$= \sin 6\pi t + \sqrt{16 - (\cos 6\pi t)^2}$$

for $t \ge 0$.

84. PHYSICS—ENGINEERING In Problem 83, find the position of the piston y when $t = 0.2$ second (to three significant digits).

 85. GEOMETRY The area of a regular n-sided polygon circumscribed about a circle of radius 1 is given by

$$A = n \tan \frac{180°}{n}$$

(A) Find A for $n = 8$, $n = 100$, $n = 1,000$, and $n = 10,000$. Compute each to five decimal places.
(B) What number does A seem to approach as $n \to \infty$? (What is the area of a circle with radius 1?)

$r = 1$

$n = 8$

 86. GEOMETRY The area of a regular n-sided polygon inscribed in a circle of radius 1 is given by

$$A = \frac{n}{2} \sin \frac{360°}{n}$$

(A) Find A for $n = 8$, $n = 100$, $n = 1,000$, and $n = 10,000$. Compute each to five decimal places.
(B) What number does A seem to approach as $n \to \infty$? (What is the area of a circle with radius 1?)

 87. ANGLE OF INCLINATION Recall (Section 2-3) the **slope** of a nonvertical line passing through points $P_1 = (x_1, y_1)$ and $P_2 = (x_2, y_2)$ is given by slope $= m = (y_2 - y_1)/(x_2 - x_1)$. The angle θ that the line L makes with the x axis, $0° \leq \theta < 180°$, is called the **angle of inclination** of the line L (see figure). So

$$\text{Slope} = m = \tan \theta, \ 0° \leq \theta < 180°$$

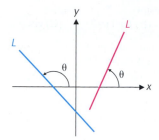

(A) Compute the slopes to two decimal places of the lines with angles of inclination 88.7° and 162.3°.
(B) Find the equation of a line passing through $(-4, 5)$ with an angle of inclination 137°. Write the answer in the form $y = mx + b$, with m and b to two decimal places.

 88. ANGLE OF INCLINATION Refer to Problem 87.
(A) Compute the slopes to two decimal places of the lines with angles of inclination 5.34° and 92.4°.
(B) Find the equation of a line passing through $(6, -4)$ with an angle of inclination 106°. Write the answer in the form $y = mx + b$, with m and b to two decimal places.

6-5 More General Trigonometric Functions and Models

> Graphing $y = A \sin Bx$ and $y = A \cos Bx$
> Graphing $y = A \sin (Bx + C)$ and $y = A \cos (Bx + C)$
> Finding an Equation from the Graph of a Simple Harmonic
> Mathematical Modeling and Data Analysis

Imagine a weight suspended from the ceiling by a spring. If the weight were pulled downward and released, then, assuming no air resistance or friction, it would move up and down with the same frequency and amplitude forever. This idealized motion is an example of **simple harmonic motion.** Simple harmonic motion can be described by functions of the form $y = A \sin (Bx + C)$ or $y = A \cos (Bx + C)$, called **simple harmonics.**

Simple harmonics are extremely important in both pure and applied mathematics. In applied mathematics, they are used in the analysis of sound waves, radio waves, X-rays, gamma rays, visible light, infrared radiation, ultraviolet radiation, seismic waves, ocean waves, electric circuits, electric generators, vibrations, bridge and building construction, spring–mass systems, bow waves of boats, sonic booms, and so on. Analysis involving simple harmonics is called *harmonic analysis.*

In Section 6-5 we study properties, graphs, and applications of simple harmonics. A brief review of graph transformations (Section 3-3) should prove helpful.

› Graphing $y = A \sin Bx$ and $y = A \cos Bx$

We visualize the graphs of functions of the form $y = A \sin Bx$ or $y = A \cos Bx$, and determine their zeros and turning points, by understanding how each of the constants A and B transforms the graph of $y = \sin x$ or $y = \cos x$.

EXAMPLE 1 | **Zeros and Turning Points**

Find the zeros and turning points of each function on the interval $[0, 2\pi]$.

(A) $y = \dfrac{1}{2} \sin x$

(B) $y = -2 \sin x$

SOLUTIONS (A) The function $y = \frac{1}{2} \sin x$ is the vertical contraction of $y = \sin x$ that is obtained by multiplying each ordinate value by $\frac{1}{2}$ (Fig. 1). Therefore its zeros on $[0, 2\pi]$ are identical to the zeros of $y = \sin x$, namely, $x = 0$, π, and 2π. Because the turning points of $y = \sin x$ are $(\pi/2, 1)$ and $(3\pi/2, -1)$, the turning points of $y = \frac{1}{2} \sin x$ are $(\pi/2, 1/2)$ and $(3\pi/2, -1/2)$.

(B) The function $y = -2 \sin x$ is the vertical expansion of $y = \sin x$ that is obtained by multiplying each ordinate value by 2, followed by a reflection in the x axis (see Fig. 1). Therefore, its zeros on $[0, 2\pi]$ are identical to the zeros of $y = \sin x$, namely $x = 0$, π, and 2π. Because the turning points of $y = \sin x$ are $(\pi/2, 1)$ and $(3\pi/2, -1)$, the turning points of $y = -2 \sin x$ are $(\pi/2, -2)$ and $(3\pi/2, 2)$.

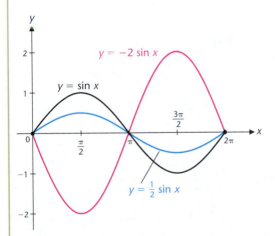

› Figure 1

MATCHED PROBLEM 1 Find the zeros and turning points of each function on the interval $[\pi/2, 5\pi/2]$.

(A) $y = -5 \cos x$

(B) $y = \dfrac{1}{3} \cos x$

As Example 1 illustrates, the graph of $y = A \sin x$ can be obtained from the graph of $y = \sin x$ by multiplying each y value of $y = \sin x$ by the constant A. The graph of $y = A \sin x$ still crosses the x axis where the graph of $y = \sin x$ crosses the x axis, because $A \cdot 0 = 0$. Because the maximum value of $\sin x$ is 1, the maximum value of $A \sin x$ is $|A| \cdot 1 = |A|$. The constant $|A|$ is called the **amplitude** of the graph of $y = A \sin x$ and indicates the maximum deviation of the graph of $y = A \sin x$ from the x axis.

The period of $y = A \sin x$ (assuming $A \neq 0$) is the same as the period of $y = \sin x$, namely 2π, because $A \sin (x + 2\pi) = A \sin x$.

EXAMPLE | **2** | **Periods**

Find the period of each function.

(A) $y = \sin 2x$
(B) $y = \sin (x/2)$

SOLUTIONS

(A) Because the function $y = \sin x$ has period 2π, the function $y = \sin 2x$ completes one cycle as $2x$ varies from

$$2x = 0 \quad \text{to} \quad 2x = 2\pi$$

or as x varies from

$$x = 0 \quad \text{to} \quad x = \pi \qquad \text{Half the period for sin x.}$$

Therefore, the period of $y = \sin 2x$ is π (Fig. 2).

(B) Because the function $y = \sin x$ has period 2π, the function $y = \sin (x/2)$ completes one cycle as $x/2$ varies from

$$\frac{x}{2} = 0 \quad \text{to} \quad \frac{x}{2} = 2\pi$$

or as x varies from

$$x = 0 \quad \text{to} \quad x = 4\pi \qquad \text{Double the period for sin x.}$$

Therefore, the period of $y = \sin (x/2)$ is 4π (see Fig. 2).

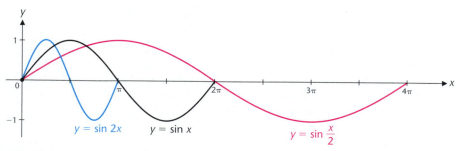

> **Figure 2**

MATCHED PROBLEM 2

Find the period of each function.

(A) $y = \cos (x/10)$
(B) $y = \cos (6\pi x)$

As Example 2 illustrates, the graph of $y = \sin Bx$, for a positive constant B, completes one cycle as Bx varies from

$$Bx = 0 \qquad \text{to} \qquad Bx = 2\pi$$

or as x varies from

$$x = 0 \qquad \text{to} \qquad x = \frac{2\pi}{B}$$

Therefore, the period of $y = \sin Bx$ is $\frac{2\pi}{B}$. Note that the amplitude of $y = \sin Bx$ is 1, the same as the amplitude of $y = \sin x$. The effect of the constant B is to compress or stretch the basic sine curve by changing the period of the function, but not its amplitude. A similar analysis applies to $y = \cos Bx$, for $B > 0$, where it can be shown that the period is also $\frac{2\pi}{B}$. We combine and summarize our results on period and amplitude as follows:

> ### PERIOD AND AMPLITUDE
>
> For $y = A \sin Bx$ or $y = A \cos Bx$, $A \neq 0$, $B > 0$:
>
> $$\text{Amplitude} = |A| \qquad \text{Period} = \frac{2\pi}{B}$$
>
> If $0 < B < 1$, the basic sine or cosine curve is stretched.
> If $B > 1$, the basic sine or cosine curve is compressed.

You can either learn the formula for the period, $\frac{2\pi}{B}$, or use the reasoning we used in deriving the formula. Recall, $\sin Bx$ or $\cos Bx$ completes one cycle as Bx varies from

$$Bx = 0 \qquad \text{to} \qquad Bx = 2\pi$$

that is, as x varies from

$$x = 0 \qquad \text{to} \qquad x = \frac{2\pi}{B}$$

EXAMPLE 3

Amplitude, Period, and Turning Points

Find the amplitude, period, and turning points of $y = -3 \cos (\pi x/2)$ on the interval $[-4, 4]$.

SOLUTION

$$\text{Amplitude} = |-3| = 3 \qquad \text{Period} = \frac{2\pi}{(\pi/2)} = 4$$

Because $y = \cos x$ has turning points at $x = 0$ and $x = \pm\pi$ (half of a complete cycle), $y = -3 \cos (\pi x/2)$ has turning points at $x = 0$ and $x = \pm 2$. So the turning points on the interval $[-4, 4]$ are $(-2, 3)$, $(0, -3)$, and $(2, 3)$. These results are confirmed by a graph of $y = -3 \cos (\pi x/2)$ (Fig. 3).

> **Figure 3**

MATCHED PROBLEM 3

Find the amplitude, period, and turning points of $y = \frac{1}{4} \sin(3\pi x)$ on the interval $[0, 1]$.

⟩⟩⟩ EXPLORE-DISCUSS 1

Find an equation of the form $y = A \cos Bx$ that produces the following graph.

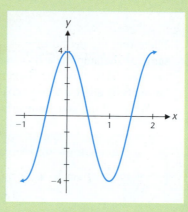

Is it possible for an equation of the form $y = A \sin Bx$ to produce the same graph? Explain.

⟩ Graphing $y = A \sin(Bx + C)$ and $y = A \cos(Bx + C)$

The graph of $y = A \sin(Bx + C)$ is a horizontal shift of the graph of the function $y = A \sin Bx$. In fact, because the period of the sine function is 2π, $y = A \sin(Bx + C)$ completes one cycle as $Bx + C$ varies from

$$Bx + C = 0 \qquad \text{to} \qquad Bx + C = 2\pi$$

or (solving for x in each equation) as x varies from

$$\underset{\text{Phase shift}}{\underbrace{x = -\frac{C}{B}}} \qquad \text{to} \qquad x = -\frac{C}{B} + \underset{\text{Period}}{\frac{2\pi}{B}}$$

We conclude that $y = A \sin(Bx + C)$ has a period of $2\pi/B$, and its graph is the graph of $y = A \sin Bx$ shifted $|-C/B|$ units to the right if $-C/B$ is positive and $|-C/B|$ units to the left if $-C/B$ is negative. The number $-C/B$ is referred to as the **phase shift.**

EXAMPLE 4

Amplitude, Period, Phase Shift, and Zeros

Find the amplitude, period, phase shift, and zeros of $y = \frac{1}{2} \cos(4x - \pi)$, and sketch the graph for $-\pi \leq x \leq \pi$.

SOLUTION

$$\text{Amplitude} = |A| = \left| \frac{1}{2} \right| = \frac{1}{2}$$

The graph completes one cycle as $4x - \pi$ varies from

$$4x - \pi = 0 \qquad \text{to} \qquad 4x - \pi = 2\pi$$

or as x varies from

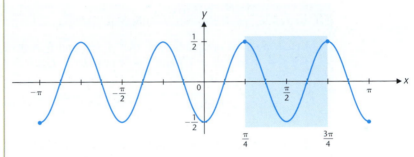

To sketch the graph, divide the interval $[\pi/4, 3\pi/4]$ into four equal parts and sketch one cycle of $y = \frac{1}{2} \cos(4x - \pi)$. Then extend the graph to cover $[-\pi, \pi]$ (Fig. 4).

> **Figure 4** $y = \frac{1}{2} \cos(4x - \pi)$, $-\pi \leq x \leq \pi$.

The zeros of $y = \frac{1}{2} \cos(4x - \pi)$ are obtained by shifting the zeros of $y = \frac{1}{2} \cos(4x)$ to the right by $\pi/4$ units. Because $x = \pi/8$ and $x = 3\pi/8$ are zeros of $y = \frac{1}{2} \cos(4x)$, $x = \pi/8 + \pi/4 = 3\pi/8$ and $x = 3\pi/8 + \pi/4 = 5\pi/8$ are zeros of $y = \frac{1}{2} \cos(4x - \pi)$. By periodicity, the zeros of $y = \frac{1}{2} \cos(4x - \pi)$ are $x = 3\pi/8 + k\pi/4$, k any integer, as confirmed by the graph.

MATCHED PROBLEM 4

Find the amplitude, period, phase shift, and zeros of $y = \frac{3}{4} \sin(2x + \pi)$, and sketch the graph for $-\pi \leq x \leq \pi$.

>>> **EXPLORE-DISCUSS 2**

Find an equation of the form $y = A \sin(Bx + C)$ that produces the following graph.

Is it possible for an equation of the form $y = A \cos(Bx + C)$ to produce the same graph? Explain.

The graphs of $y = A \sin (Bx + C) + k$ and $y = A \cos (Bx + C) + k$ are vertical shifts (up k units if $k > 0$, down k units if $k < 0$) of the graphs of $y = A \sin (Bx + C)$ and $y = A \cos (Bx + C)$, respectively.

Because $y = \sec x$ and $y = \csc x$ are unbounded functions, amplitude is not defined for functions of the form $y = A \sec (Bx + C)$ and $y = A \csc (Bx + C)$. However, because both the secant and cosecant functions have period 2π, the functions $y = A \csc (Bx + C)$ and $y = A \sec (Bx + C)$ have period $2\pi/B$ and phase shift $-C/B$.

Because $y = \tan x$ and $y = \cot x$ are unbounded functions, amplitude is not defined for functions of the form $y = A \tan (Bx + C)$ or $y = A \cot (Bx + C)$. The tangent and cotangent functions both have period π, so the functions $y = A \tan (Bx + C)$ and $y = A \cot (Bx + C)$ have period π/B and phase shift $-C/B$.

Our results on amplitude, period, and phase shift are summarized in the following box.

> **AMPLITUDE, PERIOD, AND PHASE SHIFT**

Let A, B, C be constants such that $A \neq 0$ and $B > 0$.
For $y = A \sin (Bx + C)$ and $y = A \cos (Bx + C)$:

$$\text{Amplitude} = |A| \qquad \text{Period} = \frac{2\pi}{B} \qquad \text{Phase shift} = -\frac{C}{B}$$

For $y = A \sec (Bx + C)$ and $y = A \csc (Bx + C)$:

$$\text{Period} = \frac{2\pi}{B} \qquad \text{Phase shift} = -\frac{C}{B}$$

For $y = A \tan (Bx + C)$ and $y = A \cot (Bx + C)$:

$$\text{Period} = \frac{\pi}{B} \qquad \text{Phase shift} = -\frac{C}{B}$$

Note: Amplitude is not defined for the secant, cosecant, tangent, and cotangent functions, all of which are unbounded.

> ## Finding an Equation from the Graph of a Simple Harmonic

Given the graph of a simple harmonic, it's possible to find an equation of the form $y = A \sin (Bx + C)$ or $y = A \cos (Bx + C)$ that produces the graph. Example 5 illustrates the process.

| EXAMPLE | 5 | **Finding an Equation of a Simple Harmonic Graph** |

Graph $y_1 = 3 \sin x + 4 \cos x$ using a graphing calculator, and find an equation of the form $y_2 = A \sin (Bx + C)$ that has the same graph as y_1. Find A and B exactly and C to three decimal places.

SOLUTION The graph of y_1 is shown in Figure 5. The graph appears to be a sine curve shifted to the left. The amplitude and period appear to be 5 and 2π, respectively. (We will assume this for now and check it at the end.) So $A = 5$, and because $P = 2\pi/B$, then $B = 2\pi/P = 2\pi/2\pi = 1$. Using a graphing calculator, we find that the x intercept closest to the origin, to three

> **Figure 5** $y_1 = 3 \sin x + 4 \cos x$.

decimal places, is -0.927. To find C, substitute $B = 1$ and $x = -0.927$ into the phase-shift formula $x = -C/B$ and solve for C:

$$x = -\frac{C}{B} \qquad \text{Substitute } x = -0.927, B = 1.$$

$$-0.927 = -\frac{C}{1} \qquad \text{Solve for } C.$$

$$C = 0.927$$

We now have the equation we are looking for:

$$y_2 = 5 \sin (x + 0.927)$$

CHECK Graph y_1 and y_2 in the same viewing window. If the graphs are the same, it appears that only one graph is drawn—the second graph is drawn over the first. To check further that the graphs are the same, use TRACE and switch back and forth between y_1 and y_2 at different values of x. Figure 6 shows a comparison at $x = 0$ (both graphs appear in the same viewing window).

> **Figure 6**

MATCHED PROBLEM 5

Graph $y_1 = 4 \sin x - 3 \cos x$ using a graphing calculator, and find an equation of the form $y_2 = A \sin (Bx + C)$ that has the same graph as y_1. (Find the x intercept closest to the origin to three decimal places.)

> ## Mathematical Modeling and Data Analysis

The polynomial, exponential, and logarithmic functions studied in Chapters 4 and 5 are not suitable for modeling periodic phenomena. Instead, when given a data set that indicates periodic behavior, we use a technique called **sinusoidal regression** to model the data by a function of the form $f(x) = A \sin (Bx + C) + k$.

EXAMPLE | **6** | ## Temperature Variation

The monthly average high temperatures in Fairbanks, Alaska, are given in Table 1. A sinusoidal model for the data is given by

$$y = 37.4 \sin (0.523x - 1.93) + 37.2$$

where x is time in months ($x = 1$ represents January 15, $x = 2$ represents February 15, etc.) and y is temperature in degrees Fahrenheit. Use the sinusoidal regression function to estimate the average high temperature on April 1 to one decimal place.

Table 1 Temperatures in Fairbanks, Alaska

Month	1	2	3	4	5	6	7	8	9	10	11	12
Average High (°F)	0	8	25	44	61	71	73	66	54	31	11	3
Average Low (°F)	−19	−15	−3	20	37	49	52	46	35	16	−7	−15

SOLUTION To estimate the average high temperature on April 1 we substitute $x = 3.5$:

$$y = 37.4 \sin (0.523 \cdot 3.5 - 1.93) + 37.2 \approx 33.5°$$

Technology Connections

Figure 7 shows the details of constructing the sinusoidal model of Example 6 on a graphing calculator. To observe the cyclical behavior of the data, we enter the average high temperatures for two consecutive years, from $x = 1$ to $x = 24$. The data, the sinusoidal regression function, and a plot of the data and graph of the regression function are shown in Figure 7. To estimate the average high temperature on April 1, we let $x = 3.5$ [Fig. 7(c)]. Note the slight discrepancy between the estimated high temperature (33.5°) of Example 6, and the value given in Figure 7(c) (approximately 33.6°), due to rounding the coefficients of the regression equation to three significant digits.

(a) Entering the data

(b) Finding the model

(c) Graphing the data and the model

› Figure 7

MATCHED PROBLEM 6

The monthly average low temperatures in Fairbanks, Alaska, are given in Table 1. A sinusoidal model for the data is given by

$$y = 36.7 \sin (0.524x - 2.05) + 16.4$$

where x is the time in months ($x = 1$ represents January 15, $x = 2$ represents February 15, etc.) and y is temperature in degrees Fahrenheit. Use the sinusoidal regression function to estimate the average low temperature on April 1 to one decimal place.

ANSWERS TO MATCHED PROBLEMS

1. (A) Zeros: $\pi/2, 3\pi/2, 5\pi/2$; turning points: $(\pi, 5), (2\pi, -5)$
 (B) Zeros: $\pi/2, 3\pi/2, 5\pi/2$; turning points: $(\pi, -1/3), (2\pi, 1/3)$
2. (A) 20π (B) $1/3$
3. Amplitude: $1/4$; period: $2/3$; turning points: $(1/6, 1/4), (1/2, -1/4), (5/6, 1/4)$
4. Amplitude: $3/4$; period: π; phase shift: $-\pi/2$; zeros: $k\pi/2$, k any integer

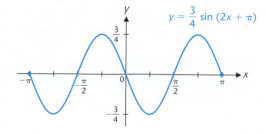

$$y = \frac{3}{4} \sin (2x + \pi)$$

5. $y_2 = 5 \sin (x - 0.644)$

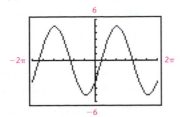

6. 8.5°F

6-5 Exercises

1. What is simple harmonic motion?

2. Describe in your own words what the graph of a simple harmonic looks like.

3. Explain the connection between the graphs of

$y = A \cos (Bx + C)$ and $y = A \cos Bx$.

4. Does every trigonometric function have an amplitude? Explain.

In Problems 5–14, find the amplitude (if applicable) and period.

5. $y = 3 \sin x$

6. $y = \frac{1}{4} \cos x$

7. $y = -\frac{1}{2} \cos x$

8. $y = -2 \sin x$

9. $y = 2 \cot 4x$

10. $y = 3 \tan 2x$

11. $y = -\frac{1}{4} \tan 8\pi x$

12. $y = -\frac{1}{2} \cot 2\pi x$

13. $y = \csc (x/2)$

14. $y = \sec \pi x$

In Problems 15–18, find the amplitude (if applicable), the period, and all zeros in the given interval.

15. $y = \sin \pi x, -2 \le x \le 2$

16. $y = \cos \pi x, -2 \le x \le 2$

17. $y = \frac{1}{2} \cot (x/2), 0 < x < 4\pi$

18. $y = \frac{1}{2} \tan (x/2), -\pi < x < 3\pi$

In Problems 19–22, find the amplitude (if applicable), the period, and all turning points in the given interval.

19. $y = 3 \cos 2x, -\pi \le x \le \pi$

20. $y = 2 \sin 4x, -\pi \le x \le \pi$

21. $y = 2 \sec \pi x, -1 \le x \le 3$

22. $y = 2 \csc (x/2), 0 < x < 8\pi$

In Problems 23–26, find the equation of the form $y = A \sin Bx$ that produces the graph shown.

23.

24.

25.

26.

89. POLLUTION The amount of sulfur dioxide pollutant from heating fuels released in the atmosphere in a city varies seasonally. Suppose the number of tons of pollutant released into the atmosphere during the nth week after January 1 for a particular city is given by

$$A(n) = 1.5 + \cos \frac{n\pi}{26} \qquad 0 \le n \le 104$$

Graph the function over the indicated interval and describe what the graph shows.

90. MEDICINE A seated normal adult breathes in and exhales about 0.82 liter of air every 4.00 seconds. The volume of air in the lungs t seconds after exhaling is approximately

$$V(t) = 0.45 - 0.37 \cos \frac{\pi t}{2} \qquad 0 \le t \le 8$$

Graph the function over the indicated interval and describe what the graph shows.

91. ELECTRICAL CIRCUIT The current in an electrical circuit is given by $I = 15 \cos (120\pi t + \pi/2)$, $0 \le t \le \frac{2}{60}$, where I is measured in amperes. State the amplitude A, period P, and phase shift. Graph the equation.

92. ELECTRICAL CIRCUIT The current in an electrical circuit is given by $I = 30 \cos (120\pi t - \pi)$, $0 \le t \le \frac{3}{60}$, where I is measured in amperes. State the amplitude A, period P, and phase shift. Graph the equation.

93. PHYSICS—ENGINEERING The thin, plastic disk shown in the figure below is rotated at 3 revolutions per second, starting at $\theta = 0$ (so at the end of t seconds, $\theta = 6\pi t$—Why?). If the disk has a radius of 3, show that the position of the shadow on the y scale from the small steel ball B is given by

$$y = 3 \sin 6\pi t$$

Graph this equation for $0 \le t \le 1$.

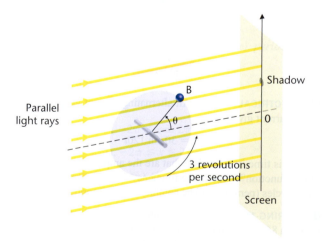

94. PHYSICS—ENGINEERING If in Problem 93 the disk started rotating at $\theta = \pi/2$, show that the position of the shadow at time t (in seconds) is given by

$$y = 3 \sin \left(6\pi t + \frac{\pi}{2} \right)$$

Graph this equation for $0 \le t \le 1$.

95. A beacon light 20 feet from a wall rotates clockwise at the rate of 1/4 revolution per second (rps) (see the figure), so that $\theta = \pi t/2$.
(A) Start counting time in seconds when the light spot is at N and write an equation for the length c of the light beam in terms of t.
(B) Graph the equation found in part A for the time interval $[0, 1]$.
(C) Describe what happens to the length c of the light beam as t goes from 0 to 1.

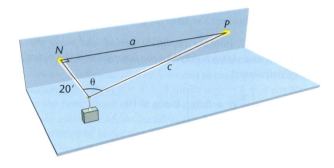

96. Refer to Problem 95.
(A) Write an equation for the distance a the light spot travels along the wall in terms of time t.
(B) Graph the equation found in part A for the time interval $[0, 1]$.
(C) Describe what happens to the distance a along the wall as t goes from 0 to 1.

97. MODELING SUNSET TIMES Sunset times for the fifth of each month over a period of 1 year were taken from a tide booklet for the San Francisco Bay to form Table 2. Daylight savings time was ignored and the times are for a 24-hour clock starting at midnight.
(A) Using 1 month as the basic unit of time, enter the data for a 2-year period in your graphing calculator and produce a scatter plot in the viewing window. Before entering Table 2 data into your graphing calculator, convert sunset times from hours and minutes to decimal hours rounded to two decimal places. Choose $15 \le y \le 20$ for the viewing window.
(B) It appears that a sine curve of the form

$$y = k + A \sin (Bx + C)$$

will closely model these data. The constants k, A, and B are easily determined from Table 2 as follows: $A = (\text{max } y - \text{min } y)/2$, $B = 2\pi/\text{Period}$, and $k = \text{min } y + A$. To estimate C, visually estimate to one decimal place the smallest positive phase shift from the plot in part A. After determining A, B, k, and C, write the resulting equation. (Your value of C may differ slightly from the answer in the back of the book.)

Table 2

x (months)	1	2	3	4	5	6	7	8	9	10	11	12
y (sunset)*	17:05	17:38	18:07	18:36	19:04	19:29	19:35	19:15	18:34	17:47	17:07	16:51

*Time on a 24-hr clock, starting at midnight.

(C) Plot the results of parts A and B in the same viewing window. (An improved fit may result by adjusting your value of C slightly.)

(D) If your graphing calculator has a sinusoidal regression feature, check your results from parts B and C by finding and plotting the regression equation.

 98. MODELING TEMPERATURE VARIATION The 30-year average monthly temperature, °F, for each month of the year for Washington, D.C., is given in Table 3 (*World Almanac*).

(A) Using 1 month as the basic unit of time, enter the data for a 2-year period in your graphing calculator and produce a scatter plot in the viewing window. Choose $0 \le y \le 80$ for the viewing window.

(B) It appears that a sine curve of the form

$$y = k + A \sin(Bx + C)$$

will closely model these data. The constants k, A, and B are easily determined from Table 3 as follows: $A = (\max\ y - \min\ y)/2$, $B = 2\pi/\text{Period}$, and $k = \min\ y + A$. To estimate C, visually estimate to one decimal place the smallest positive phase shift from the plot in part A. After determining A, B, k, and C, write the resulting equation.

(C) Plot the results of parts A and B in the same viewing window. (An improved fit may result by adjusting your value of C slightly.)

(D) If your graphing calculator has a sinusoidal regression feature, check your results from parts B and C by finding and plotting the regression equation.

Table 3

x (months)	1	2	3	4	5	6	7	8	9	10	11	12
y (temp.)	31	34	43	53	62	71	76	74	67	55	45	35

6-6 Inverse Trigonometric Functions

> Inverse Sine Function
> Inverse Cosine Function
> Inverse Tangent Function
> Summary
> Inverse Cotangent, Secant, and Cosecant Functions (Optional)

A brief review of the general concept of inverse functions discussed in Section 3-6 should prove helpful before proceeding with Section 6-6. In the box we restate a few important facts about inverse functions from Section 3-6.

> **FACTS ABOUT INVERSE FUNCTIONS**

For a one-to-one function f and its inverse f^{-1}:

1. If (a, b) is an element of f, then (b, a) is an element of f^{-1}, and conversely.

2. Range of f = Domain of f^{-1}
Domain of f = Range of f^{-1}

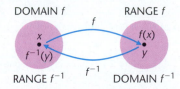

3. If $x = f^{-1}(y)$, then $y = f(x)$ for y in the domain of f^{-1} and x in the domain of f, and conversely.

4. $f(f^{-1}(y)) = y$ for y in the domain of f^{-1}
$f^{-1}(f(x)) = x$ for x in the domain of f

All trigonometric functions are periodic, so each range value can be associated with infinitely many domain values (Fig. 1). As a result, no trigonometric function is one-to-one, so, strictly speaking, no trigonometric function has an inverse. However, we can restrict the domain of each function so that it is one-to-one over the restricted domain. Then, for this restricted domain, an inverse function is guaranteed.

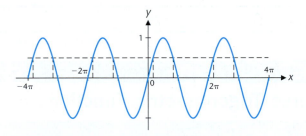

> **Figure 1** $y = \sin x$ is not one-to-one over $(-\infty, \infty)$.

Inverse trigonometric functions represent another group of basic functions that are added to our library of elementary functions. These functions are used in many applications and mathematical developments, and will be particularly useful to us when we solve trigonometric equations in Section 7-5.

> ## Inverse Sine Function

How can the domain of the sine function be restricted so that it is one-to-one? This can be done in infinitely many ways. A fairly natural and generally accepted way is illustrated in Figure 2.

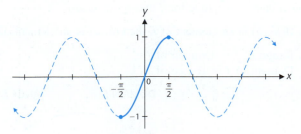

> **Figure 2** $y = \sin x$ is one-to-one over $[-\pi/2, \pi/2]$.

If the domain of the sine function is restricted to the interval $[-\pi/2, \pi/2]$, we see that the restricted function passes the horizontal line test (Section 3-6) and so is one-to-one. Note that each range value from -1 to 1 is assumed exactly once as x moves from $-\pi/2$ to $\pi/2$. We use this restricted sine function to define the *inverse sine function.*

> **DEFINITION 1** Inverse Sine Function

The **inverse sine function,** denoted by $\sin^{-1}$ or arcsin, is defined as the inverse of the restricted sine function $y = \sin x$, $-\pi/2 \le x \le \pi/2$. So

$$y = \sin^{-1} x \qquad \text{and} \qquad y = \arcsin x$$

are equivalent to

$$\sin y = x \qquad \text{where} \qquad -\pi/2 \le y \le \pi/2, \ -1 \le x \le 1$$

In words, the inverse sine of x, or the arcsine of x, is the number or angle y, $-\pi/2 \le y \le \pi/2$, whose sine is x.

To graph $y = \sin^{-1} x$, take each point on the graph of the restricted sine function and reverse the order of the coordinates. For example, because $(-\pi/2, -1)$, $(0, 0)$, and $(\pi/2, 1)$ are on the graph of the restricted sine function [Fig. 3(a)], then $(-1, -\pi/2)$, $(0, 0)$, and $(1, \pi/2)$ are on the graph of the inverse sine function, as shown in Figure 3(b). Using these three points provides us with a quick way of sketching the graph of the inverse sine function. A more accurate graph can be obtained by using a calculator.

> **Figure 3** Inverse sine function.

Restricted sine function

(a)

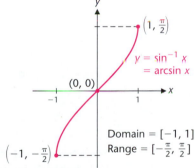

Inverse sine function

(b)

>>> **EXPLORE-DISCUSS 1**

A graphing calculator produced the graph in Figure 4 for $y_1 = \sin^{-1} x$, $-2 \le x \le 2$, and $-2 \le y \le 2$. Explain why there are no parts of the graph on the intervals $[-2, -1)$ and $(1, 2]$.

> **Figure 4**

Next we will state the important sine–inverse sine identities that follow from the general properties of inverse functions given in the box at the beginning of this section.

> ### › SINE–INVERSE SINE IDENTITIES
>
> $$\sin(\sin^{-1} x) = x \qquad -1 \le x \le 1 \qquad\qquad f(f^{-1}(x)) = x$$
>
> $$\sin^{-1}(\sin x) = x \qquad -\pi/2 \le x \le \pi/2 \qquad f^{-1}(f(x)) = x$$
>
> $\sin(\sin^{-1} 0.7) = 0.7 \qquad \sin(\sin^{-1} 1.3) \ne 1.3$
>
> $\sin^{-1}[\sin(-1.2)] = -1.2 \qquad \sin^{-1}[\sin(-2)] \ne -2$
>
> [*Note:* The number 1.3 is not in the domain of the inverse sine function, and -2 is not in the restricted domain of the sine function. Try calculating all these examples with your calculator and see what happens!]

EXAMPLE 1 **Exact Values**

Find exact values without using a calculator.

(A) $\arcsin\left(-\frac{1}{2}\right)$ (B) $\sin^{-1}(\sin 1.2)$ (C) $\cos\left[\sin^{-1}\left(\frac{2}{3}\right)\right]$

SOLUTIONS (A) $y = \arcsin\left(-\frac{1}{2}\right)$ is equivalent to

$$\sin y = -\frac{1}{2} \qquad -\frac{\pi}{2} \le y \le \frac{\pi}{2}$$

$$y = -\frac{\pi}{6} = \arcsin\left(-\frac{1}{2}\right)$$

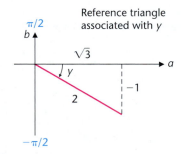

Reference triangle associated with y

[*Note:* $y \ne 11\pi/6$, even though $\sin(11\pi/6) = -\frac{1}{2}$ because y must be between $-\pi/2$ and $\pi/2$, inclusive.]

(B) $\sin^{-1}(\sin 1.2) = 1.2$ Sine–inverse sine identity, because $-\pi/2 \le 1.2 \le \pi/2$

(C) Let $y = \sin^{-1}\left(\frac{2}{3}\right)$; then $\sin y = \frac{2}{3}$, $-\pi/2 \le y \le \pi/2$. Draw the reference triangle associated with y. Then $\cos y = \cos\left[\sin^{-1}\left(\frac{2}{3}\right)\right]$ can be determined directly from the triangle (after finding the third side) without actually finding y.

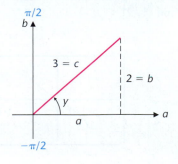

$$a^2 + b^2 = c^2$$

$$a = \sqrt{3^2 - 2^2} \qquad \text{Because } a > 0 \text{ in Quadrant I}$$

$$= \sqrt{5}$$

Therefore, $\cos\left[\sin^{-1}\left(\frac{2}{3}\right)\right] = \cos y = \sqrt{5}/3$.

MATCHED PROBLEM 1

Find exact values without using a calculator.

(A) $\arcsin(\sqrt{2}/2)$ (B) $\sin[\sin^{-1}(-0.4)]$ (C) $\tan[\sin^{-1}(-1/\sqrt{5})]$

EXAMPLE 2 **Calculator Values**

Find to four significant digits using a calculator.

(A) $\arcsin(-0.3042)$ (B) $\sin^{-1} 1.357$ (C) $\cot[\sin^{-1}(-0.1087)]$

SOLUTIONS The function keys used to represent inverse trigonometric functions vary among different brands of calculators, so read the user's manual for your calculator. Set your calculator in radian mode and follow your manual for key sequencing.

(A) $\arcsin(-0.3042) = -0.3091$

(B) $\sin^{-1} 1.357 = \text{Error}$ **1.357 is not in the domain of $\sin^{-1}$.**

(C) $\cot[\sin^{-1}(-0.1087)] = -9.145$

MATCHED PROBLEM 2

Find to four significant digits using a calculator.

(A) $\sin^{-1} 0.2903$ (B) $\arcsin(-2.305)$ (C) $\cot[\sin^{-1}(-0.3446)]$

› Inverse Cosine Function

To restrict the cosine function so that it becomes one-to-one, we choose the interval $[0, \pi]$. Over this interval the restricted function passes the horizontal line test, and each range value is assumed exactly once as x moves from 0 to π (Fig. 5). We use this restricted cosine function to define the *inverse cosine function*.

› **Figure 5** $y = \cos x$ is one-to-one over $[0, \pi]$.

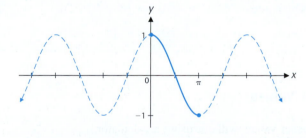

> **DEFINITION 2** Inverse Cosine Function

The **inverse cosine function,** denoted by $\cos^{-1}$ or arccos, is defined as the inverse of the restricted cosine function $y = \cos x$, $0 \le x \le \pi$. So

$$y = \cos^{-1} x \quad \text{and} \quad y = \arccos x$$

are equivalent to

$$\cos y = x \quad \text{where} \quad 0 \le y \le \pi, -1 \le x \le 1$$

In words, the inverse cosine of x, or the arccosine of x, is the number or angle y, $0 \le y \le \pi$, whose cosine is x.

Figure 6 compares the graphs of the restricted cosine function and its inverse. Notice that $(0, 1)$, $(\pi/2, 0)$, and $(\pi, -1)$ are on the restricted cosine graph. Reversing the coordinates gives us three points on the graph of the inverse cosine function.

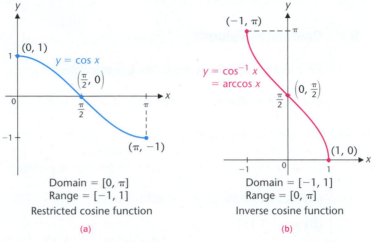

Domain $= [0, \pi]$
Range $= [-1, 1]$
Restricted cosine function

(a)

Domain $= [-1, 1]$
Range $= [0, \pi]$
Inverse cosine function

(b)

> **Figure 6** Inverse cosine function.

We complete the discussion by giving the cosine–inverse cosine identities:

> **COSINE–INVERSE COSINE IDENTITIES**

$$\cos(\cos^{-1} x) = x \qquad -1 \le x \le 1 \qquad f(f^{-1}(x)) = x$$
$$\cos^{-1}(\cos x) = x \qquad 0 \le x \le \pi \qquad f^{-1}(f(x)) = x$$

>>> **EXPLORE-DISCUSS 2**

Evaluate each of the following with a calculator. Which illustrate a cosine–inverse cosine identity and which do not? Discuss why.

(A) $\cos(\cos^{-1} 0.2)$ **(B)** $\cos[\cos^{-1}(-2)]$

(C) $\cos^{-1}(\cos 2)$ **(D)** $\cos^{-1}[\cos(-3)]$

EXAMPLE 3 Exact Values

Find exact values without using a calculator.

(A) $\arccos(-\sqrt{3}/2)$ (B) $\cos(\cos^{-1} 0.7)$ (C) $\sin[\cos^{-1}(-\tfrac{1}{3})]$

SOLUTIONS (A) $y = \arccos(-\sqrt{3}/2)$ is equivalent to

$$\cos y = -\frac{\sqrt{3}}{2} \qquad 0 \le y \le \pi$$

$$y = \frac{5\pi}{6} = \arccos\left(-\frac{\sqrt{3}}{2}\right)$$

Reference triangle associated with y

[*Note:* $y \ne -5\pi/6$, even though $\cos(-5\pi/6) = -\sqrt{3}/2$ because y must be between 0 and π, inclusive.]

(B) $\cos{(\cos^{-1}{0.7})} = 0.7$ Cosine–inverse cosine identity, because $-1 \le 0.7 \le 1$

(C) Let $y = \cos^{-1}{(-\frac{1}{3})}$; then $\cos{y} = -\frac{1}{3}$, $0 \le y \le \pi$. Draw a reference triangle associated with y. Then $\sin{y} = \sin{[\cos^{-1}{(-\frac{1}{3})}]}$ can be determined directly from the triangle (after finding the third side) without actually finding y.

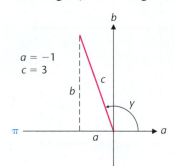

$$a^2 + b^2 = c^2$$
$$b = \sqrt{3^2 - (-1^2)} \quad \text{Because } b > 0 \text{ in Quadrant II}$$
$$= \sqrt{8}$$
$$= 2\sqrt{2}$$

Therefore, $\sin{[\cos^{-1}{(-\frac{1}{3})}]} = \sin{y} = 2\sqrt{2}/3$.

MATCHED PROBLEM 3 Find exact values without using a calculator.

(A) $\arccos{(\sqrt{2}/2)}$

(B) $\cos^{-1}{(\cos{3.05})}$

(C) $\cot{[\cos^{-1}{(-1/\sqrt{5})}]}$

EXAMPLE 4 **Calculator Values**

Find to four significant digits using a calculator.

(A) $\arccos{0.4325}$

(B) $\cos^{-1}{2.137}$

(C) $\csc{[\cos^{-1}{(-0.0349)}]}$

SOLUTIONS Set your calculator in radian mode.

(A) $\arccos{0.4325} = 1.124$

(B) $\cos^{-1}{2.137} = $ Error 2.137 is not in the domain of $\cos^{-1}$.

(C) $\csc{[\cos^{-1}{(-0.0349)}]} = 1.001$

MATCHED PROBLEM 4 Find to four significant digits using a calculator.

(A) $\cos^{-1}{(0.6773)}$ (B) $\arccos{(-1.003)}$ (C) $\cot{[\cos^{-1}{(-0.5036)}]}$

> ### Inverse Tangent Function

To restrict the tangent function so that it becomes one-to-one, we choose the interval $(-\pi/2, \pi/2)$. Over this interval the restricted function passes the horizontal line test, and each range value is assumed exactly once as x moves across this restricted domain (Fig. 7). We use this restricted tangent function to define the *inverse tangent function*.

> **Figure 7** $y = \tan x$ is one-to-one over $(-\pi/2, \pi/2)$.

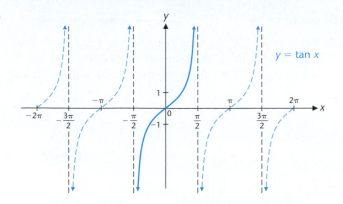

> **DEFINITION 3** Inverse Tangent Function

The **inverse tangent function,** denoted by $\tan^{-1}$ or arctan, is defined as the inverse of the restricted tangent function $y = \tan x$, $-\pi/2 < x < \pi/2$. So

$$y = \tan^{-1} x \qquad \text{and} \qquad y = \arctan x$$

are equivalent to

$$\tan y = x \qquad \text{where} \qquad -\pi/2 < y < \pi/2 \text{ and } x \text{ is a real number}$$

In words, the inverse tangent of x, or the arctangent of x, is the number or angle y, $-\pi/2 < y < \pi/2$, whose tangent is x.

Figure 8 compares the graphs of the restricted tangent function and its inverse. Notice that $(-\pi/4, -1)$, $(0, 0)$, and $(\pi/4, 1)$ are on the restricted tangent graph. Reversing the coordinates gives us three points on the graph of the inverse tangent function. Also note that the vertical asymptotes become horizontal asymptotes for the inverse function.

> **Figure 8** Inverse tangent function.

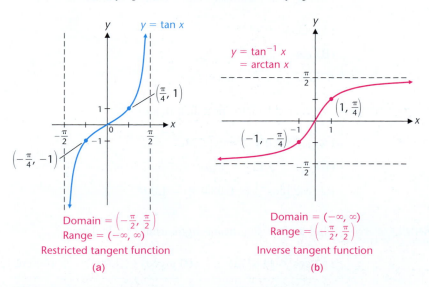

Domain $= \left(-\frac{\pi}{2}, \frac{\pi}{2}\right)$
Range $= (-\infty, \infty)$
Restricted tangent function
(a)

Domain $= (-\infty, \infty)$
Range $= \left(-\frac{\pi}{2}, \frac{\pi}{2}\right)$
Inverse tangent function
(b)

We now state the tangent–inverse tangent identities.

> **TANGENT–INVERSE TANGENT IDENTITIES**

$$\tan (\tan^{-1} x) = x \qquad -\infty < x < \infty \qquad f(f^{-1}(x)) = x$$
$$\tan^{-1} (\tan x) = x \qquad -\pi/2 < x < \pi/2 \qquad f^{-1}(f(x)) = x$$

| EXAMPLE | 5 | **Exact Values** |

Find exact values without using a calculator.

(A) $\tan^{-1}(-1/\sqrt{3})$ (B) $\tan^{-1}(\tan 0.63)$

SOLUTIONS (A) $y = \tan^{-1}(-1/\sqrt{3})$ is equivalent to

Reference triangle associated with y

$$\tan y = -\frac{1}{\sqrt{3}} \qquad -\frac{\pi}{2} < y < \frac{\pi}{2}$$

$$y = -\frac{\pi}{6} = \tan^{-1}\left(-\frac{1}{\sqrt{3}}\right)$$

[*Note:* y cannot be $11\pi/6$ because y must be between $-\pi/2$ and $\pi/2$.]

(B) $\tan^{-1}(\tan 0.63) = 0.63$ Tangent-inverse tangent identity, because $-\pi/2 \le 0.63 \le \pi/2$

MATCHED PROBLEM 5 Find exact values without using a calculator.

(A) $\arctan(-\sqrt{3})$ (B) $\tan(\tan^{-1} 43)$

› **Summary**

We summarize the definitions and graphs of the inverse trigonometric functions discussed so far for convenient reference.

› SUMMARY OF $\sin^{-1}$, $\cos^{-1}$, AND $\tan^{-1}$

$y = \sin^{-1} x$	is equivalent to	$x = \sin y$	where $-1 \le x \le 1$, $-\pi/2 \le y \le \pi/2$
$y = \cos^{-1} x$	is equivalent to	$x = \cos y$	where $-1 \le x \le 1$, $0 \le y \le \pi$
$y = \tan^{-1} x$	is equivalent to	$x = \tan y$	where $-\infty < x < \infty$, $-\pi/2 < y < \pi/2$

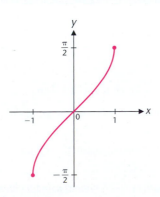

$y = \sin^{-1} x$
Domain $= [-1, 1]$
Range $= [-\frac{\pi}{2}, \frac{\pi}{2}]$

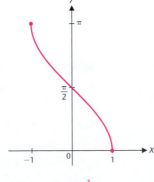

$y = \cos^{-1} x$
Domain $= [-1, 1]$
Range $= [0, \pi]$

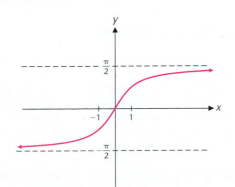

$y = \tan^{-1} x$
Domain $= (-\infty, \infty)$
Range $= \left(-\frac{\pi}{2}, \frac{\pi}{2}\right)$

> ## Inverse Cotangent, Secant, and Cosecant Functions (Optional)

For completeness, we include the definitions and graphs of the inverse cotangent, secant, and cosecant functions.

> ### DEFINITION 4 Inverse Cotangent, Secant, and Cosecant Functions

$y = \cot^{-1} x$ is equivalent to $x = \cot y$ where $0 < y < \pi$, $-\infty < x < \infty$

$y = \sec^{-1} x$ is equivalent to $x = \sec y$ where $0 \le y \le \pi$, $y \ne \pi/2$, $|x| \ge 1$

$y = \csc^{-1} x$ is equivalent to $x = \csc y$ where $-\pi/2 \le y \le \pi/2$, $y \ne 0$, $|x| \ge 1$

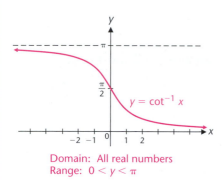

Domain: All real numbers
Range: $0 < y < \pi$

Domain: $x \le -1$ or $x \ge 1$
Range: $0 \le y \le \pi$, $y \ne \pi/2$

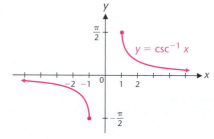

Domain: $x \le -1$ or $x \ge 1$
Range: $-\pi/2 \le y \le \pi/2$, $y \ne 0$

[*Note:* The domain restrictions used in defining $\sec^{-1}$ and $\csc^{-1}$ are not universally agreed upon.]

> ### ANSWERS TO MATCHED PROBLEMS

1. (A) $\pi/4$ (B) -0.4 (C) $-1/2$ **2.** (A) 0.2945 (B) Not defined (C) -2.724
3. (A) $\pi/4$ (B) 3.05 (C) $-1/2$ **4.** (A) 0.8267 (B) Not defined (C) -0.5829
5. (A) $-\pi/3$ (B) 43

6-6 Exercises

Unless stated to the contrary, the inverse trigonometric functions are assumed to have real number ranges (use radian mode in calculator problems). A few problems involve ranges with angles in degree measure, and these are clearly indicated (use degree mode in calculator problems).

1. Explain why the function $f(x) = \sin x$, for $0 \le x \le \pi$, has no inverse.

2. Explain why the function $f(x) = \cos x$, for $-\pi/2 \le x \le \pi/2$, has no inverse.

3. Does $\tan(\tan^{-1} x) = x$ for all real x? Explain.

4. Does $\tan^{-1}(\tan x) = x$ for all real x? Explain.

5. If a function f has an inverse, how are the graphs of f and f^{-1} related?

6. If f is increasing, is f^{-1} also increasing? Explain.

In Problems 7–18, find exact values without using a calculator.

7. $\cos^{-1} 0$ **8.** $\sin^{-1} 0$

9. $\arcsin(\sqrt{3}/2)$ **10.** $\arccos(\sqrt{3}/2)$

11. arctan $\sqrt{3}$

12. $\tan^{-1} 1$

13. $\sin^{-1}(\sqrt{2}/2)$

14. $\cos^{-1}(\frac{1}{2})$

15. arccos 1

16. arctan $(1/\sqrt{3})$

17. $\sin^{-1}(\frac{1}{2})$

18. $\tan^{-1} 0$

In Problems 19–24, evaluate to four significant digits using a calculator.

19. $\sin^{-1} 0.9103$

20. $\cos^{-1} 0.4038$

21. arctan 103.7

22. $\tan^{-1} 43.09$

23. arccos 3.051

24. arcsin 1.131

In Problems 25–30, find the exact value without using a calculator if the expression is defined.

25. arctan $(-\sqrt{3})$

26. arccos $(-\sqrt{3})$

27. $\cos^{-1}(-\sqrt{2}/2)$

28. $\sin^{-1}(-\sqrt{3}/2)$

29. arcsin (-2)

30. arctan (-1)

In Problems 31–34, evaluate to four significant digits using a calculator.

31. $\cot [\cos^{-1}(-0.7003)]$

32. $\sec [\sin^{-1}(-0.0399)]$

33. $\sqrt{5 + \cos^{-1}(1 - \sqrt{2})}$

34. $\sqrt{2} + \tan^{-1} \sqrt[3]{5}$

In Problems 35–46, find the exact value without using a calculator if the expression is defined.

35. $\tan (\tan^{-1} \sqrt{5})$

36. $\tan [\tan^{-1}(-10)]$

37. $\sin [\sin^{-1}(-1/\sqrt{3})]$

38. $\sin (\sin^{-1} \sqrt{5})$

39. $\cos [\cos^{-1}(-\sqrt{2})]$

40. $\cos [\cos^{-1}(-1)]$

41. $\sin^{-1}(\sin 1.5)$

42. $\sin^{-1}[\sin (3\pi/2)]$

43. $\cos^{-1}[\cos (-\pi)]$

44. $\cos^{-1}[\cos (-\pi/5)]$

45. $\tan^{-1}[\tan (\pi/2)]$

46. $\tan^{-1}[\tan (5\pi/4)]$

In Problems 47–52, find the exact degree measure without using a calculator if the expression is defined.

47. $\sin^{-1}(-\frac{1}{2})$

48. $\cos^{-1}(-1)$

49. $\tan^{-1}(-1)$

50. $\sin^{-1}(-\sqrt{2})$

51. arccos $(-\sqrt{3}/2)$

52. arctan $(-\sqrt{3})$

In Problems 53–56, find the degree measure of each to two decimal places using a calculator set in degree mode.

53. $\cos^{-1} 0.7253$

54. $\tan^{-1} 12.4304$

55. arcsin (-0.3662)

56. arccos (-0.9206)

57. Evaluate $\sin^{-1}(\sin 2)$ with a calculator set in radian mode, and explain why this does or does not illustrate the inverse sine–sine identity.

58. Evaluate $\cos^{-1}[\cos (-0.5)]$ with a calculator set in radian mode, and explain why this does or does not illustrate the inverse cosine–cosine identity.

In Problems 59–64, determine whether the statement is true or false. Explain.

59. None of the six trigonometric functions is one-to-one.

60. Each of the six inverse trigonometric functions is one-to-one.

61. Each of the six inverse trigonometric functions is periodic.

62. Each of the six inverse trigonometric functions is bounded.

63. The function $y = \sin^{-1} x$ is odd.

64. The function $y = \cos^{-1} x$ is even.

In Problems 65–72, graph each function over the indicated interval.

65. $y = \sin^{-1} x, -1 \le x \le 1$

66. $y = \cos^{-1} x, -1 \le x \le 1$

67. $y = \cos^{-1}(x/3), -3 \le x \le 3$

68. $y = \sin^{-1}(x/2), -2 \le x \le 2$

69. $y = \sin^{-1}(x - 2), 1 \le x \le 3$

70. $y = \cos^{-1}(x + 1), -2 \le x \le 0$

71. $y = \tan^{-1}(2x - 4), -2 \le x \le 6$

72. $y = \tan^{-1}(2x + 3), -5 \le x \le 2$

73. The identity $\cos (\cos^{-1} x) = x$ is valid for $-1 \le x \le 1$.
(A) Graph $y = \cos (\cos^{-1} x)$ for $-1 \le x \le 1$.
(B) What happens if you graph $y = \cos (\cos^{-1} x)$ over a larger interval, say $-2 \le x \le 2$? Explain.

74. The identity $\sin (\sin^{-1} x) = x$ is valid for $-1 \le x \le 1$.
(A) Graph $y = \sin (\sin^{-1} x)$ for $-1 \le x \le 1$.
(B) What happens if you graph $y = \sin (\sin^{-1} x)$ over a larger interval, say $-2 \le x \le 2$? Explain.

In Problems 75–78, write each expression as an algebraic expression in x free of trigonometric or inverse trigonometric functions.

75. $\cos (\sin^{-1} x)$

76. $\sin (\cos^{-1} x)$

77. $\cos (\arctan x)$

78. $\tan (\arcsin x)$

In Problems 79 and 80, find $f^{-1}(x)$. How must x be restricted in $f^{-1}(x)$?

79. $f(x) = 4 + 2 \cos (x - 3), 3 \le x \le (3 + \pi)$

80. $f(x) = 3 + 5 \sin (x - 1), (1 - \pi/2) \le x \le (1 + \pi/2)$

81. The identity $\cos^{-1}(\cos x) = x$ is valid for $0 \le x \le \pi$.
(A) Graph $y = \cos^{-1}(\cos x)$ for $0 \le x \le \pi$.
(B) What happens if you graph $y = \cos^{-1}(\cos x)$ over a larger interval, say $-2\pi \le x \le 2\pi$? Explain.

82. The identity $\sin^{-1}(\sin x) = x$ is valid for $-\pi/2 \le x \le \pi/2$.
(A) Graph $y = \sin^{-1}(\sin x)$ for $-\pi/2 \le x \le \pi/2$.
(B) What happens if you graph $y = \sin^{-1}(\sin x)$ over a larger interval, say $-2\pi \le x \le 2\pi$? Explain.

APPLICATIONS

83. PHOTOGRAPHY The viewing angle changes with the focal length of a camera lens. A 28-millimeter wide-angle lens has a wide viewing angle and a 300-millimeter telephoto lens has a narrow viewing angle. For a 35-millimeter format camera the viewing angle θ, in degrees, is given by

$$\theta = 2\tan^{-1}\frac{21.634}{x}$$

where x is the focal length of the lens being used. What is the viewing angle (in decimal degrees to two decimal places) of a 28-millimeter lens? Of a 100-millimeter lens?

84. PHOTOGRAPHY Referring to Problem 83, what is the viewing angle (in decimal degrees to two decimal places) of a 17-millimeter lens? Of a 70-millimeter lens?

85. (A) Graph the function in Problem 83 in a graphing calculator using degree mode. The graph should cover lenses with focal lengths from 10 millimeters to 100 millimeters.
(B) What focal-length lens, to two decimal places, would have a viewing angle of 40°? Solve by graphing $\theta = 40$ and $\theta = 2\tan^{-1}(21.634/x)$ in the same viewing window and finding the point of intersection using the INTERSECT command.

86. (A) Graph the function in Problem 83 in a graphing calculator, in degree mode, with the graph covering lenses with focal lengths from 100 millimeters to 1,000 millimeters.
(B) What focal length lens, to two decimal places, would have a viewing angle of 10°? Solve by graphing $\theta = 10$ and $\theta = \tan^{-1}(21.634/x)$ in the same viewing window and finding the point of intersection using the INTERSECT command.

87. ENGINEERING The length of the belt around the two pulleys in the figure is given by

$$L = \pi D + (d - D)\theta + 2C\sin\theta$$

where θ (in radians) is given by

$$\theta = \cos^{-1}\frac{D - d}{2C}$$

Verify these formulas, and find the length of the belt to two decimal places if $D = 4$ inches, $d = 2$ inches, and $C = 6$ inches.

88. ENGINEERING For Problem 87, find the length of the belt if $D = 6$ inches, $d = 4$ inches, and $C = 10$ inches.

89. ENGINEERING The function

$$y_1 = 4\pi - 2\cos^{-1}\frac{1}{x} + 2x\sin\left(\cos^{-1}\frac{1}{x}\right)$$

represents the length of the belt around the two pulleys in Problem 87 when the centers of the pulleys are x inches apart.
(A) Graph y_1 in a graphing calculator (in radian mode), with the graph covering pulleys with their centers from 3 to 10 inches apart.
(B) How far, to two decimal places, should the centers of the two pulleys be placed to use a belt 24 inches long? Solve by graphing y_1 and $y_2 = 24$ in the same viewing window and finding the point of intersection using the INTERSECT command.

90. ENGINEERING The function

$$y_1 = 6\pi - 2\cos^{-1}\frac{1}{x} + 2x\sin\left(\cos^{-1}\frac{1}{x}\right)$$

represents the length of the belt around the two pulleys in Problem 88 when the centers of the pulleys are x inches apart.
(A) Graph y_1 in a graphing calculator (in radian mode), with the graph covering pulleys with their centers from 3 to 20 inches apart.
(B) How far, to two decimal places, should the centers of the two pulleys be placed to use a belt 36 inches long? Solve by graphing y_1 and $y_2 = 36$ in the same viewing window and finding the point of intersection using the INTERSECT command.

91. MOTION The figure represents a circular courtyard surrounded by a high stone wall. A floodlight located at E shines into the courtyard.

(A) If a person walks x feet away from the center along DC, show that the person's shadow will move a distance given by

$$d = 2r\theta = 2r\tan^{-1}\frac{x}{r}$$

where θ is in radians. [*Hint:* Draw a line from A to C.]

(B) Find d to two decimal places if $r = 100$ feet and $x = 40$ feet.

92. MOTION In Problem 91, find d for $r = 50$ feet and $x = 25$ feet.

CHAPTER **6** Review

6-1 Angles and Their Measure

An **angle** is formed by rotating (in a plane) a ray m, called the **initial side** of the angle, around its endpoint until it coincides with a ray n, called the **terminal side** of the angle. The common endpoint of m and n is called the **vertex**. If the rotation is counterclockwise, the angle is **positive;** if clockwise, **negative**. Two angles are **coterminal** if they have the same initial and terminal sides.

An angle is in **standard position** in a rectangular coordinate system if its vertex is at the origin and its initial side is along the positive x axis. **Quadrantal angles** have their terminal sides on a coordinate axis. An angle of **1 degree** is $\frac{1}{360}$ of a complete rotation. Two positive angles are **complementary** if their sum is 90°; they are **supplementary** if their sum is 180°. An angle of **1 radian** is a central angle of a circle subtended by an arc having the same length as the radius.

$$\text{Radian measure: } \theta = \frac{s}{r}$$

$$\text{Radian–degree conversion: } \frac{\theta_{\deg}}{180°} = \frac{\theta_{\text{rad}}}{\pi \text{ radians}}$$

If a point P moves through an angle θ and arc length s, in time t, on the circumference of a circle of radius r, then the (average) **linear speed** of P is

$$v = \frac{s}{t}$$

and the (average) **angular speed** is

$$\omega = \frac{\theta}{t}$$

Because $s = r\theta$ it follows that $v = r\omega$.

6-2 Right Triangle Trigonometry

A **right triangle** is a triangle with one 90° angle. To **solve a right triangle** is to find all unknown angles and sides, given the measures of two sides or the measures of one side and an acute angle.

Trigonometric Ratios

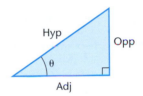

$$\sin\theta = \frac{\text{Opp}}{\text{Hyp}} \qquad \csc\theta = \frac{\text{Hyp}}{\text{Opp}}$$

$$\cos\theta = \frac{\text{Adj}}{\text{Hyp}} \qquad \sec\theta = \frac{\text{Hyp}}{\text{Adj}}$$

$$\tan\theta = \frac{\text{Opp}}{\text{Adj}} \qquad \cot\theta = \frac{\text{Adj}}{\text{Opp}}$$

Computational Accuracy

Angle to Nearest	Significant Digits for Side Measure
1°	2
10′ or 0.1°	3
1′ or 0.01°	4
10″ or 0.001°	5

6-3 Trigonometric Functions: A Unit Circle Approach

If θ is a positive angle in standard position, and P is the point of intersection of the terminal side of θ with the unit circle, then the radian measure of θ equals the length x of the arc opposite θ; and if θ is negative, the radian measure of θ equals the negative of the length of the intercepted arc. The function W that associates with each real number x the point $W(x) = P$ is called the **wrapping function,** and the point P is called a **circular point**. The function $W(x)$ can be visualized as a wrapping of the real number line, with origin at $(1, 0)$, around the unit circle—the positive real axis is wrapped counterclockwise and the negative real axis is wrapped clockwise—so that each real number is paired with a unique circular point. The function $W(x)$ is not one-to-one: for example, each of the real numbers $2\pi k$, k any integer, corresponds to the circular point $(1, 0)$.

The coordinates of key circular points in the first quadrant can be found using simple geometric facts; the coordinates of the circular point associated with any multiple of $\pi/6$ or $\pi/4$ can then be determined using symmetry properties.

Coordinates of Key Circular Points

The six **trigonometric functions—sine, cosine, tangent, cotangent, secant,** and **cosecant**—are defined in terms of the coordinates (a, b) of the circular point $W(x)$ that lies on the terminal side of the angle with radian measure x:

$$\sin x = b \qquad \cos x = \frac{1}{b} \quad b \neq 0$$

$$\cos x = a \qquad \sec x = \frac{1}{a} \quad a \neq 0$$

$$\tan x = \frac{b}{a} \quad a \neq 0 \qquad \cot x = \frac{a}{b} \quad b \neq 0$$

The trigonometric functions of any multiple of $\pi/6$ or $\pi/4$ can be determined exactly from the coordinates of the circular point. A graphing calculator can be used to graph the trigonometric functions and approximate their values for arbitrary inputs.

6-4 Properties of Trigonometric Functions

The definition of the trigonometric functions implies that the following **basic identities** hold true for all replacements of x by real numbers for which both sides of an equation are defined:

Reciprocal identities

$$\csc x = \frac{1}{\sin x} \qquad \sec x = \frac{1}{\cos x} \qquad \cot x = \frac{1}{\tan x}$$

Quotient identities

$$\tan x = \frac{\sin x}{\cos x} \qquad \cot x = \frac{\cos x}{\sin x}$$

Identities for negatives

$$\sin(-x) = -\sin x \qquad \cos(-x) = \cos x$$
$$\tan(-x) = -\tan x$$

Pythagorean identity

$$\sin^2 x + \cos^2 x = 1$$

A function f is **periodic** if there exists a positive real number p such that

$$f(x + p) = f(x)$$

for all x in the domain of f. The smallest such positive p, if it exists, is called the **fundamental period of f,** or often just the **period of f.** All the trigonometric functions are periodic.

Graph of $y = \sin x$:

Period: 2π
Domain: All real numbers
Range: $[-1, 1]$

Graph of $y = \cos x$:

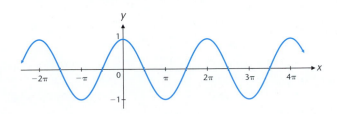

Period: 2π
Domain: All real numbers
Range: $[-1, 1]$

Graph of $y = \tan x$:

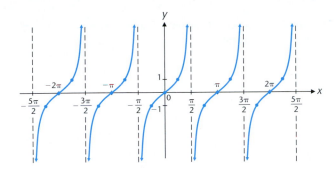

Period: π
Domain: All real numbers except $\pi/2 + k\pi$, k an integer
Range: All real numbers

Graph of $y = \cot x$:

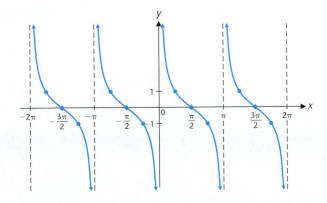

Period: π
Domain: All real numbers except $k\pi$, k an integer
Range: All real numbers

Graph of $y = \csc x$:

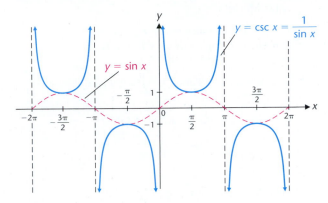

Period: 2π
Domain: All real numbers except $k\pi$, k an integer
Range: All real numbers y such that $y \le -1$ or $y \ge 1$

Graph of $y = \sec x$:

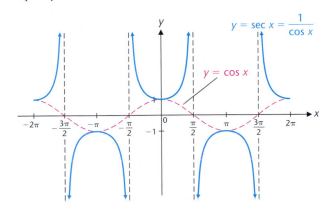

Period: 2π
Domain: All real numbers except $\pi/2 + k\pi$, k an integer
Range: All real numbers y such that $y \le -1$ or $y \ge 1$

Associated with each angle that does not terminate on a coordinate axis is a **reference triangle** for θ. The reference triangle is formed by drawing a perpendicular from point $P = (a, b)$ on the terminal side of θ to the horizontal axis. The **reference angle** α is the acute angle, always taken positive, between the terminal side of θ and the horizontal axis as indicated in the following figure.

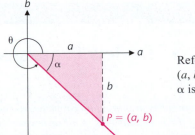

Reference Triangle
$(a, b) \ne (0, 0)$
α is always positive

6-5 More General Trigonometric Functions
 and Models

Let A, B, C be constants such that $A \neq 0$ and $B > 0$.
If $y = A \sin (Bx + C)$ or $y = A \cos (Bx + C)$:

$$\text{Amplitude} = |A| \quad \text{Period} = \frac{2\pi}{B} \quad \text{Phase shift} = -\frac{C}{B}$$

If $y = A \sec (Bx + C)$ or $y = \csc (Bx + C)$:

$$\text{Period} = \frac{2\pi}{B} \quad \text{Phase shift} = -\frac{C}{B}$$

If $y = A \tan (Bx + C)$ or $y = A \cot (Bx + C)$:

$$\text{Period} = \frac{\pi}{B} \quad \text{Phase shift} = -\frac{C}{B}$$

(Amplitude is not defined for the secant, cosecant, tangent, and cotangent functions, all of which are unbounded.)

Sinusoidal regression is used to find the function of the form $y = A \sin (Bx + C) + k$ that best fits a set of data points.

6-6 Inverse Trigonometric Functions

$y = \sin^{-1} x = \arcsin x$ if and only if $\sin y = x$,
$-\pi/2 \leq y \leq \pi/2$ and $-1 \leq x \leq 1$.

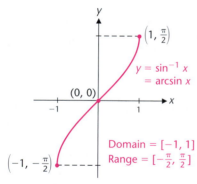

Domain = $[-1, 1]$
Range = $[-\frac{\pi}{2}, \frac{\pi}{2}]$

Inverse sine function

$y = \cos^{-1} x = \arccos x$ if and only if $\cos y = x$, $0 \leq y \leq \pi$ and $-1 \leq x \leq 1$.

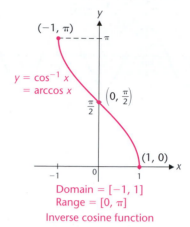

Domain = $[-1, 1]$
Range = $[0, \pi]$
Inverse cosine function

$y = \tan^{-1} x = \arctan x$ if and only if $\tan y = x$, $-\pi/2 < y < \pi/2$ and x is any real number.

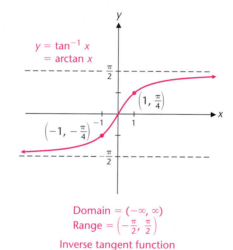

Domain = $(-\infty, \infty)$
Range = $\left(-\frac{\pi}{2}, \frac{\pi}{2}\right)$
Inverse tangent function

CHAPTER 6 Review Exercises

Work through all the problems in this chapter review and check answers in the back of the book. Answers to all review problems are there, and following each answer is a number in italics indicating the section in which that type of problem is discussed. Where weaknesses show up, review appropriate sections in the text.

1. Find the radian measure of a central angle opposite an arc 15 centimeters long on a circle of radius 6 centimeters.

2. In a circle of radius 3 centimeters, find the length of an arc opposite an angle of 2.5 radians.

3. Solve the triangle:

4. Find the reference angle associated with each angle θ.
 (A) $\theta = \pi/3$ (B) $\theta = -120°$
 (C) $\theta = -13\pi/6$ (D) $\theta = 210°$

5. In which quadrants is each negative?
 (A) $\sin \theta$ (B) $\cos \theta$ (C) $\tan \theta$

6. If $(4, -3)$ is on the terminal side of angle θ, find
 (A) $\sin \theta$ (B) $\sec \theta$ (C) $\cot \theta$

7. Complete Table 1 using exact values. Do not use a calculator.

Table 1

θ°	θ rad	sin θ	cos θ	tan θ	csc θ	sec θ	cot θ
0°				ND*			
30°							
45°	π/4		$1/\sqrt{2}$				
60°							
90°							
180°							
270°							
360°							

*ND = Not defined

8. What is the period of each of the following?
 (A) $y = \cos x$ (B) $y = \csc x$ (C) $y = \tan x$

9. Indicate the domain and range of each.
 (A) $y = \sin x$ (B) $y = \tan x$

10. Sketch a graph of $y = \sin x$, $-2\pi \le x \le 2\pi$.

11. Sketch a graph of $y = \cot x$, $-\pi < x < \pi$.

12. Verbally describe the meaning of a central angle in a circle with radian measure 0.5.

13. Describe the smallest shift of the graph of $y = \sin x$ that produces the graph of $y = \cos x$.

14. Change 1.37 radians to decimal degrees to two decimal places.

15. Solve the triangle:

16. Indicate whether the angle is a Quadrant I, II, III, or IV angle or a quadrantal angle.
 (A) $-210°$ (B) $5\pi/2$ (C) 4.2 radians

17. Which of the following angles are coterminal with 120°?
 (A) $-240°$ (B) $-7\pi/6$ (C) 840°

18. Which of the following have the same value as cos 3?
 (A) cos 3° (B) cos (3 radians) (C) cos (3 + 2π)

19. For which values of x, $0 \le x \le 2\pi$, is each of the following not defined?
 (A) $\tan x$ (B) $\cot x$ (C) $\csc x$

20. A circular point $P = (a, b)$ moves clockwise around the circumference of a unit circle starting at (1, 0) and stops after covering a distance of 8.305 units. Explain how you would find the coordinates of point P at its final position and how you would determine which quadrant P is in. Find the coordinates of P to three decimal places and the quadrant for the final position of P.

In Problems 21–36, evaluate exactly without the use of a calculator.

21. $\tan 0$ **22.** $\sec 90°$

23. $\cos^{-1} 1$ **24.** $\cos\left(-\dfrac{3\pi}{4}\right)$

25. $\sin^{-1} \dfrac{\sqrt{2}}{2}$ **26.** $\csc 300°$

27. $\arctan \sqrt{3}$ **28.** $\sin 570°$

29. $\tan^{-1}(-1)$ **30.** $\cot\left(-\dfrac{4\pi}{3}\right)$

31. $\arcsin\left(-\dfrac{1}{2}\right)$ **32.** $\cos^{-1}\left(-\dfrac{\sqrt{3}}{2}\right)$

33. $\cos(\cos^{-1} 0.33)$ **34.** $\csc[\tan^{-1}(-1)]$

35. $\sin\left[\arccos\left(-\dfrac{1}{2}\right)\right]$ **36.** $\tan\left(\sin^{-1}\dfrac{-4}{5}\right)$

Evaluate Problems 37–44 to four significant digits using a calculator.

37. $\cos 423.7°$ **38.** $\tan 93°46'17''$

39. $\sec(-2.073)$ **40.** $\sin^{-1}(-0.8277)$

41. $\arccos(-1.3281)$ **42.** $\tan^{-1} 75.14$

43. $\csc[\cos^{-1}(-0.4081)]$ **44.** $\sin^{-1}(\tan 1.345)$

45. Find the exact degree measure of each without a calculator.
 (A) $\theta = \sin^{-1}(-\tfrac{1}{2})$ (B) $\theta = \arccos(-\tfrac{1}{2})$

46. Find the degree measure of each to two decimal places using a calculator.
 (A) $\theta = \cos^{-1}(-0.8763)$ (B) $\theta = \arctan 7.3771$

47. Evaluate $\cos^{-1}[\cos(-2)]$ with a calculator set in radian mode, and explain why this does or does not illustrate the inverse cosine–cosine identity.

48. Sketch a graph of $y = -2 \cos \pi x$, $-1 \le x \le 3$. Indicate amplitude A and period P.

49. Sketch a graph of $y = -2 + 3 \sin (x/2)$, $-4\pi \le x \le 4\pi$.

50. Find the equation of the form $y = A \cos Bx$ that has the graph shown here.

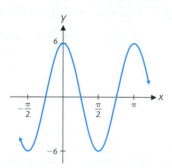

51. Find the equation of the form $y = A \sin Bx$ that has the graph shown here.

52. Describe the smallest shift and/or reflection that transforms the graph of $y = \tan x$ into the graph of $y = \cot x$.

53. Simplify each of the following using appropriate basic identities:

(A) $\sin (-x) \cot (-x)$ (B) $\dfrac{\sin^2 x}{1 - \sin^2 x}$

54. Sketch a graph of $y = 3 \sin [(x/2) + (\pi/2)]$ over the interval $-4\pi \le x \le 4\pi$.

55. Indicate the amplitude A, period P, and phase shift for the graph of $y = -2 \cos [(\pi/2) x - (\pi/4)]$. Do not graph.

56. Sketch a graph of $y = \cos^{-1} x$, and indicate the domain and range.

57. Graph $y = 1/(1 + \tan^2 x)$ in a graphing calculator that displays at least two full periods of the graph. Find an equation of the form $y = k + A \sin Bx$ or $y = k + A \cos Bx$ that has the same graph.

58. Graph each equation in a graphing calculator and find an equation of the form $y = A \tan Bx$ or $y = A \cot Bx$ that has the same graph as the given equation. Select the dimensions of the viewing window so that at least two periods are visible.

(A) $y = \dfrac{2 \sin^2 x}{\sin 2x}$ (B) $y = \dfrac{2 \cos^2 x}{\sin 2x}$

59. Determine whether each function is even, odd, or neither.

(A) $f(x) = \dfrac{1}{1 + \tan^2 x}$ (B) $g(x) = \dfrac{1}{1 + \tan x}$

In Problems 60 and 61, determine whether the statement is true or false. If true, explain why. If false, give a counterexample.

60. If α and β are the acute angles of a right triangle, then $\sin \alpha = \csc \beta$.

61. If α and β are the acute angles of a right triangle and $\alpha = \beta$, then all six trigonometric functions of α are greater than $\frac{1}{2}$ and less than $\frac{3}{2}$.

62. If in the figure the coordinates of A are $(8, 0)$ and arc length s is 20 units, find:
(A) The exact radian measure of θ
(B) The coordinates of P to three significant digits

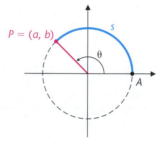

63. Find exactly the least positive real number for which
(A) $\cos x = -\frac{1}{2}$ (B) $\csc x = -\sqrt{2}$

64. Sketch a graph of $y = \sec x$, $-\pi/2 < x < 3\pi/2$.

65. Sketch a graph of $y = \tan^{-1} x$, and indicate the domain and range.

66. Indicate the period P and phase shift for the graph of $y = -5 \tan (\pi x + \pi/2)$. Do not graph.

67. Indicate the period and phase shift for the graph of $y = 3 \csc (x/2 - \pi/4)$. Do not graph.

68. Indicate whether each is symmetrical with respect to the x axis, y axis, or origin.
(A) Sine (B) Cosine (C) Tangent

69. Write as an algebraic expression in x free of trigonometric or inverse trigonometric functions:

$$\sec (\sin^{-1} x)$$

70. Try to calculate each of the following on your calculator. Explain the results.
(A) $\csc (-\pi)$ (B) $\tan (-3\pi/2)$ (C) $\sin^{-1} 2$

71. The accompanying graph is a graph of an equation of the form $y = A \sin (Bx + C)$, $-1 < -C/B < 0$. Find the equation.

 72. Graph $y = 1.2 \sin 2x + 1.6 \cos 2x$ in a graphing calculator. (Select the dimensions of the viewing window so that at least two periods are visible.) Find an equation of the form $y = A \sin (Bx + C)$ that has the same graph as the given equation. Find A and B exactly and C to three decimal places. Use the x intercept closest to the origin as the phase shift.

 73. A particular waveform is approximated by the first six terms of a Fourier series:

$$y = \frac{4}{\pi}\left(\sin x + \frac{\sin 3x}{3} + \frac{\sin 5x}{5} + \frac{\sin 7x}{7} + \frac{\sin 9x}{9} + \frac{\sin 11x}{11}\right)$$

(A) Graph this equation in a graphing calculator for $-3\pi \le x \le 3\pi$ and $-2 \le y \le 2$.

(B) The graph in part A approximates a waveform that is made up entirely of straight line segments. Sketch by hand the waveform that the Fourier series approximates.

This waveform is called a **pulse wave** or a **square wave,** and is used, for example, to test distortion and to synchronize operations in computers.

APPLICATIONS

74. ASTRONOMY A line from the sun to the Earth sweeps out an angle of how many radians in 73 days? Express the answer in terms of π.

75. GEOMETRY Find the perimeter of a square inscribed in a circle of radius 5.00 centimeters.

76. ANGULAR SPEED A wind turbine of rotor diameter 40 feet makes 80 revolutions per minute. Find the angular speed (in radians per second) and the linear speed (in feet per second) of the rotor tip.

77. ALTERNATING CURRENT The current I in alternating electrical current has an amplitude of 30 amperes and a period of $\frac{1}{60}$ second. If $I = 30$ amperes when $t = 0$, find an equation of the form $I = A \cos Bt$ that gives the current at any time $t \ge 0$.

78. RESTRICTED ACCESS A 10-foot-wide canal makes a right turn into a 15-foot-wide canal. Long narrow logs are to be floated through the canal around the right angle turn (see the figure). We are interested in finding the longest log that will go around the corner, ignoring the log's diameter.

(A) Express the length L of the line that touches the two outer sides of the canal and the inside corner in terms of θ.

(B) Complete Table 2, each to one decimal place, and estimate from the table the longest log to the nearest foot that can make it around the corner. (The longest log is the shortest distance L.)

Table 2

θ (radians)	0.4	0.5	0.6	0.7	0.8	0.9	1.0
L (feet)	42.0						

 (C) Graph the function in part A in a graphing calculator and use the MINIMUM command to find the shortest distance L to one decimal place; the length of the longest log that can make it around the corner.

(D) Explain what happens to the length L as θ approaches 0 or $\pi/2$.

79. MODELING SEASONAL BUSINESS CYCLES A soft drink company has revenues from sales over a 2-year period as shown by the accompanying graph, where $R(t)$ is revenue (in millions of dollars) for a month of sales t months after February 1.

(A) Find an equation of the form $R(t) = k + A \cos Bt$ that produces this graph, and check the result by graphing.

(B) Verbally interpret the graph

 80. MODELING TEMPERATURE VARIATION The 30-year average monthly temperature, °F, for each month of the year for Los Angeles is given in Table 3 (*World Almanac*).

(A) Using 1 month as the basic unit of time, enter the data for a 2-year period in your graphing calculator and produce a scatter plot in the viewing window. Choose $40 \le y \le 90$ for the viewing window.

(B) It appears that a sine curve of the form

$$y = k + A \sin (Bx + C)$$

will closely model these data. The constants k, A, and B are easily determined from Table 3. To estimate C, visually estimate to one decimal place the smallest positive phase shift from the plot in part A. After determining A, B, k, and C, write the resulting equation. (Your value of C may differ slightly from the answer at the back of the book.)

(C) Plot the results of parts A and B in the same viewing window. (An improved fit may result by adjusting your value of C slightly.)

(D) If your graphing calculator has a sinusoidal regression feature, check your results from parts B and C by finding and plotting the regression equation.

Table 3

x (months)	1	2	3	4	5	6	7	8	9	10	11	12
y (temperature)	58	60	61	63	66	70	74	75	74	70	63	58

CHAPTER 6

 ››› GROUP ACTIVITY A Predator–Prey Analysis Involving Mountain Lions and Deer

In some western state wilderness areas, deer and mountain lion populations are interrelated, because the mountain lions rely on the deer as a food source. The population of each species goes up and down in cycles, but out of phase with each other. A wildlife management research team estimated the respective populations in a particular region every 2 years over a 16-year period, with the results shown in Table 1.

(A) Deer Population Analysis

 1. Enter the deer population data for the time interval [0, 16] in a graphing calculator and produce a scatter plot of the data.

 2. Use sinusoidal regression to find a function of the form $y = k + A \sin(Bx + C)$ that models the data, and plot the function and the data.

 3. Write an analysis of the fluctuations and cycles of the deer population.

(B) Mountain Lion Population Analysis

 Repeat 1, 2, and 3 of part (A) for the mountain lion data.

(C) Interrelationship of the Two Populations

 Discuss the dynamics of the two interdependent populations. What causes the two populations to rise and fall, and why are they out of phase with each other?

Table 1 Mountain Lion/Deer Populations

Years	0	2	4	6	8	10	12	14	16
Deer	1,272	1,523	1,152	891	1,284	1,543	1,128	917	1,185
Mountain Lions	39	47	63	54	37	48	60	46	40

Trigonometric Identities and Conditional Equations

TRIGONOMETRIC functions are widely used in solving real-world problems and in the development of mathematics. Whatever their use, it is often of value to be able to change a trigonometric expression from one form to an equivalent more useful form. This involves the use of identities. Recall that an equation in one or more variables is said to be an identity if the left side is equal to the right side for all replacements of the variables for which both sides are defined.

For example, the equation

$$\sin^2 x + \cos^2 x = 1$$

is an identity, but the equation

$$\sin x + \cos x = 1$$

is not. The latter equation is called a conditional equation, because it holds for certain values of x (for example, $x = 0$ and $x = \pi/2$) but not for other values for which both sides are defined (for example, $x = \pi/4$). Sections 1 through 4 of Chapter 7 deal with trigonometric identities, and Section 7-5 with conditional trigonometric equations.

7-1 Basic Identities and Their Use

› Basic Identities
› Establishing Other Identities

In Section 7-1, we will review the basic identities introduced in Section 6-4 and use them to develop and verify new identities.

› Basic Identities

In the following box, we have listed the basic identities that we will use throughout this chapter. Because we will need them so often, you should make sure you're very familiar with them. All of them were established in Section 6-4 with the exception of the second and third Pythagorean identities. Those two will be addressed in both Explore-Discuss 1 and Problems 91 and 92 in the exercises.

› BASIC TRIGONOMETRIC IDENTITIES

Reciprocal identities

$$\csc x = \frac{1}{\sin x} \qquad \sec x = \frac{1}{\cos x} \qquad \cot x = \frac{1}{\tan x}$$

Quotient identities

$$\tan x = \frac{\sin x}{\cos x} \qquad \cot x = \frac{\cos x}{\sin x}$$

Identities for negatives

$$\sin(-x) = -\sin x \qquad \cos(-x) = \cos x \qquad \tan(-x) = -\tan x$$

Pythagorean identities

$$\sin^2 x + \cos^2 x = 1 \qquad \tan^2 x + 1 = \sec^2 x \qquad 1 + \cot^2 x = \csc^2 x$$

››› EXPLORE-DISCUSS 1

Divide both sides of the first Pythagorean identity by $\cos^2 x$. What is the result? Repeat for $\sin^2 x$. (These serve as a handy way to remember the second and third Pythagorean identities.)

› Establishing Other Identities

Identities are used often in the study of trigonometry to convert an expression into an equivalent form that may be more useful for a given situation. To *verify an identity* means to prove that both sides of an equation are equal when any values of the variables for which both sides are defined are substituted into that equation. Verifying identities is usually done using

basic identities or other previously verified identities, along with standard algebraic operations like multiplication, factoring, and combining and reducing fractions. In Examples 1 through 6, we will demonstrate a general procedure for verifying identities. Often, there is more than one approach that will work—the verifications presented here are just one approach. To have any chance of success in this topic, you will need to do many problems on your own.

| EXAMPLE | 1 | **Identity Verification** |

Verify the identity $\cos x \tan x = \sin x$.

VERIFICATION Generally, we will proceed by starting with the more complicated of the two sides, and transform that side into the other side in one or more steps using basic identities, algebra, or other established identities. Here we start with the left-hand side and use a quotient identity to rewrite $\tan x$:

$$\cos x \tan x = \cos x \, \frac{\sin x}{\cos x} \qquad \frac{\cos x}{\cos x} = 1$$

$$= 1 \cdot \sin x = \sin x$$

| MATCHED **PROBLEM 1** | Verify the identity $\sin x \cot x = \cos x$.

Technology Connections

Graph the left and right sides of the identity in Example 1 in a graphing calculator by letting $y_1 = \cos x \tan x$ and $y_2 = \sin x$. Use TRACE, moving back and forth between the graphs of y_1 and y_2, to compare values of y for given values of x. What does this investigation illustrate?

| EXAMPLE | 2 | **Identity Verification** |

Verify the identity $\sec(-x) = \sec x$.

VERIFICATION We start with the left-hand side and use a reciprocal identity:

$$\sec(-x) = \frac{1}{\cos(-x)} \qquad \text{Use an identity for negatives.}$$

$$= \frac{1}{\cos x} \qquad \text{Use a reciprocal identity.}$$

$$= \sec x$$

| MATCHED **PROBLEM 2** | Verify the identity $\csc(-x) = -\csc x$.

EXAMPLE **3** **Identity Verification**

Verify the identity $\cot x \cos x + \sin x = \csc x$.

VERIFICATION We start with the left-hand side and rewrite $\cot x$ in terms of $\sin x$ and $\cos x$:

$$\cot x \cos x + \sin x = \frac{\cos x}{\sin x} \cos x + \sin x \qquad \textcolor{red}{\cos x \cdot \cos x = \cos^2 x}$$

$$= \frac{\cos^2 x}{\sin x} + \sin x \qquad \textcolor{red}{\text{Rewrite } \sin x \text{ as } \frac{\sin^2 x}{\sin x} \text{ and add.}}$$

$$= \frac{\cos^2 x + \sin^2 x}{\sin x} \qquad \textcolor{red}{\text{Use } \sin^2 x + \cos^2 x = 1.}$$

$$= \frac{1}{\sin x} \qquad \textcolor{red}{\text{Use a reciprocal identity.}}$$

$$= \csc x$$

KEY ALGEBRAIC STEPS IN EXAMPLE 3

$$\frac{a}{b} a + b = \frac{a^2}{b} + b = \frac{a^2 + b^2}{b}$$

MATCHED PROBLEM 3 Verify the identity $\tan x \sin x + \cos x = \sec x$.

When verifying identities, *always* start with one side and work only on it, trying to transform it into the other side. *Never* work with both sides at the same time! Although there is no step-by-step procedure that works for all identities, the following steps serve as a general framework.

> **SUGGESTED STEPS IN VERIFYING IDENTITIES**

1. Start with the more complicated side of the identity, and transform it into the simpler side.

2. Try algebraic operations such as multiplying, factoring, combining fractions, and splitting fractions.

3. If other steps fail, express each function in terms of sine and cosine functions, and then perform appropriate algebraic operations.

4. At each step, keep the other side of the identity in mind. This often reveals what you should do to get there.

EXAMPLE **4** **Identity Verification**

Verify the identity $2 \sec x = \dfrac{1 + \sin x}{\cos x} + \dfrac{\cos x}{1 + \sin x}$.

VERIFICATION This time we'll start with the right side because it is obviously more complicated; begin by finding a common denominator and adding the two fractions.

$$\frac{1 + \sin x}{\cos x} + \frac{\cos x}{1 + \sin x} = \frac{(1 + \sin x)^2 + \cos^2 x}{\cos x\,(1 + \sin x)}$$ Multiply out the numerator.

$$= \frac{1 + 2\sin x + \sin^2 x + \cos^2 x}{\cos x\,(1 + \sin x)}$$ Use $\sin^2 x + \cos^2 x = 1$.

$$= \frac{1 + 2\sin x + 1}{\cos x\,(1 + \sin x)}$$ Simplify.

$$= \frac{2 + 2\sin x}{\cos x\,(1 + \sin x)}$$ Factor out 2.

$$= \frac{2(1 + \sin x)}{\cos x\,(1 + \sin x)}$$ $\dfrac{1 + \sin x}{1 + \sin x} = 1$.

$$= \frac{2}{\cos x}$$ Use a reciprocal identity.

$$= 2\sec x$$

KEY ALGEBRAIC STEPS IN EXAMPLE 4

$$\frac{a}{b} + \frac{b}{a} = \frac{a^2 + b^2}{ba} \qquad (1 + c)^2 = 1 + 2c + c^2 \qquad \frac{m(a + b)}{n(a + b)} = \frac{m}{n}$$

MATCHED PROBLEM 4 Verify the identity $2\csc x = \dfrac{1 + \cos x}{\sin x} + \dfrac{\sin x}{1 + \cos x}$.

EXAMPLE 5 **Identity Verification**

Verify the identity $\dfrac{\sin^2 x + 2\sin x + 1}{\cos^2 x} = \dfrac{1 + \sin x}{1 - \sin x}$.

VERIFICATION We start with the left-hand side and factor its numerator:

$$\frac{\sin^2 x + 2\sin x + 1}{\cos^2 x} = \frac{(\sin x + 1)^2}{\cos^2 x}$$ Since $\sin^2 x + \cos^2 x = 1$, $\cos^2 x = 1 - \sin^2 x$.

$$= \frac{(\sin x + 1)^2}{1 - \sin^2 x}$$ Factor the denominator as a difference of squares.

$$= \frac{(1 + \sin x)^2}{(1 - \sin x)(1 + \sin x)}$$ Divide numerator and denominator by $1 + \sin x$.

$$= \frac{1 + \sin x}{1 - \sin x}$$

KEY ALGEBRAIC STEPS IN EXAMPLE 5

$$a^2 + 2a + 1 = (a + 1)^2 \qquad 1 - b^2 = (1 - b)(1 + b)$$

MATCHED PROBLEM 5 Verify the identity $\sec^4 x - 2\sec^2 x \tan^2 x + \tan^4 x = 1$.

EXAMPLE 6 **Identity Verification**

Verify the identity $\dfrac{\tan x - \cot x}{\tan x + \cot x} = 1 - 2 \cos^2 x$.

VERIFICATION We start with the left-hand side and rewrite the tangents and cotangents in terms of sine and cosine:

$$\frac{\tan x - \cot x}{\tan x + \cot x} = \frac{\dfrac{\sin x}{\cos x} - \dfrac{\cos x}{\sin x}}{\dfrac{\sin x}{\cos x} + \dfrac{\cos x}{\sin x}}$$ Multiply numerator and denominator by $(\sin x)(\cos x)$.

$$= \frac{(\sin x)(\cos x)\left(\dfrac{\sin x}{\cos x} - \dfrac{\cos x}{\sin x}\right)}{(\sin x)(\cos x)\left(\dfrac{\sin x}{\cos x} + \dfrac{\cos x}{\sin x}\right)}$$ Distribute sin x cos x through the parentheses.

$$= \frac{\sin^2 x - \cos^2 x}{\sin^2 x + \cos^2 x}$$ Write sin² x in the numerator as $1 - \cos^2 x$; Use sin² x + cos² x = 1 in the denominator.

$$= \frac{1 - \cos^2 x - \cos^2 x}{1}$$ Simplify.

$$= 1 - 2 \cos^2 x$$

KEY ALGEBRAIC STEPS IN EXAMPLE 6

$$\frac{\dfrac{a}{b} - \dfrac{b}{a}}{\dfrac{a}{b} + \dfrac{b}{a}} = \frac{ab\left(\dfrac{a}{b} - \dfrac{b}{a}\right)}{ab\left(\dfrac{a}{b} + \dfrac{b}{a}\right)} = \frac{a^2 - b^2}{a^2 + b^2}$$

MATCHED PROBLEM 6 Verify the identity $\cot x - \tan x = \dfrac{2 \cos^2 x - 1}{\sin x \cos x}$.

Maybe more than any other topic in this book, verifying identities requires a lot of practice. You'll never get good at it by watching other people verify identities—this is *not* a spectator sport!

EXAMPLE 7 **Determining Whether an Equation is an Identity**

Determine whether each equation is an identity. If the equation is an identity, verify it. If the equation is not an identity, find a value of x for which both sides are defined but are not equal.

(A) $\tan x + 1 = (\sec x)(\cos x - \sin x)$

(B) $\tan x - 1 = (\sec x)(\sin x - \cos x)$

SOLUTIONS (A) We select several values of x (for example, $x = 0$, π, $\pi/2$, $\pi/4$, $\pi/6$) and calculate both the left and right sides of the equation

$$\tan x + 1 = (\sec x)(\cos x - \sin x)$$

Let $x = 0$.

Left side: $\tan 0 + 1 = 1$
Right side: $(\sec 0)(\cos 0 - \sin 0) = 1$

Let $x = \pi$.

Left side: $\tan \pi + 1 = 1$
Right side: $(\sec \pi)(\cos \pi - \sin \pi) = 1$

Let $x = \pi/2$.

Left side: $\tan \pi/2 + 1 =$ Undefined
Right side: $(\sec \pi/2)(\cos \pi/2 - \sin \pi/2) =$ Undefined

Let $x = \pi/4$.

Left side: $\tan \pi/4 + 1 = 2$
Right side: $(\sec \pi/4)(\cos \pi/4 - \sin \pi/4) = 0$

We have found a value of x, namely $\pi/4$, for which both sides are defined but are not equal. Therefore, the equation is *not* an identity. No further calculation is needed.

(B) We select several values of x and calculate both the left and right sides of the equation.

$$\tan x - 1 = (\sec x)(\sin x - \cos x)$$

If $x = 0$ or $x = \pi$, both sides equal -1.
If $x = \pi/2$, both sides are undefined.
If $x = \pi/4$, both sides equal 0.

If $x = \pi/6$, both sides equal $\dfrac{1}{\sqrt{3}} - 1$.

These calculations suggest that the equation is probably an identity, which we will now try to verify. We start with the right-hand side and use a quotient identity to rewrite $\sec x$:

$$(\sec x)(\sin x - \cos x) = \left(\frac{1}{\cos x}\right)(\sin x - \cos x) \qquad \text{Distribute.}$$

$$= \left(\frac{\sin x}{\cos x} - \frac{\cos x}{\cos x}\right) \qquad \frac{\sin x}{\cos x} = \tan x;\ \frac{\cos x}{\cos x} = 1$$

$$= \tan x - 1$$

MATCHED PROBLEM 7

Determine whether each equation is an identity. If the equation is an identity, verify it. If the equation is not an identity, find a value of x for which both sides are defined but are not equal.

(A) $\dfrac{\sin x}{1 - \cos^2 x} = \csc x$ (B) $\dfrac{\sin x}{1 - \cos^2 x} = \sec x$

Technology Connections

Using a graphing calculator, we can eliminate the calculations in Example 7 by comparing the graphs of each side of the given equation. Figure 1 shows the graphs of each side of the equation of Example 7(A). The equation is not an identity because the graphs do not coincide (note that when $x = \pi/4$, Y_1 has the value 2 but Y_2 has the value 0). Figure 2 shows the graphs of each side of the equation of Example 7(B). The equation appears to be an identity because the graphs coincide; to show that it is indeed an identity, it must still be verified as in the solution to Example 7(B).

> Figure 1

> Figure 2

>>> CAUTION >>>

You can *never* verify an identity by substituting in some numbers. Finding one number for which the equation is not true is enough to show that an equation is *not* an identity, but verifying that an equation *is* an identity requires the techniques of this section.

ANSWERS TO MATCHED PROBLEMS

In the following identity verifications, other correct sequences of steps are possible—the process is not unique.

1. $\sin x \cot x = \sin x \dfrac{\cos x}{\sin x} = \cos x$ 2. $\csc(-x) = \dfrac{1}{\sin(-x)} = \dfrac{1}{-\sin x} = -\csc x$

3. $\tan x \sin x + \cos x = \dfrac{\sin^2 x}{\cos x} + \cos x = \dfrac{\sin^2 x + \cos^2 x}{\cos x} = \dfrac{1}{\cos x} = \sec x$

4. $\dfrac{1 + \cos x}{\sin x} + \dfrac{\sin x}{1 + \cos x} = \dfrac{(1 + \cos x)^2 + \sin^2 x}{\sin x (1 + \cos x)} = \dfrac{1 + 2\cos x + \cos^2 x + \sin^2 x}{\sin x (1 + \cos x)}$

$$= \dfrac{2(1 + \cos x)}{\sin x (1 + \cos x)} = 2\csc x$$

5. $\sec^4 x - 2\sec^2 x \tan^2 x + \tan^4 x = (\sec^2 x - \tan^2 x)^2 = 1^2 = 1$

6. $\cot x - \tan x = \dfrac{\cos x}{\sin x} - \dfrac{\sin x}{\cos x} = \dfrac{\cos^2 x - \sin^2 x}{\sin x \cos x} = \dfrac{\cos^2 x - (1 - \cos^2 x)}{\sin x \cos x} = \dfrac{2\cos^2 x - 1}{\sin x \cos x}$

7. (A) An identity: $\dfrac{\sin x}{1 - \cos^2 x} = \dfrac{\sin x}{\sin^2 x} = \dfrac{1}{\sin x} = \csc x$

(B) Not an identity: the left side is not equal to the right side for $x = \pi/6$, for example.

7-1 Exercises

1. What does it mean to verify an identity?

2. If you are unsure of whether a given equation is or is not an identity, describe what you would do to try and decide.

3. How do you prove that an equation is *not* an identity?

4. Explain why a graphing calculator can be used to prove that an equation is not an identity, but can never be used to prove that an equation is an identity.

Verify that the equations in Problems 5–30 are identities.

5. $\sin \theta \sec \theta = \tan \theta$

6. $\cos \theta \csc \theta = \cot \theta$

7. $\cot u \sec u \sin u = 1$

8. $\tan \theta \csc \theta \cos \theta = 1$

9. $\dfrac{\sin(-x)}{\cos(-x)} = -\tan x$

10. $\cot(-x) \tan x = -1$

11. $\sin \alpha = \dfrac{\tan \alpha \cot \alpha}{\csc \alpha}$

12. $\tan \alpha = \dfrac{\cos \alpha \sec \alpha}{\cot \alpha}$

13. $\cot u + 1 = (\csc u)(\cos u + \sin u)$

14. $\tan u + 1 = (\sec u)(\sin u + \cos u)$

15. $\dfrac{\cos x - \sin x}{\sin x \cos x} = \csc x - \sec x$

16. $\dfrac{\cos^2 x - \sin^2 x}{\sin x \cos x} = \cot x - \tan x$

17. $\dfrac{\sin^2 t}{\cos t} + \cos t = \sec t$

18. $\dfrac{\cos^2 t}{\sin t} + \sin t = \csc t$

19. $\dfrac{\cos x}{1 - \sin^2 x} = \sec x$

20. $\dfrac{\sin u}{1 - \cos^2 u} = \csc u$

21. $(1 - \cos u)(1 + \cos u) = \sin^2 u$

22. $(1 - \sin t)(1 + \sin t) = \cos^2 t$

23. $\cos^2 x - \sin^2 x = 1 - 2 \sin^2 x$

24. $(\sin x + \cos x)^2 = 1 + 2 \sin x \cos x$

25. $(\sec t + 1)(\sec t - 1) = \tan^2 t$

26. $(\csc t - 1)(\csc t + 1) = \cot^2 t$

27. $\csc^2 x - \cot^2 x = 1$

28. $\sec^2 u - \tan^2 u = 1$

29. $\cot x + \sec x = \dfrac{\cos x + \tan x}{\sin x}$

30. $\sin m (\csc m - \sin m) = \cos^2 m$

In Problems 31–38, prove that the equation is not an identity by finding a value of x for which both sides are defined but are not equal.

31. $\sqrt{(x + 3)^2} = x + 3$

32. $\sqrt{x^2 + 10x + 25} = x - 5$

33. $x^2 + x = 3x^3 - x^2$

34. $|x - 4| = \sqrt{x^2 - 16}$

35. $\sin x = 1 - \cos x$

36. $\sin^2 x - \cos^2 x = -1$

37. $\sec x \tan x = \cot x$

38. $\cos(-x) = -\cos x$

In Problems 39–44, graph the two functions in the same viewing window on a graphing calculator on the interval $[-\pi, \pi]$. If the two expressions are set equal to each other, does the result appear to be an identity? Explain.

39. (A) $y = 1 + \sin x$ (B) $y = \cos x$

40. (A) $y = 3 - 3 \cos^2 x$ (B) $y = 3 \sin^2 x$

41. (A) $y = \cos x \tan x$ (B) $y = \sin x$

42. (A) $y = \sec x \cot x$ (B) $y = \sin x$

43. (A) $y = \sin x - \csc x$ (B) $y = -\cot x \csc x$

44. (A) $y = \cos x - \sec x$ (B) $y = -\sin x \tan x$

In Problems 45–52, is the equation an identity? Explain.

45. $\dfrac{x^2 - 9}{x + 3} = x - 3$

46. $\dfrac{5x}{|x|} = 5$

47. $\sqrt{x^2 + 4x + 4} = x + 2$

48. $\dfrac{1}{\sqrt{x}} = \dfrac{\sqrt{x}}{|x|}$

49. $\sin x - \cos x = 1$

50. $\sin x + \cos x = 1$

51. $\cos^3 x = \cos x - \cos x \sin^2 x$

52. $\pi - \pi \cos^2 x = \pi \sin^2 x$

Verify that the equations in Problems 53–74 are identities.

53. $\dfrac{1 - (\sin x - \cos x)^2}{\sin x} = 2 \cos x$

54. $\dfrac{1 - \cos^2 y}{(1 - \sin y)(1 + \sin y)} = \tan^2 y$

55. $\cos \theta + \sin \theta = \dfrac{\cot \theta + 1}{\csc \theta}$

56. $\sin \theta + \cos \theta = \dfrac{\tan \theta + 1}{\sec \theta}$

57. $\dfrac{1 + \cos y}{1 - \cos y} = \dfrac{\sin^2 y}{(1 - \cos y)^2}$

58. $1 - \sin y = \dfrac{\cos^2 y}{1 + \sin y}$

59. $\tan^2 x - \sin^2 x = \tan^2 x \sin^2 x$

60. $\sec^2 x + \csc^2 x = \sec^2 x \csc^2 x$

61. $\dfrac{\csc \theta}{\cot \theta + \tan \theta} = \cos \theta$

62. $\dfrac{1 + \sec \theta}{\sin \theta + \tan \theta} = \csc \theta$

63. $\ln(\tan x) = \ln(\sin x) - \ln(\cos x)$

64. $\ln(\cot x) = \ln(\cos x) - \ln(\sin x)$

65. $\dfrac{1 - \cos A}{1 + \cos A} = \dfrac{\sec A - 1}{\sec A + 1}$

66. $\dfrac{1 - \csc y}{1 + \csc y} = \dfrac{\sin y - 1}{\sin y + 1}$

67. $\sin^4 w - \cos^4 w = 1 - 2 \cos^2 w$

68. $\sin^4 x + 2 \sin^2 x \cos^2 x + \cos^4 x = 1$

69. $\sec x - \dfrac{\cos x}{1 + \sin x} = \tan x$

70. $\csc n - \dfrac{\sin n}{1 + \cos n} = \cot n$

71. $\dfrac{\cos^2 z - 3 \cos z + 2}{\sin^2 z} = \dfrac{2 - \cos z}{1 + \cos z}$

72. $\dfrac{\sin^2 t + 4 \sin t + 3}{\cos^2 t} = \dfrac{3 + \sin t}{1 - \sin t}$

73. $\dfrac{\cos^3 \theta - \sin^3 \theta}{\cos \theta - \sin \theta} = 1 + \sin \theta \cos \theta$

74. $\dfrac{\cos^3 u + \sin^3 u}{\cos u + \sin u} = 1 - \sin u \cos u$

In Problems 75–84, use a graphing calculator to test whether each of the following is an identity. If an equation appears to be an identity, verify it. If the equation does not appear to be an identity, find a value of x for which both sides are defined but are not equal.

75. $\dfrac{\sin (-x)}{\cos (-x) \tan (-x)} = -1$

76. $\dfrac{\cos (-x)}{\sin x \cot (-x)} = 1$

77. $\dfrac{\sin x}{\cos x \tan (-x)} = -1$

78. $\dfrac{\cos x}{\sin (-x) \cot (-x)} = 1$

79. $\sin x + \dfrac{\cos^2 x}{\sin x} = \sec x$

80. $\dfrac{1 - \tan^2 x}{1 - \cot^2 x} = \tan^2 x$

81. $\sin x + \dfrac{\cos^2 x}{\sin x} = \csc x$

82. $\dfrac{\tan^2 x - 1}{1 - \cot^2 x} = \tan^2 x$

83. $\dfrac{\tan x}{\sin x - 2 \tan x} = \dfrac{1}{\cos x - 2}$

84. $\dfrac{\cos x}{1 - \sin x} + \dfrac{\cos x}{1 + \sin x} = 2 \sec x$

Verify that the equations in Problems 85–90 are identities.

85. $\dfrac{2 \sin^2 x + 3 \cos x - 3}{\sin^2 x} = \dfrac{2 \cos x - 1}{1 + \cos x}$

86. $\dfrac{3 \cos^2 z + 5 \sin z - 5}{\cos^2 z} = \dfrac{3 \sin z - 2}{1 + \sin z}$

87. $\dfrac{\tan u + \sin u}{\tan u - \sin u} - \dfrac{\sec u + 1}{\sec u - 1} = 0$

88. $\dfrac{\sin x \cos y + \cos x \sin y}{\cos x \cos y - \sin x \sin y} = \dfrac{\tan x + \tan y}{1 - \tan x \tan y}$

89. $\tan \alpha + \cot \beta = \dfrac{\tan \beta + \cot \alpha}{\tan \beta \cot \alpha}$

90. $\dfrac{\cot \alpha + \cot \beta}{\cot \alpha \cot \beta - 1} = \dfrac{\tan \alpha + \tan \beta}{1 - \tan \alpha \tan \beta}$

In Problems 91 and 92, fill in the blanks citing the appropriate basic trigonometric identity.

91.

Statement	Reason
$\cot^2 x + 1 = \left(\dfrac{\cos x}{\sin x}\right)^2 + 1$	(A) _____
$= \dfrac{\cos^2 x}{\sin^2 x} + 1$	Algebra
$= \dfrac{\cos^2 x + \sin^2 x}{\sin^2 x}$	Algebra
$= \dfrac{1}{\sin^2 x}$	(B) _____
$= \left(\dfrac{1}{\sin x}\right)^2$	Algebra
$= \csc^2 x$	(C) _____

92.

Statement	Reason
$\tan^2 x + 1 = \left(\dfrac{\sin x}{\cos x}\right)^2 + 1$	(A) _____
$= \dfrac{\sin^2 x}{\cos^2 x} + 1$	Algebra
$= \dfrac{\sin^2 x + \cos^2 x}{\cos^2 x}$	Algebra
$= \dfrac{1}{\cos^2 x}$	(B) _____
$= \left(\dfrac{1}{\cos x}\right)^2$	Algebra
$= \sec^2 x$	(C) _____

Each of the equations in Problems 93–98 is an identity in certain quadrants associated with x. Indicate which quadrants.

93. $\sqrt{1 - \cos^2 x} = -\sin x$ **94.** $\sqrt{1 - \sin^2 x} = \cos x$

95. $\sqrt{1 - \sin^2 x} = |\cos x|$ **96.** $\sqrt{1 - \cos^2 x} = |\sin x|$

97. $\dfrac{\sin x}{\sqrt{1 - \sin^2 x}} = \tan x$ **98.** $\dfrac{\sin x}{\sqrt{1 - \sin^2 x}} = -\tan x$

In calculus, trigonometric substitutions provide an effective way to rationalize the radical forms $\sqrt{a^2 - u^2}$ and $\sqrt{a^2 + u^2}$, which in turn leads to the solution to an important class of problems. Problems 99–102 involve such transformations. [Recall: $\sqrt{x^2} = |x|$ for all real numbers x.]

99. In the radical form $\sqrt{a^2 - u^2}$, $a > 0$, let $u = a \sin x$, $-\pi/2 < x < \pi/2$. Simplify, using a basic identity, and write the final form free of radicals.

100. In the radical form $\sqrt{a^2 - u^2}$, $a > 0$, let $u = a \cos x$, $0 < x < \pi$. Simplify, using a basic identity, and write the final form free of radicals.

101. In the radical form $\sqrt{a^2 + u^2}$, $a > 0$, let $u = a \tan x$, $0 < x < \pi/2$. Simplify, using a basic identity, and write the final form free of radicals.

102. In the radical form $\sqrt{a^2 + u^2}$, $a > 0$, let $u = a \cot x$, $0 < x < \pi/2$. Simplify, using a basic identity, and write the final form free of radicals.

7-2 Sum, Difference, and Cofunction Identities

> ❭ **Sum and Difference Identities for Cosine**
> ❭ **Cofunction Identities**
> ❭ **Sum and Difference Identities for Sine and Tangent**
> ❭ **Summary and Use**

The basic identities discussed in Section 7-1 involved only one variable. In this section, we will consider identities that involve two variables.

❭ Sum and Difference Identities for Cosine

Without giving it too much thought, you might guess that $\cos(x - y) = \cos x - \cos y$. Unfortunately, you would be wrong. *It is extremely important in this section to keep in mind that the trigonometric functions cannot be "distributed" through parentheses.* Instead, we will use identities like the **difference identity for cosine:**

$$\cos(x - y) = \cos x \cos y + \sin x \sin y \tag{1}$$

This identity can be used to find many other useful identities. We will verify that equation (1) is an identity for the case where x and y are both in the interval $(0, 2\pi)$ and x is greater than y. We could then use basic identities and the fact that sine has period 2π to show that equation (1) is true for all real numbers.

First, we will label x and y as arcs on the unit circle [Fig. 1(a)]. Using the definitions of sine and cosine, the terminal points of x and y are $(\cos x, \sin x)$ and $(\cos y, \sin y)$, respectively. To simplify the notation, we will write $a = \cos x$, $b = \sin x$, $c = \cos y$, $d = \sin y$, and so on, as indicated.

❭ **Figure 1** Obtaining the difference identity for cosine.

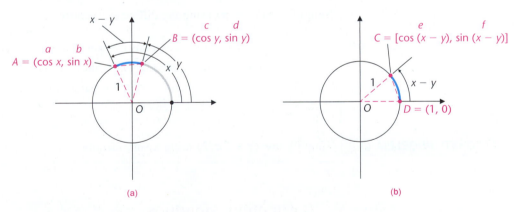

(a) (b)

Now if you rotate the triangle AOB clockwise about the origin until the terminal point B coincides with $D = (1, 0)$, then terminal point A will be at the point C, as shown in Figure 1(b). Because rotation preserves lengths,

$$d(A, B) = d(C, D) \qquad \text{Use distance formula.}$$
$$\sqrt{(c - a)^2 + (d - b)^2} = \sqrt{(1 - e)^2 + (0 - f)^2} \qquad \text{Square both sides.}$$
$$(c - a)^2 + (d - b)^2 = (1 - e)^2 + f^2 \qquad \text{Expand.}$$
$$c^2 - 2ac + a^2 + d^2 - 2db + b^2 = 1 - 2e + e^2 + f^2 \qquad \text{Rearrange.}$$
$$(c^2 + d^2) + (a^2 + b^2) - 2ac - 2db = 1 - 2e + (e^2 + f^2) \tag{2}$$

Because points A, B, and C are on unit circles, $c^2 + d^2 = 1$, $a^2 + b^2 = 1$, and $e^2 + f^2 = 1$. We make these substitutions in equation (2) and simplify:

$$1 + 1 - 2ac - 2db = 1 - 2e + 1 \qquad \text{Subtract 2 from both sides.}$$
$$-2ac - 2db = -2e \qquad \text{Multiply both sides by } -\tfrac{1}{2}.$$
$$ac + db = e$$

Therefore, switching the sides of the equation,

$$e = ac + bd \qquad (3)$$

Replacing e, a, c, b, and d with $\cos (x - y)$, $\cos x$, $\cos y$, $\sin x$, and $\sin y$, respectively (see Fig. 1), we get

$$\cos (x - y) = \cos x \cos y + \sin y \sin x$$
$$= \cos x \cos y + \sin x \sin y \qquad (4)$$

and have established the difference identity for cosine.

If we replace y with $-y$ in equation (4) and use the identities for negatives (a good exercise for you), we get

$$\cos (x + y) = \cos x \cos y - \sin x \sin y \qquad (5)$$

This is the **sum identity for cosine.**

>>> **EXPLORE-DISCUSS 1**

(A) Verify the difference identity for cosine and the sum identity for cosine if $x = \pi/2$ and $y = \pi/3$.

(B) Discuss how you would show that the equations $\cos (x - y) = \cos x - \cos y$ and $\cos (x + y) = \cos x + \cos y$ are *not* identities.

EXAMPLE 1 Using the Difference Identity

Simplify $\cos (x - \pi)$ using the difference identity.

SOLUTION
$$\cos (x - y) = \cos x \cos y + \sin x \sin y \qquad \text{Substitute } y = \pi$$
$$\cos (x - \pi) = \cos x \cos \pi + \sin x \sin \pi \qquad \cos \pi = -1,\ \sin \pi = 0$$
$$= (\cos x)(-1) + (\sin x)(0) \qquad \text{Simplify.}$$
$$= -\cos x$$

MATCHED PROBLEM 1 Simplify $\cos (x + 3\pi/2)$ using a sum identity.

› Cofunction Identities

To obtain sum and difference identities for the sine and tangent functions, we will first derive *cofunction identities* directly from the difference identity for cosine:

$$\cos (x - y) = \cos x \cos y + \sin x \sin y \qquad \text{Substitute } x = \frac{\pi}{2}.$$
$$\cos \left(\frac{\pi}{2} - y \right) = \cos \frac{\pi}{2} \cos y + \sin \frac{\pi}{2} \sin y \qquad \cos \frac{\pi}{2} = 0,\ \sin \frac{\pi}{2} = 1.$$
$$= (0)(\cos y) + (1)(\sin y) \qquad \text{Simplify.}$$
$$= \sin y$$

This is called the **cofunction identity for cosine:**

$$\cos\left(\frac{\pi}{2} - y\right) = \sin y \tag{6}$$

for y any real number or angle in radian measure. If y is in degree measure, replace $\pi/2$ with $90°$.

Now, if we replace y with $\pi/2 - x$ in equation (6), we get

$$\cos\left[\frac{\pi}{2} - \left(\frac{\pi}{2} - x\right)\right] = \sin\left(\frac{\pi}{2} - x\right) \qquad \text{Simplify left-hand side.}$$

$$\cos x = \sin\left(\frac{\pi}{2} - x\right)$$

This is the **cofunction identity for sine:**

$$\sin\left(\frac{\pi}{2} - x\right) = \cos x \tag{7}$$

where x is any real number or angle in radian measure. If x is in degree measure, replace $\pi/2$ with $90°$.

Next, we will state the **cofunction identity for tangent** (and leave its derivation to Problem 26 in Exercises 7-2):

$$\tan\left(\frac{\pi}{2} - x\right) = \cot x \tag{8}$$

for x any real number or angle in radian measure. If x is in degree measure, replace $\pi/2$ with $90°$.

> ## Sum and Difference Identities for Sine and Tangent

To derive a difference identity for sine, we first substitute $x - y$ for y in equation (6):

$$\sin(x - y) = \cos\left[\frac{\pi}{2} - (x - y)\right] \qquad \text{Use algebra.}$$

$$= \cos\left[\left(\frac{\pi}{2} - x\right) - (-y)\right] \qquad \text{Use equation (1).}$$

$$= \cos\left(\frac{\pi}{2} - x\right)\cos(-y) + \sin\left(\frac{\pi}{2} - x\right)\sin(-y) \qquad \begin{array}{l}\text{Use equations (6) and (7)}\\ \text{and identities for negatives.}\end{array}$$

$$= \sin x \cos y - \cos x \sin y$$

The same result is obtained by replacing $\pi/2$ with $90°$. So

$$\sin(x - y) = \sin x \cos y - \cos x \sin y \tag{9}$$

is the **difference identity for sine.**

Now, if we replace y in equation (9) with $-y$ (see Problem 31 in Exercises 7-2), we obtain

$$\sin(x + y) = \sin x \cos y + \cos x \sin y \tag{10}$$

which is the **sum identity for sine.**

In Problems 32 and 89–94 of Exercises 7-2, we will develop the **sum** and **difference** **identities for tangent:**

$$\tan(x - y) = \frac{\tan x - \tan y}{1 + \tan x \tan y} \tag{11}$$

for all angles or real numbers x and y for which both sides are defined. This is the **difference identity for tangent.**

If we replace y in equation (11) with $-y$ (another good exercise for you), we obtain

$$\tan (x + y) = \frac{\tan x + \tan y}{1 - \tan x \tan y} \tag{12}$$

the **sum identity for tangent.**

>>> **EXPLORE-DISCUSS 2**

(A) Verify the difference identity for tangent and the sum identity for tangent if $x = 5\pi/6$ and $y = \pi/6$.

(B) Discuss how you would show that the equations $\tan (x - y) = \tan x - \tan y$ and $\tan (x + y) = \tan x + \tan y$ are *not* identities.

› Summary and Use

Before proceeding with more examples illustrating the use of these new identities, review the list given in the box.

› **SUMMARY OF IDENTITIES**

Sum identities

$$\sin (x + y) = \sin x \cos y + \cos x \sin y$$
$$\cos (x + y) = \cos x \cos y - \sin x \sin y$$
$$\tan (x + y) = \frac{\tan x + \tan y}{1 - \tan x \tan y}$$

Difference identities

$$\sin (x - y) = \sin x \cos y - \cos x \sin y$$
$$\cos (x - y) = \cos x \cos y + \sin x \sin y$$
$$\tan (x - y) = \frac{\tan x - \tan y}{1 + \tan x \tan y}$$

Cofunction identities

(Replace $\pi/2$ with $90°$ if x is in degrees.)

$$\cos \left(\frac{\pi}{2} - x \right) = \sin x \qquad \sin \left(\frac{\pi}{2} - x \right) = \cos x \qquad \tan \left(\frac{\pi}{2} - x \right) = \cot x$$

Exact values for functions are *always* better than calculator approximations. One of the most useful aspects of the sum and difference identities is that they allow us to greatly expand the number of exact values we can find for the trigonometric functions. Example 2 illustrates one such case.

| **EXAMPLE** | **2** | **Finding Exact Values** |

Find the exact value of tan 75° in radical form.

SOLUTION Because we can write $75° = 45° + 30°$, the sum of two special angles, we can use the sum identity for tangent with $x = 45°$ and $y = 30°$:

$$\tan(x + y) = \frac{\tan x + \tan y}{1 - \tan x \tan y} \qquad \text{Substitute } x = 45°, y = 30°.$$

$$\tan(45° + 30°) = \frac{\tan 45° + \tan 30°}{1 - \tan 45° \tan 30°} \qquad \tan 45° = 1; \tan 30° = 1/\sqrt{3}.$$

$$= \frac{1 + (1/\sqrt{3})}{1 - 1(1/\sqrt{3})} \qquad \text{Multiply numerator and denominator by } \sqrt{3} \text{ and simplify.}$$

$$= \frac{\sqrt{3} + 1}{\sqrt{3} - 1} \qquad \text{Rationalize denominator and simplify.}$$

$$= 2 + \sqrt{3}$$

| **MATCHED PROBLEM 2** | Find the exact value of sin 15° in radical form. |

| **EXAMPLE** | **3** | **Finding Exact Values** |

Find the exact value of $\cos(x + y)$, given $\sin x = \frac{3}{5}$, $\cos y = \frac{4}{5}$, x is an angle in Quadrant II, and y is an angle in Quadrant I. Do not use a calculator.

SOLUTION We start with the sum identity for cosine,

$$\cos(x + y) = \cos x \cos y - \sin x \sin y$$

We know $\sin x$ and $\cos y$, but not $\cos x$ and $\sin y$. We can find the latter two using two different methods as follows (use the method that is easiest for you).

Given $\sin x = \frac{3}{5}$ and x is an angle in Quadrant II, find $\cos x$:

METHOD I. USE A REFERENCE TRIANGLE: **METHOD II. USE A UNIT CIRCLE:**

$$\cos x = \frac{a}{5} \qquad\qquad\qquad \cos x = a$$

$$a^2 + 3^2 = 5^2 \qquad\qquad a^2 + \left(\tfrac{3}{5}\right)^2 = 1$$

$$a^2 = 16 \qquad\qquad\qquad a^2 = \tfrac{16}{25}$$

$$a = \pm 4 \qquad\qquad\qquad a = \pm\tfrac{4}{5}$$

In Quadrant II, $a = -4$ In Quadrant II, $a = -\tfrac{4}{5}$

Therefore, $\cos x = -\tfrac{4}{5}$ Therefore, $\cos x = -\tfrac{4}{5}$

Given $\cos y = \frac{4}{5}$ and y is an angle in Quadrant I, find $\sin y$:

METHOD I. USE A REFERENCE TRIANGLE: **METHOD II. USE A UNIT CIRCLE:**

$$\sin y = \frac{b}{5}$$

$$4^2 + b^2 = 5^2$$

$$b^2 = 9$$

$$b = \pm 3$$

In Quadrant I, $\quad b = 3$

Therefore, $\quad \sin y = \frac{3}{5}$

$$\sin y = b$$

$$\left(\tfrac{4}{5}\right)^2 + b^2 = 1$$

$$b^2 = \tfrac{9}{25}$$

$$b = \pm \tfrac{3}{5}$$

In Quadrant I, $\quad b = \frac{3}{5}$

Therefore, $\quad \sin y = \frac{3}{5}$

We can now evaluate $\cos (x + y)$ without knowing x and y:

$$\cos (x + y) = \cos x \cos y - \sin x \sin y$$
$$= \left(-\tfrac{4}{5}\right)\left(\tfrac{4}{5}\right) - \left(\tfrac{3}{5}\right)\left(\tfrac{3}{5}\right) = -\tfrac{25}{25} = -1$$

MATCHED PROBLEM 3 | Find the exact value of $\sin (x - y)$, given $\sin x = -\frac{2}{3}$, $\cos y = \sqrt{5}/3$, x is an angle in Quadrant III, and y is an angle in Quadrant IV. Do not use a calculator.

EXAMPLE | **4** | **Identity Verification**

Verify the identity $\tan x + \cot y = \dfrac{\cos (x - y)}{\cos x \sin y}$.

VERIFICATION

We will start with the right-hand side and use the difference identity for cosine:

$$\frac{\cos (x - y)}{\cos x \sin y} = \frac{\cos x \cos y + \sin x \sin y}{\cos x \sin y} \qquad \textcolor{magenta}{\text{Write as sum of two fractions.}}$$

$$= \frac{\cos x \cos y}{\cos x \sin y} + \frac{\sin x \sin y}{\cos x \sin y} \qquad \textcolor{magenta}{\text{Simplify and use quotient identities.}}$$

$$= \cot y + \tan x \qquad \textcolor{magenta}{\text{Switch the order of terms.}}$$

$$= \tan x + \cot y$$

MATCHED PROBLEM 4 | Verify the identity $\cot y - \cot x = \dfrac{\sin (x - y)}{\sin x \sin y}$.

7-2 Exercises

1. What is a cofunction?

2. Explain how the cofunction identity $\cos(\pi/2 - x) = \sin x$ can be obtained from a difference identity.

3. Explain how each of the sum identities can be obtained from the corresponding difference identity.

4. In the sum identities, does it make a difference if x and y are given in degrees rather than in radians? Explain.

5. In the confunction identities, does it make a difference if x and y are given in degrees rather than in radians? Explain.

6. Explain why you can't use the sum identity for tangent to obtain an identity with left side $\tan(\pi/2 + x)$. How could you obtain such an identity?

In Problems 7–16, show that the equation is not an identity by finding a value of x and a value of y for which both sides are defined but are not equal.

7. $(x + y)^2 = x^2 + y^2$

8. $(x - y)^3 = x^3 - y^3$

9. $x \sin y = \sin xy$

10. $x \tan y = \tan xy$

11. $\cos(x + y) = \cos x + \cos y$

12. $\tan(x + y) = \tan x + \tan y$

13. $\tan(x - y) = \tan x - \tan y$

14. $\sin(x - y) = \sin x - \sin y$

15. $\cos(x - y) = \cos x - \cos y$

16. $\sin(x + y) = \sin x + \sin y$

In Problems 17–24, is the equation an identity? Explain, making use of the sum or difference identities.

17. $\tan(x - \pi) = \tan x$

18. $\cos(x + \pi) = \cos x$

19. $\sin(x - \pi) = \sin x$

20. $\cot(x + \pi) = \cot x$

21. $\csc(2\pi - x) = \csc x$

22. $\sec(2\pi - x) = \sec x$

23. $\sin(x - \pi/2) = -\cos x$

24. $\cos(x - \pi/2) = -\sin x$

Verify each identity in Problems 25–30 using cofunction identities for sine and cosine and basic identities discussed in Section 7-1.

25. $\cot\left(\dfrac{\pi}{2} - x\right) = \tan x$

26. $\tan\left(\dfrac{\pi}{2} - x\right) = \cot x$

27. $\csc\left(\dfrac{\pi}{2} - x\right) = \sec x$

28. $\sec\left(\dfrac{\pi}{2} - x\right) = \csc x$

29. $\cos(\pi - x) = -\cos x$

30. $\sin(\pi - x) = \sin x$

31. Replace y with $(-y)$ in the subtraction formula for sine to derive the addition formula for sine.

32. Replace y with $(-y)$ in the subtraction formula for tangent to derive the addition formula for tangent.

Convert Problems 33–38 to forms involving $\sin x$, $\cos x$, and/or $\tan x$ using sum or difference identities.

33. $\sin(30° - x)$

34. $\sin(x - 45°)$

35. $\sin(180° - x)$

36. $\cos(x + 180°)$

37. $\tan\left(x + \dfrac{\pi}{3}\right)$

38. $\tan\left(\dfrac{\pi}{4} - x\right)$

Use appropriate identities to find exact values for Problems 39–46. Do not use a calculator.

39. $\sec 75°$

40. $\sin 75°$

41. $\sin\dfrac{7\pi}{12}$ $\left[Hint: \dfrac{7\pi}{12} = \dfrac{\pi}{3} + \dfrac{\pi}{4}\right]$

42. $\cos\dfrac{\pi}{12}$ $\left[Hint: \dfrac{\pi}{12} = \dfrac{\pi}{4} - \dfrac{\pi}{6}\right]$

43. $\cos 74° \cos 44° + \sin 74° \sin 44°$

44. $\sin 22° \cos 38° + \cos 22° \sin 38°$

45. $\dfrac{\tan 27° + \tan 18°}{1 - \tan 27° \tan 18°}$ **46.** $\dfrac{\tan 110° - \tan 50°}{1 + \tan 110° \tan 50°}$

Find sin (x − y) and tan (x + y) exactly without a calculator using the information given in Problems 47–50.

47. $\sin x = -\frac{3}{5}$, $\sin y = \sqrt{8}/3$, x is a Quadrant IV angle, y is a Quadrant I angle.

48. $\sin x = \frac{2}{3}$, $\cos y = -\frac{1}{4}$, x is a Quadrant II angle, y is a Quadrant III angle.

49. $\tan x = \frac{3}{4}$, $\tan y = -\frac{1}{2}$, x is a Quadrant III angle, y is a Quadrant IV angle.

50. $\cos x = -\frac{1}{3}$, $\tan y = \frac{1}{2}$, x is a Quadrant II angle, y is a Quadrant III angle.

Verify each identity in Problems 51–64.

51. $\cos 2x = \cos^2 x - \sin^2 x$

52. $\sin 2x = 2 \sin x \cos x$

53. $\cot (x + y) = \dfrac{\cot x \cot y - 1}{\cot x + \cot y}$

54. $\cot (x - y) = \dfrac{\cot x \cot y + 1}{\cot y - \cot x}$

55. $\tan 2x = \dfrac{2 \tan x}{1 - \tan^2 x}$

56. $\cot 2x = \dfrac{\cot^2 x - 1}{2 \cot x}$

57. $\dfrac{\sin (v + u)}{\sin (v - u)} = \dfrac{\cot u + \cot v}{\cot u - \cot v}$

58. $\dfrac{\sin (u + v)}{\sin (u - v)} = \dfrac{\tan u + \tan v}{\tan u - \tan v}$

59. $\cot x - \tan y = \dfrac{\cos (x + y)}{\sin x \cos y}$

60. $\tan x - \tan y = \dfrac{\sin (x - y)}{\cos x \cos y}$

61. $\tan (x - y) = \dfrac{\cot y - \cot x}{\cot x \cot y + 1}$

62. $\tan (x + y) = \dfrac{\cot x + \cot y}{\cot x \cot y - 1}$

63. $\dfrac{\cos (x + h) - \cos x}{h} = \cos x \left(\dfrac{\cos h - 1}{h}\right) - \sin x \left(\dfrac{\sin h}{h}\right)$

64. $\dfrac{\sin (x + h) - \sin x}{h} = \sin x \left(\dfrac{\cos h - 1}{h}\right) + \cos x \left(\dfrac{\sin h}{h}\right)$

In problems 65–68, evaluate both sides of the sum identities for cosine and sine for the given values of x and y. Evaluate all functions exactly.

65. $x = \dfrac{\pi}{4}, y = \dfrac{3\pi}{4}$ **66.** $x = \dfrac{\pi}{3}, y = \dfrac{4\pi}{3}$

67. $x = \dfrac{11\pi}{6}, y = -\dfrac{5\pi}{6}$ **68.** $x = \dfrac{5\pi}{4}, y = -\dfrac{3\pi}{4}$

Evaluate both sides of the difference identity for sine and the sum identity for tangent for the values of x and y indicated in Problems 69–74. Evaluate to four significant digits using a calculator.

69. $x = 5.288, y = 1.769$ **70.** $x = 3.042, y = 2.384$

71. $x = 42.08°, y = 68.37°$ **72.** $x = 128.3°, y = 25.62°$

73. Show that $\sec (x - y) = \sec x - \sec y$ is *not* an identity.

74. Show that $\csc (x + y) = \csc x + \csc y$ is *not* an identity.

In Problems 75–86, use sum or difference identities to convert each equation to a form involving sin x, cos x, and/or tan x. Enter the original equation in a graphing calculator as y_1 and the converted form as y_2, then graph y_1 and y_2 in the same viewing window. Use TRACE to compare the two graphs.

75. $y = \sin (x + \pi/6)$ **76.** $y = \sin (x - \pi/3)$

77. $y = \cos (x - 3\pi/4)$ **78.** $y = \cos (x + 5\pi/6)$

79. $y = \tan (x + 2\pi/3)$ **80.** $y = \tan (x - \pi/4)$

In Problems 81–86, evaluate exactly as real numbers without the use of a calculator.

81. $\sin \left[\cos^{-1} \left(-\frac{4}{5}\right) + \sin^{-1} \left(-\frac{3}{5}\right)\right]$

82. $\cos \left[\sin^{-1} \left(-\frac{3}{5}\right) + \cos^{-1} \left(\frac{4}{5}\right)\right]$

83. $\sin \left[\arccos \frac{1}{2} + \arcsin (-1)\right]$

84. $\cos \left[\arccos (-\sqrt{3}/2) - \arcsin \left(-\frac{1}{2}\right)\right]$

85. Express $\sin (\sin^{-1} x + \cos^{-1} y)$ in an equivalent form free of trigonometric and inverse trigonometric functions.

86. Express $\cos (\sin^{-1} x - \cos^{-1} y)$ in an equivalent form free of trigonometric and inverse trigonometric functions.

Verify the identities in Problems 87 and 88.

87. $\cos (x + y + z) = \cos x \cos y \cos z - \sin x \sin y \cos z - \sin x \cos y \sin z - \cos x \sin y \sin z$

88. $\sin (x + y + z) = \sin x \cos y \cos z + \cos x \sin y \cos z + \cos x \cos y \sin z - \sin x \sin y \sin z$

In Problems 89–94, we will derive the subtraction formula for tangent. Begin with the expression tan (x − y) and follow the directions in each problem.

89. Rewrite $\tan (x - y)$ in terms of $\sin (x - y)$ and $\cos (x - y)$.

90. Use the subtraction formulas for both sine and cosine to expand the numerator and denominator.

91. Divide the numerator and denominator by $\cos x \cos y$.

92. Split each of the numerator and denominator into two separate fractions with denominator $\cos x \cos y$.

93. In both the numerator and denominator, simplify the two fractions using division.

94. Use the formula $\tan x = \sin x / \cos x$ to rewrite all remaining trigonometric expressions in terms of $\tan x$ or $\tan y$.

APPLICATIONS

95. ANALYTIC GEOMETRY Use the information in the figure to show that

$$\tan (\theta_2 - \theta_1) = \frac{m_2 - m_1}{1 + m_1 m_2}$$

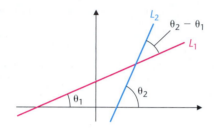

$$\tan \theta_1 = \text{Slope of } L_1 = m_1$$
$$\tan \theta_2 = \text{Slope of } L_2 = m_2$$

96. ANALYTIC GEOMETRY Find the acute angle of intersection between the two lines $y = 3x + 1$ and $y = \frac{1}{2}x - 1$. (Use the results of Problem 95.)

97. LIGHT REFRACTION Light rays passing through a plate glass window are refracted when they enter the glass and again when they leave to continue on a path parallel to the entering rays (see the figure). If the plate glass is M inches thick, the parallel displacement of the light rays is N inches, the angle of incidence is α, and the angle of refraction is β, show that

$$\tan \beta = \tan \alpha - \frac{N}{M} \sec \alpha$$

[*Hint:* First use geometric relationships to obtain

$$\frac{M}{\sec (90° - \beta)} = \frac{N}{\sin (\alpha - \beta)}$$

then use difference identities and fundamental identities to complete the derivation.]

98. LIGHT REFRACTION Use the results of Problem 97 to find β to the nearest degree if $\alpha = 43°$, $M = 0.25$ inch, and $N = 0.11$ inch.

99. SURVEYING El Capitan is a large monolithic granite peak that rises straight up from the floor of Yosemite Valley in Yosemite National Park. It attracts rock climbers worldwide. At certain times, the reflection of the peak can be seen in the Merced River that runs along the valley floor. How can the height H of El Capitan above the river be determined by using only a sextant h feet high to measure the angle of elevation, β, to the top of the peak, and the angle of depression, α, of the reflected peak top in the river? (See accompanying figure, which is not to scale.)

(A) Using right triangle relationships, show that

$$H = h \left(\frac{1 + \tan \beta \cot \alpha}{1 - \tan \beta \cot \alpha} \right)$$

(B) Using sum or difference identities, show that the result in part A can be written in the form

$$H = h \left[\frac{\sin (\alpha + \beta)}{\sin (\alpha - \beta)} \right]$$

(C) If a sextant of height 4.90 feet measures α to be 46.23° and β to be 46.15°, compute the height H of El Capitan above the Merced River to three significant digits.

7-3 Double-Angle and Half-Angle Identities

> Double-Angle Identities
> Half-Angle Identities

In Section 7-3, we will develop another useful set of identities called *double-angle* and *half-angle identities*. We can derive these identities directly from the sum and difference identities given in Section 7-2. Although the names use the word *angle,* the new identities hold for real numbers as well.

> Double-Angle Identities

Start with the sum identity for sine,

$$\sin (x + y) = \sin x \cos y + \cos x \sin y$$

and replace y with x to get

$$\sin (x + x) = \sin x \cos x + \cos x \sin x$$

On simplification, this gives

$$\sin 2x = 2 \sin x \cos x \qquad \text{Double-angle identity for sine} \qquad (1)$$

If we start with the sum identity for cosine,

$$\cos (x + y) = \cos x \cos y - \sin x \sin y$$

and replace y with x, we get

$$\cos (x + x) = \cos x \cos x - \sin x \sin x$$

On simplification, this gives

$$\cos 2x = \cos^2 x - \sin^2 x \qquad \text{First double-angle identity for cosine} \qquad (2)$$

Now, using the Pythagorean identity $\sin^2 x + \cos^2 x = 1$ in the form $\cos^2 x = 1 - \sin^2 x$ and substituting it into equation (2), we get

$$\cos 2x = 1 - \sin^2 x - \sin^2 x$$

On simplification, this gives

$$\cos 2x = 1 - 2 \sin^2 x \qquad \text{Second double-angle identity for cosine} \qquad (3)$$

Or, if we use the pythagorean identity in the form $\sin^2 x = 1 - \cos^2 x$ and substitute it into equation (2), we get

$$\cos 2x = \cos^2 x - (1 - \cos^2 x)$$

On simplification, this gives

$$\cos 2x = 2 \cos^2 x - 1 \qquad \text{Third double-angle identity for cosine} \qquad (4)$$

Double-angle identities will be established for the tangent function in Problems 85–87 of Exercise 7-3 by starting with the sum formula for tangent.

The double-angle identities are listed next for convenient reference.

> **DOUBLE-ANGLE IDENTITIES**

$$\sin 2x = 2 \sin x \cos x$$

$$\cos 2x = \cos^2 x - \sin^2 x = 1 - 2 \sin^2 x = 2 \cos^2 x - 1$$

$$\tan 2x = \frac{2 \tan x}{1 - \tan^2 x} = \frac{2 \cot x}{\cot^2 x - 1} = \frac{2}{\cot x - \tan x}$$

>>> **EXPLORE-DISCUSS 1**

(A) Verify the double-angle identities if $x = \pi/6$.

(B) Discuss how you would show that the equations $\sin 2x = 2 \sin x$, $\cos 2x = 2 \cos x$, and $\tan 2x = 2 \tan x$ are *not* identities.

EXAMPLE **1**

Identity Verification

Verify the identity $\cos 2x = \dfrac{1 - \tan^2 x}{1 + \tan^2 x}$.

VERIFICATION

We start with the right side and use quotient identities:

$$\frac{1 - \tan^2 x}{1 + \tan^2 x} = \frac{1 - \dfrac{\sin^2 x}{\cos^2 x}}{1 + \dfrac{\sin^2 x}{\cos^2 x}}$$ Multiply numerator and denominator by $\cos^2 x$.

$$= \frac{\cos^2 x \left(1 - \dfrac{\sin^2 x}{\cos^2 x}\right)}{\cos^2 x \left(1 + \dfrac{\sin^2 x}{\cos^2 x}\right)}$$ Distribute $\cos^2 x$ twice and reduce fractions,

$$= \frac{\cos^2 x - \sin^2 x}{\cos^2 x + \sin^2 x}$$ Use $\sin^2 x + \cos^2 x = 1$.

$$= \cos^2 x - \sin^2 x$$ Use a double-angle identity.

$$= \cos 2x$$

MATCHED PROBLEM 1 Verify the identity $\sin 2x = \dfrac{2 \tan x}{1 + \tan^2 x}$.

EXAMPLE **2**

Finding Exact Values

Find the exact values, without using a calculator, of $\sin 2x$ and $\cos 2x$ if $\tan x = -\frac{3}{4}$ and x is a Quadrant IV angle.

SOLUTION First draw the reference triangle for x and find any unknown sides:

$$r = \sqrt{(-3)^2 + 4^2} = 5$$

$$\sin x = -\tfrac{3}{5}$$

$$\cos x = \tfrac{4}{5}$$

Now use double-angle identities for sine and cosine:

$$\sin 2x = 2 \sin x \cos x = 2(-\tfrac{3}{5})(\tfrac{4}{5}) = -\tfrac{24}{25}$$
$$\cos 2x = 2 \cos^2 x - 1 = 2(\tfrac{4}{5})^2 - 1 = \tfrac{7}{25}$$

MATCHED PROBLEM 2 Find the exact values, without using a calculator, of $\cos 2x$ and $\tan 2x$ if $\sin x = \tfrac{4}{5}$ and x is a Quadrant II angle.

› Half-Angle Identities

Half-angle identities are simply double-angle identities stated in an alternate form. Let's start with the double-angle identity for cosine in the form

$$\cos 2m = 1 - 2 \sin^2 m$$

Now replace m with $x/2$ and solve for $\sin (x/2)$ [if $2m$ is twice m, then m is half of $2m$— think about this]:

$$\cos x = 1 - 2 \sin^2 \frac{x}{2} \qquad \text{Solve for } \sin^2 \tfrac{x}{2}.$$

$$\sin^2 \frac{x}{2} = \frac{1 - \cos x}{2} \qquad \text{Apply square root to both sides.}$$

$$\sin \frac{x}{2} = \pm \sqrt{\frac{1 - \cos x}{2}} \qquad \text{Half-angle identity for sine} \qquad (5)$$

where the choice of the sign is determined by the quadrant in which $x/2$ lies (not x!).

To obtain a half-angle identity for cosine, start with the double-angle identity for cosine in the form

$$\cos 2m = 2 \cos^2 m - 1$$

and let $m = x/2$ to obtain

$$\cos \frac{x}{2} = \pm \sqrt{\frac{1 + \cos x}{2}} \qquad \text{Half-angle identity for cosine} \qquad (6)$$

where the sign is again determined by the quadrant in which $x/2$ lies (not x!).

To obtain a *half-angle identity for tangent,* use the quotient identity and the half-angle formulas for sine and cosine:

$$\tan \frac{x}{2} = \frac{\sin \dfrac{x}{2}}{\cos \dfrac{x}{2}} = \frac{\pm \sqrt{\dfrac{1 - \cos x}{2}}}{\pm \sqrt{\dfrac{1 + \cos x}{2}}} = \pm \sqrt{\frac{1 - \cos x}{1 + \cos x}}$$

The resulting formula is

$$\tan \frac{x}{2} = \pm \sqrt{\frac{1 - \cos x}{1 + \cos x}} \qquad \text{Half-angle identity for tangent} \qquad (7)$$

where the sign is once again determined by the quadrant in which $x/2$ lies (not x!).

Simpler versions of the half-angle formula for tangent will be developed in Exercises 7-3. They are listed in the following box along with the other half-angle identities.

> **HALF-ANGLE IDENTITIES**

$$\sin\frac{x}{2} = \pm\sqrt{\frac{1-\cos x}{2}}$$

$$\cos\frac{x}{2} = \pm\sqrt{\frac{1+\cos x}{2}}$$

$$\tan\frac{x}{2} = \pm\sqrt{\frac{1-\cos x}{1+\cos x}} = \frac{\sin x}{1+\cos x} = \frac{1-\cos x}{\sin x}$$

where the sign is determined by the quadrant in which $x/2$ lies.

⟫⟫ EXPLORE-DISCUSS 2

(A) Verify the half-angle identities if $x = \pi/2$.

(B) Discuss how you would show that the equations

$$\sin\frac{x}{2} = \frac{1}{2}\sin x, \qquad \cos\frac{x}{2} = \frac{1}{2}\cos x, \qquad \text{and} \qquad \tan\frac{x}{2} = \frac{1}{2}\tan x$$

are *not* identities.

EXAMPLE **3** **Finding Exact Values**

Compute the exact value of $\sin 165°$ without a calculator using a half-angle identity.

SOLUTION

$$\sin 165° = \sin\frac{330°}{2} \qquad \text{Use half-angle identity for sine with a positive radical, because 165° is in the second quadrant.}$$

$$= \sqrt{\frac{1-\cos 330°}{2}} \qquad \cos 330° = \frac{\sqrt{3}}{2}$$

$$= \sqrt{\frac{1-(\sqrt{3}/2)}{2}} \qquad \text{Multiply numerator and denominator of radicand by 2 and simplify.}$$

$$= \frac{\sqrt{2-\sqrt{3}}}{2}$$

MATCHED PROBLEM 3 Compute the exact value of $\tan 105°$ without a calculator using a half-angle identity.

⟫⟫ CAUTION ⟫⟫

When using half-angle formulas, never include the $\pm$ in your answer. You have to choose one sign based on the quadrant that $x/2$ is in. Sine, cosine, and tangent are functions, so they need to have a unique output for any domain value.

EXAMPLE **4** **Finding Exact Values**

Find the exact values of $\cos(x/2)$ and $\cot(x/2)$ without using a calculator if $\sin x = -\frac{3}{5}$, $\pi < x < 3\pi/2$.

SOLUTION

Draw a reference triangle in the third quadrant, and find $\cos x$. Then use appropriate half-angle identities.

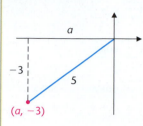

$$a = -\sqrt{5^2 - (-3)^2} = -4$$

$$\cos x = -\frac{4}{5}$$

$(a, -3)$

If $\pi < x < 3\pi/2$, then

$$\frac{\pi}{2} < \frac{x}{2} < \frac{3\pi}{4} \qquad \text{Divide each member of } \pi < x < 3\pi/2 \text{ by 2.}$$

This shows that, $x/2$ is an angle in the second quadrant where cosine and cotangent are negative, and

$$\cos\frac{x}{2} = -\sqrt{\frac{1 + \cos x}{2}} \qquad\qquad \cot\frac{x}{2} = \frac{1}{\tan (x/2)} = \frac{\sin x}{1 - \cos x}$$

$$= -\sqrt{\frac{1 + (-\frac{4}{5})}{2}} \qquad\qquad\qquad = \frac{-\frac{3}{5}}{1 - (-\frac{4}{5})} = -\frac{1}{3}$$

$$= -\sqrt{\frac{1}{10}} \text{ or } \frac{-\sqrt{10}}{10}$$

MATCHED PROBLEM 4

Find the exact values of $\sin (x/2)$ and $\tan (x/2)$ without using a calculator if $\cot x = -\frac{4}{3}$, $\pi/2 < x < \pi$.

EXAMPLE 5 **Identity Verification**

Verify the identity $\sin^2\dfrac{x}{2} = \dfrac{\tan x - \sin x}{2 \tan x}$.

VERIFICATION

We start with the left-hand side and use the half-angle identity for sine to rewrite $\sin\frac{x}{2}$:

$$\sin^2\frac{x}{2} = \left(\pm\sqrt{\frac{1 - \cos x}{2}}\right)^2 \qquad \text{Use } \sqrt{a^2} = a \text{ for any real number.}$$

$$= \frac{1 - \cos x}{2} \qquad \text{Multiply by } \frac{\tan x}{\tan x}.$$

$$= \frac{\tan x}{\tan x} \cdot \frac{1 - \cos x}{2} \qquad \text{Multiply; distribute in the numerator.}$$

$$= \frac{\tan x - \tan x \cos x}{2 \tan x} \qquad \text{Rewrite second } \tan x \text{ as } \frac{\sin x}{\cos x}.$$

$$= \frac{\tan x - \left(\dfrac{\sin x}{\cos x}\right)\cos x}{2 \tan x} \quad {}^{*} \qquad \text{Simplify.}$$

$$= \frac{\tan x - \sin x}{2 \tan x}$$

*Throughout the book, dashed boxes—called think boxes—are used to represent steps that may be performed mentally.

MATCHED PROBLEM 5

Verify the identity $\cos^2 \dfrac{x}{2} = \dfrac{\tan x + \sin x}{2 \tan x}$.

ANSWERS TO MATCHED PROBLEMS

1. $\dfrac{2 \tan x}{1 + \tan^2 x} = \dfrac{2\left(\dfrac{\sin x}{\cos x}\right)}{1 + \dfrac{\sin^2 x}{\cos^2 x}} = \dfrac{\cos^2 x \left[2\left(\dfrac{\sin x}{\cos x}\right)\right]}{\cos^2 x \left(1 + \dfrac{\sin^2 x}{\cos^2 x}\right)} = \dfrac{2 \sin x \cos x}{\cos^2 x + \sin^2 x} = 2 \sin x \cos x = \sin 2x$

2. $\cos 2x = -\frac{7}{25}$, $\tan 2x = \frac{24}{7}$ 3. $-\sqrt{3} - 2$ 4. $\sin (x/2) = 3\sqrt{10}/10$, $\tan (x/2) = 3$

5. $\cos^2 \dfrac{x}{2} = \dfrac{1 + \cos x}{2} = \dfrac{\tan x}{\tan x} \cdot \dfrac{1 + \cos x}{2} = \dfrac{\tan x + \tan x \cos x}{2 \tan x} = \dfrac{\tan x + \sin x}{2 \tan x}$

7-3 Exercises

1. Explain how the double-angle identity for sine can be obtained from the sum identity for sine.

2. Using the same technique as in Problem 1, find the identity that is obtained from the difference identity for sine.

3. Explain how the first double-angle identity for cosine can be obtained from the sum identity for cosine.

4. Using the same technique as in Problem 3, find the identity that is obtained from the difference identity for cosine.

5. How can we develop half-angle formulas using the double-angle formulas?

6. The half-angle formulas for sine and cosine both have $\pm$ at the beginning. Explain why you have to choose one or the other, and how you make that choice.

In Problems 7–12, verify each identity for the values indicated.

7. $\cos 2x = \cos^2 x - \sin^2 x$, $x = 30°$

8. $\sin 2x = 2 \sin x \cos x$, $x = 45°$

9. $\tan 2x = \dfrac{2}{\cot x - \tan x}$, $x = \dfrac{\pi}{3}$

10. $\tan 2x = \dfrac{2 \tan x}{1 - \tan^2 x}$, $x = \dfrac{\pi}{6}$

11. $\sin \dfrac{x}{2} = \pm\sqrt{\dfrac{1 - \cos x}{2}}$, $x = \pi$
 (Choose the correct sign.)

12. $\cos \dfrac{x}{2} = \pm\sqrt{\dfrac{1 + \cos x}{2}}$, $x = \dfrac{\pi}{2}$
 (Choose the correct sign.)

In Problems 13–20, find the exact value without a calculator using half-angle identities.

13. $\sin 22.5°$

14. $\tan 75°$

15. $\cos 67.5°$

16. $\tan 15°$

17. $\tan \dfrac{\pi}{8}$

18. $\cos \dfrac{\pi}{12}$

19. $\cos \dfrac{5\pi}{12}$

20. $\sin \dfrac{-7\pi}{8}$

Verify the identities in Problems 21–38.

21. $(\sin x + \cos x)^2 = 1 + \sin 2x$

22. $\sin 2x = (\tan x)(1 + \cos 2x)$

23. $\sin^2 x = \frac{1}{2}(1 - \cos 2x)$

24. $\cos^2 x = \frac{1}{2}(\cos 2x + 1)$

25. $1 - \cos 2x = \tan x \sin 2x$

26. $1 + \sin 2t = (\sin t + \cos t)^2$

27. $\sin^2 \dfrac{x}{2} = \dfrac{1 - \cos x}{2}$

28. $\cos^2 \dfrac{x}{2} = \dfrac{1 + \cos x}{2}$

29. $\cot 2x = \dfrac{1 - \tan^2 x}{2 \tan x}$

30. $\cot 2x = \dfrac{\cot x - \tan x}{2}$

31. $\cot \dfrac{\theta}{2} = \dfrac{\sin \theta}{1 - \cos \theta}$

32. $\cot \dfrac{\theta}{2} = \dfrac{1 + \cos \theta}{\sin \theta}$

33. $\cos 2u = \dfrac{1 - \tan^2 u}{1 + \tan^2 u}$

34. $\dfrac{\cos 2u}{1 - \sin 2u} = \dfrac{1 + \tan u}{1 - \tan u}$

35. $2 \csc 2x = \dfrac{1 + \tan^2 x}{\tan x}$

36. $\sec 2x = \dfrac{\sec^2 x}{2 - \sec^2 x}$

37. $\cos \alpha = \dfrac{1 - \tan^2 (\alpha/2)}{1 + \tan^2 (\alpha/2)}$

38. $\cos 2\alpha = \dfrac{\cot \alpha - \tan \alpha}{\cot \alpha + \tan \alpha}$

In Problems 39–46, show that the equation is not an identity by finding a value of x for which both sides are defined but are not equal.

39. $\tan 2x = 2 \tan x$

40. $\cos 2x = 2 \cos x$

41. $\sin \dfrac{x}{2} = \dfrac{1}{2} \sin x$

42. $\tan \dfrac{x}{2} = \dfrac{1}{2} \tan x$

43. $\cos \dfrac{x}{2} = \sqrt{\dfrac{1 + \cos x}{2}}$

44. $\sin \dfrac{x}{2} = \sqrt{\dfrac{1 - \cos x}{2}}$

45. $\tan \dfrac{x}{2} = \sqrt{\dfrac{1 - \cos x}{1 + \cos x}}$

46. $\tan 2x = \dfrac{2 \cot x}{1 - \cot^2 x}$

In Problems 47–54, is the equation an identity? Explain.

47. $\sin 2x = 2 \sin x$

48. $\sin \dfrac{x}{2} = \dfrac{1}{2} \sin x$

49. $\sin 4x = 4 \sin x \cos x$

50. $\csc 2x = 2 \csc x \sec x$

51. $\cot 2x = \dfrac{\tan x \,(\cot^2 x - 1)}{2}$

52. $\tan 4x = 4 \tan x$

53. $\cos 2x = 1 - 2 \cos^2 x$

54. $\tan 2x = \dfrac{2}{\tan x - \cot x}$

Compute the exact values of sin 2x, cos 2x, and tan 2x using the information given in Problems 55–58 and appropriate identities. Do not use a calculator.

55. $\sin x = \frac{3}{5},\ \pi/2 < x < \pi$

56. $\cos x = -\frac{4}{5},\ \pi/2 < x < \pi$

57. $\tan x = -\frac{5}{12},\ -\pi/2 < x < 0$

58. $\cot x = -\frac{5}{12},\ -\pi/2 < x < 0$

In Problems 59–62, compute the exact values of sin (x/2), cos (x/2), and tan (x/2) using the information given and appropriate identities. Do not use a calculator.

59. $\sin x = -\frac{1}{3},\ \pi < x < 3\pi/2$

60. $\cos x = -\frac{1}{4},\ \pi < x < 3\pi/2$

61. $\cot x = \frac{3}{4},\ -\pi < x < -\pi/2$

62. $\tan x = \frac{3}{4},\ -\pi < x < -\pi/2$

Suppose you are tutoring a student who is having difficulties in finding the exact values of sin θ and cos θ from the information given in Problems 63 and 64. Assuming you have worked through each problem and have identified the key steps in the solution process, proceed with your tutoring by guiding the student through the solution process using the following questions. Record the expected correct responses from the student.

(A) The angle 2θ is in what quadrant and how do you know?

(B) How can you find sin 2θ and cos 2θ? Find each.

(C) What identities relate sin θ and cos θ with either sin 2θ or cos 2θ?

(D) How would you use the identities in part C to find sin θ and cos θ exactly, including the correct sign?

(E) What are the exact values for sin θ and cos θ?

63. Find the exact values of sin θ and cos θ, given $\tan 2\theta = -\frac{4}{3}$, $0° < \theta < 90°$.

64. Find the exact values of sin θ and cos θ, given $\sec 2\theta = -\frac{5}{4}$, $0° < \theta < 90°$.

Verify each of the following identities for the value of x indicated in Problems 65–68. Compute values to five significant digits using a calculator.

(A) $\tan 2x = \dfrac{2 \tan x}{1 - \tan^2 x}$ *(B)* $\cos \dfrac{x}{2} = \pm\sqrt{\dfrac{1 + \cos x}{2}}$

(Choose the correct sign.)

65. $x = 252.06°$

66. $x = 72.358°$

67. $x = 0.934\ 57$

68. $x = 4$

In Problems 69–72, graph y_1 and y_2 in the same viewing window for $-2\pi \le x \le 2\pi$, and state the intervals for which the equation $y_1 = y_2$ is an identity.

69. $y_1 = \cos (x/2),\ y_2 = \sqrt{\dfrac{1 + \cos x}{2}}$

70. $y_1 = \cos (x/2),\ y_2 = -\sqrt{\dfrac{1 + \cos x}{2}}$

71. $y_1 = \sin (x/2),\ y_2 = -\sqrt{\dfrac{1 - \cos x}{2}}$

72. $y_1 = \sin (x/2),\ y_2 = \sqrt{\dfrac{1 - \cos x}{2}}$

Verify the identities in Problems 73–76.

73. $\cos 3x = 4 \cos^3 x - 3 \cos x$

74. $\sin 3x = 3 \sin x - 4 \sin^3 x$

75. $\cos 4x = 8 \cos^4 x - 8 \cos^2 x + 1$

76. $\sin 4x = (\cos x)(4 \sin x - 8 \sin^3 x)$

In Problems 77–82, find the exact value of each without using a calculator.

77. $\cos \left[2 \cos^{-1} \left(\frac{3}{5}\right)\right]$

78. $\sin \left[2 \cos^{-1} \left(\frac{3}{5}\right)\right]$

79. $\tan \left[2 \cos^{-1} \left(-\frac{4}{5}\right)\right]$

80. $\tan \left[2 \tan^{-1} \left(-\frac{3}{4}\right)\right]$

81. $\cos \left[\frac{1}{2} \cos^{-1} \left(-\frac{3}{5}\right)\right]$

82. $\sin \left[\frac{1}{2} \tan^{-1} \left(-\frac{4}{3}\right)\right]$

The identities in Problems 83 and 84 are useful in calculus to transfrom expressions with powers into ones without.

83. Use the formula $\cos (2x) = 1 - 2 \sin^2 x$ to show that $\sin^2 x = \dfrac{1 - \cos 2x}{2}$.

84. Use the formula $\cos (2x) = 2 \cos^2 x - 1$ to show that $\cos^2 x = \dfrac{1 + \cos 2x}{2}$.

85. Use a sum identity to derive the first double-angle formula for tangent: $\tan (2x) = \dfrac{2 \tan x}{1 - \tan^2 x}$

86. Starting with the identity in Problem 85, derive the second double-angle formula for tangent: $\tan (2x) = \dfrac{2 \cot x}{\cot^2 x - 1}$. (*Hint:* Multiply numerator and denominator by an appropriate expression.)

87. Starting with the identity in Problem 85, derive the third double-angle formula for tangent: $\tan (2x) = \dfrac{2}{\cot x - \tan x}$. (*Hint:* Multiply numerator and denominator by an appropriate expression.)

88. By looking at four cases, one for each possible quadrant that x could lie in, show that $\tan (x/2)$ and $\sin x$ always have the same sign.

89. In this problem we will derive a simpler form of the first half-angle formula for tangent. Fill in the blanks, justifying each step in the derivation.

Statement	Reason
$\left\vert \tan \dfrac{x}{2} \right\vert = \sqrt{\dfrac{1 - \cos x}{1 + \cos x}}$	(A) _____
$= \sqrt{\dfrac{1 - \cos x}{1 + \cos x} \cdot \dfrac{1 + \cos x}{1 + \cos x}}$	Multiplied by 1
$= \sqrt{\dfrac{1 - \cos^2 x}{(1 + \cos x)^2}}$	Algebra
$= \sqrt{\dfrac{\sin^2 x}{(1 + \cos x)^2}}$	(B) _____
$= \dfrac{\sqrt{\sin^2 x}}{\sqrt{(1 + \cos x)^2}}$	$\sqrt{\dfrac{a}{b}} = \dfrac{\sqrt{a}}{\sqrt{b}}$
$= \dfrac{\vert \sin x \vert}{1 + \cos x}$	(C) _____

(D) Explain why the last line above allows us to conclude that $\tan \dfrac{x}{2} = \dfrac{\sin x}{1 + \cos x}$. (*Hint:* See Problem 88.)

90. Repeat Problem 89, this time multiplying the numerator and denominator by $1 - \cos x$ to show that $\tan \dfrac{x}{2} = \dfrac{1 - \cos x}{\sin x}$.

APPLICATIONS

91. INDIRECT MEASUREMENT Find the exact value of x in the figure; then find x and θ to three decimal places. [*Hint:* Use $\cos 2\theta = 2 \cos^2 \theta - 1$.]

92. INDIRECT MEASUREMENT Find the exact value of x in the figure; then find x and θ to three decimal places. [*Hint:* Use $\tan 2\theta = (2 \tan \theta)/(1 - \tan^2 \theta)$.]

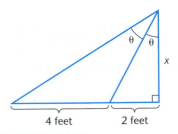

93. SPORTS—PHYSICS The theoretical distance d that a shot-putter, discus thrower, or javelin thrower can achieve on a given throw is found in physics to be given approximately by

$$d = \frac{2v_0^2 \sin \theta \cos \theta}{32 \text{ feet per second per second}}$$

where v_0 is the initial speed of the object thrown (in feet per second) and θ is the angle above the horizontal at which the object leaves the hand (see the figure).

(A) Write the formula in terms of $\sin 2\theta$ by using a suitable identity.
(B) Using the resulting equation in part A, determine the angle θ that will produce the maximum distance d for a given initial speed v_0. This result is an important consideration for shot-putters, javelin throwers, and discus throwers.

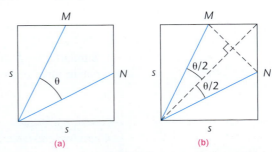

94. GEOMETRY In part (a) of the figure, M and N are the midpoints of the sides of a square. Find the exact value of $\cos \theta$. [*Hint:* The solution uses the Pythagorean theorem, the definition of sine and cosine, a half-angle identity, and some auxiliary lines as drawn in part (b) of the figure.]

(a) (b)

95. AREA An n-sided regular polygon is inscribed in a circle of radius R.
(A) Show that the area of the n-sided polygon is given by

$$A_n = \frac{1}{2} nR^2 \sin \frac{2\pi}{n}$$

[*Hint:* (Area of a triangle) = $(\frac{1}{2})$(base)(altitude). Also, a double-angle identity is useful.]

(B) For a circle of radius 1, complete Table 1, to five decimal places, using the formula in part A:

Table 1

n	10	100	1,000	10,000
A_n				

(C) What number does A_n seem to approach as n increases without bound? (What is the area of a circle of radius 1?)

(D) Will A_n exactly equal the area of the circumscribed circle for some sufficiently large n? How close can A_n be to the area of the circumscribed circle? [In calculus, the area of the circumscribed circle is called the *limit* of A_n as n increases without bound. In symbols, for a circle of radius 1, we would write $\lim_{n \to \infty} A_n = \pi$. The limit concept is the cornerstone on which calculus is constructed.]

7-4 Product–Sum and Sum–Product Identities

> Product–Sum Identities
> Sum–Product Identities

Our work with identities is concluded by developing the *product–sum* and *sum–product identities,* which are easily derived from the sum and difference identities developed in Section 7-2. These identities are used in calculus to convert product forms to more convenient sum forms. They are also used in the study of sound waves in music to convert sum forms to more convenient product forms.

> Product–Sum Identities

We will start with the sum and difference identities for sine, and add them, left side to left side and right side to right side:

$$\sin (x + y) = \sin x \cos y + \cos x \sin y$$
$$\underline{\sin (x - y) = \sin x \cos y - \cos x \sin y}$$
$$\sin (x + y) + \sin (x - y) = 2 \sin x \cos y$$

Multiplying both sides by $\frac{1}{2}$, we get a formula that turns a product into a sum:

$$\sin x \cos y = \tfrac{1}{2}[\sin (x + y) + \sin (x - y)]$$

Similarly, by adding or subtracting the appropriate sum and difference identities, we can obtain three other **product–sum identities.** These are listed in the box.

> **PRODUCT–SUM IDENTITIES**

$$\sin x \cos y = \tfrac{1}{2}[\sin (x + y) + \sin (x - y)]$$
$$\cos x \sin y = \tfrac{1}{2}[\sin (x + y) - \sin (x - y)]$$
$$\sin x \sin y = \tfrac{1}{2}[\cos (x - y) - \cos (x + y)]$$
$$\cos x \cos y = \tfrac{1}{2}[\cos (x + y) + \cos (x - y)]$$

EXAMPLE **1** **A Product as a Difference**

Write the product $\cos 3t \sin t$ as a sum or difference.

SOLUTION

$$\cos x \sin y = \tfrac{1}{2}[\sin(x+y) - \sin(x-y)] \qquad \text{Let } x = 3t \text{ and } y = t.$$
$$\cos 3t \sin t = \tfrac{1}{2}[\sin(3t+t) - \sin(3t-t)] \qquad \text{Simplify.}$$
$$= \tfrac{1}{2}\sin 4t - \tfrac{1}{2}\sin 2t$$

MATCHED PROBLEM 1 Write the product $\cos 5\theta \cos 2\theta$ as a sum or difference.

EXAMPLE **2** **Finding Exact Values**

Evaluate $\sin 105° \sin 15°$ exactly using an appropriate product–sum identity.

SOLUTION

$$\sin x \sin y = \tfrac{1}{2}[\cos(x-y) - \cos(x+y)] \qquad \text{Let } x = 105° \text{ and } y = 15°.$$
$$\sin 105° \sin 15° = \tfrac{1}{2}[\cos(105° - 15°) - \cos(105° + 15°)] \qquad \text{Simplify.}$$
$$= \tfrac{1}{2}[\cos 90° - \cos 120°] \qquad \begin{array}{l}\cos 90° = 0,\\ \cos 120° = -\tfrac{1}{2}\end{array}$$
$$= \tfrac{1}{2}[0 - (-\tfrac{1}{2})] = \tfrac{1}{4}$$

MATCHED PROBLEM 2 Evaluate $\cos 165° \sin 75°$ exactly using an appropriate product–sum identity.

› **Sum–Product Identities**

››› EXPLORE-DISCUSS 1

Starting with the identity $\sin x \cos y = \dfrac{1}{2}[\sin(x+y) + \sin(x-y)]$, replace x with $\dfrac{x+y}{2}$ and y with $\dfrac{x-y}{2}$. Then simplify and multiply both sides by 2. What does the resulting identity do?

The product–sum identities can be transformed into equivalent forms called **sum–product identities**. These identities are used to express sums and differences involving sines and cosines as products involving sines and cosines. Explore-Discuss 1 illustrates the process:

$$\sin x \cos y = \frac{1}{2}[\sin(x+y) + \sin(x-y)]$$

$$\sin \frac{x+y}{2} \cos \frac{x-y}{2} = \frac{1}{2}\left[\sin\left(\frac{x+y}{2} + \frac{x-y}{2}\right) + \sin\left(\frac{x+y}{2} - \frac{x-y}{2}\right)\right]$$

$$\sin \frac{x+y}{2} \cos \frac{x-y}{2} = \frac{1}{2}[\sin x + \sin y]$$

$$\sin x + \sin y = 2 \sin \frac{x+y}{2} \cos \frac{x-y}{2}$$

The three remaining sum–product identities are derived in a similar way from the other three product–sum identities.

> **SUM–PRODUCT IDENTITIES**

$$\sin x + \sin y = 2 \sin \frac{x + y}{2} \cos \frac{x - y}{2}$$

$$\sin x - \sin y = 2 \cos \frac{x + y}{2} \sin \frac{x - y}{2}$$

$$\cos x + \cos y = 2 \cos \frac{x + y}{2} \cos \frac{x - y}{2}$$

$$\cos x - \cos y = -2 \sin \frac{x + y}{2} \sin \frac{x - y}{2}$$

EXAMPLE 3 A Difference as a Product

Write the difference $\sin 7\theta - \sin 3\theta$ as a product.

SOLUTION

$$\sin x - \sin y = 2 \cos \frac{x + y}{2} \sin \frac{x - y}{2} \qquad \text{Let } x = 7\theta \text{ and } y = 3\theta.$$

$$\sin 7\theta - \sin 3\theta = 2 \cos \frac{7\theta + 3\theta}{2} \sin \frac{7\theta - 3\theta}{2} \qquad \text{Simplify.}$$

$$= 2 \cos 5\theta \sin 2\theta$$

MATCHED PROBLEM 3 Write the sum $\cos 3t + \cos t$ as a product.

EXAMPLE 4 Finding Exact Values

Find the exact value of $\sin 105° - \sin 15°$ using an appropriate sum–product identity.

SOLUTION

$$\sin x - \sin y = 2 \cos \frac{x + y}{2} \sin \frac{x - y}{2} \qquad \text{Let } x = 105° \text{ and } y = 15°.$$

$$\sin 105° - \sin 15° = 2 \cos \frac{105° + 15°}{2} \sin \frac{105° - 15°}{2} \qquad \text{Simplify.}$$

$$= 2 \cos 60° \sin 45° \qquad \cos 60° = \tfrac{1}{2}, \sin 45° = \tfrac{\sqrt{2}}{2}$$

$$= 2 \left(\frac{1}{2}\right)\left(\frac{\sqrt{2}}{2}\right) = \frac{\sqrt{2}}{2}$$

MATCHED PROBLEM 4 Find the exact value of $\cos 165° - \cos 75°$ using an appropriate sum–product identity.

ANSWERS TO MATCHED PROBLEMS

1. $\frac{1}{2} \cos 7\theta + \frac{1}{2} \cos 3\theta$ **2.** $(-\sqrt{3} - 2)/4$ **3.** $2 \cos 2t \cos t$ **4.** $-\sqrt{6}/2$

7-4 Exercises

1. What is a product–sum identity?

2. What is a sum–product identity?

3. Which identity is obtained if you add, left side to left side and right side to right side, the first two product–sum identities?

4. Which identity is obtained if you add, left side to left side and right side to right side, the last two product–sum identities?

5. Which identities are obtained if you substitute x for y in the product–sum identities?

6. Which identities are obtained if you substitute x for y in the sum–product identities?

In Problems 7–10, write each product as a sum or difference involving sine and cosine.

7. $\sin 3m \cos m$

8. $\cos 7A \cos 5A$

9. $\sin u \sin 3u$

10. $\cos 2\theta \sin 3\theta$

In Problems 11–14, write each difference or sum as a product involving sines and cosines.

11. $\sin 3t + \sin t$

12. $\cos 7\theta + \cos 5\theta$

13. $\cos 5w - \cos 9w$

14. $\sin u - \sin 5u$

Evaluate Problems 15–26 exactly using an appropriate identity.

15. $\sin 195° \cos 75°$

16. $\cos 75° \sin 15°$

17. $\cos 157.5° \cos 67.5°$

18. $\sin 112.5° \sin 22.5°$

19. $\cos 37.5° \sin 7.5°$

20. $\sin 262.5° \cos 52.5°$

21. $\sin \dfrac{5\pi}{8} \sin \dfrac{\pi}{8}$

22. $\cos \dfrac{3\pi}{8} \cos \dfrac{7\pi}{8}$

23. $\cos \dfrac{11\pi}{12} \cos \dfrac{\pi}{12}$

24. $\cos \dfrac{7\pi}{12} \sin \dfrac{5\pi}{12}$

25. $\sin \dfrac{13\pi}{24} \cos \dfrac{5\pi}{24}$

26. $\sin \dfrac{17\pi}{24} \sin \dfrac{\pi}{24}$

Evaluate Problems 27–34 exactly using an appropriate identity.

27. $\sin 195° - \sin 105°$

28. $\cos 105° + \cos 15°$

29. $\cos 75° - \cos 15°$

30. $\sin 165° + \sin 105°$

31. $\cos \dfrac{17\pi}{12} + \cos \dfrac{\pi}{12}$

32. $\cos \dfrac{13\pi}{12} - \cos \dfrac{5\pi}{12}$

33. $\sin \dfrac{\pi}{12} + \sin \dfrac{5\pi}{12}$

34. $\sin \dfrac{11\pi}{12} - \sin \dfrac{7\pi}{12}$

Use sum and difference identities to verify the identities in Problems 35 and 36.

35. $\cos x \cos y = \frac{1}{2}[\cos (x + y) + \cos (x - y)]$

36. $\sin x \sin y = \frac{1}{2}[\cos (x - y) - \cos (x + y)]$

37. Explain how you can transform the product–sum identity

$$\sin u \sin v = \tfrac{1}{2}[\cos (u - v) - \cos (u + v)]$$

into the sum–product identity

$$\cos x - \cos y = -2 \sin \frac{x + y}{2} \sin \frac{x - y}{2}$$

using a suitable substitution.

38. Explain how you can transform the product–sum identity

$$\cos u \cos v = \tfrac{1}{2}[\cos (u + v) + \cos (u - v)]$$

into the sum–product identity

$$\cos x + \cos y = 2 \cos \frac{x + y}{2} \cos \frac{x - y}{2}$$

using a suitable substitution.

Verify each identity in Problems 39–46.

39. $\dfrac{\sin 2t + \sin 4t}{\cos 2t - \cos 4t} = \cot t$

40. $\dfrac{\cos t - \cos 3t}{\sin t + \sin 3t} = \tan t$

41. $\dfrac{\sin x - \sin y}{\cos x - \cos y} = -\cot \dfrac{x + y}{2}$

42. $\dfrac{\sin x + \sin y}{\cos x + \cos y} = \tan \dfrac{x + y}{2}$

43. $\dfrac{\cos x + \cos y}{\sin x - \sin y} = \cot \dfrac{x - y}{2}$

44. $\dfrac{\cos x - \cos y}{\sin x + \sin y} = -\tan \dfrac{x - y}{2}$

45. $\dfrac{\cos x + \cos y}{\cos x - \cos y} = -\cot \dfrac{x + y}{2} \cot \dfrac{x - y}{2}$

46. $\dfrac{\sin x + \sin y}{\sin x - \sin y} = \dfrac{\tan [\frac{1}{2}(x + y)]}{\tan [\frac{1}{2}(x - y)]}$

In Problems 47–54, show that the equation is not an identity by finding a value of x and a value of y for which both sides are defined but are not equal.

47. $\sin x \cos y = \sin x + \cos y$

48. $\cos x \sin y = \cos x - \sin y$

49. $\sin x \sin y = \sin (x + y)$

50. $\cos x \cos y = \cos (x + y)$

51. $\cos x + \cos y = (\cos x)(\cos y)$

52. $\sin x + \sin y = (\sin x)(\sin y)$

53. $\sin x - \sin y = \cos \dfrac{x + y}{2} \sin \dfrac{x - y}{2}$

54. $\cos x - \cos y = 2 \sin \dfrac{x + y}{2} \sin \dfrac{x - y}{2}$

In Problems 55–60, is the equation an identity? Explain.

55. $\sin 3x - \sin x = 2 \cos 2x \sin x$

56. $2 \sin x \cos 2x = \sin x + \sin 3x$

57. $\cos 3x - \cos x = 2 \sin 2x \sin x$

58. $2 \cos 3x \cos 5x = \cos 8x + \cos 2x$

59. $\cos x + \cos 5x = 2 \cos 2x \cos 3x$

60. $2 \sin 4x \cos 2x = \sin 8x + \sin 2x$

Verify each of the following identities for the values of x and y indicated in Problems 61–64. Evaluate each side to five significant digits.
(A) $\cos x \sin y = \frac{1}{2}[\sin (x + y) - \sin (x - y)]$

(B) $\cos x + \cos y = 2 \cos \dfrac{x + y}{2} \cos \dfrac{x - y}{2}$

61. $x = 172.63°, y = 20.177°$

62. $x = 50.137°, y = 18.044°$

63. $x = 1.1255, y = 3.6014$

64. $x = 0.039\ 17, y = 0.61052$

In Problems 65–72, write each as a product if y is a sum or difference, or as a sum or difference if y is a product. Enter the original equation in a graphing calculator as y_1, the converted form as y_2, and graph y_1 and y_2 in the same viewing window. Use TRACE to compare the two graphs.

65. $y = \sin 2x + \sin x$ 　　　**66.** $y = \cos 3x + \cos x$

67. $y = \cos 1.7x - \cos 0.3x$ 　　**68.** $y = \sin 2.1x - \sin 0.5x$

69. $y = \sin 3x \cos x$ 　　　　**70.** $y = \cos 5x \cos 3x$

71. $y = \sin 2.3x \sin 0.7x$ 　　**72.** $y = \cos 1.9x \sin 0.5x$

Verify each identity in Problems 73 and 74.

73. $\cos x \cos y \cos z = \frac{1}{4}[\cos (x + y - z) + \cos (y + z - x)$
$+ \cos (z + x - y) + \cos (x + y + z)]$

74. $\sin x \sin y \sin z = \frac{1}{4}[\sin (x + y - z) + \sin (y + z - x)$
$+ \sin (z + x - y) - \sin (x + y + z)]$

In Problems 75–78,
(A) Graph $y_1, y_2,$ and y_3 in a graphing calculator for $0 \le x \le 1$ and $-2 \le y \le 2$.
(B) Convert y_1 to a sum or difference and repeat part A.

75. $y_1 = 2 \cos (28\pi x) \cos (2\pi x)$
$y_2 = 2 \cos (2\pi x)$
$y_3 = -2 \cos (2\pi x)$

76. $y_1 = 2 \sin (24\pi x) \sin (2\pi x)$
$y_2 = 2 \sin (2\pi x)$
$y_3 = -2 \sin (2\pi x)$

77. $y_1 = 2 \sin (20\pi x) \cos (2\pi x)$
$y_2 = 2 \cos (2\pi x)$
$y_3 = -2 \cos (2\pi x)$

78. $y_1 = 2 \cos (16\pi x) \sin (2\pi x)$
$y_2 = 2 \sin (2\pi x)$
$y_3 = -2 \sin (2\pi x)$

APPLICATIONS

*Problems 79 and 80 involve the phenomenon of sound called beats. If two tones having the same loudness and close together in pitch (frequency) are sounded, one following the other, most people have difficulty in differentiating the two tones. However, if the tones are sounded simultaneously, they will interact with each other, producing a low warbling sound called a **beat**. Musicians, when tuning an instrument with other instruments or a tuning fork, listen for these lower beat frequencies and try to eliminate them by adjusting their instruments. Problems 79 and 80 provide a visual illustration of the beat phenomenon.*

79. MUSIC—BEAT FREQUENCIES The equations $y = 0.5 \cos 128\pi t$ and $y = -0.5 \cos 144\pi t$ model sound waves with frequencies 64 and 72 hertz, respectively. If both sounds are emitted simultaneously, a *beat* frequency results.

(A) Show that
$$0.5 \cos 128\pi t - 0.5 \cos 144\pi t = \sin 8\pi t \sin 136\pi t$$

(The product form is more useful to sound engineers.)
(B) Graph each equation in a different viewing window for $0 \le t \le 0.25$:
$$y = 0.5 \cos 128\pi t$$
$$y = -0.5 \cos 144\pi t$$
$$y = 0.5 \cos 128\pi t - 0.5 \cos 144\pi t$$
$$y = \sin 8\pi t \sin 136\pi t$$

80. MUSIC—BEAT FREQUENCIES The equations $y = 0.25 \cos 256\pi t$ and $y = -0.25 \cos 288\pi t$ model sound waves with frequencies 128 and 144 hertz, respectively. If both sounds are emitted simultaneously, a *beat* frequency results.
(A) Show that
$$0.25 \cos 256\pi t - 0.25 \cos 288\pi t = 0.5 \sin 16\pi t \sin 272\pi t$$

(The product form is more useful to sound engineers.)
(B) Graph each equation in a different viewing window for $0 \le t \le 0.125$:
$$y = 0.25 \cos 256\pi t$$
$$y = -0.25 \cos 288\pi t$$
$$y = 0.25 \cos 256\pi t - 0.25 \cos 288\pi t$$
$$y = 0.5 \sin 16\pi t \sin 272\pi t$$

7-5 Trigonometric Equations

> › Solving Trigonometric Equations Using an Algebraic Approach
> › Solving Trigonometric Equations Using a Graphing Calculator

In the first four sections of this chapter, we studied trigonometric equations called **identities.** These are equations that are true for all replacements of the variable(s) for which both sides are defined. We now turn our attention to **conditional equations,** which may be true for some replacements of the variable but false for others. For example,

$$\cos x = \sin x$$

is a conditional equation, because it is true for some values of x, for example, $x = \pi/4$, and false for others, such as $x = 0$. (You should check both values.)

We will look at two approaches for solving conditional trigonometric equations: an algebraic approach and a graphing calculator approach. The algebraic approach often requires the use of algebraic manipulation, identities, and ingenuity. In some cases algebraic methods lead to exact solutions, which are very useful in certain contexts. Graphing calculator methods can be used to approximate solutions to a greater variety of trigonometric equations, but usually don't produce exact solutions. Each method has its strengths.

>>> EXPLORE-DISCUSS 1

Suppose that we need to find solutions to the equation

$$\cos x = 0.5$$

The figure shows a partial graph of the left and right sides of the equation.

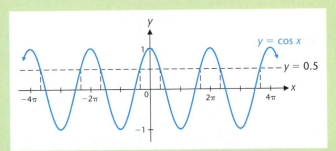

(A) How many solutions does the equation have on the interval $[0, 2\pi)$? What are they?

(B) How many solutions does the equation have on the interval $(-\infty, \infty)$? Discuss a method of writing all solutions to the equation.

› Solving Trigonometric Equations Using an Algebraic Approach

You might find the following suggestions for solving trigonometric equations using an algebraic approach useful:

> ❯ SUGGESTIONS FOR SOLVING TRIGONOMETRIC EQUATIONS ALGEBRAICALLY
>
> **1.** Treat one particular trigonometric function like a variable, and solve for it.
>
> **(A)** Consider using algebraic manipulation such as factoring, combining or separating fractions, and so on.
>
> **(B)** Consider using identities.
>
> **2.** After solving for a trigonometric function, solve for the variable.

Examples 1–5 should help make the algebraic approach clear.

EXAMPLE **1** ### Exact Solutions Using Factoring

Find all solutions exactly for $2 \cos^2 x - \cos x = 0$.

SOLUTION **Step 1.** *Solve for cos x.*

$$2 \cos^2 x - \cos x = 0 \qquad \text{Factor out cos x.}$$

$$\cos x \,(2 \cos x - 1) = 0 \qquad \text{ab = 0 only if a = 0 or b = 0}$$

$$\cos x = 0 \qquad \text{or} \qquad 2 \cos x - 1 = 0$$

$$\cos x = \tfrac{1}{2}$$

Step 2. *Solve each equation over one period $[0, 2\pi)$.* Sketch a graph of $y = \cos x$, $y = 0$, and $y = \tfrac{1}{2}$ in the same coordinate system to provide an aid to writing all solutions over one period [Fig. 1(a)], or use a unit circle diagram [Fig. 1(b)].

$$\cos x = 0 \qquad\qquad \cos x = \tfrac{1}{2}$$

$$x = \pi/2, \, 3\pi/2 \qquad x = \pi/3, \, 5\pi/3$$

❯ **Figure 1(a)**

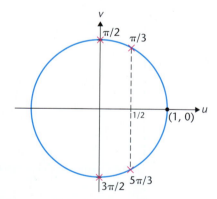

❯ **Figure 1(b)**

Step 3. *Write an expression for all solutions.* Because the cosine function is periodic with period 2π, all solutions are given by

$$x = \begin{cases} \pi/3 + 2k\pi \\ \pi/2 + 2k\pi \\ 3\pi/2 + 2k\pi \\ 5\pi/3 + 2k\pi \end{cases} \qquad k \text{ any integer}$$

MATCHED PROBLEM 1 Find all solutions exactly for $2 \sin^2 x + \sin x = 0$.

| EXAMPLE | 2 | **Approximate Solutions Using Identities and Factoring** |

Find all real solutions for $3 \cos^2 x + 8 \sin x = 7$. Compute all inverse functions to four decimal places.

SOLUTION **Step 1.** *Solve for sin x and/or cos x.* Move all nonzero terms to the left of the equal sign and use an identity so that sine is the only trig function on the left side:

$$3 \cos^2 x + 8 \sin x = 7$$
$$3 \cos^2 x + 8 \sin x - 7 = 0 \qquad \textcolor{red}{\cos^2 x = 1 - \sin^2 x}$$
$$3(1 - \sin^2 x) + 8 \sin x - 7 = 0 \qquad \textcolor{red}{\text{Distribute; multiply both sides by } -1.}$$
$$3 \sin^2 x - 8 \sin x + 4 = 0 \qquad \textcolor{red}{\text{Factor: } 3u^2 - 8u + 4 = (u - 2)(3u - 2).}$$
$$(\sin x - 2)(3 \sin x - 2) = 0 \qquad \textcolor{red}{ab = 0 \text{ only if } a = 0 \text{ or } b = 0.}$$
$$\sin x - 2 = 0 \qquad \text{or} \qquad 3 \sin x - 2 = 0$$
$$\sin x = 2 \qquad\qquad\qquad\qquad \sin x = \tfrac{2}{3}$$

Step 2. *Solve each equation over one period [0, 2π):* Sketch a graph of $y = \sin x$, $y = 2$, and $y = \tfrac{2}{3}$ in the same coordinate system to provide an aid to writing all solutions over one period [Fig. 2(a)], or use a unit circle diagram [Fig. 2(b)].
Solve the first equation:

$$\sin x = 2 \qquad \textcolor{red}{\text{No solution, because sine is never greater than 1.}}$$

Solve the second equation:

$$\sin x = \tfrac{2}{3} \qquad\qquad \textcolor{red}{\text{From the graphs we see there are solutions in}}$$
$$\qquad\qquad\qquad \textcolor{red}{\text{the first and second quadrants.}}$$
$$x = \sin^{-1} \tfrac{2}{3} = 0.7297 \qquad \textcolor{red}{\text{First quadrant solution}}$$
$$x = \pi - 0.7297 = 2.4119 \qquad \textcolor{red}{\text{Second quadrant solution}}$$

> **Figure 2(a)**

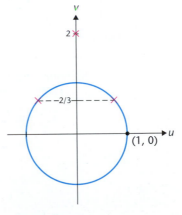

> **Figure 2(b)**

CHECK

$$\sin 0.7297 = 0.6667; \ \sin 2.4119 = 0.6666$$

(Checks may not be exact because of roundoff errors.)

Step 3. *Write an expression for all solutions.* Because the sine function is periodic with period 2π, all solutions are given by

$$x = \begin{cases} 0.7297 + 2k\pi \\ 2.4119 + 2k\pi \end{cases} \qquad k \text{ any integer}$$

MATCHED PROBLEM 2 Find all real solutions to $8 \sin^2 x = 5 - 10 \cos x$. Compute all inverse functions to four decimal places.

EXAMPLE 3 **Approximate Solutions Using Substitution**

Find θ in degree measure to three decimal places so that $5 \sin (2\theta - 5) = -3.045$, $0° \leq 2\theta - 5 \leq 360°$.

SOLUTION **Step 1.** *Make a substitution.* Let $x = 2\theta - 5$ to obtain

$$5 \sin x = -3.045, \ 0° \leq x \leq 360°$$

Step 2. *Solve for sin x.* Divide both sides by 5.

$$\sin x = \frac{-3.045}{5} = -0.609$$

Step 3. *Solve for x over $0° \leq x \leq 360°$.* Sketch a graph of $y = \sin x$ and $y = -0.609$ in the same coordinate system to provide an aid to writing all solutions over $0° \leq x \leq 360°$ [Fig. 3(a)], or use a unit circle diagram [Fig. 3(b)].

> **Figure 3(a)**

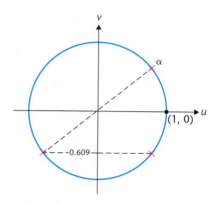

> **Figure 3(b)**

Solutions are in the third and fourth quadrants. If the reference angle is α, then $x = 180° + \alpha$ or $x = 360° - \alpha$.

$$\alpha = \sin^{-1} 0.609 = 37.517° \qquad \text{Reference angle}$$
$$x = 180° + 37.517°$$
$$= 217.517° \qquad \text{Third quadrant solution}$$
$$x = 360° - 37.517°$$
$$= 322.483° \qquad \text{Fourth quadrant solution}$$

CHECK

$$\sin 217.517° = -0.609; \ \sin 322.483° = -0.609$$

Step 4. *Now substitute $2\theta - 5$ back in for x, and solve for θ:*

$x = 217.517°$	$x = 322.483°$
$2\theta - 5 = 217.517°$	$2\theta - 5 = 322.483°$
$2\theta = 222.517°$	$2\theta = 327.483°$
$\boldsymbol{\theta = 111.259°}$	$\boldsymbol{\theta = 163.742°}$

A final check in the original equation is left to the reader.

MATCHED PROBLEM 3 Find θ in degree measure to three decimal places so that $8 \tan (6\theta + 15) = -64.328$, $-90° < 6\theta + 15 < 90°$.

EXAMPLE 4

Exact Solutions Using Identities and Factoring

Find exact solutions for $\sin^2 x = \frac{1}{2} \sin 2x$, $0 \le x \le 2\pi$.

SOLUTION The following solution includes only the key steps. Sketch graphs as appropriate on scratch paper.

$$\sin^2 x = \tfrac{1}{2} \sin 2x \qquad \text{Use double-angle identity.}$$
$$= \tfrac{1}{2}(2 \sin x \cos x) \qquad \text{Subtract sin x cos x from both sides.}$$
$$\sin^2 x - \sin x \cos x = 0 \qquad \text{Factor out sin x.}$$
$$\sin x (\sin x - \cos x) = 0 \qquad a(a - b) = 0 \text{ only if } a = 0 \text{ or } a - b = 0.$$

$$\sin x = 0 \qquad \text{or} \qquad \sin x - \cos x = 0$$
$$x = 0, \pi \qquad\qquad \sin x = \cos x \quad {\scriptstyle \text{sin x = cos x when x is } \pi/4, 5\pi/4.}$$
$$x = \pi/4, 5\pi/4$$

Combining the solutions from both equations, we have the complete set of solutions:

$$x = 0, \pi/4, \pi, 5\pi/4$$

MATCHED PROBLEM 4 Find exact solutions for $\sin 2x = \sin x$, $0 \le x \le 2\pi$.

EXAMPLE 5

Approximate Solutions Using Identities and the Quadratic Formula

Solve $\cos 2x = 4 \cos x - 2$ for all real x. Compute inverse functions to four decimal places.

SOLUTION **Step 1.** *Solve for cos x.*

$$\cos 2x = 4 \cos x - 2 \qquad \text{Use double-angle identity.}$$
$$2 \cos^2 x - 1 = 4 \cos x - 2 \qquad \text{Subtract 4 cos x − 2 from both sides.}$$
$$2 \cos^2 x - 4 \cos x + 1 = 0 \qquad {\scriptstyle \begin{array}{l}\text{Quadratic in cos x. Left side does not factor}\\\text{using integer coefficients. Solve for cos x}\\\text{using quadratic formula with } a = 2, b = -4,\\\text{and } c = 1.\end{array}}$$

$$\cos x = \frac{4 \pm \sqrt{16 - 4(2)(1)}}{2(2)}$$
$$= 1.707107 \text{ or } 0.292893$$

Step 2. *Solve each equation over one period $[0, 2\pi)$:* Sketch a graph of $y = \cos x$, $y = 1.707107$, and $y = 0.292893$ in the same coordinate system to provide an aid to writing all solutions over one period [Fig. 4(a)], or use a unit circle diagram [Fig. 4(b)].

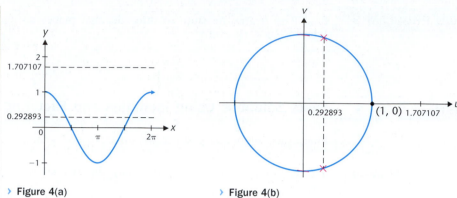

› **Figure 4(a)** › **Figure 4(b)**

Solve the first equation:

$$\cos x = 1.707107 \qquad \text{\color{red}No solution, because cosine is never greater than 1.}$$

Solve the second equation:

$$\cos x = 0.292893$$

Figures 4(a) and 4(b) indicate a first quadrant solution and a fourth quadrant solution. If the reference angle is α, then $x = \alpha$ or $x = 2\pi - \alpha$.

$$\alpha = \cos^{-1} 0.292893 = 1.2735$$
$$2\pi - \alpha = 2\pi - 1.2735 = 5.0096$$

CHECK

$$\cos 1.2735 = 0.292936; \ \cos 5.0096 = 0.292854$$

Step 3. *Write an expression for all solutions.* Because the cosine function is periodic with period 2π, all solutions are given by

$$x = \begin{cases} 1.2735 + 2k\pi \\ 5.0096 + 2k\pi \end{cases} \qquad k \text{ any integer}$$

MATCHED PROBLEM 5

Solve $\cos 2x = 2(\sin x - 1)$ for all real x. Compute inverse functions to four decimal places.

› ## Solving Trigonometric Equations Using a Graphing Calculator

The equations we've solved so far using algebraic methods can also be solved using a graphing calculator. The biggest difference is that we can almost never get exact answers using graphing calculator methods. But there are many trigonometric equations that are either extremely difficult or impossible to solve algebraically; solutions can usually be found to up to 8 or 10 decimal place accuracy with a graphing calculator.

EXAMPLE 6 ### Solution Using a Graphing Calculator

Find all real solutions to four decimal places for $2 \cos 2x = 1.35x - 2$.

SOLUTION This relatively simple trigonometric equation cannot be solved using a finite number of algebraic steps (try it!). However, it can be solved rather easily using a graphing calculator. Graph $y_1 = 2 \cos 2x$ and $y_2 = 1.35x - 2$ in the same viewing window, and find any points of intersection using the INTERSECT command. The first point of intersection is shown in Figure 5. It appears there may be more than one point of intersection, but zooming in on the portion of the graph in question shows that the two graphs do not intersect in that region (Fig. 6). The only solution is

$$x = 0.9639$$

CHECK

Left side: $2 \cos 2(0.9639) = -0.6989$ Slightly different due to rounding error.
Right side: $1.35(0.9639) - 2 = -0.6987$

> **Figure 5**

> **Figure 6**

MATCHED PROBLEM 6 Find all real solutions to four decimal places for $\sin x/2 = 0.2x - 0.5$.

EXAMPLE **7** **Solution Using a Graphing Calculator**

Find all real solutions, to four decimal places, for $\tan (x/2) = 1/x$, $-\pi < x \le 3\pi$.

SOLUTION Graph $y_1 = \tan (x/2)$ and $y_2 = 1/x$ in the same viewing window for $-\pi < x < 3\pi$ (Fig. 7). Solutions are at points of intersection; there are three at them.

Using the INTERSECT command, the three solutions are found to be

$$x = -1.3065, \ 1.3065, \ 6.5846$$

Checking these solutions is left to the reader.

> **Figure 7**

MATCHED PROBLEM 7 Find all real solutions, to four decimal places, for $0.25 \tan (x/2) = \ln x$, $0 < x < 4\pi$.

Solving trigonometric inequalities using a graphing calculator is as easy as solving trigonometric equations using a graphing calculator. Example 8 illustrates the process.

EXAMPLE 8 Solving a Trigonometric Inequality

Solve $\sin x - \cos x < 0.25x - 0.5$, using two-decimal-place accuracy.

SOLUTION Graph $y_1 = \sin x - \cos x$ and $y_2 = 0.25x - 0.5$ in the same viewing window (Fig. 8). The solution to the inequality is the intervals where the graph of y_1 is below (less than) the graph of y_2.

Finding the three points of intersection by the INTERSECT command, we see that the graph of y_1 is below the graph of y_2 on the following two intervals: $(-1.65, 0.52)$ and $(3.63, \infty)$. So the solution set to the inequality is $(-1.65, 0.52) \cup (3.63, \infty)$.

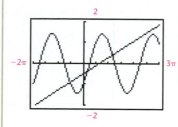

> **Figure 8**

MATCHED PROBLEM 8 Solve $\cos x - \sin x > 0.4 - 0.3x$, using two-decimal-place accuracy.

》》 EXPLORE-DISCUSS 2

How many solutions does the following equation have?

$$\sin (1/x) = 0 \qquad (1)$$

Graph $y_1 = \sin (1/x)$ and $y_2 = 0$ for each of the indicated intervals in parts A–G. From each graph estimate the number of solutions that equation (1) appears to have. What final conclusion would you be willing to make regarding the number of solutions to equation (1)? Explain.

(A) $[-20, 20]$; Can 0 be a solution? Explain.

(B) $[-2, 2]$ **(C)** $[-1, 1]$ **(D)** $[-0.1, 0.1]$ **(E)** $[-0.01, 0.01]$

(F) $[-0.001, 0.001]$ **(G)** $[-0.0001, 0.0001]$

ANSWERS TO MATCHED PROBLEMS

1. $x = \begin{cases} 0 + 2k\pi \\ \pi + 2k\pi \\ 7\pi/6 + 2k\pi \\ 11\pi/6 + 2k\pi \end{cases}$ k any integer 2. $x = \begin{cases} 1.8235 + 2k\pi \\ 4.4597 + 2k\pi \end{cases}$ k any integer

3. $-16.318°$ 4. $x = 0, \pi/3, \pi, 5\pi/3$ 5. $x = \begin{cases} 0.9665 + 2k\pi \\ 2.1751 + 2k\pi \end{cases}$ k any integer

6. $x = 5.1609$ 7. $x = 1.1828, 2.6369, 9.2004$ 8. $(-1.67, 0.64) \cup (3.46, \infty)$

7-5 Exercises

1. What is the difference between a conditional equation and an identity? What does it mean to solve a conditional equation?

2. Why do most trigonometric equations have infinitely many solutions?

3. For the equation $\sin x = 1/5$, there are two solutions between 0 and 2π: $x = 0.201$ and $x = 2.94$. Explain how you can use this information to write *all* solutions to the equation.

4. Explain how to solve a trigonometric equation using a graphing calculator.

In Problems 5–28, find exact solutions over the indicated intervals (x a real number, θ in degrees).

5. $\sin x = \dfrac{\sqrt{3}}{2}, 0 \le x \le 2\pi$

6. $\cos x = \dfrac{1}{\sqrt{2}}, 0 \le x \le 2\pi$

7. $\sin x = \dfrac{\sqrt{3}}{2}$, all real x

8. $\cos x = \dfrac{1}{\sqrt{2}}$, all real x

9. $\sin x + 1 = 0, 0 \le x < 2\pi$

10. $\cos x - 1 = 0, 0 \le x < 2\pi$

11. $\sin x + 1 = 0$, all real x

12. $\cos x - 1 = 0$, all real x

13. $\tan \theta - 1 = 0, 0° \le \theta < 360°$

14. $\tan \theta + 1 = 0, 0° \le \theta < 360°$

15. $\tan \theta - 1 = 0$, all θ

16. $\tan \theta + 1 = 0$, all θ

17. $2 \sin x - 1 = 0, 0 \le x < 2\pi$

18. $2 \cos x - 1 = 0, 0 \le x < 2\pi$

19. $2 \sin x - 1 = 0$, all real x

20. $2 \cos x - 1 = 0$, all real x

21. $2 \sin \theta + \sqrt{3} = 0, 0° \le \theta < 360°$

22. $\sqrt{2} \cos \theta - 1 = 0, 0° \le \theta < 360°$

23. $2 \sin \theta + \sqrt{3} = 0$, all θ

24. $\sqrt{2} \cos \theta - 1 = 0$, all θ

25. $\tan x - \sqrt{3} = 0, 0 \le x < 2\pi$

26. $\tan x - \sqrt{3} = 0$, all real x

27. $2 \cos x + 4 = 0$, all real x

28. $-3 \sin x - 4 = 5$, all real x

Solve the equation in Problems 29–36 to four decimal places (θ in degrees, x real).

29. $7 \cos x - 3 = 0, 0 \le x < 2\pi$

30. $5 \cos x - 2 = 0, 0 \le x < 2\pi$

31. $2 \tan \theta - 7 = 0, 0° \le \theta < 180°$

32. $4 \tan \theta + 15 = 0, 0° \le \theta < 180°$

33. $\sqrt{10} \cos 2x + 6 = 0$, all real x

34. $\dfrac{1}{5} \cos x + \dfrac{3}{5} = 0$, all real x

35. $\sqrt{5} \sin x - 2 = 0$, all real x

36. $\dfrac{11}{3} - 5 \sin x = 0$, all real x

Solve the equation in Problems 37–40 to four decimal places using a graphing calculator.

37. $1 - x = 2 \sin x$, all real x

38. $2x - \cos x = 0$, all real x

39. $\tan (x/2) = 8 - x, 0 \le x < \pi$

40. $\tan 2x = 1 + 3x, 0 \le x < \pi/4$

In Problems 41–56, find exact solutions for x real and θ in degrees.

41. $2 \sin^2 \theta + \sin 2\theta = 0$, all θ

42. $\cos^2 \theta = \frac{1}{2} \sin 2\theta$, all θ

43. $\tan x = -2 \sin x, 0 \le x < 2\pi$

44. $\cos x = \cot x, 0 \le x < 2\pi$

45. $\sec (x/2) + 2 = 0, 0 \le x < 2\pi$

46. $\tan (x/2) - 1 = 0, 0 \le x < 2\pi$

47. $2 \cos^2 \theta + 3 \sin \theta = 0, 0° \le \theta < 360°$

48. $\sin^2 \theta + 2 \cos \theta = -2, 0° \le \theta < 360°$

49. $\cos 2\theta + \cos \theta = 0, 0° \le \theta < 360°$

50. $\cos 2\theta + \sin^2 \theta = 0, 0° \le \theta < 360°$

51. $2 \sin^2 (x/2) - 3 \sin (x/2) + 1 = 0, 0 \le x \le 2\pi$

52. $4 \cos^2 2x - 4 \cos 2x + 1 = 0, 0 \le x \le 2\pi$

53. $\cos^2 x + \sin^2 x = 1, 0 \le x < 2\pi$

54. $\sin x + \cos x = 3, 0 \le x < 2\pi$

55. $2 \sin \theta = 5 + \cos \theta, 0° \le \theta < 360°$

56. $2 \cos^2 \theta = 1 + \cos 2\theta, 0° \le \theta < 360°$

Solve the equation in Problems 57–62 (x real and θ in degrees). Compute inverse functions to four significant digits.

57. $6 \sin^2 \theta + 5 \sin \theta = 6, 0° \le \theta \le 90°$

58. $4 \cos^2 \theta = 7 \cos \theta + 2, 0° \le \theta \le 180°$

59. $3 \cos^2 x - 8 \cos x = 3, 0 \le x \le \pi$

60. $8 \sin^2 x + 10 \sin x = 3, 0 \le x \le \pi/2$

61. $2 \sin x = \cos 2x, 0 \le x < 2\pi$

62. $\cos 2x + 10 \cos x = 5, 0 \le x < 2\pi$.

Solve the equation in Problems 63 and 64 for all real number solutions. Compute inverse functions to four significant digits.

63. $2 \sin^2 x = 1 - 2 \sin x$

64. $\cos^2 x = 3 - 5 \cos x$

Solve the equation in Problems 65–72 to four decimal places using a graphing calculator.

65. $2 \sin x = \cos 2x, 0 \le x < 2\pi$

66. $\cos 2x + 10 \cos x = 5, 0 \le x < 2\pi$

67. $2 \sin^2 x = 1 - 2 \sin x$, all real x

68. $\cos^2 x = 3 - 5 \cos x$, all real x

69. $\cos 2x > x^2 - 2$, all real x

70. $2 \sin (x - 2) < 3 - x^2$, all real x

71. $\cos (2x + 1) \le 0.5x - 2$, all real x

72. $\sin (3 - 2x) \ge 1 - 0.4x$, all real x

73. Explain why you can solve the inequality $2 \cos x - 3 < 0$ easily with just a bit of algebra.

74. Explain why you can solve the inequality $5 \sin x + 7 < 0$ easily with just a bit of algebra.

75. Explain the difference between evaluating $\tan^{-1} (-5.377)$ and solving the equation $\tan x = -5.377$.

76. Explain the difference between evaluating $\cos^{-1} (-0.7334)$ and solving the equation $\cos x = -0.7334$.

Find exact solutions to the equation in Problems 77–80. [Hint: Square both sides at an appropriate point, solve, then eliminate extraneous solutions at the end.]

77. $\cos x - \sin x = 1, 0 \le x < 2\pi$

78. $\sin x + \cos x = 1, 0 \le x < 2\pi$

79. $\tan x - \sec x = 1, 0 \le x < 2\pi$

80. $\sec x + \tan x = 1, 0 \le x < 2\pi$

81. Suppose that we are asked to find the zeros of $f(x) = \sin (1/x)$ for $x > 0$.

(A) Explore the graph of f over different intervals $[0.1, b]$ for various values of b, $b > 0.1$. Does the function f have a largest zero? If so, what is it (to four decimal places)? Explain what happens to the graph of f as x increases without bound. Does the graph have an asymptote? If so, what is its equation?

(B) Explore the graph of f over different intervals $(0, b]$ for various values of b, $0 < b \le 0.1$. How many zeros exist between 0 and b, for any $b > 0$, however small? Explain why this happens. Does f have a smallest positive zero? Explain.

82. Suppose that we are asked to find the zeros of $g(x) = \cos (1/x)$ for $x > 0$.

(A) Explore the graph of g over different intervals $[0.1, b]$ for various values of b, $b > 0.1$. Does the function g have a largest zero? If so, what is it (to four decimal places)? Explain what happens to the graph of g as x increases without bound. Does the graph have an asymptote? If so, what is its equation?

(B) Explore the graph of g over different intervals $(0, b]$ for various values of b, $0 < b \le 0.1$. How many zeros exist between 0 and b, for any $b > 0$, however small? Explain why this happens. Does g have a smallest positive zero? Explain.

APPLICATIONS

83. PHYSICS The equation $y = -8 \cos 2t$ represents the motion of a weight hanging from a spring after it has been pulled 8 inches below its natural length and released (neglecting air resistance and friction). The output y is the position of the weight in inches above (positive y values) or below (negative y values) the starting point after t seconds. Find the first four times when the weight returns to its starting point.

84. PHYSICS Refer to Problem 83. Find the first four times the weight is 4 inches above its starting point.

85. ELECTRIC CURRENT An alternating current generator produces a current given by the equation

$$I = 30 \sin 120\pi t$$

where t is time in seconds and I is current in amperes. Find the smallest positive t (to four significant digits) such that $I = -10$ amperes.

86. ELECTRIC CURRENT Refer to Problem 85. Find the smallest positive t (to four significant digits) such that $I = 25$ amperes.

87. OPTICS A polarizing filter for a camera contains two parallel plates of polarizing glass, one fixed and the other able to rotate. If θ is the angle of rotation from the position of maximum light transmission, then the intensity of light leaving the filter is $\cos^2 \theta$ times the intensity I of light entering the filter (see the figure).

Polarizing filter
(schematic)

Light

I

$I \cos^2 \theta$

Find the smallest positive θ (in decimal degrees to two decimal places) so that the intensity of light leaving the filter is 40% of that entering.

88. OPTICS Refer to Problem 87. Find the smallest positive θ so that the light leaving the filter is 70% of that entering.

89. ASTRONOMY The planet Mercury travels around the sun in an elliptical orbit given approximately by

$$r = \frac{3.44 \times 10^7}{1 - 0.206 \cos \theta}$$

(see the figure). Find the smallest positive θ (in decimal degrees to three significant digits) such that Mercury is 3.09×10^7 miles from the sun.

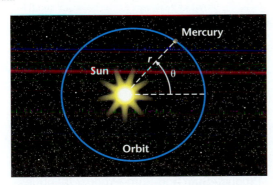

Mercury

Sun

r

θ

Orbit

90. ASTRONOMY Refer to Problem 89. Find the smallest positive θ (in decimal degrees to three significant digits) such that Mercury is 3.78×10^7 miles from the sun.

91. GEOMETRY The area of the segment of a circle in the figure is given by

$$A = \tfrac{1}{2} R^2 (\theta - \sin \theta)$$

where θ is in radian measure. Use a graphing calculator to find the radian measure, to three decimal places, of angle θ, if the radius is 8 inches and the area of the segment is 48 square inches.

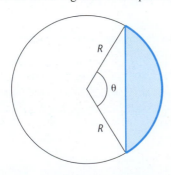

R

θ

R

92. GEOMETRY Repeat Problem 91, if the radius is 10 centimeters and the area of the segment is 40 square centimeters.

93. EYE SURGERY A surgical technique for correcting an astigmatism involves removing small pieces of tissue to change the curvature of the cornea.* In the cross section of a cornea shown in the figure, the circular arc, with radius R and central angle 2θ, represents a cross section of the surface of the cornea.
(A) If $a = 5.5$ millimeters and $b = 2.5$ millimeters, find L correct to four decimal places.
(B) Reducing the chord length $2a$ without changing the length L of the arc has the effect of pushing the cornea outward and giving it a rounder, yet still a circular, shape. With the aid of a graphing calculator in part of the solution, approximate b to four decimal places if a is reduced to 5.4 millimeters and L remains the same as it was in part A.

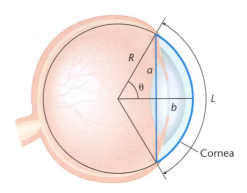

R

a

θ

b

L

Cornea

94. EYE SURGERY Refer to Problem 93.
(A) If in the figure $a = 5.4$ millimeters and $b = 2.4$ millimeters, find L correct to four decimal places.
(B) Increasing the chord length without changing the arc length L has the effect of pulling the cornea inward and giving it a flatter, yet still circular, shape. With the aid of a graphing calculator in part of the solution, approximate b to four decimal places if a is increased to 5.5 millimeters and L remains the same as it was in part A.

ANALYTIC GEOMETRY *Find simultaneous solutions for each system of equations in Problems 95 and 96 ($0° \le \theta \le 360°$). These are polar equations, which will be discussed in Chapter 8.*

95. $r = 2 \sin \theta$
$r = \sin 2\theta$

96. $r = 2 \sin \theta$
$r = 2(1 - \sin \theta)$

97. MODELING NUMBER OF DAYLIGHT HOURS The formula $y = 2.818 \sin (0.5108x - 1.605) + 12.14$ can be used to model the number of hours of daylight in Columbus, Ohio, on the 15th of each month, where x is the month, with $x = 1$ corresponding to January 15, $x = 2$ corresponding to February 15, and so on. When does Columbus have exactly 12 hours of daylight?

98. MODELING NUMBER OF DAYLIGHT HOURS The formula $y = 1.912 \sin (0.511x - 1.608) + 12.13$ can be used to model the number of hours of daylight in New Orleans on the 15th of each month, where x is the month, with $x = 1$ corresponding to January 15, $x = 2$ corresponding to February 15, and so on. When does New Orleans have exactly 14 hours of daylight?

*Based on the article "The Surgical Correction of Astigmatism" by Sheldon Rothman and Helen Strassberg in the *UMAP Journal*, Vol. v, no. 2, 1984.

CHAPTER 7 Review

7-1 Basic Identities and Their Use

The following 11 identities are basic to the process of changing trigonometric expressions to equivalent but more useful forms:

Reciprocal Identities

$$\csc x = \frac{1}{\sin x} \qquad \sec x = \frac{1}{\cos x} \qquad \cot x = \frac{1}{\tan x}$$

Quotient Identities

$$\tan x = \frac{\sin x}{\cos x} \qquad \cot x = \frac{\cos x}{\sin x}$$

Identities for Negatives

$$\sin(-x) = -\sin x \qquad \cos(-x) = \cos x$$
$$\tan(-x) = -\tan x$$

Pythagorean Identities

$$\sin^2 x + \cos^2 x = 1 \qquad \tan^2 x + 1 = \sec^2 x$$
$$1 + \cot^2 x = \csc^2 x$$

Although there is no fixed method of verification that works for all identities, the following suggested steps are helpful in many cases.

Suggested Steps in Verifying Identities

1. Start with the more complicated side of the identity, and transform it into the simpler side.

2. Try algebraic operations such as multiplying, factoring, combining fractions, and splitting fractions.

3. If other steps fail, express each function in terms of sine and cosine functions, and then perform appropriate algebraic operations.

4. At each step, keep the other side of the identity in mind. This often reveals what you should do to get there.

7-2 Sum, Difference, and Cofunction Identities

Sum Identities

$$\sin(x + y) = \sin x \cos y + \cos x \sin y$$
$$\cos(x + y) = \cos x \cos y - \sin x \sin y$$
$$\tan(x + y) = \frac{\tan x + \tan y}{1 - \tan x \tan y}$$

Difference Identities

$$\sin(x - y) = \sin x \cos y - \cos x \sin y$$
$$\cos(x - y) = \cos x \cos y + \sin x \sin y$$
$$\tan(x - y) = \frac{\tan x - \tan y}{1 + \tan x \tan y}$$

Cofunction Identities

(Replace $\pi/2$ with $90°$ if x is in degrees.)

$$\cos\left(\frac{\pi}{2} - x\right) = \sin x$$
$$\sin\left(\frac{\pi}{2} - x\right) = \cos x$$
$$\tan\left(\frac{\pi}{2} - x\right) = \cot x$$

7-3 Double-Angle and Half-Angle Identities

Double-Angle Identities

$$\sin 2x = 2 \sin x \cos x$$
$$\cos 2x = \cos^2 x - \sin^2 x = 1 - 2 \sin^2 x = 2 \cos^2 x - 1$$
$$\tan 2x = \frac{2 \tan x}{1 - \tan^2 x} = \frac{2 \cot x}{\cot^2 x - 1} = \frac{2}{\cot x - \tan x}$$

Half-Angle Identities

$$\sin\frac{x}{2} = \pm\sqrt{\frac{1 - \cos x}{2}}$$
$$\cos\frac{x}{2} = \pm\sqrt{\frac{1 + \cos x}{2}}$$
$$\tan\frac{x}{2} = \pm\sqrt{\frac{1 - \cos x}{1 + \cos x}} = \frac{\sin x}{1 + \cos x} = \frac{1 - \cos x}{\sin x}$$

7-4 Product–Sum and Sum–Product Identities

Product–Sum Identities

$$\sin x \cos y = \tfrac{1}{2}[\sin(x + y) + \sin(x - y)]$$
$$\cos x \sin y = \tfrac{1}{2}[\sin(x + y) - \sin(x - y)]$$
$$\sin x \sin y = \tfrac{1}{2}[\cos(x - y) - \cos(x + y)]$$
$$\cos x \cos y = \tfrac{1}{2}[\cos(x + y) + \cos(x - y)]$$

Sum–Product Identities

$$\sin x + \sin y = 2 \sin\frac{x + y}{2} \cos\frac{x - y}{2}$$
$$\sin x - \sin y = 2 \cos\frac{x + y}{2} \sin\frac{x - y}{2}$$
$$\cos x + \cos y = 2 \cos\frac{x + y}{2} \cos\frac{x - y}{2}$$
$$\cos x - \cos y = -2 \sin\frac{x + y}{2} \sin\frac{x - y}{2}$$

7-5 Trigonometric Equations

Sections 7-1 through 7-4 of the chapter considered trigonometric equations called **identities.** Identities are true for all replacements of the variable(s) for which both sides are defined. Section 7-5 considered **conditional equations.** Conditional equations may be true for some variable replacements, but are false for other variable replacements for which both sides are defined. The equation $\sin x = \cos x$ is an example of a conditional equation.

In *solving a trigonometric equation using an algebraic approach,* no particular rule will always lead to all solutions of every trigonometric equation you are likely to encounter. Solving trigonometric equations algebraically often requires the use of algebraic manipulation, identities, and ingenuity.

Suggestions for Solving Trigonometric Equations Algebraically

1. Treat one particular trigonometric function like a variable, and solve for it.
 (A) Consider using algebraic manipulation such as factoring, combining or separating fractions, and so on.
 (B) Consider using identities.

2. After solving for a trigonometric function, solve for the variable.

In *solving a trigonometric equation using a graphing calculator approach* you can solve a larger variety of problems than with the algebraic approach. The solutions are generally approximations (to whatever decimal accuracy desired).

CHAPTER **7** **Review Exercises**

Work through all the problems in this chapter review and check answers in the back of the book. Answers to all review problems, except verifications, are there, and following each answer is a number in italics indicating the section in which that type of problem is discussed. Where weaknesses show up, review appropriate sections in the text.

Verify each identity in Problems 1–4.

1. $\tan x + \cot x = \sec x \csc x$

2. $\sec^4 x - 2 \sec^2 x \tan^2 x + \tan^4 x = 1$

3. $\dfrac{1}{1 - \sin x} + \dfrac{1}{1 + \sin x} = 2 \sec^2 x$

4. $\cos\left(x - \dfrac{3\pi}{2}\right) = -\sin x$

In Problems 5–7, prove that the equation is not an identity by finding a value of x for which both sides are defined, but are not equal.

5. $\sin (xy) = \sin x \sin y$

6. $\cos (4x) = 4 \cos x$

7. $\sin \dfrac{x}{2} = \sqrt{\dfrac{1 - \cos x}{2}}$

8. Write as a sum: $\sin 5\alpha \cos 3\alpha$.

9. Write as a product: $\cos 7x - \cos 5x$.

10. Simplify: $\sin\left(x + \dfrac{9\pi}{2}\right)$.

Solve the equation in Problems 11 and 12 exactly (θ in degrees, x real).

11. $\sqrt{2} \cos \theta + 1 = 0$, all θ

12. $\sin x \tan x - \sin x = 0$, all real x

Solve the equation in Problems 13–16 to four decimal places (θ in degrees and x real).

13. $\sin x = 0.7088$, all real x

14. $\cos \theta = 0.2557$, all θ

15. $\cot x = -0.1692$, $-\pi/2 < x < \pi/2$

16. $3 \tan (11 - 3x) = 23.46$, $-\pi/2 < 11 - 3x < \pi/2$

17. Use a graphing calculator to test whether each of the following is an identity. If an equation appears to be an identity, verify it. If the equation does not appear to be an identity, find a value of x for which both sides are defined but are not equal.
 (A) $(\sin x + \cos x)^2 = 1 - 2 \sin x \cos x$
 (B) $\cos^2 x - \sin^2 x = 1 - 2 \sin^2 x$

Verify each identity in Problems 18–25.

18. $\dfrac{1 - 2 \cos x - 3 \cos^2 x}{\sin^2 x} = \dfrac{1 - 3 \cos x}{1 - \cos x}$

19. $(1 - \cos x)(\csc x + \cot x) = \sin x$

20. $\dfrac{1 + \sin x}{\cos x} = \dfrac{\cos x}{1 - \sin x}$

21. $\cos 2x = \dfrac{1 - \tan^2 x}{1 + \tan^2 x}$

22. $\cot x - \tan x = \dfrac{4 \cos^2 x - 2}{\sin 2x}$

23. $\left(\dfrac{1 - \cot x}{\csc x}\right)^2 = 1 - \sin 2x$

24. $\tan m + \tan n = \dfrac{\sin (m + n)}{\cos m \cos n}$

25. $\tan (x + y) = \dfrac{\cot x + \cot y}{\cot x \cot y - 1}$

26. Use a sum identity to find the exact value of $\tan 75°$.

27. Use a difference identity to find the exact value of $\cos (\pi/12)$.

28. Use a half-angle identity to find the exact value of $\sin 105°$.

29. Use a half-angle identity to find the exact value of $\cos (7\pi/8)$.

Evaluate Problems 30 and 31 exactly using appropriate sum–product or product–sum identities.

30. $\cos 195° \sin 75°$

31. $\cos 195° + \cos 105°$

32. $\sin \dfrac{11\pi}{24} \sin \dfrac{5\pi}{24}$ **33.** $\sin \dfrac{5\pi}{12} - \sin \dfrac{\pi}{12}$

In Problems 34–37, is the equation an identity? Explain.

34. $\cot^2 x = \csc^2 x + 1$

35. $\cos 3x = \cos x (\cos 2x - 2 \sin^2 x)$

36. $\sin (x + 3\pi/2) = \cos x$ **37.** $\cos (x - 3\pi/2) = \sin x$

Solve the equation in Problems 38–42 exactly (θ in degrees, x real).

38. $4 \sin^2 x - 3 = 0, 0 \le x < 2\pi$

39. $2 \sin^2 \theta + \cos \theta = 1, 0° \le \theta \le 180°$

40. $2 \sin^2 x - \sin x = 0$, all real x

41. $\sin 2x = \sqrt{3} \sin x$, all real x

42. $2 \sin^2 \theta + 5 \cos \theta + 1 = 0$, all θ

Solve the equation in Problems 43–45 to four significant digits (θ in degrees, x real).

43. $\tan \theta = 0.2557$, all θ **44.** $\sin^2 x + 2 = 4 \sin x$, all real x

45. $\tan^2 x = 2 \tan x + 1, 0 \le x < \pi$

 Use a graphing calculator to solve the equation or inequality in Problems 46–49 to four decimal places.

46. $3 \sin 2x = 2x - 2.5$, all real x

47. $3 \sin 2x > 2x - 2.5$, all real x

48. $2 \sin^2 x - \cos 2x = 1 - x^2$, all real x

49. $2 \sin^2 x - \cos 2x \le 1 - x^2$, all real x

50. Given the equation $\tan (x + y) = \tan x + \tan y$:
(A) Is $x = 0$ and $y = \pi/4$ a solution?
(B) Is the equation an identity or a conditional equation? Explain.

51. Explain the difference in evaluating $\sin^{-1} 0.3351$ and solving the equation $\sin x = 0.3351$.

 52. Use a graphing calculator to test whether each of the following is an identity. If an equation appears to be an identity, verify it. If the equation does not appear to be an identity, find a value of x for which both sides are defined but are not equal.

(A) $\dfrac{\tan x}{\sin x + 2 \tan x} = \dfrac{1}{\cos x - 2}$

(B) $\dfrac{\tan x}{\sin x - 2 \tan x} = \dfrac{1}{\cos x - 2}$

53. (A) Solve $\tan (x/2) = 2 \sin x$ exactly, $0 \le x < 2\pi$, using algebraic methods.
 (B) Solve $\tan (x/2) = 2 \sin x$, $0 \le x < 2\pi$, to four decimal places using a graphing calculator.

 54. Solve $3 \cos (x - 1) = 2 - x^2$ for all real x, to three decimal places using a graphing calculator.

Solve Problems 55–57 exactly without the use of a calculator.

55. Given $\tan x = -\frac{3}{4}, \pi/2 \le x \le \pi$, find
(A) $\sin (x/2)$ (B) $\cos 2x$

56. $\sin \left[2 \tan^{-1} \left(-\frac{3}{4}\right)\right]$

57. $\sin \left[\sin^{-1} \left(\frac{3}{5}\right) + \cos^{-1} \left(\frac{4}{5}\right)\right]$

58. (A) Solve $\cos^2 2x = \cos 2x + \sin^2 2x, 0 \le x < \pi$, exactly using algebraic methods.
 (B) Solve $\cos^2 2x = \cos 2x + \sin^2 2x, 0 \le x < \pi$, to four decimal places using a graphing calculator.

APPLICATIONS

59. INDIRECT MEASUREMENT Find the exact value of x in the figure, then find x and θ to three decimal places. [*Hint:* Use a suitable identity involving $\tan 2\theta$.]

60. ELECTRIC CURRENT An alternating current generator produces a current given by the equation

$$I = 50 \sin 120\pi(t - 0.001)$$

where t is time in seconds and I is current in amperes. Find the smallest positive t, to three significant digits, such that $I = 40$ amperes.

61. MUSIC—BEAT FREQUENCIES The equations

$$y = 0.6 \cos 184\pi t \text{ and } y = -0.6 \cos 208\pi t$$

model sound waves with frequencies 92 and 104 hertz, respectively. If both sounds are emitted simultaneously, a beat frequency results.
(A) Show that

$$0.6 \cos 184\pi t - 0.6 \cos 208\pi t = 1.2 \sin 12\pi t \sin 196\pi t$$

 (B) Graph each of the following equations in a different viewing window for $0 \le t \le 0.2$.

$$y = 0.6 \cos 184\pi t$$
$$y = -0.6 \cos 208\pi t$$
$$y = 0.6 \cos 184\pi t - 0.6 \cos 208\pi t$$
$$y = 1.2 \sin 12\pi t \sin 196\pi t$$

 62. ENGINEERING The circular arch of a bridge has an arc length of 36 feet and spans a 32-foot canal (see the figure). Determine the height of the circular arch above the water at the center of the bridge, and the radius of the circular arch, both to three decimal places. Start by drawing auxiliary lines in the figure, labeling appropriate parts, then explain how the trigonometric equation

$$\sin \theta = \frac{8}{9} \theta$$

is related to the problem. After solving the trigonometric equation for θ, the radius is easy to find and the height of the arch above the water can be found with a little ingenuity.

63. CLIMATE The equation $y = 58.1 + 24.5 \sin (0.524x + 4.1)$ can be used to model the average high temperature in degrees Fahrenheit for Pittsburgh, Pennsylvania, where x is time in months, with $x = 1$ corresponding to January 15, $x = 2$ corresponding to February 15, and so on. When is the average high temperature above 70°?

64. CLIMATE Refer to Problem 63. When is the average high temperature exactly 30°?

CHAPTER 7

››› GROUP ACTIVITY From *M* sin *Bt* + *N* cos *Bt* to *A* sin (*Bt* + *C*)— A Harmonic Analysis Tool

In solving certain kinds of more advanced applied mathematical problems—problems dealing with electrical circuits, spring-mass systems, heat flow, and so on—the solution process leads naturally to a function of the form

$$y = M \sin Bt + N \cos Bt \qquad (1)$$

Functions like this are often easier to work with if transformed into the more familiar form

$$y = A \sin (Bt + C) \qquad (2)$$

The process of finding A, B, and C, given M, N, and B, requires a little ingenuity and the use of the sum identity

$$\sin (x + y) = \sin x \cos y + \cos x \sin y \qquad (3)$$

How do we proceed? We start by trying to get the right side of equation (1) to look like the right side of identity (3). Then we use equation (3), from right to left, to obtain equation (2).
(A) *Establishing a Transformation Identity.* Show that

$$y = M \sin Bt + N \cos Bt$$
$$= \sqrt{M^2 + N^2} \sin (Bt + C) \qquad (4)$$

where C is any angle (in radians if t is real) having $P = (M, N)$ on its terminal side. [*Hint:* A first step is the following:

$$M \sin Bt + N \cos Bt = \frac{\sqrt{M^2 + N^2}}{\sqrt{M^2 + N^2}} (M \sin Bt + N \cos Bt)]$$

(B) *Use of Transformation Identity.* Use equation (4) to transform

$$y_1 = -4 \sin (t/2) + 3 \cos (t/2)$$

into the form $y_2 = A \sin (Bt + C)$, where C is chosen so that $|C|$ is minimum. Compute C to three decimal places. From the new equation, determine the amplitude, period, and phase shift.
(C) *Graphing Calculator Visualization and Verification.* Graph y_1 and y_2 from part C in the same viewing window.
(D) *Physics Application.* A weight suspended from a spring, with spring constant 64, is pulled 4 centimeters below its equilibrium position and is then given a downward thrust to produce an initial downward velocity of 24 centimeters per second. In more advanced mathematics (differential

equations) the equation of motion (neglecting air resistance and friction) is found to be given approximately by

$$y_1 = -3 \sin 8t - 4 \cos 8t$$

where y_1 is the coordinate of the bottom of the weight in Figure 1 at time t (y is in centimeters and t is in seconds). Transform the equation into the form

$$y_2 = A \sin(Bt + C)$$

and indicate the amplitude, period, and phase shift of the motion. Choose the least positive C and keep A positive.

(E) *Graphing Calculator Visualization and Verification.* Graph y_1 and y_2 from part E in the same viewing window of a graphing calculator, $0 \le t \le 6$. How many times will the bottom of the weight pass $y = 2$ in the first 6 seconds?

(F) *Solving a Trigonometric Equation.* How long, to three decimal places, will it take the bottom of the weight to reach $y = 2$ for the first time?

> **Figure 1** Spring-mass system.

Additional Topics in Trigonometry

IN Chapter 8, a number of additional topics involving trigonometry are considered. First, we return to the problem of solving triangles—not just right triangles, but any triangle. Then some of these ideas are used to develop the important concept of vector. With our knowledge of trigonometry, we introduce the *polar coordinate system*, probably the most important coordinate system after the rectangular coordinate system. After considering polar equations and their graphs, we represent complex numbers in *polar form*. Once a complex number is in polar form, it will be possible to find *n*th powers and *n*th roots of the number using an ingenious theorem established by De Moivre.

CHAPTER

8

OUTLINE

8-1 Law of Sines

> ❯ **Law of Sines Derivation**
> ❯ **Solving the ASA and AAS Cases**
> ❯ **Solving the SSA Case—Including the Ambiguous Case**

Acute triangle
(a)

Obtuse triangle
(b)

> **Figure 1** Oblique triangles.

In Chapter 6 we used trigonometric functions to solve problems concerning right triangles. We now consider analogous problems for **oblique triangles**—triangles without a right angle.

Every oblique triangle is either **acute** (all angles between 0° and 90°) or **obtuse** (one angle between 90° and 180°). Figure 1 illustrates both types of triangles.

Note how the sides and angles of the oblique triangles in Figure 1 have been labeled: Side a is opposite angle α, side b is opposite angle β, and side c is opposite angle γ. Also note that the largest side of a triangle is opposite the largest angle. Given any three of the six quantities indicated in Figure 1, we are interested in finding the remaining three, if possible. This process is called *solving the triangle.*

If only the three angles of a triangle are known, it is impossible to solve for the sides. (Why?) But if we are given two angles and a side, or two sides and an angle, or all three sides, then it is possible to determine whether a triangle having the given quantities exists, and, if so, to solve for the remaining quantities.

The basic tools for solving oblique triangles are the *law of sines,* developed in Section 8-1, and the *law of cosines,* developed in Section 8-2.

Before proceeding with specific examples, it is important to recall the rules in Table 1 regarding accuracy of angle and side measure. Table 1 is repeated inside the front cover of the text for easy reference.

Table 1 Triangles and Significant Digits

Angle to Nearest	Significant Digits for Side Measure
1°	2
10′ or 0.1°	3
1′ or 0.01°	4
10″ or 0.001°	5

> ❯ **CALCULATOR CALCULATIONS**
>
> When solving for a particular side or angle, carry out all operations within the calculator and then round to the appropriate number of significant digits (as specified in Table 1) at the end of the calculation. Your answers may still differ slightly from those in the book, depending on the order in which you solve for the sides and angles.

❯ Law of Sines Derivation

The law of sines is relatively easy to prove using the right triangle properties studied in Chapter 6. We will also use the fact that

$$\sin (180° - x) = \sin x$$

Acute triangle
(a)

180° − γ

Obtuse triangle
(b)

› **Figure 2**

which is readily obtained using a difference identity (a good exercise for you). Referring to Figure 2, we proceed as follows: Angles α and β in Figure 2(a), and also in Figure 2(b), satisfy

$$\sin \alpha = \frac{h}{b} \quad \text{and} \quad \sin \beta = \frac{h}{a}$$

Solving each equation for h, we obtain

$$h = b \sin \alpha \quad \text{and} \quad h = a \sin \beta$$

So,

$$b \sin \alpha = a \sin \beta$$
$$\frac{\sin \alpha}{a} = \frac{\sin \beta}{b} \tag{1}$$

Similarly, angles α and γ in Figure 2(a), and also in Figure 2(b), satisfy

$$\sin \alpha = \frac{m}{c} \quad \text{and} \quad \sin \gamma = \sin (180° - \gamma) = \frac{m}{a}$$

Solving each equation for m, we obtain

$$m = c \sin \alpha \quad \text{and} \quad m = a \sin \gamma$$

So,

$$c \sin \alpha = a \sin \gamma$$
$$\frac{\sin \alpha}{a} = \frac{\sin \gamma}{c} \tag{2}$$

If we combine equations (1) and (2), we obtain the **law of sines.**

> **THEOREM 1 Law of Sines**

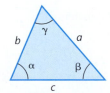

$$\frac{\sin \alpha}{a} = \frac{\sin \beta}{b} = \frac{\sin \gamma}{c}$$

In words, the ratio of the sine of an angle to its opposite side is the same as the ratio of the sine of either of the other angles to its opposite side.

Suppose that an angle of a triangle and its opposite side are known. Then the ratio of Theorem 1 can be calculated. So if one additional part of the triangle, either of the other angles or either of the other sides, is known, then the law of sines can be used to solve the triangle.

Therefore, the law of sines is used to solve triangles, given:

1. Two sides and an angle opposite one of them (SSA), or

2. Two angles and any side (ASA or AAS)

If the given information for a triangle consists of two sides and the included angle (SAS) or three sides (SSS), then the law of sines cannot be applied. The key to handling these two cases, the law of cosines, is developed in Section 8-2.

We will apply the law of sines to the easier ASA and AAS cases first, and then will turn to the more challenging SSA case.

> ### › Solving the ASA and AAS Cases

EXAMPLE **1** **Solving the ASA Case**

Solve the triangle in Figure 3.

> **Figure 3**

SOLUTION We are given two angles and the included side, which is the ASA case. Find the third angle, then solve for the other two sides using the law of sines.

First, we solve for γ:

$$\alpha + \beta + \gamma = 180°$$
$$\gamma = 180° - (\alpha + \beta)$$
$$= 180° - (28°0' + 45°20')$$
$$= 106°40'$$

Next, we solve for a:

$$\frac{\sin \alpha}{a} = \frac{\sin \gamma}{c} \qquad \text{Law of sines}$$

$$a = \frac{c \sin \alpha}{\sin \gamma}$$

$$= \frac{120 \sin 28°0'}{\sin 106°40'}$$

$$= 58.8 \text{ meters}$$

Finally, we solve for b:

$$\frac{\sin \beta}{b} = \frac{\sin \gamma}{c} \qquad \text{Law of sines}$$

$$b = \frac{c \sin \beta}{\sin \gamma}$$

$$= \frac{120 \sin 45°20'}{\sin 106°40'}$$

$$= 89.1 \text{ meters}$$

MATCHED PROBLEM 1 Solve the triangle in Figure 4.

> **Figure 4**

Note that the AAS case can always be converted to the ASA case by first solving for the third angle. For the ASA or AAS case to determine a unique triangle, the sum of the two angles must be between 0° and 180°, because the sum of all three angles in a triangle is 180° and no angle can be zero or negative.

› Solving the SSA Case—Including the Ambiguous Case

We now look at the case where we are given two sides and an angle opposite one of the sides—the SSA case. This case has several possible outcomes, depending on the measures of the two sides and the angle. Table 2 illustrates the various possibilities.

Table 2 SSA Variations

α	a $[h = b \sin \alpha]$	Number of Triangles	Figure	Case
Acute	$0 < a < h$	0		(a)
Acute	$a = h$	1		(b)
Acute	$h < a < b$	2	Ambiguous case	(c)
Acute	$a \geq b$	1		(d)
Obtuse	$0 < a \leq b$	0		(e)
Obtuse	$a > b$	1		(f)

It is unnecessary to memorize Table 2 to solve triangles in the SSA case. Instead, given sides a, b, and angle α, we use the law of sines to solve for the angle β opposite side b. The number of triangles is equal to the number of solutions β, $0° < \beta < 180°$, of the law of sines equation

$$\frac{\sin \beta}{b} = \frac{\sin \alpha}{a} \qquad (3)$$

that satisfy

$$\alpha + \beta < 180° \qquad (4)$$

In practice, we check each solution of equation (3) to determine whether inequality (4) is satisfied. If it is, we can easily solve for the remaining parts of the triangle. Examples 2–4 will make the procedure clear.

EXAMPLE 2 The SSA Case: One Triangle

Solve the triangle(s) with $a = 47$ centimeters, $b = 23$ centimeters, and $\alpha = 123°$.

SOLUTION First, we can solve for β:

$$\frac{\sin \beta}{b} = \frac{\sin \alpha}{a} \qquad \text{Law of sines}$$

$$\sin \beta = \frac{b \sin \alpha}{a} = \frac{23 \sin 123°}{47}$$

This equation has two solutions between $0°$ and $180°$:

$$\beta = \sin^{-1}\left(\frac{23 \sin 123°}{47}\right) = 24°$$

$$\beta' = 180° - 24° = 156°$$

Because

$$\alpha + \beta = 123° + 24° = 147° < 180°$$

$$\alpha + \beta' = 123° + 156° = 279° \geq 180°$$

there is only one triangle. [Note that this conclusion is consistent with Table 2. Because α is obtuse and $a > b$, we are in Case (f).]

Next, we solve for γ:

$$\alpha + \beta + \gamma = 180°$$

$$\gamma = 180° - 123° - 24° = 33°$$

Finally, we solve for c:

$$\frac{\sin \alpha}{a} = \frac{\sin \gamma}{c}$$

$$c = \frac{a \sin \gamma}{\sin \alpha} = \frac{47 \sin 33°}{\sin 123°} = 31 \text{ centimeters}$$

MATCHED PROBLEM 2 Solve the triangle(s) with $a = 88$ meters, $b = 62$ meters, and $\alpha = 81°$.

EXAMPLE 3 The SSA Case: No Triangle

Solve the triangle(s) with $a = 27$ inches, $b = 28$ inches, and $\alpha = 110°$.

SOLUTION First, we can solve for β:

$$\frac{\sin \beta}{b} = \frac{\sin \alpha}{a} \qquad \text{Law of sines}$$

$$\sin \beta = \frac{b \sin \alpha}{a} = \frac{28 \sin 110°}{27}$$

This equation has two solutions between $0°$ and $180°$:

$$\beta = \sin^{-1}\left(\frac{28 \sin 110°}{27}\right) = 77°$$

$$\beta' = 180° - 77° = 103°$$

Because

$$\alpha + \beta = 110° + 77° = 187° \geq 180°$$

$$\alpha + \beta' = 110° + 103° = 213° \geq 180°$$

there is no triangle. [Note that this conclusion is consistent with Table 2. Because α is obtuse and $a \leq b$, we are in Case (e).]

MATCHED PROBLEM 3

Solve the triangle(s) with $a = 64$ feet, $b = 79$ feet, and $\alpha = 57°$.

EXAMPLE 4

The SSA Case: Two Triangles

Solve the triangle(s) with $a = 1.0$ meters, $b = 1.8$ meters, and $\alpha = 26°$.

SOLUTION

First, we can solve for β:

$$\frac{\sin \beta}{b} = \frac{\sin \alpha}{a} \qquad \text{Law of sines}$$

$$\sin \beta = \frac{b \sin \alpha}{a} = \frac{1.8 \sin 26°}{1.0}$$

This equation has two solutions between $0°$ and $180°$:

$$\beta = \sin^{-1}\left(\frac{1.8 \sin 26°}{1.0}\right) = 52°$$

$$\beta' = 180° - 52° = 128°$$

Because

$$\alpha + \beta = 26° + 52° = 78° < 180°$$

$$\alpha + \beta' = 26° + 128° = 154° < 180°$$

there are two triangles. [Note that this conclusion is consistent with Table 2. Because α is acute and $h = b \sin \alpha < a < b$, we are in Case (c), the ambiguous case.]

Next, we solve for γ and γ':

$$\gamma = 180° - 26° - 52° = 102°$$

$$\gamma' = 180° - 26° - 128° = 26°$$

Finally, we solve for c and c':

$$c = \frac{a \sin \gamma}{\sin \alpha} = \frac{1.0 \sin 102°}{\sin 26°} = 2.2 \text{ meters}$$

$$c' = \frac{a \sin \gamma'}{\sin \alpha} = \frac{1.0 \sin 26°}{\sin 26°} = 1.0 \text{ meters}$$

In summary:

Triangle I: $\beta = 52°$ $\gamma = 102°$ $c = 2.2$ meters

Triangle II: $\beta' = 128°$ $\gamma' = 26°$ $c' = 1.0$ meters

MATCHED PROBLEM 4 Solve the triangle(s) with $a = 8$ kilometers, $b = 10$ kilometers, and $\alpha = 35°$.

⟩⟩⟩ EXPLORE-DISCUSS 1 Sides a and b and acute angle α of a triangle are given. Explain which case(s) of Table 2 could apply if, in solving the triangle, it is found that

(A) $\sin \beta > 1$ **(B)** $\sin \beta = 1$ **(C)** $\sin \beta < 1$

The law of sines is useful in many applications, as can be seen in Example 5 and the applications in Exercises 8-1.

EXAMPLE 5 **Surveying**

To measure the length d of a lake (Fig. 5), a base line AB is established and measured to be 125 meters. Angles A and B are measured to be 41.6° and 124.3°, respectively. How long is the lake?

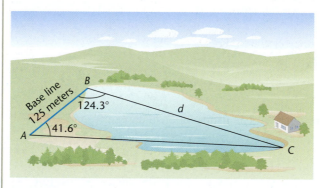

⟩ **Figure 5**

SOLUTION Find angle C and use the law of sines.

$$\text{Angle } C = 180° - (124.3° + 41.6°) \qquad \frac{\sin 14.1°}{125} = \frac{\sin 41.6°}{d}$$

$$= 14.1°$$

$$d = \frac{125 \sin 41.6°}{\sin 14.1°}$$

$$= 341 \text{ meters}$$

MATCHED PROBLEM 5 In Example 5, find the distance AC.

ANSWERS TO MATCHED PROBLEMS

1. $\gamma = 101°40'$, $b = 141$, $c = 152$ **2.** $\beta = 44°$, $\gamma = 55°$, $c = 73$ meters **3.** No solution
4. Triangle I: $\beta = 46°$, $\gamma = 99°$, $c = 14$ kilometers; Triangle II: $\beta' = 134°$, $\gamma' = 11°$,
 $c' = 2.7$ kilometers
5. 424 meters

8-1 Exercises

The labeling in the figure below is the convention we will follow in this exercise set. Your answers to some problems may differ slightly from those in the book, depending on the order in which you solve for the sides and angles of a given triangle.

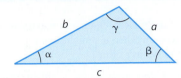

1. Explain why every oblique triangle is either acute or obtuse.

2. Explain why it is impossible to solve for the sides of a triangle if only its three angles are known.

3. Explain what the abbreviations SSA, ASA, AAS, SAS, and SSS mean in the context of solving triangles.

4. Explain how the AAS case can always be reduced to ASA.

5. Explain why the law of sines cannot be applied to the SAS or SSS cases.

6. Explain why one of the SSA variations is called the "ambiguous case."

Solve each triangle in Problems 7–14.

7. $\alpha = 73°$, $\beta = 28°$, $c = 42$ feet

8. $\alpha = 41°$, $\beta = 33°$, $c = 21$ centimeters

9. $\alpha = 122°$, $\gamma = 18°$, $b = 12$ kilometers

10. $\beta = 43°$, $\gamma = 36°$, $a = 92$ millimeters

11. $\beta = 112°$, $\gamma = 19°$, $c = 23$ yards

12. $\alpha = 52°$, $\gamma = 105°$, $c = 47$ meters

13. $\alpha = 52°$, $\gamma = 47°$, $a = 13$ centimeters

14. $\beta = 83°$, $\gamma = 77°$, $c = 25$ miles

In Problems 15–22, determine whether the information in each problem allows you to construct zero, one, or two triangles. Do not solve the triangle. Explain which case in Table 2 applies.

15. $a = 2$ inches, $b = 4$ inches, $\alpha = 30°$

16. $a = 3$ feet, $b = 6$ feet, $\alpha = 30°$

17. $a = 6$ inches, $b = 4$ inches, $\alpha = 30°$

18. $a = 8$ feet, $b = 6$ feet, $\alpha = 30°$

19. $a = 1$ inch, $b = 4$ inches, $\alpha = 30°$

20. $a = 2$ feet, $b = 6$ feet, $\alpha = 30°$

21. $a = 3$ inches, $b = 4$ inches, $\alpha = 30°$

22. $a = 5$ feet, $b = 6$ feet, $\alpha = 30°$

Solve each triangle in Problems 23–38. If a problem has no solution, say so.

23. $\alpha = 118.3°$, $\gamma = 12.2°$, $b = 17.3$ feet

24. $\beta = 27.5°$, $\gamma = 54.5°$, $a = 9.27$ inches

25. $\alpha = 67.7°$, $\beta = 54.2°$, $b = 123$ meters

26. $\alpha = 122.7°$, $\beta = 34.4°$, $b = 18.3$ kilometers

27. $\alpha = 46.5°$, $a = 7.9$ millimeters, $b = 13.1$ millimeters

28. $\alpha = 26.3°$, $a = 14.7$ inches, $b = 35.2$ inches

29. $\alpha = 15.9°$, $a = 22.4$ inches, $b = 29.6$ inches

30. $\alpha = 43.5°$, $a = 138$ centimeters, $b = 172$ centimeters

31. $\beta = 38.9°$, $a = 42.7$ inches, $b = 30.0$ inches

32. $\beta = 27.3°$, $a = 244$ centimeters, $b = 135$ centimeters

33. $\alpha = 123.2°$, $a = 101$ yards, $b = 152$ yards

34. $\alpha = 137.3°$, $a = 13.9$ meters, $b = 19.1$ meters

35. $\beta = 29°30'$, $a = 43.2$ millimeters, $b = 56.5$ millimeters

36. $\beta = 33°50'$, $a = 673$ meters, $b = 1,240$ meters

37. $\alpha = 30°$, $a = 29$ feet, $b = 58$ feet

38. $\beta = 30°$, $a = 92$ inches, $b = 46$ inches

39. Let $\alpha = 42.3°$ and $b = 25.2$ centimeters. Determine a value k so that if $0 < a < k$, there is no solution; if $a = k$, there is one solution; and if $k < a < b$, there are two solutions.

40. Let $\alpha = 37.3°$ and $b = 42.8$ centimeters. Determine a value k so that if $0 < a < k$, there is no solution; if $a = k$, there is one solution; and if $k < a < b$, there are two solutions.

41. **Mollweide's equation,**

$$(a - b) \cos \frac{\gamma}{2} = c \sin \frac{\alpha - \beta}{2}$$

is often used to check the final solution of a triangle, because all six parts of a triangle are involved in the equation. If the left side does not equal the right side after substitution, then an error has been made in solving a triangle. Use this equation to check Problem 7. (Because of rounding errors, both sides may not be exactly the same.)

42. (A) Use the law of sines and suitable identities to show that for any triangle

$$\frac{a - b}{a + b} = \frac{\tan \dfrac{\alpha - \beta}{2}}{\tan \dfrac{\alpha + \beta}{2}}$$

(B) Verify the formula with values from Problem 7.

APPLICATIONS

43. COAST GUARD Two lookout posts, A and B (10.0 miles apart), are established along a coast to watch for illegal ships coming within the 3-mile limit. If post A reports a ship S at angle $BAS = 37°30'$ and post B reports the same ship at angle $ABS = 20°0'$, how far is the ship from post A? How far is the ship from the shore (assuming the shore is along the line joining the two observation posts)?

44. FIRE LOOKOUT A fire at F is spotted from two fire lookout stations, A and B, which are 10.0 miles apart. If station B reports the fire at angle $ABF = 53°0'$ and station A reports the fire at angle $BAF = 28°30'$, how far is the fire from station A? From station B?

45. NATURAL SCIENCE The tallest trees in the world grow in Redwood National Park in California; they are taller than a football field is long. Find the height of one of these trees, given the information in the figure. (The 100-foot measurement is accurate to three significant digits.)

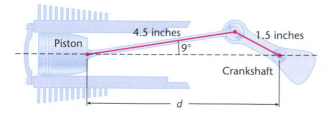

37°10'
44°0'
100 feet

46. SURVEYING To measure the height of Mt. Whitney in California, surveyors used a scheme like the one shown in the figure in Problem 45. They set up a horizontal base line 2,000 feet long at the foot of the mountain and found the angle nearest the mountain to be 43°5'; the angle farthest from the mountain was found to be 38°0'. If the base line was 5,000 feet above sea level, how high is Mt. Whitney above sea level?

47. ENGINEERING A 4.5-inch piston rod joins a piston to a 1.5-inch crankshaft (see the figure). How far is the base of the piston from the center of the crankshaft (distance d) when the rod makes an angle of 9° with the centerline? There are two answers to the problem.

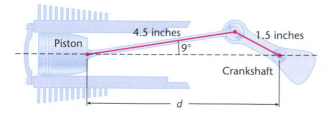

Piston 4.5 inches 9° 1.5 inches
Crankshaft
d

48. ENGINEERING Repeat Problem 47 if the piston rod is 6.3 inches, the crankshaft is 1.7 inches, and the angle is 11°.

49. ASTRONOMY The orbits of the Earth and Venus are approximately circular, with the sun at the center. A sighting of Venus is made from Earth, and the angle α is found to be 18°40'. If the radius of the orbit of the Earth is 1.495×10^8 kilometers and the radius of the orbit of Venus is 1.085×10^8 kilometers, what are the possible distances from the Earth to Venus? (See the figure.)

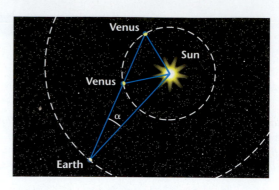

Venus
Sun
Venus
α
Earth

50. ASTRONOMY In Problem 49, find the maximum angle α. [*Hint:* The angle is maximum when a straight line joining the Earth and Venus is tangent to Venus's orbit.]

51. SURVEYING A tree growing on a hillside casts a 102-foot shadow straight down the hill (see the figure). Find the vertical height of the tree if, relative to the horizontal, the hill slopes 15.0° and the angle of elevation of the sun is 62.0°.

52. SURVEYING Find the height of the tree in Problem 51 if the shadow length is 157 feet and, relative to the horizontal, the hill slopes 11.0° and the angle of elevation of the sun is 42.0°.

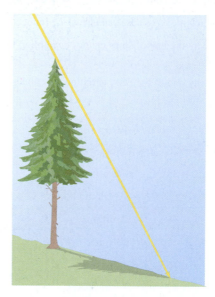

53. LIFE SCIENCE A cross-section of the cornea of an eye, a circular arc, is shown in the figure. Find the arc radius R and the arc length s, given the chord length $C = 11.8$ millimeters and the central angle $\theta = 98.9°$.

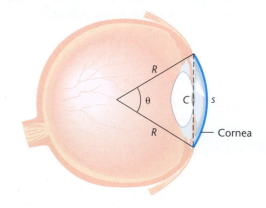

R
θ C s
R
Cornea

54. LIFE SCIENCE Referring to the preceding figure, find the arc radius R and the arc length s, given the chord length $C = 10.2$ millimeters and the central angle $\theta = 63.2°$.

55. SURVEYING The procedure illustrated in Problems 45 and 46 is used to determine an inaccessible height h when a base line d on a line perpendicular to h can be established (see the figure) and the angles α and β can be measured. Show that

$$h = d\left[\frac{\sin \alpha \sin \beta}{\sin (\beta - \alpha)}\right]$$

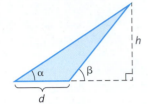

56. SURVEYING The layout in the figure is used to determine an inaccessible height h when a base line d in a plane perpendicular to h can be established and the angles α, β, and γ can be measured. Show that

$$h = d \sin \alpha \csc (\alpha + \beta) \tan \gamma$$

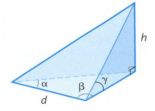

8-2 Law of Cosines

> Law of Cosines Derivation

> Solving the SAS Case

> Solving the SSS Case

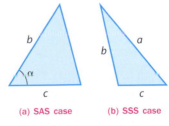

(a) SAS case (b) SSS case

> Figure 1

If in a triangle two sides and the included angle are given (SAS), or three sides are given (SSS), the law of sines cannot be used to solve the triangle—neither case involves an angle and its opposite side (Fig. 1). Both cases can be solved starting with the *law of cosines,* which is the subject matter for Section 8-2.

> ## Law of Cosines Derivation

Theorem 1 states the **law of cosines.**

> **THEOREM 1 Law of Cosines**

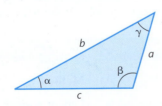

$$a^2 = b^2 + c^2 - 2bc \cos \alpha$$
$$b^2 = a^2 + c^2 - 2ac \cos \beta$$
$$c^2 = a^2 + b^2 - 2ab \cos \gamma$$

All three equations say essentially the same thing.

The law of cosines is used to solve triangles, given:

1. Two sides and the included angle (SAS), or

2. Three sides (SSS)

We will establish the first equation in Theorem 1. The other two equations then can be obtained from this one simply by relabeling the figure. We start by locating a triangle in a rectangular coordinate system. Figure 2 shows three typical triangles.

For an arbitrary triangle located as in Figure 2, the distance formula is used to obtain

$$a = \sqrt{(h - c)^2 + (k - 0)^2} \qquad \text{Square both sides.}$$

$$a^2 = (h - c)^2 + k^2 \qquad \text{Expand.} \tag{1}$$

$$= h^2 - 2hc + c^2 + k^2$$

› **Figure 2** Three representative triangles.

(a) (b) (c)

From Figure 2, we note that

$$b^2 = h^2 + k^2$$

Substituting b^2 for $h^2 + k^2$ in equation (1), we obtain

$$a^2 = b^2 + c^2 - 2hc \tag{2}$$

But

$$\cos \alpha = \frac{h}{b}$$

$$h = b \cos \alpha$$

By replacing h in equation (2) with $b \cos \alpha$, we reach our objective:

$$a^2 = b^2 + c^2 - 2bc \cos \alpha$$

[*Note:* If α is acute, then $\cos \alpha$ is positive; if α is obtuse, then $\cos \alpha$ is negative.]

› Solving the SAS Case

For the SAS case, start by using the law of cosines to find the side opposite the given angle. Then use either the law of cosines or the law of sines to find a second angle. Because of the simpler computations, the law of sines will generally be used to find the second angle. This gives the following strategy for solving the SAS case.

› **STRATEGY FOR SOLVING THE SAS CASE**

Step	Find	Method
1	Side opposite given angle	Law of cosines
2	Second angle (Find the angle opposite the shorter of the two given sides—this angle will always be acute.)	Law of sines
3	Third angle	Subtract the sum of the measures of the given angle and the angle found in step 2 from 180°.

EXAMPLE	1	**Solving the SAS Case**

Solve the triangle in Figure 3.

> **Figure 3**

SOLUTION

First, we solve for b using the law of cosines:

$$b^2 = a^2 + c^2 - 2ac \cos \beta \qquad\qquad \textcolor{red}{\text{Law of cosines}}$$
$$b = \sqrt{a^2 + c^2 - 2ac \cos \beta}$$
$$= \sqrt{(10.3)^2 + (6.45)^2 - 2(10.3)(6.45) \cos 32.4°}$$
$$= 5.96 \text{ cm}$$

Next, we solve for γ (the angle opposite the shorter side) using the law of sines:

$$\frac{\sin \gamma}{c} = \frac{\sin \beta}{b} \qquad\qquad \textcolor{red}{\text{Law of sines}}$$

$$\sin \gamma = \frac{c \sin \beta}{b} \qquad\qquad \textcolor{red}{\text{Solve for } \gamma.}$$

$$\gamma = \sin^{-1}\left(\frac{c \sin \beta}{b}\right) \qquad \textcolor{red}{\text{Because } \gamma \text{ is acute, the inverse sine function gives us } \gamma \text{ directly.}}$$

$$= \sin^{-1}\left(\frac{6.45 \sin 32.4°}{5.96}\right)$$

$$= 35.4°$$

Finally, we solve for α:

$$\alpha = 180° - (\beta + \gamma)$$
$$= 180° - (32.4° + 35.4°) = 112.2°$$

MATCHED PROBLEM 1	Solve the triangle with $\alpha = 77.5°$, $b = 10.4$ feet, and $c = 17.7$ feet.

> ## Solving the SSS Case

Starting with three sides of a triangle, the problem is to find the three angles. Subsequent calculations are simplified if we solve for the obtuse angle first, if present. The law of cosines is used for this purpose. A second angle, which must be acute, can be found using either law, although computations are usually simpler with the law of sines.

>>> **EXPLORE-DISCUSS 1**

(A) Starting with $a^2 = b^2 + c^2 - 2bc \cos \alpha$, show that

$$\alpha = \cos^{-1}\left(\frac{a^2 - b^2 - c^2}{-2bc}\right) \qquad\qquad (3)$$

(B) Does equation (3) give us the correct angle α regardless of whether α is acute or obtuse? Explain.

The preceding discussion leads to the following strategy for solving the SSS case:

> **STRATEGY FOR SOLVING THE SSS CASE**

Step	Find	Method
1	Angle opposite longest side—this will be the obtuse angle, if there is one.	Law of cosines
2	Either of the remaining angles, which will be acute. (Why?)	Law of sines
3	Third angle	Subtract the sum of the measures of the angles found in steps 1 and 2 from 180°.

EXAMPLE 2 Solving the SSS Case

Solve the triangle with $a = 27.3$ meters, $b = 17.8$ meters, and $c = 35.2$ meters.

SOLUTION Three sides of the triangle are given and we are to find the three angles. This is the SSS case.

Sketch the triangle (Fig. 4) and use the law of cosines to find the largest angle, then use the law of sines to find one of the two remaining acute angles.

First, we solve for γ using the law of cosines:

$$c^2 = a^2 + b^2 - 2ab \cos \gamma \qquad \text{Law of cosines}$$

$$\cos \gamma = \frac{a^2 + b^2 - c^2}{2ab} \qquad \text{Solve for } \gamma.$$

$$\gamma = \cos^{-1}\left(\frac{a^2 + b^2 - c^2}{2ab}\right)$$

$$= \cos^{-1}\left[\frac{(27.3)^2 + (17.8)^2 - (35.2)^2}{2(27.3)(17.8)}\right]$$

$$= 100.5°$$

Next, we solve for α using the law of sines:

$$\frac{\sin \alpha}{a} = \frac{\sin \gamma}{c} \qquad \text{Law of sines}$$

$$\sin \alpha = \frac{a \sin \gamma}{c} = \frac{27.3 \sin 100.5°}{35.2} \qquad \text{Solve for } \alpha.$$

$$\alpha = \sin^{-1}\left(\frac{27.3 \sin 100.5°}{35.2}\right)$$

$$= 49.7° \qquad \alpha \text{ is acute.}$$

Finally, we solve for β:

$$\alpha + \beta + \gamma = 180°$$

$$\beta = 180° - (\alpha + \gamma)$$

$$= 180° - (49.7° + 100.5°)$$

$$= 29.8°$$

17.8 m γ 27.3 m

α β

35.2 m

> **Figure 4**

MATCHED PROBLEM 2 Solve the triangle with $a = 1.25$ yards, $b = 2.05$ yards, and $c = 1.52$ yards.

EXAMPLE 3 ### Finding the Side of a Regular Polygon

If a seven-sided regular polygon is inscribed in a circle of radius 22.8 centimeters, find the length of one side of the polygon.

SOLUTION Sketch a figure (Fig. 5) and use the law of cosines.

$$d^2 = 22.8^2 + 22.8^2 - 2(22.8)(22.8) \cos \frac{360°}{7}$$

$$d = \sqrt{2(22.8)^2 - 2(22.8)^2 \cos \frac{360°}{7}}$$

$$= 19.8 \text{ centimeters}$$

Actually, you only need to sketch the triangle:

22.8 $\frac{360°}{7}$ 22.8

d

> **Figure 5**

MATCHED PROBLEM 3 If an 11-sided regular polygon is inscribed in a circle with radius 4.63 inches, find the length of one side of the polygon.

⟩⟩⟩ EXPLORE-DISCUSS 2 **(A)** The area of a rectangle is its length times its width. Use this fact to explain why the area of a right triangle is one-half the product of its legs (the legs are the sides adjacent to the right angle).

(B) Explain why the area A of any triangle is given by $A = \frac{1}{2} bh$ [*Hint*: Use right triangles.]

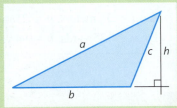

a c h

b

(C) The law of cosines can be used to derive the formula given by Heron of Alexandria (A.D. 75) for the area of any triangle in terms of the lengths a, b, and c of its sides:

$$A = \sqrt{s(s - a)(s - b)(s - c)} \qquad \text{where} \qquad s = \frac{a + b + c}{2}$$

is the **semiperimeter** of the triangle. Verify that Heron's formula gives the correct area for any equilateral triangle.

| EXAMPLE | 4 | **Finding the Area of a Triangle** |

Find the area of each triangle (to the same number of significant digits as the side with the least number of significant digits):

(A) $a = 4.2$ inches, $b = 8.7$ inches, $\gamma = 25°$
(B) $a = 3.52$ inches, $b = 2.91$ inches, $c = 4.67$ inches

SOLUTIONS

(A) We are given two sides and the included angle. Either of the known sides can be used as the base; we will use side a. The height h is the perpendicular distance to the base (extend the base if necessary), from the vertex that is not on the base (Fig. 6). Therefore,

> Figure 6

$$\frac{h}{8.7} = \sin 25° \qquad \text{Multiply both sides by 8.7.}$$

$$h = 8.7 \sin 25°$$

The area of any triangle (see Explore-Discuss 2) is given by

$$A = \tfrac{1}{2}(\text{base})(\text{height}) \qquad \text{Base} = 4.2, \text{ height} = 8.7 \sin 25°$$
$$= \tfrac{1}{2}(4.2)(8.7 \sin 25°)$$
$$= 7.7 \text{ square inches} \qquad \text{To two significant digits}$$

(B) Because we are given the lengths of all three sides, we can calculate the semiperimeter s and then use Heron's formula (see Explore-Discuss 2):

$$s = \frac{a + b + c}{2}$$
$$= \frac{3.52 + 2.91 + 4.67}{2}$$
$$= 5.55$$

$$A = \sqrt{s(s - a)(s - b)(s - c)} \qquad \text{Substitute values of } s, a, b, \text{ and } c.$$
$$= \sqrt{5.55(5.55 - 3.52)(5.55 - 2.91)(5.55 - 4.67)} \qquad \text{Calculate.}$$
$$= 5.12 \text{ square inches} \qquad \text{To three significant digits}$$

MATCHED PROBLEM 4

Find the area of each triangle (to the same number of significant digits as the side with the least number of significant digits):

(A) $a = 38$ meters, $b = 25$ meters, $\gamma = 74°$
(B) $a = 135$ yards, $b = 94$ yards, $c = 172$ yards

>>> CAUTION >>>

The formula for the area of a triangle,

$$A = \tfrac{1}{2}(\text{base})(\text{height}), \text{ is often written as } A = \tfrac{1}{2}bh.$$

Don't confuse the letter b in the formula, which stands for base, with the letters a, b, and c that we use to label the sides of a triangle. Remember that any known side of a triangle can serve as the base, provided the height h is calculated with respect to that base (see Example 4A).

8-2 Exercises

The labeling in the figure shown here is the convention we will
follow in this exercise set. Your answers to some problems may
differ slightly from those in the book, depending on the order in
which you solve for the sides and angles of a given triangle.

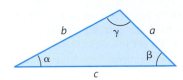

1. Explain how to solve a triangle given two of its sides and the
included angle.

2. Explain how to solve a triangle given all three of its sides.

3. Explain how the law of cosines simplifies if $\gamma = 90°$.

4. Explain how the law of cosines simplifies if $a = b = c$.

5. In Euclidean geometry, the SAS congruence theorem says that
if two sides and the included angle are congruent, respectively,
to two sides and the included angle of a second triangle, then
the two triangles are congruent. Is there an ASA congruence
theorem? An AAA congruence theorem? Explain.

6. Refer to Problem 5. Is there an SSA congruence theorem? An
SSS congruence theorem? Explain.

7. Referring to the figure, if $\alpha = 47.3°$, $b = 11.7$ centimeters, and
$c = 6.04$ centimeters, which of the two angles, β or γ, can you
say for certain is acute and why?

8. Referring to the figure, if $\alpha = 93.5°$, $b = 5.34$ inches, and
$c = 8.77$ inches, which of the two angles, β or γ, can you say
for certain is acute and why?

Solve each triangle in Problems 9–12.

9. $\alpha = 71.2°$, $b = 5.32$ yards, $c = 5.03$ yards

10. $\beta = 57.3°$, $a = 6.08$ centimeters, $c = 5.25$ centimeters

11. $\gamma = 120°20'$, $a = 5.73$ millimeters, $b = 10.2$ millimeters

12. $\alpha = 135°50'$, $b = 8.44$ inches, $c = 20.3$ inches

13. Referring to the figure at the beginning of the exercise set, if
$a = 13.5$ feet, $b = 20.8$ feet, and $c = 8.09$ feet, then if the trian-
gle has an obtuse angle, which angle must it be and why?

14. Suppose you are told that a triangle has sides $a = 12.5$ cen-
timeters, $b = 25.3$ centimeters, and $c = 10.7$ centimeters.
Explain why the triangle has no solution.

Solve each triangle in Problems 15–18 if the triangle has a
solution. Use decimal degrees for angle measure.

15. $a = 4.00$ meters, $b = 10.2$ meters, $c = 9.05$ meters

16. $a = 10.5$ miles, $b = 20.7$ miles, $c = 12.2$ miles

17. $a = 6.00$ kilometers, $b = 5.30$ kilometers,
$c = 5.52$ kilometers

18. $a = 31.5$ meters, $b = 29.4$ meters, $c = 33.7$ meters

Problems 19–34 represent a variety of problems involving both the
law of sines and the law of cosines. Solve each triangle. If a
problem does not have a solution, say so.

19. $\alpha = 94.5°$, $\gamma = 88.3°$, $b = 23.7$ centimeters

20. $\beta = 85.6°$, $\gamma = 97.3°$, $a = 14.3$ millimeters

21. $\beta = 104.5°$, $a = 17.2$ inches, $c = 11.7$ inches

22. $\beta = 27.3°$, $a = 13.7$ yards, $c = 20.1$ yards

23. $\alpha = 57.2°$, $\gamma = 112.0°$, $c = 24.8$ meters

24. $\beta = 132.4°$, $\gamma = 17.3°$, $b = 67.6$ kilometers

25. $\beta = 38.4°$, $a = 11.5$ inches, $b = 14.0$ inches

26. $\gamma = 66.4°$, $b = 25.5$ meters, $c = 25.5$ meters

27. $a = 32.9$ meters, $b = 42.4$ meters, $c = 20.4$ meters

28. $a = 10.5$ centimeters, $b = 5.23$ centimeters,
$c = 8.66$ centimeters

29. $\gamma = 58.4°$, $b = 7.23$ meters, $c = 6.54$ meters

30. $\alpha = 46.7°$, $a = 18.1$ meters, $b = 22.6$ meters

31. $\beta = 39.8°$, $a = 12.5$ inches, $b = 7.31$ inches

32. $\gamma = 47.9°$, $b = 35.2$ inches, $c = 25.5$ inches

33. $\beta = 13.6°$, $b = 21.6$ meters, $c = 58.4$ meters

34. $\beta = 25.1°$, $b = 53.7$ meters, $c = 98.5$ meters

In Problems 35–44, find the area of each triangle (to the same
number of significant digits as the side with the least number of
significant digits).

35. $a = 33$ yards, $b = 28$ yards, $\alpha = 90°$

36. $a = 542$ yards, $b = 167$ yards, $\alpha = 90°$

37. $a = 75$ meters, $b = 14$ meters, $\gamma = 37°$

38. $a = 183$ meters, $b = 10.1$ meters, $\gamma = 49.3°$

39. $a = 152$ feet, $b = 363$ feet, $\gamma = 112.5°$

40. $a = 42$ feet, $b = 210$ feet, $\gamma = 139°$

41. $\alpha = 72°$, $\beta = 48°$, $c = 2.6$ meters

42. $\alpha = 41°$, $\beta = 113°$, $c = 9.5$ meters

43. $a = 237$ yards, $b = 513$ yards, $c = 455$ yards

44. $a = 95$ yards, $b = 19$ yards, $c = 104$ yards

45. Show, using the law of cosines, that if $\gamma = 90°$, then $c^2 = a^2 + b^2$ (the Pythagorean theorem).

46. Show, using the law of cosines, that if $c^2 = a^2 + b^2$, then $\gamma = 90°$.

47. Show that for any triangle,

$$\frac{a^2 + b^2 + c^2}{2abc} = \frac{\cos \alpha}{a} + \frac{\cos \beta}{b} + \frac{\cos \gamma}{c}$$

48. Show that for any triangle,

$$a = b \cos \gamma + c \cos \beta$$

49. A triangle has sides $a = 1$, $b = \sqrt{3}$, and included angle $\gamma = 30°$. A student uses the law of cosines to find that $c = 1$, and then uses the law of sines to find that $\sin \beta = \sqrt{3}/2$. He concludes that $\beta = 60°$, so the third angle $\alpha = 90°$. But no right triangle has sides 1, $\sqrt{3}$, 1. Explain what is wrong with his strategy, and solve the triangle correctly.

50. A triangle has sides $a = 1$, $b = \sqrt{3}$, and $c = 1$. A student uses the law of cosines to find that $\alpha = 30°$, and then uses the law of sines to find that $\sin \beta = \sqrt{3}/2$. She concludes that $\beta = 60°$, so the third angle $\gamma = 90°$. But no right triangle has sides 1, $\sqrt{3}$, 1. Explain what is wrong with her strategy, and solve the triangle correctly.

APPLICATIONS

51. SURVEYING To find the length AB of a small lake, a surveyor measured angle ACB to be $96°$, AC to be 91 yards, and BC to be 71 yards. What is the approximate length of the lake?

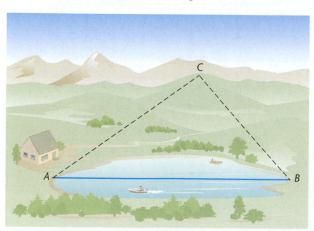

52. SURVEYING Refer to Problem 51. If a surveyor finds $\angle ACB = 110°$, $AC = 85$ meters, and $BC = 73$ meters, what is the approximate length of the lake?

53. GEOMETRY Find the measure in decimal degrees of a central angle subtended by a chord of length 112 millimeters in a circle of radius 72.8 millimeters.

54. GEOMETRY Find the measure in decimal degrees of a central angle subtended by a chord of length 13.8 feet in a circle of radius 8.26 feet.

55. GEOMETRY Two adjacent sides of a parallelogram meet at an angle of $35°10'$ and have lengths of 3 and 8 feet. What is the length of the shorter diagonal of the parallelogram (to three significant digits)?

56. GEOMETRY What is the length of the longer diagonal of the parallelogram in Problem 55 (to three significant digits)?

57. NAVIGATION Los Angeles and Las Vegas are approximately 200 miles apart. A pilot 80 miles from Los Angeles finds that she is $6°20'$ off course relative to her start in Los Angeles. How far is she from Las Vegas at this time? (Compute the answer to three significant digits.)

58. SEARCH AND RESCUE At noon, two search planes set out from San Francisco to find a downed plane in the ocean. Plane A travels due west at 400 miles per hour, and plane B flies northwest at 500 miles per hour. At 2 P.M. plane A spots the survivors of the downed plane and radios plane B to come and assist in the rescue. How far is plane B from plane A at this time (to three significant digits)?

59. GEOMETRY Find the perimeter of a pentagon inscribed in a circle of radius 12.6 meters.

60. GEOMETRY Find the perimeter of a nine-sided regular polygon inscribed in a circle of radius 7.09 centimeters.

61. ANALYTIC GEOMETRY If point A in the figure has coordinates $(3, 4)$ and point B has coordinates $(4, 3)$, find the radian measure of angle θ to three decimal places.

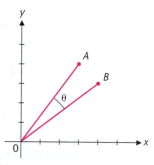

62. ANALYTIC GEOMETRY If point A has coordinates $(4, 3)$ and point B has coordinates $(5, 1)$, find the radian measure of $\angle AOB$ to three decimal places.

63. ENGINEERING Three circles of radius 2.03, 5.00, and 8.20 centimeters are tangent to one another (see the figure). Find the three angles formed by the lines joining their centers (to the nearest $10'$).

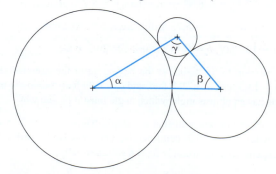

64. ENGINEERING Three circles of radius 2.00, 5.00, and 8.00 inches are tangent to each other (see the figure for Problem 63). Find the three angles formed by the lines joining their centers (to the nearest 10′).

65. GEOMETRY A rectangular solid has sides as indicated in the figure. Find ∠ CAB to the nearest degree.

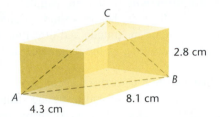

66. GEOMETRY Referring to Problem 65, find ∠ ACB to the nearest degree.

67. SPACE SCIENCE For communications between a space shuttle and the White Sands tracking station in southern New Mexico, two satellites are placed in geostationary orbit, 130° apart relative to the center of the Earth, and 22,300 miles above the surface of the Earth (see the figure). (A satellite in **geostationary** orbit remains stationary above a fixed point on the surface of the Earth.) Radio signals are sent from the tracking station by way of the satellites to the shuttle, and vice versa. How far to the nearest 100 miles is one of the geostationary satellites from the White Sands tracking station, W? The radius of the Earth is 3,964 miles.

68. SPACE SCIENCE A satellite S, in circular orbit around the Earth, is sighted by a tracking station T (see the figure). The distance TS is determined by radar to be 1,034 miles, and the angle of elevation above the horizon is 32.4°. How high is the satellite above the Earth at the time of the sighting? The radius of the Earth is 3,964 miles.

8-3 Vectors in the Plane

> Vectors
> Vector Addition and Scalar Multiplication
> Unit Vectors
> Velocity Vectors
> Force Vectors

Many physical quantities such as length, area, or volume can be completely specified by a single real number. They are called *scalar quantities*. Other quantities such as directed distances, velocities, and forces require both a magnitude and direction. They are *vector quantities*. Vector quantities have wide application in many areas of science and engineering.

In Section 8-3, we introduce the concept of a vector and study its applications. Although we restrict our discussion to vectors in the plane, the methods we introduce can be readily generalized to vectors in three-dimensional or higher-dimensional spaces.

> ## Vectors

A **vector v** is a quantity that has both magnitude and direction. We picture a vector as an arrow from an initial point O to a terminal point P with this provision: arrows that have the same length (magnitude) and direction represent the same vector (Fig. 1).

> **Figure 1** All three arrows represent the same vector v.

The vector **v** of Figure 1 is also denoted by $\overrightarrow{O_1P_1}$ (or $\overrightarrow{O_2P_2}$ or $\overrightarrow{O_3P_3}$). We use boldface letters such as **v** to denote vectors. But because it is difficult to write boldface by hand, we suggest that you use $\vec{v}$ as a substitute for **v** when you want to denote a vector by a single letter.

The **magnitude** of the vector $\mathbf{v} = \overrightarrow{OP}$, denoted by $|\mathbf{v}|$, $|\vec{v}|$, or $|\overrightarrow{OP}|$, is the length of the line segment OP. Two vectors have the **same direction** if they are parallel and point in the same direction. Two vectors have **opposite directions** if they are parallel and point in opposite directions. The **zero vector**, denoted by **0** or $\vec{0}$, has magnitude 0 and arbitrary direction. Two vectors are **equal** if they have the same magnitude and direction. So a vector can be translated from one location to another as long as the magnitude and direction do not change.

Any vector $\overrightarrow{AB}$ in a rectangular coordinate system can be translated so that its initial point is the origin O. The vector $\overrightarrow{OP}$ such that $\overrightarrow{OP} = \overrightarrow{AB}$ is said to be the **standard vector** for $\overrightarrow{AB}$ (Fig. 2). Note that $\overrightarrow{OP}$ is the standard vector for infinitely many vectors—all vectors with the same magnitude and direction as $\overrightarrow{OP}$.

Given the coordinates of the endpoints of vector $\overrightarrow{AB}$, how do we find its corresponding standard vector $\overrightarrow{OP}$? The coordinates of the origin O, the initial point of OP, are always $(0, 0)$. The coordinates of P, the terminal point of $\overrightarrow{OP}$, are given by

$$(x_p, y_p) = (x_b - x_a, y_b - y_a)$$

where $A = (x_a, y_a)$ and $B = (x_b, y_b)$.

> **Figure 2** $\overrightarrow{OP}$ is the standard vector for $\overrightarrow{AB}$.

EXAMPLE	1

Finding a Standard Vector for a Given Vector

Given the geometric vector $\overrightarrow{AB}$ with initial point $A = (3, 4)$ and terminal point $B = (7, -1)$, find the coordinates of the point P such that $\overrightarrow{OP} = \overrightarrow{AB}$.

SOLUTION The coordinates of P are given by

$$\begin{aligned}
(x_p, y_p) &= (x_b - x_a, y_b - y_a) \\
&= (7 - 3, -1 - 4) \\
&= (4, -5)
\end{aligned}$$

Note in Figure 3 that if we start at A, then move to the right four units and down five units, we will be at B. If we start at the origin, then move to the right four units and down five units, we will be at P.

> **Figure 3**

MATCHED PROBLEM 1 Given the geometric vector $\overrightarrow{AB}$ with initial point $A = (8, -3)$ and terminal point $B = (4, 5)$, find the standard vector $\overrightarrow{OP}$ for $\overrightarrow{AB}$.

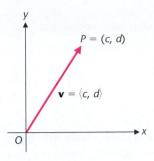

> **Figure 4** Vector v = ⟨c, d⟩.

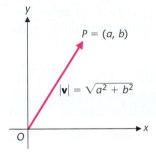

> **Figure 5** Magnitude of v = ⟨a, b⟩.

Example 1 suggests that there is a one-to-one correspondence between vectors in a rectangular coordinate system and points in the system. Any vector $\overrightarrow{AB}$ is completely specified by the point $P = (x_p, y_p)$ such that $\overrightarrow{OP} = \overrightarrow{AB}$ (we are not concerned that $\overrightarrow{OP}$ has a different position than $\overrightarrow{AB}$; we are free to translate a vector anywhere we please). Conversely, any point P of the system corresponds to the vector $\overrightarrow{OP}$.

A vector can be denoted by an ordered pair of real numbers. To avoid confusion, we use ⟨c, d⟩ to denote the vector $\overrightarrow{OP}$ with initial point (0, 0) and terminal point (c, d) (Fig. 4). The real numbers c and d are called the **scalar components** of the vector ⟨c, d⟩. Two vectors **u** = ⟨a, b⟩ and **v** = ⟨c, d⟩ are equal if their corresponding components are equal, that is, if $a = c$ and $b = d$. The zero vector is **0** = ⟨0, 0⟩. The magnitude of the vector ⟨a, b⟩ is the length of the line segment from (0, 0) to (a, b) [Fig. 5].

> **MAGNITUDE OF A VECTOR**
>
> The **magnitude,** or **norm,** of a vector **v** = ⟨a, b⟩ is denoted by |**v**| and given by
>
> $$|\mathbf{v}| = \sqrt{a^2 + b^2}$$

EXAMPLE	2	**Finding the Magnitude of a Vector**

Find the magnitude of the vector **v** = ⟨3, −5⟩.

SOLUTION
$$|\mathbf{v}| = \sqrt{3^2 + (-5)^2} = \sqrt{34}$$

MATCHED PROBLEM 2 Find the magnitude of the vector **v** = ⟨−2, 4⟩.

> **Figure 6** Vector addition: tail-to-tip rule.

› Vector Addition and Scalar Multiplication

The sum **u** + **v** of two vectors **u** and **v** is defined by the **tail-to-tip rule:** Translate **v** so that its tail (initial point) is at the tip (terminal point) of **u**. Then, the vector from the tail of **u** to the tip of **v** is the sum, denoted **u** + **v**, of the vectors **u** and **v** (Fig. 6).

If **u** and **v** are not parallel, the **parallelogram rule** gives an alternative description of **u** + **v**: The sum of two nonparallel vectors **u** and **v** is the diagonal of the parallelogram formed using **u** and **v** as adjacent sides (Fig. 7).

> **Figure 7** Vector addition: parallelogram rule.

The vector **u** + **v** is also called the **resultant** of the two vectors **u** and **v,** and **u** and **v** are called **vector components** of **u** + **v.**

>>> EXPLORE-DISCUSS 1

Use the tail-to-tip rule and/or the parallelogram rule to explain why vector addition is *commutative,* that is, for any vectors **u** and **v,**

$$\mathbf{u} + \mathbf{v} = \mathbf{v} + \mathbf{u}$$

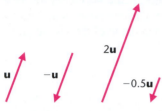

> **Figure 8** Scalar multiplication.

The **scalar product** $k\mathbf{u}$ of a scalar (real number) k and a vector **u** is the vector with magnitude $|k||\mathbf{u}|$ that has the same direction as **u** if k is positive and the opposite direction if k is negative. For example, 2**u** has twice the magnitude of **u** and the same direction. Similarly, $-0.5\mathbf{u}$ has half the magnitude of **u** and the opposite direction (Fig. 8).

Both the sum **u** + **v** and the scalar product $k\mathbf{u}$ are easy to calculate if the scalar components of **u** and **v** are given: for the sum, just add corresponding components; for the scalar product, multiply each component by the scalar.

› **VECTOR ADDITION AND SCALAR MULTIPLICATION**

If $\mathbf{u} = \langle a, b \rangle$ and $\mathbf{v} = \langle c, d \rangle$ are vectors and k is a scalar (real number), then

$$\mathbf{u} + \mathbf{v} = \langle a + c, b + d \rangle \qquad \textbf{Vector addition}$$
$$k\mathbf{u} = \langle ka, kb \rangle \qquad \textbf{Scalar multiplication}$$

EXAMPLE 3 **Vector Addition and Scalar Multiplication**

Let $\mathbf{u} = \langle 4, -3 \rangle$, $\mathbf{v} = \langle 2, 3 \rangle$, and $\mathbf{w} = \langle 0, -5 \rangle$. Find

(A) $\mathbf{u} + \mathbf{v}$ (B) $-2\mathbf{u}$ (C) $2\mathbf{u} - 3\mathbf{v}$ (D) $3\mathbf{u} + 2\mathbf{v} - \mathbf{w}$

SOLUTIONS

(A) $\mathbf{u} + \mathbf{v} = \langle 4, -3 \rangle + \langle 2, 3 \rangle = \langle 6, 0 \rangle$

(B) $-2\mathbf{u} = -2\langle 4, -3 \rangle = \langle -8, 6 \rangle$

(C) $2\mathbf{u} - 3\mathbf{v} = 2\langle 4, -3 \rangle - 3\langle 2, 3 \rangle$
$$= \langle 8, -6 \rangle + \langle -6, -9 \rangle = \langle 2, -15 \rangle$$

(D) $3\mathbf{u} + 2\mathbf{v} - \mathbf{w} = 3\langle 4, -3 \rangle + 2\langle 2, 3 \rangle - \langle 0, -5 \rangle$
$$= \langle 12, -9 \rangle + \langle 4, 6 \rangle + \langle 0, 5 \rangle$$
$$= \langle 16, 2 \rangle$$

MATCHED PROBLEM 3

Let $\mathbf{u} = \langle -5, 3 \rangle$, $\mathbf{v} = \langle 4, -6 \rangle$, and $\mathbf{w} = \langle -2, 0 \rangle$. Find

(A) $\mathbf{u} + \mathbf{v}$ (B) $-3\mathbf{u}$ (C) $3\mathbf{u} - 2\mathbf{v}$ (D) $2\mathbf{u} - \mathbf{v} + 3\mathbf{w}$

Vector addition and scalar multiplication possess algebraic properties similar to the real numbers. These properties enable us to manipulate symbols representing vectors and scalars in much the same way we manipulate symbols that represent real numbers in algebra. The algebraic properties are listed here for convenient reference.

> **ALGEBRAIC PROPERTIES OF VECTORS**

A. Addition Properties. For all vectors **u**, **v**, and **w**,

1. $\mathbf{u} + \mathbf{v} = \mathbf{v} + \mathbf{u}$	**Commutative Property**
2. $\mathbf{u} + (\mathbf{v} + \mathbf{w}) = (\mathbf{u} + \mathbf{v}) + \mathbf{w}$	**Associative Property**
3. $\mathbf{u} + \mathbf{0} = \mathbf{0} + \mathbf{u} = \mathbf{u}$	**Additive Identity**
4. $\mathbf{u} + (-\mathbf{u}) = (-\mathbf{u}) + \mathbf{u} = \mathbf{0}$	**Additive Inverse**

B. Scalar Multiplication Properties. For all vectors **u** and **v** and all scalars m and n:

1. $m(n\mathbf{u}) = (mn)\mathbf{u}$	**Associative Property**
2. $m(\mathbf{u} + \mathbf{v}) = m\mathbf{u} + m\mathbf{v}$	**Distributive Property**
3. $(m + n)\mathbf{u} = m\mathbf{u} + n\mathbf{u}$	**Distributive Property**
4. $1\mathbf{u} = \mathbf{u}$	**Multiplicative Identity**

> ## Unit Vectors

Any vector that has magnitude 1 is called a **unit vector.** If **v** is an arbitrary nonzero vector and k is a scalar, then the scalar product $k\mathbf{v}$ has magnitude $|k||\mathbf{v}|$. Therefore, by choosing k to be $1/|\mathbf{v}|$, the scalar product $k\mathbf{v}$ will be a unit vector with the same direction as **v.**

> **A UNIT VECTOR WITH THE SAME DIRECTION AS v**

If **v** is a nonzero vector, then

$$\mathbf{u} = \frac{1}{|\mathbf{v}|}\mathbf{v}$$

is a unit vector with the same direction as **v.**

EXAMPLE 4 | **Finding a Unit Vector with the Same Direction as a Given Vector**

Given a vector $\mathbf{v} = \langle 1, -2 \rangle$, find a unit vector **u** with the same direction as **v.**

SOLUTION

$$|\mathbf{v}| = \sqrt{1^2 + (-2)^2} = \sqrt{5}$$

$$\mathbf{u} = \frac{1}{|\mathbf{v}|}\mathbf{v} = \frac{1}{\sqrt{5}}\langle 1, -2 \rangle$$

$$= \left\langle \frac{1}{\sqrt{5}}, -\frac{2}{\sqrt{5}} \right\rangle$$

CHECK.

$$|\mathbf{u}| = \sqrt{\left(\frac{1}{\sqrt{5}}\right)^2 + \left(-\frac{2}{\sqrt{5}}\right)^2} = \sqrt{\frac{1}{5} + \frac{4}{5}} = \sqrt{1} = 1$$

We can see that **u** is a unit vector with the same direction as **v.**

MATCHED PROBLEM 4 | Given a vector $\mathbf{v} = \langle 3, 1 \rangle$, find a unit vector **u** with the same direction as **v.**

The unit vectors in the directions of the positive x axis and the positive y axis are denoted by **i** and **j,** respectively.

> **THE i AND j UNIT VECTORS**

$$\mathbf{i} = \langle 1, 0 \rangle$$
$$\mathbf{j} = \langle 0, 1 \rangle$$

Why are the **i** and **j** unit vectors so important? Any vector $\mathbf{v} = \langle a, b \rangle$ can be expressed as a linear combination of these two vectors; that is, as $a\mathbf{i} + b\mathbf{j}.$

$$\mathbf{v} = \langle a, b \rangle = \langle a, 0 \rangle + \langle 0, b \rangle$$
$$= a \langle 1, 0 \rangle + b \langle 0, 1 \rangle = a\mathbf{i} + b\mathbf{j}$$

EXAMPLE 5

Expressing a Vector in Terms of the i and j Unit Vectors

Express each vector as a linear combination of the **i** and **j** unit vectors.

(A) $\langle -2, 4 \rangle$ (B) $\langle 2, 0 \rangle$ (C) $\langle 0, -7 \rangle$

SOLUTIONS

(A) $\langle -2, 4 \rangle = -2\mathbf{i} + 4\mathbf{j}$

(B) $\langle 2, 0 \rangle = 2\mathbf{i} + 0\mathbf{j} = 2\mathbf{i}$

(C) $\langle 0, -7 \rangle = 0\mathbf{i} - 7\mathbf{j} = -7\mathbf{j}$

MATCHED PROBLEM 5

Express each vector as a linear combination of the **i** and **j** unit vectors.

(A) $\langle 5, -3 \rangle$ (B) $\langle -9, 0 \rangle$ (C) $\langle 0, 6 \rangle$

EXAMPLE 6

Algebraic Operations on Vectors Expressed in Terms of the i and j Unit Vectors

For $\mathbf{u} = \mathbf{i} - 2\mathbf{j}$ and $\mathbf{v} = 5\mathbf{i} + 2\mathbf{j},$ compute each of the following:

(A) $\mathbf{u} + \mathbf{v}$ (B) $\mathbf{u} - \mathbf{v}$ (C) $2\mathbf{u} - 3\mathbf{v}$

SOLUTIONS

(A) $\mathbf{u} + \mathbf{v} = (\mathbf{i} - 2\mathbf{j}) + (5\mathbf{i} + 2\mathbf{j}) = \mathbf{i} - 2\mathbf{j} + 5\mathbf{i} + 2\mathbf{j} = 6\mathbf{i} + 0\mathbf{j} = 6\mathbf{i}$

(B) $\mathbf{u} - \mathbf{v} = (\mathbf{i} - 2\mathbf{j}) - (5\mathbf{i} + 2\mathbf{j}) = \mathbf{i} - 2\mathbf{j} - 5\mathbf{i} - 2\mathbf{j} = -4\mathbf{i} - 4\mathbf{j}$

(C) $2\mathbf{u} + 3\mathbf{v} = 2(\mathbf{i} - 2\mathbf{j}) + 3(5\mathbf{i} + 2\mathbf{j}) = 2\mathbf{i} - 4\mathbf{j} + 15\mathbf{i} + 6\mathbf{j} = 17\mathbf{i} + 2\mathbf{j}$

MATCHED PROBLEM 6

For $\mathbf{u} = 2\mathbf{i} - \mathbf{j}$ and $\mathbf{v} = 4\mathbf{i} + 5\mathbf{j},$ compute each of the following:

(A) $\mathbf{u} + \mathbf{v}$ (B) $\mathbf{u} - \mathbf{v}$ (C) $3\mathbf{u} - 2\mathbf{v}$

› **Velocity Vectors**

A vector that represents the direction and speed of an object in motion is called a **velocity vector.** Problems involving objects in motion often can be analyzed using vector methods. Many of these problems involve the use of a **navigational compass,** which is marked clockwise in degrees starting at north as indicated in Figure 9.

Navigational compass

› **Figure 9**

EXAMPLE	7

Apparent and Actual Velocity

An airplane has a compass heading (the direction the plane is pointing) of 85° and an airspeed (relative to the air) of 140 miles per hour. The wind is blowing from north to south at 66 miles per hour. The velocity of a plane relative to the air is called **apparent velocity,** and the velocity relative to the ground is called **resultant,** or **actual, velocity.** The resultant velocity is the vector sum of the apparent velocity and the wind velocity. Find the resultant velocity; that is, find the actual speed and direction of the airplane relative to the ground. Directions are given to the nearest degree and magnitudes to two significant digits.

SOLUTION Vectors [Fig. 10(a)] are used to represent the apparent velocity and the wind velocity. Add the two vectors using the tail-to-tip method to obtain the resultant (actual) velocity vector [Fig. 10(b)]. From the vector diagram [Fig. 10(b)], we obtain the triangle in Figure 11 and solve for γ, c, and α.

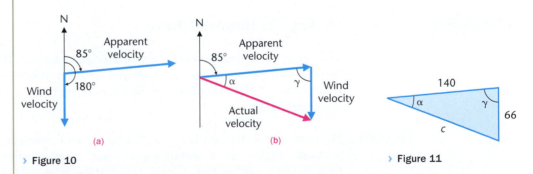

› **Figure 10** › **Figure 11**

SOLVE FOR γ

Because the wind velocity vector is parallel to the north–south line, $\gamma = 85°$ [alternate interior angles of two parallel lines cut by a transversal are equal—see Fig. 10(b)].

SOLVE FOR c

Use the law of cosines:

$$c^2 = a^2 + b^2 - 2ab \cos \gamma$$
$$c = \sqrt{a^2 + b^2 - 2ab \cos \gamma}$$
$$= \sqrt{66^2 + 140^2 - 2(66)(140) \cos 85°}$$
$$= 150 \text{ miles per hour} \qquad \textcolor{red}{\textit{Speed relative to the ground}}$$

SOLVE FOR α

Use the law of sines:

$$\frac{\sin \alpha}{a} = \frac{\sin \gamma}{c}$$

$$\alpha = \sin^{-1}\left(\frac{a \sin \gamma}{c}\right)$$

$$= \sin^{-1}\left(\frac{66 \sin 85}{150}\right) = 26°$$

Actual heading $= 85° + \alpha = 85° + 26° = 111°$. So the magnitude and direction of the resultant velocity vector are 150 miles per hour and 111°, respectively. That is, the plane, relative to the ground, is traveling at 150 miles per hour in a direction of 111°.

MATCHED PROBLEM 7

A river is flowing southwest (225°) at 3.0 miles per hour. A boat crosses the river with a compass heading of 90°. If the speedometer on the boat reads 5.0 miles per hour (the boat's speed relative to the water), what is the resultant velocity? That is, what is the boat's actual speed and direction relative to the ground? Directions are to the nearest degree, and magnitudes are to two significant digits.

› **Force Vectors**

A vector that represents the direction and magnitude of an applied force is called a **force vector.** If an object is subjected to two forces, then the sum of these two forces, the **resultant force,** is a single force. If the resultant force replaced the original two forces, it would act on the object in the same way as the two original forces taken together. In physics it is shown that the resultant force vector can be obtained using vector addition to add the two individual force vectors. It seems natural to use the parallelogram rule for adding force vectors, as is illustrated in Example 8.

EXAMPLE 8 Finding the Resultant Force

Two forces of 30 and 70 pounds act on a point in a plane. If the angle between the force vectors is 40°, what are the magnitude and direction (relative to the 70-pound force) of the resultant force? The magnitudes of the forces are to two significant digits and the angles to the nearest degree.

SOLUTION We start with a diagram (Fig. 12), letting vectors represent the various forces. Because adjacent angles in a parallelogram are supplementary, the measure of angle $OCB = 180° - 40° = 140°$. We can now find the magnitude of the resultant vector **R** using the law of cosines (Fig. 13).

› Figure 12 › Figure 13

$$|\mathbf{R}|^2 = 30^2 + 70^2 - 2(30)(70)\cos 140°$$

$$|\mathbf{R}| = \sqrt{30^2 + 70^2 - 2(30)(70)\cos 140°}$$

$$= 95 \text{ pounds}$$

To find θ, the direction of **R,** we use the law of sines (Fig. 14).

$$\frac{\sin \theta}{30} = \frac{\sin 140°}{95}$$

$$\sin \theta = \frac{30 \sin 140°}{95}$$

$$\theta = \sin^{-1}\left(\frac{30 \sin 140°}{95}\right) = 12°$$

> **Figure 14**

The two given forces are equivalent to a single force of 95 pounds in the direction of 12° (relative to the 70-pound force).

MATCHED PROBLEM 8 Repeat Example 8 using an angle of 100° between the two forces.

Instead of adding vectors, many problems require the resolution of vectors into components. As we indicated earlier, whenever a vector is expressed as the sum or resultant of two vectors, the two vectors are called *vector components* of the given vector. Example 9 illustrates an application of the process of resolving a vector into vector components.

EXAMPLE 9 ## Resolving a Force Vector into Components

A car weighing 3,210 pounds is on a driveway inclined 20.2° to the horizontal. Neglecting friction, find the magnitude of the force parallel to the driveway that will keep the car from rolling down the hill.

SOLUTION We start by drawing a vector diagram (Fig. 15).

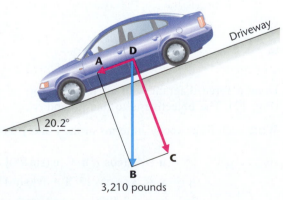

> **Figure 15**

The force vector $\overrightarrow{DB}$ acts in a downward direction and represents the weight of the car. Note that $\overrightarrow{DB} = \overrightarrow{DC} + \overrightarrow{DA}$, where $\overrightarrow{DC}$ is the perpendicular component of $\overrightarrow{DB}$ relative to the driveway and $\overrightarrow{DA}$ is the parallel component of $\overrightarrow{DB}$ relative to the driveway.

To keep the car at D from rolling down the hill, we need a force with the magnitude of $\overrightarrow{DA}$ but oppositely directed. To find $|\overrightarrow{DA}|$, we note that $\angle ABD = 20.2°$. This is true because $\angle ABD$ and the driveway angle have the same complement, $\angle ADB$.

$$\sin 20.2° = \frac{|\overrightarrow{DA}|}{3{,}210}$$
$$|\overrightarrow{DA}| = 3{,}210 \sin 20.2°$$
$$= 1{,}110 \text{ pounds}$$

MATCHED PROBLEM 9 Find the magnitude of the perpendicular component of $\overrightarrow{DB}$ in Example 9.

An object at rest is said to be in **static equilibrium.** Example 10 illustrates how important physics and engineering problems can be solved using the condition for static equilibrium: For an object to remain in static equilibrium, the sum of all the force vectors acting on the object must be the zero vector.

EXAMPLE 10 **Tension in Cables**

A cable car, used to ferry people and supplies across a river, weighs 2,500 pounds fully loaded. The car stops when partway across and deflects the cable relative to the horizontal, as indicated in Figure 16. What is the tension in each part of the cable running to each tower? (The *tension* in a cable is the magnitude of the force it exerts in the direction parallel to the cable.)

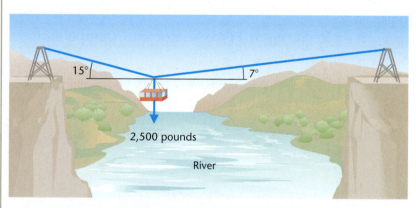

> Figure 16

SOLUTION **Step 1.** Draw a force diagram with all force vectors in standard position at the origin (Fig. 17). The objective is to find $|\mathbf{u}|$ and $|\mathbf{v}|$.

Step 2. Write each force vector in terms of the $\mathbf{i}$ and $\mathbf{j}$ unit vectors:

$$\mathbf{u} = |\mathbf{u}|(\cos 7°)\mathbf{i} + |\mathbf{u}|(\sin 7°)\mathbf{j}$$
$$\mathbf{v} = |\mathbf{v}|(-\cos 15°)\mathbf{i} + |\mathbf{v}|(\sin 15°)\mathbf{j}$$
$$\mathbf{w} = -2{,}500\mathbf{j}$$

Step 3. For the system to be in static equilibrium, the sum of the force vectors must be the zero vector. That is,

$$\mathbf{u} + \mathbf{v} + \mathbf{w} = \mathbf{0}$$

$|\mathbf{w}| = 2{,}500$ pounds

> Figure 17

Replacing vectors **u, v,** and **w** from step 2, we obtain

$$[|\mathbf{u}|(\cos 7°)\mathbf{i} + |\mathbf{u}|(\sin 7°)\mathbf{j}] + [|\mathbf{v}|(-\cos 15°)\mathbf{i} + |\mathbf{v}|(\sin 15°)\mathbf{j}] - 2{,}500\mathbf{j} = 0\mathbf{i} + 0\mathbf{j}$$

which, upon combining **i** and **j** vectors, becomes

$$[|\mathbf{u}|(\cos 7°) + |\mathbf{v}|(-\cos 15°)]\mathbf{i} + [|\mathbf{u}|(\sin 7°) + |\mathbf{v}|(\sin 15°) - 2{,}500]\mathbf{j} = 0\mathbf{i} + 0\mathbf{j}$$

Because two vectors are equal if and only if their corresponding components are equal, we are led to the following system of two equations in the two variables $|\mathbf{u}|$ and $|\mathbf{v}|$:

$$(\cos 7°)|\mathbf{u}| + (-\cos 15°)|\mathbf{v}| \qquad = 0$$
$$(\sin 7°)|\mathbf{u}| + (\sin 15°)|\mathbf{v}| - 2{,}500 = 0$$

Solving this system by standard methods, we find that

$$|\mathbf{u}| = 6{,}400 \text{ pounds} \qquad \text{and} \qquad |\mathbf{v}| = 6{,}600 \text{ pounds}$$

Did you expect that the tension in each part of the cable is more than the weight hanging from the cable?

MATCHED PROBLEM 10

Repeat Example 10 with 15° replaced with 13°, 7° replaced with 9°, and the 2,500 pounds replaced with 1,900 pounds.

ANSWERS TO MATCHED PROBLEMS

1. $\overrightarrow{OP} = \langle -4, 8 \rangle$ 2. $2\sqrt{5}$
3. (A) $\langle -1, -3 \rangle$ (B) $\langle 15, -9 \rangle$ (C) $\langle -23, 21 \rangle$ (D) $\langle -20, 12 \rangle$
4. $\mathbf{u} = \langle 3/\sqrt{10}, 1/\sqrt{10} \rangle$ 5. (A) $5\mathbf{i} - 3\mathbf{j}$ (B) $-9\mathbf{i}$ (C) $6\mathbf{j}$
6. (A) $6\mathbf{i} + 4\mathbf{j}$ (B) $-2\mathbf{i} - 6\mathbf{j}$ (C) $-2\mathbf{i} - 13\mathbf{j}$
7. Resultant velocity: magnitude = 3.6 miles per hour, direction = 126°
8. $|\mathbf{R}| = 71$ pounds, $\theta = 25°$ 9. $|\overrightarrow{DC}| = 3{,}010$ pounds
10. $|\mathbf{u}| = 4{,}900$ pounds, $|\mathbf{v}| = 5{,}000$ pounds

8-3 Exercises

Express all angle measures in decimal degrees. In navigation problems, refer to the figure of a navigational compass.

Navigational compass

1. What is a vector?

2. Explain how to add two vectors using the parallelogram rule.

3. Explain the difference between a vector and a scalar.

4. What is a unit vector?

5. Explain how the unit vector *i* is different from the complex number *i*.

6. Explain the difference between apparent velocity and actual velocity.

In Problems 7–14, find the standard vector $\overrightarrow{OP}$ for each vector $\overrightarrow{AB}$.

7. $A = (4, 6)$; $B = (10, 11)$

8. $A = (2, 7)$; $B = (3, 15)$

9. $A = (3, -9)$; $B = (-4, 5)$

10. $A = (-5, 2)$; $B = (8, -1)$

11. $A = (0, 0)$; $B = (-6, 7)$

12. $A = (9, -7)$; $B = (0, 0)$

13. $A = (5, 8)$; $B = (0, 0)$

14. $A = (0, 0)$; $B = (7, 1)$

In Problems 15–22, find the magnitude of each vector.

15. $\langle -10, 0 \rangle$ **16.** $\langle 0, 8 \rangle$

17. $\langle 5, -12 \rangle$ **18.** $\langle -4, 3 \rangle$

19. $\langle -24, 7 \rangle$ **20.** $\langle 10, -10 \rangle$

21. $\mathbf{i} + \mathbf{j}$ **22.** $2\mathbf{i} - \mathbf{j}$

Problems 23–30 refer to figure (a) and (b) showing vector addition for vectors $\mathbf{u}$ and $\mathbf{v}$.

Tail-to-tip rule	Parallelogram rule
(a)	(b)

In Problems 23–26, find $|\mathbf{u} + \mathbf{v}|$ and α given $|\mathbf{u}|$, $|\mathbf{v}|$, and θ in figures (a) and (b).

23. $|\mathbf{u}| = 66$ grams, $|\mathbf{v}| = 22$ grams, $\theta = 68°$

24. $|\mathbf{u}| = 120$ grams, $|\mathbf{v}| = 84$ grams, $\theta = 44°$

25. $|\mathbf{u}| = 21$ knots, $|\mathbf{v}| = 3.2$ knots, $\theta = 53°$

26. $|\mathbf{u}| = 8.0$ knots, $|\mathbf{v}| = 2.0$ knots, $\theta = 64°$

In Problems 27–30, find $|\mathbf{u}|$ and $|\mathbf{v}|$, given $|\mathbf{u} + \mathbf{v}|$, α and θ in figures (a) and (b).

27. $|\mathbf{u} + \mathbf{v}| = 14$ kilograms, $\alpha = 25°$, $\theta = 79°$

28. $|\mathbf{u} + \mathbf{v}| = 33$ kilograms, $\alpha = 17°$, $\theta = 43°$

29. $|\mathbf{u} + \mathbf{v}| = 223$ miles per hour, $\alpha = 42.3°$, $\theta = 69.4°$

30. $|\mathbf{u} + \mathbf{v}| = 437$ miles per hour, $\alpha = 17.8°$, $\theta = 50.5°$

In Problems 31–34, find:
(A) $\mathbf{u} + \mathbf{v}$ (B) $\mathbf{u} - \mathbf{v}$ (C) $2\mathbf{u} - \mathbf{v} + 3\mathbf{w}$

31. $\mathbf{u} = \langle 2, 1 \rangle$, $\mathbf{v} = \langle -1, 3 \rangle$, $\mathbf{w} = \langle 3, 0 \rangle$

32. $\mathbf{u} = \langle -1, 2 \rangle$, $\mathbf{v} = \langle 3, -2 \rangle$, $\mathbf{w} = \langle 0, -2 \rangle$

33. $\mathbf{u} = \langle -4, -1 \rangle$, $\mathbf{v} = \langle 2, 2 \rangle$ $\mathbf{w} = \langle 0, 1 \rangle$

34. $\mathbf{u} = \langle -3, 2 \rangle$, $\mathbf{v} = \langle -2, 2 \rangle$, $\mathbf{w} = \langle -3, 0 \rangle$

In Problems 35–40, express $\mathbf{v}$ in terms of the $\mathbf{i}$ and $\mathbf{j}$ unit vectors.

35. $\mathbf{v} = \langle -3, 4 \rangle$ **36.** $\mathbf{v} = \langle 2, -5 \rangle$

37. $\mathbf{v} = \langle 3, 0 \rangle$ **38.** $\mathbf{v} = \langle 0, -27 \rangle$

39. $\mathbf{v} = \overrightarrow{AB}$, where $A = (2, 3)$ and $B = (-3, 1)$

40. $\mathbf{v} = \overrightarrow{AB}$, where $A = (-2, -1)$ and $B = (0, 2)$

In Problems 41–46, let $\mathbf{u} = 3\mathbf{i} - 2\mathbf{j}$, $\mathbf{v} = 2\mathbf{i} + 4\mathbf{j}$, and $\mathbf{w} = 2\mathbf{i}$, and perform the indicated operations.

41. $\mathbf{u} + \mathbf{v}$ **42.** $\mathbf{u} - \mathbf{v}$

43. $2\mathbf{u} - 3\mathbf{v}$ **44.** $3\mathbf{u} + 2\mathbf{v}$

45. $2\mathbf{u} - \mathbf{v} - 2\mathbf{w}$ **46.** $\mathbf{u} - 3\mathbf{v} + 2\mathbf{w}$

In Problems 47–54, find a unit vector with the same direction as $\mathbf{v}$.

47. $\mathbf{v} = \langle 4, 3 \rangle$ **48.** $\mathbf{v} = \langle 5, 12 \rangle$

49. $\mathbf{v} = \langle -1, 1 \rangle$ **50.** $\mathbf{v} = \langle 2, -3 \rangle$

51. $\mathbf{v} = \langle -8, 0 \rangle$ **52.** $\mathbf{v} = \langle 0, -17 \rangle$

53. $\mathbf{v} = 5\mathbf{i} + \sqrt{11}\mathbf{j}$ **54.** $\mathbf{v} = -\sqrt{2}\mathbf{i} - \sqrt{7}\mathbf{j}$

In Problems 55–62, determine whether the statement is true or false. If true, explain why. If false, give a counterexample.

55. If two vectors have the same magnitude, then they are equal.

56. Every vector has the same magnitude as its standard vector.

57. If a vector has the same initial and terminal points, then it is the zero vector.

58. The only unit vectors in the plane are $\mathbf{i}$ and $\mathbf{j}$.

59. Every vector $\mathbf{v}$ has the same magnitude as $\mathbf{v} + \mathbf{v}$.

60. Every vector $\mathbf{v}$ has the same direction as $\mathbf{v} + \mathbf{v}$.

61. The magnitude of every vector is a positive real number.

62. If $\mathbf{u}$ and $\mathbf{v}$ are unit vectors, then $\mathbf{u} + \mathbf{v}$ is a unit vector.

In Problems 63–70, let $\mathbf{u} = \langle a, b \rangle$, $\mathbf{v} = \langle c, d \rangle$, and $\mathbf{w} = \langle e, f \rangle$ be vectors and m and n be scalars. Prove each of the following vector properties using appropriate properties of real numbers and the definitions of vector addition and scalar multiplication.

63. $\mathbf{u} + (\mathbf{v} + \mathbf{w}) = (\mathbf{u} + \mathbf{v}) + \mathbf{w}$

64. $\mathbf{u} + \mathbf{v} = \mathbf{v} + \mathbf{u}$ **65.** $\mathbf{u} + \mathbf{0} = \mathbf{u}$

66. $\mathbf{u} + (-\mathbf{u}) = \mathbf{0}$ **67.** $(m + n)\mathbf{u} = m\mathbf{u} + n\mathbf{u}$

68. $m(\mathbf{u} + \mathbf{v}) = m\mathbf{u} + m\mathbf{v}$

69. $m(n\mathbf{u}) = (mn)\mathbf{u}$ **70.** $1\mathbf{u} = \mathbf{u}$

APPLICATIONS

In Problems 71–74, assume the north, east, south, and west directions are exact.

71. NAVIGATION An airplane is flying with a compass heading of 285° and an airspeed of 230 miles per hour. A steady wind of 35 miles per hour is blowing in the direction of 260°. What is the plane's actual velocity; that is, what is its speed and direction relative to the ground?

72. NAVIGATION A power boat crossing a wide river has a compass heading of 25° and speed relative to the water of 15 miles per hour. The river is flowing in the direction of 135° at 3.9 miles per hour. What is the boat's actual velocity; that is, what is its speed and direction relative to the ground?

73. NAVIGATION Two docks are directly opposite each other on a southward-flowing river. A boat pilot needs to go in a straight line from the east dock to the west dock in a ferryboat with a cruising speed in still water of 8.0 knots. If the river's current is 2.5 knots, what compass heading should be maintained while crossing the river? What is the actual speed of the boat relative to the land?

74. NAVIGATION An airplane can cruise at 255 miles per hour in still air. If a steady wind of 46.0 miles per hour is blowing from the west, what compass heading should the pilot fly for the course of the plane relative to the ground to be north (0°)? Compute the ground speed for this course.

75. RESULTANT FORCE A large ship has gone aground in a harbor and two tugs, with cables attached, attempt to pull it free. If one tug pulls with a compass course of 52° and a force of 2,300 pounds and a second tug pulls with a compass course of 97° and a force of 1,900 pounds, what is the compass direction and the magnitude of the resultant force?

76. RESULTANT FORCE Repeat Problem 75 if one tug pulls with a compass direction of 161° and a force of 2,900 kilograms and a second tug pulls with a compass direction of 192° and a force of 3,600 kilograms.

77. RESOLUTION OF FORCES An automobile weighing 4,050 pounds is standing on a driveway inclined 5.5° with the horizontal.
(A) Find the magnitude of the force parallel to the driveway necessary to keep the car from rolling down the hill.
(B) Find the magnitude of the force perpendicular to the driveway.

78. RESOLUTION OF FORCES Repeat Problem 77 for a car weighing 2,500 pounds parked on a hill inclined at 15° to the horizontal.

79. RESOLUTION OF FORCES If two weights are fastened together and placed on inclined planes, as shown in the figure, neglecting friction, which way will they slide?

80. RESOLUTION OF FORCES If two weights are fastened together and placed on inclined planes as indicated in the figure, neglecting friction, which way will they slide?

In Problems 81–88, compute all answers to three significant digits.

81. STATIC EQUILIBRIUM A unicyclist at a certain point on a tightrope deflects the rope as indicated in the figure. If the total weight of the cyclist and the unicycle is 155 pounds, how much tension is in each part of the cable?

82. STATIC EQUILIBRIUM Repeat Problem 81 with the left angle 4.2°, the right angle 5.3°, and the total weight 112 pounds.

83. STATIC EQUILIBRIUM A weight of 1,000 pounds is suspended from two cables as shown in the figure. What is the tension in each cable?

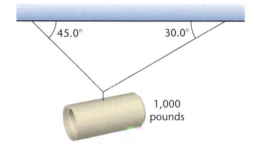

84. STATIC EQUILIBRIUM A weight of 500 pounds is supported by two cables as illustrated. What is the tension in each cable?

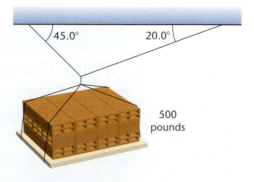

85. STATIC EQUILIBRIUM A 400-pound sign is suspended as shown in figure (a) on the next page. The corresponding force diagram (b) is formed by observing the following: Member *AB* is "pushing" at *B* and is under compression. This "pushing" force also can be thought of as the force vector **a** "pulling" to the right at *B*. The force vector **b** reflects the fact that member *CB* is under tension— that is, it is "pulling" at *B*. The force vector **c** corresponds to the weight of the sign "pulling" down at *B*. Find the magnitudes of the

forces in the rigid supporting members; that is, find |**a**| and |**b**| in the force diagram (b).

(a)

(b)

86. STATIC EQUILIBRIUM A weight of 1,000 kilograms is supported as shown in the figure. What are the magnitudes of the forces on the members *AB* and *BC*?

87. STATIC EQUILIBRIUM A 1,250-pound weight is hanging from a hoist as indicated in the figure. What are the magnitudes of the forces on the members *AB* and *BC*?

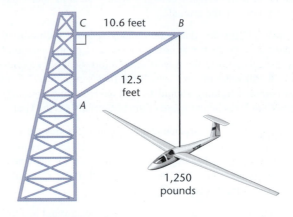

88. STATIC EQUILIBRIUM A weight of 5,000 kilograms is supported as shown in the figure. What are the magnitudes of the forces on the members *AB* and *BC*?

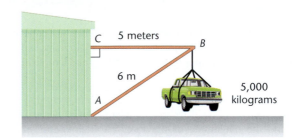

8-4 Polar Coordinates and Graphs

› Polar Coordinate System
› Converting from Polar to Rectangular Form, and Vice Versa
› Graphing Polar Equations
› Some Standard Polar Curves
› Application

Up until now we have used only the rectangular coordinate system. Other coordinate systems have particular advantages in certain situations. Of the many that are possible, the *polar coordinate system* ranks second in importance to the rectangular coordinate system and is the subject matter of this section.

› Polar Coordinate System

To form a **polar coordinate system** in a plane (Fig. 1), start with a fixed point O and call it the **pole, or origin.** From this point draw a half-line, or ray (usually horizontal and to the right), and call this line the **polar axis.**

If P is an arbitrary point in a plane, then associate polar coordinates (r, θ) (Fig. 1) with it as follows: Starting with the polar axis as the initial side of an angle, rotate the terminal side until it, or the extension of it through the pole, passes through the point. The θ coordinate in (r, θ) is this angle, in degree or radian measure. The angle θ is positive if the rotation is counterclockwise and negative if the rotation is clockwise. The r coordinate in (r, θ) is the directed distance from the pole to the point P. It is positive if measured from the pole along the terminal side of θ and negative if measured along the terminal side extended through the pole.

Figure 2 illustrates a point P with three different sets of polar coordinates. Study this figure carefully. The pole has polar coordinates $(0, \theta)$ for arbitrary θ. For example, $(0, 0°)$, $(0, \pi/3)$, and $(0, -371°)$ are all coordinates of the pole.

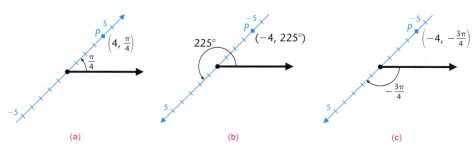

> **Figure 1** Polar coordinate system.

> **Figure 2** Polar coordinates of a point.

We now see a distinct difference between rectangular and polar coordinates for the given point. For a given point in a rectangular coordinate system, there exists exactly one set of rectangular coordinates. On the other hand, in a polar coordinate system, a point has infinitely many sets of polar coordinates.

Just as graph paper with a rectangular grid is readily available for plotting rectangular coordinates, polar graph paper is available for plotting polar coordinates.

| EXAMPLE | 1 | **Plotting Points in a Polar Coordinate System** |

Plot the following points in a polar coordinate system:

(A) $A = (3, 30°)$, $B = (-8, 180°)$, $C = (5, -135°)$, $D = (-10, -45°)$

(B) $A = (5, \pi/3)$, $B = (-6, 5\pi/6)$, $C = (7, -\pi/2)$, $D = (-4, -\pi/6)$

SOLUTIONS

(A)

(B)

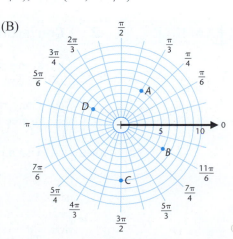

MATCHED PROBLEM 1

Plot the following points in a polar coordinate system:

(A) $A = (8, 45°)$, $B = (-5, 150°)$, $C = (4, -210°)$, $D = (-6, -90°)$

(B) $A = (9, \pi/6)$, $B = (-3, -\pi)$, $C = (-7, 7\pi/4)$, $D = (5, -5\pi/6)$

>>> EXPLORE-DISCUSS 1

A point in a polar coordinate system has coordinates $(5, 30°)$. How many other polar coordinates does the point have for θ restricted to $-360° \le \theta \le 360°$? Find the other coordinates of the point and explain how they are found.

› Converting from Polar to Rectangular Form, and Vice Versa

Often, it is necessary to transform coordinates or equations in rectangular form into polar form, or vice versa. The following polar–rectangular relationships are useful in this regard:

> ### › POLAR–RECTANGULAR RELATIONSHIPS
>
> We have the following relationships between rectangular coordinates (x, y) and polar coordinates (r, θ):
>
>
>
> $$r^2 = x^2 + y^2$$
>
> $$\sin \theta = \frac{y}{r} \quad \text{or} \quad y = r \sin \theta$$
>
> $$\cos \theta = \frac{x}{r} \quad \text{or} \quad x = r \cos \theta$$
>
> $$\tan \theta = \frac{y}{x}$$
>
> [*Note:* The signs of x and y determine the quadrant for θ. The angle θ is chosen so that $-\pi < \theta \le \pi$ or $-180° < \theta \le 180°$, unless directed otherwise.]

Many calculators can automatically convert rectangular coordinates to polar form, and vice versa. (Read the manual for your particular calculator.) Example 2 illustrates calculator conversions in both directions.

EXAMPLE 2

Converting from Polar to Rectangular Form, and Vice Versa

(A) Convert the polar coordinates $(-4, 1.077)$ to rectangular coordinates to three decimal places.

(B) Convert the rectangular coordinates $(-3.207, -5.719)$ to polar coordinates with θ in degree measure, $-180° < \theta \le 180°$ and $r \ge 0$.

SOLUTIONS

(A) Use a calculator set in radian mode.

$$(r, \theta) = (-4, 1.077)$$
$$x = r \cos \theta = (-4) \cos 1.077 = -1.896$$
$$y = r \sin \theta = (-4) \sin 1.077 = -3.522$$

The rectangular coordinates are $(-1.896, -3.522)$.

> Figure 3

> Figure 4

Figure 3 shows the same conversion done in a graphing calculator with a built-in conversion routine.

(B) Use a calculator set in degree mode.

$$(x, y) = (-3.207, -5.719)$$
$$r = \sqrt{x^2 + y^2} = \sqrt{(-3.207)^2 + (-5.719)^2} = 6.557$$
$$\tan \theta = \frac{y}{x} = \frac{-5.719}{-3.207}$$

θ is a third-quadrant angle and is to be chosen so that $-180° < \theta \le 180°$.

$$\theta = -180° + \tan^{-1}\left(\frac{-5.719}{-3.207}\right) = -119.28°$$

The polar coordinates are $(6.557, -199.28°)$.

Figure 4 shows the same conversion done in a graphing calculator with a built-in conversion routine.

MATCHED PROBLEM 2

(A) Convert the polar coordinates $(8.677, -1.385)$ to rectangular coordinates to three decimal places.

(B) Convert the rectangular coordinates $(-6.434, 4.023)$ to polar coordinates with θ in degree measure, $-180° < \theta \le 180°$ and $r \ge 0$.

Generally, a more important use of the polar–rectangular relationships is in the conversion of equations in rectangular form to polar form, and vice versa.

EXAMPLE 3 **Converting an Equation from Rectangular Form to Polar Form**

Change $x^2 + y^2 - 4y = 0$ to polar form.

SOLUTION Use $r^2 = x^2 + y^2$ and $y = r \sin \theta$.

$$x^2 + y^2 - 4y = 0$$
$$r^2 - 4r \sin \theta = 0$$
$$r(r - 4 \sin \theta) = 0$$
$$r = 0 \quad \text{or} \quad r - 4 \sin \theta = 0$$

The graph of $r = 0$ is the pole. Because the pole is included in the graph of $r - 4 \sin \theta = 0$ (let $\theta = 0$), we can discard $r = 0$ and keep only

$$r - 4 \sin \theta = 0$$

or

$$r = 4 \sin \theta \qquad \text{The polar form of } x^2 + y^2 - 4y = 0.$$

MATCHED PROBLEM 3 Change $x^2 + y^2 - 6x = 0$ to polar form.

| EXAMPLE | 4 | **Converting an Equation from Polar Form to Rectangular Form** |

Change $r = -3 \cos \theta$ to rectangular form.

SOLUTION

The transformation of this equation as it stands into rectangular form is fairly difficult. With a little trick, however, it becomes easy. We multiply both sides by r, which simply adds the pole to the graph. But the pole is already part of the graph of $r = -3 \cos \theta$ (let $\theta = \pi/2$), so we haven't actually changed anything.

$$r = -3 \cos \theta \qquad \text{Multiply both sides by } r.$$
$$r^2 = -3r \cos \theta \qquad r^2 = x^2 + y^2 \text{ and } r \cos \theta = x$$
$$x^2 + y^2 = -3x$$
$$x^2 + y^2 + 3x = 0$$

| MATCHED PROBLEM 4 |

Change $r + 2 \sin \theta = 0$ to rectangular form.

› Graphing Polar Equations

We now turn to graphing polar equations. The **graph** of a polar equation, such as $r = 3\theta$ or $r = 6 \cos \theta$, in a polar coordinate system is the set of all points having coordinates that satisfy the polar equation. Certain curves have simpler representations in polar coordinates, and other curves have simpler representations in rectangular coordinates.

To establish fundamentals in graphing polar equations, we start with a point-by-point graph. We then consider a more rapid way of making rough sketches of certain polar curves. And, finally, we show how polar curves are graphed in a graphing utility.

To plot a polar equation using **point-by-point plotting,** just as in rectangular coordinates, make a table of values that satisfy the equation, plot these points, then join them with a smooth curve. Example 5 illustrates the process.

| EXAMPLE | 5 | **Point-by-Point Plotting** |

(A) Graph $r = 8 \cos \theta$ with θ in radians.

(B) Convert the polar equation in part A to rectangular form, and identify the graph.

SOLUTIONS

(A) We construct a table using multiples of $\pi/6$, plot these points, then join the points with a smooth curve (Fig. 5).

θ	r
0	8.0
$\pi/6$	6.9
$\pi/3$	4.0
$\pi/2$	0.0
$2\pi/3$	-4.0
$5\pi/6$	-6.9
π	8.0
Graph repeats	

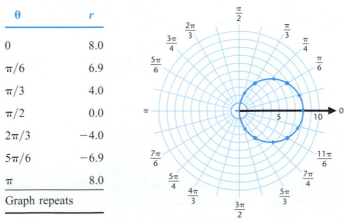

› Figure 5

(B)
$$r = 8 \cos \theta \qquad \text{Multiply both sides by } r.$$
$$r^2 = 8r \cos \theta \qquad \text{Change to rectangular form.}$$
$$x^2 + y^2 = 8x$$
$$x^2 - 8x + y^2 = 0 \qquad \text{Complete the square on the left side.}$$
$$x^2 - 8x + 16 + y^2 = 16 \qquad \text{Write in standard form.}$$
$$(x - 4)^2 + y^2 = 4^2 \qquad \text{Standard equation of a circle.}$$

The graph in part A is a circle with center at $(4, 0)$ and radius 4 (see Section 2-2).

MATCHED PROBLEM 5

(A) Graph $r = 8 \sin \theta$ with θ in degrees.

(B) Convert the polar equation in part A to rectangular form, and identify the graph.

If only a rough sketch of a polar equation involving $\sin \theta$ or $\cos \theta$ is desired, you can speed up the point-by-point graphing process by taking advantage of the uniform variation of $\sin \theta$ and $\cos \theta$ as θ moves around a unit circle. This process is referred to as **rapid polar sketching.** It is convenient to visualize Figure 6 in the process. With a little practice most of the table work in rapid sketching can be done mentally and a rough sketch can be made directly from the equation.

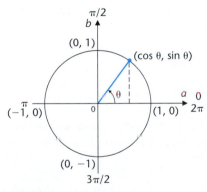

> Figure 6

EXAMPLE 6 **Rapid Polar Sketching**

Sketch $r = 4 + 4 \cos \theta$ using rapid sketching techniques with θ in radians.

SOLUTION We set up a table that indicates how r varies as we let θ vary through each set of quadrant values:

θ Varies from	$\cos \theta$ Varies from	$4 \cos \theta$ Varies from	$r = 4 + 4 \cos \theta$ Varies from
0 to $\pi/2$	1 to 0	4 to 0	8 to 4
$\pi/2$ to π	0 to -1	0 to -4	4 to 0
π to $3\pi/2$	-1 to 0	-4 to 0	0 to 4
$3\pi/2$ to 2π	0 to 1	0 to 4	4 to 8

Notice that as θ increases from 0 to $\pi/2$, cos θ decreases from 1 to 0, 4 cos θ decreases from 4 to 0, and $r = 4 + 4$ cos θ decreases from 8 to 4, and so on. Sketching these values, we obtain the graph in Figure 7, called a **cardioid.**

$r = 4 + 4 \cos \theta$
Cardioid

› **Figure 7**

MATCHED PROBLEM 6

Sketch $r = 5 + 5 \sin \theta$ using rapid sketching techniques with θ in radians.

EXAMPLE **7** **Rapid Polar Sketching**

Sketch $r = 8 \cos 2\theta$ with θ in radians.

SOLUTION

Start by letting 2θ (instead of θ) range through each set of quadrant values. That is, start with values for 2θ in the second column of the table, fill in the table to the right, and then fill in the first column for θ.

⌐ **Start with the second column**

θ **Varies from**	2θ **Varies from**	cos 2θ **Varies from**	$r = 8$ cos 2θ **Varies from**
0 to $\pi/4$	0 to $\pi/2$	1 to 0	8 to 0
$\pi/4$ to $\pi/2$	$\pi/2$ to π	0 to -1	0 to -8
$\pi/2$ to $3\pi/4$	π to $3\pi/2$	-1 to 0	-8 to 0
$3\pi/4$ to π	$3\pi/2$ to 2π	0 to 1	0 to 8
π to $5\pi/4$	2π to $5\pi/2$	1 to 0	8 to 0
$5\pi/4$ to $3\pi/2$	$5\pi/2$ to 3π	0 to -1	0 to -8
$3\pi/2$ to $7\pi/4$	3π to $7\pi/2$	-1 to 0	-8 to 0
$7\pi/4$ to 2π	$7\pi/2$ to 4π	0 to 1	0 to 8

As 2θ increases from 0 to $\pi/2$, θ increases from 0 to $\pi/4$, and r decreases from 8 to 0. As 2θ increases from $\pi/2$ to π, θ increases from $\pi/4$ to $\pi/2$, and r decreases from 0 to -8,

and so on. Continue until the graph starts to repeat. Plotting the values, we obtain the graph in Figure 8, called a **four-leafed rose:**

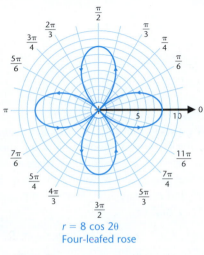

$r = 8 \cos 2\theta$
Four-leafed rose

> **Figure 8**

MATCHED PROBLEM 7

Sketch $r = 6 \sin 2\theta$ with θ in radians.

We now turn to **graphing polar equations in a graphing calculator.** Example 8 illustrates the process.

EXAMPLE 8 Graphing in a Graphing Calculator

Graph each of the following polar equations in a graphing calculator (parts B and C are from Examples 6 and 7).

(A) $r = 3\theta$, $0 \leq \theta \leq 3\pi/2$ (Archimedes' spiral)

(B) $r = 4 + 4 \cos \theta$ (cardioid)

(C) $r = 8 \cos 2\theta$ (four-leafed rose)

SOLUTIONS Set the graphing calculator in polar mode and select polar coordinates and radian measure. Adjust window values to accommodate the whole graph. A squared graph is often desirable in showing the true shape of the curve, and is used here. Many graphing calculators, including the one used here, do not show a polar grid. When using TRACE, many graphing calculators offer a choice between polar coordinates and rectangular coordinates for points on the polar curve. The graphs of the preceding equations are shown in Figure 9.

(A) $r = 3\theta$, $0 \leq \theta \leq 3\pi/2$

(B) $r = 4 + 4 \cos \theta$

(C) $r = 8 \cos 2\theta$

> **Figure 9**

MATCHED PROBLEM 8

Graph each of the following polar equations in a graphing calculator.

(A) $r = 2\theta$, $0 \leq \theta \leq 2\pi$

(B) $r = 5 + 5 \sin \theta$

(C) $r = 6 \sin 2\theta$

>>> EXPLORE-DISCUSS 2

(A) Graph $r_1 = 10 \sin \theta$ and $r_2 = 10 \cos \theta$ in the same viewing window. Use TRACE on r_1 and estimate the polar coordinates where the two graphs intersect. Do the same thing for r_2. Which intersection point appears to have the same polar coordinates on each curve and consequently represents a simultaneous solution to both equations? Which intersection point appears to have different polar coordinates on each curve and consequently does not represent a simultaneous solution? Solve the system for r and θ.

(B) Explain how rectangular coordinate systems differ from polar coordinate systems relative to intersection points and simultaneous solutions of systems of equations in the respective systems.

› Some Standard Polar Curves

In a rectangular coordinate system the simplest types of equations to graph are found by setting the rectangular variables x and y equal to constants:

$$x = a \qquad \text{and} \qquad y = b$$

The graphs are straight lines: The graph of $x = a$ is a vertical line, and the graph of $y = b$ is a horizontal line. A glance at Table 1 shows that horizontal and vertical lines do not have simple equations in polar coordinates.

Table 1 Standard Polar Graphs

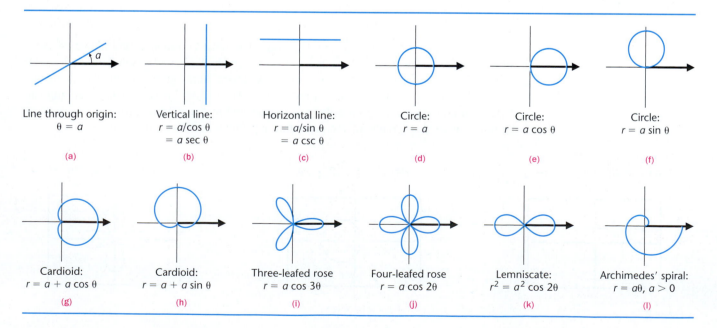

Line through origin:
$\theta = a$
(a)

Vertical line:
$r = a/\cos \theta$
$= a \sec \theta$
(b)

Horizontal line:
$r = a/\sin \theta$
$= a \csc \theta$
(c)

Circle:
$r = a$
(d)

Circle:
$r = a \cos \theta$
(e)

Circle:
$r = a \sin \theta$
(f)

Cardioid:
$r = a + a \cos \theta$
(g)

Cardioid:
$r = a + a \sin \theta$
(h)

Three-leafed rose:
$r = a \cos 3\theta$
(i)

Four-leafed rose:
$r = a \cos 2\theta$
(j)

Lemniscate:
$r^2 = a^2 \cos 2\theta$
(k)

Archimedes' spiral:
$r = a\theta$, $a > 0$
(l)

Two of the simplest types of polar equations to graph in a polar coordinate system are found by setting the polar variables r and θ equal to constants:

$$r = a \quad \text{and} \quad \theta = b$$

Figure 10 illustrates the graphs of $\theta = \pi/4$ and $r = 5$.

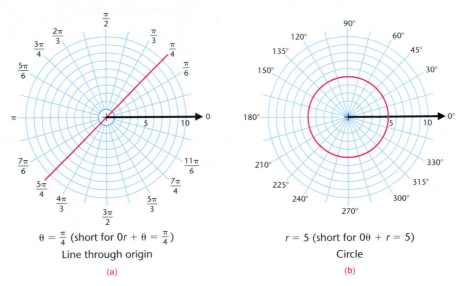

$\theta = \frac{\pi}{4}$ (short for $0r + \theta = \frac{\pi}{4}$)
Line through origin
(a)

$r = 5$ (short for $0\theta + r = 5$)
Circle
(b)

> **Figure 10**

Table 1 illustrates a number of standard polar graphs and their equations. Polar graphing is often made easier if you have some idea of the final form.

› Application

Serious sailboat racers make polar plots of boat speeds at various angles to the wind with various sail combinations at different wind speeds. With many polar plots for different sizes and types of sails at different wind speeds, they are able to accurately choose a sail for the optimum performance for different points of sail relative to any given wind strength. Figure 11 illustrates one such polar plot, where the maximum speed appears to be about 7.5 knots at 105° off the wind (with spinnaker sail set).

> **Figure 11** Polar diagram showing optimum sailing speed at different sailing angles to the wind.

ANSWERS TO MATCHED PROBLEMS

1. (A)

(B)

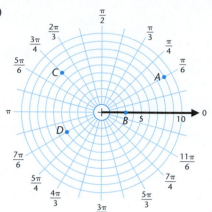

2. (A) (1.603, −8.528) (B) (7.588, 147.98°) **3.** $r = 6 \cos \theta$ **4.** $x^2 + y^2 + 2y = 0$

5. (A)

θ	r
0°	0.0
30°	4.0
60°	6.9
90°	8.0
120°	6.9
150°	4.0
180°	0.0
Graph Repeats	

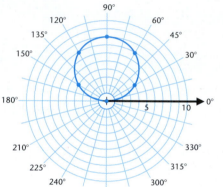

Circle:
$r = 8 \sin \theta$

(B) $x^2 + (y - 4)^2 = 4^2$, a circle with center at (0, 4) and radius 4

6. $r = 5 + 5 \sin \theta$, cardioid

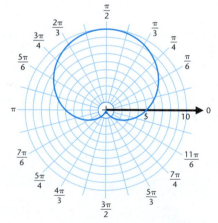

7. $r = 6 \sin 2\theta$, four-leafed rose

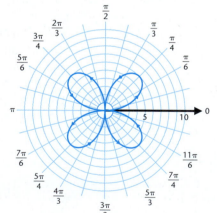

8. (A) $r = 2\theta$, $0 \le \theta \le 2\pi$

(B) $r = 5 + 5 \sin \theta$

(C) $r = 6 \sin 2\theta$

8-4 Exercises

1. If a point has polar coordinates (r, θ), explain the significance of r.

2. If a point has polar coordinates (r, θ), explain the significance of θ.

3. Explain why the point with rectangular coordinates $(1, 0)$ has more than one set of polar coordinates.

4. If you are given the rectangular coordinates of a point, explain how you can find a set of polar coordinates of the same point.

5. If you are given the polar coordinates of a point, explain how you can find the rectangular coordinates of the same point.

6. Explain the difference between point-by-point plotting and rapid polar sketching.

Plot A, B, and C in Problems 7–14 in a polar coordinate system.

7. $A = (4, 0°)$, $B = (7, 180°)$, $C = (9, 45°)$

8. $A = (8, 0°)$, $B = (5, 90°)$, $C = (6, 30°)$

9. $A = (-4, 0°)$, $B = (-7, 180°)$, $C = (-9, 45°)$

10. $A = (-8, 0°)$, $B = (-5, 90°)$, $C = (-6, 30°)$

11. $A = (8, -\pi/3)$, $B = (4, -\pi/4)$, $C = (10, -\pi/6)$

12. $A = (6, -\pi/6)$, $B = (5, -\pi/2)$, $C = (8, -\pi/4)$

13. $A = (-6, -\pi/6)$, $B = (-5, -\pi/2)$, $C = (-8, -\pi/4)$

14. $A = (-6, -\pi/2)$, $B = (-5, -\pi/3)$, $C = (-8, -\pi/4)$

15. A point in a polar coordinate system has coordinates $(-5, 3\pi/4)$. Find all other polar coordinates for the point, $-2\pi \le \theta \le 2\pi$, and verbally describe how the coordinates are associated with the point.

16. A point in a polar coordinate system has coordinates $(6, -30°)$. Find all other polar coordinates for the point, $-360° \le \theta \le 360°$, and verbally describe how the coordinates are associated with the point.

Graph the polar equation in Problems 17 and 18 in a polar coordinate system using point-by-point plotting and the special values 0, $\pi/6$, $\pi/4$, $\pi/3$, $\pi/2$, $2\pi/3$, $3\pi/4$, $5\pi/6$, and π for θ.

17. $r = 10 \sin \theta$ **18.** $r = 10 \cos \theta$

Graph the polar equation in Problems 19–22 in a polar coordinate system.

19. $r = 8$ **20.** $r = 5$

21. $\theta = \pi/3$ **22.** $\theta = \pi/6$

In Problems 23–28, convert the polar coordinates to rectangular coordinates to three decimal places.

23. $(6, \pi/6)$ **24.** $(7, 2\pi/3)$

25. $(-2, 7\pi/8)$ **26.** $(3, -3\pi/7)$

27. $(-4.233, -2.084)$ **28.** $(-9.028, -0.663)$

In Problems 29–34, convert the rectangular coordinates to polar coordinates with θ in degree measure, $-180° < \theta \le 180°$, and $r \ge 0$.

29. $(3.5, 7.1)$ **30.** $(6.9, 4.7)$

31. $(22, -14)$ **32.** $(16, -27)$

33. $(-7.33, -2.04)$ **34.** $(-8.33, 4.29)$

In Problems 35–44, use rapid graphing techniques to sketch the graph of each polar equation.

35. $r = 4 \sin \theta$ **36.** $r = 4 \cos \theta$

37. $r = 10 \sin 2\theta$ **38.** $r = 8 \cos 2\theta$

39. $r = 5 \cos 3\theta$ **40.** $r = 6 \sin 3\theta$

41. $r = 2 + 2 \sin \theta$ **42.** $r = 3 + 3 \cos \theta$

43. $r = 2 + 4 \sin \theta$ **44.** $r = 2 + 4 \cos \theta$

 Problems 45–50 are exploratory problems requiring the use of a graphing calculator.

45. Graph each polar equation in its own viewing window: $r = 2 + 2 \sin \theta$, $r = 4 + 2 \sin \theta$, $r = 2 + 4 \sin \theta$.

46. Graph each polar equation in its own viewing window: $r = 2 + 2 \cos \theta$, $r = 4 + 2 \cos \theta$, $r = 2 + 4 \cos \theta$.

47. (A) Graph each polar equation in its own viewing window: $r = 4 \sin \theta$, $r = 4 \sin 3\theta$, $r = 4 \sin 5\theta$.
(B) What would you guess to be the number of leaves for $r = 4 \sin 7\theta$?
(C) What would you guess to be the number of leaves for $r = a \sin n\theta$, $a > 0$ and n odd?

48. (A) Graph each polar equation in its own viewing window: $r = 4 \cos \theta$, $r = 4 \cos 3\theta$, $r = 4 \cos 5\theta$.
(B) What would you guess to be the number of leaves for $r = 4 \cos 7\theta$?
(C) What would you guess to be the number of leaves for $r = a \cos n\theta$, $a > 0$ and n odd?

49. (A) Graph each polar equation in its own viewing window: $r = 4 \sin 2\theta$, $r = 4 \sin 4\theta$, $r = 4 \sin 6\theta$.
(B) What would you guess to be the number of leaves for $r = 4 \sin 8\theta$?
(C) What would you guess to be the number of leaves for $r = a \sin n\theta$, $a > 0$ and n even?

50. (A) Graph each polar equation in its own viewing window: $r = 4 \cos 2\theta$, $r = 4 \cos 4\theta$, $r = 4 \cos 6\theta$.
(B) What would you guess to be the number of leaves for $r = 4 \cos 8\theta$?
(C) What would you guess to be the number of leaves for $r = a \cos n\theta$, $a > 0$ and n even?

In Problems 51–56, change each rectangular equation to polar form.

51. $y^2 = 5y - x^2$ **52.** $6x - x^2 = y^2$

53. $y = x$ **54.** $x^2 + y^2 = 9$

55. $y^2 = 4x$ **56.** $2xy = 1$

In Problems 57–62, change each polar equation to rectangular form.

57. $r(3 \cos \theta - 4 \sin \theta) = -1$ **58.** $r(2 \cos \theta + \sin \theta) = 4$

59. $r = -2 \sin \theta$ **60.** $r = 8 \cos \theta$

61. $\theta = \pi/4$ **62.** $r = 4$

 Problems 63 and 64 are exploratory problems requiring the use of a graphing calculator.

63. Graph $r = 1 + 2 \sin (n\theta)$ for various values of n, n a natural number. Describe how n is related to the number of large petals and the number of small petals on the graph and how the large and small petals are related to each other relative to n.

64. Graph $r = 1 + 2 \cos (n\theta)$ for various values of n, n a natural number. Describe how n is related to the number of large petals and the number of small petals on the graph and how the large and small petals are related to each other relative to n.

 In Problems 65–68 graph each system of equations on the same set of polar coordinate axes. Then solve the system simultaneously. [Note: Any solution (r_1, θ_1) to the system must satisfy each equation in the system and therefore identifies a point of intersection of the two graphs. However, there may be other points of intersection of the two graphs that do not have any coordinates that satisfy both equations. This represents a major difference between the rectangular coordinate system and the polar coordinate system.]

65. $r = 4 \cos \theta$ **66.** $r = 2 \cos \theta$
 $r = -4 \sin \theta$ $r = 2 \sin \theta$
 $0 \le \theta \le \pi$ $0 \le \theta \le \pi$

67. $r = 6 \cos \theta$ **68.** $r = 8 \sin \theta$
 $r = 6 \sin 2\theta$ $r = 8 \cos 2\theta$
 $0° \le \theta \le 360°$ $0° \le \theta \le 360°$

APPLICATIONS

69. ANALYTIC GEOMETRY A distance formula for the distance between two points in a polar coordinate system follows directly from the law of cosines:

$$d^2 = r_1^2 + r_2^2 - 2r_1r_2 \cos (\theta_2 - \theta_1)$$
$$d = \sqrt{r_1^2 + r_2^2 - 2r_1r_2 \cos (\theta_2 - \theta_1)}$$

Find the distance (to three decimal places) between the two points $P_1 = (4, \pi/4)$ and $P_2 = (1, \pi/2)$.

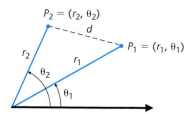

70. ANALYTIC GEOMETRY Refer to Problem 69. Find the distance (to three decimal places) between the two points $P_1 = (2, 30°)$ and $P_2 = (3, 60°)$.

Problems 71–72 refer to the polar diagram in the figure. Polar diagrams of this type are used extensively by serious sailboat racers, and this polar diagram represents speeds in knots of a high-performance sailboat sailing at various angles to a wind blowing at 20 knots.

71. SAILBOAT RACING Referring to the figure, estimate to the nearest knot the speed of the sailboat sailing at the following angles to the wind: 30°, 75°, 135°, and 180°.

72. SAILBOAT RACING Referring to the figure, estimate to the nearest knot the speed of the sailboat sailing at the following angles to the wind: 45°, 90°, 120°, and 150°.

 73. CONIC SECTIONS Using a graphing calculator, graph the equation

$$r = \frac{8}{1 - e \cos \theta}$$

for the following values of e (called the **eccentricity** of the conic) and identify each curve as a hyperbola, an ellipse, or a parabola.
(A) $e = 0.4$ (B) $e = 1$ (C) $e = 1.6$

(It is instructive to explore the graph for other positive values of e. See the Chapter 8 Group Activity for information on parabola, ellipse, and hyperbola.)

 74. CONIC SECTIONS Using a graphing calculator, graph the equation

$$r = \frac{8}{1 - e \cos \theta}$$

for the following values of e and identify each curve as a hyperbola, an ellipse, or a parabola.
(A) $e = 0.6$ (B) $e = 1$ (C) $e = 2$

 75. ASTRONOMY (A) The planet Mercury travels around the sun in an elliptical orbit given approximately by

$$r = \frac{3.442 \times 10^7}{1 - 0.206 \cos \theta}$$

where r is measured in miles and the sun is at the pole. Graph the orbit. Use TRACE to find the distance from Mercury to the sun at **aphelion** (greatest distance from the sun) and at **perihelion** (shortest distance from the sun).
(B) Johannes Kepler (1571–1630) showed that a line joining a planet to the sun sweeps out equal areas in space in equal intervals in time (see the figure). Use this information to determine whether a planet travels faster or slower at aphelion than at perihelion. Explain your answer.

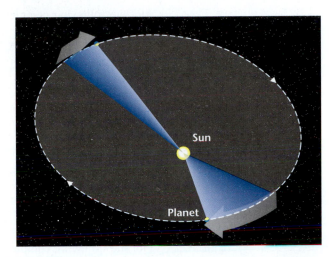

8-5 Complex Numbers and De Moivre's Theorem

> **Rectangular Form**
> **Polar Form**
> **Multiplication and Division**
> **Powers—De Moivre's Theorem**
> **Roots**
> **Historical Note**

Utilizing polar concepts studied in Section 8-4, we now show how complex numbers can be written in polar form. A brief review of Section 1-4 on complex numbers should prove helpful before proceeding further.

> Rectangular Form

Recall from Section 1-4 that a complex number is any number that can be written in the form

$$x + yi$$

where x and y are real numbers and i is the imaginary unit. (We use $x + yi$ and $x + iy$ interchangeably; each has its advantages in keeping notation simple.) So associated with each complex number $x + yi$ is a unique ordered pair of real numbers (x, y), and vice versa. For example,

$$5 + 3i \qquad \text{corresponds to} \qquad (5, 3)$$

Associating these ordered pairs of real numbers with points in a rectangular coordinate system, we obtain a **complex plane** (Fig. 1). When complex numbers are associated with points in a rectangular coordinate system, we refer to the x axis as the **real axis** and the y axis as the **imaginary axis.** The complex number $x + yi$ is said to be in **rectangular form.**

> **Figure 1** Complex plane.

EXAMPLE 1 **Plotting in the Complex Plane**

Plot the following complex numbers in a complex plane:

$$A = 2 + 3i \qquad B = -3 + 5i \qquad C = -4 \qquad D = -3i$$

SOLUTION

MATCHED PROBLEM 1 Plot the following complex numbers in a complex plane:

$$A = 4 + 2i \qquad B = 2 - 3i \qquad C = -5 \qquad D = 4i$$

> ## Polar Form

Each point (x, y) of the plane corresponds to a unique complex number z, namely, in *rectangular form*, $z = x + iy$. But the point (x, y) can also be specified by polar coordinates. Therefore, the complex number z can be given a polar form that depends on r and θ. The **polar form** of z is written $z = re^{i\theta}$. (When convenient, we write $re^{\theta i}$ in place of $re^{i\theta}$.)

	Points	Complex numbers
Rectangular form	(x, y)	$x + iy$
Polar form	(r, θ)	$re^{i\theta}$

The point with rectangular coordinates $(1, 1)$ has polar coordinates $(\sqrt{2}, \pi/4)$. (Why?) Therefore, the complex number $z = 1 + i$ has the polar form $z = \sqrt{2}e^{i\pi/4}$. A graphing calculator can convert a complex number in rectangular form to polar form and vice versa (see Fig. 2, where θ is in radians and numbers are displayed to two decimal places).

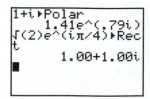

> **Figure 2** $1 + i = \sqrt{2}e^{i\pi/4}$.

The polar–rectangular relationships of Section 8-4 lead to the following connections between the rectangular and polar forms of a complex number.

> **POLAR–RECTANGULAR RELATIONSHIPS FOR COMPLEX NUMBERS**

If $x + iy = re^{i\theta}$, then

$$x = r \cos \theta \qquad r = \sqrt{x^2 + y^2}$$

$$y = r \sin \theta \qquad \tan \theta = \frac{y}{x}, \qquad x \neq 0$$

Therefore: $x + iy = r(\cos \theta + i \sin \theta) = re^{i\theta}$ and $e^{i\theta} = \cos \theta + i \sin \theta$

If $z = re^{i\theta}$, then the number r is called the **modulus,** or **absolute value,** of z and is denoted by **mod** z or $|z|$. The angle θ (in radians or degrees) is called the **argument** of z and is denoted by **arg** z. Recall that (r, θ) and $(r, \theta + 2\pi)$ represent the same point in polar coordinates. Therefore, $z = re^{i\theta} = re^{i(\theta + 2\pi)}$. So the argument of a complex number is not unique. But we usually choose the argument θ so that $-\pi < \theta \leq \pi$ (or $-180° < \theta \leq 180°$).

EXAMPLE	2

From Rectangular to Polar Form

Write each complex number in parts A–C in polar form, θ in radians, $-\pi < \theta \leq \pi$. Compute the modulus and arguments for parts A and B exactly; compute the modulus and argument for part C to two decimal places.

(A) $z_1 = 1 - i$ (B) $z_2 = -\sqrt{3} + i$ (C) $z = -5 - 2i$

SOLUTIONS Locate each point in a complex plane first; then if x and y are associated with special angles, r and θ can often be determined by inspection.

(A) A sketch shows that z_1 is associated with a special 45° triangle (Fig. 3). So by inspection, $r = \sqrt{2}$, $\theta = -\pi/4$ (not $7\pi/4$), and

$$z_1 = \sqrt{2}e^{i(-\pi/4)}$$

(B) A sketch shows that z_2 is associated with a special 30°–60° triangle (Fig. 4). So by inspection, $r = 2$, $\theta = 5\pi/6$, and

$$z_2 = 2e^{i(5\pi/6)}$$

(C) A sketch shows that z_3 is not associated with a special triangle (see Figure 5 on the next page) on the next page. So we proceed as follows:

$$r = \sqrt{(-5)^2 + (-2)^2} = 5.39 \qquad \text{To two decimal places}$$

$$\theta = -\pi + \tan^{-1}\left(\tfrac{2}{5}\right) = -2.76 \qquad \text{To two decimal places}$$

This gives us

$$z_3 = 5.39e^{(-2.76)i} \qquad \text{To two decimal places}$$

> Figure 3

> Figure 4

> **Figure 5**

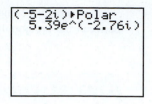

> **Figure 6**
$(-5 - 2i) = 5.39e^{(-2.76)i}$

Figure 6 shows the same conversion done by a graphing calculator with a built-in conversion routine (with numbers displayed to two decimal places).

MATCHED PROBLEM 2 Write each complex number in parts A–C in polar form, θ in radians, $-\pi < \theta \le \pi$. Compute the modulus and arguments for parts A and B exactly; compute the modulus and argument for part C to two decimal places.

(A) $-1 + i$ (B) $1 + i\sqrt{3}$ (C) $-3 - 7i$

EXAMPLE 3 **From Polar to Rectangular Form**

Write each complex number in parts A–C in rectangular form. Compute the exact values for parts A and B; for part C, compute a and b for $a + bi$ to two decimal places.

(A) $z_1 = 2e^{(5\pi/6)i}$ (B) $z_2 = 3e^{(-60°)i}$ (C) $z_3 = 7.19e^{(-2.13)i}$

SOLUTIONS (A) $x + iy = 2e^{(5\pi/6)i}$

$\qquad = 2[\cos(5\pi/6) + i\sin(5\pi/6)]$ $\cos\dfrac{5\pi}{6} = -\dfrac{\sqrt{3}}{2}, \sin\dfrac{5\pi}{6} = \dfrac{1}{2}$

$\qquad = 2\left(-\dfrac{\sqrt{3}}{2}\right) + i2\left(\dfrac{1}{2}\right)$ Simplify.

$\qquad = -\sqrt{3} + i$

(B) $x + iy = 3e^{(-60°)i}$

$\qquad = 3[\cos(-60°) + i\sin(-60°)]$ $\cos(-60°) = \dfrac{1}{2}, \sin(-60°) = -\dfrac{\sqrt{3}}{2}$

$\qquad = 3\left(\dfrac{1}{2}\right) + i3\left(-\dfrac{\sqrt{3}}{2}\right)$ Simplify.

$\qquad = \dfrac{3}{2} - \dfrac{3\sqrt{3}}{2}i$

(C) $x + iy = 7.19e^{(-2.13)i}$

$\qquad = 7.19[\cos(-2.13) + i\sin(-2.13)]$

$\qquad = -3.81 - 6.09i$

Figure 7 shows the same conversion done by a graphing calculator with a built-in conversion routine.

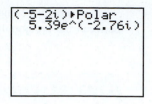

> **Figure 7**
$7.19e^{(-2.13)i} = -3.81 - 6.09i.$

MATCHED PROBLEM 3 Write each complex number in parts A–C in rectangular form. Compute the exact values for parts A and B; for part C compute a and b for $a + bi$ to two decimal places.

(A) $z_1 = \sqrt{2}e^{(-\pi/2)i}$ (B) $z_2 = 3e^{120°i}$ (C) $z_3 = 6.49e^{(-2.08)i}$

>>> **EXPLORE-DISCUSS 1**

Let $z_1 = \sqrt{3} + i$ and $z_2 = 1 + i\sqrt{3}$.

(A) Find $z_1 z_2$ and z_1/z_2 using the rectangular forms of z_1 and z_2.

(B) Find $z_1 z_2$ and z_1/z_2 using the polar forms of z_1 and z_2, θ in degrees. (Assume the product and quotient exponent laws hold for $e^{i\theta}$.)

(C) Convert the results from part B back to rectangular form and compare with the results in part A.

> ## Multiplication and Division

There is a particular advantage in representing complex numbers in polar form: multiplication and division become very easy. Theorem 1 provides the reason. (The polar form of a complex number obeys the product and quotient rules for exponents: $b^m b^n = b^{m+n}$ and $b^m/b^n = b^{m-n}$.)

> ### THEOREM 1 Products and Quotients in Polar Form

If $z_1 = r_1 e^{i\theta_1}$ and $z_2 = r_2 e^{i\theta_2}$, then

1. $z_1 z_2 = r_1 e^{i\theta_1} r_2 e^{i\theta_2} = r_1 r_2 e^{i(\theta_1 + \theta_2)}$

2. $\dfrac{z_1}{z_2} = \dfrac{r_1 e^{i\theta_1}}{r_2 e^{i\theta_2}} = \dfrac{r_1}{r_2} e^{i(\theta_1 - \theta_2)}$

Theorem 1 says that to multiply two complex numbers, you multiply their moduli and add their arguments. Similarly, to divide two complex numbers, you divide their moduli and subtract their arguments.

We will establish the multiplication property and leave the quotient property for Problem 66 in Exercises 8-5.

$$z_1 z_2 = r_1 e^{i\theta_1} r_2 e^{i\theta_2} \qquad \text{Use polar–rectangular relationships.}$$

$$= r_1 r_2 (\cos\theta_1 + i\sin\theta_1)(\cos\theta_2 + i\sin\theta_2) \qquad \text{Multiply.}$$

$$= r_1 r_2 (\cos\theta_1 \cos\theta_2 + i\cos\theta_1 \sin\theta_2 \qquad \text{Group real parts and imaginary parts.}$$
$$+ i\sin\theta_1 \cos\theta_2 - \sin\theta_1 \sin\theta_2)$$

$$= r_1 r_2 [(\cos\theta_1 \cos\theta_2 - \sin\theta_1 \sin\theta_2) \qquad \text{Use sum identities.}$$
$$+ i(\cos\theta_1 \sin\theta_2 + \sin\theta_1 \cos\theta_2)]$$

$$= r_1 r_2 [\cos(\theta_1 + \theta_2) + i\sin(\theta_1 + \theta_2)] \qquad \text{Use polar–rectangular relationships.}$$

$$= r_1 r_2 e^{i(\theta_1 + \theta_2)}$$

| EXAMPLE | 4 | **Products and Quotients** |

If $z_1 = 8e^{45°i}$ and $z_2 = 2e^{30°i}$, find

(A) $z_1 z_2$ **(B)** z_1/z_2

SOLUTIONS **(A)** $z_1 z_2 = 8e^{45°i} \cdot 2e^{30°i} \quad \boxed{= 8 \cdot 2e^{i(45° + 30°)}}^{*} = 16e^{75°i}$

(B) $\dfrac{z_1}{z_2} = \dfrac{8e^{45°i}}{2e^{30°i}} \quad \boxed{= \tfrac{8}{2}e^{i(45° - 30°)}} = 4e^{15°i}$

*Throughout the book, dashed boxes—called think boxes—are used to represent steps that may be performed mentally.

MATCHED PROBLEM 4

If $z_1 = 9e^{165°i}$ and $z_2 = 3e^{55°i}$, find

(A) $z_1 z_2$ (B) z_1/z_2

› Powers—De Moivre's Theorem

Abraham De Moivre (1667–1754), of French birth, spent most of his life in London tutoring, writing, and publishing mathematics. He belonged to many prestigious professional societies in England, Germany, and France. He was a close friend of Isaac Newton. The theorem that bears his name gives a formula for raising any complex number to the power n where n is a natural number.

> **› THEOREM 2 De Moivre's Theorem**
>
> If $z = re^{i\theta}$ and n is a natural number, then $z^n = r^n e^{i(n\theta)}$.

De Moivre's theorem follows from the formula for the product of complex numbers in polar form. If $n = 2$ and $z = re^{i\theta}$, then

$$z^2 = re^{i\theta}re^{i\theta} = r^2 e^{i(2\theta)}$$

In other words, to square a complex number, you square the modulus and double the argument. Similarly, to cube a complex number you cube the modulus and triple the argument. De Moivre's theorem says that to raise a complex number to the power n, you raise the modulus to the power n and multiply the argument by n.

EXAMPLE 5

The Natural Number Power of a Complex Number

Use De Moivre's theorem to find $(1 + i)^{10}$. Write the answer in exact rectangular form.

SOLUTION

First note that the polar form of $1 + i$ is $\sqrt{2}e^{45°i}$. Therefore,

$$\begin{aligned}
(1 + i)^{10} &= (\sqrt{2}e^{45°i})^{10} & & \text{Use De Moivre's theorem.} \\
&= (\sqrt{2})^{10}e^{(10 \cdot 45°)i} & & \text{Simplify.} \\
&= 32e^{450°i} & & \text{Change to rectangular form.} \\
&= 32(\cos 450° + i \sin 450°) & & \text{cos } 450° = 0, \text{ sin } 450° = 1 \\
&= 32(0 + i) & & \text{Simplify.} \\
&= 32i & & \text{Rectangular form}
\end{aligned}$$

MATCHED PROBLEM 5

Use De Moivre's theorem to find $(1 + i\sqrt{3})^5$. Write the answer in exact polar and rectangular forms.

EXAMPLE 6

The Natural Number Power of a Complex Number

Use De Moivre's theorem to find $(-\sqrt{3} + i)^6$. Write the answer in exact rectangular form.

SOLUTION First note that the polar form of $\sqrt{3} + i$ is $2e^{150°i}$. Therefore,

$$(-\sqrt{3} + i)^6 = (2e^{150°i})^6 \qquad \text{Use De Moivre's theorem.}$$

$$= 2^6 e^{(6 \cdot 150°)i} \qquad \text{Simplify.}$$

$$= 64 e^{900°i} \qquad \text{Change to rectangular form.}$$

$$= 64(\cos 900° + i \sin 900°) \qquad \cos 900° = -1, \sin 900° = 0$$

$$= 64(-1 + i0) \qquad \text{Simplify.}$$

$$= -64 \qquad \text{Rectangular form}$$

[*Note:* $-\sqrt{3} + i$ must be a sixth root of -64, because $(-\sqrt{3} + i)^6 = -64$.]

MATCHED PROBLEM 6 Use De Moivre's theorem to find $(1 - i\sqrt{3})^4$. Write the answer in exact polar and rectangular forms.

› Roots

Let $n > 1$ be an integer. A complex number w is an **nth root of z** if $w^n = z$. For example, 2 and -2 are square roots (second roots) of 4 because $2^2 = 4$ and $(-2)^2 = 4$. Similarly, $3i$ and $-3i$ are square roots of -9 because $(3i)^2 = -9$ and $(-3i)^2 = -9$. The nth root theorem gives a formula for all of the nth roots of any nonzero complex number.

> ### › THEOREM 3 nth Root Theorem
>
> Let $n > 1$ be an integer and let $z = re^{i\theta}$ be a nonzero complex number. Then z has n distinct nth roots given by
>
> $$r^{1/n} e^{i(\theta/n + k360°/n)} \qquad k = 0, 1, \ldots, n - 1$$

The proof of Theorem 3 is left to Problems 67 and 68 in Exercises 8-5. The nth root theorem implies that every nonzero complex number z has two square roots, three cube roots, four fourth roots, and so on. Furthermore, all n of the nth roots of z have the same modulus, so they all lie on the same circle centered at the origin, and they are equally spaced around that circle.

EXAMPLE 7 **Finding All Sixth Roots of a Complex Number**

Find six distinct sixth roots of $-1 + i\sqrt{3}$, and plot them in a complex plane.

SOLUTION First write $-1 + i\sqrt{3}$ in polar form:

$$-1 + i\sqrt{3} = 2e^{120°i}$$

Using the nth-root theorem, all six roots are given by

$$2^{1/6} e^{(120°/6 + k360°/6)i} = 2^{1/6} e^{(20° + k60°)i} \qquad k = 0, 1, 2, 3, 4, 5$$

So

$$w_1 = 2^{1/6}e^{(20° + 0 \cdot 60)i} = 2^{1/6}e^{20°i}$$
$$w_2 = 2^{1/6}e^{(20° + 1 \cdot 60)i} = 2^{1/6}e^{80°i}$$
$$w_3 = 2^{1/6}e^{(20° + 2 \cdot 60)i} = 2^{1/6}e^{140°i}$$
$$w_4 = 2^{1/6}e^{(20° + 3 \cdot 60)i} = 2^{1/6}e^{200°i}$$
$$w_5 = 2^{1/6}e^{(20° + 4 \cdot 60)i} = 2^{1/6}e^{260°i}$$
$$w_6 = 2^{1/6}e^{(20° + 5 \cdot 60)i} = 2^{1/6}e^{320°i}$$

> Figure 8

All roots are easily graphed in the complex plane after the first root is located. The root points are equally spaced around a circle of radius $2^{1/6}$ at an angular increment of 60° from one root to the next (Fig. 8).

MATCHED PROBLEM 7 Find five distinct fifth roots of $1 + i$. Leave the answers in polar form and plot them in a complex plane.

EXAMPLE 8 Solving a Cubic Equation

Solve $x^3 + 1 = 0$. Write final answers in rectangular form, and plot them in a complex plane.

SOLUTION

$$x^3 + 1 = 0$$
$$x^3 = -1$$

We see that x is a cube root of -1, and there are a total of three roots. To find the three roots, we first write -1 in polar form:

$$-1 = 1e^{180°i}$$

Using the nth-root theorem, all three cube roots of -1 are given by

$$1^{1/3}e^{(180°/3 + k360°/3)i} = 1e^{(60° + k120°)i} \qquad k = 0, 1, 2$$

So

$$w_1 = 1e^{60°i} = \cos 60° + i \sin 60° = \frac{1}{2} + i\frac{\sqrt{3}}{2}$$
$$w_2 = 1e^{180°i} = \cos 180° + i \sin 180° = -1$$
$$w_3 = 1e^{300°i} = \cos 300° + i \sin 300° = \frac{1}{2} - i\frac{\sqrt{3}}{2}$$

(*Note:* This problem can also be solved using factoring and the quadratic formula—try it.)

> Figure 9

The three roots are graphed in Figure 9.

MATCHED PROBLEM 8 Solve $x^3 - 1 = 0$. Write final answers in rectangular form, and plot them in a complex plane.

> **Historical Note**

There is hardly an area in mathematics that does not have some imprint of the famous Swiss mathematician Leonhard Euler (1707–1783), who spent most of his productive life at the New St. Petersburg Academy in Russia and the Prussian Academy in Berlin. One of the

most prolific writers in the history of the subject, he is credited with making the following familiar notations standard:

$$f(x) \quad \text{function notation}$$
$$e \quad \text{natural logarithmic base}$$
$$i \quad \text{imaginary unit, } \sqrt{-1}$$

For our immediate interest, he is also responsible for the extraordinary relationship

$$e^{i\theta} = \cos \theta + i \sin \theta$$

If we let $\theta = \pi$, we obtain an equation that relates five of the most important numbers in the history of mathematics:

$$e^{i\pi} + 1 = 0$$

ANSWERS TO MATCHED PROBLEMS

1.

2. (A) $\sqrt{2}e^{i(3\pi/4)}$
(B) $2e^{i(\pi/3)}$
(C) $7.62e^{(-1.98)i}$

3. (A) $-i\sqrt{2}$ (B) $-\dfrac{3}{2} + \dfrac{3\sqrt{3}}{2}i$ (C) $-3.16 - 5.67i$

4. (A) $z_1 z_2 = 27e^{220°i}$ (B) $z_1/z_2 = 3e^{110°i}$

5. $32e^{300°i} = 16 - i16\sqrt{3}$ **6.** $16e^{(-240°)i} = -8 + i8\sqrt{3}$

7. $w_1 = 2^{1/10}e^{9°i}$, $w_2 = 2^{1/10}e^{81°i}$,
$w_3 = 2^{1/10}e^{153°i}$, $w_4 = 2^{1/10}e^{225°i}$, $w_5 = 2^{1/10}e^{297°i}$ **8.** $1, -\dfrac{1}{2} + i\dfrac{\sqrt{3}}{2}, -\dfrac{1}{2} - i\dfrac{\sqrt{3}}{2}$

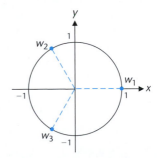

8-5 Exercises

1. What is the modulus of a complex number?

2. What is the argument of a complex nummber?

3. Explain how to locate the product of two complex numbers that lie on the unit circle.

4. Explain how to locate the quotient of two complex numbers that lie on the unit circle.

5. Explain how to locate the cube of a complex number that lies on the unit circle.

6. Explain how to locate the cube root of a complex number that lies on the unit circle.

In Problems 7–14, plot each set of complex numbers in a complex plane.

7. $A = 3 + 4i, B = -2 - i, C = 2i$

8. $A = 4 + i, B = -3 + 2i, C = -3i$

9. $A = 3 - 3i, B = 4, C = -2 + 3i$

10. $A = -3, B = -2 - i, C = 4 + 4i$

11. $A = 2e^{(\pi/3)i}, B = \sqrt{2}e^{(\pi/4)i}, C = 4e^{(\pi/2)i}$

12. $A = 2e^{(\pi/6)i}, B = 4e^{\pi i}, C = \sqrt{2}e^{(3\pi/4)i}$

13. $A = 4e^{(-150°)i}, B = 3e^{20°i}, C = 5e^{(-90°)i}$

14. $A = 2e^{150°i}, B = 3e^{(-50°)i}, C = 4e^{75°i}$

In Problems 15–18, convert to the polar form $re^{i\theta}$. For Problems 15 and 16, choose θ in degrees, $-180° < \theta \le 180°$; for Problems 17 and 18 choose θ in radians, $-\pi < \theta \le \pi$. Compute the modulus and arguments for parts A and B exactly; compute the modulus and argument for part C to two decimal places.

15. (A) $\sqrt{3} + i$ (B) $-1 - i$ (C) $5 - 6i$

16. (A) $-1 + i\sqrt{3}$ (B) $-3i$ (C) $-7 - 4i$

17. (A) $-i\sqrt{3}$ (B) $-\sqrt{3} - i$ (C) $-8 + 5i$

18. (A) $\sqrt{3} - i$ (B) $-2 + 2i$ (C) $6 - 5i$

In Problems 19–22, change the complex number in parts A–C to rectangular form. Compute the exact values for parts A and B; for part C compute a and b for $a + bi$ to two decimal places.

19. (A) $2e^{(\pi/3)i}$ (B) $\sqrt{2}e^{(-45°)i}$ (C) $3.08e^{2.44i}$

20. (A) $2e^{30°i}$ (B) $\sqrt{2}e^{(-3\pi/4)i}$ (C) $5.71e^{(-0.48)i}$

21. (A) $6e^{(\pi/6)i}$ (B) $\sqrt{7}e^{(-90°)i}$ (C) $4.09e^{(-122.88°)i}$

22. (A) $\sqrt{3}e^{(-\pi/2)i}$ (B) $\sqrt{2}e^{135°i}$ (C) $6.83e^{(-108.82°)i}$

In Problems 23–28, find z_1z_2 and z_1/z_2 in the polar form $re^{i\theta}$.

23. $z_1 = 7e^{82°i}, z_2 = 2e^{31°i}$

24. $z_1 = 6e^{132°i}, z_2 = 3e^{93°i}$

25. $z_1 = 5e^{52°i}, z_2 = 2e^{83°i}$

26. $z_1 = 3e^{67°i}, z_2 = 2e^{97°i}$

27. $z_1 = 3.05e^{1.76i}, z_2 = 11.94e^{2.59i}$

28. $z_1 = 7.11e^{0.79i}, z_2 = 2.66e^{1.07i}$

In Problems 29–34, use De Moivre's theorem to evaluate each. Leave answers in polar form.

29. $(2e^{30°i})^3$

30. $(5e^{15°i})^3$

31. $(\sqrt{2}e^{10°i})^6$

32. $(\sqrt{2}e^{15°i})^8$

33. $(1 + i\sqrt{3})^3$

34. $(\sqrt{3} + i)^8$

In Problems 35–40, find the value of each expression and write the final answer in exact rectangular form. (Verify the results in Problems 35–40 by evaluating each directly on a calculator.)

35. $(-\sqrt{3} - i)^4$ **36.** $(-1 + i)^4$

37. $(1 - i)^8$ **38.** $(-\sqrt{3} + i)^5$

39. $\left(-\dfrac{1}{2} + \dfrac{\sqrt{3}}{2}i\right)^3$ **40.** $\left(-\dfrac{1}{2} - \dfrac{\sqrt{3}}{2}i\right)^3$

For n and z as indicated in Problems 41–46, find all nth roots of z. Leave answers in the polar form $re^{i\theta}$.

41. $z = 8e^{30°i}, n = 3$

42. $z = 8e^{45°i}, n = 3$

43. $z = 81e^{60°i}, n = 4$

44. $z = 16e^{90°i}, n = 4$

45. $z = 1 - i, n = 5$

46. $z = -1 + i, n = 3$

For n and z as indicated in Problems 47–52, find all nth roots of z. Write answers in the polar form $re^{i\theta}$ and plot in a complex plane.

47. $z = 8, n = 3$ **48.** $z = 1, n = 4$

49. $z = -16, n = 4$ **50.** $z = -8, n = 3$

51. $z = i, n = 6$ **52.** $z = -i, n = 5$

53. (A) Show that $1 + i$ is a root of $x^4 + 4 = 0$. How many other roots does the equation have?
(B) The root $1 + i$ is located on a circle of radius $\sqrt{2}$ in the complex plane as indicated in the figure. Locate the other three roots of $x^4 + 4 = 0$ on the figure and explain geometrically how you found their location.
(C) Verify that each complex number found in part B is a root of $x^4 + 4 = 0$.

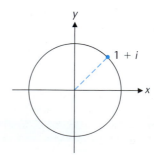

54. (A) Show that -2 is a root of $x^3 + 8 = 0$. How many other roots does the equation have?
(B) The root -2 is located on a circle of radius 2 in the complex plane as indicated in the figure on the next page. Locate the other two roots of $x^3 + 8 = 0$ on the figure and explain geometrically how you found their location.
(C) Verify that each complex number found in part B is a root of $x^3 + 8 = 0$.

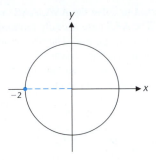

In Problems 55–58, solve each equation for all roots. Write final answers in the polar form $re^{i\theta}$ and exact rectangular form.

55. $x^3 + 64 = 0$ **56.** $x^3 - 64 = 0$

57. $x^3 - 27 = 0$

58. $x^3 + 27 = 0$

In Problems 59–64, determine whether the statement is true or false. If true, explain why. If false, give a counterexample.

59. If two numbers lie on the real axis, then their product lies on the real axis.

60. If two numbers lie on the imaginary axis, then their quotient lies on the imaginary axis.

61. If z is a positive real number, then all of the fourth roots of z are real.

62. If z is a positive real number, then all of the square roots of z are real.

63. If w is a square root of 1, then w is a sixth root of 1.

64. If w is a sixth root of 1, then w is a square root of 1.

65. Suppose that z is a complex number that is not real. Explain why none of the nth roots of z lies on the x axis.

66. Prove

$$\frac{z_1}{z_2} = \frac{r_1 e^{i\theta_1}}{r_2 e^{i\theta_2}} = \frac{r_1}{r_2} e^{i(\theta_1 - \theta_2)}$$

67. Show that

$$\left[r^{1/n} e^{(\theta/n + k360°/n)i} \right]^n = re^{i\theta}$$

for any natural number n and any integer k.

68. Show that

$$r^{1/n} e^{(\theta/n + k360°/n)i}$$

is the same number for $k = 0$ and $k = n$.

In Problems 69–72, write answers in the polar form $re^{i\theta}$.

69. Find all complex zeros for $P(x) = x^5 - 32$.

70. Find all complex zeros for $P(x) = x^6 + 1$.

71. Solve $x^5 + 1 = 0$ in the set of complex numbers.

72. Solve $x^3 - i = 0$ in the set of complex numbers.

In Problems 73 and 74, write answers using exact rectangular forms.

73. Write $P(x) = x^6 + 64$ as a product of linear factors.

74. Write $P(x) = x^6 - 1$ as a product of linear factors.

CHAPTER **8** **Review**

8-1 Law of Sines

An **oblique triangle** is a triangle without a right angle. An oblique triangle is **acute** if all angles are between 0° and 90° and **obtuse** if one angle is between 90° and 180°. The labeling convention shown in these figures is followed in Chapter 8.

Acute triangle

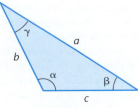
Obtuse triangle

The objective in Sections 8-1 and 8-2 is to solve an oblique triangle given any three of the six quantities indicated in either figure, if a solution exists. The law of sines, discussed in Section 8-1, and the law of cosines, discussed in Section 8-2, are used for this purpose. Accuracy in computation is governed by Table 1.

Table 1 Triangles and Significant Digits

Angle to Nearest	Significant Digits for Side Measure
1°	2
10' or 0.1°	3
1' or 0.01°	4
10" or 0.001°	5

The **law of sines** is given as

$$\frac{\sin \alpha}{a} = \frac{\sin \beta}{b} = \frac{\sin \gamma}{c}$$

and is generally used to solve the ASA, AAS, and SSA cases for oblique triangles. The AAS case is easily reduced to the ASA case by solving for the third angle first. The SSA case has a number of variations, including the ambiguous case. These variations are summarized in Table 2. Note that the ambiguous case always results in two triangles, one obtuse and one acute.

Table 2 SSA Variations

α	$a[h = b \sin \alpha]$	Number of Triangles	Figure
Acute	$0 < a < h$	0	
Acute	$a = h$	1	
Acute	$h < a < b$	2	Ambiguous case
Acute	$a \geq b$	1	
Obtuse	$0 < a \leq b$	0	
Obtuse	$a > b$	1	

8-2 Law of Cosines

The **law of cosines** is given as

$$a^2 = b^2 + c^2 - 2bc \cos \alpha$$
$$b^2 = a^2 + c^2 - 2ac \cos \beta$$
$$c^2 = a^2 + b^2 - 2ab \cos \gamma$$

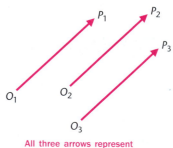

All three arrows represent the same vector **v**

and is generally used as the first step in solving the SAS and SSS cases for oblique triangles. After a side or angle is found using the law of cosines, it is usually easier to continue the solving process with the law of sines.

8-3 Vectors in the Plane

A **vector v** is a quantity that has both magnitude and direction. We picture a vector as an arrow from an initial point O to a terminal point P with this provision: arrows that have the same length (magnitude) and direction represent the same vector (see the figure).

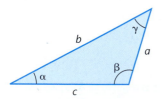

The vector **v** of the figure is also denoted by $\overrightarrow{O_1P_1}$ (or $\overrightarrow{O_2P_2}$ or $\overrightarrow{O_3P_3}$).

The **magnitude** of the vector $\mathbf{v} = \overrightarrow{OP}$, denoted by $|\mathbf{v}|$, $|\vec{v}|$, or $|\overrightarrow{OP}|$, is the length of the line segment OP. Two vectors have the **same direction** if they are parallel and point in the same direction. Two vectors have **opposite directions** if they are parallel and point in opposite directions. The **zero vector,** denoted by **0** or $\vec{0}$, has magnitude 0 and arbitrary direction. Two vectors are **equal** if they have the same magnitude and direction.

Any vector $\overrightarrow{AB}$ in a rectangular coordinate system can be translated so that its initial point is the origin O. The vector $\overrightarrow{OP}$ such that $\overrightarrow{OP} = \overrightarrow{AB}$ is said to be the **standard vector** for $\overrightarrow{AB}$, as shown in the figure on the next page.

OP is the standard vector for AB

Referring to the figure, if $A = (x_a, y_a)$ and $B = (x_b, y_b)$, then the coordinates of the point P are given by $(x_p, y_p) = (x_b - x_a, y_b - y_a)$.

There is a one-to-one correspondence between vectors in a rectangular coordinate system and points in the system: Any vector $\overrightarrow{AB}$ is associated with the point $P = (x_p, y_p)$ such that $\overrightarrow{OP} = \overrightarrow{AB}$ and any point P is associated with the vector $\overrightarrow{OP}$. A vector can therefore be denoted by an ordered pair of real numbers. We use $\langle c, d \rangle$ to denote the vector with initial point $(0, 0)$ and terminal point (c, d). The real numbers c and d are called the **scalar components** of the vector $\langle c, d \rangle$. Two vectors $\mathbf{u} = \langle a, b \rangle$ and $\mathbf{v} = \langle c, d \rangle$ are equal if their corresponding components are equal, that is, if $a = c$ and $b = d$. The zero vector is $\mathbf{0} = \langle 0, 0 \rangle$. The magnitude of the vector $\mathbf{u} = \langle a, b \rangle$ is given by

$$|\mathbf{u}| = \sqrt{a^2 + b^2}$$

The sum $\mathbf{u} + \mathbf{v}$ of two vectors $\mathbf{u}$ and $\mathbf{v}$ is defined by the **tail-to-tip rule:** Translate $\mathbf{v}$ so that its tail (initial point) is at the tip (terminal point) of $\mathbf{u}$. Then, the vector from the tail of $\mathbf{u}$ to the tip of $\mathbf{v}$ is the sum, denoted $\mathbf{u} + \mathbf{v}$, of the vectors $\mathbf{u}$ and $\mathbf{v}$.

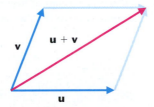

Vector addition: tail-to-tip rule

If $\mathbf{u}$ and $\mathbf{v}$ are not parallel, the **parallelogram rule** gives an alternative description of $\mathbf{u} + \mathbf{v}$: The sum of two nonparallel vectors $\mathbf{u}$ and $\mathbf{v}$ is the diagonal formed using $\mathbf{u}$ and $\mathbf{v}$ as adjacent sides.

Vector addition: parallelogram rule

The vector $\mathbf{u} + \mathbf{v}$ is also called the **resultant** of the two vectors $\mathbf{u}$ and $\mathbf{v}$, and $\mathbf{u}$ and $\mathbf{v}$ are called **vector components** of $\mathbf{u} + \mathbf{v}$.

The **scalar product** $k\mathbf{u}$ of a scalar (real number) k and a vector $\mathbf{u}$ is the vector with magnitude $|k||\mathbf{u}|$ that has the same direction as $\mathbf{u}$ if k is positive and the opposite direction if k is negative.

Both the sum $\mathbf{u} + \mathbf{v}$ and the scalar product $k\mathbf{u}$ are easy to calculate if the scalar components of $\mathbf{u}$ and $\mathbf{v}$ are given. If $\mathbf{u} = \langle a, b \rangle$, $\mathbf{v} = \langle c, d \rangle$, and k is a scalar (real number), then

$$\mathbf{u} + \mathbf{v} = \langle a + c, b + d \rangle \qquad \text{Vector addition}$$

$$k\mathbf{u} = \langle ka, kb \rangle \qquad \text{Scalar multiplication}$$

Any vector that has magnitude 1 is called a **unit vector.** If $\mathbf{v}$ is an arbitrary nonzero vector, then $u = (1/|\mathbf{v}|)\mathbf{v}$, is a unit vector with the same direction as $\mathbf{v}$. The unit vectors in the directions of the positive x axis and the positive y axis are denoted by $\mathbf{i}$ and $\mathbf{j}$, respectively.

Every vector can be expressed in terms of the $\mathbf{i}$ and $\mathbf{j}$ unit vectors: $\langle a, b \rangle = a\mathbf{i} + b\mathbf{j}$.

The following algebraic properties of vector addition and scalar multiplication enable us to manipulate symbols representing vectors and scalars in much the same way we manipulate symbols that represent real numbers in algebra.

> **ALGEBRAIC PROPERTIES OF VECTORS**
>
> **A. Addition Properties.** For all vectors $\mathbf{u}$, $\mathbf{v}$, and $\mathbf{w}$:
>
> **1.** $\mathbf{u} + \mathbf{v} = \mathbf{v} + \mathbf{u}$ **Commutative Property**
>
> **2.** $\mathbf{u} + (\mathbf{v} + \mathbf{w}) = (\mathbf{u} + \mathbf{v}) + \mathbf{w}$ **Associative Property**
>
> **3.** $\mathbf{u} + \mathbf{0} = \mathbf{0} + \mathbf{u} = \mathbf{u}$ **Additive Identity**
>
> **4.** $\mathbf{u} + (-\mathbf{u}) = (-\mathbf{u}) + \mathbf{u} = \mathbf{0}$ **Additive Inverse**
>
> **B. Scalar Multiplication Properties.** For all vectors $\mathbf{u}$ and $\mathbf{v}$ and all scalars m and n:
>
> **1.** $m(n\mathbf{u}) = (mn)\mathbf{u}$ **Associative Property**
>
> **2.** $m(\mathbf{u} + \mathbf{v}) = m\mathbf{u} + m\mathbf{v}$ **Distributive Property**
>
> **3.** $(m + n)\mathbf{u} = m\mathbf{u} + n\mathbf{u}$ **Distributive Property**
>
> **4.** $1\mathbf{u} = \mathbf{u}$ **Multiplicative Identity**

A vector that represents the direction and speed of an object in motion is called a **velocity vector.** The velocity of an airplane relative to the air is called the **apparent velocity,** and the velocity relative to the ground is called the **resultant,** or **actual, velocity.** The resultant velocity is the vector sum of the apparent velocity and wind velocity. Similar statements apply to objects in water subject to currents.

A vector that represents the direction and magnitude of an applied force is called a **force vector.** If an object is subjected to two forces, then the sum of these two forces, the **resultant force,** is a single force acting on the object in the same way as the two original forces taken together. An object at rest is said to be in **static equilibrium.** For an object to remain in static equilibrium, the sum of all the force vectors acting on the object must be the zero vector.

8-4 Polar Coordinates and Graphs

The figure illustrates a **polar coordinate system.** The fixed point O is called the **pole** or **origin,** and the horizontal arrow is called the **polar axis.** We have the following **relationships between rectangular coordinates (x, y) and polar coordinates (r, θ):**

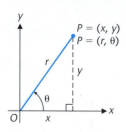

$$r^2 = x^2 + y^2$$

$$\sin \theta = \frac{y}{r} \quad \text{or} \quad y = r \sin \theta$$

$$\cos \theta = \frac{x}{r} \quad \text{or} \quad x = r \cos \theta$$

$$\tan \theta = \frac{y}{x}$$

[*Note:* The signs of x and y determine the quadrant for θ. The angle θ is chosen so that $-\pi < \theta \leq \pi$ or $-180° < \theta \leq 180°$, unless directed otherwise.]

Polar graphs can be obtained by **point-by-point plotting** much in the same way graphs in rectangular coordinates are formed. Make a table of values that satisfy the polar equation, plot these points, then join them with a smooth curve.

Graphs can also be obtained by **rapid polar sketching.** If only a rough sketch of a polar equation involving $\sin \theta$ or $\cos \theta$ is desired, we can speed up the point-by-point graphing process by taking advantage of the uniform variation of $\sin \theta$ and $\cos \theta$ as θ moves through each set of quadrant values. Graphing calculators can produce polar graphs almost instantly.

The table shows some standard polar curves with their equations:

Standard Polar Graphs

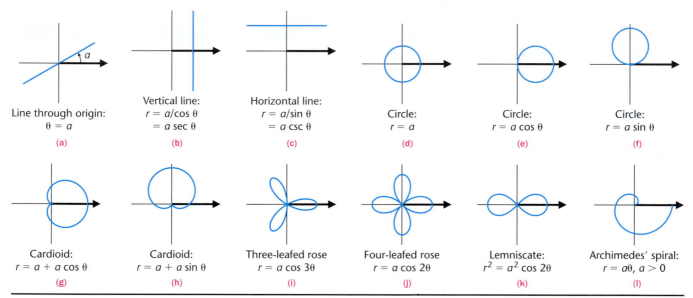

Line through origin:
$\theta = a$
(a)

Vertical line:
$r = a/\cos \theta$
$= a \sec \theta$
(b)

Horizontal line:
$r = a/\sin \theta$
$= a \csc \theta$
(c)

Circle:
$r = a$
(d)

Circle:
$r = a \cos \theta$
(e)

Circle:
$r = a \sin \theta$
(f)

Cardioid:
$r = a + a \cos \theta$
(g)

Cardioid:
$r = a + a \sin \theta$
(h)

Three-leafed rose:
$r = a \cos 3\theta$
(i)

Four-leafed rose:
$r = a \cos 2\theta$
(j)

Lemniscate:
$r^2 = a^2 \cos 2\theta$
(k)

Archimedes' spiral:
$r = a\theta, a > 0$
(l)

8-5 Complex Numbers and De Moivre's Theorem

Each point (x, y) of the plane corresponds to a unique complex number z. The **rectangular form** of z is written $z = x + iy$. The point (x, y) can also be specified by polar coordinates. Therefore, the complex number z can be given a polar form that depends on r and θ. The **polar form** of z is written $z = re^{i\theta}$.

	Points	Complex numbers
Rectangular form	(x, y)	$x + iy$
Polar form	(r, θ)	$re^{i\theta}$

In advanced mathematics the notation $e^{i\theta}$ is used to represent a value of a generalized version of the natural exponential function $f(x) = e^x$ with base $e \approx 2.718$. Our use of the notation $re^{i\theta}$ does not depend on that interpretation. Instead, $re^{i\theta}$ simply denotes the complex number that corresponds to the point with polar coordinates (r, θ). As is the case with polar coordinates, θ can be given in either radians or degrees. We assume, however, that r and θ are chosen so that r is nonnegative.

The polar–rectangular relationships of Section 8-4 lead to the following connections between the rectangular and polar forms of a complex number $z = x + iy = re^{i\theta}$:

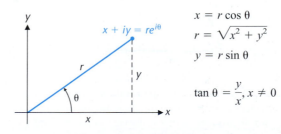

$$x = r \cos \theta$$
$$r = \sqrt{x^2 + y^2}$$
$$y = r \sin \theta$$
$$\tan \theta = \frac{y}{x}, x \neq 0$$

$x + iy = r(\cos \theta + i \sin \theta) = re^{i\theta}$ and $e^{i\theta} = \cos \theta + i \sin \theta$

If $z = re^{i\theta}$, then the number r is called the **modulus,** or **absolute value,** of z and is denoted by **mod** z or $|z|$. The angle θ (in radians or degrees) is called the **argument** of z and is denoted by **arg** z. The argument of a complex number is not unique, but we usually choose the argument θ so that $-\pi < \theta \leq \pi$ (or $-180° < \theta \leq 180°$).

Products and **quotients** of complex numbers are easily calculated from their polar forms:

$$r_1 e^{i\theta_1} r_2 e^{i\theta_2} = r_1 r_2 e^{i(\theta_1 + \theta_2)} \qquad \textbf{Product}$$

$$\frac{r_1 e^{i\theta_1}}{r_2 e^{i\theta_2}} = \frac{r_1}{r_2} e^{i(\theta_1 - \theta_2)} \qquad \textbf{Quotient}$$

De Moivre's theorem gives a formula for raising any complex number to the power n where n is a natural number: If $z = re^{i\theta}$, then $z^n = r^n e^{i(n\theta)}$.

Let $n > 1$ be an integer. A complex number w is an **nth root of z** if $w^n = z$. The **nth root theorem** gives a formula for all of the nth roots of any nonzero complex number: If $z = re^{i\theta}$ is a nonzero complex number, then z has n distinct nth roots given by

$$r^{1/n} e^{i(\theta/n + k360°/n)} \qquad k = 0, 1, \ldots, n - 1$$

Because all n of the nth roots of z have the same modulus, they all lie on the same circle centered at the origin, and they are equally spaced around that circle.

CHAPTER **8** Review Exercises

Work through all the problems in this chapter review and check answers in the back of the book. Answers to all review problems are there, and following each answer is a number in italics indicating the section in which that type of problem is discussed. Where weaknesses show up, review appropriate sections in the text.

Problems in these exercises use the following labeling of sides and angles:

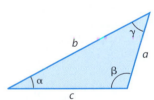

In Problems 1–3, determine whether the information in each problem allows you to construct 0, 1, or 2 triangles. Do not solve the triangle.

1. $a = 11$ meters, $b = 3.7$ meters, $\alpha = 67°$

2. $c = 15$ centimeters, $\alpha = 97°$, $\beta = 84°$

3. $a = 18$ feet, $b = 22$ feet, $\alpha = 54°$

4. Referring to the figure at the beginning of the exercises, if $\alpha = 52.6°$, $b = 57.1$ centimeters, and $c = 79.5$ centimeters, which of the two angles, β or γ, can you say for certain is acute and why?

In Problems 5–7, solve each triangle, given the indicated information.

5. $\alpha = 67°$, $\beta = 38°$, and $c = 49$ meters

6. $\alpha = 15°$, $b = 9.1$ feet, and $c = 12$ feet

7. $\gamma = 121°$, $c = 11$ centimeters, and $b = 4.2$ centimeters

8. Given geometric vectors **u** and **v** as indicated in the figure, find $|\mathbf{u} + \mathbf{v}|$ and θ, given $|\mathbf{u}| = 160$ miles per hour and $|\mathbf{v}| = 55$ miles per hour.

9. Write the algebraic vector $\langle a, b \rangle$ corresponding to the geometric vector $\overrightarrow{AB}$ with endpoints $A = (2, 6)$ and $B = (5, -1)$.

10. Find the magnitude of the vector $\langle -3, -5 \rangle$.

11. Sketch a graph of $\theta = \pi/6$ in a polar coordinate system.

12. Sketch a graph of $r = 6$ in a polar coordinate system.

13. Plot in a complex plane: $A = 3 + 5i$, $B = -1 - i$, $C = -3i$.

14. A point in a polar coordinate system has coordinates $(10, -30°)$. Find all other polar coordinates for the point, $-360° \le \theta \le 360°$, and verbally describe how the coordinates are associated with the point.

15. Plot in a complex plane: $A = 5e^{30°i}$, $B = 10e^{(-\pi/2)i}$, $C = 7e^{(3\pi/4)i}$.

16. (A) Change $1 - i\sqrt{3}$ to the polar form $re^{i\theta}$, $r \ge 0$, $-180° < \theta \le 180°$.
(B) Change $4e^{(-30°)i}$ to exact rectangular form.

17. (A) Find $[(-1/2) - (\sqrt{3}/2)i]^3$ using De Moivre's theorem. Write the final answer in exact rectangular form.
(B) Verify the results in part A with a calculator.

18. Find $(2e^{15°i})^4$ using De Moivre's theorem, and write the final answer in exact rectangular form.

19. Referring to the figure at the beginning of the exercises, if $a = 434$ meters, $b = 302$ meters, and $c = 197$ meters, then if the triangle has an obtuse angle which angle must it be and why?

In Problems 20–23, solve each triangle. If a problem does not have a solution, say so. If a triangle has two solutions, say so, and solve the obtuse case.

20. $\beta = 115.4°$, $a = 5.32$ centimeters, $c = 7.05$ centimeters

21. $\alpha = 63.2°$, $a = 179$ millimeters, $b = 205$ millimeters

22. $\alpha = 26.4°$, $a = 52.2$ kilometers, $b = 84.6$ kilometers

23. $a = 19.0$ inches, $b = 27.8$ inches, $c = 26.1$ inches

24. If four nonzero force vectors with different magnitudes and directions are acting on an object at rest, what must the sum of all four vectors be for the object to remain at rest?

25. Given geometric vectors **u** and **v** as indicated in the figure, find $|\mathbf{u} + \mathbf{v}|$ and α, given $|\mathbf{u}| = 75.2$ kilograms, $|\mathbf{v}| = 34.2$ kilograms, and $\theta = 57.2°$.

26. Express each vector in terms of **i** and **j** unit vectors:
(A) $\mathbf{u} = \langle -3, 9 \rangle$ (B) $\mathbf{v} = \langle 0, -2 \rangle$

For the indicated vectors in Problems 27 and 28, find
(A) $\mathbf{u} - \mathbf{v}$ (B) $3\mathbf{u} - \mathbf{v} + 2\mathbf{w}$

27. $\mathbf{u} = \langle -2, 3 \rangle$, $\mathbf{v} = \langle 2, -4 \rangle$, $\mathbf{w} = \langle -3, 0 \rangle$

28. $\mathbf{u} = \mathbf{i} - 2\mathbf{j}$, $\mathbf{v} = 3\mathbf{i} + 2\mathbf{j}$, $\mathbf{w} = -\mathbf{j}$

29. Find a unit vector **u** with the same direction as $\mathbf{v} = \langle -1, -3 \rangle$.

In Problems 30–33, use rapid sketching techniques to sketch each graph in a polar coordinate system.

30. $r = 6 + 4 \cos \theta$

31. $r = 8 + 8 \sin \theta$

32. $r = 10 \cos 2\theta$

33. $r = 8 \sin 3\theta$

34. Graph $r = 6 \cos \dfrac{\theta}{7}$ for $0 \le \theta \le 7\pi$.

35. Graph $r = 6 \cos \dfrac{\theta}{9}$ for $0 \le \theta \le 9\pi$.

36. Graph $r = 8 (\sin \theta)^{2n}$, for $n = 1, 2$, and 3. How many leaves do you expect the graph will have for arbitrary n?

37. Graph $r = 3/(1 - e \cos \theta)$ for the following values of e and identify each curve as an ellipse, a parabola, or a hyperbola:
(A) $e = 0.55$ (B) $e = 1$ (C) $e = 1.7$

38. Convert $x^2 + y^2 = 6x$ to polar form.

39. Convert $r = 5 \cos \theta$ to rectangular form.

40. Change the following complex numbers to the polar form $re^{i\theta}$, $r \ge 0$, $-180° < \theta \le 180°$: $z_1 = -1 + i$, $z_2 = -1 - i\sqrt{3}$, $z_3 = 5$.

41. Change the following complex numbers to exact rectangular form: $z_1 = \sqrt{2}e^{(\pi/4)i}$, $z_2 = 3e^{210°i}$, $z_3 = 2e^{(-2\pi/3)i}$.

42. If $z_1 = 8e^{25°i}$ and $z_2 = 4e^{19°i}$, find
(A) $z_1 z_2$ (B) z_1/z_2

Leave answers in the polar form $re^{i\theta}$.

43. (A) Write $(1 + i\sqrt{3})^4$ in exact rectangular form. Use De Moivre's theorem.
(B) Verify part A by evaluating $(1 + i\sqrt{3})^4$ directly on a calculator.

44. Find all cube roots of i. Write final answers in exact rectangular form, and locate the roots on a circle in the complex plane.

45. Find all cube roots of $-4\sqrt{3} + 4i$ exactly. Leave answers in the polar form $re^{i\theta}$.

46. Show that $4e^{15°i}$ is a square root of $8\sqrt{3} + 8i$.

47. Change the rectangular coordinates $(5.17, -2.53)$ to polar coordinates to two decimal places, $r \ge 0$, $-180° < \theta \le 180°$.

48. Change the polar coordinates $(5.81, -2.72)$ to rectangular coordinates to two decimal places.

49. Change the complex number $-3.18 + 4.19i$ to the polar form $re^{i\theta}$ to two decimal places, $r \ge 0$, $-180° < \theta \le 180°$.

50. Change the complex number $7.63e^{(-162.27°)i}$ to rectangular form $a + bi$, where a and b are computed to two decimal places.

51. (A) The cube root of a complex number is shown in the figure. Geometrically locate all other cube roots of the number on the figure, and explain how they were located.
(B) Determine geometrically the other cube roots of the number in exact rectangular form.
(C) Cube each cube root from parts A and B.

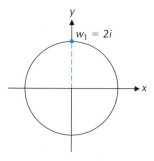

52. For an oblique triangle with $\alpha = 23.4°$, $b = 44.6$ millimeters, and a the side opposite angle α, determine a value k so that if $0 < a < k$, there is no solution; if $a = k$, there is one solution; and if $k < a < b$, there are two solutions.

53. Show that for any triangle
$$\frac{a^2 + b^2 + c^2}{2abc} = \frac{\cos \alpha}{a} + \frac{\cos \beta}{b} + \frac{\cos \gamma}{c}$$

54. Let $\mathbf{u} = \langle a, b \rangle$ and $\mathbf{v} = \langle c, d \rangle$ be vectors and m a scalar; prove
(A) $(\mathbf{u} + \mathbf{v}) = (\mathbf{v} + \mathbf{u})$
(B) $m(\mathbf{u} + \mathbf{v}) = m\mathbf{u} + m\mathbf{v}$

55. Given the polar equation $r = 4 + 4 \cos(\theta/2)$.
 (A) Sketch a graph of the equation using rapid graphing techniques.
 (B) Verify the graph in part A on a graphing calculator.

 56. (A) Graph $r = -8 \sin \theta$ and $r = 8 \cos \theta$, $0 \le \theta \le \pi$, in the same viewing window. Use TRACE to determine which intersection point has coordinates that satisfy both equations simultaneously.
 (B) Solve the equations simultaneously to verify the results in part A.
 (C) Explain why the pole is not a simultaneous solution, even though the two curves intersect at the pole.

57. Find all solutions, real and imaginary, for $x^8 - 1 = 0$. Write roots in exact rectangular form.

58. Write $P(x) = x^3 - 8i$ as a product of linear factors.

APPLICATIONS

For Problems 59–61, use the navigational compass shown. Assume directions given in terms of north, east, south, and west are exact.

Navigational compass

59. NAVIGATION An airplane flies east at 256 miles per hour, and another airplane flies southeast at 304 miles per hour. After 2 hours, how far apart are the two planes?

60. NAVIGATION An airplane flies with an airspeed of 450 miles per hour and a compass heading of 75°. If the wind is blowing at 65 miles per hour out of the north (from north to south), what is the plane's actual direction and speed relative to the ground? Compute direction to the nearest degree and speed to the nearest mile per hour.

61. NAVIGATION An airplane that can cruise at 500 miles per hour in still air is to fly due east. If the wind is blowing from the northeast at 50 miles per hour, what compass heading should the pilot choose? What will be the actual speed of the plane relative to the ground? Compute direction to the nearest degree and speed to the nearest mile per hour.

62. COASTAL NAVIGATION The owner of a pleasure boat cruising along a coast wants to pass a rocky point at a safe distance (see the figure). Sightings of the rocky point are made at A and at B, 1.0 mile apart. If the boat continues on the same course, how close will it come to the point? That is, find d in the figure to the nearest tenth of a mile.

63. FORCES Two forces **u** and **v** are acting on an object as indicated in the figure. Find the direction and magnitude of the resultant force **u** + **v** relative to force **v**.

64. STATIC EQUILIBRIUM Two forces **u** and **v** are acting on an object as indicated in the figure. What third force **w** must be added to achieve static equilibrium? Give direction relative to **u**.

65. ENGINEERING A cable car weighing 1,000 pounds is used to cross a river (see the figure). What is the tension in each half of the cable when the car is located as indicated? Compute the answer to three significant digits.

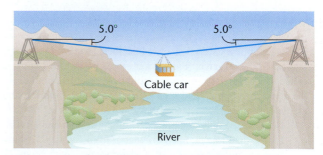

66. ASTRONOMY.
(A) The planet Mars travels around the sun in an elliptical orbit given approximately by

$$r = \frac{1.41 \times 10^8}{1 - 0.0934 \cos \theta} \qquad (1)$$

where r is measured in miles and the sun is at the pole. Graph the orbit. Use TRACE to find the distance (to three significant digits) from Mars to the sun at **aphelion** (greatest distance from the sun) and at **perihelion** (shortest distance from the sun).
(B) Referring to equation (1), r is maximum when the denominator is minimum, and r is minimum when the denominator is maximum. Use this information to find the distance from Mars to the sun at aphelion and at perihelion.

CHAPTER **8**

››› GROUP ACTIVITY Polar Equations of Conic Sections*

A **conic section** is the cross section obtained by intersecting a right circular cone[†] and a plane (Fig. 1). A **circle** is obtained by cutting the cone by a plane perpendicular to the axis of the cone. Tilt the plane slightly to obtain an **ellipse.** If the plane is parallel to an edge of the cone, the cross section is a **parabola.** If the plane is titled more to the vertical, then the cross section cuts both nappes and is a **hyperbola** with two branches.

Circle Ellipse Parabola Hyperbola

› **Figure 1** Conic sections.

A second way to define the conic sections is in terms of *eccentricity.* Let F be a fixed point, called the **focus,** and let d be a fixed line called the **directrix** (Fig. 2). Let e be a positive real number called the **eccentricity.** A **conic section** is the set of all points P in the plane containing F and d such that the distance from P to the focus F is e times the distance from P to the directrix d.

1. Let F be the pole in a polar coordinate system and let d be the vertical line that is a distance p from F (Fig. 2). Show that the polar equation of a conic section with focus F, directrix d, and eccentrictiy e is given by

$$r = \frac{ep}{1 - e \cos \theta} \qquad (1)$$

 2. Use a graphing calculator to explore the graph of equation (1) if $0 < e < 1$ and p is a positive number. Summarize the results of holding e fixed and changing p, and of holding p fixed and changing e. Which conic section is produced?

 3. Repeat Problem 2 if $e > 1$.

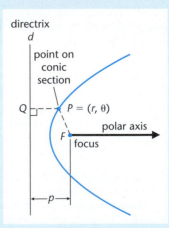

› **Figure 2** Conic sections.

 4. Repeat Problem 2 if $e = 1$ (summarize the results of changing p).

*The conic sections, and their applications to astronomy, architecture, medicine, and engineering, are discussed in detail in Chapter 9.
[†]Starting with a fixed line L and a fixed point V on L, the surface formed by all straight lines through V making a constant angle θ with L is called a **right circular cone.** The fixed line L is called the **axis** of the cone, and V is its **vertex.** The two parts of the cone separated by the vertex are called **nappes.**

Additional Topics in Analytic Geometry

ANALYTIC geometry is the study of geometric objects using algebraic techniques. René Descartes (1596–1650), the French philosopher and mathematician, is generally recognized as the founder of the subject. We used analytic geometry in Chapter 2 to obtain equations of lines and circles. In Chapter 9, we take a similar approach to the study of parabolas, ellipses, and hyperbolas. Each of these geometric objects is a *conic section,* that is, the intersection of a plane and a cone. We will derive equations for the conic sections and explore a wealth of applications in architecture, communications, engineering, medicine, optics, and space science.

OUTLINE

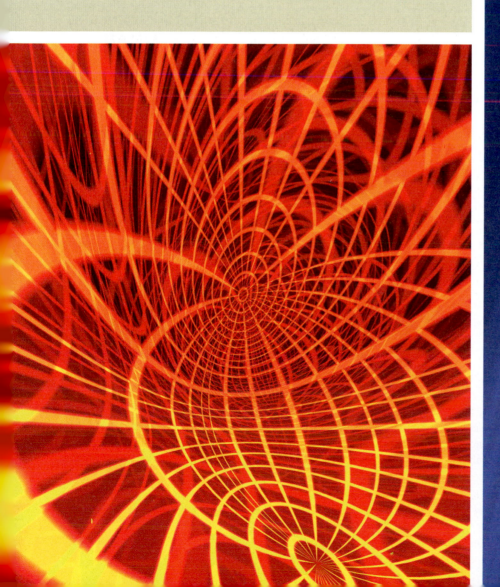

9-1 Conic Sections; Parabola

> Conic Sections
> Definition of a Parabola
> Drawing a Parabola
> Standard Equations and Their Graphs
> Applications

In Section 9-1 we introduce the general concept of a conic section and then discuss the particular conic section called a *parabola*. In Sections 9-2 and 9-3 we will discuss two other conic sections called *ellipses* and *hyperbolas*.

> Conic Sections

In Section 2-3 we found that the graph of a first-degree equation in two variables,

$$Ax + By = C \tag{1}$$

where A and B are not both 0, is a straight line, and every straight line in a rectangular coordinate system has an equation of this form. What kind of graph will a second-degree equation in two variables,

$$Ax^2 + Bxy + Cy^2 + Dx + Ey + F = 0 \tag{2}$$

where A, B, and C are not all 0, yield for different sets of values of the coefficients? The graphs of equation (2) for various choices of the coefficients are plane curves obtainable by intersecting a cone* with a plane, as shown in Figure 1. These curves are called **conic sections.**

> **Figure 1** Conic sections.

Circle Ellipse Parabola Hyperbola

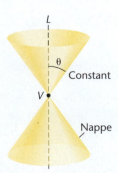

If a plane cuts clear through one nappe, then the intersection curve is called a **circle** if the plane is perpendicular to the axis and an **ellipse** if the plane is not perpendicular to the axis. If a plane cuts only one nappe, but does not cut clear through, then the intersection curve is called

*Starting with a fixed line L and a fixed point V on L, the surface formed by all straight lines through V making a constant angle θ with L is called a **right circular cone.** The fixed line L is called the **axis** of the cone, and V is its **vertex.** The two parts of the cone separated by the vertex are called **nappes.**

a **parabola.** Finally, if a plane cuts through both nappes, but not through the vertex, the resulting intersection curve is called a **hyperbola.** A plane passing through the vertex of the cone produces a **degenerate conic**—a point, a line, or a pair of lines.

Conic sections are very useful and are readily observed in your immediate surroundings: wheels (circle), the path of water from a garden hose (parabola), some serving platters (ellipses), and the shadow on a wall from a light surrounded by a cylindrical or conical lamp shade (hyperbola) are some examples (Fig. 2). We will discuss many applications of conics throughout the remainder of this chapter.

> **Figure 2** Examples of conics.

Wheel (circle)	Water from garden hose (parabola)	Serving platter (ellipse)	Lamp light shadow (hyperbola)
(a)	(b)	(c)	(d)

› Definition of a Parabola

The following definition of a parabola is a **coordinate-free definition.** It does not depend on the coordinates of points in any coordinate system.

> **DEFINITION 1** Parabola

A **parabola** is the set of all points in a plane equidistant from a fixed point F and a fixed line L (not containing F) in the plane. The fixed point F is called the **focus,** and the fixed line L is called the **directrix.** A line through the focus perpendicular to the directrix is called the **axis of symmetry,** and the point on the axis of symmetry halfway between the directrix and focus is called the **vertex.**

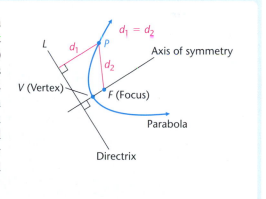

› Drawing a Parabola

Using Definition 1, we can draw a parabola with fairly simple equipment—a straightedge, a right-angle drawing triangle, a piece of string, a thumbtack, and a pencil. Referring to Figure 3 on the next page tape the straightedge along the line AB and place the thumbtack above the line AB. Place one leg of the triangle along the straightedge as indicated, then take a piece of string the same length as the other leg, tie one end to the thumbtack, and fasten the other end with tape at C on the triangle. Now press the string to the edge of the triangle, and keeping the string taut, slide the triangle along the straightedge. Because DE will always equal DF, the resulting curve will be part of a parabola with directrix AB lying along the straightedge and focus F at the thumbtack.

› **Figure 3** Drawing a parabola.

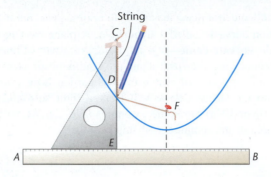

The line through the focus F that is perpendicular to the axis of symmetry of a parabola intersects the parabola in two points G and H. Explain why the distance from G to H is twice the distance from F to the directrix of the parabola.

› Standard Equations and Their Graphs

Using the definition of a parabola and the distance formula

$$d = \sqrt{(x_2 - x_1)^2 + (y_2 - y_1)^2} \tag{3}$$

we can derive simple standard equations for a parabola located in a rectangular coordinate system with its vertex at the origin and its axis of symmetry along a coordinate axis. We start with the axis of symmetry of the parabola along the x axis and the focus at $F = (a, 0)$. We locate the parabola in a coordinate system as in Figure 4 and label key lines and points. This is an important step in finding an equation of a geometric figure in a coordinate system. Note that the parabola opens to the right if $a > 0$ and to the left if $a < 0$. The vertex is at the origin, the directrix is $x = -a$, and the coordinates of M are $(-a, y)$.

› **Figure 4** Parabola with vertex at the origin and axis of symmetry the x axis.

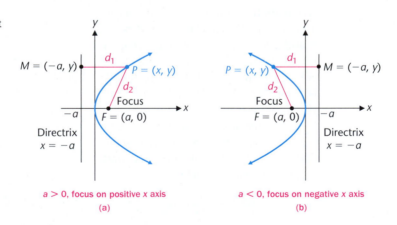

$a > 0$, focus on positive x axis
(a)

$a < 0$, focus on negative x axis
(b)

The point $P = (x, y)$ is a point on the parabola if and only if

$$d_1 = d_2$$
$$d(P, M) = d(P, F) \qquad \text{Use equation (3).}$$
$$\sqrt{(x + a)^2 + (y - y)^2} = \sqrt{(x - a)^2 + (y - 0)^2} \qquad \text{Square both sides.}$$
$$(x + a)^2 = (x - a)^2 + y^2 \qquad \text{Simplify.}$$
$$x^2 + 2ax + a^2 = x^2 - 2ax + a^2 + y^2$$
$$y^2 = 4ax \tag{4}$$

Equation (4) is the standard equation of a parabola with vertex at the origin, axis of symmetry the x axis, and focus at $(a, 0)$.

By a similar derivation (see Problem 57 in Exercises 9-1), the standard equation of a parabola with vertex at the origin, axis of symmetry the y axis, and focus at $(0, a)$ is given by equation (5).

$$x^2 = 4ay \qquad (5)$$

Looking at Figure 5, note that the parabola opens upward if $a > 0$ and downward if $a < 0$.

> **Figure 5** Parabola with vertex at the origin and axis of symmetry the y axis.

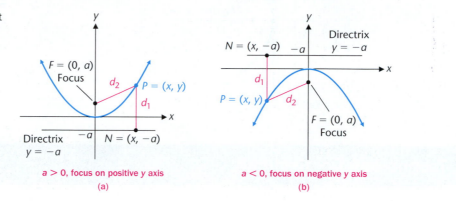

$a > 0$, focus on positive y axis

(a)

$a < 0$, focus on negative y axis

(b)

We summarize these results for easy reference in Theorem 1.

> **THEOREM 1** Standard Equations of a Parabola with Vertex at (0, 0)

1. $y^2 = 4ax$
 Vertex: $(0, 0)$
 Focus: $(a, 0)$
 Directrix: $x = -a$
 Symmetric with respect to the x axis
 Axis of symmetry the x axis

2. $x^2 = 4ay$
 Vertex: $(0, 0)$
 Focus: $(0, a)$
 Directrix: $y = -a$
 Symmetric with respect to the y axis
 Axis of symmetry the y axis

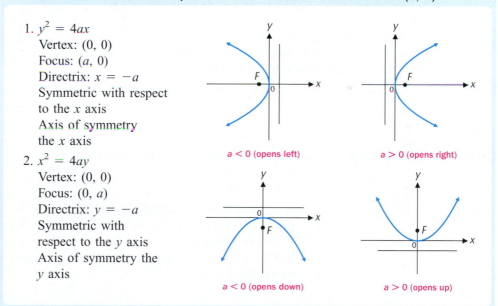

$a < 0$ (opens left)

$a > 0$ (opens right)

$a < 0$ (opens down)

$a > 0$ (opens up)

EXAMPLE **1** **Graphing a Parabola**

Locate the focus and directrix and sketch the graph of $y^2 = 16x$.

SOLUTION The equation $y^2 = 16x$ has the form $y^2 = 4ax$ with $4a = 16$, so $a = 4$. Therefore, the focus is $(4, 0)$ and the directrix is the line $x = -4$. To sketch the graph, we choose some values of x that make the right side of the equation a perfect square and solve for y.

x	0	1	4
y	0	± 4	± 8

Note that x must be greater than or equal to 0 for y to be a real number. Then we plot the resulting points. Because $a > 0$, the parabola opens to the right (Fig. 6).

> Figure 6

Technology Connections

To graph $y^2 = 16x$ on a graphing calculator, we solve the equation for y.

$$y^2 = 16x \qquad \text{Take square roots.}$$
$$y = \pm 4\sqrt{x}$$

This results in two functions, $y = 4\sqrt{x}$ and $y = -4\sqrt{x}$. Entering these functions in a graphing calculator (Fig. 7) and graphing in a standard viewing window produces the graph of the parabola (Fig. 8).

> Figure 7

> Figure 8

MATCHED PROBLEM 1 Graph $y^2 = -8x$, and locate the focus and directrix.

⟫⟫ CAUTION ⟫⟫ A common error in making a quick sketch of $y^2 = 4ax$ or $x^2 = 4ay$ is to sketch the first with the y axis as its axis of symmetry and the second with the x axis as its axis of symmetry. The graph of $y^2 = 4ax$ is symmetric with respect to the x axis, and the graph of $x^2 = 4ay$ is symmetric with respect to the y axis, as a quick symmetry check will reveal.

EXAMPLE **2** **Finding the Equation of a Parabola**

(A) Find the equation of a parabola having the origin as its vertex, the y axis as its axis of symmetry, and $(-10, -5)$ on its graph.

(B) Find the coordinates of its focus and the equation of its directrix.

SOLUTIONS (A) Because the axis of symmetry of the parabola is the y axis, the parabola has an equation of the form $x^2 = 4ay$. Because $(-10, -5)$ is on the graph, we have

$$x^2 = 4ay \qquad \text{Substitute } x = -10 \text{ and } y = -5.$$
$$(-10)^2 = 4a(-5) \qquad \text{Simplify.}$$
$$100 = -20a \qquad \text{Divide both sides by } -20.$$
$$a = -5$$

Therefore the equation of the parabola is

$$x^2 = 4(-5)y$$
$$x^2 = -20y$$

(B) Focus: $F = (0, a) = (0, -5)$
 Directrix: $y = -a$
 $$y = 5$$

MATCHED PROBLEM 2 (A) Find the equation of a parabola having the origin as its vertex, the x axis as its axis of symmetry, and $(4, -8)$ on its graph.

(B) Find the coordinates of its focus and the equation of its directrix.

› Applications

If you are observant, you will find many applications of parabolas in the physical world. Parabolas are key to the design of suspension bridges, arch bridges, microphones, symphony shells, satellite antennas, radio and optical telescopes, radar equipment, solar furnaces, and searchlights.

Figure 9(a) illustrates a parabolic reflector used in all reflecting telescopes—from 3- to 6-inch home types to the 200-inch research instrument on Mount Palomar in California. Parallel light rays from distant celestial bodies are reflected to the focus off a parabolic mirror. If the light source is the sun, then the parallel rays are focused at F and we have a solar furnace. Temperatures of over 6,000°C have been achieved by such furnaces. If we locate a light source at F, then the rays in Figure 9(a) reverse, and we have a spotlight or a searchlight. Automobile headlights can use parabolic reflectors with special lenses over the light to diffuse the rays into useful patterns.

Figure 9(b) shows a suspension bridge, such as the Golden Gate Bridge in San Francisco. The suspension cable is a parabola. It is interesting to note that a free-hanging cable, such as a telephone line, does not form a parabola. It forms another curve called a *catenary*.

Figure 9(c) shows a concrete arch bridge. If all the loads on the arch are to be compression loads (concrete works very well under compression), then using physics and advanced mathematics, it can be shown that the arch must be parabolic.

Parabolic reflector

(a)

Suspension bridge

(b)

Arch bridge

(c)

› **Figure 9** Uses of parabolic forms.

| EXAMPLE | 3 | **Parabolic Reflector** |

A **paraboloid** is formed by revolving a parabola about its axis of symmetry. A spotlight in the form of a paraboloid 5 inches deep has its focus 2 inches from the vertex. Find, to one decimal place, the radius R of the opening of the spotlight.

SOLUTION **Step 1.** Locate a parabolic cross section containing the axis of symmetry in a rectangular coordinate system, and label all known parts and parts to be found. This is a very important step and can be done in infinitely many ways. We can make things simpler for ourselves by locating the vertex at the origin and choosing a coordinate axis as the axis of symmetry. We choose the y axis as the axis of symmetry of the parabola with the parabola opening upward (Fig. 10).

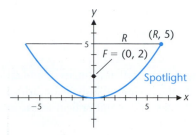

> **Figure 10**

Step 2. Find the equation of the parabola in the figure. Because the parabola has the y axis as its axis of symmetry and the vertex at the origin, the equation is of the form

$$x^2 = 4ay$$

We are given $F = (0, a) = (0, 2)$; so $a = 2$, and the equation of the parabola is

$$x^2 = 8y$$

Step 3. Use the equation found in step 2 to find the radius R of the opening. Because $(R, 5)$ is on the parabola, we have

$$R^2 = 8(5)$$
$$R = \sqrt{40} \approx 6.3 \text{ inches}$$

MATCHED PROBLEM 3 Repeat Example 3 with a paraboloid 12 inches deep and a focus 9 inches from the vertex.

ANSWERS TO MATCHED PROBLEMS

1. Focus: $(-2, 0)$
Directrix: $x = 2$

x	0	-2
y	0	± 4

2. (A) $y^2 = 16x$ (B) Focus: $(4, 0)$; Directrix: $x = -4$
3. $R = 20.8$ inches

9-1 Exercises

1. List the seven different types of conic sections.

2. Explain how each of the seven types of conic sections can be obtained as the intersection of a cone and a plane.

3. What is a degenerate conic?

4. Give a coordinate-free definition of a parabola in your own words.

5. What happens to light rays that are parallel to the axis of a parabolic mirror when they hit the mirror?

6. What happens to light rays that are emitted from the focus of a parabolic mirror when they hit the mirror?

In Problems 7–10, a parabola has its vertex at the origin and the given directrix. Find the coordinates of the focus.

7. $x = 8$ **8.** $x = -5$

9. $y = -10$ **10.** $y = 6$

In Problems 11–14, a parabola has its vertex at the origin and the given focus. Find the equation of the directrix.

11. $(0, -15)$ **12.** $(0, 9)$

13. $(25, 0)$ **14.** $(-21, 0)$

In Problems 15–24, graph each equation, and locate the focus and directrix.

15. $y^2 = 4x$ **16.** $y^2 = 8x$

17. $x^2 = 8y$ **18.** $x^2 = 4y$

19. $y^2 = -12x$ **20.** $y^2 = -4x$

21. $x^2 = -4y$ **22.** $x^2 = -8y$

23. $y^2 = -20x$ **24.** $x^2 = -24y$

In Problems 25–30, find the coordinates to two decimal places of the focus of the parabola.

25. $y^2 = 39x$ **26.** $x^2 = 58y$

27. $x^2 = -105y$ **28.** $y^2 = -93x$

29. $y^2 = -77x$ **30.** $x^2 = -205y$

In Problems 31–38, find the equation of a parabola with vertex at the origin, axis of symmetry the x or y axis, and

31. Directrix $y = -3$ **32.** Directrix $y = 4$

33. Focus $(0, -7)$ **34.** Focus $(0, 5)$

35. Directrix $x = 6$ **36.** Directrix $x = -9$

37. Focus $(2, 0)$ **38.** Focus $(-4, 0)$

In Problems 39–44, find the equation of the parabola having its vertex at the origin, its axis of symmetry as indicated, and passing through the indicated point.

39. y axis; $(4, 2)$ **40.** x axis; $(4, 8)$

41. x axis; $(-3, 6)$ **42.** y axis; $(-5, 10)$

43. y axis; $(-6, -9)$ **44.** x axis; $(-6, -12)$

In Problems 45–48, find the first-quadrant points of intersection for each pair of parabolas to three decimal places.

45. $x^2 = 4y$
$\quad\ y^2 = 4x$ **46.** $y^2 = 3x$
$\qquad\qquad\qquad\quad\ x^2 = 3y$

47. $y^2 = 6x$
$\quad\ x^2 = 5y$ **48.** $x^2 = 7y$
$\qquad\qquad\qquad\quad\ y^2 = 2x$

49. Consider the parabola with equation $x^2 = 4ay$.
 (A) How many lines through $(0, 0)$ intersect the parabola in exactly one point? Find their equations.
 (B) Find the coordinates of all points of intersection of the parabola with the line through $(0, 0)$ having slope $m \neq 0$.

50. Find the coordinates of all points of intersection of the parabola with equation $x^2 = 4ay$ and the parabola with equation $y^2 = 4bx$.

51. The line segment AB through the focus in the figure is called a **focal chord** of the parabola. Find the coordinates of A and B.

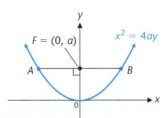

52. The line segment AB through the focus in the figure is called a **focal chord** of the parabola. Find the coordinates of A and B.

In Problems 53–56, use the definition of a parabola and the distance formula to find the equation of a parabola with

53. Directrix $y = -4$ and focus $(2, 2)$

54. Directrix $y = 2$ and focus $(-3, 6)$

55. Directrix $x = 2$ and focus $(6, -4)$

56. Directrix $x = -3$ and focus $(1, 4)$

57. Use the definition of a parabola and the distance formula to derive the equation of a parabola with focus $F = (0, a)$ and directrix $y = -a$ for $a \neq 0$.

58. Let F be a fixed point and let L be a fixed line in the plane that contains F. Describe the set of all points in the plane that are equidistant from F and L.

APPLICATIONS

59. ENGINEERING The parabolic arch in the concrete bridge in the figure must have a clearance of 50 feet above the water and span a distance of 200 feet. Find the equation of the parabola after inserting a coordinate system with the origin at the vertex of the parabola and the vertical y axis (pointing upward) along the axis of symmetry of the parabola.

60. ASTRONOMY The cross section of a parabolic reflector with 6-inch diameter is ground so that its vertex is 0.15 inch below the rim (see the figure).

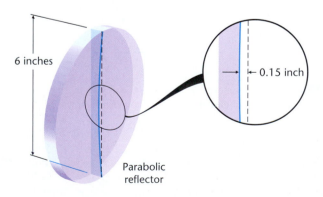

6 inches

← 0.15 inch

Parabolic reflector

(A) Find the equation of the parabola after inserting an xy coordinate system with the vertex at the origin and the y axis (pointing upward) the axis of symmetry of the parabola.
(B) How far is the focus from the vertex?

61. SPACE SCIENCE A designer of a 200-foot-diameter parabolic electromagnetic antenna for tracking space probes wants to place the focus 100 feet above the vertex (see the figure).

200 ft

Focus

100 ft

Radio telescope

(A) Find the equation of the parabola using the axis of symmetry of the parabola as the y axis (up positive) and vertex at the origin.
(B) Determine the depth of the parabolic reflector.

62. SIGNAL LIGHT A signal light on a ship is a spotlight with parallel reflected light rays (see the figure). Suppose the parabolic reflector is 12 inches in diameter and the light source is located at the focus, which is 1.5 inches from the vertex.

Signal light

Focus

(A) Find the equation of the parabola using the axis of symmetry of the parabola as the x axis (right positive) and vertex at the origin.
(B) Determine the depth of the parabolic reflector.

9-2 Ellipse

> Definition of an Ellipse
> Drawing an Ellipse
> Standard Equations of Ellipses and Their Graphs
> Applications

We start our discussion of the ellipse with a coordinate-free definition. Using this definition, we show how an ellipse can be drawn and we derive standard equations for ellipses specially located in a rectangular coordinate system.

> Definition of an Ellipse

The following is a coordinate-free definition of an ellipse:

> **DEFINITION 1** Ellipse

An **ellipse** is the set of all points P in a plane such that the sum of the distances from P to two fixed points in the plane is a constant (the constant is required to be greater than the distance between the two fixed points). Each of the fixed points, F' and F, is called a **focus,** and together they are called **foci.** Referring to the figure, the line segment $V'V$ through the foci is the **major axis.** The perpendicular bisector $B'B$ of the major axis is the **minor axis.** Each end of the major axis, V' and V, is called a **vertex.** The midpoint of the line segment $F'F$ is called the **center** of the ellipse.

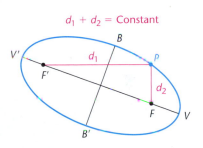

> Drawing an Ellipse

An ellipse is easy to draw. All you need is a piece of string, two thumbtacks, and a pencil or pen (see Figure 1 on the next page.) Place the two thumbtacks in a piece of cardboard. These form the foci of the ellipse. Take a piece of string longer than the distance between the two thumbtacks—this represents the constant in the definition—and tie each end to a thumbtack. Finally, catch the tip of a pencil under the string and move it while keeping the string taut. The resulting figure is by definition an ellipse. Ellipses of different shapes result, depending on the placement of thumbtacks and the length of the string joining them.

Note that $d_1 + d_2$ always adds up to the length of the string, which does not change.

> **Figure 1** Drawing an ellipse.

> Standard Equations of Ellipses and Their Graphs

Using the definition of an ellipse and the distance formula, we can derive standard equations for an ellipse located in a rectangular coordinate system. We start by placing an ellipse in the coordinate system with the foci on the x axis at $F' = (-c, 0)$ and $F = (c, 0)$ with $c > 0$ (Fig. 2). By definition 1 the constant sum $d_1 + d_2$ is required to be greater than $2c$ (the distance between F and F'). Therefore, the ellipse intersects the x axis at points $V' = (-a, 0)$ and $V = (a, 0)$ with $a > c > 0$, and it intersects the y axis at points $B' = (-b, 0)$ and $B = (b, 0)$ with $b > 0$.

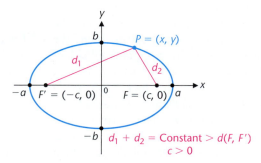

> **Figure 2** Ellipse with foci on x axis.

Study Figure 2: Note first that if $P = (a, 0)$, then $d_1 + d_2 = 2a$. (Why?) Therefore, the constant sum $d_1 + d_2$ is equal to the distance between the vertices. Second, if $P = (0, b)$, then $d_1 = d_2 = a$ and $a^2 = b^2 + c^2$ by the Pythagorean theorem; in particular, $a > b$.

Referring again to Figure 2, the point $P = (x, y)$ is on the ellipse if and only if

$$d_1 + d_2 = 2a$$

Using the distance formula for d_1 and d_2, eliminating radicals, and simplifying (see Problem 49 in Exercises 9-2), we obtain the equation of the ellipse pictured in Figure 2:

$$\frac{x^2}{a^2} + \frac{y^2}{b^2} = 1$$

By similar reasoning (see Problem 50 in Exercises 9-2) we obtain the equation of an ellipse centered at the origin with foci on the y axis. Both cases are summarized in Theorem 1.

> **THEOREM 1** Standard Equations of an Ellipse with Center at (0, 0)

1. $\dfrac{x^2}{a^2} + \dfrac{y^2}{b^2} = 1$ $a > b > 0$

 x intercepts: $\pm a$ (vertices)
 y intercepts: $\pm b$
 Foci: $F' = (-c, 0)$, $F = (c, 0)$

 $$c^2 = a^2 - b^2$$

 Major axis length $= 2a$
 Minor axis length $= 2b$

2. $\dfrac{x^2}{b^2} + \dfrac{y^2}{a^2} = 1$ $a > b > 0$

 x intercepts: $\pm b$
 y intercepts: $\pm a$ (vertices)
 Foci: $F' = (0, -c)$, $F = (0, c)$

 $$c^2 = a^2 - b^2$$

 Major axis length $= 2a$
 Minor axis length $= 2b$

 [*Note:* Both graphs are symmetric with respect to the x axis, y axis, and origin. Also, the major axis is always longer than the minor axis.]

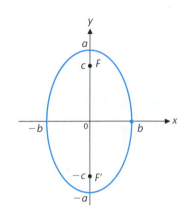

>>> **EXPLORE-DISCUSS 1**

The line through a focus F of an ellipse that is perpendicular to the major axis intersects the ellipse in two points G and H. For each of the two standard equations of an ellipse with center $(0, 0)$, find an expression in terms of a and b for the distance from G to H.

EXAMPLE 1 **Graphing an Ellipse**

Find the coordinates of the foci, find the lengths of the major and minor axes, and graph the following equation:

$$9x^2 + 16y^2 = 144$$

SOLUTION First, write the equation in standard form by dividing both sides by 144 and determine a and b:

$$9x^2 + 16y^2 = 144 \qquad \text{Divide both sides by 144.}$$

$$\dfrac{9x^2}{144} + \dfrac{16y^2}{144} = \dfrac{144}{144} \qquad \text{Simplify.}$$

$$\dfrac{x^2}{16} + \dfrac{y^2}{9} = 1$$

$$a = 4 \qquad \text{and} \qquad b = 3$$

*Throughout the book, dashed boxes—called think boxes—are used to represent steps that may be performed mentally.

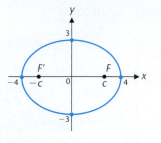

> **Figure 3**

x intercepts: ± 4 Major axis length: $2(4) = 8$

y intercepts: ± 3 Minor axis length: $2(3) = 6$

$$\text{Foci: } c^2 = a^2 - b^2 \quad \text{Substitute } a = 4 \text{ and } b = 3.$$
$$= 16 - 9$$
$$= 7$$
$$c = \sqrt{7} \quad \text{c must be positive.}$$

So the foci are $F' = (-\sqrt{7}, 0)$ and $F = (\sqrt{7}, 0)$.

Plot the foci and intercepts and sketch the ellipse (Fig. 3).

MATCHED PROBLEM 1

Find the coordinates of the foci, find the lengths of the major and minor axes, and graph the following equation:

$$x^2 + 4y^2 = 4$$

EXAMPLE 2 Graphing an Ellipse

Find the coordinates of the foci, find the lengths of the major and minor axes, and graph the following equation:

$$2x^2 + y^2 = 10$$

SOLUTION First, write the equation in standard form by dividing both sides by 10 and determine a and b:

$$2x^2 + y^2 = 10 \quad \text{Divide both sides by 10.}$$

$$\boxed{\frac{2x^2}{10} + \frac{y^2}{10} = \frac{10}{10}} \quad \text{Simplify.}$$

$$\frac{x^2}{5} + \frac{y^2}{10} = 1$$

$$a = \sqrt{10} \quad \text{and} \quad b = \sqrt{5}$$

y intercepts: $\pm\sqrt{10} \approx \pm 3.16$ Major axis length: $2\sqrt{10} \approx 6.32$

x intercepts: $\pm\sqrt{5} \approx \pm 2.24$ Minor axis length: $2\sqrt{5} \approx 4.47$

$$\text{Foci: } c^2 = a^2 - b^2 \quad \text{Substitute } a = \sqrt{10}, \ b = \sqrt{5}.$$
$$= 10 - 5$$
$$= 5$$
$$c = \sqrt{5} \quad \text{c must be positive.}$$

So the foci are $F' = (0, -\sqrt{5})$ and $F = (0, \sqrt{5})$.

Plot the foci and intercepts and sketch the ellipse (Fig. 4).

> **Figure 4**

Technology Connections

To graph the ellipse of Example 2 on a graphing calculator, solve the original equation for y:

$2x^2 + y^2 = 10$ Subtract $2x^2$ from both sides.

$\quad y^2 = 10 - 2x^2$ Take square roots of both sides.

$\quad y = \pm\sqrt{10 - 2x^2}$

This produces two functions, $y_1 = \sqrt{10 - 2x^2}$ and $y_2 = -\sqrt{10 - 2x^2}$, which are graphed in Figure 5. Notice that we used a squared viewing window to avoid distorting the shape of the ellipse. Also note the gaps in the graph

near the x intercepts; they are due to the relatively low resolution of the graphing calculator screen.

> **Figure 5**

MATCHED PROBLEM 2

Find the coordinates of the foci, find the lengths of the major and minor axes, and graph the following equation:

$$3x^2 + y^2 = 18$$

EXAMPLE 3

Finding the Equation of an Ellipse

Find an equation of an ellipse in the form

$$\frac{x^2}{M} + \frac{y^2}{N} = 1 \qquad M, N > 0$$

if the center is at the origin, the major axis is along the y axis, and

(A) Length of major axis = 20 (B) Length of major axis = 10
Length of minor axis = 12 Distance of foci from center = 4

SOLUTIONS (A) Compute x and y intercepts and make a rough sketch of the ellipse, as shown in Figure 6.

$$\frac{x^2}{b^2} + \frac{y^2}{a^2} = 1$$

$$a = \frac{20}{2} = 10 \qquad b = \frac{12}{2} = 6$$

$$\frac{x^2}{36} + \frac{y^2}{100} = 1$$

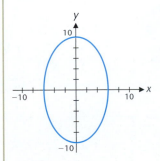

> **Figure 6**

(B) Make a rough sketch of the ellipse, as shown in Figure 7; locate the foci and y intercepts, then determine the x intercepts using the fact that $a^2 = b^2 + c^2$:

$$\frac{x^2}{b^2} + \frac{y^2}{a^2} = 1$$

$$a = \frac{10}{2} = 5 \qquad b^2 = 5^2 - 4^2 = 25 - 16 = 9$$
$$b = 3$$

$$\frac{x^2}{9} + \frac{y^2}{25} = 1$$

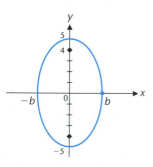

> **Figure 7**

MATCHED PROBLEM 3

Find an equation of an ellipse in the form

$$\frac{x^2}{M} + \frac{y^2}{N} = 1 \qquad M, N > 0$$

if the center is at the origin, the major axis is along the x axis, and

(A) Length of major axis = 50 (B) Length of minor axis = 16
Length of minor axis = 30 Distance of foci from center = 6

› Applications

Ellipses have many applications: orbits of satellites, planets, and comets; shapes of galaxies; gears and cams, some airplane wings, boat keels, and rudders; tabletops; public fountains; and domes in buildings are a few examples (Fig. 8).

Planetary motion
(a)

Elliptical gears
(b)

Elliptical dome
(c)

> **Figure 8** Uses of elliptical forms.

Johannes Kepler (1571–1630), a German astronomer, discovered that planets move in elliptical orbits, with the sun at a focus, and not in circular orbits as had been thought before [Fig. 8(a)]. Figure 8(b) shows a pair of elliptical gears with pivot points at foci. Such gears transfer constant rotational speed to variable rotational speed, and vice versa. Figure 8(c) shows an elliptical dome. An interesting property of such a dome is that a sound or light source at one focus will reflect off the dome and pass through the other focus. One of the chambers in the Capitol Building in Washington, D.C., has such a dome, and is referred to as a whispering room because a whispered sound at one focus can be easily heard at the other focus.

A fairly recent application in medicine is the use of elliptical reflectors and ultrasound to break up kidney stones. A device called a lithotripter is used to generate intense sound waves that break up the stone from outside the body, eliminating the need for surgery. To be certain that the waves do not damage other parts of the body, the reflecting property of the ellipse is used to design and correctly position the lithotripter.

| EXAMPLE | 4 | **Medicinal Lithotripsy** |

A lithotripter is formed by rotating the portion of an ellipse below the minor axis around the major axis (Fig. 9). The lithotripter is 20 centimeters wide and 16 centimeters deep. If the ultrasound source is positioned at one focus of the ellipse and the kidney stone at the other, then all the sound waves will pass through the kidney stone. How far from the kidney stone should the point V on the base of the lithotripter be positioned to focus the sound waves on the kidney stone? Round the answer to one decimal place.

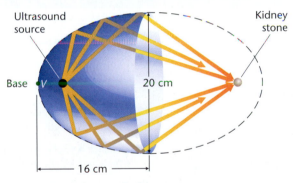

> **Figure 9** Lithotripter.

SOLUTION From Figure 9 we see that $a = 16$ and $b = 10$ for the ellipse used to form the lithotripter. So the distance c from the center to either the kidney stone or the ultrasound source is given by

$$c = \sqrt{a^2 - b^2} = \sqrt{16^2 - 10^2} = \sqrt{156} \approx 12.5$$

and the distance from the base of the lithotripter to the kidney stone is $16 + 12.5 = 28.5$ centimeters.

| MATCHED PROBLEM 4 | Because lithotripsy is an external procedure, the lithotripter described in Example 4 can be used only on stones within 12.5 centimeters of the surface of the body. Suppose a kidney stone is located 14 centimeters from the surface. If the diameter is kept fixed at 20 centimeters, how deep must a lithotripter be to focus on this kidney stone? Round answer to one decimal place.

ANSWERS TO MATCHED PROBLEMS

1.

Foci: $F' = (-\sqrt{3}, 0)$, $F = (\sqrt{3}, 0)$
Major axis length $= 4$
Minor axis length $= 2$

2.

Foci: $F' = (0, -\sqrt{12})$, $F = (0, \sqrt{12})$
Major axis length $= 2\sqrt{18} \approx 8.49$
Minor axis length $= 2\sqrt{6} \approx 4.90$

3. (A) $\dfrac{x^2}{625} + \dfrac{y^2}{225} = 1$ (B) $\dfrac{x^2}{100} + \dfrac{y^2}{64} = 1$ **4.** 17.2 centimeters

9-2 Exercises

1. Give a coordinate-free definition of an ellipse in your own words.

2. Explain how the major axis of an ellipse differs from the minor axis.

3. Given the major axis of an ellipse and the foci, describe a procedure for drawing the ellipse.

4. Is the graph of an ellipse the graph of a function? Explain.

5. Is a circle an ellipse? Explain.

6. Using the definition of an ellipse, explain why the minor axis is shorter than the major axis.

In Problems 7–10, find the distance between the foci of the ellipse.

7. Major axis length $= 10$
Minor axis length $= 8$

8. Major axis length $= 26$
Minor axis length $= 10$

9. Major axis length $= 2$
Minor axis length $= 1$

10. Major axis length $= 4$
Minor axis length $= 3$

In Problems 11–14, find the length of the major axis of the ellipse.

11. Distance between foci $= 14$
Minor axis length $= 48$

12. Distance between foci $= 10$
Minor axis length $= 1$

13. Distance between foci $= 5$
Minor axis length $= 5$

14. Distance between foci $= 3$
Minor axis length $= 3\sqrt{3}$

In Problems 15–20, sketch a graph of each equation, find the coordinates of the foci, and find the lengths of the major and minor axes.

15. $\dfrac{x^2}{25} + \dfrac{y^2}{4} = 1$ **16.** $\dfrac{x^2}{9} + \dfrac{y^2}{4} = 1$ **17.** $\dfrac{x^2}{4} + \dfrac{y^2}{25} = 1$

18. $\dfrac{x^2}{4} + \dfrac{y^2}{9} = 1$ **19.** $x^2 + 9y^2 = 9$ **20.** $4x^2 + y^2 = 4$

In Problems 21–24, match each equation with one of graphs (a)–(d).

21. $9x^2 + 16y^2 = 144$ **22.** $16x^2 + 9y^2 = 144$

jor axis o
foot shor

23. $4x^2 + y^2 = 16$ **24.** $x^2 + 4y^2 = 16$

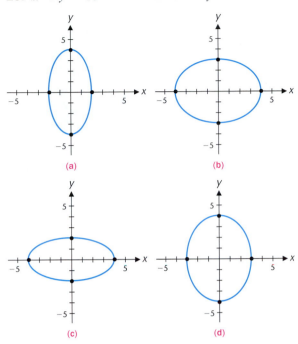

(a) (b)

(c) (d)

In Problems 25–30, sketch a graph of each equation, find the coordinates of the foci, and find the lengths of the major and minor axes.

25. $25x^2 + 9y^2 = 225$ **26.** $16x^2 + 25y^2 = 400$

27. $2x^2 + y^2 = 12$ **28.** $4x^2 + 3y^2 = 24$

29. $4x^2 + 7y^2 = 28$ **30.** $3x^2 + 2y^2 = 24$

In Problems 31–42, find an equation of an ellipse in the form

$$\frac{x^2}{M} + \frac{y^2}{N} = 1 \qquad M, N > 0$$

if the center is at the origin, and

31. The graph is

32. The graph is

33. The graph is

34. The graph is

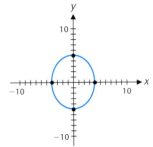

35. Major axis on x axis
Major axis length $= 10$
Minor axis length $= 6$

36. Major axis on x axis
Major axis length $= 14$
Minor axis length $= 10$

37. Major axis on y axis
Major axis length $= 22$
Minor axis length $= 16$

38. Major axis on y axis
Major axis length $= 24$
Minor axis length $= 18$

39. Major axis on x axis
Major axis length $= 16$
Distance of foci from center $= 6$

40. Major axis on y axis
Major axis length $= 24$
Distance of foci from center $= 10$

41. Major axis on y axis
Minor axis length $= 20$
Distance of foci from center $= \sqrt{70}$

42. Major axis on x axis
Minor axis length $= 14$
Distance of foci from center $= \sqrt{200}$

43. Explain why an equation whose graph is an ellipse does not define a function.

44. Consider all ellipses having $(0, \pm 1)$ as the ends of the minor axis. Describe the connection between the elongation of the ellipse and the distance from a focus to the origin.

45. Find an equation of the set of points in a plane, each of whose distance from $(2, 0)$ is one-half its distance from the line $x = 8$. Identify the geometric figure.

46. Find an equation of the set of points in a plane, each of whose distance from $(0, 9)$ is three-fourths its distance from the line $y = 16$. Identify the geometric figure.

47. Let
star
tha
con

48. Let
suc
the
is e

49. Stu
elli
(0,
tion
tha

$\sqrt{(x +}$

$(x +$

$\sqrt{(x -}$

$(x -$

$\left(1 - \dfrac{c^{\cdot}}{a^{\cdot}}\right.$

50. Stu
elli
cep
equ
pag

$\sqrt{x^2 +}$

$x^2 +$

$\sqrt{x^2 +}$

$x^2 +$

$x^2 + \Big(1$

APPL

51. ENG
the figur
a distanc
a coordi
the majc
points tc
feet fron

› Drawing a Hyperbola

Thumbtacks, a straightedge, string, and a pencil are all that are needed to draw a hyperbola (Fig. 1). Place two thumbtacks in a piece of cardboard—these form the foci of the hyperbola. Rest one corner of the straightedge at the focus F' so that it is free to rotate about this point. Cut a piece of string shorter than the length of the straightedge, and fasten one end to the straightedge corner A and the other end to the thumbtack at F. Now push the string with a pencil up against the straightedge at B. Keeping the string taut, rotate the straightedge about F', keeping the corner at F'. The resulting curve will be part of a hyperbola. Other parts of the hyperbola can be drawn by changing the position of the straightedge and string. To see that the resulting curve meets the conditions of the definition, note that the difference of the distances BF' and BF is

$$BF' - BF = BF' + BA - BF - BA$$
$$= AF' - (BF + BA)$$
$$= \left(\begin{array}{c}\text{Straightedge} \\ \text{length}\end{array}\right) - \left(\begin{array}{c}\text{String} \\ \text{length}\end{array}\right)$$
$$= \text{Constant}$$

› **Figure 1** Drawing a hyperbola.

› Standard Equations of Hyperbolas and Their Graphs

Using the definition of a hyperbola and the distance formula, we can derive standard equations for a hyperbola located in a rectangular coordinate system. We start by placing a hyperbola in the coordinate system with the foci on the x axis at $F' = (-c, 0)$ and $F = (c, 0)$ with $c > 0$ (Fig. 2). By definition 1, the constant difference $|d_1 - d_2|$ is

› **Figure 2** Hyperbola with foci on the x axis.

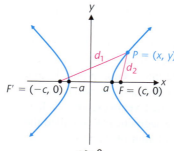

$c > 0$
$|d_1 - d_2| = \text{Positive constant} < d(F, F')$

required to be less than $2c$ (the distance between F and F'). Therefore, the hyperbola intersects the x axis at points $V' = (-a, 0)$ and $V = (a, 0)$ with $c > a > 0$. The hyperbola does not intersect the y axis, because the constant difference $|d_1 - d_2|$ is required to be positive by definition 1.

Study Figure 2: Note that if $P = (a, 0)$, then $|d_1 - d_2| = 2a$. (Why?) Therefore, the constant $|d_1 - d_2|$ is equal to the distance between the vertices.

It is convenient to let $b = \sqrt{c^2 - a^2}$, so that $c^2 = a^2 + b^2$. (Unlike the situation for ellipses, b may be greater than or equal to a.)

Referring again to Figure 2, the point $P = (x, y)$ is on the hyperbola if and only if

$$|d_1 - d_2| = 2a$$

Using the distance formula for d_1 and d_2, eliminating radicals, and simplifying (see Problem 57 in Exercises 9-3), we obtain the equation of the hyperbola pictured in Figure 2:

$$\frac{x^2}{a^2} - \frac{y^2}{b^2} = 1$$

Although the hyperbola does not intersect the y axis, the points $(0, b)$ and $(0, -b)$ are significant; the line segment joining them is called the **conjugate axis** of the hyperbola. Note that the conjugate axis is perpendicular to the transverse axis, that is, the line segment joining the vertices $(a, 0)$ and $(-a, 0)$. The rectangle with corners (a, b), $(a, -b)$, $(-a, -b)$, and $(-a, b)$ is called the **asymptote rectangle** because its extended diagonals are asymptotes for the hyperbola (Fig. 3). In other words, the hyperbola approaches the lines $y = \pm\frac{b}{a}x$ as $|x|$ becomes larger (see Problems 53 and 54 in Exercises 9-3). As a result, it is helpful to include the asymptote rectangle and its extended diagonals when sketching the graph of a hyperbola.

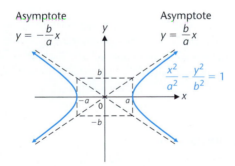

> **Figure 3** Asymptotes.

Note that the four corners of the asymptote rectangle (Fig. 3) are equidistant from the origin, at distance $\sqrt{a^2 + b^2} = c$. Therefore,

A circle, with center at the origin, that passes through all four corners of the asymptote rectangle of a hyperbola also passes through its foci.

By similar reasoning (see Problem 58 in Exercises 9-3) we obtain the equation of a hyperbola centered at the origin with foci on the y axis. Both cases are summarized in Theorem 1.

> **THEOREM 1** Standard Equations of a Hyperbola with Center at (0, 0)

1. $\dfrac{x^2}{a^2} - \dfrac{y^2}{b^2} = 1$

 x intercepts: $\pm a$ (vertices)
 y intercepts: none
 Foci: $F' = (-c, 0)$, $F = (c, 0)$

 $$c^2 = a^2 + b^2$$

 Transverse axis length $= 2a$
 Conjugate axis length $= 2b$

 Asymptotes: $y = \pm\dfrac{b}{a}x$

2. $\dfrac{y^2}{a^2} - \dfrac{x^2}{b^2} = 1$

 x intercepts: none
 y intercepts: $\pm a$ (vertices)
 Foci: $F' = (0, -c)$, $F = (0, c)$

 $$c^2 = a^2 + b^2$$

 Transverse axis length $= 2a$
 Conjugate axis length $= 2b$

 Asymptotes: $y = \pm\dfrac{a}{b}x$

[*Note:* Both graphs are symmetric with respect to the x axis, y axis, and origin.]

>>> **EXPLORE-DISCUSS 1**

The line through a focus F of a hyperbola that is perpendicular to the transverse axis intersects the hyperbola in two points G and H. For each of the two standard equations of a hyperbola with center $(0, 0)$, find an expression in terms of a and b for the distance from G to H.

EXAMPLE 1 **Graphing Hyperbolas**

Find the coordinates of the foci, find the lengths of the transverse and conjugate axes, find the equations of the asymptotes, and graph the following equation:

$$9x^2 - 16y^2 = 144$$

SOLUTION First, write the equation in standard form by dividing both sides by 144 and determine a and b:

$$9x^2 - 16y^2 = 144 \qquad \text{Divide both sides by 144.}$$

$$\frac{9x^2}{144} - \frac{16y^2}{144} = \frac{144}{144} \qquad \text{Simplify.}$$

$$\frac{x^2}{16} - \frac{y^2}{9} = 1$$

$$a = 4 \qquad \text{and} \qquad b = 3$$

x intercepts: ± 4 Transverse axis length $= 2(4) = 8$

y intercepts: none Conjugate axis length $= 2(3) = 6$

$$\text{Foci: } c^2 = a^2 + b^2 \quad \textcolor{magenta}{\text{Substitute } a = 4 \text{ and } b = 3.}$$
$$= 16 + 9$$
$$= 25$$
$$c = 5$$

So the foci are $F' = (-5, 0)$ and $F = (5, 0)$.

Plot the foci and x intercepts, sketch the asymptote rectangle and the asymptotes, then sketch the hyperbola (Fig. 4). The equations of the asymptotes are $y = \pm\frac{3}{4}x$ (note that the diagonals of the asymptote rectangle have slope $\pm\frac{3}{4}$).

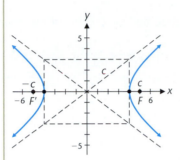

> **Figure 4**

MATCHED PROBLEM 1

Find the coordinates of the foci, find the lengths of the transverse and conjugate axes, and graph the following equation:

$$16x^2 - 25y^2 = 400$$

EXAMPLE 2 **Graphing Hyperbolas**

Find the coordinates of the foci, find the lengths of the transverse and conjugate axes, find the equations of the asymptotes, and graph the following equation:

$$16y^2 - 9x^2 = 144$$

SOLUTION Write the equation in standard form:

$$16y^2 - 9x^2 = 144 \quad \textcolor{magenta}{\text{Divide both sides by 144.}}$$
$$\frac{y^2}{9} - \frac{x^2}{16} = 1$$
$$a = 3 \quad \text{and} \quad b = 4$$

y intercepts: ± 3 Transverse axis length $= 2(3) = 6$

x intercepts: none Conjugate axis length $= 2(4) = 8$

$$\text{Foci: } c^2 = a^2 + b^2 \quad \textcolor{magenta}{\text{Substitute } a = 3 \text{ and } b = 4.}$$
$$= 9 + 16$$
$$= 25$$
$$c = 5$$

So the foci are $F' = (0, -5)$ and $F = (0, 5)$.

Plot the foci and y intercepts, sketch the asymptote rectangle and the asymptotes, then sketch the hyperbola (Fig. 5). The equations of the asymptotes are $y = \pm\frac{3}{4}x$ (note that the diagonals of the asymptote rectangle have slope $\pm\frac{3}{4}$).

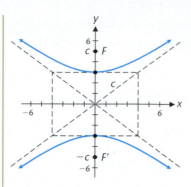

› **Figure 5**

MATCHED PROBLEM 2 Find the coordinates of the foci, find the lengths of the transverse and conjugate axes, and graph the following equation:

$$25y^2 - 16x^2 = 400$$

Two hyperbolas of the form

$$\frac{x^2}{M} - \frac{y^2}{N} = 1 \quad \text{and} \quad \frac{y^2}{N} - \frac{x^2}{M} = 1 \quad M, N > 0$$

are called **conjugate hyperbolas.** In Examples 1 and 2 and in Matched Problems 1 and 2, the hyperbolas are conjugate hyperbolas—they share the same asymptotes.

››› **CAUTION** ››› When making a quick sketch of a hyperbola, it is a common error to have the hyperbola opening up and down when it should open left and right, or vice versa. The mistake can be avoided if you first locate the intercepts accurately.

EXAMPLE 3 Graphing Hyperbolas

Find the coordinates of the foci, find the lengths of the transverse and conjugate axes, and graph the following equation:

$$2x^2 - y^2 = 10$$

SOLUTION

$$2x^2 - y^2 = 10 \quad \text{Divide both sides by 10.}$$

$$\frac{x^2}{5} - \frac{y^2}{10} = 1$$

$$a = \sqrt{5} \quad \text{and} \quad b = \sqrt{10}$$

x intercepts: $\pm\sqrt{5}$ Transverse axis length $= 2\sqrt{5} \approx 4.47$

y intercepts: none Conjugate axis length $= 2\sqrt{10} \approx 6.32$

Foci: $c^2 = a^2 + b^2$ Substitute $a = \sqrt{5}$ and $b = \sqrt{10}$.

$$= 5 + 10$$

$$= 15$$

$$c = \sqrt{15}$$

So the foci are $F' = (-\sqrt{15}, 0)$ and $F = (\sqrt{15}, 0)$.

Plot the foci and x intercepts, sketch the asymptote rectangle and the asymptotes, then sketch the hyperbola (Fig. 6).

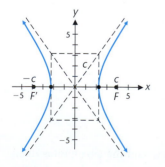

› **Figure 6**

MATCHED PROBLEM 3

Find the coordinates of the foci, find the lengths of the transverse and conjugate axes, and graph the following equation:

$$y^2 - 3x^2 = 12$$

EXAMPLE 4 Finding the Equation of a Hyperbola

Find an equation of a hyperbola in the form

$$\frac{y^2}{M} - \frac{x^2}{N} = 1 \qquad M, N > 0$$

if the center is at the origin, and:

(A) Length of transverse axis is 12 (B) Length of transverse axis is 6
 Length of conjugate axis is 20 Distance of foci from center is 5

SOLUTIONS (A) Start with

$$\frac{y^2}{a^2} - \frac{x^2}{b^2} = 1$$

and find a and b:

$$a = \frac{12}{2} = 6 \qquad \text{and} \qquad b = \frac{20}{2} = 10$$

So the equation is

$$\frac{y^2}{36} - \frac{x^2}{100} = 1$$

(B) Start with

$$\frac{y^2}{a^2} - \frac{x^2}{b^2} = 1$$

and find a and b:

$$a = \frac{6}{2} = 3$$

To find b, sketch the asymptote rectangle (Fig. 7), label known parts, and use the Pythagorean theorem:

$$b^2 = 5^2 - 3^2$$
$$= 16$$
$$b = 4$$

So the equation is

$$\frac{y^2}{9} - \frac{x^2}{16} = 1$$

> **Figure 7** Asymptote rectangle.

MATCHED PROBLEM 4

Find an equation of a hyperbola in the form

$$\frac{x^2}{M} - \frac{y^2}{N} = 1 \qquad M, N > 0$$

if the center is at the origin, and:

(A) Length of transverse axis is 50
 Length of conjugate axis is 30

(B) Length of conjugate axis is 12
 Distance of foci from center is 9

⟩⟩⟩ EXPLORE-DISCUSS 2

(A) Does the line with equation $y = x$ intersect the hyperbola with equation $x^2 - (y^2/4) = 1$? If so, find the coordinates of all intersection points.

(B) Does the line with equation $y = 3x$ intersect the hyperbola with equation $x^2 - (y^2/4) = 1$? If so, find the coordinates of all intersection points.

(C) For which values of m does the line with equation $y = mx$ intersect the hyperbola $\dfrac{x^2}{a^2} - \dfrac{y^2}{b^2} = 1$? Find the coordinates of all intersection points.

⟩ Applications

You may not be aware of the many important uses of hyperbolic forms. They are encountered in the study of comets; the loran system of navigation for pleasure boats, ships, and aircraft; sundials; capillary action; nuclear reactor cooling towers; optical and radio telescopes; and contemporary architectural structures. The TWA building at Kennedy Airport is a *hyperbolic paraboloid,* and the St. Louis Science Center Planetarium is a *hyperboloid.* With such structures, thin concrete shells can span large spaces [Fig. 8(a)]. Some comets from outer space occasionally enter the sun's gravitational field, follow a hyperbolic path around the sun (with the sun as a focus), and then leave, never to be seen again [Fig. 8(b)]. Example 5 illustrates the use of hyperbolas in navigation.

⟩ **Figure 8** Uses of hyperbolic forms.

St. Louis Planetarium

(a)

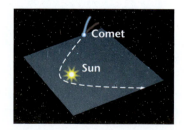

Comet around sun

(b)

EXAMPLE 5

Navigation

A ship is traveling on a course parallel to and 60 miles from a straight shoreline. Two transmitting stations, S_1 and S_2, are located 200 miles apart on the shoreline (Fig. 9). By timing radio signals from the stations, the ship's navigator determines that the ship is between the two stations and 50 miles closer to S_2 than to S_1. Find the distance from the ship to each station. Round answers to one decimal place.

> **Figure 9** $d_1 - d_2 = 50$.

SOLUTION If d_1 and d_2 are the distances from the ship to S_1 and S_2, respectively, then $d_1 - d_2 = 50$ and the ship must be on the hyperbola with foci at S_1 and S_2 and fixed difference 50, as illustrated in Figure 10. In the derivation of the equation of a hyperbola, we represented the fixed difference as $2a$. So for the hyperbola in Figure 10 we have

$$c = 100$$
$$a = \tfrac{1}{2}(50) = 25$$
$$b = \sqrt{100^2 - 25^2} = \sqrt{9{,}375}$$

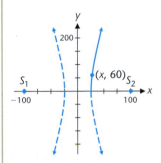

> **Figure 10**

The equation for this hyperbola is

$$\frac{x^2}{625} - \frac{y^2}{9{,}375} = 1$$

Substitute $y = 60$ and solve for x (see Fig. 10):

$$\frac{x^2}{625} - \frac{60^2}{9{,}375} = 1 \qquad \text{Add } \tfrac{60^2}{9{,}375} \text{ to both sides.}$$

$$\frac{x^2}{625} = \frac{3{,}600}{9{,}375} + 1 \qquad \text{Multiply both sides by 625.}$$

$$x^2 = 625\,\frac{3{,}600 + 9{,}375}{9{,}375} \qquad \text{Simplify.}$$

$$= 865$$

So $x = \sqrt{865} \approx 29.41$ (The negative square root is discarded, because the ship is closer to S_2 than to S_1.)

Distance from ship to S_1 Distance from ship to S_2

$$d_1 = \sqrt{(29.41 + 100)^2 + 60^2} \qquad\qquad d_2 = \sqrt{(29.41 - 100)^2 + 60^2}$$
$$= \sqrt{20{,}346.9841} \qquad\qquad\qquad\qquad = \sqrt{8{,}582.9841}$$
$$\approx 142.6 \text{ miles} \qquad\qquad\qquad\qquad\quad \approx 92.6 \text{ miles}$$

Notice that the difference between these two distances is 50, as it should be.

MATCHED PROBLEM 5 Repeat Example 5 if the ship is 80 miles closer to S_2 than to S_1.

Example 5 illustrates a simplified form of the loran (LOng RAnge Navigation) system. In practice, three transmitting stations are used to send out signals simultaneously (Fig. 11), instead of the two used in Example 5. A computer onboard a ship will record these signals and use them to determine the differences of the distances that the ship is to S_1 and S_2, and to S_2 and S_3. Plotting all points so that these distances remain constant produces two branches, p_1 and p_2, of a hyperbola with foci S_1 and S_2, and two branches, q_1 and q_2, of a hyperbola with foci S_2 and S_3. It is easy to tell which branches the ship is on by comparing the signals from each station. The intersection of a branch of each hyperbola locates the ship and the computer expresses this in terms of longitude and latitude.

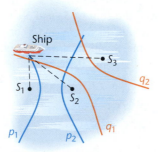

❯ **Figure 11** Loran navigation.

ANSWERS TO MATCHED PROBLEMS

1.

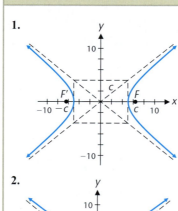

$\dfrac{x^2}{25} - \dfrac{y^2}{16} = 1$

Foci: $F' = (-\sqrt{41}, 0)$, $F = (\sqrt{41}, 0)$

Transverse axis length = 10

Conjugate axis length = 8

2.

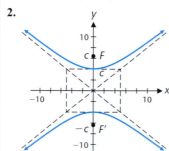

$\dfrac{y^2}{16} - \dfrac{x^2}{25} = 1$

Foci: $F' = (0, -\sqrt{41})$, $F = (0, \sqrt{41})$

Transverse axis length = 8

Conjugate axis length = 10

3.

$\dfrac{y^2}{12} - \dfrac{x^2}{4} = 1$

Foci: $F' = (0, -4)$, $F = (0, 4)$

Transverse axis length = $2\sqrt{12} \approx 6.93$

Conjugate axis length = 4

4. (A) $\dfrac{x^2}{625} - \dfrac{y^2}{225} = 1$ **(B)** $\dfrac{x^2}{45} - \dfrac{y^2}{36} = 1$ **5.** $d_1 = 159.5$ miles, $d_2 = 79.5$ miles

9-3 Exercises

1. Give a coordinate-free definition of a hyperbola in your own words.

2. Explain how the transverse axis of a hyperbola differs from the conjugate axis.

3. Given the transverse axis and foci of a hyperbola, describe a procedure for drawing the hyperbola.

4. Is the graph of a hyperbola the graph of a function? Explain.

5. Is the conjugate axis of a hyperbola always shorter then the transverse axis? Explain.

6. Explain what an asymptote rectangle is, and how it is related to the graph of a hyperbola.

In Problems 7–10, find the distance between the foci of the hyperbola.

7. Transverse axis length = 24
Conjugate axis length = 18

8. Transverse axis length = 25
Conjugate axis length = 60

9. Transverse axis length = 1
Conjugate axis length = 3

10. Transverse axis length = 7
Conjugate axis length = 1

In Problems 11–14, match each equation with one of graphs (a)–(d).

11. $x^2 - y^2 = 1$

12. $y^2 - x^2 = 1$

13. $y^2 - x^2 = 4$

14. $x^2 - y^2 = 4$

(a)

(c)

(b)

(d)

Sketch a graph of each equation in Problems 15–26, find the coordinates of the foci, and find the lengths of the transverse and conjugate axes.

15. $\dfrac{x^2}{9} - \dfrac{y^2}{4} = 1$

16. $\dfrac{x^2}{9} - \dfrac{y^2}{25} = 1$

17. $\dfrac{y^2}{4} - \dfrac{x^2}{9} = 1$

18. $\dfrac{y^2}{25} - \dfrac{x^2}{9} = 1$

19. $4x^2 - y^2 = 16$

20. $x^2 - 9y^2 = 9$

21. $9y^2 - 16x^2 = 144$

22. $4y^2 - 25x^2 = 100$

23. $3x^2 - 2y^2 = 12$

24. $3x^2 - 4y^2 = 24$

25. $7y^2 - 4x^2 = 28$

26. $3y^2 - 2x^2 = 24$

In Problems 27–38, find an equation of a hyperbola in the form

$$\frac{x^2}{M} - \frac{y^2}{N} = 1 \quad \text{or} \quad \frac{y^2}{N} - \frac{x^2}{M} = 1 \quad M, N > 0$$

if the center is at the origin, and:

27. The graph is

28. The graph is

29. The graph is

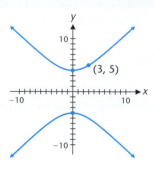

(3, 5)

30. The graph is

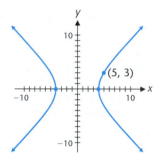

(5, 3)

31. Transverse axis on x axis
Transverse axis length $= 14$
Conjugate axis length $= 10$

32. Transverse axis on x axis
Transverse axis length $= 8$
Conjugate axis length $= 6$

33. Transverse axis on y axis
Transverse axis length $= 24$
Conjugate axis length $= 18$

34. Transverse axis on y axis
Transverse axis length $= 16$
Conjugate axis length $= 22$

35. Transverse axis on x axis
Transverse axis length $= 18$
Distance of foci from center $= 11$

36. Transverse axis on x axis
Transverse axis length $= 16$
Distance of foci from center $= 10$

37. Conjugate axis on x axis
Conjugate axis length $= 14$
Distance of foci from center $= \sqrt{200}$

38. Conjugate axis on x axis
Conjugate axis length $= 10$
Distance of foci from center $= \sqrt{70}$

In Problems 39–46, find the equations of the asymptotes of each hyperbola.

39. $\dfrac{x^2}{25} - \dfrac{y^2}{4} = 1$ **40.** $\dfrac{x^2}{16} - \dfrac{y^2}{36} = 1$

41. $\dfrac{y^2}{4} - \dfrac{x^2}{16} = 1$ **42.** $\dfrac{y^2}{9} - \dfrac{x^2}{25} = 1$

43. $9x^2 - y^2 = 9$ **44.** $x^2 - 4y^2 = 4$

45. $2y^2 - 3x^2 = 1$

46. $5y^2 - 6x^2 = 1$

47. (A) How many hyperbolas have center at $(0, 0)$ and a focus at $(1, 0)$? Find their equations.
(B) How many ellipses have center at $(0, 0)$ and a focus at $(1, 0)$? Find their equations.
(C) How many parabolas have center at $(0, 0)$ and focus at $(1, 0)$? Find their equations.

48. How many hyperbolas have the lines $y = \pm 2x$ as asymptotes? Find their equations.

49. Find all intersection points of the graph of the hyperbola $x^2 - y^2 = 1$ with the graph of each of the following lines:
(A) $y = 0.5x$
(B) $y = 2x$

For what values of m will the graph of the hyperbola and the graph of the line $y = mx$ intersect? Find the coordinates of these intersection points.

50. Find all intersection points of the graph of the hyperbola $y^2 - x^2 = 1$ with the graph of each of the following lines:
(A) $y = 0.5x$
(B) $y = 2x$

For what values of m will the graph of the hyperbola and the graph of the line $y = mx$ intersect? Find the coordinates of these intersection points.

51. Find all intersection points of the graph of the hyperbola $y^2 - 4x^2 = 1$ with the graph of each of the following lines:
(A) $y = x$
(B) $y = 3x$

For what values of m will the graph of the hyperbola and the graph of the line $y = mx$ intersect? Find the coordinates of these intersection points.

52. Find all intersection points of the graph of the hyperbola $4x^2 - y^2 = 1$ with the graph of each of the following lines:
(A) $y = x$
(B) $y = 3x$

For what values of m will the graph of the hyperbola and the graph of the line $y = mx$ intersect? Find the coordinates of these intersection points.

53. Consider the hyperbola with equation

$$\frac{x^2}{a^2} - \frac{y^2}{b^2} = 1$$

(A) Show that $y = \pm \frac{b}{a}x\sqrt{1 - \frac{a^2}{x^2}}$.

(B) Explain why the hyperbola approaches the lines $y = \pm \frac{b}{a}x$ as $|x|$ becomes larger.
(C) Does the hyperbola approach its asymptotes from above or below? Explain.

54. Consider the hyperbola with equation

$$\frac{y^2}{a^2} - \frac{x^2}{b^2} = 1$$

(A) Show that $y = \pm\frac{a}{b}x\sqrt{1 + \frac{b^2}{x^2}}$.

(B) Explain why the hyperbola approaches the lines $y = \pm\frac{a}{b}x$ as $|x|$ becomes larger.

(C) Does the hyperbola approach its asymptotes from above or below? Explain.

55. Let F and F' be two points in the plane and let c be a constant such that $c > d(F, F')$. Describe the set of all points P in the plane such that the absolute value of the difference of the distances from P to F and F' is equal to the constant c.

56. Let F and F' be two points in the plane and let c denote the constant $d(F, F')$. Describe the set of all points P in the plane such that the absolute value of the difference of the distances from P to F and F' is equal to the constant c.

57. Study the following derivation of the standard equation of a hyperbola with foci $(\pm c, 0)$, x intercepts $(\pm a, 0)$, and endpoints of the conjugate axis $(0, \pm b)$. Explain why each equation follows from the equation that precedes it. [*Hint:* Recall that $c^2 = a^2 + b^2$.]

$$|d_1 - d_2| = 2a$$

$$\sqrt{(x + c)^2 + y^2} = \pm 2a + \sqrt{(x - c)^2 + y^2}$$

$$(x + c)^2 + y^2 = 4a^2 \pm 4a\sqrt{(x - c)^2 + y^2} + (x - c)^2 + y^2$$

$$\pm\sqrt{(x - c)^2 + y^2} = a - \frac{cx}{a}$$

$$(x - c)^2 + y^2 = a^2 - 2cx + \frac{c^2x^2}{a^2}$$

$$\left(1 - \frac{c^2}{a^2}\right)x^2 + y^2 = a^2 - c^2$$

$$\frac{x^2}{a^2} - \frac{y^2}{b^2} = 1$$

58. Study the following derivation of the standard equation of a hyperbola with foci $(0, \pm c)$, y intercepts $(0, \pm a)$, and endpoints of the conjugate axis $(\pm b, 0)$. Explain why each equation follows from the equation that precedes it. [*Hint:* Recall that $c^2 = a^2 + b^2$.]

$$|d_1 - d_2| = 2a$$

$$\sqrt{x^2 + (y + c)^2} = \pm 2a + \sqrt{x^2 + (y - c)^2}$$

$$x^2 + (y + c)^2 = 4a^2 \pm 4a\sqrt{x^2 + (y - c)^2} + x^2 + (y - c)^2$$

$$\pm\sqrt{x^2 + (y - c)^2} = a - \frac{cy}{a}$$

$$x^2 + (y - c)^2 = a^2 - 2cy + \frac{c^2y^2}{a^2}$$

$$x^2 + \left(1 - \frac{c^2}{a^2}\right)y^2 = a^2 - c^2$$

$$\frac{y^2}{a^2} - \frac{x^2}{b^2} = 1$$

ECCENTRICITY *Problems 59 and 60 (and Problems 45 and 46 in Exercises 9-2) are related to a property of conics called **eccentricity**, which is denoted by a positive real number E. Parabolas, ellipses, and hyperbolas all can be defined in terms of E, a fixed point called a focus, and a fixed line not containing the focus called a directrix as follows: The set of points in a plane each of whose distance from a fixed point is E times its distance from a fixed line is an ellipse if $0 < E < 1$, a parabola if $E = 1$, and a hyperbola if $E > 1$.*

59. Find an equation of the set of points in a plane each of whose distance from $(3, 0)$ is three-halves its distance from the line $x = \frac{4}{3}$. Identify the geometric figure.

60. Find an equation of the set of points in a plane each of whose distance from $(0, 4)$ is four-thirds its distance from the line $y = \frac{9}{4}$. Identify the geometric figure.

APPLICATIONS

61. ARCHITECTURE An architect is interested in designing a thin-shelled dome in the shape of a hyperbolic paraboloid, as shown in Figure (a). Find the equation of the hyperbola located in a coordinate system [Fig. (b)] satisfying the indicated conditions. How far is the hyperbola above the vertex 6 feet to the right of the vertex? Compute the answer to two decimal places.

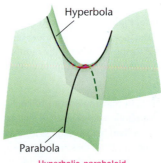

Hyperbola

Parabola

Hyperbolic paraboloid
(a)

(8, 12)

Hyperbola part of dome
(b)

62. NUCLEAR POWER A nuclear reactor cooling tower is a **hyperboloid**, that is, a hyperbola rotated around its conjugate axis, as shown in Figure (a) on page 604. The equation of the hyperbola in Figure (b) used to generate the hyperboloid is

$$\frac{x^2}{100^2} - \frac{y^2}{150^2} = 1$$

Nuclear reactor cooling tower
(a)

Hyperbola part of dome
(b)

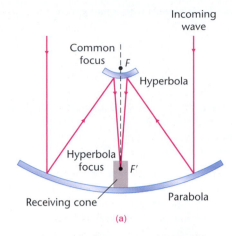

Incoming wave

Common focus F

Hyperbola

Hyperbola focus F'

Receiving cone

Parabola

(a)

Radio telescope

(b)

If the tower is 500 feet tall, the top is 150 feet above the center of the hyperbola, and the base is 350 feet below the center, what is the radius of the top and the base? What is the radius of the smallest circular cross section in the tower? Compute answers to three significant digits.

63. SPACE SCIENCE In tracking space probes to the outer planets, NASA uses large parabolic reflectors with diameters equal to two-thirds the length of a football field. Needless to say, many design problems are created by the weight of these reflectors. One weight problem is solved by using a hyperbolic reflector sharing the parabola's focus to reflect the incoming electromagnetic waves to the other focus of the hyperbola where receiving equipment is installed (see the figure).

For the receiving antenna shown in the figure, the common focus F is located 120 feet above the vertex of the parabola, and focus F' (for the hyperbola) is 20 feet above the vertex. The vertex of the reflecting hyperbola is 110 feet above the vertex for the parabola. Introduce a coordinate system by using the axis of the parabola as the y axis (up positive), and let the x axis pass through the center of the hyperbola (right positive). What is the equation of the reflecting hyperbola? Write y in terms of x.

9-4 Translation and Rotation of Axes

> **Translation of Axes**
> **Translation Used in Graphing**
> **Rotation of Axes**
> **Rotation Used in Graphing**
> **Identifying Conics**

In Sections 9-1, 9-2, and 9-3 we found standard equations for parabolas, ellipses, and hyperbolas with axes on the coordinate axes and centered relative to the origin. Each of those standard equations was a special case of the equation

$$Ax^2 + Bxy + Cy^2 + Dx + Ey + F = 0 \qquad (1)$$

for appropriate constants A, B, C, D, E, and F. In this section, we show that *every* equation of the form (1) has a graph that is either a conic, a degenerate conic (that is, a point, a line, or a pair of lines), or the empty set. The difficulty is that a conic with an equation of form (1) might not be centered at the origin, and might have axes that are skewed with respect to the coordinate axes. To overcome the difficulty we use two basic mathematical tools: *translation of axes* and *rotation of axes.* With these tools we will be able to choose a new coordinate system (that depends on the constants A, B, C, D, E, and F) in which the equation has an especially transparent and useful form.

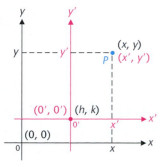

> **Figure 1** Translation of coordinates.

› Translation of Axes

If you move a sheet of paper on a desk top, without rotating the paper and without flipping it over, you *translate* the paper to its new position. Similarly, a **translation of coordinate axes** occurs when the new coordinate axes have the same direction as and are parallel to the original coordinate axes. To see how coordinates in the original system are changed when moving to the translated system, and vice versa, refer to Figure 1.

A point P in the plane has two sets of coordinates: (x, y) in the original system and (x', y') in the translated system. If the coordinates of the origin of the translated system are (h, k) relative to the original system, then the old and new coordinates are related as given in Theorem 1.`

> **THEOREM 1** Translation Formulas

1. $x = x' + h$ $\qquad$ 2. $x' = x - h$
 $y = y' + k$ $\qquad\qquad$ $y' = y - k$

It can be shown that these formulas hold for (h, k) located anywhere in the original coordinate system.

EXAMPLE $\quad$ 1 $\quad$ | $\quad$ **Equation of a Curve in a Translated System**

A curve has the equation

$$(x - 4)^2 + (y + 1)^2 = 36$$

If the origin is translated to $(4, -1)$, find the equation of the curve in the translated system and identify the curve.

SOLUTION $\quad$ Because $(h, k) = (4, -1)$, use translation formulas

$$x' = x - h = x - 4$$
$$y' = y - k = y + 1$$

to obtain, after substitution,

$$x'^2 + y'^2 = 36$$

This is the equation of a circle of radius 6 with center at the new origin. The coordinates of the new origin in the original coordinate system are $(4, -1)$ (see Figure 2 on the next page.) Note that this result agrees with our general treatment of the circle in Section 2-2.

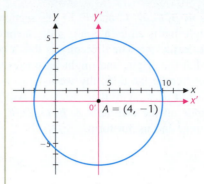

> **Figure 2** $(x - 4)^2 + (y + 1)^2 = 36$.

MATCHED PROBLEM 1 A curve has the equation $(y + 2)^2 = 8(x - 3)$. If the origin is translated to $(3, -2)$, find an equation of the curve in the translated system and identify the curve.

Suppose the coordinate axes in the xy system have been translated to (h, k), as in Figure 1 on page 605. Then, as illustrated by Example 1, the circle $x'^2 + y'^2 = r^2$ has the equation $(x - h)^2 + (y - k)^2 = r^2$ in the original xy system. In a similar manner we use the standard equations for the parabola, ellipse, and hyperbola centered at the origin to obtain more general standard equations for conics centered at the point (h, k) (see Table 1). Note that when $h = 0$ and $k = 0$ the standard equations of Table 1 are exactly the standard equations obtained in Sections 9-1, 9-2, and 9-3.

Table 1 Standard Equations for Conics

<p align="center">Parabolas</p>

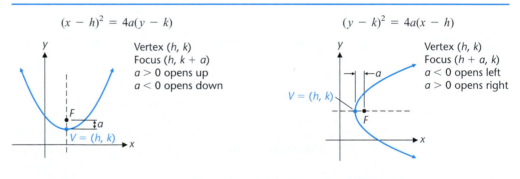

$(x - h)^2 = 4a(y - k)$

Vertex (h, k)
Focus $(h, k + a)$
$a > 0$ opens up
$a < 0$ opens down

$(y - k)^2 = 4a(x - h)$

Vertex (h, k)
Focus $(h + a, k)$
$a < 0$ opens left
$a > 0$ opens right

<p align="center">Circles</p>

$(x - h)^2 + (y - k)^2 = r^2$

Center (h, k)
Radius r

Table 1 Continued

Ellipses

$$\frac{(x-h)^2}{a^2} + \frac{(y-k)^2}{b^2} = 1 \qquad a > b > 0 \qquad \frac{(x-h)^2}{b^2} + \frac{(y-k)^2}{a^2} = 1$$

Center (h, k)
Major axis $2a$
Minor axis $2b$

Center (h, k)
Major axis $2a$
Minor axis $2b$

Hyperbolas

$$\frac{(x-h)^2}{a^2} - \frac{(y-k)^2}{b^2} = 1 \qquad\qquad \frac{(y-k)^2}{a^2} - \frac{(x-h)^2}{b^2} = 1$$

Center (h, k)
Transverse axis $2a$
Conjugate axis $2b$

Center (h, k)
Transverse axis $2a$
Conjugate axis $2b$

› **Translation Used in Graphing**

Any equation of the form

$$Ax^2 + Cy^2 + Dx + Ey + F = 0 \tag{2}$$

has a graph that is a conic, a degenerate conic, or the empty set [note that equation (2) is the same as equation (1) on page 604 with $B = 0$]. To see this, we use the technique of completing the square discussed in Section 1-5. If we can transform equation (2) into one of the standard forms of Table 1, then we will be able to identify its graph and sketch it rather quickly. Some examples should help make the process clear.

EXAMPLE 2 **Graphing a Conic**

Given the equation

$$y^2 - 6y - 4x + 1 = 0 \tag{3}$$

(A) Transform the equation into one of the standard forms in Table 1 and identify the conic.

(B) Find the equation in the translated system.

(C) Graph the conic.

SOLUTIONS (A) Complete the square in equation (3) relative to each variable that is squared—in this case y:

$$y^2 - 6y - 4x + 1 = 0 \qquad \text{\textcolor{red}{Add } }4x - 1\text{ \textcolor{red}{to both sides.}}$$

$$y^2 - 6y \qquad\qquad = 4x - 1 \qquad \text{\textcolor{red}{Add 9 to both sides to complete the square on the left side.}}$$

$$y^2 - 6y \qquad + 9 = 4x + 8 \qquad \text{\textcolor{red}{Factor.}}$$

$$(y - 3)^2 = 4(x + 2) \tag{4}$$

From Table 1 we recognize equation (4) as an equation of a parabola opening to the right with vertex at $(h, k) = (-2, 3)$.

(B) Find the equation of the parabola in the translated system with origin $0'$ at $(h, k) = (-2, 3)$. The equations of translation are read directly from equation (4):

$$x' = x + 2$$
$$y' = y - 3$$

Making these substitutions in equation (4) we obtain

$$y'^2 = 4x' \tag{5}$$

the equation of the parabola in the $x'y'$ system.

(C) Graph equation (5) in the $x'y'$ system following the process discussed in Section 9-1. The resulting graph is the graph of the original equation relative to the original xy coordinate system (Fig. 3).

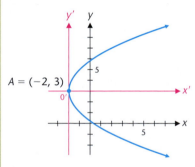

> **Figure 3**

MATCHED PROBLEM 2

Repeat Example 2 for the equation $x^2 + 4x + 4y - 12 = 0$.

EXAMPLE **3** **Graphing a Conic**

Given the equation

$$9x^2 - 4y^2 - 36x - 24y - 36 = 0$$

(A) Transform the equation into one of the standard forms in Table 1 and identify the conic.

(B) Find the equation in the translated system.

(C) Graph the conic.

(D) Find the coordinates of any foci relative to the original system.

SOLUTIONS

(A) Complete the square relative to both x and y.

$$9x^2 - 4y^2 - 36x - 24y - 36 = 0 \qquad \text{Add 36 to both sides.}$$

$$9x^2 - 36x \qquad - 4y^2 - 24y \qquad = 36 \qquad \text{Factor out coefficients of } x^2 \text{ and } y^2.$$

$$9(x^2 - 4x \qquad) - 4(y^2 + 6y \qquad) = 36 \qquad \text{Complete squares.}$$

$$9(x^2 - 4x + 4) - 4(y^2 + 6y + 9) = 36 + 36 - 36 \qquad \text{Factor.}$$

$$9(x - 2)^2 - 4(y + 3)^2 = 36 \qquad \text{Divide both sides by 36.}$$

$$\frac{(x - 2)^2}{4} - \frac{(y + 3)^2}{9} = 1$$

From Table 1 we recognize the last equation as an equation of a hyperbola opening left and right with center at $(h, k) = (2, -3)$.

(B) Find the equation of the hyperbola in the translated system with origin $0'$ at $(h, k) = (2, -3)$. The equations of translation are read directly from the last equation in part A:

$$x' = x - 2$$
$$y' = y + 3$$

Making these substitutions, we obtain

$$\frac{x'^2}{4} - \frac{y'^2}{9} = 1$$

the equation of the hyperbola in the $x'y'$ system.

(C) Graph the equation obtained in part B in the $x'y'$ system following the process discussed in Section 9-3. The resulting graph is the graph of the original equation relative to the original xy coordinate system (Fig. 4).

(D) Find the coordinates of the foci. To find the coordinates of the foci in the original system, first find the coordinates in the translated system:

$$c'^2 = 2^2 + 3^2 = 13$$
$$c' = \sqrt{13}$$
$$-c' = -\sqrt{13}$$

So the coordinates in the translated system are

$$F' = (-\sqrt{13}, 0) \qquad \text{and} \qquad F = (\sqrt{13}, 0)$$

Now, use

$$x = x' + h = x' + 2$$
$$y = y' + k = y' - 3$$

to obtain

$$F' = (-\sqrt{13} + 2, -3) \qquad \text{and} \qquad F = (\sqrt{13} + 2, -3)$$

as the coordinates of the foci in the original system.

> **Figure 4**

Technology Connections

To graph the equation of Example 1 on a graphing calculator, write it as a quadratic equation in the variable y, and use the quadratic formula to solve for y.

$$9x^2 - 4y^2 - 36x - 24y - 36 = 0$$

Write in the form $ay^2 + by + c = 0$.

$$4y^2 + 24y + (-9x^2 + 36x + 36) = 0$$

Use the quadratic formula with $a = 4$, $b = 24$, and $c = -9x^2 + 36x + 36$

$$y = \frac{-24 \pm \sqrt{24^2 - 4(4)(-9x^2 + 36x + 36)}}{8}$$

$$= -3 \pm 1.5\sqrt{x^2 - 4x} \qquad (6)$$

The two functions determined by equation (6) are graphed in Figure 5.

> **Figure 5**

MATCHED PROBLEM 3

Repeat Example 3 for the equation

$$9x^2 + 16y^2 + 36x - 32y - 92 = 0$$

⟩⟩⟩ EXPLORE-DISCUSS 1

If $A \neq 0$ and $C \neq 0$, show that the translation of axes $x' = x + \dfrac{D}{2A}$, $y' = y + \dfrac{E}{2C}$ transforms the equation $Ax^2 + Cy^2 + Dx + Ey + F = 0$ into an equation of the form $Ax'^2 + Cy'^2 = K$.

EXAMPLE **4** ## Finding the Equation of a Conic

Find the equation of a hyperbola with vertices on the line $x = -4$, conjugate axis on the line $y = 3$, length of the transverse axis $= 4$, and length of the conjugate axis $= 6$.

SOLUTION Locate the vertices, asymptote rectangle, and asymptotes in the original coordinate system [Fig. 6(a)], then sketch the hyperbola and translate the origin to the center of the hyperbola [Fig. 6(b)].

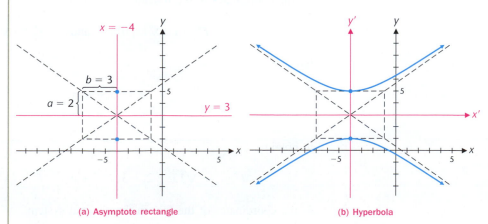

(a) Asymptote rectangle (b) Hyperbola

> **Figure 6**

Next write the equation of the hyperbola in the translated system:

$$\frac{y'^2}{4} - \frac{x'^2}{9} = 1$$

The origin in the translated system is at $(h, k) = (-4, 3)$, and the translation formulas are

$$x' = x - h = x - (-4) = x + 4$$
$$y' = y - k = y - 3$$

So the equation of the hyperbola in the original system is

$$\frac{(y - 3)^2}{4} - \frac{(x + 4)^2}{9} = 1$$

or, after simplifying and writing in the form of equation (1) on page 604,

$$4x^2 - 9y^2 + 32x + 54y + 19 = 0$$

MATCHED PROBLEM 4

Find the equation of an ellipse with foci on the line $x = 4$, minor axis on the line $y = -3$, length of the major axis $= 8$, and length of the minor axis $= 4$.

>>> EXPLORE-DISCUSS 2

Use the strategy of completing the square to transform each equation to an equation in an $x'y'$ coordinate system. Note that the equation you obtain is not one of the standard forms in Table 1; instead, it is either the equation of a degenerate conic or the equation has no solution. If the solution set of the equation is not empty, graph it and identify the graph (a point, a line, two parallel lines, or two intersecting lines).

(A) $x^2 + 2y^2 - 2x + 16y + 33 = 0$

(B) $4x^2 - y^2 - 24x - 2y + 35 = 0$

(C) $y^2 - 2y - 15 = 0$

(D) $5x^2 + y^2 + 12y + 40 = 0$

(E) $x^2 - 18x + 81 = 0$

› Rotation of Axes

To handle the general equation of the form

$$Ax^2 + Bxy + Cy^2 + Dx + Ey + F = 0 \tag{1}$$

when $B \neq 0$, we need to be able to *rotate,* not just translate, coordinate axes. If you hold a sheet of paper to a desk top with a pencil point, and move the paper without moving the pencil point, you *rotate* the paper. Similarly, a **rotation of coordinate axes** occurs when the origin is kept fixed and the x' and y' axes are obtained by rotating the x and y axes counterclockwise through an angle θ, as shown in Figure 7 on the next page.

› **Figure 7**

Referring to Figure 7 and using trigonometry, we have

$$x' = r \cos \alpha \qquad y' = r \sin \alpha \qquad (7)$$

and

$$x = r \cos (\theta + \alpha) \qquad y = r \sin (\theta + \alpha) \qquad (8)$$

Using sum identities from trigonometry for the equations in (8), we obtain

$$
\begin{aligned}
x &= r \cos (\theta + \alpha) && \text{Use sum identity for cosine.} \\
&= r (\cos \theta \cos \alpha - \sin \theta \sin \alpha) && \text{Distribute } r. \\
&= r \cos \theta \cos \alpha - r \sin \theta \sin \alpha && \text{Associate.} \\
&= (r \cos \alpha) \cos \theta - (r \sin \alpha) \sin \theta && \text{Substitute } x' = r \cos \alpha \text{ and } y' = r \sin \alpha. \\
&= x' \cos \theta - y' \sin \theta && (9)
\end{aligned}
$$

$$
\begin{aligned}
y &= r \sin (\theta + \alpha) && \text{Use sum identity for sine.} \\
&= r(\sin \theta \cos \alpha + \cos \theta \sin \alpha) && \text{Distribute } r. \\
&= r \sin \theta \cos \alpha + r \cos \theta \sin \alpha && \text{Associate.} \\
&= (r \cos \alpha) \sin \theta + (r \sin \alpha) \cos \theta && \text{Substitute } x' = r \cos \alpha \text{ and } y' = r \sin \alpha. \\
&= x' \sin \theta + y' \cos \theta && (10)
\end{aligned}
$$

So equations (9) and (10) together transform the xy coordinate system into the $x'y'$ coordinate system.

Equations (9) and (10) can be solved for x' and y' in terms of x and y to produce formulas that transform the $x'y'$ coordinate system back into the xy coordinate system. Omitting the details, the formulas for the transformation in the reverse direction are

$$x' = x \cos \theta + y \sin \theta \qquad y' = -x \sin \theta + y \cos \theta \qquad (11)$$

These results are summarized in Theorem 2.

> ### › THEOREM 2 Rotation Formulas
>
> If the xy coordinate axes are rotated counterclockwise through an angle θ, then the xy and $x'y'$ coordinates of a point P are related by:
>
> 1. $x = x' \cos \theta - y' \sin \theta$ 2. $x' = x \cos \theta + y \sin \theta$
> $y = x' \sin \theta + y' \cos \theta$ $y' = -x \sin \theta + y \cos \theta$

> ## Rotation Used in Graphing

We now investigate how rotation formulas are used in graphing.

EXAMPLE	5

Using the Rotation of Axes Formulas

Transform the equation $xy = -2$ using a rotation of axes through $45°$. Graph the new equation and identify the curve.

SOLUTION Use the rotation formulas:

$$x = x'\cos 45° - y'\sin 45° = \frac{\sqrt{2}}{2}(x' - y')$$

$$y = x'\sin 45° + y'\cos 45° = \frac{\sqrt{2}}{2}(x' + y')$$

$$xy = -2 \qquad \text{Substitute for } x \text{ and } y.$$

$$\frac{\sqrt{2}}{2}(x' - y')\frac{\sqrt{2}}{2}(x' + y') = -2 \qquad \text{Simplify.}$$

$$\frac{1}{2}(x'^2 - y'^2) = -2 \qquad \text{Distribute } \tfrac{1}{2}.$$

$$\frac{x'^2}{2} - \frac{y'^2}{2} = -2 \qquad \text{Divide both sides by } -2.$$

$$\frac{y'^2}{4} - \frac{x'^2}{4} = 1$$

This is a standard equation for a hyperbola. Summarizing, the graph of $xy = -2$ in the $x'y'$ coordinate system is a hyperbola with equation

$$\frac{y'^2}{4} - \frac{x'^2}{4} = 1$$

> **Figure 8**

as shown in Figure 8.

Notice that the asymptotes in the rotated system are the x and y axes in the original system.

MATCHED PROBLEM 5

Transform the equation $2xy = 1$ using a rotation of axes through $45°$. Graph the new equation and identify the curve. Check by graphing on a graphing calculator.

In Example 5, a $45°$ rotation transformed the original equation into one with no $x'y'$ term. This made it easy to recognize that the graph of the transformed equation was a hyperbola. In general, how do we determine the angle of rotation that will transform an equation with an xy term into one with no $x'y'$ term? To find out, we substitute

$$x = x'\cos \theta - y'\sin \theta \qquad \text{and} \qquad y = x'\sin \theta + y'\cos \theta$$

into equation (1) on page 604 to obtain

$$A(x'\cos\theta - y'\sin\theta)^2 + B(x'\cos\theta - y'\sin\theta)(x'\sin\theta + y'\cos\theta)$$
$$+ C(x'\sin\theta + y'\cos\theta)^2 + D(x'\cos\theta - y'\sin\theta)$$
$$+ E(x'\sin\theta + y'\cos\theta) + F = 0$$

After multiplying and collecting terms, we have

$$A'x'^2 + B'x'y' + C'y'^2 + D'x' + E'y' + F = 0 \tag{12}$$

where

$$B' = 2(C - A)\sin\theta\cos\theta + B(\cos^2\theta - \sin^2\theta) \tag{13}$$

For the $x'y'$ term in equation (12) to drop out, B' must be 0. We won't worry about A', C', D', and E' at this point; they will automatically be determined once we find θ so that $B' = 0$. We set the right side of equation (13) equal to 0 and solve for θ:

$$2(C - A)\sin\theta\cos\theta + B(\cos^2\theta - \sin^2\theta) = 0$$

Using the double-angle identities from trigonometry, $\sin 2\theta = 2\sin\theta\cos\theta$ and $\cos 2\theta = \cos^2\theta - \sin^2\theta$, we obtain

$$(C - A)\sin 2\theta + B\cos 2\theta = 0 \qquad \text{Add } (A - C)\sin 2\theta \text{ to both sides.}$$

$$B\cos 2\theta = (A - C)\sin 2\theta \qquad \text{Divide both sides by } B\sin 2\theta.$$

$$\frac{\cos 2\theta}{\sin 2\theta} = \frac{A - C}{B} \qquad \text{Use quotient identity.}$$

$$\cot 2\theta = \frac{A - C}{B} \tag{14}$$

So if we choose θ so that $\cot 2\theta = (A - C)/B$, then $B' = 0$ and the $x'y'$ term in equation (12) will drop out. There is always an angle θ between $0°$ and $90°$ that solves equation (14), because the range of $y = \cot 2\theta$ for $0° < \theta < 90°$ is the set of all real numbers (Fig. 9).

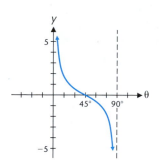

> **Figure 9**

> **THEOREM 3** Angle of Rotation to Eliminate the $x'y'$ Term

To transform the equation

$$Ax^2 + Bxy + Cy^2 + Dx + Ey + F = 0$$

into an equation in x' and y' with no $x'y'$ term, find θ so that

$$\cot 2\theta = \frac{A - C}{B} \qquad \text{and} \qquad 0° < \theta < 90°$$

and use the rotation formulas in Theorem 2.

EXAMPLE 6 **Identifying and Graphing an Equation with an xy Term**

Given the equation $17x^2 - 6xy + 9y^2 = 72$, find the angle of rotation so that the transformed equation will have no $x'y'$ term. Sketch and identify the graph.

SOLUTION

$$17x^2 - 6xy + 9y^2 = 72 \tag{15}$$

$$\cot 2\theta = \frac{A - C}{B} = \frac{17 - 9}{-6} = -\frac{4}{3}$$

So 2θ is a quadrant II angle, and using the reference triangle in the figure, we can see that $\cos 2\theta = -\frac{4}{5}$. We can find the rotation formulas exactly by the use of the half-angle identities

$$\sin \theta = \sqrt{\frac{1 - \cos 2\theta}{2}}$$

and

$$\cos \theta = \sqrt{\frac{1 + \cos 2\theta}{2}}$$

Using these identities, we obtain

$$\sin \theta = \sqrt{\frac{1 - (-\frac{4}{5})}{2}} = \frac{3}{\sqrt{10}}$$

and

$$\cos \theta = \sqrt{\frac{1 + (-\frac{4}{5})}{2}} = \frac{1}{\sqrt{10}}$$

So the rotation formulas are

$$x = \frac{1}{\sqrt{10}}x' - \frac{3}{\sqrt{10}}y'$$

and (16)

$$y = \frac{3}{\sqrt{10}}x' + \frac{1}{\sqrt{10}}y'$$

Substituting equations (16) into equation (15), we have

$$17\left(\frac{1}{\sqrt{10}}x' - \frac{3}{\sqrt{10}}y'\right)^2 - 6\left(\frac{1}{\sqrt{10}}x' - \frac{3}{\sqrt{10}}y'\right)\left(\frac{3}{\sqrt{10}}x' + \frac{1}{\sqrt{10}}y'\right)$$
$$+ 9\left(\frac{3}{\sqrt{10}}x' + \frac{1}{\sqrt{10}}y'\right)^2 = 72$$

$$\frac{17}{10}(x' - 3y')^2 - \frac{6}{10}(x' - 3y')(3x' + y') + \frac{9}{10}(3x' + y')^2 = 72$$

Further simplification leads to

$$\frac{x'^2}{9} + \frac{y'^2}{4} = 1$$

which is a standard equation for an ellipse. To graph, we rotate the original axes through an angle θ determined as follows:

$$\cot 2\theta = -\frac{4}{3}$$

$$2\theta \approx 143.1301°$$

$$\theta \approx 71.57°$$

> Figure 10

We could also use either

$$\sin \theta = \frac{3}{\sqrt{10}} \quad \text{or} \quad \cos \theta = \frac{1}{\sqrt{10}}$$

to determine the angle of rotation. Summarizing these results, the graph of $17x^2 - 6xy + 9y^2 = 72$ in the $x'y'$ coordinate system formed by a rotation of $71.57°$ is an ellipse with equation

$$\frac{x'^2}{9} + \frac{y'^2}{4} = 1$$

as shown in Figure 10.

MATCHED PROBLEM 6 Given the equation $3x^2 + 26\sqrt{3}xy - 23y^2 = 144$, find the angle of rotation so that the transformed equation will have no $x'y'$ term. Sketch and identify the graph. Check by graphing on a graphing calculator.

> Identifying Conics

The **discriminant** of the general second-degree equation in two variables [equation (1)] is $B^2 - 4AC$. It can be shown that the value of this expression does not change when the axes are translated or rotated. This forms the basis for Theorem 4.

> **THEOREM 4 Identifying Conics**

The graph of the equation

$$Ax^2 + Bxy + Cy^2 + Dx + Ey + F = 0 \qquad (1)$$

is, excluding degenerate cases,

1. A hyperbola if $B^2 - 4AC > 0$
2. A parabola if $B^2 - 4AC = 0$
3. An ellipse if $B^2 - 4AC < 0$

The proof of Theorem 4 is beyond the scope of this book. Its use is best illustrated by example.

EXAMPLE 7 **Identifying Conics**

Identify the following conics.

(A) $x^2 - xy + y^2 = 5$
(B) $x^2 - xy - y^2 = 5$
(C) $x^2 - 4xy + 4y^2 + x = 5$

SOLUTIONS | (A) The discriminant is

$$B^2 - 4AC = (-1)^2 - 4(1)(1) = -3 < 0$$

so by Theorem 4 the conic is an ellipse.

(B) The discriminant is

$$B^2 - 4AC = (-1)^2 - 4(1)(-1) = 5 > 0$$

so by Theorem 4 the conic is a hyperbola.

(C) The discriminant is

$$B^2 - 4AC = (-4)^2 - 4(1)(4) = 0$$

so by Theorem 4 the conic is a parabola.

Technology Connections

Each of the equations in Example 7 can be graphed by the method illustrated in Example 6, or, as an alternative, by a graphing calculator. For example, to graph the equation $x^2 - xy + y^2 = 5$ using a graphing calculator, first write the equation as a quadratic in the variable y, then use the quadratic formula to solve for y:

$x^2 - xy + y^2 = 5$ Write as a quadratic in y.

$y^2 - xy + x^2 - 5 = 0$ Use the quadratic formula with $a = 1$, $b = -x$, and $c = x^2 - 5$.

$$y = \frac{x \pm \sqrt{(-x)^2 - 4(1)(x^2 - 5)}}{2}$$ Simplify.

$$= \frac{x \pm \sqrt{20 - 3x^2}}{2}$$

Graphing

$$y_1 = \frac{x + \sqrt{20 - 3x^2}}{2} \quad \text{and} \quad y_2 = \frac{x - \sqrt{20 - 3x^2}}{2}$$

produces the ellipse of Example 7(A) [Fig. 11].

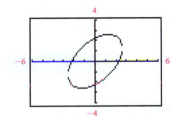

❯ **Figure 11**

MATCHED PROBLEM 7 Identify the following conics.

(A) $x^2 + xy + 2y^2 = 10$

(B) $x^2 + xy - 2y^2 = 10$

(C) $x^2 - 2xy + y^2 - x = 10$

ANSWERS TO MATCHED PROBLEMS

1. $y'^2 = 8x'$; a parabola **2.** (A) $(x + 2)^2 = -4(y - 4)$; a parabola (B) $x'^2 = -4y'$

(C)

3. (A) $\dfrac{(x+2)^2}{16} + \dfrac{(y-1)^2}{9} = 1$; ellipse (B) $\dfrac{x'^2}{16} + \dfrac{y'^2}{9} = 1$

(C) (D) Foci: $F' = (-\sqrt{7}-2, 1), F = (\sqrt{7}-2, 1)$

4. $\dfrac{(x-4)^2}{4} + \dfrac{(y+3)^2}{16} = 1$, or $4x^2 + y^2 - 32x + 6y + 57 = 0$

5. $x'^2 - y'^2 = 1$; hyperbola **6.** $\dfrac{x'^2}{9} - \dfrac{y'^2}{4} = 1$; $\theta = 30°$; hyperbola

7. (A) Ellipse (B) Hyperbola (C) Parabola

9-4 Exercises

1. Explain what a translation is in your own words.

2. Explain what a rotation is in your own words.

3. What is the discriminant of the equation $Ax^2 + Bxy + Cy^2 + Dx + Ey + F = 0$?

4. Explain how the discriminant can be used to determine whether the graph of a second-degree equation in two variables is a parabola, ellipse, or hyperbola.

In Problems 5–12:
(A) Find translation formulas that translate the origin to the indicated point (h, k).
(B) Write the equation of the curve for the translated system.
(C) Identify the curve.

5. $(x-3)^2 + (y-5)^2 = 81$; $(3, 5)$

6. $(x-3)^2 = 8(y+2)$; $(3, -2)$

7. $\dfrac{(x+7)^2}{9} + \dfrac{(y-4)^2}{16} = 1$; $(-7, 4)$

8. $(x+2)^2 + (y+6)^2 = 36$; $(-2, -6)$

9. $(y+9)^2 = 16(x-4)$; $(4, -9)$

10. $\dfrac{(y-9)^2}{10} - \dfrac{(x+5)^2}{6} = 1$; $(-5, 9)$

11. $\dfrac{(x+8)^2}{12} + \dfrac{(y+3)^2}{8} = 1$; $(-8, -3)$

12. $\dfrac{(x+7)^2}{25} - \dfrac{(y-8)^2}{50} = 1$; $(-7, 8)$

In Problems 13–18:
(A) Write each equation in one of the standard forms listed in Table 1.
(B) Identify the curve.

13. $16(x-3)^2 - 9(y+2)^2 = 144$

14. $(y+2)^2 - 12(x-3) = 0$

15. $6(x+5)^2 + 5(y+7)^2 = 30$

16. $12(y-5)^2 - 8(x-3)^2 = 24$

17. $(x+6)^2 + 24(y-4) = 0$

18. $4(x-7)^2 + 7(y-3)^2 = 28$

In Problems 19–22, find the x'y' coordinates of the given points if the coordinate axes are rotated through the indicated angle.

19. $(1, 0), (0, 1), (1, -1), (-3, 4), \theta = 30°$

20. $(1, 0), (0, 1), (-1, 2), (-2, 5), \theta = 60°$

21. $(1, 0), (0, 1), (-1, -2), (1, -3), \theta = 45°$

22. $(1, 1), (-1, -1), (1, -1), (-1, 1), \theta = 90°$

In Problems 23–26, find the equations of the x' and y' axes in terms of x and y if the xy coordinate axes are rotated through the indicated angle.

23. $\theta = 30°$ **24.** $\theta = 60°$

25. $\theta = 45°$ **26.** $\theta = 90°$

In Problems 27–34, transform each equation into one of the standard forms in Table 1. Identify the curve and graph it.

27. $4x^2 + 9y^2 - 16x - 36y + 16 = 0$

28. $16x^2 + 9y^2 + 64x + 54y + 1 = 0$

29. $x^2 + 8x + 8y = 0$

30. $y^2 + 12x + 4y - 32 = 0$

31. $x^2 + y^2 + 12x + 10y + 45 = 0$

32. $x^2 + y^2 - 8x - 6y = 0$

33. $-9x^2 + 16y^2 - 72x - 96y - 144 = 0$

34. $16x^2 - 25y^2 - 160x = 0$

In Problems 35–40, find the coordinates of any foci relative to the original coordinate system.

35. Problem 27 **36.** Problem 28 **37.** Problem 29

38. Problem 30 **39.** Problem 33 **40.** Problem 34

In Problems 41–44, complete the square in each equation, identify the transformed equation, and graph.

41. $x^2 - 2x + y^2 + 4y + 5 = 0$

42. $x^2 - 6x + 2y^2 + 4y + 11 = 0$

43. $x^2 + 8x - 4y^2 + 8y + 12 = 0$

44. $x^2 + 4x - y^2 + 6y - 5 = 0$

45. If $A \neq 0$, $C = 0$, and $E \neq 0$, find h and k so that the translation of axes $x = x' + h$, $y = y' + k$ transforms the equation $Ax^2 + Cy^2 + Dx + Ey + F = 0$ into one of the standard forms of Table 1.

46. If $A = 0$, $C \neq 0$, and $D \neq 0$, find h and k so that the translation of axes $x = x' + h$, $y = y' + k$ transforms the equation $Ax^2 + Cy^2 + Dx + Ey + F = 0$ into one of the standard forms of Table 1.

In Problems 47–50, find the transformed equation when the axes are rotated through the indicated angle. Sketch and identify the graph.

47. $x^2 + y^2 = 49, \theta = 45°$ **48.** $x^2 + y^2 = 25, \theta = 60°$

49. $2x^2 + \sqrt{3}xy + y^2 - 10 = 0, \theta = 30°$

50. $x^2 + 8xy + y^2 - 75 = 0, \theta = 45°$

In Problems 51–56, find the angle of rotation so that the transformed equation will have no x'y' term. Sketch and identify the graph.

51. $x^2 - 4xy + y^2 = 12$ **52.** $x^2 + xy + y^2 = 6$

53. $8x^2 - 4xy + 5y^2 = 36$ **54.** $5x^2 - 4xy + 8y^2 = 36$

55. $x^2 - 2\sqrt{3}xy + 3y^2 - 16\sqrt{3}x - 16y = 0$

56. $x^2 + 2\sqrt{3}xy + 3y^2 + 8\sqrt{3}x - 8y = 0$

In Problems 57–66, find the equations (in the original xy coordinate system) of the asymptotes of each hyperbola.

57. $(x - 3)^2 - (y + 2)^2 = 1$ **58.** $(x + 1)^2 - (y - 4)^2 = 1$

59. $\dfrac{x^4}{4} - \dfrac{(y + 1)^2}{25} = 1$ **60.** $\dfrac{(x - 5)^2}{36} - \dfrac{y^2}{4} = 1$

61. $9(y - 5)^2 - 16(x + 2)^2 = 144$

62. $25(y + 3)^2 - 9(x - 1)^2 = 225$

63. $3(y + 4)^2 - x^2 = 1$

64. $y^2 - 5(x + 2)^2 = 1$

65. $xy - 9 = 0$

66. $4xy + 1 = 0$

In Problems 67–78, use the given information to find the equation of each conic. Express the answer in the form $Ax^2 + Cy^2 + Dx + Ey + F = 0$ with integer coefficients and $A > 0$.

67. A parabola with vertex at $(2, 5)$, axis the line $x = 2$, and passing through the point $(-2, 1)$.

68. A parabola with vertex at $(4, -1)$, axis the line $y = -1$, and passing through the point $(2, 3)$.

69. An ellipse with major axis on the line $y = -3$, minor axis on the line $x = -2$, length of major axis $= 8$, and length of minor axis $= 4$.

70. An ellipse with major axis on the line $x = -4$, minor axis on the line $y = 1$, length of major axis $= 4$, and length of minor axis $= 2$.

71. An ellipse with vertices $(4, -7)$ and $(4, 3)$ and foci $(4, -6)$ and $(4, 2)$.

72. An ellipse with vertices $(-3, 1)$ and $(7, 1)$ and foci $(-1, 1)$ and $(5, 1)$.

73. A hyperbola with transverse axis on the line $x = 2$, length of transverse axis $= 4$, conjugate axis on the line $y = 3$, and length of conjugate axis $= 2$.

74. A hyperbola with transverse axis on the line $y = -5$, length of transverse axis $= 6$, conjugate axis on the line $x = 2$, and length of conjugate axis $= 6$.

75. An ellipse with the following graph:

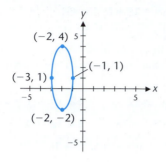

76. An ellipse with the following graph:

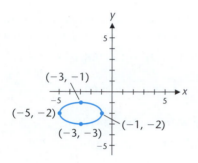

77. A hyperbola with the following graph:

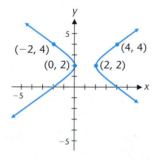

78. A hyperbola with the following graph:

In Problems 79–84, use the discriminant to identify each graph. Graph on a graphing calculator.

79. $13x^2 + 10xy + 13y^2 - 72 = 0$

80. $3x^2 - 10xy + 3y^2 + 8 = 0$

81. $x^2 - 6\sqrt{3}xy - 5y^2 - 8 = 0$

82. $16x^2 + 24xy + 9y^2 + 15x - 20y = 0$

83. $16x^2 - 24xy + 9y^2 + 60x + 80y = 0$

84. $7x^2 - 6\sqrt{3}xy + 13y^2 - 16 = 0$

In Problems 85 and 86, use a rotation followed by a translation to transform each equation into a standard form. Sketch and identify the curve.

85. $x^2 + 2\sqrt{3}xy + 3y^2 - 8\sqrt{3}x - 8y - 4 = 0$

86. $73x^2 + 72xy + 52y^2 - 260x - 320y + 400 = 0$

CHAPTER 9 Review

9-1 Conic Sections; Parabola

The plane curves obtained by intersecting a right circular cone with a plane are called **conic sections.** If the plane cuts clear through one nappe, then the intersection curve is called a **circle** if the plane is perpendicular to the axis and an **ellipse** if the plane is not perpendicular to the axis. If a plane cuts only one nappe, but does not cut clear through, then the intersection curve is called a **parabola.** If a plane cuts through both nappes, but not through the vertex, the resulting intersection curve is called a **hyperbola.** A plane passing through the vertex of the cone produces a **degenerate conic**—a point, a line, or a pair of lines. The figure illustrates the four nondegenerate conics.

Circle Ellipse Parabola Hyperbola

The graph of

$$Ax^2 + Bxy + Cy^2 + Dx + Ey + F = 0$$

is a conic, a degenerate conic, or the empty set.

The following is a coordinate-free definition of a parabola:

Parabola

A **parabola** is the set of all points in a plane equidistant from a fixed point F and a fixed line L (not containing F) in the plane. The fixed point F is called the **focus**, and the fixed line L is called the **directrix**. A line through the focus perpendicular to the directrix is called the **axis of symmetry**, and the point on the axis halfway between the directrix and focus is called the **vertex**.

From the definition of a parabola, we can obtain the following standard equations:

Standard Equations of a Parabola with Vertex at (0, 0)

1. $y^2 = 4ax$
Vertex: $(0, 0)$
Focus: $(a, 0)$
Directrix: $x = -a$
Symmetric with respect to the x axis
Axis of symmetry the x axis

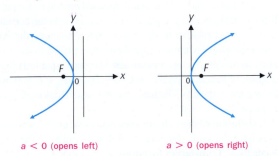

a < 0 (opens left) a > 0 (opens right)

2. $x^2 = 4ay$
Vertex: $(0, 0)$
Focus: $(0, a)$
Directrix: $y = -a$
Symmetric with respect to the y axis
Axis of symmetry the y axis

a < 0 (opens down) a > 0 (opens up)

9-2 Ellipse

The following is a coordinate-free definition of an ellipse:

Ellipse

An **ellipse** is the set of all points P in a plane such that the sum of the distances from P to two fixed points in the plane is a constant (the constant is required to be greater than the distance between the two fixed points). Each of the fixed points, F' and F, is called a **focus**, and together they are called **foci**. Referring to the figure, the line segment $V'V$ through the foci is the **major axis**. The perpendicular bisector $B'B$ of the major axis is the **minor axis**. Each end of the major axis, V' and V, is called a **vertex**. The midpoint of the line segment $F'F$ is called the **center** of the ellipse.

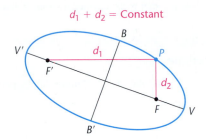

From the definition of an ellipse, we can obtain the following standard equations:

Standard Equations of an Ellipse with Center at (0, 0)

1. $\dfrac{x^2}{a^2} + \dfrac{y^2}{b^2} = 1 \qquad a > b > 0$

x intercepts: $\pm a$ (vertices)
y intercepts: $\pm b$
Foci: $F' = (-c, 0),\ F = (c, 0) \qquad c^2 = a^2 - b^2$
Major axis length $= 2a$
Minor axis length $= 2b$

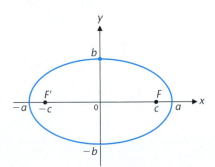

2. $\dfrac{x^2}{b^2} + \dfrac{y^2}{a^2} = 1 \qquad a > b > 0$

x intercepts: $\pm b$
y intercepts: $\pm a$ (vertices)
Foci: $F' = (0, -c),\ F = (0, c) \qquad c^2 = a^2 - b^2$
Major axis length $= 2a$
Minor axis length $= 2b$

[*Note:* Both graphs are symmetric with respect to the x axis, y axis, and origin. Also, the major axis is always longer than the minor axis.]

9-3 Hyperbola

The following is a coordinate-free definition of a hyperbola:

Hyperbola

A **hyperbola** is the set of all points P in a plane such that the absolute value of the difference of the distances from P to two fixed points in the plane is a positive constant (the constant is required to be less than the distance between the two fixed points). Each of the fixed points, F' and F, is called a **focus.** The intersection points V' and V of the line through the foci and the two branches of the hyperbola are called **vertices,** and each is called a **vertex.** The line segment $V'V$ is called the **transverse axis.** The midpoint of the transverse axis is the **center** of the hyperbola. The line segment perpendicular to the transverse axis through the center is called the **conjugate axis.**

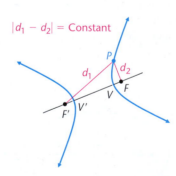

From the definition of a hyperbola, we can obtain the following standard equations:

Standard Equations of a Hyperbola with Center at (0, 0)

1. $\dfrac{x^2}{a^2} - \dfrac{y^2}{b^2} = 1$

 x intercepts: $\pm a$ (vertices)
 y intercepts: none
 Foci: $F' = (-c, 0)$, $F = (c, 0)$ $c^2 = a^2 + b^2$
 Transverse axis length $= 2a$
 Conjugate axis length $= 2b$

 Asymptotes: $y = \pm\dfrac{b}{a}x$

2. $\dfrac{y^2}{a^2} - \dfrac{x^2}{b^2} = 1$

 x intercepts: none
 y intercepts: $\pm a$ (vertices)
 Foci: $F' = (0, -c)$, $F = (0, c)$ $c^2 = a^2 + b^2$
 Transverse axis length $= 2a$
 Conjugate axis length $= 2b$

 Asymptotes: $y = \pm\dfrac{a}{b}x$

[*Note:* Both graphs are symmetric with respect to the x axis, y axis, and origin.]

9-4 Translation and Rotation of Axes

In Sections 9-1, 9-2, and 9-3 we found standard equations for parabolas, ellipses, and hyperbolas located with their axes on the coordinate axes and centered relative to the origin. We now move the conics away from the origin while keeping their axes parallel to the coordinate axes. In this process we obtain new standard equations that are special cases of the equation $Ax^2 + Cy^2 + Dx + Ey + F = 0$, where A and C are not both zero. The basic mathematical tool used is *translation of axes.*

A **translation of coordinate axes** occurs when the new coordinate axes have the same direction as and are parallel to the original coordinate axes. **Translation formulas** are as follows:

1. $x = x' + h$ **2.** $x' = x - h$
 $y = y' + k$ $y' = y - k$

where (h, k) are the coordinates of the origin $0'$ relative to the original system.

Table 1 on page 606 lists the standard equations for conics. If the xy coordinate axes are rotated counterclockwise through an angle θ into the $x'y'$ coordinate axes, then the xy and $x'y'$ coordinate systems are related by the **rotation formulas:**

1. $x = x' \cos \theta - y' \sin \theta$ **2.** $x' = x \cos \theta + y \sin \theta$

$\, y = x' \sin \theta + y' \cos \theta$ $x' = -x \sin \theta + y \cos \theta$

To transform the general quadratic equation

$$Ax^2 + Bxy + Cy^2 + Dx + Ey + F = 0$$

into an equation in x' and y' with no $x'y'$ term, choose the *angle of rotation* θ to satisfy $\cot 2\theta = (A - C)/B$ and $0° < \theta < 90°$. The **discriminant** of the general second-degree equation in two variables is $B^2 - 4AC$ and the graph is

1. A hyperbola if $B^2 - 4AC > 0$

2. A parabola if $B^2 - 4AC = 0$

3. An ellipse if $B^2 - 4AC < 0$

CHAPTER **9** **Review Exercises**

Work through all the problems in this chapter review and check answers in the back of the book. Answers to all review problems are there, and following each answer is a number in italics indicating the section in which that type of problem is discussed. Where weaknesses show up, review appropriate sections in the text.

In Problems 1–6, graph each equation and locate foci. Locate the directrix for any parabolas. Find the lengths of major, minor, transverse, and conjugate axes where applicable.

1. $9x^2 + 25y^2 = 225$ **2.** $x^2 = -12y$

3. $25y^2 - 9x^2 = 225$ **4.** $x^2 - y^2 = 16$

5. $y^2 = 8x$ **6.** $2x^2 + y^2 = 8$

In Problems 7–9:

(A) Write each equation in one of the standard forms listed in Table 1 on page 606.

(B) Identify the curve.

7. $4(y + 2)^2 - 25(x - 4)^2 = 100$

8. $(x + 5)^2 + 12(y + 4) = 0$

9. $16(x - 6)^2 + 9(y - 4)^2 = 144$

10. Find the $x'y'$ coordinates of the point $(3, 4)$ when the axes are rotated through

 (A) 30° (B) 45° (C) 60°

11. Find the equations of the x' and y' axes in terms of x and y if the axes are rotated through an angle of 75°.

12. Find the equation of the parabola having its vertex at the origin, its axis of symmetry the x axis, and $(-4, -2)$ on its graph.

In Problems 13 and 14, find the equation of the ellipse in the form

$$\frac{x^2}{M} + \frac{y^2}{N} = 1 \qquad M, N > 0$$

if the center is at the origin, and:

13. Major axis on x axis

 Major axis length = 12

 Minor axis length = 10

14. Major axis on y axis

 Minor axis length = 12

 Distance between foci = 16

In Problems 15 and 16, find the equation of the hyperbola in the form

$$\frac{x^2}{M} - \frac{y^2}{N} = 1 \qquad or \qquad \frac{y^2}{M} - \frac{x^2}{N} = 1 \qquad M, N > 0$$

if the center is at the origin, and:

15. Transverse axis on y axis

 Conjugate axis length = 6

 Distance between foci = 8

16. Transverse axis on x axis

 Transverse axis length = 14

 Conjugate axis length = 16

17. Find the equation of the parabola having directrix $y = 5$ and focus $(0, -5)$.

18. Find the foci of the ellipse through the point $(-6, 0)$ if the center is at the origin, the major axis is on the x axis, and the major axis has twice the length of the minor axis.

19. Find the y intercepts of a hyperbola if the center is at the origin, the conjugate axis is on the x axis and has length 4, and $(0, -3)$ is a focus.

20. Find the directrix of a parabola having its vertex at the origin and focus $(-4, 0)$.

21. Find the points of intersection of the parabolas $x^2 = 8y$ and $y^2 = -x$.

22. Find the x intercepts of an ellipse if the center is at the origin, the major axis is on the y axis and has length 14, and $(0, -1)$ is a focus.

23. Find the foci of the hyperbola through the point $(0, -4)$ if the center is at the origin, the transverse axis is on the y axis, and the conjugate axis has twice the length of the transverse axis.

In Problems 24–26, transform each equation into one of the standard forms in Table 1 on page 606. Identify the curve and graph it.

24. $16x^2 + 4y^2 + 96x - 16y + 96 = 0$

25. $x^2 - 4x - 8y - 20 = 0$

26. $4x^2 - 9y^2 + 24x - 36y - 36 = 0$

27. Given the equation $x^2 - \sqrt{3}xy + 2y^2 - 10 = 0$, find the transformed equation when the axes are rotated through $30°$. Sketch and identify the graph.

28. Given the equation $5x^2 + 26xy + 5y^2 + 72 = 0$, find the angle of rotation so that the transformed equation will have no $x'y'$ term. Sketch and identify the graph.

29. Given the equation $3x^2 + 4xy + 2y^2 - 20 = 0$, identify the curve.

30. Use the definition of a parabola and the distance formula to find the equation of a parabola with directrix $x = 6$ and focus at $(2, 4)$.

31. Find an equation of the set of points in a plane each of whose distance from $(4, 0)$ is twice its distance from the line $x = 1$. Identify the geometric figure.

32. Find an equation of the set of points in a plane each of whose distance from $(4, 0)$ is two-thirds its distance from the line $x = 9$. Identify the geometric figure.

In Problems 33–35, find the coordinates of any foci relative to the original coordinate system.

33. Problem 24 **34.** Problem 25 **35.** Problem 26

In Problems 36–39, find the equations of the asymptotes of each hyperbola.

36. $\dfrac{x^2}{49} - \dfrac{y^2}{25} = 1$ **37.** $\dfrac{y^2}{64} - \dfrac{x^2}{4} = 1$

38. $4x^2 - y^2 = 1$ **39.** $xy - 1 = 0$

APPLICATIONS

40. COMMUNICATIONS A parabolic satellite television antenna has a diameter of 8 feet and is 1 foot deep. How far is the focus from the vertex?

41. ENGINEERING An elliptical gear is to have foci 8 centimeters apart and a major axis 10 centimeters long. Letting the x axis lie along the major axis (right positive) and the y axis lie along the minor axis (up positive), write the equation of the ellipse in the standard form

$$\frac{x^2}{a^2} + \frac{y^2}{b^2} = 1$$

42. SPACE SCIENCE A hyperbolic reflector for a radio telescope (such as that illustrated in Problem 63, Exercises 9-3) has the equation

$$\frac{y^2}{40^2} - \frac{x^2}{30^2} = 1$$

If the reflector has a diameter of 30 feet, how deep is it? Compute the answer to three significant digits.

CHAPTER 9

››› GROUP ACTIVITY Focal Chords

Many of the applications of the conic sections are based on their reflective or focal properties. One of the interesting algebraic properties of the conic sections concerns their focal chords.

If a line through a focus F contains two points G and H of a conic section, then the line segment GH is called a **focal chord**. Let $G = (x_1, y_1)$ and $H = (x_2, y_2)$ be points on the graph of $x^2 = 4ay$ such that GH is a focal chord. Let u denote the length of GF and v the length of FH (Fig. 1).

> **Figure 1** Focal chord GH of the parabola $x^2 = 4ay$.

(A) Use the distance formula to show that $u = y_1 + a$.

(B) Show that G and H lie on the line $y - a = mx$, where $m = (y_2 - y_1)/(x_2 - x_1)$.

(C) Solve $y - a = mx$ for x and substitute in $x^2 = 4ay$, obtaining a quadratic equation in y. Explain why $y_1 y_2 = a^2$.

(D) Show that $\dfrac{1}{u} + \dfrac{1}{v} = \dfrac{1}{a}$.

(E) Show that $u + v - 4a = \dfrac{(u - 2a)^2}{u - a}$. Explain why this implies that $u + v \geq 4a$, with equality if and only if $u = v = 2a$.

(F) Which focal chord is the shortest? Is there a longest focal chord?

(G) Is $\dfrac{1}{u} + \dfrac{1}{v}$ a constant for focal chords of the ellipse? For focal chords of the hyperbola? Obtain evidence for your answers by considering specific examples.

(H) The conic section with focus at the origin, directrix the line $x = D > 0$, and eccentricity $E > 0$ has the polar equation

$$r = \frac{DE}{1 + E \cos \theta}.$$ Explain how this polar equation makes it easy to show that $\dfrac{1}{u} + \dfrac{1}{v} = \dfrac{1}{a}$ for a parabola. Use the polar equation to determine the sum $\dfrac{1}{u} + \dfrac{1}{v}$ for a focal chord of an ellipse or hyperbola.

Systems of Equations and Matrices

↧

WE have seen many real-world situations where solving an equation is valuable. But the world is a very complicated place, and many more situations lead to more than one variable. In that case, solving a system of equations becomes important. In this chapter, we will study a variety of methods for solving systems of equations. We will begin with linear systems with two or three variables using algebraic techniques similar to those we used for solving individual equations. Then we will introduce a variety of matrix methods for solving linear systems. These methods can be applied to very large systems that model very complicated real-world problems.

10-1 Systems of Linear Equations

> › Systems of Equations
> › Solving by Graphing
> › Solving by Substitution
> › Solving Using Elimination by Addition
> › Applications

We have seen a wide variety of real-world problems that can be solved by writing and solving an equation. But a lot of problems have extra conditions that makes writing a single equation impractical. In this case, two or more equations might be needed to model the situation. In this section, we'll examine how to solve two or more equations together, then see how to apply what we learn.

› Systems of Equations

To illustrate the basic concepts, we'll use a simple example. At one campus coffee shop, muffins cost $2 each, and lattes are $3 each. If a total of seven items are sold for $18, how many of each item were sold?

There are two natural variables in the problem: the number of muffins, which we'll call x, and the number of lattes, which we'll call y. Then

$$x + y = 7 \quad \text{Seven items total}$$
$$2x + 3y = 18 \quad \text{Total cost is \$18.}$$

This is called a **system of linear equations in two variables**. The solution to the problem is found by finding all pairs of numbers x and y that make both equations true.

In general, we will study solving linear systems of the type

$$ax + by = h$$
$$cx + dy = k \quad \text{System of two linear equations in two variables}$$

where x and y are variables, a, b, c, and d are real numbers called the **coefficients** of x and y, and h and k are real numbers called the **constant terms** in the equations. A pair of numbers $x = x_0$ and $y = y_0$ is a **solution** of this system if each equation is satisfied by the pair. The set of all such pairs of numbers is called the **solution set** for the system. To **solve** a system is to find its solution set.

› Solving by Graphing

Recall that the graph of a linear equation is the line consisting of all ordered pairs that satisfy the equation. To solve the coffee shop problem by graphing, we will graph both equations in the same coordinate system. The coordinates of any points that the lines have in common must be solutions to the system, because they must satisfy both equations.

| EXAMPLE | 1 | Solving a System by Graphing |

Solve the coffee shop problem by graphing:
$$x + y = 7$$
$$2x + 3y = 18$$

SOLUTION | Find the x and y intercepts for each line.

$$x + y = 7 \qquad\qquad 2x + 3y = 18$$

x	y
0	7
7	0

x	y
0	6
9	0

Plot these points, graph the two lines, estimate the intersection point visually (Fig. 1), and check the estimate.

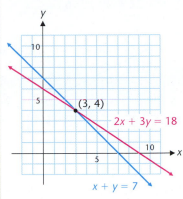

(3, 4)

$2x + 3y = 18$

$x + y = 7$

› **Figure 1**

$$x = 3 \qquad \text{Muffins*}$$
$$y = 4 \qquad \text{Lattes}$$

CHECK

$$x + y = 7 \qquad\qquad 2x + 3y = 18$$
$$3 + 4 \overset{?}{=} 7 \qquad\qquad 2(3) + 3(4) \overset{?}{=} 18$$
$$7 \overset{\checkmark}{=} 7 \qquad\qquad 18 \overset{\checkmark}{=} 18$$

MATCHED PROBLEM 1 | Solve by graphing: $\quad x - y = 3$
$ x + 2y = -3$

Technology Connections

To solve Example 1 with a graphing calculator, first solve each equation for y:

$x + y = 7$ Subtract x from both sides.

$\quad y = 7 - x$

$2x + 3y = 18$ Subtract $2x$ from both sides.

$\quad 3y = 18 - 2x$ Divide both sides by 3.

$\quad y = 6 - \frac{2}{3}x$

Next, enter these functions in the equation editor of a graphing calculator (Fig. 2) and use the intersect command to find the intersection point (Fig. 3).

From Figure 3, we see that the solution is

$$x = 3 \qquad \text{Muffins}$$
$$y = 4 \qquad \text{Lattes}$$

› **Figure 2**

› **Figure 3**

*When the solution set for a linear system is a single point, we will follow the common practice of writing the solution as (3, 4) or as $x = 3$, $y = 4$, rather than the more formal expression $\{(3, 4)\}$.

It is clear that Example 1 has exactly one solution, because the lines have exactly one point of intersection. In general, lines in a rectangular coordinate system are related to each other in one of three ways, as illustrated in Example 2.

EXAMPLE 2 Determining the Nature of Solutions

Match each of the following systems with one of the graphs in Figure 4 and discuss the nature of the solutions:

(A) $2x - 3y = 2$
 $x + 2y = 8$

(B) $4x + 6y = 12$
 $2x + 3y = -6$

(C) $2x - 3y = -6$
 $-x + \frac{3}{2}y = 3$

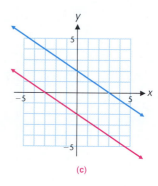

(a) (b) (c)

> **Figure 4**

SOLUTIONS (A) Write each equation in slope–intercept form:

$$2x - 3y = 2 \qquad\qquad x + 2y = 8$$
$$-3y = -2x + 2 \qquad\qquad 2y = -x + 8$$
$$y = \tfrac{2}{3}x - \tfrac{2}{3} \qquad\qquad y = -\tfrac{1}{2}x + 4 \qquad \textcolor{red}{\text{One positive slope, one negative}}$$

The graphs of these two lines match graph (b). There is exactly one solution: $x = 4$, $y = 2$.

(B) $4x + 6y = 12 \qquad\qquad 2x + 3y = -6$
$$6y = -4x + 12 \qquad\qquad 3y = -2x - 6 \qquad \textcolor{red}{\text{Slopes are equal.}}$$
$$y = -\tfrac{2}{3}x + 2 \qquad\qquad y = -\tfrac{2}{3}x - 2$$

The graphs of these parallel lines match graph (c). There is no solution.

(C) $2x - 3y = -6 \qquad\qquad -x + \tfrac{3}{2}y = 3$
$$-3y = -2x - 6 \qquad\qquad \tfrac{3}{2}y = x + 3$$
$$y = \tfrac{2}{3}x + 2 \qquad\qquad y = \tfrac{2}{3}x + 2 \qquad \textcolor{red}{\text{Same line!}}$$

The graph of these identical lines match graph (a). There are an infinite number of solutions.

MATCHED PROBLEM 2 Solve each of the following systems by graphing:

(A) $2x + 3y = 12$
 $x - 3y = -3$

(B) $x - 3y = -3$
 $-2x + 6y = 12$

(C) $2x - 3y = 12$
 $-x + \frac{3}{2}y = -6$

Next, we'll define some terms that can be used to describe the different types of solutions to systems of equations illustrated in Example 2.

> **SYSTEMS OF LINEAR EQUATIONS: BASIC TERMS**

A system of linear equations is **consistent** if it has one or more solutions and **inconsistent** if no solutions exist. Furthermore, a consistent system is said to be **independent** if it has exactly one solution (often referred to as the **unique solution**) and **dependent** if it has more than one solution.

Referring to the three systems in Example 2, the system in part A [Fig. 4(b)] is consistent and independent, with the unique solution $x = 4$ and $y = 2$. The system in part B [Fig. 4(c)] is inconsistent, with no solution. And the system in part C [Fig. 4(a)] is consistent and dependent, with an infinite number of solutions: all the points on the two coinciding lines.

>>> **EXPLORE-DISCUSS 1**

Can a consistent and dependent linear system have exactly two solutions? Exactly three solutions? Explain.

In general, any two lines in a rectangular coordinate plane either intersect in exactly one point, or are parallel, or coincide (have identical graphs). So, the systems in Example 2 illustrate the only three possible types of solutions for systems of two linear equations in two variables. These ideas are summarized in Theorem 1.

> **THEOREM 1 Possible Solutions to a Linear System**

A system of linear equations must have

1. Exactly one solution Consistent and independent

 or

2. No solution Inconsistent

 or

3. Infinitely many solutions Consistent and dependent

Note: While the geometric discussion presented here only applies to systems of equations with two variables, the same three possibilities remain for systems of linear equations with more than two variables.

> ## Solving by Substitution

The accuracy of solutions found by graphing depends a lot on how accurate the graph is when the graphs are drawn by hand. If the solutions are found using a graphing calculator, you will likely get very accurate solutions, but they probably won't be exact. Worse still, the solutions can be very difficult to find, depending on the window settings that you choose. Also, for systems with more than two variables, the geometry gets extremely complicated. For all of these reasons, we will next turn our attention to solving systems algebraically. There are a number of different techniques that can be used. One of the simplest is the **substitution method**.

We will return to the coffee shop problem from page 626 to illustrate the substitution method.

EXAMPLE 3 **Solving a System by Substitution**

Use substitution to solve the coffee shop problem: $x + y = 7$
$2x + 3y = 18$

SOLUTION **Step 1:** Solve either equation for one variable. It will be easy to solve the first equation for y in terms of x:

$$x + y = 7 \qquad \text{Solve the first equation for } y \text{ in terms of } x.$$
$$y = 7 - x \qquad \text{Substitute into the second equation.}$$

Step 2: Substitute $7 - x$ for y in the second equation.

$$2x + 3y = 18 \qquad y = 7 - x, \text{ so replace } y \text{ with } 7 - x.$$
$$2x + 3(7 - x) = 18 \qquad \text{Multiply out parentheses.}$$
$$2x + 21 - 3x = 18 \qquad \text{Collect } x \text{ terms on the left and constant terms on the right.}$$
$$-x = -3 \qquad \text{Multiply both sides by } -1.$$
$$x = 3$$

Step 3: Replace x with 3 in $y = 7 - x$:

$$y = 7 - x$$
$$y = 7 - 3$$
$$y = 4$$

The solution is 3 muffins and 4 lattes, as we found and checked earlier.

MATCHED PROBLEM 3 Solve by substitution and check: $x - y = 3$
$x + 2y = -3$

The following box summarizes the steps for solving a system using the substitution method.

> **› SOLVING SYSTEMS OF TWO LINEAR EQUATIONS IN TWO VARIABLES: THE SUBSTITUTION METHOD**
>
> **1.** Choose one of the two equations and solve it for one of the two variables. (Make a choice that avoids fractions, if possible.)
>
> **2.** Substitute the result of step 1 into the equation that was not used in step 1 and solve the resulting linear equation in one variable.
>
> **3.** Substitute the result of step 2 into the expression obtained in step 1 to find the value of the second variable.

››› EXPLORE-DISCUSS 2 Use substitution to solve each of the following systems. Discuss the nature of the solution sets you obtain.

$$x + 3y = 4 \qquad x + 3y = 4$$
$$2x + 6y = 7 \qquad 2x + 6y = 8$$

› Solving Using Elimination by Addition

Now we turn to **elimination by addition.** This is probably the most important method of solution, since it is readily generalized to larger systems. The method involves the replacement of systems of equations with simpler *equivalent systems,* by performing appropriate operations, until we obtain a system with an obvious solution. **Equivalent systems** of equations are, as you would expect, systems that have exactly the same solution set. Theorem 2 lists operations that produce equivalent systems.

> **THEOREM 2** Elementary Equation Operations Producing Equivalent Systems

A system of linear equations is transformed into an equivalent system if:

1. Two equations are interchanged.

2. An equation is multiplied by a nonzero constant.

3. A constant multiple of another equation is added to a given equation.

We'll return one more time to the coffee shop problem to illustrate why elimination by addition works so well. The system of equations was

$$x + y = 7$$
$$2x + 3y = 18$$

Notice that if we use the third operation in Theorem 2, adding -2 times the first equation to the second one, we get

$$-2x - 2y = -14$$
$$\underline{2x + 3y = 18}$$
$$y = 4$$

This eliminated x, and left behind an equation with only y. We could then easily substitute back in to find x.

We will rely mostly on operations 2 and 3 for now, but operation 1 will come in especially handy later in the section. Examples 4 and 5 illustrate the use of elimination by addition on two and three variable systems.

| EXAMPLE | 4 | Solving a System Using Elimination by Addition |

Solve using elimination by addition: $3x - 2y = 8$
$2x + 5y = -1$

SOLUTION We will use Theorem 2 to eliminate one of the variables and get an easy equation with one variable.

$$3x - 2y = 8$$
$$2x + 5y = -1$$

If we multiply the top equation by 5, the bottom by 2, and then add, we can eliminate y.

$$15x - 10y = 40$$
$$\underline{4x + 10y = -2}$$
$$19x = 38$$

Now solve for x.

$$x = 2$$

The equation $x = 2$ paired with either of the two original equations produces an equivalent system. So, we can substitute $x = 2$ back into either of the two original equations to solve for y. We choose the second equation.

$$2(2) + 5y = -1$$
$$5y = -5$$
$$y = -1$$

SOLUTION $x = 2, y = -1,$ or $(2, -1).$

CHECK

$$\begin{array}{cc} 3x - 2y = 8 & 2x + 5y = -1 \\ 3(2) - 2(-1) \overset{?}{=} 8 & 2(2) + 5(-1) \overset{?}{=} -1 \\ 8 \overset{\checkmark}{=} 8 & -1 \overset{\checkmark}{=} -1 \end{array}$$

MATCHED PROBLEM 4

Solve using elimination by addition: $6x + 3y = 3$
$5x + 4y = 7$

When a system has three equations, we will use elimination to reduce to a system with two equations and two variables, then solve like we did in Example 4. To help you follow a solution, we will number the equations as E_1, E_2, and so on.

EXAMPLE 5 Solution Using Elimination by Addition

$$\begin{array}{ll} x + 2y + 3z = 2 & E_1 \\ 3x - 5y - 4z = 15 & E_2 \\ -2x - 3y + 2z = 2 & E_3 \end{array}$$

SOLUTION Since the coefficient of x in E_1 is 1, our calculations will be simplified if we use E_1 to eliminate x from the other equations. First we eliminate x from E_2 by multiplying E_1 by -3 and adding the result to E_2.

Equivalent System

$$\begin{array}{lll} -3x - 6y - 9z = -6 & -3E_1 \\ \underline{3x - 5y - 4z = 15} & E_2 \\ -11y - 13z = 9 & E_4 \end{array} \qquad \begin{array}{ll} x + 2y + 3z = 2 & E_1 \\ -11y - 13z = 9 & E_4 \\ -2x - 3y + 2z = 2 & E_3 \end{array}$$

Now we use E_1 to eliminate x (the same variable eliminated above) from E_3 by multiplying E_1 by 2 and adding the result to E_3.

Equivalent System

$$\begin{array}{lll} 2x + 4y + 6z = 4 & 2E_1 \\ \underline{-2x - 3y + 2z = 2} & E_3 \\ y + 8z = 6 & E_5 \end{array} \qquad \begin{array}{ll} x + 2y + 3z = 2 & E_1 \\ -11y - 13z = 9 & E_4 \\ y + 8z = 6 & E_5 \end{array}$$

Notice that E_4 and E_5 form a system of two equations with two variables. Next we use E_5 to eliminate y from E_4 and replace E_4 with the result.

$$\begin{array}{ll} 11y + 88z = 66 & 11E_5 \\ \underline{-11y - 13z = 9} & E_4 \\ 75z = 75 & E_6 \end{array}$$

Now we can easily solve for z.

$$75z = 75 \qquad E_6$$
$$z = 1$$

Next substitute $z = 1$ in E_4 or E_5 and solve for y.

$$y + 8z = 6 \qquad E_5$$
$$y + 8(1) = 6$$
$$y = -2$$

Finally, substitute $y = -2$ and $z = 1$ in any of E_1, E_2, or E_3 and solve for x.

$$x + 2y + 3z = 2 \qquad E_1$$
$$x + 2(-2) + 3(1) = 2$$
$$x = 3$$

The solution to the original system is $(3, -2, 1)$ or $x = 3$, $y = -2$, $z = 1$.

CHECK To check the solution, we must check each equation in the original system:

$$
\begin{array}{lll}
x + 2y + 3z = 2 \quad E_1 & 3x - 5y - 4z = 15 \quad E_2 & -2x - 3y + 2z = 2 \quad E_3 \\
3 + 2(-2) + 3(1) \stackrel{?}{=} 2 & 3(3) - 5(-2) - 4(1) \stackrel{?}{=} 15 & -2(3) - 3(-2) + 2(1) \stackrel{?}{=} 2 \\
2 \stackrel{\checkmark}{=} 2 & 15 \stackrel{\checkmark}{=} 15 & 2 \stackrel{\checkmark}{=} 2
\end{array}
$$

MATCHED PROBLEM 5	Solve:

$$2x + 3y - 5z = -12$$
$$3x - 2y + 2z = 1$$
$$4x - 5y - 4z = -12$$

Let's see what happens in the solution process when a system either has no solution or has infinitely many solutions. Consider the solutions to the following system:

$$2x + 6y = -3$$
$$x + 3y = 2$$

Solution by Substitution

Solve the second equation for x and substitute in the first equation.

$$x = 2 - 3y$$
$$2(2 - 3y) + 6y = -3$$
$$4 - 6y + 6y = -3$$
$$4 = -3$$

Solution by Elimination

Multiply the second equation by -2 and add to the first equation.

$$
\begin{array}{r}
2x + 6y = -3 \\
-2x - 6y = -4 \\
\hline
0 = -7
\end{array}
$$

Both methods of solution lead to a contradiction (a statement that is false). An assumption that the original system has solutions must be false. This tells us that the system has no solution. The graphs of the equations are parallel and the system is inconsistent.

Now consider the system

$$x - \tfrac{1}{2}y = 4$$
$$-2x + y = -8$$

Solution by Substitution	**Solution by Elimination**
Solve the first equation for x and substitute in the second equation.	Multiply the first equation by 2 and add to the second equation.

$$x = \tfrac{1}{2}y + 4$$
$$-2\left(\tfrac{1}{2}y + 4\right) + y = -8$$
$$-y - 8 + y = -8$$
$$-8 = -8$$

$$\begin{aligned} 2x - y &= \ \ 8 \\ -2x + y &= -8 \\ \hline 0 &= \ \ 0 \end{aligned}$$

This time both solution methods lead to a statement that is always true. This means that the two original equations are equivalent. That is, their graphs coincide. The system is dependent and has an infinite number of solutions. There are many different ways to represent this infinite solution set. For example,

$$S_1 = \{(x, y) \mid y = 2x - 8,\ x \text{ any real number}\}$$

and

$$S_2 = \{(x, y) \mid x = \tfrac{1}{2}y + 4,\ y \text{ any real number}\}$$

both represent the solutions to this system. For reasons that will become apparent later, it is customary to introduce a new variable, called a **parameter,** and express both variables in terms of this new variable. If we let $x = s$ and $y = 2s - 8$ in S_1, we can express the solution set as

$$\{(s, 2s - 8) \mid s \text{ any real number}\}$$

Some **particular solutions** to this system are obtained by choosing particular values for the parameter.

$s = -1$	$s = 2$	$s = 5$	$s = 9.4$
$(-1, -10)$	$(2, -4)$	$(5, 2)$	$(9.4, 10.8)$

EXAMPLE	**6**	**Using Elimination by Addition**

Solve:

$$\begin{aligned} x + \ y + \ \ z &= 3 & E_1 \\ x - \ y - 5z &= 1 & E_2 \\ 2x + 3y + 5z &= 6 & E_3 \end{aligned}$$

SOLUTION Use E_1 to eliminate z from E_2 and replace E_2 with the result.

Equivalent System

$$\begin{aligned} 5x + 5y + 5z &= 15 & 5E_1 \\ x - \ y - 5z &= \ \ 1 & E_2 \\ \hline 6x + 4y \ \ \ \ \ \ &= 16 & E_4 \end{aligned}$$

$$\begin{aligned} x + \ y + \ z &= \ \ 3 & E_1 \\ 6x + 4y \ \ \ \ \ \ &= 16 & E_4 \\ 2x + 3y + 5z &= \ \ 6 & E_3 \end{aligned}$$

Use E_1 to eliminate z from E_3 and replace E_3 with the result.

Equivalent System

$$\begin{aligned} -5x - 5y - 5z &= -15 & -5E_1 \\ 2x + 3y + 5z &= \ \ 6 & E_3 \\ \hline -3x - 2y \ \ \ \ \ \ &= \ \ -9 & E_5 \end{aligned}$$

$$\begin{aligned} x + \ y + z &= \ \ 3 & E_1 \\ 6x + 4y \ \ \ \ \ &= 16 & E_4 \\ -3x - 2y \ \ \ \ \ &= -9 & E_5 \end{aligned}$$

Now treat E_4 and E_5 as a system of two equations, and eliminate y.

$$
\begin{array}{rrl}
6x + 4y = & 16 & E_4 \\
-6x - 4y = & -18 & 2E_5 \\
\hline
0 = & -2 & E_6
\end{array}
$$

Stop! We have obtained a contradiction. The original system is inconsistent and has no solution. (*Note:* It's impossible to check in this case.)

MATCHED PROBLEM 6

Solve:

$$
\begin{aligned}
2x + 3y - 5z &= 3 \\
3x - 2y + 2z &= 2 \\
x - 5y + 7z &= 1
\end{aligned}
$$

EXAMPLE 7 **Using Elimination by Addition**

Solve:

$$
\begin{array}{rl}
x + y + z = 1 & E_1 \\
2x + y - z = 3 & E_2 \\
3x + y - 3z = 5 & E_3
\end{array}
$$

SOLUTION Use E_1 to eliminate y from E_2 and replace E_2 with the result.

Equivalent System

$$
\begin{array}{rrl}
-x - y - z = & -1 & -E_1 \\
2x + y - z = & 3 & E_2 \\
\hline
x \quad\ - 2z = & 2 & E_4
\end{array}
\qquad
\begin{array}{rl}
x + y + z = 1 & E_1 \\
x \quad\ - 2z = 2 & E_4 \\
3x + y - 3z = 5 & E_3
\end{array}
$$

Use E_1 to eliminate y from E_3 and replace E_3 with the result.

Equivalent System

$$
\begin{array}{rrl}
-x - y - z = & -1 & -E_1 \\
3x + y - 3z = & 5 & E_3 \\
\hline
2x \quad\ - 4z = & 4 & E_5
\end{array}
\qquad
\begin{array}{rl}
x + y + z = 1 & E_1 \\
x \quad\ - 2z = 2 & E_4 \\
2x \quad\ - 4z = 4 & E_5
\end{array}
$$

Use E_4 to eliminate z from E_5 and replace E_5 with the result.

Equivalent System

$$
\begin{array}{rrl}
-2x \quad\ + 4z = & -4 & -2E_4 \\
2x \quad\ - 4z = & 4 & E_5 \\
\hline
0 = & 0 & E_6
\end{array}
\qquad
\begin{array}{rl}
x + y + z = 1 & E_1 \\
x \quad\ - 2z = 2 & E_4
\end{array}
$$

Since E_6 is true for all x, y, and z, it provides no information about the systems' solution set and can be discarded. The solutions to the last equivalent system can be described by introducing a parameter. If we let $z = s$, then, using E_4, we can write $x = 2s + 2$. Substituting for x and z in E_1 and solving for y, we have

$$
\begin{array}{rl}
x + y + z = 1 & E_1 \\
2s + 2 + y + s = 1 & \\
y = -3s - 1 &
\end{array}
$$

The solution set is given by

$$\{(2s + 2, -3s - 1, s)|s \text{ any real number}\}$$

The check is left to the reader.

MATCHED PROBLEM 7

Solve:

$$
\begin{aligned}
3x + 2y + 4z &= 5 \\
2x + y + 5z &= 2 \\
x \quad\quad + 6z &= -1
\end{aligned}
$$

>>> EXPLORE-DISCUSS 3

Refer to the solution to Example 7. The given representation of the solution set is not the only one. Which of the following is a representation of the solution set? Justify your answer.

(A) $\{(t, 2 - 1.5t, 0.5t - 1) \mid t \text{ any real number}\}$

(B) $\{(2u + 4, -2u - 3, u) \mid u \text{ any real number}\}$

Let $y = v$, where v is any real number, express x and z in terms of v, and find another representation of the solution set for Example 7.

› Applications

Examples 8–10 illustrate the advantages of using systems of equations in solving word problems.

EXAMPLE 8

Airspeed

An airplane makes the 2,400-mile trip from Washington, D.C. to San Francisco in 7.5 hours and makes the return trip in 6 hours. Assuming that the plane travels at a constant airspeed and that the wind blows at a constant rate from west to east, find the plane's airspeed and the wind rate.

SOLUTION

Let x represent the airspeed of the plane and let y represent the rate at which the wind is blowing (both in miles per hour). The plane's speed relative to the ground is determined by combining these two rates; that is,

$$x - y = \text{Ground speed flying east to west (airspeed} - \text{wind)}$$
$$x + y = \text{Ground speed flying west to east (airspeed} + \text{wind)}$$

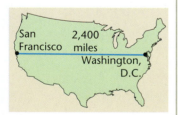

Applying the familiar formula $D = RT$ to each leg of the trip leads to the following system of equations:

$$2,400 = 7.5(x - y) \qquad \text{Washington to San Francisco: 7.5 hr, 2,400 mi}$$
$$2,400 = 6(x + y) \qquad \text{San Francisco to Washington: 6 hr, 2,400 mi}$$

After simplification, we have

$$
\begin{aligned}
x - y &= 320 \\
x + y &= 400
\end{aligned}
$$

Add these two equations to eliminate y:

$$2x = 720$$

$$x = 360 \text{ mph} \qquad \text{Airspeed}$$

Substitute for x in the second equation:

$$x + y = 400$$
$$360 + y = 400$$
$$y = \textbf{40 mph} \quad \text{Wind rate}$$

CHECK
$$2{,}400 = 7.5(x - y) \qquad\qquad 2{,}400 = 6(x + y)$$
$$2{,}400 \stackrel{?}{=} 7.5(360 - 40) \qquad 2{,}400 \stackrel{?}{=} 6(360 + 40)$$
$$2{,}400 \stackrel{\checkmark}{=} 2{,}400 \qquad\qquad 2{,}400 \stackrel{\checkmark}{=} 2{,}400$$

MATCHED PROBLEM 8

A boat takes 8 hours to travel 80 miles upstream and 5 hours to return to its starting point. Find the speed of the boat in still water and the speed of the current.

The quantity of a product that people are willing to buy (known as the *demand*) during some period of time depends on its price. Generally, the higher the price, the less the demand; the lower the price, the greater the demand. Similarly, the quantity of a product that a supplier is willing to sell during some period of time (known as the *supply*) also depends on the price. Generally, a supplier will be willing to supply more of a product at higher prices and less of a product at lower prices. The simplest supply and demand model is a linear model.

If the demand for a product is greater than the supply, the price tends to rise. If the demand is less than the supply, the price tends to fall. So the price tends to stabilize at an **equilibrium price;** at that price, the supply and demand are equal, and that common quantity is called the **equilibrium quantity.** Example 9 illustrates the basic concepts of supply and demand.

EXAMPLE 9

Supply and Demand

Using collected data and regression analysis, an analyst arrives at the following price–demand and price–supply equations for the sale of cherries each day in a major urban area.

$$p = -0.2q + 5.6 \quad \text{Demand equation (consumer)}$$
$$p = 0.1q + 1.7 \quad \text{Supply equation (supplier)}$$

where q represents the quantity of cherries in thousands of pounds and p represents the price in dollars per pound. For example, we see (Fig. 5) that consumers will purchase 11 thousand pounds ($q = 11$) when the price is $p = -0.2(11) + 5.6 = \$3.40$ per pound. On the other hand, suppliers will be willing to supply 17 thousand pounds of cherries at $3.40 per pound (solve $3.4 = 0.1q + 1.7$ for q). So, at $3.40 per pound the suppliers are willing to supply more cherries than the consumers are willing to purchase. The supply exceeds the demand at that price, and the price will come down. Find the equilibrium quantity and the equilibrium price.

> Figure 5

SOLUTION To find the equilibrium quantity, we solve the linear system

$$p = -0.2q + 5.6 \qquad \text{Demand equation (consumer)}$$
$$p = \ \ \ 0.1q + 1.7 \qquad \text{Supply equation (supplier)}$$

using substitution (substituting $p = -0.2q + 5.6$ into the second equation).

$$p = 0.1q + 1.7 \qquad \text{Substitute } p = -0.2q + 5.6.$$
$$-0.2q + 5.6 = 0.1q + 1.7 \qquad \text{Add } 0.2q \text{ to both sides.}$$
$$5.6 = 0.3q + 1.7 \qquad \text{Subtract 1.7 from both sides.}$$
$$3.9 = 0.3q \qquad \text{Divide both sides by 0.3.}$$
$$\boldsymbol{q = 13 \text{ thousand pounds}} \qquad \text{Equilibrium quantity}$$

Now substitute $q = 13$ back into either of the original equations in the system and solve for p (we choose the second equation):

$$p = 0.1(13) + 1.7$$
$$\boldsymbol{p = \$3 \text{ per pound}} \qquad \text{Equilibrium price}$$

So if the price of cherries is $3 per pound, then the supplier would supply 13,000 pounds of cherries and the consumer would demand (purchase) 13,000 pounds of cherries. In other words, the market would be in equilibrium (see Fig. 6).

> Figure 6

MATCHED PROBLEM 9 The price–demand and price–supply equations for strawberries in a certain city are

$$p = -0.2q + 4 \qquad \text{Demand equation}$$
$$p = 0.04q + 1.84 \qquad \text{Supply equation}$$

where q represents the quantity in thousands of pounds and p represents the price in dollars. Find the equilibrium quantity and the equilibrium price.

EXAMPLE 10 **Production Scheduling**

A garment industry manufactures three shirt styles. Each style shirt requires the services of three departments as listed in the table. The cutting, sewing, and packaging departments have available a maximum of 1,160, 1,560, and 480 labor-hours per week, respectively. How many of each style shirt must be produced each week for the plant to operate at full capacity?

	Style A	Style B	Style C	Time Available
Cutting department	0.2 hr	0.4 hr	0.3 hr	1,160 hr
Sewing department	0.3 hr	0.5 hr	0.4 hr	1,560 hr
Packaging department	0.1 hr	0.2 hr	0.1 hr	480 hr

SOLUTION Let

$$x = \text{Number of style A shirts produced per week}$$
$$y = \text{Number of style B shirts produced per week}$$
$$z = \text{Number of style C shirts produced per week}$$

Then

$$0.2x + 0.4y + 0.3z = 1{,}160 \qquad \text{Cutting department}$$
$$0.3x + 0.5y + 0.4z = 1{,}560 \qquad \text{Sewing department}$$
$$0.1x + 0.2y + 0.1z = \ \ \ 480 \qquad \text{Packaging department}$$

We can clear the system of decimals by multiplying each side of each equation by 10:

$$2x + 4y + 3z = 11{,}600 \qquad E_1$$
$$3x + 5y + 4z = 15{,}600 \qquad E_2$$
$$x + 2y + z = 4{,}800 \qquad E_3$$

Use E_3 to eliminate z from E_1 and replace E_1 with the result.

<center>**Equivalent System**</center>

$$
\begin{array}{ll}
2x + 4y + 3z = 11{,}600 & E_1 \\
-3x - 6y - 3z = -14{,}400 & -3E_3 \\
\hline
-x - 2y = -2{,}800 & E_4
\end{array}
\qquad
\begin{array}{ll}
-x - 2y = -2{,}800 & E_4 \\
3x + 5y + 4z = 15{,}600 & E_2 \\
x + 2y + z = 4{,}800 & E_3
\end{array}
$$

Use E_3 to eliminate z from E_2 and replace E_2 with the result.

<center>**Equivalent System**</center>

$$
\begin{array}{ll}
3x + 5y + 4z = 15{,}600 & E_2 \\
-4x - 8y - 4z = -19{,}200 & -4E_3 \\
\hline
-x - 3y = -3{,}600 & E_5
\end{array}
\qquad
\begin{array}{ll}
-x - 2y = -2{,}800 & E_4 \\
-x - 3y = -3{,}600 & E_5 \\
x + 2y + z = 4{,}800 & E_3
\end{array}
$$

Now treat E_4 and E_5 as a system of two equations; eliminate x.

$$
\begin{array}{ll}
x + 2y = 2{,}800 & -E_4 \\
-x - 3y = -3{,}600 & E_5 \\
\hline
-y = {-800} & E_6
\end{array}
$$

From E_6 we see that

$$y = 800$$

Substitute $y = 800$ in E_4 or E_5 and solve for x.

$$
\begin{array}{ll}
-x - 2y = -2{,}800 & E_4 \\
-x - 2(800) = -2{,}800 & \\
x = 1{,}200 &
\end{array}
$$

Substitute $x = 1{,}200$ and $y = 800$ in E_1, E_2, or E_3 and solve for z.

$$
\begin{array}{ll}
x + 2y + z = 4{,}800 & E_3 \\
1{,}200 + 2(800) + z = 4{,}800 & \\
z = 2{,}000 &
\end{array}
$$

Each week, the company should produce 1,200 style A shirts, 800 style B shirts, and 2,000 style C shirts to operate at full capacity. You should check this solution.

MATCHED PROBLEM 10

Repeat Example 10 with the cutting, sewing, and packaging departments having available a maximum of 1,180, 1,560, and 510 labor-hours per week, respectively.

ANSWERS TO MATCHED PROBLEMS

1.

$x = 1, y = -2$

Check:
$$x - y = 3$$
$$1 - (-2) \overset{?}{=} 3$$
$$3 \overset{\checkmark}{=} 3$$
$$x + 2y = -3$$
$$1 + 2(-2) \overset{?}{=} -3$$
$$-3 \overset{\checkmark}{=} -3$$

2. (A) (3, 2) or $x = 3$ and $y = 2$ (B) No solutions (C) Infinite number of solutions

3. $x = 1, y = -2$

4. $x = -1, y = 3$

5. $(-1, 0, 2)$ or $x = -1, y = 0, z = 2$

6. Inconsistent system with no solution

7. $\{(-6s - 1, 7s + 4, s) \mid s$ any real number$\}$

8. Boat: 13 mph; current: 3 mph

9. Equilibrium quantity = 9 thousand pounds; Equilibrium price = \$2.20 per pound

10. Each week, the company should produce 900 style A shirts, 1,300 style B shirts, and 1,600 style C shirts to operate at full capacity.

10-1 Exercises

1. Explain in your own words how to solve a system of two linear equations by graphing.

2. Explain in your own words how to solve a system of two linear equations by substitution.

3. Explain in your own words how to solve a system of two linear equations using elimination by addition.

4. Which of the three solving techniques is the best choice for a system of three equations? Why?

5. Can a system of two linear equations have exactly two solutions? Explain.

6. Describe how the solution sets differ for systems of linear equations that are consistent, inconsistent, and dependent.

Match each system in Problems 7–10 with one of the following graphs, and use the graph to solve the system.

7. $2x - 4y = 8$
 $x - 2y = 0$

8. $x + y = 3$
 $x - 2y = 0$

9. $2x - y = 5$
 $3x + 2y = -3$

10. $4x - 2y = 10$
 $2x - y = 5$

(a)

(b)

(c)

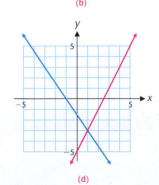

(d)

Solve the system of equations in Problems 11–46.

11. $x + y = 7$
$x - y = 3$

12. $x - y = 2$
$x + y = 4$

13. $3x - 2y = 12$
$7x + 2y = 8$

14. $3x - y = 2$
$x + 2y = 10$

15. $3u + 5v = 15$
$6u + 10v = -30$

16. $m + 2n = 4$
$2m + 4n = -8$

17. $3x - y = -2$
$-9x + 3y = 6$

18. $2x - 8y = 10$
$8x - 32y = 40$

19. $x - y = 4$
$x + 3y = 12$

20. $3x - y = 7$
$2x + 3y = 1$

21. $4x + 3y = 26$
$3x - 11y = -7$

22. $9x - 3y = 24$
$11x + 2y = 1$

23. $7m + 12n = -1$
$5m - 3n = 7$

24. $3p + 8q = 4$
$15p + 10q = -10$

25. $y = 0.08x$
$y = 100 + 0.04x$

26. $0.2u - 0.5v = 0.07$
$0.8u - 0.3v = 0.79$

27. $\frac{2}{5}x + \frac{3}{2}y = 2$
$\frac{7}{3}x - \frac{5}{4}y = -5$

28. $5x - 2y = 8$
$2x + 3y = -10$

29. $-2.3y + 4.1z = -14.21$
$10.1y - 2.9z = 26.15$

30. $5.4x + 4.2y = -12.9$
$3.7x + 6.4y = -4.5$

31. $-2x = 2$
$x - 3y = 2$
$-x + 2y + 3z = -7$

32. $2y + z = -4$
$x - 3y + 2z = 9$
$-y = 3$

33. $2y - z = 2$
$-4y + 2z = 1$
$x - 2y + 3z = 0$

34. $x + y - z = 3$
$x - 2z = 1$
$y + z = 2$

35. $x - 3y = 2$
$2y + z = -1$
$x - y + z = 1$

36. $-4x + 3y = 1$
$8x - 6y = 4$
$2x - 4y + 3z = 6$

37. $2x + z = -5$
$x - 3z = -6$
$4x + 2y - z = -9$

38. $x - 3y + z = 4$
$-x + 4y - 4z = 1$
$2x - y + 5z = -3$

39. $x - y + z = 1$
$2x + y + z = 6$
$7x - y + 5z = 15$

40. $2x - y + 3z = 7$
$x + 2y - z = -3$
$3x + y + 2z = 2$

41. $2a + 4b + 3c = -6$
$a - 3b + 2c = -15$
$-a + 2b - c = 9$

42. $3u - 2v + 3w = 11$
$2u + 3v - 2w = -5$
$u + 4v - w = -5$

43. $2x - 3y + 3z = -5$
$3x + 2y - 5z = 34$
$5x - 4y - 2z = 23$

44. $x + 2y + z = 2$
$-2x + 3y - 2z = -3$
$x - 5y + z = 2$

45. $-x + 2y - z = -4$
$2x + 5y - 4z = -16$
$x + y - z = -4$

46. $x - 8y + 2z = -1$
$x - 3y + z = 1$
$2x - 11y + 3z = 2$

In Problems 47 and 48, solve each system for p and q in terms of x and y. Explain how you could check your solution and then perform the check.

47. $x = 2 + p - 2q$
$y = 3 - p + 3q$

48. $x = -1 + 2p - q$
$y = 4 - p + q$

Problems 49 and 50 refer to the system

$$ax + by = h$$
$$cx + dy = k$$

where x and y are variables and a, b, c, d, h, and k are real constants.

49. Solve the system for x and y in terms of the constants a, b, c, d, h, and k. Clearly state any assumptions you must make about the constants during the solution process.

50. Discuss the nature of solutions to systems that do not satisfy the assumptions you made in Problem 49.

APPLICATIONS

51. **AIRSPEED** It takes a private airplane 8.75 hours to make the 2,100-mile flight from Atlanta to Los Angeles and 5 hours to make the return trip. Assuming that the wind blows at a constant rate from Los Angeles to Atlanta, find the airspeed of the plane and the wind rate.

52. **AIRSPEED** A plane carries enough fuel for 20 hours of flight at an airspeed of 150 miles per hour. How far can it fly into a 30 mph headwind and still have enough fuel to return to its starting point? (This distance is called the *point of no return.*)

53. **RATE–TIME** A crew of eight can row 20 kilometers per hour in still water. The crew rows upstream and then returns to its starting point in 15 minutes. If the river is flowing at 2 km/h, how far upstream did the crew row?

54. **RATE–TIME** It takes a boat 2 hours to travel 20 miles down a river and 3 hours to return upstream to its starting point. What is the rate of the current in the river?

55. **BUSINESS** A company that supplies bulk candy to bakeries has one batch of chocolate chips that are 50% dark chocolate and 50% milk chocolate. They have another batch that is 80% dark chocolate and 20% milk chocolate. One of their customers sends in a rush order for 100 lb of a mix that is 68% dark chocolate. How many pounds from each batch should be mixed to meet this order?

56. **BUSINESS** A jeweler has two bars of gold alloy in stock, one of 12 carats and the other of 18 carats (24-carat gold is pure gold, 12-carat is $\frac{12}{24}$ pure, 18-carat gold is $\frac{18}{24}$ pure, and so on). How many grams of each alloy must be mixed to obtain 10 grams of 14-carat gold?

57. BREAK-EVEN ANALYSIS It costs a small recording company $17,680 to prepare a compact disc. This is a one-time fixed cost that covers recording, package design, and so on. Variable costs, including such things as manufacturing, marketing, and royalties, are $4.60 per CD. If the CD is sold to music shops for $8 each, how many must be sold for the company to break even?

58. FINANCE Suppose you have $12,000 to invest. If part is invested at 10% and the rest at 15%, how much should be invested at each rate to yield 12% on the total amount invested?

59. PRODUCTION A supplier for the electronics industry manufactures keyboards and screens for graphing calculators at plants in Mexico and Taiwan. The hourly production rates at each plant are given in the table. How many hours should each plant be operated to fill an order for exactly 4,000 keyboards and exactly 4,000 screens?

Plant	Keyboards	Screens
Mexico	40	32
Taiwan	20	32

60. PRODUCTION A company produces Italian sausages and bratwursts at plants in Green Bay and Sheboygan. The hourly production rates at each plant are given in the table. How many hours should each plant be operated to exactly fill an order for 62,250 Italian sausages and 76,500 bratwursts?

Plant	Italian Sausage	Bratwurst
Green Bay	800	800
Sheboygan	500	1,000

61. SUPPLY AND DEMAND Suppose the supply and demand equations for printed T-shirts in a resort town for a particular week are

$$p = 0.007q + 3 \quad \text{Supply equation}$$
$$p = -0.018q + 15 \quad \text{Demand equation}$$

where p is the price in dollars and q is the quantity.
(A) Find the supply and the demand (to the nearest unit) if T-shirts are priced at $4 each. Discuss the stability of the T-shirt market at this price level.
(B) Find the supply and the demand (to the nearest unit) if T-shirts are priced at $8 each. Discuss the stability of the T-shirt market at this price level.
(C) Find the equilibrium price and quantity.
(D) Graph the two equations in the same coordinate system and identify the equilibrium point, supply curve, and demand curve.

62. SUPPLY AND DEMAND Suppose the supply and demand equations for printed baseball caps in a resort town for a particular week are

$$p = 0.006q + 2 \quad \text{Supply equation}$$
$$p = -0.014q + 13 \quad \text{Demand equation}$$

where p is the price in dollars and q is the quantity in hundreds.
(A) Find the supply and the demand (to the nearest unit) if baseball caps are priced at $4 each. Discuss the stability of the baseball cap market at this price level.

(B) Find the supply and the demand (to the nearest unit) if baseball caps are priced at $8 each. Discuss the stability of the baseball cap market at this price level.
(C) Find the equilibrium price and quantity.
(D) Graph the two equations in the same coordinate system and identify the equilibrium point, supply curve, and demand curve.

63. SUPPLY AND DEMAND At $0.60 per bushel, the daily supply for wheat is 450 bushels and the daily demand is 645 bushels. When the price is raised to $0.90 per bushel, the daily supply increases to 750 bushels and the daily demand decreases to 495 bushels. Assume that the supply and demand equations are linear.
(A) Find the supply equation.
(B) Find the demand equation.
(C) Find the equilibrium price and quantity.

64. SUPPLY AND DEMAND At $1.40 per bushel, the daily supply for soybeans is 1,075 bushels and the daily demand is 580 bushels. When the price falls to $1.20 per bushel, the daily supply decreases to 575 bushels and the daily demand increases to 980 bushels. Assume that the supply and demand equations are linear.
(A) Find the supply equation.
(B) Find the demand equation.
(C) Find the equilibrium price and quantity.

65. EARTH SCIENCE An earthquake emits a primary wave and a secondary wave. Near the surface of the Earth the primary wave travels at about 5 miles per second and the secondary wave at about 3 miles per second. From the time lag between the two waves arriving at a given receiving station, it is possible to estimate the distance to the quake. (The *epicenter* can be located by obtaining distance bearings at three or more stations.) Suppose a station measured a time difference of 16 seconds between the arrival of the two waves. How long did each wave travel, and how far was the earthquake from the station?

66. EARTH SCIENCE A ship using sound-sensing devices above and below water recorded a surface explosion 6 seconds sooner by its underwater device than its above-water device. Sound travels in air at about 1,100 feet per second and in seawater at about 5,000 feet per second.
(A) How long did it take each sound wave to reach the ship?
(B) How far was the explosion from the ship?

67. PRODUCTION SCHEDULING A company manufactures three products; lawn mowers, snowblowers, and chain saws. The labor, material, and shipping costs for manufacturing one unit of each product are given in the table. The weekly allocations for labor, materials, and shipping are $35,000, $50,000, and $20,000, respectively. How many of each type of product should be manufactured each week in order to exactly use the weekly allocations?

Product	Labor	Materials	Shipping
Lawn mower	$20	$35	$15
Snowblower	$30	$50	$25
Chain saw	$45	$40	$10

68. PRODUCTION SCHEDULING A company manufactures three products; desk chairs, file cabinets, and printer stands. The labor, material, and shipping costs for manufacturing one unit of each product are given in the table. The weekly allocations for labor, materials, and shipping are $21,100, $31,500, and $11,900, respec-

tively. How many of each type of product should be manufactured each week in order to exactly use the weekly allocations?

Product	Desk Chair	File Cabinet	Printer Stand
Labor	$30	$35	$40
Materials	$45	$60	$55
Shipping	$25	$20	$15

69. PRODUCTION SCHEDULING A company has plants located in Michigan, New York, and Ohio where it manufactures laptop computers, desktop computers, and servers. The number of units of each product that can be produced per day at each plant are given in the table below. The company has orders for 2,150 laptop computers, 2,300 desktop computers, and 2,500 servers. How many days should the company operate each plant in order to exactly fill these orders?

Plant	Michigan	New York	Ohio
Laptop	10	70	60
Desktop	20	50	80
Server	40	30	90

70. PRODUCTION SCHEDULING A company has plants located in Maine, Utah, and Oregon where it manufactures stoves, refrigerators, and dishwashers. The number of units of each product that can be produced per day at each plant are given in the table. The company has orders for 1,500 stoves, 2,350 refrigerators, and 2,400 dishwashers.

How many days should the company operate each plant in order to exactly fill these orders? Set up a system of equations whose solution would answer this question and solve the system.

Plant	Stoves	Refrigerators	Dishwashers
Maine	30	70	60
Utah	20	50	50
Oregon	40	30	40

71. INVESTMENT Due to recent volatility in the stock market, Catalina's financial advisor suggests that she reallocate $70,000 of her retirement fund to bonds. He recommends a mix of treasury bonds earning 4% annually, municipal bonds earning 3.5% annually, and corporate bonds earning 4.5% annually. For tax reasons, he also recommends that the amount invested in treasury bonds should be equal to the sum of the amount invested in the other categories. If Catalina follows these recommendations, and the goal is to produce $2,900 in annual interest income, how much will she invest in each of the three types of bonds?

72. INVESTMENT When the real estate market begins to rebound, Catalina (see Problem 71) decides to reallocate her investment mix. At this point, her investment has grown to $76,000. She'll leave some money in treasury and corporate bonds, but will replace municipal bonds with a real estate investment trust that guarantees a 6.5% annual return. If she plans to leave as much in treasury bonds as the sum of the other two investments, how much should she invest in each to reach her new goal of earning an annual interest income of $3,600?

10-2	**Solving Systems of Linear Equations Using Gauss–Jordan Elimination**

> **Matrices and Row Operations**
> **Reduced Matrices**
> **Solving Systems by Gauss–Jordan Elimination**
> **Application**

In this section, we introduce *Gauss–Jordan elimination,* a step-by-step procedure for solving systems of linear equations. This procedure works for any system of linear equations and is easily implemented on a computer. In fact, the TI-84 has a built-in procedure for performing Gauss–Jordan elimination.

> ## Matrices and Row Operations

In solving systems of equations using elimination by addition, the coefficients of the variables and the constant terms played a central role. The process can be made more efficient

by the introduction of a mathematical form called a *matrix*. A **matrix** (plural **matrices**) is a rectangular array of numbers written within brackets. Two examples are

$$A = \begin{bmatrix} 1 & -3 & 7 \\ 5 & 0 & -4 \end{bmatrix} \qquad B = \begin{bmatrix} -5 & 4 & 11 \\ 0 & 1 & 6 \\ -2 & 12 & 8 \\ -3 & 0 & -1 \end{bmatrix} \tag{1}$$

Each number in a matrix is called an **element** of the matrix. Matrix A has six elements arranged in two rows and three columns. Matrix B has 12 elements arranged in four rows and three columns. If a matrix has m rows and n columns, it is called an **$m \times n$ matrix** (read "m by n matrix"). The expression $m \times n$ is called the **size** of the matrix, and the numbers m and n are called the **dimensions** of the matrix. It is important to note that the number of rows is always given first. Referring to equations (1), A is a 2×3 matrix and B is a 4×3 matrix. A matrix with n rows and n columns is called a **square matrix of order n.** A matrix with only one column is called a **column matrix,** and a matrix with only one row is called a **row matrix.** These definitions are illustrated by the following:

<div align="center">

3 × 3 **4 × 1** **1 × 4**

$$\begin{bmatrix} 0.5 & 0.2 & 1.0 \\ 0.0 & 0.3 & 0.5 \\ 0.7 & 0.0 & 0.2 \end{bmatrix} \qquad \begin{bmatrix} 3 \\ -2 \\ 1 \\ 0 \end{bmatrix} \qquad \begin{bmatrix} 2 & \frac{1}{2} & 0 & -\frac{2}{3} \end{bmatrix}$$

Square matrix of order 3 Column matrix Row matrix

</div>

The **position** of an element in a matrix is the row and column containing the element. This is usually denoted using **double subscript notation** a_{ij}, where i is the row and j is the column containing the element a_{ij}, as illustrated next:

$$A = \begin{bmatrix} 1 & 5 & -3 & 4 \\ 6 & 0 & -4 & 1 \\ -2 & 3 & 4 & 7 \end{bmatrix} \qquad \begin{array}{l} a_{11} = 1, a_{12} = 5, a_{13} = -3, a_{14} = 4 \\ a_{21} = 6, a_{22} = 0, a_{23} = -4, a_{24} = 1 \\ a_{31} = -2, a_{32} = 3, a_{33} = 4, a_{34} = 7 \end{array}$$

Note that a_{12} is read "a sub one two," not "a sub twelve." The elements $a_{11} = 1$, $a_{22} = 0$, and $a_{33} = 4$ make up the *principal diagonal* of A. In general, the **principal diagonal** of a matrix A consists of the elements a_{ii}, $i = 1, 2, \ldots n$.

Technology Connections

Most graphing calculators are capable of storing and manipulating matrices. Figure 1 shows matrix A displayed in the matrix editing screen of a TI-84 graphing calculator. The size of the matrix is given at the top of the screen, and the position of the currently selected element is given at the bottom. Notice that a comma is used in the notation for the position. This is common practice on graphing calculators but it's almost never written or typed that way.

> **Figure 1** Matrix notation on a TI-84 graphing calculator.*

*The onscreen display of A was too large to fit on the screen of a TI-84, so we pasted together two screen shots to form Figure 1. When this happens on your graphing calculator, you will have to scroll left and right and/or up and down to see the entire matrix.

Now we turn our attention to the connection between matrices and systems of equations. Consider the system of equations

$$\begin{aligned} x + 5y - 3z &= 4 \\ 6x \qquad - 4z &= 1 \\ -2x + 3y + 4z &= 7 \end{aligned} \qquad (2)$$

If we remove the variables and leave behind the numbers, we can think of the result as a matrix:

$$\left[\begin{array}{ccc|c} 1 & 5 & -3 & 4 \\ 6 & 0 & -4 & 1 \\ -2 & 3 & 4 & 7 \end{array}\right]$$

Coefficient matrix

Constant matrix

$$\left[\begin{array}{ccc} 1 & 5 & -3 \\ 6 & 0 & -4 \\ -2 & 3 & 4 \end{array}\right] \qquad \left[\begin{array}{c} 4 \\ 1 \\ 7 \end{array}\right]$$

> **Figure 2**

This is known as the **augmented coefficient matrix** for the system. We can also define the **coefficient matrix** and the **constant matrix** for the system, as shown in Figure 2. The augmented coefficient matrix contains all of the information about the system needed to solve it. Note that we put in a coefficient of zero for the missing y in the second equation, and that we drew a vertical bar to separate the coefficients from the constants. (Matrices displayed on a graphing calculator won't have that line.)

Since we would like to be able to use matrices to solve large systems with many variables, moving forward we will use x_1, x_2, x_3, and the like, rather than x, y, z, and so on. In this notation, we will rewrite system (2) as

$$\begin{aligned} x_1 + 5x_2 - 3x_3 &= 4 \\ 6x_1 \qquad - 4x_3 &= 1 \\ -2x_1 + 3x_2 + 4x_3 &= 7 \end{aligned}$$

In Section 10-1, we used E_i to denote the equations in a linear system. Now we use R_i to denote the rows and C_i to denote the columns, respectively, in a matrix, as illustrated below for system (2).

$$\begin{array}{cccc} C_1 & C_2 & C_3 & C_4 \end{array}$$
$$\left[\begin{array}{ccc|c} 1 & 5 & -3 & 4 \\ 6 & 0 & -4 & 1 \\ -2 & 3 & 4 & 7 \end{array}\right] \begin{array}{c} R_1 \\ R_2 \\ R_3 \end{array} \qquad (3)$$

Our goal will be to learn how to perform the basic steps we used to solve systems using elimination by addition, but on an augmented matrix. This enables us to focus on the numbers without being concerned about algebraic manipulations.

EXAMPLE 1 **Writing an Augmented Coefficient Matrix**

Write the augmented coefficient matrix corresponding to each of the following systems.

(A) $\begin{aligned} 2x_1 - 4x_2 &= 5 \\ -3x_1 + x_2 &= -6 \end{aligned}$
(B) $\begin{aligned} -3x_1 \qquad + 2x_3 &= -4 \\ 7x_1 - 5x_2 + 3x_3 &= 0 \end{aligned}$
(C) $\begin{aligned} 2x_1 - x_2 \qquad &= 4 \\ 3x_1 \qquad - 5x_3 &= 6 \\ -2x_2 + x_3 &= -3 \end{aligned}$

SOLUTIONS

(A) $\left[\begin{array}{cc|c} 2 & -4 & 5 \\ -3 & 1 & -6 \end{array}\right]$
(B) $\left[\begin{array}{ccc|c} -3 & 0 & 2 & -4 \\ 7 & -5 & 3 & 0 \end{array}\right]$
(C) $\left[\begin{array}{ccc|c} 2 & -1 & 0 & 4 \\ 3 & 0 & -5 & 6 \\ 0 & -2 & 1 & -3 \end{array}\right]$

MATCHED PROBLEM 1

Write the augmented coefficient matrix corresponding to each of the following systems.

(A) $\begin{aligned} -x_1 + 2x_2 &= -3 \\ 3x_1 - 5x_2 &= 8 \end{aligned}$
(B) $\begin{aligned} -2x_2 + 2x_3 &= -4 \\ 7x_1 - 5x_2 + 3x_3 &= 0 \end{aligned}$
(C) $\begin{aligned} 2x_1 - x_2 + x_3 &= 4 \\ 3x_1 + 4x_2 \qquad &= 6 \\ x_1 \qquad + 5x_3 &= -3 \end{aligned}$

Recall that two linear systems are said to be equivalent if they have the same solution set. In Theorem 2, Section 10-1, we used the operations listed next to transform linear systems into equivalent systems:

(A) Two equations are interchanged.

(B) An equation is multiplied by a nonzero constant.

(C) A constant multiple of one equation is added to another equation.

Paralleling this approach, we now say that two augmented matrices are **row-equivalent,** denoted by the symbol $\sim$ between the two matrices, if they are augmented matrices of equivalent systems of equations. How do we transform augmented matrices into row-equivalent matrices? We use Theorem 1, which gives the matrix analogs of operations (A), (B), and (C).

> **THEOREM 1** Elementary Row Operations Producing Row-Equivalent Matrices

An augmented matrix is transformed into a row-equivalent matrix if any of the following **row operations** is performed:

1. Two rows are interchanged ($R_i \leftrightarrow R_j$).
2. A row is multiplied by a nonzero constant ($kR_i \rightarrow R_i$).
3. A constant multiple of one row is added to another row ($kR_j + R_i \rightarrow R_i$).

[*Note:* The arrow means "replaces."]

EXAMPLE 2 **Row Operations**

Perform each of the indicated row operations on the following augmented coefficient matrix.

$$\begin{bmatrix} 1 & -4 & | & 3 \\ 2 & 4 & | & -8 \end{bmatrix}$$

(A) $R_1 \leftrightarrow R_2$ (B) $\frac{1}{2}R_2 \rightarrow R_2$ (C) $(-2)R_1 + R_2 \rightarrow R_2$

SOLUTIONS

(A) $\begin{bmatrix} 2 & 4 & | & -8 \\ 1 & -4 & | & 3 \end{bmatrix}$ (B) $\begin{bmatrix} 1 & -4 & | & 3 \\ 1 & 2 & | & -4 \end{bmatrix}$ (C) $\begin{bmatrix} 1 & -4 & | & 3 \\ 0 & 12 & | & -14 \end{bmatrix}$

MATCHED PROBLEM 2 Perform each of the indicated row operations on the following augmented coefficient matrix.

$$\begin{bmatrix} 1 & -2 & | & 3 \\ 3 & -6 & | & -3 \end{bmatrix}$$

(A) $R_1 \leftrightarrow R_2$ (B) $\frac{1}{3}R_2 \rightarrow R_2$ (C) $(-3)R_1 + R_2 \rightarrow R_2$

> ## Reduced Matrices

The goal of the elimination process is to transform a system of equations into an equivalent system whose solution is easy to find. Now our goal is to use a sequence of matrix row operations to transform an augmented coefficient matrix into a simpler equivalent matrix that corresponds to a system with an obvious solution. Example 3 illustrates the process of interpreting the solution of a system given its augmented coefficient matrix.

EXAMPLE 3 **Interpreting an Augmented Coefficient Matrix**

Write the system corresponding to each of the following augmented coefficient matrices and find its solution.

$$(A) \begin{bmatrix} 1 & 0 & 0 & | & -4 \\ 0 & 1 & 0 & | & 6 \\ 0 & 0 & 1 & | & 0 \end{bmatrix} \quad (B) \begin{bmatrix} 1 & 0 & 2 & | & -4 \\ 0 & 1 & -3 & | & 6 \\ 0 & 0 & 0 & | & 1 \end{bmatrix} \quad (C) \begin{bmatrix} 1 & 0 & 2 & | & -4 \\ 0 & 1 & -3 & | & 6 \\ 0 & 0 & 0 & | & 0 \end{bmatrix}$$

SOLUTIONS (A) The corresponding system is

$$\begin{aligned} x_1 &= -4 \qquad & 1 \cdot x_1 + 0 \cdot x_2 + 0 \cdot x_3 = x_1 \\ x_2 &= 6 \\ x_3 &= 0 \end{aligned}$$

and $(-4, 6, 0)$ is the solution.

(B) The corresponding system is

$$\begin{aligned} x_1 \qquad + \quad 2x_3 &= -4 \\ x_2 - \quad 3x_3 &= 6 \\ 0 \cdot x_1 + 0 \cdot x_2 + 0 \cdot x_3 &= 1 \end{aligned}$$

The third equation, $0 = 1$, is a contradiction, so the system has no solutions.

(C) The first two rows of this augmented coefficient matrix correspond to the system

$$\begin{aligned} x_1 \quad + 2x_3 &= -4 \\ x_2 - 3x_3 &= 6 \end{aligned}$$

The third row corresponds to the equation $0 = 0$, which is always true and can be discarded.

This is a dependent system with an infinite number of solutions. Introducing a parameter s, we can write

$$\begin{aligned} x_1 \quad + 2s &= -4 \qquad & x_1 &= -2s - 4 \\ x_2 - 3s &= 6 \quad \text{or} \quad & x_2 &= 3s + 6 \\ x_3 &= s & x_3 &= s \end{aligned}$$

So the solution set is

$$\{(-2s - 4, 3s + 6, s) \mid s \text{ any real number}\}$$

MATCHED PROBLEM 3 Write the system corresponding to each of the following augmented coefficient matrices and find its solution.

$$(A) \begin{bmatrix} 1 & 0 & 0 & | & 5 \\ 0 & 1 & 0 & | & -7 \\ 0 & 0 & 1 & | & 0 \end{bmatrix} \quad (B) \begin{bmatrix} 1 & 0 & -3 & | & 5 \\ 0 & 1 & 4 & | & -7 \\ 0 & 0 & 0 & | & 0 \end{bmatrix} \quad (C) \begin{bmatrix} 1 & 0 & -3 & | & 5 \\ 0 & 1 & 4 & | & -7 \\ 0 & 0 & 0 & | & 1 \end{bmatrix}$$

⟩⟩⟩ EXPLORE-DISCUSS 1 If an augmented coefficient matrix contains a row where every element on the left of the vertical line is 0 and the single element on the right is a nonzero number, what can you say about the solution of the corresponding system?

Next, we will define a particular matrix form that makes it simple to find solutions of the corresponding system.

> **DEFINITION 1** Reduced Matrix

A matrix is in **reduced form*** if:

1. Each row consisting entirely of 0's is below any row having at least one nonzero element.

2. The leftmost nonzero element in each row is 1.

3. The column containing the leftmost 1 of a given row has 0's above and below the 1.

4. The leftmost 1 in any row is to the right of the leftmost 1 in the preceding row.

For example, each of the following matrices is in reduced form. Before moving on, you should verify that each matrix satisfies all four conditions in Definition 1.

$$\left[\begin{array}{cc|c} 1 & 0 & 2 \\ 0 & 1 & -3 \end{array}\right] \left[\begin{array}{ccc|c} 1 & 0 & 0 & 2 \\ 0 & 1 & 0 & -1 \\ 0 & 0 & 1 & 3 \end{array}\right] \left[\begin{array}{cc|c} 1 & 0 & 3 \\ 0 & 1 & -1 \\ 0 & 0 & 0 \end{array}\right] \left[\begin{array}{cccc|c} 1 & 4 & 0 & 0 & 3 \\ 0 & 0 & 1 & 0 & 2 \\ 0 & 0 & 0 & 1 & 6 \end{array}\right] \left[\begin{array}{ccc|c} 1 & 0 & 4 & 0 \\ 0 & 1 & 3 & 0 \\ 0 & 0 & 0 & 1 \end{array}\right]$$

EXAMPLE 4 Reduced Forms

The matrices shown next are not in reduced form. Indicate which condition in the definition is violated for each matrix. State the row operation(s) required to transform the matrix to reduced form, and find the reduced form.

(A) $\left[\begin{array}{cc|c} 0 & 1 & -2 \\ 1 & 0 & 3 \end{array}\right]$ (B) $\left[\begin{array}{ccc|c} 1 & 2 & -2 & 3 \\ 0 & 0 & 1 & -1 \end{array}\right]$

(C) $\left[\begin{array}{cc|c} 1 & 0 & -3 \\ 0 & 0 & 0 \\ 0 & 1 & -2 \end{array}\right]$ (D) $\left[\begin{array}{ccc|c} 1 & 0 & 0 & -1 \\ 0 & 2 & 0 & 3 \\ 0 & 0 & 1 & -5 \end{array}\right]$

SOLUTIONS (A) Condition 4 is violated: The leftmost 1 in row 2 is not to the right of the leftmost 1 in row 1. Perform the row operation $R_1 \leftrightarrow R_2$ to obtain the reduced form:

$$\left[\begin{array}{cc|c} 0 & 1 & -2 \\ 1 & 0 & 3 \end{array}\right] \xrightarrow{R_1 \leftrightarrow R_2} \left[\begin{array}{cc|c} 1 & 0 & 3 \\ 0 & 1 & -2 \end{array}\right]$$

(B) Condition 3 is violated: The column containing the leftmost 1 in row 2 does not have a zero above the 1. Perform the row operation $2R_2 + R_1 \to R_1$ to obtain the reduced form:

$$\left[\begin{array}{ccc|c} 1 & 2 & -2 & 3 \\ 0 & 0 & 1 & -1 \end{array}\right] \xrightarrow{2R_2 + R_1 \to R_1} \left[\begin{array}{ccc|c} 1 & 2 & 0 & 1 \\ 0 & 0 & 1 & -1 \end{array}\right]$$

(C) Condition 1 is violated: The second row contains all zeros, and it is not below any row having at least one nonzero element. Perform the row operation $R_2 \leftrightarrow R_3$ to obtain the reduced form:

$$\left[\begin{array}{cc|c} 1 & 0 & -3 \\ 0 & 0 & 0 \\ 0 & 1 & -2 \end{array}\right] \xrightarrow{R_2 \leftrightarrow R_3} \left[\begin{array}{cc|c} 1 & 0 & -3 \\ 0 & 1 & -2 \\ 0 & 0 & 0 \end{array}\right]$$

*The reduced form we have defined here is sometimes called the **reduced row echelon form,** and most graphing calculators use the abbreviation **rref** to refer to it. There are other reduced forms that can be used to solve systems of equations, but we will use the term "reduced form" for simplicity.

(D) Condition 2 is violated: The leftmost nonzero element in row 2 is not a 1. Perform the row operation $\frac{1}{2}R_2 \to R_2$ to obtain the reduced form:

$$\begin{bmatrix} 1 & 0 & 0 & | & -1 \\ 0 & 2 & 0 & | & 3 \\ 0 & 0 & 1 & | & -5 \end{bmatrix} \quad \frac{1}{2}R_2 \to R_2 \quad \begin{bmatrix} 1 & 0 & 0 & | & -1 \\ 0 & 1 & 0 & | & \frac{3}{2} \\ 0 & 0 & 1 & | & -5 \end{bmatrix}$$

MATCHED PROBLEM 4

The matrices below are not in reduced form. Indicate which condition in the definition is violated for each matrix. State the row operation(s) required to transform the matrix to reduced form and find the reduced form.

(A) $\begin{bmatrix} 1 & 0 & | & 2 \\ 0 & 3 & | & -6 \end{bmatrix}$ (B) $\begin{bmatrix} 1 & 5 & 4 & | & 3 \\ 0 & 1 & 2 & | & -1 \\ 0 & 0 & 0 & | & 0 \end{bmatrix}$

(C) $\begin{bmatrix} 0 & 1 & 0 & | & -3 \\ 1 & 0 & 0 & | & 0 \\ 0 & 0 & 1 & | & 2 \end{bmatrix}$ (D) $\begin{bmatrix} 1 & 2 & 0 & | & 3 \\ 0 & 0 & 0 & | & 0 \\ 0 & 0 & 1 & | & 4 \end{bmatrix}$

› Solving Systems by Gauss–Jordan Elimination

We are now ready to outline the Gauss–Jordan elimination method for solving systems of linear equations. The method systematically transforms an augmented matrix into a reduced form. The system corresponding to a reduced augmented coefficient matrix is called a **reduced system.** As we will see, reduced systems are easy to solve.

The Gauss–Jordan elimination method is named after the German mathematician Carl Friedrich Gauss (1777–1855) and the German geodesist Wilhelm Jordan (1842–1899). Gauss, one of the greatest mathematicians of all time, used a method of solving systems of equations that was later generalized by Jordan to solve problems in large-scale surveying.

EXAMPLE 5 Solving a System Using Gauss–Jordan Elimination

Solve by Gauss–Jordan elimination: $2x_1 - 2x_2 + x_3 = 3$
$$3x_1 + x_2 - x_3 = 7$$
$$x_1 - 3x_2 + 2x_3 = 0$$

SOLUTION Write the augmented matrix and follow the steps indicated at the right to produce a reduced form.

$$\begin{bmatrix} 2 & -2 & 1 & | & 3 \\ 3 & 1 & -1 & | & 7 \\ 1 & -3 & 2 & | & 0 \end{bmatrix} \quad R_1 \leftrightarrow R_3$$

Need a 1 here.

Step 1: Choose the leftmost nonzero column and get a 1 at the top.

$$\sim \begin{bmatrix} 1 & -3 & 2 & | & 0 \\ 3 & 1 & -1 & | & 7 \\ 2 & -2 & 1 & | & 3 \end{bmatrix} \quad \begin{array}{l} (-3)R_1 + R_2 \to R_2 \\ (-2)R_1 + R_3 \to R_3 \end{array}$$

Need 0's here.

Step 2: Use multiples of the row containing the 1 from step 1 to get zeros in all remaining places in the column containing this 1.

$$\sim \begin{bmatrix} 1 & -3 & 2 & | & 0 \\ 0 & 10 & -7 & | & 7 \\ 0 & 4 & -3 & | & 3 \end{bmatrix} \quad 0.1R_2 \to R_2$$

Need a 1 here.

Step 3: Repeat step 1 with the *submatrix* formed by (mentally) deleting the top (shaded) row.

$$\sim \begin{bmatrix} 1 & -3 & 2 & | & 0 \\ 0 & 1 & -0.7 & | & 0.7 \\ 0 & 4 & -3 & | & 3 \end{bmatrix} \quad \begin{aligned} 3R_2 + R_1 \to R_1 \\ \\ (-4)R_2 + R_3 \to R_3 \end{aligned}$$

Need 0's here.

Step 4: Repeat step 2 with the *entire matrix.*

$$\sim \begin{bmatrix} 1 & 0 & -0.1 & | & 2.1 \\ 0 & 1 & -0.7 & | & 0.7 \\ 0 & 0 & -0.2 & | & 0.2 \end{bmatrix} \quad (-5)R_3 \to R_3$$

Need a 1 here.

Step 3: Repeat step 1 with the *submatrix* formed by (mentally) deleting the top two (shaded) rows.

$$\sim \begin{bmatrix} 1 & 0 & -0.1 & | & 2.1 \\ 0 & 1 & -0.7 & | & 0.7 \\ 0 & 0 & 1 & | & -1 \end{bmatrix} \quad \begin{aligned} 0.1R_3 + R_1 \to R_1 \\ \\ 0.7R_3 + R_2 \to R_2 \end{aligned}$$

Need 0's here.

Step 4: Repeat step 2 with the *entire matrix.*

$$\sim \begin{bmatrix} 1 & 0 & 0 & | & 2 \\ 0 & 1 & 0 & | & 0 \\ 0 & 0 & 1 & | & -1 \end{bmatrix}$$

The matrix is now in reduced form, and we can proceed to solve the corresponding reduced system.

$$\begin{aligned} x_1 & = 2 \\ x_2 & = 0 \\ x_3 & = -1 \end{aligned}$$

The solution to this system is $x_1 = 2$, $x_2 = 0$, $x_3 = -1$. You should check this solution in the original system.

> **GAUSS–JORDAN ELIMINATION**

Step 1. Choose the leftmost nonzero column and use appropriate row operations to get a 1 at the top.

Step 2. Use multiples of the row containing the 1 from step 1 to get zeros in all remaining places in the column containing this 1.

Step 3. Repeat step 1 with the **submatrix** formed by (mentally) deleting the row used in step 2 and all rows above this row.

Step 4. Repeat step 2 with the **entire matrix,** including the mentally deleted rows. Continue this process until the entire matrix is in reduced form.

[*Note:* If at any point in this process we obtain a row with all zeros to the left of the vertical line and a nonzero number to the right, we can stop, since we will have a contradiction: $0 = n$, $n \ne 0$. We can then conclude that the system has no solution.]

MATCHED PROBLEM 5

Solve by Gauss–Jordan elimination:
$$\begin{aligned} 3x_1 + x_2 - 2x_3 & = 2 \\ x_1 - 2x_2 + x_3 & = 3 \\ 2x_1 - x_2 - 3x_3 & = 3 \end{aligned}$$

Technology Connections

Figure 3 illustrates the solution of Example 5 on a TI-84 graphing calculator using the built-in rref (reduced row-echelon form) routine for finding reduced forms. Notice that in row 2 and column 4 of the reduced form the graphing calculator has displayed the very small number -3.5E-13 instead of the exact value 0. This is a common occurrence caused by rounding error on a graphing calculator and causes no problems. Just replace any very small numbers displayed in scientific notation with 0.

```
[A]
  [[2 -2 1   3]
   [3  1 -1   7]
   [1 -3 2   0]]
rref([A])
  [[1 0 0 2     ]
   [0 1 0 -3.5E-13]
   [0 0 1 -1    ]]
```

> **Figure 3** Using rref on a TI-84 graphing calculator.

EXAMPLE 6 **Solving a System Using Gauss–Jordan Elimination**

Solve by Gauss–Jordan elimination:
$$2x_1 - 4x_2 + x_3 = -4$$
$$4x_1 - 8x_2 + 7x_3 = 2$$
$$-2x_1 + 4x_2 - 3x_3 = 5$$

SOLUTION

$$\begin{bmatrix} 2 & -4 & 1 & | & -4 \\ 4 & -8 & 7 & | & 2 \\ -2 & 4 & -3 & | & 5 \end{bmatrix}$$

$0.5R_1 \rightarrow R_1$ (To get 1 in upper left corner)

$$\sim \begin{bmatrix} 1 & -2 & 0.5 & | & -2 \\ 4 & -8 & 7 & | & 2 \\ -2 & 4 & -3 & | & 5 \end{bmatrix}$$

(Next, get zeros below that 1.)

$(-4)R_1 + R_2 \rightarrow R_2$

$2R_1 + R_3 \rightarrow R_3$

$$\sim \begin{bmatrix} 1 & -2 & 0.5 & | & -2 \\ 0 & 0 & 5 & | & 10 \\ 0 & 0 & -2 & | & 1 \end{bmatrix}$$

$0.2R_2 \rightarrow R_2$ Note that column 3 is the leftmost nonzero column in this submatrix.

$$\sim \begin{bmatrix} 1 & -2 & 0.5 & | & -2 \\ 0 & 0 & 1 & | & 2 \\ 0 & 0 & -2 & | & 1 \end{bmatrix}$$

$(-0.5)R_2 + R_1 \rightarrow R_1$

$2R_2 + R_3 \rightarrow R_3$

$$\sim \begin{bmatrix} 1 & -2 & 0 & | & -3 \\ 0 & 0 & 1 & | & 2 \\ 0 & 0 & 0 & | & 5 \end{bmatrix}$$

We stop the Gauss–Jordan elimination, even though the matrix is not in reduced form, since the last row produces a contradiction

The system is inconsistent and has no solution.

MATCHED PROBLEM 6

Solve by Gauss–Jordan elimination:
$$2x_1 - 4x_2 - x_3 = -8$$
$$4x_1 - 8x_2 + 3x_3 = 4$$
$$-2x_1 + 4x_2 + x_3 = 11$$

Note that if we were to use **rref** on a graphing calculator for Example 6, it would continue reducing further. But the final reduced form would still show a contradiction.

EXAMPLE **7** **Solving a System Using Gauss–Jordan Elimination**

Solve by Gauss–Jordan elimination:
$$\begin{aligned} 3x_1 + 6x_2 - 9x_3 &= 15 \\ 2x_1 + 4x_2 - 6x_3 &= 10 \\ -2x_1 - 3x_2 + 4x_3 &= -6 \end{aligned}$$

SOLUTION

$$\begin{bmatrix} 3 & 6 & -9 & | & 15 \\ 2 & 4 & -6 & | & 10 \\ -2 & -3 & 4 & | & -6 \end{bmatrix}$$ $\frac{1}{3}R_1 \to R_1$

$$\sim \begin{bmatrix} 1 & 2 & -3 & | & 5 \\ 2 & 4 & -6 & | & 10 \\ -2 & -3 & 4 & | & -6 \end{bmatrix}$$ $(-2)R_1 + R_2 \to R_2$
 $2R_1 + R_3 \to R_3$

$$\sim \begin{bmatrix} 1 & 2 & -3 & | & 5 \\ 0 & 0 & 0 & | & 0 \\ 0 & 1 & -2 & | & 4 \end{bmatrix}$$ $R_2 \leftrightarrow R_3$ Note that we must interchange rows 2 and 3 to obtain a nonzero entry at the top of the second column of this submatrix.

$$\sim \begin{bmatrix} 1 & 2 & -3 & | & 5 \\ 0 & 1 & -2 & | & 4 \\ 0 & 0 & 0 & | & 0 \end{bmatrix}$$ $(-2)R_2 + R_1 \to R_1$

$$\sim \begin{bmatrix} 1 & 0 & 1 & | & -3 \\ 0 & 1 & -2 & | & 4 \\ 0 & 0 & 0 & | & 0 \end{bmatrix}$$ This matrix is now in reduced form. Write the corresponding reduced system and solve.

$$\begin{aligned} x_1 \quad + \quad x_3 &= -3 \\ x_2 - 2x_3 &= \quad 4 \end{aligned}$$ We discard the equation corresponding to the third (all 0) row in the reduced form, since it is satisfied by all values of x_1, x_2, and x_3.

Note that the leftmost variable in each equation appears in one and only one equation. We solve for the leftmost variables x_1 and x_2 in terms of the remaining variable x_3:

$$\begin{aligned} x_1 &= -x_3 - 3 \\ x_2 &= \quad 2x_3 + 4 \end{aligned}$$

This dependent system has an infinite number of solutions. We will use a parameter to represent all the solutions. If we let $x_3 = t$, then for any real number t,

$$\begin{aligned} x_1 &= -t - 3 \\ x_2 &= 2t + 4 \\ x_3 &= t \end{aligned}$$

is a solution. You should check that $(-t - 3, 2t + 4, t)$ is a solution of the original system for any real number t. Some particular solutions are

$t = 0$	$t = -2$	$t = 3.5$
$(-3, 4, 0)$	$(-1, 0, -2)$	$(-6.5, 11, 3.5)$

MATCHED PROBLEM 7 Solve by Gauss–Jordan elimination:
$$\begin{aligned} 2x_1 - 2x_2 - 4x_3 &= -2 \\ 3x_1 - 3x_2 - 6x_3 &= -3 \\ -2x_1 + 3x_2 + x_3 &= 7 \end{aligned}$$

In general,

If the number of leftmost 1's in a reduced augmented coefficient matrix is less than the number of variables in the system and there are no contradictions, then the system is dependent and has infinitely many solutions.

There are many different ways to use the reduced augmented coefficient matrix to describe the infinite number of solutions of a dependent system. We will always proceed as follows: Solve each equation in a reduced system for its leftmost variable and then introduce a different parameter for each remaining variable. Example 8 illustrates a dependent system where two parameters are required to describe the solution.

EXAMPLE | **8** | **Solving a System Using Gauss–Jordan Elimination**

Solve by Gauss–Jordan elimination:

$$\begin{aligned} x_1 + 2x_2 + 4x_3 + x_4 - x_5 &= 1 \\ 2x_1 + 4x_2 + 8x_3 + 3x_4 - 4x_5 &= 2 \\ x_1 + 3x_2 + 7x_3 + \qquad 3x_5 &= -2 \end{aligned}$$

SOLUTION

$$\begin{bmatrix} 1 & 2 & 4 & 1 & -1 & | & 1 \\ 2 & 4 & 8 & 3 & -4 & | & 2 \\ 1 & 3 & 7 & 0 & 3 & | & -2 \end{bmatrix} \quad \begin{array}{l} (-2)R_1 + R_2 \to R_2 \\ (-1)R_1 + R_3 \to R_3 \end{array}$$

$$\sim \begin{bmatrix} 1 & 2 & 4 & 1 & -1 & | & 1 \\ 0 & 0 & 0 & 1 & -2 & | & 0 \\ 0 & 1 & 3 & -1 & 4 & | & -3 \end{bmatrix} \quad R_2 \leftrightarrow R_3$$

$$\sim \begin{bmatrix} 1 & 2 & 4 & 1 & -1 & | & 1 \\ 0 & 1 & 3 & -1 & 4 & | & -3 \\ 0 & 0 & 0 & 1 & -2 & | & 0 \end{bmatrix} \quad (-2)R_2 + R_1 \to R_1$$

$$\sim \begin{bmatrix} 1 & 0 & -2 & 3 & -9 & | & 7 \\ 0 & 1 & 3 & -1 & 4 & | & -3 \\ 0 & 0 & 0 & 1 & -2 & | & 0 \end{bmatrix} \quad \begin{array}{l} (-3)R_3 + R_1 \to R_1 \\ R_3 + R_2 \to R_2 \end{array}$$

$$\sim \begin{bmatrix} 1 & 0 & -2 & 0 & -3 & | & 7 \\ 0 & 1 & 3 & 0 & 2 & | & -3 \\ 0 & 0 & 0 & 1 & -2 & | & 0 \end{bmatrix} \quad \text{Matrix is in reduced form.}$$

$$\begin{aligned} x_1 \qquad - 2x_3 \qquad - 3x_5 &= 7 \\ x_2 + 3x_3 \qquad + 2x_5 &= -3 \\ x_4 - 2x_5 &= 0 \end{aligned}$$

Solve for the leftmost variables x_1, x_2, and x_4 in terms of the remaining variables x_3 and x_5:

$$\begin{aligned} x_1 &= 2x_3 + 3x_5 + 7 \\ x_2 &= -3x_3 - 2x_5 - 3 \\ x_4 &= 2x_5 \end{aligned}$$

If we let $x_3 = s$ and $x_5 = t$, then for any real numbers s and t,

$$\begin{aligned} x_1 &= 2s + 3t + 7 \\ x_2 &= -3s - 2t - 3 \\ x_3 &= s \\ x_4 &= 2t \\ x_5 &= t \end{aligned}$$

is a solution. The check is left for you to perform.

MATCHED PROBLEM 8

Solve by Gauss–Jordan elimination:

$$\begin{aligned} x_1 - x_2 + 2x_3 \qquad - 2x_5 &= 3 \\ -2x_1 + 2x_2 - 4x_3 - x_4 + x_5 &= -5 \\ 3x_1 - 3x_2 + 7x_3 + x_4 - 4x_5 &= 6 \end{aligned}$$

› Application

Dependent systems probably seem very abstract to you—a solution like the one in Example 8 doesn't seem like it would apply to any real–world situations. But in Example 9, we will solve a problem where a dependent system leads to real solutions.

EXAMPLE **9**	**Purchasing**

A chemical manufacturer plans to purchase a fleet of 24 railroad tank cars with a combined carrying capacity of 250,000 gallons. Tank cars with three different carrying capacities are available: 6,000 gallons, 8,000 gallons, and 18,000 gallons. How many of each type of tank car should be purchased?

SOLUTION Let

$$x_1 = \text{Number of 6,000-gallon tank cars}$$
$$x_2 = \text{Number of 8,000-gallon tank cars}$$
$$x_3 = \text{Number of 18,000-gallon tank cars}$$

Then

$$x_1 + x_2 + x_3 = 24 \qquad \text{Total number of tank cars}$$
$$6{,}000x_1 + 8{,}000x_2 + 18{,}000x_3 = 250{,}000 \qquad \text{Total carrying capacity}$$

Now we can form the augmented matrix of the system and solve by using Gauss–Jordan elimination:

$$\begin{bmatrix} 1 & 1 & 1 & | & 24 \\ 6{,}000 & 8{,}000 & 18{,}000 & | & 250{,}000 \end{bmatrix} \qquad \tfrac{1}{1{,}000}R_2 \rightarrow R_2 \text{ (simplify } R_2)$$

$$\sim \begin{bmatrix} 1 & 1 & 1 & | & 24 \\ 6 & 8 & 18 & | & 250 \end{bmatrix} \qquad (-6)R_1 + R_2 \rightarrow R_2$$

$$\sim \begin{bmatrix} 1 & 1 & 1 & | & 24 \\ 0 & 2 & 12 & | & 106 \end{bmatrix} \qquad \tfrac{1}{2}R_2 \rightarrow R_2$$

$$\sim \begin{bmatrix} 1 & 1 & 1 & | & 24 \\ 0 & 1 & 6 & | & 53 \end{bmatrix} \qquad (-1)R_2 + R_1 \rightarrow R_1$$

$$\sim \begin{bmatrix} 1 & 0 & -5 & | & -29 \\ 0 & 1 & 6 & | & 53 \end{bmatrix} \qquad \text{Matrix is in reduced form.}$$

$$x_1 - 5x_3 = -29 \qquad \text{or} \qquad x_1 = 5x_3 - 29$$
$$x_2 + 6x_3 = 53 \qquad \text{or} \qquad x_2 = -6x_3 + 53$$

Let $x_3 = t$. Then for t any real number,

$$x_1 = 5t - 29$$
$$x_2 = -6t + 53$$
$$x_3 = t$$

is a solution—or is it? Since the variables in this system represent the number of tank cars purchased, the values of x_1, x_2, and x_3 must be nonnegative integers. The third equation requires that t must be a nonnegative integer. The first equation requires that $5t - 29 \geq 0$, so t must be at least 6. The middle equation requires that $-6t + 53 \geq 0$, so t can be no larger than 8.

So, 6, 7, and 8 are the only possible values for t. There are three different possible combinations that meet the company's specifications of 24 tank cars with a total carrying capacity of 250,000 gallons, as shown in Table 1:

Table 1

t	6,000-Gallon Tank Cars x_1	8,000-Gallon Tank Cars x_2	18,000-Gallon Tank Cars x_3
6	1	17	6
7	6	11	7
8	11	5	8

The final choice would probably be influenced by other factors. For example, the company might want to minimize the cost of the 24 tank cars.

MATCHED PROBLEM 9

A commuter airline plans to purchase a fleet of 30 airplanes with a combined carrying capacity of 960 passengers. The three available types of planes carry 18, 24, and 42 passengers, respectively. How many of each type of plane should be purchased?

ANSWERS TO MATCHED PROBLEMS

1. (A) $\begin{bmatrix} -1 & 2 & | & -3 \\ 3 & -5 & | & 8 \end{bmatrix}$ (B) $\begin{bmatrix} 0 & -2 & 2 & | & -4 \\ 7 & -5 & 3 & | & 0 \end{bmatrix}$ (C) $\begin{bmatrix} 2 & -1 & 1 & | & 4 \\ 3 & 4 & 0 & | & 6 \\ 1 & 0 & 5 & | & -3 \end{bmatrix}$

2. (A) $\begin{bmatrix} 3 & -6 & | & -3 \\ 1 & -2 & | & 3 \end{bmatrix}$ (B) $\begin{bmatrix} 1 & -2 & | & 3 \\ 1 & -2 & | & -1 \end{bmatrix}$ (C) $\begin{bmatrix} 1 & -2 & | & 3 \\ 0 & 0 & | & -12 \end{bmatrix}$

3. (A) $x_1 = 5, x_2 = -7, x_3 = 0$ or $(5, -7, 0)$
 (B) $x_1 = 3s + 5, x_2 = -4s - 7, x_3 = s$, s any real number; or
 $\{(3s + 5, -4s - 7, s) \mid s$ any real number$\}$ (C) No solution

4. (A) Condition 2 is violated: The 3 in row 2 and column 2 should be a 1. Perform the operation $\frac{1}{3}R_2 \to R_2$ to obtain:

$$\begin{bmatrix} 1 & 0 & | & 2 \\ 0 & 1 & | & -2 \end{bmatrix}$$

(B) Condition 3 is violated: The 5 in row 1 and column 2 should be a 0. Perform the operation $(-5)R_2 + R_1 \to R_1$ to obtain:

$$\begin{bmatrix} 1 & 0 & -6 & | & 8 \\ 0 & 1 & 2 & | & -1 \\ 0 & 0 & 0 & | & 0 \end{bmatrix}$$

(C) Condition 4 is violated: The leftmost 1 in the second row is not to the right of the leftmost 1 in the first row. Perform the operation $R_1 \leftrightarrow R_2$ to obtain:

$$\begin{bmatrix} 1 & 0 & 0 & | & 0 \\ 0 & 1 & 0 & | & -3 \\ 0 & 0 & 1 & | & 2 \end{bmatrix}$$

(D) Condition 1 is violated: The all-zero second row should be at the bottom. Perform the operation $R_2 \leftrightarrow R_3$ to obtain:

$$\begin{bmatrix} 1 & 2 & 0 & | & 3 \\ 0 & 0 & 1 & | & 4 \\ 0 & 0 & 0 & | & 0 \end{bmatrix}$$

5. $x_1 = 1$, $x_2 = -1$, $x_3 = 0$ or $(1, -1, 0)$

6. No solution

7. $x_1 = 5t + 4$, $x_2 = 3t + 5$, $x_3 = t$, t any real number; or $\{(5t + 4, 3t + 5, t) \mid t \text{ any real number}\}$

8. $x_1 = s + 7$, $x_2 = s$, $x_3 = t - 2$, $x_4 = -3t - 1$, $x_5 = t$, s and t any real numbers; or $\{(s + 7, s, t - 2, -3t - 1, t) \mid s \text{ and } t \text{ any real numbers}\}$

9.

t	18-Passenger Planes x_1	24-Passenger Planes x_2	42-Passenger Planes x_3
14	2	14	14
15	5	10	15
16	8	6	16
17	11	2	17

10-2 Exercises

1. What is the size of a matrix?

2. What is a row matrix? What is its size?

3. What is a column matrix? What is its size?

4. What is a square matrix?

5. What does a_{ij} mean?

6. What is the principal diagonal of a matrix?

7. What is an augmented coefficient matrix?

8. What operations can you perform on an augmented coefficient matrix to produce a row-equivalent matrix?

9. What is a reduced matrix and how is it used to solve a system of linear equations?

10. Describe the Gauss–Jordan elimination process in your own words.

In Problems 11–18, indicate whether each matrix is in reduced form.

11. $\begin{bmatrix} 1 & 0 & | & -1 \\ 0 & 2 & | & 6 \end{bmatrix}$

12. $\begin{bmatrix} 1 & 0 & | & 5 \\ 0 & 1 & | & -3 \end{bmatrix}$

13. $\begin{bmatrix} 0 & 1 & -2 & | & 0 \\ 0 & 0 & 0 & | & 1 \\ 0 & 0 & 0 & | & 0 \end{bmatrix}$

14. $\begin{bmatrix} 1 & -1 & 4 & | & 0 \\ 0 & 0 & 0 & | & 0 \\ 0 & 0 & 0 & | & 1 \end{bmatrix}$

15. $\begin{bmatrix} 0 & 0 & 1 & | & 2 \\ 0 & 1 & 0 & | & -5 \\ 1 & 0 & 0 & | & 4 \end{bmatrix}$

16. $\begin{bmatrix} 1 & -2 & 4 & | & 1 \\ 0 & 0 & 1 & | & -3 \\ 0 & 0 & 0 & | & 0 \end{bmatrix}$

17. $\begin{bmatrix} 0 & 1 & 6 & 0 & | & -8 \\ 0 & 0 & 0 & 1 & | & 1 \end{bmatrix}$

18. $\begin{bmatrix} 0 & 0 & 1 & | & 0 \\ 0 & 0 & 0 & | & 0 \end{bmatrix}$

In Problems 19–26, write the linear system corresponding to each reduced augmented matrix and solve.

19. $\begin{bmatrix} 1 & 0 & 0 & | & -2 \\ 0 & 1 & 0 & | & 3 \\ 0 & 0 & 1 & | & 0 \end{bmatrix}$

20. $\begin{bmatrix} 1 & 0 & 0 & 0 & | & -2 \\ 0 & 1 & 0 & 0 & | & 0 \\ 0 & 0 & 1 & 0 & | & 1 \\ 0 & 0 & 0 & 1 & | & 3 \end{bmatrix}$

21. $\begin{bmatrix} 1 & 0 & -2 & | & 3 \\ 0 & 1 & 1 & | & -5 \\ 0 & 0 & 0 & | & 0 \end{bmatrix}$

22. $\begin{bmatrix} 1 & -2 & 0 & | & -3 \\ 0 & 0 & 1 & | & 5 \\ 0 & 0 & 0 & | & 0 \end{bmatrix}$

23. $\begin{bmatrix} 1 & 0 & | & 0 \\ 0 & 1 & | & 0 \\ 0 & 0 & | & 1 \end{bmatrix}$

24. $\begin{bmatrix} 1 & 0 & | & 5 \\ 0 & 1 & | & -3 \\ 0 & 0 & | & 0 \end{bmatrix}$

25. $\begin{bmatrix} 1 & -2 & 0 & -3 & | & -5 \\ 0 & 0 & 1 & 3 & | & 2 \end{bmatrix}$

26. $\begin{bmatrix} 1 & 0 & -2 & 3 & | & 4 \\ 0 & 1 & -1 & 2 & | & -1 \end{bmatrix}$

Perform each of the row operations indicated in Problems 27–38 on the following matrix:

$$\begin{bmatrix} 1 & -3 & | & 2 \\ 4 & -6 & | & -8 \end{bmatrix}$$

27. $R_1 \leftrightarrow R_2$

28. $\frac{1}{2}R_2 \rightarrow R_2$

29. $-4R_1 \rightarrow R_1$

30. $-2R_1 \rightarrow R_1$

31. $2R_2 \rightarrow R_2$

32. $-1R_2 \rightarrow R_2$

33. $(-4)R_1 + R_2 \rightarrow R_2$

34. $(-\frac{1}{2})R_2 + R_1 \rightarrow R_1$

35. $(-2)R_1 + R_2 \rightarrow R_2$

36. $(-3)R_1 + R_2 \rightarrow R_2$

37. $(-1)R_1 + R_2 \rightarrow R_2$

38. $1R_1 + R_2 \rightarrow R_2$

Use row operations to change each matrix in Problems 39–44 to reduced form.

39. $\begin{bmatrix} 1 & 2 & | & -1 \\ 0 & 1 & | & 3 \end{bmatrix}$

40. $\begin{bmatrix} 1 & 3 & | & 1 \\ 0 & 2 & | & -4 \end{bmatrix}$

41. $\begin{bmatrix} 1 & 0 & -3 & | & 1 \\ 0 & 1 & 2 & | & 0 \\ 0 & 0 & 3 & | & -6 \end{bmatrix}$

42. $\begin{bmatrix} 1 & 0 & 4 & | & 0 \\ 0 & 1 & -3 & | & -1 \\ 0 & 0 & -2 & | & 2 \end{bmatrix}$

43. $\begin{bmatrix} 1 & 2 & -2 & | & -1 \\ 0 & 3 & -6 & | & 1 \\ 0 & -1 & 2 & | & -\frac{1}{3} \end{bmatrix}$

44. $\begin{bmatrix} 0 & -2 & 8 & | & 1 \\ 2 & -2 & 6 & | & -4 \\ 0 & -1 & 4 & | & \frac{1}{2} \end{bmatrix}$

Solve Problems 45–70 using Gauss–Jordan elimination.

45. $\begin{aligned} x_1 - 4x_2 &= -2 \\ -2x_1 + x_2 &= -3 \end{aligned}$

46. $\begin{aligned} x_1 - 3x_2 &= -5 \\ -3x_1 - x_2 &= 5 \end{aligned}$

47. $\begin{aligned} x_1 + 2x_2 &= 4 \\ 2x_1 + 4x_2 &= -8 \end{aligned}$

48. $\begin{aligned} 2x_1 - 3x_2 &= -2 \\ -4x_1 + 6x_2 &= 7 \end{aligned}$

49. $\begin{aligned} 3x_1 - 6x_2 &= -9 \\ -2x_1 + 4x_2 &= 6 \end{aligned}$

50. $\begin{aligned} 2x_1 - 4x_2 &= -2 \\ -3x_1 + 6x_2 &= 3 \end{aligned}$

51. $\begin{aligned} 2x_1 + 4x_2 - 10x_3 &= -2 \\ 3x_1 + 9x_2 - 21x_3 &= 0 \\ x_1 + 5x_2 - 12x_3 &= 1 \end{aligned}$

52. $\begin{aligned} 3x_1 + 5x_2 - x_3 &= -7 \\ x_1 + x_2 + x_3 &= -1 \\ 2x_1 + 11x_3 &= 7 \end{aligned}$

53. $\begin{aligned} 3x_1 + 8x_2 - x_3 &= -18 \\ 2x_1 + x_2 + 5x_3 &= 8 \\ 2x_1 + 4x_2 + 2x_3 &= -4 \end{aligned}$

54. $\begin{aligned} 2x_1 + 7x_2 + 15x_3 &= -12 \\ 4x_1 + 7x_2 + 13x_3 &= -10 \\ 3x_1 + 6x_2 + 12x_3 &= -9 \end{aligned}$

55. $\begin{aligned} 2x_1 - x_2 - 3x_3 &= 8 \\ x_1 - 2x_2 &= 7 \end{aligned}$

56. $\begin{aligned} 2x_1 + 4x_2 - 6x_3 &= 10 \\ 3x_1 + 3x_2 - 3x_3 &= 6 \end{aligned}$

57. $\begin{aligned} 2x_1 - x_2 &= 0 \\ 3x_1 + 2x_2 &= 7 \\ x_1 - x_2 &= -1 \end{aligned}$

58. $\begin{aligned} 2x_1 - x_2 &= 0 \\ 3x_1 + 2x_2 &= 7 \\ x_1 - x_2 &= -2 \end{aligned}$

59. $\begin{aligned} 3x_1 - 4x_2 - x_3 &= 1 \\ 2x_1 - 3x_2 + x_3 &= 1 \\ x_1 - 2x_2 + 3x_3 &= 2 \end{aligned}$

60. $\begin{aligned} -2x_1 + x_2 + 3x_3 &= -7 \\ x_1 - 4x_2 + 2x_3 &= 0 \\ x_1 - 3x_2 + x_3 &= 1 \end{aligned}$

61. $\begin{aligned} 2x_1 - 2x_2 - 4x_3 &= -2 \\ -3x_1 + 3x_2 + 6x_3 &= 3 \end{aligned}$

62. $\begin{aligned} 4x_1 - x_2 + 2x_3 &= 3 \\ -4x_1 + x_2 - 3x_3 &= -10 \\ 8x_1 - 2x_2 + 9x_3 &= -1 \end{aligned}$

63. $\begin{aligned} 2x_1 - 5x_2 - 3x_3 &= 7 \\ -4x_1 + 10x_2 + 2x_3 &= 6 \\ 6x_1 - 15x_2 - x_3 &= -19 \end{aligned}$

64. $\begin{aligned} 5x_1 - 3x_2 + 2x_3 &= 13 \\ 2x_1 - x_2 - 3x_3 &= 1 \\ 4x_1 - 2x_2 + 4x_3 &= 12 \end{aligned}$

65. $\begin{aligned} x_1 + 2x_2 - 4x_3 - x_4 &= 7 \\ 2x_1 + 5x_2 - 9x_3 - 4x_4 &= 16 \\ x_1 + 5x_2 - 7x_3 - 7x_4 &= 13 \end{aligned}$

66. $\begin{aligned} 2x_1 + 4x_2 + 5x_3 + 4x_4 &= 8 \\ x_1 + 2x_2 + 2x_3 + x_4 &= 3 \end{aligned}$

67. $\begin{aligned} x_1 - x_2 + 3x_3 - 2x_4 &= 1 \\ -2x_1 + 4x_2 - 3x_3 + x_4 &= 0.5 \\ 3x_1 - x_2 + 10x_3 - 4x_4 &= 2.9 \\ 4x_1 - 3x_2 + 8x_3 - 2x_4 &= 0.6 \end{aligned}$

68. $\begin{aligned} x_1 + x_2 + 4x_3 + x_4 &= 1.3 \\ -x_1 + x_2 - x_3 &= 1.1 \\ 2x_1 + x_3 + 3x_4 &= -4.4 \\ 2x_1 + 5x_2 + 11x_3 + 3x_4 &= 5.6 \end{aligned}$

69. $\begin{aligned} x_1 - 2x_2 + x_3 + x_4 + 2x_5 &= 2 \\ -2x_1 + 4x_2 + 2x_3 + 2x_4 - 2x_5 &= 0 \\ 3x_1 - 6x_2 + x_3 + x_4 + 5x_5 &= 4 \\ -x_1 + 2x_2 + 3x_3 + x_4 + x_5 &= 3 \end{aligned}$

70. $\begin{aligned} x_1 - 3x_2 + x_3 + x_4 + 2x_5 &= 2 \\ -x_1 + 5x_2 + 2x_3 + 2x_4 - 2x_5 &= 0 \\ 2x_1 - 6x_2 + 2x_3 + 2x_4 + 4x_5 &= 4 \\ -x_1 + 3x_2 - x_3 - x_5 &= -3 \end{aligned}$

71. Consider a consistent system of three linear equations in three variables. Discuss the nature of the solution set for the system if the reduced form of the augmented coefficient matrix has
(A) One leftmost 1
(B) Two leftmost 1's
(C) Three leftmost 1's
(D) Four leftmost 1's

72. Consider a system of three linear equations in three variables. Give examples of two reduced forms that are not row equivalent if the system is
(A) Consistent and dependent
(B) Inconsistent

APPLICATIONS

73. BUYING Suppose that you have a $129 credit on your account at Amazon.com, and you want to spend it all on sale CDs at $10 each, sale DVDs at $12 each, and sale books at $7 each. If you buy 13 items total, how many will you buy of each?

74. PETTY CRIME Shady Grady finds a parking meter with a broken lock and scoops out the change inside. The meter accepts nickels, dimes, and quarters, and there were 32 coins inside with a total value of $6.80. How many of each type of coin did Grady get?

75. CHEMISTRY A chemist has two solutions of sulfuric acid: a 20% solution and an 80% solution. How much of each should be used to obtain 100 liters of a 62% solution?

76. CHEMISTRY A chemist has two solutions: one containing 40% alcohol and another containing 70% alcohol. How much of each should be used to obtain 80 liters of a 49% solution?

77. GEOMETRY Find a, b, and c so that the graph of the parabola with equation $y = a + bx + cx^2$ passes through the points $(-2, 3)$, $(-1, 2)$, and $(1, 6)$.

78. GEOMETRY Find a, b, and c so that the graph of the parabola with equation $y = a + bx + cx^2$ passes through the points $(1, 3)$, $(2, 2)$, and $(3, 5)$.

79. PRODUCTION SCHEDULING A small manufacturing plant makes three types of inflatable boats: one-person, two-person, and four-person models. Each boat requires the services of three departments, as listed in the table. The cutting, assembly, and packaging departments have available a maximum of 380, 330, and 120 labor-hours per week, respectively. How many boats of each type must be produced each week for the plant to operate at full capacity?

	One-Person Boat	Two-Person Boat	Four-Person Boat
Cutting department	0.5 h	1.0 h	1.5 h
Assembly department	0.6 h	0.9 h	1.2 h
Packaging department	0.2 h	0.3 h	0.5 h

80. PRODUCTION SCHEDULING Repeat Problem 79 assuming the cutting, assembly, and packaging departments have available a maximum of 350, 330, and 115 labor-hours per week, respectively.

81. PRODUCTION SCHEDULING Rework Problem 79 assuming the packaging department is no longer used.

82. PRODUCTION SCHEDULING Rework Problem 80 assuming the packaging department is no longer used.

83. PRODUCTION SCHEDULING Rework Problem 79 assuming the four-person boat is no longer produced.

84. PRODUCTION SCHEDULING Rework Problem 80 assuming the four-person boat is no longer produced.

85. NUTRITION A dietitian in a hospital is to arrange a special diet using three basic foods. The diet is to include exactly 340 units of calcium, 180 units of iron, and 220 units of vitamin A. The number of units per ounce of each special ingredient for each of the foods is indicated in the table. How many ounces of each food must be used to meet the diet requirements?

	Units per Ounce		
	Food A	Food B	Food C
Calcium	30	10	20
Iron	10	10	20
Vitamin A	10	30	20

86. NUTRITION Repeat Problem 85 if the diet is to include exactly 400 units of calcium, 160 units of iron, and 240 units of vitamin A.

87. NUTRITION Solve Problem 85 with the assumption that food C is no longer available.

88. NUTRITION Solve Problem 86 with the assumption that food C is no longer available.

89. NUTRITION Solve Problem 85 assuming the vitamin A requirement is deleted.

90. NUTRITION Solve Problem 86 assuming the vitamin A requirement is deleted.

91. SOCIOLOGY Two sociologists have grant money to study school busing in a particular city. They wish to conduct an opinion survey using 600 telephone contacts and 400 house contacts. Survey company A has personnel to do 30 telephone and 10 house contacts per hour; survey company B can handle 20 telephone and 20 house contacts per hour. How many hours should be scheduled for each firm to produce exactly the number of contacts needed?

92. SOCIOLOGY Repeat Problem 91 if 650 telephone contacts and 350 house contacts are needed.

93. DELIVERY CHARGES United Express, a nationwide package delivery service, charges a base price for overnight delivery of packages weighing 1 pound or less and a surcharge for each additional pound (or fraction thereof). A customer is billed $27.75 for shipping a 5-pound package and $64.50 for shipping a 20-pound package. Find the base price and the surcharge for each additional pound.

94. DELIVERY CHARGES Refer to Problem 93. Federated Shipping, a competing overnight delivery service, informs the customer in Problem 93 that it would ship the 5-pound package for $29.95 and the 20-pound package for $59.20.
(A) If Federated Shipping computes its cost in the same manner as United Express, find the base price and the surcharge for Federated Shipping.
(B) Devise a simple rule that the customer can use to choose the cheaper of the two services for each package shipped. Justify your answer.

95. RESOURCE ALLOCATION A coffee manufacturer uses Colombian and Brazilian coffee beans to produce two blends, robust and mild. A pound of the robust blend requires 12 ounces of Colombian beans and 4 ounces of Brazilian beans. A pound of the mild blend requires 6 ounces of Colombian beans and 10 ounces of Brazilian beans. Coffee is shipped in 132-pound burlap bags. The company has 50 bags of Colombian beans and 40 bags of Brazilian beans on hand. How many pounds of each blend should it produce in order to use all the available beans?

96. RESOURCE ALLOCATION Refer to Problem 95.
(A) If the company decides to discontinue production of the robust blend and only produce the mild blend, how many pounds of the mild blend can it produce and how many beans of each type will it use? Are there any beans that are not used?
(B) Repeat part A if the company decides to discontinue production of the mild blend and only produce the robust blend.

10-3 Matrix Operations

> Adding and Subtracting Matrices
> Multiplying a Matrix by a Number
> Finding the Product of Two Matrices

In Section 10-2, we introduced basic matrix terminology and solved systems of equations by performing row operations on augmented coefficient matrices. Matrices have many other useful applications and possess an interesting mathematical structure in their own right. As we will see, matrix addition and multiplication are similar to real number addition and multiplication in many respects, but there are some important differences.

> Adding and Subtracting Matrices

Before we can discuss arithmetic operations for matrices, we have to define equality for matrices. Two matrices are **equal** if they have the same size and their corresponding elements are equal. For example,

$$\underset{2 \times 3}{\begin{bmatrix} a & b & c \\ d & e & f \end{bmatrix}} = \underset{2 \times 3}{\begin{bmatrix} u & v & w \\ x & y & z \end{bmatrix}} \qquad \text{if and only if} \qquad \begin{matrix} a = u & b = v & c = w \\ d = x & e = y & f = z \end{matrix}$$

The **sum of two matrices** of the same size is a matrix with elements that are the sums of the corresponding elements of the two given matrices.

Addition is not defined for matrices of different sizes.

EXAMPLE 1 **Matrix Addition**

Add:

(A) $\begin{bmatrix} 2 & -3 & 0 \\ 1 & 2 & -5 \end{bmatrix} + \begin{bmatrix} 3 & 1 & 2 \\ -3 & 2 & 5 \end{bmatrix}$ (B) $\begin{bmatrix} 2 & 1 & 4 \\ 3 & 2 & -3 \end{bmatrix} + \begin{bmatrix} 0 & 2 \\ -3 & 5 \\ -1 & 4 \end{bmatrix}$

SOLUTIONS

(A) $\begin{bmatrix} 2 & -3 & 0 \\ 1 & 2 & -5 \end{bmatrix} + \begin{bmatrix} 3 & 1 & 2 \\ -3 & 2 & 5 \end{bmatrix} = \begin{bmatrix} (2+3) & (-3+1) & (0+2) \\ (1-3) & (2+2) & (-5+5) \end{bmatrix}$ *

$= \begin{bmatrix} 5 & -2 & 2 \\ -2 & 4 & 0 \end{bmatrix}$

(B) $\begin{bmatrix} 2 & 1 & 4 \\ 3 & 2 & -3 \end{bmatrix} + \begin{bmatrix} 0 & 2 \\ -3 & 5 \\ -1 & 4 \end{bmatrix}$

Because the first matrix is 2×3 and the second is 3×2, this sum is not defined.

*Throughout the book, dashed boxes—called think boxes—are used to represent steps that may be performed mentally.

MATCHED PROBLEM 1 Add:

(A) $\begin{bmatrix} 3 & 2 \\ -1 & -1 \\ 0 & 3 \end{bmatrix} + \begin{bmatrix} -2 & 3 \\ 1 & -1 \\ 2 & -2 \end{bmatrix}$ (B) $\begin{bmatrix} 1 & -2 & 7 \end{bmatrix} + \begin{bmatrix} -2 & 4 & 3 & -1 \end{bmatrix}$

Technology Connections

Graphing calculators can be used to solve problems involving matrix operations. Figure 1 illustrates the solutions to Example 1A and 1B on a graphing calculator.

```
[A]
          [[2 -3 0 ]
           [1 2  -5]]
[B]
          [[3  1 2]
           [-3 2 5]]
[A]+[B]
          [[5  -2 2]
           [-2 4  0]]
```

(a) Example 1A

```
[A]
          [[2 1 4 ]
           [3 2 -3]]
[B]
          [[0  2]
           [-3 5]
           [-1 4]]
[A]+[B]
ERR:DIM MISMATCH
1:Quit
2:Goto
```

(b) Example 1B

> **Figure 1** Matrix addition on a graphing calculator.

Because we add two matrices by adding their corresponding elements (which are real numbers), it follows from the properties of real numbers that matrices of the same size are commutative and associative relative to addition. That is, if A, B, and C are matrices of the same size, then

$$A + B = B + A \qquad \text{Commutative}$$
$$(A + B) + C = A + (B + C) \qquad \text{Associative}$$

$\begin{bmatrix} 0 & 0 & 0 & 0 \\ 0 & 0 & 0 & 0 \\ 0 & 0 & 0 & 0 \end{bmatrix}$

$\begin{bmatrix} 0 \\ 0 \\ 0 \\ 0 \end{bmatrix}$

$\begin{bmatrix} 0 & 0 \\ 0 & 0 \end{bmatrix}$

$\begin{bmatrix} 0 & 0 & 0 \end{bmatrix}$

> **Figure 2** Zero matrices.

A matrix with elements that are all 0's is called a **zero matrix.** Examples of zero matrices are shown in Figure 2.

[*Note:* "0" is often used to denote the zero matrix of any size.]

The **negative of a matrix M,** denoted by $-M$, is a matrix with elements that are the negatives of the elements in M. So if

$$M = \begin{bmatrix} a & b \\ c & d \end{bmatrix}$$

then

$$-M = \begin{bmatrix} -a & -b \\ -c & -d \end{bmatrix}$$

Based on our definition of addition, $M + (-M) = 0$ (a zero matrix).

If A and B are matrices of the same size, then we define **subtraction** as follows.

$$A - B = A + (-B)$$

To subtract matrix B from matrix A, we subtract corresponding elements.

EXAMPLE 2 **Matrix Subtraction**

Subtract: $\begin{bmatrix} 3 & -2 \\ 5 & 0 \end{bmatrix} - \begin{bmatrix} -2 & 2 \\ 3 & 4 \end{bmatrix}$

SOLUTION

$$\begin{bmatrix} 3 & -2 \\ 5 & 0 \end{bmatrix} - \begin{bmatrix} -2 & 2 \\ 3 & 4 \end{bmatrix} = \begin{bmatrix} 3 - (-2) & -2 - 2 \\ 5 - 3 & 0 - 4 \end{bmatrix} = \begin{bmatrix} 5 & -4 \\ 2 & -4 \end{bmatrix}$$

MATCHED PROBLEM 2 Subtract: $[2 \quad -3 \quad 5] - [3 \quad -2 \quad 1]$

EXAMPLE 3 Matrix Equations

Find a, b, c, and d so that

$$\begin{bmatrix} a & b \\ c & d \end{bmatrix} - \begin{bmatrix} 2 & -1 \\ -5 & 6 \end{bmatrix} = \begin{bmatrix} 4 & 3 \\ -2 & 4 \end{bmatrix}$$

SOLUTION

$$\begin{bmatrix} a & b \\ c & d \end{bmatrix} - \begin{bmatrix} 2 & -1 \\ -5 & 6 \end{bmatrix} = \begin{bmatrix} 4 & 3 \\ -2 & 4 \end{bmatrix}$$ Subtract the matrices on the left side.

$$\begin{bmatrix} a - 2 & b - (-1) \\ c - (-5) & d - 6 \end{bmatrix} = \begin{bmatrix} 4 & 3 \\ -2 & 4 \end{bmatrix}$$ Simplify.

$$\begin{bmatrix} a - 2 & b + 1 \\ c + 5 & d - 6 \end{bmatrix} = \begin{bmatrix} 4 & 3 \\ -2 & 4 \end{bmatrix}$$ Set corresponding elements equal to each other.

$$a - 2 = 4 \qquad b + 1 = 3 \qquad c + 5 = -2 \qquad d - 6 = 4$$
$$a = 6 \qquad\quad b = 2 \qquad\quad c = -7 \qquad\quad d = 10$$

MATCHED PROBLEM 3 Find a, b, c, and d so that

$$\begin{bmatrix} a & b \\ c & d \end{bmatrix} - \begin{bmatrix} -4 & 2 \\ 1 & -3 \end{bmatrix} = \begin{bmatrix} -2 & 5 \\ 8 & 2 \end{bmatrix}$$

› **Multiplying a Matrix by a Number**

The **product of a number k and a matrix M,** denoted by kM, is a matrix formed by multiplying each element of M by k.

EXAMPLE 4 Multiplying a Matrix by a Number

Multiply: $-2 \begin{bmatrix} 3 & -1 & 0 \\ -2 & 1 & 3 \\ 0 & -1 & -2 \end{bmatrix}$

SOLUTION

$$-2 \begin{bmatrix} 3 & -1 & 0 \\ -2 & 1 & 3 \\ 0 & -1 & -2 \end{bmatrix} = \begin{bmatrix} -6 & 2 & 0 \\ 4 & -2 & -6 \\ 0 & 2 & 4 \end{bmatrix}$$

MATCHED PROBLEM 4 Multiply: $10 \begin{bmatrix} 1.3 \\ 0.2 \\ 3.5 \end{bmatrix}$

Multiplication of two numbers can be interpreted as repeated addition if one of the numbers is a positive integer. That is,

$$2a = a + a \qquad 3a = a + a + a \qquad 4a = a + a + a + a$$

and so on. How does this apply to multiplication of a matrix by a number?

Matrix operations have many applications, particularly in business.

EXAMPLE 5

Sales and Commissions

Ms. Fong and Mr. Petris are salespeople for a new car agency that sells only two models. August was the last month for this year's models, and next year's models were introduced in September. Gross dollar sales for each month are given in the following matrices:

	AUGUST SALES	
	Compact	Luxury
Fong	$36,000	$72,000
Petris	$72,000	$0

$= A$

	SEPTEMBER SALES	
	Compact	Luxury
Fong	$144,000	$288,000
Petris	$180,000	$216,000

$= B$

For example, Ms. Fong had $36,000 in compact sales in August and Mr. Petris had $216,000 in luxury car sales in September.

(A) What were the combined dollar sales in August and September for each salesperson and each model?

(B) What was the increase in dollar sales from August to September?

(C) If both salespeople receive a 3% commission on gross dollar sales, compute the commission for each salesperson for each model sold in September.

SOLUTIONS

We use matrix addition for part A, matrix subtraction for part B, and multiplication of a matrix by a number for part C.

(A) $A + B = \begin{bmatrix} \$180,000 & \$360,000 \\ \$252,000 & \$216,000 \end{bmatrix}$ Fong Sum of sales for August and September
 Petris

Compact Luxury (above first matrix)

(B) $B - A = \begin{bmatrix} \$108,000 & \$216,000 \\ \$108,000 & \$216,000 \end{bmatrix}$ Fong September sales − August sales
 Petris

(C) $0.03B = \begin{bmatrix} (0.03)(\$144,000) & (0.03)(\$288,000) \\ (0.03)(\$180,000) & (0.03)(\$216,000) \end{bmatrix}$ 3% of September sales

Compact Luxury (above matrix)

$= \begin{bmatrix} \$4,320 & \$8,640 \\ \$5,400 & \$6,480 \end{bmatrix}$ Fong
 Petris

MATCHED PROBLEM 5

Repeat Example 5 with

$$A = \begin{bmatrix} \$72,000 & \$72,000 \\ \$36,000 & \$72,000 \end{bmatrix} \qquad \text{and} \qquad B = \begin{bmatrix} \$180,000 & \$216,000 \\ \$144,000 & \$216,000 \end{bmatrix}$$

Example 5 involved an agency with only two salespeople and two models. A more realistic problem might involve 20 salespeople and 15 models. Problems of this size are often solved using spreadsheets on a computer. Figure 3 illustrates a spreadsheet solution to Example 5.

	A	B	C	D	E	F	G
1		Compact	Luxury	Compact	Luxury	Compact	Luxury
2		August Sales		September Sales		September Commissions	
3	Fong	$36,000	$72,000	$144,000	$288,000	$4,320	$8,640
4	Petris	$72,000	$0	$180,000	$216,000	$5,400	$6,480
5		Combined Sales		Sales Increases			
6	Fong	$180,000	$360,000	$108,000	$216,000		
7	Petris	$252,000	$216,000	$108,000	$216,000		

> Figure 3

> Finding the Product of Two Matrices

Next we will define a way to multiply two matrices. It will probably seem strange to you at first; eventually you will see examples of why it is useful in many problems. In particular, matrix multiplication will help us to develop an alternative method for solving linear systems that have the same number of variables and equations.

We start by defining the product of two special matrices, a row matrix and a column matrix.

> ### DEFINITION 1 Product of a Row Matrix and a Column Matrix

The **product** of a $1 \times n$ row matrix and an $n \times 1$ column matrix is a 1×1 matrix given by

$$\underset{1 \times n}{[a_1 \ a_2 \cdots a_n]} \overset{n \times 1}{\begin{bmatrix} b_1 \\ b_2 \\ \vdots \\ b_n \end{bmatrix}} = [a_1 b_1 + a_2 b_2 + \cdots + a_n b_n]$$

Note that the number of elements in the row matrix and in the column matrix must be the same for the product to be defined.

| EXAMPLE | 6 | Product of a Row Matrix and a Column Matrix |

Multiply: $[2 \quad -3 \quad 0]\begin{bmatrix} -5 \\ 2 \\ -2 \end{bmatrix}$

SOLUTION

$$[2 \quad -3 \quad 0]\begin{bmatrix} -5 \\ 2 \\ -2 \end{bmatrix} = [(2)(-5) + (-3)(2) + (0)(-2)]$$

$$= [-10 - 6 + 0] = [-16]$$

MATCHED PROBLEM 6

Multiply: $[-1 \quad 0 \quad 3 \quad 2]\begin{bmatrix} 2 \\ 3 \\ 4 \\ -1 \end{bmatrix}$

The answer to Example 6 is a 1×1 matrix, which we represented with $[-16]$. From now on, if the result of a calculation is a 1×1 matrix, we'll usually omit the brackets and write the answer as a real number.

| EXAMPLE | 7 | **Production Scheduling** |

A factory produces a slalom water ski that requires 4 labor-hours in the fabricating department and 1 labor-hour in the finishing department. Fabricating personnel receive $10 per hour, and finishing personnel receive $8 per hour. Find the total labor cost per ski.

SOLUTION Total labor cost per ski is given by the product

$$[4 \ \ 1]\begin{bmatrix} 10 \\ 8 \end{bmatrix} = [(4)(10) + (1)(8)] = [40 + 8] = [48] \text{ or \$48 per ski}$$

MATCHED PROBLEM 7 If the factory in Example 7 also produces a trick water ski that requires 6 labor-hours in the fabricating department and 1.5 labor-hours in the finishing department. Find the total-labor cost per ski by multiplying an appropriate row matrix and column matrix.

We will now use the product of a $1 \times n$ row matrix and an $n \times 1$ column matrix to extend the definition of matrix product to more general matrices.

> **DEFINITION 2** Matrix Product

If A is an $m \times p$ matrix and B is a $p \times n$ matrix, then the **matrix product** of A and B, denoted AB, is an $m \times n$ matrix whose element in the ith row and jth column is the real number obtained from the product of the ith row of A and the jth column of B. If the number of columns in A does not equal the number of rows in B, then the matrix product AB is **not defined**.

Must be
the same
$(b = c)$

Size of
product

$\underset{a \times b}{\underbrace{}} \ \underset{c \times d}{\underbrace{}}$ $a \times d$

$A \cdot B = AB$

> Figure 4

It is important to check sizes before starting the multiplication process. If A is an $a \times b$ matrix and B is a $c \times d$ matrix, then if $b = c$, the product AB will exist and will be an $a \times d$ matrix (see Fig. 4). If $b \neq c$, then the product AB does not exist.

The definition is not as complicated as it looks. An example should help clarify the process. For

$$A = \begin{bmatrix} 2 & 3 & -1 \\ -2 & 1 & 2 \end{bmatrix} \quad \text{and} \quad B = \begin{bmatrix} 1 & 3 \\ 2 & 0 \\ -1 & 2 \end{bmatrix}$$

A is 2×3, B is 3×2, and so AB is 2×2. To find the first row of AB, we take the product of the first row of A with every column of B and write each result as a real number, not a 1×1 matrix. The second row of AB is computed in the same manner. The four products of row and column matrices used to produce the four elements in AB are shown in the dashed box below. These products are usually calculated mentally, or with the aid of a calculator, and need not be written out. The shaded portions highlight the steps involved in computing the element in the first row and second column of AB.

<div align="center">

2×3 3×2

$$\begin{bmatrix} 2 & 3 & -1 \\ -2 & 1 & 2 \end{bmatrix} \begin{bmatrix} 1 & 3 \\ 2 & 0 \\ -1 & 2 \end{bmatrix} = \begin{bmatrix} \begin{bmatrix} 2 & 3 & -1 \end{bmatrix}\begin{bmatrix} 1 \\ 2 \\ -1 \end{bmatrix} & \begin{bmatrix} 2 & 3 & -1 \end{bmatrix}\begin{bmatrix} 3 \\ 0 \\ 2 \end{bmatrix} \\ \begin{bmatrix} -2 & 1 & 2 \end{bmatrix}\begin{bmatrix} 1 \\ 2 \\ -1 \end{bmatrix} & \begin{bmatrix} -2 & 1 & 2 \end{bmatrix}\begin{bmatrix} 3 \\ 0 \\ 2 \end{bmatrix} \end{bmatrix}$$

$$= \begin{bmatrix} (2)(1) + (3)(2) + (-1)(-1) & (2)(3) + (3)(0) + (-1)(2) \\ (-2)(1) + (1)(2) + (2)(-1) & (-2)(3) + (1)(0) + (2)(2) \end{bmatrix}$$

2×2

$$= \begin{bmatrix} 9 & 4 \\ -2 & -2 \end{bmatrix}$$

</div>

EXAMPLE	**8**	**Matrix Multiplication**

Given

$$A = \begin{bmatrix} 2 & 1 \\ 1 & 0 \\ -1 & 2 \end{bmatrix} \qquad B = \begin{bmatrix} 1 & -1 & 0 & 1 \\ 2 & 1 & 2 & 0 \end{bmatrix} \qquad C = \begin{bmatrix} 2 & 6 \\ -1 & -3 \end{bmatrix} \qquad D = \begin{bmatrix} 1 & 2 \\ 3 & 6 \end{bmatrix}$$

Find each product that is defined:

(A) AB (B) BA (C) CD (D) DC

SOLUTIONS

(A) $AB = \overset{3 \times 2}{\begin{bmatrix} 2 & 1 \\ 1 & 0 \\ -1 & 2 \end{bmatrix}} \overset{2 \times 4}{\begin{bmatrix} 1 & -1 & 0 & 1 \\ 2 & 1 & 2 & 0 \end{bmatrix}}$

$$= \begin{bmatrix} (2)(1) + (1)(2) & (2)(-1) + (1)(1) & (2)(0) + (1)(2) & (2)(1) + (1)(0) \\ (1)(1) + (0)(2) & (1)(-1) + (0)(1) & (1)(0) + (0)(2) & (1)(1) + (0)(0) \\ (-1)(1) + (2)(2) & (-1)(-1) + (2)(1) & (-1)(0) + (2)(2) & (-1)(1) + (2)(0) \end{bmatrix}$$

3×4

$$= \begin{bmatrix} 4 & -1 & 2 & 2 \\ 1 & -1 & 0 & 1 \\ 3 & 3 & 4 & -1 \end{bmatrix}$$

(B) $BA = \overset{2 \times 4}{\begin{bmatrix} 1 & -1 & 0 & 1 \\ 2 & 1 & 2 & 0 \end{bmatrix}} \overset{3 \times 2}{\begin{bmatrix} 2 & 1 \\ 1 & 0 \\ -1 & 2 \end{bmatrix}}$

Product is not defined.

(C) $CD = \overset{2 \times 2}{\begin{bmatrix} 2 & 6 \\ -1 & -3 \end{bmatrix}} \overset{2 \times 2}{\begin{bmatrix} 1 & 2 \\ 3 & 6 \end{bmatrix}} = \begin{bmatrix} (2)(1) + (6)(3) & (2)(2) + (6)(6) \\ (-1)(1) + (-3)(3) & (-1)(2) + (-3)(6) \end{bmatrix}$

2×2

$$= \begin{bmatrix} 20 & 40 \\ -10 & -20 \end{bmatrix}$$

$$(D)\ DC = \begin{bmatrix} 1 & 2 \\ 3 & 6 \end{bmatrix}^{2\times 2} \begin{bmatrix} 2 & 6 \\ -1 & -3 \end{bmatrix}^{2\times 2} = \begin{bmatrix} (1)(2) + (2)(-1) & (1)(6) + (2)(-3) \\ (3)(2) + (6)(-1) & (3)(6) + (6)(-3) \end{bmatrix} = \begin{bmatrix} 0 & 0 \\ 0 & 0 \end{bmatrix}^{2\times 2}$$

MATCHED PROBLEM 8

Find each product, if it is defined:

$$(A)\ \begin{bmatrix} -1 & 0 & 3 & -2 \\ 1 & 2 & 2 & 0 \end{bmatrix} \begin{bmatrix} -1 & 1 \\ 2 & 3 \\ 1 & 0 \end{bmatrix} \qquad (B)\ \begin{bmatrix} -1 & 1 \\ 2 & 3 \\ 1 & 0 \end{bmatrix} \begin{bmatrix} -1 & 0 & 3 & -2 \\ 1 & 2 & 2 & 0 \end{bmatrix}$$

$$(C)\ \begin{bmatrix} 1 & 2 \\ -1 & -2 \end{bmatrix} \begin{bmatrix} -2 & 4 \\ 1 & -2 \end{bmatrix} \qquad (D)\ \begin{bmatrix} -2 & 4 \\ 1 & -2 \end{bmatrix} \begin{bmatrix} 1 & 2 \\ -1 & -2 \end{bmatrix}$$

In the arithmetic of real numbers, it doesn't matter in which order we multiply; for example, $5 \times 7 = 7 \times 5$. In matrix multiplication, however, it does make a difference. That is, AB does not always equal BA, even if both multiplications are defined and both products are the same size (see Examples 8C and 8D). In other words,

Matrix multiplication is not commutative.

Also, AB may be zero with neither A nor B equal to zero (see Example 8D). That is,

The zero property does not hold for matrix multiplication.

(See Section R-1 for a discussion of the zero property for real numbers.)

Just as we used the familiar algebraic notation AB to represent the product of matrices A and B, we use the notation A^2 for AA (the product of A with itself), A^3 for AAA, and so on.

⟩⟩⟩ EXPLORE-DISCUSS 2

In addition to the commutative and zero properties, there are other significant differences between real number multiplication and matrix multiplication.

(A) In real number multiplication, the only real number whose square is 0 is the real number 0 ($0^2 = 0$). Find at least one 2×2 matrix A with all elements nonzero such that $A^2 = 0$, where 0 is the 2×2 zero matrix.

(B) In real number multiplication, the only nonzero real number that is equal to its square is the real number 1 ($1^2 = 1$). Find at least one 2×2 matrix A with all elements nonzero such that $A^2 = A$.

We'll return to our study of the properties of matrix multiplication in Section 10-4. We will conclude this section with an application of matrix multiplication.

EXAMPLE 9 Labor Costs

If we combine the time requirements for making slalom and trick water skis discussed in Example 7 and Matched Problem 7, we get

Labor-hours per ski

	Assembly department	Finishing department
Trick ski	6 h	1.5 h
Slalom ski	4 h	1 h

$$= L$$

Now suppose that the company has two manufacturing plants, X and Y, in different parts of the country and that the hourly rates for each department are given in the following matrix:

Hourly Wages

	Plant X	Plant Y	
Assembly department	$10	$12	$= H$
Finishing department	$ 8	$10	

Find the matrix products HL and LH, and decide if either matrix has a meaningful interpretation in terms of ski production.

SOLUTION Since H and L are both 2×2 matrices, we can find the product of H and L in either order and the result will be a 2×2 matrix:

$$HL = \begin{bmatrix} 10 & 12 \\ 8 & 10 \end{bmatrix} \begin{bmatrix} 6 & 1.5 \\ 4 & 1 \end{bmatrix} = \begin{bmatrix} 108 & 27 \\ 88 & 22 \end{bmatrix}$$

$$LH = \begin{bmatrix} 6 & 1.5 \\ 4 & 1 \end{bmatrix} \begin{bmatrix} 10 & 12 \\ 8 & 10 \end{bmatrix} = \begin{bmatrix} 72 & 87 \\ 48 & 58 \end{bmatrix}$$

How can we interpret the elements in these products? Let's begin with the product HL. The element 108 in the first row and first column of HL is the product of the first row matrix of H and the first column matrix of L:

Plant Plant
X Y

$$[10 \quad 12] \begin{bmatrix} 6 \\ 4 \end{bmatrix} \begin{array}{l} \text{Trick} \\ \text{Slalom} \end{array} = 10(6) + 12(4) = 60 + 48 = 108$$

Notice that $60 is the labor cost for assembling a trick ski at Plant X and $48 is the labor cost for assembling a slalom ski at Plant Y. Although both numbers represent labor costs, it makes no sense to add them together. They do not pertain to the same type of ski or to the same plant. So, even though the product HL happens to be defined mathematically, it has no useful interpretation in this problem.

Now let's consider the product LH. The element 72 in the first row and first column of LH is given by the following product:

Assembly Finishing

$$[6 \quad 1.5] \begin{bmatrix} 10 \\ 8 \end{bmatrix} \begin{array}{l} \text{Assembly} \\ \text{Finishing} \end{array} = 6(10) + 1.5(8) = 60 + 12 = 72$$

where $60 is the labor cost for assembling a trick ski at Plant X and $12 is the labor cost for finishing a trick ski at Plant X. The sum is the total labor cost for producing a trick ski at Plant X. The other elements in LH also represent total labor costs, as indicated by the row and column labels shown below:

Labor costs per ski

	Plant X	Plant Y	
$LH =$	$72	$87	Trick ski
	$48	$58	Slalom ski

MATCHED PROBLEM 9 Refer to Example 9. The company wants to know how many hours to schedule in each department in order to produce 1,000 trick skis and 2,000 slalom skis. These production requirements can be represented by either of the following matrices:

Trick skis Slalom skis

$$P = [1{,}000 \quad 2{,}000] \qquad Q = \begin{bmatrix} 1{,}000 \\ 2{,}000 \end{bmatrix} \begin{array}{l} \text{Trick skis} \\ \text{Slalom skis} \end{array}$$

Using the labor-hour matrix L from Example 9, find PL or LQ, whichever has a meaningful interpretation for this problem, and label the rows and columns accordingly.

>>> **CAUTION** >>>

Example 9 and Matched Problem 9 illustrate an important point about matrix multiplication. Even if you are using a graphing calculator to perform the calculations in a matrix product, you will still need to know the definition of matrix multiplication so that you can interpret the results correctly.

ANSWERS TO MATCHED PROBLEMS

1. (A) $\begin{bmatrix} 1 & 5 \\ 0 & -2 \\ 2 & 1 \end{bmatrix}$ (B) Not defined 2. $[-1 \quad -1 \quad 4]$ 3. $a = -6, b = 7, c = 9, d = -1$

4. $\begin{bmatrix} 13 \\ 2 \\ 35 \end{bmatrix}$ 5. (A) $\begin{bmatrix} \$252,000 & \$288,000 \\ \$180,000 & \$288,000 \end{bmatrix}$ (B) $\begin{bmatrix} \$108,000 & \$144,000 \\ \$108,000 & \$144,000 \end{bmatrix}$ (C) $\begin{bmatrix} \$5,400 & \$6,480 \\ \$4,320 & \$6,480 \end{bmatrix}$

6. $[8]$ 7. $[6 \quad 1.5]\begin{bmatrix} 10 \\ 8 \end{bmatrix} = [72]$ or \$72

8. (A) Not defined (B) $\begin{bmatrix} 2 & 2 & -1 & 2 \\ 1 & 6 & 12 & -4 \\ -1 & 0 & 3 & -2 \end{bmatrix}$ (C) $\begin{bmatrix} 0 & 0 \\ 0 & 0 \end{bmatrix}$ (D) $\begin{bmatrix} -6 & -12 \\ 3 & 6 \end{bmatrix}$

9.
Assembly Finishing
$PL = [14,000 \quad 3,500]$ Labor hours

10-3 Exercises

1. What conditions must matrices A and B satisfy so that $A + B$ exists?

2. What conditions must matrices A and B satisfy so that AB exists?

3. What conditions must matrices A and B satisfy so that BA exists?

4. What conditions must matrices A and B satisfy so that both AB and BA exist?

5. What is the negative of a matrix?

6. How do you subtract two matrices?

7. How do you multiply a matrix by a number?

8. If A is a $1 \times n$ matrix and B is an $n \times 1$ matrix, how do you find the product AB? What is the size of AB?

9. If A is a $1 \times n$ matrix and B is an $n \times 1$ matrix, how do you find the product BA? What is the size of BA?

10. Describe the operation of matrix multiplication in your own words.

Perform the indicated operations in Problems 11–24, if possible.

11. $\begin{bmatrix} 5 & -2 \\ 3 & 0 \end{bmatrix} + \begin{bmatrix} -3 & 7 \\ 1 & -6 \end{bmatrix}$ **12.** $\begin{bmatrix} 0 & 8 \\ 2 & -1 \end{bmatrix} + \begin{bmatrix} 9 & -4 \\ 7 & 5 \end{bmatrix}$

13. $\begin{bmatrix} 4 & 0 \\ -2 & 3 \\ 8 & 1 \end{bmatrix} + \begin{bmatrix} -1 & 2 \\ 0 & 5 \\ 4 & -6 \end{bmatrix}$

14. $\begin{bmatrix} 6 & -2 & 3 \\ 4 & -8 & -7 \end{bmatrix} + \begin{bmatrix} 3 & 9 & -1 \\ 6 & -2 & 4 \end{bmatrix}$

15. $\begin{bmatrix} 4 & 0 \\ -2 & 3 \\ 8 & 1 \end{bmatrix} + \begin{bmatrix} -1 & 0 & 4 \\ 2 & 5 & -6 \end{bmatrix}$

16. $\begin{bmatrix} 6 & -2 & 3 \\ 4 & -8 & -7 \end{bmatrix} + \begin{bmatrix} 3 & 6 \\ 9 & -2 \\ -1 & 4 \end{bmatrix}$

17. $\begin{bmatrix} 5 & -1 & 0 \\ 4 & 6 & 3 \end{bmatrix} - \begin{bmatrix} 2 & 4 & -6 \\ 3 & 5 & -5 \end{bmatrix}$

18. $\begin{bmatrix} 6 & 2 \\ -4 & 1 \\ 3 & 0 \end{bmatrix} - \begin{bmatrix} 0 & 5 \\ -7 & 2 \\ -1 & 0 \end{bmatrix}$

19. $\begin{bmatrix} 4 & -7 \\ 10 & 11 \\ -13 & -9 \end{bmatrix} - \begin{bmatrix} 4 & 10 & -13 \\ -7 & 11 & -9 \end{bmatrix}$

20. $\begin{bmatrix} \frac{1}{2} & -\frac{3}{4} \\ \frac{5}{4} & -\frac{3}{2} \end{bmatrix} - \begin{bmatrix} \frac{9}{2} & \frac{1}{4} \\ -\frac{7}{4} & \frac{1}{2} \end{bmatrix}$

21. $\begin{bmatrix} 2.4 & -2.8 & 3.9 \\ -1.6 & 0 & 4.2 \end{bmatrix} - \begin{bmatrix} 7 & -2.2 & -2.2 \\ -3.2 & -3.2 & 1 \end{bmatrix}$

22. $\begin{bmatrix} 10 \\ 20 \end{bmatrix} - [20 \quad 10]$ **23.** $4\begin{bmatrix} 3 & -4 & 7 \\ -2 & 9 & 5 \end{bmatrix}$

24. $5\begin{bmatrix} -7 & 3 & 0 & 9 \\ 4 & -5 & 6 & 2 \end{bmatrix}$

Find the products in Problems 25–38.

25. $[5 \quad 3]\begin{bmatrix} 4 \\ 7 \end{bmatrix}$ **26.** $[-2 \quad 4]\begin{bmatrix} 3 \\ -8 \end{bmatrix}$

27. $\begin{bmatrix} -5 \\ -3 \end{bmatrix}[4 \quad -2]$ **28.** $\begin{bmatrix} 3 \\ -4 \end{bmatrix}[2 \quad -1]$

29. $[3 \quad -2 \quad -4]\begin{bmatrix} 1 \\ 2 \\ -3 \end{bmatrix}$ **30.** $[1 \quad -2 \quad 2]\begin{bmatrix} 2 \\ -1 \\ 1 \end{bmatrix}$

31. $\begin{bmatrix} 1 \\ 2 \\ -3 \end{bmatrix}[3 \quad -2 \quad -4]$ **32.** $\begin{bmatrix} 2 \\ -1 \\ 1 \end{bmatrix}[1 \quad -2 \quad 2]$

33. $\begin{bmatrix} -6 & 3 \\ 2 & -5 \end{bmatrix}\begin{bmatrix} 1 \\ 3 \end{bmatrix}$ **34.** $\begin{bmatrix} 3 & 7 \\ -1 & -9 \end{bmatrix}\begin{bmatrix} 4 \\ -1 \end{bmatrix}$

35. $\begin{bmatrix} 5 & 1 \\ 4 & 6 \end{bmatrix}\begin{bmatrix} 2 & 0 \\ 3 & 8 \end{bmatrix}$ **36.** $\begin{bmatrix} -2 & 7 \\ 3 & -1 \end{bmatrix}\begin{bmatrix} 4 & -1 \\ 0 & 5 \end{bmatrix}$

37. $\begin{bmatrix} 8 & -3 \\ -5 & 3 \end{bmatrix}\begin{bmatrix} 2 & 0 \\ 0 & 6 \end{bmatrix}$ **38.** $\begin{bmatrix} 7 & 0 \\ 0 & 3 \end{bmatrix}\begin{bmatrix} 9 & -2 \\ -4 & -1 \end{bmatrix}$

Problems 39–56 refer to the following matrices.

$$A = \begin{bmatrix} 2 & -1 & 3 \\ 0 & 4 & -2 \end{bmatrix} \quad B = \begin{bmatrix} -3 & 1 \\ 2 & 5 \end{bmatrix}$$

$$C = \begin{bmatrix} -1 & 0 & 2 \\ 4 & -3 & 1 \\ -2 & 3 & 5 \end{bmatrix} \quad D = \begin{bmatrix} 3 & -2 \\ 0 & -1 \\ 1 & 2 \end{bmatrix}$$

Perform the indicated operations, if possible.

39. CA **40.** AC **41.** BA

42. AB **43.** C^2 **44.** B^2

45. $C + DA$ **46.** $B + AD$ **47.** $0.2CD$

48. $0.1DB$ **49.** $2DB + 5CD$ **50.** $3BA + 4AC$

51. $(-1)AC + 3DB$ **52.** $(-2)BA + 6CD$

53. CDA **54.** ACD

55. DBA **56.** BAD

In Problems 57 and 58, use a graphing calculator to calculate B, B^2, B^3, ... and AB, AB^2, AB^3, Describe any patterns you observe in each sequence of matrices.

57. $A = [0.3 \quad 0.7]$ and $B = \begin{bmatrix} 0.4 & 0.6 \\ 0.2 & 0.8 \end{bmatrix}$

58. $A = [0.4 \quad 0.6]$ and $B = \begin{bmatrix} 0.9 & 0.1 \\ 0.3 & 0.7 \end{bmatrix}$

59. Find a, b, c, and d so that

$$\begin{bmatrix} a & b \\ c & d \end{bmatrix} + \begin{bmatrix} 2 & -3 \\ 0 & 1 \end{bmatrix} = \begin{bmatrix} 1 & -2 \\ 3 & -4 \end{bmatrix}$$

60. Find x and y so that

$$\begin{bmatrix} 3x & 5 \\ -1 & 4x \end{bmatrix} + \begin{bmatrix} 2y & -3 \\ -6 & -y \end{bmatrix} = \begin{bmatrix} 7 & 2 \\ -7 & 2 \end{bmatrix}$$

61. Find w, x, y, and z so that

$$\begin{bmatrix} 3 & 0 \\ -7 & -11 \end{bmatrix} + \begin{bmatrix} w & x \\ y & z \end{bmatrix} = \begin{bmatrix} 9 & 1 \\ 4 & 6 \end{bmatrix}$$

62. Find x and y so that

$$\begin{bmatrix} x & -1 \\ -2 & y \end{bmatrix} - \begin{bmatrix} 4y & 4 \\ 5 & 3x \end{bmatrix} = \begin{bmatrix} 12 & -5 \\ -4 & -14 \end{bmatrix}$$

In Problems 63 and 64, let a, b, and c be any nonzero real numbers, and let

$$A = \begin{bmatrix} a & b \\ c & -a \end{bmatrix} \text{ and } I = \begin{bmatrix} 1 & 0 \\ 0 & 1 \end{bmatrix}$$

63. If $A^2 = 0$, how are a, b, and c related? Use this relationship to provide several examples of 2×2 matrices with no zero entries whose square is the zero matrix.

64. If $A^2 = I$, how are a, b, and c related? Use this relationship to provide several examples of 2×2 matrices with no zero entries whose square is the matrix I.

Problems 65 and 66 refer to the matrices

$$A = \begin{bmatrix} a & b \\ c & d \end{bmatrix} \text{ and } B = \begin{bmatrix} 1 & 1 \\ 1 & 1 \end{bmatrix}$$

65. If $AB = 0$, how are a, b, c, and d related? Use this relationship to provide several examples of 2×2 matrices A with no zero entries that satisfy $AB = 0$.

66. If $BA = 0$, how are a, b, c, and d related? Use this relationship to provide several examples of 2×2 matrices A with no zero entries that satisfy $BA = 0$.

67. Find x and y so that

$$\begin{bmatrix} 1 & 3 \\ -2 & -2 \end{bmatrix}\begin{bmatrix} x & 1 \\ 3 & 2 \end{bmatrix} = \begin{bmatrix} y & 7 \\ y & -6 \end{bmatrix}$$

68. Find x and y so that

$$\begin{bmatrix} x & -1 \\ 1 & 0 \end{bmatrix}\begin{bmatrix} 2 & 1 \\ 4 & 1 \end{bmatrix} = \begin{bmatrix} y & y \\ 2 & 1 \end{bmatrix}$$

69. Find a, b, c, and d so that

$$\begin{bmatrix} 1 & 3 \\ 1 & 4 \end{bmatrix}\begin{bmatrix} a & b \\ c & d \end{bmatrix} = \begin{bmatrix} 6 & -5 \\ 7 & -7 \end{bmatrix}$$

70. Find a, b, c, and d so that

$$\begin{bmatrix} 1 & -2 \\ 2 & -3 \end{bmatrix}\begin{bmatrix} a & b \\ c & d \end{bmatrix} = \begin{bmatrix} 1 & 0 \\ 3 & 2 \end{bmatrix}$$

71. A square matrix is a **diagonal matrix** if all elements not on the principal diagonal are zero. So a 2×2 diagonal matrix has the form

$$A = \begin{bmatrix} a & 0 \\ 0 & d \end{bmatrix}$$

where a and d are any real numbers. Discuss the validity of each of the following statements. If the statement is always true, explain why. If not, give examples.

(A) If A and B are 2×2 diagonal matrices, then $A + B$ is a 2×2 diagonal matrix.

(B) If A and B are 2×2 diagonal matrices, then $A + B = B + A$.

(C) If A and B are 2×2 diagonal matrices, then AB is a 2×2 diagonal matrix.

(D) If A and B are 2×2 diagonal matrices, then $AB = BA$.

72. A square matrix is an **upper triangular matrix** if all elements below the principal diagonal are zero. So a 2×2 upper triangular matrix has the form

$$A = \begin{bmatrix} a & b \\ 0 & d \end{bmatrix}$$

where a, b, and d are any real numbers. Discuss the validity of each of the following statements. If the statement is always true, explain why. If not, give examples.

(A) If A and B are 2×2 upper triangular matrices, then $A + B$ is a 2×2 upper triangular matrix.

(B) If A and B are 2×2 upper triangular matrices, then $A + B = B + A$.

(C) If A and B are 2×2 upper triangular matrices, then AB is a 2×2 upper triangular matrix.

(D) If A and B are 2×2 upper triangular matrices, then $AB = BA$.

73. A company with two different plants makes satellite radios and GPS units. The production costs for each item are given in the following matrices:

	Plant X Radio	Plant X GPS		Plant Y Radio	Plant Y GPS	
Materials	$30	$25	$= A$	$36	$27	$= B$
Labor	$60	$80		$54	$74	

Find the matrix $\frac{1}{2}(A + B)$, and explain what information it provides.

74. Suppose that the company in Problem 73 experiences an increase in the cost of both labor and materials at plant X. Find the matrix $\frac{1}{2}(1.2A + B)$. If it provides the average cost of production for the two plants, by how much were the costs at plant X increased?

75. MARKUP An import car dealer sells three models of a car. Current dealer invoice price (cost) and the retail price for the basic models and the indicated options are given in the following two matrices (where "Air" means air conditioning):

Dealer Invoice Price

	Basic Car	Air	CD changer	Cruise Control	
Model A	$10,400	$682	$215	$182	$= M$
Model B	$12,500	$721	$295	$182	
Model C	$16,400	$827	$443	$192	

Retail Price

	Basic Car	Air	CD changer	Cruise Control	
Model A	$13,900	$783	$263	$215	$= N$
Model B	$15,000	$838	$395	$236	
Model C	$18,300	$967	$573	$248	

We define the markup matrix to be $N - M$ (**markup** is the difference between the retail price and the dealer invoice price). Suppose the value of the dollar has had a sharp decline and the dealer invoice price is to have an across-the-board 15% increase next year. To stay competitive with domestic cars, the dealer increases the retail prices only 10%. Calculate a markup matrix for next year's models and the indicated options. (Compute results to the nearest dollar.)

76. MARKUP Referring to Problem 75, what is the markup matrix resulting from a 20% increase in dealer invoice prices and an increase in retail prices of 15%? (Compute results to the nearest dollar.)

77. LABOR COSTS A company with manufacturing plants located in different parts of the country has labor-hour and wage requirements for the manufacturing of three types of inflatable boats as given in the following two matrices:

Labor-Hours per Boat

	Cutting Department	Assembly Department	Packaging Department	
$M =$	0.6 h	0.6 h	0.2 h	One-person boat
	1.0 h	0.9 h	0.3 h	Two-person boat
	1.5 h	1.2 h	0.4 h	Four-person boat

Hourly Wages

	Plant I	Plant II	
$N =$	$8	$9	Cutting department
	$10	$12	Assembly department
	$5	$6	Packaging department

(A) Find the labor costs for a one-person boat manufactured at plant I.

(B) Find the labor costs for a four-person boat manufactured at plant II.

(C) Discuss possible interpretations of the elements in the matrix products MN and NM.

(D) If either of the products MN or NM has a meaningful interpretation, find the product and label its rows and columns.

78. INVENTORY VALUE A personal computer retail company sells five different computer models through three stores located in a large metropolitan area. The inventory of each model on hand in each store is summarized in matrix M. Wholesale (W) and retail (R) values of each model computer are summarized in matrix N.

Model

	A	B	C	D	E	
$M =$	4	2	3	7	1	Store 1
	2	3	5	0	6	Store 2
	10	4	3	4	3	Store 3

	W	R	
$N =$	$700	$840	A
	$1,400	$1,800	B
	$1,800	$2,400	C
	$2,700	$3,300	D
	$3,500	$4,900	E

(A) What is the retail value of the inventory at store 2?

(B) What is the wholesale value of the inventory at store 3?

(C) Discuss possible interpretations of the elements in the matrix products MN and NM.

(D) If either of the products *MN* or *NM* has a meaningful interpretation, find the product and label its rows and columns.

(E) Discuss methods of matrix multiplication that can be used to find the total inventory of each model on hand at all three stores. State the matrices that can be used, and perform the necessary operations.

(F) Discuss methods of matrix multiplication that can be used to find the total inventory of all five models at each store. State the matrices that can be used, and perform the necessary operations.

79. AIRFREIGHT A nationwide airfreight service has connecting flights between five cities, as illustrated in the figure. To represent this schedule in matrix form, we construct a 5×5 **incidence matrix** *A*, where the rows represent the origins of each flight and the columns represent the destinations. We place a 1 in the *i*th row and *j*th column of this matrix if there is a connecting flight from the *i*th city to the *j*th city and a 0 otherwise. We also place 0s on the principal diagonal, because a connecting flight with the same origin and destination does not make sense.

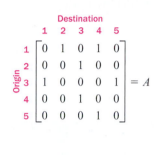

$$ \begin{array}{c} \\ \text{Origin} \\ \\ \end{array} \begin{array}{c} 1 \\ 2 \\ 3 \\ 4 \\ 5 \end{array} \begin{bmatrix} 0 & 1 & 0 & 1 & 0 \\ 0 & 0 & 1 & 0 & 0 \\ 1 & 0 & 0 & 0 & 1 \\ 0 & 0 & 1 & 0 & 0 \\ 0 & 0 & 0 & 1 & 0 \end{bmatrix} = A $$

Now that the schedule has been represented in the mathematical form of a matrix, we can perform operations on this matrix to obtain information about the schedule.

(A) Find A^2. What does the 1 in row 2 and column 1 of A^2 indicate about the schedule? What does the 2 in row 1 and column 3 indicate about the schedule? In general, how would you interpret each element off the principal diagonal of A^2? [*Hint:* Examine the diagram for possible connections between the *i*th city and the *j*th city.]

(B) Find A^3. What does the 1 in row 4 and column 2 of A^3 indicate about the schedule? What does the 2 in row 1 and column 5 indicate about the schedule? In general, how would you interpret each element off the principal diagonal of A^3?

(C) Compute $A, A + A^2, A + A^2 + A^3, \dots$, until you obtain a matrix with no zero elements (except possibly on the principal diagonal), and interpret.

80. AIRFREIGHT Refer to Problem 79. Find the incidence matrix *A* for the flight schedule illustrated in the figure. Compute $A, A + A^2, A + A^2 + A^3, \dots$, until you obtain a matrix with no zero elements (except possibly on the principal diagonal), and interpret.

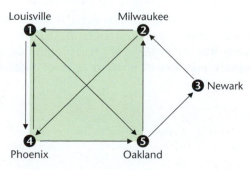

81. POLITICS In a local election, a group hired a public relations firm to promote its candidate in three ways: telephone, house calls, and direct mail. The cost per contact is given in matrix *M*:

$$ \begin{matrix} & \text{Cost per} \\ & \text{Contact} \end{matrix} $$

$$ M = \begin{bmatrix} \$0.80 \\ \$1.50 \\ \$0.40 \end{bmatrix} \begin{matrix} \text{Telephone} \\ \text{House Call} \\ \text{Mail} \end{matrix} $$

The number of contacts of each type made in two adjacent cities is given in matrix *N*:

$$ N = \begin{matrix} \text{Telephone} & \text{House Call} & \text{Mail} \\ \begin{bmatrix} 1{,}000 & 500 & 5{,}000 \\ 2{,}000 & 800 & 8{,}000 \end{bmatrix} & & \begin{matrix} \text{Berkeley} \\ \text{Oakland} \end{matrix} \end{matrix} $$

(A) Find the total amount spent in Berkeley.

(B) Find the total amount spent in Oakland.

(C) Discuss possible interpretations of the elements in the matrix products *MN* and *NM*.

(D) If either of the products *MN* or *NM* has a meaningful interpretation, find the product and label its rows and columns.

(E) Discuss methods of matrix multiplication that can be used to find the total number of telephone calls, house calls, and letters. State the matrices that can be used, and perform the necessary operations.

(F) Discuss methods of matrix multiplication that can be used to find the total number of contacts in Berkeley and in Oakland. State the matrices that can be used, and perform the necessary operations.

82. NUTRITION A nutritionist for a cereal company blends two cereals in different mixes. The amounts of protein, carbohydrate, and fat (in grams per ounce) in each cereal are given by matrix *M*. The amounts of each cereal used in the three mixes are given by matrix *N*.

$$ M = \begin{matrix} \text{Cereal} & \text{Cereal} \\ A & B \end{matrix} \\ \begin{bmatrix} 4 \text{ g/oz} & 2 \text{ g/oz} \\ 20 \text{ g/oz} & 16 \text{ g/oz} \\ 3 \text{ g/oz} & 1 \text{ g/oz} \end{bmatrix} \begin{matrix} \text{Protein} \\ \text{Carbohydrate} \\ \text{Fat} \end{matrix} $$

$$ N = \begin{matrix} \text{Mix X} & \text{Mix Y} & \text{Mix Z} \\ \begin{bmatrix} 15 \text{ oz} & 10 \text{ oz} & 5 \text{ oz} \\ 5 \text{ oz} & 10 \text{ oz} & 15 \text{ oz} \end{bmatrix} & & \begin{matrix} \text{Cereal A} \\ \text{Cereal B} \end{matrix} \end{matrix} $$

(A) Find the amount of protein in mix *X*.

(B) Find the amount of fat in mix *Z*.

(C) Discuss possible interpretations of the elements in the matrix products *MN* and *NM*.

(D) If either of the products *MN* or *NM* has a meaningful interpretation, find the product and label its rows and columns.

83. TOURNAMENT SEEDING To rank players for an upcoming tennis tournament, a club decides to have each player play one set with every other player. The results are given in the table.

Player	Defeated
1. Aaron	Charles, Dan, Elvis
2. Bart	Aaron, Dan, Elvis
3. Charles	Bart, Dan
4. Dan	Frank
5. Elvis	Charles, Dan, Frank
6. Frank	Aaron, Bart, Charles

Player	Defeated
1. Anne	Diane
2. Bridget	Anne, Carol, Diane
3. Carol	Anne
4. Diane	Carol, Erlene
5. Erlene	Anne, Bridget, Carol

(A) Express the outcomes as an incidence matrix A by placing a 1 in the ith row and jth column of A if player i defeated player j, and a 0 otherwise (see Problem 79).

(B) Compute the matrix $B = A + A^2$

(C) Discuss matrix multiplication methods that can be used to find the sum of each of the rows in B. State the matrices that can be used and perform the necessary operations.

(D) Rank the players from strongest to weakest. Explain the reasoning behind your ranking.

84. PLAYER RANKING Each member of a chess team plays one match with every other player. The results are given in the table.

(A) Express the outcomes as an incidence matrix A by placing a 1 in the ith row and jth column of A if player i defeated player j, and a 0 otherwise (see Problem 79).

(B) Compute the matrix $B = A + A^2$

(C) Discuss matrix multiplication methods that can be used to find the sum of each of the rows in B. State the matrices that can be used and perform the necessary operations.

(D) Rank the players from strongest to weakest. Explain the reasoning behind your ranking.

10-4 Solving Systems of Linear Equations Using Matrix Inverse Methods

> **The Identity Matrix for Multiplication**
> **Finding the Inverse of a Square Matrix**
> **Matrix Equations**
> **Matrix Equations and Systems of Linear Equations**
> **Application: Cryptography**

Now that we know a bit about matrix multiplication, we will see how it can be used to solve certain systems of equations.

> The Identity Matrix for Multiplication

We know that for any real number a, $1 \cdot a = a \cdot 1 = a$. The number 1 is called the *identity* for real number multiplication. Is there a matrix analog? That is, if M is an arbitrary matrix, is there a matrix I with the property that $IM = MI = M$? It turns out that, in general, the answer is no. But the set of **square matrices of order n** (matrices with n rows and n columns) does have an identity.

(A) Pick any 2×2 matrix you like, and multiply it by the following matrix in both possible orders.

$$\begin{bmatrix} 1 & 0 \\ 0 & 1 \end{bmatrix}$$

(B) Repeat (A) for any 3×3 matrix you like, but multiply by the matrix

$$\begin{bmatrix} 1 & 0 & 0 \\ 0 & 1 & 0 \\ 0 & 0 & 1 \end{bmatrix}$$

What can you conclude?

> **DEFINITION 1** Identity Matrix

The **identity matrix for multiplication** for the set of all square matrices of order n is the square matrix of order n, denoted by I, with 1's along the principal diagonal (from upper left corner to lower right corner) and 0's elsewhere.

In Explore-Discuss 1, we saw that

$$\begin{bmatrix} 1 & 0 \\ 0 & 1 \end{bmatrix} \quad \text{and} \quad \begin{bmatrix} 1 & 0 & 0 \\ 0 & 1 & 0 \\ 0 & 0 & 1 \end{bmatrix}$$

are the identity matrices for square matrices of order 2 and 3, respectively.

We will show in Exercises 10-4 that if M is any square matrix of order n and I is the identity matrix of order n, then

$$IM = MI = M$$

Note: If M is an $m \times n$ matrix that is not square ($m \neq n$), then it is still possible to multiply M on the left and on the right by an identity matrix, but not with the same-size identity matrix. To avoid the complications involved with associating two different identity matrices with each nonsquare matrix, we will restrict our attention in this section to square matrices.

> ## Finding the Inverse of a Square Matrix

In the set of real numbers, we know that for each real number a, except 0, there exists a real number a^{-1} such that

$$a^{-1}a = 1$$

The number a^{-1} is called the *inverse* of the number a relative to multiplication, or the *multiplicative inverse* of a. For example, 2^{-1} is the *multiplicative inverse* of 2, since $2^{-1}(2) = 1$. We will use this idea to define the *inverse of a square matrix*.

> › **DEFINITION 2** Inverse of a Square Matrix
>
> If A is a square matrix of order n and if there exists a matrix A^{-1} (read "A inverse") such that
>
> $$A^{-1}A = AA^{-1} = I$$
>
> then A^{-1} is called the **multiplicative inverse of A** or, more simply, the **inverse of A.** If no such matrix exists, then A is said to be a **singular matrix.**

>>> EXPLORE-DISCUSS 2

Let $A = \begin{bmatrix} 4 & 2 \\ 2 & 2 \end{bmatrix}$ $B = \begin{bmatrix} \frac{1}{4} & \frac{1}{2} \\ \frac{1}{2} & \frac{1}{2} \end{bmatrix}$ $C = \begin{bmatrix} \frac{1}{2} & -\frac{1}{2} \\ -\frac{1}{2} & 1 \end{bmatrix}$

(A) How are the entries in A and B related?

(B) Find AB. Is B the inverse of A?

(C) Find AC. Is C the inverse of A?

The multiplicative inverse of a nonzero real number a also can be written as $1/a$, but this notation is never used for matrix inverses.

Let's use Definition 2 to find A^{-1}, if it exists, for

$$A = \begin{bmatrix} 2 & 3 \\ 1 & 2 \end{bmatrix}$$

We are looking for a matrix

$$A^{-1} = \begin{bmatrix} a & c \\ b & d \end{bmatrix}$$

such that

$$AA^{-1} = A^{-1}A = I$$

We can write

$$\underset{A}{\begin{bmatrix} 2 & 3 \\ 1 & 2 \end{bmatrix}} \underset{A^{-1}}{\begin{bmatrix} a & c \\ b & d \end{bmatrix}} = \underset{I}{\begin{bmatrix} 1 & 0 \\ 0 & 1 \end{bmatrix}}$$

and try to find a, b, c, and d so that the product of A and A^{-1} is the identity matrix I. Multiplying A and A^{-1} on the left side, we get

$$\begin{bmatrix} (2a + 3b) & (2c + 3d) \\ (a + 2b) & (c + 2d) \end{bmatrix} = \begin{bmatrix} 1 & 0 \\ 0 & 1 \end{bmatrix}$$

which is true only if

$2a + 3b = 1$ $\qquad\qquad$ $2c + 3d = 0$ $\qquad$ **Use Gauss–Jordan**
$a + 2b = 0$ $\qquad\qquad\quad$ $c + 2d = 1$ $\qquad$ **elimination to solve each system.**

$\begin{bmatrix} 2 & 3 & | & 1 \\ 1 & 2 & | & 0 \end{bmatrix}$ $\quad R_1 \leftrightarrow R_2 \quad$ $\begin{bmatrix} 2 & 3 & | & 0 \\ 1 & 2 & | & 1 \end{bmatrix}$ $\quad R_1 \leftrightarrow R_2$

$\begin{bmatrix} 1 & 2 & | & 0 \\ 2 & 3 & | & 1 \end{bmatrix}$ $\quad -2R_1 + R_2 \rightarrow R_2 \quad$ $\begin{bmatrix} 1 & 2 & | & 1 \\ 2 & 3 & | & 0 \end{bmatrix}$ $\quad -2R_1 + R_2 \rightarrow R_2$

$$\begin{bmatrix} 1 & 2 & | & 0 \\ 0 & -1 & | & 1 \end{bmatrix} \quad (-1)R_2 \to R_2 \qquad\qquad \begin{bmatrix} 1 & 2 & | & 1 \\ 0 & -1 & | & -2 \end{bmatrix} \quad (-1)R_2 \to R_2$$

$$\begin{bmatrix} 1 & 2 & | & 0 \\ 0 & 1 & | & -1 \end{bmatrix} \quad (-1)R_2 + R_1 \to R_1 \qquad\qquad \begin{bmatrix} 1 & 2 & | & 1 \\ 0 & 1 & | & 2 \end{bmatrix} \quad (-1)R_2 + R_1 \to R_1$$

$$\begin{bmatrix} 1 & 0 & | & 2 \\ 0 & 1 & | & -1 \end{bmatrix} \qquad\qquad\qquad\qquad \begin{bmatrix} 1 & 0 & | & -3 \\ 0 & 1 & | & 2 \end{bmatrix}$$

$$a = 2,\, b = -1 \qquad\qquad\qquad\qquad c = -3,\, d = 2$$

$$A^{-1} = \begin{bmatrix} a & b \\ c & d \end{bmatrix} = \begin{bmatrix} 2 & -3 \\ -1 & 2 \end{bmatrix}$$

CHECK

$$\underset{A}{\begin{bmatrix} 2 & 3 \\ 1 & 2 \end{bmatrix}} \underset{A^{-1}}{\begin{bmatrix} 2 & -3 \\ -1 & 2 \end{bmatrix}} = \underset{I}{\begin{bmatrix} 1 & 0 \\ 0 & 1 \end{bmatrix}} = \underset{A^{-1}}{\begin{bmatrix} 2 & -3 \\ -1 & 2 \end{bmatrix}} \underset{A}{\begin{bmatrix} 2 & 3 \\ 1 & 2 \end{bmatrix}}$$

Unlike nonzero real numbers, inverses do not always exist for nonzero square matrices. For example, if

$$B = \begin{bmatrix} 2 & 1 \\ 4 & 2 \end{bmatrix}$$

then, proceeding as before, we are led to the systems

$$2a + b = 1 \qquad\qquad\qquad 2c + d = 0 \qquad \text{Use Gauss–Jordan}$$
$$4a + 2b = 0 \qquad\qquad\qquad 4c + 2d = 1 \qquad \text{elimination to solve each system.}$$

$$\begin{bmatrix} 2 & 1 & | & 1 \\ 4 & 2 & | & 0 \end{bmatrix} \quad (-2)R_1 + R_2 \qquad\qquad \begin{bmatrix} 2 & 1 & | & 0 \\ 4 & 2 & | & 1 \end{bmatrix} \quad (-2)R_1 + R_2$$

$$\begin{bmatrix} 2 & 1 & | & 0 \\ 0 & 0 & | & -2 \end{bmatrix} \qquad\qquad\qquad\qquad \begin{bmatrix} 2 & 1 & | & 0 \\ 0 & 0 & | & 1 \end{bmatrix}$$

The last row of each augmented coefficient matrix contains a contradiction. So each system is inconsistent and has no solution. We conclude that B^{-1} does not exist and B is a singular matrix.

Technology Connections

Most graphing calculators can find matrix inverses and can identify singular matrices. Figure 1 shows the calculation of A^{-1} for the matrix A discussed earlier. Figure 2 shows the error message that results when the inverse operation is applied to the singular matrix B discussed earlier.

Note that the inverse operation is performed by pressing the x^{-1} key. Entering [A]^(−1) results in an error message (Fig. 3).

› Figure 1

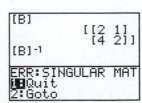

› Figure 2

› Figure 3

Being able to find inverses, when they exist, leads to direct and simple solutions to many practical problems.

The algebraic method outlined for finding the inverse, if it exists, gets very involved for matrices of order larger than 2. Now that we know what we are looking for, we can use augmented matrices, as in Section 10-2, to make the process more efficient. Details are illustrated in Example 1.

EXAMPLE 1 **Finding an Inverse**

Find the inverse, if it exists, of

$$A = \begin{bmatrix} 1 & -1 & 1 \\ 0 & 2 & -1 \\ 2 & 3 & 0 \end{bmatrix}$$

SOLUTION We start as before and write

$$\overset{A}{\begin{bmatrix} 1 & -1 & 1 \\ 0 & 2 & -1 \\ 2 & 3 & 0 \end{bmatrix}} \overset{A^{-1}}{\begin{bmatrix} a & d & g \\ b & e & h \\ c & f & i \end{bmatrix}} = \overset{I}{\begin{bmatrix} 1 & 0 & 0 \\ 0 & 1 & 0 \\ 0 & 0 & 1 \end{bmatrix}}$$

Equating corresponding terms, we see that this is true only if

$$\begin{array}{ccc} a - b + c = 1 & d - e + f = 0 & g - h + i = 0 \\ 2b - c = 0 & 2e - f = 1 & 2h - i = 0 \\ 2a + 3b \quad\;\; = 0 & 2d + 3e \quad\;\; = 0 & 2g + 3h \quad\;\; = 1 \end{array}$$

Now we write augmented matrices for each of the three systems:

First
$$\begin{bmatrix} 1 & -1 & 1 & | & 1 \\ 0 & 2 & -1 & | & 0 \\ 2 & 3 & 0 & | & 0 \end{bmatrix}$$
Second
$$\begin{bmatrix} 1 & -1 & 1 & | & 0 \\ 0 & 2 & -1 & | & 1 \\ 2 & 3 & 0 & | & 0 \end{bmatrix}$$
Third
$$\begin{bmatrix} 1 & -1 & 1 & | & 0 \\ 0 & 2 & -1 & | & 0 \\ 2 & 3 & 0 & | & 1 \end{bmatrix}$$

If you look carefully at the side-by-side solutions on pages 674 and 675, you will see that the exact same row operations were performed on each augmented matrix. The same would happen here; all three preceding augmented matrices have the same coefficient matrix. To save time, we'll combine all three into one, as shown next.

$$\begin{bmatrix} 1 & -1 & 1 & | & 1 & 0 & 0 \\ 0 & 2 & -1 & | & 0 & 1 & 0 \\ 2 & 3 & 0 & | & 0 & 0 & 1 \end{bmatrix} = [A \,|\, I] \qquad (1)$$

We now try to perform row operations on matrix (1) until we obtain a row-equivalent matrix that looks like matrix (2):

$$\overset{I \qquad\qquad\quad B}{\begin{bmatrix} 1 & 0 & 0 & | & a & d & g \\ 0 & 1 & 0 & | & b & e & h \\ 0 & 0 & 1 & | & c & f & i \end{bmatrix}} = [I \,|\, B] \qquad (2)$$

If this can be done, then the new matrix to the right of the vertical bar is A^{-1}! Now let's try to transform matrix (1) into a form like that of matrix (2). We follow the same

sequence of steps as in the solution of linear systems by Gauss–Jordan elimination (see Section 10-2):

$$
\begin{array}{c}
\underset{A}{} \qquad \underset{I}{}
\end{array}
$$

$$
\left[\begin{array}{ccc|ccc}
1 & -1 & 1 & 1 & 0 & 0 \\
0 & 2 & -1 & 0 & 1 & 0 \\
2 & 3 & 0 & 0 & 0 & 1
\end{array}\right] \quad (-2)R_1 + R_3 \to R_3
$$

$$
\sim \left[\begin{array}{ccc|ccc}
1 & -1 & 1 & 1 & 0 & 0 \\
0 & 2 & -1 & 0 & 1 & 0 \\
0 & 5 & -2 & -2 & 0 & 1
\end{array}\right] \quad \tfrac{1}{2}R_2 \to R_2
$$

$$
\sim \left[\begin{array}{ccc|ccc}
1 & -1 & 1 & 1 & 0 & 0 \\
0 & 1 & -\tfrac{1}{2} & 0 & \tfrac{1}{2} & 0 \\
0 & 5 & -2 & -2 & 0 & 1
\end{array}\right] \quad \begin{array}{l} R_2 + R_1 \to R_1 \\[6pt] (-5)R_2 + R_3 \to R_3 \end{array}
$$

$$
\sim \left[\begin{array}{ccc|ccc}
1 & 0 & \tfrac{1}{2} & 1 & \tfrac{1}{2} & 0 \\
0 & 1 & -\tfrac{1}{2} & 0 & \tfrac{1}{2} & 0 \\
0 & 0 & \tfrac{1}{2} & -2 & -\tfrac{5}{2} & 1
\end{array}\right] \quad 2R_3 \to R_3
$$

$$
\sim \left[\begin{array}{ccc|ccc}
1 & 0 & \tfrac{1}{2} & 1 & \tfrac{1}{2} & 0 \\
0 & 1 & -\tfrac{1}{2} & 0 & \tfrac{1}{2} & 0 \\
0 & 0 & 1 & -4 & -5 & 2
\end{array}\right] \quad \begin{array}{l} (-\tfrac{1}{2})R_3 + R_1 \to R_1 \\[6pt] \tfrac{1}{2}R_3 + R_2 \to R_2 \end{array}
$$

$$
\sim \left[\begin{array}{ccc|ccc}
1 & 0 & 0 & 3 & 3 & -1 \\
0 & 1 & 0 & -2 & -2 & 1 \\
0 & 0 & 1 & -4 & -5 & 2
\end{array}\right] = [\,I\,|\,B\,]
$$

We suspect that matrix B is actually A^{-1}, but we should check.

CHECK Because the definition of matrix inverse requires that

$$
A^{-1}A = I \qquad \text{and} \qquad AA^{-1} = I \tag{3}
$$

it appears that we must compute both $A^{-1}A$ and AA^{-1} to check our work. However, it can be shown that if one of the equations in (3) is satisfied, then the other is also satisfied. So, for checking purposes it's enough to compute either $A^{-1}A$ or AA^{-1}—we don't need to do both.

$$
A^{-1}A = \left[\begin{array}{ccc}
3 & 3 & -1 \\
-2 & -2 & 1 \\
-4 & -5 & 2
\end{array}\right]\left[\begin{array}{ccc}
1 & -1 & 1 \\
0 & 2 & -1 \\
2 & 3 & 0
\end{array}\right] = \left[\begin{array}{ccc}
1 & 0 & 0 \\
0 & 1 & 0 \\
0 & 0 & 1
\end{array}\right] = I
$$

MATCHED PROBLEM 1

Let $A = \left[\begin{array}{ccc} 3 & -1 & 1 \\ -1 & 1 & 0 \\ 1 & 0 & 1 \end{array}\right]$

(A) Form the augmented matrix $[A \,|\, I]$.

(B) Use row operations to transform $[A \,|\, I]$ into $[I \,|\, B]$.

(C) Verify by multiplication that $B = A^{-1}$.

The procedure used in Example 1 can be used to find the inverse of any square matrix if the inverse exists, and will also indicate when the inverse does not exist. These ideas are summarized in Theorem 1.

> **THEOREM 1** Inverse of a Square Matrix A

If $[A\,|\,I]$ is transformed by row operations into $[I\,|\,B]$, then the resulting matrix B is A^{-1}. If, however, we obtain all 0s in one or more rows to the left of the vertical line, then A^{-1} does not exist.

EXAMPLE 2 **Finding a Matrix Inverse**

Find A^{-1}, given $A = \begin{bmatrix} 4 & -1 \\ -6 & 2 \end{bmatrix}$

SOLUTION

$$\begin{bmatrix} 4 & -1 & | & 1 & 0 \\ -6 & 2 & | & 0 & 1 \end{bmatrix} \quad \tfrac{1}{4}R_1 \to R_1$$

$$\sim \begin{bmatrix} 1 & \tfrac{1}{4} & | & \tfrac{1}{4} & 0 \\ -6 & 2 & | & 0 & 1 \end{bmatrix} \quad 6R_1 + R_2 \to R_2$$

$$\sim \begin{bmatrix} 1 & -\tfrac{1}{4} & | & \tfrac{1}{4} & 0 \\ 0 & \tfrac{1}{2} & | & \tfrac{3}{2} & 1 \end{bmatrix} \quad 2R_2 \to R_2$$

$$\sim \begin{bmatrix} 1 & -\tfrac{1}{4} & | & \tfrac{1}{4} & 0 \\ 0 & 1 & | & 3 & 2 \end{bmatrix} \quad \tfrac{1}{4}R_2 + R_1 \to R_1$$

$$\sim \begin{bmatrix} 1 & 0 & | & 1 & \tfrac{1}{2} \\ 0 & 1 & | & 3 & 2 \end{bmatrix}$$

$$A^{-1} = \begin{bmatrix} 1 & \tfrac{1}{2} \\ 3 & 2 \end{bmatrix}$$

You should check our work by showing that $A^{-1}A = I$.

MATCHED PROBLEM 2 Find A^{-1}, given $A = \begin{bmatrix} 2 & -6 \\ 1 & -2 \end{bmatrix}$

EXAMPLE 3 **Finding an Inverse**

Find B^{-1}, if it exists, given $B = \begin{bmatrix} 10 & -2 \\ -5 & 1 \end{bmatrix}$

SOLUTION

$$\begin{bmatrix} 10 & -2 & | & 1 & 0 \\ -5 & 1 & | & 0 & 1 \end{bmatrix} \sim \begin{bmatrix} 1 & -\tfrac{1}{5} & | & \tfrac{1}{10} & 0 \\ -5 & 1 & | & 0 & 1 \end{bmatrix}$$

$$\sim \begin{bmatrix} 1 & -\tfrac{1}{5} & | & \tfrac{1}{10} & 0 \\ 0 & 0 & | & \tfrac{1}{2} & 1 \end{bmatrix}$$

We have all 0s in the second row to the left of the vertical line. Therefore, B^{-1} does not exist.

MATCHED PROBLEM 3 Find B^{-1}, if it exists, given $B = \begin{bmatrix} 6 & -3 \\ -2 & 1 \end{bmatrix}$

› Matrix Equations

Before we discuss the solution of matrix equations, you might find it helpful to briefly review the basic properties of real numbers discussed in Section R-1.

››› EXPLORE-DISCUSS 3

Let a, b, and c be real numbers, with $a \neq 0$. Solve each equation for x.

(A) $ax = b$ **(B)** $ax + b = c$

Solving simple matrix equations follows very much the same procedures used in solving real number equations. We have, however, less freedom with matrix equations, because matrix multiplication is not commutative. In solving matrix equations, we will be guided by the properties of matrices summarized in Theorem 2. (Some of these properties were introduced previously.)

> **› THEOREM 2 Basic Properties of Matrices**
>
> Assuming all products and sums are defined for the indicated matrices A, B, C, I, and 0, then
>
> **Addition Properties**
> **Associative:** $(A + B) + C = A + (B + C)$
> **Commutative:** $A + B = B + A$
> **Additive Identity:** $A + 0 = 0 + A = A$
> **Additive Inverse:** $A + (-A) = (-A) + A = 0$
>
> **Multiplication Properties**
> **Associative Property:** $A(BC) = (AB)C$
> **Multiplicative Identity:** $AI = IA = A$
> **Multiplicative Inverse:** If A is a square matrix and A^{-1} exists,
> then $AA^{-1} = A^{-1}A = I$.
>
> **Combined Properties**
> **Left Distributive:** $A(B + C) = AB + AC$
> **Right Distributive:** $(B + C)A = BA + CA$
>
> **Equality**
> **Addition:** If $A = B$, then $A + C = B + C$.
> **Left Multiplication:** If $A = B$, then $CA = CB$.
> **Right Multiplication:** If $A = B$, then $AC = BC$.

The process of solving certain types of simple matrix equations is best illustrated by an example.

EXAMPLE **4** Solving a Matrix Equation

Given an $n \times n$ matrix A and $n \times 1$ column matrices B and X, solve $AX = B$ for X. Assume all necessary inverses exist.

SOLUTION We are interested in finding a column matrix X that satisfies the matrix equation $AX = B$. To solve this equation, we multiply both sides, on the left, by A^{-1}, assuming it exists, to isolate X on the left side.

$$AX = B \qquad \text{Use the left multiplication property.}$$
$$A^{-1}(AX) = A^{-1}B \qquad \text{Associative property}$$
$$(A^{-1}A)X = A^{-1}B \qquad A^{-1}A = I$$
$$IX = A^{-1}B \qquad IX = X$$
$$X = A^{-1}B$$

>>> **CAUTION** >>>

1. Do not mix the left multiplication property and the right multiplication property. If $AX = B$, then

$$A^{-1}(AX) \neq BA^{-1}$$

2. Matrix division is not defined. If a, b, and x are real numbers, then the solution of $ax = b$ can be written either as $x = a^{-1}b$ or as $x = \frac{b}{a}$. But if A, B, and X are matrices, the solution of $AX = B$ must be written as $X = A^{-1}B$. The expression $\frac{B}{A}$ is not defined for matrices.

MATCHED PROBLEM 4 Given an $n \times n$ matrix A and $n \times 1$ column matrices B, C, and X, solve $AX + C = B$ for X. Assume all necessary inverses exist.

> ## Matrix Equations and Systems of Linear Equations

We will now show how independent systems of linear equations with the same number of variables as equations can be solved by first converting the system into a matrix equation of the form $AX = B$ and using $X = A^{-1}B$, as obtained in Example 4.

EXAMPLE 5 **Using Inverses to Solve Systems of Equations**

Use matrix inverse methods to solve the system

$$\begin{aligned} x_1 - x_2 + x_3 &= 1 \\ 2x_2 - x_3 &= 1 \\ 2x_1 + 3x_2 \quad &= 1 \end{aligned} \qquad (4)$$

SOLUTION First, we will convert the system of equations (4) into a matrix equation:

$$\overset{A}{\begin{bmatrix} 1 & -1 & 1 \\ 0 & 2 & -1 \\ 2 & 3 & 0 \end{bmatrix}} \overset{X}{\begin{bmatrix} x_1 \\ x_2 \\ x_3 \end{bmatrix}} = \overset{B}{\begin{bmatrix} 1 \\ 1 \\ 1 \end{bmatrix}} \qquad (5)$$

You should check that the matrix equation (5) is equivalent to the original system of equations (4) by performing the multiplication on the left side, and then equating corresponding elements.

If we can find the column matrix X, it will provide a solution to the system. In Example 4, we found that if $AX = B$ and A^{-1} exists, then $X = A^{-1}B$. So our job is to find A^{-1}

and multiply it by the constant matrix B on the left. In Example 1, we found that the inverse of matrix A is

$$A^{-1} = \begin{bmatrix} 3 & 3 & -1 \\ -2 & -2 & 1 \\ -4 & -5 & 2 \end{bmatrix}$$

So the equation $X = A^{-1}B$ is

$$\underset{X}{\begin{bmatrix} x_1 \\ x_2 \\ x_3 \end{bmatrix}} = \underset{A^{-1}}{\begin{bmatrix} 3 & 3 & -1 \\ -2 & -2 & 1 \\ -4 & -5 & 2 \end{bmatrix}} \underset{B}{\begin{bmatrix} 1 \\ 1 \\ 1 \end{bmatrix}} = \begin{bmatrix} 5 \\ -3 \\ -7 \end{bmatrix}$$

and we can conclude that $x_1 = 5$, $x_2 = -3$, and $x_3 = -7$. Check this result in system (4). ◉

MATCHED PROBLEM 5

Use matrix inverse methods to solve the system:

$$\begin{aligned} 3x_1 - x_2 + x_3 &= 1 \\ -x_1 + x_2 \quad\quad &= 3 \\ x_1 \quad\quad + x_3 &= 2 \end{aligned}$$

[*Note*: The inverse of the coefficient matrix was found in Matched Problem 1.]

> **USING INVERSE METHODS TO SOLVE SYSTEMS OF EQUATIONS**

If the number of equations in a system equals the number of variables and the coefficient matrix has an inverse, then the system will always have a unique solution that can be found by using the inverse of the coefficient matrix to solve the corresponding matrix equation.

Matrix equation	Solution
$AX = B$	$X = A^{-1}B$

At first, matrix inverse methods don't seem any better than Gauss–Jordan elimination— both require applying row operations to an augmented matrix. The advantage of the inverse method becomes apparent when solving a number of systems with a common coefficient matrix, as in Example 6.

EXAMPLE 6

Using Inverses to Solve Systems of Equations

Use matrix inverse methods to solve each of the following systems:

(A) $\begin{aligned} x_1 - x_2 + x_3 &= 3 \\ 2x_2 - x_3 &= 1 \\ 2x_1 + 3x_2 \quad\quad &= 4 \end{aligned}$ (B) $\begin{aligned} x_1 - x_2 + x_3 &= -5 \\ 2x_2 - x_3 &= 2 \\ 2x_1 + 3x_2 \quad\quad &= -3 \end{aligned}$

SOLUTIONS

Notice that both systems have the same coefficient matrix A as system (4) in Example 5. Only the constant terms have been changed. So we can use A^{-1} to solve these systems just as we did in Example 5.

(A) X A^{-1} B

$$\begin{bmatrix} x_1 \\ x_2 \\ x_3 \end{bmatrix} = \begin{bmatrix} 3 & 3 & -1 \\ -2 & -2 & 1 \\ -4 & -5 & 2 \end{bmatrix} \begin{bmatrix} 3 \\ 1 \\ 4 \end{bmatrix} = \begin{bmatrix} 8 \\ -4 \\ -9 \end{bmatrix}$$

The solution is, $x_1 = 8$, $x_2 = -4$, and $x_3 = -9$

(B) X A^{-1} B

$$\begin{bmatrix} x_1 \\ x_2 \\ x_3 \end{bmatrix} = \begin{bmatrix} 3 & 3 & -1 \\ -2 & -2 & 1 \\ -4 & -5 & 2 \end{bmatrix} \begin{bmatrix} -5 \\ 2 \\ -3 \end{bmatrix} = \begin{bmatrix} -6 \\ 3 \\ 4 \end{bmatrix}$$

The solution is, $x_1 = -6$, $x_2 = 3$, and $x_3 = 4$

MATCHED PROBLEM 6 Use matrix inverse methods to solve each of the following systems (see Matched Problem 5):

(A) $3x_1 - x_2 + x_3 = 3$ (B) $3x_1 - x_2 + x_3 = -5$
 $-x_1 + x_2 \quad\quad = -3$ $-x_1 + x_2 \quad\quad = 1$
 $x_1 \quad\quad + x_3 = 2$ $x_1 \quad\quad + x_3 = -4$

As Examples 5 and 6 illustrate, inverse methods are very convenient for hand calculations because once the inverse is found, it can be used to solve any new system formed by changing only the constant terms. Since most graphing calculators can compute the inverse of a matrix, this method also adapts readily to graphing calculator solutions. However, if your graphing calculator also has a built-in procedure for finding the reduced form of an augmented coefficient matrix, then it is just as convenient to use Gauss–Jordan elimination. Furthermore, Gauss–Jordan elimination can be used in all cases and, as noted previously, matrix inverse methods cannot always be used.

The application in Example 7 illustrates the usefulness of matrix inverses.

EXAMPLE 7 Investment Allocation

An investment adviser currently has two types of investments available for clients: an investment A that pays 4% per year and an investment B of higher risk that pays 8% per year. Clients may divide their investments between the two to achieve any total return desired between 4 and 8%. However, the higher the desired return, the higher the risk. How should each client listed in the table invest to achieve the indicated return?

	Client			
	1	**2**	**3**	**k**
Total investment	$20,000	$50,000	$10,000	k_1
Annual return desired	$1,200	$3,750	$500	k_2
	(6%)	(7.5%)	(5%)	

SOLUTION We will first solve the problem for an arbitrary client k using inverses, and then apply the result to the three specific clients.

Let

$$x_1 = \text{Amount invested in } A$$
$$x_2 = \text{Amount invested in } B$$

Then

$$x_1 + x_2 = k_1 \qquad \text{\textcolor{magenta}{Total invested}}$$
$$0.04x_1 + 0.08x_2 = k_2 \qquad \text{\textcolor{magenta}{Total annual return (4\% of } X_1 \text{ + 8\% of } X_2)}$$

Write as a matrix equation:

$$\underset{A}{\begin{bmatrix} 1 & 1 \\ 0.04 & 0.08 \end{bmatrix}} \underset{X}{\begin{bmatrix} x_1 \\ x_2 \end{bmatrix}} = \underset{B}{\begin{bmatrix} k_1 \\ k_2 \end{bmatrix}}$$

We now find A^{-1} by starting with $[A \mid I]$ and proceeding as discussed earlier.

$$\begin{bmatrix} 1 & 1 & | & 1 & 0 \\ 0.04 & 0.08 & | & 0 & 1 \end{bmatrix} \qquad \text{\textcolor{magenta}{100 } } R_2 \rightarrow R_2 \text{ \textcolor{magenta}{(To eliminate decimals)}}$$

$$\sim \begin{bmatrix} 1 & 1 & | & 1 & 0 \\ 4 & 8 & | & 0 & 100 \end{bmatrix} \qquad \text{\textcolor{magenta}{$-4R_1 + R_2 \rightarrow R_2$}}$$

$$\sim \begin{bmatrix} 1 & 1 & | & 1 & 0 \\ 0 & 4 & | & -4 & 100 \end{bmatrix} \qquad \text{\textcolor{magenta}{$\frac{1}{4} R_2 \rightarrow R_2$}}$$

$$\sim \begin{bmatrix} 1 & 1 & | & 1 & 0 \\ 0 & 1 & | & -1 & 25 \end{bmatrix} \qquad \text{\textcolor{magenta}{$(-1)R_2 + R_1 \rightarrow R_1$}}$$

$$\sim \begin{bmatrix} 1 & 0 & | & 2 & -25 \\ 0 & 1 & | & -1 & 25 \end{bmatrix}$$

So A has an inverse, and

$$A^{-1} = \begin{bmatrix} 2 & -25 \\ -1 & 25 \end{bmatrix}$$

CHECK

$$\underset{A^{-1}}{\begin{bmatrix} 2 & -25 \\ -1 & 25 \end{bmatrix}} \underset{A}{\begin{bmatrix} 1 & 1 \\ 0.04 & 0.08 \end{bmatrix}} = \underset{I}{\begin{bmatrix} 1 & 0 \\ 0 & 1 \end{bmatrix}}$$

Also,

$$\underset{X}{\begin{bmatrix} x_1 \\ x_2 \end{bmatrix}} = \underset{A^{-1}}{\begin{bmatrix} 2 & -25 \\ -1 & 25 \end{bmatrix}} \underset{B}{\begin{bmatrix} k_1 \\ k_2 \end{bmatrix}}$$

To solve each client's investment problem, we replace k_1 and k_2 with appropriate values from the table and multiply by A^{-1}.

\textcolor{magenta}{Client 1}

$$\begin{bmatrix} x_1 \\ x_2 \end{bmatrix} = \begin{bmatrix} 2 & -25 \\ -1 & 25 \end{bmatrix} \begin{bmatrix} 20,000 \\ 1,200 \end{bmatrix} = \begin{bmatrix} 10,000 \\ 10,000 \end{bmatrix}$$

To draw \$1,200 interest, invest \$10,000 at 4% and \$10,000 at 8%.

\textcolor{magenta}{Client 2}

$$\begin{bmatrix} x_1 \\ x_2 \end{bmatrix} = \begin{bmatrix} 2 & -25 \\ -1 & 25 \end{bmatrix} \begin{bmatrix} 50,000 \\ 3,750 \end{bmatrix} = \begin{bmatrix} 6,250 \\ 43,750 \end{bmatrix}$$

To draw \$3,750 interest, invest \$6,250 at 4% and \$43,750 at 8%.

\textcolor{magenta}{Client 3}

$$\begin{bmatrix} x_1 \\ x_2 \end{bmatrix} = \begin{bmatrix} 2 & -25 \\ -1 & 25 \end{bmatrix} \begin{bmatrix} 10,000 \\ 500 \end{bmatrix} = \begin{bmatrix} 7,500 \\ 2,500 \end{bmatrix}$$

To draw \$500 interest, invest \$7,500 at 4% and \$2,500 at 8%.

MATCHED PROBLEM 7 Repeat Example 7 with investment A paying 5% and investment B paying 9%.

> ## Application: Cryptography

Matrix inverses can be used to provide a simple and effective procedure for encoding and decoding messages. To begin, we assign the numbers 1 to 26 to the letters in the alphabet, as shown. We also assign the number 27 to a blank to provide for space between words. (A more sophisticated code could include both uppercase and lowercase letters and punctuation symbols.)

A	B	C	D	E	F	G	H	I	J	K	L	M	N
1	2	3	4	5	6	7	8	9	10	11	12	13	14

O	P	Q	R	S	T	U	V	W	X	Y	Z	Blank
15	16	17	18	19	20	21	22	23	24	25	26	27

The message SPRING BREAK corresponds to the sequence

$$19 \quad 16 \quad 18 \quad 9 \quad 14 \quad 7 \quad 27 \quad 2 \quad 18 \quad 5 \quad 1 \quad 11$$

Any matrix whose elements are positive integers and whose inverse exists can be used as an **encoding matrix.** For example, to use the 2×2 matrix

$$A = \begin{bmatrix} 4 & 3 \\ 5 & 4 \end{bmatrix}$$

to encode the preceding message, first we divide the numbers in the sequence into groups of 2 and use these groups as the columns of a matrix with 2 rows. (We would have added an extra blank in the last entry if the last column had an empty space.) Then we multiply this matrix on the left by A:

$$\begin{bmatrix} 4 & 3 \\ 5 & 4 \end{bmatrix} \begin{bmatrix} 19 & 18 & 14 & 27 & 18 & 1 \\ 16 & 9 & 7 & 2 & 5 & 11 \end{bmatrix} = \begin{bmatrix} 124 & 99 & 77 & 114 & 87 & 37 \\ 159 & 126 & 98 & 143 & 110 & 49 \end{bmatrix}$$

The coded message is

$$124 \quad 159 \quad 99 \quad 126 \quad 77 \quad 98 \quad 114 \quad 143 \quad 87 \quad 110 \quad 37 \quad 49$$

This message can be decoded simply by putting it back into matrix form and multiplying on the left by the **decoding matrix** A^{-1}. Since A^{-1} is easily determined if A is known, the encoding matrix A is the only key needed to decode messages encoded in this manner. Although simple in concept, codes of this type can be very difficult to crack.

EXAMPLE 8 Cryptography

The message

$$31 \quad 54 \quad 69 \quad 37 \quad 64 \quad 82 \quad 23 \quad 50 \quad 66 \quad 51 \quad 69 \quad 75 \quad 23 \quad 30 \quad 36 \quad 65 \quad 84 \quad 84$$

was encoded with the matrix A shown next. Use a graphing calculator to decode this message.

$$A = \begin{bmatrix} 0 & 2 & 1 \\ 1 & 2 & 1 \\ 2 & 1 & 1 \end{bmatrix}$$

SOLUTION We begin by entering the 3×3 encoding matrix A (Fig. 4). Then we enter the coded message in the columns of a matrix C with three rows (Fig. 4). If B is the matrix containing the uncoded message, then B and C are related by $C = AB$. To find B, we multiply both sides of the equation $C = AB$ by A^{-1} (Fig. 5).

> Figure 4

> Figure 5

Writing the numbers in the columns of this matrix in sequence and using the correspondence between numbers and letters noted earlier produces the decoded message:

23 8 15 27 9 19 27 11 1 18 12 27 7 1 21 19 19 27

W H O I S K A R L G A U S S

The answer to this question can be found somewhere in this chapter.

MATCHED PROBLEM 8

The message

46 84 85 55 101 100 59 95 132 25 42 53 52 91 90 43 71 83 19 37 25

was encoded with the matrix A shown here. Decode this message.

$$A = \begin{bmatrix} 1 & 1 & 1 \\ 2 & 1 & 2 \\ 2 & 3 & 1 \end{bmatrix}$$

ANSWERS TO MATCHED PROBLEMS

1. (A) $\left[\begin{array}{ccc|ccc} 3 & -1 & 1 & 1 & 0 & 0 \\ -1 & 1 & 0 & 0 & 1 & 0 \\ 1 & 0 & 1 & 0 & 0 & 1 \end{array}\right]$ (B) $\left[\begin{array}{ccc|ccc} 1 & 0 & 0 & 1 & 1 & -1 \\ 0 & 1 & 0 & 1 & 2 & -1 \\ 0 & 0 & 1 & -1 & -1 & 2 \end{array}\right]$

(C) $\begin{bmatrix} 1 & 1 & -1 \\ 1 & 2 & -1 \\ -1 & -1 & 2 \end{bmatrix}\begin{bmatrix} 3 & -1 & 1 \\ -1 & 1 & 0 \\ 1 & 0 & 1 \end{bmatrix} = \begin{bmatrix} 1 & 0 & 0 \\ 0 & 1 & 0 \\ 0 & 0 & 1 \end{bmatrix}$

2. $\begin{bmatrix} -1 & 3 \\ -\frac{1}{2} & 1 \end{bmatrix}$ **3.** Does not exit

4. $AX + C = B$

$(AX + C) - C = B - C$
$AX + (C - C) = B - C$
$AX + 0 = B - C$

$AX = B - C$

$A^{-1}(AX) = A^{-1}(B - C)$
$(A^{-1}A)X = A^{-1}(B - C)$
$IX = A^{-1}(B - C)$

$X = A^{-1}(B - C)$

5. $x_1 = 2, x_2 = 5, x_3 = 0$

6. (A) $x_1 = -2, x_2 = -5, x_3 = 4$ (B) $x_1 = 0, x_2 = 1, x_3 = -4$

7. $A^{-1} = \begin{bmatrix} 2.25 & -25 \\ -1.25 & 25 \end{bmatrix}$; Client 1: \$15,000 in A and \$5,000 in B; Client 2: \$18,750 in A and \$31,250 in B; Client 3: \$10,000 in A

8. WHO IS WILHELM JORDAN

10-4 Exercises

1. What is an identity matrix?

2. What is the (multiplicative) inverse of a real number? Does every real number have an inverse?

3. What is the (multiplicative) inverse of a matrix? Does every matrix have an inverse?

4. What is a singular matrix?

5. Describe the process for finding the inverse of a matrix by hand.

6. Explain how inverse matrices can be used to solve systems of linear equations by hand.

7. Explain how inverse matrices can be used to solve systems of linear equations on a graphing calculator.

8. How would you solve a linear system that has more variables than equations?

9. How would you solve a linear system that has fewer variables than equations?

10. How would you solve a linear system if the number of variables and the number of equations are equal?

Perform the indicated operations in Problems 11–14.

11. $\begin{bmatrix} 1 & 0 \\ 0 & 1 \end{bmatrix}\begin{bmatrix} 2 & -3 \\ 4 & 5 \end{bmatrix}$ **12.** $\begin{bmatrix} 2 & -3 \\ 4 & 5 \end{bmatrix}\begin{bmatrix} 1 & 0 \\ 0 & 1 \end{bmatrix}$

13. $\begin{bmatrix} 1 & 0 & 0 \\ 0 & 1 & 0 \\ 0 & 0 & 1 \end{bmatrix}\begin{bmatrix} -2 & 1 & 3 \\ 2 & 4 & -2 \\ 5 & 1 & 0 \end{bmatrix}$

14. $\begin{bmatrix} -2 & 1 & 3 \\ 2 & 4 & -2 \\ 5 & 1 & 0 \end{bmatrix}\begin{bmatrix} 1 & 0 & 0 \\ 0 & 1 & 0 \\ 0 & 0 & 1 \end{bmatrix}$

In Problems 15–24, examine the product of the two matrices to determine if each is the inverse of the other.

15. $\begin{bmatrix} 3 & -4 \\ -2 & 3 \end{bmatrix}; \begin{bmatrix} 3 & 4 \\ 2 & 3 \end{bmatrix}$ **16.** $\begin{bmatrix} -2 & -1 \\ -4 & 2 \end{bmatrix}; \begin{bmatrix} 1 & -1 \\ 2 & -2 \end{bmatrix}$

17. $\begin{bmatrix} 2 & 2 \\ -1 & -1 \end{bmatrix}; \begin{bmatrix} 1 & 1 \\ -1 & -1 \end{bmatrix}$ **18.** $\begin{bmatrix} 5 & -7 \\ -2 & 3 \end{bmatrix}; \begin{bmatrix} 3 & 7 \\ 2 & 5 \end{bmatrix}$

19. $\begin{bmatrix} -5 & 2 \\ -8 & 3 \end{bmatrix}; \begin{bmatrix} 3 & -2 \\ 8 & -5 \end{bmatrix}$ **20.** $\begin{bmatrix} 7 & 4 \\ -5 & -3 \end{bmatrix}; \begin{bmatrix} 3 & 4 \\ -5 & -7 \end{bmatrix}$

21. $\begin{bmatrix} 1 & 2 & 0 \\ 0 & 1 & 0 \\ -1 & -1 & 1 \end{bmatrix}; \begin{bmatrix} 1 & -2 & 0 \\ 0 & 1 & 0 \\ 1 & -1 & 0 \end{bmatrix}$

22. $\begin{bmatrix} 1 & 0 & 1 \\ -3 & 1 & -2 \\ 0 & 0 & 1 \end{bmatrix}; \begin{bmatrix} 1 & 0 & -1 \\ 3 & 1 & -1 \\ 0 & 0 & 1 \end{bmatrix}$

23. $\begin{bmatrix} 1 & -1 & 1 \\ 0 & 2 & -1 \\ 2 & 3 & 0 \end{bmatrix}; \begin{bmatrix} 3 & 3 & -1 \\ -2 & -2 & 1 \\ -4 & -5 & 2 \end{bmatrix}$

24. $\begin{bmatrix} 1 & 0 & -1 \\ 3 & 1 & -1 \\ 0 & 0 & 0 \end{bmatrix}; \begin{bmatrix} 1 & 0 & -1 \\ -3 & 1 & -2 \\ 0 & 0 & 1 \end{bmatrix}$

Write the matrix equations in Problems 25–28 as systems of linear equations without matrices.

25. $\begin{bmatrix} 2 & -1 \\ 1 & 3 \end{bmatrix}\begin{bmatrix} x_1 \\ x_2 \end{bmatrix} = \begin{bmatrix} 3 \\ -2 \end{bmatrix}$ **26.** $\begin{bmatrix} -3 & 1 \\ -1 & 2 \end{bmatrix}\begin{bmatrix} x_1 \\ x_2 \end{bmatrix} = \begin{bmatrix} -2 \\ 5 \end{bmatrix}$

27. $\begin{bmatrix} -2 & 0 & 1 \\ 1 & 2 & 1 \\ 0 & 1 & -1 \end{bmatrix}\begin{bmatrix} x_1 \\ x_2 \\ x_3 \end{bmatrix} = \begin{bmatrix} 3 \\ -4 \\ 2 \end{bmatrix}$

28. $\begin{bmatrix} 1 & -2 & 0 \\ -3 & 1 & -1 \\ 2 & 0 & 4 \end{bmatrix}\begin{bmatrix} x_1 \\ x_2 \\ x_3 \end{bmatrix} = \begin{bmatrix} 3 \\ -2 \\ 5 \end{bmatrix}$

Write each system in Problems 29–32 as a matrix equation of the form AX = B.

29. $\begin{aligned} 4x_1 - 3x_2 &= 2 \\ x_1 + 2x_2 &= 1 \end{aligned}$ **30.** $\begin{aligned} x_1 - 2x_2 &= 7 \\ -3x_1 + x_2 &= -3 \end{aligned}$

31. $\begin{aligned} x_1 - 2x_2 + x_3 &= -1 \\ -x_1 + x_2 &= 2 \\ 2x_1 + 3x_2 + x_3 &= -3 \end{aligned}$ **32.** $\begin{aligned} 2x_1 + 3x_3 &= 5 \\ x_1 - 2x_2 + x_3 &= -4 \\ -x_1 + 3x_2 &= 2 \end{aligned}$

In Problems 33–40, find x_1 and x_2.

33. $\begin{bmatrix} x_1 \\ x_2 \end{bmatrix} = \begin{bmatrix} 3 & -2 \\ 1 & 4 \end{bmatrix}\begin{bmatrix} -2 \\ 1 \end{bmatrix}$ **34.** $\begin{bmatrix} x_1 \\ x_2 \end{bmatrix} = \begin{bmatrix} -2 & 1 \\ -1 & 2 \end{bmatrix}\begin{bmatrix} 3 \\ -2 \end{bmatrix}$

35. $\begin{bmatrix} x_1 \\ x_2 \end{bmatrix} = \begin{bmatrix} -2 & 3 \\ 2 & -1 \end{bmatrix}\begin{bmatrix} 3 \\ 2 \end{bmatrix}$ **36.** $\begin{bmatrix} x_1 \\ x_2 \end{bmatrix} = \begin{bmatrix} 3 & -1 \\ 0 & 2 \end{bmatrix}\begin{bmatrix} -2 \\ 1 \end{bmatrix}$

37. $\begin{bmatrix} 1 & -1 \\ 1 & -2 \end{bmatrix}\begin{bmatrix} x_1 \\ x_2 \end{bmatrix} = \begin{bmatrix} 5 \\ 7 \end{bmatrix}$ **38.** $\begin{bmatrix} 1 & 3 \\ 1 & 4 \end{bmatrix}\begin{bmatrix} x_1 \\ x_2 \end{bmatrix} = \begin{bmatrix} 9 \\ 6 \end{bmatrix}$

39. $\begin{bmatrix} 1 & 1 \\ 2 & -3 \end{bmatrix}\begin{bmatrix} x_1 \\ x_2 \end{bmatrix} = \begin{bmatrix} 15 \\ 10 \end{bmatrix}$ **40.** $\begin{bmatrix} 1 & 1 \\ 3 & -2 \end{bmatrix}\begin{bmatrix} x_1 \\ x_2 \end{bmatrix} = \begin{bmatrix} 10 \\ 20 \end{bmatrix}$

In Problems 41–60, given A, find A^{-1}, if it exists. Check each inverse by showing $A^{-1}A = I$.

41. $\begin{bmatrix} 1 & 9 \\ 0 & 1 \end{bmatrix}$ **42.** $\begin{bmatrix} 0 & -1 \\ -1 & 3 \end{bmatrix}$ **43.** $\begin{bmatrix} -1 & -2 \\ 2 & 5 \end{bmatrix}$

44. $\begin{bmatrix} 3 & -4 \\ -2 & 3 \end{bmatrix}$ **45.** $\begin{bmatrix} -5 & 7 \\ 2 & -3 \end{bmatrix}$ **46.** $\begin{bmatrix} 11 & 4 \\ 3 & 1 \end{bmatrix}$

47. $\begin{bmatrix} 3 & 9 \\ 2 & 6 \end{bmatrix}$

48. $\begin{bmatrix} 2 & -4 \\ -3 & 6 \end{bmatrix}$

49. $\begin{bmatrix} 2 & 3 \\ 3 & 5 \end{bmatrix}$

50. $\begin{bmatrix} -5 & 4 \\ 4 & -3 \end{bmatrix}$

51. $\begin{bmatrix} 1 & -1 & 0 \\ -1 & 1 & -1 \\ 0 & -1 & 1 \end{bmatrix}$

52. $\begin{bmatrix} 2 & -1 & 0 \\ 0 & 1 & 1 \\ 1 & 0 & 1 \end{bmatrix}$

53. $\begin{bmatrix} 1 & 2 & 5 \\ 3 & 5 & 9 \\ 1 & 1 & -2 \end{bmatrix}$

54. $\begin{bmatrix} 1 & -1 & 1 \\ -2 & 3 & 2 \\ 3 & -3 & 2 \end{bmatrix}$

55. $\begin{bmatrix} 2 & 2 & -1 \\ 0 & 4 & -1 \\ -1 & -2 & 1 \end{bmatrix}$

56. $\begin{bmatrix} 4 & 2 & -1 \\ 1 & 1 & -1 \\ -3 & -1 & 1 \end{bmatrix}$

57. $\begin{bmatrix} 2 & 1 & 1 \\ 1 & 1 & 0 \\ -1 & -1 & 0 \end{bmatrix}$

58. $\begin{bmatrix} 1 & -1 & 0 \\ 2 & -1 & 1 \\ 0 & 1 & 1 \end{bmatrix}$

59. $\begin{bmatrix} 1 & 5 & 10 \\ 0 & 1 & 4 \\ 1 & 6 & 15 \end{bmatrix}$

60. $\begin{bmatrix} 1 & -5 & -10 \\ 0 & 1 & 6 \\ 1 & -4 & -3 \end{bmatrix}$

Write each system in Problems 61–68 as a matrix equation and solve using inverses. [Note: the inverse of each coefficient matrix was found earlier in this exercise set in the indicated problem.]

61. $-x_1 - 2x_2 = k_1$
$2x_1 + 5x_2 = k_2$
(A) $k_1 = 2, k_2 = 5$
(B) $k_1 = -4, k_2 = 1$
(C) $k_1 = -3, k_2 = -2$
(see Problem 43.)

62. $3x_1 - 4x_2 = k_1$
$-2x_1 + 3x_2 = k_2$
(A) $k_1 = 3, k_2 = -1$
(B) $k_1 = 6, k_2 = 5$
(C) $k_1 = 0, k_2 = -4$
(see Problem 44.)

63. $-5x_1 + 7x_2 = k_1$
$2x_1 - 3x_2 = k_2$
(A) $k_1 = -5, k_2 = 1$
(B) $k_1 = 8, k_2 = -4$
(C) $k_1 = 6, k_2 = 0$
(see Problem 45.)

64. $11x_1 + 4x_2 = k_1$
$3x_1 + x_2 = k_2$
(A) $k_1 = -2, k_2 = -3$
(B) $k_1 = -1, k_2 = 9$
(C) $k_1 = 4, k_2 = 5$
(see Problem 46.)

65. $x_1 - x_2 \quad\quad = k_1$
$-x_1 + x_2 - x_3 = k_2$
$\quad\quad - x_2 + x_3 = k_3$
(A) $k_1 = 1, k_2 = 1, k_3 = 2$
(B) $k_1 = -1, k_2 = 0, k_3 = -4$
(C) $k_1 = 3, k_2 = -2, k_3 = 0$
(see Problem 51.)

66. $2x_1 - x_2 \quad\quad = k_1$
$\quad\quad x_2 + x_3 = k_2$
$x_1 \quad\quad + x_3 = k_3$
(A) $k_1 = -2, k_2 = 4, k_3 = -1$
(B) $k_1 = 2, k_2 = -3, k_3 = 1$
(C) $k_1 = -1, k_2 = 2, k_3 = -5$
(see Problem 52.)

67. $x_1 + 2x_2 + 5x_3 = k_1$
$3x_1 + 5x_2 + 9x_3 = k_2$
$x_1 + x_2 - 2x_3 = k_3$
(A) $k_1 = 0, k_2 = 1, k_3 = 4$
(B) $k_1 = 5, k_2 = -1, k_3 = 0$
(C) $k_1 = -6, k_2 = 0, k_3 = 2$
(see Problem 53.)

68. $x_1 - x_2 + x_3 = k_1$
$-2x_1 + 3x_2 + 2x_3 = k_2$
$3x_1 - 3x_2 + 2x_3 = k_3$
(A) $k_1 = 3, k_2 = -1, k_3 = 0$
(B) $k_1 = 0, k_2 = 4, k_3 = 5$
(C) $k_1 = -2, k_2 = 0, k_3 = 1$
(see Problem 54.)

For $n \times n$ matrices A and B and $n \times 1$ matrices C, D, and X, solve each matrix equation in Problems 69–74 for X. Assume all necessary inverses exist.

69. $AX = BX + C$

70. $AX + BX = C + D$

71. $X = AX + C$

72. $X + C = AX - BX$

73. $AX + C = 3X$

74. $AX + C = BX - 7X + D$

75. Discuss the existence of A^{-1} for 2×2 diagonal matrices of the form

$$A = \begin{bmatrix} a & 0 \\ 0 & d \end{bmatrix}$$

76. Discuss the existence of A^{-1} for 2×2 upper triangular matrices of the form

$$A = \begin{bmatrix} a & b \\ 0 & d \end{bmatrix}$$

77. Find A^{-1} and A^2 for each of the following matrices.

(A) $A = \begin{bmatrix} 3 & 2 \\ -4 & -3 \end{bmatrix}$ (B) $A = \begin{bmatrix} -2 & -1 \\ 3 & 2 \end{bmatrix}$

78. Based on your observations in Problem 77, if $A = A^{-1}$ for a square matrix A, what is A^2? Give a mathematical argument to support your conclusion.

79. Find $(A^{-1})^{-1}$ for each of the following matrices.

(A) $A = \begin{bmatrix} 4 & 2 \\ 1 & 3 \end{bmatrix}$ (B) $A = \begin{bmatrix} 5 & 5 \\ -1 & 3 \end{bmatrix}$

80. Based on your observations in Problem 79, if A^{-1} exists for a square matrix A, what is $(A^{-1})^{-1}$? Give a mathematical argument to support your conclusion.

81. Find $(AB)^{-1}$, $A^{-1}B^{-1}$, and $B^{-1}A^{-1}$ for each of the following pairs of matrices.

(A) $A = \begin{bmatrix} 3 & 4 \\ 2 & 3 \end{bmatrix}$ and $B = \begin{bmatrix} 3 & 7 \\ 2 & 5 \end{bmatrix}$

(B) $A = \begin{bmatrix} 1 & -1 \\ 2 & 3 \end{bmatrix}$ and $B = \begin{bmatrix} 6 & 2 \\ 2 & 1 \end{bmatrix}$

82. Based on your observations in Problems 81, which of the following is a true statement? Give a mathematical argument to support your conclusion.

(A) $(AB)^{-1} = A^{-1}B^{-1}$

(B) $(AB)^{-1} = B^{-1}A^{-1}$

APPLICATIONS

Problems 83–86 refer to the encoding matrix $A = \begin{bmatrix} 3 & 5 \\ 1 & 2 \end{bmatrix}$

83. CRYPTOGRAPHY Encode the message LEBRON JAMES with the matrix A.

84. CRYPTOGRAPHY Encode the message KOBE BRYANT with the matrix A.

85. CRYPTOGRAPHY The following message was encoded with the matrix A. Decode the message.

 31 12 150 55 57 20 150 59 103 39
 160 61 61 22 192 73

86. CRYPTOGRAPHY The following message was encoded with the matrix A. Decode the message.

 49 18 103 41 159 62 61 22 47 18
 105 41

Problems 87–90 require the use of a graphing calculator. To use the 5×5 encoding matrix B given below, form a matrix with 5 rows and as many columns as necessary to accommodate each message.

$$B = \begin{bmatrix} 1 & 0 & 1 & 0 & 1 \\ 0 & 1 & 1 & 0 & 3 \\ 2 & 1 & 1 & 1 & 1 \\ 0 & 0 & 1 & 0 & 2 \\ 1 & 1 & 1 & 2 & 1 \end{bmatrix}$$

87. CRYPTOGRAPHY Encode the message NEW ENGLAND PATRIOTS with the matrix B.

88. CRYPTOGRAPHY Encode the message PITTSBURGH STEELERS with the matrix B.

89. CRYPTOGRAPHY The following message was encoded with the matrix B. Decode the message.

 32 25 55 19 41 51 64 103 39 100 62
 109 114 62 92 58 115 105 73 113 39
 110 85 65 111

90. CRYPTOGRAPHY The following message was encoded with the matrix B. Decode the message.

 44 45 88 29 82 51 61 86 45 84 35 63
 74 37 77 46 108 61 72 65

Solve Problems 91–97 using systems of equations and matrix inverses.

91. RESOURCE ALLOCATION A concert hall has 10,000 seats. If tickets are $20 and $30, how many of each type of ticket should be sold (assuming that all seats can be sold) to bring in each of the returns indicated in the table? Use decimals in computing the inverse.

	Concert		
	1	2	3
Tickets sold	10,000	10,000	10,000
Return required	$240,000	$250,000	$270,000

92. PRODUCTION SCHEDULING Labor and material costs for manufacturing two guitar models are given in the following table:

Guitar Model	Labor Cost	Material Cost
A	$30	$20
B	$40	$30

If a total of $3,000 a week is allowed for labor and material, how many of each model should be produced each week to exactly use each of the allocations of the $3,000 indicated in the following table? Use decimals in computing the inverse.

	Weekly Allocation		
	1	2	3
Labor	$1,800	$1,750	$1,720
Material	$1,200	$1,250	$1,280

93. CIRCUIT ANALYSIS A direct current electric circuit consisting of conductors (wires), resistors, and batteries is diagrammed in the figure.

If I_1, I_2, and I_3 are the currents (in amperes) in the three branches of the circuit and V_1 and V_2 are the voltages (in volts) of the two batteries, then *Kirchhoff's* laws* can be used to show that the currents satisfy the following system of equations:

$$I_1 - I_2 + I_3 = 0$$
$$I_1 + I_2 = V_1$$
$$I_2 + 2I_3 = V_2$$

Solve this system for:

(A) $V_1 = 10$ volts, $V_2 = 10$ volts

(B) $V_1 = 10$ volts, $V_2 = 15$ volts

(C) $V_1 = 15$ volts, $V_2 = 10$ volts

*Gustav Kirchhoff (1824–1887), a German physicist, was among the first to apply theoretical mathematics to physics. He is best known for his development of certain properties of electric circuits, which are now known as **Kirchhoff's laws.**

94. CIRCUIT ANALYSIS Repeat Problem 93 for the electric circuit shown in the figure.

$$I_1 - I_2 + I_3 = 0$$
$$I_1 + 2I_2 = V_1$$
$$2I_2 + 2I_3 = V_2$$

95. GEOMETRY The graph of $f(x) = ax^2 + bx + c$ passes through the points $(1, k_1)$, $(2, k_2)$, and $(3, k_3)$. Determine a, b, and c for:
(A) $k_1 = -2, k_2 = 1, k_3 = 6$
(B) $k_1 = 4, k_2 = 3, k_3 = -2$
(C) $k_1 = 8, k_2 = -5, k_3 = 4$

96. GEOMETRY Repeat Problem 95 if the graph passes through the points $(-1, k_1)$, $(0, k_2)$, and $(1, k_3)$.

97. DIETS A biologist has available two commercial food mixes with the following percentages of protein and fat:

Mix	Protein (%)	Fat (%)
A	20	2
B	10	6

How many ounces of each mix should be used to prepare each of the diets listed in the following table?

	Diet		
	1	2	3
Protein	20 oz	10 oz	10 oz
Fat	6 oz	4 oz	6 oz

10-5 Determinants and Cramer's Rule

> **Defining First- and Second-Order Determinants**
> **Evaluating Third-Order Determinants**
> **Using Cramer's Rule to Solve Systems of Equations**

In this section, we'll study one more method for solving systems of linear equations using matrices. Like the inverse method, it works only for systems with the same number of equations and variables. The biggest advantage is that it's purely computational—it requires very little symbol manipulation. The method is based on *determinants*.

› **Defining First- and Second-Order Determinants**

For any square matrix A, the **determinant** of A is a real number denoted by **det (A)** or $|A|$*. If A is a square matrix of order n, then det (A) is called a **determinant of order n**. If $A = [a_{11}]$ is a square matrix of order 1, then

$$\det (A) = a_{11}$$

is a **first-order determinant.** Now we proceed to define determinants of higher order.

Given a second-order square matrix $A = \begin{bmatrix} a_{11} & a_{12} \\ a_{21} & a_{22} \end{bmatrix}$, the **second-order determinant** of A is

$$\det (A) = \begin{vmatrix} a_{11} & a_{12} \\ a_{21} & a_{22} \end{vmatrix} = a_{11}a_{22} - a_{21}a_{12} \tag{1}$$

*The absolute value notation will now have two interpretations: the absolute value of a real number or the determinant of a square matrix. These concepts are not the same. You must always interpret $|A|$ in terms of the context in which it is used.

Formula (1) is easily remembered if you notice that the expression on the right is the product of the elements on the **principal diagonal,** from upper left to lower right, minus the product of the elements on the **secondary diagonal,** from lower left to upper right.

EXAMPLE 1 **Evaluating a Second-Order Determinant**

Find $\begin{vmatrix} -1 & 2 \\ -3 & -4 \end{vmatrix}$.

SOLUTIONS

$$\det (A) = \begin{vmatrix} -1 & 2 \\ -3 & -4 \end{vmatrix}$$
$$= (-1)(-4) - (-3)(2)$$
$$= 4 - (-6)$$
$$= 10$$

MATCHED PROBLEM 1 Find $\begin{vmatrix} 3 & -5 \\ 4 & -2 \end{vmatrix}$.

≫ CAUTION ≫

Remember that $A = \begin{bmatrix} 3 & -5 \\ 4 & -2 \end{bmatrix}$ is a matrix, but $\begin{vmatrix} 3 & -5 \\ 4 & -2 \end{vmatrix}$ represents a real number, the determinant of A. We will often refer to $\begin{vmatrix} 3 & -5 \\ 4 & -2 \end{vmatrix}$ as a determinant, and refer to the process of finding the real number it represents as "evaluating the determinant."

Technology Connections

Most graphing calculators have a command to calculate determinants. On the TI-84, it is on the MATRIX-MATH menu. In Figure 1, the determinant from Example 1 is calculated.

```
[A]
        [[-1  2 ]
         [-3 -4]]
det([A])
              10
```

▸ Figure 1

▸ Evaluating Third-Order Determinants

Given the matrix $A = \begin{bmatrix} a_{11} & a_{12} & a_{13} \\ a_{21} & a_{22} & a_{23} \\ a_{31} & a_{32} & a_{33} \end{bmatrix}$, **the third-order determinant** of A is

$$\det (A) = \begin{vmatrix} a_{11} & a_{12} & a_{13} \\ a_{21} & a_{22} & a_{23} \\ a_{31} & a_{32} & a_{33} \end{vmatrix} = a_{11}a_{22}a_{33} - a_{11}a_{32}a_{23} + a_{21}a_{32}a_{13} - a_{21}a_{12}a_{33} \qquad (2)$$
$$+ a_{31}a_{12}a_{23} - a_{31}a_{22}a_{13}$$

Don't panic! You don't need to memorize formula (2). After we introduce the ideas of *minor* and *cofactor* below, we will state a theorem that can be used to obtain the same result with much less trouble.

The **minor of an element** in a third-order determinant is a second-order determinant obtained by deleting the row and column that contains the element. For example, in the determinant in formula (2),

Deletions are usually done mentally.

$$\text{Minor of } a_{23} = \begin{vmatrix} a_{11} & a_{12} \\ a_{31} & a_{32} \end{vmatrix} \quad \begin{vmatrix} a_{11} & a_{12} & a_{13} \\ a_{21} & a_{22} & a_{23} \\ a_{31} & a_{32} & a_{33} \end{vmatrix} = a_{11}a_{32} - a_{31}a_{12}$$

$$\text{Minor of } a_{32} = \begin{vmatrix} a_{11} & a_{13} \\ a_{21} & a_{23} \end{vmatrix} \quad \begin{vmatrix} a_{11} & a_{12} & a_{13} \\ a_{21} & a_{22} & a_{23} \\ a_{31} & a_{32} & a_{33} \end{vmatrix} = a_{11}a_{23} - a_{21}a_{13}$$

⟫⟫ EXPLORE-DISCUSS 1 Write the minors of the other seven elements in the determinant in formula (2).

A quantity closely associated with the minor of an element is the **cofactor of an element** a_{ij} (from the ith row and jth column), which is defined as the product of the minor of a_{ij} and $(-1)^{i+j}$.

> **DEFINITION 1 Cofactor**
>
> Cofactor of $a_{ij} = (-1)^{i+j}$ (Minor of a_{ij})

So a cofactor is just a minor with either a positive or negative sign. The sign is determined by raising -1 to a power that is the sum of the numbers indicating the row and column in which the element appears. Note that $(-1)^{i+j}$ is 1 if $i + j$ is even and -1 if $i + j$ is odd. So if we are given the determinant

$$\begin{vmatrix} a_{11} & a_{12} & a_{13} \\ a_{21} & a_{22} & a_{23} \\ a_{31} & a_{32} & a_{33} \end{vmatrix}$$

then

$$\text{Cofactor of } a_{23} = (-1)^{2+3} \begin{vmatrix} a_{11} & a_{12} \\ a_{31} & a_{32} \end{vmatrix} = -\begin{vmatrix} a_{11} & a_{12} \\ a_{31} & a_{32} \end{vmatrix} = -(a_{11}a_{32} - a_{31}a_{12})$$

$$\text{Cofactor of } a_{11} = (-1)^{1+1} \begin{vmatrix} a_{22} & a_{23} \\ a_{32} & a_{33} \end{vmatrix} = \begin{vmatrix} a_{22} & a_{23} \\ a_{32} & a_{33} \end{vmatrix} = a_{22}a_{33} - a_{32}a_{23}$$

EXAMPLE 2 Finding Cofactors

Find the cofactors of -2 and 5 in the determinant

$$\begin{vmatrix} -2 & 0 & 3 \\ 1 & -6 & 5 \\ -1 & 2 & 0 \end{vmatrix}$$

SOLUTION

$$\text{Cofactor of } -2 = (-1)^{1+1}\begin{vmatrix} -6 & 5 \\ 2 & 0 \end{vmatrix} = \begin{vmatrix} -6 & 5 \\ 2 & 0 \end{vmatrix} \qquad \color{red}{-2 \text{ is } a_{11}}$$

$$= (-6)(0) - (2)(5) = -10$$

$$\text{Cofactor of } 5 = (-1)^{2+3}\begin{vmatrix} -2 & 0 \\ -1 & 2 \end{vmatrix} = -\begin{vmatrix} -2 & 0 \\ -1 & 2 \end{vmatrix} \qquad \color{red}{5 \text{ is } a_{23}}$$

$$= -[(-2)(2) - (-1)(0)] = 4$$

MATCHED PROBLEM 2 Find the cofactors of 2 and 3 in the determinant in Example 2.

[*Note:* The sign in front of the minor, $(-1)^{i+j}$, can be determined rather mechanically by using a checkerboard pattern of $+$ and $-$ signs over the determinant, starting with $+$ in the upper left-hand corner:

$$\begin{matrix} + & - & + \\ - & + & - \\ + & - & + \end{matrix}$$

Use either the checkerboard or the exponent method—whichever is easier for you—to determine the sign in front of the minor.]

Theorem 1 will give us a step-by-step procedure for finding third-order determinants without having to memorize formula (2).

> **THEOREM 1 Computing a Third-Order Determinant**
>
> The value of a determinant of order 3 is the sum of three products obtained by multiplying each element in any row or any column by its cofactor. This is called *expanding along a row or column*.

Proving Theorem 1 requires six different calculations: expanding an arbitrary third-order determinant along each of the rows and columns, and showing that the result matches formula (2). You will be asked to complete a couple of those cases in the exercises.

EXAMPLE 3 Evaluating a Third-Order Determinant

Evaluate $\begin{vmatrix} 2 & -2 & 0 \\ -3 & 1 & 2 \\ 1 & -3 & -1 \end{vmatrix}$

SOLUTION We can choose any row or column to expand along. We will choose the first row because of the zero: we won't need to find that cofactor because it will be multiplied by zero.

$$\begin{vmatrix} 2 & -2 & 0 \\ -3 & 1 & 2 \\ 1 & -3 & -1 \end{vmatrix} = a_{11}\begin{pmatrix} \text{Cofactor} \\ \text{of } a_{11} \end{pmatrix} + a_{12}\begin{pmatrix} \text{Cofactor} \\ \text{of } a_{12} \end{pmatrix} + a_{13}\begin{pmatrix} \text{Cofactor} \\ \text{of } a_{13} \end{pmatrix}$$

$$= 2\left[(-1)^{1+1}\begin{vmatrix} 1 & 2 \\ -3 & -1 \end{vmatrix}\right] + (-2)\left[(-1)^{1+2}\begin{vmatrix} -3 & 2 \\ 1 & -1 \end{vmatrix}\right] + 0$$

$$= (2)(1)[(1)(-1) - (-3)(2)] + (-2)(-1)[(-3)(-1) - (1)(2)]$$

$$= (2)(5) + (2)(1) = 12$$

MATCHED PROBLEM 3

Evaluate $\begin{vmatrix} 2 & 1 & -1 \\ -2 & -3 & 0 \\ -1 & 2 & 1 \end{vmatrix}$

It's important to note that the determinant will work out the same regardless of which row or column you choose to expand along. So if possible, you should choose a row or column with one or more zeros to minimize the number of computations.

› ## Using Cramer's Rule to Solve Systems of Equations

Now we will see how determinants can be used to solve systems of equations. We'll start by investigating two equations in two variables, and then extend our results to three equations in three variables.

Instead of thinking of each system of linear equations in two variables as a different problem, let's see what happens when we attempt to solve the general system

$$a_{11}x + a_{12}y = k_1 \tag{3A}$$
$$a_{21}x + a_{22}y = k_2 \tag{3B}$$

once and for all, in terms of the unspecified real constants a_{11}, a_{12}, a_{21}, a_{22}, k_1, and k_2.

We proceed by multiplying equations (3A) and (3B) by suitable constants so that when the resulting equations are added, left side to left side and right side to right side, one of the variables drops out. Suppose we choose to eliminate y. What constant should we use to make the coefficients of y the same except for the signs? Multiply equation (3A) by a_{22} and (3B) by $-a_{12}$; then add:

a_{22}(3A): $a_{11}a_{22}x + a_{12}a_{22}y = k_1a_{22}$

$-a_{12}$(3B): $-a_{21}a_{12}x - a_{12}a_{22}y = -k_2a_{12}$

$\overline{a_{11}a_{22}x - a_{21}a_{12}x + 0y = k_1a_{22} - k_2a_{12}}$ *y* is eliminated. **Factor out *x*.**

$(a_{11}a_{22} - a_{21}a_{12})x = k_1a_{22} - k_2a_{12}$ Solve for *x*.

$$x = \frac{k_1a_{22} - k_2a_{12}}{a_{11}a_{22} - a_{21}a_{12}}$$ $a_{11}a_{22} - a_{21}a_{12} \neq 0$

At this point, the numerator and denominator might remind you of second-order determinants. In fact, the value of x can be written as

$$x = \frac{\begin{vmatrix} k_1 & a_{12} \\ k_2 & a_{22} \end{vmatrix}}{\begin{vmatrix} a_{11} & a_{12} \\ a_{21} & a_{22} \end{vmatrix}}$$

Similarly, starting with system (3A) and (3B) and eliminating x (this is left as an exercise), we obtain

$$y = \frac{\begin{vmatrix} a_{11} & k_1 \\ a_{21} & k_2 \end{vmatrix}}{\begin{vmatrix} a_{11} & a_{12} \\ a_{21} & a_{22} \end{vmatrix}}$$

These results are summarized in Theorem 2, **Cramer's rule**, which is named after the Swiss mathematician Gabriel Cramer (1704–1752).

> **THEOREM 2** Cramer's Rule for Two Equations and Two Variables

Given the system

$$a_{11}x + a_{12}y = k_1$$
$$a_{21}x + a_{22}y = k_2 \qquad \text{with} \qquad D = \begin{vmatrix} a_{11} & a_{12} \\ a_{21} & a_{22} \end{vmatrix} \neq 0$$

then

$$x = \frac{\begin{vmatrix} k_1 & a_{12} \\ k_2 & a_{22} \end{vmatrix}}{D} \qquad \text{and} \qquad y = \frac{\begin{vmatrix} a_{11} & k_1 \\ a_{21} & k_2 \end{vmatrix}}{D}$$

The determinant D is called the *coefficient determinant*. If $D \neq 0$, then the system has exactly one solution, which is given by Cramer's rule. If, on the other hand, $D = 0$, then it can be shown that the system is either inconsistent and has no solutions or is dependent and has an infinite number of solutions. In that case, we would need to use other methods to determine the exact nature of the solutions.

EXAMPLE **4** **Solving a Two-Variable System with Cramer's Rule**

Solve using Cramer's rule: $3x - 5y = 2$
$$-4x + 3y = -1$$

SOLUTIONS First find the determinant of the coefficient matrix:

$$D = \begin{vmatrix} 3 & -5 \\ -4 & 3 \end{vmatrix} = 9 - 20 = -11$$

Now replace the x column with the constants and find the determinant, then divide by -11.

$$x = \frac{\begin{vmatrix} 2 & -5 \\ -1 & 3 \end{vmatrix}}{-11} = \frac{6 - 5}{-11} = -\frac{1}{11}$$

Now repeat, this time replacing the y column with the constants.

$$y = \frac{\begin{vmatrix} 3 & 2 \\ -4 & -1 \end{vmatrix}}{-11} = \frac{-3 - (-8)}{-11} = -\frac{5}{11}$$

The solution to the system is $x = -\dfrac{1}{11}, y = -\dfrac{5}{11}$.

MATCHED PROBLEM 4 Solve using Cramer's rule: $3x + 2y = -4$
$$-4x + 3y = -10$$

Cramer's rule can be generalized completely for any size linear system that has the *same number of variables as equations*. However, it cannot be used to solve systems where the number of variables is not equal to the number of equations. In Theorem 3 we state without proof Cramer's rule for three equations in three variables.

> **THEOREM 3 Cramer's Rule for Three Equations in Three Variables**

Given the system

$$\begin{aligned} a_{11}x + a_{12}y + a_{13}z &= k_1 \\ a_{21}x + a_{22}y + a_{23}z &= k_2 \\ a_{31}x + a_{32}y + a_{33}z &= k_3 \end{aligned} \quad \text{with} \quad D = \begin{vmatrix} a_{11} & a_{12} & a_{13} \\ a_{21} & a_{22} & a_{23} \\ a_{31} & a_{32} & a_{33} \end{vmatrix} \neq 0$$

then

$$x = \frac{\begin{vmatrix} k_1 & a_{12} & a_{13} \\ k_2 & a_{22} & a_{23} \\ k_3 & a_{32} & a_{33} \end{vmatrix}}{D} \qquad y = \frac{\begin{vmatrix} a_{11} & k_1 & a_{13} \\ a_{21} & k_2 & a_{23} \\ a_{31} & k_3 & a_{33} \end{vmatrix}}{D} \qquad z = \frac{\begin{vmatrix} a_{11} & a_{12} & k_1 \\ a_{21} & a_{22} & k_2 \\ a_{31} & a_{32} & k_3 \end{vmatrix}}{D}$$

You can easily remember these determinant formulas for x, y, and z if you observe the following:

1. Determinant D is formed from the coefficients of x, y, and z, keeping the same relative position in the determinant as found in the system of equations.
2. Determinant D appears in the denominators for x, y, and z.
3. The numerator for x can be obtained from D by replacing the coefficients of x ($a_{11}, a_{21},$ and a_{31}) with the constants k_1, k_2, and k_3, respectively. Similar statements can be made for the numerators for y and z.

| EXAMPLE | 5 | **Solving a Three-Variable System with Cramer's Rule**

Solve using Cramer's rule:
$$\begin{aligned} x + y \quad\;\;\; &= 2 \\ 3y - z &= -4 \\ x \quad\;\; + z &= 3 \end{aligned}$$

SOLUTION

$$D = \begin{vmatrix} 1 & 1 & 0 \\ 0 & 3 & -1 \\ 1 & 0 & 1 \end{vmatrix} = 2$$

$$x = \frac{\begin{vmatrix} 2 & 1 & 0 \\ -4 & 3 & -1 \\ 3 & 0 & 1 \end{vmatrix}}{2} = \frac{7}{2}$$

$$y = \frac{\begin{vmatrix} 1 & 2 & 0 \\ 0 & -4 & -1 \\ 1 & 3 & 1 \end{vmatrix}}{2} = -\frac{3}{2}$$

$$z = \frac{\begin{vmatrix} 1 & 1 & 2 \\ 0 & 3 & -4 \\ 1 & 0 & 3 \end{vmatrix}}{2} = -\frac{1}{2}$$

MATCHED PROBLEM 5

Solve using Cramer's rule:
$$\begin{aligned} 3x \quad\quad - z &= 5 \\ x - y + z &= 0 \\ x + y \quad\quad &= 1 \end{aligned}$$

Cofactor expansion can be used to find determinants of orders higher than 3, so Cramer's rule can be used for systems with more than three variables. For large systems, however, the Gauss-Jordan method, which involves fewer arithmetic operations than Cramer's Rule, is a more practical choice.

ANSWERS TO MATCHED PROBLEMS

1. 14 **2.** Cofactor of 2 = 13; cofactor of 3 = -4 **3.** 3
4. $x = \frac{8}{17}, y = -\frac{46}{17}$ **5.** $x = \frac{6}{5}, y = -\frac{1}{5}, z = -\frac{7}{5}$

10-5 Exercises

1. Explain the difference between $\begin{bmatrix} a_{11} & a_{12} \\ a_{21} & a_{22} \end{bmatrix}$ and $\begin{vmatrix} a_{11} & a_{12} \\ a_{21} & a_{22} \end{vmatrix}$.

2. Explain the difference between a matrix and a minor.

3. Explain the difference between a minor and a cofactor.

4. How do you evaluate a third-order determinant?

5. If A is the 2×2 coefficient matrix for a linear system and det $(A) = 0$, what can you conclude about the solution set for the system?

6. Can you use Cramer's rule to solve a linear system with a 3×2 coefficient matrix? Explain.

7. Can you use Cramer's rule to solve a linear system with a 4×4 coefficient matrix? Explain.

8. List all the possible solution methods for linear systems that we have discussed in this chapter. Which is your favorite and why?

Evaluate each second-order determinant in Problems 9–14.

9. $\begin{vmatrix} 5 & 4 \\ 2 & 3 \end{vmatrix}$ **10.** $\begin{vmatrix} 8 & -3 \\ 4 & 1 \end{vmatrix}$

11. $\begin{vmatrix} 3 & -7 \\ -5 & 6 \end{vmatrix}$ **12.** $\begin{vmatrix} 9 & -2 \\ 4 & 0 \end{vmatrix}$

13. $\begin{vmatrix} 4.3 & -1.2 \\ -5.1 & 3.7 \end{vmatrix}$ **14.** $\begin{vmatrix} -0.7 & -2.3 \\ 1.9 & -4.8 \end{vmatrix}$

Solve the system in Problems 15–22 using Cramer's rule.

15. $\begin{aligned} x + 2y &= 1 \\ x + 3y &= -1 \end{aligned}$ **16.** $\begin{aligned} x + 2y &= 3 \\ x + 3y &= 5 \end{aligned}$

17. $\begin{aligned} 2x + y &= 1 \\ 5x + 3y &= 2 \end{aligned}$ **18.** $\begin{aligned} x + 3y &= 1 \\ 2x + 8y &= 0 \end{aligned}$

19. $\begin{aligned} 2x - y &= -3 \\ -x + 3y &= 3 \end{aligned}$ **20.** $\begin{aligned} -3x + 2y &= 1 \\ 2x - 3y &= -3 \end{aligned}$

21. $\begin{aligned} 4x - 3y &= 4 \\ 3x + 2y &= -2 \end{aligned}$ **22.** $\begin{aligned} 5x + 2y &= -1 \\ 2x - 3y &= 2 \end{aligned}$

Problems 23–30 pertain to the following determinant:

$$\begin{vmatrix} 5 & -1 & -3 \\ 3 & 4 & 6 \\ 0 & -2 & 8 \end{vmatrix}$$

Write the minor of each element given in Problems 23–26. Leave the answer in determinant form.

23. a_{11} **24.** a_{33} **25.** a_{23} **26.** a_{12}

Write the cofactor of each element given in Problems 27–30, and evaluate each.

27. a_{11} **28.** a_{33} **29.** a_{23} **30.** a_{12}

Evaluate the determinant in Problems 31–40 using cofactors.

31. $\begin{vmatrix} 1 & 0 & 0 \\ -2 & 4 & 3 \\ 5 & -2 & 1 \end{vmatrix}$ **32.** $\begin{vmatrix} 2 & -3 & 5 \\ 0 & -3 & 1 \\ 0 & 6 & 2 \end{vmatrix}$

33. $\begin{vmatrix} 0 & 1 & 5 \\ 3 & -7 & 6 \\ 0 & -2 & -3 \end{vmatrix}$ **34.** $\begin{vmatrix} 4 & -2 & 0 \\ 9 & 5 & 4 \\ 1 & 2 & 0 \end{vmatrix}$

35. $\begin{vmatrix} -1 & 2 & -3 \\ -2 & 0 & -6 \\ 4 & -3 & 2 \end{vmatrix}$ **36.** $\begin{vmatrix} 0 & 2 & -1 \\ -6 & 3 & 1 \\ 7 & -9 & -2 \end{vmatrix}$

37. $\begin{vmatrix} 1 & 4 & 1 \\ 1 & 1 & -2 \\ 2 & 1 & -1 \end{vmatrix}$ **38.** $\begin{vmatrix} 3 & 2 & 1 \\ -1 & 5 & 1 \\ 2 & 3 & 1 \end{vmatrix}$

39. $\begin{vmatrix} 1 & 4 & 3 \\ 2 & 1 & 6 \\ 3 & -2 & 9 \end{vmatrix}$ **40.** $\begin{vmatrix} 4 & -6 & 3 \\ -1 & 4 & 1 \\ 5 & -6 & 3 \end{vmatrix}$

Solve Problems 41–44 to two significant digits using Cramer's rule.

41. $0.9925x - 0.9659y = \quad 0$
$\quad 0.1219x + 0.2588y = 2,500$

42. $0.9877x - 0.9744y = \quad 0$
$\quad 0.1564x + 0.2250y = 1,900$

43. $0.9954x - 0.9942y = \quad 0$
$\quad 0.0958x + 0.1080y = 155$

44. $0.9973x - 0.9957y = \quad 0$
$\quad 0.0732x + 0.0924y = 112$

Solve Problems 45–52 using Cramer's rule:

45. $\begin{aligned} x + y &= 0 \\ 2y + z &= -5 \\ -x \quad + z &= -3 \end{aligned}$ **46.** $\begin{aligned} x + y &= -4 \\ 2y + z &= 0 \\ -x \quad + z &= 5 \end{aligned}$

47. $\begin{aligned} x + y &= 1 \\ 2y + z &= 0 \\ -y + z &= 1 \end{aligned}$ **48.** $\begin{aligned} x + 3y &= -3 \\ 2y + z &= 3 \\ -x \quad + 3z &= 7 \end{aligned}$

49. $\begin{aligned} 3y + z &= -1 \\ x \quad + 2z &= 3 \\ x - 3y &= -2 \end{aligned}$ **50.** $\begin{aligned} x \quad - z &= 3 \\ 2x - y &= -3 \\ x + y + z &= 1 \end{aligned}$

51. $\begin{aligned} 2y - z &= -3 \\ x - y - z &= 2 \\ x - y + 2z &= 4 \end{aligned}$ **52.** $\begin{aligned} 2x + y &= 2 \\ x - y + z &= -1 \\ x + y + z &= 2 \end{aligned}$

Discuss the number of solutions for the systems in Problems 53 and 54 where a and b are real numbers. Use Cramer's rule where appropriate and Gauss–Jordan elimination otherwise.

53. $\begin{aligned} ax + 3y &= b \\ 2x + 4y &= 5 \end{aligned}$ **54.** $\begin{aligned} 2x + ay &= b \\ 3x + 4y &= 7 \end{aligned}$

In Problems 55 and 56, use Cramer's rule to solve for x only.

55. $\begin{aligned} 2x - 3y + z &= -3 \\ -4x + 3y + 2z &= -11 \\ x - y - z &= 3 \end{aligned}$ **56.** $\begin{aligned} x + 4y - 3z &= 25 \\ 3x + y - z &= 2 \\ -4x + y + 2z &= 1 \end{aligned}$

In Problems 57 and 58, use Cramer's rule to solve for y only.

57. $\begin{aligned} 12x - 14y + 11z &= 5 \\ 15x + 7y - 9z &= -13 \\ 5x - 3y + 2z &= 0 \end{aligned}$ **58.** $\begin{aligned} 2x - y + 4z &= 15 \\ -x + y + 2z &= 5 \\ 3x + 4y - 2z &= 4 \end{aligned}$

In Problems 59 and 60, use Cramer's rule to solve for z only.

59. $\begin{aligned} 3x - 4y + 5z &= 18 \\ -9x + 8y + 7z &= -13 \\ 5x - 7y + 10z &= 33 \end{aligned}$ **60.** $\begin{aligned} 13x + 11y + 10z &= 2 \\ 10x + 8y + 7z &= 1 \\ 8x + 5y + 4z &= 4 \end{aligned}$

If A is a 3 × 3 matrix, det A can be evaluated by the following **diagonal expansion.** *Form a 3 × 5 matrix by augmenting A on the right with its first two columns, and compute the diagonal products* $p_1, p_2, \ldots, p_6$ *indicated by the arrows:*

Diagonal **expansion formula**

The determinant of A is given by [compare with formula (2)]

$$\det A = p_1 + p_2 + p_3 - p_4 - p_5 - p_6$$
$$= a_{11}a_{22}a_{33} + a_{12}a_{23}a_{31} + a_{13}a_{21}a_{32} - a_{13}a_{22}a_{31} - a_{11}a_{23}a_{32} - a_{12}a_{21}a_{33}$$

[Caution: The diagonal expansion procedure works only for 3 × 3 matrices. Do not apply it to matrices of any other size.]

Use the diagonal expansion formula to evaluate the determinants in Problems 61 and 62.

61. $\begin{vmatrix} 2 & 6 & -1 \\ 5 & 3 & -7 \\ -4 & -2 & 1 \end{vmatrix}$ **62.** $\begin{vmatrix} 4 & 1 & -5 \\ 1 & 2 & -6 \\ -3 & -1 & 7 \end{vmatrix}$

A square matrix is called an **upper triangular matrix** *if all elements below the principal diagonal are zero. In Problems 63–66, determine whether the statement is true or false. If true, explain why. If false, give a counterexample.*

63. If the determinant of an upper triangular matrix is 0, then the elements on the principal diagonal are all 0.

64. If A and B are upper triangular matrices, then $\det(A + B) = \det A + \det B$.

65. The determinant of an upper triangular matrix is the product of the elements on the principal diagonal.

66. If A and B are upper triangular matrices, then $\det(AB) = (\det A)(\det B)$.

67. Show that the expansion of the determinant

$$\begin{vmatrix} a_{11} & a_{12} & a_{13} \\ a_{21} & a_{22} & a_{23} \\ a_{31} & a_{32} & a_{33} \end{vmatrix}$$

by the first column is the same as its expansion by the third row, and that both match formula (2).

68. Repeat Problem 67, using the second row and the third column.

69. If

$$A = \begin{bmatrix} 2 & 3 \\ 1 & -2 \end{bmatrix} \quad \text{and} \quad B = \begin{bmatrix} -1 & 3 \\ 2 & 1 \end{bmatrix}$$

show that $\det(AB) = (\det A)(\det B)$.

70. If

$$A = \begin{bmatrix} a & b \\ c & d \end{bmatrix} \quad \text{and} \quad B = \begin{bmatrix} w & x \\ y & z \end{bmatrix}$$

show that $\det(AB) = (\det A)(\det B)$.

It is clear that $x = 0$, $y = 0$, $z = 0$ is a solution to each of the systems given in Problem 71. Use Cramer's rule to determine whether this solution is unique. [Hint: If $D \neq 0$, what can you conclude? If $D = 0$, what can you conclude?]

71. (a). $\begin{aligned} x - 4y + 9z &= 0 \\ 4x - y + 6z &= 0 \\ x - y + 3z &= 0 \end{aligned}$ **(b).** $\begin{aligned} 3x - y + 3z &= 0 \\ 5x + 5y - 9z &= 0 \\ -2x + y - 3z &= 0 \end{aligned}$

72. Prove Theorem 2 for y.

APPLICATIONS

73. REVENUE ANALYSIS A supermarket sells two brands of coffee: brand A at $\$p$ per pound and brand B at $\$q$ per pound. The daily demand equations for brands A and B are, respectively,

$$\begin{aligned} x &= 200 - 6p + 4q \\ y &= 300 + 2p - 3q \end{aligned} \qquad (1)$$

(both in pounds). The daily revenue R is given by

$$R = xp + yq$$

(A) To analyze the effect of price changes on the daily revenue, an economist wants to express the daily revenue R in terms of p and q only. Use system (1) to eliminate x and y in the equation for R, expressing the daily revenue in terms of p and q.

(B) To analyze the effect of changes in demand on the daily revenue, the economist now wants to express the daily revenue in terms of x and y only. Use Cramer's rule to solve system (1) for p and q in terms of x and y and then express the daily revenue R in terms of x and y.

74. REVENUE ANALYSIS A company manufactures ten-speed and three-speed bicycles. The weekly demand equations are

$$\begin{aligned} p &= 230 - 10x + 5y \\ q &= 130 + 4x - 4y \end{aligned} \qquad (2)$$

where $\$p$ is the price of a ten-speed bicycle, $\$q$ is the price of a three-speed bicycle, x is the weekly demand for ten-speed bicycles, and y is the weekly demand for three-speed bicycles. The weekly revenue R is given by

$$R = xp + yq$$

(A) Use system (2) to express the daily revenue in terms of x and y only.

(B) Use Cramer's rule to solve system (2) for x and y in terms of p and q, and then express the daily revenue R in terms of p and q only.

CHAPTER **10** Review

10-1 Systems of Linear Equations

A system of two linear equations in two variables is a system of the form

$$\begin{aligned} ax + by &= h \\ cx + dy &= k \end{aligned} \qquad (1)$$

where x and y are variables; a, b, c, and d are real numbers called the **coefficients** of x and y, and h and k are real numbers called the **constant terms** in the equations. The ordered pair of numbers (x_0, y_0) is a **solution** to system (1) if each equation is satisfied by the pair. The set of all such ordered pairs of numbers is called the **solution set** for the system. To **solve** a system is to find its solution set.

In general, a system of linear equations has exactly one solution, no solution, or infinitely many solutions. A system of linear equations is **consistent** if it has one or more solutions and **inconsistent** if no solutions exist. A consistent system is said to be **independent** if it has exactly one solution and **dependent** if it has more than one solution.

To solve a system by **substitution,** solve either equation for either variable, substitute in the other equation, solve the resulting linear equation in one variable, and then substitute this value into the expression obtained in the first step to find the other variable.

Two systems of equations are **equivalent** if both have the same solution set. To solve a system of equations using **elimination by addition,** use Theorem 2 to find a simpler equivalent system whose solution is obvious.

As stated in Theorem 2, a system of linear equations is transformed into an equivalent system if:

1. Two equations are interchanged.

2. An equation is multiplied by a nonzero constant.

3. A constant multiple of another equation is added to a given equation.

The solution set S of a dependent system is often expressed in terms of a **parameter.** Any element in S is called a **particular solution.**

Any equation that can be written in the form

$$ax + by + cz = k$$

where a, b, c, and k are constants (not all a, b, and c zero) is called a **linear equation in three variables.** The method of elimination by addition can be used for systems of linear equations in three variables.

10-2 Solving Systems of Linear Equations Using Gauss–Jordan Elimination

The method of solution using elimination by addition can be transformed into a more efficient method for larger-scale systems by the introduction of an *augmented matrix*. A **matrix** is a rectangular array of numbers written within brackets. Each number in a matrix is called an **element** of the matrix. If a matrix has *m* rows and *n* columns, it is called an **m × n matrix** (read "*m* by *n* matrix"). The expression *m × n* is called the **size** of the matrix, and the numbers *m* and *n* are called the **dimensions** of the matrix. A matrix with *n* rows and *n* columns is called a **square matrix of order *n*.** A matrix with only one column is called a **column matrix,** and a matrix with only one row is called a **row matrix.** The **position** of an element in a matrix is the row and column containing the element. This is usually denoted using **double subscript notation** a_{ij}, where *i* is the row and *j* is the column containing the element a_{ij}. The **principal diagonal** of a matrix A consists of the elements a_{ii}, $i = 1, 2, \dots, n$. Rather than using x, y, and z to denote variables, we will use subscript notation x_1, x_2, and x_3.

Related to the system

$$x_1 + 5x_2 - 3x_3 = 4$$
$$6x_1 \qquad - 4x_3 = 1$$
$$-2x_1 + 3x_2 + 4x_3 = 7$$

are the following matrices:

Coefficient matrix	Constant matrix	Augmented coefficient matrix	
$\begin{bmatrix} 1 & 5 & -3 \\ 6 & 0 & -4 \\ -2 & 3 & 4 \end{bmatrix}$	$\begin{bmatrix} 4 \\ 1 \\ 7 \end{bmatrix}$	$\left[\begin{array}{ccc	c} 1 & 5 & -3 & 4 \\ 6 & 0 & -4 & 1 \\ -2 & 3 & 4 & 7 \end{array}\right]$

Two augmented matrices are **row-equivalent,** denoted by the symbol ~ between the two matrices, if they are augmented matrices of equivalent systems of equations. An augmented matrix is transformed into a row-equivalent matrix if any of the following **row operations** is performed:

1. Two rows are interchanged.

2. A row is multiplied by a nonzero constant.

3. A constant multiple of another row is added to a given row.

These correspond to the operations on equations from Theorem 2 in Section 10–1. The following symbols are used to describe these row operations:

1. $R_i \leftrightarrow R_j$ means "interchange row *i* with row *j*."

2. $kR_i \rightarrow R_i$ means "multiply row *i* by the constant *k*."

3. $kR_j + R_i \rightarrow R_i$ means "multiply row *j* by the constant *k* and add to row *i*."

As before, our objective is to start with the augmented matrix of a linear system and transform it using row operations into a simple form where the solution can be found easily. The simple form, called the **reduced form,** is achieved if:

1. Each row consisting entirely of 0's is below any row having at least one nonzero element.

2. The leftmost nonzero element in each row is 1.

3. The column containing the leftmost 1 of a given row has 0's above and below the 1.

4. The leftmost 1 in any row is to the right of the leftmost 1 in the preceding row.

A **reduced system** is a system of linear equations that corresponds to a reduced augmented matrix. When a reduced system has more variables than equations and contains no contradictions, the system is dependent and has infinitely many solutions.

The **Gauss–Jordan elimination** procedure for solving a system of linear equations is given in step-by-step form as follows:

Step 1. Choose the leftmost nonzero column, and use appropriate row operations to get a 1 at the top.

Step 2. Use multiples of the row containing the 1 from step 1 to get zeros in all remaining places in the column containing this 1.

Step 3. Repeat step 1 with the **submatrix** formed by (mentally) deleting the row used in step 2 and all rows above this row.

Step 4. Repeat step 2 with the **entire matrix,** including the mentally deleted rows. Continue this process until the entire matrix is in reduced form.

If at any point in the preceding process we obtain a row with all 0's to the left of the vertical line and a nonzero number *n* to the right, we can stop, since we have a contradiction: $0 = n$, $n \neq 0$. We can then conclude that the system has no solution. If this does not happen and we obtain an augmented matrix in reduced form without any contradictions, the solution can be found by converting back to equation form.

10-3 Matrix Operations

Two matrices are **equal** if they are the same size and their corresponding elements are equal. The **sum of two matrices** of the same size is a matrix with elements that are the sums of the corresponding elements of the two given matrices. Matrix addition is **commutative** and **associative.** A matrix with all zero elements is called the **zero matrix.** The **negative of a matrix *M*,** denoted $-M$, is a matrix with elements that are the negatives of the elements in M. If A and B are matrices of the same size, then we define **subtraction** as follows: $A - B = A + (-B)$. The **product of a number *k* and a matrix *M*,** denoted by kM, is a matrix formed by multiplying each element of M by k. The **product** of a $1 \times n$ row matrix and an $n \times 1$ column matrix is a 1×1 matrix given by

$$\underset{1 \times n}{[a_1 \quad a_2 \quad \cdots \quad a_n]} \overset{n \times 1}{\begin{bmatrix} b_1 \\ b_2 \\ \vdots \\ b_n \end{bmatrix}} = \underset{1 \times 1}{[a_1 b_1 + a_2 b_2 + \cdots + a_n b_n]}$$

If A is an $m \times p$ matrix and B is a $p \times n$ matrix, then the **matrix product** of A and B, denoted AB, is an $m \times n$ matrix whose element in the *i*th row and *j*th column is the real number obtained from the product of the *i*th row of A and the *j*th column of B. If the number of columns in A does not equal the number of rows in B, then the matrix product AB is **not defined. Matrix multiplication is not commutative,** and the **zero property does not hold for matrix multiplication.** That is, for matrices A and B, the matrix product AB can be zero without either A or B being the zero matrix.

10-4 Solving Systems of Linear Equations Using Matrix Inverse Methods

The **identity matrix for multiplication** for the set of all square matrices of order n is the square matrix of order n, denoted by I, with 1's along the **principal diagonal** (from upper left corner to lower right corner) and 0's elsewhere. If M is a square matrix of order n and I is the identity matrix of order n, then

$$IM = MI = M$$

If M is a **square matrix of order n** and if there exists a matrix M^{-1} (read "M inverse") such that

$$M^{-1}M = MM^{-1} = I$$

then M^{-1} is called the **multiplicative inverse of M** or, more simply, the **inverse of M.** If the augmented matrix $[M\,|\,I]$ is transformed by row operations into $[I\,|\,B]$, then the resulting matrix B is M^{-1}. If, however, we obtain all 0's in one or more rows to the left of the vertical line, then M^{-1} does not exist and M is called a **singular matrix.**

A system of linear equations with the same number of variables as equations such as

$$a_{11}x_1 + a_{12}x_2 + a_{13}x_3 = k_1$$
$$a_{21}x_1 + a_{22}x_2 + a_{23}x_3 = k_2$$
$$a_{31}x_1 + a_{32}x_2 + a_{33}x_3 = k_3$$

can be written as the matrix equation

$$\underset{A}{\begin{bmatrix} a_{11} & a_{12} & a_{13} \\ a_{21} & a_{22} & a_{23} \\ a_{31} & a_{32} & a_{33} \end{bmatrix}} \underset{X}{\begin{bmatrix} x_1 \\ x_2 \\ x_3 \end{bmatrix}} = \underset{B}{\begin{bmatrix} k_1 \\ k_2 \\ k_3 \end{bmatrix}}$$

If the inverse of A exists, then the matrix equation has a unique solution given by

$$X = A^{-1}B$$

After multiplying B by A^{-1} *on the left*, it is easy to read the solution to the original system of equations.

10-5 Determinants and Cramer's Rule

Associated with each square matrix A is a real number called the **determinant** of the matrix. The determinant of A is denoted by **det A**, or simply by writing the array of elements in A using vertical lines in place of square brackets. For example,

$$\det \begin{bmatrix} a_{11} & a_{12} \\ a_{21} & a_{22} \end{bmatrix} = \begin{vmatrix} a_{11} & a_{12} \\ a_{21} & a_{22} \end{vmatrix}$$

A **determinant of order n** is a determinant with n rows and n columns.

The **value of a second-order determinant** is the real number given by

$$\begin{vmatrix} a_{11} & a_{12} \\ a_{21} & a_{22} \end{vmatrix} = a_{11}a_{22} - a_{21}a_{12}$$

The **value of a third-order determinant** is the sum of three products obtained by multiplying each element of any one row (or each element of any one column) by its cofactor. The **cofactor of an element a_{ij}** (from the ith row and jth column) is the product of the minor of a_{ij} and $(-1)^{i+j}$. The **minor of an element a_{ij}** is the determinant remaining after deleting the ith row and jth column.

Systems of equations having the same number of variables as equations can also be solved using determinants and Cramer's rule. **Cramer's rule for three equations and three variables** is as follows: Given the system

$$a_{11}x + a_{12}y + a_{13}z = k_1$$
$$a_{21}x + a_{22}y + a_{23}z = k_2 \quad \text{with} \quad D = \begin{vmatrix} a_{11} & a_{12} & a_{13} \\ a_{21} & a_{22} & a_{23} \\ a_{31} & a_{32} & a_{33} \end{vmatrix} \neq 0$$
$$a_{31}x + a_{32}y + a_{33}z = k_3$$

then

$$x = \frac{\begin{vmatrix} k_1 & a_{12} & a_{13} \\ k_2 & a_{22} & a_{23} \\ k_3 & a_{32} & a_{33} \end{vmatrix}}{D} \quad y = \frac{\begin{vmatrix} a_{11} & k_1 & a_{13} \\ a_{21} & k_2 & a_{23} \\ a_{31} & k_3 & a_{33} \end{vmatrix}}{D} \quad z = \frac{\begin{vmatrix} a_{11} & a_{12} & k_1 \\ a_{21} & a_{22} & k_2 \\ a_{31} & a_{32} & k_3 \end{vmatrix}}{D}$$

Cramer's rule can be generalized completely for any size linear system that has the same number of variables as equations. The formulas are easily remembered if you observe the following:

1. Determinant D is formed from the coefficients of x, y, and z, keeping the same relative position in the determinant as found in the system of equations.

2. Determinant D appears in the denominators for x, y, and z.

3. The numerator for x can be obtained from D by replacing the coefficients of x (a_{11}, a_{21}, and a_{31}) with the constants k_1, k_2, and k_3, respectively. Similar statements can be made for the numerators for y and z.

Cramer's rule is rarely used to solve systems of order higher than 3 by hand, because more efficient methods are available. Cramer's rule, however, is a valuable tool in more advanced theoretical and applied mathematics.

CHAPTER 10 Review Exercises

Work through all the problems in this chapter review and check answers in the back of the book. Answers to all review problems are there, and following each answer is a number in italics indicating the section in which that type of problem is discussed. Where weaknesses show up, review appropriate sections in the text.

Solve the system in Problems 1–5 using substitution or elimination by addition.

1. $2x + y = 7$
$3x - 2y = 0$

2. $3x - 6y = 5$
$-2x + 4y = 1$

3. $4x - 3y = -8$
$-2x + \frac{3}{2}y = 4$

4. $x - 3y + z = 4$
$-x + 4y - 4z = 1$
$2x - y + 5z = -3$

5. $2x + y - z = 5$
$x - 2y - 2z = 4$
$3x + 4y + 3z = 3$

6. Solve the system by graphing.
$3x - 2y = 8$
$x + 3y = -1$

Perform each of the row operations indicated in Problems 10–9 on the following augmented matrix:

$$\begin{bmatrix} 1 & -4 & | & 5 \\ 3 & -6 & | & 12 \end{bmatrix}$$

7. $R_1 \leftrightarrow R_2$

8. $\frac{1}{3}R_2 \to R_2$

9. $(-3)R_1 + R_2 \to R_2$

In Problems 10–12, write the linear system corresponding to each reduced augmented matrix and solve.

10. $\begin{bmatrix} 1 & 0 & | & 4 \\ 0 & 1 & | & -7 \end{bmatrix}$

11. $\begin{bmatrix} 1 & -1 & | & 4 \\ 0 & 0 & | & 1 \end{bmatrix}$

12. $\begin{bmatrix} 1 & -1 & | & 4 \\ 0 & 0 & | & 0 \end{bmatrix}$

In Problems 13–21, perform the operations that are defined, given the following matrices:

$$A = \begin{bmatrix} 4 & -2 \\ 0 & 3 \end{bmatrix} \quad B = \begin{bmatrix} -1 & 5 \\ -4 & 6 \end{bmatrix} \quad C = \begin{bmatrix} -1 & 4 \end{bmatrix} \quad D = \begin{bmatrix} 3 \\ -2 \end{bmatrix}$$

13. AB

14. CD

15. CB

16. AD

17. $A + B$

18. $C + D$

19. $A + C$

20. $2A - 5B$

21. $CA + C$

22. Find the inverse of

$$A = \begin{bmatrix} 4 & 7 \\ -1 & -2 \end{bmatrix}$$

Show that $A^{-1}A = I$.

23. Write the system

$$3x_1 + 2x_2 = k_1$$
$$4x_1 + 3x_2 = k_2$$

as a matrix equation, and solve using matrix inverse methods for:
(A) $k_1 = 3, k_2 = 5$ (B) $k_1 = 7, k_2 = 10$
(C) $k_1 = 4, k_2 = 2$

Evaluate the determinants in Problems 24 and 25.

24. $\begin{vmatrix} 2 & -3 \\ -5 & -1 \end{vmatrix}$

25. $\begin{vmatrix} 2 & 3 & -4 \\ 0 & 5 & 0 \\ 1 & -4 & -2 \end{vmatrix}$

26. Solve the system using Cramer's rule:

$$3x - 2y = 8$$
$$x + 3y = -1$$

27. Use Gauss–Jordan elimination to solve the system

$$x_1 - x_2 = 4$$
$$2x_1 + x_2 = 2$$

Then write the linear system represented by each augmented matrix in your solution, and solve each of these systems graphically. Discuss the relationship between the solutions of these systems.

 28. Use an intersection routine on a graphing calculator to approximate the solution of the following system to two decimal places:

$$x + 3y = 9$$
$$-2x + 7y = 10$$

Solve the system in Problems 29–34 using Gauss–Jordan elimination.

29. $3x_1 + 2x_2 = 3$
$x_1 + 3x_2 = 8$

30. $x_1 + x_2 = 1$
$x_1 - x_3 = -2$
$x_2 + 2x_3 = 4$

31. $x_1 + 2x_2 + 3x_3 = 1$
$2x_1 + 3x_2 + 4x_3 = 3$
$x_1 + 2x_2 + x_3 = 3$

32. $x_1 + 2x_2 - x_3 = 2$
$2x_1 + 3x_2 + x_3 = -3$
$3x_1 + 5x_2 = -1$

33. $x_1 - 2x_2 = 1$
$2x_1 - x_2 = 0$
$x_1 - 3x_2 = -2$

34. $x_1 + 2x_2 - x_3 = 2$
$3x_1 - x_2 + 2x_3 = -3$

In Problems 35–40, perform the operations that are defined, given the following matrices:

$$A = \begin{bmatrix} 1 & 2 \\ 4 & 5 \\ -3 & -1 \end{bmatrix} \quad B = \begin{bmatrix} 6 \\ 0 \\ -4 \end{bmatrix} \quad C = \begin{bmatrix} 2 & 4 & -1 \end{bmatrix}$$

$$D = \begin{bmatrix} 7 & 0 & -5 \\ 0 & 8 & -2 \end{bmatrix} \quad E = \begin{bmatrix} 9 & -3 \\ -6 & 2 \end{bmatrix}$$

35. AD **36.** DA **37.** BC

38. CB **39.** DE **40.** ED

41. Find the inverse of

$$A = \begin{bmatrix} 1 & 0 & 4 \\ -2 & 1 & 0 \\ 4 & -1 & 4 \end{bmatrix}$$

Show that $AA^{-1} = I$.

42. Write the system

$$x_1 + 2x_2 + 3x_3 = k_1$$
$$2x_1 + 3x_2 + 4x_3 = k_2$$
$$x_1 + 2x_2 + x_3 = k_3$$

as a matrix equation, and solve using matrix inverse methods for:
(A) $k_1 = 1, k_2 = 3, k_3 = 3$
(B) $k_1 = 0, k_2 = 0, k_3 = -2$
(C) $k_1 = -3, k_2 = -4, k_3 = 1$

Evaluate the determinants in Problems 43 and 44.

43. $\begin{vmatrix} -\frac{1}{4} & \frac{3}{2} \\ \frac{1}{2} & \frac{2}{3} \end{vmatrix}$

44. $\begin{vmatrix} 2 & -1 & 1 \\ -3 & 5 & 2 \\ 1 & -2 & 4 \end{vmatrix}$

45. Solve for y only using Cramer's rule:

$$\begin{aligned} x - 2y + z &= -6 \\ y - z &= 4 \\ 2x + 2y + z &= 2 \end{aligned}$$

46. Solve using Gauss–Jordan elimination:

$$\begin{aligned} x_1 + x_2 + x_3 &= 7{,}000 \\ 0.04x_1 + 0.05x_2 + 0.06x_3 &= 360 \\ 0.04x_1 + 0.05x_2 - 0.06x_3 &= 120 \end{aligned}$$

47. Show that

$$\begin{vmatrix} u & v \\ w & x \end{vmatrix} = \begin{vmatrix} u + kv & v \\ w + kx & x \end{vmatrix}$$

48. Discuss the number of solutions for the system corresponding to the reduced form shown here if
(A) $m \neq 0$
(B) $m = 0$ and $n \neq 0$
(C) $m = 0$ and $n = 0$

$$\begin{bmatrix} 1 & 0 & -3 & | & 4 \\ 0 & 1 & 2 & | & 5 \\ 0 & 0 & m & | & n \end{bmatrix}$$

49. Discuss the number of solutions for a system of n equations in n variables if the coefficient matrix:
(A) Has an inverse.
(B) Does not have an inverse.

50. If A is a nonzero square matrix of order n satisfying $A^2 = 0$, can A^{-1} exist? Explain.

51. For $n \times n$ matrices A and C and $n \times 1$ column matrices B and X, solve for X assuming all necessary inverses exist:

$$AX - B = CX$$

52. Find the inverse of

$$A = \begin{bmatrix} 4 & 5 & 6 \\ 4 & 5 & -6 \\ 1 & 1 & 1 \end{bmatrix}$$

Show that $A^{-1}A = I$.

53. Clear the decimals in the system

$$\begin{aligned} 0.04x_1 + 0.05x_2 + 0.06x_3 &= 360 \\ 0.04x_1 + 0.05x_2 - 0.06x_3 &= 120 \\ x_1 + x_2 + x_3 &= 7{,}000 \end{aligned}$$

by multiplying the first two equations by 100. Then write the resulting system as a matrix equation and solve using the inverse found in Problem 52.

APPLICATIONS

54. BUSINESS A container holds 120 packages. Some of the packages weigh $\frac{1}{2}$ pound each, and the rest weigh $\frac{1}{3}$ pound each. If the total contents of the container weigh 48 pounds, how many are there of each type of package?

55. DIET A laboratory assistant needs a food mix that contains, among other things, 27 grams of protein, 5.4 grams of fat, and 19 grams of moisture. He has available mixes A, B, and C with the compositions listed in the table. How many grams of each mix should be used to get the desired diet mix? Set up a system of equations and solve using Gauss–Jordan elimination.

Mix	Protein (%)	Fat (%)	Moisture (%)
A	30	3	10
B	20	5	20
C	10	4	10

56. RESOURCE ALLOCATION A Colorado mining company operates mines at Big Bend and Saw Pit. The Big Bend mine produces ore that is 5% nickel and 7% copper. The Saw Pit mine produces ore that is 3% nickel and 4% copper. How many tons of ore should be produced at each mine to obtain the amounts of nickel and copper listed in the table? Set up a matrix equation and solve using matrix inverses.

	Nickel	Copper
(A)	3.6 tons	5 tons
(B)	3 tons	4.1 tons
(C)	3.2 tons	4.4 tons

57. LABOR COSTS A company with manufacturing plants in North and South Carolina has labor-hour and wage requirements for the manufacturing of computer desks and printer stands as given in matrices L and H:

Labor-hour requirements

	Fabricating department	Assembly department	Packaging department	
$L =$	1.7 h	2.4 h	0.8 h	Desk
	0.9 h	1.8 h	0.6 h	Stand

Hourly wages

	North Carolina plant	South Carolina plant	
$H =$	\$11.50	\$10.00	Fabricating department
	\$9.50	\$8.50	Assembly department
	\$5.00	\$4.50	Packaging department

(A) Find the labor cost for producing one printer stand at the South Carolina plant.
(B) Discuss possible interpretations of the elements in the matrix products HL and LH.
(C) If either of the products HL or LH has a meaningful interpretation, find the product and label its rows and columns.

58. LABOR COSTS The monthly production of computer desks and printer stands for the company in Problem 57 for the months of January and February are given in matrices J and F:

January production
North Carolina plant South Carolina plant

$$J = \begin{bmatrix} 1{,}500 & 1{,}650 \\ 850 & 700 \end{bmatrix} \begin{matrix} \text{Desks} \\ \text{Stands} \end{matrix}$$

February production
North Carolina plant South Carolina plant

$$F = \begin{bmatrix} 1{,}700 & 1{,}810 \\ 930 & 740 \end{bmatrix} \begin{matrix} \text{Desks} \\ \text{Stands} \end{matrix}$$

(A) Find the average monthly production for the months of January and February.

(B) Find the increase in production from January to February.

(C) Find $J \begin{bmatrix} 1 \\ 1 \end{bmatrix}$ and interpret.

59. CRYPTOGRAPHY The following message was encoded with the matrix B shown below. Decode the message:

21 21 27 30 28 31 29 34 50
46 35 62 19 21 39 52 52 79

$$B = \begin{bmatrix} 1 & 1 & 0 \\ 1 & 0 & 1 \\ 1 & 1 & 1 \end{bmatrix}$$

60. PUZZLE A piggy bank contains 30 coins worth $1.90.
(A) If the bank contains only nickels and dimes, how many coins of each type does it contain?
(B) If the bank contains nickels, dimes, and quarters, how many coins of each type does it contain?

CHAPTER **10**

››› GROUP ACTIVITY Modeling with Systems of Linear Equations

In this group activity, we will consider two real-world problems that can be solved using systems of linear equations: heat conduction and traffic flow. Both problems involve using a grid and a basic assumption to construct the model (the system of equations). Gauss–Jordan elimination is then used to solve the model. In the heat conduction problem, the solution of the model is easily interpreted in terms of the original problem. The system in the second problem is dependent, and the solution requires a more careful interpretation.

I HEAT CONDUCTION

A metal grid consists of four thin metal bars. The end of each bar of the grid is kept at a constant temperature, as shown in Figure 1. We assume that the temperature at each intersection point in the grid is the average of the temperatures at the four

adjacent points in the grid (adjacent points are either other intersection points or ends of bars). So the temperature x_1 at the intersection point in the upper left-hand corner of the grid must satisfy

$$x_1 = \tfrac{1}{4}(\underset{\text{Left}}{40} + \underset{\text{Above}}{0} + \underset{\text{Right}}{x_2} + \underset{\text{Below}}{x_3})$$

Find equations for the temperature at the other three intersection points, and solve the resulting system to find the temperature at each intersection point in the grid.

II TRAFFIC FLOW

The rush-hour traffic flow for a network of four one-way streets in a city is shown in Figure 2 on page 704. The numbers next to each street indicate the number of vehicles per hour that enter and leave the network on that street. The variables $x_1, x_2, x_3,$ and x_4 represent the flow of traffic between the four intersections in the network. For a smooth flow of traffic, we assume that the number of vehicles entering each intersection should always equal the number leaving. For example, since 1,500 vehicles enter the intersection of 5th Street and Washington Avenue each hour and $x_1 + x_4$ vehicles leave this intersection, we see that $x_1 + x_4 = 1{,}500$.
(A) Find the equations determined by the traffic flow at each of the other three intersections.
(B) Find the solution to the system in part A.

> **Figure 1**

› **Figure 2**

(C) What is the maximum number of vehicles that can travel from Washington Avenue to Lincoln Avenue on 5th Street? What is the minimum number?

(D) If traffic lights are adjusted so that 1,000 vehicles per hour travel from Washington Avenue to Lincoln Avenue on 5th Street, determine the flow around the rest of the network.

Sequences, Induction, and Probability

THE lists

$$1, 4, 9, 16, 25, 36, 49, 64, \ldots$$

and

$$3, 6, 3, 1, 4, 2, 1, 4, \ldots$$

are examples of sequences. In the first sequence, a pattern is noticeable: You probably recognize it as the sequence of perfect squares. Its terms are increasing, and as we will see, the differences between terms form a clear pattern. You probably don't recognize the second sequence because the terms don't suggest an obvious pattern. In fact, we obtained the second sequence by recording the results of repeatedly tossing a single die. Sequences, and the related concept of *series,* are useful tools in almost all areas of mathematics. In this chapter, they will play roles in the development of several topics: a method of proof called *mathematical induction,* techniques for counting, and probability.

11-1 Sequences and Series

> Defining Sequences
> Defining Series

In this section, we introduce special notation and formulas for representing and generating sequences and sums of sequences.

> Defining Sequences

Consider the following list of numbers: 1, 3, 5, 7, 9, This is an example of a **sequence,** which can be defind informally as a list of numbers in a specific order. This particular sequence is the sequence of positive odd integers.

Now consider the function f given by

$$f(n) = 2n - 1 \qquad\qquad (1)$$

where the domain of f is $\{1, 2, 3, \ldots\}$ (that is, the set of natural numbers N). Note that

$$f(1) = 2(1) - 1 = 1$$
$$f(2) = 2(2) - 1 = 3$$
$$f(3) = 2(3) - 1 = 5$$

The outputs of the function f form the same list of odd positive integers that we started with above. This provides an alternative (and more precise) definition of sequence: A **sequence** is a function whose domain is a set of successive integers.

While the function f above is a perfectly good way to describe a sequence, a special notation for describing sequences with formulas has evolved over the years. Our first order of business should be to become familiar with this notation.

To start, the range value $f(n)$ is usually symbolized more compactly with a symbol such as a_n. So in place of equation (1) we write

$$a_n = 2n - 1$$

The domain is understood to be the set of natural numbers N unless stated to the contrary or the context indicates otherwise. The elements in the range are called **terms of the sequence:** a_1 is the first term, a_2 the second term, and a_n the nth term, or the **general term:**

$$a_1 = 2(1) - 1 = 1 \qquad \text{First term}$$
$$a_2 = 2(2) - 1 = 3 \qquad \text{Second term}$$
$$a_3 = 2(3) - 1 = 5 \qquad \text{Third term}$$
$$\vdots \qquad\qquad \vdots$$

The ordered list of elements

$$1, 3, 5, \ldots, 2n - 1, \ldots$$

in which the terms of a sequence are written in their natural order with respect to the domain values, is often informally referred to as a sequence. A sequence is also represented in the abbreviated form $\{a_n\}$, where a symbol for the nth term is placed between braces. For example, we can refer to the sequence

$$1, 3, 5, \ldots, 2n - 1, \ldots$$

as the sequence $\{2n - 1\}$.

If the domain of a function is a finite set of successive integers, then the sequence is called a **finite sequence.** If the domain is an infinite set of successive integers, then the sequence is called an **infinite sequence.** The preceding sequence $\{2n - 1\}$ is an example of an infinite sequence.

Technology Connections

Some graphing calculators have a special sequence mode that can be useful when studying sequences. Figure 1(a) shows the sequence $\{2n - 1\}$ entered in the sequence editor.

Figure 1(b) shows the graph of the sequence. Figure 1(c) displays the sequence in a table.

(a)　　　　　　　　　　(b)　　　　　　　　　　(c)

> **Figure 1**

Some sequences are specified by a **recursion formula**—that is, a formula that defines each term in terms of one or more preceding terms. The sequence we have chosen to illustrate a recursion formula is a very famous sequence in the history of mathematics called the **Fibonacci sequence.** It is named after the most celebrated mathematician of the thirteenth century, Leonardo Fibonacci from Italy (1180?–1250?).

| EXAMPLE | 1 | **Fibonacci Sequence** |

List the first seven terms of the sequence specified by

$$a_1 = 1$$
$$a_2 = 1$$
$$a_n = a_{n-2} + a_{n-1} \qquad n \geq 3$$

SOLUTION

$$a_1 \;\;= 1$$
$$a_2 \;\;= 1$$
$$a_3 \;\;= a_1 + a_2 = 1 + 1 \;\;= 2 \quad *$$
$$a_4 \;\;= a_2 + a_3 = 1 + 2 \;\;= 3$$
$$a_5 \;\;= a_3 + a_4 = 2 + 3 \;\;= 5$$
$$a_6 \;\;= a_4 + a_5 = 3 + 5 \;\;= 8$$
$$a_7 \;\;= a_5 + a_6 = 5 + 8 \;\;= 13$$

*Throughout the book, dashed boxes—called think boxes—are used to represent steps that may be performed mentally.

MATCHED PROBLEM 1

List the first seven terms of the sequence specified by

$$a_1 = 1$$
$$a_2 = 1$$
$$a_n = a_{n-2} - a_{n-1} \qquad n \geq 3$$

>>> EXPLORE-DISCUSS 1

A multiple-choice test question asked for the next term in the sequence:

$$1, 3, 9, \ldots$$

and gave the following choices:

(A) 16 **(B)** 19 **(C)** 27

Which is the correct answer?

Compare the first four terms of the following sequences:

(A) $a_n = 3^{n-1}$ **(B)** $b_n = 1 + 2(n-1)^2$ **(C)** $c_n = 8n + \dfrac{12}{n} - 19$

Now which of the choices appears to be correct?

Now we consider the reverse problem. That is, can a sequence be defined just by listing the first three or four terms of the sequence? And can we then use these initial terms to find a formula for the nth term? In general, without other information, the answer to the first question is no. As Explore-Discuss 1 illustrates, many different sequences may start off with the same terms. Simply listing the first three terms, or any other finite number of terms, does not specify a particular sequence. In fact, it can be shown that given any list of m numbers, there are an infinite number of sequences whose first m terms agree with these given numbers.

What about the second question? That is, given a few terms, can we find the general formula for at least one sequence whose first few terms agree with the given terms? The answer to this question is a qualified yes. If we can observe a simple pattern in the given terms, then we may be able to construct a general term that will produce the pattern. Example 2 illustrates this approach.

EXAMPLE 2

Finding the General Term of a Sequence

Find the general term of a sequence whose first four terms are

(A) 5, 6, 7, 8, ... (B) 2, −4, 8, −16, ...

SOLUTIONS

(A) Because these terms are consecutive integers, one solution is $a_n = n$, $n \geq 5$. If we want the domain of the sequence to be all natural numbers, then another solution is $b_n = n + 4$.

(B) Each of these terms can be written as the product of a power of 2 and a power of −1:

$$2 = (-1)^0 2^1$$
$$-4 = (-1)^1 2^2$$
$$8 = (-1)^2 2^3$$
$$-16 = (-1)^3 2^4$$

If we choose the domain to be all natural numbers, then a solution is

$$a_n = (-1)^{n-1} 2^n$$

| MATCHED PROBLEM 2 | Find the general term of a sequence whose first four terms are |

(A) $2, 4, 6, 8, \ldots$ (B) $1, -\frac{1}{2}, \frac{1}{4}, -\frac{1}{8}, \ldots$

In general, there is usually more than one way of representing the nth term of a given sequence. This was seen in the solution of Example 2, part A. However, unless stated to the contrary, we assume the domain of the sequence is the set of natural numbers N.

>>> **EXPLORE-DISCUSS 2**

The sequence with general term $b_n = \dfrac{\sqrt{5}}{5}\left(\dfrac{1 + \sqrt{5}}{2}\right)^n$ is closely related to the Fibonacci sequence. Compute the first 20 terms of both sequences and discuss the relationship. [The first seven values of b_n are shown in Fig. 2(b)].

(a)

(b)

> **Figure 2**

> ## Defining Series

If $a_1, a_2, a_3, \ldots, a_n, \ldots$ is a sequence, then the expression

$$a_1 + a_2 + a_3 + \cdots + a_n + \cdots$$

is called a **series.** If the sequence is finite, the corresponding series is a **finite series.** If the sequence is infinite, the corresponding series is an **infinite series.** For example,

$$1, 2, 4, 8, 16 \qquad \text{Finite sequence}$$
$$1 + 2 + 4 + 8 + 16 \qquad \text{Finite series}$$

We will restrict our discussion to finite series in this section.

Series are often represented in a compact form called **summation notation** using the symbol $\sum$, which is a stylized version of the Greek letter sigma. Consider the following examples:

$$\sum_{k=1}^{4} a_k = a_1 + a_2 + a_3 + a_4$$

$$\sum_{k=3}^{7} b_k = b_3 + b_4 + b_5 + b_6 + b_7$$

$$\sum_{k=0}^{n} c_k = c_0 + c_1 + c_2 + \cdots + c_n \qquad \text{Domain is the set of integers } k \text{ satisfying } 0 \le k \le n.$$

The terms on the right are obtained from the expression on the left by successively replacing the **summing index** k with integers, starting with the first number indicated below $\sum$ and ending with the number that appears above $\sum$. For example, if we are given the sequence

$$\frac{1}{2}, \frac{1}{4}, \frac{1}{8}, \cdots, \frac{1}{2^n}$$

the corresponding series is

$$\sum_{k=1}^{n} \frac{1}{2^k} = \frac{1}{2} + \frac{1}{4} + \frac{1}{8} + \cdots + \frac{1}{2^n}$$

| EXAMPLE | 3 | **Writing the Terms of a Series** |

Write without summation notation: $\displaystyle\sum_{k=1}^{5} \frac{k-1}{k}$

SOLUTION

$$\sum_{k=1}^{5} \frac{k-1}{k} = \frac{1-1}{1} + \frac{2-1}{2} + \frac{3-1}{3} + \frac{4-1}{4} + \frac{5-1}{5}$$

$$= 0 + \frac{1}{2} + \frac{2}{3} + \frac{3}{4} + \frac{4}{5}$$

MATCHED PROBLEM 3 Write without summation notation: $\displaystyle\sum_{k=0}^{5} \frac{(-1)^k}{2k+1}$

If the terms of a series are alternately positive and negative, it is called an **alternating series**. Example 4 deals with the representation of such a series.

| EXAMPLE | 4 | **Writing a Series in Summation Notation** |

Write the following series using summation notation:

$$1 - \frac{1}{2} + \frac{1}{3} - \frac{1}{4} + \frac{1}{5} - \frac{1}{6}$$

(A) Start the summing index at $k = 1$.

(B) Start the summing index at $k = 0$.

SOLUTIONS

(A) $(-1)^{k-1}$ provides the alternation of sign, and $1/k$ provides the other part of each term. So we can write

$$\sum_{k=1}^{6} \frac{(-1)^{k-1}}{k}$$

as can be easily checked.

(B) $(-1)^k$ provides the alternation of sign, and $1/(k+1)$ provides the other part of each term. We write the series as

$$\sum_{k=0}^{5} \frac{(-1)^k}{k+1}$$

as can be checked.

MATCHED PROBLEM 4 Write the following series using summation notation:

$$1 - \frac{2}{3} + \frac{4}{9} - \frac{8}{27} + \frac{16}{81}$$

(A) Start with $k = 1$. (B) Start with $k = 0$.

1. $1, 1, 0, 1, -1, 2, -3$ 2. (A) $a_n = 2n$ (B) $a_n = (-1)^{n-1}\left(\dfrac{1}{2}\right)^{n-1}$

3. $1 - \dfrac{1}{3} + \dfrac{1}{5} - \dfrac{1}{7} + \dfrac{1}{9} - \dfrac{1}{11}$ 4. (A) $\displaystyle\sum_{k=1}^{5} (-1)^{k-1}\left(\dfrac{2}{3}\right)^{k-1}$ (B) $\displaystyle\sum_{k=0}^{4} (-1)^{k}\left(\dfrac{2}{3}\right)^{k}$

11-1 Exercises

1. Explain the difference between a sequence and a series.

2. What is a recursion formula?

3. Explain how the Fibonacci sequence can be defined by means of a recursion formula.

4. Explain summation notation.

5. Explain why the following statement is not true: The general term of the sequence $1, 3, 7, \ldots$ is $2^n - 1$.

6. Explain why at least one term must be provided when defining a sequence recursively.

Write the first four terms for each sequence in Problems 7–12.

7. $a_n = n - 2$

8. $a_n = n + 3$

9. $a_n = \dfrac{n-1}{n+1}$

10. $a_n = \left(1 + \dfrac{1}{n}\right)^n$

11. $a_n = (-2)^{n+1}$

12. $a_n = \dfrac{(-1)^{n+1}}{n^2}$

13. Write the eighth term in the sequence in Problem 7.

14. Write the tenth term in the sequence in Problem 8.

15. Write the one-hundredth term in the sequence in Problem 9.

16. Write the two-hundredth term in the sequence in Problem 10.

In Problems 17–22, write each series in expanded form without summation notation.

17. $\displaystyle\sum_{k=1}^{5} k$

18. $\displaystyle\sum_{k=1}^{4} k^2$

19. $\displaystyle\sum_{k=1}^{3} \dfrac{1}{10^k}$

20. $\displaystyle\sum_{k=1}^{5} \left(\dfrac{1}{3}\right)^k$

21. $\displaystyle\sum_{k=1}^{4} (-1)^k$

22. $\displaystyle\sum_{k=1}^{6} (-1)^{k+1} k$

Write the first five terms of each sequence in Problems 23–32.

23. $a_n = (-1)^{n+1} n^2$

24. $a_n = (-1)^{n+1}\left(\dfrac{1}{2^n}\right)$

25. $a_n = \dfrac{1}{3}\left(1 - \dfrac{1}{10^n}\right)$

26. $a_n = n[1 - (-1)^n]$

27. $a_n = (-\tfrac{1}{2})^{n-1}$

28. $a_n = (-\tfrac{3}{2})^{n-1}$

29. $a_1 = 7;\ a_n = a_{n-1} - 4,\ n \geq 2$

30. $a_1 = 3;\ a_n = a_{n-1} + 5,\ n \geq 2$

31. $a_1 = 4;\ a_n = \tfrac{1}{4} a_{n-1},\ n \geq 2$

32. $a_1 = 2;\ a_n = 2a_{n-1},\ n \geq 2$

In Problems 33–36, write the first seven terms of each sequence.

33. $a_1 = 1, a_2 = 2, a_n = a_{n-2} + 2a_{n-1}, n \geq 3$

34. $a_1 = 1, a_2 = -1, a_n = a_{n-2} - a_{n-1}, n \geq 3$

35. $a_1 = -1, a_2 = 2, a_n = 2a_{n-2} + a_{n-1}, n \geq 3$

36. $a_1 = 2, a_2 = 1, a_n = -a_{n-2} + a_{n-1}, n \geq 3$

In Problems 37–48, find a general term a_n for the given sequence $a_1, a_2, a_3, a_4, \ldots$

37. $-2, -1, 0, 1, \ldots$

38. $10, 11, 12, 13, \ldots$

39. $5, 7, 9, 11, \ldots$

40. $1, -1, -3, -5, \ldots$

41. $-1, 1, -1, 1, \ldots$

42. $1, -\tfrac{1}{2}, \tfrac{1}{3}, -\tfrac{1}{4}, \ldots$

43. $2, \tfrac{3}{2}, \tfrac{4}{3}, \tfrac{5}{4}, \ldots$

44. $\tfrac{1}{3}, \tfrac{2}{4}, \tfrac{3}{5}, \tfrac{4}{6}, \ldots$

45. $-3, 9, -27, 81, \ldots$

46. $5, 25, 125, 625, \ldots$

47. $x, \dfrac{x^2}{2}, \dfrac{x^3}{3}, \dfrac{x^4}{4}, \ldots$

48. $x, -x^3, x^5, -x^7, \ldots$

In Problems 49–54:

(A) Find the first four terms of the sequence.
(B) Find a general term b_n for a different sequence that has the same first three terms as the given sequence.

49. $a_n = n^2 - n + 2$

50. $a_n = 9n^2 - 21n + 14$

51. $a_n = 6n^2 - 11n + 6$

52. $a_n = 25n^2 - 60n + 36$

53. $a_n = 2n^2 - 8n + 7$

54. $a_n = -4n^2 + 15n - 12$

In Problems 55–58, use a graphing calculator to graph the first 20 terms of each sequence.

55. $a_n = 1/n$

56. $a_n = 2 + \pi n$

57. $a_n = (-0.9)^n$

58. $a_1 = -1, a_n = \tfrac{2}{3} a_{n-1} + \tfrac{1}{2}$

In Problems 59–64, write each series in expanded form without summation notation.

59. $\displaystyle\sum_{k=1}^{4} \frac{(-2)^{k+1}}{k}$

60. $\displaystyle\sum_{k=1}^{5} (-1)^{k+1}(2k-1)^2$

61. $\displaystyle\sum_{k=1}^{3} \frac{1}{k}x^{k+1}$

62. $\displaystyle\sum_{k=1}^{5} x^{k-1}$

63. $\displaystyle\sum_{k=1}^{5} \frac{(-1)^{k+1}}{k}x^k$

64. $\displaystyle\sum_{k=0}^{4} \frac{(-1)^k x^{2k+1}}{2k+1}$

In Problems 65–72, write each series using summation notation with the summing index k starting at k = 1.

65. $1^2 + 2^2 + 3^2 + 4^2$

66. $2 + 3 + 4 + 5 + 6$

67. $\dfrac{1}{2} + \dfrac{1}{2^2} + \dfrac{1}{2^3} + \dfrac{1}{2^4} + \dfrac{1}{2^5}$

68. $1 - \dfrac{1}{2} + \dfrac{1}{3} - \dfrac{1}{4}$

69. $1 + \dfrac{1}{2^2} + \dfrac{1}{3^2} + \cdots + \dfrac{1}{n^2}$

70. $2 + \dfrac{3}{2} + \dfrac{4}{3} + \cdots + \dfrac{n+1}{n}$

71. $1 - 4 + 9 - \cdots + (-1)^{n+1}n^2$

72. $\dfrac{1}{2} - \dfrac{1}{4} + \dfrac{1}{8} - \cdots + \dfrac{(-1)^{n+1}}{2^n}$

The sequence

$$a_n = \frac{a_{n-1}^2 + M}{2a_{n-1}} \qquad n \geq 2, \ M \ a \ positive \ real \ number$$

can be used to find $\sqrt{M}$ to any decimal-place accuracy desired. To start the sequence, choose a_1 arbitrarily from the positive real numbers. Problems 73 and 74 are related to this sequence.

73. (A) Find the first four terms of the sequence

$$a_1 = 3 \qquad a_n = \frac{a_{n-1}^2 + 2}{2a_{n-1}} \qquad n \geq 2$$

(B) Compare the terms with $\sqrt{2}$ from a calculator.
(C) Repeat parts A and B letting a_1 be any other positive number, say 1.

74. (A) Find the first four terms of the sequence

$$a_1 = 2 \qquad a_n = \frac{a_{n-1}^2 + 5}{2a_{n-1}} \qquad n \geq 2$$

(B) Find $\sqrt{5}$ with a calculator, and compare with the results of part A.
(C) Repeat parts A and B letting a_1 be any other positive number, say 3.

75. Let $\{a_n\}$ denote the Fibonacci sequence and let $\{b_n\}$ denote the sequence defined by $b_1 = 1$, $b_2 = 3$, and $b_n = b_{n-1} + b_{n-2}$ for $n \geq 3$. Compute 10 terms of the sequence $\{c_n\}$, where $c_n = b_n/a_n$. Describe the terms of $\{c_n\}$ for large values of n.

76. Define sequences $\{u_n\}$ and $\{v_n\}$ by $u_1 = 1$, $v_1 = 0$, $u_n = u_{n-1} + v_{n-1}$ and $v_n = u_{n-1}$ for $n \geq 2$. Find the first 10 terms of each sequence, and explain their relationship to the Fibonacci sequence.

In calculus, it can be shown that

$$e^x = \sum_{k=0}^{\infty} \frac{x^k}{k!} \approx 1 + \frac{x}{1!} + \frac{x^2}{2!} + \frac{x^3}{3!} + \cdots + \frac{x^n}{n!}$$

where the larger n is, the better the approximation. Problems 77 and 78 refer to this series. Note that n!, read "n factorial," is defined by $0! = 1$ and $n! = 1 \cdot 2 \cdot 3 \cdot \cdots \cdot n$ for $n \in N$.

77. Approximate $e^{0.2}$ using the first five terms of the series. Compare this approximation with your calculator evaluation of $e^{0.2}$.

78. Approximate $e^{-0.5}$ using the first five terms of the series. Compare this approximation with your calculator evaluation of $e^{-0.5}$.

79. Show that $\displaystyle\sum_{k=1}^{n} ca_k = c\sum_{k=1}^{n} a_k$

80. Show that $\displaystyle\sum_{k=1}^{n} (a_k + b_k) = \sum_{k=1}^{n} a_k + \sum_{k=1}^{n} b_k$

APPLICATIONS

81. PHYSICS Suppose that a rubber ball is dropped from a height of 20 feet. If it bounces 10 times, with each bounce going half as high as the one before, the heights of these bounces can be described by the sequence $a_n = 10(\frac{1}{2})^{n-1}$ ($1 \leq n \leq 10$).
(A) How high is the fifth bounce? The tenth?
(B) Find the value of the series $\displaystyle\sum_{n=1}^{10} a_n$. What does this number represent?

82. PHYSICS A bungee jumper dives off a bridge that is 300 feet above the ground. He bounces back 100 feet on the first bounce, then continues to bounce nine more times before coming to rest, with each bounce 1/3 as high as the previous. The heights of these bounces can be described by the sequence $a_n = 100(\frac{1}{3})^{n-1}$ ($1 \leq n \leq 10$).
(A) How high is the fifth bounce? The tenth?
(B) Find the value of the series $\displaystyle\sum_{n=1}^{10} a_n$. What does this number represent?

83. SALARY INCREMENT Suppose that you are offered a job with a starting annual salary of $40,000 and annual increases of 4% of the current salary.
(A) Write out the first six terms of a sequence a_n whose terms describe your salary in the first 6 years on this job.
(B) Write the general term of the sequence in part A.
(C) Find the value of the series $\displaystyle\sum_{n=1}^{6} a_n$. What does this number represent?

84. SALARY INCREMENT A marketing firm is advertising entry-level positions with a starting annual salary of $24,000 and annual increments of 3% of the current salary.
(A) Write out the first six terms of a sequence a_n whose terms describe the salary for this position in the first 6 years on this job.
(B) Write the general term of the sequence in part A.
(C) Find the value of the series $\displaystyle\sum_{n=1}^{6} a_n$. What does this number represent?

11-2 Mathematical Induction

> Using Counterexamples
> Using Mathematical Induction
> Additional Examples of Mathematical Induction
> Three Famous Problems

Many of the most important facts and formulas in this book have been stated as theorems. But a theorem is not a theorem until it has been proved, and proving theorems is one of the most challenging tasks in mathematics. There is a big difference between being pretty sure that a statement is true, and *proving* that statement. Let's look at an example.

Suppose that we are interested in the sum of the first n consecutive odd integers, where n is a positive integer. We can begin by writing the sums for the first few values of n to see if we can observe a pattern:

$$1 = 1 \quad \text{\textit{n} = 1}$$
$$1 + 3 = 4 \quad \text{\textit{n} = 2}$$
$$1 + 3 + 5 = 9 \quad \text{\textit{n} = 3}$$
$$1 + 3 + 5 + 7 = 16 \quad \text{\textit{n} = 4}$$
$$1 + 3 + 5 + 7 + 9 = 25 \quad \text{\textit{n} = 5}$$

Is there any pattern to the sums 1, 4, 9, 16, and 25? You most likely noticed that each is a perfect square and, in fact, each is the square of the number of terms in the sum. So the following *conjecture** seems reasonable:

CONJECTURE P: For each positive integer n,

$$1 + 3 + 5 + \cdots + (2n - 1) = n^2$$

(Recall that the general term $2n - 1$ was used to list the odd positive integers in the last section.)

At this point, you may be pretty sure that our conjecture is true. You might even look at the previous five calculations and think that we have proved our conjecture. But in actuality, all we have proved is that the conjecture is true for $n = 1, 2, 3, 4,$ and 5. We are trying to prove that it is true for *every* positive integer, not just those five! With that in mind, continuing to check the conjecture for specific n's like 6, 7, 8, . . . is pointless: You can keep trying for the rest of your life, but you will never be able to check *every* positive integer. Instead, in this section, we will use a much more powerful tool called *mathematical induction* to prove conjectures. Before we learn about this method of proof, we first consider how to prove that a conjecture is false.

> Using Counterexamples

Consider the following conjecture:

CONJECTURE Q: For each positive integer n, the number $n^2 - n + 41$ is a prime number.

Since the conjecture states that this fact is true for *every* positive integer n, if we can find even one positive integer n for which it is false, then the conjecture will be proved false.

A single case or example for which a conjecture fails is called a **counterexample.** We checked the conjecture for a few particular cases in Table 1. From the table, it certainly appears

Table 1

n	$n^2 - n + 41$	Prime?
1	41	Yes
2	43	Yes
3	47	Yes
4	53	Yes
5	61	Yes

*A **conjecture** is a statement that is believed to be true, but has not been proved.

that conjecture Q has a good chance of being true. You may want to check a few more cases. If you persist, you will find that conjecture Q is true for n up to 40.

Most students would guess that the statement is always true long before getting to $n = 41$. But then something interesting happens at $n = 41$:

$$41^2 - 41 + 41 = 41^2$$

which is not prime. Because $n = 41$ provides a counterexample, conjecture Q is false. Here we see the danger of generalizing without proof from a few special cases, even if that "few" is 40 cases!

This example was discovered by Euler (1701–1783), the same mathematician that introduced the number e as the base of the natural exponential function.

EXAMPLE 1 **Finding a Counterexample**

Prove that the following conjecture is false by finding a counterexample: For every positive integer $n \geq 2$, at least half of the positive integers less than or equal to n are prime.

SOLUTION We will check the conjecture for positive integer values of n starting at 2.

n	Primes less than or equal to n	Fraction of positive integers less than or equal to n that are prime	True or false
2	2	1/2	True
3	2, 3	2/3	True
4	2, 3	2/4	True
5	2, 3, 5	3/5	True
6	2, 3, 5	3/6	True
7	2, 3, 5, 7	4/7	True
8	2, 3, 5, 7	4/8	True
9	2, 3, 5, 7	4/9	False

Since $n = 9$ provides a counterexample, the conjecture is false.

MATCHED PROBLEM 1 Prove that the following conjecture is false by finding a counterexample: For every positive integer n, the last digit of n^3 is less than 9.

› **Using Mathematical Induction**

To begin our study of proving conjectures, we will state the *principle of mathematical induction,* which forms the basis for all of our work in this section.

› **THEOREM 1 Principle of Mathematical Induction**

Let P_n be a statement associated with each positive integer n, and suppose the following conditions are satisfied:

1. P_1 is true.

2. For any positive integer k, if P_k is true, then P_{k+1} is also true.

Then the statement P_n is true for all positive integers n.

Theorem 1 must be read very carefully. At first glance, it seems to say that if we assume a statement is true, then it is true. But that is not the case at all. If the two conditions in Theorem 1 are satisfied, then we can reason as follows:

P_1 is true. Condition 1

P_2 is true, because P_1 is true. Condition 2

P_3 is true, because P_2 is true. Condition 2

P_4 is true, because P_3 is true. Condition 2

$\vdots$ $\vdots$

Because this chain of implications never ends, we will eventually reach P_n for any positive integer n.

This is *not* the same as checking each case separately: The truth of *any* case follows from knowing that the previous one is true once we have established condition 2.

To help visualize this process, picture a row of dominoes that goes on forever (Fig. 1) and interpret the conditions in Theorem 1 as follows: Condition 1 says that the first domino can be pushed over. Condition 2 says that if the kth domino falls, then so does the $(k + 1)$st domino. Together, these two conditions imply that all the dominoes must fall.

Condition 1: The first domino can be pushed over.
(a)

Condition 2: If the kth domino falls, then so does the $(k + 1)$st.
(b)

Conclusion: All the dominoes will fall.
(c)

> **Figure 1** Interpreting mathematical induction.

In Example 2 we illustrate proof by mathematical induction by returning to our conjecture P from the beginning of the section.

EXAMPLE 2 **Proving a Conjecture Using Induction**

Prove that for all positive integers n,

$$1 + 3 + 5 + \cdots + (2n - 1) = n^2$$

SOLUTION State P_n:

$$P_n:\quad 1 + 3 + 5 + \cdots + (2n - 1) = n^2$$

CONDITION 1 Show that P_1 is true.

$$P_1:\quad 1 = 1^2$$

CONDITION 2 Show that if P_k is true, then P_{k+1} must be true.

It's a good idea to always write out both P_k and P_{k+1} at the beginning of this step to see what we can use, and what we need to prove.

$P_k:\quad 1 + 3 + 5 + \cdots + (2k - 1) = k^2$ We assume this is a true statement.

$P_{k+1}:\quad 1 + 3 + 5 + \cdots + (2k - 1) + [2(k + 1) - 1] = (k + 1)^2$ We need to show that this is also true.

Note that P_{k+1} can be simplified a bit:

$$P_{k+1}:\quad 1 + 3 + 5 + \cdots + (2k - 1) + (2k + 1) = (k + 1)^2$$

We will perform algebraic operations on the equation P_k (which we know is true) with a goal of obtaining P_{k+1}. Note that the left side of P_{k+1} is the left side of P_k plus the addition term $2k + 1$.

$$1 + 3 + 5 + \cdots + (2k - 1) = k^2 \qquad \text{Add } 2k + 1 \text{ to both sides.}$$

$$1 + 3 + 5 + \cdots + (2k - 1) + (2k + 1) = k^2 + 2k + 1 \qquad \text{Factor the right side.}$$

$$1 + 3 + 5 + \cdots + (2k - 1) + (2k + 1) = (k + 1)^2 \qquad \text{This is } P_{k+1}!$$

P_{k+1} was obtained by adding the same number to both sides of P_k, so if P_k is true, then P_{k+1} must be as well.

CONCLUSION

Both conditions of Theorem 1 are satisfied, so P_n is true for all positive integers n.

MATCHED PROBLEM 2 Prove that for all positive integers n

$$1 + 2 + 3 + \cdots + n = \frac{n(n + 1)}{2}$$

› Additional Examples of Mathematical Induction

Now we will consider some additional examples of proof by induction. The first is another summation formula. Mathematical induction is the primary tool for proving that formulas of this type are true.

EXAMPLE 3 Proving a Summation Formula

Prove that for all positive integers n

$$\frac{1}{2} + \frac{1}{4} + \frac{1}{8} + \cdots + \frac{1}{2^n} = \frac{2^n - 1}{2^n}$$

PROOF State P_n:

$$P_n: \quad \frac{1}{2} + \frac{1}{4} + \frac{1}{8} + \cdots + \frac{1}{2^n} = \frac{2^n - 1}{2^n}$$

PART 1 Show that P_1 is true.

$$P_1: \quad \frac{1}{2} = \frac{2^1 - 1}{2^1}$$
$$= \frac{1}{2}$$

So P_1 is true.

PART 2 Show that if P_k is true, then P_{k+1} is true. Again, it is a good idea to always write out both P_k and P_{k+1} at the beginning of any induction proof to see what is assumed and what must be proved:

$$P_k: \quad \frac{1}{2} + \frac{1}{4} + \frac{1}{8} + \cdots + \frac{1}{2^k} = \frac{2^k - 1}{2^k} \qquad \text{We assume } P_k \text{ is true.}$$

$$P_{k+1}: \quad \frac{1}{2} + \frac{1}{4} + \frac{1}{8} + \cdots + \frac{1}{2^k} + \frac{1}{2^{k+1}} = \frac{2^{k+1} - 1}{2^{k+1}} \qquad \text{We must show that } P_{k+1} \text{ follows from } P_k.$$

We start with the true statement P_k, add $1/2^{k+1}$ to both sides, and simplify the right side:

$$\frac{1}{2} + \frac{1}{4} + \frac{1}{8} + \cdots + \frac{1}{2^k} = \frac{2^k - 1}{2^k}$$ Add $\frac{1}{2^{k+1}}$ to both sides.

$$\frac{1}{2} + \frac{1}{4} + \frac{1}{8} + \cdots + \frac{1}{2^k} + \frac{1}{2^{k+1}} = \frac{2^k - 1}{2^k} + \frac{1}{2^{k+1}}$$ Find common denominator for right-hand side.

$$= \frac{2^k - 1}{2^k} \cdot \frac{2}{2} + \frac{1}{2^{k+1}}$$ Write as single fraction.

$$= \frac{2^{k+1} - 2 + 1}{2^{k+1}}$$ Simplify.

$$= \frac{2^{k+1} - 1}{2^{k+1}}$$

So

$$\frac{1}{2} + \frac{1}{4} + \frac{1}{8} + \cdots + \frac{1}{2^k} + \frac{1}{2^{k+1}} = \frac{2^{k+1} - 1}{2^{k+1}}$$ P_{k+1}

and we have shown that if P_k is true, then P_{k+1} is true.

CONCLUSION

Both conditions in Theorem 1 are satisfied. Therefore, P_n is true for all positive integers n. ◉

MATCHED PROBLEM 3 Prove that for all positive integers n

$$\frac{2}{3} + \frac{2}{9} + \frac{2}{27} + \cdots + \frac{2}{3^n} = \frac{3^n - 1}{3^n}$$

Example 4 provides a proof of a law of exponents that previously we had to assume was true. First we redefine a^n for n a positive integer, using a recursion formula:

> **DEFINITION 1** Recursive Definition of a^n

For n a positive integer

$$a^1 = a$$
$$a^{n+1} = a^n a \qquad n > 1$$

EXAMPLE 4 **Proving a Law of Exponents**

Prove that $(xy)^n = x^n y^n$ for all positive integers n.

PROOF State P_n:

$$P_n: (xy)^n = x^n y^n$$

PART 1 Show that P_1 is true.

$$(xy)^1 = xy \qquad \text{Definition 1}$$
$$= x^1 y^1 \qquad \text{Definition 1}$$

So P_1 is true.

PART 2 Show that if P_k is true, then P_{k+1} is true.

$$P_k: \quad (xy)^k = x^k y^k \qquad \text{Assume } P_k \text{ is true.}$$
$$P_{k+1}: \quad (xy)^{k+1} = x^{k+1} y^{k+1} \qquad \text{Show that } P_{k+1} \text{ follows from } P_k.$$

Here we start with the left side of P_{k+1} and use P_k to find the right side of P_{k+1}:

$$\begin{aligned} (xy)^{k+1} &= (xy)^k (xy)^1 & \text{Use } P_k: (xy)^k = x^k y^k \\ &= x^k y^k xy & \text{Use properties of real numbers.} \\ &= (x^k x)(y^k y) & \text{Use Definition 1 twice.} \\ &= x^{k+1} y^{k+1} \end{aligned}$$

So $(xy)^{k+1} = x^{k+1} y^{k+1}$, and we have shown that if P_k is true, then P_{k+1} is true.

CONCLUSION

Both conditions in Theorem 1 are satisfied. Therefore, P_n is true for all positive integers n. ◉

MATCHED PROBLEM 4

Prove that $(x/y)^n = x^n/y^n$ for all positive integers n.

◉

Example 5 deals with factors of integers. Before we start, recall that an integer p is *divisible* by an integer q if $p = qr$ for some integer r.

EXAMPLE 5 Proving a Divisibility Property

Prove that $4^{2n} - 1$ is divisible by 5 for all positive integers n.

PROOF Use the definition of divisibility to state P_n as follows:

$$P_n: \quad 4^{2n} - 1 = 5r \qquad \text{for some integer } r$$

PART 1 Show that P_1 is true.

$$P_1: \quad 4^2 - 1 = 15 = 5 \cdot 3$$

So P_1 is true.

PART 2 Show that if P_k is true, then P_{k+1} is true.

$$P_k: \quad 4^{2k} - 1 = 5r \quad \text{for some integer } r \qquad \text{Assume } P_k \text{ is true.}$$
$$P_{k+1}: \quad 4^{2(k+1)} - 1 = 5s \quad \text{for some integer } s \qquad \text{Show that } P_{k+1} \text{ must follow.}$$

As before, we start with the true statement P_k:

$$\begin{aligned} 4^{2k} - 1 &= 5r & \text{Multiply both sides by } 4^2. \\ 4^2(4^{2k} - 1) &= 4^2(5r) & \text{Simplify.} \\ 4^{2k+2} - 16 &= 80r & \text{Add 15 to both sides.} \\ 4^{2(k+1)} - 1 &= 80r + 15 & \text{Factor out 5.} \\ &= 5(16r + 3) \end{aligned}$$

So

$$4^{2(k+1)} - 1 = 5s \qquad P_{k+1}$$

where $s = 16r + 3$ is an integer, and we have shown that if P_k is true, then P_{k+1} is true.

CONCLUSION

Both conditions in Theorem 1 are satisfied. Therefore, P_n is true for all positive integers n. ◉

MATCHED PROBLEM 5 Prove that $8^n - 1$ is divisible by 7 for all positive integers n.

In some cases, a conjecture may be true only for $n \geq m$, where m is a positive integer, rather than for all $n \geq 0$. For example, see Problems 53 and 54 in Exercises 11-2. The principle of mathematical induction can be extended to cover cases like this as follows:

> ### › THEOREM 2 Extended Principle of Mathematical Induction

> Let m be a positive integer, let P_n be a statement associated with each integer $n \geq m$, and suppose the following conditions are satisfied:
>
> 1. P_m is true.
> 2. For any integer $k \geq m$, if P_k is true, then P_{k+1} is also true.
>
> Then the statement P_n is true for all integers $n \geq m$.

› Three Famous Problems

The problem of determining whether a certain statement about the positive integers is true may be extremely difficult. Proofs may require remarkable insight and ingenuity and the development of techniques far more advanced than mathematical induction. Consider, for example, the famous problems of proving the following statements:

1. **Lagrange's Four Square Theorem, 1772:** Each positive integer can be expressed as the sum of four or fewer squares of positive integers.

2. **Fermat's Last Theorem, 1637:** For $n > 2$, $x^n + y^n = z^n$ does not have solutions in the natural numbers.

3. **Goldbach's Conjecture, 1742:** Every positive even integer greater than 2 is the sum of two prime numbers.

The first statement was considered by the early Greeks and finally proved in 1772 by Lagrange. Fermat's last theorem, defying the best mathematical minds for over 350 years, finally succumbed to a 200-page proof by Professor Andrew Wiles of Princeton University in 1993. To this date no one has been able to prove or disprove Goldbach's conjecture.

››› EXPLORE-DISCUSS 1

(A) Explain the difference between a theorem and a conjecture.

(B) Why is "Fermat's last theorem" a misnomer? Suggest more accurate names for the result.

ANSWERS TO MATCHED PROBLEMS

1. The last digit of $9^3 = 729$ is greater than 8.
2. Sketch of proof.

$$P_n: 1 + 2 + 3 + \cdots + n = \frac{n(n + 1)}{2}$$

Condition 1. $1 = \dfrac{1(1 + 1)}{2}$. P_1 is true.

Condition 2. Show that if P_k is true, then P_{k+1} is true.

$$1 + 2 + 3 + \cdots + k = \frac{k(k+1)}{2} \qquad P_k$$

$$1 + 2 + 3 + \cdots + k + (k+1) = \frac{k(k+1)}{2} + (k+1)$$

$$= \frac{(k+1)(k+2)}{2} \qquad P_{k+1}$$

Conclusion: P_n is true for all positive integers n.

3. Sketch of proof. P_n: $\dfrac{2}{3} + \dfrac{2}{9} + \dfrac{2}{27} + \cdots + \dfrac{2}{3^n} = \dfrac{3^n - 1}{3^n}$

Part 1. $\dfrac{2}{3} = \dfrac{3^1 - 1}{3}$. P_1 is true.

Part 2. Show that if P_k is true, then P_{k+1} is true.

$$\frac{2}{3} + \frac{2}{9} + \frac{2}{27} + \cdots + \frac{2}{3^k} = \frac{3^k - 1}{3^k} \qquad P_k$$

$$\frac{2}{3} + \frac{2}{9} + \frac{2}{27} + \cdots + \frac{2}{3^k} + \frac{2}{3^{k+1}} = \frac{3^k - 1}{3^k} + \frac{2}{3^{k+1}}$$

$$= \frac{3^{k+1} - 1}{3^k} \qquad P_{k+1}$$

Conclusion: P_n is true for all positive integers n.

4. Sketch of proof. P_n: $\left(\dfrac{x}{y}\right)^n = \dfrac{x^n}{y^n}$

Part 1. $\left(\dfrac{x}{y}\right)^1 = \dfrac{x}{y} = \dfrac{x^1}{y^1}$. P_1 is true.

Part 2. Show that if P_k is true, then P_{k+1} is true.

$$\left(\frac{x}{y}\right)^{k+1} = \left(\frac{x}{y}\right)^k \left(\frac{x}{y}\right) = \frac{x^k}{y^k}\left(\frac{x}{y}\right) = \frac{x^k x}{y^k y} = \frac{x^{k+1}}{y^{k+1}}$$

Conclusion: P_n is true for all positive integers n.

5. Sketch of proof. P_n: $8^n - 1 = 7r$ for some integer r

Part 1. $8^1 - 1 = 7 = 7 \cdot 1$. P_1 is true.

Part 2. Show that if P_k is true, then P_{k+1} is true.

$$8^k - 1 = 7r \qquad P_k$$

$$8(8^k - 1) = 8(7r)$$

$$8^{k+1} - 1 = 56r + 7 = 7(8r + 1) = 7s \qquad P_{k+1}$$

Conclusion: P_n is true for all positive integers n.

11-2 Exercises

1. What is a counterexample?

2. Explain how falling dominoes can be compared to the principle of mathematical induction.

3. In Theorem 1 (principle of mathematical induction), what do P_k and P_{k+1} represent?

4. The number $n^2 - n + 41$ is prime for $n = 1, 2, \ldots, 40$. Does this prove that $n^2 - n + 41$ is prime for every natural number n? Explain.

In Problems 5–8, find the first positive integer n that causes the statement to fail.

5. $(3 + 5)^n = 3^n + 5^n$

6. $n < 10$

7. $n^2 = 3n - 2$

8. $n^3 + 11n = 6n^2 + 6$

Verify each statement P_n in Problems 9–14 for $n = 1, 2,$ and 3.

9. P_n: $2 + 6 + 10 + \cdots + (4n - 2) = 2n^2$

10. P_n: $4 + 8 + 12 + \cdots + 4n = 2n(n + 1)$

11. P_n: $a^5 a^n = a^{5+n}$ **12.** P_n: $(a^5)^n = a^{5n}$

13. P_n: $9^n - 1$ is divisible by 4

14. P_n: $4^n - 1$ is divisible by 3

Write P_k and P_{k+1} for P_n as indicated in Problems 15–20.

15. P_n in Problem 9 **16.** P_n in Problem 10

17. P_n in Problem 11 **18.** P_n in Problem 12

19. P_n in Problem 13 **20.** P_n in Problem 14

In Problems 21–26, use mathematical induction to prove that each P_n holds for all positive integers n.

21. P_n in Problem 9 **22.** P_n in Problem 10

23. P_n in Problem 11 **24.** P_n in Problem 12

25. P_n in Problem 13 **26.** P_n in Problem 14

In Problems 27–30, prove the statement is false by finding a counterexample.

27. If $n > 2$, then any polynomial of degree n has at least one real zero.

28. Any positive integer $n > 7$ can be written as the sum of three or fewer squares of positive integers.

29. If n is a positive integer, then there is at least one prime number p such that $n < p < n + 6$.

30. If $a, b, c,$ and d are positive integers such that $a^2 + b^2 = c^2 + d^2$, then $a = c$ or $a = d$.

In Problems 31–46, use mathematical induction to prove each proposition for all positive integers n, unless restricted otherwise.

31. $2 + 2^2 + 2^3 + \cdots + 2^n = 2^{n+1} - 2$

32. $\dfrac{1}{2} + \dfrac{1}{4} + \dfrac{1}{8} + \cdots + \dfrac{1}{2^n} = 1 - \left(\dfrac{1}{2}\right)^n$

33. $1^2 + 3^2 + 5^2 + \cdots + (2n - 1)^2 = \frac{1}{3}(4n^3 - n)$

34. $1 + 8 + 16 + \cdots + 8(n - 1) = (2n - 1)^2$; $n > 1$

35. $1^2 + 2^2 + 3^2 + \cdots + n^2 = \dfrac{n(n + 1)(2n + 1)}{6}$

36. $1 \cdot 2 + 2 \cdot 3 + 3 \cdot 4 + \cdots + n(n + 1) = \dfrac{n(n + 1)(n + 2)}{3}$

37. $\dfrac{a^n}{a^3} = a^{n-3}$; $n > 3$ **38.** $\dfrac{a^5}{a^n} = \dfrac{1}{a^{n-5}}$; $n > 5$

39. $a^m a^n = a^{m+n}$; $m, n \in N$ [*Hint:* Choose m as an arbitrary element of N, and then use induction on n.]

40. $(a^n)^m = a^{mn}$; $m, n \in N$

41. $x^n - 1$ is divisible by $x - 1$, $x \neq 1$ [*Hint:* Divisible means that $x^n - 1 = (x - 1)Q(x)$ for some polynomial $Q(x)$.]

42. $x^n - y^n$ is divisible by $x - y$; $x \neq y$

43. $x^{2n} - 1$ is divisible by $x - 1$; $x \neq 1$

44. $x^{2n} - 1$ is divisible by $x + 1$; $x \neq -1$

45. $1^3 + 2^3 + 3^3 + \cdots + n^3 = (1 + 2 + 3 + \cdots + n)^2$ [*Hint:* See Matched Problem 2 following Example 2.]

46. $\dfrac{1}{1 \cdot 2 \cdot 3} + \dfrac{1}{2 \cdot 3 \cdot 4} + \dfrac{1}{3 \cdot 4 \cdot 5} + \cdots$
$$+ \dfrac{1}{n(n + 1)(n + 2)} = \dfrac{n(n + 3)}{4(n + 1)(n + 2)}$$

In Problems 47–50, suggest a formula for each expression, and prove your conjecture using mathematical induction, $n \in N$.

47. $2 + 4 + 6 + \cdots + 2n$

48. $\dfrac{1}{1 \cdot 2} + \dfrac{1}{2 \cdot 3} + \dfrac{1}{3 \cdot 4} + \cdots + \dfrac{1}{n(n + 1)}$

49. The number of lines determined by n points in a plane, no three of which are collinear

50. The number of diagonals in a polygon with n sides

Prove Problems 51–54 true for all integers n as specified.

51. If $a > 1$, then $a^n > 1$; $n \in N$

52. If $0 < a < 1$, then $0 < a^n < 1$; $n \in N$

53. $n^2 > 2n$; $n \geq 3$

54. $2^n > n^2$; $n \geq 5$

In Problems 55–58, determine whether the statement is true or false. If true, prove using mathematical induction. If false, find a counterexample.

55. If n is a positive integer, then
$$1 - 2 + 3 - \cdots + (2n - 1) = n$$
(that is, the alternating sum of the first $2n - 1$ positive integers is equal to n).

56. If n is a positive integer, then
$$1^2 - 2^2 + 3^2 - \cdots + (-1)^{n+1}n^2 = \dfrac{(-1)^{n+1}n(n + 1)}{2}$$

57. If n is a positive integer, then
$$3^{n+1} + 4^{n+1} + \cdots + (n + 3)^{n+1} = (n + 4)^{n+1}$$

58. If n is a positive integer, then $n^2 + 21n + 1$ is a prime number.

If $\{a_n\}$ and $\{b_n\}$ are two sequences, we write $\{a_n\} = \{b_n\}$ if and only if $a_n = b_n$ for all $n \in N$. In Problems 59–62, use mathematical induction to show that $\{a_n\} = \{b_n\}$.

59. $a_1 = 1$, $a_n = a_{n-1} + 2$; $b_n = 2n - 1$

60. $a_1 = 2$, $a_n = a_{n-1} + 2$; $b_n = 2n$

61. $a_1 = 2$, $a_n = 2^2 a_{n-1}$; $b_n = 2^{2n-1}$

62. $a_1 = 2$, $a_n = 3a_{n-1}$; $b_n = 2 \cdot 3^{n-1}$

11-3 Arithmetic and Geometric Sequences

› Arithmetic and Geometric Sequences
› Developing *n*th-Term Formulas
› Developing Sum Formulas for Finite Arithmetic Series
› Developing Sum Formulas for Finite Geometric Series
› Developing a Sum Formula for Infinite Geometric Series

For most sequences, it is difficult to add up an arbitrary number of terms of the sequence without adding the terms one at a time. In this section, we will study two special types of sequences, *arithmetic sequences* and *geometric sequences*. One of the things that make them special is that we can develop formulas for the sum of the corresponding series.

› Arithmetic and Geometric Sequences

Consider the sequence defined by the general term $a_n = 5 + 2(n - 1)$, $n \geq 1$. The first five terms are 5, 7, 9, 11, and 13. It's not hard to see that after starting at 5, every term is obtained by adding 2 to the previous term. This is an example of an arithmetic sequence.

› **DEFINITION 1** Arithmetic Sequence

A sequence

$$a_1, a_2, a_3, \ldots, a_n, \ldots$$

is called an **arithmetic sequence,** or **arithmetic progression,** if there exists a constant d, called the **common difference,** such that

$$a_n - a_{n-1} = d$$

That is,

$$a_n = a_{n-1} + d \qquad \text{for every } n > 1$$

In short, a sequence is arithmetic when every term is obtained by adding some fixed number to the previous term. This fixed number is called the *common difference,* and is usually represented by the letter d.

Now consider the sequence with general term $a_n = 5(2)^{n-1}$. The first five terms are 5, 10, 20, 40, and 80. It also starts at 5, but this time every term is obtained by *multiplying* the previous term by 2. This is an example of a geometric sequence.

> **DEFINITION 2 Geometric Sequence**

A sequence

$$a_1, a_2, a_3, \ldots, a_n, \ldots$$

is called a **geometric sequence,** or **geometric progression,** if there exists a nonzero constant r, called the **common ratio,** such that

$$\frac{a_n}{a_{n-1}} = r$$

That is,

$$a_n = ra_{n-1} \qquad \text{for every } n > 1$$

In short, a sequence is geometric when every term is obtained by multiplying the previous term by some fixed number. This fixed number is called the *common ratio,* and is usually represented by the letter r.

>>> EXPLORE-DISCUSS 1

(A) Graph the arithmetic sequence 5, 7, 9,
Describe the graphs of all arithmetic sequences with common difference 2.

(B) Graph the geometric sequence 5, 10, 20,
Describe the graphs of all geometric sequences with common ratio 2.

EXAMPLE 1 **Recognizing Arithmetic and Geometric Sequences**

Which of the following can be the first four terms of an arithmetic sequence? Of a geometric sequence?

(A) 1, 2, 3, 5, . . . (B) −1, 3, −9, 27, . . .

(C) 3, 3, 3, 3, . . . (D) 10, 8.5, 7, 5.5, . . .

SOLUTIONS (A) Because $2 - 1 \neq 5 - 3$, there is no common difference, so the sequence is not an arithmetic sequence. Because $\frac{2}{1} \neq \frac{3}{2}$, there is no common ratio, so the sequence is not geometric either.

(B) The sequence is geometric with common ratio −3, but it is not arithmetic.

(C) The sequence is arithmetic with common difference 0 and it is also geometric with common ratio 1.

(D) The sequence is arithmetic with common difference −1.5, but it is not geometric.

MATCHED PROBLEM 1 Which of the following can be the first four terms of an arithmetic sequence? Of a geometric sequence?

(A) 8, 2, 0.5, 0.125, . . . (B) −7, −2, 3, 8, . . . (C) 1, 5, 25, 100, . . .

› Developing *nth*-Term Formulas

If $\{a_n\}$ is an arithmetic sequence with common difference d, then

$$a_2 = a_1 + d$$
$$a_3 = a_2 + d = a_1 + 2d$$
$$a_4 = a_3 + d = a_1 + 3d$$

This suggests Theorem 1, which can be proved by mathematical induction (see Problem 67 in Exercises 11-3).

> **THEOREM 1** The *nth* Term of an Arithmetic Sequence

$$a_n = a_1 + (n - 1)d \qquad \text{for every } n > 1$$

Similarly, if $\{a_n\}$ is a geometric sequence with common ratio r, then

$$a_2 = a_1 r$$
$$a_3 = a_2 r = a_1 r^2$$
$$a_4 = a_3 r = a_1 r^3$$

This suggests Theorem 2, which can also be proved by mathematical induction (see Problem 71 in Exercises 11-3).

> **THEOREM 2** The *nth* Term of a Geometric Sequence

$$a_n = a_1 r^{n-1} \qquad \text{for every } n > 1$$

EXAMPLE **2** **Finding Terms in Arithmetic and Geometric Sequences**

(A) If the first and tenth terms of an arithmetic sequence are 3 and 30, respectively, find the fiftieth term of the sequence.

(B) If the first and tenth terms of a geometric sequence are 1 and 4, find the seventeenth term to three decimal places.

SOLUTIONS (A) First use Theorem 1 with $a_1 = 3$ and $a_{10} = 30$ to find d:

$$a_n = a_1 + (n - 1)d \qquad \text{Substitute } n = 10.$$
$$a_{10} = a_1 + (10 - 1)d \qquad \text{Substitute } a_{10} = 30 \text{ and } a_1 = 3.$$
$$30 = 3 + 9d \qquad \text{Solve for } d.$$
$$d = 3$$

Now find a_{50}:

$$a_{50} = a_1 + (50 - 1)3 \qquad \text{Substitute } a_1 = 3.$$
$$= 3 + 49 \cdot 3 \qquad \text{Simplify.}$$
$$= 150$$

(B) First let $n = 10$, $a_1 = 1$, $a_{10} = 4$ and use Theorem 2 to find r.

$$a_n = a_1 r^{n-1} \qquad \text{Substitute } n = 10, a_{10} = 4, \text{ and } a_1 = 1.$$
$$4 = 1r^{10-1} \qquad \text{Solve for } r.$$
$$r = 4^{1/9}$$

Now use Theorem 2 again, this time with $n = 17$.

$$a_{17} = a_1 r^{16} = 1(4^{1/9})^{16} = 4^{16/9} \approx 11.758$$

MATCHED PROBLEM 2

(A) If the first and fifteenth terms of an arithmetic sequence are -5 and 23, respectively, find the seventy-third term of the sequence.

(B) Find the eighth term of the geometric sequence $\dfrac{1}{64}, -\dfrac{1}{32}, \dfrac{1}{16}, \ldots$.

› Developing Sum Formulas for Finite Arithmetic Series

If $a_1, a_2, a_3, \ldots, a_n$ is a finite arithmetic sequence, then the corresponding series $a_1 + a_2 + a_3 + \cdots + a_n$ is called an *arithmetic series*. We will derive two simple and very useful formulas for the sum of an arithmetic series. Let d be the common difference of the arithmetic sequence $a_1, a_2, a_3, \ldots, a_n$ and let S_n denote the sum of the series $a_1 + a_2 + a_3 + \cdots + a_n$.

Then

$$S_n = a_1 + (a_1 + d) + \cdots + [a_1 + (n - 2)d] + [a_1 + (n - 1)d]$$

Reversing the order of the sum, we obtain

$$S_n = [a_1 + (n - 1)d] + [a_1 + (n - 2)d] + \cdots + (a_1 + d) + a_1$$

Adding the left sides of these two equations and corresponding elements of the right sides, we see that

$$\begin{aligned} 2S_n &= [2a_1 + (n - 1)d] + [2a_1 + (n - 1)d] + \cdots + [2a_1 + (n - 1)d] \\ &= n[2a_1 + (n - 1)d] \end{aligned}$$

This can be restated as in Theorem 3.

> **THEOREM 3** Sum of an Arithmetic Series—First Form

$$S_n = \frac{n}{2}[2a_1 + (n - 1)d]$$

By replacing $a_1 + (n - 1)d$ with a_n, we obtain a second useful formula for the sum.

> **THEOREM 4** Sum of an Arithmetic Series—Second Form

$$S_n = \frac{n}{2}(a_1 + a_n)$$

The proof of the first sum formula by mathematical induction is left as an exercise (see Problem 68 in Exercises 11-3).

EXAMPLE 3 Finding the Sum of an Arithmetic Series

Find the sum of the first 26 terms of an arithmetic series if the first term is -7 and $d = 3$.

SOLUTION Let $n = 26$, $a_1 = -7$, $d = 3$, and use Theorem 3.

$$S_n = \frac{n}{2}[2a_1 + (n-1)d] \qquad \text{Substitute } n = 26, a_1 = -7, \text{ and } d = 3.$$

$$S_{26} = \frac{26}{2}[2(-7) + (26-1)3] \qquad \text{Simplify.}$$

$$= 793$$

MATCHED PROBLEM 3 Find the sum of the first 52 terms of an arithmetic series if the first term is 23 and $d = -2$.

EXAMPLE 4 Finding the Sum of an Arithmetic Series

Find the sum of all the odd numbers between 51 and 99, inclusive.

SOLUTION First, use $a_1 = 51$, $a_n = 99$, and Theorem 1 to find n:

$$a_n = a_1 + (n-1)d \qquad \text{Substitute } a_n = 99, a_1 = 51, \text{ and } d = 2.$$

$$99 = 51 + (n-1)2 \qquad \text{Solve for } n.$$

$$n = 25$$

Now use Theorem 4 to find S_{25}:

$$S_n = \frac{n}{2}(a_1 + a_n) \qquad \text{Substitute } n = 25, a_1 = 51, \text{ and } a_n = 99.$$

$$S_{25} = \frac{25}{2}(51 + 99)$$

$$= 1,875$$

MATCHED PROBLEM 4 Find the sum of all the even numbers between -22 and 52, inclusive.

EXAMPLE 5 Prize Money

A 16-team bowling league has $8,000 to be awarded as prize money. If the last-place team is awarded $275 in prize money and the award increases by the same amount for each successive finishing place, how much will the first-place team receive?

SOLUTION If a_1 is the award for the first-place team, a_2 is the award for the second-place team, and so on, then the prize money awards form an arithmetic sequence with $n = 16$, $a_{16} = 275$, and $S_{16} = 8,000$. Use Theorem 4 to find a_1.

$$S_n = \frac{n}{2}(a_1 + a_n) \qquad \text{Substitute } n = 16, S_{16} = 8,000, a_{16} = 275.$$

$$8,000 = \frac{16}{2}(a_1 + 275) \qquad \text{Solve for } a_1.$$

$$a_1 = 725$$

The first-place team receives $725.

MATCHED PROBLEM 5 Refer to Example 5. How much prize money is awarded to the second-place team?

› **Developing Sum Formulas for Finite Geometric Series**

If $a_1, a_2, a_3, \ldots, a_n$ is a finite geometric sequence, then the corresponding series $a_1 + a_2 + a_3 + \cdots + a_n$ is called a *geometric series*. As with arithmetic series, we can derive two simple and very useful formulas for the sum of a geometric series. Let r be the common ratio of the geometric sequence $a_1, a_2, a_3, \ldots, a_n$ and let S_n denote the sum of the series $a_1 + a_2 + a_3 + \cdots + a_n$. Then

$$S_n = a_1 + a_1 r + a_1 r^2 + a_1 r^3 + \cdots + a_1 r^{n-2} + a_1 r^{n-1}$$

Multiply both sides of this equation by r to obtain

$$r S_n = a_1 r + a_1 r^2 + a_1 r^3 + \cdots + a_1 r^{n-1} + a_1 r^n$$

Now subtract the left side of the second equation from the left side of the first, and the right side of the second equation from the right side of the first to obtain

$$S_n - r S_n = a_1 - a_1 r^n \qquad \text{\color{red}Factor out } S_n$$

$$S_n(1 - r) = a_1 - a_1 r^n$$

Solving for S_n, we obtain the following formula for the sum of a geometric series:

› **THEOREM 5 Sum of a Geometric Series—First Form**

$$S_n = \frac{a_1 - a_1 r^n}{1 - r} \qquad r \neq 1$$

Because $a_n = a_1 r^{n-1}$, or $r a_n = a_1 r^n$, the sum formula also can be written in the following form:

› **THEOREM 6 Sum of a Geometric Series—Second Form**

$$S_n = \frac{a_1 - r a_n}{1 - r} \qquad r \neq 1$$

The proof of the first sum formula (Theorem 5) by mathematical induction is left as an exercise (see Problem 72, Exercises 11-3).

If $r = 1$, then

$$S_n = a_1 + a_1(1) + a_1(1^2) + \cdots + a_1(1^{n-1}) = n a_1$$

| **EXAMPLE** | **6** | **Finding the Sum of a Geometric Series** |

Find the sum of the first 20 terms of a geometric series if the first term is 1 and $r = 2$.

SOLUTION Let $n = 20$, $a_1 = 1$, $r = 2$, and use Theorem 5.

$$S_n = \frac{a_1 - a_1 r^n}{1 - r} \qquad \text{\color{red}Substitute } n = 20,\ a_1 = 1,\ \text{and } r = 2.$$

$$= \frac{1 - 1 \cdot 2^{20}}{1 - 2} = 1{,}048{,}575$$

Technology Connections

To calculate the sum of a series with a graphing calculator, first generate the sequence using the sequence command, then find its sum using the sum command. Figure 1 shows the solution to Example 6.

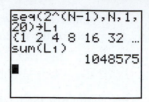

> Figure 1

MATCHED PROBLEM 6 Find the sum, to two decimal places, of the first 14 terms of a geometric series if the first term is $\frac{1}{64}$ and $r = -2$.

> Developing a Sum Formula for Infinite Geometric Series

Consider a geometric series with $a_1 = 5$ and $r = \frac{1}{2}$. What happens to the sum S_n as n increases? To answer this question, we first write the sum formula in the more convenient form

$$S_n = \frac{a_1 - a_1 r^n}{1 - r} = \frac{a_1}{1 - r} - \frac{a_1 r^n}{1 - r} \tag{1}$$

For $a_1 = 5$ and $r = \frac{1}{2}$,

$$S_n = 10 - 10\left(\frac{1}{2}\right)^n$$

Let's look at some of the S_ns:

$$S_2 = 10 - 10\left(\frac{1}{4}\right) = 7.5$$

$$S_3 = 10 - 10\left(\frac{1}{8}\right) = 8.75$$

$$S_4 = 10 - 10\left(\frac{1}{16}\right) = 9.375$$

$$\vdots$$

$$S_{20} = 10 - 10\left(\frac{1}{1,048,576}\right) = 9.999990\ldots$$

It appears that $\left(\frac{1}{2}\right)^n$ becomes smaller and smaller as n increases and that the sum gets closer and closer to 10.

In general, it is possible to show that, if $|r| < 1$, then r^n will get closer and closer to 0 as n increases. Symbolically, $r^n \to 0$ as $n \to \infty$. So the term

$$\frac{a_1 r^n}{1 - r}$$

in equation (1) will tend to 0 as n increases, and S_n will tend to

$$\frac{a_1}{1 - r}$$

In other words, if $|r| < 1$, then S_n can be made as close to

$$\frac{a_1}{1 - r}$$

as we wish by taking n sufficiently large. So we can define the **sum of an infinite geometric series** by the following formula:

> › **DEFINITION 3** Sum of an Infinite Geometric Series
>
> $$S_\infty = \frac{a_1}{1 - r} \qquad |r| < 1$$

If $|r| \geq 1$, an infinite geometric series has no sum.

EXAMPLE 7

Expressing a Repeating Decimal as a Fraction

Represent the repeating decimal $0.454\ 545 \cdots = 0.\overline{45}$ as the quotient of two integers. Recall that a repeating decimal names a rational number and that any rational number can be represented as the quotient of two integers.

SOLUTION

$$0.\overline{45} = 0.45 + 0.0045 + 0.000\ 045 + \cdots$$

The right side of the equation is an infinite geometric series with $a_1 = 0.45$ and $r = 0.01$. The sum is

$$S_\infty = \frac{a_1}{1 - r} = \frac{0.45}{1 - 0.01} = \frac{0.45}{0.99} = \frac{5}{11}$$

This shows that, $0.\overline{45}$ and $\frac{5}{11}$ name the same rational number. You can check the result by dividing 5 by 11.

MATCHED PROBLEM 7

Repeat Example 7 for $0.818\ 181 \cdots = 0.\overline{81}$.

EXAMPLE 8

Economy Stimulation

A state government uses proceeds from a lottery to provide a tax rebate for property owners. Suppose an individual receives a $500 rebate and spends 80% of this, and each of the recipients of the money spent by this individual also spends 80% of what he or she receives, and this process continues without end. According to the **multiplier doctrine** in economics, the effect of the original $500 tax rebate on the economy is multiplied many times. What is the total amount spent if the process continues as indicated?

SOLUTION

The individual receives $500 and spends $0.8(500) = \$400$. The recipients of this $400 spend $0.8(400) = \$320$, the recipients of this $320 spend $0.8(320) = \$256$, and so on. The total spending generated by the $500 rebate is

$$400 + 320 + 256 + \cdots = 400 + 0.8(400) + (0.8)^2(400) + \cdots$$

which we recognize as an infinite geometric series with $a_1 = 400$ and $r = 0.8$. The total amount spent is

$$S_\infty = \frac{a_1}{1-r} = \frac{400}{1-0.8} = \frac{400}{0.2} = \$2{,}000$$

MATCHED PROBLEM 8 Repeat Example 8 if the tax rebate is \$1,000 and the percentage spent by all recipients is 90%.

>>> EXPLORE-DISCUSS 2

(A) Find an infinite geometric series with $a_1 = 10$ whose sum is 1,000.

(B) Find an infinite geometric series with $a_1 = 10$ whose sum is 6.

(C) Suppose that an infinite geometric series with $a_1 = 10$ has a sum. Explain why that sum must be greater than 5.

ANSWERS TO MATCHED PROBLEMS

1. (A) The sequence is geometric with $r = \frac{1}{4}$, but not arithmetic.
(B) The sequence is arithmetic with $d = 5$, but not geometric.
(C) The sequence is neither arithmetic nor geometric.
2. (A) 139 (B) -2 **3.** $-1{,}456$ **4.** 570 **5.** \$695 **6.** -85.33
7. $\frac{9}{11}$ **8.** \$9,000

11-3 Exercises

1. What is an arithmetic sequence?

2. What is a geometric sequence?

3. Explain the terms "common difference" and "common ratio."

4. Explain how a repeating decimal can be viewed as a geometric series.

5. Which infinite arithmetic series have a sum?

6. Which infinite geometric series have a sum?

In Problems 7 and 8, determine whether the following can be the first three terms of an arithmetic or geometric sequence, and, if so, find the common difference or common ratio and the next two terms of the sequence.

7. (A) $-11, -16, -21, \ldots$ (B) $2, -4, 8, \ldots$
(C) $1, 4, 9, \ldots$ (D) $\frac{1}{2}, \frac{1}{6}, \frac{1}{18}, \ldots$

8. (A) $5, 20, 100, \ldots$ (B) $-5, -5, -5, \ldots$
(C) $7, 6.5, 6, \ldots$ (D) $512, 256, 128, \ldots$

Let $a_1, a_2, a_3, \ldots, a_n, \ldots$ be an arithmetic sequence. In Problems 9–16, find the indicated quantities.

9. $a_1 = -5, d = 4; a_2 = ?, a_3 = ?, a_4 = ?$

10. $a_1 = -18, d = 3; a_2 = ?, a_3 = ?, a_4 = ?$

11. $a_1 = -3, d = 5; a_{15} = ?, S_{11} = ?$

12. $a_1 = 3, d = 4; a_{22} = ?, S_{21} = ?$

13. $a_1 = 1, a_2 = 5; S_{21} = ?$

14. $a_1 = 5, a_2 = 11; S_{11} = ?$

15. $a_1 = 7, a_2 = 5; a_{15} = ?$

16. $a_1 = -3, d = -4; a_{10} = ?$

Let $a_1, a_2, a_3, \ldots, a_n, \ldots$ be a geometric sequence. In Problems 17–22, find each of the indicated quantities.

17. $a_1 = -6, r = -\frac{1}{2}; a_2 = ?, a_3 = ?, a_4 = ?$

18. $a_1 = 12, r = \frac{2}{3}; a_2 = ?, a_3 = ?, a_4 = ?$

19. $a_1 = 81, r = \frac{1}{3}; a_{10} = ?$

20. $a_1 = 64, r = \frac{1}{2}; a_{13} = ?$

21. $a_1 = 3, a_7 = 2{,}187, r = 3; S_7 = ?$

22. $a_1 = 1, a_7 = 729, r = -3; S_7 = ?$

Let $a_1, a_2, a_3, \ldots, a_n, \ldots$ be an arithmetic sequence. In Problems 23–30, find the indicated quantities.

23. $a_1 = 3, a_{20} = 117; d = ?, a_{101} = ?$

24. $a_1 = 7, a_8 = 28; d = ?, a_{25} = ?$

25. $a_1 = -12, a_{40} = 22; S_{40} = ?$

26. $a_1 = 24, a_{24} = -28; S_{24} = ?$

27. $a_1 = \frac{1}{3}, a_2 = \frac{1}{2}; a_{11} = ?, S_{11} = ?$

28. $a_1 = \frac{1}{6}, a_2 = \frac{1}{4}; a_{19} = ?, S_{19} = ?$

29. $a_3 = 13, a_{10} = 55; a_1 = ?$

30. $a_9 = -12, a_{13} = 3; a_1 = ?$

Let $a_1, a_2, a_3, \ldots, a_n, \ldots$ be a geometric sequence. Find each of the indicated quantities in Problems 31–42.

31. $a_1 = 8, a_2 = 2; r = ?$

32. $a_1 = -6, a_2 = 2; r = ?$

33. $a_1 = 120, a_4 = -15; r = ?$

34. $a_1 = \sqrt{2}, a_6 = 8; r = ?$

35. $a_1 = 9, r = \frac{2}{3}; S_{10} = ?$

36. $a_1 = 3, r = 5; S_9 = ?$

37. $a_1 = 1, a_8 = 2{,}187; S_8 = ?$

38. $a_1 = \frac{1}{2}, a_{12} = 1{,}024; S_{12} = ?$

39. $a_3 = 72, a_6 = -243; a_1 = ?$

40. $a_4 = 8, a_5 = 6; a_1 = ?$

41. $a_1 = 1, a_4 = -1; a_{100} = ?$

42. $a_1 = -1, a_8 = 1; a_{99} = ?$

43. $S_{51} = \displaystyle\sum_{k=1}^{51}(3k + 3) = ?$

44. $S_{40} = \displaystyle\sum_{k=1}^{40}(2k - 3) = ?$

45. $S_7 = \displaystyle\sum_{k=1}^{7}(-3)^{k-1} = ?$

46. $S_7 = \displaystyle\sum_{k=1}^{7} 3^k = ?$

47. Find the sum of all the even integers between 21 and 135.

48. Find the sum of all the odd integers between 100 and 500.

49. Show that the sum of the first n odd natural numbers is n^2, using appropriate formulas from Section 11-3.

50. Show that the sum of the first n even natural numbers is $n + n^2$, using appropriate formulas from Section 11-3.

In Problems 51–60, find the sum of each infinite geometric series that has a sum.

51. $2 + \frac{1}{2} + \frac{1}{8} + \cdots$

52. $6 + 2 + \frac{2}{3} + \cdots$

53. $3 - 1 + \frac{1}{3} - \cdots$

54. $1 + \frac{4}{3} + \frac{16}{9} + \cdots$

55. $1 + 0.1 + 0.01 + \cdots$

56. $10 - 2 + 0.4 - \cdots$

57. $-1 + \frac{1}{2} - \frac{1}{4} + \cdots$

58. $-6 + 4 - \frac{8}{3} + \cdots$

59. $1 - 1 + 1 - \cdots$

60. $-100 - 80 - 64 - \cdots$

In Problems 61–66, represent each repeating decimal as the quotient of two integers.

61. $0.\overline{7} = 0.7777\cdots$ **62.** $0.\overline{5} = 0.5555\cdots$

63. $0.\overline{54} = 0.545\,454\cdots$ **64.** $0.\overline{27} = 0.272\,727\cdots$

65. $3.\overline{216} = 3.216\,216\,216\cdots$

66. $5.\overline{63} = 5.636\,363\cdots$

67. Prove, using mathematical induction, that if $\{a_n\}$ is an arithmetic sequence, then

$$a_n = a_1 + (n - 1)d \qquad \text{for every } n > 1$$

68. Prove, using mathematical induction, that if $\{a_n\}$ is an arithmetic sequence, then

$$S_n = \frac{n}{2}[2a_1 + (n - 1)d]$$

69. If in a given sequence, $a_1 = -2$ and $a_n = -3a_{n-1}, n > 1$, find a_n in terms of n.

70. For the sequence in Problem 69, find $S_n = \displaystyle\sum_{k=1}^{n} a_k$ in terms of n.

71. Prove, using mathematical induction, that if $\{a_n\}$ is a geometric sequence, then

$$a_n = a_1 r^{n-1} \qquad n \in N$$

72. Prove, using mathematical induction, that if $\{a_n\}$ is a geometric sequence, then

$$S_n = \frac{a_1 - a_1 r^n}{1 - r} \qquad n \in N, r \neq 1$$

73. Is there an arithmetic sequence that is also geometric? Explain.

74. Is there an infinite geometric sequence with $a_1 = 1$ that has sum equal to $\frac{1}{2}$? Explain.

APPLICATIONS

75. BUSINESS In investigating different job opportunities, you find that firm A will start you at $25,000 per year and guarantee you a raise of $1,200 each year whereas firm B will start you at $28,000 per year but will guarantee you a raise of only $800 each year. Over a period of 15 years, how much would you receive from each firm?

76. BUSINESS In Problem 75, what would be your annual salary at each firm for the tenth year?

77. ECONOMICS The government, through a subsidy program, distributes $1,000,000. If we assume that each individual or agency spends 0.8 of what is received, and 0.8 of this is spent, and so on, how much total increase in spending results from this government action?

78. ECONOMICS Because of reduced taxes, an individual has an extra $600 in spendable income. If we assume that the individual spends 70% of this on consumer goods, that the producers of these goods in turn spend 70% of what they receive on consumer goods, and that this process continues indefinitely, what is the total amount spent on consumer goods?

79. BUSINESS If $$P$ is invested at r% compounded annually, the amount A present after n years forms a geometric sequence with a common ratio $1 + r$. Write a formula for the amount present after n years. How long will it take a sum of money P to double if invested at 6% interest compounded annually?

80. POPULATION GROWTH If a population of A_0 people grows at the constant rate of r% per year, the population after t years forms a geometric sequence with a common ratio $1 + r$. Write a formula for the total population after t years. If the world's population is increasing at the rate of 2% per year, how long will it take to double?

81. FINANCE Eleven years ago an investment earned $7,000 for the year. Last year the investment earned $14,000. If the earnings from the investment have increased the same amount each year, what is the yearly increase and how much income has accrued from the investment over the past 11 years?

82. AIR TEMPERATURE As dry air moves upward, it expands. In so doing, it cools at the rate of about 5°F for each 1,000-foot rise. This is known as the **adiabatic process.**
(A) Temperatures at altitudes that are multiples of 1,000 feet form what kind of a sequence?
(B) If the ground temperature is 80°F, write a formula for the temperature T_n in terms of n, if n is in thousands of feet.

83. ENGINEERING A rotating flywheel coming to rest rotates 300 revolutions the first minute (see the figure). If in each subsequent minute it rotates two-thirds as many times as in the preceding minute, how many revolutions will the wheel make before coming to rest?

84. PHYSICS The first swing of a bob on a pendulum is 10 inches. If on each subsequent swing it travels 0.9 as far as on the preceding swing, how far will the bob travel before coming to rest?

85. FOOD CHAIN A plant is eaten by an insect, an insect by a trout, a trout by a salmon, a salmon by a bear, and the bear is eaten by you. If only 20% of the energy is transformed from one stage to the next, how many calories must be supplied by plant food to provide you with 2,000 calories from the bear meat?

86. GENEALOGY If there are 30 years in a generation, how many direct ancestors did each of us have 600 years ago? By *direct* ancestors we mean parents, grandparents, great-grandparents, and so on.

87. PHYSICS An object falling from rest in a vacuum near the surface of the Earth falls 16 feet during the first second, 48 feet during the second second, 80 feet during the third second, and so on.
(A) How far will the object fall during the eleventh second?
(B) How far will the object fall in 11 seconds?
(C) How far will the object fall in t seconds?

88. PHYSICS In Problem 87, how far will the object fall during:
(A) The twentieth second? (B) The tth second?

89. BACTERIA GROWTH A single cholera bacterium divides every $\frac{1}{2}$ hour to produce two complete cholera bacteria. If we start with a colony of A_0 bacteria, how many bacteria will we have in t hours, assuming adequate food supply?

90. CELL DIVISION One leukemic cell injected into a healthy mouse will divide into two cells in about $\frac{1}{2}$ day. At the end of the day these two cells will divide again, with the doubling process continuing each $\frac{1}{2}$ day until there are 1 billion cells, at which time the mouse dies. On which day after the experiment is started does this happen?

91. ASTRONOMY Ever since the time of the Greek astronomer Hipparchus, second century B.C., the brightness of stars has been measured in terms of magnitude. The brightest stars, excluding the sun, are classed as magnitude 1, and the dimmest visible to the eye are classed as magnitude 6. In 1856, the English astronomer N. R. Pogson showed that first-magnitude stars are 100 times brighter than sixth-magnitude stars. If the ratio of brightness between consecutive magnitudes is constant, find this ratio. [*Hint:* If b_n is the brightness of an nth-magnitude star, find r for the geometric sequence $b_1, b_2, b_3, \ldots$, given $b_1 = 100b_6$.]

92. PUZZLE If a sheet of very thin paper 0.001-inch thick is torn in half, and each half is again torn in half, and this process is repeated for a total of 32 times, how high will the stack of paper be if the pieces are placed one on top of the other? Give the answer to the nearest mile.

93. PUZZLE If you place 1¢ on the first square of a chessboard, 2¢ on the second square, 4¢ on the third, and so on, continuing to double the amount until all 64 squares are covered, how much money will be on the sixty-fourth square? How much money will there be on the whole board?

94. MUSIC The notes on a piano, as measured in cycles per second, form a geometric sequence.
(A) If A is 400 cycles per second and A′, 12 notes higher, is 800 cycles per second, find the constant ratio r.
(B) Find the cycles per second for C, three notes higher than A.

95. ATMOSPHERIC PRESSURE If atmospheric pressure decreases roughly by a factor of 10 for each 10-mile increase in altitude up to 60 miles, and if the pressure is 15 pounds per square inch at sea level, what will the pressure be 40 miles up?

96. ZENO'S PARADOX Visualize a hypothetical 440-yard oval racetrack that has tapes stretched across the track at the halfway point and at each point that marks the halfway point of each remaining distance thereafter. A runner running around the track has to break the first tape before the second, the second before the third, and so on. From this point of view it appears that he will never finish the race. This famous paradox is attributed to the Greek philosopher Zeno (495–435 B.C.). If we assume the runner runs at 440 yards per minute, the times between tape breakings form an infinite geometric sequence. What is the sum of this sequence?

97. GEOMETRY If the midpoints of the sides of an equilateral triangle are joined by straight lines, the new figure will be an equilateral triangle with a perimeter equal to half the original. If we start with an equilateral triangle with perimeter 1 and form a sequence of "nested" equilateral triangles proceeding as described, what will be the total perimeter of all the triangles that can be formed in this way?

98. PHOTOGRAPHY The shutter speeds and f-stops on a camera are given as follows:

Shutter speeds: $1, \frac{1}{2}, \frac{1}{4}, \frac{1}{8}, \frac{1}{15}, \frac{1}{30}, \frac{1}{60}, \frac{1}{125}, \frac{1}{250}, \frac{1}{500}$
f-stops: $1.4, 2, 2.8, 4, 5.6, 8, 11, 16, 22$

These are very close to being geometric sequences. Estimate their common ratios.

99. GEOMETRY We know that the sum of the interior angles of a triangle is 180°. Show that the sums of the interior angles of polygons with 3, 4, 5, 6, . . . sides form an arithmetic sequence. Find the sum of the interior angles for a 21-sided polygon.

11-4 Multiplication Principle, Permutations, and Combinations

> **Counting with the Multiplication Principle**
> **Using Factorial Notation**
> **Counting Permutations**
> **Counting Combinations**

Section 11-4 introduces some new mathematical tools that are usually referred to as *counting techniques*. In general, a **counting technique** is a mathematical method of determining the number of objects in a set without actually enumerating the objects in the set as 1, 2, 3, For example, we can count the number of squares in a checkerboard

> **Figure 1**

(Fig. 1) by counting 1, 2, 3, ..., 64. This is enumeration. Or we can note that there are eight rows with eight squares in each row. So the total number of squares must be $8 \times 8 = 64$. This is a very simple counting technique.

Now consider the problem of assigning telephone numbers. How many different seven-digit telephone numbers can be formed? As we will soon see, the answer is $10^7 = 10,000,000$, a number that is much too large to obtain by enumeration. This shows that counting techniques are essential tools if the number of elements in a set is very large. The techniques developed in this section will be applied to a brief introduction to probability theory in Section 11-5, and to a famous algebraic formula in Section 11-6.

> ## Counting with the Multiplication Principle

We start with an example.

| EXAMPLE | 1 | **Combined Outcomes** |

Suppose we flip a coin and then throw a single die (Fig. 2). What are the possible combined outcomes?

SOLUTION

One way to solve this problem is to use a **tree diagram:**

Heads Tails

Coin outcomes

Die outcomes

> **Figure 2** Coin and die outcomes.

There are 12 possible combined outcomes—two ways in which the coin can come up followed by six ways in which the die can come up.

MATCHED PROBLEM 1

Use a tree diagram to determine the number of possible outcomes of throwing a single die followed by flipping a coin.

Now suppose you are asked, "From the 26 letters in the alphabet, how many ways can 3 letters appear in a row on a license plate if no letter is repeated?" To try to count the possibilities using a tree diagram would be extremely tedious, to say the least. The following **multiplication principle,** also called the **fundamental counting principle,** enables us to solve this problem easily. In addition, it forms the basis for several other counting techniques developed later in this section.

> **MULTIPLICATION PRINCIPLE**

1. If two operations O_1 and O_2 are performed in order with N_1 possible outcomes for the first operation and N_2 possible outcomes for the second operation, then there are

$$N_1 \cdot N_2$$

possible combined outcomes of the first operation followed by the second.

2. In general, if n operations $O_1, O_2, \ldots, O_n$, are performed in order, with possible number of outcomes $N_1, N_2, \ldots, N_n$, respectively, then there are

$$N_1 \cdot N_2 \cdot \cdots \cdot N_n$$

possible combined outcomes of the operations performed in the given order.

In Example 1, we see that there are two possible outcomes from the first operation of flipping a coin and six possible outcomes from the second operation of throwing a die. So by the multiplication principle, there are $2 \cdot 6 = 12$ possible combined outcomes of flipping a coin followed by throwing a die. (Now try using the multiplication principle to solve Matched Problem 1.)

To answer the license plate question, we reason as follows: There are 26 ways the first letter can be chosen. After a first letter is chosen, 25 letters remain, so there are 25 ways a second letter can be chosen. And after 2 letters are chosen, there are 24 ways a third letter can be chosen. Using the multiplication principle, there are $26 \cdot 25 \cdot 24 = 15,600$ possible ways 3 letters can be chosen from the alphabet without allowing any letter to repeat. By not allowing any letter to repeat, earlier selections affect the choice of subsequent selections. If we allow letters to repeat, then earlier selections do not affect the choice in subsequent selections, and there are 26 possible choices for each of the 3 letters. So, if we allow letters to repeat, there are $26 \cdot 26 \cdot 26 = 26^3 = 17,576$ possible ways the 3 letters can be chosen from the alphabet.

| EXAMPLE | 2 | **Computer-Generated Tests** |

Many universities and colleges are now using computer-assisted testing procedures. Suppose a screening test is to consist of five questions, and a computer stores five equivalent questions for the first test question, eight equivalent questions for the second, six for the third, five for the fourth, and ten for the fifth. How many different five-question tests can the computer select? Two tests are considered different if they differ in one or more questions.

SOLUTION

O_1: Select the first question N_1: five ways
O_2: Select the second question N_2: eight ways
O_3: Select the third question N_3: six ways
O_4: Select the fourth question N_4: five ways
O_5: Select the fifth question N_5: ten ways

The computer can generate

$$5 \cdot 8 \cdot 6 \cdot 5 \cdot 10 = 12,000 \text{ different tests}$$

MATCHED PROBLEM 2

Each question on a multiple-choice test has five choices. If there are five such questions on a test, how many different response sheets are possible if only one choice is marked for each question?

EXAMPLE | **3** | **Counting Code Words**

How many three-letter code words are possible using the first eight letters of the alphabet if:

(A) No letter can be repeated? (B) Letters can be repeated?

(C) Adjacent letters cannot be alike?

SOLUTIONS

(A) No letter can be repeated.

O_1: Select first letter N_1: eight ways
O_2: Select second letter N_2: seven ways Because one letter has been used
O_3: Select third letter N_3: six ways Because two letters have been used

There are

$$8 \cdot 7 \cdot 6 = 336 \text{ possible code words}$$

(B) Letters can be repeated.

O_1: Select first letter N_1: eight ways
O_2: Select second letter N_2: eight ways Repeats are allowed.
O_3: Select third letter N_3: eight ways Repeats are allowed.

There are

$$8 \cdot 8 \cdot 8 = 8^3 = 512 \text{ possible code words}$$

(C) Adjacent letters cannot be alike.

O_1: Select first letter N_1: eight ways
O_2: Select second letter N_2: seven ways Cannot be the same as the first
O_3: Select third letter N_3: seven ways Cannot be the same as the second, but can be the same as the first

There are

$$8 \cdot 7 \cdot 7 = 392 \text{ possible code words}$$

MATCHED PROBLEM 3

How many four-letter code words are possible using the first ten letters of the alphabet under the three conditions stated in Example 3?

>>> EXPLORE-DISCUSS 1

The postal service of a developing country is choosing a five-character postal code consisting of letters (of the English alphabet) and digits. At least a half a million postal codes must be accommodated. Which format would you recommend to make the codes easy to remember?

The multiplication principle can be used to develop two additional counting techniques that are extremely useful in more complicated counting problems. Both of these methods use factorial notation, which we introduce next.

› Using Factorial Notation

For n a natural number, n **factorial**—denoted by $n!$—is the product of the first n natural numbers. **Zero factorial** is defined to be 1.

> **DEFINITION 1** *n* Factorial

For *n* a natural number

$$n! = n(n-1) \cdot \cdots \cdot 2 \cdot 1$$
$$1! = 1$$
$$0! = 1$$

It is also useful to note that

> **THEOREM 1** Recursion Formula for *n* Factorial

$$n! = n \cdot (n-1)!$$

EXAMPLE 4 **Evaluating Factorials**

Evaluate each expression:

(A) 4! (B) 5! (C) $\dfrac{7!}{6!}$ (D) $\dfrac{8!}{5!}$ (E) $\dfrac{9!}{6!3!}$

SOLUTIONS

(A) $4! = 4 \cdot 3 \cdot 2 \cdot 1 = 24$ (B) $5! = 5 \cdot 4 \cdot 3 \cdot 2 \cdot 1 = 120$ (C) $\dfrac{7!}{6!} = \dfrac{7 \cdot 6!}{6!} = 7$

(D) $\dfrac{8!}{5!} = \dfrac{8 \cdot 7 \cdot 6 \cdot 5!}{5!} = 336$ (E) $\dfrac{9!}{6!3!} = \dfrac{\overset{3}{9} \cdot \overset{4}{8} \cdot 7 \cdot 6!}{6! \, 3 \cdot 2 \cdot 1} = 84$

MATCHED PROBLEM 4

Find (A) 6! (B) $\dfrac{6!}{5!}$ (C) $\dfrac{9!}{6!}$ (D) $\dfrac{10!}{7!3!}$

⟫ CAUTION ⟫

When reducing fractions involving factorials, don't confuse the single integer *n* with the symbol *n*!, which represents the product of *n* consecutive integers.

$$\frac{6!}{3!} \neq 2! \qquad \frac{6!}{3!} = \frac{6 \cdot 5 \cdot 4 \cdot 3!}{3!} = 6 \cdot 5 \cdot 4 = 120$$

⟫⟫ EXPLORE-DISCUSS 2

A student used a calculator* to solve Matched Problem 4, as shown in Figure 3. Check these answers. If any are incorrect, explain why and find a correct calculator solution.

```
6!
               720
6!/5!
                 6
9!/6!
               504
10!/7!3!
              4320
```

> **Figure 3**

*The factorial symbol ! and related symbols can be found under the MATH-PROB menus on a TI-84 or TI-86.

It is interesting and useful to note that $n!$ grows very rapidly. Compare the following:

$$5! = 120 \qquad 10! = 3{,}628{,}800 \qquad 15! = 1{,}307{,}674{,}368{,}000$$

If $n!$ is too large for a calculator to store and display, an error message is displayed. Find the value of n such that your calculator will evaluate $n!$, but not $(n + 1)!$.

› Counting Permutations

Suppose four pictures are to be arranged from left to right on one wall of an art gallery. How many arrangements are possible? Using the multiplication principle, there are four ways of selecting the first picture. After the first picture is selected, there are three ways of selecting the second picture. After the first two pictures are selected, there are two ways of selecting the third picture. And after the first three pictures are selected, there is only one way to select the fourth. So, the number of arrangements possible for the four pictures is

$$4 \cdot 3 \cdot 2 \cdot 1 = 4! \qquad \text{or} \qquad 24$$

In general, we refer to a particular arrangement, or **ordering,** of n objects without repetition as a **permutation** of the n objects. How many permutations of n objects are there? From the preceding reasoning, there are n ways in which the first object can be chosen, there are $n - 1$ ways in which the second object can be chosen, and so on. Applying the multiplication principle, we have Theorem 2.

> ### › THEOREM 2 Permutations of n Objects
>
> The number of permutations of n objects, denoted by $P_{n,n}$, is given by
>
> $$P_{n,n} = n \cdot (n - 1) \cdot \cdots \cdot 1 = n!$$

Now suppose the director of the art gallery decides to use only two of the four available pictures on the wall, arranged from left to right. How many arrangements of two pictures can be formed from the four? There are four ways the first picture can be selected. After selecting the first picture, there are three ways the second picture can be selected. So, the number of arrangements of two pictures from four pictures, denoted by $P_{4,2}$, is given by

$$P_{4,2} = 4 \cdot 3 = 12$$

Or, in terms of factorials, multiplying $4 \cdot 3$ by 1 in the form $2!/2!$, we have

$$P_{4,2} = 4 \cdot 3 = \frac{4 \cdot 3 \cdot 2!}{2!} = \frac{4!}{2!}$$

This last form gives $P_{4,2}$ in terms of factorials, which is useful in some cases.

A **permutation of a set of n objects taken r at a time** is an arrangement of the r objects in a specific order. So, reasoning in the same way as in the preceding example, we find that the number of permutations of n objects taken r at a time, $0 \le r \le n$, denoted by $P_{n,r}$, is given by

$$P_{n,r} = n(n - 1)(n - 2) \cdot \cdots \cdot (n - r + 1)$$

Multiplying the right side of this equation by 1 in the form $(n - r)!/(n - r)!$, we obtain a factorial form for $P_{n,r}$:

$$P_{n,r} = n(n - 1)(n - 2) \cdot \cdots \cdot (n - r + 1) \frac{(n - r)!}{(n - r)!}$$

But

$$n(n - 1)(n - 2) \cdot \cdots \cdot (n - r + 1)(n - r)! = n!$$

We have developed Theorem 3.

> ### › THEOREM 3 Permutation of *n* Objects Taken *r* at a Time
>
> The number of permutations of *n* objects taken *r* at a time is given by
>
> $$P_{n,r} = \underbrace{n(n - 1)(n - 2) \cdot \cdots \cdot (n - r + 1)}_{r \text{ factors}}$$
>
> or
>
> $$P_{n,r} = \frac{n!}{(n - r)!} \qquad 0 \le r \le n$$

Note that if $r = n$, then the number of permutations of *n* objects taken *n* at a time is

$$P_{n,n} = \frac{n!}{(n - n)!} = \frac{n!}{0!} = n! \qquad \text{Recall, 0! = 1.}$$

which agrees with Theorem 2, as it should.

The permutation symbol $P_{n,r}$ *also can be denoted by* P_r^n, $_nP_r$, *or* $P(n, r)$. Many calculators use $_nP_r$ to denote the function that evaluates the permutation symbol.

EXAMPLE 5 ### Selecting Officers

From a committee of eight people, in how many ways can we choose a chair and a vice-chair, assuming one person cannot hold more than one position?

SOLUTION We are actually asking for the number of permutations of eight objects taken two at a time—that is, $P_{8,2}$:

$$P_{8,2} = \frac{8!}{(8 - 2)!} = \frac{8!}{6!} = \frac{8 \cdot 7 \cdot 6!}{6!} = 56$$

MATCHED PROBLEM 5 From a committee of ten people, in how many ways can we choose a chair, vice-chair, and secretary, assuming one person cannot hold more than one position?

EXAMPLE 6 ### Evaluating $P_{n,r}$

Find the number of permutations of 25 objects taken

(A) Two at a time

(B) Four at a time

(C) Eight at a time

SOLUTION | Figure 4 shows the solution on a calculator.

```
25 nPr 2
              600
25 nPr 4
           303600
25 nPr 8
     4.3609104E10
```

› **Figure 4**

MATCHED PROBLEM 6 | Find the number of permutations of 30 objects taken

(A) Two at a time　　　(B) Four at a time　　　(C) Six at a time

› Counting Combinations

Now suppose that an art museum owns eight paintings by a given artist and another art museum hopes to borrow three of these paintings for a special show. How many ways can three paintings be selected for shipment out of the eight available? Here, the order of the items selected doesn't matter. What we are actually interested in is how many subsets of three objects can be formed from a set of eight objects. We call such a subset a **combination** of eight objects taken three at a time. The total number of combinations is denoted by the symbol

$$C_{8,3} \qquad \text{or} \qquad \binom{8}{3}$$

To find the number of combinations of eight objects taken three at a time, $C_{8,3}$, we make use of the formula for $P_{n,r}$ and the multiplication principle. We know that the number of permutations of eight objects taken three at a time is given by $P_{8,3}$, and we have a formula for computing this quantity. Now suppose we think of $P_{8,3}$ in terms of two operations:

O_1:　Select a subset of three objects (paintings)

N_1:　$C_{8,3}$ ways

O_2:　Arrange the subset in a given order

N_2:　3! ways

The combined operation, O_1 followed by O_2, produces a permutation of eight objects taken three at a time. So,

$$P_{8,3} = C_{8,3} \cdot 3!$$

To find $C_{8,3}$, we replace $P_{8,3}$ in the preceding equation with $8!/(8-3)!$ and solve for $C_{8,3}$:

$$\frac{8!}{(8-3)!} = C_{8,3} \cdot 3!$$

$$C_{8,3} = \frac{8!}{3!(8-3)!} = \frac{8 \cdot 7 \cdot 6 \cdot 5!}{3 \cdot 2 \cdot 1 \cdot 5!} = 56$$

The museum can make 56 different selections of three paintings from the eight available.

A **combination of a set of n objects taken r at a time** is an r-element subset of the n objects. Reasoning in the same way as in the example, the number of combinations of n

objects taken r at a time, $0 \leq r \leq n$, denoted by $C_{n,r}$, can be obtained by solving for $C_{n,r}$ in the relationship

$$P_{n,r} = C_{n,r} \cdot r!$$

$$C_{n,r} = \frac{P_{n,r}}{r!}$$

$$= \frac{n!}{r!(n-r)!} \qquad P_{n,r} = \frac{n!}{(n-r)!}$$

> **THEOREM 4 Combination of n Objects Taken r at a Time**
>
> The number of combinations of n objects taken r at a time is given by
>
> $$C_{n,r} = \binom{n}{r} = \frac{P_{n,r}}{r!} = \frac{n!}{r!(n-r)!} \qquad 0 \leq r \leq n$$

The combination symbols $C_{n,r}$ and $\binom{n}{r}$ also can be denoted by C_r^n, $_nC_r$, or $C(n, r)$.

EXAMPLE 7 **Selecting Subcommittees**

From a committee of eight people, in how many ways can we choose a subcommittee of two people?

SOLUTION Notice how this example differs from Example 5, where we wanted to know how many ways a chair and a vice-chair can be chosen from a committee of eight people. In Example 5, ordering matters. In choosing a subcommittee of two people, the ordering does not matter. So, we are actually asking for the number of combinations of eight objects taken two at a time. The number is given by

$$C_{8,2} = \binom{8}{2} = \frac{8!}{2!(8-2)!} = \frac{8 \cdot 7 \cdot 6!}{2 \cdot 1 \cdot 6!} = 28$$

MATCHED PROBLEM 7 How many subcommittees of three people can be chosen from a committee of eight people?

EXAMPLE 8 **Evaluating $C_{n,r}$**

Find the number of combinations of 25 objects taken

(A) Two at a time (B) Four at a time (C) Eight at a time

SOLUTION Figure 5 shows the solution on a calculator. Compare these results with Example 6.

```
25 nCr 2
              300
25 nCr 4
            12650
25 nCr 8
         1081575
```

> **Figure 5**

MATCHED PROBLEM 8

Find the number of combinations of 30 objects taken

(A) Two at a time (B) Four at a time (C) Six at a time

Remember: In a permutation, order counts. In a combination, order does not count.

To determine whether a permutation or combination is needed, decide whether rearranging the collection or listing makes a difference. If so, use permutations. If not, use combinations.

>>> EXPLORE-DISCUSS 3

Each of the following is a selection without repetition. Would you consider the selection to be a combination? A permutation? Discuss your reasoning.

(A) A student checks out three books from the library.

(B) A baseball manager names his starting lineup.

(C) The newly elected president names his cabinet members.

(D) The president selects a delegation of three cabinet members to attend the funeral of a head of state.

(E) An orchestra conductor chooses three pieces of music for a symphony program.

> **Figure 6** A standard deck of cards.

A **standard deck** of 52 cards (Fig. 6) has four 13-card suits: diamonds, hearts, clubs, and spades. Each 13-card suit contains cards numbered from 2 to 10, a jack, a queen, a king, and an ace. The jack, queen, and king are called **face cards.** Depending on the game, the ace may be counted as the lowest and/or the highest card in the suit. Example 9, as well as other examples and exercises in Chapter 11, refer to this standard deck.

EXAMPLE 9 Counting Card Hands

Out of a standard 52-card deck, how many 5-card hands will have three aces and two kings?

SOLUTION

O_1: Choose three aces out of four possible Order is not important.
N_1: $C_{4,3}$
O_2: Choose two kings out of four possible Order is not important.
N_2: $C_{4,2}$

Using the multiplication principle, we have

$$\text{Number of hands} = C_{4,3} \cdot C_{4,2} = 4 \cdot 6 = 24$$

MATCHED PROBLEM 9

From a standard 52-card deck, how many 5-card hands will have three hearts and two spades?

EXAMPLE 10 Counting Serial Numbers

Serial numbers for a product are to be made using two letters followed by three numbers. If the letters are to be taken from the first eight letters of the alphabet with no repeats and the numbers from the 10 digits 0 through 9 with no repeats, how many serial numbers are possible?

SOLUTION

O_1: Choose two letters out of eight available Order is important.

N_1: $P_{8,2}$

O_2: Choose three numbers out of ten available Order is important.

N_2: $P_{10,3}$

Using the multiplication principle, we have

$$\text{Number of serial numbers} = P_{8,2} \cdot P_{10,3} = 40{,}320$$

MATCHED PROBLEM 10

Repeat Example 10 under the same conditions, except the serial numbers are now to have three letters followed by two digits with no repeats.

ANSWERS TO MATCHED PROBLEMS

1.

HT	HT	HT	HT	HT	HT
1	2	3	4	5	6

Start

There are 12 outcomes. **2.** 5^5, or 3,125

3. (A) $10 \cdot 9 \cdot 8 \cdot 7 = 5{,}040$ (B) $10 \cdot 10 \cdot 10 \cdot 10 = 10{,}000$
 (C) $10 \cdot 9 \cdot 9 \cdot 9 = 7{,}290$ **4.** (A) 720 (B) 6 (C) 504 (D) 120

5. $P_{10,3} = \dfrac{10!}{(10-3)!} = 720$ **6.** (A) 870 (B) 657,720 (C) 427,518,000

7. $C_{8,3} = \dfrac{8!}{3!(8-3)!} = 56$ **8.** (A) 435 (B) 27,405 (C) 593,775

9. $C_{13,3} \cdot C_{13,2} = 22{,}308$ **10.** $P_{8,3} \cdot P_{10,2} = 30{,}240$

11-4 Exercises

1. What is a permutation?

2. What is a combination?

3. Explain how $n!$ can be defined by means of a recursion formula.

4. State the multiplication principle for counting in your own words.

5. Explain how permutations and combinations differ with respect to order.

6. Explain how permutations and combinations are alike with respect to repetition.

Evaluate the expression in Problems 7–16:

7. $9!$ **8.** $10!$ **9.** $11!$

10. $12!$ **11.** $\dfrac{11!}{8!}$ **12.** $\dfrac{14!}{12!}$

13. $\dfrac{5!}{2!3!}$ **14.** $\dfrac{6!}{4!2!}$ **15.** $\dfrac{7!}{4!(7-4)!}$

16. $\dfrac{8!}{3!(8-3)!}$

17. The figure shows calculator solutions to Problems 11, 13, and 15. Check these answers. If any are incorrect, explain why and find a correct calculator solution.

```
11 nPr 8
           6652800
5!/2!3!
               360
7 nCr 4
                35
■
```

18. The figure shows calculator solutions to Problems 12, 14, and 16. Check these answers. If any are incorrect, explain why and find a correct calculator solution.

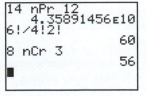

```
14 nPr 12
     4.35891456E10
6!/4!2!
                60
8 nCr 3
                56
■
```

In Problems 19–26, evaluate.

19. $P_{13,4}$ **20.** $C_{20,10}$ **21.** $P_{13,9}$

22. $C_{20,4}$ **23.** $C_{15,8}$ **24.** $P_{11,3}$

25. $C_{15,12}$ **26.** $P_{11,8}$

In Problems 27 and 28, would you consider the selection to be a combination or a permutation? Explain your reasoning.

27. (A) The recently elected chief executive officer (CEO) of a company named three new vice-presidents, of marketing, research, and manufacturing.
 (B) The CEO selected three of her vice-presidents to attend the dedication ceremony of a new plant.

28. (A) An individual rented four DVDs from a rental store to watch over a weekend.
 (B) The same individual did some holiday shopping by buying four DVDs, one for his father, one for his mother, one for his younger sister, and one for his older brother.

29. A particular new car model is available with five choices of color, three choices of transmission, four types of interior, and two types of engine. How many different variations of this model car are possible?

30. A deli serves sandwiches with the following options: three kinds of bread, five kinds of meat, and lettuce or sprouts. How many different sandwiches are possible, assuming one item is used out of each category?

31. In a horse race, how many different finishes among the first three places are possible for a 10-horse race? Exclude ties.

32. In a long-distance foot race, how many different finishes among the first five places are possible for a 50-person race? Exclude ties.

33. How many ways can a subcommittee of three people be selected from a committee of seven people? How many ways can a president, vice-president, and secretary be chosen from a committee of seven people?

34. Suppose nine cards are numbered with the nine digits from 1 to 9. A three-card hand is dealt, one card at a time. How many hands are possible where:
 (A) Order is taken into consideration?
 (B) Order is not taken into consideration?

35. There are 10 teams in a league. If each team is to play every other team exactly once, how many games must be scheduled?

36. Given seven points, no three of which are on a straight line, how many lines can be drawn joining two points at a time?

37. How many four-letter code words are possible from the first six letters of the alphabet, with no letter repeated? Allowing letters to repeat?

38. How many five-letter code words are possible from the first seven letters of the alphabet, with no letter repeated? Allowing letters to repeat?

39. A combination lock has five wheels, each labeled with the 10 digits from 0 to 9. How many opening combinations of five numbers are possible, assuming no digit is repeated? Assuming digits can be repeated?

40. A small combination lock on a suitcase has three wheels, each labeled with digits from 0 to 9. How many opening combinations of three numbers are possible, assuming no digit is repeated? Assuming digits can be repeated?

41. From a standard 52-card deck, how many 5-card hands will have all hearts?

42. From a standard 52-card deck, how many 5-card hands will have all face cards? All face cards, but no kings? Consider only jacks, queens, and kings to be face cards.

43. How many different license plates are possible if each contains three letters followed by three digits? How many of these license plates contain no repeated letters and no repeated digits?

44. How may five-digit zip codes are possible? How many of these codes contain no repeated digits?

45. From a standard 52-card deck, how many 7-card hands have exactly five spades and two hearts?

46. From a standard 52-card deck, how many 5-card hands will have two clubs and three hearts?

47. A catering service offers eight appetizers, ten main courses, and seven desserts. A banquet chairperson is to select three appetizers, four main courses, and two desserts for a banquet. How many ways can this be done?

48. Three research departments have 12, 15, and 18 members, respectively. If each department is to select a delegate and an alternate to represent the department at a conference, how many ways can this be done?

 49. (A) Use a graphing calculator to display the sequences $P_{10,0}$, $P_{10,1}$, ..., $P_{10,10}$ and 0!, 1!, ..., 10! in table form, and show that $P_{10,r} \geq r!$ for $r = 0, 1, \ldots, 10$.
 (B) Find all values of r such that $P_{10,r} = r!$
 (C) Explain why $P_{n,r} \geq r!$ whenever $0 \leq r \leq n$.

50. (A) How are the sequences $\dfrac{P_{10,0}}{0!}$, $\dfrac{P_{10,1}}{1!}$, ..., $\dfrac{P_{10,10}}{10!}$ and $C_{10,0}$, $C_{10,1}$, ..., $C_{10,10}$ related?
 (B) Use a graphing calculator to graph each sequence and confirm the relationship of part A.

51. A sporting goods store has 12 pairs of ski gloves of 12 different brands thrown loosely in a bin. The gloves are all the same size. In how many ways can a left-hand glove and a right-hand glove be selected that do not match relative to brand?

52. A sporting goods store has six pairs of running shoes of six different styles thrown loosely in a basket. The shoes are all the same size. In how many ways can a left shoe and a right shoe be selected that do not match?

53. Eight distinct points are selected on the circumference of a circle.
 (A) How many chords can be drawn by joining the points in all possible ways?
 (B) How many triangles can be drawn using these eight points as vertices?
 (C) How many quadrilaterals can be drawn using these eight points as vertices?

54. Five distinct points are selected on the circumference of a circle.
 (A) How many chords can be drawn by joining the points in all possible ways?
 (B) How many triangles can be drawn using these five points as vertices?

55. How many ways can two people be seated in a row of five chairs? Three people? Four people? Five people?

56. Each of two countries sends five delegates to a negotiating conference. A rectangular table is used with five chairs on each long side. If each country is assigned a long side of the table, how many seating arrangements are possible? [*Hint:* Operation 1 is assigning a long side of the table to each country.]

57. A basketball team has five distinct positions. Out of eight players, how many starting teams are possible if
 (A) The distinct positions are taken into consideration?
 (B) The distinct positions are not taken into consideration?
 (C) The distinct positions are not taken into consideration, but either Mike or Ken, but not both, must start?

58. How many committees of four people are possible from a group of nine people if
 (A) There are no restrictions?
 (B) Both Juan and Mary must be on the committee?
 (C) Either Juan or Mary, but not both, must be on the committee?

59. A 5-card hand is dealt from a standard 52-card deck. Which is more likely: the hand contains exactly one king or the hand contains no hearts?

60. A 10-card hand is dealt from a standard 52-card deck. Which is more likely: all cards in the hand are red or the hand contains all four aces?

11-5 Sample Spaces and Probability

> Sample Spaces and Events
> Finding the Probability of an Event
> Making Equally Likely Assumptions
> Finding or Approximating Empirical Probability

This section provides an introduction to probability. It's going to need to be a relatively brief one, because probability is a topic to which entire books and courses are devoted. Probability involves many subtle notions, and care must be taken at the beginning to understand the fundamental concepts on which the subject is based. Our development of probability, because of space limitations, must be somewhat informal. More formal and precise treatments can be found in books on probability.

> Sample Spaces and Events

Our first step in constructing a mathematical model for probability studies is to describe the type of experiments on which probability studies are based. Some types of experiments do not yield the same results, no matter how carefully they are repeated under the same conditions. These experiments are called **random experiments.** Some standard examples of random experiments are flipping coins, rolling dice, observing the frequency of defective items from an assembly line, or observing the frequency of deaths in a certain age group.

Probability theory is a branch of mathematics that has been developed to deal with outcomes of random experiments. In the work that follows, the word **experiment** will be used to mean a random experiment.

The outcomes of experiments are typically described in terms of *sample spaces* and *events*. Our second step in constructing a mathematical model for probability studies is to define these two terms.

Consider the experiment, "A single six-sided die is rolled." What outcomes might we observe? We might be interested in the number of dots facing up, or whether the number of dots facing up is an even number, or whether the number of dots facing up is divisible by 3, and so on. The list of possible outcomes appears endless. In general, there is no unique method of analyzing all possible outcomes of an experiment. Therefore, before conducting an experiment, it is important to decide just what outcomes are of interest.

In the die experiment, suppose we limit our interest to the number of dots facing up when the die comes to rest. Having decided what to observe, we make a list of outcomes of the experiment, called *simple events,* such that in each trial of the experiment, one and only one of the results on the list will occur. The set of simple events for the experiment is called a **sample space** for the experiment. The sample space S we have chosen for the die-rolling experiment is

$$S = \{1, 2, 3, 4, 5, 6\}$$

Now consider the outcome, "The number of dots facing up is an even number." This outcome is not a simple event, because it will occur whenever 2, 4, or 6 dots appear, that is, whenever an element in the subset

$$E = \{2, 4, 6\}$$

occurs. Subset E is called a *compound event.* In general, we have the following definition:

> **DEFINITION 1** Event
>
> Given a sample space S for an experiment, we define an **event E** to be any subset of S. If an event E has only one element in it, it is called a **simple event.** If event E has more than one element, it is called a **compound event.** We say that **an event E occurs** if any of the simple events in E occurs.

| EXAMPLE | 1 | **Choosing a Sample Space** |

A nickel and a dime are tossed. How will we identify a sample space for this experiment?

SOLUTIONS There are a number of possibilities, depending on our interest. We will consider three.

(A) If we are interested in whether each coin falls heads (H) or tails (T), then, using a tree diagram, we can easily determine an appropriate sample space for the experiment:

Nickel Outcomes	Dime Outcomes	Combined Outcomes
H	H	HH
	T	HT
T	H	TH
	T	TT

Start

The sample space is

$$S_1 = \{HH, HT, TH, TT\}$$

and there are four simple events in the sample space.

(B) If we are interested only in the number of heads that appear on a single toss of the two coins, then we can let

$$S_2 = \{0, 1, 2\}$$

and there are three simple events in the sample space.

(C) If we are interested in whether the coins match (M) or don't match (D), then we can let

$$S_3 = \{M, D\}$$

and there are only two simple events in the sample space.

MATCHED PROBLEM 1

An experiment consists of recording the boy–girl composition of families with two children.

(A) What is an appropriate sample space if we are interested in the gender of each child in the order of their births? Draw a tree diagram.

(B) What is an appropriate sample space if we are interested only in the number of girls in a family?

(C) What is an appropriate sample space if we are interested only in whether the genders are alike (A) or different (D)?

(D) What is an appropriate sample space for all three interests expressed above?

In Example 1, sample space S_1 contains more information than either S_2 or S_3. If we know which outcome has occurred in S_1, then we know which outcome has occurred in S_2 and S_3. However, the reverse is not true. In this sense, we say that S_1 is a more **fundamental sample space** than either S_2 or S_3.

> **Important Remark: There is no one correct sample space for a given experiment. When specifying a sample space for an experiment, we include as much detail as necessary to answer *all* questions of interest regarding the outcomes of the experiment. If in doubt, include more elements in the sample space rather than fewer.**

Now let's return to the two-coin problem in Example 1 and the sample space

$$S_1 = \{HH, HT, TH, TT\}$$

Suppose we are interested in the outcome, "Exactly 1 head is up." Looking at S_1, we find that it occurs if either of the two simple events HT or TH occurs.* So, to say that the event, "Exactly 1 head is up" occurs is the same as saying the experiment has an outcome in the set

$$E = \{HT, TH\}$$

This is a subset of the sample space S_1. The event E is a compound event.

*Technically, we should write {HT} and {TH}, because there is a logical distinction between an element of a set and a subset consisting of only that element. But we will just keep this in mind and drop the braces for simple events to simplify the notation.

EXAMPLE 2

Rolling Two Dice

Consider an experiment of rolling two dice. A convenient sample space that will enable us to answer many questions about events of interest is shown in Figure 1. Let S be the set of all ordered pairs listed in the figure. Note that the simple event (3, 2) is to be distinguished from the simple event (2, 3). The former indicates a 3 turned up on the first die and a 2 on the second, whereas the latter indicates a 2 turned up on the first die and a 3 on the second. What is the event that corresponds to each of the following outcomes?

(A) A sum of 7 turns up. (B) A sum of 11 turns up.

(C) A sum less than 4 turns up. (D) A sum of 12 turns up.

SECOND DIE

	⚀	⚁	⚂	⚃	⚄	⚅
⚀	(1, 1)	(1, 2)	(1, 3)	(1, 4)	(1, 5)	(1, 6)
⚁	(2, 1)	(2, 2)	(2, 3)	(2, 4)	(2, 5)	(2, 6)
⚂	(3, 1)	(3, 2)	(3, 3)	(3, 4)	(3, 5)	(3, 6)
⚃	(4, 1)	(4, 2)	(4, 3)	(4, 4)	(4, 5)	(4, 6)
⚄	(5, 1)	(5, 2)	(5, 3)	(5, 4)	(5, 5)	(5, 6)
⚅	(6, 1)	(6, 2)	(6, 3)	(6, 4)	(6, 5)	(6, 6)

FIRST DIE

› **Figure 1** A sample space for rolling two dice.

SOLUTIONS

(A) By "A sum of 7 turns up," we mean that the sum of all dots on both turned-up faces is 7. This outcome corresponds to the event

$$\{(6, 1), (5, 2), (4, 3), (3, 4), (2, 5), (1, 6)\}$$

(B) "A sum of 11 turns up" corresponds to the event

$$\{(6, 5), (5, 6)\}$$

(C) "A sum less than 4 turns up" corresponds to the event

$$\{(1, 1), (2, 1), (1, 2)\}$$

(D) "A sum of 12 turns up" corresponds to the event

$$\{(6, 6)\}$$

MATCHED PROBLEM 2

Refer to the sample space in Example 2 (Fig. 1). What is the event that corresponds to each of the following outcomes?

(A) A sum of 5 turns up.

(B) A sum that is a prime number greater than 7 turns up.

Informally, to facilitate discussion, we often use the terms *event* and *outcome of an experiment* interchangeably. So, in Example 2 we might say "the event 'A sum of 11 turns up'" in place of "the outcome 'A sum of 11 turns up,'" or even write

$$E = \text{A sum of 11 turns up} = \{(6, 5), (5, 6)\}$$

Technically speaking, an event is the mathematical counterpart of an outcome of an experiment.

› Finding the Probability of an Event

The next step in developing our mathematical model for probability studies is the introduction of a *probability function*. This is a function that assigns to an arbitrary event associated with a sample space a real number between 0 and 1, inclusive. We start by discussing ways in which probabilities are assigned to simple events in S.

> › **DEFINITION 2 Probabilities for Simple Events**
>
> Given a sample space
>
> $$S = \{e_1, e_2, \ldots, e_n\}$$
>
> with n simple events, to each simple event e_i we assign a real number, denoted by $P(e_i)$, that is called the **probability of the event e_i.** These numbers may be assigned in an arbitrary manner as long as the following two conditions are satisfied:
>
> **1.** $0 \le P(e_i) \le 1$
>
> **2.** $P(e_1) + P(e_2) + \cdots + P(e_n) = 1$ The sum of the probabilities of all simple events in the sample space is 1.
>
> Any probability assignment that meets conditions 1 and 2 is said to be an **acceptable probability assignment.**

Our mathematical theory does not explain how acceptable probabilities are assigned to simple events. These assignments are generally based on the expected or actual percentage of times a simple event occurs when an experiment is repeated a large number of times. Assignments based on this principle are called *reasonable*.

Let an experiment be the flipping of a single coin, and let us choose a sample space S to be

$$S = \{H, T\}$$

If a coin appears to be fair, we are inclined to assign probabilities to the simple events in S as follows:

$$P(H) = \tfrac{1}{2} \quad \text{and} \quad P(T) = \tfrac{1}{2}$$

These assignments are based on reasoning that, because there are two ways a coin can land, in the long run a head will turn up half the time and a tail will turn up half the time. These probability assignments are acceptable, because both of the conditions for acceptable probability assignments in Definition 2 are satisfied:

1. $0 \le P(H) \le 1, 0 \le P(T) \le 1$

2. $P(H) + P(T) = \tfrac{1}{2} + \tfrac{1}{2} = 1$

But there are other acceptable assignments. Maybe after flipping a coin 1,000 times we find that the head turns up 376 times and the tail turns up 624 times. With this result, we might suspect that the coin is not fair and assign the simple events in the sample space S the probabilities

$$P(H) = .376 \quad \text{and} \quad P(T) = .624$$

This is also an acceptable assignment. But the probability assignment

$$P(H) = 1 \quad \text{and} \quad P(T) = 0$$

though acceptable, is not reasonable, unless the coin has two heads. The assignment

$$P(H) = .6 \quad \text{and} \quad P(T) = .8$$

is not acceptable, because $.6 + .8 = 1.4$, which violates condition 2 in Definition 2.

In probability studies, the 0 to the left of the decimal is usually omitted; we write .8 and not 0.8.

It is important to keep in mind that out of the infinitely many possible acceptable probability assignments to simple events in a sample space, we are generally inclined to choose one assignment over another based on reasoning or experimental results.

Given an acceptable probability assignment for simple events in a sample space S, how do we define the probability of an arbitrary event E associated with S?

> **DEFINITION 3** Probability of an Event E

Given an acceptable probability assignment for the simple events in a sample space S, we define the **probability of an arbitrary event E,** denoted by $P(E)$, as follows:

1. If E is the empty set, then $P(E) = 0$.

2. If E is a simple event, then $P(E)$ has already been assigned.

3. If E is a compound event, then $P(E)$ is the sum of the probabilities of all the simple events in E.

4. If E is the sample space S, then $P(E) = P(S) = 1$. This is a special case of 3.

EXAMPLE 3 **Finding Probabilities of Events**

Let's return to Example 1, the tossing of a nickel and dime, and the sample space

$$S = \{HH, HT, TH, TT\}$$

Because there are four simple outcomes and the coins are assumed to be fair, it appears that each outcome should occur in the long run 25% of the time. Let's assign the same probability of $\frac{1}{4}$ to each simple event in S:

Simple event, e_i	HH	HT	TH	TT
$P(e_i)$	$\frac{1}{4}$	$\frac{1}{4}$	$\frac{1}{4}$	$\frac{1}{4}$

This is an acceptable assignment according to Definition 2 and a reasonable assignment for ideal coins that are perfectly balanced or coins close to ideal.

(A) What is the probability of getting exactly one head?

(B) What is the probability of getting at least one head?

(C) What is the probability of getting a head or a tail?

(D) What is the probability of getting three heads?

SOLUTIONS (A) $E_1 =$ Getting one head $= \{HT, TH\}$

Because E_1 is a compound event, we use item 3 in Definition 3 and find $P(E_1)$ by adding the probabilities of the simple events in E_1.

$$P(E_1) = P(HT) + P(TH) = \tfrac{1}{4} + \tfrac{1}{4} = \tfrac{1}{2}$$

(B) E_2 = Getting at least 1 head = {HH, HT, TH}

$\quad P(E_2) = P(HH) + P(HT) + P(TH)$

$\qquad = \frac{1}{4} + \frac{1}{4} + \frac{1}{4} = \frac{3}{4}$

(C) E_3 = {HH, HT, TH, TT} = S

$\quad P(E_3) = P(S) = 1$ $\frac{1}{4} + \frac{1}{4} + \frac{1}{4} + \frac{1}{4} = 1$

(D) E_3 = Getting three heads = $\varnothing$ **Empty set**

$\quad P(\varnothing) = 0$

> ### STEPS FOR FINDING PROBABILITIES OF EVENTS
>
> **Step 1.** Set up an appropriate sample space S for the experiment.
>
> **Step 2.** Assign acceptable probabilities to the simple events in S.
>
> **Step 3.** To obtain the probability of an arbitrary event E, add the probabilities of the simple events in E.

The function P defined in steps 2 and 3 is called a **probability function.** The domain of this function is all possible events in the sample space S, and the range is a set of real numbers between 0 and 1, inclusive.

MATCHED PROBLEM 3

Return to Matched Problem 1, recording the boy–girl composition of families with two children and the sample space

$$S = \{BB, BG, GB, GG\}$$

Statistics from the U.S. Census Bureau indicate that an acceptable and reasonable probability for this sample space is

Simple event, e_i	BB	BG	GB	GG
$P(e_i)$	.26	.25	.25	.24

Find the probabilities of the following events:

(A) E_1 = Having at least one girl in the family

(B) E_2 = Having at most one girl in the family

(C) E_3 = Having two children of the same sex in the family

› Making Equally Likely Assumptions

In tossing a nickel and dime (Example 3), we assigned the same probability, $\frac{1}{4}$, to each simple event in the sample space S = {HH, HT, TH, TT}. By assigning the same probability to each simple event in S, we are actually making the assumption that each simple event is as likely to occur as any other. We refer to this as an **equally likely assumption.** In general, we have Definition 4.

> **DEFINITION 4** Probability of a Simple Event Under an Equally Likely Assumption

If, in a sample space

$$S = \{e_1, e_2, \cdots, e_n\}$$

with n elements, we assume each simple event e_i is as likely to occur as any other, then we assign the probability $1/n$ to each. That is,

$$P(e_i) = \frac{1}{n}$$

Under an equally likely assumption, we can develop a very useful formula for finding probabilities of arbitrary events associated with a sample space S. Consider the following example.

If a single die is rolled and we assume each face is as likely to come up as any other, then for the sample space

$$S = \{1, 2, 3, 4, 5, 6\}$$

we assign a probability of $\frac{1}{6}$ to each simple event, because there are six simple events. Then the probability of

$$E = \text{Rolling a prime number} = \{2, 3, 5\}$$

is

$$P(E) = P(2) + P(3) + P(5) = \tfrac{1}{6} + \tfrac{1}{6} + \tfrac{1}{6} = \tfrac{3}{6} = \tfrac{1}{2}$$

So, under the assumption that each simple event is as likely to occur as any other, the computation of the probability of the occurrence of any event E in a sample space S is the number of elements in E divided by the number of elements in S.

> **THEOREM 1** Probability of an Arbitrary Event Under an Equally Likely Assumption

If we assume each simple event in sample space S is as likely to occur as any other, then the probability of an arbitrary event E in S is given by

$$P(E) = \frac{\text{Number of elements in } E}{\text{Number of elements in } S} = \frac{n(E)}{n(S)}$$

EXAMPLE 4 **Finding Probabilities of Events**

If in rolling two dice we assume each simple event in the sample space shown in Figure 1 on page 748 is as likely as any other, find the probabilities of the following events:

(A) E_1 = A sum of 7 turns up (B) E_2 = A sum of 11 turns up

(C) E_3 = A sum less than 4 turns up (D) E_4 = A sum of 12 turns up

SOLUTIONS Referring to Figure 1, we see that:

(A) $P(E_1) = \dfrac{n(E_1)}{n(S)} = \dfrac{6}{36} = \dfrac{1}{6}$ (B) $P(E_2) = \dfrac{n(E_2)}{n(S)} = \dfrac{2}{36} = \dfrac{1}{18}$

(C) $P(E_3) = \dfrac{n(E_3)}{n(S)} = \dfrac{3}{36} = \dfrac{1}{12}$ (D) $P(E_4) = \dfrac{n(E_4)}{n(S)} = \dfrac{1}{36}$

MATCHED PROBLEM 4 Under the conditions in Example 4, find the probabilities of the following events:

(A) E_5 = A sum of 5 turns up

(B) E_6 = A sum that is a prime number greater than 7 turns up

>>> EXPLORE-DISCUSS 1 A box contains four red balls and seven green balls. A ball is drawn at random and then, without replacing the first ball, a second ball is drawn. Discuss whether or not the equally likely assumption would be appropriate for the sample space $S = \{RR, RG, GR, GG\}$.

We now turn to some examples that make use of the counting techniques developed in Section 11-4.

EXAMPLE 5 ## Drawing Cards

In drawing 5 cards from a 52-card deck without replacement, what is the probability of getting five spades?

SOLUTION Let the sample space S be the set of all 5-card hands from a 52-card deck. Because the order in a hand does not matter, $n(S) = C_{52,5}$. The event we seek is

$$E = \text{Set of all 5-card hands from 13 spades}$$

Again, the order does not matter and $n(E) = C_{13,5}$. Assuming that each 5-card hand is as likely as any other,

$$P(E) = \frac{n(E)}{n(S)} = \frac{C_{13,5}}{C_{52,5}} = \frac{13!/5!8!}{52!/5!47!} = \frac{13!}{5!8!} \cdot \frac{5!47!}{52!} \approx .0005$$

MATCHED PROBLEM 5 In drawing 7 cards from a 52-card deck without replacement, what is the probability of getting seven hearts?

EXAMPLE 6 ## Selecting Committees

The board of regents of a university is made up of 12 men and 16 women. If a committee of six is chosen at random, what is the probability that it will contain three men and three women?

SOLUTION Let S = Set of all 6-person committees out of 28 people:

$$n(S) = C_{28,6}$$

Let E = Set of all 6-person committees with 3 men and 3 women. To find $n(E)$, we use the multiplication principle and the following two operations:

O_1: Select 3 men out of the 12 available N_1: $C_{12,3}$
O_2: Select 3 women out of the 16 available N_2: $C_{16,3}$

So

$$n(E) = C_{12,3} \cdot C_{16,3}$$

and

$$P(E) = \frac{n(E)}{n(S)} = \frac{C_{12,3} \cdot C_{16,3}}{C_{28,6}} \approx .327$$

MATCHED PROBLEM 6

What is the probability that the committee in Example 6 will have four men and two women?

› Finding or Approximating Empirical Probability

In the earlier examples in this section we made a reasonable assumption about an experiment and used deductive reasoning to assign probabilities. For example, it is reasonable to assume that an ordinary coin will come up heads about as often as it will come up tails. Probabilities determined in this manner are called **theoretical probabilities.** No experiments are ever conducted. But what if the theoretical probabilities are not obvious? Then we assign probabilities to simple events based on the results of actual experiments. Probabilities determined from the results of actually performing an experiment are called **empirical probabilities.** As an experiment is repeated over and over, the percentage of times an event occurs may get closer and closer to a single fixed number. If so, this single fixed number is generally called the **actual probability** of the event.

››› EXPLORE-DISCUSS 2

Like a coin, a thumbtack tossed into the air will land in one of two positions, point up or point down [Fig. 2(a)]. Unlike a coin, we would not expect both events to occur with the same frequency. Indeed, the frequencies of landing point up and point down may well vary from one thumbtack to another [Fig. 2(b)]. Find two thumbtacks of different sizes and guess which one is likely to land point up more frequently. Then toss each tack 100 times and record the number of times each lands point up. Did the experiment confirm your initial guess?

(a) Point up or point down (b) Two different tacks

› **Figure 2**

Suppose when tossing one of the thumbtacks in Explore-Discuss 2, we observe that the tack lands point up 43 times and point down 57 times. Based on this experiment, it seems reasonable to say that for this particular thumbtack

$$P(\text{Point up}) = \frac{43}{100} = .43$$

$$P(\text{Point down}) = \frac{57}{100} = .57$$

Probability assignments based on the results of repeated trials of an experiment are called **approximate empirical probabilities.**

In general, if we conduct an experiment n times and an event E occurs with **frequency** $f(E)$, then the ratio $f(E)/n$ is called the **relative frequency** of the occurrence of event E in n trials. We define the **empirical probability** of E, denoted by $P(E)$, by the number, if it exists, that the relative frequency $f(E)/n$ approaches as n gets larger and larger. Of course, for any particular n, the relative frequency $f(E)/n$ is generally only approximately equal to $P(E)$. However, as n increases, we expect the approximation to improve.

> ### › DEFINITION 5 Empirical Probability
>
> If $f(E)$ is the frequency of event E in n trials, then
>
> $$P(E) \approx \frac{\text{Frequency of occurrence of } E}{\text{Total number of trials}} = \frac{f(E)}{n}$$

If we can also deduce theoretical probabilities for an experiment, then we expect the approximate empirical probabilities to approach the theoretical probabilities. If this does not happen, then we should begin to suspect the manner in which the theoretical probabilities were computed. If $P(E)$ is the theoretical probability of an event E and the experiment is performed n times, then the **expected frequency** of the occurrence of E is $n \cdot P(E)$.

EXAMPLE	7	**Finding Approximate Empirical and Theoretical Probabilities**

Two coins are tossed 500 times with the following frequencies of outcomes:

Two heads: 121

One head: 262

Zero heads: 117

(A) Compute the approximate empirical probability for each outcome.

(B) Compute the theoretical probability for each outcome.

(C) Compute the expected frequency for each outcome.

SOLUTIONS

(A) $P(\text{two heads}) \approx \dfrac{121}{500} = .242$

$P(\text{one head}) \approx \dfrac{262}{500} = .524$

$P(\text{zero heads}) \approx \dfrac{117}{500} = .234$

(B) A sample space of equally likely simple events is $S = \{HH, HT, TH, TT\}$. Let

$$E_1 = \text{two heads} = \{HH\}$$
$$E_2 = \text{one head} = \{HT, TH\}$$
$$E_3 = \text{zero heads} = \{TT\}$$

Then

$$P(E_1) = \frac{n(E_1)}{n(S)} = \frac{1}{4} = .25$$

$$P(E_2) = \frac{n(E_2)}{n(S)} = \frac{2}{4} = .50$$

$$P(E_3) = \frac{n(E_3)}{n(S)} = \frac{1}{4} = .25$$

(C) The expected frequencies are

$$E_1: 500(.25) = 125$$
$$E_2: 500(.5) \;\; = 250$$
$$E_3: 500(.25) = 125$$

The actual frequencies obtained from performing the experiment are reasonably close to the expected frequencies. Increasing the number of trials of the experiment would most likely produce even better approximations.

MATCHED PROBLEM 7 One die is rolled 500 times with the following frequencies of outcomes:

Outcome	1	2	3	4	5	6
Frequency	89	83	77	91	72	88

(A) Compute the approximate empirical probability for each outcome.

(B) Compute the theoretical probability for each outcome.

(C) Compute the expected frequency for each outcome.

Technology Connections

The data in Example 7 were *not* generated by tossing two coins 500 times. Instead, the experiment was simulated by a random number generator on a graphing calculator. The command randint (0, 1, 500) produces a random sequence of 500 terms; each term is 0 or 1 with equal liklihood. Thinking of 1 as heads and 0 as tails, such a sequence represents 500 tosses of a single coin. Adding two such sequences together produces a sequence of 500 terms in which each term represents the number of heads in a toss of two coins

[see Fig. 3(a)]. We determine the frequency of each outcome (0, 1, or 2 heads) in 500 tosses of two coins as follows: first, we construct a histogram [Figs. 3(b) and 3(c)], then we use the TRACE command to read off the frequencies [Figs. 3(d), 3(e), and 3(f)]. Compare with the data of Example 7.

If you perform the same simulation on your graphing calculator, you are not likely to get exactly the same results. But the approximate empirical probabilities you obtain will be close to the theoretical probabilities.

(a) Generating the random numbers

(b) Setting up the histogram

(c) Selecting the window variables

(d) 0 heads: 117

(e) 1 head: 262

(f) 2 heads: 121

› **Figure 3** Simulating 500 tosses of two coins.

EXAMPLE 8 Empirical Probabilities for an Insurance Company

An insurance company selected 1,000 drivers at random in a particular city to determine a relationship between age and accidents. The data obtained are listed in Table 1. Compute the approximate empirical probabilities of the following events for a driver chosen at random in the city:

(A) E_1: being under 20 years old *and* having exactly three accidents in 1 year

(B) E_2: being 30–39 years old *and* having one or more accidents in 1 year

(C) E_3: having no accidents in 1 year

(D) E_4: being under 20 years old *or** having exactly three accidents in 1 year

Table 1

Age	Accidents in 1 Year				
	0	**1**	**2**	**3**	**Over 3**
Under 20	50	62	53	35	20
20–29	64	93	67	40	36
30–39	82	68	32	14	4
40–49	38	32	20	7	3
Over 49	43	50	35	28	24

SOLUTIONS

(A) $P(E_1) \approx \dfrac{35}{1,000} = .035$

(B) $P(E_2) \approx \dfrac{68 + 32 + 14 + 4}{1,000} = .118$

(C) $P(E_3) \approx \dfrac{50 + 64 + 82 + 38 + 43}{1,000} = .277$

(D) $P(E_4) \approx \dfrac{50 + 62 + 53 + 35 + 20 + 40 + 14 + 7 + 28}{1,000} = .309$

Notice that in this type of problem, which is typical of many realistic problems, approximate empirical probabilities are the only type we can compute.

MATCHED PROBLEM 8

Referring to Table 1 in Example 8, compute the approximate empirical probabilities of the following events for a driver chosen at random in the city:

(A) E_1: being under 20 years old with no accidents in 1 year

(B) E_2: being 20–29 years old and having fewer than two accidents in 1 year

(C) E_3: not being over 49 years old

Approximate empirical probabilities are often used to test theoretical probabilities. Equally likely assumptions may not be justified in reality. In addition to this use, there are many situations in which it is either very difficult or impossible to compute the theoretical

*Interpret "*or*" in its inclusive sense, as customary in mathematics (a driver who is both under 20 and has three accidents must be counted once in the frequency of E_4).

probabilities for given events. For example, insurance companies use past experience to establish approximate empirical probabilities to predict future accident rates; baseball teams use batting averages, which are approximate empirical probabilities based on past experience, to predict the future performance of a player; and pollsters use approximate empirical probabilities to predict outcomes of elections.

ANSWERS TO MATCHED PROBLEMS

1. (A) $S_1 = \{BB, BG, GB, GG\}$;

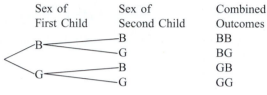

Sex of First Child	Sex of Second Child	Combined Outcomes
B	B	BB
	G	BG
G	B	GB
	G	GG

(B) $S_2 = \{0, 1, 2\}$ (C) $S_3 = \{A, D\}$ (D) The sample space in part A.

2. (A) $\{(4, 1), (3, 2), (2, 3), (1, 4)\}$ (B) $\{(6, 5), (5, 6)\}$

3. (A) .74 (B) .76 (C) .5 **4.** (A) $P(E_5) = \frac{1}{9}$ (B) $P(E_6) = \frac{1}{18}$

5. $C_{13,7}/C_{52,7} \approx .000013$ **6.** $(C_{12,4} \cdot C_{16,2})/C_{28,6} \approx .158$

7. (A) $P(E_1) \approx .178, P(E_2) \approx .166, P(E_3) \approx .154, P(E_4) \approx .182, P(E_5) \approx .144, P(E_6) \approx .176$
(B) $\frac{1}{6} \approx .167$ for each (C) 83.3 for each

8. (A) $P(E_1) \approx .05$ (B) $P(E_2) \approx .157$ (C) $P(E_3) \approx .82$

11-5 Exercises

1. What is a sample space?

2. Explain in your own words how the probability of an event is defined in terms of probabilities of simple events.

3. Explain the difference between a theoretical probability and an empirical probability.

4. What is an equally likely assumption?

5. A single fair die is rolled. What is the probability of getting a one or a six?

6. A single fair die is rolled. What is the probability of getting a number greater than three?

7. A single card is drawn from a standard 52-card deck. What is the probability of getting a red card?

8. A single card is drawn from a standard 52-card deck. What is the probability of getting a club?

9. A fair coin is tossed twice. What is the probability of getting two heads?

10. A fair coin is tossed twice. What is the probability of getting at least one head?

11. Two fair dice are rolled. What is the probability of getting doubles?

12. Two fair dice are rolled. What is the probability of getting double sixes?

13. A single card is drawn from a standard 52-card deck. What is the probability of getting a king or a queen?

14. A single card is drawn from a standard 52-card deck. What is the probability of getting a numbered card (that is, a two through ten)?

15. A fair coin is tossed three times. What is the probability of getting exactly two tails?

16. A fair coin is tossed three times. What is the probability of getting three tails?

17. How would you interpret $P(E) = 1$?

18. How would you interpret $P(E) = 0$?

19. A spinner can land on four different colors: red (R), green (G), yellow (Y), and blue (B). If we do not assume each color is as likely to turn up as any other, which of the following probability assignments have to be rejected, and why?
(A) $P(R) = .15, P(G) = -.35, P(Y) = .50, P(B) = .70$
(B) $P(R) = .32, P(G) = .28, P(Y) = .24, P(B) = .30$
(C) $P(R) = .26, P(G) = .14, P(Y) = .30, P(B) = .30$

20. Under the probability assignments in Problem 19, part C, what is the probability that the spinner will not land on blue?

21. Under the probability assignments in Problem 19, part C, what is the probability that the spinner will land on red or yellow?

22. Under the probability assignments in Problem 19, part C, what is the probability that the spinner will not land on red or yellow?

23. A ski jumper has jumped over 300 feet in 25 out of 250 jumps. What is the approximate empirical probability of the next jump being over 300 feet?

24. In a certain city there are 4,000 youths between 16 and 20 years old who drive cars. If 560 of them were involved in accidents last year, what is the approximate empirical probability of a youth in this age group being involved in an accident this year?

25. Out of 420 times at bat, a baseball player gets 189 hits. What is the approximate empirical probability that the player will get a hit next time at bat?

26. In a medical experiment, a new drug is found to help 2,400 out of 3,000 people. If a doctor prescribes the drug for a particular patient, what is the approximate empirical probability that the patient will be helped?

27. A small combination lock on a suitcase has three wheels, each labeled with the 10 digits from 0 to 9. If an opening combination is a particular sequence of three digits with no repeats, what is the probability of a person guessing the right combination?

28. A combination lock has five wheels, each labeled with the 10 digits from 0 to 9. If an opening combination is a particular sequence of five digits with no repeats, what is the probability of a person guessing the right combination?

Problems 29–34 involve an experiment consisting of dealing 5 cards from a standard 52-card deck. In Problems 29–32, what is the probability of being dealt:

29. Five black cards

30. Five hearts

31. Five face cards if an ace is considered to be a face card.

32. Five nonface cards if an ace is considered to be a one and not a face card.

33. If we are interested in the number of aces in a 5-card hand, would $S = \{0, 1, 2, 3, 4\}$ be an acceptable sample space? Would it be an equally-likely sample space? Explain.

34. If we are interested in the number of black cards in a 5-card hand, would $S = \{0, 1, 2, 3, 4, 5\}$ be an acceptable sample space? Would it be an equally-likely sample space? Explain.

35. If four-digit numbers less than 5,000 are randomly formed from the digits 1, 3, 5, 7, and 9, what is the probability of forming a number divisible by 5? Digits may be repeated; for example, 1,355 is acceptable.

36. If code words of four letters are generated at random using the letters A, B, C, D, E, and F, what is the probability of forming a word without a vowel in it? Letters may be repeated.

37. Suppose five thank-you notes are written and five envelopes are addressed. Accidentally, the notes are randomly inserted into the envelopes and mailed without checking the addresses. What is the probability that all five notes will be inserted into the correct envelopes?

38. Suppose six people check their coats in a checkroom. If all claim checks are lost and the six coats are randomly returned, what is the probability that all six people will get their own coats back?

In Problems 39–50, an experiment consists of rolling two fair dice. Let a and b denote the numbers of dots on the two sides facing up. Use the sample space shown in Figure 1 on page 748 to find the probability of each event.

39. The sum of a and b is 3.

40. The sum of a and b is 5.

41. The sum of a and b is greater than 9.

42. The sum of a and b is less than 6.

43. The product of a and b is 12.

44. The product of a and b is 6.

45. The product of a and b is less than 5.

46. The product of a and b is greater than 15.

47. $a = b$

48. $a \neq b$

49. At least one of a or b is a 6.

50. Exactly one of a or b is a 6.

51. Five thousand people work in a large auto plant. An individual is selected at random and his or her birthday (month and day, not year) is recorded. Set up an appropriate sample space for this experiment and assign acceptable probabilities to the simple events.

52. In a hotly contested three-way race for governor of Minnesota, the leading candidates are running neck-and-neck while the third candidate is receiving half the support of either of the others. Registered voters are chosen at random and are asked for which of the three they are most likely to vote. Set up an appropriate sample space for the random survey experiment and assign acceptable probabilities to the simple events.

53. A pair of dice is rolled 500 times with the following frequencies:

Sum	2	3	4	5	6	7	8	9	10	11	12
Frequency	11	35	44	50	71	89	72	52	36	26	14

(A) Compute the approximate empirical probability for each outcome.

(B) Compute the theoretical probability for each outcome, assuming fair dice.

(C) Compute the expected frequency of each outcome.

(D) Describe how a random number generator could be used to simulate this experiment. If your graphing calculator has a random number generator, use it to simulate 500 tosses of a pair of dice and compare your results with part C.

54. Three coins are flipped 500 times with the following frequencies of outcomes:

Three heads: 58 Two heads: 198
One head: 190 Zero heads: 54

(A) Compute the approximate empirical probability for each outcome.

(B) Compute the theoretical probability for each outcome, assuming fair coins.

(C) Compute the expected frequency of each outcome.

(D) Describe how a random number generator could be used to simulate this experiment. If your graphing calculator has a random number generator, use it to simulate 500 tosses of three coins and compare your results with part C.

55. (A) Is it possible to get 29 heads in 30 flips of a fair coin? Explain.
(B) If you flip a coin 50 times and get 42 heads, would you suspect that the coin was unfair? Why or why not? If you suspect an unfair coin, what empirical probabilities would you assign to the simple events of the sample space?

56. (A) Is it possible to get nine double sixes in 12 rolls of a pair of fair dice? Explain.
(B) If you roll a pair of dice 40 times and get 14 double sixes, would you suspect that the dice were unfair? Why or why not? If you suspect unfair dice, what empirical probability would you assign to the event of rolling a double six?

An experiment consists of tossing three fair coins, but one of the three coins has a head on both sides. Compute the probabilities of obtaining the indicated results in Problems 57–62.

57. One head
58. Two heads
59. Three heads
60. Zero heads
61. More than one head
62. More than one tail

An experiment consists of rolling two fair dice and adding the dots on the two sides facing up. Each die has one dot on two opposite faces, two dots on two opposite faces, and three dots on two opposite faces. Compute the probabilities of obtaining the indicated sums in Problems 63–70.

63. 2 **64.** 3 **65.** 4 **66.** 5 **67.** 6

68. 7 **69.** An odd sum **70.** An even sum

An experiment consists of dealing 5 cards from a standard 52-card deck. In Problems 71–78, what is the probability of being dealt the following cards?

71. Five cards, jacks through aces

72. Five cards, 2 through 10

73. Four aces

74. Four of a kind

75. Straight flush, ace high; that is, 10, jack, queen, king, ace in one suit

76. Straight flush, starting with 2; that is, 2, 3, 4, 5, 6 in one suit

77. Two aces and three queens

78. Two kings and three aces

APPLICATIONS

79. MARKET ANALYSIS A company selected 1,000 households at random and surveyed them to determine a relationship between income level and the number of television sets in a home. The information gathered is listed in the table:

	Televisions per Household				
Yearly Income ($)	0	1	2	3	Above 3
Less than 12,000	0	40	51	11	0
12,000–19,999	0	70	80	15	1
20,000–39,999	2	112	130	80	12
40,000–59,999	10	90	80	60	21
60,000 or more	30	32	28	25	20

Compute the approximate empirical probabilities:
(A) Of a household earning $12,000–$19,999 per year *and* owning exactly three television sets
(B) Of a household earning $20,000–$39,999 per year *and* owning more than one television set
(C) Of a household earning $60,000 or more per year *or* owning more than three television sets
(D) Of a household not owning zero television sets

80. MARKET ANALYSIS Use the sample results in Problem 79 to compute the approximate empirical probabilities:
(A) Of a household earning $40,000–$59,999 per year *and* owning zero television sets
(B) Of a household earning $12,000–$39,999 per year *and* owning more than two television sets
(C) Of a household earning less than $20,000 per year *or* owning exactly two television sets
(D) Of a household not owning more than three television sets

11-6 The Binomial Formula

> Using Pascal's Triangle
> The Binomial Formula
> Proving the Binomial Formula

In a surprising number of areas in math, it turns out to be useful to expand expressions of the form $(a + b)^n$, where n is a natural number. This is known as a *binomial expansion*. Expanding a binomial is pretty straightforward for small values of n, but gets hard very

quickly as n increases. The good news is that it turns out that the coefficients in such an expansion are related to counting techniques that we have already learned about.

› Using Pascal's Triangle

Let's begin by expanding $(a + b)^n$ for the first few values of n. We include $n = 0$, which is not a natural number, for reasons of completeness that will become apparent later.

$$\begin{aligned}(a + b)^0 &= 1\\(a + b)^1 &= a + b\\(a + b)^2 &= a^2 + 2ab + b^2\\(a + b)^3 &= a^3 + 3a^2b + 3ab^2 + b^3\end{aligned}\qquad(1)$$

››› EXPLORE-DISCUSS 1

Based on the expansions in equations (1), how many terms would you expect $(a + b)^n$ to have? What is the first term? What is the last term?

> **Figure 1** Pascal's triangle.

Now let's examine just the coefficients of the expansions in equations (1) arranged in a form that is usually referred to as **Pascal's triangle** (Fig. 1).

It is convenient to refer to the top row of Pascal's triangle (containing a single 1) as row 0. Then row 1 is "1 1," row 2 is "1 2 1," and row 3 is "1 3 3 1." For n a natural number, the first two entries of row n are 1 and n.

››› EXPLORE-DISCUSS 2

Refer to Figure 1.

(A) How is the middle element of row 2 related to the elements in the row above?

(B) How are the two inner elements of row 3 related to the elements in the row above?

(C) Based on your observations in parts A and B, conjecture the entries of row 4 and row 5. Check your conjecture by expanding $(a + b)^4$ and $(a + b)^5$.

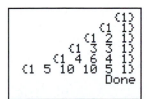

> **Figure 2**

Many students find Pascal's triangle a useful tool for determining the coefficients in the expansion of $(a + b)^n$, especially for small values of n. Figure 2 contains output from a program called PASCAL.* You should recognize the output in the table—it is the first six lines of Pascal's triangle. The major drawback of using this triangle, whether it is generated by hand or on a graphing calculator, is that to find the elements in a given row, you must write out all the preceding rows. It would be useful to find a formula that gives the coefficients for a binomial expansion directly. Fortunately, such a formula exists—the combination formula $C_{n,r}$ introduced in Section 11-4.

› The Binomial Formula

When working with binomial expansions, it is customary to use the notation $\binom{n}{r}$ for $C_{n,r}$. Recall the combination formula from Section 11-4.

*Programs for TI-84 and TI-86 graphing calculators can be found at the website for this book.

COMBINATION FORMULA For nonnegative integers r and n, $0 \leq r \leq n$,

$$\binom{n}{r} = C_{n,r} = \frac{n!}{r!(n-r)!}$$

Theorem 1 establishes that the coefficients in a binomial expansion can always be expressed in terms of the combination formula. This is a very important theoretical result and a very practical tool. As we will see, using this theorem in conjunction with a graphing calculator provides a very efficient method for expanding binomials.

> **THEOREM 1** Binomial Formula

For n a positive integer

$$(a + b)^n = \sum_{k=0}^{n} \binom{n}{k} a^{n-k} b^k$$

We defer the proof of Theorem 1 until the end of this section. Because the values of the combination formula are the coefficients in a binomial expansion, it is natural to call them **binomial coefficients.**

EXAMPLE 1 Using the Binomial Formula

Use the binomial formula to expand $(x + y)^6$.

$$(x + y)^6 = \sum_{k=0}^{6} \binom{6}{k} x^{6-k} y^k$$

$$= \binom{6}{0} x^6 y^0 + \binom{6}{1} x^5 y + \binom{6}{2} x^4 y^2 + \binom{6}{3} x^3 y^3 + \binom{6}{4} x^2 y^4 + \binom{6}{5} xy^5 + \binom{6}{6} x^0 y^6$$

$$= x^6 + 6x^5 y + 15x^4 y^2 + 20x^3 y^3 + 15x^2 y^4 + 6xy^5 + y^6$$

Note that the coefficients (1, 6, 15, 20, 15, 6, 1) are the entries of row 6 of Pascal's triangle.

MATCHED PROBLEM 1 Use the binomial formula to expand $(x + 1)^5$.

EXAMPLE 2 Using the Binomial Formula

Use the binomial formula to expand $(3p - 2q)^4$.

SOLUTION

$$(3p - 2q)^4 = [(3p) + (-2q)]^4$$

$$= \sum_{k=0}^{4} \binom{4}{k} (3p)^{4-k} (-2q)^k$$

$$= \sum_{k=0}^{4} \binom{4}{k} 3^{4-k} (-2)^k p^{4-k} q^k$$

$$= 1(3)^4(-2)^0 p^4 q^0 + 4(3)^3(-2)p^3 q + 6(3)^2(-2)^2 p^2 q^2$$
$$+ 4(3)(-2)^3 pq^3 + 1(3)^0(-2)^4 p^0 q^4$$

$$= 81p^4 - 216p^3 q + 216p^2 q^2 - 96pq^3 + 16q^4$$

Note that the coefficients (81, −216, 216, −96, 16) are formed by multiplying the entries in row 4 of Pascal's triangle (1, 4, 6, 4, 1) by the appropriate powers of 3 and −2.

Technology Connections

The table feature on a graphing calculator provides an efficient alternative to calculating the coefficients of Example 2 one by one (Fig. 3).

> Figure 3 $y_1 = C_{4,x} 3^{4-x}(-2)^x$.

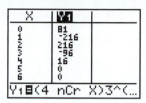

MATCHED PROBLEM 2 Use the binomial formula to expand $(2m - 5n)^3$.

>>> EXPLORE-DISCUSS 3

(A) Compute each term and also the sum of the alternating series

$$\binom{6}{0} - \binom{6}{1} + \binom{6}{2} - \cdots + \binom{6}{6}$$

(B) What result about an alternating series can be deduced by letting $a = 1$ and $b = -1$ in the binomial formula?

EXAMPLE 3 **Using the Binomial Formula**

Find the term containing x^9 in the expansion of $(x + 3)^{14}$.

SOLUTION In the expansion

$$(x + 3)^{14} = \sum_{k=0}^{14} \binom{14}{k} x^{14-k} 3^k$$

the exponent of x is 9 when $k = 5$. So the term containing x^9 is

$$\binom{14}{5} x^9 3^5 = (2{,}002)(243)x^9 = 486{,}486x^9$$

MATCHED PROBLEM 3 Find the term containing y^8 in the expansion of $(2 + y)^{14}$.

EXAMPLE 4 **Using the Binomial Formula**

If the terms in the expansion of $(x - 2)^{20}$ are arranged in decreasing powers of x, find the fourth term and the sixteenth term.

SOLUTION In the expansion of $(a + b)^n$, the exponent of b in the rth term is $r - 1$ and the exponent of a is $n - (r - 1)$. Therefore

Fourth term:

$$\binom{20}{3} x^{17} (-2)^3$$

$$= \frac{20 \cdot 19 \cdot 18}{3 \cdot 2 \cdot 1} x^{17}(-8)$$

$$= -9{,}120x^{17}$$

Sixteenth term:

$$\binom{20}{15} x^5 (-2)^{15}$$

$$= \frac{20 \cdot 19 \cdot 18 \cdot 17 \cdot 16}{5 \cdot 4 \cdot 3 \cdot 2 \cdot 1} x^5(-32{,}768)$$

$$= -508{,}035{,}072x^5$$

MATCHED PROBLEM 4 If the terms in the expansion of $(u - 1)^{18}$ are arranged in decreasing powers of u, find the fifth term and the twelfth term.

› Proving the Binomial Formula

We now prove that the binomial formula holds for all natural numbers n using mathematical induction.

PROOF State the conjecture.

$$P_n: (a + b)^n = \sum_{j=0}^{n} \binom{n}{j} a^{n-j} b^j$$

PART 1 Show that P_1 is true.

$$\sum_{j=0}^{1} \binom{1}{j} a^{1-j} b^j = \binom{1}{0} a + \binom{1}{1} b = a + b = (a + b)^1$$

P_1 is true.

PART 2 Show that if P_k is true, then P_{k+1} is true.

$$P_k: (a + b)^k = \sum_{j=0}^{k} \binom{k}{j} a^{k-j} b^j \qquad \text{Assume } P_k \text{ is true.}$$

$$P_{k+1}: (a + b)^{k+1} = \sum_{j=0}^{k+1} \binom{k+1}{j} a^{k+1-j} b^j \qquad \text{Show } P_{k+1} \text{ is true.}$$

We begin by multiplying both sides of P_k by $(a + b)$:

$$(a + b)^k (a + b) = \left[\sum_{j=0}^{k} \binom{k}{j} a^{k-j} b^j \right] (a + b)$$

The left side of this equation is the left side of P_{k+1}. Now we multiply out the right side of the equation and try to obtain the right side of P_{k+1}:

$$(a + b)^{k+1} = \left[\binom{k}{0} a^k + \binom{k}{1} a^{k-1} b + \binom{k}{2} a^{k-2} b^2 + \cdots + \binom{k}{k} b^k \right] (a + b) \qquad \text{Use the distributive property.}$$

$$= \left[\binom{k}{0} a^{k+1} + \binom{k}{1} a^k b + \binom{k}{2} a^{k-1} b^2 + \cdots + \binom{k}{k} a b^k \right]$$

$$+ \left[\binom{k}{0} a^k b + \binom{k}{1} a^{k-1} b^2 + \cdots + \binom{k}{k-1} a b^k + \binom{k}{k} b^{k+1} \right] \qquad \text{Combine like terms.}$$

$$= \binom{k}{0} a^{k+1} + \left[\binom{k}{0} + \binom{k}{1} \right] a^k b + \left[\binom{k}{1} + \binom{k}{2} \right] a^{k-1} b^2 + \cdots$$

$$+ \left[\binom{k}{k-1} + \binom{k}{k} \right] a b^k + \binom{k}{k} b^{k+1}$$

We now use the following facts (the proofs are left as exercises; see Problems 63–65, Exercises 11-6).

$$\binom{k}{r-1} + \binom{k}{r} = \binom{k+1}{r} \qquad \binom{k}{0} = \binom{k+1}{0} \qquad \binom{k}{k} = \binom{k+1}{k+1}$$

to rewrite the right side as

$$\binom{k+1}{0} a^{k+1} + \binom{k+1}{1} a^k b + \binom{k+1}{2} a^{k-1} b^2 + \cdots$$

$$+ \binom{k+1}{k} a b^k + \binom{k+1}{k+1} b^{k+1} = \sum_{j=0}^{k+1} \binom{k+1}{j} a^{k+1-j} b^j$$

Because the right side of the last equation is the right side of P_{k+1}, we have shown that P_{k+1} follows from P_k.

CONCLUSION

P_n is true. That is, the binomial formula holds for all positive integers n.

ANSWERS TO MATCHED PROBLEMS

1. $x^5 + 5x^4 + 10x^3 + 10x^2 + 5x + 1$ 2. $8m^3 - 60m^2n + 150mn^2 - 125n^3$
3. $192,192y^8$ 4. $3,060u^{14}; -31,824u^7$

11-6 Exercises

1. What is a binomial?

2. What is a binomial coefficient?

3. Explain how the entries in Pascal's triangle are generated.

4. How can Pascal's triangle be used to expand $(a + b)^5$?

In Problems 5–12, use Pascal's triangle to evaluate each expression.

5. $\dbinom{8}{3}$ 6. $\dbinom{8}{4}$ 7. $\dbinom{9}{6}$ 8. $\dbinom{9}{7}$

9. $C_{7,5}$ 10. $C_{7,3}$ 11. $C_{9,0}$ 12. $C_{10,10}$

In Problems 13–20, evaluate each expression.

13. $\dbinom{13}{3}$ 14. $\dbinom{13}{9}$ 15. $\dbinom{12}{4}$ 16. $\dbinom{12}{11}$

17. $C_{52,3}$ 18. $C_{52,4}$ 19. $C_{12,6}$ 20. $C_{12,11}$

Expand Problems 21–32 using the binomial formula.

21. $(m + n)^3$ 22. $(x + 2)^3$ 23. $(2x - 3y)^3$

24. $(3u + 2v)^3$ 25. $(x - 2)^4$ 26. $(x - y)^4$

27. $(m + 3n)^4$ 28. $(3p - q)^4$ 29. $(2x - y)^5$

30. $(2x - 1)^5$ 31. $(m + 2n)^6$ 32. $(2x - y)^6$

In Problems 33–42, find the term of the binomial expansion containing the given power of x.

33. $(x + 1)^7; x^4$ 34. $(x + 1)^8; x^5$ 35. $(2x - 1)^{11}; x^6$

36. $(3x + 1)^{12}; x^7$ 37. $(2x + 3)^{18}; x^{14}$ 38. $(3x - 2)^{17}; x^5$

39. $(x^2 - 1)^6; x^8$ 40. $(x^2 - 1)^9; x^7$ 41. $(x^2 + 1)^9; x^{11}$

42. $(x^2 + 1)^{10}; x^{14}$

In Problems 43–50, find the indicated term in each expansion if the terms of the expansion are arranged in decreasing powers of the first term in the binomial.

43. $(u + v)^{15}$; seventh term 44. $(a + b)^{12}$; fifth term

45. $(2m + n)^{12}$; eleventh term 46. $(x + 2y)^{20}$; third term

47. $[(w/2) - 2]^{12}$; seventh term 48. $(x - 3)^{10}$; fourth term

49. $(3x - 2y)^8$; sixth term 50. $(2p - 3q)^7$; fourth term

In Problems 51–54, use the binomial formula to expand and simplify the difference quotient

$$\frac{f(x + h) - f(x)}{h}$$

for the indicated function f. Discuss the behavior of the simplified form as h approaches 0.

51. $f(x) = x^3$ 52. $f(x) = x^4$

53. $f(x) = x^5$ 54. $f(x) = x^6$

In Problems 55–58, use a graphing calculator to graph each sequence and to display it in table form.

55. Find the number of terms of the sequence

$$\dbinom{20}{0}, \dbinom{20}{1}, \dbinom{20}{2}, \dots, \dbinom{20}{20}$$

that are greater than one-half of the largest term.

56. Find the number of terms of the sequence

$$\dbinom{40}{0}, \dbinom{40}{1}, \dbinom{40}{2}, \dots, \dbinom{40}{40}$$

that are greater than one-half of the largest term.

57. (A) Find the largest term of the sequence $a_0, a_1, a_2, \dots, a_{10}$ to three decimal places, where

$$a_k = \dbinom{10}{k}(0.6)^{10-k}(0.4)^k$$

(B) According to the binomial formula, what is the sum of the series $a_0 + a_1 + a_2 + \cdots + a_{10}$?

58. (A) Find the largest term of the sequence $a_0, a_1, a_2, \dots, a_{10}$ to three decimal places, where

$$a_k = \dbinom{10}{k}(0.3)^{10-k}(0.7)^k$$

(B) According to the binomial formula, what is the sum of the series $a_0 + a_1 + a_2 + \cdots + a_{10}$?

59. Evaluate $(1.01)^{10}$ to four decimal places, using the binomial formula. [*Hint:* Let $1.01 = 1 + 0.01$.]

60. Evaluate $(0.99)^6$ to four decimal places, using the binomial formula.

61. Show that: $\dbinom{n}{r} = \dbinom{n}{n-r}$

62. Show that: $\dbinom{n}{0} = \dbinom{n}{n}$

63. Show that: $\dbinom{k}{r-1} + \dbinom{k}{r} = \dbinom{k+1}{r}$

64. Show that: $\dbinom{k}{0} = \dbinom{k+1}{0}$

65. Show that: $\dbinom{k}{k} = \dbinom{k+1}{k+1}$

66. Show that: $\dbinom{n}{r}$ is given by the recursion formula

$$\dbinom{n}{r} = \frac{n-r+1}{r}\dbinom{n}{r-1}$$

where $\dbinom{n}{0} = 1$.

67. Write $2^n = (1+1)^n$ and expand, using the binomial formula to obtain

$$2^n = \dbinom{n}{0} + \dbinom{n}{1} + \dbinom{n}{2} + \cdots + \dbinom{n}{n}$$

68. Write $0 = (1-1)^n$ and expand, using the binomial formula, to obtain

$$0 = \dbinom{n}{0} - \dbinom{n}{1} + \dbinom{n}{2} - \cdots + (-1)^n\dbinom{n}{n}$$

CHAPTER 11 Review

11-1 Sequences and Series

A **sequence** is a function with the domain a set of successive integers. The symbol a_n, called the **nth term,** or **general term,** represents the range value associated with the domain value n. Unless specified otherwise, the domain is understood to be the set of natural numbers. A **finite sequence** has a finite domain, and an **infinite sequence** has an infinite domain. A **recursion formula** defines each term of a sequence in terms of one or more of the preceding terms. For example, the **Fibonacci sequence** is defined by $a_n = a_{n-1} + a_{n-2}$ for $n \geq 3$, where $a_1 = a_2 = 1$. If $a_1, a_2, \ldots, a_n, \ldots$ is a sequence, then the expression $a_1 + a_2 + \cdots + a_n + \cdots$ is called a **series.** A finite sequence produces a **finite series,** and an infinite sequence produces an **infinite series.** Series can be represented using the **summation notation:**

$$\sum_{k=m}^{n} a_k = a_m + a_{m+1} + \cdots + a_n$$

where k is called the **summing index.** If the terms in the series are alternately positive and negative, the series is called an **alternating series.**

11-2 Mathematical Induction

A wide variety of statements can be proven using the **principle of mathematical induction:** Let P_n be a statement associated with each positive integer n and suppose the following conditions are satisfied:

1. P_1 is true.

2. For any positive integer k, if P_k is true, then P_{k+1} is also true.

Then the statement P_n is true for all positive integers n.

To use mathematical induction to prove statements involving laws of exponents, it is convenient to state a **recursive definition of a^n:**

$$a^1 = a \quad \text{and} \quad a^{n+1} = a^n a \quad \text{for any integer } n > 1$$

To deal with conjectures that may be true only for $n \geq m$, where m is a positive integer, we use the **extended principle of mathematical induction:** Let m be a positive integer, let P_n be a statement associated with each integer $n \geq m$, and suppose the following conditions are satisfied:

1. P_m is true.

2. For any integer $k \geq m$, if P_k is true, then P_{k+1} is also true.

Then the statement P_n is true for all integers $n \geq m$.

11-3 Arithmetic and Geometric Sequences

A sequence is called an **arithmetic sequence,** or **arithmetic progression,** if there exists a constant d, called the **common difference,** such that

$$a_n - a_{n-1} = d \quad \text{or} \quad a_n = a_{n-1} + d$$
$$\text{for every } n > 1$$

The following formulas are useful when working with arithmetic sequences and their corresponding series:

$$a_n = a_1 + (n-1)d \qquad \textbf{nth-Term Formula}$$

$$S_n = \frac{n}{2}[2a_1 + (n-1)d] \qquad \textbf{Sum Formula—First Form}$$

$$S_n = \frac{n}{2}(a_1 + a_n) \qquad \textbf{Sum Formula—Second Form}$$

A sequence is called a **geometric sequence,** or a **geometric progression,** if there exists a nonzero constant r, called the **common ratio,** such that

$$\frac{a_n}{a_{n-1}} = r \quad \text{or} \quad a_n = ra_{n-1} \quad \text{for every } n > 1$$

The following formulas are useful when working with geometric sequences and their corresponding series:

$$a_n = a_1 r^{n-1} \qquad \textbf{\textit{n}th-Term Formula}$$

$$S_n = \frac{a_1 - a_1 r^n}{1 - r} \quad r \neq 1 \qquad \textbf{Sum Formula—First Form}$$

$$S_n = \frac{a_1 - ra_n}{1 - r} \quad r \neq 1 \qquad \textbf{Sum Formula—Second Form}$$

$$S_\infty = \frac{a_1}{1 - r} \quad |r| < 1 \qquad \textbf{Sum of an Infinite Geometric Series}$$

11-4 Multiplication Principle, Permutations, and Combinations

A **counting technique** is a mathematical method of determining the number of objects in a set without actually enumerating them. Given a sequence of operations, **tree diagrams** are often used to list all the possible combined outcomes. To count the number of combined outcomes without listing them, we use the **multiplication principle** (also called the **fundamental counting principle**):

1. If operations O_1 and O_2 are performed in order with N_1 possible outcomes for the first operation and N_2 possible outcomes for the second operation, then there are

$$N_1 \cdot N_2$$

possible outcomes of the first operation followed by the second.

2. In general, if n operations $O_1, O_2, \ldots, O_n$ are performed in order, with possible number of outcomes $N_1, N_2, \ldots, N_n$, respectively, then there are

$$N_1 \cdot N_2 \cdot \cdots \cdot N_n$$

possible combined outcomes of the operations performed in the given order.

The symbol $\boldsymbol{n!}$ is read $\boldsymbol{n}$ **factorial** and 0! is defined to be 1.

A particular arrangement or ordering of n objects without repetition is called a **permutation.** The number of permutations of n objects is given by

$$P_{n,n} = n \cdot (n - 1) \cdot \cdots \cdot 1 = n!$$

A **permutation of a set of n objects taken r at a time** is an arrangement of the r objects in a specific order. The number of permutations of n objects taken r at a time is given by

$$P_{n,r} = \frac{n!}{(n - r)!} \qquad 0 \leq r \leq n$$

A **combination of a set of n objects taken r at a time** is an r-element subset of the n objects. The number of combinations of n objects taken r at a time is given by

$$C_{n,r} = \binom{n}{r} = \frac{P_{n,r}}{r!} = \frac{n!}{r!(n - r)!} \qquad 0 \leq r \leq n$$

In a permutation, order is important. In a combination, order is not important.

11-5 Sample Spaces and Probability

The outcomes of an experiment are called **simple events** if one and only one of these results will occur in each trial of the experiment. The set of all simple events is called the **sample space.** Any subset of the sample space is called an **event.** An event is a **simple event** if it has only one element in it and a **compound event** if it has more than one element in it. We say that **an event E occurs** if any of the simple events in E occurs. A sample space S_1 is **more fundamental** than a second sample space S_2 if knowledge of which event occurs in S_1 tells us which event in S_2 occurs, but not conversely.

Given a sample space $S = \{e_1, e_2, \ldots, e_n\}$ with n simple events, to each simple event e_i we assign a real number denoted by $P(e_i)$, that is called the **probability of the event e_i** and satisfies:

1. $0 \leq P(e_i) \leq 1$

2. $P(e_1) + P(e_2) + \cdots + P(e_n) = 1$

Any probability assignment that meets conditions 1 and 2 is said to be an **acceptable probability assignment.**

Given an acceptable probability assignment for the simple events in a sample space S, the **probability of an arbitrary event E** is defined as follows:

1. If E is the empty set, then $P(E) = 0$.

2. If E is a simple event, then $P(E)$ has already been assigned.

3. If E is a compound event, then $P(E)$ is the sum of the probabilities of all the simple events in E.

4. If E is the sample space S, then $P(E) = P(S) = 1$.

If each of the simple events in a sample space $S = \{e_1, e_2, \ldots, e_n\}$ with n simple events is **equally likely** to occur, then we assign the probability $1/n$ to each. If E is an arbitrary event in S, then

$$P(E) = \frac{\text{Number of elements in } E}{\text{Number of elements in } S} = \frac{n(E)}{n(S)}$$

If we conduct an experiment n times and event E occurs with **frequency** $f(E)$, then the ratio $f(E)/n$ is called the **relative frequency** of the occurrence of event E in n trials. As n increases, $f(E)/n$ usually approaches a number that is called the **empirical probability** $P(E)$. So $f(E)/n$ is used as an **approximate empirical probability** for $P(E)$.

If $P(E)$ is the theoretical probability of an event E and the experiment is performed n times, then the **expected frequency** of the occurrence of E is $n \cdot P(E)$.

11-6 Binomial Formula

Pascal's triangle is a triangular array of coefficients for the expansion of the binomial $(a + b)^n$, where n is a positive integer. Notation for the combination formula is

$$\binom{n}{r} = C_{n,r} = \frac{n!}{r!(n - r)!}$$

For n a positive integer, the **binomial formula** is

$$(a + b)^n = \sum_{k=0}^{n} \binom{n}{k} a^{n-k} b^k$$

The numbers $\binom{n}{k}$, $0 \leq k \leq n$, are called **binomial coefficients.**

CHAPTER 11 Review Exercises

Work through all the problems in this chapter review and check answers in the back of the book. Answers to all review problems are there, and following each answer is a number in italics indicating the section in which that type of problem is discussed. Where weaknesses show up, review appropriate sections in the text.

1. Determine whether each of the following can be the first three terms of a geometric sequence, an arithmetic sequence, or neither.
 (A) $16, -8, 4, \ldots$ (B) $5, 7, 9, \ldots$
 (C) $-8, -5, -2, \ldots$ (D) $2, 3, 5, \ldots$
 (E) $-1, 2, -4, \ldots$

In Problems 2–5:
(A) Write the first four terms of each sequence.
(B) Find a_{10}. *(C) Find S_{10}.*

2. $a_n = 2n + 3$ 3. $a_n = 32(\tfrac{1}{2})^n$

4. $a_1 = -8; a_n = a_{n-1} + 3, n \geq 2$

5. $a_1 = -1, a_n = (-2)a_{n-1}, n \geq 2$

6. Find S_∞ in Problem 3.

Evaluate the expression in Problems 7–10.

7. $6!$ 8. $\dfrac{22!}{19!}$

9. $\dfrac{7!}{2!(7-2)!}$ 10. $C_{6,2}$ and $P_{6,2}$

11. A single die is rolled and a coin is flipped. How many combined outcomes are possible? Solve
 (A) By using a tree diagram
 (B) By using the multiplication principle

12. How many seating arrangements are possible with six people and six chairs in a row? Solve by using the multiplication principle.

13. Solve Problem 12 using permutations or combinations, whichever is applicable.

14. In a single deal of 5 cards from a standard 52-card deck, what is the probability of being dealt five clubs?

15. Betty and Bill are members of a 15-person ski club. If the president and treasurer are selected by lottery, what is the probability that Betty will be president and Bill will be treasurer? A person cannot hold more than one office.

16. A drug has side effects for 50 out of 1,000 people in a test. What is the approximate empirical probability that a person using the drug will have side effects?

Verify the statement P_n in Problems 17–19 for $n = 1, 2,$ and 3.

17. P_n: $5 + 7 + 9 + \cdots + (2n + 3) = n^2 + 4n$

18. P_n: $2 + 4 + 8 + \cdots + 2^n = 2^{n+1} - 2$

19. P_n: $49^n - 1$ is divisible by 6

In Problems 20–22, write P_k and P_{k+1}.

20. For P_n in Problem 17 21. For P_n in Problem 18

22. For P_n in Problem 19

23. Either prove the statement is true or prove it is false by finding a counterexample: If n is a positive integer, then the sum of the series $1 + \dfrac{1}{2} + \dfrac{1}{3} + \cdots + \dfrac{1}{n}$ is less than 4.

Write Problems 24 and 25 without summation notation, and find the sum.

24. $S_{10} = \displaystyle\sum_{k=1}^{10} (2k - 8)$

25. $S_7 = \displaystyle\sum_{k=1}^{7} \dfrac{16}{2^k}$

26. $S_\infty = 27 - 18 + 12 + \cdots = ?$

27. Write
$$S_n = \frac{1}{3} - \frac{1}{9} + \frac{1}{27} + \cdots + \frac{(-1)^{n+1}}{3^n}$$
using summation notation, and find S_∞.

28. Someone tells you that the following approximate empirical probabilities apply to the sample space $\{e_1, e_2, e_3, e_4\}$: $P(e_1) \approx .1$, $P(e_2) \approx -.2$, $P(e_3) \approx .6$, $P(e_4) \approx 2$. There are three reasons why P cannot be a probability function. Name them.

29. Six distinct points are selected on the circumference of a circle. How many triangles can be formed using these points as vertices?

30. In an arithmetic sequence, $a_1 = 13$ and $a_7 = 31$. Find the common difference d and the fifth term a_5.

31. How many three-letter code words are possible using the first eight letters of the alphabet if no letter can be repeated? If letters can be repeated? If adjacent letters cannot be alike?

32. Two coins are flipped 1,000 times with the following frequencies:

 Two heads: 210
 One head: 480
 Zero heads: 310

 (A) Compute the empirical probability for each outcome.
 (B) Compute the theoretical probability for each outcome.
 (C) Using the theoretical probabilities computed in part B, compute the expected frequency of each outcome, assuming fair coins.

33. From a standard deck of 52 cards, what is the probability of obtaining a 5-card hand:
(A) Of all diamonds?
(B) Of three diamonds and two spades?

Write answers in terms of $C_{n,r}$ or $P_{n,r}$, as appropriate. Do not evaluate.

34. A group of 10 people includes one married couple. If four people are selected at random, what is the probability that the married couple is selected?

35. A spinning device has three numbers, 1, 2, 3, each as likely to turn up as the other. If the device is spun twice, what is the probability that:
(A) The same number turns up both times?
(B) The sum of the numbers turning up is 5?

36. Use the formula for the sum of an infinite geometric series to write $0.727\,272 \cdots = 0.\overline{72}$ as the quotient of two integers.

37. Solve the following problems using $P_{n,r}$ or $C_{n,r}$, as appropriate:
(A) How many three-digit opening combinations are possible on a combination lock with six digits if the digits cannot be repeated?
(B) Suppose five tennis players have made the finals. If each of the five players is to play every other player exactly once, how many games must be scheduled?

Evaluate Problems 38–40.

38. $\dfrac{20!}{18!(20 - 18)!}$ **39.** $\dbinom{16}{12}$ **40.** $\dbinom{11}{11}$

41. Expand $(x - y)^5$ using the binomial formula.

42. Find the term containing x^6 in the expansion of $(x + 2)^9$.

43. If the terms in the expansion of $(2x - y)^{12}$ are arranged in descending powers of x, find the tenth term.

Establish each statement in Problems 44–46 for all natural numbers using mathematical induction.

44. P_n in Problem 17 **45.** P_n in Problem 18

46. P_n in Problem 19

 In Problems 47 and 48, find the smallest positive integer n such that $a_n < b_n$ by graphing the sequences $\{a_n\}$ and $\{b_n\}$ with a graphing calculator. Check your answer by using a graphing calculator to display both sequences in table form.

47. $a_n = C_{50,n}, b_n = 3^n$

48. $a_1 = 100, a_n = 0.99a_{n-1} + 5, b_n = 9 + 7n$

49. How many different families with five children are possible, excluding multiple births, where the sex of each child in the order of their birth is taken into consideration? How many families are possible if the order pattern is not taken into account?

50. A free-falling body travels $g/2$ feet in the first second, $3g/2$ feet during the next second, $5g/2$ feet the next, and so on. Find the distance fallen during the twenty-fifth second and the total distance fallen from the start to the end of the twenty-fifth second.

51. How many ways can two people be seated in a row of four chairs?

52. Expand $(x + i)^6$, where i is the imaginary unit, using the binomial formula.

53. If three people are selected from a group of seven men and three women, what is the probability that at least one woman is selected?

54. Three fair coins are tossed 1,000 times with the following frequencies of outcomes:

Number of heads	0	1	2	3
Frequency	120	360	350	170

(A) What is the approximate empirical probability of obtaining two heads?
(B) What is the theoretical probability of obtaining two heads?
(C) What is the expected frequency of obtaining two heads?

Prove that each statement in Problems 55–59 holds for all positive integers using mathematical induction.

55. $\displaystyle\sum_{k=1}^{n} k^3 = \left(\sum_{k=1}^{n} k\right)^2$

56. $x^{2n} - y^{2n}$ is divisible by $x - y$, $x \neq y$

57. $\dfrac{a^n}{a^m} = a^{n-m}$; $n > m$; n, m positive integers

58. $\{a_n\} = \{b_n\}$, where $a_n = a_{n-1} + 2, a_1 = -3, b_n = -5 + 2n$

59. $(1!)1 + (2!)2 + (3!)3 + \cdots + (n!)n = (n + 1)! - 1$ (From U.S.S.R. Mathematical Olympiads, 1955–1956, Grade 10.)

APPLICATIONS

60. LOAN REPAYMENT You borrow $7,200 and agree to pay 1% of the unpaid balance each month for interest. If you decide to pay an additional $300 each month to reduce the unpaid balance, how much interest will you pay over the 24 months it will take to repay this loan?

61. ECONOMICS Due to reduced taxes, an individual has an extra $2,400 in spendable income. If we assume that the individual spends 75% of this on consumer goods, and the producers of those consumer goods in turn spend 75% on consumer goods, and that this process continues indefinitely, what is the total amount (to the nearest dollar) spent on consumer goods?

62. COMPOUND INTEREST If $500 is invested at 6% compounded annually, the amount A present after n years forms a geometric sequence with common ratio $1 + 0.06 = 1.06$. Use a geometric sequence formula to find the amount A in the account (to the nearest cent) after 10 years; after 20 years.

63. TRANSPORTATION A distribution center A wishes to distribute its products to five different retail stores, B, C, D, E, and F, in a city. How many different route plans can be constructed so that a single truck can start from A, deliver to each store exactly once, and then return to the center?

64. MARKET ANALYSIS A DVD distributor selected 1,000 persons at random and surveyed them to determine a relationship between age of purchaser and annual DVD purchases. The results are given in the table on page 770.

CHAPTERS 1–3 Cumulative Review Exercises

Work through all the problems in this cumulative review and check answers in the back of the book. Answers to all review problems are there, and following each answer is a number in italics indicating the section in which that type of problem is discussed. Where weaknesses show up, review appropriate sections in the text.

1. Solve for x: $\dfrac{7x}{5} - \dfrac{3 + 2x}{2} = \dfrac{x - 10}{3} + 2$

In Problems 2–4, solve and graph the inequality.

2. $2(3 - y) + 4 \le 5 - y$ **3.** $|x - 2| < 7$

4. $x^2 + 3x \ge 10$

5. Perform the indicated operations and write the answer in standard form:
(A) $(2 - 3i) - (-5 + 7i)$
(B) $(1 + 4i)(3 - 5i)$
(C) $\dfrac{5 + i}{2 + 3i}$

In Problems 6–9, solve the equation.

6. $3x^2 = -12x$ **7.** $4x^2 - 20 = 0$

8. $x^2 - 6x + 2 = 0$ **9.** $x - \sqrt{12 - x} = 0$

10. Given the points $A = (3, 2)$ and $B = (5, 6)$, find:
(A) Distance between A and B.
(B) Slope of the line through A and B.
(C) Slope of a line perpendicular to the line through A and B.

11. Find the equation of the circle with radius $\sqrt{2}$ and center:
(A) $(0, 0)$ (B) $(-3, 1)$

12. Graph $2x - 3y = 6$ and indicate its slope and intercepts.

13. Indicate whether each set defines a function. Find the domain and range of each function.
(A) $\{(1, 1), (2, 1), (3, 1)\}$
(B) $\{(1, 1), (1, 2), (1, 3)\}$
(C) $\{(-2, 2), (-1, -1), (0, 0), (1, -1), (2, 2)\}$

14. For $f(x) = x^2 - 2x + 5$ and $g(x) = 3x - 2$, find:
(A) $f(-2) + g(3)$
(B) $(f + g)(x)$
(C) $(f \circ g)(x)$
(D) $\dfrac{f(a + h) - f(a)}{h}$

15. How are the graphs of the following related to the graph of $y = |x|$?
(A) $y = 2|x|$
(B) $y = |x - 2|$
(C) $y = |x| - 2$

Problems 16–18 refer to the function f given by the graph:

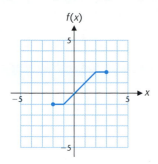

16. Find the domain and range of f. Express answers in interval notation.

17. Is f an even function, an odd function, or neither? Explain.

18. Use the graph of f to sketch a graph of the following:
(A) $y = -f(x + 1)$ (B) $y = 2f(x) - 2$

In Problems 19–21, solve the equation.

19. $\dfrac{x + 3}{2x + 2} + \dfrac{5x + 2}{3x + 3} = \dfrac{5}{6}$ **20.** $\dfrac{3}{x} = \dfrac{6}{x + 1} - \dfrac{1}{x - 1}$

21. $2x + 1 = 3\sqrt{2x - 1}$

In Problems 22–24, solve and graph the inequality.

22. $|4x - 9| > 3$ **23.** $\sqrt{(3m - 4)^2} \le 2$

24. $\dfrac{x + 1}{2} > x - 2$

25. For what real values of x does the following expression represent a real number?
$$\dfrac{\sqrt{x - 2}}{x - 4}$$

26. Perform the indicated operations and write the final answers in standard form:
(A) $(2 - 3i)^2 - (4 - 5i)(2 - 3i) - (2 + 10i)$
(B) $\dfrac{3}{5} + \dfrac{4}{5}i + \dfrac{1}{\frac{3}{5} + \frac{4}{5}i}$ (C) i^{35}

27. Convert to $a + bi$ form, perform the indicated operations, and write the final answers in standard form:
(A) $(5 + 2\sqrt{-9}) - (2 - 3\sqrt{-16})$
(B) $\dfrac{2 + 7\sqrt{-25}}{3 - \sqrt{-1}}$ (C) $\dfrac{12 - \sqrt{-64}}{\sqrt{-4}}$

In Problems 28–31, solve the equation.

28. $1 + \dfrac{14}{y^2} = \dfrac{6}{y}$ **29.** $4x^{2/3} - 4x^{1/3} - 3 = 0$

30. $u^4 + u^2 - 12 = 0$

31. $\sqrt{8t - 2} - 2\sqrt{t} = 1$

Use a calculator to solve the equation or inequality in Problems 32 and 33. Compute answers to two decimal places.

32. $-3.45 < 1.86 - 0.33x \le 7.92$

33. $2.35x^2 + 10.44x - 16.47 = 0$

34. Solve for y in terms of x:

$$\frac{x - 2}{x + 1} = \frac{2y + 1}{y - 2}$$

35. Find each of the following for the function f given by the graph shown in the figure.
(A) The domain of f
(B) The range of f
(C) $f(-3) + f(-2) + f(2)$
(D) The intervals over which f is increasing
(E) The x coordinates of any points of discontinuity

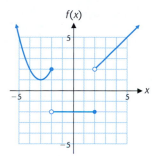

$f(x)$

36. Write equations of the lines
(A) Parallel to
(B) Perpendicular to
the line $3x + 2y = 12$ and passing through the point $(-6, 1)$. Write the final answers in the slope–intercept form $y = mx + b$.

37. Find the domain of $g(x) = \sqrt{x + 4}$.

38. Graph $f(x) = x^2 - 2x - 8$. Show the axis of symmetry and vertex, and find the range, intercepts, and maximum or minimum value of $f(x)$.

39. Given $f(x) = 1/(x - 2)$ and $g(x) = (x + 3)/x$, find $f \circ g$. What is the domain of $f \circ g$?

40. Find $f^{-1}(x)$ for $f(x) = 2x + 5$.

41. Graph, finding the domain, range, and any points of discontinuity:

$$f(x) = \begin{cases} x - 1 & \text{if } x < 0 \\ x^2 + 1 & \text{if } x \ge 0 \end{cases}$$

42. Graph:
(A) $y = 2\sqrt{x} + 1$
(B) $y = -\sqrt{x + 1}$

43. The graph in the figure is the result of applying a sequence of transformations to the graph of $y = |x|$. Describe the transformations verbally and write an equation for the graph in the figure.

44. Let $f(x) = \sqrt{x + 4}$.
(A) Find $f^{-1}(x)$.
(B) Find the domain and range of f and f^{-1}.
(C) Graph f, f^{-1}, and $y = x$ on the same coordinate system. Check by graphing f, f^{-1}, and $y = x$ in a squared window on a graphing calculator.

45. Find the center and radius of the circle given by the equation $x^2 - 6x + y^2 + 2y = 0$. Graph the circle and show the center and the radius.

46. Discuss symmetry with respect to the x axis, y axis, and the origin for the equation

$$xy + |xy| = 5$$

47. Write an equation for the graph in the figure in the form $y = a(x - h)^2 + k$, where a is either -1 or $+1$ and h and k are integers.

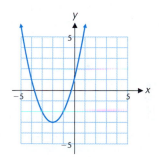

48. Solve for y in terms of x:

$$\frac{x + y}{y - \dfrac{x + y}{x - y}} = 1$$

49. Find all roots: $3x^2 = 2\sqrt{2x} - 1$.

50. Consider the quadratic equation

$$x^2 + bx + 1 = 0$$

where b is a real number. Discuss the relationship between the values of b and the three types of roots listed in Table 1 in Section 1-5.

51. Find all solutions: $\sqrt{3 - 2x} - \sqrt{x + 7} = \sqrt{x + 4}$.

52. Write in standard form: $\dfrac{a + bi}{a - bi}$, $a, b \ne 0$.

53. Given $f(x) = x^2$ and $g(x) = \sqrt{4 - x^2}$, find:
(A) Domain of g
(B) f/g and its domain
(C) $f \circ g$ and its domain

 54. Let $f(x) = x^2 - 2x - 3, x \geq 1$.
(A) Find $f^{-1}(x)$.
(B) Find the domain and range of f^{-1}.
(C) Graph f, f^{-1}, and $y = x$ on the same coordinate system. Check by graphing f, f^{-1}, and $y = x$ in a squared window on a graphing calculator.

APPLICATIONS

55. NUMBERS Find a number such that the number exceeds its reciprocal by $\frac{3}{2}$.

56. RATE–TIME A boat travels upstream for 35 miles and then returns to its starting point. If the round-trip took 4.8 hours and the boat's speed in still water is 15 miles per hour, what is the speed of the current?

57. CHEMISTRY How many gallons of distilled water must be mixed with 24 gallons of a 90% sulfuric acid solution to obtain a 60% solution?

58. BREAK-EVEN ANALYSIS The publisher's fixed costs for the production of a new study guide are $41,800. Variable costs are $4.90 per book. If the book is sold to bookstores for $9.65, how many must be sold for the publisher to break even?

59. FINANCE An investor instructs a broker to buy a certain stock whenever the price per share p of the stock is within $10 of $200. Express this instruction as an absolute value inequality.

60. PRICE AND DEMAND The weekly demand for mouthwash in a chain of drugstores is 1,160 bottles at a price of $3.79 each. If the price is lowered to $3.59, the weekly demand increases to 1,340 bottles. Assuming that the relationship between the weekly demand x and the price per bottle p is linear, express x as a function of p. How many bottles would the store sell each week if the price were lowered to $3.29?

61. BUSINESS—PRICING A telephone company begins a new pricing plan that charges customers for local calls as follows: The first 60 calls each month are 6 cents each, the next 90 are 5 cents each, the next 150 are 4 cents each, and any additional calls are 3 cents each. If C is the cost, in dollars, of placing x calls per month, write a piecewise definition of C as a function of x and graph.

62. CONSTRUCTION A homeowner has 80 feet of chain-link fencing to be used to construct a dog pen adjacent to a house (see the figure).
(A) Express the area $A(x)$ enclosed by the pen as a function of the width x.
(B) From physical considerations, what is the domain of the function A?
(C) Graph A and determine the dimensions of the pen that will make the area maximum.

63. COMPUTER SCIENCE Let $f(x) = x - 2[\![x/2]\!]$. This function can be used to determine if an integer is odd or even.
(A) Find $f(1), f(2), f(3)$, and $f(4)$.
(B) Find $f(n)$ for any integer n. [*Hint:* Consider two cases, $n = 2k$ and $n = 2k + 1$, k an integer.]

64. DEPRECIATION Office equipment was purchased for $20,000 and is assumed to depreciate linearly to a scrap value of $4,000 after 8 years.
(A) Find a linear function $v = d(t)$ that relates value v in dollars to time t in years.
(B) Find $t = d^{-1}(v)$.

65. PROFIT AND LOSS ANALYSIS At a price of $\$p$ per unit, the marketing department at a company estimates that the weekly cost C and the weekly revenue R, in thousands of dollars, will be given by the equations

$$C = 88 - 12p \qquad \text{Cost equation}$$
$$R = 15p - 2p^2 \qquad \text{Revenue equation}$$

Find the prices for which the company has:
(A) A profit (B) A loss

66. SHIPPING A ship leaves port A, sails east to port B, and then north to port C, a total distance of 115 miles. The next day the ship sails directly from port C back to port A, a distance of 85 miles. Find the distance between ports A and B and between ports B and C.

67. PHYSICS The distance s above the ground (in feet) of an object dropped from a hot-air balloon t seconds after it is released is given by

$$s = a + bt^2$$

where a and b are constants. Suppose the object is 2,100 feet above the ground 5 seconds after its release and 900 feet above the ground 10 seconds after its release.
(A) Find the constants a and b.
(B) How high is the balloon?
(C) How long does the object fall?

68. PRICE AND DEMAND The demand for barley q (in thousands of bushels) and the corresponding price p (in cents) at a midwestern grain exchange are shown in the figure.

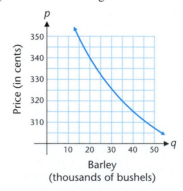

(A) What is the demand (to the nearest thousand bushels) when the price is 325 cents per bushel?
(B) Does the demand increase or decrease if the price is increased to 340 cents per bushel? By how much?
(C) Does the demand increase or decrease if the price is decreased to 315 cents per bushel? By how much?
(D) Write a brief description of the relationship between price and demand illustrated by this graph.

(E) Use the graph to estimate the price (to the nearest cent) when the demand is 20, 25, 30, 35, and 40 thousand bushels. Use these data to find a quadratic regression model for the price of barley using the demand as the independent variable.

69. STOPPING DISTANCE Table 1 contains data related to the length of the skid marks left by an automobile when making an emergency stop. A model for the skid mark length L (in feet) is

$$L = f(s) = 0.05s^2 - 0.2s + 6.5, s \geq 20$$

where s is speed in miles per hour.
(A) Graph $L = f(s)$ and the data for skid mark length on the same axes.
(B) Find $s = f^{-1}(L)$ and find its domain and range.
(C) An insurance investigator finds skid marks 220 feet long at the scene of an accident involving this automobile. How fast (to the nearest mile per hour) was the automobile traveling when it made these skid marks?

Table 1 Skid Marks

Speed (mph)	Length of Skid Marks (feet)
20	24
30	48
40	77
50	115
60	187
70	246
80	312

CHAPTERS 4–5 Cumulative Review Exercises

Work through all the problems in this cumulative review and check answers in the back of the book. Answers to all review problems are there, and following each answer is a number in italics indicating the section in which that type of problem is discussed. Where weaknesses show up, review appropriate sections in the text.

1. Let $P(x)$ be the polynomial whose graph is shown in the figure.
(A) Assuming that $P(x)$ has integer zeros and leading coefficient 1, find the lowest-degree equation that could produce this graph.
(B) Describe the left and right behavior of $P(x)$.

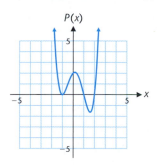

2. Match each equation with the graph of f, g, m, or n in the figure.
(A) $y = (\frac{3}{4})^x$
(B) $y = (\frac{4}{3})^x$
(C) $y = (\frac{3}{4})^x + (\frac{4}{3})^x$
(D) $y = (\frac{4}{3})^x - (\frac{3}{4})^x$

3. For $P(x) = 3x^3 + 5x^2 - 18x - 3$ and $D(x) = x + 3$, use synthetic division to divide $P(x)$ by $D(x)$, and write the answer in the form $P(x) = D(x)Q(x) + R$.

4. Let $P(x) = 2(x + 2)(x - 3)(x - 5)$. What are the zeros of $P(x)$?

5. Let $P(x) = 4x^3 - 5x^2 - 3x - 1$. How do you know that $P(x)$ has at least one real zero between 1 and 2?

6. Let $P(x) = x^3 + x^2 - 10x + 8$. Find all rational zeros for $P(x)$.

7. Solve for x.
(A) $y = 10^x$
(B) $y = \ln x$

8. Simplify.
(A) $(2e^x)^3$
(B) $\dfrac{e^{3x}}{e^{-2x}}$

9. Solve for x exactly. Do not use a calculator or a table.
(A) $\log_3 x = 2$
(B) $\log_3 81 = x$
(C) $\log_x 4 = -2$

10. Solve for x to three significant digits.
(A) $10^x = 2.35$
(B) $e^x = 87,500$
(C) $\log x = -1.25$
(D) $\ln x = 2.75$

In Problems 11 and 12, translate each statement into an equation using k as the constant of proportionality.

11. E varies directly as p and inversely as the cube of x.

12. F is jointly proportional to q_1 and q_2 and inversely proportional to the square of r.

13. Explain why the graph in the figure is not the graph of a polynomial function.

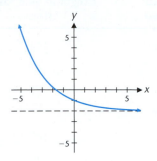

14. Explain why the graph in the figure is not the graph of a rational function.

15. The function f subtracts the square root of the domain element from three times the natural log of the domain element. Write an algebraic definition of f.

16. Write a verbal description of the function $f(x) = 100e^{0.5x} - 50$.

17. Let $f(x) = \dfrac{2x + 8}{x + 2}$.
 (A) Find the domain and the intercepts for f.
 (B) Find the vertical and horizontal asymptotes for f.
 (C) Sketch the graph of f. Draw vertical and horizontal asymptotes with dashed lines.

18. Find all zeros of $P(x) = (x^3 + 4x)(x + 4)$, and specify those zeros that are x intercepts.

19. Solve $(x^3 + 4x)(x + 4) \le 0$.

20. If $P(x) = 2x^3 - 5x^2 + 3x + 2$, find $P(\frac{1}{2})$ using the remainder theorem and synthetic division.

21. Which of the following is a factor of $P(x)$?
$$P(x) = x^{25} - x^{20} + x^{15} + x^{10} - x^5 + 1$$
 (A) $x - 1$ (B) $x + 1$

22. Let $P(x) = x^4 - 8x^2 + 3$.
 (A) Graph $P(x)$ and describe the graph verbally, including the number of x intercepts, the number of turning points, and the left and right behavior.
 (B) Approximate the largest x intercept to two decimal places.

23. Let $P(x) = x^5 - 8x^4 + 17x^3 + 2x^2 - 20x - 8$.
 (A) Approximate the zeros of $P(x)$ to two decimal places and state the multiplicity of each zero.
 (B) Can any of these zeros be approximated with the bisection method? The MAXIMUM or MINIMUM commands? Explain.

24. Let $P(x) = x^4 + 2x^3 - 20x^2 - 30$.
 (A) Find the smallest positive and largest negative integers that, by Theorem 1 in Section 4-2, are upper and lower bounds, respectively, for the real zeros of $P(x)$.
 (B) If $(k, k+1)$, k an integer, is the interval containing the largest real zero of $P(x)$, determine how many additional intervals are required in the bisection method to approximate this zero to one decimal place.
 (C) Approximate the real zeros of $P(x)$ to two decimal places.

25. Find all zeros (rational, irrational, and imaginary) exactly for $P(x) = 4x^3 - 20x^2 + 29x - 15$.

26. Final all zeros (rational, irrational, and imaginary) exactly for $P(x) = x^4 + 5x^3 + x^2 - 15x - 12$, and factor $P(x)$ into linear factors.

In Problems 27–36, solve for x exactly. Do not use a calculator or a table.

27. $2^{x^2} = 4^{x+4}$ **28.** $2x^2 e^{-x} + xe^{-x} = e^{-x}$

29. $e^{\ln x} = 2.5$ **30.** $\log_x 10^4 = 4$

31. $\log_9 x = -\frac{3}{2}$

32. $\ln(x + 4) - \ln(x - 4) = 2 \ln 3$

33. $\ln(2x^2 + 2) = 2 \ln(2x - 4)$

34. $\log x + \log(x + 15) = 2$

35. $\log(\ln x) = -1$ **36.** $4(\ln x)^2 = \ln x^2$

In Problems 37–41, solve for x to three significant digits.

37. $x = \log_3 41$ **38.** $\ln x = 1.45$

39. $4(2^x) = 20$ **40.** $10e^{-0.5x} = 1.6$

41. $\dfrac{e^x - e^{-x}}{e^x + e^{-x}} = \dfrac{1}{2}$

42. G is directly proportional to the square of x. If $G = 10$ when $x = 5$, find G when $x = 7$.

43. H varies inversely as the cube of r. If $H = 162$ when $r = 2$, find H when $r = 3$.

In Problems 44–50, find the domain, range, and the equations of any horizontal or vertical asymptotes.

44. $f(x) = 3 + 2^x$

45. $f(x) = 2 - \log_3(x - 1)$

46. $f(x) = 5 - 4x^3$

47. $f(x) = 3 + 2x^4$

48. $f(x) = \dfrac{5}{x + 3}$

49. $f(x) = 20e^{-x} - 15$

50. $f(x) = 8 + \ln(x + 2)$

51. If the graph of $y = \ln x$ is reflected in the line $y = x$, the graph of the function $y = e^x$ is obtained. Discuss the functions that are obtained by reflecting the graph of $y = \ln x$ in the x axis and in the y axis.

52. (A) Explain why the equation $e^{-x} = \ln x$ has exactly one solution.
 (B) Approximate the solution of the equation to two decimal places.

In Problems 53 and 54, factor each polynomial in two ways:
(A) As a product of linear factors (with real coefficients) and
quadratic factors (with real coefficients and imaginary zeros).
(B) As a product of linear factors with complex coefficients.

53. $P(x) = x^4 + 9x^2 + 18$

54. $P(x) = x^4 - 23x^2 - 50$

55. Graph f and indicate any horizontal, vertical, or oblique asymptotes with dashed lines:

$$f(x) = \frac{x^2 + 4x + 8}{x + 2}$$

56. Let $P(x) = x^4 - 28x^3 + 262x^2 - 922x + 1.083$. Approximate (to two decimal places) the x intercepts and the local extrema.

57. Find a polynomial of lowest degree with leading coefficient 1 that has zeros -1 (multiplicity 2), 0 (multiplicity 3), $3 + 5i$, and $3 - 5i$. Leave the answer in factored form. What is the degree of the polynomial?

58. If $P(x)$ is a fourth-degree polynomial with integer coefficients and if i is a zero of $P(x)$, can $P(x)$ have any irrational zeros? Explain.

59. Let $P(x) = x^4 + 9x^3 - 500x^2 + 20,000$.
 (A) Find the smallest positive integer multiple of 10 and the largest negative integer multiple of 10 that, by Theorem 1 in Section 4-2, are upper and lower bounds, respectively, for the real zeros of $P(x)$.
 (B) Approximate the real zeros of $P(x)$ to two decimal places.

60. Find all zeros (rational, irrational, and imaginary) exactly for

$$P(x) = x^5 - 4x^4 + 3x^3 + 10x^2 - 10x - 12$$

and factor $P(x)$ into linear factors.

61. Find rational roots exactly and irrational roots to two decimal places for

$$P(x) = x^5 + 4x^4 + x^3 - 11x^2 - 8x + 4$$

62. Give an example of a rational function $f(x)$ that satisfies the following conditions: the real zeros of f are 5 and 8; $x = 1$ is the only vertical asymptote; and the line $y = 3$ is a horizontal asymptote.

63. Use natural logarithms to solve for n.

$$A = P \frac{(1 + i)^n - 1}{i}$$

64. Solve $\ln y = 5x + \ln A$ for y. Express the answer in a form that is free of logarithms.

65. Solve for x.

$$y = \frac{e^x - 2e^{-x}}{2}$$

66. Solve $\dfrac{x^3 - x}{x^3 - 8} \geq 0$.

 67. Solve (to three decimal places)

$$\frac{4x}{x^2 - 1} < 3$$

APPLICATIONS

68. SHIPPING A mailing service provides customers with rectangular shipping containers. The length plus the girth of one of these containers is 10 feet (see the figure). If the end of the container is square and the volume is 8 cubic feet, find the side length of the end. Find solutions exactly; round irrational solutions to two decimal places.

69. GEOMETRY The diagonal of a rectangle is 2 feet longer than one of the sides, and the area of the rectangle is 6 square feet. Find the dimensions of the rectangle to two decimal places.

70. POPULATION GROWTH If the Democratic Republic of the Congo has a population of about 60 million people and a doubling time of 23 years, find the population in
(A) 5 years (B) 30 years

Compute answers to three significant digits.

71. COMPOUND INTEREST How long will it take money invested in an account earning 7% compounded annually to double? Use the annual compounding growth model $P = P_0(1 + r)^t$, and compute the answer to three significant digits.

72. COMPOUND INTEREST Repeat Problem 71 using the continuous compound interest model $P = P_0 e^{rt}$.

73. EARTHQUAKES If the 1906 and 1989 San Francisco earthquakes registered 8.3 and 7.1, respectively, on the Richter scale, how many times more powerful was the 1906 earthquake than the 1989 earthquake? Use the formula $M = \frac{2}{3} \log (E/E_0)$, where $E_0 = 10^{4.40}$ joules, and compute the answer to one decimal place.

74. SOUND If the decibel level at a rock concert is 88, find the intensity of the sound at the concert. Use the formula $D = 10 \log (I/I_0)$, where $I_0 = 10^{-12}$ watts per square meter, and compute the answer to two significant digits.

75. ASTRONOMY The square of the time t required for a planet to make one orbit around the sun varies directly as the cube of its mean (average) distance d from the sun. Write the equation of variation, using k as the constant of variation.

76. PHYSICS Atoms and molecules that make up the air constantly fly about like microscopic missiles. The velocity v of a particular particle at a fixed temperature varies inversely as the square root of its molecular weight w. If an oxygen molecule in air at room temperature has an average velocity of 0.3 mile/second, what will be the average velocity of a hydrogen molecule, given that the hydrogen molecule is one-sixteenth as heavy as the oxygen molecule?

Problems 77 and 78 require a graphing calculator or a computer that can calculate linear, quadratic, cubic, and exponential regression models for a given data set.

77. Table 1 shows the life expectancy (in years) at birth for residents of the United States from 1970 to 1995. Let x represent years since 1970. Use the indicated regression model to estimate the life expectancy (to the nearest tenth of a year) for a U.S. resident born in 2010.
(A) Linear regression (B) Quadratic regression
(C) Cubic regression (D) Exponential regression

Table 1

Year	Life Expectancy
1970	70.8
1975	72.6
1980	73.7
1985	74.7
1990	75.4
1995	75.9
2000	77.0
2005	77.7

Source: U.S. Census Bureau

78. Refer to Problem 77. The Census Bureau projected the life expectancy for a U.S. resident born in 2010 to be 77.9 years. Which of the models in Problem 77 is closest to the Census Bureau projection?

CHAPTERS **6–8** **Cumulative Review Exercises**

Work through all the problems in this cumulative review and check answers in the back of the book. Answers to all review problems, except verifications, are there, and following each answer is a number in italics indicating the section in which that type of problem is discussed. Where weaknesses show up, review appropriate sections in the text.

1. In a circle of radius 6 meters, find the length of an arc opposite an angle of 0.31 radians.

2. Solve the triangle.

3. In which quadrants is each positive?
(A) $\sin \theta$ (B) $\cos \theta$ (C) $\tan \theta$

4. If $(-3, 4)$ is on the terminal side of an angle θ, find
(A) $\cos \theta$ (B) $\csc \theta$ (C) $\tan \theta$

5. Find the reference angle associated with each angle θ:
(A) $-3\pi/4$ (B) $245°$ (C) $-30°$

6. Indicate the domain, range, and period of each.
(A) $y = \sin x$ (B) $y = \cos x$ (C) $y = \tan x$

7. Sketch a graph of $y = \cos x$, $-\pi/2 \leq x \leq 5\pi/2$.

8. Sketch a graph of $y = \tan x$, $-\pi/2 < x < 3\pi/2$.

9. Describe the meaning of a central angle in a circle with radian measure 2.

10. Describe the smallest shift of the graph of $y = \cos x$ to produce the graph of $y = \sin x$.

Verify each identity in Problems 11–14.

11. $\cot \theta \sec \theta = \csc \theta$

12. $\sec x - \cos x = \tan x \sin x$

13. $\sin (x - \pi/2) = -\cos x$

14. $\csc 2x = \frac{1}{2} \csc x \sec x$

15. Use a graphing calculator to test whether each of the following is an identity. If an equation appears to be an identity, verify it. If the equation does not appear to be an identity, find a value of x for which both sides are defined but are not equal.
(A) $\dfrac{\sin^2 x}{\cos x} + \cos x = \csc x$

(B) $\dfrac{\sin^2 x}{\cos x} + \cos x = \sec x$

16. If in a triangle, $a = 32.5$ feet, $c = 77.2$ feet, and $\beta = 61.3°$, without solving the triangle or drawing any pictures, which of the two angles, α or γ, can you say for certain is acute and why?

Solve the equation Problems 17 and 18 to four decimal places.

17. $\sin x = 0.3188, 0 \le x \le 2\pi$

18. $\tan \theta = -4.076, -90° < \theta < 90°$

19. Solve the triangle.

20. Write the algebraic vector $\langle a, b \rangle$ corresponding to the geometric vector $\overrightarrow{AB}$ with endpoints $A = (-3, 2)$ and $B = (3, -1)$.

21. A point in a polar coordinate system has coordinates $(-5, 150°)$. Find all other polar coordinates for the point, $-360° \le \theta \le 360°$, and verbally describe how the coordinates are associated with the point.

22. Sketch a graph of $r = 6 \cos \theta$ in a polar coordinate system.

23. Plot in a complex plane: $A = -3 + 4i$ and $B = 4e^{60°i}$.

24. Find $(2e^{10°i})^3$. Write the final answer in exact rectangular form.

25. Which of the following angles are coterminal with $150°$: $30°$, $-7\pi/6$, $870°$?

26. Change 1.31 radians to decimal degrees to two decimal places.

27. Which of the following have the same value as $\cos 8$?
 (A) $\cos (8 \text{ rad})$ (B) $\cos 8°$ (C) $\cos (8 - 4\pi)$

Evaluate the expression in Problems 28–35 exactly without a calculator. If the function is not defined at the value, say so.

28. $\sin (-5\pi/6)$ **29.** $\sec 330°$

30. $\cos^{-1} (-1)$ **31.** $\sin^{-1} 1.5$

32. $\arccos \left(-\frac{1}{2}\right)$ **33.** $\sin (\sin^{-1} 0.55)$

34. $\cos \left[\sin^{-1} \left(-\frac{4}{5}\right) \right]$ **35.** $\cos \left[\tan^{-1} (-2) \right]$

36. Evaluate to four significant digits using a calculator. If a function is not defined, say so.
 (A) $\tan 84°12'55''$ (B) $\sec (-1.8409)$
 (C) $\tan^{-1} (-84.32)$ (D) $\cos^{-1} (\tan 2.314)$

37. Sketch a graph of $y = 2 - 2 \cos (\pi x/2), -1 \le x \le 5$.

38. (A) Find the exact degree measure of $\theta = \cos^{-1} (-\sqrt{3}/2)$ without a calculator.
 (B) Find the degree measure of $\theta = \sin^{-1} (-0.338)$ to three decimal places using a calculator.

39. Evaluate $\sin^{-1} (\sin 3)$ with a calculator set in radian mode, and explain why this does or does not illustrate a sine–inverse sine identity.

40. A circular point $P = (a, b)$ moves counterclockwise around the circumference of a unit circle starting at $(1, 0)$ and stops after covering a distance of 11.205 units. Explain how you would find the coordinates of point P at its final position and how you would determine which quadrant P is in. Find the coordinates of P to three decimal places and the quadrant for the final position of P.

41. Explain the difference in solving the equation $\tan x = -24.5$ and evaluating $\tan^{-1} (-24.5)$.

42. Find an equation of the form $y = k + a \sin Bx$ that produces the graph shown.

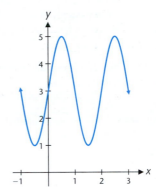

43. Sketch a graph of $y = 3 \sin (2x - \pi), -\pi \le x \le 2\pi$. Indicate amplitude A, period P, and phase shift.

44. Sketch a graph of $y = 2 \tan (\pi x/2 - \pi/2), 0 < x < 4$. Indicate the period P and phase shift.

45. Sketch a graph of $y = \sin x$ and $y = \csc x$ in the same coordinate system.

46. Describe the smallest left shift and/or reflection that transforms the graph of $y = \cot x$ into the graph of $y = \tan x$.

47. Given the equation $\sin 2x = 2 \sin x$,
 (A) Are $x = 0$ and $x = \pi$ solutions?
 (B) Is the equation an identity or a conditional equation? Explain.

Verify each identity in Problems 48–53.

48. $\dfrac{\sin u}{1 + \cos u} + \cot u = \csc u$

49. $\sec x + \tan x = \dfrac{\cos x}{1 - \sin x}$

50. $\tan \dfrac{x}{2} = \csc x - \cot x$

51. $\csc^2 \dfrac{x}{2} = 2 \csc x (\csc x + \cot x)$

52. $\dfrac{2}{1 + \cos 2x} = \sec^2 x$

53. $\dfrac{\cos x + \cos y}{\sin x - \sin y} = \cot \dfrac{x - y}{2}$

 [*Hint:* Use sum–product identities.]

54. Find $\cos(x - y)$ exactly without a calculator given $\sin x = (-2/\sqrt{5})$, $\cos y = (-2/\sqrt{5})$, x a Quadrant IV angle, and y a Quadrant III angle.

55. Compute the exact value of $\sin 2x$ and $\cos (x/2)$ without a calculator, given $\sin x = \frac{3}{5}$, $\pi/2 \leq x \leq \pi$.

Solve Problems 56 and 57 exactly without a calculator, θ in degrees and x real.

56. $2 \sin^2 \theta + \sin \theta = 1$, $0 \leq \theta < 360°$

57. $\sin 2x = \sin x$, all real solutions

58. (A) Solve $\cot x = -2 \cos x$ exactly, $0 \leq x \leq 2\pi$.
 (B) Solve $\cot x = -2 \cos x$ to three decimal places using a graphing calculator, $0 \leq x \leq 2\pi$.

59. Solve $2 \cos x = x - \cos 2x$ to three decimal places for all real solutions using a graphing calculator.

In Problems 60–62, solve each triangle labeled as in the figure. If a problem does not have a solution, say so. If a triangle has two solutions, solve the obtuse case.

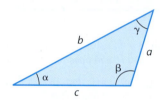

60. $a = 21.3$ meters, $b = 37.4$ meters, $c = 48.2$ meters

61. $\alpha = 125.4°$, $b = 25.4$ millimeters, $a = 20.3$ millimeters

62. $\alpha = 52.9°$, $b = 37.1$ inches, $a = 34.4$ inches

63. Assume in a triangle that γ is acute, $a = 92.5$ centimeters, and $b = 43.4$ centimeters. Which of the angles, α or β, can you say for certain is acute and why?

64. Given vectors as indicated in the figures, find $|\mathbf{u} + \mathbf{v}|$ and α, given $|\mathbf{u}| = 25.3$ pounds, $|\mathbf{v}| = 13.4$ pounds, and $\theta = 48.3°$.

Tail-to-tip rule Parallelogram rule

(a) (b)

65. Find $2\mathbf{u} - \mathbf{v} + 3\mathbf{w}$ for,
 (A) $\mathbf{u} = \langle -1, 2 \rangle$, $\mathbf{v} = \langle 0, -2 \rangle$, $\mathbf{w} = \langle 1, -1 \rangle$
 (B) $\mathbf{u} = 2\mathbf{i} - \mathbf{j}$, $\mathbf{v} = \mathbf{i} + 3\mathbf{j}$, $\mathbf{w} = 2\mathbf{j}$

66. Convert to polar form: $x^2 + y^2 = 8y$.

67. Convert $r = -4 \cos \theta$ to rectangular form.

Use rapid sketching techniques to graph the equation in Problems 68 and 69 in a polar coordinate system.

68. $r = 4 + 4 \cos \theta$ **69.** $r = 6 \sin 3\theta$

70. Graph $r = 5(\cos 2\theta)^{2n}$, for $n = 1, 2$, and 3. How many leaves do you expect the graph will have for arbitrary n?

71. Graph $r = e^{(\cos \theta)} - 2 \cos (4\theta)$ using a squared window and 0.05 for a step size for θ. The resulting curve is often referred to as *a butterfly curve*.

72. Change the rectangular coordinates $(-2.78, -3.19)$ to polar coordinates to two decimal places, $r \geq 0$, $-180° < \theta \leq 180°$.

73. Change the polar coordinates $(6.22, -4.08)$ to rectangular coordinates to two decimal places.

74. Change $2e^{(-\pi/6)i}$ to exact rectangular form.

75. Change $z = -1 + i\sqrt{3}$ to the polar form $re^{i\theta}$, θ in degrees.

76. Compute $(1 - i\sqrt{3})^6$ using De Moivre's theorem and write the final answer in $a + bi$ form.

77. Find all cube roots of $-i$ exactly. Write final answers in the form $a + bi$, and locate the roots on a circle in the complex plane.

78. Change the complex number $-4.88 - 3.17i$ to the polar form $re^{i\theta}$ to two decimal places, $r \geq 0$, $-180° < \theta \leq 180°$.

79. Change the complex number $6.97e^{163.87°i}$ to rectangular form $a + bi$, where a and b are computed to two decimal places.

80. (A) The fourth root of a complex number is shown in the figure. Geometrically locate all other fourth roots of the number on the figure, and explain how they were located.
 (B) Determine geometrically the other fourth roots of the number in exact rectangular form.
 (C) Raise each fourth root from parts A and B to the fourth power.

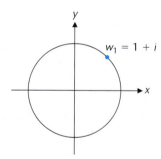

81. If, in the figure, the coordinates of A are $(1, 0)$ and arc length s is 1.2 units, find the coordinates of P to three significant digits.

82. Sketch a graph of $y = 1 + \sec x$, $-3\pi/2 < x < 3\pi/2$.

83. The accompanying graph is a graph of an equation of the form $y = A \cos (Bx + C)$, $0 < -B/C < 1$. Find the equation by finding A, B, and C exactly. What are the period, amplitude, and phase shift?

 84. Graph $1.6 \sin 2x - 1.2 \cos 2x$ in a graphing calculator. (Select the dimensions of a viewing window so that at least two periods are visible.) Find an equation of the form $y = A \sin (Bx + C)$ that has the same graph as the given equation. Find A and B exactly and C to three decimal places. Use the x intercept closest to the origin as the phase shift. To check your results graph both equations in the same viewing window and use TRACE while shifting back and forth between the two graphs.

85. Write $\csc (\cos^{-1} x)$ as an algebraic expression in x free of trigonometric or inverse trigonometric functions.

Solve Problems 86 and 87 without a calculator.

86. $\sin \left[2 \cot^{-1} \left(\frac{3}{4} \right) \right] = ?$

87. Given $\sec x = -5/3$, $\pi/2 \le x \le \pi$, find
 (A) $\sin (x/2)$ (B) $\cos 2x$

88. (A) Solve $2 \sin^2 x = 3 \cos x$ exactly for all real solutions, $0 \le x \le 2\pi$.
 (B) Solve $2 \sin^2 x = 3 \cos x$ to four decimal places using a graphing calculator, $0 \le x \le 2\pi$.

89. (A) Use rapid sketching techniques to sketch a graph of the polar equation $r^2 = 36 \cos 2\theta$.
 (B) Verify the graph in part A using a graphing calculator.

 90. (A) Graph $r_1 = 2 + 2 \cos \theta$ and $r_2 = 6 \cos \theta$ in the same viewing window, $0 \le \theta \le 2\pi$.
 (B) Use TRACE to determine how many times the graph of r_2 crosses the graph of r_1 as θ goes from 0 to 2π.
 (C) Solve the two equations simultaneously to find the exact solutions for $0 \le \theta \le 2\pi$.
 (D) Explain why the number of solutions found in part C does not agree with the number of times r_1 crosses r_2, $0 \le \theta \le 2\pi$.

APPLICATIONS

91. ASTRONOMY A line from the sun to the Earth sweeps out an angle of how many radians in 5 days?

92. METEOROLOGY A weather balloon is released and rises vertically. Two weather stations C and D in the same vertical plane as the balloon and 1,000 meters apart sight the balloon at the same time and record the information given in the figure. At the time of sighting, how high was the balloon to the nearest meter?

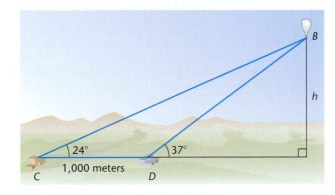

93. GEOMETRY Find the length to two decimal places of one side of a regular pentagon inscribed in a circle with radius 5 inches.

94. GEOMETRY Find $\angle ABC$ to the nearest degree in the rectangular solid shown in the figure.

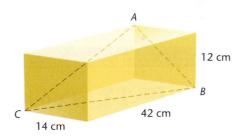

95. ELECTRICAL CIRCUIT The current I in an alternating electrical circuit has an amplitude of 50 amperes and a period of $\frac{1}{110}$ second. If $I = 50$ amperes when $t = 0$, find an equation of the form $I = A \cos Bt$ that gives the current at time $t \ge 0$.

96. NAVIGATION An airplane flies with an airspeed of 260 miles per hour and a compass heading of 110°. If a 36 mile per hour wind is blowing out of the north, what is the plane's actual heading and ground speed? Compute direction to the nearest degree and ground speed to the nearest mile per hour.

97. ENGINEERING A 65-pound child glides across a small river on a homemade cable trolley (see the figure). What is the tension on each half of the support cable when the child is in the center? Compute your answer to nearest pound.

98. GEOMETRY A circular arc of 10 centimeters has a chord of 8 centimeters as shown in the figure.

(A) Explain how the radius is given by the equation

$$\sin \frac{5}{R} = \frac{4}{R}$$

(B) What difficulties do you encounter in trying to solve the equation in part A exactly using algebraic and trigonometric methods?

(C) Show on a graphing calculator how to approximate the radius of the circle R, and find R to three decimal places.

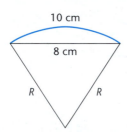

10 cm

8 cm

R R

99. MODELING TEMPERATURE VARIATION The 30-year average monthly temperature, in degrees Fahrenheit, for each month of the year for Washington, D.C., is given in Table 1 (from the *World Almanac*).

(A) Using 1 month as the basic unit of time, enter the data for a 2-year period in your graphing calculator and produce a scatter plot in the viewing window. Choose $25 \le y \le 80$ for the viewing window.

(B) It appears that a sine curve of the form

$$y = k + A \sin (Bx + C)$$

will closely model these data. The constants k, A, and B are easily determined from Table 1. To estimate C, visually estimate to one decimal place the smallest positive phase shift from the plot in part A. After determining A, B, k, and C, write the resulting equation. (Your value of C may differ slightly from the answer in the book.)

(C) Plot the results of parts A and B in the same viewing window. (An improved fit may result by adjusting your value of C a little.)

(D) If your graphing calculator has a sinusoidal regression feature, check your results from parts B and C by finding and plotting the regression equation.

Table 1 Monthly Average Temperatures, Washington, D.C.

x (months)	1	2	3	4	5	6	7	8	9	10	11	12
y (temperature)	31	34	43	53	62	71	76	74	67	55	45	35

CHAPTERS 9–11 Cumulative Review Exercises

Work through all the problems in this cumulative review and check answers in the back of the book. Answers to all review problems are there, and following each answer is a number in italics indicating the section in which that type of problem is discussed. Where weaknesses show up, review appropriate sections in the text. Note that Problems 4, 15, 16, 40, 41, 48, 49, and 88 are from sections that appear online.

1. Solve using substitution or elimination by addition:

$$3x - 5y = 11$$
$$2x + 3y = 1$$

2. Solve by graphing: $2x - y = -4$
$$3x + y = -1$$

3. Solve by substitution or elimination by addition:

$$-6x + 3y = 2$$
$$2x - y = 1$$

4. Solve by graphing: $3x + 5y \le 15$
$$x, y \ge 0$$

5. Determine whether each of the following can be the first three terms of an arithmetic sequence, a geometric sequence, or neither.

(A) 20, 15, 10, . . . (B) 5, 25, 125, . . .
(C) 5, 25, 50, . . . (D) 27, −9, 3, . . .
(E) −9, −6, −3, . . .

In Problems 6–8:
(A) Write the first four terms of each sequence.
(B) Find a_8. (C) Find S_8.

6. $a_n = 2 \cdot 5^n$ **7.** $a_n = 3n - 1$

8. $a_1 = 100; a_n = a_{n-1} - 6, n \ge 2$

9. Evaluate each of the following:

(A) $8!$ (B) $\dfrac{32!}{30!}$ (C) $\dfrac{9!}{3!(9-3)!}$

10. Evaluate each of the following:

(A) $\dbinom{7}{2}$ (B) $C_{7,2}$ (C) $P_{7,2}$

In Problems 11–13, graph each equation and locate foci. Locate the directrix for any parabolas. Find the lengths of major, minor, transverse, and conjugate axes where applicable.

11. $25x^2 - 36y^2 = 900$ **12.** $25x^2 + 36y^2 = 900$

13. $25x^2 - 36y = 0$

14. Find each determinant:

(A) $\begin{vmatrix} -3 & 5 \\ 2 & -2 \end{vmatrix}$ (B) $\begin{vmatrix} 5 & 3 \\ -5 & -3 \end{vmatrix}$

15. Solve $x^2 + y^2 = 2$
$$2x - y = 1$$

16. Find the maximum and minimum value of $z = 2x + 3y$ over the feasible region S:

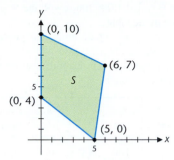

17. Perform the operations that are defined, given the following matrices:

$$M = \begin{bmatrix} 2 & 1 \\ 1 & -3 \end{bmatrix} \qquad N = \begin{bmatrix} 1 & 2 \\ -1 & 3 \end{bmatrix}$$

$$P = [1 \quad 2] \qquad Q = \begin{bmatrix} -1 \\ 2 \end{bmatrix}$$

(A) $M - 2N$ (B) $P + Q$ (C) PQ
(D) MN (E) PN (F) QM

18. A coin is flipped three times. How many combined outcomes are possible? Solve
(A) By using a tree diagram
(B) By using the multiplication principle

19. How many ways can four distinct books be arranged on a shelf? Solve
(A) By using the multiplication principle
(B) By using permutations or combinations, whichever is applicable

20. In a single deal of 3 cards from a standard 52-card deck, what is the probability of being dealt three diamonds?

21. Each of the 10 digits 0 through 9 is printed on 1 of 10 different cards. Four of these cards are drawn in succession without replacement. What is the probability of drawing the digits 4, 5, 6, and 7 by drawing 4 on the first draw, 5 on the second draw, 6 on the third draw, and 7 on the fourth draw? What is the probability of drawing the digits 4, 5, 6, and 7 in any order?

22. A thumbtack lands point down in 38 out of 100 tosses. What is the approximate empirical probability of the tack landing point up?

23. Write the linear system corresponding to each augmented matrix and solve:

(A) $\begin{bmatrix} 1 & 0 & | & 3 \\ 0 & 1 & | & -4 \end{bmatrix}$ (B) $\begin{bmatrix} 1 & -2 & | & 3 \\ 0 & 0 & | & 0 \end{bmatrix}$

(C) $\begin{bmatrix} 1 & -2 & | & 3 \\ 0 & 0 & | & 1 \end{bmatrix}$

24. Given the system: $x_1 + x_2 = 3$
 $-x_1 + x_2 = 5$

(A) Write the augmented matrix for the system.
(B) Transform the augmented matrix into reduced form.
(C) Write the solution to the system.

25. Given the system: $x_1 - 3x_2 = k_1$
 $2x_1 - 5x_2 = k_2$

(A) Write the system as a matrix equation of the form $AX = B$.
(B) Find the inverse of the coefficient matrix A.
(C) Use A^{-1} to find the solution for $k_1 = -2$ and $k_2 = 1$.
(D) Use A^{-1} to find the solution for $k_1 = 1$ and $k_2 = -2$.

26. Use Gauss–Jordan elimination to solve the system

$$x_1 + 3x_2 = 10$$
$$2x_1 - x_2 = -1$$

Then write the linear system represented by each augmented matrix in your solution, and solve each of these systems graphically. Discuss the relationship between the solutions of these systems.

27. Solve graphically to two decimal places:

$$-2x + 3y = 7$$
$$3x + 4y = 18$$

Verify the statement P_n in Problems 28 and 29 for $n = 1, 2,$ and 3.

28. P_n: $1 + 5 + 9 + \cdots + (4n - 3) = n(2n - 1)$

29. P_n: $n^2 + n + 2$ is divisible by 2

In Problems 30 and 31, write P_k and P_{k+1}.

30. For P_n in Problem 28 **31.** For P_n in Problem 29

32. Find the equation of the parabola having its vertex at the origin, its axis the y axis, and $(2, -8)$ on its graph.

33. Find an equation of an ellipse in the form

$$\frac{x^2}{M} + \frac{y^2}{N} = 1 \qquad M, N > 0$$

if the center is at the origin, the major axis is the x axis, the major axis length is 10, and the distance of the foci from the center is 3.

34. Find an equation of a hyperbola in the form

$$\frac{x^2}{M} - \frac{y^2}{N} = 1 \qquad M, N > 0$$

if the center is at the origin, the transverse axis length is 16, and the distance of the foci from the center is $\sqrt{89}$.

Solve Problems 35–37 using Gauss–Jordan elimination.

35. $x_1 + 2x_2 - x_3 = 3$
 $x_2 + x_3 = -2$
 $2x_1 + 3x_2 + x_3 = 0$

36. $x_1 + x_2 - x_3 = 2$
 $4x_2 + 6x_3 = -1$
 $6x_2 + 9x_3 = 0$

37. $x_1 - 2x_2 + x_3 = 1$
 $3x_1 - 2x_2 - x_3 = -5$

38. Given $M = [1 \quad 2 \quad -1]$ and $N = \begin{bmatrix} 1 \\ -1 \\ 2 \end{bmatrix}$. Find:

(A) MN (B) NM

39. Given

$$L = \begin{bmatrix} 2 & -1 & 0 \\ 1 & 2 & 1 \end{bmatrix} \quad M = \begin{bmatrix} 1 & 2 \\ -1 & 0 \\ 1 & 1 \end{bmatrix} \quad N = \begin{bmatrix} 2 & 1 \\ -1 & 0 \end{bmatrix}$$

Find, if defined: (A) $LM - 2N$ (B) $ML + N$

In Problems 40 and 41, solve the system.

40. $x^2 - 3xy + 3y^2 = 1$
　　　$xy = 1$

41. $x^2 - 3xy + y^2 = -1$
　　　$x^2 - xy = 0$

In Problems 42 and 43, find the determinant.

42. $\begin{vmatrix} 1 & 0 & 4 \\ 2 & 5 & -1 \\ 3 & 0 & -6 \end{vmatrix}$　　**43.** $\begin{vmatrix} -4 & 5 & -6 \\ 3 & -2 & -1 \\ 2 & 4 & 6 \end{vmatrix}$

44. Find all real solutions to two decimal places

$$x^2 + 2xy - y^2 = 1$$
$$9x^2 + 4xy + y^2 = 15$$

45. Write $\sum_{k=1}^{5} k^k$ without summation notation and find the sum.

46. Write the series $\dfrac{2}{2!} - \dfrac{2^2}{3!} + \dfrac{2^3}{4!} - \dfrac{2^4}{5!} + \dfrac{2^5}{6!} - \dfrac{2^6}{7!}$ using summation notation with the summation index k starting at $k = 1$.

47. Find S_∞ for the geometric series $108 - 36 + 12 - 4 + \cdots$.

48. Graph the solution region and indicate whether the solution region is bounded or unbounded. Find the coordinates of each corner point.

$$3x + 2y \geq 12$$
$$x + 2y \geq 8$$
$$x, y \geq 0$$

49. Solve the linear programming problem:

Maximize　$z = 4x + 9y$
Subject to　$x + 2y \leq 14$
　　　　　　$2x + y \leq 16$
　　　　　　$x, y \geq 0$

50. Given the system: $x_1 + 4x_2 + 2x_3 = k_1$
　　　　　　$2x_1 + 6x_2 + 3x_3 = k_2$
　　　　　　$2x_1 + 5x_2 + 2x_3 = k_3$
(A) Write the system as a matrix equation of the form $AX = B$.
(B) Find the inverse of the coefficient matrix A.
(C) Use A^{-1} to solve the system when $k_1 = -1$, $k_2 = 2$, and $k_3 = 1$.
(D) Use A^{-1} to solve the system when $k_1 = 2$, $k_2 = 0$, and $k_3 = -1$.

51. How many four-letter code words are possible using the first six letters of the alphabet if no letter can be repeated? If letters can be repeated? If adjacent letters cannot be alike?

52. A basketball team with 12 members has two centers. If 5 players are selected at random, what is the probability that both

centers are selected? Express the answer in terms of $C_{n,r}$ or $P_{n,r}$, as appropriate, and evaluate.

53. A single die is rolled 1,000 times with the frequencies of outcomes shown in the table.
(A) What is the approximate empirical probability that the number of dots showing is divisible by 3?
(B) What is the theoretical probability that the number of dots showing is divisible by 3?

Number of dots facing up	1	2	3	4	5	6
Frequency	160	155	195	180	140	170

54. Let $a_n = 100(0.9)^n$ and $b_n = 10 + 0.03n$. Find the least positive integer n such that $a_n < b_n$ by graphing the sequences $\{a_n\}$ and $\{b_n\}$ with a graphing calculator. Check your answer by using a graphing calculator to display both sequences in table form.

55. Evaluate each of the following:

(A) $P_{25,5}$　　(B) $C(25, 5)$　　(C) $\binom{25}{20}$

56. Expand $(a + \frac{1}{2}b)^6$ using the binomial formula.

57. Find the fifth and the eighth terms in the expansion of $(3x - y)^{10}$.

Prove each statement in Problems 58 and 59 for all positive integers using mathematical induction.

58. P_n in Problem 28　　**59.** P_n in Problem 29

60. Find the sum of all the odd integers between 50 and 500.

61. Use the formula for the sum of an infinite geometric series to write $2.\overline{45} = 2.454\,545\cdots$ as the quotient of two integers.

62. Let $a_k = \binom{30}{k}(0.1)^{30-k}(0.9)^k$ for $k = 0, 1, \cdots, 30$. Use a graphing calculator to find the largest term of the sequence $\{a_k\}$ and the number of terms that are greater than 0.01.

63. Use Cramer's rule to solve the system for x only:

$$-2x \qquad + 3z = -13$$
$$x - 6y + 5z = -16$$
$$-x + 2y = -1$$

64. Use Cramer's rule to solve the system in Problem 63 for y.

65. Use Cramer's rule to solve the system in Problem 63 for z.

66. How many nine-digit zip codes are possible? How many of these have no repeated digits?

67. Use mathematical induction to prove that the following statement holds for all positive integers:

$$P_n: \frac{1}{1 \cdot 3} + \frac{1}{3 \cdot 5} + \frac{1}{5 \cdot 7} + \cdots$$
$$+ \frac{1}{(2n-1)(2n+1)} = \frac{n}{2n+1}$$

68. Three-digit numbers are randomly formed from the digits 1, 2, 3, 4, and 5. What is the probability of forming an even number if digits cannot be repeated? If digits can be repeated?

69. Discuss the number of solutions for the system corresponding to the reduced form shown below if
(A) $m = 0$ and $n = 0$ (B) $m = 0$ and $n \neq 0$
(C) $m \neq 0$

$$\begin{bmatrix} 1 & 0 & -5 & | & 2 \\ 0 & 1 & 3 & | & 6 \\ 0 & 0 & m & | & n \end{bmatrix}$$

70. If a square matrix A satisfies the equation $A^2 = A$, find A. Assume that A^{-1} exists.

71. Which of the following augmented matrices are in reduced form?

$$L = \begin{bmatrix} 1 & 0 & 0 & | & 2 \\ 0 & 1 & 0 & | & 0 \\ 0 & 0 & 1 & | & -1 \end{bmatrix} \quad M = \begin{bmatrix} 1 & 0 & 3 & | & 3 \\ 0 & 1 & -2 & | & 2 \\ 0 & 0 & 0 & | & 0 \end{bmatrix}$$

$$N = \begin{bmatrix} 0 & 0 & | & 0 \\ 1 & 0 & | & 2 \\ 0 & 1 & | & -3 \end{bmatrix} \quad P = \begin{bmatrix} 1 & 2 & 0 & 2 & | & -2 \\ 0 & 0 & 1 & 3 & | & 1 \end{bmatrix}$$

*Recall that a square matrix is called **upper triangular** if all elements below the principal diagonal are zero, and it is called **diagonal** if all elements not on the principal diagonal are zero. A square matrix is called **lower triangular** if all elements above the principal diagonal are zero. In Problems 72–77, determine whether the statement is true or false. If true, explain why. If false, give a counterexample.*

72. The sum of two upper triangular matrices is upper triangular.

73. The product of two lower triangular matrices is lower triangular.

74. The sum of an upper triangular matrix and a lower triangular matrix is a diagonal matrix.

75. The product of an upper triangular matrix and a lower triangular matrix is a diagonal matrix.

76. A matrix that is both upper triangular and lower triangular is a diagonal matrix.

77. If a diagonal matrix has no zero elements on the principal diagonal, then it has an inverse.

78. Use the binomial formula to expand $(x - 2i)^6$, where i is the imaginary unit.

79. Use the definition of a parabola and the distance formula to find the equation of a parabola with directrix $y = 3$ and focus $(6, 1)$.

80. An ellipse has vertices $(\pm 4, 0)$ and foci $(\pm 2, 0)$. Find the y intercepts.

81. A hyperbola has vertices $(2, \pm 3)$ and foci $(2, \pm 5)$. Find the length of the conjugate axis.

82. Seven distinct points are selected on the circumference of a circle. How many triangles can be formed using these seven points as vertices?

83. Use mathematical induction to prove that $2^n < n!$ for all integers $n > 3$.

84. Use mathematical induction to show that $\{a_n\} = \{b_n\}$, where $a_1 = 3$, $a_n = 2a_{n-1} - 1$ for $n > 1$, and $b_n = 2^n + 1$, $n \geq 1$.

85. Find an equation of the set of points in the plane each of whose distance from $(1, 4)$ is three times its distance from the x axis. Write the equation in the form $Ax^2 + Cy^2 + Dx + Ey + F = 0$, and identify the curve.

86. A box of 12 lightbulbs contains 4 defective bulbs. If three bulbs are selected at random, what is the probability of selecting at least one defective bulb?

APPLICATIONS

87. ECONOMICS The government, through a subsidy program, distributes $2,000,000. If we assume that each individual or agency spends 75% of what it receives, and 75% of this is spent, and so on, how much total increase in spending results from this government action?

88. GEOMETRY Find the dimensions of a rectangle with perimeter 24 meters and area 32 square meters.

89. ENGINEERING An automobile headlight contains a parabolic reflector with a diameter of 8 inches. If the light source is located at the focus, which is 1 inch from the vertex, how deep is the reflector?

90. ARCHITECTURE A sound whispered at one focus of a whispering chamber can be easily heard at the other focus. Suppose that a cross section of this chamber is a semielliptical arch that is 80 feet wide and 24 feet high (see the figure). How far is each focus from the center of the arch? How high is the arch above each focus?

91. FINANCE An investor has $12,000 to invest. If part is invested at 8% and the rest in a higher-risk investment at 14%, how much should be invested at each rate to produce the same yield as if all had been invested at 10%?

92. DIET In an experiment involving mice, a zoologist needs a food mix that contains, among other things, 23 grams of protein, 6.2 grams of fat, and 16 grams of moisture. She has on hand mixes of the following compositions: Mix A contains 20% protein, 2% fat, and 15% moisture, mix B contains 10% protein, 6% fat, and 10% moisture; and mix C contains 15% protein, 5% fat, and 5% moisture. How many grams of each mix should be used to get the desired diet mix?

93. PURCHASING A soft-drink distributor has budgeted $300,000 for the purchase of 12 new delivery trucks. If a model A truck costs $18,000, a model B truck costs $22,000, and a model C truck costs $30,000, how many trucks of each model should the distributor purchase to use exactly all the budgeted funds?

B-1 Scientific Notation and Significant Digits

> Significant Digits
> Rounding Convention

› Significant Digits

Most calculations involving problems in the real world deal with numbers that are only approximate. It therefore seems reasonable to assume that a final answer should not be any more accurate than the least accurate number used in the calculation. This is an important point, because calculators tend to give the impression that greater accuracy is achieved than is warranted.

Suppose we want to compute the length of the diagonal of a rectangular field from measurements of its sides of 237.8 meters and 61.3 meters. Using the Pythagorean theorem and a calculator, we find

$$d = \sqrt{237.8^2 + 61.3^2}$$
$$= 245.573\ 878\ldots$$

61.3 meters

237.8 meters

The calculator answer suggests an accuracy that is not justified. What accuracy is justified? To answer this question, we introduce the idea of *significant digits*.

Whenever we write a measurement such as 61.3 meters, we assume that the measurement is accurate to the last digit written. So the measurement 61.3 meters indicates that the measurement was made to the nearest tenth of a meter. That is, the actual width is between 61.25 meters and 61.35 meters. In general, the digits in a number that indicate the accuracy of the number are called **significant digits.** If all the digits in a number are nonzero, then they are all significant. So the measurement 61.3 meters has three significant digits, and the measurement 237.8 meters has four significant digits.

What are the significant digits in the number 7,800? The accuracy of this number is not clear. It could represent a measurement with any of the following accuracies:

Between 7,750 and 7,850	Correct to the hundreds place
Between 7,795 and 7,805	Correct to the tens place
Between 7,799.5 and 7,800.5	Correct to the units place

To give a precise definition of significant digits that resolves this ambiguity, we use scientific notation.

> ### › DEFINITION 1 Significant Digits
>
> If a number x is written in scientific notation as
>
> $$x = a \times 10^n \qquad 1 \le a < 10,\ n \text{ an integer}$$
>
> then the number of significant digits in x is the number of digits in a.

Using this definition,

$$7.8 \times 10^3 \qquad \text{has two significant digits}$$
$$7.80 \times 10^3 \qquad \text{has three significant digits}$$
$$7.800 \times 10^3 \qquad \text{has four significant digits}$$

All three of these measurements have the same decimal representation (7,800), but each represents a different accuracy.

Definition 1 tells us how to write a number so that the number of significant digits is clear, but it does not tell us how to interpret the accuracy of a number that is not written in scientific notation. We will use the following convention for numbers that are written as decimal fractions:

> **SIGNIFICANT DIGITS IN DECIMAL FRACTIONS**
>
> The number of significant digits in a number with no decimal point is found by counting the digits from left to right, starting with the first digit and ending with the last *nonzero* digit.
>
> The number of significant digits in a number containing a decimal point is found by counting the digits from left to right, starting with the first *nonzero* digit and ending with the last digit.

Applying this rule to the number 7,800, we conclude that this number has two significant digits. If we want to indicate that it has three or four significant digits, we must use scientific notation.

EXAMPLE 1

Significant Digits in Decimal Fractions

Underline the significant digits in the following numbers:

(A) 70,007 (B) 82,000 (C) 5.600 (D) 0.0008 (E) 0.000 830

SOLUTIONS (A) <u>70,007</u> (B) <u>82</u>,000 (C) <u>5.600</u> (D) 0.000<u>8</u> (E) 0.000 <u>830</u>

MATCHED PROBLEM 1 Underline the significant digits in the following numbers:

(A) 5,009 (B) 12,300 (C) 23.4000 (D) 0.00050 (E) 0.0012

› Rounding Convention

In calculations involving multiplication, division, powers, and roots, we adopt the following convention:

> **› ROUNDING CALCULATED VALUES**
>
> The result of a calculation is rounded to the same number of significant digits as the number used in the calculation that has the least number of significant digits.

So, in computing the length of the diagonal of the rectangular field shown earlier, we write the answer rounded to three significant digits because the width has three significant digits and the length has four significant digits:

$$d = 246 \text{ meters} \qquad \text{Three significant digits}$$

One Final Note: In rounding a number that is exactly halfway between a larger and a smaller number, we use the convention of making the final result even.

EXAMPLE 2 **Rounding Numbers**

Round each number to three significant digits.

(A) 43.0690 (B) 48.05 (C) 48.15 (D) $8.017\ 632 \times 10^{-3}$

SOLUTIONS (A) 43.1

(B) 48.0

(C) 48.2 Use the convention of making the digit before the 5 even if it is odd, or leaving it alone if it is even.

(D) 8.02×10^{-3}

MATCHED PROBLEM 2 Round each number to three significant digits.

(A) 3.1495 (B) 0.004 135 (C) 32,450 (D) $4.314\ 764\ 09 \times 10^{12}$

ANSWERS TO MATCHED PROBLEMS

1. (A) <u>5,009</u> (B) <u>12,3</u>00 (C) <u>23.4000</u> (D) 0.000<u>50</u> (E) 0.00<u>12</u>
2. (A) 3.15 (B) 0.004 14 (C) 32,400 (D) 4.31×10^{12}

B-1 Exercises

In Problems 1–12, underline the significant digits in each number.

1. 123,005 2. 3,400,002 3. 20,040
4. 300,600 5. 6.0 6. 7.00
7. 80.000 8. 900.0000 9. 0.012
10. 0.0015 11. 0.000 960 12. 0.000 700

In Problems 13–22, round each number to three significant digits.

13. 3.0780 14. 4.0240 15. 924,300
16. 643,820 17. 23.65 18. 23.75

19. $2.816\ 743 \times 10^3$
20. 56.114×10^4
21. $6.782\ 045 \times 10^{-4}$
22. $5.248\ 102 \times 10^{-3}$

In Problems 23 and 24, find the diagonal of the rectangle with the indicated side measurements. Round answers to the number of significant digits appropriate for the given measurements.

23. 25 feet by 20 feet
24. 2,900 yards by 1,570 yards

B-2

Partial Fractions

> Basic Theorems
> Partial Fraction Decomposition

You have now had considerable experience combining two or more rational expressions into a single rational expression. For example, problems such as

$$\frac{2}{x + 5} + \frac{3}{x - 4} = \frac{2(x - 4) + 3(x + 5)}{(x + 5)(x - 4)} = \frac{5x + 7}{(x + 5)(x - 4)}$$

should seem routine. Frequently in more advanced courses, particularly in calculus, it is useful to be able to reverse this process—that is, to be able to express a rational expression as the sum of two or more simpler rational expressions called **partial fractions.** As is often the case with reverse processes, the process of decomposing a rational expression into partial fractions is more difficult than combining rational expressions. Basic to the process is the factoring of polynomials, so many of the topics discussed in Chapter 4 can be put to effective use. Partial fraction decomposition is usually accomplished by solving a related system of linear equations. If you are familiar with basic techniques for solving linear systems discussed earlier in this book, such as Gauss–Jordan elimination, inverse matrix solutions, or Cramer's rule, you may use these as you see fit. However, all of the linear systems encountered in this section can also be solved by some special techniques developed here. Mathematically equivalent to the techniques mentioned, these special techniques are generally easier to use in partial fraction decomposition problems.

We confine our attention to rational expressions of the form $P(x)/D(x)$, where $P(x)$ and $D(x)$ are polynomials with real coefficients. In addition, we assume that the degree of $P(x)$ is less than the degree of $D(x)$. If the degree of $P(x)$ is greater than or equal to that of $D(x)$, we have only to divide $P(x)$ by $D(x)$ to obtain

$$\frac{P(x)}{D(x)} = Q(x) + \frac{R(x)}{D(x)}$$

where the degree of $R(x)$ is less than that of $D(x)$. For example,

$$\frac{x^4 - 3x^3 + 2x^2 - 5x + 1}{x^2 - 2x + 1} = x^2 - x - 1 + \frac{-6x + 2}{x^2 - 2x + 1}$$

If the degree of $P(x)$ is less than that of $D(x)$, then $P(x)/D(x)$ is called a **proper fraction.**

> ## Basic Theorems

Our task now is to establish a systematic way to decompose a proper fraction into the sum of two or more partial fractions. Theorems 1, 2, and 3 take care of the problem completely.

> ### THEOREM 1 Equal Polynomials

Two polynomials are equal to each other if and only if the coefficients of terms of like degree are equal.

For example, if

then

$$B = -3$$ Substitute $B = -3$ into the second equation to solve for A.

$$A + 2B = 5$$

$$A + 2(-3) = 5$$

$$A = 11$$

>>> EXPLORE-DISCUSS 1

If

$$x + 5 = A(x + 1) + B(x - 3) \qquad (1)$$

is a polynomial identity (that is, both sides represent the same polynomial), then equating coefficients produces the system

$$1 = A + B$$ Equating coefficients of x

$$5 = A - 3B$$ Equating constant terms

(A) Solve this system graphically.

(B) For an alternate method of solution, substitute $x = 3$ in equation (1) to find A and then substitute $x = -1$ in equation (1) to find B. Explain why this method is valid.

The Linear and Quadratic Factors Theorem from Chapter 4 (page 290) underlies the technique of decomposing a rational function into partial fractions. We restate the theorem here.

> **THEOREM 2** Linear and Quadratic Factors Theorem

For a polynomial of degree $n > 0$ with real coefficients, there always exists a factorization involving only linear and/or quadratic factors with real coefficients in which the quadratic factors have imaginary zeros.

The quadratic formula can be used to determine easily whether a given quadratic factor $ax^2 + bx + c$, with real coefficients, has imaginary zeros. If $b^2 - 4ac < 0$, then $ax^2 + bx + c$ has imaginary zeros. Otherwise its zeros are real. Therefore, $ax^2 + bx + c$ has imaginary zeros if and only if it cannot be factored as a product of linear factors with real coefficients.

> ## Partial Fraction Decomposition

We are now ready to state Theorem 3, which forms the basis for partial fraction decomposition.

> **THEOREM 3** Partial Fraction Decomposition

Any proper fraction $P(x)/D(x)$ reduced to lowest terms can be decomposed into the sum of partial fractions as follows:

1. If $D(x)$ has a nonrepeating linear factor of the form $ax + b$, then the partial fraction decomposition of $P(x)/D(x)$ contains a term of the form

$$\frac{A}{ax + b} \qquad A \text{ a constant}$$

2. If $D(x)$ has a k-repeating linear factor of the form $(ax + b)^k$, then the partial fraction decomposition of $P(x)/D(x)$ contains k terms of the form

$$\frac{A_1}{ax + b} + \frac{A_2}{(ax + b)^2} + \cdots + \frac{A_k}{(ax + b)^k} \qquad A_1, A_2, \ldots, A_k \text{ constants}$$

3. If $D(x)$ has a nonrepeating quadratic factor of the form $ax^2 + bx + c$ that has imaginary zeros, then the partial fraction decomposition of $P(x)/D(x)$ contains a term of the form

$$\frac{Ax + B}{ax^2 + bx + c} \qquad A, B \text{ constants}$$

4. If $D(x)$ has a k-repeating quadratic factor of the form $(ax^2 + bx + c)^k$, where $ax^2 + bx + c$ has imaginary zeros, then the partial fraction decomposition of $P(x)/D(x)$ contains k terms of the form

$$\frac{A_1x + B_1}{ax^2 + bx + c} + \frac{A_2x + B_2}{(ax^2 + bx + c)^2} + \cdots + \frac{A_kx + B_k}{(ax^2 + bx + c)^k}$$

$$A_1, \ldots, A_k, B_1, \ldots, B_k \text{ constants}$$

Let's see how the theorem is used to obtain partial fraction decompositions in several examples.

EXAMPLE 1 Nonrepeating Linear Factors

Decompose into partial fractions: $\dfrac{5x + 7}{x^2 + 2x - 3}$.

SOLUTION We first try to factor the denominator. If it can't be factored in the real numbers, then we can't go any further. In this example, the denominator factors, so we apply part 1 from Theorem 3:

$$\frac{5x + 7}{(x - 1)(x + 3)} = \frac{A}{x - 1} + \frac{B}{x + 3} \qquad (2)$$

To find the constants A and B, we combine the fractions on the right side of equation (2) to obtain

$$\frac{5x + 7}{(x - 1)(x + 3)} = \frac{A(x + 3) + B(x - 1)}{(x - 1)(x + 3)}$$

Because these fractions have the same denominator, their numerators must be equal. So

$$5x + 7 = A(x + 3) + B(x - 1) \qquad (3)$$

We could multiply the right side and find A and B by using Theorem 1, but in this case it is easier to take advantage of the fact that equation (3) is an identity—that is, it must hold for all values of x. In particular, we note that if we let $x = 1$, then the second term of the right side drops out and we can solve for A:

$$5 \cdot 1 + 7 = A(1 + 3) + B(1 - 1)$$
$$12 = 4A$$
$$A = 3$$

Similarly, if we let $x = -3$, the first term drops out and we find

$$-8 = -4B$$
$$B = 2$$

Now we have the decomposition:

$$\frac{5x + 7}{x^2 + 2x - 3} = \frac{3}{x - 1} + \frac{2}{x + 3} \tag{4}$$

as can easily be checked by adding the two fractions on the right.

MATCHED PROBLEM 1 Decompose into partial fractions: $\dfrac{7x + 6}{x^2 + x - 6}$.

Technology Connections

A graphing calculator can also be used to check a partial fraction decomposition. To check Example 1, we graph the left and right sides of equation (4) in a graphing calculator (Fig. 1). Discuss how the TRACE command on the graphing calculator can be used to check that the graphing calculator is displaying two identical graphs.

> Figure 1

EXAMPLE 2 Repeating Linear Factors

Decompose into partial fractions: $\dfrac{6x^2 - 14x - 27}{(x + 2)(x - 3)^2}$.

SOLUTION Using parts 1 and 2 from Theorem 3, we write

$$\frac{6x^2 - 14x - 27}{(x + 2)(x - 3)^2} = \frac{A}{x + 2} + \frac{B}{x - 3} + \frac{C}{(x - 3)^2}$$
$$= \frac{A(x - 3)^2 + B(x + 2)(x - 3) + C(x + 2)}{(x + 2)(x - 3)^2}$$

So for all x,

$$6x^2 - 14x - 27 = A(x - 3)^2 + B(x + 2)(x - 3) + C(x + 2)$$

If $x = 3$, then If $x = -2$, then

$$-15 = 5C \qquad 25 = 25A$$
$$C = -3 \qquad\ A = 1$$

There are no other values of x that will cause terms on the right to drop out. Because any value of x can be substituted to produce an equation relating A, B, and C, we let $x = 0$ and obtain

$$-27 = 9A - 6B + 2C \qquad \text{Substitute } A = 1 \text{ and } C = -3.$$
$$-27 = 9 - 6B - 6$$
$$B = 5$$

Therefore,

$$\frac{6x^2 - 14x - 27}{(x + 2)(x - 3)^2} = \frac{1}{x + 2} + \frac{5}{x - 3} - \frac{3}{(x - 3)^2}$$

MATCHED PROBLEM 2

Decompose into partial fractions: $\dfrac{x^2 + 11x + 15}{(x - 1)(x + 2)^2}$.

EXAMPLE 3

Nonrepeating Linear and Quadratic Factors

Decompose into partial fractions: $\dfrac{5x^2 - 8x + 5}{(x - 2)(x^2 - x + 1)}$.

SOLUTION First, we see that the quadratic in the denominator can't be factored further in the real numbers. Then, we use parts 1 and 3 from Theorem 3 to write

$$\frac{5x^2 - 8x + 5}{(x - 2)(x^2 - x + 1)} = \frac{A}{x - 2} + \frac{Bx + C}{x^2 - x + 1}$$

$$= \frac{A(x^2 - x + 1) + (Bx + C)(x - 2)}{(x - 2)(x^2 - x + 1)}$$

So for all x,

$$5x^2 - 8x + 5 = A(x^2 - x + 1) + (Bx + C)(x - 2)$$

If $x = 2$, then

$$9 = 3A$$
$$A = 3$$

If $x = 0$, then, using $A = 3$, we have

$$5 = 3 - 2C$$
$$C = -1$$

If $x = 1$, then, using $A = 3$ and $C = -1$, we have

$$2 = 3 + (B - 1)(-1)$$
$$B = 2$$

Therefore,

$$\frac{5x^2 - 8x + 5}{(x - 2)(x^2 - x + 1)} = \frac{3}{x - 2} + \frac{2x - 1}{x^2 - x + 1}$$

MATCHED PROBLEM 3 Decompose into partial fractions: $\dfrac{7x^2 - 11x + 6}{(x - 1)(2x^2 - 3x + 2)}$.

EXAMPLE 4 **Repeating Quadratic Factors**

Decompose into partial fractions: $\dfrac{x^3 - 4x^2 + 9x - 5}{(x^2 - 2x + 3)^2}$.

SOLUTION Because $x^2 - 2x + 3$ can't be factored further in the real numbers, we proceed to use part 4 from Theorem 3 to write

$$\frac{x^3 - 4x^2 + 9x - 5}{(x^2 - 2x + 3)^2} = \frac{Ax + B}{x^2 - 2x + 3} + \frac{Cx + D}{(x^2 - 2x + 3)^2}$$

$$= \frac{(Ax + B)(x^2 - 2x + 3) + Cx + D}{(x^2 - 2x + 3)^2}$$

So for all x,

$$x^3 - 4x^2 + 9x - 5 = (Ax + B)(x^2 - 2x + 3) + Cx + D$$

Because the substitution of carefully chosen values of x doesn't lead to the immediate determination of A, B, C, or D, we multiply and rearrange the right side to obtain

$$x^3 - 4x^2 + 9x - 5 = Ax^3 + (B - 2A)x^2 + (3A - 2B + C)x + (3B + D)$$

Now we use Theorem 1 to equate coefficients of terms of like degree:

$$A = 1$$
$$B - 2A = -4$$
$$3A - 2B + C = 9$$
$$3B + D = -5$$

From these equations we easily find that $A = 1$, $B = -2$, $C = 2$, and $D = 1$. Now we can write

$$\frac{x^3 - 4x^2 + 9x - 5}{(x^2 - 2x + 3)^2} = \frac{x - 2}{x^2 - 2x + 3} + \frac{2x + 1}{(x^2 - 2x + 3)^2}$$

MATCHED PROBLEM 4 Decompose into partial fractions: $\dfrac{3x^3 - 6x^2 + 7x - 2}{(x^2 - 2x + 2)^2}$.

ANSWERS TO MATCHED PROBLEMS

1. $\dfrac{4}{x - 2} + \dfrac{3}{x + 3}$ 2. $\dfrac{3}{x - 1} - \dfrac{2}{x + 2} + \dfrac{1}{(x + 2)^2}$

3. $\dfrac{2}{x - 1} + \dfrac{3x - 2}{2x^2 - 3x + 2}$ 4. $\dfrac{3x}{x^2 - 2x + 2} + \dfrac{x - 2}{(x^2 - 2x + 2)^2}$

B-2 Exercises

In Problems 1–4, find A and B so that the right side is equal to the left. After cross-multiplying to produce a polynomial equation, solve each problem two ways (see Explore-Discuss 1). First, equate the coefficients of both sides to determine a linear system for A and B and solve this system. Second, solve for A and B by evaluating both sides for selected values of x.

1. $\dfrac{7x - 14}{(x - 4)(x + 3)} = \dfrac{A}{x - 4} + \dfrac{B}{x + 3}$

2. $\dfrac{9x + 21}{(x + 5)(x - 3)} = \dfrac{A}{x + 5} + \dfrac{B}{x - 3}$

3. $\dfrac{17x - 1}{(2x - 3)(3x - 1)} = \dfrac{A}{2x - 3} + \dfrac{B}{3x - 1}$

4. $\dfrac{x - 11}{(3x + 2)(2x - 1)} = \dfrac{A}{3x + 2} + \dfrac{B}{2x - 1}$

In Problems 5–10, find A, B, C, and D, so that the right side is equal to the left.

5. $\dfrac{3x^2 + 7x + 1}{x(x + 1)^2} = \dfrac{A}{x} + \dfrac{B}{x + 1} + \dfrac{C}{(x + 1)^2}$

6. $\dfrac{x^2 - 6x + 11}{(x + 1)(x - 2)^2} = \dfrac{A}{x + 1} + \dfrac{B}{x - 2} + \dfrac{C}{(x - 2)^2}$

7. $\dfrac{3x^2 + x}{(x - 2)(x^2 + 3)} = \dfrac{A}{x - 2} + \dfrac{Bx + C}{x^2 + 3}$

8. $\dfrac{5x^2 - 9x + 19}{(x - 4)(x^2 + 5)} = \dfrac{A}{x - 4} + \dfrac{Bx + C}{x^2 + 5}$

9. $\dfrac{2x^2 + 4x - 1}{(x^2 + x + 1)^2} = \dfrac{Ax + B}{x^2 + x + 1} + \dfrac{Cx + D}{(x^2 + x + 1)^2}$

10. $\dfrac{3x^3 - 3x^2 + 10x - 4}{(x^2 - x + 3)^2} = \dfrac{Ax + B}{x^2 - x + 3} + \dfrac{Cx + D}{(x^2 - x + 3)^2}$

In Problems 11–30, decompose into partial fractions.

11. $\dfrac{-x + 22}{x^2 - 2x - 8}$

12. $\dfrac{-x - 21}{x^2 + 2x - 15}$

13. $\dfrac{3x - 13}{6x^2 - x - 12}$

14. $\dfrac{11x - 11}{6x^2 + 7x - 3}$

15. $\dfrac{x^2 - 12x + 18}{x^3 - 6x^2 + 9x}$

16. $\dfrac{5x^2 - 36x + 48}{x(x - 4)^2}$

17. $\dfrac{5x^2 + 3x + 6}{x^3 + 2x^2 + 3x}$

18. $\dfrac{6x^2 - 15x + 16}{x^3 - 3x^2 + 4x}$

19. $\dfrac{2x^3 + 7x + 5}{x^4 + 4x^2 + 4}$

20. $\dfrac{-5x^2 + 7x - 18}{x^4 + 6x^2 + 9}$

21. $\dfrac{x^3 - 7x^2 + 17x - 17}{x^2 - 5x + 6}$

22. $\dfrac{x^3 + x^2 - 13x + 11}{x^2 + 2x - 15}$

23. $\dfrac{4x^2 + 5x - 9}{x^3 - 6x - 9}$

24. $\dfrac{4x^2 - 8x + 1}{x^3 - x + 6}$

25. $\dfrac{x^2 + 16x + 18}{x^3 + 2x^2 - 15x - 36}$

26. $\dfrac{5x^2 - 18x + 1}{x^3 - x^2 - 8x + 12}$

27. $\dfrac{-x^2 + x - 7}{x^4 - 5x^3 + 9x^2 - 8x + 4}$

28. $\dfrac{-2x^3 + 12x^2 - 20x - 10}{x^4 - 7x^3 + 17x^2 - 21x + 18}$

29. $\dfrac{4x^5 + 12x^4 - x^3 + 7x^2 - 4x + 2}{4x^4 + 4x^3 - 5x^2 + 5x - 2}$

30. $\dfrac{6x^5 - 13x^4 + x^3 - 8x^2 + 2x}{6x^4 - 7x^3 + x^2 + x - 1}$

B-3 Parametric Equations

> Parametric Equations and Plane Curves
> Parametric Equations and Conic Sections
> Projectile Motion
> Cycloid

> ### Parametric Equations and Plane Curves

Consider the two equations

$$\begin{aligned} x &= t + 1 \\ y &= t^2 - 2t \end{aligned} \qquad -\infty < t < \infty \qquad (1)$$

> **Figure 1** Graph of $x = t + 1$, $y = t^2 - 2t$, $-\infty < t < \infty$.

Each value of t determines a value of x, a value of y, and therefore, an ordered pair (x, y). To graph the set of ordered pairs (x, y) determined by letting t assume all real values, we construct Table 1 listing selected values of t and the corresponding values of x and y. Then we plot the ordered pairs (x, y) and connect them with a continuous curve, as shown in Figure 1. The variable t is called a *parameter* and does not appear on the graph. Equations (1) are called *parametric equations* because both x and y are expressed in terms of the parameter t. The graph of the ordered pairs (x, y) is called a *plane curve*.

Table 1

t	0	1	2	3	4	-1	-2
x	1	2	3	4	5	0	-1
y	0	-1	0	3	8	3	8

Technology Connections

Parametric equations can also be graphed on a graphing calculator. Figure 2(a) shows the Parametric mode selected on a Texas Instruments TI-84 calculator. Figure 2(b) shows the equation editor with the parametric equations in (1) en-tered as x_{1T} and y_{1T}. In Figure 2(c), notice that there are three new window variables, Tmin, Tmax, and Tstep, that must be entered by the user.

(a)

(b)

(c)

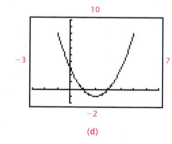

(d)

> **Figure 2** Graphing parametric equations on a graphing calculator.

In some cases, it is possible to eliminate the parameter by solving one of the equations for t and substituting into the other. In the example just considered, solving the first equation for t in terms of x, we have

$$t = x - 1$$

Then, substituting the result into the second equation, we obtain

$$y = (x - 1)^2 - 2(x - 1)$$
$$= x^2 - 4x + 3$$

We recognize this as the equation of a parabola, as we would guess from Figure 1.

In other cases, it may not be easy or possible to eliminate the parameter to obtain an equation in just x and y. For example, for

$$\begin{aligned} x &= t + \log t \\ y &= t - e^t \end{aligned} \qquad t > 0$$

you will not find it possible to solve either equation for t in terms of functions we have considered.

Is there more than one parametric representation for a plane curve? The answer is yes. In fact, there is an unlimited number of parametric representations for the same plane curve. The following are two additional representations of the parabola in Figure 1.

$$x = t + 3$$
$$y = t^2 + 2t \qquad -\infty < t < \infty \tag{2}$$

$$x = t$$
$$y = t^2 - 4t + 3 \qquad -\infty < t < \infty \tag{3}$$

The concepts introduced in the preceding discussion are summarized in Definition 1.

> **DEFINITION 1** Parametric Equations and Plane Curves

A **plane curve** is the set of points (x, y) determined by the **parametric equations**

$$x = f(t)$$
$$y = g(t)$$

where the **parameter** t varies over an interval I and the functions f and g are both defined on the interval I.

Why are we interested in parametric representations of plane curves? It turns out that this approach is more general than using equations with two variables as we have been doing. In addition, the approach generalizes to curves in three- and higher-dimensional spaces. Other important reasons for using parametric representations of plane curves will be brought out in the discussion and examples that follow.

EXAMPLE 1 **Eliminating the Parameter**

Eliminate the parameter and identify the plane curve given parametrically by

$$x = \sqrt{t}$$
$$y = \sqrt{9 - t} \qquad 0 \le t \le 9 \tag{4}$$

SOLUTION To eliminate the parameter t, we solve each equation (4) for t:

$$x = \sqrt{t} \qquad y = \sqrt{9 - t}$$
$$x^2 = t \qquad y^2 = 9 - t$$
$$t = 9 - y^2$$

Equating the last two equations, we have

$$x^2 = 9 - y^2$$
$$x^2 + y^2 = 9 \qquad \text{A circle of radius 3 centered at (0, 0)}$$

As the parameter t increases from 0 to 9, x will increase from 0 to 3 and y will decrease from 3 to 0.

So the graph of the parametric equations in (4) is the quarter of the circle of radius 3 centered at the origin that lies in the first quadrant (Fig. 3).

> Figure 3

MATCHED PROBLEM 1 Eliminate the parameter and identify the plane curve given parametrically by $x = \sqrt{4 - t}$, $y = -\sqrt{t}, 0 \le t \le 4$.

› **Parametric Equations and Conic Sections**

Trigonometric functions provide very effective representations for many conic sections. Examples 2 and 3 illustrate the basic concepts.

| EXAMPLE | 2 | **Identifying a Conic Section in Parametric Form** |

Eliminate the parameter θ and identify the plane curve given by

$$x = 8 \cos \theta$$
$$y = 4 \sin \theta \qquad 0 \leq \theta \leq 2\pi \qquad (5)$$

SOLUTION To eliminate the parameter θ, we solve the first equation in (5) for $\cos \theta$, the second for $\sin \theta$, and substitute into the Pythagorean identity $\cos^2 \theta + \sin^2 \theta = 1$:

$$\cos \theta = \frac{x}{8} \qquad \text{and} \qquad \sin \theta = \frac{y}{4}$$

$$\cos^2 \theta + \sin^2 \theta = 1$$

$$\left(\frac{x}{8}\right)^2 + \left(\frac{y}{4}\right)^2 = 1$$

$$\frac{x^2}{64} + \frac{y^2}{16} = 1$$

The graph is an ellipse (Fig. 4).

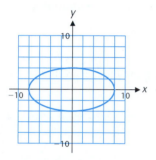

› **Figure 4** Graph of $x = 8 \cos \theta$, $y = 4 \sin \theta$, $0 \leq \theta \leq 2\pi$.

| MATCHED PROBLEM 2 | Eliminate the parameter θ and identify the plane curve given by $x = 4 \cos \theta$, $y = 4 \sin \theta$, $0 \leq \theta \leq 2\pi$. |

››› **EXPLORE-DISCUSS 1**

Identify each of the three plane curves given parametrically by

$$x_1 = 5 \cos \theta \qquad x_2 = 2 \cos \theta \qquad x_3 = 5 \cos \theta$$
$$y_1 = 5 \sin \theta \qquad y_2 = 2 \sin \theta \qquad y_3 = 2 \sin \theta$$

Graph one period of each curve ($0 \leq \theta \leq 2\pi$). What happens if you graph less than one period? More than one period?

| EXAMPLE | 3 | **Parametric Equations for Conic Sections** |

Find parametric equations for the conic section with the given equation:

(A) $25x^2 + 9y^2 - 100x + 54y - 44 = 0$ (B) $x^2 - 16y^2 - 10x + 32y - 7 = 0$

SOLUTIONS (A) By completing the square in x and y we obtain the standard form $\frac{(x - 2)^2}{9} + \frac{(y + 3)^2}{25} = 1$. So the graph is an ellipse with center $(2, -3)$ and major axis on the line $x = 2$. Because $\cos^2 \theta + \sin^2 \theta = 1$, a parametric representation with parameter θ is obtained by letting $\frac{x - 2}{3} = \cos \theta$, $\frac{y + 3}{5} = \sin \theta$:

$$x = 2 + 3 \cos \theta$$
$$y = -3 + 5 \sin \theta$$

> **Figure 5** $x = 2 + 3 \cos \theta$, $y = -3 + 5 \sin \theta$, $0 \le \theta \le 2\pi$.

Because $\sin \theta$ and $\cos \theta$ have period 2π, graphing these equations for $0 \le \theta \le 2\pi$ will produce a complete graph of the ellipse (Fig. 5).

(B) By completing the square in x and y we obtain the standard form $\dfrac{(x-5)^2}{16} - (y-1)^2 = 1$. So the graph is a hyperbola with center $(5, 1)$ and transverse axis on the line $y = 1$. Because $\sec^2 \theta - \tan^2 \theta = 1$, a parametric representation with parameter θ is obtained by letting $\dfrac{x-5}{4} = \sec \theta$, $y - 1 = \tan \theta$:

$$x = 5 + 4 \sec \theta$$
$$y = 1 + \tan \theta$$

The period of $\tan \theta$ is π, but the period of $\sec \theta$ is 2π, so we have to use $0 \le \theta \le 2\pi$ to produce a complete graph of the hyperbola (Fig. 6). To be precise, we should exclude $\theta = \pi/2$ and $3\pi/2$, because the tangent function is not defined at these values.

> **Figure 6** $x = 5 + 4 \sec \theta$, $y = 1 + \tan \theta$, $0 \le \theta \le 2\pi$, $\theta \ne \dfrac{\pi}{2}, \dfrac{3\pi}{2}$.

MATCHED PROBLEM 3

Find parametric equations for the conic section with the given equation:

(A) $36x^2 + 16y^2 + 504x - 96y + 1{,}332 = 0$

(B) $16y^2 - 9x^2 - 36x + 128y + 76 = 0$

> ## Projectile Motion

Newton's laws and advanced mathematics can be used to determine the *path of a projectile*. If v_0 is the initial speed of the projectile, at an angle α with the horizontal and a_0 is the initial altitude of the projectile (Fig. 7), then, neglecting air resistance, the **path of the projectile** is given by

$$
\begin{aligned}
x &= (v_0 \cos \alpha)t \\
y &= a_0 + (v_0 \sin \alpha)t - 4.9t^2
\end{aligned}
\qquad 0 \le t \le b \qquad (6)
$$

> **Figure 7** Projectile motion.

The parameter t represents time in seconds, and x and y are distances measured in meters. Solving the first equation in equations (6) for t in terms of x, substituting into the second equation, and simplifying produces the following equation:

$$y = a_0 + (\tan \alpha)x - \frac{4.9}{v_0^2 \cos^2 \alpha} x^2 \tag{7}$$

You should verify this by supplying the omitted details.

We recognize equation (7) as a parabola. This equation in x and y describes the path the projectile follows but tells us little else about its flight. On the other hand, the parametric equations (6) not only determine the path of the projectile but also tell us where it is at any time t. Furthermore, using concepts from physics and calculus, the parametric equations can be used to determine the velocity and acceleration of the projectile at any time t. This illustrates another advantage of using parametric representations of plane curves.

EXAMPLE 4 Projectile Motion

An automobile drives off a 50-meter cliff traveling at 25 meters per second (Fig. 8). When (to the nearest tenth of a second) will the automobile strike the ground? How far (to the nearest meter) from the base of the cliff is the point of impact?

SOLUTION At the instant the automobile leaves the cliff, the velocity is 25 meters per second, the angle with the horizontal is 0, and the altitude is 50 meters. Substituting these values in equations (6), the parametric equations for the path of the automobile are

$$x = 25t$$
$$y = 50 - 4.9t^2$$

The automobile strikes the ground when $y = 0$. Using the parametric equation for y, we have

$$y = 50 - 4.9t^2 = 0$$
$$-4.9t^2 = -50$$
$$t = \sqrt{\frac{-50}{-4.9}} \approx 3.2 \text{ seconds}$$

The distance from the base of the cliff is the same as the value of x. Substituting $t = 3.2$ in the first parametric equation, the distance from the base of the cliff at the point of impact is $x = 25(3.2) = 80$ meters.

> Figure 8

MATCHED PROBLEM 3 A gardener is holding a hose in a horizontal position 1.5 meters above the ground. Water is leaving the hose at a speed of 5 meters per second. What is the distance (to the nearest tenth of a meter) from the gardener's feet to the point where the water hits the ground?

The **range of a projectile** at an altitude $a_0 = 0$ is the distance from the point of firing to the point of impact. If we keep the initial speed v_0 of the projectile constant and vary the angle α in Figure 7, we obtain different parabolic paths followed by the projectile and different ranges. The maximum range is obtained when $\alpha = 45°$. Furthermore, assuming that the projectile always stays in the same vertical plane, then there are points in the air and on the ground that the projectile cannot reach, regardless of the angle α used, $0° \le \alpha \le 180°$. Using more advanced mathematics, it can be shown that the reachable region is separated from the nonreachable region by a parabola called an **envelope** of the other parabolas (Fig. 9).

> **Figure 9** Reachable region of a projectile.

Envelope

> **Cycloid**

We now consider an unusual curve called a *cycloid,* which has a fairly simple parametric representation and a very complicated representation in terms of *x* and *y* only. The path traced by a point on the rim of a circle that rolls along a line is called a **cycloid.** To derive parametric equations for a cycloid we roll a circle of radius *a* along the *x* axis with the tracing point *P* on the rim starting at the origin (Fig. 10).

> **Figure 10** Cycloid.

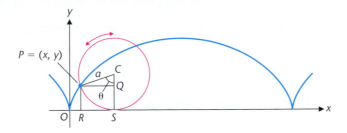

Because the circle rolls along the *x* axis without slipping (see Fig. 10), we see that

$$
\begin{aligned}
d(O, S) &= \text{arc } PS \\
&= a\theta
\end{aligned} \qquad \theta \text{ in radians} \tag{8}
$$

where *S* is the point of contact between the circle and the *x* axis. Referring to triangle *CPQ*, we see that

$$
d(P, Q) = a \sin \theta \qquad 0 \le \theta \le \pi/2 \tag{9}
$$
$$
d(Q, C) = a \cos \theta \qquad 0 \le \theta \le \pi/2 \tag{10}
$$

Using these results, we have

$$
\begin{aligned}
x &= d(O, R) \\
&= d(O, S) - d(R, S) \\
&= (\text{arc } PS) - d(P, Q) \\
&= a\theta - a \sin \theta \qquad \text{Use equations (8) and (9).} \\
y &= d(R, P) \\
&= d(S, C) - d(Q, C) \\
&= a - a \cos \theta \qquad \text{Use equation (10) and the fact that } d(S, C) = a.
\end{aligned}
$$

Although θ in equations (9) and (10) was restricted so that $0 \le \theta \le \pi/2$, it can be shown that the derived parametric equations generate the whole cycloid for $-\infty < \theta < \infty$. The graph specifies a periodic function with period $2\pi a$. So in general, we have Theorem 1.

> **THEOREM 1** Parametric Equations for a Cycloid

For a circle of radius a rolled along the x axis, the resulting cycloid generated by a point on the rim starting at the origin is given by

$$x = a\theta - a \sin \theta$$
$$y = a - a \cos \theta \qquad -\infty < \theta < \infty$$

> **Figure 11** Cycloid path.

The cycloid is a good example of a curve that is very difficult to represent without the use of a parameter. A cycloid has a very interesting physical property. An object sliding without friction from a point P to a point Q lower than P, but not on the same vertical line as P, will arrive at Q in a shorter time traveling along a cycloid than on any other path (Fig. 11).

⟩⟩⟩ **EXPLORE-DISCUSS 2**

Let Q be a point b units from the center of a wheel of radius a, where $0 < b < a$. If the wheel rolls along the x axis with the tracing point Q starting at $(0, a - b)$, explain why parametric equations for the path of Q are given by

$$x = a\theta - b \sin \theta$$
$$y = a - b \cos \theta$$

ANSWERS TO MATCHED PROBLEMS

1. The quarter of the circle of radius 2 centered at the origin that lies in the fourth quadrant.
2. $x^2 + y^2 = 16$, circle of radius 4 centered at $(0, 0)$
3. (A) Ellipse: $x = -7 + 4 \cos \theta$, $y = 3 + 6 \sin \theta$, $0 \le \theta \le 2\pi$
 (B) Hyperbola: $x = -2 + 4 \tan \theta$, $y = -4 + 3 \sec \theta$, $0 \le \theta \le 2\pi$, $\theta \ne \dfrac{\pi}{2}, \dfrac{3\pi}{2}$
4. 2.8 meters

B-3 Exercises

1. If $x = t^2$ and $y = t^2 - 2$, then $y = x - 2$. Discuss the differences between the graph of the parametric equations and the graph of the line $y = x - 2$.

2. If $x = t^2$ and $y = t^4 - 2$, then $y = x^2 - 2$. Discuss the differences between the graph of the parametric equations and the graph of the parabola $y = x^2 - 2$.

In Problems 3–12, the interval for the parameter is the whole real line. For each pair of parametric equations, eliminate the parameter t and find an equation for the curve in terms of x and y. Identify and graph the curve.

3. $x = -t, y = 2t - 2$ **4.** $x = t, y = t + 1$

5. $x = -t^2, y = 2t^2 - 2$ **6.** $x = t^2, y = t^2 + 1$

7. $x = 3t, y = -2t$ **8.** $x = 2t, y = t$

9. $x = \frac{1}{4}t^2, y = t$ **10.** $x = 2t, y = t^2$

11. $x = \frac{1}{4}t^4, y = t^2$ **12.** $x = 2t^2, y = t^4$

In Problems 13–24, obtain an equation in x and y by eliminating the parameter. Identify the curve.

13. $x = t - 2, y = 4 - 2t$

14. $x = t - 1, y = 2t + 2$

15. $x = t - 1, y = \sqrt{t}, t \ge 0$

16. $x = \sqrt{t}, y = t + 1, t \ge 0$

17. $x = \sqrt{t}, y = 2\sqrt{16 - t}, 0 \le t \le 16$

18. $x = -3\sqrt{t}, y = \sqrt{25 - t}, 0 \le t \le 25$

19. $x = -\sqrt{t + 1}, y = -\sqrt{t - 1}, t \ge 1$

20. $x = \sqrt{2 - t}, y = -\sqrt{4 - t}, t \le 2$

21. $x = 3 \sin \theta, y = 4 \cos \theta, 0 \le \theta \le 2\pi$

22. $x = 3 \sin \theta, y = 3 \cos \theta, 0 \le \theta \le 2\pi$

23. $x = 2 + 2 \sin \theta, y = 3 + 2 \cos \theta, 0 \le \theta \le 2\pi$

24. $x = 3 + 4 \sin \theta, y = 2 + 2 \cos \theta, 0 \le \theta \le 2\pi$

25. If $A \ne 0$, $C = 0$, and $E \ne 0$, find parametric equations for $Ax^2 + Cy^2 + Dx + Ey + F = 0$. Identify the curve.

26. If $A = 0$, $C \ne 0$, and $D \ne 0$, find parametric equations for $Ax^2 + Cy^2 + Dx + Ey + F = 0$. Identify the curve.

In Problems 27–30, eliminate the parameter and find the standard equation for the curve. Name the curve and find its center.

27. $x = 3 + 6 \cos t, y = 2 + 4 \sin t, 0 \le t \le 2\pi$

28. $x = 1 + 3 \sec t, y = -2 + 2 \tan t, 0 \le t \le 2\pi, t \ne \dfrac{\pi}{2}, \dfrac{3\pi}{2}$

29. $x = -3 + 2 \tan t, y = -1 + 5 \sec t, 0 \le t \le 2\pi, t \ne \dfrac{\pi}{2}, \dfrac{3\pi}{2}$

30. $x = -4 + 5 \cos t, y = 1 + 8 \sin t, 0 \le t \le 2\pi$

In Problems 31–36, the interval for the parameter is the entire real line. Obtain an equation in x and y by eliminating the parameter and identify the curve.

31. $x = \sqrt{t^2 + 1}, y = \sqrt{t^2 + 9}$

32. $x = \sqrt{t^2 + 4}, y = \sqrt{t^2 + 1}$

33. $x = \dfrac{2}{\sqrt{t^2 + 1}}, y = \dfrac{2t}{\sqrt{t^2 + 1}}$

34. $x = \dfrac{3t}{\sqrt{t^2 + 1}}, y = \dfrac{3}{\sqrt{t^2 + 1}}$

35. $x = \dfrac{8}{t^2 + 4}, y = \dfrac{4t}{t^2 + 4}$

36. $x = \dfrac{4t}{t^2 + 1}, y = \dfrac{4t^2}{t^2 + 1}$

In Problems 37–40, find the standard form of each equation. Name the curve and find its center. Then use trigonometric functions to find parametric equations for the curve.

37. $25x^2 - 200x - 9y^2 - 18y + 616 = 0$

38. $36x^2 + 360x + 4y^2 - 8y + 760 = 0$

39. $4x^2 - 24x + 49y^2 + 392y + 624 = 0$

40. $16x^2 + 32x - 9y^2 - 36y - 164 = 0$

 41. Consider the following two pairs of parametric equations:

$$x_1 = t, y_1 = e^t, -\infty < t < \infty$$
$$x_2 = e^t, y_2 = t, -\infty < t < \infty$$

(A) Graph both pairs of parametric equations in a squared viewing window and discuss the relationship between the graphs.

(B) Eliminate the parameter and express each equation as a function of x. How are these functions related?

42. Consider the following two pairs of parametric equations:

$$x_1 = t, y_1 = \log t, t > 0$$
$$x_2 = \log t, y_2 = t, t > 0$$

(A) Graph both pairs of parametric equations in a squared viewing window and discuss the relationship between the graphs.

(B) Eliminate the parameter and express each equation as a function of x. How are these functions related?

APPLICATIONS

43. PROJECTILE MOTION An airplane flying at an altitude of 1,000 meters is dropping medical supplies to hurricane victims on an island. The path of the plane is horizontal, the speed is 125 meters per second, and the supplies are dropped at the instant the plane crosses the shoreline. How far inland (to the nearest meter) will the supplies land?

44. PROJECTILE MOTION One stone is dropped vertically from the top of a tower 40 meters high. A second stone is thrown horizontally from the top of the tower with a speed of 30 meters per second. How far apart (to the nearest tenth of a meter) are the stones when they land?

45. PROJECTILE MOTION A projectile is fired with an initial speed of 300 meters per second at an angle of 45° to the horizontal. Neglecting air resistance, find

(A) The time of impact

(B) The horizontal distance covered (range) in meters and kilometers at time of impact

(C) The maximum height in meters of the projectile

Compute all answers to three decimal places.

46. PROJECTILE MOTION Repeat Problem 45 if the same projectile is fired at 40° to the horizontal instead of 45°.

› Similar Triangles

(A) Two triangles are similar if two angles of one triangle have the same measure as two angles of the other.

(B) If two triangles are similar, their corresponding sides are proportional:

$$\frac{a}{a'} = \frac{b}{b'} = \frac{c}{c'}$$

› Pythagorean Theorem

$$c^2 = a^2 + b^2$$

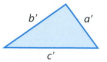

› Rectangle

$A = ab$ **Area**

$P = 2a + 2b$ **Perimeter**

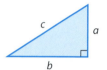

› Parallelogram

$h =$ Height

$A = ah = ab \sin\theta$ **Area**

$P = 2a + 2b$ **Perimeter**

› Triangle

$h =$ Height

$A = \frac{1}{2}hc$ **Area**

$P = a + b + c$ **Perimeter**

$s = \frac{1}{2}(a + b + c)$ **Semiperimeter**

$A = \sqrt{s(s - a)(s - b)(s - c)}$ **Area—Heron's formula**

› Trapezoid

Base a is parallel to base b.

$h =$ Height

$A = \frac{1}{2}(a + b)h$ **Area**

› Circle

R = Radius

D = Diameter

$D = 2R$

$A = \pi R^2 = \frac{1}{4}\pi D^2$ Area

$C = 2\pi R = \pi D$ Circumference

$\dfrac{C}{D} = \pi$ For all circles

$\pi \approx 3.141\ 59$

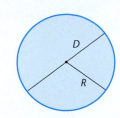

› Rectangular Solid

$V = abc$ Volume

$T = 2ab + 2ac + 2bc$ Total surface area

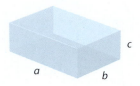

› Right Circular Cylinder

R = Radius of base

h = Height

$V = \pi R^2 h$ Volume

$S = 2\pi R h$ Lateral surface area

$T = 2\pi R(R + h)$ Total surface area

› Right Circular Cone

R = Radius of base

h = Height

s = Slant height

$V = \frac{1}{3}\pi R^2 h$ Volume

$S = \pi Rs = \pi R\sqrt{R^2 + h^2}$ Lateral surface area

$T = \pi R(R + s) = \pi R(R + \sqrt{R^2 + h^2})$ Total surface area

› Sphere

R = Radius

D = Diameter

$D = 2R$

$V = \frac{4}{3}\pi R^3 = \frac{1}{6}\pi D^3$ Volume

$S = 4\pi R^2 = \pi D^2$ Surface area

STUDENT ANSWER APPENDIX

CHAPTER R Exercises R-1

35. (A) $\{1, \sqrt{144}\}$ (B) $\{-3, 0, 1, \sqrt{144}\}$ (C) $\{-3, -\frac{2}{3}, 0, 1, \frac{9}{5}, \sqrt{144}\}$ (D) $\{\sqrt{3}\}$

37. (A) $0.888\,888\ldots$; repeating; repeated digit: 8
(B) $0.272\,727\ldots$; repeating; repeated digits: 27 (C) $2.236\,067\,977\ldots$; nonrepeating and nonterminating (D) 1.375; terminating

Exercises R-2

61. $\dfrac{n^8}{m^{12}}$ **67.** $-2\sqrt{3}$ **69.** $6\sqrt[3]{5} - \sqrt[3]{25}$ **71.** $5\sqrt[3]{2}$ **83.** $\sqrt{2}/2$ or $\frac{1}{2}\sqrt{2}$

Exercises R-4

59. $\dfrac{-x(x+y)}{y}$

CHAPTER 1 Exercises 1-2

5. $-8 \le x \le 7$ x

7. $-6 \le x < 6$ x

9. $x \ge -6$ x

11. $(-2, 6]$ x

13. $(-7, 8)$ x

15. $(-\infty, -2]$ x

29. $x < 5; (-\infty, 5)$ x

31. $y \ge 2$ y

33. $N < -8; (-\infty, -8)$ N

35. $t > 2; (2, \infty)$ t

37. $m > 3; (3, \infty)$ m

39. $B \ge -4; [-4, \infty)$ B

41. $-2 < t \le 3; (-2, 3]$ t

43. $(-5, 7]$ x

45. $(2, 4)$ x

47. $(-\infty, \infty)$ x

49. $(-\infty, -1) \cup [3, 7)$ x

51. $(1, 5)$ x

53. $(-\infty, 6]$ x

55. $q < -14; (-\infty, -14)$ q

57. $x \ge 4.5; [4.5, \infty)$ x

59. $-20 \le x \le 20; [-20, 20]$ x

61. $6 < x < 12$ x

63. $-8 \le x < -3; [-8, -3)$ x

65. $-42 \le x < 30$ x

67. $x < 10; (-\infty, 10)$ x

69. $x \ge 8; [8, \infty)$ x

77. (A) and (C) $a > 0$ and $b > 0$, or $a < 0$ and $b < 0$ (B) and (D) $a > 0$ and $b < 0$, or $a < 0$ and $b > 0$

Exercises 1-3

31. y is 3 units from 5; $y = 2, 8$ y

33. y is less than 3 units from 5; $2 < y < 8; (2, 8)$ y

35. y is more than 3 units from 5; $y < 2$ or $y > 8; (-\infty, 2) \cup (8, \infty)$ y

37. u is 3 units from -8; $u = -11, -5$ u

39. u is no more than 3 units from -8; $-11 \le u \le -5; [-11, -5]$ u

41. u is at least 3 units from -8; $u \le -11$ or $u \ge -5; (-\infty, -11] \cup [-5, \infty)$ u

51. $u \le -11$ or $u \ge -6; (-\infty, -11] \cup [-6, \infty)$ **53.** $-35 < C < -\frac{5}{9}; (-35, -\frac{5}{9})$ **55.** $-2 < x < 2; (-2, 2)$ **57.** $-\frac{1}{3} \le t \le 1; [-\frac{1}{3}, 1]$

65. The distance from x to 3 is between zero and 0.1; $(2.9, 3) \cup (3, 3.1)$; x

67. The distance from x to a is between 0 and 1/10; $\left(a - \dfrac{1}{10}, a\right) \cup \left(a, a + \dfrac{1}{10}\right)$ x

Exercises 1-4

9. (A) $-\dfrac{3}{2}$ (B) $\dfrac{5}{6}i$ (C) $-\dfrac{3}{2}-\dfrac{5}{6}i$ 11. (A) 6.5 (B) 2.1i (C) $6.5-2.1i$ 13. (A) 0 (B) πi (C) $-\pi i$

15. (A) 4π (B) 0 (C) 4π

Exercises 1-5

19. $z=\pm 4\sqrt{2}/5$ 25. Two real roots: $x=1\pm\sqrt{2}$ 27. No real roots: $x=1\pm i\sqrt{2}$ 29. No real roots: $t=(3\pm i\sqrt{7})/2$

31. Two real roots: $t=(3\pm\sqrt{7})/2$ 33. $x=2\pm\sqrt{5}$ 35. $r=(-5\pm\sqrt{3})/2$ 37. $u=(-2\pm i\sqrt{11})/2$ 43. $y=(3\pm\sqrt{5})/2$

45. $x=(3\pm\sqrt{13})/2$

Exercises 1-6

27. $2u^2-4u=0, u=x^{-3}$ 29. Not of quadratic type 31. $\dfrac{10}{9}+4u-7u^2=0, u=\dfrac{1}{x^2}$ 35. $m=\pm\sqrt{3},\pm i\sqrt{5}$ 39. $y=-64,\frac{27}{8}$

51. $y=\dfrac{1}{3}\pm\dfrac{i\sqrt{2}}{3}$ 53. $t=\pm\dfrac{\sqrt{2}}{2},\pm\sqrt{2}$ 63. $x=\pm\sqrt{\dfrac{5\pm\sqrt{13}}{6}}$ (four roots)

Chapter 1 Review Exercises

5. $-14<y<-4;(-14,-4)$ *(1-3)* 9. $x=\pm\sqrt{\frac{7}{2}}$ or $\pm\frac{1}{2}\sqrt{14}$ *(1-5)* 13. $m=-\frac{1}{2}\pm(\sqrt{3}/2)i$ *(1-5)*

19. $x\geq-19;[-19,\infty)$ *(1-2)* 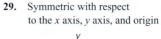 27. $x=(1\pm\sqrt{43})/3$ *(1-5)* 37. $I=(E\pm\sqrt{E^2-4PR})/(2R)$ *(1-5)*

CHAPTER 2 Exercises 2-1

15. 17.

19. Points: $A=(2,4), B=(3,-1), C=(-4,0), D=(-5,2)$
 Reflections: $A'=(-2,4), B'=(-3,-1), C'=(4,0), D'=(5,2)$

21. Points: $A=(-3,-3), B=(0,4), C=(-3,2), D=(5,-1)$
 Reflections: $A'=(3,3), B'=(0,-4), C'=(3,-2), D'=(-5,1)$

23. No symmetry with respect to x axis, y axis, or origin 25. Symmetric with respect to the origin 27. Symmetric with respect to the x axis 29. Symmetric with respect to the x axis, y axis, and origin

35. (A) (B) (C) (D)

47. Symmetric with respect to the x axis

49. Symmetric with respect to the y axis

51. Symmetric with respect to the x axis, y axis, and origin

53. Symmetric with respect to the origin

55. Symmetric with respect to the y axis

57. Symmetric with respect to the y axis

59.

61.

63. $y = \pm\sqrt{3 - 2x}$

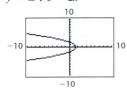

65. $y = -1 \pm\sqrt{x^2 - 4}$

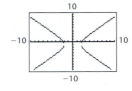

67. Symmetric with respect to the y axis

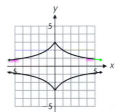

69. Symmetric with respect to the origin

71. No symmetry with respect to the x axis, y axis, or origin

73. Symmetric with respect to the x axis, y axis, and origin

75. Symmetric with respect to the x axis, y axis, and origin

81.

83. (A) 3,000 cases (B) Demand decreases by 400 cases (C) Demand increases by 600 cases

87. (A)

Exercises 2-2

21. $x^2 + y^2 = 4$

23. $(x - 1)^2 + y^2 = 1$

25. $(x + 2)^2 + (y - 1)^2 = 9$

33. The set of all points that are two units from the point $(0, 2)$.
$x^2 + (y - 2)^2 = 4$

35. The set of all points that are four units from the point $(1, 1)$.
$(x - 1)^2 + (y - 1)^2 = 16$

43. Center: $(0, -2)$; radius: 3 **45.** Center: $(-4, 2)$; radius: $\sqrt{7}$ **47.** Center: $(-3, 0)$; radius: 5 **49.** Center: $(3, 2)$; radius: 7

51. Center: $(-4, 3)$;
radius: $\sqrt{17}$

53. $y = \pm\sqrt{3 - x^2}$

55. $y = -1 \pm \sqrt{2 - (x + 3)^2}$

73. (A) $A = (0, 0), B = (0, 13.5), C = (0, 27), D = (60, 27), E = (78, 27), F = (78, 13.5), G = (78, 0)$ (B) 62 feet, 79 feet

77. (A) $(x + 12)^2 + (y + 5)^2 = 26^2$; center: $(-12, -5)$; radius: 26
(B) 13.5 miles

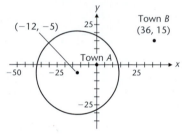

Exercises 2-3

19. Slope $= -\frac{3}{5}$ **21.** Slope $= -\frac{3}{4}$ **23.** Slope $= -2$ **25.** Slope $= \frac{4}{5}$

 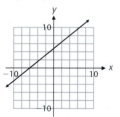

27. Slope $= 2$ **29.** Slope not defined **31.** Slope $= 0$

41. $y = \frac{3}{2}x + \frac{23}{2}$ **47.** $y = -\frac{2}{5}x + 2$ **67.** slope $AB = -\frac{3}{4} =$ slope DC
69. (slope AB)(slope BC) $= \left(-\frac{3}{4}\right)\left(\frac{4}{3}\right) = -1$

75. $3x + 4y = 25$

77. $x - y = 10$

79. $232 = 5x - 12y$

81. (A)

x	0	5,000	10,000	15,000	20,000	25,000	30,000
B	212	203	194	185	176	167	158

 (B) The boiling point drops 9°F for each 5,000-ft increase in altitude.

87. (A) $F = \frac{9}{5}C + 32$ (B) 68°F, 30°C

Exercises 2-4

5. (A) $C = 2{,}147 + 75x$
 (B) The rate of change of cost with respect to production is $75.
 (C) Increasing production by 1 unit increases cost by $75

7. (A) The rate of change of height with respect to DBH is 4.06 feet per inch.
 (B) Increasing DBH by 1 inch increases height by 4.06 feet.
 (C) 73 feet
 (D) 19 inches

9. (A) Robinson: The rate of change of weight with respect to height is 3.7 pounds per inch.
 Miller: The rate of change of weight with respect to height is 3 pounds per inch.
 (B) Robinson: 130.2 pounds; Miller: 135 pounds
 (C) Robinson: 5′9″; Miller: 5′8″

11. $s = 0.75t + 717$; speed increases 0.75 mph for each 1°F change in temperature.

15. (A) $V = 142{,}000 - 7{,}500t$
 (B) The tractor's value is decreasing at the rate of $7,500 per year.
 (C) $97,000

17. (A) $R = 1.4C - 7$
 (B) The slope is 1.4; this is the rate of change of retail price with respect to cost.
 (C) $137

23. (A)

 (B) 0.97 million
 (C) 1.3 million

Chapter 2 Review Exercises

1. *(2-1)*

3. (A) Symmetric with respect to the origin (B) No symmetry with respect to the x axis, y axis, or origin

(C) Symmetric with respect to the *y* axis (D) Symmetric with respect to the *x* axis *(2-1)*

9. (A) *(2-2)*

(B) $d(A, C) = 2\sqrt{10}, d(B, C) = \sqrt{10}, d(A, B) = \sqrt{50}$, perimeter $= 16.56$
(C) $d(A, C)^2 + d(B, C)^2 = d(A, B)^2$; right triangle
(D) Midpoint of side $AC = (0, 1)$, of side $BC = (2.5, 3.5)$, and of side $AB = (1.5, 0.5)$

11. Slope $= -\frac{3}{2}$ *(2-3)* **15.** Symmetric with respect to the *y* axis *(2-1)*

17. Symmetric with respect to **27.** $y = -x + 7$ *(1-5, 2-2)*
the *x* axis, *y* axis, and origin *(2-1)*

29. (A) *(2-1)* (B) (C) (D)

 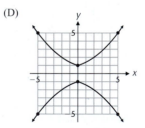

35. (A) The rate of change of body surface area with respect to weight is 0.3433.
(B) Body surface area increases by 34.33 cm².
(C) 6,470.5 cm² *(2-4)*

37. (A) $H = 0.7(220 - A)$ (B) $H = 140$ beats per minute (C) $A = 40$ years old *(2-4)*

CHAPTER 3 Exercises 3-1

39. Not a function; for example, when $x = 0, y = \pm 2$ **41.** A function with domain all real numbers
43. Not a function; for example, when $x = 0, y = \pm 7$ **45.** A function with domain all real numbers
59. $[-4, 1) \cup (1, \infty); -4 \le x < 1$ or $x > 1$ **67.** Function f multiplies the square of the domain element by 2 then adds 5 to the result.
69. Function z divides the sum of four times the domain element and 5 by the square root of the domain element.

79. (A) $-8x + 3 - 4h$ (B) $-4x - 4a + 3$ **81.** (A) $\dfrac{1}{\sqrt{x + h + 2} + \sqrt{x + 2}}$ (B) $\dfrac{1}{\sqrt{x + 2} + \sqrt{a + 2}}$

83. (A) $\dfrac{-4}{x(x + h)}$ (B) $\dfrac{-4}{ax}$ **91.** The cost is a flat $17 per month, plus $2.40 for each hour of airtime.

93. (A) $s(0) = 0, s(1) = 16, s(2) = 64, s(3) = 144$ (B) $64 + 16h$
(C) Let $q(h) = [s(2 + h) - s(2)]/h$

h	-1	-0.1	-0.01	-0.001	0.001	0.01	0.1	1
$q(h)$	48	62.4	63.84	63.984	64.016	64.16	65.6	80

(D) $q(h)$, the average velocity from 2 to $2 + h$ seconds, approaches 64 feet per second

97. $F = 8x + (250/x) - 12$;

x	4	5	6	7
F	82.5	78	77.7	79.7

Exercises 3-2

9. (A) $[-4, 4]$ (B) $[-3, 3)$ (C) 0 (D) 0 (E) $[-4, 4]$ (F) None (G) None (H) None
11. (A) $(-\infty, \infty)$ (B) $[-4, \infty]$ (C) $-3, 1$ (D) -3 (E) $[-1, \infty)$ (F) $(-\infty, -1]$ (G) None (H) None
13. (A) $(-\infty, 2) \cup (2, \infty)$ (B) $(-\infty, -1) \cup [1, \infty)$ (C) None (D) 1 (E) None (F) $(-\infty, -2], (2, \infty)$
(G) $[-2, 2)$ (H) $x = 2$

21. One possible answer:

23. One possible answer:

25. One possible answer:

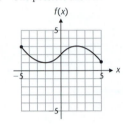

27. Slope $= 2$,
x intercept $= -2$,
y intercept $= 4$

29. Slope $= -\frac{1}{2}$,
x intercept $= -\frac{10}{3}$,
y intercept $= -\frac{5}{3}$

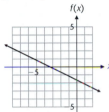

31. Slope $= -2.3$,
x intercept $= 3.1$,
y intercept $= 7.1$

37. Domain: $\{x \mid x \neq -2\}$; x intercept: 4; y intercept: -3
39. Domain: $\{x \mid x \neq \frac{5}{4}\}$; x intercept: $\frac{2}{3}$; y intercept: $\frac{2}{5}$
41. Domain: $\{x \mid x \neq 2\}$; x intercept: 0; y intercept: 0
43. Domain: $\{x \mid x \neq -3, 3\}$; x intercept: ± 4; y intercept: $\frac{16}{9}$
45. Domain: $\{x \mid x \neq -5, 5\}$; no x intercept; y intercept: $-\frac{7}{25}$
47. (A) $f(-1) = 0, f(0) = 1, f(1) = 0$
(B)

(C) Domain: $[-1, 1]$; range: $[0, 1]$; continuous on its domain

49. (A) $f(-2) = -2, f(-1)$ is not defined, $f(2) = 4$
(B)

(C) Domain: $[-3, -1) \cup (-1, 2]$; range: $\{-2, 4\}$; discontinuous at $x = -1$

51. (A) $f(-2) = 0, f(-1)$ is not defined, $f(0) = -2$ (B)

(C) Domain: $(-\infty, -1) \cup (-1, \infty)$; range: R;
discontinuous at $x = -1$

53. (A) $f(-3) = 0, f(-2) = -2, f(0) = -2, f(3) = -2, f(4) = 4$
(B)

(C) Domain: R; range: $[-2, \infty)$; continuous on its domain

55. (A) $f(-3) = -\frac{3}{2}, f(-2) = 1, f(0) = 1, f(3) = 1, f(4) = \frac{5}{2}$
(B)

(C) Domain: R; range: R; continuous on its domain

57. (A) $f(-1) = \frac{10}{3}, f(0)$ is not defined, $f(1) = \frac{5}{2}$,
$f(2)$ is not defined, $f(3) = -\frac{3}{2}$
(B)

(C) Domain: $(-\infty, 0) \cup (0, 2) \cup (2, \infty)$; range: $(-\infty, 4)$;
discontinuous at $x = 0$ and $x = 2$

63. $f(x) = \begin{cases} 3 & \text{if } x < -2 \\ -2x - 2 & \text{if } -2 < x < 1 \\ -1 & \text{if } x > 1 \end{cases}$

65. $f(x) = \begin{cases} 1 - x & \text{if } x < 0 \\ 1 + x & \text{if } x \geq 0 \end{cases}$

Domain: R; range: $[1, \infty)$;
continuous on its domain

67. $f(x) = \begin{cases} -x + 2 & \text{if } x < 2 \\ x - 2 & \text{if } x \geq 2 \end{cases}$

Domain: R; range: $[0, \infty)$;
continuous on its domain

69. (A) One possible answer:

(B) The graph must cross the
x axis exactly once.

71. (A) One possible answer:

(B) The graph must cross the x axis at least twice.
There is no upper limit on the number of
times it can cross the x axis.

73. Graphs of f and g

Graph of m

Graph of n

75. Graphs of f and g

Graph of m

Graph of n

77. Graphs of f and g Graph of m Graph of n

81. $R(x) = \begin{cases} 32 & \text{if } 0 \le x \le 100 \\ 16 + 0.16x & \text{if } x > 100 \end{cases}$

83. $E(x) = \begin{cases} 200 & \text{if } 0 \le x \le 3{,}000 \\ 80 + 0.04x & \text{if } 3{,}000 < x < 8{,}000 \\ 180 + 0.04x & \text{if } x \ge 8{,}000 \end{cases}$

Discontinuous at $x = 8{,}000$

$E(5{,}750) = \$310$, $E(9{,}200) = \$548$

85.

x	4	-4	6	-6	24	25	247	-243	-245	-246
$f(x)$	0	0	10	-10	20	30	250	-240	-240	-250

; f rounds numbers to the tens place.

89. (A) $C(x) = \begin{cases} 15 & 0 < x \le 1 \\ 18 & 1 < x \le 2 \\ 21 & 2 < x \le 3 \\ 24 & 3 < x \le 4 \\ 27 & 4 < x \le 5 \\ 30 & 5 < x \le 6 \end{cases}$

(B) No, since $f(x) \ne C(x)$ at $x = 1, 2, 3, 4, 5,$ or 6

91. $T(x) = \begin{cases} 0.03x & \text{if } 0 \le x \le 10{,}000 \\ 0.05x - 200 & \text{if } x > 10{,}000 \end{cases}$

93. $T(x) = \begin{cases} 0.0535x & 0 \le x \le 19{,}890 \\ 0.0705x - 338.25 & 19{,}890 < x \le 65{,}330 \\ 0.0785x - 860.41 & x > 65{,}330 \end{cases}$

$T(10{,}000) = \$535$

$T(30{,}000) = \$1{,}776.75$

$T(100{,}000) = \$6{,}989.60$

Exercises 3-3

5. Domain: $[0, \infty)$; Range: $(-\infty, 0]$

11. Domain: $[-2, 2]$; range: $[0, 4]$

13. Domain: $[-2, 2]$; range: $[1, 3]$

15. Domain: $[0, 4]$; range: $[-2, 2]$

17. Domain: $[-4, 0]$; range: $[-1, 1]$

19. Domain: $[-2, 2]$; range: $[-2, 2]$

21. Domain: $[-2, 2]$; range: $[-2, 2]$

23. Domain: $[-1, 1]$; range: $[-1, 1]$

25. Domain: $[-2, 2]$; range: $[-2, 2]$

45.

47.

49.

51.

53.

55.

57.

59.

61.

79. (A) f is a horizontal shrink of $y = \sqrt[3]{x}$ by a factor of $1/8$. g is a vertical stretch of $y = \sqrt[3]{x}$ by a factor of 2. (B) The graphs are identical.
(C) $f(x) = \sqrt[3]{8x} = \sqrt[3]{8} \cdot \sqrt[3]{x} = 2\sqrt[3]{x}$

81. (A) The graphs are different; order is significant. (B) i. $f(x) = -(x^2 - 5)$ ii. $f(x) = -x^2 - 5$

91.

$f(x)$

93.

$g(x)$

95. Conclusion: any function can be written as the sum of two other functions, one even and the other odd.

97.

99. Each graph is a vertical translation of the graph of $y = 0.004(x - 10)^3$.

101. Each graph is a portion of the graph of a horizontal translation followed by a vertical shrink (except for $C = 8$) of the graph of $y = t^2$. Larger values of C correspond to a smaller opening.

Exercises 3-4

7. Vertex: $(-3, -4)$; axis: $x = -3$

9. Vertex: $\left(\frac{3}{2}, -5\right)$; axis: $x = \frac{3}{2}$

11. Vertex: $(-10, 20)$; axis: $x = -10$

13. The graph is shifted 2 units right and 1 unit up. **15.** The graph is reflected in the x axis, then shifted 1 unit left.

17. The graph is shifted 2 units right and 3 units down.

25. $f(x) = (x - 2)^2 + 1$; vertex: $(2, 1)$; axis: $x = 2$

27. $h(x) = -(x + 1)^2 - 2$; vertex: $(-1, -2)$; axis: $x = -1$

29. $m(x) = 2(x - 3)^2 + 4$; vertex: $(3, 4)$; axis: $x = 3$

31. $f(x) = \frac{1}{2}(x + 3)^2 - 8$; vertex: $(-3, -8)$; axis: $x = -3$

33. $f(x) = 2(x - 6)^2 + 18$; vertex: $(6, 18)$; axis: $x = 6$

35. Vertex: $(-4, -8)$; The graph is symmetric about the axis, $x = -4$. It decreases until reaching a minimum at $(-4, -8)$, then increases. The range is $[-8, \infty)$.

37. Vertex: $\left(-\frac{7}{2}, \frac{65}{4}\right)$; The graph is symmetric about the axis, $x = -\frac{7}{2}$. It increases until reaching a maximum at $\left(-\frac{7}{2}, \frac{65}{4}\right)$, then decreases. The range is $\left(-\infty, \frac{65}{4}\right]$.

39. Vertex: $\left(\frac{9}{4}, \frac{19}{4}\right)$; The graph is symmetric about the axis, $x = \frac{9}{4}$. It decreases until reaching a minimum at $\left(\frac{9}{4}, \frac{19}{4}\right)$, then increases. The range is $\left[\frac{19}{4}, \infty\right)$.

41. Vertex: $\left(\frac{5}{2}, \frac{149}{2}\right)$; The graph is symmetric about the axis, $x = \frac{5}{2}$. It increases until reaching a maximum at $\left(\frac{5}{2}, \frac{149}{2}\right)$, then decreases. The range is $\left(-\infty, \frac{149}{2}\right]$.

43. Vertex: $\left(-\frac{3}{2}, -\frac{9}{4}\right)$; axis of symmetry: $x = 0$; domain: $(-\infty, \infty)$; range: $\left[-\frac{9}{4}, \infty\right)$; $\min f(x) = f\left(-\frac{3}{2}\right) = -\frac{9}{4}$; decreasing on $\left(-\infty, -\frac{3}{2}\right)$; increasing on $\left(-\frac{3}{2}, \infty\right)$

45. Vertex: $(2, -9)$; axis of symmetry: $x = 2$; domain: $(-\infty, \infty)$; range: $[-9, \infty)$; $\min f(x) = f(2) = -9$; decreasing on $(-\infty, 2)$; increasing on $(2, \infty)$

81. The minimum product is -225 for the numbers 15 and -15. There is no maximum product.

83. 26 employees; \$322,800 **85.** (A) 2003 (B) The domain values should be whole numbers.
(C) 56 mph

97. (B)

105. (A) $R(x) = 3.5x - 0.00007x^2$; domain: $[0, 50{,}000]$; $C(x) = 24{,}500 + 0.35x$, domain: $[0, \infty)$
(B) $x = 10{,}000$ and $x = 35{,}000$

(C) The company makes a profit for those sales levels for which the graph of the revenue function is above the graph of the cost function, that is, if the sales are between 10,000 and 35,000 gallons. The company suffers a loss for those sales levels for which the graph of the revenue function is below the graph of the cost function, that is, if the sales are between 0 and 10,000 gallons or between 35,000 and 50,000 gallons.
(D) The maximum profit is \$10,937.50 when 22,500 gallons are sold at a price of \$1.92 per gallon.

Exercises 3-5

7.

x	-3	-2	-1	0	1	2	3
$(f + g)(x)$	3	3	1	-1	-3	-3	-3

9.

x	-3	-2	-1	0	1	2	3
$(fg)(x)$	2	0	-2	-2	0	2	2

27. $(f \circ g)(-7) = 3$; $(f \circ g)(0) = 9$; $(f \circ g)(4) = -10$

29. $(f + g)(x) = 5x + 1$; $(f - g)(x) = 3x - 1$; $(fg)(x) = 4x^2 + 4x$;

$\left(\dfrac{f}{g}\right)(x) = \dfrac{4x}{x + 1}$; domain $f + g$, $f - g$, $fg = (-\infty, \infty)$; domain of $f/g = (-\infty, -1) \cup (-1, \infty)$

31. $(f + g)(x) = 3x^2 + 1$; $(f - g)(x) = x^2 - 1$; $(fg)(x) = 2x^4 + 2x^2$;

$\left(\dfrac{f}{g}\right)(x) = \dfrac{2x^2}{x^2 + 1}$; domain of each function: $(-\infty, \infty)$

33. $(f + g)(x) = x^2 + 3x + 4$; $(f - g)(x) = -x^2 + 3x + 6$; $(fg)(x) = 3x^3 + 5x^2 - 3x - 5$;

$\left(\dfrac{f}{g}\right)(x) = \dfrac{3x + 5}{x^2 - 1}$; domain $f + g$, $f - g$, fg: $(-\infty, \infty)$; domain of f/g: $(-\infty, -1) \cup (-1, 1) \cup (1, \infty)$

35. $(f + g)(x) = \sqrt{2 - x} + \sqrt{x + 3}$; $(f - g)(x) = \sqrt{2 - x} - \sqrt{x + 3}$; $(fg)(x) = \sqrt{6 - x - x^2}$;

$\left(\dfrac{f}{g}\right)(x) = \sqrt{\dfrac{2 - x}{x + 3}}$. The domain of the functions $f + g$, $f - g$, and fg is $[-3, 2]$. The domain of $\dfrac{f}{g}$ is $(-3, 2]$.

37. $(f + g)(x) = 2\sqrt{x} - 2$; $(f - g)(x) = 6$; $(fg)(x) = x - 2\sqrt{x} - 8$;

$\left(\dfrac{f}{g}\right)(x) = \dfrac{\sqrt{x} + 2}{\sqrt{x} - 4}$. The domain of $f + g$, $f - g$, and fg is $[0, \infty)$. Domain of $\dfrac{f}{g} = [0, 16) \cup (16, \infty)$.

39. $(f + g)(x) = \sqrt{x^2 + x - 6} + \sqrt{7 + 6x - x^2}$; $(f - g)(x) = \sqrt{x^2 + x - 6} - \sqrt{7 + 6x - x^2}$; $(fg)(x) = \sqrt{-x^4 + 5x^3 + 19x^2 - 29x - 42}$;

$\left(\dfrac{f}{g}\right)(x) = \sqrt{\dfrac{x^2 + x - 6}{7 + 6x - x^2}}$. The domain of the functions $f + g$, $f - g$, and fg is $[2, 7]$. The domain of $\dfrac{f}{g}$ is $[2, 7)$.

41. $(f + g)(x) = 2x$; $(f - g)(x) = \dfrac{2}{x}$; $(fg)(x) = x^2 - \dfrac{1}{x^2}$; $\left(\dfrac{f}{g}\right)(x) = \dfrac{x^2 + 1}{x^2 - 1}$ The domain of $f + g$, $f - g$, and fg is $(-\infty, 0) \cup (0, \infty)$.

The domain of $\dfrac{f}{g}$ is $(-\infty, -1) \cup (-1, 0) \cup (0, 1) \cup (1, \infty)$.

43. $(f \circ g)(x) = (x^2 - x + 1)^3$; domain: $(-\infty, \infty)$; $(g \circ f)(x) = x^6 - x^3 + 1$; domain: $(-\infty, \infty)$

45. $(f \circ g)(x) = |2x + 4|$; domain: $(-\infty, \infty)$; $(g \circ f)(x) = 2|x + 1| + 3$; domain: $(-\infty, \infty)$

47. $(f \circ g)(x) = (2x^3 + 4)^{1/3}$; domain: $(-\infty, \infty)$; $(g \circ f)(x) = 2x + 4$; domain: $(-\infty, \infty)$

49. $(f \circ g)(x) = \sqrt{x - 4}$; domain: $[4, \infty)$; $(g \circ f)(x) = \sqrt{x} - 4$; domain: $[0, \infty)$

51. $(f \circ g)(x) = \dfrac{1}{x} + 2$; domain: $(-\infty, 0) \cup (0, \infty)$; $(g \circ f)(x) = \dfrac{1}{x + 2}$; domain: $(-\infty, -2) \cup (-2, \infty)$

53. $(f \circ g)(x) = \sqrt{4 - x^2}$; domain of $f \circ g$ is $[-2, 2]$; $(g \circ f)(x) = 4 - x$; domain of $g \circ f$ is $(-\infty, 4]$.

55. $(f \circ g)(x) = \dfrac{6x - 10}{x}$; domain of $f \circ g$ is $(-\infty, 0) \cup (0, 2) \cup (2, \infty)$; $(g \circ f)(x) = \dfrac{x + 5}{5 - x}$; domain of $g \circ f$ is $(-\infty, 0) \cup (0, 5) \cup (5, \infty)$.

57. $(f \circ g)(x) = x$; domain: $(-\infty, 2) \cup (2, \infty)$; $(g \circ f)(x) = x$; domain: $(-\infty, 0) \cup (0, \infty)$

59. $(f \circ g)(x) = \sqrt{16 - x^2}$; domain of $f \circ g$ is $[-4, 4]$; $(g \circ f)(x) = \sqrt{34 - x^2}$; domain of $g \circ f$ is $[-5, 5]$.

65. $(f \circ g)(x) = (g \circ f)(x) = x$; the graphs of f and g are symmetric with respect to the line $y = x$.

67. $(f \circ g)(x) = (g \circ f)(x) = x$; the graphs of f and g are symmetric with respect to the line $y = x$.

69. $(f \circ g)(x) = x$, $(g \circ f)(x) = x$; the graphs of f and g are symmetric with respect to the line $y = x$.

71. $(f \circ g)(x) = x$, $(g \circ f)(x) = x$; the graphs of f and g are symmetric with respect to the line $y = x$.

73. $g(x) = 2x - 7$; $f(x) = x^4$; $h(x) = (f \circ g)(x)$

77. $f(x) = x^7$; $g(x) = 3x - 5$; $h(x) = (g \circ f)(x)$

85. $(f + g)(x) = 2x$; $(f - g)(x) = \dfrac{2}{x}$; $(fg)(x) = x^2 - \dfrac{1}{x^2}$;

$\dfrac{f}{g}(x) = \dfrac{x^2 + 1}{x^2 - 1}$ The domain of $f + g$, $f - g$, and fg is

$(-\infty, 0) \cup (0, \infty)$. The domain of $\dfrac{f}{g}$ is

$(-\infty, -1) \cup (-1, 0) \cup (0, 1) \cup (1, \infty)$.

87. $(f + g)(x) = 2$; $(f - g)(x) = \dfrac{-2x}{|x|}$; $(fg)(x) = 0$; $\left(\dfrac{f}{g}\right)(x) = 0$

The domain of $f + g$, $f - g$, and fg is $(-\infty, 0) \cup (0, \infty)$.

Domain of $\dfrac{f}{g}$ is $(0, \infty)$.

75. $g(x) = 4 + 2x$; $f(x) = x^{1/2}$; $h(x) = (f \circ g)(x)$

79. $f(x) = x^{-1/2}$; $g(x) = 4x + 3$; $h(x) = (g \circ f)(x)$

Exercises 3-6

7. The original set and the reversed set are both one-to-one functions.

9. The original set is a function. The reversed set is not a function.

11. Neither set is a function.

41. Domain of $f = [-4, 4]$
range of $f = [1, 5]$
domain of $f^{-1} = [1, 5]$
range of $f^{-1} = [-4, 4]$

43. Domain of $f = [-5, 3]$
range of $f = [-3, 5]$
domain of $f^{-1} = [-3, 5]$
range of $f^{-1} = [-5, 3]$

45. $f^{-1}(x) = \frac{1}{3}x$
domain of $f = (-\infty, \infty)$
range of $f = (-\infty, \infty)$
domain of $f^{-1} = (-\infty, \infty)$
range of $f^{-1} = (-\infty, \infty)$

47. $f^{-1}(x) = (x + 3)/4$
domain of $f = (-\infty, \infty)$
range of $f = (-\infty, \infty)$
domain of $f^{-1} = (-\infty, \infty)$
range of $f^{-1} = (-\infty, \infty)$

49. $f^{-1}(x) = 5x - 2$
domain of $f = (-\infty, \infty)$
range of $f = (-\infty, \infty)$
domain of $f^{-1} = (-\infty, \infty)$
range of $f^{-1} = (-\infty, \infty)$

51. $f^{-1}(x) = (x - 3)^2, x \geq 3$
domain of $f = [0, \infty)$
range of $f = [3, \infty)$
domain of $f^{-1} = [3, \infty)$
range of $f^{-1} = [0, \infty)$

53. $f^{-1}(x) = 16 - 4x^2, x \geq 0$
domain of $f = (-\infty, 16]$
range of $f = [0, \infty)$
domain of $f^{-1} = [0, \infty)$
range of $f^{-1} = (-\infty, 16]$

55. $f^{-1}(x) = (3 - x)^2 + 1, x \leq 3$
domain of $f = [1, \infty)$
range of $f = (-\infty, 3]$
domain of $f^{-1} = (-\infty, 3]$
range of $f^{-1} = [1, \infty)$

57. $f^{-1}(x) = \sqrt{x - 5}$
domain of $f = [0, \infty)$
range of $f = [5, \infty)$
domain of $f^{-1} = [5, \infty)$
range of $f^{-1} = [0, \infty)$

59. $f^{-1}(x) = -\sqrt{4 - x}$
domain of $f = (-\infty, 0]$
range of $f = (-\infty, 4]$
domain of $f^{-1} = (-\infty, 4]$
range of $f^{-1} = (-\infty, 0]$

61. $f^{-1}(x) = \sqrt{x + 16} - 4$
domain of $f = [-4, \infty)$
range of $f = [-16, \infty)$
domain of $f^{-1} = [-16, \infty)$
range of $f^{-1} = [-4, \infty)$

63. $f^{-1}(x) = 2 - \sqrt{x}$
domain of $f = (-\infty, 2]$
range of $f = [0, \infty)$
domain of $f^{-1} = [0, \infty)$
range of $f^{-1} = (-\infty, 2]$

65. $f^{-1}(x) = 1 + \sqrt{x - 2}$
domain of $f = [1, \infty)$
range of $f = [2, \infty)$
domain of $f^{-1} = [2, \infty)$
range of $f^{-1} = [1, \infty)$

67. $f^{-1}(x) = -\sqrt{x + 3} - 1$
domain of $f = (-\infty, -1]$
range of $f = [-3, \infty)$
domain of $f^{-1} = [-3, \infty)$
range of $f^{-1} = (-\infty, -1]$

69. $f^{-1}(x) = \sqrt{9 - x^2}$
domain of $f = [0, 3]$
range of $f = [-3, 0]$
domain of $f^{-1} = [-3, 0]$
range of $f^{-1} = [0, 3]$

71. $f^{-1}(x) = -\sqrt{9 - x^2}$
domain of $f = [-3, 0]$
range of $f = [0, 3]$
domain of $f^{-1} = [0, 3]$
range of $f^{-1} = [-3, 0]$

73. $f^{-1}(x) = -\sqrt{2x - x^2}$
domain of $f = [-1, 0]$
range of $f = [0, 1]$
domain of $f^{-1} = [0, 1]$
range of $f^{-1} = [-1, 0]$

75. $f^{-1}(x) = \dfrac{2}{3 - x}$ **77.** $f^{-1}(x) = \dfrac{2 + x}{x}$ **79.** $f^{-1}(x) = \dfrac{x}{2 - x}$ **81.** $f^{-1}(x) = \dfrac{4x + 5}{3x - 2}$ **83.** $f^{-1}(x) = (4 - x)^5 - 2$

85. The x intercept of f is the y intercept of f^{-1} and the y intercept of f is the x intercept of f^{-1}.

89. One possible answer: domain $x \leq 2, f^{-1}(x) = 2 - \sqrt{x}$ **91.** One possible answer: domain $0 \leq x \leq 2, f^{-1}(x) = 2 - \sqrt{4 - x^2}$

95. (A) $[200, 1{,}000]$ (B) $d^{-1}(q) = \dfrac{15{,}000}{q} - 5$; domain: $[200, 1{,}000]$; range: $[10, 70]$

97. (A) $r = m(w) = 1.25w + 3$; domain: $[0, \infty)$; range: $[3, \infty)$ (B) $w = m^{-1}(r) = 0.8r - 2.4$; domain: $[3, \infty)$; range: $[0, \infty)$

99. $s = f^{-1}(L) = 10 + \sqrt{\dfrac{50}{3}(L - 20)}$; domain: $[20, \infty)$; range: $[10, \infty)$

Chapter 3 Review Exercises

1. (A) Function (B) Function (C) Not a function *(3-1)*

3. If there is at least one team that has won more than one Super Bowl, then the correspondence is not a function because one input (team) will correspond with more than one output (year). There are several teams that have won at least two Super Bowls, so it is not a function. *(3-1)*

23.

x	-3	-2	-1	0	1	2	3	*(3-5)*
$(fg)(x)$	-12	-6	-2	0	0	2	6	

31. *(3-3)*

33. *(3-3)*

35. *(3-3)*

39. (A) $(f/g)(x) = (x^2 - 4)/(x + 3)$; domain of $f/g = (-\infty, -3) \cup (-3, \infty)$ (B) $(g/f)(x) = (x + 3)/(x^2 - 4)$; domain of $g/f = (-\infty, -2) \cup (-2, 2) \cup (2, \infty)$ (C) $(f \circ g)(x) = x^2 + 6x + 5$; domain of $f \circ g = (-\infty, \infty)$
(D) $(g \circ f)(x) = x^2 - 1$; domain of $g \circ f = (-\infty, \infty)$ *(3-5)*

49. The function f multiplies the square of the domain element by 3, adds 4 times the domain element, and then subtracts 6. *(3-1)*

51. This equation does not define a function. For example, the ordered pairs $(2, 2)$ and $(2, -2)$ both satisfy the equation. *(3-1)*

53. Domain: $[0, \infty)$; y intercept: 2; no x intercepts *(3-1, 3-2)* 55. Domain: $(-\infty, 3)$; y intercept: 0; x intercept: 0 *(3-1, 3-2)*

57. Domain: $[0, 16) \cup (16, \infty)$; y intercept: $\frac{1}{4}$; no x intercepts *(3-1, 3-2)*

61. (A) $(f \circ g)(x) = \sqrt{|x|} - 8$, $(g \circ f)(x) = |\sqrt{x} - 8|$ (B) Domain of $f \circ g = (-\infty, \infty)$, domain of $g \circ f = [0, \infty)$ *(3-5)*

67. $g(x) = 5 - 3|x - 2|$ *(3-3)*

69. The graph of $y = \sqrt[3]{x}$ is vertically stretched by a factor of 2, reflected through the x axis, shifted 1 unit left and 1 unit down. Equation: $y = -2\sqrt[3]{x + 1} - 1$ *(3-3)*

73. $t(x) = 0.25x^2 + x - 3$ *(3-3)*

75. *(3-3)*

77. *(3-3)*

79. *(3-3)*

81. *(3-3)*

83. $x < -2$ or $x > 6$; $(-\infty, -2) \cup (6, \infty)$ (3-4)

85. (A) $x^2\sqrt{1-x}$, domain $= (-\infty, 1]$ (B) $x^2/\sqrt{1-x}$, domain $= (-\infty, 1)$ (C) $1 - x$, domain $= (-\infty, 1]$
 (D) $\sqrt{1-x^2}$, domain $= [-1, 1]$ (3-5)

87. (A) $f^{-1}(x) = x^2 + 1$ (B) Domain of $f = [1, \infty) =$ Range of f^{-1} Range of $f = [0, \infty) =$ Domain of f^{-1} (3-6)
 (C)

89. (A) (B) (3-3)

91. (A) $E(x) = \begin{cases} 120 & \text{if } 0 \le x \le 2{,}000 \\ 0.1x - 80 & \text{if } 2{,}000 < x \le 5{,}000 \\ 0.1x + 170 & \text{if } x > 5{,}000 \end{cases}$

93. (A) $f = f(c) = 1.6c$ (B) \$168 (C) $c = f^{-1}(r) = 0.625r$; domain: $[16, \infty)$; range: $[10, \infty)$ (D) \$24.99 (3-6)

95. (A) $[1, 3]$ (B) $q = g^{-1}(p) = \dfrac{4{,}500}{p} - 500$; domain: $[1, 3]$; range: $[1{,}000, 4{,}000]$ (C) $R(p) = 4{,}500 - 500p$

 (D) $R(q) = 9q/(1 + 0.002q)$ (3-6)

97. (A) $A(x) = 60x - \frac{3}{2}x^2$ (B) $0 < x < 40$ (C) $x = 20, y = 15$ (3-4)

99. $T(x) = \begin{cases} 0.02x & \text{if } \quad 0 \le x \le 3{,}000 \\ 0.03x - 30 & \text{if } \quad 3{,}000 < x \le 5{,}000 \\ 0.05x - 130 & \text{if } \quad 5{,}000 < x \le 17{,}000 \\ 0.0575x - 257.5 & \text{if } 17{,}000 \le x \end{cases}$

x	\$2,000	\$4,000	\$10,000	\$30,000
$T(x)$	\$40	\$90.00	\$370	\$1,467.50

(3-2)

CHAPTER 4 Exercises 4-1

37. $\dfrac{4x^2 + 10x - 9}{x + 3} = 4x - 2 - \dfrac{3}{x + 3}$ **39.** $\dfrac{2x^3 - 3x + 1}{x - 2} = 2x^2 + 4x + 5 + \dfrac{11}{x - 2}$

63. $P(x) \to \infty$ as $x \to \infty$ and $P(x) \to -\infty$ as $x \to -\infty$; three intercepts and two local extrema

65. $P(x) \to -\infty$ as $x \to \infty$ and $P(x) \to \infty$ as $x \to -\infty$; three intercepts and two local extrema

67. $P(x) \to \infty$ as $x \to \infty$ and as $x \to -\infty$; four intercepts and three local extrema

77. x intercepts: $-12.69, -0.72, 4.41$; local maximum: $P(2.07) \approx 96.07$; local minimum: $P(-8.07) \approx -424.07$

79. x intercepts: $-16.06, 0.50, 15.56$; local maximum: $P(-9.13) \approx 65.86$; local minimum: $P(9.13) \approx -55.86$

81. x intercepts: $-16.15, -2.53, 1.56, 14.12$; local minimum: $P(-11.68) \approx -1{,}395.99$; local maximum: $P(-0.50) \approx 95.72$; local minimum: $P(9.92) \approx -1{,}140.27$

93. (A)
```
CubicReg
y=ax³+bx²+cx+d
a=.0210939441
b=-.1287915774
c=6.192857497
d=20.84904986
```
 (B) \$4,062 billion

95. (A)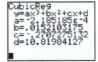
```
CubicReg
y=ax³+bx²+cx+d
a=-2.185185ε-4
b=.0152103175
c=-.2707275132
d=10.81984127
```
 (B) -3.6 (implausible estimate)

Exercises 4-2

35. (A) Upper bound: 2; lower bound: -2 (B) 1.4 (or -1.4)

Exercises 4-3

9. 0 (multiplicity 3), $-\frac{1}{2}$ (multiplicity 2); degree of $P(x)$ is 5

11. $2i$ (multiplicity 3); $-2i$ (multiplicity 4); -2 (multiplicity 5); 2 (multiplicity 5); degree of $P(x)$ is 17

15. $P(x) = (x + 7)^3[x - (-3 + \sqrt{2})][x - (-3 - \sqrt{2})]$; degree 5

17. $P(x) = [x - (2 - 3i)][x - (2 - 3i)](x + 4)^2$; degree 4
87. (A) $-\frac{1}{2} + \frac{\sqrt{3}}{2}i, -\frac{1}{2} - \frac{\sqrt{3}}{2}i$ (B) 3
91. No, because $P(x)$ is not a polynomial with real coefficients (the coefficient of x is the imaginary number $2i$).

Exercises 4-4

15. Domain: all real numbers except 0; x intercept: 3 **17.** Domain: all real numbers except ±2; x intercept: -6
19. Domain: all real numbers; x intercepts: $-4, 1$ **21.** Domain: all real numbers except ±6; x intercepts: none
23. Vertical asymptote: $x = -2$; horizontal asymptote: $y = 5$ **25.** Vertical asymptotes: $x = -4, x = 4$; horizontal asymptote: $y = 0$
27. Vertical asymptote: $x = 0$; horizontal asymptote: none **29.** Vertical asymptotes: $x = -3, x = 0$; horizontal asymptote: $y = \dfrac{3}{2}$
37. The graph of f is the same as the graph of g except that f has a hole at $(-2, \frac{1}{6})$.

39.

41.

43.

45.

47.

49.

51.

53. $f(x) = \dfrac{3(x^2 - 1)(x^2 - 4)}{x^4 + 1}$ **55.** $f(x) = \dfrac{(2x + 5)(x - 10) + 100}{x - 10}$ **73.** Vertical asymptote: $x = 1$; oblique asymptote: $y = 2x + 2$
77. Vertical asymptote: $x = 0$; oblique asymptote: $y = 2x - 3$

79.

81.

83.

Vertical asymptote: $x = 0$ Vertical asymptote: $x = 2$ Vertical asymptote: $x = 0$
Oblique asymptote: $y = x$ Oblique asymptote: $y = \frac{1}{2}x - 1$ Oblique asymptote: $y = -\frac{1}{4}x$

85. Domain: $x \neq 2$, or $(-\infty, 2) \cup (2, \infty)$; $f(x) = x + 2$ **87.** Domain: $x \neq 2, -2$ or $(-\infty, -2) \cup (-2, 2) \cup (2, \infty)$; $r(x) = \dfrac{1}{x - 2}$

Vertical asymptote: $x = 2$
Horizontal asymptote: $y = 0$

89. As $t \to \infty, N \to 50$ **91.** As $t \to \infty, N \to 5$

93. (A) $\overline{C}(n) = 25n + 175 + \dfrac{2,500}{n}$ **95.** (A) $L(x) = 2x + \dfrac{450}{x}$

(B) 10 yr (B) $(0, \infty)$

(C) (C) 15 ft by 15 ft

 (D)

Chapter 4 Review Exercise

1. Zeros: $-1, 3$; turning points: $(-1, 0), (1, 2), (3, 0)$; $P(x) \to \infty$ as $x \to \infty$ and $P(x) \to \infty$ as $x \to -\infty$ *(4-1)*

9. $\pm 1, \pm 3, \pm 5, \pm 15, \pm\dfrac{1}{5}, \pm\dfrac{3}{5}$ *(4-3)*

11. (A) Domain: all real numbers except 5; x intercept: 0 (B) Domain: all real numbers except -4 and 2; x intercept: $-\dfrac{3}{7}$ *(4-4)*

13. The graph does not increase or decrease without bound as $x \to \infty$ and as $x \to -\infty$ *(4-1)*

29. $(x + 1)(2x - 1)\left(x - \dfrac{1 + i\sqrt{3}}{2}\right)\left(x - \dfrac{1 - i\sqrt{3}}{2}\right)$ *(4-3)*

33. (A) Upper bound: 7; lower bound: -5 (B) Four intervals (C) $-4.67, 6.62$ *(4-2)*

47. (A) 3 (B) $-\dfrac{3}{2} \pm \dfrac{3i\sqrt{3}}{2}$ *(4-3)* **49.** *(4-4)*

53. 3: None of the candidates for rational zeros ($\pm 1, \pm 2,$ and ± 4) are actually zeros. *(4-3)* **59.** $v = k\dfrac{\sqrt{T}}{\sqrt{w}}$ *(4-5)*

63. (A)

(B) 1,915

CHAPTER 5 Exercises 5-1

25. The graph of g is the same as the graph of f stretched vertically by a factor of 3; g is increasing; horizontal asymptote: $y = 0$

27. The graph of g is the same as the graph of f reflected through the y axis and shrunk vertically by a factor of $\frac{1}{3}$; g is decreasing; horizontal asymptote: $y = 0$

29. The graph of g is the same as the graph of f shifted upward 2 units; g is increasing; horizontal asymptote: $y = 2$

31. The graph of g is the same as the graph of f shifted 2 units to the left; g is increasing; horizontal asymptote: $y = 0$

53. In every case, $y = 1$. The function $y = 1^x$ is simply the constant function $y = 1$.

55. The graph of g is the same as the graph of f reflected through the x axis; g is increasing; horizontal asymptote: $y = 0$

57. The graph of g is the same as the graph of f stretched horizontally by a factor of 2 and shifted upward 3 units; g is decreasing; horizontal asymptote: $y = 3$

59. The graph of g is the same as the graph of f stretched vertically by a factor of 500; g is increasing; horizontal asymptote: $y = 0$

61. The graph of g is the same as the graph of f shifted 3 units to the right, stretched vertically by a factor of 2, and shifted upward 1 unit; g is increasing; horizontal asymptote: $y = 1$

63. The graph of g is the same as the graph of f shifted 2 units to the right, reflected through the origin, stretched vertically by a factor of 4, and shifted upward 3 units; g is increasing; horizontal asymptote: $y = 3$

65. $\dfrac{e^{-2x}(-2x - 3)}{x^4}$ **69.** No local extrema; no x intercept; y intercept: 2.14; horizontal asymptote: $y = 2$

71. Local maximum: $s(0) = 1$; no x intercepts: y intercept: 1: horizontal asymptote: x axis

73. No local extrema; no x intercept; y intercept: 50; horizontal asymptotes: x axis and $y = 200$

75. Local minimum: $f(0) = 1$; no x intercepts; no horizontal asymptotes

79. $2^{1.4} = 2.6390$; $2^{1.41} = 2.6574$; $2^{1.414} = 2.6648$; $2^{1.4142} = 2.6651$; $2^{1.41421} = 2.6651$; $2^{1.414214} = 2.6651$; $2^{\sqrt{2}} \approx 2.6651$

81. **83.**

85. As $x \to \infty$, $f_n(x) \to 0$; the line $y = 0$ is a horizontal asymptote.
As $x \to -\infty$, $f_1(x) \to -\infty$ and $f_3(x) \to -\infty$, while $f_2(x) \to \infty$.
As $x \to -\infty$, $f_n(x) \to \infty$ if n is even and $f_n(x) \to -\infty$ if n is odd.

97. Flagstar: \$5,488.61; UmbrellaBank.com: \$5,470.85; Allied First Bank: \$5,463.71

Exercises 5-2

13.

25.

33. q approaches 0.000 9 coulombs, the upper limit for the charge on the capacitor.

35. (C) A approaches 100 deer, the upper limit for the number of deer the island can support.

37. $y = 14{,}910(0.8163)^x$; estimated purchase price: \$14,910; estimated value after 10 years: \$1,959

39. (A) $y = \dfrac{906}{1 + 2.27e^{-0.169x}}$ (B) 2010: 893.3 billion 2020: 903.6 billion

Exercises 5-3

19.

x	$y = 3^x$
-3	$\frac{1}{27}$
-2	$\frac{1}{9}$
-1	$\frac{1}{3}$
0	1
1	3
2	9
3	27

x	$y = \log_3 x$
$\frac{1}{27}$	-3
$\frac{1}{9}$	-2
$\frac{1}{3}$	-1
1	0
3	1
9	2
27	3

21.

x	$y = \left(\frac{2}{3}\right)^x$
-3	$\frac{27}{8}$
-2	$\frac{9}{4}$
-1	$\frac{3}{2}$
0	1
1	$\frac{2}{3}$
2	$\frac{4}{9}$
3	$\frac{8}{27}$

x	$y = \log_{2/3} x$
$\frac{27}{8}$	-3
$\frac{9}{4}$	-2
$\frac{3}{2}$	-1
1	0
$\frac{2}{3}$	1
$\frac{4}{9}$	2
$\frac{8}{27}$	3

61. b is any positive real number except 1. **79.** $\log x - \log y$ **81.** $4 \log x + 3 \log y$ **83.** $\ln\left(\dfrac{x}{y}\right)$ **85.** $\ln\left(\dfrac{x^2 y^5}{z}\right)$

91. The graph of g is the same as the graph of f shifted upward 3 units; g is increasing. Domain: $(0, \infty)$; vertical asymptote: $x = 0$

93. The graph of g is the same as the graph of f shifted 2 units to the right; g is decreasing. Domain: $(2, \infty)$; vertical asymptote: $x = 2$

95. The graph of g is the same as the graph of f reflected through the x axis and shifted downward 1 unit; g is decreasing. Domain: $(0, \infty)$; vertical asymptote: $x = 0$

97. The graph of g is the same as the graph of f reflected through the x axis, stretched vertically by a factor of 3, and shifted upward 5 units; g is decreasing. Domain: $(0, \infty)$; vertical asymptote: $x = 0$

103. (A) $f^{-1}(x) = 2 - 3^x$
(B) (C)

107.

109.

Exercises 5-4

25. (A) $y = 11.9 + 24.1 \ln x$; 2008: 73.7%; 2015: 84.1% (B) No; the predicted percentage goes over 100 sometime around 2034.

Exercises 5-5

87. (A) 7.94×10^{14} joules

Chapter 5 Review Exercises

55. The graph of g is the same as the graph of f stretched vertically by a factor of 2 and shifted downward 4 units; g is increasing.
Domain: all real numbers
Horizontal asymptote: $y = -4$

(5-1)

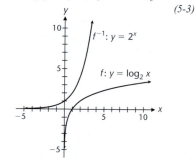

57. The graph of g is the same as the graph of f stretched vertically by a factor of 2 and shifted upward 1 unit; g is decreasing.
Domain: $(0, \infty)$;
Vertical asymptote: $x = 0$

(5-3)

67. Domain $f = (0, \infty) =$ Range f^{-1}
Range $f = (-\infty, \infty) =$ Domain f^{-1}

(5-3)

$f^{-1}: y = 2^x$

$f: y = \log_2 x$

75. (A)

81. (A) $y = 43.3(1.09)^x$; 2010: $574 billion; 2020: $1,360 billion

CHAPTER 6 Exercises 6-1

69. A central angle of radian measure 1 is an angle subtended by an arc of the same length as the radius of the circle.

81. The 7.5° angle and θ have a common side. (An extended vertical pole in Alexandria will pass through the center of the Earth.) The sun's rays are essentially parallel when they arrive at the Earth. So the other two sides of the angles are parallel, because a sun ray to the bottom of the well, when extended, will pass through the center of the Earth. From geometry we know that the alternate interior angles made by a line intersecting two parallel lines are equal. Therefore, θ = 7.5°.

Exercises 6-2

33. $\beta = 54.6°$ or $54°40'$, $\alpha = 35°20'$, $c = 10.4$

57. (A) As θ approaches 90°, $OA = \cos θ$ approaches 0. (B) As θ approaches 90°, $DE = \cot θ$ approaches 0.
(C) As θ approaches 90°, $OC = \sec θ$ increases without bound.

59. (A) As θ approaches 0°, $AD = \sin\theta$ approaches 0. (B) As θ approaches 0°, $CD = \tan\theta$ approaches 0.
(C) As θ approaches 0°, $OE = \csc\theta$ increases without bound.
73. (B)

θ	$C(\theta)$
10°	$368,222
20°	$363,435
30°	$360,622
40°	$360,146
50°	$360,050

Exercises 6-3

59. Zeros: none; turning points: $(\pi, -1)$, $(2\pi, 1)$, $(3\pi, -1)$

77. $W(x)$ is the coordinates of a point on a unit circle that is $|x|$ units from $(1, 0)$, in a counterclockwise direction if x is positive and in a clockwise direction if x is negative. $W(x + 4\pi)$ has the same coordinates as $W(x)$, because we return to the same point every time we go around the unit circle any integer multiple of 2π units (the circumference of the circle) in either direction.

81. $\cos x < 0$ in Quadrants II and III; $\sec x > 0$ in Quadrants I and IV; therefore, it is not possible to have both true for the same value of x.

95. $a_1 = 0.5$, $a_2 = 1.377583$, $a_3 = 1.569596$, $a_4 = 1.570796$, $a_5 = 1.570796$; $\dfrac{\pi}{2} \approx 1.570796$

Exercises 6-4

11. (A) $-2\pi, -\pi, 0, \pi, 2\pi$ (B) $-3\pi/2, -\pi/2, \pi/2, 3\pi/2$ (C) No x intercepts
13. (A) None (B) $-3\pi/2, -\pi/2, \pi/2, 3\pi/2$ (C) $-2\pi, -\pi, 0, \pi, 2\pi$
15. (A) No vertical asymptotes (B) $-3\pi/2, -\pi/2, \pi/2, 3\pi/2$ (C) $-2\pi, -\pi, 0, \pi, 2\pi$
17. (A) A shift of $\pi/2$ to the left will transform the cosecant graph into the secant graph. [The answer is not unique—see part B.]
(B) The graph of $y = -\csc(x - \pi/2)$ is a $\pi/2$ shift to the right and a reflection in the x axis of the graph of $y = \csc x$. The result is the graph of $y = \sec x$.
31. $\sin\theta = \frac{4}{5}$, $\csc\theta = \frac{5}{4}$, $\cos\theta = \frac{3}{5}$, $\sec\theta = \frac{5}{3}$, $\tan\theta = \frac{4}{3}$, $\cot\theta = \frac{3}{4}$
33. $\sin\theta = \dfrac{\sqrt{3}}{2}$, $\csc\theta = \dfrac{2}{\sqrt{3}}$, $\cos\theta = -\dfrac{1}{2}$, $\sec\theta = -2$, $\tan\theta = -\sqrt{3}$, $\cot\theta = -\dfrac{1}{\sqrt{3}}$
41. $120°$ or $\dfrac{2\pi}{3}$ radians **43.** $210°$ or $\dfrac{7\pi}{6}$ radians **45.** $240°$ or $\dfrac{4\pi}{3}$ radians
49. $\sin\theta = -\dfrac{2}{3}$, $\sec\theta = -\dfrac{3}{\sqrt{5}}$, $\tan\theta = \dfrac{2}{\sqrt{5}}$, $\cot\theta = \dfrac{\sqrt{5}}{2}$, $\csc\theta = -\dfrac{3}{2}$
51. Tangent and secant, because $\tan\theta = b/a$ and $\sec\theta = r/a$ and $a = 0$ if $P = (a, b)$ is on the vertical axis (division by zero is not defined).
67. (A)

$y = -3\cos x$ $y = \cos x$

$y = 2\cos x$

(B) No
(C) 1 unit; 2 units; 3 units
(D) The deviation of the graph from the x axis is changed by changing A. The deviation appears to be $|A|$.

69. (A)

$y = \sin 2x$ $y = \sin 3x$ $y = \sin x$

(B) 1; 2; 3
(C) n

71. (A)

$y = \cos(x + \pi/2)$ $y = \cos x$ $y = \cos(x - \pi/2)$

(B) The graph of $y = \cos x$ is shifted $|C|$ units to the right if $C < 0$ and $|C|$ units to the left if $C > 0$.

73. For each case, the number is not in the domain of the function and an error message of some type will appear.
75. (A) Both graphs are almost indistinguishable the closer the x is to the origin.
(B)

x	-0.3	-0.2	-0.1	0.0	0.1	0.2	0.3
$\sin x$	-0.296	-0.199	-0.100	0.000	0.100	0.199	0.296

85. (A) 3.31371, 3.14263, 3.14160, 3.14159 (B) $\pi = 3.1415926\ldots$

Exercises 6-5

5. $A = 3, P = 2\pi$ **7.** $A = \frac{1}{2}, P = 2\pi$ **9.** Period $= \dfrac{\pi}{4}$ **11.** Period $= \dfrac{1}{8}$ **13.** Period $= 4\pi$

15. $A = 1, P = \dfrac{2\pi}{\pi} = 2$; zeros: $-2, -1, 0, 1, 2$ **17.** Period $= 2\pi$; zeros: $\pi, 3\pi$

19. $A = 3, P = \dfrac{2\pi}{\pi} = \pi$; turning points: $\left(-\dfrac{\pi}{2}, -3\right), (0, 3), \left(\dfrac{\pi}{2}, -3\right)$ **21.** Period $= 2$; turning points: $(0, 2), (1, -2), (2, 2)$

31. Amplitude: 4; period: 2π; phase shift: 0 **33.** Amplitude: $\frac{1}{2}$; period: 2π; phase shift: $-\dfrac{\pi}{4}$ **35.** Period: π; phase shift

37. Period: $\dfrac{\pi}{2}$; phase shift: 0 **39.** Amplitude: 2π; period: 4; phase shift: 0 **41.** Amplitude: 3; period: 1; phase shift: $-\frac{1}{2}$

43. Period: 2π; phase shift: $-\pi$ **45.** Period: 2; phase shift: 0 **57.** $y = 2 \cot 2x$

 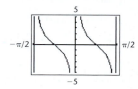

59. $y = \cot (x/2)$ **61.** $y = \csc 3x$ **63.** $y = \tan 2x$

65. $y = -4 \sin\left(\dfrac{\pi}{2}x - \dfrac{\pi}{2}\right)$ **67.** $y = \dfrac{1}{2} \cos\left(\dfrac{1}{4}x - \dfrac{3\pi}{4}\right)$

69. $A = 3.5, P = 4$, phase shift $= -0.5$ **71.** $A = 50, P = 1$, phase shift $= 0.25$

79.

The amplitude is decreasing with time. This is often referred to as a *damped sine wave*. Examples are a car's vertical motion, which is damped by the suspension system after the car goes over a bump, and the slowing down of a pendulum that is released away from the vertical line of suspension (air resistance and friction).

81.

The amplitude is increasing with time. In physical and electrical systems this is referred to as *resonance*. Some examples are the swinging of a bridge during high winds and the movement of tall buildings during an earthquake. Some bridges and buildings are destroyed when the resonance reaches the elastic limits of the structure.

83.

85. $A = \frac{1}{3}$, $P = \frac{2\pi}{8} = \frac{\pi}{4}$

89. The graph shows the seasonal changes of sulfur dioxide pollutant in the atmosphere; more is produced during winter months because of increased heating.

91. $A = 15$, $P = \frac{1}{60}$, phase shift $= -\frac{1}{240}$ **93.** $A = 3$, $P = \frac{1}{3}$

95. (A) $c = 20 \sec(\pi t/2)$, $[0, 1)$ (B)

(C) The length of the light beam starts at 20 feet and increases slowly at first, then increases rapidly without end.

97. (A)

(B) $y = 18.22 + 1.37 \sin\left(\frac{\pi x}{6} - 1.75\right)$ (C)

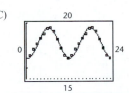

Exercises 6-6

57. $\sin^{-1}(\sin 2) = 1.416 \neq 2$. For the identity $\sin^{-1}(\sin x) = x$ to hold, x must be in the restricted domain of the sine function; that is, $-\frac{\pi}{2} \le x \le \frac{\pi}{2}$. The number 2 is not in the restricted domain.

65. **67.** **69.** **71.**

73. (A)

(B) The domain of $\cos^{-1}$ is restricted to $-1 \le x \le 1$; hence no graph will appear for other x.

79. $f^{-1}(x) = 3 + \cos^{-1}\dfrac{x-4}{2}; 2 \le x \le 6$

81. (A)

(B) The domain for $\cos x$ is $(-\infty, \infty)$ and the range is $[-1, 1]$, which is the domain for $\cos^{-1} x$. Thus, $y = \cos^{-1}(\cos x)$ has a graph over the interval $(-\infty, \infty)$, but $\cos^{-1}(\cos x) = x$ only on the restricted domain of $\cos x$, $[0, \pi]$.

85. (A)

(B) 59.44 mm

89. (A)

(B) 7.22 inches

91. (B) 76.10 feet

Chapter 6 Review Exercises

7.

$\theta°$	θ rad	$\sin \theta$	$\cos \theta$	$\tan \theta$	$\csc \theta$	$\sec \theta$	$\cot \theta$	*(6-1, 6-3)*
0°	0	0	1	0	ND*	1	ND	
30°	$\pi/6$	1/2	$\sqrt{3}/2$	$1/\sqrt{3}$	2	$2/\sqrt{3}$	$\sqrt{3}$	
45°	$\pi/4$	$1/\sqrt{2}$	$1/\sqrt{2}$	1	$\sqrt{2}$	$\sqrt{2}$	1	
60°	$\pi/3$	$\sqrt{3}/2$	1/2	$\sqrt{3}$	$2/\sqrt{3}$	2	$1/\sqrt{3}$	
90°	$\pi/2$	1	0	ND	1	ND	0	
180°	π	0	-1	0	ND	-1	ND	
270°	$3\pi/2$	-1	0	ND	-1	ND	0	
360°	2π	0	1	0	ND	1	ND	

*ND = not defined

9. (A) Domain $= (-\infty, \infty)$, range $= [-1, 1]$

(B) Domain is set of all real numbers except $x = \dfrac{2k+1}{2}\pi$, k an integer, range $=$ all real numbers *(6-4)*

11. *(6-4)*

13. If the graph of $y = \sin x$ is shifted $\dfrac{\pi}{2}$ units to the left, the result will be the graph of $y = \cos x$. *(6-4)*

47. $\cos^{-1}[\cos(-2)] = 2$. For the identity $\cos^{-1}(\cos x) = x$ to hold, x must be in the restricted domain of the cosine function; that is, $0 \le x \le \pi$. The number -2 is not in the restricted domain. *(6-6)*

49. *(6-5)*

57. $y = \frac{1}{2}\cos 2x + \frac{1}{2}$ *(6-5)*

65.

Domain = all real numbers; range $= \left(-\dfrac{\pi}{2}, \dfrac{\pi}{2}\right)$ *(6-6)*

73. (A)

(B) *(6-5)*

79. (A) $R(t) = 4 - 3\cos\dfrac{\pi}{6}t$. (B) The graph shows the seasonal changes in soft drink consumption. Most is consumed in August and the least in February. *(6-5)*

CHAPTER 7 Exercises 7-1

39.

Not an identity

41.

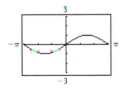

Appears to be an identity

43.

Not an identity

Exercises 7-2

75. $y_1 = \sin(x + \pi/6); y_2 = \dfrac{\sqrt{3}}{2}\sin x + \dfrac{1}{2}\cos x$

77. $y_1 = \cos(x - 3\pi/4); y_2 = -\dfrac{\sqrt{2}}{2}\cos x + \dfrac{\sqrt{2}}{2}\sin x$

79. $y_1 = \tan(x + 2\pi/3); y_2 = \dfrac{\tan x - \sqrt{3}}{1 + \sqrt{3}\tan x}$

89. $\tan(x - y) = \dfrac{\sin(x - y)}{\cos(x - y)}$

91. $\dfrac{\sin x \cos y - \cos x \sin y}{\cos x \cos y + \sin x \sin y} = \dfrac{\dfrac{\sin x \cos y - \cos x \sin y}{\cos x \cos y}}{\dfrac{\cos x \cos y + \sin x \sin y}{\cos x \cos y}}$

93. $\dfrac{\dfrac{\sin x \cos y}{\cos x \cos y} - \dfrac{\cos x \sin y}{\cos x \cos y}}{\dfrac{\cos x \cos y}{\cos x \cos y} + \dfrac{\sin x \sin y}{\cos x \cos y}} = \dfrac{\dfrac{\sin x}{\cos x} - \dfrac{\sin y}{\cos y}}{1 + \dfrac{\sin x \sin y}{\cos x \cos y}}$

Exercises 7-3

59. $\sin\dfrac{x}{2} = \sqrt{\dfrac{3 + 2\sqrt{2}}{6}}, \cos\dfrac{x}{2} = -\sqrt{\dfrac{3 - 2\sqrt{2}}{6}}, \tan\dfrac{x}{2} = -3 - 2\sqrt{2}$ **61.** $\sin\dfrac{x}{2} = -\dfrac{2\sqrt{5}}{5}, \cos\dfrac{x}{2} = \dfrac{\sqrt{5}}{5}, \tan\dfrac{x}{2} = -2$

63. (A) 2θ is a second quadrant angle, because θ is a first quadrant angle and $\tan 2\theta$ is negative for 2θ in the second quadrant and not for 2θ in the first.
(B) Construct a reference triangle for 2θ in the second quadrant with $(a, b) = (-3, 4)$. Use the Pythagorean theorem to find $r = 5$. Then $\sin 2\theta = 4/5$
and $\cos 2\theta = -3/5$. (C) The double-angle identities $\cos 2\theta = 1 - 2\sin^2\theta$ and $\cos 2\theta = 2\cos^2\theta - 1$.
(D) Use the identities in part C in the form
$$\sin\theta = \sqrt{\dfrac{1 - \cos 2\theta}{2}} \quad \text{and} \quad \cos\theta = \sqrt{\dfrac{1 + \cos 2\theta}{2}}$$
The positive radicals are used because θ is in quadrant one.
(E) $\sin\theta = 2\sqrt{5}/5; \cos\theta = \sqrt{5}/5$

65. (A) $-0.72335 = -0.72335$ (B) $-0.58821 = -0.58821$
69. $y_1 = y_2$ for $[-\pi, \pi]$ **71.** $y_1 = y_2$ for $[-2\pi, 0]$

89. (A) Since $\tan\dfrac{x}{2} = \pm\sqrt{\dfrac{1 - \cos x}{1 + \cos x}}$, $\tan\dfrac{x}{2}$ and $\sqrt{\dfrac{1 - \cos x}{1 + \cos x}}$ are always equal other than possibly their sign.
(B) $1 - \cos^2 x = \sin^2 x$; Pythagorean identity
(C) $\sqrt{\sin^2 x} = |\sin x|$ because $\sqrt{a^2} = |a|$ for any real number a; $\sqrt{(1 + \cos x)^2} = 1 + \cos x$ because $1 + \cos x$ is never negative.
(D) Since $\tan(x/2)$ and $\sin x$ always have the same sign, and since $1 + \cos x$ is never negative, $\tan(x/2)$ and $\sin x/(1 + \cos x)$ always have the same sign for any x.

95. (B) Table 1

n	10	100	1,000	10,000
A_n	2.93893	3.13953	3.14157	3.14159

(C) A_n appears to approach π, the area of the circle with radius 1.
(D) A_n will not exactly equal the area of the circumscribing circle for any n no matter how large n is chosen; however, A_n can be made as close to the area of the circumscribing circle as we like by making n sufficiently large.

Exercises 7-4

37. Let $x = u + v$ and $y = u - v$ and solve the resulting system for u and v in terms of x and y, then substitute the results into the first identity. The second identity will result after a small amount of algebraic manipulation.

65. $y_2 = 2\sin\dfrac{3x}{2}\cos\dfrac{x}{2}$ **67.** $y_2 = -2\sin x \sin 0.7x$ **69.** $y_2 = \frac{1}{2}(\sin 4x + \sin 2x)$

71. $y_2 = \frac{1}{2}(\cos 1.6x - \cos 3x)$

75. (A)

(B) $y_1 = \cos(30\pi x) + \cos(26\pi x)$; graph same as part A

77. (A)

(B) $y_1 = \sin(22\pi x) + \sin(18\pi x)$; graph same as part A

79. (B)

Exercises 7-5

73. Isolating $\cos x$, we get $\cos x < \frac{3}{2}$, which is true for all

75. $\tan^{-1}(-5.377)$ has exactly one value, -1.387; the equation $\tan x = -5.377$ has infinitely many solutions, which are found by adding $k\pi$, k any integer, to each solution in one period of $\tan x$.

81. (A) The largest zero for f is 0.3183. As x increases without bound, $1/x$ tends to 0 through positive numbers, and $\sin(1/x)$ tends to 0 through positive numbers. $y = 0$ is a horizontal asymptote for the graph of f. (B) Infinitely many zeros exist between 0 and b, for any b, however small. The exploration graphs suggest this conclusion, which is reinforced by the following reasoning. Note that for each interval $(0, b]$, however small, as x tends to zero through positive numbers, $1/x$ increases without bound, and as $1/x$ increases without bound, $\sin(1/x)$ will cross the x axis an unlimited number of times. The function f does not have a smallest zero, because, between 0 and b, no matter how small b is, there is always an unlimited number of zeros.

83. After 0.785 sec, 2.36 sec, 3.93 sec, and 5.50 sec

Chapter 7 Review Exercises

13. $x = \begin{cases} 0.7878 + 2k\pi \\ 2.3538 + 2k\pi \end{cases}$ k any integer *(7-5)* **27.** $\dfrac{\sqrt{2} + \sqrt{6}}{4}$ *(7-2)* **29.** $-\dfrac{\sqrt{2} + \sqrt{2}}{2}$ *(7-3)*

41. $x = 0 + 2k\pi, x = \pi + 2k\pi, x = \dfrac{\pi}{6} + 2k\pi, x = \dfrac{11\pi}{6} + 2k\pi$, k any integer. The first two can also be written together as $x = k\pi$, k any integer. *(7-5)*

51. $\sin^{-1} 0.3351$ has exactly one value, whereas the equation $\sin x = 0.3351$ has infinitely many solutions. *(7-5)*

55. (A) $\dfrac{3}{\sqrt{10}}$ or $\dfrac{3\sqrt{10}}{10}$ (B) $\frac{7}{25}$ *(7-3)*

61. (B) $y = 0.6\cos 184\pi t$ $y = -0.6\cos 208\pi t$ $y = 0.6\cos 184\pi t - 0.6\cos 208\pi t$ $y = 1.2\sin 12\pi t \sin 196\pi t$

(7-4)

CHAPTER 8 Exercises 8-1

29. Triangle I: $\beta = 158.8°$, $\gamma = 5.3°$, $c = 7.55$ inches; triangle II: $\beta = 21.2°$, $\gamma = 142.9°$, $c = 49.3$ inches

31. Triangle I: $\alpha = 116.6°$, $\gamma = 24.5°$, $c = 19.8$ inches; triangle II: $\alpha = 63.4°$, $\gamma = 77.7°$, $c = 46.7$ inches

Exercises 8-2

7. Angle γ is acute. A triangle can have at most one obtuse angle. Because α is acute, then, if the triangle has an obtuse angle it must be the angle opposite the longer of the two sides, b and c. So γ, the angle opposite the shorter of the two sides, c, must be acute.

13. If the triangle has an obtuse angle, then it must be the angle opposite the longest side; in this case, β.

29. Triangle I: $\beta = 109.7°$, $\alpha = 11.9°$, $a = 1.58$ meters; triangle II: $\beta = 70.3°$, $\alpha = 51.3°$, $a = 5.99$ meters

33. Triangle I: $\gamma = 140.5°$, $\alpha = 25.9°$, $a = 40.1$ meters; triangle II: $\gamma = 39.5°$, $\alpha = 126.9°$, $a = 73.5$ meters

Exercises 8-3

49. $\left\langle -\dfrac{1}{\sqrt{2}}, \dfrac{1}{\sqrt{2}} \right\rangle$

Exercises 8-4

7.

9.

11.

13.

15. $(5, -\pi/4)$: The polar axis is rotated $\pi/4$ radians clockwise (negative direction) and the point is located five units from the pole along the positive polar axis. $(5, 7\pi/4)$: The polar axis is rotated $7\pi/4$ radians counterclockwise (positive direction) and the point is located five units from the pole along the positive polar axis. $(-5, -5\pi/4)$: The polar axis is rotated $5\pi/4$ radians clockwise (negative direction) and the point is located five units from the pole along the negative polar axis.

17.

19.

21.

35.

37.

39.

41.

43.

45.

47. (A)

(B) 7 (C) *n*

49. (A)

 (B) 16 (C) 2*n*

63. For each *n*, there are *n* large petals and *n* small petals. For *n* odd, the small petals are within the large petals; for *n* even, the small petals are between the large petals.

65.

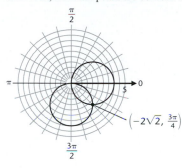

$(r, \theta) = \left(-2\sqrt{2}, \dfrac{3\pi}{4}\right)$ [*Note:* $(0, 0)$ is not a solution

of the system even though the graphs cross at the origin.]

67.

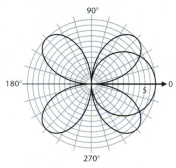

$(r, \theta) = (0, 90°), (0, 270°), (3\sqrt{3}, 30°), (-3\sqrt{3}, 150°)$
[*Note:* $(0, 0)$ is not a solution of the system even
though the graphs cross at the origin.]

73. (A) Ellipse

(B) Parabola

(C) Hyperbola

75. (A) Aphelion: 4.34×10^7 miles; perihelion: 2.85×10^7 miles

(B) Faster at perihelion. Because the distance from the sun to Mercury is less at perihelion than at aphelion, the planet must move faster near perihelion for the line joining Mercury to the sun to sweep out equal areas in equal intervals of time.

Exercises 8-5

7.

9.

11.

13.

45. $w_1 = 2^{1/10} e^{(-9°)i}, w_2 = 2^{1/10} e^{63°i}, w_3 = 2^{1/10} e^{135°i}, w_4 = 2^{1/10} e^{207°i}, w_5 = 2^{1/10} e^{279°i}$

47. $w_1 = 2e^{0°i}, w_2 = 2e^{120°i},$ **49.** $w_1 = 2e^{45°i}, w_2 = 2e^{135°i},$ **51.** $w_1 = 1e^{15°i}, w_2 = 1e^{75°i}, w_3 = 1e^{135°i}$
$w_3 = 2e^{240°i}$ $w_3 = 2e^{225°i}, w_4 = 2e^{315°i}$ $w_4 = 1e^{195°i}, w_5 = 1e^{255°i}, w_6 = 1e^{315°i}$

53. (A) $(1 + i)^4 + 4 = -4 + 4 = 0.$ There are three other roots.

(B) The four roots are equally spaced around the circle. Because there are four roots, the angle between successive roots on the circle is $360°/4 = 90°.$

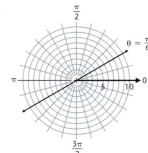

(C) $(-1 + i)^4 + 4 = -4 + 4 = 0;$
$(-1 - i)^4 + 4 = -4 + 4 = 0; (1 - i)^4 + 4 = -4 + 4 = 0$

55. $x_1 = 4e^{60°i} = 2 + 2\sqrt{3}i, x_2 = 4e^{180°i} = -4, x_3 = 4e^{300°i} = 2 - 2\sqrt{3}i$

57. $x_1 = 3e^{0°i} = 3, x_2 = 3e^{120°i} = -\dfrac{3}{2} + \dfrac{3\sqrt{3}}{2}i, x_3 = 3e^{240°i} = -\dfrac{3}{2} - \dfrac{3\sqrt{3}}{2}i$

73. $P(x) = (x - 2i)(x + 2i)[x - (-\sqrt{3} + i)][x - (-\sqrt{3} - i)][x - (\sqrt{3} + i)][x - (\sqrt{3} - i)]$

Chapter 8 Review Exercises

11. *(8-4)* **13.** *(8-5)*

15. *(8-5)* **31.** *(8-4)* **33.** *(8-4)*

35. *(8-4)*

37. (A) Ellipse

(B) Parabola

(C) Hyperbola

(8-4)

51. (A) There are a total of three cube roots and they are spaced equally around a circle of radius 2.

(B) $w_2 = -\sqrt{3} - i, w_3 = \sqrt{3} - i$

(C) The cube of each cube root is $-8i$. *(8-5)*

55. (A)

(B)

(8-4)

57. $1, -1, i, -i, \dfrac{\sqrt{2}}{2} + i\dfrac{\sqrt{2}}{2}, \dfrac{\sqrt{2}}{2} - i\dfrac{\sqrt{2}}{2}, -\dfrac{\sqrt{2}}{2} + i\dfrac{\sqrt{2}}{2}, -\dfrac{\sqrt{2}}{2} - i\dfrac{\sqrt{2}}{2}$ *(8-5)*

CHAPTER 9 Exercises 9-1

15.

Directrix $x = -1$ $F = (1, 0)$

17.

$F = (0, 2)$ Directrix $y = -2$

19.

Directrix $x = 3$ $F = (-3, 0)$

21.

$F = (0, -1)$ Directrix $y = 1$

23.

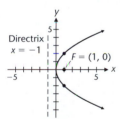

Directrix $x = 5$ $F = (-5, 0)$

45.

$(4, 4)$ $(0, 0)$

47.

$(5.313, 5.646)$ $(0, 0)$

57. $\sqrt{(x - x)^2 + (y + a)^2} = \sqrt{(x - 0)^2 + (y - a)^2}$

$(y + a)^2 = x^2 + (y - a)^2$

$y^2 + 2ay + a^2 = x^2 + y^2 - 2ay + a^2$

$x^2 = 4ay$

Exercises 9-2

15. Foci: $F' = (-\sqrt{21}, 0)$, $F = (\sqrt{21}, 0)$;
major axis length $= 10$;
minor axis length $= 4$

17. Foci: $F' = (0, -\sqrt{21})$, $F = (0, \sqrt{21})$;
major axis length $= 10$;
minor axis length $= 4$

19. Foci: $F' = (-\sqrt{8}, 0)$, $F = (\sqrt{8}, 0)$;
major axis length $= 6$;
minor axis length $= 2$

25. Foci: $F' = (0, -4)$, $F = (0, 4)$;
major axis length $= 10$;
minor axis length $= 6$

27. Foci: $F' = (0, -\sqrt{6})$, $F = (0, \sqrt{6})$;
minor axis length $= 2\sqrt{12} \approx 6.93$;
major axis length $= 2\sqrt{6} \approx 4.90$

29. Foci: $F' = (-\sqrt{3}, 0)$, $F = (\sqrt{3}, 0)$;
major axis length $= 2\sqrt{7} \approx 5.29$;
minor axis length $= 4$

35. $\dfrac{x^2}{25} + \dfrac{y^2}{9} = 1$ **37.** $\dfrac{x^2}{64} + \dfrac{y^2}{121} = 1$

51. $\dfrac{x^2}{400} + \dfrac{y^2}{144} = 1$; 7.94 feet approximately **53.** (A) $\dfrac{x^2}{576} + \dfrac{y^2}{15.9} = 1$ (B) 5.13 feet

Exercises 9-3

15. Foci: $F' = (-\sqrt{13}, 0)$, $F = (\sqrt{13}, 0)$;
transverse axis length $= 6$;
conjugate axis length $= 4$

17. Foci: $F' = (0, -\sqrt{13})$, $F = (0, \sqrt{13})$;
transverse axis length $= 4$;
conjugate axis length $= 6$

19. Foci: $F' = (-\sqrt{20}, 0)$, $F = (\sqrt{20}, 0)$;
transverse axis length $= 4$;
conjugate axis length $= 8$

21. Foci: $F' = (0, -5)$, $F = (0, 5)$;
transverse axis length $= 8$;
conjugate axis length $= 6$

23. Foci: $F' = (-\sqrt{10}, 0)$, $F = (\sqrt{10}, 0)$;
transverse axis length $= 4$;
conjugate axis length $= 2\sqrt{6} \approx 4.90$

25. Foci: $F' = (0, -\sqrt{11})$, $F = (0, \sqrt{11})$;
transverse axis length $= 4$;
conjugate axis length $= 2\sqrt{7} \approx 5.29$

45. $y = \pm\dfrac{\sqrt{3}}{\sqrt{2}}x$

47. (A) Infinitely many; $\dfrac{x^2}{a^2} - \dfrac{y^2}{1 - a^2} = 1 \; (0 < a < 1)$ (B) Infinitely many; $\dfrac{x^2}{a^2} + \dfrac{y^2}{a^2 - 1} = 1 \; (a > 1)$ (C) One; $y^2 = 4x$

49. (A) $(2/\sqrt{3}, 1/\sqrt{3})$, $(-2/\sqrt{3}, -1/\sqrt{3})$
(B) No intersection points
The graphs intersect at $x = \pm1/(\sqrt{1 - m^2})$ and $y = \pm m/(\sqrt{1 - m^2})$ for $-1 < m < 1$.

51. (A) No intersection points

(B) $(1/\sqrt{5}, 3/\sqrt{5}), (-1/\sqrt{5}, -3/\sqrt{5})$
The graphs intersect at $x = \pm 1/(\sqrt{m^2 - 4})$ and $y = \pm m/(\sqrt{m^2 - 4})$ for $m < -2$ or $m > 2$.

59. $\dfrac{x^2}{4} - \dfrac{y^2}{5} = 1$; hyperbola

Exercises 9-4

7. (A) $x' = x + 7, y' = y - 4$ (B) $\dfrac{x'^2}{9} + \dfrac{y'^2}{16} = 1$ (C) Ellipse

11. (A) $x' = x + 8, y' = y + 3$ (B) $\dfrac{x'^2}{12} + \dfrac{y'^2}{8} = 1$ (C) Ellipse **13.** (A) $\dfrac{(x - 3)^2}{9} - \dfrac{(y + 2)^2}{16} = 1$ (B) Hyperbola

15. (A) $\dfrac{(x + 5)^2}{5} + \dfrac{(y + 7)^2}{6} = 1$ (B) Ellipse

19. $(\sqrt{3}/2, -1/2), (1/2, \sqrt{3}/2), ((\sqrt{3} - 1)/2, (-1 - \sqrt{3})/2), ((-3\sqrt{3} + 4)/2, (3 + 4\sqrt{3})/2)$

21. $(\sqrt{2}/2, -\sqrt{2}/2), (\sqrt{2}/2, \sqrt{2}/2), (-3\sqrt{2}/2, -\sqrt{2}/2), (-\sqrt{2}, -2\sqrt{2})$ **23.** y' axis: $y = -\sqrt{3}x$; x' axis: $y = \dfrac{1}{\sqrt{3}}x$

27. $\dfrac{(x - 2)^2}{9} + \dfrac{(y - 2)^2}{4} = 1$; ellipse **29.** $(x + 4)^2 = -8(y - 2)$; parabola **31.** $(x + 6)^2 + (y + 5)^2 = 16$; circle

33. $\dfrac{(y - 3)^2}{9} - \dfrac{(x + 4)^2}{16} = 1$; hyperbola **35.** $F' = (-\sqrt{5} + 2, 2)$ and $F = (\sqrt{5} + 2, 2)$

41. $(x - 1)^2 + (y + 2)^2 = 0$; the point $(1, -2)$
(a degenerate circle)

43. $(x + 4)^2 - 4(y - 1)^2 = 0$; the lines $y = 0.5x + 3$ and $y = -0.5x - 1$,
intersecting at $(-4, 1)$ (a degenerate hyperbola)

47. $x'^2 + y'^2 = 49$; circle **49.** $\dfrac{x'^2}{4} + \dfrac{y'^2}{20} = 1$; ellipse **51.** $\dfrac{y'^2}{4} - \dfrac{x'^2}{12} = 1$; hyperbola **53.** $\dfrac{x'^2}{9} + \dfrac{y'^2}{4} = 1$; ellipse,
$\theta = 45°$ $\theta = 63.43°$

55. $y'^2 = 8x'$; parabola
$\theta = 30°$

63. $y + 4 = \pm\dfrac{1}{\sqrt{3}}x$

79. Ellipse

81. Hyperbola

83. Parabola

85. $\theta = 60°$; $x'^2 - 2\sqrt{3}x' + 2y' - 1 = 0$;
translate $0'$ to $(\sqrt{3}, 2)$; $x''^2 = -2y''$; parabola

Chapter 9 Review Exercises

1. Foci: $F' = (-4, 0)$, $F = (4, 0)$;
major axis length $= 10$;
minor axis length $= 6$ *(9-2)*

3. Foci: $F' = (0, -\sqrt{34})$, $F = (0, \sqrt{34})$;
transverse axis length $= 6$;
conjugate axis length $= 10$ *(9-3)*

5. *(9-1)*

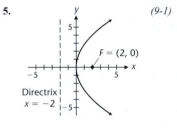

7. (A) $\dfrac{(y + 2)^2}{25} - \dfrac{(x - 4)^2}{4} = 1$ (B) Hyperbola *(9-4)*

25. $(x - 2)^2 = 4(2)(y + 3)$; parabola *(9-4)*

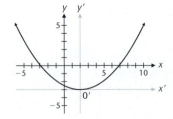

27. $\dfrac{x'^2}{20} + \dfrac{y'^2}{4} = 1$; ellipse *(9-4)*

29. Ellipse *(9-4)* **33.** $F' = (-3, -\sqrt{12} + 2)$ and $F = (-3, \sqrt{12} + 2)$ *(9-4)* **35.** $F' = (-\sqrt{13} - 3, -2)$ and $F = (\sqrt{13} - 3, -2)$ *(9-4)*

CHAPTER 10 Exercises 10-1

35. $\{(3s + 2, s, -2s - 1) \mid s \text{ any real number}\}$ **39.** $\{(-2s + 5, s, 3s - 4) \mid s \text{ any real number}\}$

45. $\{(\frac{1}{3}s - \frac{4}{3}, \frac{2}{3}s - \frac{8}{3}, s) \mid s \text{ any real number}\}$ **49.** $x = \dfrac{dh - bk}{ad - bc}, y = \dfrac{ak - ch}{ad - bc}, ad - bc \neq 0$

61. (A) Supply: 143 T-shirts; demand: 611 T-shirts
 (B) Supply: 714 T-shirts; demand: 389 T-shirts
 (C) Equilibrium price: $6.36; equilibrium quantity: 480 T-shirts
 (D)

71. $35,000 treasury bonds; $7,500 municipal bonds; $27,500 corporate bonds

Exercises 10-2

21. $x_1 = 2t + 3, x_2 = -t - 5, x_3 = t, t$ any real number

27. $\begin{bmatrix} 4 & -6 & -8 \\ 1 & -3 & 2 \end{bmatrix}$ **29.** $\begin{bmatrix} -4 & 12 & -8 \\ 4 & -6 & -8 \end{bmatrix}$ **31.** $\begin{bmatrix} 1 & -3 & 2 \\ 8 & -12 & -16 \end{bmatrix}$ **33.** $\begin{bmatrix} 1 & -3 & 2 \\ 0 & 6 & -16 \end{bmatrix}$ **35.** $\begin{bmatrix} 1 & -3 & 2 \\ 2 & 0 & -12 \end{bmatrix}$

41. $\begin{bmatrix} 1 & 0 & 0 & -5 \\ 0 & 1 & 0 & 4 \\ 0 & 0 & 1 & -2 \end{bmatrix}$ **43.** $\begin{bmatrix} 1 & 0 & 2 & -\frac{5}{3} \\ 0 & 1 & -2 & \frac{1}{3} \\ 0 & 0 & 0 & 0 \end{bmatrix}$

49. Infinitely many solutions; for any real number $s, x_2 = s, x_1 = 2s - 3$
73. Either 11 CDs, 1 DVD and 1 book; 6 CDs, 4 DVDs, and 3 books; or 1 CD, 7 DVDs, and 5 books
81. One-person boats: $t - 80$; two-person boats: $-2t + 420$; four-person boats: $t, 80 \le t \le 210, t$ an integer
83. No solution; no production schedule will use all the labor-hours in all departments.

Exercises 10-3

11. $\begin{bmatrix} 2 & 5 \\ 4 & -6 \end{bmatrix}$ **31.** $\begin{bmatrix} 3 & -2 & -4 \\ 6 & -4 & -8 \\ -9 & 6 & 12 \end{bmatrix}$ **41.** $\begin{bmatrix} -6 & 7 & -11 \\ 4 & 18 & -4 \end{bmatrix}$ **43.** $\begin{bmatrix} -3 & 6 & 8 \\ -18 & 12 & 10 \\ 4 & 6 & 24 \end{bmatrix}$ **45.** $\begin{bmatrix} 5 & -11 & 15 \\ 4 & -7 & 3 \\ 0 & 10 & 4 \end{bmatrix}$

47. $\begin{bmatrix} -0.2 & 1.2 \\ 2.6 & -0.6 \\ -0.2 & 2.2 \end{bmatrix}$ **49.** $\begin{bmatrix} -31 & 16 \\ 61 & -25 \\ -3 & 77 \end{bmatrix}$ **53.** $\begin{bmatrix} -2 & 25 & -15 \\ 26 & -25 & 45 \\ -2 & 45 & -25 \end{bmatrix}$ **55.** $\begin{bmatrix} -26 & -15 & -25 \\ -4 & -18 & 4 \\ 2 & 43 & -19 \end{bmatrix}$

73. $\frac{1}{2}(A + B) = \begin{bmatrix} \$33 & \$26 \\ \$57 & \$77 \end{bmatrix}$

This is the average cost of materials and labor for each product at the two plants.

75.

	Basic car	Air	Markup AM/FM radio	Cruise control
Model A	$3,330	$77	$42	$27
Model B	$2,125	$93	$95	$50
Model C	$1,270	$113	$121	$52

77. (A) $11.80 (B) $30.30
 (C) MN gives the labor costs per boat at each plant.
 (D)

	Plant I	Plant II	
$MN =$	$11.80	$13.80	One-person boat
	$18.50	$21.60	Two-person boat
	$26.00	$30.30	Four-person boat

79. (A) $A^2 = \begin{bmatrix} 0 & 0 & 2 & 0 & 0 \\ 1 & 0 & 0 & 0 & 1 \\ 0 & 1 & 0 & 2 & 0 \\ 1 & 0 & 0 & 0 & 1 \\ 0 & 0 & 1 & 0 & 0 \end{bmatrix}$

There is one way to travel from Baltimore to Atlanta with one intermediate connection; there are two ways to travel from Atlanta to Chicago with one intermediate connection. In general, the elements in A^2 indicate the number of different ways to travel from the ith city to the jth city with one intermediate connection.

(B) $A^3 = \begin{bmatrix} 2 & 0 & 0 & 0 & 2 \\ 0 & 1 & 0 & 2 & 0 \\ 0 & 0 & 3 & 0 & 0 \\ 0 & 1 & 0 & 2 & 0 \\ 1 & 0 & 0 & 0 & 1 \end{bmatrix}$

There is one way to travel from Denver to Baltimore with two intermediate connections; there are two ways to travel from Atlanta to El Paso with two intermediate connections. In general, the elements in A^3 indicate the number of different ways to travel from the ith city to the jth city with two intermediate connections.

$$(C) \quad A + A^2 + A^3 + A^4 = \begin{bmatrix} 2 & 3 & 2 & 5 & 2 \\ 1 & 1 & 4 & 2 & 1 \\ 4 & 1 & 3 & 2 & 4 \\ 1 & 1 & 4 & 2 & 1 \\ 1 & 1 & 1 & 3 & 1 \end{bmatrix}$$

It is possible to travel from any origin to any destination with at most three intermediate connections.

81. (A) \$3,550 (B) \$6,000 (C) *NM* gives the total cost per town.

(D) $NM = \begin{bmatrix} \$3,550 \\ \$6,000 \end{bmatrix} \begin{matrix} \text{Berkeley} \\ \text{Oakland} \end{matrix}$

(E) $[1 \quad 1]N = [3{,}000 \quad 1{,}300 \quad 13{,}000]$

where the columns are labeled Cost/town, Telephone call, House call, Letter, and Total contacts.

(F) $N\begin{bmatrix} 1 \\ 1 \\ 1 \end{bmatrix} = \begin{bmatrix} 6{,}500 \\ 10{,}800 \end{bmatrix} \begin{matrix} \text{Berkeley} \\ \text{Oakland} \end{matrix}$

83. (A) $\begin{bmatrix} 0 & 0 & 1 & 1 & 1 & 0 \\ 1 & 0 & 0 & 1 & 1 & 0 \\ 0 & 1 & 0 & 1 & 0 & 0 \\ 0 & 0 & 0 & 0 & 0 & 1 \\ 0 & 0 & 1 & 1 & 0 & 1 \\ 1 & 1 & 1 & 0 & 0 & 0 \end{bmatrix}$ (B) $\begin{bmatrix} 0 & 1 & 2 & 3 & 1 & 2 \\ 1 & 0 & 2 & 3 & 2 & 2 \\ 1 & 1 & 0 & 2 & 1 & 1 \\ 1 & 1 & 1 & 0 & 0 & 1 \\ 1 & 2 & 2 & 2 & 0 & 2 \\ 2 & 2 & 2 & 3 & 2 & 0 \end{bmatrix}$ (C) $BC = \begin{bmatrix} 9 \\ 10 \\ 6 \\ 4 \\ 9 \\ 11 \end{bmatrix}$ where $C = \begin{bmatrix} 1 \\ 1 \\ 1 \\ 1 \\ 1 \\ 1 \end{bmatrix}$

(D) Frank, Bart, Aaron and Elvis (tie), Charles, Dan

Exercises 10-4

11. $\begin{bmatrix} 2 & -3 \\ 4 & 5 \end{bmatrix}$ **25.** $\begin{matrix} 2x_1 - x_2 = 3 \\ x_1 + 3x_2 = -2 \end{matrix}$ **29.** $\begin{bmatrix} 4 & -3 \\ 1 & 2 \end{bmatrix}\begin{bmatrix} x_1 \\ x_2 \end{bmatrix} = \begin{bmatrix} 2 \\ 1 \end{bmatrix}$ **31.** $\begin{bmatrix} 1 & -2 & 1 \\ -1 & 1 & 0 \\ 2 & 3 & 1 \end{bmatrix}\begin{bmatrix} x_1 \\ x_2 \\ x_3 \end{bmatrix} = \begin{bmatrix} -1 \\ 2 \\ -3 \end{bmatrix}$ **41.** $\begin{bmatrix} 1 & -9 \\ 0 & 1 \end{bmatrix}$

43. $\begin{bmatrix} -5 & -2 \\ 2 & 1 \end{bmatrix}$ **45.** $\begin{bmatrix} -3 & -7 \\ -2 & -5 \end{bmatrix}$ **51.** $\begin{bmatrix} 0 & -1 & -1 \\ -1 & -1 & -1 \\ -1 & -1 & 0 \end{bmatrix}$ **53.** $\begin{bmatrix} -19 & 9 & -7 \\ 15 & -7 & 6 \\ -2 & 1 & -1 \end{bmatrix}$ **55.** $\begin{bmatrix} 1 & 0 & 1 \\ \frac{1}{2} & \frac{1}{2} & 1 \\ 2 & 1 & 4 \end{bmatrix}$ **59.** $\begin{bmatrix} -9 & -15 & 10 \\ 4 & 5 & -4 \\ -1 & -1 & 1 \end{bmatrix}$

81. (A) $(AB)^{-1} = \begin{bmatrix} 29 & -41 \\ -12 & 17 \end{bmatrix}, A^{-1}B^{-1} = \begin{bmatrix} 23 & -33 \\ -16 & 23 \end{bmatrix},$

$B^{-1}A^{-1} = \begin{bmatrix} 29 & -41 \\ -12 & 17 \end{bmatrix}$

83. 61 22 96 38 115 43 131 47 68 27 110 43 **85.** BEYONCE KNOWLES
87. 42 43 88 33 101 40 61 62 40 49 40 103 72 56 69 52 81 99 53 101 **89.** RAIDERS OF THE LOST ARK
91. Concert 1: 6,000 \$20 tickets, 4,000 \$30 tickets
Concert 2: 5,000 \$20 tickets, 5,000 \$30 tickets
Concert 3: 3,000 \$20 tickets, 7,000 \$30 tickets

Exercises 10-5

19. $x = -\frac{6}{5}, y = \frac{3}{5}$ **21.** $x = \frac{2}{17}, y = -\frac{20}{17}$ **23.** $\begin{vmatrix} 4 & 6 \\ -2 & 8 \end{vmatrix}$ **25.** $\begin{vmatrix} 5 & -1 \\ 0 & -2 \end{vmatrix}$ **27.** $(-1)^{1+1}\begin{vmatrix} 4 & 6 \\ -2 & 8 \end{vmatrix} = 44$ **29.** $(-1)^{2+3}\begin{vmatrix} 5 & -1 \\ 0 & -2 \end{vmatrix} = 10$

47. $x = \frac{4}{3}, y = -\frac{1}{3}, z = \frac{2}{3}$ **49.** $x = -9, y = -\frac{7}{3}, z = 6$ **51.** $x = \frac{3}{2}, y = -\frac{7}{6}, z = \frac{2}{3}$

53. If $a = \frac{3}{2}$ and $b = \frac{15}{4}$, there are an infinite number of solutions. If $a = \frac{3}{2}$ and $b \neq \frac{15}{4}$, there are no solutions. If $a \neq \frac{8}{3}$, there is one solution.

71. (A) Since $D = 0$, the system either has no solution or infinitely many. Since $x = 0, y = 0, z = 0$ is a solution, the second case must hold.
(B) Since $D \neq 0$, by Cramer's rule, $x = 0, y = 0, z = 0$ is the only solution.

73. (A) $R = 200p + 300q - 6p^2 + 6pq - 3q^2$
(B) $p = -0.3x - 0.4y + 180, q = -0.2x - 0.6y + 220, R = 180x + 220y - 0.3x^2 - 0.6xy - 0.6y^2$

Chapter 10 Review Exercises

3. Infinitely many solutions $[t, (4t + 8)/3]$, for any real number t *(10-1)*

7. $\begin{bmatrix} 3 & -6 & | & 12 \\ 1 & -4 & | & 5 \end{bmatrix}$ *(10-2)* **13.** $\begin{bmatrix} 4 & 8 \\ -12 & 18 \end{bmatrix}$ *(10-3)* **17.** $\begin{bmatrix} 3 & 3 \\ -4 & 9 \end{bmatrix}$ *(10-3)*

23. (A) $x_1 = -1, x_2 = 3$ (B) $x_1 = 1, x_2 = 2$ (C) $x_1 = 8, x_2 = -10$ *(10-4)*

27. $x_1 = 2, x_2 = -2$; each pair of lines has the same intersection point. *(10-1, 10-2)*

$x_1 - x_2 = 4$	$x_1 - x_2 = 4$	$x_1 - x_2 = 4$	$x_1 = 2$
$2x_1 + x_2 = 2$	$3x_2 = -6$	$x_2 = -2$	$x_2 = -2$

35. $\begin{bmatrix} 7 & 16 & -9 \\ 28 & 40 & -30 \\ -21 & -8 & 17 \end{bmatrix}$ *(10-3)* **37.** $\begin{bmatrix} 12 & 24 & -6 \\ 0 & 0 & 0 \\ -8 & -16 & 4 \end{bmatrix}$ *(10-3)*

57. (A) \$27 (B) Elements in LH give the total cost of manufacturing each product at each plant.

(C)

	North Carolina	South Carolina	
$LH = $	\$46.35	\$41.00	Desks
	\$30.45	\$27.00	Stands

(10-3)

CHAPTER 11 Exercises 11-1

15. $\frac{99}{101}$ **19.** $\frac{1}{10} + \frac{1}{100} + \frac{1}{1,000}$ **25.** 0.3, 0.33, 0.333, 0.3333, 0.33333

55.

57.

59. $\frac{4}{1} - \frac{8}{2} + \frac{16}{3} - \frac{32}{4}$ **73.** (A) 3, 1.83, 1.46, 1.415 (B) Calculator $\sqrt{2} = 1.4142135\ldots$ (C) $a_1 = 1$; 1, 1.5, 1.417, 1.414

81. (A) 0.625 ft; 0.02 ft (B) 19.98

83. (A) 40,000, 41,600, 43,264, 44,998.56, 46,794.34, 48,666.12 (B) $40,000(1.04)^{n-1}$ (C) 265,319.02

Exercises 11-2

11. $P_1: a^5 a^1 = a^{5+1}$; $P_2: a^5 a^2 = a^5(a^1 a) = (a^5 a)a = a^6 a = a^7 = a^{5+2}$; $P_3: a^5 a^3 = a^5(a^2 a) = a^5(a^1 a)a = [(a^5 a)a]a = a^8 = a^{5+3}$

13. $P_1: 9^1 - 1 = 8$ is divisible by 4; $P_2: 9^2 - 1 = 80$ is divisible by 4; $P_3: 9^3 - 1 = 728$ is divisible by 4

15. $P_k: 2 + 6 + 10 + \cdots + (4k - 2) = 2k^2$; $P_{k+1}: 2 + 6 + 10 + \cdots + (4k - 2) + (4k + 2) = 2(k + 1)^2$

17. $P_k: a^5 a^k = a^{5+k}$; $P_{k+1}: a^5 a^{k+1} = a^{5+k+1}$

49. $1 + 2 + 3 + \cdots + (n - 1) = \dfrac{n(n - 1)}{2}, n \geq 2$

Exercises 11-3

7. (A) Arithmetic with $d = -5$; $-26, -31$ (B) Geometric with $r = -2$; $-16, 32$ (C) Neither (D) Geometric with $r = \frac{1}{3}, \frac{1}{54}, \frac{1}{162}$

Exercises 11-4

39. No repeats: $10 \cdot 9 \cdot 8 \cdot 7 \cdot 6 = 30{,}240$; with repeats: $10 \cdot 10 \cdot 10 \cdot 10 \cdot 10 = 100{,}000$

43. $26 \cdot 26 \cdot 26 \cdot 10 \cdot 10 \cdot 10 = 17{,}576{,}000$ possible license plates; no repeats: $26 \cdot 25 \cdot 24 \cdot 10 \cdot 9 \cdot 8 = 11{,}232{,}000$

49. (B) $r = 0, 10$ (C) Each is the product of r consecutive integers, the largest of which is n for $P_{n,r}$ and r for $r!$

55. Two people: $5 \cdot 4 = 20$; three people: $5 \cdot 4 \cdot 3 = 60$; four people: $5 \cdot 4 \cdot 3 \cdot 2 = 120$; five people: $5 \cdot 4 \cdot 3 \cdot 2 \cdot 1 = 120$

57. (A) $P_{8,5} = 6{,}720$ (B) $C_{8,5} = 56$ (C) $C_{2,1} \cdot C_{6,4} = 30$

59. There are $C_{4,1} \cdot C_{48,4} = 778{,}320$ hands that contain exactly one king, and $C_{39,5} = 575{,}757$ hands containing no hearts, so the former is more likely.

Exercises 11-5

19. (A) No probability can be negative (B) $P(R) + P(G) + P(Y) + P(B) \neq 1$ (C) Is an acceptable probability assignment.

31. $\dfrac{C_{16,5}}{C_{52,5}} \approx .0017$

53. (A) $P(2) \approx .022$, $P(3) \approx .07$, $P(4) \approx .088$, $P(5) \approx .1$, $P(6) \approx .142$, $P(7) \approx .178$, $P(8) \approx .144$, $P(9) \approx .104$, $P(10) \approx .072$,
$P(11) \approx .052$, $P(12) \approx .028$

(B) $P(2) = \frac{1}{36}$, $P(3) = \frac{2}{36}$, $P(4) = \frac{3}{36}$, $P(5) = \frac{4}{36}$, $P(6) = \frac{5}{36}$, $P(7) = \frac{6}{36}$, $P(8) = \frac{5}{36}$, $P(9) = \frac{4}{36}$, $P(10) = \frac{3}{36}$, $P(11) = \frac{2}{36}$, $P(12) = \frac{1}{36}$

(C)

Sum	Expected frequency	Sum	Expected frequency
2	13.9	8	69.4
3	27.8	9	55.6
4	41.7	10	41.7
5	55.6	11	27.8
6	69.4	12	13.9
7	83.3		

Exercises 11-6

21. $m^3 + 3m^2n + 3mn^2 + n^3$ **23.** $8x^3 - 36x^2y + 54xy^2 - 27y^3$ **25.** $x^4 - 8x^3 + 24x^2 - 32x + 16$
27. $m^4 + 12m^3n + 54m^2n^2 + 108mn^3 + 81n^4$ **29.** $32x^5 - 80x^4y + 80x^3y^2 - 40x^2y^3 + 10xy^4 - y^5$
31. $m^6 + 12m^5n + 60m^4n^2 + 160m^3n^3 + 240m^2n^4 + 192mn^5 + 64n^6$ **51.** $3x^2 + 3xh + h^2$; approaches $3x^2$
53. $5x^4 + 10x^3h + 10x^2h^2 + 5xh^3 + h^4$; approaches $5x^4$

Chapter 11 Review Exercises

1. (A) Geometric (B) Arithmetic (C) Arithmetic (D) Neither (E) Geometric *(11-1, 11-3)*
3. (A) $16, 8, 4, 2$ (B) $a_{10} = \frac{1}{32}$ (C) $S_{10} = 31\frac{31}{32}$ *(11-1, 11-3)* **9.** 21 *(11-4)*
11. (A) 12 combined outcomes: (B) $6 \cdot 2 = 12$ *(11-5)*

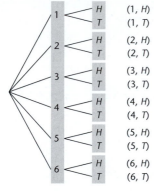

1	H	(1, H)
	T	(1, T)
2	H	(2, H)
	T	(2, T)
3	H	(3, H)
	T	(3, T)
4	H	(4, H)
	T	(4, T)
5	H	(5, H)
	T	(5, T)
6	H	(6, H)
	T	(6, T)

17. $P_1: 5 = 1^2 + 4 \cdot 1 = 5$; $P_2: 5 + 7 = 2^2 + 4 \cdot 2$; $P_3: 5 + 7 + 9 = 3^2 + 4 \cdot 3$ *(11-2)*
21. $P_k: 2 + 4 + 8 + \cdots + 2^k = 2^{k+1} - 2$; $P_{k+1}: 2 + 4 + 8 + \cdots + 2^k + 2^{k+1} = 2^{k+2} - 2$ *(11-2)*
27. $S_n = \sum_{k=1}^{n} \frac{(-1)^{k+1}}{3^k}$; $S_\infty = \frac{1}{4}$ *(11-3)* **33.** (A) $\dfrac{C_{13,5}}{C_{52,5}}$ (B) $\dfrac{C_{13,3} \cdot C_{13,2}}{C_{52,5}}$ *(11-5)*

APPENDIX A Cumulative Review Exercise for Chapters 1–3

3. $-5 < x < 9$ *(1-3)*
$(-5, 9)$

13. (A) Function; domain: $\{1, 2, 3\}$; range: $\{1\}$ (B) Not a function (C) Function; domain: $\{-2, -1, 0, 1, 2\}$; range: $\{-1, 0, 2\}$ *(3-1)*
23. $\frac{2}{3} \le m \le 2$ *(1-3)* **25.** $x \ge 2, x \ne 4$ *(1-2)*
$[\frac{2}{3}, 2]$ $[2, 4) \cup (4, \infty)$

35. (A) All real numbers (B) $\{-2\} \cup [1, \infty)$ (C) 1 (D) $[-3, -2]$ and $[2, \infty)$ (E) $-2, 2$ *(3-1, 3-2)*
39. $(f \circ g)(x) = \dfrac{x}{3 - x}$; Domain: $x \ne 0, 3$ *(3-5)*

41. Domain: $(-\infty, \infty)$ *(3-2)*
Range: $(-\infty, -1) \cup [1, \infty)$
Discontinuous at $x = 0$

45. Center: $(3, -1)$; radius: $\sqrt{10}$ *(2-2)*

53. (A) Domain g: $[-2, 2]$ (B) $\left(\dfrac{f}{g}\right)(x) = \dfrac{x^2}{\sqrt{4 - x^2}}$; Domain $\left(\dfrac{f}{g}\right)$: $(-2, 2)$ (C) $(f \circ g)(x) = 4 - x^2$; Domain $(f \circ g)$: $[-2, 2]$ *(3-5)*

61. $C(x)\begin{cases} 0.06x & \text{if } 0 \le x \le 60 \\ 0.05x + 0.6 & \text{if } 60 < x \le 150 \\ 0.04x + 2.1 & \text{if } 150 < x \le 300 \\ 0.03x + 5.1 & \text{if } 300 < x \end{cases}$

(3-2)

63. (A) $f(1) = f(3) = 1, f(2) = f(4) = 0$ (B) $f(n) = \begin{cases} 1 & \text{if } n \text{ is an odd integer} \\ 0 & \text{if } n \text{ is an even integer} \end{cases}$ *(3-2)*

65. (A) Profit: $\$5.5 < p < \8 or $(\$5.5, \$8)$ (B) Loss: $\$0 \le p < 5.5$ or $p > \$8$ or $[\$0, \$5.5) \cup (\$8, \infty)$ *(3-4)*

69. (A) *(3-4)* (B) $s = f^{-1}(L) = 2 + \sqrt{20L - 126}$; domain: $[22.5, \infty)$; range: $[20, \infty)$ (C) 67 mph

Cumulative Review Exercises for Chapters 4 and 5

17. (A) Domain: $x \ne -2$; x intercept: $x = -4$; y intercept: $y = 4$ (B) Vertical asymptote: $x = -2$; horizontal asymptote: $y = 2$
(C) *(4-4)*

23. (A) -0.56 (double zero); 2 (simple zero); 3.56 (double zero) (B) -0.56 can be approximated with a maximum routine; 2 can be approximated with the bisection: 3.56 can be approximated with a minimum routine *(4-2)*

51. A reflection through the x axis transforms the graph of $y = \ln x$ into the graph of $y = -\ln x$. A reflection through the y axis transforms the graph of $y = \ln x$ into the graph of $y = \ln(-x)$. *(5-3)*

55. Vertical asymptote: $x = -2$; *(4-4)*
oblique asymptote: $y = x + 2$

Cumulative Review Exercises for Chapters 6–8

7. *(6-4)*

21. (5, −30°): The polar axis is rotated 30° clockwise (negative direction) and the point is located five units from the pole along the positive polar axis. (−5, −210°): The polar axis is rotated 210° clockwise (negative direction) and the point is located five units from the pole along the negative polar axis. (5, 330°): The polar axis is rotated 330° counterclockwise (positive direction) and the point is located five units from the pole along the positive polar axis. *(8-4)*

23. *(8-5)* **29.** $\dfrac{2}{\sqrt{3}}$ *(6-2)* **37.** *(6-5)*

39. $\sin^{-1}(\sin 3) = 0.142$. For the identity $\sin^{-1}(\sin x) = x$ to hold, x must be in the restricted domain of the sine function; that is, $-\pi/2 \le x \le \pi/2$. The number 3 is not in the restricted domain. *(6-6)*

41. The equation has infinitely many solutions $[x = \tan^{-1}(-24.5) + k\pi, k$ any integer$]$; $\tan^{-1}(-24.5)$ has a unique value (-1.530 to three decimal places). *(6-6)*

43. $A = 3, P = \pi$, phase shift $= \dfrac{\pi}{2}$ *(6-5)* **45.** *(6-4)*

55. $\sin 2x = -\dfrac{24}{25}, \cos\dfrac{x}{2} = \sqrt{\dfrac{1}{10}}$ or $\dfrac{\sqrt{10}}{10}$ *(7-3)* **57.** $x = k\pi, \dfrac{\pi}{3} + 2k\pi, -\dfrac{\pi}{3} + 2k\pi, k$ any integer *(7-5)*

63. β must be acute. A triangle can have at most one obtuse angle, and because γ is acute, the obtuse angle, if present, must be opposite the longer of the two sides a and b. *(8-2)*

69. *(8-4)* **71.** *(8-4)*

77. $w_1 = \dfrac{\sqrt{3}}{2} - \dfrac{1}{2}i, w_2 = i, w_3 = -\dfrac{\sqrt{3}}{2} - \dfrac{1}{2}i$ *(8-5)*

83. $y = 3\cos(2\pi x - \pi/4)$; amplitude $= 3$, period $= 1$, phase shift $= 1/8$ *(6-5)*

89. (A) (B) *(8-4)*

99. (A) (B) $y = 53.5 + 22.5 \sin(\pi x/6 - 2.1)$ (C) *(6-5)*

Cumulative Review Exercises for Chapters 9–11

5. (A) Arithmetic (B) Geometric (C) Neither (D) Geometric (E) Arithmetic *(11-3)*

7. (A) 2, 5, 8, 11 (B) $a_8 = 23$ (C) $S_8 = 100$ *(11-3)*

11. Foci: $F' = (-\sqrt{61}, 0), F = (\sqrt{61}, 0)$;
transverse axis length $= 12$;
conjugate axis length $= 10$ *(9-3)*

13. *(9-1)*

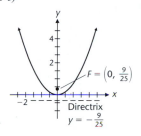

$F = \left(0, \frac{9}{25}\right)$

Directrix $y = -\frac{9}{25}$

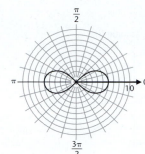

17. (A) $\begin{bmatrix} 0 & -3 \\ 3 & -9 \end{bmatrix}$ (B) Not defined (C) [3] (D) $\begin{bmatrix} 1 & 7 \\ 4 & -7 \end{bmatrix}$ (E) $[-1, 8]$ (F) Not defined *(10-4)*

23. (A) $x_1 = 3, x_2 = -4$ (B) $x_1 = 2t + 3, x_2 = t, t$ any real number. (C) No solution *(10-3)*

25. (A) $\begin{bmatrix} 1 & -3 \\ 2 & -5 \end{bmatrix}\begin{bmatrix} x_1 \\ x_2 \end{bmatrix} = \begin{bmatrix} k_1 \\ k_2 \end{bmatrix}$ (B) $A^{-1} = \begin{bmatrix} -5 & 3 \\ -2 & 1 \end{bmatrix}$ (C) $x_1 = 13, x_2 = 5$ (D) $x_1 = -11, x_2 = -4$ *(10-5)*

31. $P_k: k^2 + k + 2 = 2r$ for some integer r; $P_{k+1}: (k+1)^2 + (k+1) + 2 = 2s$ for some integer s *(11-2)*

39. (A) $\begin{bmatrix} -1 & 2 \\ 2 & 3 \end{bmatrix}$ (B) Not defined *(10-4)* **41.** $(0, i), (0, -i), (1, 1), (-1, -1)$ *(10-6)*

69. (A) Infinite number of solutions (B) No solution (C) Unique solution *(10-3)*

93. 1 model A truck, 6 model B trucks, and 5 model C trucks; or 3 model A trucks, 3 model B trucks, and 6 model C trucks; or 5 model A trucks and 7 model C trucks. *(10-3)*

95. (A) $M\begin{bmatrix} 0.25 \\ 0.25 \\ 0.25 \\ 0.25 \end{bmatrix} = \begin{bmatrix} 82.25 \\ 83 \\ 92 \\ 83.75 \\ 82 \end{bmatrix} \begin{matrix} \text{Ann} \\ \text{Bob} \\ \text{Carol} \\ \text{Dan} \\ \text{Eric} \end{matrix}$ (B) $M\begin{bmatrix} 0.2 \\ 0.2 \\ 0.2 \\ 0.4 \end{bmatrix} = \begin{bmatrix} 83 \\ 84.8 \\ 91.8 \\ 85.2 \\ 80.8 \end{bmatrix} \begin{matrix} \text{Ann} \\ \text{Bob} \\ \text{Carol} \\ \text{Dan} \\ \text{Eric} \end{matrix}$

(C) Class averages
 Test 1 Test 2 Test 3 Test 4
$[0.2 \quad 0.2 \quad 0.2 \quad 0.2 \quad 0.2]M = [84.4 \quad 81.8 \quad 85 \quad 87.2]$ *(10-4)*

APPENDIX B Exercises B-2

11. $\dfrac{-4}{x+2} + \dfrac{3}{x-4}$ **13.** $\dfrac{3}{3x+4} - \dfrac{1}{2x-3}$ **15.** $\dfrac{2}{x} - \dfrac{1}{x-3} - \dfrac{3}{(x-3)^2}$ **17.** $\dfrac{2}{x} + \dfrac{3x-1}{x^2+2x+3}$ **19.** $\dfrac{2x}{x^2+2} + \dfrac{3x+5}{(x^2+2)^2}$

21. $x - 2 + \dfrac{3}{x-2} - \dfrac{2}{x-3}$ **23.** $\dfrac{2}{x-3} + \dfrac{2x+5}{x^2+3x+3}$ **25.** $\dfrac{2}{x-4} - \dfrac{1}{x+3} + \dfrac{3}{(x+3)^2}$ **29.** $x + 2 - \dfrac{2}{x+2} + \dfrac{1}{2x-1} + \dfrac{x-1}{2x^2-x+1}$

Exercises B-3

3. $y = -2x - 2$; straight line

5. $y = -2x - 2, x \leq 0$; a ray (part of a straight line)

7. $y = -\frac{2}{3}x$; straight line

9. $y^2 = 4x$; parabola

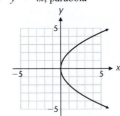

11. $y^2 = 4x, y \geq 0$; parabola (upper half)

25. $x = t, y = \dfrac{At^2 + Dt + F}{-E}, -\infty < t < \infty$; parabola

27. $\dfrac{(x-3)^2}{36} + \dfrac{(y-2)^2}{16} = 1$; ellipse with center $(3, 2)$

29. $\dfrac{(y+1)^2}{25} - \dfrac{(x+3)^2}{4} = 1$; hyperbola with center $(-3, -1)$

37. $\dfrac{(y+1)^2}{25} - \dfrac{(x-4)^2}{9} = 1$; hyperbola with center $(4, -1)$; $x = 4 + 3 \tan t, y = -1 + 5 \sec t, 0 \leq t \leq 2\pi, t \neq \dfrac{\pi}{2}, \dfrac{3\pi}{2}$

39. $\dfrac{(x-3)^2}{49} + \dfrac{(y+4)^2}{4} = 1$; ellipse with center $(3, -4)$; $x = 3 + 7 \cos t, y = -4 + 2 \sin t, 0 \leq t \leq 2\pi$

41. (A) The graphs are symmetric about to the line $y = x$.
 (B) 1. $y = e^x$
 2. $x = e^y$ or $y = \ln x$
 Function 2 is the inverse of function 1.

45. (A) 43.292 seconds (B) 9,183.620 meters, 9.184 kilometers (C) 2,295.918 meters

SUBJECT INDEX